Minor Minerals, Major Implications: Using Key Mineral Phases to Unravel the Formation and Evolution of Earth's Crust

The scientific and production quality of the Geological Society's books matches that of its journals. Since 1997, all book proposals are reviewed by two individual experts and the Society's Books Editorial Committee. Proposals are only accepted once any identified weaknesses are addressed.

The Geological Society of London is signed up to the Committee on Publication Ethics (COPE) and follows the highest standards of publication ethics. Once a book has been accepted, the volume editors agree to follow the Society's Code of Publication Ethics and facilitate a peer review process involving two independent reviewers. This is overseen by the Society Book Editors who ensure these standards are adhered to.

Geological Society books are timely volumes in topics of current interest. Proposals are often devised by editors around a specific theme or they may arise from meetings. Irrespective of origin, editors seek additional contributions throughout the editing process to ensure that the volume is balanced and representative of the current state of the field.

Submitting a book proposal
More information about submitting a proposal and producing a book for the Society can be found at https://www.geolsoc.org.uk/proposals.

It is recommended that reference to all or part of this book should be made in one of the following ways:

van Schijndel, V., Cutts, K., Pereira, I., Guitreau, M., Volante, S. and Tedeschi, M. (eds) 2024. *Minor Minerals, Major Implications: Using Key Mineral Phases to Unravel the Formation and Evolution of Earth's Crust*. Geological Society, London, Special Publications, **537**, https://doi.org/10.1144/SP537

Mako, C. A., Caddick, M. J., Law, R. D. and Thigpen, J. R. 2024 Monazite–xenotime thermometry: a review of best practices and an example from the Caledonides of northern Scotland. *Geological Society, London, Special Publications*, **537**, 185–208, https://doi.org/10.1144/SP537-2022-246

Geological Society Special Publication No. 537

Minor Minerals, Major Implications: Using Key Mineral Phases to Unravel the Formation and Evolution of Earth's Crust

Edited by

V. van Schijndel
University of Potsdam, Germany

K. Cutts
Geological Survey of Finland, Finland

I. Pereira
University of Coimbra, Portugal

M. Guitreau
Université Clermont Auvergne, France

S. Volante
Ruhr-Universität Bochum, Germany

and

M. Tedeschi
Universidade Federal de Minas Gerais, Brazil

2024
Published by
The Geological Society
London

The Geological Society of London

The Geological Society of London is a not-for-profit organization, and a registered charity (no. 210161). Our aims are to improve knowledge and understanding of the Earth, to promote Earth science education and awareness, and to promote professional excellence and ethical standards in the work of Earth scientists, for the public good. Founded in 1807, we are the oldest geological society in the world. Today, we are a world-leading communicator of Earth science – through scholarly publishing, library and information services, cutting-edge scientific conferences, education activities and outreach to the general public. We also provide impartial scientific information and evidence to support policy-making and public debate about the challenges facing humanity. For more about the Society, please go to https://www.geolsoc.org.uk/

The Geological Society Publishing House (Bath, UK) produces the Society's international journals and books, and acts as European distributor for selected publications of the American Association of Petroleum Geologists (AAPG), the Geological Society of America (GSA), the Society for Sedimentary Geology (SEPM) and the Geologists' Association (GA). GSL Fellows may purchase these societies' publications at a discount. The Society's online bookshop is at https://www.geolsoc.org.uk/bookshop

To find out about joining the Society and benefiting from substantial discounts on publications of GSL and other Societies go to https://www.geolsoc.org.uk/membership or contact the Fellowship Department at: The Geological Society, Burlington House, Piccadilly, London W1J 0BG: Tel. +44 (0)20 7434 9944; Fax +44 (0)20 7439 8975; E-mail: enquiries@geolsoc.org.uk.

For information about the Society's meetings, go to https://www.geolsoc.org.uk/events. To find out more about the Society's Corporate Patrons Scheme visit https://www.geolsoc.org.uk/patrons

Proposing a book
If you are interested in proposing a book then please visit: https://www.geolsoc.org.uk/proposals

Published by The Geological Society from:
The Geological Society Publishing House, Unit 7, Brassmill Enterprise Centre, Brassmill Lane, Bath BA1 3JN, UK

The Lyell Collection: www.lyellcollection.org
Online bookshop: www.geolsoc.org.uk/bookshop
Orders: Tel. +44 (0)1225 445046, Fax +44 (0)1225 442836

British Library Cataloguing in Publication Data

A catalogue record for this book is available from the British Library.
ISBN 978-1-78620-594-0
ISSN 0305-8719

Distributors
For details of international agents and distributors see:
https://www.geolsoc.org.uk/agentsdistributors

Typeset by Nova Techset Private Limited, Bengaluru & Chennai, India
Printed and bound by CPI Group (UK) Ltd, Croydon CR0 4YY, UK

Contents

Magmatism

Minor minerals, major implications: using key mineral phases to unravel the formation and evolution of Earth's crust

Valby van Schijndel[1]*, Kathryn A. Cutts[2], Inês Pereira[3], Martin Guitreau[4], Silvia Volante[5] and Mahyra Tedeschi[6]

[1]GFZ German Research Centre for Geosciences, 14473 Potsdam, Germany

[2]Geological Survey of Finland, P.O. Box 96, FI-02151 Espoo, Finland

[3]Universidade de Coimbra, Centro de Geociências, Coimbra, Portugal

[4]Université Clermont Auvergne, Laboratoire Magmas et Volcans, Aubière, France

[5]Structural Geology and Tectonics Group, Geological Institute, Department of Earth Sciences, ETH Zürich

[6]UFMG-Federal University of Minas Gerais, Belo Horizonte, Brazil

VVS, 0000-0002-2823-8200; KAC, 0000-0002-7190-1944; IP, 0000-0001-9028-2483; MG, 0000-0001-7156-6536; SV, 0000-0001-8807-4087; MT, 0000-0001-6573-0846
*Correspondence: valby@gfz-potsdam.de

Abstract: The investigation of key minerals including zircon, apatite, titanite, rutile, monazite, xenotime, allanite, baddeleyite and garnet can retain critical information about petrogenetic and geodynamic processes and may be utilized to understand complex geological histories and the dynamic evolution of the continental crust. They act as small but often robust petrochronological capsules and provide information about crustal evolution, from local processes to plate tectonics and supercontinent cycles. They offer us insights into processes of magmatism, sedimentation, metamorphism and alteration, even when the original protolith is not preserved. *In situ* techniques have enabled a more in-depth understanding of trace element behaviour in these minerals within their textural context. This has led to more meaningful ages for many stages of geological events. New developments of analytical procedures have further allowed us to expand our petrochronological toolbox while improving precision and accuracy. Combining multiple proxies with multiple minerals has contributed to new interpretations of the crustal history of our planet.

This Special Publication aims at showcasing recent developments and applications using key minerals such as zircon, apatite, titanite, rutile, monazite, xenotime, allanite, baddeleyite and garnet in igneous, metamorphic and detrital studies, with an emphasis on solving key scientific questions related to the formation and secular evolution of the Earth's crust. Several questions regarding the composition of Earth's first crust, the evolution of the continental crust composition through time, the geodynamic processes involved and the timing of continental crust growth and reworking remain open. Improvements in analytical techniques, such as *in situ* geochronology and geochemistry of minor minerals, novel developments of isotopic systems and trace element systematics are crucial to overcome many of the issues with precise and accurate dating of these phases.

The investigation of these key mineral phases has provided breakthroughs in our understanding about the continental crust composition and evolution of the Earth's crust, petrogenetic and geodynamic processes related to crustal growth and reworking, as well as the kinematics and timing of activation and reactivation of old crustal-scale structures (e.g. Iizuka *et al.* 2005, 2010; Condie and Aster 2010; Dhuime *et al.* 2012; Hawkesworth *et al.* 2013; Mulder and Cawood 2022). These minerals work like repositories, recording chemical and chronological information, and can thus contribute in understanding petrogenetic and geodynamic processes.

In this Special Publication, we have included the more commonly used accessory minerals and garnet which are prized for their petrochronologic value; from the traditionally most used and studied accessory mineral, zircon, to relatively more challenging isotopic and chemical systems of phases such as allanite and xenotime. Garnet, although a major mineral in many rocks, does also occur as an accessory or minor mineral in a detrital context as well as in migmatites and magmatic rocks. Due to recent analytical developments, it can be used for *in situ*

From: van Schijndel, V., Cutts, K., Pereira, I., Guitreau, M., Volante, S. and Tedeschi, M. (eds) 2024. *Minor Minerals, Major Implications: Using Key Mineral Phases to Unravel the Formation and Evolution of Earth's Crust.*
Geological Society, London, Special Publications, **537**, 1–7.
First published online October 18, 2023, https://doi.org/10.1144/SP537-2023-110

Lu-Hf (e.g. Simpson *et al.* 2021, 2023) and U–Pb (e.g. Seman *et al.* 2017; Millonig *et al.* 2020; Schannor *et al.* 2021) dating. Therefore, the continuing analytical advancements, including improvement of scientific equipment, data handling, processing and interpretation allow us to better understand complex geological processes. This Special Publication comprises contributions reviewing the multi-tools and applications of these key minerals, recent technique developments and new approaches and applications using focused case studies investigating igneous, metamorphic and detrital rocks that helps us put together the continental crust evolution puzzle.

The following seventeen contributions in this Special Publication have been arranged into four themes: reviews, novel approaches, metamorphic and magmatic applications (recognizing that a number of papers straddle multiple themes).

Reviews

The previous review volumes contemplating Petrochronology such as Kohn *et al.* (2017) and Ferrero *et al.* (2019) remain excellent resources. In the process of producing this volume, however, it became clear that the use of accessory minerals in petrochronology is quite a fast-moving area of research. Thus, novel mineral tools and methods recently developed are not covered in these volumes. In order to briefly introduce the current state and highlight future potential in the areas of sedimentary (detrital), igneous and metamorphic petrology two new review papers are included in this volume.

Pereira *et al.* (2023) revised the use and application of petrochronology applied to some detrital heavy minerals to unravel the growth and reworking of continental crust throughout geological time. The authors focus on heavy minerals that are widespread in detrital sediments and with suitable compositions for routine geochemical analysis, including isotopic dating: apatite, garnet, monazite, rutile, titanite and zircon. The aim of this paper is to provide a combined overview of sedimentary processes affecting the occurrence and distribution of these key detrital heavy minerals with both traditional and cutting-edge *in situ* analytical techniques. These techniques allow for examination of mineral, elemental and isotopic geochemistry, as well as geochronology, while preserving the textural context. The implications of these techniques are discussed in the study of crustal processes, provenance and source-to-sink identification. The authors emphasize the importance of combining multiple heavy minerals in these types of studies and review the different minerals that can be used by considering their strengths, limitations and potential future developments in the fields of petrology, sedimentology and geochemistry.

Volante *et al.* (2023) provide a review on the petrogeochemical and thermobarometric tools applied to garnet, zircon, titanite, rutile, allanite, monazite, xenotime and apatite. The authors review each mineral's chemical and textural features, what geochemical information, both elemental and isotopic, can be extracted from them, their geochronological applications and limitations and how they can be utilized to obtain thermobarometric constraints. Additionally, this contribution offers insights about how these tools have been applied in metamorphic and igneous applications. For example, in metamorphic processes, they illustrate and discuss the interplay between some of these different minerals during metamorphism, and how their changing chemistries can be used to trace and define P-T-t loops. As for magmatic applications, they show and discuss examples where these key mineral phases have been used to untangle complicated processes, such as magma mingling and mixing, and which minerals and techniques are best applied to obtain information about parental melts and source composition, to discuss crustal evolution processes. The authors conclude by highlighting recent analytical advances and developments, which have allowed the field to move forward in the past decades, and how we can best combine and apply these and other novel tools, driving the field forward.

Novel approaches

Continuous development and improvement of the investigation of key minerals for petrochronological research to better understand the evolution of the continental crust is crucial. This requires not only the development of new analytical tools, methods and techniques but also the development of new reference materials, data reduction schemes, imaging and modelling tools and machine learning to process datasets. New advancements in methods and applications ranging from using experimental research to the development of new *in situ* dating applications are highlighted by the following papers.

Joachim-Mrosko *et al.* (2023) present new experimental data tackling the diffusion and solubility of Al, Fe and H in natural rutile and in a high pressure TiO_2 polymorph. These experiments were run in a box furnace, using piston cylinder apparatus at $P \leq 3.5$ GPa and in a multi-anvil press at $P > 3.5$ GPa, considering the desired targeted pressure conditions (0.0001 up to 7 GPa), at 1273 and 1373 K and CCO-buffer conditions. In this study, the authors find Al diffusivities to be 8 to 9 orders of magnitude faster than in previous studies. This can potentially be explained by a combination of oxygen fugacity variations, hydrolytic weakening and enhanced Al-diffusion through extended defects in natural rutile. Their experiments also reveal that Fe is affected by the phase transition to an α-PbO_2-type TiO_2 polymorph from 6 to 7 GPa, matched by an increase in water concentration. These results need to be considered in studies tackling water transport and cycling at high-pressure conditions and in the development of Al-in-rutile as a geothermobarometer.

Denys *et al.* (2023) discuss the fluid induced alteration of allanite under low temperature conditions based on experiments with natural grains. Allanite is an important host of trace elements in the continental crust. This contribution evaluates the chemical processes, temperature and Ph conditions acting in the alteration of allanite and the associated mineral assemblages and microstructures. The paper reveals that the driving force behind the alteration of allanite resides in the precipitation at a secondary REE-mineral reactive front, causing a local disequilibrium close to the reaction interface between the fluid and the solid.

Glorie *et al.* (2023) test the robustness of the Lu-Hf system in apatite by applying the method to four case-studies in combination with the U–Pb system in zircon and apatite. In all investigated cases, the U–Pb system in apatite records a secondary overprint. In non-foliated granites, the Lu-Hf system in apatite preserves primary crystallization ages, whereas in foliated granites the Lu-Hf system records the timing of recrystallization. Apatite Lu-Hf ages from a lower crustal xenoliths also preserve primary ages despite being erupted with young alkali basalts, showing that the Lu-Hf system in apatite can retain primary ages despite exposure to magmatic temperatures for relatively short durations (and large grain size of >800 µm). This makes the Lu-Hf system in apatite a useful tool for evaluating the age of mantle or lower crustal xenoliths, which may not contain zircon. The results of this work suggest that the Lu-Hf system in apatite has a higher closure temperature than U–Pb, with a closure temperature between 660 and 730°C for typical grain sizes (100–300 µm). These observations suggest that the Lu-Hf system applied to apatite is a useful addition to traditional U–Pb dating of zircon or monazite, particularly if zircon and monazite are absent, or difficult to interpret due to inheritance or open system behaviour. The ease of the *in situ* laser-ablation based method also allows for campaign-style studies to be conducted, creating new opportunities for magmatic and metamorphic studies.

Mako *et al.* (2023) review the best practices for monazite-xenotime thermometry using examples from the Caledonides of northern Scotland. Currently five different thermometric calibrations are available, which produce significantly different results. These different thermometric calibrations are evaluated using a compilation of published compositional data, which have independently determined pressure and temperature conditions. It was found that the Pyle *et al.* (2001) calibration of the Heinrich *et al.* (1997) data provides the most meaningful temperature estimates. Equilibrium between monazite and xenotime is essential for temperature calculation, and it is possible to evaluate this using major and trace element composition of both monazite and xenotime. Following the evaluation of the calibrations, monazite-xenotime thermometry was applied to samples from the Northern Highland Terrane of Scotland. Thermometry in combination with U–Pb data indicates initially slow regional cooling across the Scandian thrust nappes followed by rapid cooling and exhumation to the surface. Quartz c-axis deformation thermometry in the lower nappes suggests that monazite and xenotime record the timing and temperature of deformation and indicate that ductile deformation occurred as late as 415–410 Ma.

Cutts *et al.* (2023) present an *in situ* application of Pb-Pb dating conducted via LA-MC-ICPMS to a single large garnet grain revealing growth during two metamorphic events at *c.* 3.45 and 3.2 Ga. The same garnet grain was used to model the P-T conditions of both events. Results suggest the two metamorphic events occurred at moderate pressure conditions of 0.7 GPa for the *c.* 3.45 Ga event and 0.8–0.9 GPa for the 3.2 Ga event. The temperature was estimated at 700°C for both events. The earlier event has a significantly younger age than the adjacent TTG pluton (*c.* 3.5 Ga) and partial convective overturn has no mechanism to return the rock to the surface following >20 km burial. The simplest explanation is that both of these metamorphic events are due to lateral collision events in the Archean. This is consistent with the structural setting of the Barberton Granite-Greenstone Belt. This study adds to the growing number of works that utilize *in situ* garnet geochronology and highlights the utility of obtaining age and P–T conditions from one crystal in order to unravel Archean polymetamorphism and gain insights about Archean tectonism.

Fumes *et al.* (2023) evaluate under which P–T conditions resetting of trace elements in detrital/inherited rutile takes place. Particular attention is given to the resetting of Zr contents and the application of the Zr-in-rutile geothermometer. They show that Zr contents in detrital rutile may re-equilibrate in quartzite, but commonly at higher temperatures than for rutile in neighbouring metapelites. Rutile from lower amphibolite facies quartzite (580°C, 0.9 GPa) show a large spread in Zr concentration and highest concentrations of Zr yield temperatures that are too high to relate to the metamorphic overprint and therefore are interpreted as an inherited detrital signature. A narrower spread in Zr concentration is observed in rutile from upper amphibolite facies quartzite (630°C for 1.4 GPa), indicating that the Zr content in the rutile re-equilibrated at these conditions. Therefore, these conditions are taken as the minimum conditions for inter-grain resetting of Zr in rutile for the quartzites in this study. The results indicate that rutile from high-temperature facies quartzite may indeed be used to calculate metamorphic peak-temperature conditions. Re-equilibration of rutile grains in quartzites takes place at about 50°C higher than in metapelites, and studies targeting detrital Zr-in-rutile thermometry should target quartzites instead of metapelites.

Metamorphism

These contributions display the wealth of information that can be provided by a more conventional approach using either a regional campaign style with comparison of results from multiple samples or multiple methods on one or two samples.

Schulz and Krause (2023) utilize a campaign style case study from the Saxothuringian Zone in the eastern Variscan Orogen to showcase the powerful tool of electron probe microanalyser (EPMA) geochronology in monazite. The strength of Th-U–Pb monazite dating using EPMA lies in being an *in situ* non-destructive microanalytical technique, where textural context of the targeted mineral is preserved and can be used to distinguished different age populations. An EPMA petrochronology campaign on monazite-rich metamorphic rock samples from the Saxothuringian Zone combined with geothermobarometry of garnet-bearing mineral assemblages distinguishes successive tectono-magmatic and metamorphic events key to reconstruct the Variscan tectono-metamorphic sequence of events in this region.

Skuzovatov *et al.* (2023) present a study that reports on the geochemical behaviour of zircon during subduction using trace element concentrations, U–Pb ages, Lu-Hf systematics and $\delta^{18}O$ signatures in two orthogneisses and one paragneiss that are juxtaposed with mafic eclogites within the North Muya block (Neoproterozoic Baikalides, northern Central Asian Orogenic belt). Petrological data indicate that this crustal package was buried along a single subduction P-T-t path during which zircon recrystallization was in the absence of external fluids. The authors provide evidence that this metamorphic transformation was really significant only in a H_2O-enriched protolith of the paragneiss. They further conclude that the formation of ^{18}O-depleted metamorphic overgrowths occurred during a distinct metamorphic stage without or with minor localized fluid infiltration. These rims probably formed during peak collision-driven heating, and are likely coeval with recrystallized high-grade zircon and Mn-enrichment in garnet rims.

Blereau and Spencer (2023) re-visit the metamorphic evolution of the Natal Metamorphic Province in order to better understand the relationship between the granulite facies Mzumbe and Margate Terranes during metamorphism and to re-evaluate the peak metamorphic conditions experienced. They investigate, through monazite LA–ICP–MS U–Pb geochronology, the age of peak metamorphism and argue that the thermal peak of metamorphism in the Mzumbe Terrane (987 ± 8 Ma) postdates the thermal peak metamorphism in the Margate Terrane (1033 ± 5 Ma) by ~40 myr. This indicates that the metamorphic evolution of the Natal Belt is more complex than previously thought. The Natal Metamorphic Province is part of the ~1.1 Ga Namaqua-Natal Belt and this provides new information on the early history of Rodinia as the Natal Metamorphic Province comprises several terranes that accreted onto the Kaapvaal Craton in the late Mesoproterozoic during to the assembly of Rodinia.

Alam *et al.* (2023) present new geochemical data (zircon U–Pb ages, whole-rock Sm–Nd and elemental data) on mafic granulites from the Central Indian Tectonic Zone to constraint their origin, timing of metamorphism, and relationship with the potential fragmentation of the supercontinent Columbia. The authors compiled available tectonic, metamorphic and petrogeochemical data to compare with their new data. They found that the studied mafic granulites have tholeiitic affinity and suggest that depletion in key elements such as Nb, P, Zr and Ti combined with positive enrichment in Ba, U and Pb are indicative of a variably enriched subcontinental lithospheric mantle (SCLM) source for the mafic granulite protolith magmas. Zircon U–Pb data gave ages between 1564 ± 8 Ma and 1598 ± 9 Ma which are interpreted as dating the granulite facies metamorphism. The authors further interpret whole-rock TDM model ages of 2.9–3.4 Ga as corresponding to the timing of extraction of the mafic granulite protolith magmas from the mantle. Finally, the authors show that a mineral isochron age of ~1.0 Ga points to the fact that these mafic granulites underwent another more recent event during the early Neoproterozoic.

Magmatism

Many studies apply zircon U–Pb ages with Hf-O isotopic or trace element data to shed light on the evolution of magmatic systems. Several of our contributions have employed additional methods such as U–Pb in baddeleyite, titanite or monazite, Rb–Sr in mica or feldspar and stress the importance of a multimineral approach or highlight issues when dealing with zircon alone such as the presence of antecrysts.

Gumsley *et al.* (2023) provide a detailed study based on geochemistry and U–Pb ID-TIMS baddeleyite data from the Mutare and Fingeren dyke swarms in southern Africa and East Antarctica. The dykes form a newly discovered Large Igneous Province (LIP) on the eastern Kalahari Craton, are dated at *c.* 724–712 Ma and have an EMORB-like geochemistry. The study reviews their possible palaeogeographical positions with regard to the enigmatic Kalahari Craton. The EMORB-like geochemistry of the dykes suggests an asthenosphere mantle source associated with rifting for the LIP. The authors also discuss additional implications of this new LIP and its possible involvement in the onset of the Sturtian Snowball Earth, which may have been driven by a large low-shear-velocity province (LLSVP) or superplume beneath Rodinia as it broke up.

Arzamastsev *et al.* (2023) discuss the formation of magmatic v. metamorphic titanite based on U–Pb and trace element data – corroborated by the dating of zircon inclusions in titanite – from the Kola alkaline granite massif, northeastern Fennoscandian shield. The titanite data was also used to elucidate the Neoarchaean tectonothermal history of the shield. The U–Pb and trace element data allowed the identification of different populations of titanite related to crystallization from the alkaline granite melt, metamorphic reactions involving the breakdown of Zr-bearing minerals and solid-state *in situ* recrystallization triggered by fluids. Titanite dating also revealed the presence of inherited titanite grains in peralkaline granites, suggesting that the U–Pb system can remain closed to temperatures up to 800°C.

Lim *et al.* (2023) propose a model to explore the role of lower crustal mush zones in the formation and evolution of arc magmas. They present new U–Pb and Lu-Hf isotope data on zircon crystals from mafic microgranular enclaves and the associated Satkatbong diorite from Korea. These data exhibit a significant range of ages and Hf isotope signatures that the authors interpret as evidence for prolonged growth in a hot reservoir, which is the justification of their model. The authors conclude that a realistic reservoir for the origin of the magmas would be a low-degree deep mush zone that would be frequently re-activated by the input of new mafic magma and, hence, maintained high temperatures for more than 30 My. One important outcome of this model is that the time of intrusion cannot be inferred from zircon since crystals are antecrysts for the most part.

Baggott *et al.* (2023) use samples of the Cape Woolamai S-type granite in SE Australia to constrain the timing of magmatic and hydrothermal fluids and their disruptive influence on isotopic systematics. To investigate this disruptive behaviour following fluid-rock interaction in magmatic systems, the authors use multiple geochronometers to compare and distinguish the different fluid-related events. In this view, U–Pb geochronology of zircon and monazite, and Rb–Sr geochronology of feldspar and micas was carried out, which indicated the existence of successive magmatic and hydrothermal events. The results of this study show that many events were not recorded in the most commonly used geochronometer, zircon. This contribution emphasizes the strength of integrating multiple geochronometers to untangle complex fluid-related histories in magmatic rocks which cannot be done using a single-mineral geochronological approach.

Lai *et al.* (2023) present a synthesis of zircon ages and Hf isotope signatures at the scale of the island of Sumatra and compare this to the Sunda arc to infer the tempo of magma formation, eruption and volcanic history of this region. Using new and published ages, the authors identified two major magmatic stages that were active from the Paleocene to the Early Eocene (66–48 Ma) and from the Early Miocene to the present (23–0 Ma) with a long-lasting magmatic rest of ~25 My in between. Most isotope signatures indicate juvenile sources (mildly to strongly depleted mantle) although some sharply negative εHf(t) values (down to −13) can be found around the Toba volcano. This is consistent with tapping of a mantle wedge with little influence of subducted material except around Toba, as indicated by the authors. The final conclusions are that the first magmatic stage ceased due to the closure of the Neo-Tethyan Ocean and the second magmatic stage progressively started at various locations across Sumatra after 25 My of quiescence to develop the modern Sunda subduction.

Acknowledgements We are grateful to all the contributors to this Special Publication and the reviewers who provided valuable and positive feed-back that allowed the authors to improve their manuscripts significantly. This Special Publication features a snap shot of the current state of the use of key minerals in metamorphic, magmatic and detrital studies to increase our understanding of the overall evolution of the continental crust. Special thanks to the editorial board of the Geological Society of London for supporting the production of this Special Publication.

Competing interests The authors declare that they have no known competing financial interests or personal relationships that could have appeared to influence the work reported in this paper.

Author contributions **VVS**: conceptualization (lead), writing – original draft (lead), writing – review & editing (lead); **KAC**: conceptualization (equal), writing – original draft (equal), writing – review & editing (equal); **IP**: conceptualization (equal), writing – original draft (equal), writing – review & editing (equal); **MG**: conceptualization (equal), writing – original draft (equal), writing – review & editing (equal); **SV**: conceptualization (equal), writing – original draft (equal), writing – review & editing (equal); **MT**: conceptualization (equal), writing – original draft (equal), writing – review & editing (equal).

Funding This research received no specific grant from any funding agency in the public, commercial or not-for profit sectors.

Data availability Data sharing is not applicable to this article.

References

Alam, M., Kailuna, T.V., Varma, R.R. and Ahmad, T. 2023. Zircon U–Pb geochronology, Nd isotopes and geochemistry of mafic granulites from the Central Indian Tectonic Zone (CITZ): isotopic constraints on Proterozoic crustal evolution. *Geological Society, London, Special Publications*, **537**, https://doi.org/10.1144/SP537-2022-135

Arzamastsev, A.A., Belyatsky, B.V., Rodionov, N.V., Antonov, A.V., Lepekhina, E.N. and Sergeev, S.A. 2023. Evolution of the neoarchaean Kola alkaline granites, northeastern Fennoscandian Shield: insights from SHRIMP-II titanite and zircon U–Pb isotope and rare earth elements data. *Geological Society, London, Special Publications*, **537**, https://doi.org/10.1144/SP537-2022-233

Baggott, K., Jacobsen, Y. *et al.* 2023. On the virtues and pitfalls of combined *laser ablation* Rb–Sr biotite and U–Pb monazite- zircon geochronology: an example from the isotopically disturbed Cape Woolamai Granite, SE Australia. *Geological Society, London, Special Publications*, **537**, https://doi.org/10.1144/SP537-2022-320

Blereau, E. and Spencer, C . 2023. Re-evaluating metamorphism in the southern Natal Province, South Africa. *Geological Society, London, Special Publications*, **537**, https://doi.org/10.1144/SP537-2022-222

Condie, K.C. and Aster, R.C. 2010. Episodic zircon age spectra of orogenic granitoids: the supercontinent connection and continental growth. *Precambrian Research*, **180**, 227–236, https://doi.org/10.1016/j.precamres.2010.03.008

Cutts, K., Lana, C., Stevens, G. and Buick, I. 2023. *In situ* Pb-Pb garnet geochronology as a tool for resolving polymetamorphism: a case for Paleoarchaean lateral tectonic thickening. *Geological Society, London, Special Publications*, **537**, https://doi.org/10.1144/SP537-2022-339

Denys, A., Auzende, A.L., Janots, E., Montes-Hernandez, G., Findling, N., Lanari, P. and Magnin, V. 2023. Experimental alteration of allanite at 200°C: the role of pH and aqueous ligands. *Geological Society, London, Special Publications*, **537**, https://doi.org/10.1144/SP537-2023-21

Dhuime, B., Hawkesworth, C.J., Cawood, P.A. and Storey, C.D. 2012. A change in the geodynamics of continental growth 3 billion years ago. *Science (New York, NY)*, **335**, 1334–1336, https://doi.org/10.1126/science.1216066

Ferrero, S., Lanari, P., Goncalves, P. and Grosch, E.G. 2019. *Metamorphic Geology: Microscale to Mountain Belts*. Geological Society, London, Special Publications, **478,** https://doi.org/10.1144/SP478

Fumes, R.A., Luvizotto, G.L., Pereira, I. and Moraes, R. 2023. Trace element changes in rutile geochemistry in quartzite through increasing *P–T* from lower amphibolite to eclogite facies conditions. *Geological Society, London, Special Publications*, **537**, https://doi.org/10.1144/SP537-2022-207

Glorie, S., Hand, M. *et al.* 2023. Robust laser ablation Lu-Hf dating of apatite: an empirical evaluation. *Geological Society, London, Special Publications*, **537**, https://doi.org/10.1144/SP537-2022-205

Gumsley, A.P., de Kock, M. *et al.* 2023. The Mutare–Fingeren dyke swarm: the enigma of the Kalahari Craton's exit from supercontinent Rodinia. *Geological Society, London, Special Publications*, **537**, https://doi.org/10.1144/SP537-2022-206

Hawkesworth, C., Cawood, P. and Dhuime, B. 2013. Continental growth and the crustal record. *Tectonophysics*, **609**, 651–660, https://doi.org/10.1016/j.tecto.2013.08.013

Heinrich, W., Andrehs, G. and Franz, G. 1997. Monazite – xenotime miscibility gap thermometry. I. An empirical calibration. *Journal of Metamorphic Geology*, **15**, 3–16, https://doi.org/10.1111/j.1525-1314.1997.t01-1-00052

Joachim-Mrosko, B., Konzett, J., Ludwig, T., Griffiths, T., Habler, G., Libowitzky, E. and Stalder, R. 2023. Al and H incorporation and Al-diffusion in natural rutile and its high-pressure polymorph TiO_2 (II). *Geological Society, London, Special Publications*, **537**, https://doi.org/SP537-2022-187

Iizuka, T., Hirata, T., Komiya, T., Rino, S., Katayama, I., Motoki, A. and Maruyama, S. 2005. U–Pb and Lu-Hf isotope systematics of zircons from the Mississippi River sand: Implications for reworking and growth of continental crust. *Geology*, **33**, 485–488, https://doi.org/10.1130/G21427.1

Iizuka, T., Komiya, T., Rino, S., Maruyama, S. and Hirata, T. 2010. Detrital zircon evidence for Hf isotopic evolution of granitoid 2175 crust and continental growth. *Geochimica et Cosmochimica Acta*, **74**, 2450–2472, https://doi.org/10.1016/j.gca.2010.01.023

Kohn, M.J., Engi, M. and Lanari, P. 2017. *Petrochronology: Methods and Applications*. Berlin, Boston, De Gruyter, https://doi.org/10.1515/9783110561890

Lai, Y.M., Liu, P.P. *et al.* 2023. Zircon U–Pb geochronology and Hf isotopic compositions of igneous rocks from Sumatra: implications for the Cenozoic magmatic evolution of the western Sunda Arc. *Geological Society, London, Special Publications*, **537**, https://doi.org/10.1144/SP537-2022-199

Lim, H., Nebel, O. *et al.* 2023. Lower crustal hot zones as zircon incubators? Inherited zircon antecryts in diorites from a mafic mush reservoir. *Geological Society, London, Special Publications*, **537**, https://doi.org/10.1144/SP537-2021-195

Mako, C.A., Caddick, M.J., Law, R.D. and Thigpen, J.R. 2023. Monazite-xenotime thermometry: a review of best practices and an example from the Caledonides of northern Scotland. *Geological Society, London, Special Publications*, **537**, https://doi.org/10.1144/SP537-2022-246

Millonig, L.J., Albert, R., Gerdes, A., Avigad, D. and Dietsch, C. 2020. Exploring laser ablation U–Pb dating of regional metamorphic garnet–the Straits Schist, Connecticut, USA. *Earth and Planetary Science Letters*, **552**, 116589, https://doi.org/10.1016/j.epsl.2020.116589

Mulder, J.A. and Cawood, P.A. 2022. Evaluating preservation bias in the continental growth record against the monazite archive. *Geology*, **50**, 243–247, https://doi.org/10.1130/G49416.1

Pereira, I., van Schijndel, V., Tedeschi, M., Cutts, K. and Guitreau, M. 2023. A review of detrital heavy mineral contributions to furthering our understanding of continental crust formation and evolution. *Geological Society, London, Special Publications*, **537**, https://doi.org/10.1144/SP537-2022-250

Pyle, J.M., Spear, F.S., Rudnick, R.L. and Mcdonough, W.F. 2001. Monazite – xenotime – garnet equilibrium in metapelites and a new monazite – garnet thermometer. *Journal of Petrology*, **42**, 2083–2107, https://doi.org/10.1093/petrology/42.11.2083

Schannor, M., Lana, C., Nicoli, G., Cutts, K., Buick, I., Gerdes, A. and Hecht, L. 2021. Reconstructing the metamorphic evolution of the Araçuaí orogen (SE Brazil) using in situ U–Pb garnet dating and P–T modelling. *Journal of Metamorphic Geology*, **39**, 1145–1171, https://doi.org/10.1111/jmg.12605

Schulz, B. and Krause, J. 2023. Electron probe petrochronology of monazite and garnet bearing metamorphic rocks in the Saxothuringian Allochthonous Domains (Erzgebirge, Granulite and Münchberg Massifs). *Geological Society, London, Special Publications*, **537**, https://doi.org/10.1144/SP537-2022-195

Seman, S., Stockli, D.F. and McLean, N.M. 2017. U–Pb geochronology of grossular-andradite garnet. *Chemical Geology*, **460**, 106–116, https://doi.org/10.1016/j.chemgeo.2017.04.020

Simpson, A., Gilbert, S. *et al.* 2021. In-situ Lu-Hf geochronology of garnet, apatite and xenotime by LA ICP MS/MS. *Chemical Geology*, **577**, 120299, https://doi.org/10.1016/j.chemgeo.2021.120299

Simpson, A., Glorie, S., Hand, M., Spandler, C. and Gilbert, S. 2023. Garnet Lu-Hf speed dating: a novel method to rapidly resolve polymetamorphic histories. *Gondwana Research*, **121**, 215–234, https://doi.org/10.1016/j.gr.2023.04.011

Skuzovatov, S., Wang, K.L., Li, X.H., Iizuka, Y. and Shatsky, V. 2023. Zircon trace-element and isotopes (U–Pb, Lu-Hf, $\delta^{18}O$) response to fluid-deficient metamorphism of a subducted continental terrane (North Muya, Eastern Siberia). *Geological Society, London, Special Publications*, **537**, https://doi.org/10.1144/SP537-2022-309

Volante, S., Blereau, E., Guitreau, M., Tedeschi, M., van Schijndel, V. and Cutts, K. 2023. Current applications using key mineral phases in igneous and metamorphic geology: perspectives for the future. *Geological Society, London, Special Publications*, **537**, https://doi.org/10.1144/SP537-2022-254

A review of detrital heavy mineral contributions to furthering our understanding of continental crust formation and evolution

Inês Pereira[1,2]*, Valby van Schijndel[3], Mahyra Tedeschi[4,5], Kathryn Cutts[6] and Martin Guitreau[1]

[1]Laboratoire Magmas et Volcans, CNRS-UMR6524, IRD-UMR163, OPGC, Université Clermont Auvergne, F-63178 Aubière, France

[2]Universidade de Coimbra, Centro de Geociências, Departamento de Ciências da Terra, Coimbra, Portugal

[3]Institute of Geosciences, University of Potsdam, 14476 Potsdam, Germany

[4]UFMG-Federal University of Minas Gerais, Belo Horizonte, Brazil

[5]Institute of Geological Sciences, University of Bern, 3012 Bern, Switzerland

[6]Geological Survey of Finland, P.O. Box 96, FI-02151 Espoo, Finland

IP, 0000-0001-9028-2483; VvS, 0000-0002-2823-8200; MT, 0000-0001-6573-0846; KC, 0000-0002-7190-1944; MG, 0000-0001-7156-6536
*Correspondence: ines.pereira@dct.uc.pt

Abstract: Detrital heavy minerals have helped address geologically complex issues such as the nature and origin of the early terrestrial crust, the growth and evolution of the continental crust, and the onset of plate tectonics, together with palaeogeographic and supercontinent cycles reconstructions. With the advent of *in situ* analytical techniques and a more complete understanding of trace element behaviour in rock-forming and accessory minerals, we have now at our disposal a powerful suite of tools that we can apply to multiple proxies found as detrital minerals. These can be *in situ* dating, trace element or isotopic tracing applied to both mineral hosts and their inclusions. We opted to showcase minerals that occur as primary minerals in a wide range of rock compositions and that can provide reliable age information. Additionally, over recent decades their chemistries have been tested as proxies to understand crustal processes. These are zircon, garnet, apatite, monazite, rutile and titanite. We include an overview and provide some approaches to overcome common biases that specifically affect these minerals. This review brings together petrological, sedimentological and geochemical considerations related to the application of these detrital minerals in crustal evolution studies, highlighting their strengths, limitations and possible future developments.

Heavy minerals in detrital studies

The evolution of the continental crust involves the formation and consumption of lithological units in refined metamorphic and magmatic reworking cycles. Therefore, a significant volume of crustal rocks can become unavailable to direct observation, although often they can stay preserved in the detrital sedimentary record. The formation and evolution of the continental crust over time can, in fact, be stored in igneous, metamorphic and detrital records. Investigations on detrital minerals have provided extremely relevant information to help unravel petrological and tectonic processes (Pereira *et al.* 2020, 2021; Arboit *et al.* 2021; Schönig *et al.* 2022), track mineral deposits (Grütter *et al.* 2004; Che *et al.* 2013; Forbes *et al.* 2015; Mao *et al.* 2016; Hardman *et al.* 2018) and crustal recycling (Baldwin *et al.* 2021; Andersen *et al.* 2022). Detrital mineral studies have helped geoscientists address key geological questions, such as the emergence and nature of the early crust, major crust forming episodes, major crustal composition and the onset of subduction, palaeogeography and supercontinent cycles, in both regional to more global studies (e.g. Cawood *et al.* 2012, 2013; Dhuime *et al.* 2012; Nebel *et al.* 2014; Cavosie *et al.* 2019; Pereira *et al.* 2021). Igneous and metamorphic mineral phases carry distinctive elemental signatures that can be used as a genetic toolbox to help decode their origin once found in a detrital context. Of particular interest are those heavy minerals (HMs, minerals with density >2.9 g cm^{-3}; Garzanti and Andò 2019) that are widespread as detrital sediments and have compositions suitable for routine geochemical analysis, i.e. zircon (e.g. Roberts and Spencer 2015), garnet (e.g. Morton and Hallsworth 2007), apatite (e.g. O'Sullivan *et al.* 2020), monazite (e.g. Gaschnig

From: van Schijndel, V., Cutts, K., Pereira, I., Guitreau, M., Volante, S. and Tedeschi, M. (eds) 2024. *Minor Minerals, Major Implications: Using Key Mineral Phases to Unravel the Formation and Evolution of Earth's Crust.*
Geological Society, London, Special Publications, **537**, 9–55.
First published online July 7, 2023, https://doi.org/10.1144/SP537-2022-250

2019), rutile (e.g. Triebold *et al.* 2012) and titanite (e.g. Scibiorski and Cawood 2022). In this review article we focus on these minerals since they can be routinely analysed and dated by *in situ* methods, now including garnet, and hence offer more provenance information.

The main analytical methods applied in provenance research are based on petrographic and geochemical (elemental and isotopic) data (Mange and Morton 2007). Petrographic characterization includes morphological parameters such as colour, shape, habit and zoning patterns (Fig. 1). For instance, the determination of the shape and elongation of zircon grains are often included in provenance studies (Corfu *et al.* 2003; Markwitz and Kirkland 2018; Makuluni *et al.* 2019; Gärtner *et al.* 2022).

Detailed petrographic and geochemical studies of some detrital minerals, especially those that can be dated, have helped decipher the provenance, tectonic setting, environment of deposition, and diagenetic processes of the sedimentary units that encapsulate them. The coupling of classical petrographic analyses on HMs with geochronological, geochemical and inclusions investigations enables minerals to be used as tracers of crustal growth, reworking and recycling. Since zircon, apatite, monazite, rutile, titanite and garnet are formed during different magmatic and metamorphic stages and in different conditions, they represent valuable independent tracers in provenance analysis (e.g. Guo *et al.* 2020). Other HMs such as tourmaline or spinel may be used to improve provenance analysis (Bónová *et al.* 2018; Qasim *et al.* 2020) but do not directly provide age constraints.

Differences in how the above listed minerals record source rock information, owing to variable closure temperatures and solubilities, are advantageous. Using a multi-mineral approach enhances

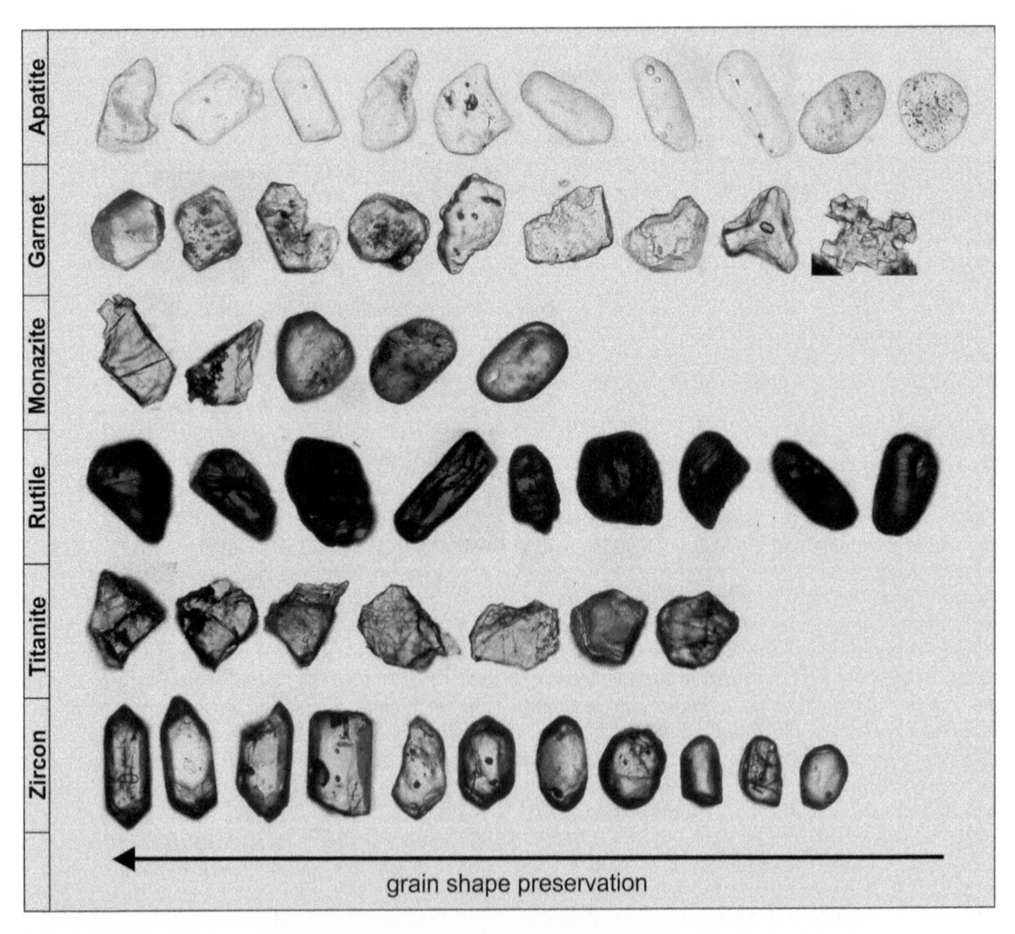

Fig. 1. Transmitted light photomicrographs of the most common heavy minerals employed in provenance investigations and herein discussed. The grains are presented from the most preserved on the left to more corroded/rounded on the right. Source: the image is courtesy of Laura Stutenbecker.

the ability to fully characterize crustal processes, from lower to higher temperatures, variable rock compositions and petrogenetic settings (O'Sullivan *et al.* 2016; Liu *et al.* 2017; Gaschnig 2019; Chew *et al.* 2020; Guo *et al.* 2020; Joshi *et al.* 2021). As a complement to bulk sediment geochemistry, each of these HMs constitutes comprehensive archives that can be deconvoluted by different techniques to unravel the deformation history through microstructures, the timings and rates of processes, and the conditions under which these petrological processes took place (Caby *et al.* 2008; Garçon *et al.* 2017; Altrin *et al.* 2018; Gillespie *et al.* 2018; Papapavlou *et al.* 2018*b*; Ansberque *et al.* 2019; Baldwin *et al.* 2021).

In this contribution, we present a review of processes influencing the presence and abundance of HMs in sedimentary rocks and an overview of the different approaches that have been used to extract precious information encapsulated in the detrital record using zircon, garnet, apatite, monazite, rutile and titanite. The strengths and weaknesses for tracing source to sink mechanisms, deciphering sediment provenance and crustal processes are introduced for this set of minerals. We list the tools that can be applied to study these minerals in order to decipher the evolution of the continental crust (e.g. composition, age, thermal history), including a brief evaluation of their inherent biases. Our purpose here is to reconcile key information arising from the different fields of sedimentology, petrology and geochemistry, which are now key in detrital mineral studies. Finally, we evaluate perspectives for future crustal evolution studies targeting the detrital record.

Heavy mineral distribution in sediments: some considerations

It has long been established that HM concentrations in sediments are mainly dependent on the chemistry and the crustal level of rock exposures within the source terranes (Garzanti and Andò 2007*a*). The tectonic setting also seems to exert an influence on the concentration of heavy minerals in sediments (Fig. 2a), as it works as a major control on weathering rates and on the availability of rock types in a catchment area, namely in their composition (and density), and thus the heavy mineral suites available during weathering (Garzanti and Andò 2007*b*).

Sediments draining cratonic continental blocks (e.g. Africa in Fig. 2) yield low heavy mineral concentrations (HMCs), often lower than 1% (Fig. 2a), and a significant proportion of these correspond to zircon, tourmaline and/or rutile (ZTR; 35%) and garnet (11%) (Fig. 2b).

The magmatic arc crust, on the other hand, supplies very important sources of HMs to sediments (22–25% HMC), either the upper crustal basaltic to rhyolitic volcanic rocks in undissected volcanic arcs or the medium to lower crust, gabbroic to calc-alkaline batholiths in dissected arcs (Fig. 2a). Dissected arc batholiths (e.g. in the Himalaya) provide greater proportions of amphiboles, low-grade minerals (epidote-group), pyroxenes and important amounts of titanite (Fig. 2b; Garzanti and Andò 2007*b*). This is in stark contrast with undissected volcanic arcs (e.g. Aegean), where sediments are dominated by augite and hypersthene, with minor amphiboles (Fig. 3b; Garzanti and Andò 2007*a*). Finally, when sediments are sourced from continental collision provinces, variable proportions of HMC can be retrieved (Fig. 2a); if a significant proportion of obducted ophiolites are exposed (e.g. Kizildag in Fig. 2), they will contribute to higher concentrations of pyroxenes, olivine and spinel (Fig. 2b), whereas if the crust experienced high-pressure and/or high-grade metamorphism, hornblende, epidote and garnet become predominant in the sediments, followed by ZTR and clinopyroxene (such as omphacite) (Fig. 2b).

Interestingly, Garzanti *et al.* (2003) demonstrated that despite a close 'source to sink' relationship between modern sedimentary systems in the Arabian Gulf and ophiolite source rocks in Oman, an important sediment recycling process is evidenced by the higher proportions of light mineral concentrations, low plagioclase/feldspar and low HMCs, with HMs exhibiting strongly rounded shapes. While in the rift sequences (Gulf of Aden; Fig. 2b), they found a higher proportion of Ca-amphiboles (hornblende) and clino- and orthopyroxenes in the HMC, indicating proximity to mafic magmatic source rocks; with recycling these tend to decrease in respect to epidote, garnet and the ultrastable minerals, ZTR (Garzanti *et al.* 2013).

Indeed, multiple factors interplay in the abundance of detrital heavy minerals, including provenance-specific factors, sedimentary processes and post-deposition dissolution phenomena (e.g. Morton and Hallsworth 1999; Garzanti and Andò 2007*a*). Density-sorting during transport and deposition can segregate these minerals within the sedimentary environment, imposing variabilities in the sediment's isotopic signatures (e.g. Garçon *et al.* 2013). A dichotomy in the proportion of ultrastable (ZTR) and less stable heavy minerals (amphiboles, pyroxenes) between modern sands and detrital sedimentary rocks can be traced to selective leaching of unstable mineral species during diagenesis (e.g. Morton 1985; Fig. 3). This process may significantly alter initial light and heavy minerals ratios and even ratios between different heavy mineral species, which usually offer valuable provenance information (such

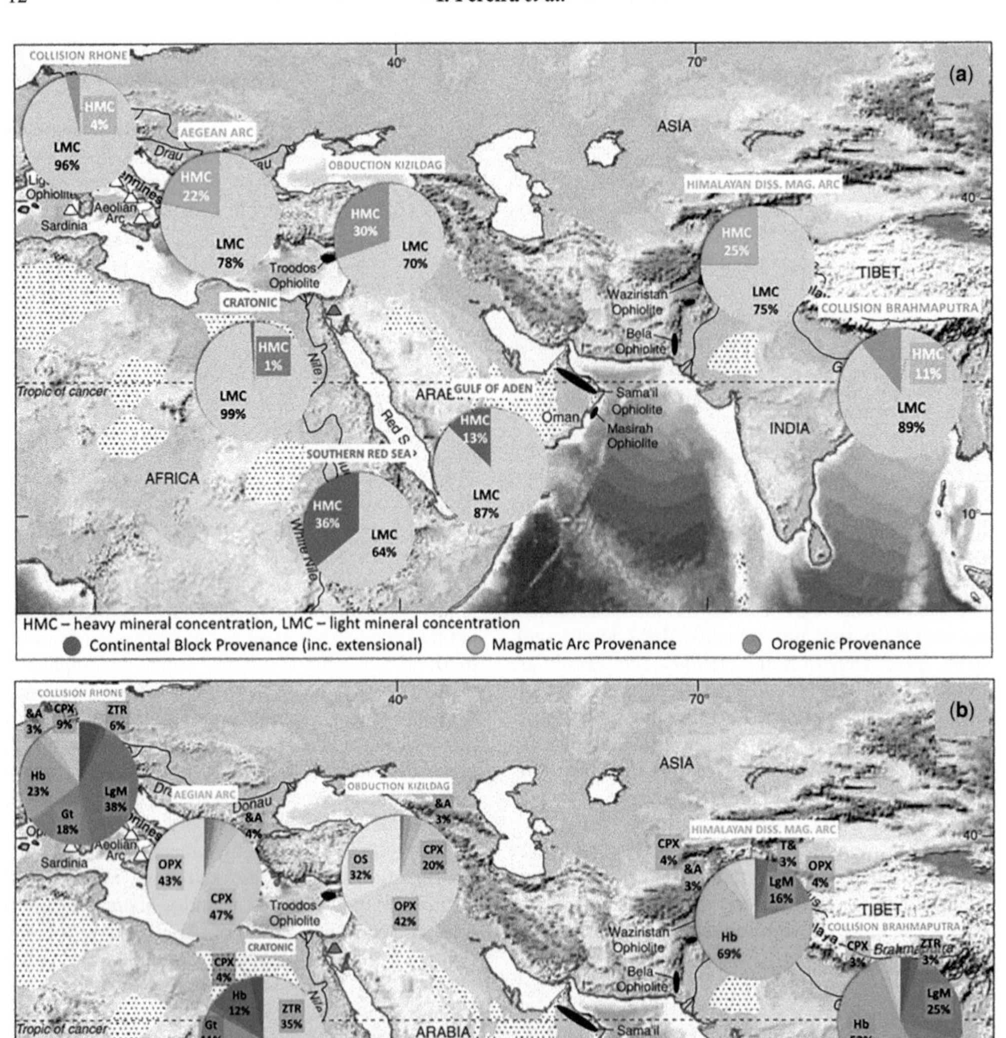

Fig. 2. Heavy mineral distribution in river sediments from different tectonic contexts, including evolved continental crust (stable or in extension) associated with magmatic arcs or from collisional orogenic domains. (**a**) Distribution of heavy ($d > 2.9$ g cm^{-3}) and light mineral concentrations; and (**b**) distribution of different heavy mineral species. Only heavy mineral groups higher than 2% are depicted in the pie charts. Source: data are from Garzanti and Andò (2007*b*), based on 95 representative samples, and projected on a map modified after Garzanti and Andò (2007*b*).

as the depositional tectonic settings as discussed above; Garzanti and Andò 2007*b*).

This implies that often these provenance-specific heavy minerals are lost after weathering and diagenesis, leading to a relative increase of other, more robust heavy minerals (e.g. zircon and rutile). When studying sedimentary rocks, and especially so when they have endured even very-low-grade

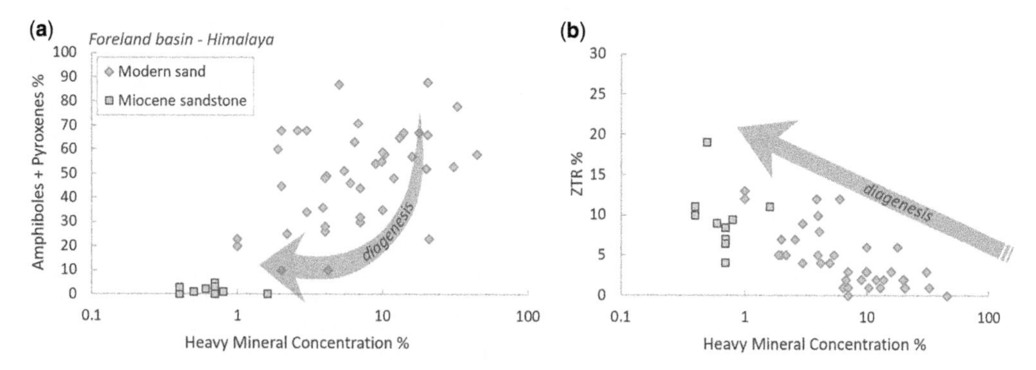

Fig. 3. Heavy mineral concentration and suites between modern sands and sandstones from the Himalayan foreland. Heavy mineral concentration correlation with (**a**) amphiboles and pyroxenes proportions and (**b**) Zircon, Tourmaline and Rutile index, to assess the effects of diagenesis on these mineral suites. Source: data are from Garzanti *et al.* (2004, 2005) (modern sands; 115 major tributaries of the Indus, Ganga, and Brahmaputra rivers) and Najman *et al.* (2003) (Miocene sandstones; 74 samples from 16 different stratigraphic units from Pakistan to Nepal). Modified from Garzanti and Andò (2007*a*).

metamorphism, these other heavy mineral species are the most reliable evidence left of sediment provenance. This is partially the reason why researchers have invested their time and efforts in the continuing development of microanalytical textural and geochemical analyses applied to these suites of heavy minerals; all the information that we are able to extract may provide the missing evidence to solve a puzzling scientific question.

Detrital heavy minerals: from source to sink

While sediments and sedimentary rocks are testimonies of the crust, the degree to which sediments truly represent the composition of the continental crust and of its lithologies is highly debatable and the focus of many studies (McLennan *et al.* 1980; Allègre and Rousseau 1984; Garzanti *et al.* 2010, 2021; Carpentier *et al.* 2013; Garçon *et al.* 2013; Greber *et al.* 2017; Klaver *et al.* 2021). Several different factors impact both the heavy minerals species and their proportion during weathering and deposition and even after deposition. In the following, we will briefly describe some of the main factors that control and impact the budget of certain heavy minerals in their host rocks, from source to sink.

Source

Source fertility: some considerations. Apatite, garnet, monazite, rutile, titanite and zircon are commonly found in different magmatic, metamorphic and detrital sedimentary rocks. However, bulk composition, conditions, as well as pressure (*P*) and temperature (*T*) or fluid conditions can influence the appearance of certain minerals in a rock, and thus given rock types can be more or less fertile to certain mineral species, impacting provenance studies' outcomes (Morton and Hallsworth 1999; Moecher and Samson 2006; Flowerdew *et al.* 2019; Chew *et al.* 2020).

Apatite and zircon fertilities in the Alps have been shown to differ by more than 3 orders of magnitude (Malusà *et al.* 2016). Moreover, if the source area is dominated by sedimentary rocks, because *P*–*T* conditions are not appropriate to produce most of these HMs, the sedimentary formations will not have an expression in a sediment derived from them (Malusà *et al.* 2016). Conversely, apatite fertility has been shown to reach its maximum in high-grade metamorphic rocks, while in zircon fertility is higher with migmatite and granitoid sources (Malusà *et al.* 2016). For a more in-depth overview of the common rock types and petrogenetic conditions that favour the growth and occurrence of these minerals, see Volante *et al.* (2022).

The decoupling of fertility can be used as an opportunity if a multiproxy mineral approach is applied in provenance studies. Flowerdew *et al.* (2019) illustrate well this source fertility bias, by finding a detrital rutile population older than detrital zircon from the same unit. Since Pb in rutile is more prone to reset than in zircon, an older detrital rutile population indicates the existence of a rutile-bearing source with no new grown zircon. Such sources could be eclogite, ultra-high pressure (UHP) rocks or high pressure (HP) amphibolites (Flowerdew *et al.* 2019).

In mafic plutonic rocks, apatite is also often present (Chew *et al.* 2020), becoming a valid alternative proxy in the absence of zircon (Moecher and Samson 2006). In the case of Si undersaturation, baddeleyite is also an alternative, although it tends to crystallize in very small crystals (<20 μm along the *c*-axis) and is therefore unlikely to be retrieved later as a detrital mineral. As reported in Volante *et al.* (2022), titanite fertility in magmatic and metamorphic rocks is controlled by bulk-rock composition (Ca/Al values) and by reactions of major rock-forming minerals, and thus typically we can find titanite in low temperature, high pressure (LT–HP) and medium temperature, medium pressure metamafic rocks, I-type granitoids and calc-silicate rocks, and scarcely in peraluminous granites, especially with higher oxygen fugacities (Frost *et al.* 2001; Kohn 2017). This considerably restricts the sources that detrital titanite may be used for. For a more detailed overview of source fertility biases in detrital studies we refer to Flowerdew *et al.* (2019).

Mineral textural arrangements. If sediments or sedimentary rocks are going to reflect the mineralogy of the source area rocks, these heavy minerals need to be released during weathering from their lithoclasts. The efficiency of this process is going to depend on the transport distance, but also on grain sizes, textures and structures of these minerals. For instance, inherited zircon has often been found to be present as an inclusion in garnet, pyroxene or amphibole in gneisses, eclogite or amphibolites (Rubatto 2017) and in biotite in granites. These zircon populations may then be missing in sediments, because they are locked in the other minerals, even though they were present in the source area rocks.

Typically, minerals such as zircon, rutile, titanite apatite and monazite are found in rocks at grain sizes smaller than 500 µm, although exceptions occur. Indeed, in low-grade metamorphic rocks, apatite can be on average 20 to 150 µm in size (Nutman 2007) and titanite between 30 and 100 µm (Cave *et al.* 2015), while monazite even at higher grades can be as small as 20 to 40 µm in diameter, even though it tends to coarsen (Bea and Montero 1999; Schulz 2021). Monazite at greenschist facies is commonly xenoblastic and may form irregular grain agglomerates, while at granulite facies will form discrete hipidioblastic to idioblastic grains (Schulz 2021). Titanite can yield larger grain sizes, commonly bigger than 100 µm and up to 1 mm (Cox *et al.* 1998; Kylander-Clark *et al.* 2008). These minerals may sit at intergranular sites or, as previously mentioned, be included by minerals that grew at a later stage during petrogenesis. It is therefore very common to find them as inclusions in other rock-forming or other accessory minerals (zircon, apatite or monazite in garnet; rutile in garnet, clinopyroxene or biotite; titanite in amphibole or plagioclase; apatite in zircon; Fig. 4a–d). Some of these grain inclusions are only a few micrometres in size (20–40 µm; e.g. Şengün and Zack 2016; Antoine *et al.* 2020), yielding different compositions and ages than the equivalent matrix minerals (e.g. Hart *et al.* 2018; Antoine *et al.* 2020). Indeed, the inherent source-dependent grain size has been shown to create a relevant impact in the detrital budget of apatite (Malusà *et al.* 2016). Apatites from plutonic rocks in the Alps are of very fine grain sizes and mostly included in feldspars (Malusà *et al.* 2016), precluding their proper release during weathering. Garnet, on the other hand, is typically larger than 0.5 mm (Fig. 4a) and frequently up to several millimetres across, and it generally occupies intergrain sites, facilitating its release as a detritus to the sedimentary environment.

As rocks break down owing to weathering, grain boundaries are usually the weakest planes through which fragmentation takes place. In the presence of mineral textural arrangements, where grain boundaries are well defined (e.g. granoblastic textures, euhedral grains in magmatic rocks), an efficient mineral segregation is possible. However, there are cases when these grain boundaries may not be the most efficient pathways. For instance, poikiloblastic garnet (Fig. 4e) or xenoblastic titanite (Fig. 4f) are often present in metamorphic rocks. As these metamorphic rocks are weathered down into smaller fragments, these grains may break and become crystal fragments of very small sizes. As this process takes place, it may be harder (1) for these minerals to survive weathering and (2) to retrieve their original source-distinctive features (e.g. garnet chemistry zoning). Thus, as these rocks and minerals are weathered into sediments, some of these populations, such as mineral inclusions or very fine-grained minerals, might be missing from the fractions commonly analysed in provenance or detrital studies.

Coarsening of metamorphic accessory minerals during increasing metamorphic grade can also impact provenance studies. Apatite under chlorite zone conditions is often <40 µm in size, while in migmatites or plutonic rocks it can reach sizes of 300–1000 µm (Nutman 2007; Marfin *et al.* 2020). This can result in undersampling of low-grade apatite in respect to higher-grade or plutonic apatite.

Weathering, transport and deposition

Selective mineral sorting of HMs during transport can take place owing to their variable densities, grain sizes and shapes. Climate may exert controls on the survivability of given HMs. Temperature and precipitation are two of the main factors that may accelerate weathering, by means of exploring weaknesses in the rocks. Low-grade (phyllosilicate-rich) metamorphic rocks and volcanic rocks show moderate weathering rates in comparison with plutonic or gneiss (quartz- and feldspar-rich) rocks, which yield much lower weathering rates (Nagy 1995; White 1995; Le Pera *et al.* 2001). In a more humid climate, increased precipitation not only accelerates water percolation, but also promotes vegetation growth and soil development, and weathering (White *et al.*

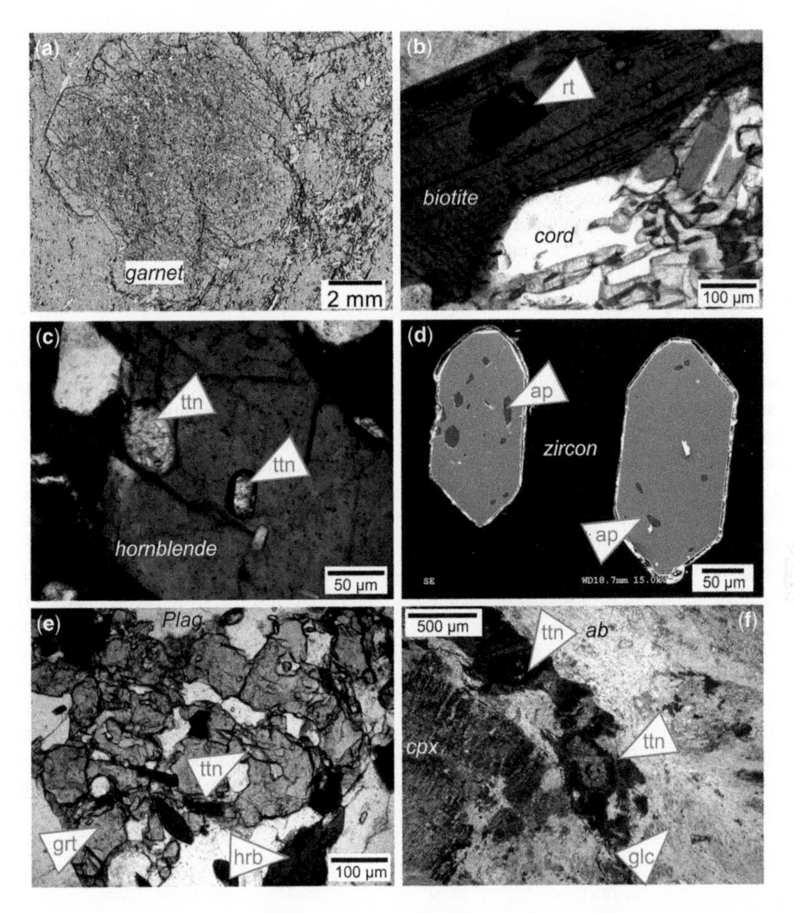

Fig. 4. Microphotographs of accessory minerals and garnet textural arrangements. (**a**) Garnet full of mineral inclusions from Zermatt; (**b**) rutile as an inclusion in biotite in LP granulite (from Madagascar); (**c**) titanite inclusions in green hornblende from *HP* amphibolite (from South Africa); (**d**) back-scattered electron image of apatite inclusions in zircon crystals from high Ba–Sr granites from Scotland; (**e**) titanite inclusions in poikiloblastic garnet in HP amphibolite (from South Africa); and (**f**) xenoblastic titanite in blueschist (from Western Alps, France). Source: (a) courtesy of T. Markmann; (d) courtesy of E. Bruand.

1999). Chemical weathering may impart strong mineralogical variations in the sediments produced, altering mineral ratios such as quartz and feldspar (Morton and Hallsworth 1999; von Eynatten *et al.* 2016). Variations in the soil pH in transport-limited erosion can lead to differential weathering of certain HMs (e.g. Morton and Hallsworth 1999). Zircon or rutile are not greatly affected by these processes, in stark contrast to minerals such as garnet, apatite and monazite, which may dissolve more efficiently under humid climates (Caracciolo 2020).

Mineral sorting by unidirectional flows: continental sandstones. In their study of the Blue Nile, Garzanti and Andò (2007*a*, *b*) showed that large modern rivers may carry large amounts of HMs, normally higher than 1%, and frequently between 10 and 40% (HMC index, Garzanti and Andò 2007*a*). Depending on the crustal exposure level of the source areas, amphiboles frequently dominate the HM budget, often followed by clinopyroxenes, while ultrastable minerals – zircon, tourmaline and rutile – are minor (ZTR index frequently <10% HMC). Such is the case of the Blue Nile River Sands (Garzanti and Andò 2007*b*), where the HM concentration in these sands increases with decreasing grain size fraction of both the bedload (transported along the bed; point-bars) and suspended sediment (found in levee deposits).

Heavy minerals are also usually more concentrated in bedloads than in the suspended sediments in the same site owing to their hydraulic behaviour (Garzanti and Andò 2007*a*). Minerals with greater potential to provide petrological information about their sources, such as amphiboles and pyroxenes, exhibit a different behaviour: clinopyroxene concentration increases with fining of sediment grain size, but usually bedloads are less concentrated in these than in suspended loads, while amphiboles are usually more concentrated in the coarser tail of both suspended and bedload sediments (Garzanti and Andò 2007*a*). Denser minerals will be concentrated in the finer sand grain-size fractions of either transport mode. Ultradense minerals such as zircon, rutile and tourmaline are even more concentrated in bedloads. In the case of the Blue Nile River, Garzanti and Andò (2007*b*) interpret these variations as a consequence of strong hydraulic behaviour-related fractionation induced by tractive currents. The proportion of clinopyroxenes, amphiboles and ZTR may then change owing to their hydraulic behaviour with very little bearing on source mineral fertility (Garzanti and Andò 2007*a*). For instance, this fractionation has been shown to have a bearing on the age distributions of detrital zircon grains in the Río Ornoco Delta, Venezuela (Ibañez-Mejia *et al.* 2018), with multiple zircon grain fractions yielding statistically dissimilar age distributions.

Mineral sorting by waves: beach sandstones. In coastal areas, sediments are transported not by a unidirectional flow, but by oscillatory flows, impacting the HM distribution in these sedimentary environments. The occurrence of HM placers at the swash zone on beaches has been ascertained to result from mineral entrainment and bedload transport, followed by efficient grain sorting through the action of surf and breaking waves (Komar 2007). Indeed, sand provenance and redistribution and HM proportions found in beaches are primarily constrained by the available sources, but also by the flow dynamics and sand transport by currents and waves. These factors tend to strongly sort minerals in a beach by density, grain-shape and size, not owing to settling velocities, but by lower entrainment of small-grained HMs (e.g. Komar and Wang 1984; Komar 2007). This mineral entrainment sorting may be efficient enough to induce higher concentration factors of ilmenite and zircon, followed by garnet and then epidote, augite and hornblende in the same sandy beach face (Komar and Wang 1984).

The interplay between source fertility, heavy mineral textures and abundances in the source rocks with weathering and hydraulic mineral sorting will result in a distribution of a suite of HMs that may change downstream (Fig. 5). One of the first aspects is HM dilution, with certain rock types providing proportionally more of some given HMs than others, which concurs with different HM hydraulic behaviour. This results in the dilution of some minor HMs (such as monazite and apatite) in respect to others (amphiboles, pyroxenes, epidotes, garnet, zircon). This implies that the information retrieved from one sample averaging an entire catchment

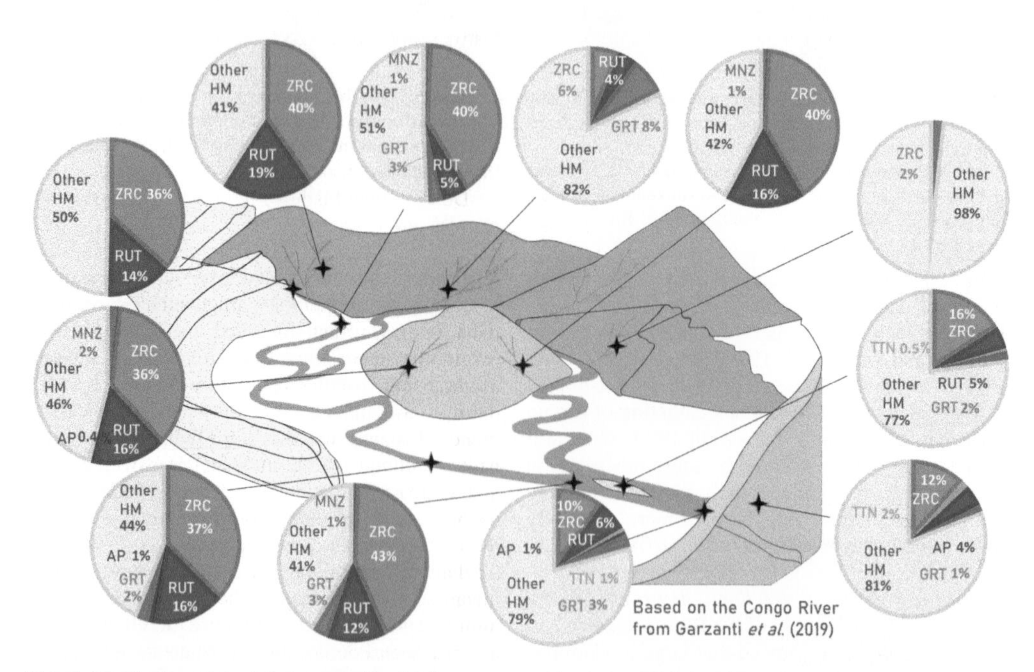

Fig. 5. Distribution of detrital zircon, rutile, titanite, garnet, apatite and monazite against other heavy minerals proportions in stream and river sediments schematics inspired by the Congo River and its tributaries. Note the changes in the heavy mineral (HM) suite distributions downstream. Source: data are from Garzanti *et al.* (2019).

area may not be entirely representative of the exposed basement and interpretations based on them may miss or emphasize particular sources. This can be illustrated by the HM species proportion distribution along the Congo River and its tributaries (Fig. 5) as reported by Garzanti *et al.* (2019).

Post-deposition: burial diagenesis, metamorphism and fluids

It has been pointed out that a relative scarcity of heavy mineral concentration in sedimentary rocks (HMC index <1) should be attributed to post-deposition alteration (Garzanti and Andò 2007*a*, *b*). Indeed, with decreasing HM proportion in sedimentary rocks, these authors observed decreasing proportions of minerals such as garnet, epidote and other high-grade minerals (staurolite, andalusite, kyanite and sillimanite) with a relative increase of the ZTR index. These differences do not reflect a change in source provenance but a direct effect of diagenesis on the stability of these minerals. Changes in the amount or proportion of HMs or the disappearance of certain minerals are influenced by fluid circulation. Therefore, to ensure proper assessments of HM suites, it is important to consider how easily fluids can flow through sediments and sedimentary rocks (Morton and Hallsworth 1999). The ability of HMs to withstand weathering and diagenesis differs, with varying degrees of resistance. The susceptibility to diagenesis is highest in titanite, moderate in garnet and lower in monazite and apatite, which are more resistant to diagenesis than

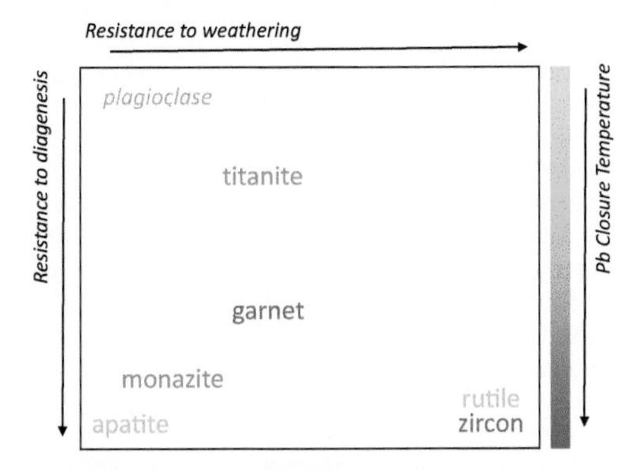

Fig. 6. Resistance to weathering combines both chemical and mechanical weathering, while resistance to diagenesis encompasses mostly burial diagenesis phenomena. Plagioclase is shown for reference. For Pb closure temperatures see caption of Figure 7. Source: Modified from Flowerdew *et al.* (2019); relative resistance data are from Pettijohn (1957), Le Pera *et al.* (2001), Morton and Hallsworth (2007) and Turner and Morton (2007).

weathering. Rutile and zircon are the most durable HMs and can withstand both processes (Fig. 6).

In the following, we briefly overview the stability of some of these detrital minerals, titanite, garnet, rutile, monazite, apatite and zircon, during post-deposition, diagenetic, metamorphic or fluid-related processes.

Titanite stability is largely influenced by bulk-rock composition (such as Ca/Al values; Force 1991). In Ca-rich rocks, ilmenite in the presence of a Ca-bearing mineral reacts to form titanite, while in Ca-poor lithologies, titanite reacts with any aluminosilicates present to form ilmenite (Frost *et al.* 2001). For this reason, titanite is more stable in metamafic rocks than in some metapelitic rocks, and more stable in metaluminous, intermediate-SiO_2 magmatic rocks than in peraluminous, high-SiO_2 granites. Frost *et al.* (2001) noted that titanite is usually present in more oxidized and hydrous conditions. This implies that titanite will be highly prone to recrystallization even at relatively low metamorphic grade (starting in the chlorite metamorphic zone) under oxidizing conditions. Indeed, titanite is so reactive in the presence of fluids that it has been used to constrain shear zone evolution histories owing to continued dissolution and reprecipitation (Papapavlou *et al.* 2018*a*; Moser *et al.* 2022). Detrital titanite and new titanite have been identified in metagreywackes at the chlorite and biotite metamorphic zone (Grapes *et al.* 2001), but generally detrital titanite grains begin to dissolve during diagenesis (Turner and Morton 2007) and new titanite tends to appear at the chlorite metamorphic zone (Schmidt *et al.* 1997) at the expense of other detrital minerals such as ilmenite and Ca amphiboles. This suggests that limited amounts of titanite will survive above these metamorphic conditions.

Detrital garnet is a common HM in modern sediments derived from a metamorphosed source terrane (Garzanti *et al.* 2014; Krippner *et al.* 2016). Owing to the generally large grain sizes in metamorphic and igneous rocks, garnet is easily found in most grain size fractions in sediments. While garnet has moderate resistance to mechanical weathering and transport, studies have reported its dissolution during burial diagenesis (Morton *et al.* 1989). This process has been linked to garnet's grossular component (Morton 1987; Morton and Hallsworth 1999) and to reactivity with organic acids produced by decomposing organic matter (Hansley 1987). Most detrital garnet will disappear during the later stages of diagenesis and the earliest stages of prograde metamorphism, reacting to produce chlorite (Adetunji and Alao-Daniel 2020). Yet some detrital garnet has been documented in Miocene sedimentary rocks from the Qaidam Basin (Tibet), attesting to its survivability, even if limited (Hong *et al.* 2020). In this case, a possible reason for the preservation of detrital garnet in these sandstones may be linked to the relatively dry climate in Tibet during deposition of the Qaidam sediments (Jian *et al.* 2014). Preservation of detrital garnet upon metamorphism is rare, but detrital garnet cores overgrown by new garnet have been reported from HP metaconglomerates from the western Alps (Manzotti and Ballèvre 2013).

Detrital rutile can break down under sub-chlorite zone conditions (Cave *et al.* 2015) but will commonly break down under upper chlorite to biotite metamorphic zone conditions. In quartz-rich protoliths, such as quartz-arenites, it may be stable at higher temperatures (>500°C) (Luvizotto *et al.* 2009; Fumes *et al.*, in progress, this volume; Pereira and Storey 2023). While Ti is generally insoluble, it can be mobilized in halogen-bearing aqueous fluids (e.g. Ayers and Watson 1993; Rapp *et al.* 2010). This implies that in the presence of such fluids, detrital rutile may dissolve at lower metamorphic grades, such as sub-chlorite to chlorite metamorphic zones, especially in the dependence of faults that may act as fluid conduits (Zeh *et al.* 2018; Pereira *et al.* 2019; Pereira and Storey 2023).

Detrital monazite is not very robust to weathering (Papoulis *et al.* 2004), even if it can be found as a detrital mineral in modern sediments (e.g. Guo *et al.* 2020; Zhou *et al.* 2020). During prograde metamorphism, under biotite zone conditions, monazite can be replaced by light rare earth element (LREE)-bearing minerals such as allanite, xenotime or apatite (Bons 1988; Rasmussen and Muhling 2007, 2009; Janots *et al.* 2008; Krenn *et al.* 2008; Volante *et al.* 2023). New monazite commonly appears after the staurolite zone, following the breakdown of LREE-bearing minerals (e.g. Akers *et al.* 1993). Although there are not many reports of detrital monazite present in metasediments under chlorite to biotite zone grades, reports of detrital monazite surviving at higher grades, such as amphibole facies, exist (Parrish 1990; Suzuki *et al.* 1994; Krenn *et al.* 2008; Iizuka *et al.* 2011).

While temperature, pressure and bulk-rock composition play an important role in the persistence and stability of monazite during metamorphism (e.g. Rasmussen and Muhling 2009), the presence of fluids imparts a major control on the stability of detrital monazite. Experiments carried out to test monazite stability in the presence of NaOH-rich fluids show monazite dissolution at 300°C, a process that becomes more efficient at increasing temperatures (Grand'Homme *et al.* 2018). At 400°C, these authors reported precipitation of new monazite, replacing a primary monazite grain after dissolution–precipitation processes. Indeed, similar textures and overgrowths have been reported in originally detrital monazite found in the low-grade Witwatersrand metaclastic rocks (Rasmussen and Muhling 2009).

While most provenance studies using modern sediments report the presence of detrital apatite (e.g. O'Sullivan *et al.* 2016; Deng *et al.* 2018; Garzanti *et al.* 2021), only a few studies have reported detrital apatite from older sedimentary rocks (>Mesozoic). In one of these studies, detrital apatite was reported from Meso-Neoproteorozoic and Cambrian un-metamorphosed to very weakly metamorphosed sedimentary successions from Siberia (Gillespie *et al.* 2018), another one from the un- to very weakly metamorphosed Mesoproterozoic Stoer Group from NW Scotland (Kenny *et al.* 2019) and another from Ordovician sandstones from Ireland (Sullivan and Chew 2020). Strongly heterogeneous Nd isotopic compositions of apatite grains from metapelitic rocks until 500–550°C support the persistence of detrital apatite during low-grade metamorphism (Hammerli *et al.* 2014), typically up to upper greenschist facies (Nutman 2007; Henrichs *et al.* 2018), upon which it fully recrystallizes as new metamorphic apatite.

Mesozoic sandstones of the North Sea show interesting apatite textures, with corroded both detrital apatite grains and authigenic apatite grains (Bouch *et al.* 2002). These authors interpreted the higher degrees of grain corrosion as a response to the presence of acidic waters, stored in the sediments of fluvial floodplains. Strong rare earth element (REE) enrichments of new authigenic apatite (>2 wt%) indicate the dissolution of other REE-bearing minerals such as garnet during its growth (Bouch *et al.* 2002), implying dissolution–reprecipitation during burial diagenesis (Morton and Hallsworth 1999; Bouch *et al.* 2002; Porten *et al.* 2015). This chemical process is enhanced by acidic fluids during low-grade metamorphism, leading to the presence/growth of different apatite generations in metasandstones.

Zircon is one of the most physiochemically robust detrital heavy minerals, which is why it is one of the few minerals that can survive multiple sedimentation cycles (Morton and Hallsworth 2007; Andersen *et al.* 2016, 2022). Additionally, zircon has been found to withstand extreme metamorphic and magmatic processes, with magmatic zircon preserved even at eclogite facies (Rubatto and Hermann 2003) or during partial melting events, with detrital/inherited zircon found both in migmatites and in their derived S-type leucogranites (e.g. Richter *et al.* 2016; Ferreira *et al.* 2019, 2022).

Yet zircon can be affected by low-temperature fluid alteration (<100°C, such as during seafloor alteration, leading to pore and micro-inclusion development (Spandler *et al.* 2004) or by late magmatic hydrothermal/aqueous fluids (Park *et al.* 2016). Through dissolution–reprecipitation, multiple generations of zircon may be identified, under either metamorphic HP (Tomaschek *et al.* 2003; Rubatto *et al.* 2008) or HT conditions (Park *et al.* 2016).

Dissolution–reprecipitation has been observed to affect detrital zircon from the very famous Jack Hills metaconglomerate (Hoskin 2005). Mushy texture overgrowths on primary magmatic zircon and highly variable and enriched trace element compositions, with low Sm/La and elevated La concentrations, enabled these to be distinguished from magmatic-type zircon. These fluid-alteration processes have been interpreted to be more pervasive in the presence of metamict, amorphous zircon (Tole 1985). While the nature of the fluids in the case of the Hadean zircon grains of Jack Hills is thought to result from magma degassing (Hoskin 2005), no constraints on the temperature of these hydrothermal fluids have been made, even though they may range from high to intermediate (such as in Park *et al.* 2016).

Now that we have reviewed the main processes that influence the presence and stability of some HMs of interest to this review, we will continue to present and discuss which tools we have at our disposal to probe them and that enhance our understanding about their source rocks and thus of the continental crust from which they derive.

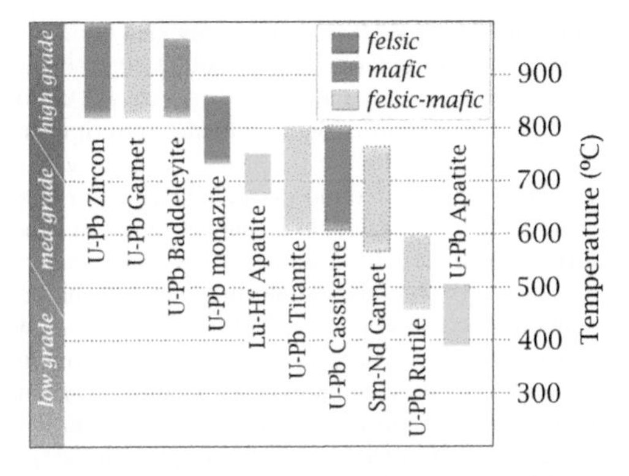

Fig. 7. Pb closure temperatures summaries for zircon, garnet, baddeleyite, titanite, cassiterite, monazite, rutile and apatite, Hf for apatite and Nd for garnet. Bars are coloured according to the higher fertility of felsic, mafic or both compositional sources for each mineral. Source: modified after Chew and Spikings (2015), Kohn *et al.* (2017), Zhang *et al.* (2017), Flowerdew *et al.* (2019), Li *et al.* (2019) and Maneiro *et al.* (2022).

What is inside your detrital petrochronology toolbox?

Detrital minerals can provide important information regarding the age and evolution of the crust they derived from, either igneous or metamorphic, including clues about the composition of these source rocks.

Different detrital minerals are produced in different rock types (i.e. felsic rocks are likely to have zircon and monazite whereas mafic rocks contain titanite and rutile). For example, the U–Pb ages obtained from detrital zircon may be different from those obtained from monazite or rutile in the same rock (Fig. 7).

Zircon has a high closure temperature that enables dating high-temperature metamorphic events or magmatic events (Cherniak and Watson 2003; Harley and Kelly 2007; Fig. 7). In contrast, titanite or rutile have much lower closure temperatures, so a lower amphibolite-facies metamorphic event will more likely be recorded by titanite and rutile (Fig. 7). Attempting to correlate terranes that have experienced moderate metamorphism targeting monazite and rutile, rather than zircon, will be more beneficial. It also means that these different accessory minerals provide complementary information to each other. Zircon provides clues about the magmatic/high-grade metamorphic history of the source terrane, whereas monazite/rutile/titanite/apatite will indicate high- to medium- or low-grade metamorphic events and the cooling history or the source region.

The following subsections outline the utility of various heavy mineral species and the strengths of using a multimineral approach when the goal is to gain greater insight about the source terrane. Zircon and garnet are reported separately from the other suite of HMs discussed in this work because they have been used extensively in provenance studies, the former owing to its physicochemical and geochronometer robustness, while the latter as a reliable tracer of sources.

Detrital zircon: the advent of in-situ U–Pb, Hf and O isotopes

Applications. Zircon is the detrital mineral that has received most of the attention in detrital studies. First, U–Pb dating was originally developed in zircon in the 1950s by Esper S. Larsen and George Tilton (e.g. Larsen *et al.* 1952; Tilton *et al.* 1955) because this mineral can incorporate significant amounts of U with virtually no Pb owing to its crystalline structure (e.g. Larsen *et al.* 1952; Finch and Hanchar 2003). This makes zircon very straightforward to date compared with other common Pb-bearing minerals (e.g. Schoene 2014). Also, zircon can survive most geological processes (e.g. Poldervaart and Eckelmann 1955; Hinton and Long 1979) because it has a high crystallization temperature and a high closure temperature for most elements and isotope systematics, which arise from very slow diffusion of most elements within its lattice (e.g. Lee *et al.* 1997; Cherniak and Watson 2003; Harley and Kelly 2007). Lastly, zircon crystallizes in a large variety of magmatic rocks, especially in granitoids that are relevant to continental crust formation and evolution (e.g. Pupin 1980; Pettingill and Patchett 1981; Amelin *et al.* 2000; Iizuka *et al.* 2005; Roberts and Spencer 2015).

The study of zircon in detrital systems has long been the focus of sedimentary studies aimed at tackling provenance issues, understanding sedimentary dispersal systems and supporting tectonic reconstructions (e.g. Tyler 1931; Twenhofel 1941; Dodson *et al.* 1988; Gehrels *et al.* 1995; Fedo *et al.* 2003). However, continental crust studies have also used detrital zircons as natural samplers of continental areas (e.g. Froude *et al.* 1983; Wilde *et al.* 2001; Iizuka *et al.* 2005; Belousova *et al.* 2010; Voice *et al.* 2011; Roberts and Spencer 2015). The widespread use of detrital zircons took place much later than zircon U–Pb development itself because of technical and interpretative issues that arose from the fact that zircon had to be analysed as individual crystals and/or batches in solution mode (e.g. Davis *et al.* 2003), hence without textural control. Another reason is that a very significant part of analysed zircons exhibited discordant ages, the origins of which were highly debated at the time (e.g. Tilton *et al.* 1957; Tilton 1960; Silver and Deutsch 1963; Wetherill 1963; Pidgeon *et al.* 1966; Steiger and Wasserburg 1966; Allègre *et al.* 1974). Some early studies of detrital zircons were nevertheless conducted to obtain average formation ages of some crustal segments (e.g. Ledent *et al.* 1964; Tatsumoto and Patterson 1964) and then discriminate populations (e.g. (Gaudette *et al.* 1981; Schärer and Allègre 1982; Davis *et al.* 1989). These issues were drastically reduced with the advent of *in-situ* analytical techniques, especially by secondary ionization mass spectrometry (SIMS/SHRIMP; e.g. Froude *et al.* 1983; Morton *et al.* 1996) and later on by laser-ablation inductively coupled plasma mass spectrometry (LA–ICP–MS/LA–MC–ICP–MS; e.g. Hirata and Nesbitt 1995; Machado and Gauthier 1996; Košler *et al.* 2002) that made it possible to analyse zircon within individual growth zones targeted using CL (cathodoluminescence) images that revealed internal textures (Tedeschi *et al.* 2023). This approach reduces ambiguities in data interpretations, especially concerning whether zircon domains crystallized under magmatic conditions, formed by metamorphism or were affected by secondary alteration (e.g. Vavra 1990; Hanchar and Miller 1993; Corfu *et al.* 2003). Currently, most detrital zircon studies are done using LA–ICP–MS because it is much more cost and time effective compared with any other U–Pb dating technique.

Dating detrital zircon crystals provides age information about the continental catchment basin, allowing us to understand the nature of detrital sediment sources and reconstruct the continental crust ages on a global scale for large river systems. However, such age interpretations have their limits since zircon can form in granitoids of various origins which, in turn, presents different meanings for crustal evolution (e.g. Stevenson and Patchett 1990; Barbarin 1999; Hawkesworth and Kemp 2006*a*, *b*). The Lu–Hf isotope system is a powerful source tracer that can help discriminate zircon formed from granitic magmas that originate from crustal reworking from those corresponding to juvenile addition (mantle extraction; e.g. Patchett *et al.* 1981; Amelin *et al.* 2000). Consequently, the Lu–Hf isotope system has rapidly been coupled with U–Pb dating of detrital zircon to discuss the origin of these zircon crystals (e.g. Stevenson and Patchett 1990; Amelin *et al.* 1999; Iizuka *et al.* 2005; Belousova *et al.* 2010). This has led to a boom in studies involving coupled U–Pb and Lu–Hf measurements in zircon that started in the early 2000s and coincided with the development of LA–MC–ICP–MS measurement of Lu–Hf isotopes in zircon (e.g. Griffin *et al.* 2000; Machado and Simonetti 2001). This approach is particularly useful for detrital zircon studies because they require a very large number of crystals to be analysed for the sake of representativeness and robustness of statistics (Dodson *et al.* 1988; Vermeesch 2004; Iizuka *et al.* 2013). Since the 2000s, analytical and methodological improvements have made Lu–Hf isotope data in zircon more reliable and their downstream interpretations more robust (e.g. Gerdes and Zeh 2009; Fisher *et al.* 2011, 2014; Guitreau and Blichert-Toft 2014; Spencer *et al.* 2020).

Hafnium isotopes alone can only provide a first-order estimation of the nature of the source from which granitoids formed (i.e. mantle or crust, e.g. Roberts and Spencer 2015; Payne *et al.* 2016). Therefore, stable O isotopes are often used to complement U–Pb and Lu–Hf isotopes to place further constraints on granite origin (e.g. Kemp *et al.* 2007; Dhuime *et al.* 2012; Iizuka *et al.* 2013) because O isotopes essentially allow discrimination between juvenile magmas and those that interacted with or were derived from partial melting of sediments formed by low-temperature aqueous alteration of crustal lithologies (e.g. clay-rich; Valley 2003). Although some ambiguities can remain in data interpretations (Couzinié *et al.* 2016), this three-fold approach makes data interpretation more reliable and supposedly converge towards realistic geological hypotheses. More information about data interpretation in this context can be seen in Volante *et al.* (2022) and the following section.

Zircon can incorporate large amounts of trace elements (e.g. Hoskin and Schaltegger 2003) that provide information on various aspects of zircon petrogenesis and preservation over time. The most common type of information targeted by trace element investigation is (1) the origin of zircon (i.e.

magmatic, metamorphic or hydrothermal), (2) the parental rock type (mafic, felsic or strongly alkaline), (3) the parental granite type (A-, S-, I-(Anorogenic, Sedimentary, Igneous) or tonalite, trondhjemite and granodiorite (TTG), sometimes referred to as M), (4) the pristineness (alteration, weathering or primary feature), (5) the oxidation state of parental magmas and (6) the crystallization temperature. More details about the use of trace elements and its limitations are given in Volante *et al.* (2022). The first point can be tackled using Th/U ratios, trace element patterns and absolute concentrations (e.g. Hoskin and Ireland 2000; Rubatto 2002; Hoskin 2005; Guitreau *et al.* 2019). The second and third points are commonly addressed using trace element ratios and concentrations (e.g. Grimes *et al.* 2007, 2015; Bouvier *et al.* 2012; Tang *et al.* 2021; Guitreau *et al.* 2022). The fourth point is commonly tackled using REE patterns and absolute elemental concentrations (e.g. Bouvier *et al.* 2012; Zeh *et al.* 2014; Pidgeon *et al.* 2019; Ranjan *et al.* 2020; Guitreau *et al.* 2022). The fifth point is achieved by Ce anomaly determination (e.g. Trail *et al.* 2012; Smythe and Brenan 2016). The final point is done using Ti-in-zircon thermometry (e.g. Watson *et al.* 2006; Cherniak and Watson 2007; Schiller and Finger 2019).

Over the last two decades, detrital zircon U–Pb, Lu–Hf and/or O isotopes, as well as trace element concentrations have shed new light on the formation and evolution of continental crust with unprecedented temporal resolution. For instance, they allow the sizes of catchment basins of major rivers to be assessed (e.g. Iizuka *et al.* 2010), give a first order vision of crustal segment evolutions (e.g. Kemp *et al.* 2006), provide an inventory of felsic lithologies exposed in and around crustal terranes (e.g. Zeh *et al.* 2008), and, more importantly, allow crustal growth to be assessed globally (e.g. Belousova *et al.* 2010). Such studies have also been useful to trace different types of orogens (Collins *et al.* 2011), to determine the tectonic setting of detrital sediment deposition (Cawood *et al.* 2012) as well as to date the onset of modern plate tectonics (Dhuime *et al.* 2012). The transcription of the global detrital zircon age record into crustal growth is very challenging because it is episodic. The actual origin and meaning of this episodicity is greatly debated between a primary feature and a preservation bias, both with many subtleties (Campbell and Allen 2008; Hawkesworth *et al.* 2009; Condie and Aster 2010; Guitreau *et al.* 2012; Parman 2015; Spencer 2020).

Detrital zircons are also our only window into the Hadean Eon with zircon providing extremely useful information about this time period (Froude *et al.* 1983; Compston and Pidgeon 1986; Mojzsis *et al.* 2001; Wilde *et al.* 2001; Hopkins *et al.* 2008; Kemp *et al.* 2010; Bell *et al.* 2014; Cavosie *et al.* 2019; Harrison 2020). The major conclusions about the Hadean gleaned from zircon is the presence of granites (*sensu largo*) and, probably, an Hadean protocrust *c.* 150 myr after the formation of the Solar System. Detrital Hadean zircons have various affinities and are sometimes interpreted differently by various authors (e.g. Mojzsis *et al.* 2001; Wilde *et al.* 2001; Cavosie *et al.* 2004; Watson *et al.* 2006; Grimes *et al.* 2007; Bouvier *et al.* 2012; Burnham and Berry 2017; Laurent *et al.* 2022; Borisova *et al.* 2021), but most studies are consistent with the existence of hydrous superficial cycles that involved liquid water before 4.3 Ga. These conclusions sharply contrast with the hellish vision based on the name of this Eon and considerations about energy involved in Earth's accretion.

Some 'thorns' in detrital zircon data acquisition and handling. Biases that are common to all the HMs in this work are reviewed in the section 'The main biases affecting detrital heavy minerals'. In the following we focus on those biases that are more inherent to zircon. Most detrital zircon studies date a large number of zircon grains only once and then apply a discordance filter (e.g. Belousova *et al.* 2010; Dhuime *et al.* 2012). These studies commonly conduct texture-controlled measurements to avoid zircon crystals that do not look igneous and pristine, arguably assuming that they will probably provide concordant ages (e.g. Connelly 2001). However, when dealing with ancient crystals, textures revealed by CL and/or BSE images are not always a reliable test for U–Pb concordance (e.g. Guitreau *et al.* 2018). A simple methodological approach by Guitreau and Blichert-Toft (2014) suggests analysing zircon multiple times instead of once to avoid discarding populations because of discordant data. Trace element data are useful complements to textures to determine whether or not a zircon is likely to have remained pristine after its growth (e.g. Hoskin and Schaltegger 2003; Bell 2016; Pidgeon *et al.* 2017; Guitreau *et al.* 2022; Tedeschi *et al.* 2023).

Most studies focused on detrital zircon involve isotopic analyses of U–Pb, Lu–Hf and O isotope systems to discuss the formation timings and sources of crustal rocks. Some limitations in data interpretation exist owing to the out-of-context nature of detrital zircon, which will be discussed in the following paragraphs. Figure 8 illustrates the most typical ways that Hf isotope data can be interpreted together with U–Pb ages to make quantitative estimates regarding crustal growth. Once a zircon is dated and Hf isotopes corrected for radiogenic ingrowth, we obtain the green point in Figure 9, which is the initial ε_{Hf} that the zircon recorded when it formed, hence, at the age given by U–Pb isotopes. If we assume that the age is perfectly concordant the Hf

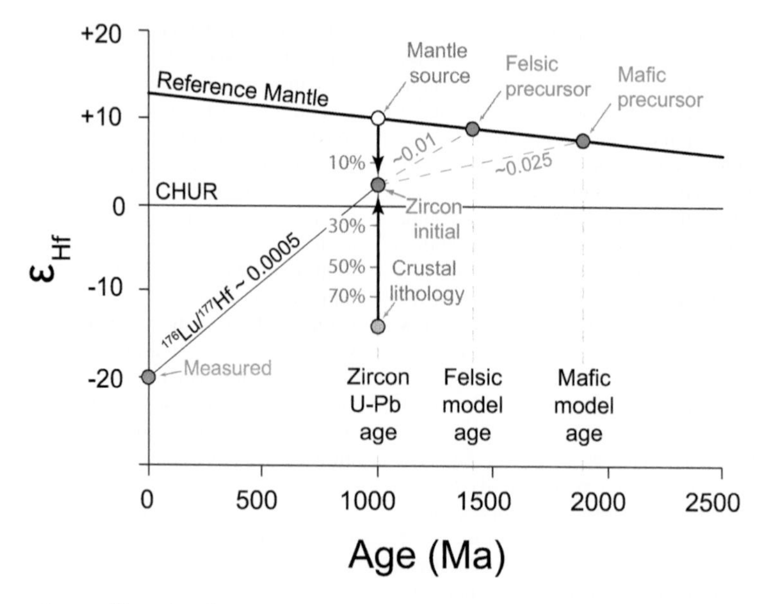

Fig. 8. ε_{Hf} v. age diagram illustrating the three main ways to interpret the zircon Hf isotope signature. The zircon Lu–Hf isotope measurement provides present-day information (blue spot) that has to be corrected for radiogenic ingrowth to obtain the initial Hf isotope signature (green spot) at the age of 1000 Ma obtained using U–Pb geochronology. This correction is done using the measured $^{176}Lu/^{177}Hf$, which is commonly very low and close to 0.0005 (e.g. Amelin *et al.* 2000; Guitreau *et al.* 2012). Percentages in grey represent the mass proportion of the crustal lithology in a mixture between a mantle-derived melt and a crustal lithology. Values in red correspond to the $^{176}Lu/^{177}Hf$ typical of felsic and mafic lithologies although they commonly exhibit a range that can vary significantly. See text for more details.

isotope signature is reliable. There are three ways to interpret this signature:

(1) The zircon ε_{Hf} is the same as that of its source and because it is positive, it can reflect a juvenile addition to the crust. Consequently, the entire mass of the zircon parental magma was extracted from the mantle and added to the crust.

(2) This ε_{Hf} signature is the consequence of mixing between a mantle-derived melt (white spot in Fig. 8) and a crustal lithology (grey spot in Fig. 8) either through contamination or actual

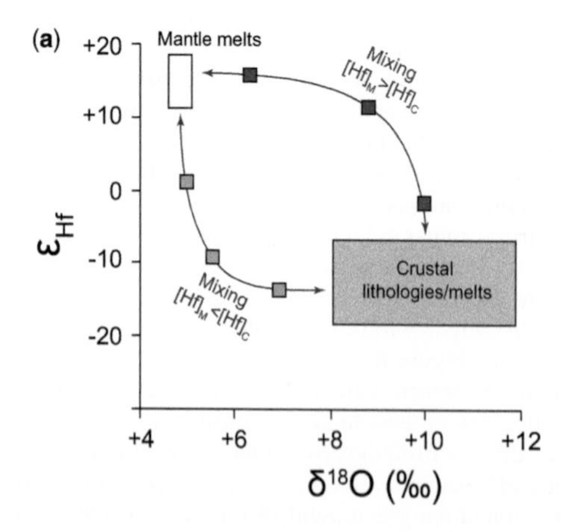

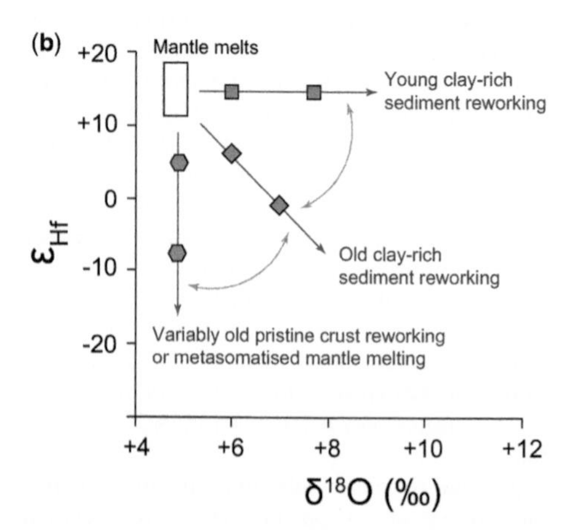

Fig. 9. ε_{Hf} v. $\delta^{18}O$ diagram illustrating two contrasting ways to interpret zircon data. The inset (a) corresponds to the approach commonly adopted for igneous crystals and (b) for detrital ones. $[Hf]_M$ and $[Hf]_C$ in inset (a) correspond to the concentration of Hf in mantle-derived melts and the crustal lithologies/melts, respectively.

magma mixing. In this case, the ε_{Hf} reflects a hybrid source and not all of the zircon parental magma is extracted from the mantle and added to the crust, more likely 80% in this example.

(3) The zircon ε_{Hf} signature reflects the reworking of a pre-existing crustal lithology, named precursor, that was extracted earlier on from the mantle and evolved within the crust for some time before being partially melted to give rise to the zircon parental magma. In this case, the entire mass of zircon parental magma is considered as redistribution of matter within the crust with nothing extracted from the mantle.

Thus, this granite-forming event is not a crustal growth event. Depending on whether this precursor is mafic or felsic, the crustal residence time required to reach the same isotopic composition would vary. This is illustrated in Figure 8 with the two red spots that show that a mafic precursor would require more time than a felsic precursor for its ε_{Hf} signature to drift from that of its source (reference mantle). The precursor reworking idea corresponds to the concept of two-stage model ages (e.g. Hawkesworth and Kemp 2006*b*) that has been applied to most detrital zircon studies and coupled U–Pb-Lu–Hf isotopes in zircon studies. We would like to stress here that calculating two-stage model ages is therefore making a strong petrological assumption on the source of zircon parental magma which, in this case, is a single lithological type with a single age. Although possible, this is a one-sided vision of granite formation that does not encompass all known granite formation processes and sources (Nédélec and Bouchez 2015) and, thus, most likely results in ages that have no geological meaning (e.g. Arndt and Goldstein 1987; Vervoort and Kemp 2016).

To further illustrate the interpretative biases that may affect crustal growth reconstructions from the detrital record, we present in Figure 9 how one can interpret coupled Hf and O isotope signatures in igneous and detrital zircon.

Figure 9a shows a schematic distribution of datapoints for magma mixing or crustal contamination depending on the respective concentrations of Hf between a mantle-derived melt and a crustal lithology/crustal-derived melt. Zircon ε_{Hf} and $\delta^{18}O$ data should define hyperbolas. There are numerous examples of hybrid source involvement in granite formation that exhibit such signatures or similar mixing hyperbola between isotopic systems (e.g. Depaolo 1981; Foland and Allen 1991; Kemp *et al.* 2007; Gaschnig *et al.* 2011; Couzinié *et al.* 2016). Figure 9b presents the way coupled ε_{Hf} and $\delta^{18}O$ are interpreted in detrital zircon. Note the often sharply different interpretations that could arise from considering data points in Figure 9a but interpreted using the approach in Figure 9b. For instance, if the two pale green points close to the mantle end member in Figure 9a are interpreted using the approach of Figure 9b, one could conclude that zircon data provide evidence for reworking of ancient crust, implying no addition to the crust, when the data could in fact indicate a major growth event with a minor crustal involvement. Unfortunately, the out-of-context nature of detrital zircon renders accurate and objective interpretations complicated to make, and thus one is left with first-order interpretations. Despite the above discussed potential sources of bias, the detrital zircon record is still very rich in information to uncover and a major source of improvement for the future is very likely to come from the use of combined mineral records (e.g. zircon, apatite rutile and monazite) to reliably unravel continental crust evolution.

Detrital garnet; your favourite metamorphic mineral as sand

Applications. The use of garnet in provenance investigations started with Morton (1985) and has significantly increased in the past 20 years. This special interest in detrital garnet derives not only from its relative abundance in sediments derived from metamorphic belts, presenting relative stability to weathering and diagenesis (Morton and Hallsworth 2007) but also, and foremost, because of its observable characteristic variations in major and trace element compositions. These variations reflect the bulk-rock compositions and therefore they can bring hints about the composition and *P–T* evolution of the source area, allowing us to unravel tectonic settings (Spear and Kohn 1996; Mange and Morton 2007; Baxter *et al.* 2013). Thus, detrital garnet is the object of several investigations, used as a proxy in provenance studies but also as a mineral pathfinder, such as in the exploration for potential diamond-bearing rocks, where diamond is usually found in association with pyrope (Dawson and Stephens 1975; Schulze 2003; Grütter *et al.* 2004; Hardman *et al.* 2018). In the following, we are going to explore why and how detrital garnet can be part of detrital studies tackling the evolution of the crust.

Garnet has been considered the ‘ultimate petrochronometer’ (Baxter *et al.* 2017) owing to the wide range of solid solution compositions, which in turn reflects the variable thermodynamic conditions in which they have been formed. Garnet's ‘superpower’ in igneous and metamorphic rock studies derives from the possibility to correlate its petrological context to time constraints, either directly by isotopic dating (e.g. Millonig *et al.* 2020; Simpson *et al.* 2021) or by correlating garnet growth with time constraints from accessory minerals via REE composition (e.g. monazite, Rocha *et al.* 2017;

zircon, Tedeschi *et al.* 2017). Notwithstanding, garnet usually hosts mineral inclusions (porphyroblastic crystals) and preserves chemical zoning (Baxter *et al.* 2017 and references therein), which can be used to constrain *P–T–t* information from one single mineral, making garnet particularly useful for when it is found as a detrital mineral in sedimentary rocks.

Discrimination diagrams based on garnet chemistry. Most detrital garnet studies employ the use of geochemistry-based discrimination schemes in order to assign the garnet grains to groups of host rocks (Morton *et al.* 2004; Tolosana-Delgado *et al.* 2018). Such discrimination schemes mostly rely on ternary diagrams used to classify among the garnet solid solution end members – almandine, pyrope, spessartine and grossular (e.g. Aubrecht *et al.* 2009; Stutenbecker *et al.* 2017). Besides the advances from the use of CaO- and Cr_2O_3-based discrimination for mantle-derived garnets (Grütter *et al.* 2004) and the step-wise classification from Suggate and Hall (2014), which incorporates more chemical elements and garnet endmembers, some issues remained overlooked. These were mainly the overlap of discrimination fields in the classification plots, as well as the limited use of variables and 'sensible multivariate statistics'. Propelled by these issues, Tolosana-Delgado *et al.* (2018) proposed a multivariant discrimination scheme based on a comprehensive database, comprising different garnet-bearing rocks, and by applying sensible multivariate statistics that considered the compositional nature of garnet analyses. The outputs, in turn, reflect the probabilities of association with a specific host-rock group instead of strict values/boundaries. The oxides SiO_2, Al_2O_3, MgO, CaO, MnO and FeO, together with TiO_2 and Cr_2O_3, are used to discriminate among eclogite-facies (A), amphibolite-facies (B), granulite-facies (C), ultramafic (D) and other igneous (E) host rock groups. In the classification, the first step uses Cr_2O_3 to distinguish garnet crystals from ultramafic sources, followed by the discrimination of garnets from felsic plutonic rocks. These discriminations yield a >90% classification success rate. However, the discrimination becomes less robust for discriminating garnet grains from different metamorphic facies with a 50–90% success rate, depending on the facies considered. These can be explained by two main factors. The first is the difficulty of accurately discriminating different types of garnet derived from mafic protoliths, owing to the compositional similarities between their protoliths (e.g. basalt). Better results are obtained for metamorphosed sedimentary rocks owing to their compositional variability in comparison with mafic igneous rocks (Krippner *et al.* 2014; Tolosana-Delgado *et al.* 2018). The second is that garnet chemistry is sensitive to the *P–T* path described by a rock, and because it often crosses different metamorphic facies, such as eclogite, amphibolite and granulite, during its evolution (e.g. Ganade de Araujo *et al.* 2014; Borges *et al.* 2023), it becomes more difficult to use garnet chemistry. The observed issues in determining metamorphic facies highlight the need to use probabilities rather than sharp discriminations between the metamorphic proto-sources (Tolosana-Delgado *et al.* 2018). More robust results, with correlations between 84 and 95%, were obtained by Schönig *et al.* (2021*a*, *b*) using a supervised random forest machine-learning based scheme. The discrimination scheme enables the host-rock settings, metamorphic facies and compositions to be distinguished. In order to improve the classification, their work employed a robust dataset of *c.* 13 000 garnet data considering also host-rock types and compositions that had not been considered in previous discrimination schemes. For instance, Tolosana-Delgado *et al.* (2018) considered igneous and ultramafic rocks, and amphibolite-, granulite- and eclogite-facies metamorphic rocks, while Schönig *et al.* (2021*a*, *b*) also included metasomatic, blueschist/greenschist-facies, metasedimentary, alkaline and calc-silicate rocks.

Čopjaková *et al.* (2005) introduced the use of trace elements analysed with LA–ICP–MS in garnet, by comparing them with the major elements, including X-ray compositional maps. Trace elements were successfully employed to distinguish garnet grains derived from granitoids and metapelites of different metamorphic degrees, with implications for the tectonic evolution of the Qaidam basin (Hong *et al.* 2020).

The use of garnet geochemistry discrimination may further provide indirect time constraints on sedimentation when applied as a proxy of proto-sources whose ages are well known. Stutenbecker *et al.* (2019) used unusual grossular and spessartine contents in foreland basin deposits to discriminate among Alpine source rocks. Detrital garnet composition allowed those authors to identify an unnoticed younger provenance shift related to the erosion of granites from the external crystalline massifs.

Detrital garnet dating: the awakening to a higher throughput. Because garnet usually grows at temperatures between *c.* 400 and 700°C (Caddick and Kohn 2013; Baxter *et al.* 2017), but persists up to *c.* 1000°C (e.g. Rocha *et al.* 2017; Motta *et al.* 2021), it can provide complementary chronological information to that obtained from zircon or monazite (see sections 'Detrital Zircon: the advent of *in-situ* U-Pb, Hf and O isotopes' and 'Calling out a task force: using Monazite, Titanite, Rutile and Apatite in detrital studies'). As pointed out by Maneiro *et al.* (2022), garnet maintains a record of peak and prograde metamorphism, depending on the subsequent *P–T* conditions (Caddick *et al.* 2010). Indeed, the Pb diffusion temperature is comparable with that of zircon, while the Nd closure temperature is slightly lower, but comparable with the titanite or monazite Pb closure temperatures (Dutch and Hand 2010; Fig. 7 in the section 'What is inside your detrital petrochronology toolbox?').

Despite representing a significant advance in provenance investigations, direct detrital garnet dating remains minimally employed (Oliver *et al.* 2000; Maneiro *et al.* 2019, 2022). This is mainly due to the complexity of applying the Sm–Nd method, the detrital garnet most commonly employed to date. The dating process is time consuming owing to: (1) the small size and zoning of the sample; (2) the process (dissolution) required to use thermal ionization mass spectrometry; and (3) the difficulty of determining the whole-rock composition to obtain the second point in the isochron regression (Maneiro *et al.* 2019, 2022). Regarding this last point, Oliver *et al.* (2000) used the sedimentary whole rock, whereas Maneiro *et al.* (2019) used mineral inclusions within garnet assuming them to be a proxy of the isotopic composition of the source. Since the sediment contains detritus from several sources and it can be affected by diagenesis, it is less likely to represent the source. Instead, the method used by Maneiro *et al.* (2019) allowed the identification of metamorphic ages in the French Broad River (USA) younger than those recorded by monazite and similar to ages recorded by some zircon rims, providing new constraints to the region. The recent developments and advances in *in-situ* Lu–Hf and U–Pb garnet techniques (Millonig *et al.* 2020; Simpson *et al.* 2021) might reveal a promising future for garnet-based detrital investigations. Indeed, we are now witnessing the first successful results of applying U–Pb and Lu–Hf single grain detrital garnet dating using laser ablation, single and triple quadrupole ICP-MS, respectively (Mark and Stutenbecker 2022).

Detrital garnet – a geological pandora's box. When it comes to the investigation of inclusions in detrital heavy minerals, garnet has a leading role (Schönig *et al.* 2018, 2022). As a common phase in high- to ultrahigh-pressure rocks, inclusions in garnet may provide information regarding deep crustal levels in source areas (Fig. 10), which are of extreme importance to geodynamic interpretations, but often no longer exposed at the surface (e.g. Schönig *et al.* 2018; Baldwin *et al.* 2021). Mineral inclusions in garnet have mainly been investigated using Raman spectroscopy (see Schönig *et al.* 2022 for a review). One application is to robustly identify coesite or diamond, which may offer insights about metamorphic conditions (Schönig *et al.* 2022). Alone, these minerals are not definite evidence of UHP metamorphism and do not indicate unambiguously the protolith of garnet; detrital Cr-poor garnets originating in the mantle can also contain coesite or diamond and be misinterpreted as UHP relicts. In such cases, a multivariant discrimination scheme is paramount (Hardman *et al.* 2018). Monomineralic coesite inclusions in detrital garnet were first reported by Schönig *et al.* (2018) in sandstones from the Western Gneiss Region of Norway, as well as inclusions of kyanite, omphacite, diopside, enstatite, amphibole and epidote. The inclusions combined with garnet geochemistry demonstrated that garnet can be used as a proxy for (U)HP provenance (see an example in Fig. 10). A similar approach was employed by Baldwin *et al.* (2021) using beach sands from Papua New Guinea to reinforce evidence from the youngest UHP metamorphism on Earth. These authors further applied elastic thermobarometry on quartz and zircon inclusions in garnet to reconstruct the rocks' burial trajectories. Furthermore, the presence of graphite, carbonates and CO_2 inclusions in garnet are interpreted as evidence of deep carbon recycling. In the Saxonian Erzgebirge area, Schönig *et al.* (2019, 2020, 2021*b*) identified monomineralic coesite, bimineralic coesite + quartz and diamond inclusions. The diamond occurs as monomineralic or part of polyphasic inclusions with quartz, plagioclase, graphite, rutile, apatite and phyllosilicates. Schönig *et al.* (2020) describes 'melt' inclusions with cristobalite, kokchetavite and kumdykolite, which is in agreement with the nanogranitoid inclusions described by Stöckhert *et al.* (2001) and Borghini *et al.* (2018, 2020) in the Erzgebirge and Saxon granulite massifs. This highlights the capacity of garnet to retain melt inclusions even after weathering from its protolith. Nevertheless, the presence of numerous inclusions also plays a role in the grain-size distribution of detrital garnet (Schönig *et al.* 2021*b*). With increasing mineral inclusions, garnet tends to become more mechanically fragile and to fracture. This, in turn, favours fluid percolation, and the transformation of mineral inclusions, such as coesite to quartz, promoting garnet disintegration and reduced weathering survivability.

Caveats specific to detrital garnet applications. Schönig *et al.* (2018) highlighted that a given garnet composition can correspond to more than one *P–T* stage (cf. Lanari *et al.* 2017) and that two main processes that affect garnet chemistry should be acknowledged: changes in the bulk rock owing to dissolution and crystallization of garnet (e.g. Marmo *et al.* 2002; Lanari and Engi 2017) and diffusion in garnet when it undergoes high-temperatures (e.g. Caddick *et al.* 2010; Carlson 2012; Kohn and Penniston-Dorland 2017). Therefore, care must be taken when using detrital garnet compositions to infer *P–T* conditions. Furthermore,

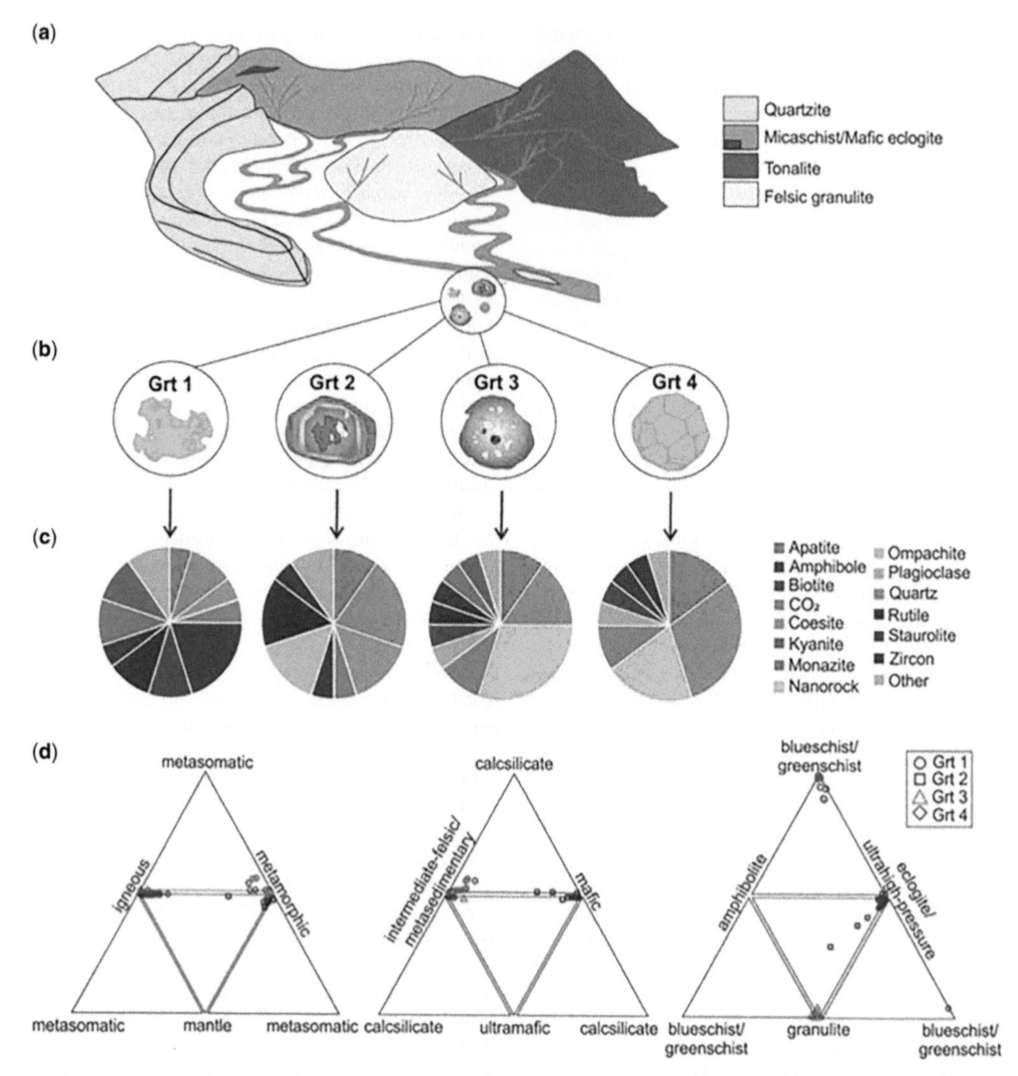

Fig. 10. Theoretical example of a detrital garnet investigation workflow. A river sediment (**a**) sample contains four different garnet populations (**b**). The investigation of the inclusions (**c**) content together with the machine learning (**d**) approach applied to the major element compositions from Schönig *et al.* (2021*a*) enables the different settings, rock composition and metamorphic facies to be distinguished. Correlations with proto-sources in the area identify the probable sources. In this example, the area is related to an active margin evolution, with subduction under ultra-high pressure conditions, the formation of blueschists and eclogites and a continental magmatic arc.

despite recent advances, discrimination diagrams show significant misclassification (lower estimates) for intermediate to felsic rocks and metasediments metamorphosed under eclogite to UHP conditions (Schönig *et al.* 2018). Thus, correlations with UHP rocks should consider both mineral chemistry and inclusion compositions and/or raman barometry applied to quartz (Enami *et al.* 2007; Zhong *et al.* 2019; Alvaro *et al.* 2020) and now also zircon (Zhong *et al.* 2019; Campomenosi *et al.* 2021) as an independent pressure estimate. While inclusions allow good results for Sm–Nd isochron production, owing to the characteristic chemical zoning and diversity in inclusion distribution, establishing the mineral assemblages in equilibrium might be challenging and, therefore, it represents a potential source of uncertainty. The recent advances in Lu–Hf and U–Pb *in-situ* garnet dating may allow those assumptions to be overcome, but, in turn, might introduce new issues, such as mixed ages or poor age resolution (Millonig *et al.* 2020). It is worth highlighting that *in-situ* dating methods with higher uncertainties can complicate distinction between close-in-age metamorphic events, as well as determination of the rates of processes (e.g. subduction, exhumation). However, this might not constitute a major issue for detrital applications.

Calling out a task force: using monazite, titanite, rutile and apatite in detrital studies

As seen earlier in this section, other U-bearing accessory minerals can be targeted for detrital studies because they can be dated using U–Th–Pb isotopes and provenance information may be obtained from their trace element compositions. Monazite, titanite, rutile and apatite can provide insight into both magmatic and metamorphic processes, ranging from high- to low-temperature and from low- to high-pressure conditions. Therefore, evaluating detrital populations of these minerals should lead to a more complete reconstruction of the composition and tectonometamorphic evolution of crustal terranes and even of the continental crust on a global scale. This is well demonstrated in O'Sullivan *et al.* (2016), where the lack of late Paleozoic detrital zircon ages in riverbed sediments from the French Broad River (USA) is shown, although sediments were drained from an area affected by significant tectonometamorphism at that time. This study shows that detrital zircon may be 'blind' to significant geological events that affected source areas and, thus, does not provide a complete record of past geological events. Despite a lack of detrital zircon recording the latest, magmatic-poor tectonic events, abundant detrital apatite and rutile found in the same sediments revealed very few older inherited ages and very clear Paleozoic detrital populations that match ages related to the three main Paleozoic tectonometamorphic events recorded by the source basement. Unlike zircon, detrital monazite collected from the same stream sediments (Hietpas *et al.* 2010) revealed late Paleozoic ages, even though proportionally less frequent than early Paleozoic ages. This testifies to the fact that even though more monazite may be produced within magmatic rocks, some metamorphic monazite was also produced during the late Paleozoic tectonometamorphic events. Consequently, monazite is also a very useful tracer of metamorphic events. Indeed, early work on detrital monazite already combining single grain U–Pb and Sm–Nd isotopic systematics was employed to better trace sediment provenance (Ross *et al.* 1991).

How can we use these suites of minerals to better understand the geodynamic evolution of the continental crust? Which tools to use and what caveats are there when using these other proxies? In the following paragraphs, we summarize the existing tools that have been applied to apatite, monazite, rutile and titanite.

How can detrital apatite, monazite, rutile and titanite ages untangle geological complexities? Many recent provenance studies rely on the acquisition of U–Pb ages from multiple minerals to reconstruct the geology of the source area and place stronger constraints on sediment provenance (O'Sullivan *et al.* 2016; Gaschnig 2019; Guo *et al.* 2020; Zhou *et al.* 2020; Arboit *et al.* 2021). Ages provide information about magma crystallization or crustal reworking through metamorphic events. As we saw earlier in this section, different HMs provide ages that may correspond to different petrogenetic stages (early or late crystallization; prograde, peak or retrograde/cooling) and thus, together, they provide a clearer and more complete scenario. In Figure 11 we explore these, by looking at detrital ages retrieved from sediments collected from two rivers in the Himalaya, one draining an area in Lhasa and another close to Everest – Pumchu (data are from Guo *et al.* 2020).

Without any additional information (e.g. trace elements and mineral inclusions) interpretations about source area can only be made on the basis of detrital ages and by inferring the probable nature of the source rock based on accessory mineral fertilities (as discussed in the section 'Source'). In this example from Guo *et al.* (2020), the surface geology is well exposed (Fig. 11a–c), and a nearly direct comparison can be made between the detrital record and the associated bedrock geology. The Lhasa Terrane preserves a different evolution than that of the Tethyan and Greater Himalaya Sequences. In Lhasa, zircon and titanite main age peaks reflect the onset of the Himalayan collision (Fig. 11d). These correspond to the ages of the Gangdese arc batholiths that are widespread in the source area (Guo *et al.* 2020). Zircon also yields other older age peaks that probably correspond to detrital zircon grains hosted in the Paleozoic metasedimentary rocks (Guo *et al.* 2020). Monazite ages record a Meso-Cenozoic evolution, yielding a main age peak younger than 20 Ma (Fig. 11d) that can imply either magmatic or metamorphic sources. Interestingly, detrital rutile ages reveal the existence of protracted metamorphism from modern (Himalayan collision) to late Paleoproterozoic (Fig. 11d) that may be indicative of metamorphism in the source area and/or recycling of older detrital rutile components hosted in the Paleozoic metasedimentary rocks (Guo *et al.* 2020).

Conversely, in the Pumchu area these detrital minerals record different main events/bedrock geology (Guo *et al.* 2020). Zircon reveals a more protracted evolution, with ages spreading between the Paleozoic and late Archean (Fig. 11e). Detrital rutile indicates a nearly bimodal age distribution between a recent metamorphic event (<50 Ma) and a late Neoproterozoic to Cambrian event (Fig. 11e). Detrital monazite yields a main age peak at *c.* 30 Ma, similar

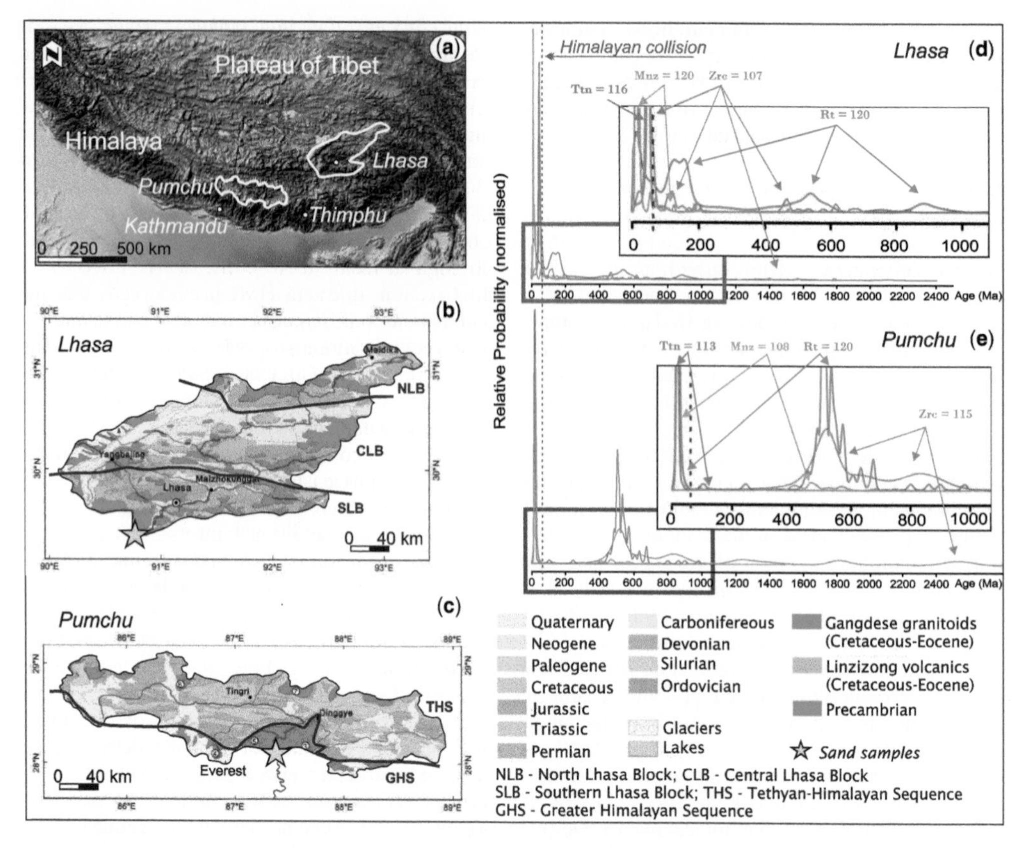

Fig. 11. Provenance analysis using detrital titanite, monazite, rutile and zircon ages from (**a**) two areas in the Himalayas, Lhasa and Pumchu; (**b**) and (**c**) basement geology of the Lhasa and Pumchu covering Phanerozoic and Precambrian ages, where two samples were collected; (**d**) and (**e**) relative probability distribution age plots of detrital minerals from the Lhasa and Pumchu, respectively. Source: elevation map is from National Oceanic and Atmospheric Administration database resources, geological maps and U–Pb data are modified from Guo *et al.* (2020).

to titanite (Fig. 11e). Together, these seem to indicate a recent metamorphic basement source for titanite, rutile and monazite, but necessarily at high grade (potentially migmatites and S-type leucogranites) coming from the Greater Himalayan Sequence (Guo *et al.* 2020). Looking at the surface proportion between these rocks and the sedimentary to low-grade metasedimentary rocks from the Tethyan–Himalayan sequence in the drainage basin, it is clear that detrital mineral age distributions do not correspond to these (Fig. 11c, e). In Pumchu, pre-Mesozoic rutile and zircon result from a second cycle of sedimentation and do not indicate currently exposed pre-Mesozoic basement in this area (Guo *et al.* 2020). However, together they can be used to constrain the palaeosource of the late Paleozoic–early Mesozoic sedimentary rocks to a Gondwana-like terrane that endured significant magmatism and metamorphism between the late Neoproterozoic and Cambrian.

As seen in this example, ages alone may lead to equivocal source types and detrital age distributions may be biased owing to rock-mineral fertility and resistance to weathering. By using a multi-proxy approach, some of these events may be better untangled.

Trace-element based provenance tools and other isotopic systematics

Apatite. Together with the incorporation of U and Th, the REEs, Y, Mn and Sr are the most common substituting elements in apatite, all of which substitute for Ca. REEs, Y and Sr are the elements most effective for discriminating magma types from each other (Bruand *et al.* 2020). Belousova *et al.* (2001), Jennings *et al.* (2011) and Bruand *et al.* (2014) have shown that there is a strong correlation between Sr in apatite and in the corresponding whole rock. Bruand *et al.* (2020) confirm that apatite Sr content can allow discrimination between HP and

LP-TTG (low pressure). The high Sr in HP-TTG corresponds to a deep source in equilibrium with garnet and rutile but no plagioclase (Moyen and Martin 2012). In addition, correlations between the unusually high Sr content of monazite and HP conditions have also been previously reported in the Bohemian Massif peraluminous rocks (Finger and Krenn 2007). Metamorphic apatite from low- to medium-grade metapelites and metabasites can be identified by very low Th and LREE concentrations and are often also low in U (Spear and Pyle 2002; Henrichs *et al.* 2018, 2019). Even though the REE and U abundances are diagnostic of metamorphic grade in metapelitic apatite, it is not always possible to date low-grade apatite grains owing to low U and high common Pb contents (Henrichs *et al.* 2019). The behaviour of trace elements in apatite is usually controlled by both the degree of fractionation and the composition of the host rock from which a detrital apatite grain was derived (e.g. Belousova *et al.* 2002; Bruand *et al.* 2016; Henrichs *et al.* 2018; Sullivan and Chew 2020). Therefore, they can be used to trace the environment in which individual apatite grains formed. Still in the early 2000s the main discrimination diagrams available were focused on igneous apatite, but over the last decade datasets of apatite trace element compositions from metamorphic rocks and ultramafic and alkali-rich igneous rocks have been published, making apatite composition a more reliable provenance tool. The Sr/Y v. ΣLREE biplot proposed by O'Sullivan *et al.* (2020) is an easy tool to discriminate the broad provenance of apatite with six lithology categories: low- and medium-grade metamorphic apatite; high-grade metamorphic apatite; felsic granitoids; mafic granitoids and mafic igneous apatite; alkali-rich igneous apatite; and ultramafic–carbonatitic apatite (see Volante *et al.* 2022). Isotopic tracing of the Sm–Nd system in apatite is another provenance technique in apatite that is now used more frequently (e.g. Foster and Carter 2007; Carter and Foster 2009), also in combination with trace elements (Malusà *et al.* 2017). As a detrital tracer, apatite is able to record both the age (e.g. Carrapa *et al.* 2009) and the lithology of its source (e.g. Abdullin *et al.* 2016; Sullivan and Chew 2020). Thus, by combining geochronological data and the geochemistry of detrital apatite, it is possible to get an extremely useful provenance tool (e.g. O'Sullivan *et al.* 2018). The significant increase of apatite trace element data in the literature makes it possible to create more specific provenance interpretations or new discrimination diagrams (Eu/Y–Ce discrimination diagram; Zhou *et al.* 2022) by using large dataset analysis techniques.

Monazite. The REEs are an important part of monazite mineral structure, and this phosphate mineral can contain up to 60% REE oxides. Monazite is mainly a LREE, Th and U phase, while xenotime contains many heavy REEs and yttrium. Light REEs, Eu and some heavy REEs are important discriminators that reflect elemental fractionation during magmatism and/or metamorphism but also the compositions of the metamorphic protoliths and primary magmas (Itano *et al.* 2020). The combination of geochemical parameters such as $[Gd/Lu]_N$, Eu/Eu* and $[Th/U]_N$ (Itano *et al.* 2016; Gaschnig 2019) and (Sm/Nd)N v. (La/Yb)N (Guo *et al.* 2020) can be used to discriminate between magmatic and metamorphic monazite. Metamorphic monazites tend to have $[Gd/Lu]_N > 500$ or $Eu/Eu^* < 0.2$ and $[Th/U]_N > 20$ if $[Gd/Lu]_N < 500$ while magmatic monazites tend to have $[Gd/Lu]_N < 500$, and similar low Eu/Eu* and $[Th/U]_N > 20$.

Sm–Nd isotope systematics may also be used to decipher the provenance and recrystallization history of monazite, which, similar to Hf isotopes in zircon, reflects the isotope composition of the bulk rock (Volante *et al.* 2022). This indicates that monazite inherits its Sm–Nd isotopic composition from its parental rock and hence detrital monazite is potentially a useful tool for studying continental growth (Iizuka *et al.* 2011; Liu *et al.* 2017).

Rutile. Nb–Cr applied to detrital rutile allows discrimination of metafelsic (log (Nb/Cr) > 1 and Nb > 8 ppm) from metamafic (log (Nb/Cr) < 1) sources (Zack *et al.* 2004; Triebold *et al.* 2007; Meinhold *et al.* 2008) and it is one of the most widely applied tools in detrital rutile studies. Rutile grains derived from amphibolite-facies *P–T* conditions occasionally will not be properly discriminated (Meinhold *et al.* 2008). An analysis based on a dataset of >1100 rutile grains from different metapelitic and metamafic rocks found that we should not expect this inconsistency to propagate, on average, to more than 10% in a detrital rutile population (Pereira *et al.* 2021). Evaluating the protolith of detrital rutile is key to minimizing uncertainties in applying the Zr-in-rutile geothermometer to metapelitic-derived grains (Triebold *et al.* 2012; Pereira *et al.* 2021). Because rutile can precipitate from hydrothermal/aqueous fluids and also from differentiated magmas, additional tools have been tested to differentiate these from metamorphic-type rutile. Extremely low Nb/Ta and Zr/Hf compositions are typical of rutile precipitated from hydrothermal fluids, while very high Nb/Ta values are typical of rutile derived from granitic rocks (Pereira *et al.* 2019). Additionally, Sc, Sb and W enrichments have been linked to ore forming processes, although W-rich rutile can also be associated with granite-derived rutile (Plavsa *et al.* 2018; Agangi *et al.* 2019; Porter *et al.* 2020). These different trace element tools can be used to filter detrital rutile according to the aims of a given study. As detailed in Volante *et al.*

(2023), Zr concentrations in rutile are temperature dependent (with a minor pressure dependency), and calibrations of Zr-in-rutile (e.g. Zack *et al.* 2002; Tomkins *et al.* 2007; Kohn 2020) allow estimates of metamorphic temperatures from detrital rutile grains to be made. Natural scatter of T_{Zr} in rutile found in metamorphic rocks can range from 50 up to 100°C (Pereira *et al.* 2021), implying that a given detrital rutile grain can deviate from the rock equilibrated temperatures by that much. Because pressure is not easily constrained from detrital grains, opting for one pressure within a reasonable range known where rutile is stable (i.e. 0.7–2.0 GPa) can incur in a 50–75°C deviation from the true temperature (Pereira and Storey 2023). Even so, this is a helpful tool, and one that combined with U–Pb dating can provide further constraints on the tectonothermal evolution of metamorphic sources and their compositions. Additionally, mineral inclusions, such as glaucophane, omphacite, coesite or kyanite, can offer further insights into metamorphic grades (Hart *et al.* 2016, 2018). Less often utilized but potentially very useful is the application of *in-situ* Hf isotopic geochemistry applied to rutile. It has only been occasionally attempted owing to analytical challenges (Ewing *et al.* 2011, 2014; Li *et al.* 2015; Wang *et al.* 2020; Zhang *et al.* 2020). Ewing *et al.* (2014) demonstrated that Hf in rutile is a robust tool, not prone to resetting during retrogression after high-grade, granulite facies and yielding the same isotopic compositions as zircon grown in the same metamorphic event. As the analytical methods improve, and smaller spot sizes can be used to measure Hf in rutile, these can in turn be used to trace the nature of detrital rutile protoliths (mafic juvenile or pelitic and more evolved) in a similar way to that used in Wang *et al.* (2020).

Titanite. Molar Al/Fe values in titanite have been used to discriminate metamorphic from igneous titanite (Aleinikoff *et al.* 2002; Gaschnig 2019; Scibiorski and Cawood 2022), while high Th/U values (>1) have been usually attributed to igneous titanite (Aleinikoff *et al.* 2002). Sr compositions of titanite have been shown to reflect the degree of magma differentiation, with higher SiO_2 magmas corresponding to lower Sr in titanite (Bruand *et al.* 2014) and low Zr/Y (Scibiorski and Cawood 2022). Moreover, TTG titanite shows higher Y (in Y–LREE–Sr systematics) than calc-alkaline type titanite (Bruand *et al.* 2020). Titanite chemistry is very sensitive to bulk-rock chemistry and to main rock-forming mineral reactions and crystallization. All of these factors make the use of single trace element tools to discriminate metamorphic titanite complicated. However, the Dy/Yb values in titanite seem to be coherent with the presence (Dy/Yb > 2) or absence (Dy/Yb < 2) of garnet in metamorphic rocks regardless of their grade or composition (Scibiorski *et al.* 2019). Combining these trace element tools with U–Pb dating enables us to differentiate igneous- from metamorphic-type titanite, and to go further in attempting to distinguish magmatic settings and metamorphic conditions through time. For the latter, the Zr-in-titanite geothermometer (Hayden *et al.* 2008) can be applied, even if it relies on estimating approximate pressure, alpha-quartz and rutile activities. Additionally, *in-situ* O isotopes can be accurately measured, and reveal the source-type magmas that titanite crystallized from, including any participation of post-magmatic fluids (Bruand *et al.* 2019). However, Bruand *et al.* (2019) show that in sector zoned titanite crystals, the brightest domains in BSE do not provide reliable oxygen isotopic compositions, and therefore careful crystal zone targeting is required. O isotopes can be combined with *in-situ* Nd isotope geochemistry (Hammerli *et al.* 2019; Fisher *et al.* 2020), which provides information regarding mantle extraction and crustal recycling processes in the source of the titanite-hosting magmas. These isotopic systematics can help better understand the evolution of the crust during the early stages of crustal growth (Fisher *et al.* 2020). However, Nd in titanite can be disturbed during high-grade metamorphism, and usually a more reliable interpretation of Nd isotopic data can only be achieved when it is coupled with other mineral isotopic systematics (such as Hf in cogenetic zircon; Fisher *et al.* 2020). Regardless of complex data handling, Zhou *et al.* (2020) successfully applied Nd isotopic systematics to detrital titanite from modern sediments, illustrating the path to novel approaches applied to detrital studies aiming to place further constraints on our understanding about crustal evolution.

Applications to detrital studies: understanding the continental crust. The combination of trace element-based tools and U–Pb dating applied to the above-mentioned HMs is a very powerful approach that can arguably help constrain the composition and structure of the continental crust and its tectonothermal evolution. In the following paragraphs, we highlight a few examples of recent studies that successfully applied this approach.

Gaschnig (2019) performed U–Pb dating and applied trace element systematics to detrital zircon, monazite, titanite and rutile from an alluvial sample of the Merrimack River in New England (USA) to infer a more complete crustal evolution of the catchment area. The author argued that by conducting such a comprehensive multimineral, multiproxy study he was able to untangle more than just an Ordovician magmatic event that peaked during the Lower Devonian, followed by younger Jurassic and Cretaceous magmatic events, as recorded by detrital zircon ages. The author used (Fe/Al)pfu values to distinguish igneous from metamorphic titanite, Th/U and Ce/Yb signatures to discriminate igneous from metamorphic zircon, the Eu anomaly and Gd/Lu_N values in monazite to trace igneous from metamorphic grains, and Cr–Nb and Zr-in-rutile to recognize protolith types and metamorphic grades. As an example, Gaschnig (2019) identifies a younger Permian regional metamorphism event, totally unseen by zircon, but recorded by monazite and rutile, reaching amphibolite-facies. Additionally, the difference between their age peaks (*c.* 27 Ma) combined with their different Pb diffusivities enabled Gaschnig (2019) to infer that peak metamorphism was followed by slow cooling rates. This study illustrates how a combination of petrogeochemical and geochronological tools applied to multiple minerals can offer a better portrayal of the geology of a given region from a single sample. Yet it does not dwell much on the uncertainties of applying those discrimination diagrams nor other potential limitations and biases.

Going another step further, Zhou *et al.* (2020) investigated the detrital record of zircon, monazite, titanite, apatite and rutile in terms of U–Pb, Sm–Nd and Lu–Hf compositions, using modern sediments from Northern China. By applying these isotope pairs, the authors were able to discriminate different types of source rocks, but also determine probable model ages and continental crust growth events. Zircon Hf and monazite and apatite Nd model ages indicate a crustal growth episode between 2.9 and 2.7 Ga in the North China Craton, and a more recent episode, at 2.5 Ga, only inferred by detrital monazite. Titanite Nd isotopic model ages and compositions indicate multiple crustal growth episodes, including in more recent times, during the Paleo- and Mesoproterozoic. Together, the authors discuss consistent crustal growth in northern China since the Mesoarchean, through several episodes of crustal growth. Apatite and rutile indicate a protracted metamorphic event from 1.9 to 1.6 Ga and then another one between 0.5 and 0.26 Ga. Metamorphic temperatures recorded by rutile are equivalent to eclogite and granulite temperature ranges, hence, metamorphic conditions that are found in the source area rocks are associated with accretion and subduction evolution of a previous ocean. Altogether, by integrating different proxies and petrogeochemical tools, the authors were able to identify new crustal growth episodes that were not clearly recorded by the basement geology and confirm that these tools can be applied elsewhere to better probe the geological record.

Using a different approach, Pereira *et al.* (2021) recently combined trace element-based discrimination diagrams to filter metamorphic rutile from other rutile-types, with an assessment of rutile stability and Zr retention after crystallization. By taking such steps, Zr-in-rutile thermometry can be applied with a higher degree of confidence to test the evolution of metamorphic conditions through time. These authors demonstrated that the signal of low T/P metamorphism, characteristic of cold subduction, is clearly recorded by detrital rutile grains from late Neoproterozoic and Paleozoic times, but also during the late Mesoproterozoic and late Paleoproterozoic. This study relies on the assumption, based on current thermodynamic and experimental data, that only HP metamorphic rutile (s.s) is stable at temperatures lower than 550°C ($P \geq 1.3 \pm 0.1$ GPa). Regardless, this approach illustrates how detrital metamorphic minerals may be used to help elucidate the evolution of metamorphic gradients through time. This, in turn, would help in better understanding the secular evolution of plate tectonics and of the continental crust.

The preservation bias of the continental crust discussion, briefly introduced earlier in this section, has recently been revisited by Mulder and Cawood (2022) using detrital monazite. The authors compared a dataset of over 100 000 monazite grains ages, which unambiguously record the continental collision stages during orogenesis, against a global compilation of detrital zircon ages. By comparing the detrital age records of these two different minerals, and finding they are statistically very similar, the authors argue that the detrital zircon record has been biased by a selective preservation process related to continental collision during the multiple supercontinent assemblies. This is a rather novel approach contributing to this overarching and contentious issue behind the record of continental crust growth.

Applications and implications of heavy minerals in detrital studies

The main biases affecting detrital heavy minerals

The interplay of different biases affecting detrital sediments limits our capabilities to understand provenance and crustal evolution. These biases may be geological, technical/methodological and interpretative or a combination of these.

The varying abundance of HMs in different crustal rocks may result in the over-representation, or under-representation, of some crustal components in the sedimentary record. As an example, studies that use coupled U–Pb and Lu–Hf isotope systematics applied to zircon to reconstruct the continental growth in a quantitative way (i.e. establishment of growth curves through time) used either recent sediments (e.g. Iizuka *et al.* 2005; Dhuime *et al.* 2012), that are possibly biased by ancient crust

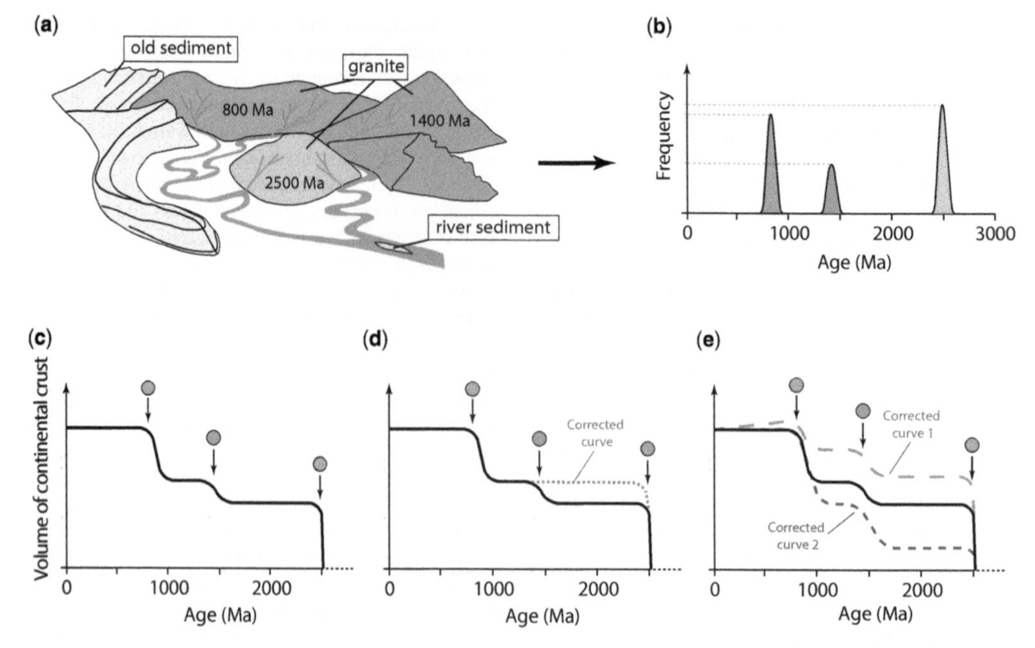

Fig. 12. Schematic illustration of the use of detrital minerals to reconstruct crustal growth curves. (**a**) A catchment basin made of zircon-bearing granitoid bodies formed 800, 1400 and 2500 My ago with the respective colours dark pink, green and orange. An old sediment is also present but not sampled. (**b**) A hypothetical histogram of U–Pb ages for zircons that would be recovered from the river sediment with the same colour code as in (a). Insets (**c**)–(**e**) show a cumulative histogram obtained from (b) with coloured circles indicating granite formation as in (a). Insets (d) and (e) contain dashed 'corrected' curves which correspond to adjustments made from the cumulative histogram depending on the biases that affect the detrital zircon record. These corrected curves are therefore true crustal growth curves in our examples. See text for details.

disappearance (e.g. Iizuka *et al.* 2017), or a combination of detrital and igneous zircons from various ages and origins (Belousova *et al.* 2010), thereby distorting quantitative information. Figure 12 illustrates the use of detrital minerals to reconstruct crustal growth and the effect of above-mentioned issues. Crystals released by host-granite erosion are transported to a deposition site (river sediment in Fig. 12a) from which the catchment basin age information can be retrieved by dating zircon using U–Pb geochronology (Fig. 12b). The common approach is to assume that the amount of zircon at a given age is proportional to the volume or mass of granite formed at this age (e.g. Iizuka *et al.* 2005; Belousova *et al.* 2010; Condie and Aster 2010; Dhuime *et al.* 2012). By doing so, the histogram in Figure 12b can be transformed into a cumulative histogram that directly gives a crustal growth curve for the studied crustal province, hence preserving both qualitative (age) and quantitative (volume) information (Fig. 12c). This approach makes sense on the first order but can actually be questioned for several reasons. Among those, we emphasize the effect of granite source, of erosion/recycling, and of mineral fertility in granite.

Figure 12d shows that if the 1400 Ma old granite (green) is a crust-reworking product, it does not mark a growth event and the growth curve should be corrected accordingly. If erosion/recycling modified the original abundances of granitoids in the catchment basin, the true growth curve would be different from the cumulative histogram (Fig. 12e), hence resulting in either an over-estimation of the amount of continental crust formed early on in the growth history (corrected curve 2 in Fig. 12e) or an under-estimation (corrected curve 1 in Fig. 12e). Similarly, if catchment basin granites contain variable amounts of zircon, referred to as the fertility issue (e.g. Dickinson 2008; Flowerdew *et al.* 2019), the amount of these later found in the detrital record no longer represents volumes of granite, hence, the cumulative curve has to be corrected because of over- or under-estimation issues. This fertility issue can be extended to all mineral types because protolith composition and/or the *P–T* evolution of a terrane will condition the minerals that grow, and thus may become available in the detrital record.

Another source of bias comes from mineral transport from source to sink and their different hydraulic behaviours. Depending on the energy of transport, medium-sized crystals of various ages can be fractionated from each other. Despite these minerals being all heavy ($d > 2.9$ g cm^{-3}), their densities differ slightly between them (apatite 3.1–3.35, garnet 3.13–4.32, titanite 3.4–3.55, rutile 4.18–4.25, zircon 4.6–4.7 and monazite 5.0–5.3 g cm^{-3}; Klein 2003; Deer *et al.* 2013). This fractionation can even take place within the same mineral type between different fractions (e.g. Lawrence *et al.* 2011). Mineral sorting during transport may affect the representativeness of the mineral record for the catchment basin (e.g. Fedo *et al.* 2003; Lawrence *et al.* 2011; Garçon and Chauvel 2014; Roberts and Spencer 2015). Furthermore, grain shapes may also impact their proportions, as discussed before, such as in the case of xenomorphic or anhedral crystals (Deal *et al.* 2023). Additionally, garnet tends to form equidimensional grains, while apatite and titanite, for example, form more elongated and tabular grains, yielding a different hydraulic behaviour. Accessory and minor minerals in igneous and metamorphic rocks have different grain size distributions. While these differences may be subtle, with grain size variations within 100 or 250 µm between accessory and minor minerals, these are enhanced by variations in HM densities. Krippner *et al.* (2016) found that MgO-rich almandine garnet was more abundant in the sediment coarser grain fractions, while MnO-rich almandine garnet was more concentrated in the finer grained fraction. This shows that small composition variations can lead to detrital garnet fractionation. All of these factors combined may fractionate these HMs during transport and deposition, generating a bias when only one single sample/fraction is taken.

Source to sink links that geologists use for tectonic and crustal reconstructions assume a simple single-stage event between erosion of catchment basins and the deposition site, which is probably not the case in most sediments (e.g. Andersen *et al.* 2016). This is commonly linked to the stability of some HMs during weathering, transport and diagenesis. As introduced earlier, garnet, titanite and to a lesser extent monazite and apatite, are more prone to mechanical and chemical weathering than rutile and zircon. This impacts the survivability of these other minerals with respect to rutile and zircon and suggests that phenomena that take place during weathering and then during post deposition can promote dissolution heterogeneously across these different heavy mineral species. Thus, zircon and rutile may be more easily recyclable than the others, and their detrital age distributions may be linked to second-cycle sedimentation (e.g. Be'eri-Shlevin *et al.* 2018). Apatite and monazite can have very small grain sizes at relatively low to moderate metamorphic conditions (under 50 µm). These grains, after weathering and diagenesis, may disappear more easily than monazite and apatite derived from gneisses or migmatites, which can have larger grain sizes. This in turn can generate a bias, favouring the identification of higher-grade metamorphic and magmatic events. This can be further enhanced if some given mineral compositions are more prone to react during diagenesis than others, which seems to be the case for garnets with high grossular component (Morton and Hallsworth 2007).

Monazite and zircon are quite rich in U, making them easily amenable for U–Pb dating, but it is not the case for apatite, rutile, titanite and garnet, which often reach less than 2 ppm (Kohn 2017; Millonig *et al.* 2020; Pereira *et al.* 2020). These phases can incorporate variable amounts of Pb at the time of crystallization, resulting in more imprecise and potentially less accurate U–Pb dates that are discarded from final datasets. Very low U-bearing grains may reflect a particular source, i.e. (meta) mafic sources, and thus we can easily bias our understanding of the source area. A similar bias can be created using a single mineral type in which the amount of U correlates with another compositional parameter, in either the mineral or parental rock (e.g. garnet whose increasing U concentrations can be correlated to increased Ca content; Seman *et al.* 2017).

It has been demonstrated over the years that detrital HM studies include a human-derived bias during sample preparation and mineral picking (Dröllner *et al.* 2021). Using a density shaking table has been shown to produce a bias where part of the smaller grains is lost (Sláma and Koler 2012). Mineral picking has been shown to represent a source of bias towards better known mineral colours and grain shapes (Dröllner *et al.* 2021), also favouring larger crystals. Magnetic separation can induce bias by removing populations with variable Fe content, plus grains which have Fe-bearing inclusions.

The main analytical bias during LA–ICP–MS U–Pb dating is caused by the $^{206}Pb/^{238}U$ laser-induced elemental fractionation owing to different matrix behaviours between unknowns and the reference materials used for standardization (e.g. Košler and Sylvester 2003). These differences between unknowns and the reference materials are mainly caused by variable composition, age and radiation damage of the unknowns, which are most common in zircon. In fact, the state of preservation of the zircon lattice, which is a direct consequence of the U and Th contents (e.g. Holland and Gottfried 1955; Murakami *et al.* 1991; Ewing *et al.* 2003), can influence the establishment of crystallization ages and even crystal selection for further analysis because of the appearance of zircon grains (e.g. presence of cracks, crystal morphology, colour, transparency). Therefore, some zircon populations in the detrital record can be partly or entirely missed, or discarded, depending on their appearance and/or their 'disturbed' U–Pb isotope systematics.

Solutions to overcome inherent biases

Variations in morphology, grain size and density can impact HM hydraulic behaviours, and hence distribution along a stream path. This sort of bias may be overcome by sampling multiple, equivalent detrital sediments, and processing not only the 125–350 μm grain size fractions, but also the finer-grained fractions where some of the heavy minerals may be concentrated (e.g. some lower-grade apatite and monazite). This approach should minimize uncertainties related to mineral transport and, in turn, strengthen interpretations based on data collected from those samples.

A sampling strategy to minimize operator-related biases includes grid sampling. In non-modern settings, sampling should also consider lateral facies variations and therefore be firmly supported by the stratigraphy. Sample preparation and mineral separation procedures (shaking tables and magnetic separator) have also been shown to introduce bias. It is advisable to also treat the medium density fraction after density separation and check the HMs in all magnetically separated fractions. To overcome mineral picking bias, sample preparation for geochronology studies that avoids picking and uses bulk-mounting can be advantageous (Dröllner *et al.* 2021). While this may impact polishing quality owing to variable grain sizes and mineral softness, it decreases operator inflicted bias and reduces final mineral preparation time. However, by avoiding picking and standard mounting, minerals with different responses to electron imaging will be mounted side-by-side, impacting the quality and efficiency of electron imaging and of microanalytical routines. Isozaki *et al.* (2018) developed and reported an automatic zircon separator, which considerably reduces this human-derived bias without impacting zircon preparation for further microanalysis, by combining imaging processing with an automated picking and mounting tool. While this is a very useful development, it is expensive and difficult to implement in most labs.

Selecting grains for analysis by automated grain determinations such as scanning electron microscopy (SEM)–energy-dispersive spectrometry (Vermeesch *et al.* 2017), Raman spectroscopy (Lünsdorf *et al.* 2019) or X-ray micro-computed tomography (Cooperdock *et al.* 2022) can help reduce the bias caused by mineral separation techniques. However, care needs to be taken with these techniques as they might create new biases, since the classification relies on the quality of the underlying mineral databases (Chew *et al.* 2020).

One of the biggest challenges of petrochronology applied to detrital heavy minerals is the assessment of preservation of the original features/signatures for each grain. While we have focused on the diffusion of Pb to discuss U–Pb ages, other important elements that are measured in these different minerals – O, Ti, Zr, Y-REE, Sr, Cr, Al, etc. – yield different diffusivities (Cherniak 2000, 2010, 2015; Cherniak *et al.* 2007; Farver 2010). Because of the complex evolution of Earth's lithosphere, with multiple supercontinent assemblies and break-ups, terranes may experience several cycles of tectonic, magmatic and metamorphic activity, especially with increasing age. These events may affect the source area rocks or the sedimentary rocks where detrital minerals are stored and promote elemental diffusion. This can lead to potential element decoupling signatures (Xiong *et al.* 2012; Kohn *et al.* 2016; Gordon *et al.* 2021; Pitra *et al.* 2022) and misleading interpretations. There are no given recipes on how to easily overcome uncertainties related to this potential decoupling. Depending on the heavy accessory minerals considered, the metamorphic grade affecting the sedimentary units and the presence of fluids needs to be properly checked and stated. Additionally, characterizing mineral inclusions can help corroborate mineral trace element chemistry and interpretations. Another strategy is to subsample the detrital mineral population and conduct spot profiling or grain elemental mapping to verify zoning/diffusion profiles (e.g. Petrus *et al.* 2017; Paul *et al.* 2019). In any case, a safe approach in detrital mineral studies is to analyse minerals several times instead of doing a single spot, as this may reveal whether a crystal is homogeneous or not (e.g. Guitreau and Blichert-Toft 2014; Guitreau *et al.* 2018). Finally, and particularly for studies aiming to characterize crustal sources and their evolution using multiple minerals, it may be advisable to start with more robust isotopes. Lu–Hf in zircon may provide a clue to the amount of possible crustal sources for the detrital population. Based on this data the crustal structure and evolution can be built up for the provenance of the sample.

Finally, we give a brief consideration regarding the recycling of detrital material that can disrupt the source to sink path of sediments. Zircon is very prone to survive through multiple sedimentary cycles. A way to assess if zircon grains are not the result of a first sedimentary cycle is by conducting a systematic recovery of detrital zircon core-rim U–Pb ages (Liu *et al.* 2022) or core-rim geochemistry (isotopes or trace elements), combined with a grain morphology characterization of multiple detrital minerals with different physical properties (such as in Zoleikhaei *et al.* 2021). Discrimination of first cycle and multi-cycle detrital zircon may also be possible by calculating a source-normalized α-dose for the detrital zircon, constraining the degree of grain transport and recycling (Dröllner *et al.* 2022).

Other minerals are not as easily recycled as zircon, except for rutile (Be'eri-Shlevin *et al.* 2018).

In general, to overcome most biases and get a more complete picture of the tectonothermal evolution of a given terrane/region, it is most advantageous to sample diverse lithologies (Zhou *et al.* 2020) and use a multi-mineral plus multi-proxy approach (e.g. Gaschnig 2019).

Summary and perspectives

The present and future of detrital heavy mineral research

Studying detrital sediments to infer global crustal information is not limited to heavy mineral studies (e.g. Allègre and Rousseau 1984; Taylor and McLennan 1985), but in contrast to whole-rock approaches that average source and timing, the study of individual minerals such as zircon allows for the reconstruction of magma extraction (sources and timing), as well as metamorphic events, that led to the formation and evolution of continents.

Owing to its reliability and robustness, zircon has played the main role in detrital investigations and is the basis of many advances in tectonic interpretations. However, with a better comprehension of the biases inherent to fertility (e.g. Moecher and Samson 2006; Cawood *et al.* 2013), data treatment (e.g. Vermeesch 2012; Andersen *et al.* 2019; Barham *et al.* 2022), metamorphism (e.g. Tedeschi *et al.* 2023) and others, the employment of complementary approaches is necessary in order to obtain a more complete and reliable picture from the detrital record. Because rocks usually have a complex history, from their igneous origin to their recycling during sedimentary or metamorphic cycles, often passing through different pressure, temperatures and fluid/melt conditions, a multi-mineral, multi-proxy approach constitutes the best way to conduct these studies. The use of different minerals allows for reconstructing the different stages a rock went through; for instance, garnet can record prograde to peak conditions (Maneiro *et al.* 2022), while zircon is more likely to record the high temperature to the cooling path stage (Kelsey *et al.* 2008) and rutile can record both peak and cooling temperatures (Zack and Kooijman 2017). Therefore, together, these minerals offer much more insights into the evolution of terranes and of the continental crust that was eroded. Recent analytical advances have opened a broad range of options for detrital mineral dating, and also the potential to correlate these time constraints with petrological conditions (metamorphic grades, magma chemistries, etc., e.g. Bruand *et al.* 2020; O'Sullivan *et al.* 2020; Glorie *et al.* 2023). These advances are extremely useful, as the detrital record (i.e. (meta)sedimentary rocks) may be the only evidence of past basement rocks (Kellett *et al.* 2018). Indeed, these detrital minerals are the only records we have of the early evolution of planet Earth, since the oldest preserved crustal record are detrital zircon grains preserved in sedimentary rocks (e.g. Cavosie *et al.* 2019).

Zircon has paved the way towards the generalized use of detrital minerals in the study of catchment basin lithologies, crustal growth reconstructions and sedimentary process dynamics owing to its refractory nature and the wealth of techniques and tools that can be applied to it (Fig. 13). Zircon has numerous strengths, but it also has limitations that, notably, pertain to elemental and isotopic systematics interpretations that are mostly made on a first-order level, as well as the fact that zircon formation is essentially linked to felsic igneous rocks (SiO_2 >58 wt%). Consequently, inferences made on the nature of crustal magmas that characterize the continental crust can sometimes be poorly constrained. Improvements in zircon elemental and isotopic systematics and their tracing to the parental magma compositions will inevitably produce better constraints on our understanding about the nature of crustal magmas. This may be attained by improvement or development of new isotope systematics in zircon, such as stable Si and Zr isotopes that can inform on the nature of the parental granitoids (e.g. Guitreau *et al.* 2022) and/or on the differentiation of parental magmas (Zhu *et al.* 2023). The same is true for zircon mineral inclusion studies as well as geochemical and isotopic signatures of those mineral inclusions (e.g. Emo *et al.* 2018; Antoine *et al.* 2020).

Garnet plays a growing and significant role in provenance studies. This is mainly due to its versatility, which is derived from its varied elemental and isotopic composition, the preservation of zoning and the capacity of hosting inclusions (Fig. 13). Recent advances have enhanced the discrimination schemes about host rock, bulk-rock composition and metamorphic facies grain derivations (Schönig *et al.* 2021*a*), and have enabled *in-situ* garnet dating (Millonig *et al.* 2020; Simpson *et al.* 2021), which can now be applied to detrital garnet (Mark and Stutenbecker 2022). In addition, the use of trace-element compositions has room to increase, particularly the use of trace-element mapping, following the advances in metamorphic petrology investigations (e.g. Raimondo *et al.* 2017).

Detrital apatite can be used for identifying mid to deep crustal processes, since it is a good tracer of both metamorphic and magmatic processes, and it may be used as a U–Pb thermochronology tool in the range of ~350–570°C (Chew and Spikings 2015, 2021 and references therein). Next to zircon, apatite is probably one of most used and versatile accessory minerals for petrochronology in detrital

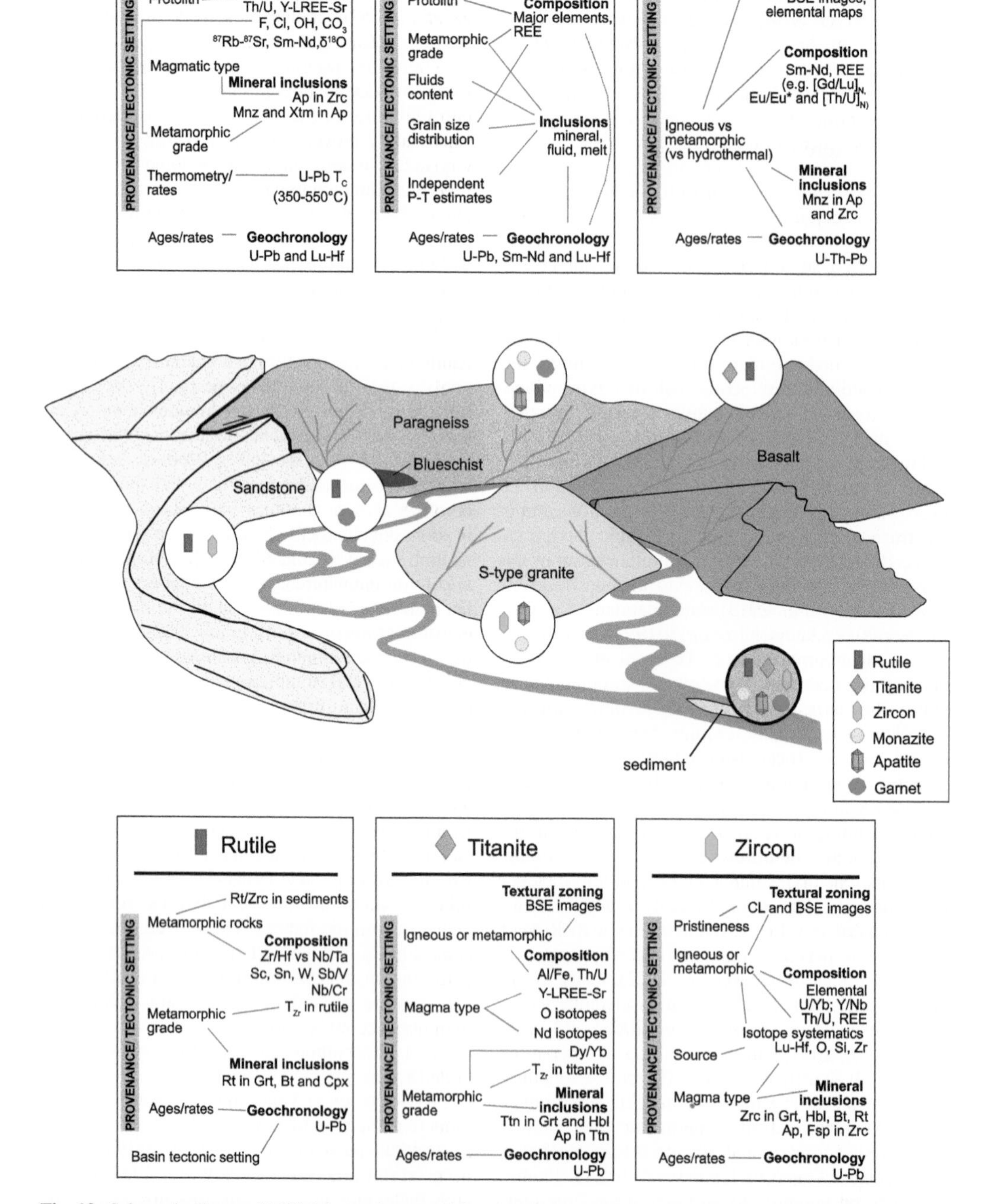

Fig. 13. Schematic diagram highlighting and summarizing the main tools that we can apply to detrital apatite, garnet, monazite, rutile, titanite and zircon to better characterize the nature of source areas and their geodynamic evolution through time and space.

studies (Fig. 13). Although U–Th–Pb dating of apatite by LA–ICP–MS is now routinely possible, it still has some difficulties owing to the presence of common Pb, its intermediate closure temperature and/or age-resetting by metamorphic/metasomatic processes (Glorie *et al.* 2022). The Lu–Hf isotope system in apatite has a higher closure temperature (*c.* 675–730°C; Glorie *et al.* 2023), which makes it an important dating method that can expand the apatite petrochronological toolbox (Fig. 13).

Monazite can be extremely useful for identifying metamorphic events in source terrains of stream sediments because monazite grows from low metamorphic grades up to granulite facies. Targeting monazite in a stream sediment sample, therefore, gives a complete picture of the metamorphic evolution of the source terrane (Fig. 13). This could be used in conjunction with trace element geochemistry to separate metamorphic and magmatic monazite grains. More recently, Sm–Nd isotope systematics have been applied to monazite to determine provenance and the isotopic composition of the bulk rock to trace crustal evolution processes (Guo *et al.* 2020; Itano *et al.* 2020; Fig. 13). Thus, monazite is a very valuable complement to zircon in provenance and crustal composition studies.

Recent advances are pushing the applications of detrital rutile beyond provenance and determination of age-temperature, attempting to utilize rutile to infer tectonic processes (Pereira *et al.* 2020, 2021). To improve our confidence in such approaches, there is still a need to improve trace element-based source discrimination diagrams, both between metamorphic and other types of rutile (Agangi *et al.* 2019; Pereira *et al.* 2019), and between different metamorphic rock types. By improving these tools, we may be able to derive metamorphic grades (*P–T*) from single detrital rutile grains more precisely and accurately. Together with crystal chemistry, mineral inclusions can offer such insight, but this implies applying systematic mineral inclusion characterization to detrital grains (Fig. 13). Automated SEM feature mapping may facilitate such tasks.

The application of robust textural and chemical composition of titanite allows the discrimination of magmatic from metamorphic or hydrothermal events (Garber *et al.* 2017; Olierook *et al.* 2019; Scibiorski *et al.* 2019), which when coupled to U–Pb dating (Storey *et al.* 2007; Kirkland *et al.* 2020), provides a unique tool to track the continental crust through time (Fig. 13). Oxygen and Nd isotopic and trace element geochemistry of igneous titanite can help discriminate different magma types and sources (Bruand *et al.* 2014, 2019, 2020; Hammerli *et al.* 2014). The Zr-in-titanite thermometry (Hayden *et al.* 2008) has rejuvenated the interest in using titanite to track metamorphic conditions, as this tool coupled with trace-element discrimination tools can assist in such tasks. Because titanite is usually found with mineral inclusions, advances in utilizing mineral inclusions to determine entrapment pressures (such as are now applied to zircon; Gonzalez *et al.* 2021) may be interesting to determine magma depths and metamorphic grades. Since many titanite grains have low Nd concentrations, which prevent them from being routinely analysed with the current generation of mass spectrometers, analytical improvements in Nd isotopic analyses will be key to determining the isotopic composition of detrital titanite populations.

The multiverse of detrital minerals: application of multiple proxies to current issues in crustal evolution

Continental crust evolution is a complex topic that involves the evaluation of compositional changes and differentiation processes, as well as changes in its volume, as a consequence continental crust growth over time. This has led to discussions about continent freeboard, supercontinent cycles, Wilson cycles and the onset of plate tectonics at a global scale, driven by subduction and the forces exerted on the subducting plate (e.g. Cawood *et al.* 2018). While much of what is currently known about the genesis and evolution of the continental crust comes from the magmatic and metamorphic rock records, it is undisputed that the detrital sedimentary record has contributed to and will surely continue to contribute to the debate. The role of detrital minerals goes beyond regional correlations and tectonic setting reconstructions and has extended the understanding of secular changes in crustal formation (e.g. Voice *et al.* 2011; Pereira *et al.* 2021; Mulder and Cawood 2022). This assessment is particularly important because detrital minerals often represent the only record of ancient periods and processes that have evolved through time. They represent snapshots of ancient continental areas that have now disappeared. As acknowledged in the present manuscript, there are biases that affect the detrital record and, thus, complicate our ability to properly interpret it and draw robust conclusions. Yet the detrital mineral record has arguably allowed us to expand our ability to probe the Earth, by assessing the eroded crust.

From this perspective, zircon has by far been the mineral from this suite of detrital HMs that has deserved the most attention and that has contributed the most to these discussions. Only in recent years have we witnessed an expansion to include the other detrital HMs to tackle some of the crustal evolution overarching questions (e.g. Sullivan and Chew 2020; Pereira *et al.* 2021; Mulder and Cawood 2022) to which zircon could not bring more firm constraints.

Detrital minerals and the detrital mineral record are strongly influenced by the emergence of continents (i.e. rise above sea-level). In this way, continental freeboard and the development of detrital sedimentary systems have influenced the continental crust itself by reworking and recycling processes, and thus constitute a hallmark on the continental crust evolution. Detrital zircon age distributions have been used to demonstrate the presence of continents well above the global ocean level as early as the Meso- to Neoarchean transition (Reimink *et al.* 2021), consistent with Sr isotope data from marine carbonate interpretations (e.g. Shields and Veizer 2002). This approach was taken by using a very large detrital zircon age dataset (>450 000 zircon grains) and evaluating the presence of uni- or bimodal detrital zircon age distributions, typical of sediments in a juvenile restricted crustal block, against more multimodal detrital age distributions, typical of sediments derived from larger continental masses and their interiors (Reimink *et al.* 2021). Similar findings are reported by Gaschnig *et al.* (2022) using Hf compositions in zircon from diamictites. Diamictites, as glaciation products, constitute a robust record of sediment sources. These authors have looked at diamictites from three different continents and four different global-scale glaciation periods to assess the nature of the exposed crusts at each time interval. The very scarce record of Mesoarchean or older detrital components (>3.6 Ga) in all samples, but mostly in older deposits, is used as supporting evidence of a very limited exposure of continental crust with Paleo- and Eoarchean ages, but more of Mesoarchean age (Gaschnig *et al.* 2022).

Continent freeboard is tighly linked to the composition of the continental crust, as it is the dichotomy between densities and therefore buoyancies of the mafic oceanic crust and the more felsic continental crust that is responsible for variations in the hypsometric curve. There is a debate about the nature of the early crust, and how and when we went from an earlier more mafic crust to a more differentiated and felsic continental crust (Kamber 2015; Tang *et al.* 2016; Greber *et al.* 2017; Rollinson 2017). As referred to in the section ‘Detrital Zircon: the advent of *in-situ* U-Pb, Hf and O isotopes’, the nature of the oldest crust can only be assessed through the mineral record, namely by detrital zircon grains (e.g. Wilde *et al.* 2001; Cavosie *et al.* 2019) and from a limited suite of Eoarchean rocks (e.g. Roth *et al.* 2014; Guitreau *et al.* 2019). The combination of U–Pb, Hf and O isotopes shows that Hadean zircon grains have chemistries that are supportive of the existence of a Hadean felsic crust (Cavosie *et al.* 2019), but at the same time, mineral inclusions and zircon trace element data equally suggest the existence of an heterogeneous crust, probably made from high-temperature, primitive sources and low-temperature, more evolved TTG-like magma sources, together with hydrated low-temperature crustal reworking products (e.g. Harrison 2009; Kemp *et al.* 2010; Drabon *et al.* 2021; Laurent *et al.* 2022). The existence of a felsic crust does not, however, indicate a predominance of a felsic Hadean crust and these Hadean grains may not even be representative of the Hadean crust globally. Since these are very precious samples, it is the combination of other useful proxies and techniques that can provide more insight into the type of crustal rocks existing at the time. One of the ways is to explore mineral inclusions in those zircons and their chemistries. Apatite, as we have seen, is a highly versatile mineral and commonly found as an inclusion in zircon. However, it has been recently shown that apatite abundance in zircon is inversely correlated with the increase in whole-rock SiO_2 content (Bell *et al.* 2018). With this, these authors have looked at mineral inclusion suites in Hadean and Eoarchean detrital zircon, discarding inclusions that were deemed secondary or affected by fluids. They further suggest that the Hadean detrital zircon grains from Western Australia (Jack Hills) seem to be derived from felsic granitoids while Eoarchean detrital zircon grains from Canada (Nuvvuagittuq supracrustal units) seem to be derived from intermediate to felsic-type granitoids. Additionally, apatite inclusions in detrital zircon can be targeted for F and Cl studies, to determine the concentration of these halogens in the granitic magmas in which zircon crystallized (Kendall-Langley *et al.* 2021), as well as their trace element and isotopic geochemistry to better constrain the composition and nature of those original source rocks (Emo *et al.* 2018; Bruand *et al.* 2020; O’Sullivan *et al.* 2020). The combination of these proxies can provide more robust data to assess the felsic v. intermediate or more mafic nature of the early crust, and the geodynamic settings behind those magma sources.

The processes behind continental crust differentiation are also a heated source of debate, as different petrogenetic contexts and geodynamic scenarios can lead to very similar products. Consequently, discussion about the onset of plate tectonics arises, where mechanisms driven by dominant vertical ‘sagduction’ and horizontal tectonics in the Archean are proposed, and a transient stage from ancient tectonics to modern subduction processes is commonly identified during the Proterozoic (e.g. Cawood *et al.* 2018 and references therein). The global record of detrital zircon has been extensively used to explore the links between zircon production, continental crust growth and the operation of modern-like subduction and of plate tectonics (e.g. Dhuime *et al.* 2012; Roberts and Spencer 2015). As we discussed in the section ‘The main biases affecting detrital heavy minerals’, interpretations based on the detrital zircon record regarding continental crust growth and estimated volumes over time are highly controversial (e.g. Payne *et al.* 2016; Rollinson 2017). This implies that we may need to apply other proxies and tools to circumvent current issues, and use alternative ways to infer crustal thickness and magma generation depths (e.g. Tang *et al.* 2021; Moreira *et al.* 2023). While with detrital zircon the main target is the magmatic crust response to geodynamic changes, other proxies are now applied to test changes in the metamorphic record. Pereira *et al.* (2021) employed U–Pb–T_{Zr} and trace element discrimination schemes to detrital rutile to demonstrate the persistence of rutile grains with features typical of rutile derived from low geothermal gradients, typical of modern subduction settings, as old as 2.1 Ga. This study confirms that the record of LT–HP rocks of late Paleoproterozoic age (e.g. François *et al.* 2018; Brown and Johnson 2019) is more widespread than the preserved metamorphic rock record and can also be recognized in the sedimentary record. A similar approach may be tested using garnet or titanite, as argued in the sections 'What is inside your detrital petrochronology toolbox?' and 'Applications and implications of heavy minerals in detrital studies'.

Recognizing the operation of these different tectonic processes is also valuable to the understanding of the mechanisms that have ruled continental crust formation and evolution in the past. While nowadays we acknowledge that continental crust growth and reworking comprise a continuous process, driven by plate tectonics, as we go back in time the equivocal record, often weighed on the global detrital zircon record, can serve as an indication of a periodicity in the growth of new continental crust (e.g. Condie and Aster 2010; Arndt and Davaille 2013) or just a reflection of a preservation bias, intrinsically related to crustal block amalgamation during supercontinent formation (e.g. Hawkesworth *et al.* 2009; Spencer *et al.* 2020). Comparing the ages of zircon and monazite in the same detrital record adds a new layer to this discussion (Mulder and Cawood 2022), as monazite forms in higher amounts under collisional conditions and thus better records these collisional stages. These authors have argued that the similarities between their detrital records can be reconciled by means of a selective preservation process affecting the detrital zircon record. This explanation also applies to similarities between the detrital zircon and rutile records (Pereira and Storey 2023).

The combination of multiple proxies (e.g. elemental and isotope geochemistry, including geochronology) applied to various minerals in provenance studies with automated HM determinations employing highly efficient sample preparation protocols could become the norm soon (e.g. Chew *et al.* 2020). Such an approach that combines the strengths of different detrital HMs, as depicted in the present contribution, will surely enable unprecedented reliability in data interpretation, and therefore, bring exciting new insights and breakthroughs in continental crust evolution.

Acknowledgements This is contribution no. 593 of the ClerVolc program of the International Research Center for Disaster Sciences and Sustainable Development of the University of Clermont Auvergne. We would like to thank Silvia Volante for the efficient editorial handling, as well as the invaluable comments received from Avishai Abbo and two other anonymous reviewers. We also take this opportunity to extend our thanks to collaborators, mentors, and colleagues who have been working on these topics and have contributed to the advancement of detrital heavy mineral applications in crustal evolution and provenance studies. We acknowledge that we were unable to include all the available literature on these topics and apologise in advance for any omissions.

Competing interests Nothing to declare.

Author contributions **IP**: conceptualization (lead), formal analysis (lead), project administration (lead), writing – original draft (lead), writing – review & editing (lead); **VVS**: conceptualization (equal), formal analysis (equal), writing – original draft (equal), writing – review & editing (equal); **MT**: conceptualization (equal), formal analysis (equal), writing – original draft (equal), writing – review & editing (equal); **KC**: conceptualization (supporting), formal analysis (supporting), writing – original draft (supporting), writing – review & editing (equal); **MG**: conceptualization (supporting), formal analysis (equal), writing – original draft (equal), writing – review & editing (equal).

Funding This research was partially financed by the Portuguese National Science Foundation I.P. (Portugal), with a fellowship contract to I. Pereira under the Individual Call to Scientific Employment Stimulus (2021.01616.CEECIND) and grants UIDB/00073/2020 and UIDP/00073/2020 awarded to Centro de Geociências (Universidade de Coimbra – Portugal). Additional support was given by the French Government Laboratory of Excellence initiative no. ANR-10-LABX-0006, the Region Auvergne and the European Regional Development Fund (I.Pereira and M.Guitreau). M. Tedeschi is thankful to the National Council for Scientific and Technological Development and the Brazilian Research Council for financial support.

Data availability Data sharing is not applicable to this article as no datasets were generated or analysed during the current study.

References

Abdullin, F., Solé, J., Solari, L., Shchepetilnikova, V., Meneses-Rocha, J.J., Pavlinova, N. and Rodríguez-Trejo, A. 2016. Single-grain apatite geochemistry of

Permian–Triassic granitoids and Mesozoic and Eocene sandstones from Chiapas, southeast Mexico: implications for sediment provenance. *International Geology Review*, **58**, 1132–1157, https://doi.org/10.1080/00206814.2016.1150212

Adetunji, A. and Alao-Daniel, A.B. 2020. Low temperature grain-scale retrograde alteration of detrital minerals in Ajali formation from Benin Flank of Anambra Basin, Nigeria. *Ife Journal of Science*, **22**, 167–174, https://doi.org/10.4314/ijs.v22i3.14

Agangi, A., Reddy, S.M., Plavsa, D., Fougerouse, D., Clark, C., Roberts, M. and Johnson, T.E. 2019. Antimony in rutile as a pathfinder for orogenic gold deposits. *Ore Geology Reviews*, **106**, 1–11, https://doi.org/10.1016/j.oregeorev.2019.01.018

Akers, W.T., Grove, M., Harrison, T.M. and Ryerson, F.J. 1993. The instability of rhabdophane and its unimportance in monazite paragenesis. *Chemical Geology*, **110**, 169–176, https://doi.org/10.1016/0009-2541(93)90252-E

Aleinikoff, J.N., Wintsch, R.P., Fanning, C.M. and Dorais, M.J. 2002. U–Pb geochronology of zircon and polygenetic titanite from the Glastonbury Complex, Connecticut, USA: an integrated SEM, EMPA, TIMS, and SHRIMP study. *Chemical Geology*, **188**, 125–147, https://doi.org/10.1016/S0009-2541(02)00076-1

Allègre, C.J. and Rousseau, D. 1984. The growth of the continent through geological time studied by Nd isotope analysis of shales. *Earth and Planetary Science Letters*, **67**, 19–34, https://doi.org/10.1016/0012-821X(84)90035-9

Allègre, C.J., Albarède, F., Grünenfelder, M. and Köppel, V. 1974. ^{238}U/^{206}Pb–^{235}U/^{207}Pb–^{232}Th/^{208}Pb zircon geochronology in alpine and non-alpine environment. *Contributions to Mineralogy and Petrology*, **43**, 163–194, https://doi.org/10.1007/BF01134835

Altrin, J.S.A., García, P.C.M. and León, A.C.Z. 2018. Provenance discrimination between Atasta and Alvarado beach sands, western Gulf of Mexico, Mexico: constraints from detrital zircon chemistry and U–Pb geochronology. *Geological Journal*, **53**, 2824–2848, https://doi.org/10.1002/gj.3122

Alvaro, M., Mazzucchelli, M.L. *et al.* 2020. Fossil subduction recorded by quartz from the coesite stability field. *Geology*, **48**, 24–28, https://doi.org/10.1130/G46617.1

Amelin, Y., Lee, D., Halliday, A.N., Pidgeon, R.T. and Hills, J. 1999. Nature of the Earth's earliest crust from hafnium isotopes in single detrital zircons. *Nature*, **399**, 252–255, https://doi.org/10.1038/20426

Amelin, Y., Lee, D.C. and Halliday, A.N. 2000. Early-middle Archaean crustal evolution deduced from Lu–Hf and U–Pb isotopic studies of single zircon grains. *Geochimica et Cosmochimica Acta*, **64**, 4205–4225, https://doi.org/10.1016/S0016-7037(00)00493-2

Andersen, T., Kristoffersen, M. and Elburg, M.A. 2016. How far can we trust provenance and crustal evolution information from detrital zircons? A South African case study. *Gondwana Research*, **34**, 129–148, https://doi.org/10.1016/j.gr.2016.03.003

Andersen, T., Elburg, M.A. and Magwaza, B.N. 2019. Sources of bias in detrital zircon geochronology: discordance, concealed lead loss and common lead correction. *Earth-Science Reviews*, **197**, 102899, https://doi.org/10.1016/j.earscirev.2019.102899

Andersen, T., van Niekerk, H. and Elburg, M.A. 2022. Detrital zircon in an active sedimentary recycling system: challenging the 'source-to-sink' approach to zircon-based provenance analysis. *Sedimentology*, **69**, 2436–2462, https://doi.org/10.1111/sed.12996

Ansberque, C., Mark, C., Caul, J.T. and Chew, D.M. 2019. Combined in-situ determination of halogen (F, Cl) content in igneous and detrital apatite by SEM-EDS and LA-Q–ICPMS: a potential new provenance tool. *Chemical Geology*, **524**, 406–420, https://doi.org/10.1016/j.chemgeo.2019.07.012

Antoine, C., Bruand, E., Guitreau, M. and Devidal, J.L. 2020. Understanding preservation of primary signatures in apatite by comparing matrix and zircon-hosted crystals from the Eoarchean Acasta Gneiss Complex (Canada). *Geochemistry, Geophysics, Geosystems*, **21**, https://doi.org/10.1029/2020GC008923

Arboit, F., Min, M., Chew, D., Mitchell, A., Drost, K., Badenszki, E. and Daly, J.S. 2021. Constraining the links between the Himalayan belt and the Central Myanmar Basins during the Cenozoic: an integrated multi-proxy detrital geochronology and trace-element geochemistry study. *Geoscience Frontiers*, **12**, 657–676, https://doi.org/10.1016/j.gsf.2020.05.024

Arndt, N. and Davaille, A. 2013. Episodic Earth evolution. *Tectonophysics*, **609**, 661–674, https://doi.org/10.1016/j.tecto.2013.07.002

Arndt, N.T. and Goldstein, S.L. 1987. Use and abuse of crust-formation ages. *Geology*, **15**, 893–895, https://doi.org/10.1130/0091-7613(1987)15<893

Aubrecht, R., Méres, Š., Sýkora, M. and Mikuš, T. 2009. Provenance of the detrital garnets and spinels from the Albian sediments of the Czorsztyn Unit (Pieniny Klippen Belt, Western Carpathians, Slovakia). *Geologica Carpathica*, **60**, 463–483, https://doi.org/10.2478/v10096-009-0034-z

Ayers, J.C. and Watson, E.B. 1993. Rutile solubility and mobility in supercritical aqueous fluids. *Contributions to Mineralogy and Petrology*, **114**, 321–330, https://doi.org/10.1007/BF01046535

Baldwin, S.L., Schönig, J., Gonzalez, J.P., Davies, H. and von Eynatten, H. 2021. Garnet sand reveals rock recycling processes in the youngest exhumed high- and ultrahigh-pressure terrane on Earth. *Proceedings of the National Academy of Sciences of the United States of America*, **118**, https://doi.org/10.1073/pnas.2017231118

Barbarin, B. 1999. A review of the relationships between granitoid types, their origins and their geodynamic environments. *Lithos*, **46**, 605–626, https://doi.org/10.1016/S0024-4937(98)00085-1

Barham, M., Kirkland, C.L. and Handoko, A.D. 2022. Understanding ancient tectonic settings through detrital zircon analysis. *Earth and Planetary Science Letters*, **583**, 117425, https://doi.org/10.1016/j.epsl.2022.117425

Baxter, E.F., Caddick, M.J. and Ague, J.J. 2013. Garnet: Common mineral, uncommonly useful. *Elements*, **9**, 415–419, https://doi.org/10.2113/gselements.9.6.415

Baxter, E.F., Caddick, M.J. and Dragovic, B. 2017. Garnet: a rock-forming mineral petrochronometer. *Reviews in Mineralogy and Geochemistry*, **83**, 469–533, https://doi.org/10.2138/rmg.2017.83.15

Bea, F. and Montero, P. 1999. Behavior of accessory phases and redistribution of Zr, REE, Y, Th, and U during metamorphism and partial melting of metapelites in the lower crust: an example from the Kinzigite Formation of Ivrea–Verbano, NW Italy. *Geochimica et Cosmochimica Acta*, **63**, 1133–1153, https://doi.org/10.1016/S0016-7037(98)00292-0

Be'eri-Shlevin, Y., Avigad, D. and Gerdes, A. 2018. The White Nile as a source for Nile sediments: assessment using U–Pb geochronology of detrital rutile and monazite. *Journal of African Earth Sciences*, **140**, 1–8, https://doi.org/10.1016/j.jafrearsci.2017.12.032

Bell, E.A. 2016. Preservation of primary mineral inclusions and secondary mineralization in igneous zircon: a case study in orthogneiss from the Blue Ridge, Virginia. *Contributions to Mineralogy and Petrology*, **171**, 1–15, https://doi.org/10.1007/s00410-016-1236-x

Bell, E.A., Harrison, T.M., Kohl, I.E. and Young, E.D. 2014. Eoarchean crustal evolution of the Jack Hills zircon source and loss of Hadean crust. *Geochimica et Cosmochimica Acta*, **146**, 27–42, https://doi.org/10.1016/j.gca.2014.09.028

Bell, E.A., Boehnke, P., Harrison, T.M. and Wielicki, M.M. 2018. Mineral inclusion assemblage and detrital zircon provenance. *Chemical Geology*, **477**, 151–160, https://doi.org/10.1016/j.chemgeo.2017.12.024

Belousova, E.A., Walters, S., Griffin, W.L. and O'Reilly, S.Y. 2001. Trace-element signatures of apatites in granitoids from the Mt Isa Inlier, Northwestern Queensland. *Australian Journal of Earth Sciences*, **48**, 603–619, https://doi.org/10.1046/j.1440-0952.2001.00879.x

Belousova, E.A., Griffin, W.L., O'Reilly, S.Y. and Fisher, N.I. 2002. Igneous zircon: trace element composition as an indicator of source rock type. *Contributions to Mineralogy and Petrology*, **143**, 602–622, https://doi.org/10.1007/s00410-002-0364-7

Belousova, E.A., Kostitsyn, Y.A., Griffin, W.L., Begg, G.C., O'Reilly, S.Y. and Pearson, N.J. 2010. The growth of the continental crust: constraints from zircon Hf-isotope data. *Lithos*, **119**, 457–466, https://doi.org/10.1016/j.lithos.2010.07.024

Bónová, K., Mikuš, T. and Bóna, J. 2018. Is Cr-spinel geochemistry enough for solving the provenance dilemma? Case study from the palaeogene sandstones of the western Carpathians (eastern Slovakia). *Minerals*, **8**, https://doi.org/10.3390/min8120543

Bons, A.J. 1988. *Intracrystalline Deformation and Slaty Cleavage Development in Very Low-grade Slates from the Central Pyrenees*. Geologica Ultraiectina.

Borges, N., dos Santos, T.J.S., Tedeschi, M., Galante, D. and Luvizotto, G.L. 2023. *P–T–t* reconstruction of a coesite-bearing retroeclogite reveals a new UHP occurrence in the Western Gondwana margin (NE-Brazil). *Lithos*, **446–447**, 107138, https://doi.org/10.1016/J.LITHOS.2023.107138

Borghini, A., Ferrero, S., Wunder, B., Laurent, O., O'Brien, P.J. and Ziemann, M.A. 2018. Granitoid melt inclusions in orogenic peridotite and the origin of garnet clinopyroxenite. *Geology*, **46**, 1007–1010, https://doi.org/10.1130/G45316.1

Borghini, A., Ferrero, S., O'Brien, P.J., Laurent, O., Günter, C. and Ziemann, M.A. 2020. Cryptic metasomatic agent measured in situ in Variscan mantle rocks: Melt inclusions in garnet of eclogite, Granulitgebirge, Germany. *Journal of Metamorphic Geology*, **38**, 207–234, https://doi.org/10.1111/jmg.12519

Borisova, A.Y., Nédélec, A. *et al.* 2021. Hadean zircon formed due to hydrated ultramafic protocrust melting. *Geology*, **50**, 300–304, https://doi.org/10.1130/G49354.1

Bouch, J.E., Hole, M.J., Trewin, N.H., Chenery, S. and Morton, A.C. 2002. Authigenic apatite in a fluvial sandstone sequence: evidence for rare-earth element mobility during diagenesis and a tool for diagenetic correlation. *Journal of Sedimentary Research*, **72**, 59–67, https://doi.org/10.1306/040301720059

Bouvier, A., Takayuki, U., Kozdon, R., Valley, J.W., Jack, S.Á. and Ttg, H.Á. 2012. Li isotopes and trace elements as a petrogenetic tracer in zircon: insights from Archean TTGs and sanukitoids. *Contributions to Mineralogy and Petrology*, **163**, 745–768, https://doi.org/10.1007/s00410-011-0697-1

Brown, M. and Johnson, T. 2019. Time's arrow, time's cycle: granulite metamorphism and geodynamics. *Mineralogical Magazine*, **83**, 323–338, https://doi.org/10.1180/mgm.2019.19

Bruand, E., Storey, C. and Fowler, M. 2014. Accessory mineral chemistry of high Ba–Sr granites from Northern Scotland: constraints on petrogenesis and records of whole-rock Signature. *Journal of Petrology*, **55**, 1619–1651, https://doi.org/10.1093/petrology/egu037

Bruand, E., Storey, C. and Fowler, M. 2016. An apatite for progress: inclusions in zircon and titanite constrain petrogenesis and provenance. *Geology*, **44**, 91–94, https://doi.org/10.1130/G37301.1

Bruand, E., Storey, C., Fowler, M. and Heilimo, E. 2019. Oxygen isotopes in titanite and apatite, and their potential for crustal evolution research. *Geochimica et Cosmochimica Acta*, **255**, 144–162, https://doi.org/10.1016/j.gca.2019.04.002

Bruand, E., Fowler, M. *et al.* 2020. Accessory mineral constraints on crustal evolution: elemental fingerprints for magma discrimination. *Geochemical Perspectives Letters*, **13**, 7–12, https://doi.org/10.7185/geochemlet.2006

Burnham, A.D. and Berry, A.J. 2017. Formation of Hadean granites by melting of igneous crust. *Nature Geoscience*, **10**, 457–461.

Caby, R., Buscail, F., Dembélé, D., Diakité, S., Sacko, S. and Bal, M. 2008. Neoproterozoic garnet–glaucophanites and eclogites: new insights for subduction metamorphism of the Gourma fold and thrust belt (eastern Mali). *Geological Society, London, Special Publications*, **297**, 203–216, https://doi.org/10.1144/SP297.9

Caddick, M.J. and Kohn, M.J. 2013. Garnet: witness to the evolution of destructive plate boundaries. *Elements*, **9**, 427–432, https://doi.org/10.2113/gselements.9.6.427

Caddick, M.J., Konopásek, J. and Thompson, A.B. 2010. Preservation of garnet growth zoning and the duration of prograde metamorphism. *Journal of Petrology*, **51**, 2327–2347, https://doi.org/10.1093/petrology/egq059

Campbell, I.H. and Allen, C.M. 2008. Formation of supercontinents linked to increases in atmospheric oxygen. *Nature Geoscience*, **1**, 554–558, https://doi.org/10.1038/ngeo259

Campomenosi, N., Scambelluri, M. *et al.* 2021. Using the elastic properties of zircon–garnet host-inclusion pairs for thermobarometry of the ultrahigh-pressure Dora–Maira whiteschists: problems and perspectives. *Contributions to Mineralogy and Petrology*, **176**, 1–17, https://doi.org/10.1007/s00410-021-01793-6

Caracciolo, L. 2020. Sediment generation and sediment routing systems from a quantitative provenance analysis perspective: review, application and future development. *Earth-Science Reviews*, **209**, 103226, https://doi.org/10.1016/j.earscirev.2020.103226

Carlson, W.D. 2012. Rates and mechanism of Y, REE, and Cr diffusion in garnet. *American Mineralogist*, **97**, 1598–1618, https://doi.org/10.2138/am.2012.4108

Carpentier, M., Weis, D. and Chauvel, C. 2013. Large U loss during weathering of upper continental crust: the sedimentary record. *Chemical Geology*, **340**, 91–104, https://doi.org/10.1016/j.chemgeo.2012.12.016

Carrapa, B., DeCelles, P.G., Reiners, P.W., Gehrels, G.E. and Sudo, M. 2009. Apatite triple dating and white mica $^{40}Ar/^{39}Ar$ thermochronology of syntectonic detritus in the Central Andes: a multiphase tectonothermal history. *Geology*, **37**, 407–410, https://doi.org/10.1130/G25698A.1

Carter, A. and Foster, G.L. 2009. Improving constraints on apatite provenance: Nd measurement on fission-track-dated grains. *Geological Society, London, Special Publications*, **324**, 57–72, https://doi.org/10.1144/SP324.5

Cave, B.J., Stepanov, A.S., Large, R.R., Halpin, J.A. and Thompson, J. 2015. Release of trace elements through the sub-greenschist facies breakdown of detrital rutile to metamorphic titanite in the Otago Schist, New Zealand. *Canadian Mineralogist*, **53**, 379–400, https://doi.org/10.3749/canmin.1400097

Cavosie, J.A., Wilde, S.A., Liu, D, Weiblen, P.W. and Valley, J.W. 2004. Internal zoning and U-Th-Pb chemistry of Jack Hills detrital zircons: a mineral record of early Archean to Mesoproterozoic (4348-1576 Ma) magmatism. *Precambrian Research*, **135**, 251–279.

Cavosie, A.J., Valley, J.W. and Wilde, S.A. 2019. Chapter 12 – The oldest terrestrial mineral record: thirty years of research on Hadean Zircon from Jack Hills, Western Australia. *In*: Martin J. Van Kranendonk, Vickie C. Bennett, and Elis Hoffmann, J. (eds) *Earth's Oldest Rocks* (2nd edn), Elsevier, 255–278, https://doi.org/10.1016/b978-0-444-63901-1.00012-5

Cawood, P.A., Hawkesworth, C.J., Pisarevsky, S.A., Dhuime, B., Capitanio, F.A. and Nebel, O. 2018. Geological archive of the onset of plate tectonics. *Philosophical Transactions of the Royal Society, A*, **376**, 20170405, https://doi.org/10.1098/rsta.2017.0405

Cawood, P.A., Hawkesworth, C.J. and Dhuime, B. 2012. Detrital zircon record and tectonic setting. *Geology*, **40**, 875–878, https://doi.org/10.1130/G32945.1

Cawood, P.A., Hawkesworth, C.J. and Dhuime, B. 2013. The continental record and the generation of continental crust. *Bulletin of the Geological Society of America*, **125**, 14–32, https://doi.org/10.1130/B30722.1

Che, X.D., Linnen, R.L., Wang, R.C., Groat, L.A. and Br, A.A. 2013. Distribution of trace and rare earth elements in titanite from tungsten and molybdenum deposits in Yukon and British Columbia, Canada. *Canadian Mineralogist*, **51**, 415–438, https://doi.org/10.3749/canmin.51.3.415

Cherniak, D.J. 2000. Rare earth element diffusion in apatite. *Geochimica et Cosmochimica Acta*, **64**, 3871–3885, https://doi.org/10.1016/S0016-7037(00)00467-1

Cherniak, D.J. 2010. Diffusion in accessory minerals: zircon, titanite, apatite, monazite and xenotime. *Reviews in Mineralogy and Geochemistry*, **72**, 827–869, https://doi.org/10.2138/rmg.2010.72.18

Cherniak, D.J. 2015. Nb and Ta diffusion in titanite. *Chemical Geology*, **413**, 44–50, https://doi.org/10.1016/j.chemgeo.2015.08.010

Cherniak, D.J. and Watson, E.B. 2003. Diffusion in Zircon history – a brief review of bulk-release and early lower-resolution diffusion measurements. *Reviews in Mineralogy & Geochemistry*, **53**, 113–143, https://doi.org/10.2113/0530089

Cherniak, D.J. and Watson, E.B. 2007. Ti diffusion in zircon. *Chemical Geology*, **242**, 470–483, https://doi.org/10.1016/j.chemgeo.2007.05.005

Cherniak, D.J., Manchester, J. and Watson, E.B. 2007. Zr and Hf diffusion in rutile. *Earth and Planetary Science Letters*, **261**, 267–279, https://doi.org/10.1016/j.epsl.2007.06.027

Chew, D.M. and Spikings, R.A. 2015. Geochronology and thermochronology using apatite: time and temperature, lower crust to surface. *Elements*, **11**, 189–194, https://doi.org/10.2113/gselements.11.3.189

Chew, D.M. and Spikings, R.A. 2021. Apatite U–Pb thermochronology: a review. *Minerals*, **11**, 1095, https://doi.org/10.3390/min11101095

Chew, D., O'Sullivan, G., Caracciolo, L., Mark, C. and Tyrrell, S. 2020. Sourcing the sand: accessory mineral fertility, analytical and other biases in detrital U–Pb provenance analysis. *Earth-Science Reviews*, **202**, 103093, https://doi.org/10.1016/j.earscirev.2020.103093

Collins, W.J., Belousova, E.A., Kemp, A.I.S. and Murphy, J.B. 2011. Two contrasting Phanerozoic orogenic systems revealed by hafnium isotope data. *Nature Geoscience*, **4**, 333–337, https://doi.org/10.1038/ngeo1127

Compston, W. and Pidgeon, R.T. 1986. Jack Hills, evidence of more very old detrital zircons in Western Australia. *Nature*, **321**, 766–769, https://doi.org/10.1038/321766a0

Condie, K.C. and Aster, R.C. 2010. Episodic zircon age spectra of orogenic granitoids: the supercontinent connection and continental growth. *Precambrian Research*, **180**, 227–236, https://doi.org/10.1016/j.precamres.2010.03.008

Connelly, J.N. 2001. Degree of preservation in igneous zonation in zircon as a signpost for concordancy in U/Pb geochronology. *Chemical Geology*, **172**, 25–39, https://doi.org/10.1016/S0009-2541(00)00234-5

Cooperdock, E.H.G., Hofmann, F., Tibbetts, R., Carrera, A., Takase, A. and Celestian, A.J. 2022. Rapid phase identification of apatite and zircon grains for geochronology using X-ray micro-computed tomography. *Geochronology*, **4**, 501–515, https://doi.org/10.5194/gchron-4-501-2022

Čopjaková, R., Sulovský, P. and Paterson, B.A. 2005. Major and trace elements in pyrope–almandine garnets as sediment provenance indicators of the Lower Carboniferous Culm sediments, Drahany Uplands, Bohemian Massif. *Lithos*, **82**, 51–70, https://doi.org/10.1016/j.lithos.2004.12.006

Corfu, F., Hanchar, J.M., Hoskin, P.W.O. and Kinny, P. 2003. Atlas of zircon textures. *Reviews in Mineralogy and Geochemistry*, **53**, 469–500, https://doi.org/10.2113/0530469

Couzinié, S., Laurent, O., Moyen, J.F., Zeh, A., Bouilhol, P. and Villaros, A. 2016. Post-collisional magmatism: crustal growth not identified by zircon Hf–O isotopes. *Earth and Planetary Science Letters*, **456**, 182–195, https://doi.org/10.1016/j.epsl.2016.09.033

Cox, R.A., Dunning, G.R. and Indares, A. 1998. Petrology and U–Pb geochronology of mafic, high-pressure, metamorphic coronites from the Tshenukutish domain, eastern Grenville Province. *Precambrian Research*, **90**, 59–83, https://doi.org/10.1016/s0301-9268(98)00033-3

Davis, D.W., Poulsen, K.H. and Kamo, S.L. 1989. New insights into Archean crustal development from geochronology in the Rainy Lake Area, Superior Province, Canada. *The Journal of Geology*, **97**, 379–398, https://doi.org/10.1086/629318

Davis, D.W., Williams, I.S. and Krogh, T.E. 2003. Historical development of zircon geochronology. *Reviews in Mineralogy and Geochemistry*, **53**, 145–181, https://doi.org/10.1515/9781501509322-009

Dawson, J.B. and Stephens, W.E. 1975. Statistical classification of garnets from kimberlite and associated xenoliths. *The Journal of Geology*, **83**, 589–607, https://doi.org/10.1086/628143

Deal, E., Venditti, J.G., Benavides, S.J., Bradley, R., Zhang, Q., Kamrin, K. and Perron, J.T. 2023. Grain shape effects in bed load sediment transport. *Nature*, **613**, 298–302, https://doi.org/10.1038/s41586-022-05564-6

Deer, W.A., Howie, R.A. and Zussman, J. 2013. *An Introduction to the Rock-forming Minerals*. Mineralogical Society of Great Britain and Ireland, https://doi.org/10.1180/dhz

Deng, B., Chew, D., Jiang, L., Mark, C., Cogné, N., Wang, Z. and Liu, S. 2018. Heavy mineral analysis and detrital U–Pb ages of the intracontinental Paleo-Yangzte basin: implications for a transcontinental source-to-sink system during Late Cretaceous time. *Bulletin of the Geological Society of America*, **130**, 2087–2109, https://doi.org/10.1130/B32037.1

Depaolo, D.J. 1981. A neodymium and strontium isotopic study of the Mesozoic calc-alkaline granitic batholiths of the Sierra Nevada and Peninsular Ranges, California. *Journal of Geophysical Research*, **86**, 10470–10488, https://doi.org/10.1029/JB086iB11p10470

Dhuime, B., Hawkesworth, C.J., Cawood, P.A. and Storey, C.D. 2012. A change in the geodynamics of continental growth 3 billion years ago. *Science (New York)*, **335**, 1334–1337, https://doi.org/10.1126/science.1216066

Dickinson, W.R. 2008. Impact of differential zircon fertility of granitoid basement rocks in North America on age populations of detrital zircons and implications for granite petrogenesis. *Earth and Planetary Sciences Letters*, **275** (1), 80–92, https://doi.org/10.1016/j.epsl.2008.08.003

Dodson, M.H., Compston, W., Williams, I.S. and Wilson, J.F. 1988. A search for ancient detrital zircons in Zimbabwean sediments. *Journal of the Geological Society*, **145**, 977–983, https://doi.org/10.1144/gsjgs.145.6.0977

Drabon, N., Byerly, B.L., Byerly, G.R., Wooden, J.L., Brenhin Keller, C. and Lowe, D.R. 2021. Heterogeneous Hadean crust with ambient mantle affinity recorded in detrital zircons of the Green Sandstone Bed, South Africa. *Proceedings of the National Academy of Sciences of the United States of America*, **118**, 1–9, https://doi.org/10.1073/pnas.2004370118

Dröllner, M., Barham, M., Kirkland, C.L. and Ware, B. 2021. Every zircon deserves a date: selection bias in detrital geochronology. *Geological Magazine*, **158**, 1135–1142, https://doi.org/10.1017/S0016756821000145

Dröllner, M., Barham, M. and Kirkland, C.L. 2022. Gaining from loss: detrital zircon source-normalized α-dose discriminates first- v. multi-cycle grain histories. *Earth and Planetary Science Letters*, **579**, 117346, https://doi.org/10.1016/j.epsl.2021.117346

Dutch, R. and Hand, M. 2010. Retention of Sm–Nd isotopic ages in garnets subjected to high-grade thermal reworking: implications for diffusion rates of major and rare earth elements and the Sm–Nd closure temperature in garnet. *Contributions to Mineralogy and Petrology*, **159**, 93–112, https://doi.org/10.1007/s00410-009-0418-1

Emo, R.B., Smit, M.A. *et al.* 2018. Evidence for evolved Hadean crust from Sr isotopes in apatite within Eoarchean zircon from the Acasta Gneiss Complex. *Geochimica et Cosmochimica Acta*, **235**, 450–462, https://doi.org/10.1016/j.gca.2018.05.028

Enami, M., Nishiyama, T. and Mouri, T. 2007. Laser Raman microspectrometry of metamorphic quartz: a simple method for comparison of metamorphic pressures. *American Mineralogist*, **92**, 1303–1315, https://doi.org/10.2138/am.2007.2438

Ewing, R.C., Meldrum, A., Wang, L., Weber, W.J. and Corrales, L.R. 2003. Radiation effects in zircon. *Reviews in Mineralogy and Geochemistry*, **53**, 387–425, https://doi.org/10.2113/0530387

Ewing, T.A., Rubatto, D., Eggins, S.M. and Hermann, J. 2011. In situ measurement of hafnium isotopes in rutile by LA–MC–ICPMS: protocol and applications. *Chemical Geology*, **281**, 72–82, https://doi.org/10.1016/j.chemgeo.2010.11.029

Ewing, T.A., Rubatto, D. and Hermann, J. 2014. Hafnium isotopes and Zr/Hf of rutile and zircon from lower crustal metapelites (Ivrea–Verbano Zone, Italy): implications for chemical differentiation of the crust. *Earth and Planetary Science Letters*, **389**, 106–118, https://doi.org/10.1016/j.epsl.2013.12.029

Farver, J.R. 2010. Oxygen and hydrogen diffusion in minerals. *Reviews in Mineralogy and Geochemistry*, **72**, 447–507, https://doi.org/10.2138/rmg.2010.72.10

Fedo, C.M., Sircombe, K.N. and Rainbird, R.H. 2003. Detrital zircon analysis of the sedimentary record. *Reviews in Mineralogy and Geochemistry*, **53**, 277–303, https://doi.org/10.1515/9781501509322-013

Ferreira, J.A., Bento dos Santos, T., Pereira, I. and Mata, J. 2019. Tectonically assisted exhumation and cooling of Variscan granites in an anatectic complex of the Central Iberian Zone, Portugal: constraints from LA–ICP–MS zircon and apatite U–Pb ages. *International Journal of Earth Sciences*, **108**, 2153–2175, https://doi.org/10.1007/s00531-019-01755-1

Ferreira, J.A., Pereira, I., Bento dos Santos, T. and Mata, J. 2022. U–Pb age constraints on the protolith, cooling and exhumation of a Variscan middle crust migmatite complex from the Central Iberian Zone: insights into the Variscan metamorphic evolution and Ediacaran paleogeographic implications. *Journal of the Geological Society*, **179**, https://doi.org/10.1144/jgs2021-072

Finch, R.J. and Hanchar, J.M. 2003. Structure and chemistry of zircon and zircon-group minerals. *Reviews in Mineralogy and Geochemistry*, **53**, 1–25, https://doi.org/10.2113/0530001

Finger, F. and Krenn, E. 2007. Three metamorphic monazite generations in a high-pressure rock from the Bohemian Massif and the potentially important role of apatite in stimulating polyphase monazite growth along a *PT* loop. *Lithos*, **95**, 103–115, https://doi.org/10.1016/j.lithos.2006.06.003

Fisher, C.M., Hanchar, J.M., Samson, S.D., Dhuime, B., Blichert-Toft, J., Vervoort, J.D. and Lam, R. 2011. Synthetic zircon doped with hafnium and rare earth elements: a reference material for in situ hafnium isotope analysis. *Chemical Geology*, **286**, 32–47, https://doi.org/10.1016/j.chemgeo.2011.04.013

Fisher, C.M., Vervoort, J.D. and Hanchar, J.M. 2014. Guidelines for reporting zircon Hf isotopic data by LA–MC–ICPMS and potential pitfalls in the interpretation of these data. *Chemical Geology*, **363**, 125–133, https://doi.org/10.1016/j.chemgeo.2013.10.019

Fisher, C.M., Bauer, A.M. and Vervoort, J.D. 2020. Disturbances in the Sm–Nd isotope system of the Acasta Gneiss Complex – implications for the Nd isotope record of the early Earth. *Earth and Planetary Science Letters*, **530**, https://doi.org/10.1016/j.epsl.2019.115900

Flowerdew, M.J., Fleming, E.J., Morton, A.C., Frei, D., Chew, D.M. and Daly, J.S. 2019. Assessing mineral fertility and bias in sedimentary provenance studies: examples from the Barents Shelf. *Geological Society, London, Special Publications*, **484**, 255–274, https://doi.org/10.1144/sp484.11

Foland, K.A. and Allen, J.C. 1991. Magma sources for Mesozoic anorogenic granites of the White Mountain magma series, New England, USA. *Contributions to Mineralogy and Petrology*, **109**, 195–211, https://doi.org/10.1007/BF00306479

Forbes, C., Giles, D., Freeman, H., Sawyer, M. and Normington, V. 2015. Glacial dispersion of hydrothermal monazite in the Prominent Hill deposit: an exploration tool. *Journal of Geochemical Exploration*, **156**, 10–33, https://doi.org/10.1016/j.gexplo.2015.04.011

Force, E.R. 1991. Geology of Titanium Mineral Deposits, Special Paper. *Geological Society of America*, **259**, https://doi.org/10.1130/SPE259-p1

Foster, G.L. and Carter, A. 2007. Insights into the patterns and locations of erosion in the Himalaya – a combined fission-track and in situ Sm–Nd isotopic study of detrital apatite. *Earth and Planetary Science Letters*, **257**, 407–418, https://doi.org/10.1016/j.epsl.2007.02.044

François, C., Debaille, V., Paquette, J.L., Baudet, D. and Javaux, E.J. 2018. The earliest evidence for modern-style plate tectonics recorded by HP–LT metamorphism in the Paleoproterozoic of the Democratic Republic of the Congo. *Scientific Reports*, **8**, 1–10, https://doi.org/10.1038/s41598-018-33823-y

Frost, B.R., Chamberlain, K.R. and Schumacher, J.C. 2001. Sphene (titanite): phase relations and role as a geochronometer. *Chemical Geology*, **172**, 131–148, https://doi.org/10.1016/S0009-2541(00)00240-0

Froude, D.O., Ireland, T.R., Kinny, P.D., Williams, I.S., Compston, W., Williams, I.R. and Myers, J.S. 1983. Ion microprobe identification of 4100–4200 Myr-old terrestrial zircons. *Nature*, **304**, 616–618, https://doi.org/10.1038/304616a0

Fumes, R.A., Luvizotto, G.L., Pereira, I. and Moraes, R. In review. Trace element changes in rutile geochemistry in quartzite through increasing P-T from lower amphibolite to eclogite facies conditions. *Geological Society, London, Special Publications*, **537**, https://doi.org/10.1144/SP537-2022-207

Ganade de Araujo, C.E., Rubatto, D., Hermann, J., Cordani, U.G., Caby, R. and Basei, M.A.S. 2014. Ediacaran 2500-km-long synchronous deep continental subduction in the West Gondwana Orogen. *Nature Communications*, **5**, 1–8, https://doi.org/10.1038/ncomms6198

Garber, J.M., Hacker, B.R., Kylander-Clark, A.R.C., Stearns, M. and Seward, G. 2017. Controls on trace element uptake in metamorphic titanite: implications for petrochronology. *Journal of Petrology*, **58**, 1031–1057, https://doi.org/10.1093/petrology/egx046

Garçon, M. and Chauvel, C. 2014. Where is basalt in river sediments, and why does it matter? *Earth and Planetary Science Letters*, **407**, 61–69, https://doi.org/10.1016/j.epsl.2014.09.033

Garçon, M., Chauvel, C., France-Lanord, C., Huyghe, P. and Lavé, J.Ô. 2013. Continental sedimentary processes decouple Nd and Hf isotopes. *Geochimica et Cosmochimica Acta*, **121**, 177–195, https://doi.org/10.1016/j.gca.2013.07.027

Garçon, M., Carlson, R.W., Shirey, S.B., Arndt, N.T., Horan, M.F. and Mock, T.D. 2017. Erosion of Archean continents: the Sm–Nd and Lu–Hf isotopic record of Barberton sedimentary rocks. *Geochimica et Cosmochimica Acta*, **206**, 216–235, https://doi.org/10.1016/j.gca.2017.03.006

Gärtner, A., Hofmann, M. *et al.* 2022. Implications for sedimentary transport processes in southwestern Africa: a combined zircon morphology and age study including extensive geochronology databases. *International Journal of Earth Sciences*, **111**, 767–788, https://doi.org/10.1007/s00531-021-02146-1

Garzanti, E. and Andò, S. 2007*a*. Heavy mineral concentration in modern sands: implications for provenance interpretation. *Developments in Sedimentology*, **58**, 517–545, https://doi.org/10.1016/S0070-4571(07)58020-9

Garzanti, E. and Andò, S. 2007*b*. Plate tectonics and heavy mineral suites of modern sands. *Developments in Sedimentology*, **58**, 741–763, https://doi.org/10.1016/S0070-4571(07)58029-5

Garzanti, E. and Andò, S. 2019. Heavy minerals for junior woodchucks. *Minerals*, **9**, 1–25, https://doi.org/10.3390/min9030148

Garzanti, E., Ando, S., Vezzoli, G. and Dell'era, D. 2003. From rifted margins to foreland basins: investigating provenance and sediment dispersal across desert Arabia (Oman, UAE). *Journal of Sedimentary Research*, **73**, 572–588.

Garzanti, E., Vezzoli, G. *et al.* 2004. Collision – Orogen Provenance (Western Alps): Detrital Signatures and Unroofing Trends. *The Journal of Geology*, **1123**, 2, https://doi.org/10.1086/381655

Garzanti, E., Vezzoli, G., Andò, S., Paparella, P. and Clift, P.D. 2005. Petrology of Indus River sands: a key to interpret erosion history of the Western Himalayan Syntaxis. *Earth and Planetary Science Letters*, **229**, 287–302, https://doi.org/10.1016/j.epsl.2004.11.008

Garzanti, E., Andò, S., France-Lanord, C., Vezzoli, G., Censi, P., Galy, V. and Najman, Y. 2010. Mineralogical and chemical variability of fluvial sediments. 1. Bedload sand (Ganga–Brahmaputra, Bangladesh). *Earth and Planetary Science Letters*, **299**, 368–381, https://doi.org/10.1016/j.epsl.2010.09.017

Garzanti, E., Padoan, M., Andò, S., Resentini, A., Vezzoli, G. and Lustrino, M. 2013. Weathering and relative durability of detrital minerals in equatorial climate: sand petrology and geochemistry in the east African rift. *Journal of Geology*, **121**, 547–580, https://doi.org/10.1086/673259

Garzanti, E., Vermeesch, P., Andò, S., Lustrino, M., Padoan, M. and Vezzoli, G. 2014. Ultra-long distance littoral transport of Orange sand and provenance of the Skeleton Coast Erg (Namibia). *Marine Geology*, **357**, 25–36, https://doi.org/10.1016/j.margeo.2014.07.005

Garzanti, E., Vermeesch, P. *et al.* 2019. Congo River sand and the equatorial quartz factory. *Earth-Science Reviews*, **197**, 102918, https://doi.org/10.1016/j.earscirev.2019.102918

Garzanti, E., Bayon, G., Dennielou, B., Barbarano, M., Limonta, M. and Vezzoli, G. 2021. The Congo deep-sea fan: mineralogical, REE, and ND-isotope variability in Quartzose passive-margin sand. *Journal of Sedimentary Research*, **91**, 433–450, https://doi.org/10.2110/JSR.2020.100

Gaschnig, R.M. 2019. Benefits of a multiproxy approach to detrital mineral provenance analysis: an example from the Merrimack River, New England, USA. *Geochemistry, Geophysics, Geosystems*, **20**, 1557–1573, https://doi.org/10.1029/2018GC008005

Gaschnig, R.M., Vervoort, J.D., Lewis, R.S. and Tikoff, B. 2011. Isotopic evolution of the idaho batholith and Challis intrusive province, Northern US Cordillera. *Journal of Petrology*, **52**, 2397–2429, https://doi.org/10.1093/petrology/egr050

Gaschnig, R.M., Horan, M.F., Rudnick, R.L., Vervoort, J.D. and Fisher, C.M. 2022. History of crustal growth in Africa and the Americas from detrital zircon and Nd isotopes in glacial diamictites. *Precambrian Research*, **373**, 106641, https://doi.org/10.1016/j.precamres.2022.106641

Gaudette, H.E., Vitrac-Michard, A. and Allègre, C.J. 1981. North American Precambrian history recorded in a single sample: high-resolution UPb systematics of the Potsdam sandstone detrital zircons, New York State. *Earth and Planetary Science Letters*, **54**, 248–260, https://doi.org/10.1016/0012-821X(81)90008-X

Gehrels, G.E., Dickinson, W.R., Ross, G.M., Stewart, J.H. and Howell, D.G. 1995. Detrital zircon reference for Cambrian to Triassic miogeoclinal strata of western North America. *Geology*, **23**, 831–834, https://doi.org/10.1130/0091-7613(1995)023<0831:DZRFCT>2.3.CO;2

Gerdes, A. and Zeh, A. 2009. Zircon formation v. zircon alteration – new insights from combined U–Pb and Lu–Hf in-situ LA–ICP–MS analyses, and consequences for the interpretation of Archean zircon from the Central Zone of the Limpopo Belt. *Chemical Geology*, **261**, 230–243, https://doi.org/10.1016/j.chemgeo.2008.03.005

Gillespie, J., Glorie, S., Khudoley, A. and Collins, A.S. 2018. Detrital apatite U–Pb and trace element analysis as a provenance tool: insights from the Yenisey Ridge (Siberia). *Lithos*, **314–315**, 140–155, https://doi.org/10.1016/j.lithos.2018.05.026

Glorie, S., Gillespie, J. *et al.* 2022. Detrital apatite Lu–Hf and U–Pb geochronology applied to the southwestern Siberian margin. *Terra Nova*, **34**, 201–209, https://doi.org/10.1111/ter.12580

Glorie, S., Hand, M. *et al.* 2023. Robust laser ablation Lu–Hf dating of apatite: an empirical evaluation. *Geological Society, London, Special Publications*, **537**, https://doi.org/10.1144/SP537-2022-205

Gonzalez, J.P., Mazzucchelli, M.L., Angel, R.J. and Alvaro, M. 2021. Elastic geobarometry for anisotropic inclusions in anisotropic host minerals: quartz-in-zircon. *Journal of Geophysical Research: Solid Earth*, **126**, https://doi.org/10.1029/2021JB022080

Gordon, S.M., Kirkland, C.L. *et al.* 2021. Deformation-enhanced recrystallization of titanite drives decoupling between U–Pb and trace elements. *Earth and Planetary Science Letters*, **560**, 116810, https://doi.org/10.1016/j.epsl.2021.116810

Grand'Homme, A., Janots, E. *et al.* 2018. Mass transport and fractionation during monazite alteration by anisotropic replacement. *Chemical Geology*, **484**, 51–68, https://doi.org/10.1016/j.chemgeo.2017.10.008

Grapes, R., Roser, B. and Kashai, K. 2001. Composition of monocrystalline detrital and authigenic minerals, metamorphic grade, and provenance of Torlesse and Waipapa graywacke, central North Island, New Zealand. *International Geology Review*, **43**, 139–175, https://doi.org/10.1080/00206810109465005

Greber, N.D., Dauphas, N., Bekker, A., Ptáček, M.P., Bindeman, I.N. and Hofmann, A. 2017. Titanium isotopic evidence for felsic crust and plate tectonics 3.5 billion years ago. *Science (New York)*, **357**, 1271–1274, https://doi.org/10.1126/science.aan8086

Griffin, W.L., Pearson, N.J., Belousova, E., Jackson, S.E., Van Achterbergh, E., O'Reilly, S.Y. and Shee, S.R. 2000. The Hf isotope composition of cratonic mantle: LAM–MC–ICPMS analysis of zircon megacrysts in kimberlites. *Geochimica et Cosmochimica Acta*, **64**, 133– 147, https://doi.org/10.1016/S0016-7037(99)00343-9

Grimes, C.B., John, B.E. *et al.* 2007. Trace element chemistry of zircons from oceanic crust: a method for distinguishing detrital zircon provenance. *Geology*, **35**, 643–646, https://doi.org/10.1130/G23603A.1

Grimes, C.B., Wooden, J.L., Cheadle, M.J. and John, B.E. 2015. 'Fingerprinting' tectono-magmatic provenance using trace elements in igneous zircon. *Contributions to Mineralogy and Petrology*, **170**, 1–26, https://doi.org/10.1007/s00410-015-1199-3

Grütter, H.S., Gurney, J.J., Menzies, A.H. and Winter, F. 2004. An updated classification scheme for mantle-derived garnet, for use by diamond explorers. *Lithos*, **77**, 841–857, https://doi.org/10.1016/j.lithos.2004.04.012

Guitreau, M. and Blichert-Toft, J. 2014. Implications of discordant U–Pb ages on Hf isotope studies of detrital zircons. *Chemical Geology*, **385**, 17–25, https://doi.org/10.1016/j.chemgeo.2014.07.014

Guitreau, M., Blichert-Toft, J., Martin, H., Mojzsis, S.J. and Albarède, F. 2012. Hafnium isotope evidence from Archean granitic rocks for deep-mantle origin of continental crust. *Earth and Planetary Science Letters*, **337–338**, 211–223, https://doi.org/10.1016/j.epsl.2012.05.029

Guitreau, M., Mora, N. and Paquette, J.L. 2018. Crystallization and disturbance histories of single zircon crystals from Hadean-Eoarchean Acasta Gneisses examined by LA–ICP–MS U–Pb traverses. *Geochemistry, Geophysics, Geosystems*, **19**, 272–291, https://doi.org/10.1002/2017GC007310

Guitreau, M., Boyet, M. *et al.* 2019. Hadean protocrust reworking at the origin of the Archean Napier Complex (Antarctica). *Geochemical Perspectives Letters*, **12**, 7–11, https://doi.org/10.7185/GEOCHEMLET.1927

Guitreau, M., Gannoun, A., Deng, Z., Chaussidon, M., Moynier, F., Barbarin, B. and Marin-Carbonne, J. 2022. Stable isotope geochemistry of silicon in granitoid zircon. *Geochimica et Cosmochimica Acta*, **316**, 273–294, https://doi.org/10.1016/j.gca.2021.09.029

Guo, R., Hu, X., Garzanti, E., Lai, W., Yan, B. and Mark, C. 2020. How faithfully do the geochronological and geochemical signatures of detrital zircon, titanite, rutile and monazite record magmatic and metamorphic events? A case study from the Himalaya and Tibet. *Earth-Science Reviews*, **201**, 103082, https://doi.org/10.1016/j.earscirev.2020.103082

Hammerli, J., Kemp, A.I.S. and Spandler, C. 2014. Neodymium isotope equilibration during crustal metamorphism revealed by in situ microanalysis of REE-rich accessory minerals. *Earth and Planetary Science Letters*, **392**, 133–142, https://doi.org/10.1016/j.epsl.2014.02.018

Hammerli, J., Kemp, A.I.S. and Whitehouse, M.J. 2019. In situ trace element and Sm–Nd isotope analysis of accessory minerals in an Eoarchean tonalitic gneiss from Greenland: implications for Hf and Nd isotope decoupling in Earth's ancient rocks. *Chemical Geology*, **524**, 394–405, https://doi.org/10.1016/j.chemgeo.2019.06.025

Hanchar, J.M. and Miller, C.F. 1993. Zircon zonation patterns as revealed by cathodoluminescence and backscattered electron images: implications for interpretation of complex crustal histories. *Chemical Geology*, **110**, 1–13, https://doi.org/10.1016/0009-2541(93)90244-D

Hansley, P.L. 1987. Petrologic and experimental evidence for the etching of garnets by organic acids in the upper Jurassic Morrison formation, Northwestern New Mexico. *Journal of Sedimentary Petrology*, **57**, 666–681.

Hardman, M.F., Pearson, D.G., Stachel, T. and Sweeney, R.J. 2018. Statistical approaches to the discrimination of crust- and mantle-derived low-Cr garnet – major-element-based methods and their application in diamond exploration. *Journal of Geochemical Exploration*, **186**, 24–35, https://doi.org/10.1016/j.gexplo.2017.11.012

Harley, S.L. and Kelly, N.M. 2007. Zircon: tiny but timely. *Elements*, **3**, 13–18, https://doi.org/10.2113/gselements.3.1.13

Harrison, T.M. 2009. The Hadean crust: evidence from >4 ga zircons. *Annual Review of Earth and Planetary Sciences*, **37**, 479–505, https://doi.org/10.1146/annurev.earth.031208.100151

Harrison, T.M. 2020. *Hadean Earth*. Springer, https://doi.org/10.1007/978-3-030-46687-9

Hart, E., Storey, C., Bruand, E., Schertl, H.P. and Alexander, B.D. 2016. Mineral inclusions in rutile: a novel recorder of HP–UHP metamorphism. *Earth and Planetary Science Letters*, **446**, 137–148, https://doi.org/10.1016/j.epsl.2016.04.035

Hart, E., Storey, C., Harley, S.L. and Fowler, M. 2018. A window into the lower crust: trace element systematics and the occurrence of inclusions/intergrowths in granulite-facies rutile. *Gondwana Research*, **59**, 76–86, https://doi.org/10.1016/j.gr.2018.02.021

Hawkesworth, C.J. and Kemp, A.I.S. 2006*a*. The differentiation and rates of generation of the continental crust. *Chemical Geology*, **226**, 134–143, https://doi.org/10.1016/j.chemgeo.2005.09.017

Hawkesworth, C.J. and Kemp, A.I.S. 2006*b*. Using hafnium and oxygen isotopes in zircons to unravel the record of crustal evolution. *Chemical Geology*, **226**, 144–162, https://doi.org/10.1016/j.chemgeo.2005.09.018

Hawkesworth, C., Cawood, P., Kemp, T., Storey, C. and Dhuime, B. 2009. A matter of preservation. *Science (New York)*, **323**, 49–50, https://doi.org/10.1126/science.1168549

Hayden, L.A., Watson, E.B. and Wark, D.A. 2008. A thermobarometer for sphene (titanite). *Contribution*, **155**, 529–540, https://doi.org/10.1007/s00410-007-0256-y

Henrichs, I.A., O'Sullivan, G., Chew, D.M., Mark, C., Babechuk, M.G., McKenna, C. and Emo, R. 2018. The trace element and U–Pb systematics of metamorphic apatite. *Chemical Geology*, **483**, 218–238, https://doi.org/10.1016/j.chemgeo.2017.12.031

Henrichs, I.A., Chew, D.M., O'Sullivan, G.J., Mark, C., McKenna, C. and Guyett, P. 2019. Trace element (Mn–Sr–Y–Th–REE) and U–Pb isotope systematics of metapelitic apatite during progressive greenschist- to amphibolite-facies barrovian metamorphism. *Geochemistry, Geophysics, Geosystems*, **20**, 4103–4129, https://doi.org/10.1029/2019GC008359

Hietpas, J., Samson, S., Moecher, D. and Schmitt, A.K. 2010. Recovering tectonic events from the sedimentary record: detrital monazite plays in high fidelity. *Geology*, **38**, 167–170, https://doi.org/10.1130/G30265.1

Hinton, R.W. and Long, J.V.P. 1979. High-resolution ion-microprobe measurement of lead isotopes: variations within single zircons from Lac Seul, northwestern Ontario. *Earth and Planetary Science Letters*, **45**, 309–325, https://doi.org/10.1016/0012-821X(79)90132-8

Hirata, T. and Nesbitt, R.W. 1995. U–Pb isotope geochronology of zircon: evaluation of the laser probe–inductively coupled plasma mass spectrometry technique. *Geochimica et Cosmochimica Acta*, **59**,

2491–2500, https://doi.org/10.1016/0016-7037(95)00144-1

Holland, H.D. and Gottfried, D. 1955. The effect of nuclear radiation on the structure of zircon. *Acta Crystallographica*, **8**, 291–300, https://doi.org/10.1107/S0365110X55000947.

Hong, D., Jian, X., Fu, L. and Zhang, W. 2020. Garnet trace element geochemistry as a sediment provenance indicator: an example from the Qaidam basin, northern Tibet. *Marine and Petroleum Geology*, **116**, 104316, https://doi.org/10.1016/j.marpetgeo.2020.104316

Hopkins, M., Harrison, T.M. and Manning, C.E. 2008. Low heat flow inferred from >4 Gyr zircons suggests Hadean plate boundary interactions. *Nature*, **456**, 493–496, https://doi.org/10.1038/nature07465

Hoskin, P.W.O. 2005. Trace-element composition of hydrothermal zircon and the alteration of Hadean zircon from the Jack Hills, Australia. *Geochimica et Cosmochimica Acta*, **69**, 637–648, https://doi.org/10.1016/j.gca.2004.07.006

Hoskin, P.W.O. and Ireland, T.R. 2000. Rare earth element chemistry of zircon and its use as a provenance indicator. *Geology*, **28**, 627–630, https://doi.org/10.1130/0091-7613(2000)028<0627:REECOZ>2.3.CO;2

Hoskin, P.W.O. and Schaltegger, U. 2003. The composition of zircon and igneous and metamorphic petrogenesis. *Reviews in Mineralogy & Geochemistry*, **53**, 27–62, https://doi.org/10.1515/9781501509322-005

Ibañez-Mejia, M., Pullen, A., Pepper, M., Urbani, F., Ghoshal, G. and Ibañez-Mejia, J.C. 2018. Use and abuse of detrital zircon U–Pb geochronology-a case from the Río Orinoco delta, eastern Venezuela. *Geology*, **46**, 1019–1022, https://doi.org/10.1130/G45596.1

Iizuka, T., Hirata, T., Komiya, T., Rino, S., Katayama, I., Motoki, A. and Maruyama, S. 2005. U–Pb and Lu–Hf isotope systematics of zircons from the Mississippi River sand: implications for reworking and growth of continental crust. *Geology*, **33**, 485–488, https://doi.org/10.1130/G21427.1

Iizuka, T., Komiya, T., Rino, S., Maruyama, S. and Hirata, T. 2010. Detrital zircon evidence for Hf isotopic evolution of granitoid crust and continental growth. *Geochimica et Cosmochimica Acta*, **74**, 2450–2472, https://doi.org/10.1016/j.gca.2010.01.023

Iizuka, T., Nebel, O. and McCulloch, M.T. 2011. Tracing the provenance and recrystallization processes of the Earth's oldest detritus at Mt. Narryer and Jack Hills, Western Australia: an in situ Sm–Nd isotopic study of monazite. *Earth and Planetary Science Letters*, **308**, 350–358, https://doi.org/10.1016/j.epsl.2011.06.006

Iizuka, T., Campbell, I.H., Allen, C.M., Gill, J.B., Maruyama, S. and Makoka, F. 2013. Evolution of the African continental crust as recorded by U–Pb, Lu–Hf and O isotopes in detrital zircons from modern rivers. *Geochimica et Cosmochimica Acta*, **107**, 96–120, https://doi.org/10.1016/j.gca.2012.12.028

Iizuka, T., Yamaguchi, T., Itano, K., Hibiya, Y. and Suzuki, K. 2017. What Hf isotopes in zircon tell us about crust–mantle evolution. *Lithos*, **274–275**, 304–327, https://doi.org/10.1016/j.lithos.2017.01.006

Isozaki, Y., Yamamoto, S. *et al.* 2018. High-reliability zircon separation for hunting the oldest material on Earth: an automatic zircon separator with image-processing/microtweezers-manipulating system and double-step dating. *Geoscience Frontiers*, **9**, 1073–1083, https://doi.org/10.1016/j.gsf.2017.04.010

Itano, K., Iizuka, T., Chang, Q., Kimura, J.I. and Maruyama, S. 2016. U–Pb chronology and geochemistry of detrital monazites from major African rivers: constraints on the timing and nature of the Pan-African Orogeny. *Precambrian Research*, **282**, 139–156, https://doi.org/10.1016/j.precamres.2016.07.008

Itano, K., Ueki, K., Iizuka, T. and Kuwatani, T. 2020. Geochemical discrimination of monazite source rock based on machine learning techniques and multinomial logistic regression analysis. *Geosciences (Switzerland)*, **10**, https://doi.org/10.3390/geosciences10020063

Janots, E., Engi, M., Berger, A., Allaz, J., Schwarz, J. and Spandler, C. 2008. Prograde metamorphic sequence of REE minerals in pelitic rocks of the Central Alps: implications for allanite–monazite–xenotime phase relations from 250 to 610°C. *Journal of Metamorphic Geology*, **26**, 509–526, https://doi.org/10.1111/j.1525-1314.2008.00774.x

Jennings, E.S., Marschall, H.R., Hawkesworth, C.J. and Storey, C.D. 2011. Characterization of magma from inclusions in zircon: apatite and biotite work well, feldspar less so. *Geology*, **39**, 863–866, https://doi.org/10.1130/G32037.1

Jian, X., Guan, P., Fu, S.T., Zhang, D.W., Zhang, W. and Zhang, Y.S. 2014. Miocene sedimentary environment and climate change in the northwestern Qaidam basin, northeastern Tibetan Plateau: facies, biomarker and stable isotopic evidences. *Palaeogeography, Palaeoclimatology, Palaeoecology*, **414**, 320–331, https://doi.org/10.1016/j.palaeo.2014.09.011

Joshi, K.B., Banerji, U.S., Dubey, C.P. and Oliveira, E.P. 2021. Heavy minerals in provenance studies: an overview. *Arabian Journal of Geosciences*, **14**, https://doi.org/10.1007/s12517-021-07687-y

Kamber, B.S. 2015. The evolving nature of terrestrial crust from the Hadean, through the Archaean, into the Proterozoic. *Precambrian Research*, **258**, 48–82, https://doi.org/10.1016/j.precamres.2014.12.007

Kellett, D.A., Weller, O.M., Zagorevski, A. and Regis, D. 2018. A petrochronological approach for the detrital record: tracking mm-sized eclogite clasts in the northern Canadian Cordillera. *Earth and Planetary Science Letters*, **494**, 23–31, https://doi.org/10.1016/j.epsl.2018.04.036

Kelsey, D.E., Clark, C. and Hand, M. 2008. Thermobarometric modelling of zircon and monazite growth in melt-bearing systems: examples using model metapelitic and metapsammitic granulites. *Journal of Metamorphic Geology*, **26**, 199–212, https://doi.org/10.1111/j.1525-1314.2007.00757.x

Kemp, A.I.S., Hawkesworth, C.J., Paterson, B.A. and Kinny, P.D. 2006. Episodic growth of the Gondwana supercontinent from hafnium and oxygen isotopes in zircon. *Nature*, **439**, 580–583.

Kemp, A.I.S., Hawkesworth, C.J. *et al.* 2007. Magmatic and crustal differentiation history of granitic rocks from Hf–O isotopes in zircon. *Science (New York)*, **315**, 980–983, https://doi.org/10.1126/science.1136154

Kemp, A.I.S., Wilde, S.A. *et al.* 2010. Hadean crustal evolution revisited: new constraints from Pb–Hf isotope systematics of the Jack Hills zircons. *Earth and*

Planetary Science Letters, **296**, 45–56, https://doi.org/10.1016/j.epsl.2010.04.043

Kendall-Langley, L.A., Kemp, A.I.S., Hawkesworth, C.J., Craven, J., Talavera, C., Hinton, R. and Roberts, M.P. 2021. Quantifying F and Cl concentrations in granitic melts from apatite inclusions in zircon. *Contributions to Mineralogy and Petrology*, **176**, https://doi.org/10.1007/s00410-021-01813-5

Kenny, G.G., O'Sullivan, G.J., Alexander, S., Simms, M.J., Chew, D.M. and Kamber, B.S. 2019. On the track of a Scottish impact structure: a detrital zircon and apatite provenance study of the Stac Fada Member and wider Stoer Group, NW Scotland. *Geological Magazine*, **156**, 1863–1876, https://doi.org/10.1017/S0016756819000220

Kirkland, C.L., Yakymchuk, C., Gardiner, N.J., Szilas, K., Hollis, J., Olierook, H. and Steenfelt, A. 2020. Titanite petrochronology linked to phase equilibrium modelling constrains tectono-thermal events in the Akia Terrane, West Greenland. *Chemical Geology*, **536**, 119467, https://doi.org/10.1016/j.chemgeo.2020.119467

Klaver, M., MacLennan, S.A., Ibañez-Mejia, M., Tissot, F.L.H., Vroon, P.Z. and Millet, M.A. 2021. Reliability of detrital marine sediments as proxy for continental crust composition: the effects of hydrodynamic sorting on Ti and Zr isotope systematics. *Geochimica et Cosmochimica Acta*, **310**, 221–239, https://doi.org/10.1016/j.gca.2021.05.030

Klein, E.M. 2003. Geochemistry of the igneous oceanic crust. *In*: Holland, H.D. and Turenkian, K.K. (eds) *Teatrise on Geochemistry*. Pergamon, 433–463, https://doi.org/10.1016/B0-08-043751-6/03030-9

Kohn, M.J. 2017. Titanite petrochronology. *Reviews in Mineralogy and Geochemistry*, **83**, 419–441, https://doi.org/10.2138/rmg.2017.83.13

Kohn, M.J. 2020. A refined zirconium-in-rutile thermometer. *American Mineralogist*, **105**, 963–971, https://doi.org/10.2138/am-2020-7091

Kohn, M.J. and Penniston-Dorland, S.C. 2017. Diffusion: obstacles and opportunities in petrochronology. *Reviews in Mineralogy and Geochemistry*, **83**, 103–152, https://doi.org/10.2138/rmg.2017.83.4

Kohn, M.J., Penniston-Dorland, S.C. and Ferreira, J.C.S. 2016. Implications of near-rim compositional zoning in rutile for geothermometry, geospeedometry, and trace element equilibration. *Contributions to Mineralogy and Petrology*, **171**, 1–15, https://doi.org/10.1007/s00410-016-1285-1

Kohn, M.J., Engi, M. and Lanari, P. 2017. Petrochronology. Methods and Applications. *Mineralogical Society of America Reviews in Mineralogy and Geochemistry*, **83**, 575.

Komar, P.D. 2007. The entrainment, transport and sorting of heavy minerals by waves and currents. *Developments in Sedimentology*, **58**, 3–48, https://doi.org/10.1016/S0070-4571(07)58001-5

Komar, P.D. and Wang, C. 1984. Processes of selective grain transport and the formation of placers on beaches. *Journal of Geology*, **92**, 637–655, https://doi.org/10.1086/628903

Košler, J. and Sylvester, P.J. 2003. Present trends and the future of zircon in Geochronology: Laser ablation ICPMS. *Reviews in Mineralogy and Geochemistry*, **53**, 243–275.

Košler, J., Fonneland, H., Sylvester, P., Tubrett, M. and Pedersen, R.B. 2002. U–Pb dating of detrital zircons for sediment provenance studies – a comparison of laser ablation ICPMS and SIMS techniques. *Chemical Geology*, **182**, 605–618, https://doi.org/10.1016/S0009-2541(01)00341-2

Krenn, E., Ustaszewski, K. and Finger, F. 2008. Detrital and newly formed metamorphic monazite in amphibolite-facies metapelites from the Motajica Massif, Bosnia. *Chemical Geology*, **254**, 164–174, https://doi.org/10.1016/j.chemgeo.2008.03.012

Krippner, A., Meinhold, G., Morton, A.C. and Von Eynatten, H. 2014. Evaluation of garnet discrimination diagrams using geochemical data of garnets derived from various host rocks. *Sedimentary Geology*, **306**, 36–52, https://doi.org/10.1016/j.sedgeo.2014.03.004

Krippner, A., Meinhold, G., Morton, A.C. and Von Eynatten, H. 2016. Heavy-mineral and garnet compositions of stream sediments and HP–UHP basement rocks from the Western Gneiss Region, SW Norway. *Norwegian Journal of Geology*, **96**, 7–17, https://doi.org/10.17850/njg96-1-02

Kylander-Clark, A.R.C., Hacker, B.R. and Mattinson, J.M. 2008. Slow exhumation of UHP terranes: titanite and rutile ages of the Western Gneiss Region, Norway. *Earth and Planetary Science Letters*, **272**, 531–540, https://doi.org/10.1016/j.epsl.2008.05.019

Lanari, P. and Engi, M. 2017. Local bulk composition effects on metamorphic mineral assemblages. *Reviews in Mineralogy and Geochemistry*, **83**, 55–102, https://doi.org/10.2138/rmg.2017.83.3

Lanari, P., Giuntoli, F., Loury, C., Burn, M. and Engi, M. 2017. An inverse modeling approach to obtain *P–T* conditions of metamorphic stages involving garnet growth and resorption. *European Journal of Mineralogy*, **29**, 181–199, https://doi.org/10.1127/ejm/2017/0029-2597

Larsen, J.S., Keevil, N.B. and Harrison, H.C. 1952. Method for determining the age of igneous rocks using the accessory minerals. *Bulletin of the Geological Society of America*, **63**, 1045–1052, https://doi.org/10.1130/0016-7606(1952)63[1045:MFDTAO]2.0.CO;2

Laurent, O., Moyen, J.F., Wotzlaw, J.F., Björnsen, J. and Bachmann, O. 2022. Early Earth zircons formed in residual granitic melts produced by tonalite differentiation. *Geology*, **50**, 437–441, https://doi.org/10.1130/G49232.1

Lawrence, R.L., Cox, R., Mapes, R.W. and Coleman, D.S. 2011. Hydrodynamic fractionation of zircon age populations. *Bulletin of the Geological Society of America*, **123**, 295–305, https://doi.org/10.1130/B30151.1

Ledent, D., Patterson, C. and Tilton, G.R. 1964. Ages of zircon and feldspar concentrates from North American beach and river sands. *The Journal of Geology*, **72**, 112–122, https://doi.org/10.1086/626967

Lee, J.K.W., Williams, I.S. and Ellis, D.J. 1997. Pb, U and Th diffusion in natural zircon. *Nature*, **390**, 159–162, https://doi.org/10.1038/36554

Le Pera, E., Critelli, S. and Sorriso-Valvo, M. 2001. Weathering of gneiss in Calabria, southern Italy. *Catena*, **42**, 1–15, https://doi.org/10.1016/S0341-8162(00)00117-X

Li, D., Tan, C., Miao, F., Liu, Q., Zhang, Y. and Sun, X. 2019. Initiation of Zn–Pb mineralization in the Pingbao Pb–Zn skarn district, South China: constraints from

U–Pb dating of grossular-rich garnet. *Ore Geology Reviews*, **107**, 587–599, https://doi.org/10.1016/j.oregeorev.2019.03.011

Li, Y., Yang, Y.H., Jiao, S.J., Wu, F.Y., Yang, J.H., Xie, L.W. and Huang, C. 2015. In situ determination of hafnium isotopes from rutile using LA–MC–ICP–MS. *Science China Earth Sciences*, **58**, 2134–2144, https://doi.org/10.1007/s11430-015-5215-2

Liu, L., Stockli, D.F., Lawton, T.F., Xu, J., Stockli, L.D., Fan, M. and Nadon, G.C. 2022. Reconstructing source-to-sink systems from detrital zircon core and rim ages. *Geology*, **50**, 691–696, https://doi.org/10.1130/G49904.1

Liu, X.C., Wu, Y.B., Fisher, C.M., Hanchar, J.M., Beranek, L., Gao, S. and Wang, H. 2017. Tracing crustal evolution by U–Th–Pb, Sm–Nd, and Lu–Hf isotopes in detrital monazite and zircon from modern rivers. *Geology*, **45**, 103–106, https://doi.org/10.1130/G38720.1

Lünsdorf, N.K., Kalies, J., Ahlers, P., Dunkl, I. and von Eynatten, H. 2019. Semi-automated heavy-mineral analysis by Raman spectroscopy. *Minerals*, **9**, 385, https://doi.org/10.3390/min9070385

Luvizotto, G.L., Zack, T., Triebold, S. and Von Eynatten, H. 2009. Rutile occurrence and trace element behavior in medium-grade metasedimentary rocks: example from the Erzgebirge, Germany. *Mineralogy and Petrology*, **97**, 233–249, https://doi.org/10.1007/s00710-009-0092-z

Machado, N. and Gauthier, G. 1996. Determination of $^{207}Pb/^{206}Pb$ ages on zircon and monazite by laser-ablation ICPMS and application to a study of sedimentary provenance and metamorphism in southeastern Brazil. *Geochimica et Cosmochimica Acta*, **60**, 5063–5073, https://doi.org/10.1016/S0016-7037(96)00287-6

Machado, N. and Simonetti, A. 2001. U–Pb dating and Hf isotopic composition of zircon by laser-ablation–MC–ICP–MS. *In*: Sylvester, P. (ed) *Laser Ablation–ICPMS in the Earth Sciences: Principles and Applications*, Wiley, Canada, **29**, 121–146.

Makuluni, P., Kirkland, C.L. and Barham, M. 2019. Zircon grain shape holds provenance information: a case study from southwestern Australia. *Geological Journal*, **54**, 1279–1293, https://doi.org/10.1002/gj.3225

Malusà, M.G., Resentini, A. and Garzanti, E. 2016. Hydraulic sorting and mineral fertility bias in detrital geochronology. *Gondwana Research*, **31**, 1–19, https://doi.org/10.1016/j.gr.2015.09.002

Malusà, M.G., Wang, J., Garzanti, E., Liu, Z.C., Villa, I.M., and Wittmann, H. 2017. Trace-element and Nd-isotope systematics in detrital apatite of the Po river catchment: Implications for provenance discrimination and the lag-time approach to detrital thermochronology. *Lithos*, **290**, 48–59, http://doi.org/10.1016/j.lithos.2017.08.006

Maneiro, K.A., Baxter, E.F., Samson, S.D., Marschall, H.R. and Hietpas, J. 2019. Detrital garnet geochronology: application in tributaries of the French broad river, southern appalachian mountains, USA. *Geology*, **47**, 1189–1192, https://doi.org/10.1130/G46840.1

Maneiro, K.A., Jordan, M.K. and Baxter, E.F. 2022. Detrital garnet geochronology: a new window into ancient tectonics and sedimentary provenance. *In*: Sims, K.W.W., Maher, K. and Schrag, D.P. (eds) *Isotopic Constraints on Earth System Processes*. American Geophysical Union, https://doi.org/10.1002/9781119595007.ch09

Mange, M.A. and Morton, A.C. 2007. Geochemistry of heavy minerals. *Developments in Sedimentology*, **58**, 345–391, https://doi.org/10.1016/S0070-4571(07)58013-1

Manzotti, P. and Ballèvre, M. 2013. Multistage garnet in high-pressure metasediments: alpine overgrowths on variscan detrital grains. *Geology*, **41**, 1151–1154, https://doi.org/10.1130/G34741.1

Mao, M., Rukhlov, A.S., Rowins, S.M., Spence, J. and Coogan, L.A. 2016. Apatite trace element compositions: a robust new tool for mineral exploration. *Economic Geology*, **111**, 1187–1222, https://doi.org/10.2113/econgeo.111.5.1187

Marfin, A.E., Ivanov, A.V., Kamenetsky, V.S., Abersteiner, A., Yakich, T.Y. and Dudkin, T.V. 2020. Contact metamorphic and metasomatic processes at the Kharaelakh Intrusion, Oktyabrsk Deposit, Norilsk-Talnakh Ore District: application of LA–ICP–MS dating of perovskite, apatite, garnet, and titanite. *Economic Geology*, **115**, 1213–1226, https://doi.org/10.5382/ECONGEO.4744

Mark, C. and Stutenbecker, L. 2022. Detrital garnet Lu-Hf and U-Pb geochronometry coupled with compositional analysis: Possibilities and limitations as a sediment provenance indicator. *EGU22, 24th EGU General Assembly, 23–27 May, 2022, Vienna, Austria*, https://doi.org/10.5194/egusphere-egu22-6405

Markwitz, V. and Kirkland, C.L. 2018. Source to sink zircon grain shape: constraints on selective preservation and significance for Western Australian Proterozoic basin provenance. *Geoscience Frontiers*, **9**, 415–430, https://doi.org/10.1016/j.gsf.2017.04.004

Marmo, B.A., Clarke, G.L. and Powell, R. 2002. Fractionation of bulk rock composition due to porphyroblast growth: effects on eclogite facies mineral equilibria, Pam Peninsula, New Caledonia. *Journal of Metamorphic Geology*, **20**, 151–165, https://doi.org/10.1046/j.0263-4929.2001.00346.x

McLennan, S.M., Nance, W.B. and Taylor, S.R. 1980. Rare earth element–thorium correlations in sedimentary rocks, and the composition of the continental crust. *Geochimica et Cosmochimica Acta*, **44**, 1833–1839, https://doi.org/10.1016/0016-7037(80)90232-X

Meinhold, G., Anders, B., Kostopoulos, D. and Reischmann, T. 2008. Rutile chemistry and thermometry as provenance indicator: an example from Chios Island, Greece. *Sedimentary Geology*, **203**, 98–111, https://doi.org/10.1016/j.sedgeo.2007.11.004

Millonig, L.J., Albert, R., Gerdes, A., Avigad, D. and Dietsch, C. 2020. Exploring laser ablation U–Pb dating of regional metamorphic garnet – the Straits Schist, Connecticut, USA. *Earth and Planetary Science Letters*, **552**, 116589, https://doi.org/10.1016/j.epsl.2020.116589

Moecher, D.P. and Samson, S.D. 2006. Differential zircon fertility of source terranes and natural bias in the detrital zircon record: implications for sedimentary provenance analysis. *Earth and Planetary Science Letters*, **247**, 252–266, https://doi.org/10.1016/j.epsl.2006.04.035

Mojzsis, S.J., Harrison, T.M. and Pidgeon, R.T. 2001. Oxygen-isotope evidence from ancient zircons for liquid water at the Earth's surface 4300 Myr ago sources and may therefore provide information about the nature. *Nature*, **409**, 178–181, https://doi.org/10.1038/35051557

Moreira, H., Buzenchi, A., Hawkesworth, C.J. and Dhuime, B. 2023. Plumbing the depths of magma crystallization using 176 Lu/177 Hf in zircon as a pressure proxy. *Geology*, **XX**, 1–5, https://doi.org/10.1130/G50659.1/5761913/g50659.pdf

Morton, A.C. 1985. A new approach to provenance studies: electron microprobe analysis of detrital garnets from Middle Jurassic sandstones of the northern North Sea. *Sedimentology*, **32**, 553–566, https://doi.org/10.1111/j.1365-3091.1985.tb00470.x

Morton, A.C. 1987. Influences of provenance and diagenesis on detrital garnet suites in the Paleocene Forties sandstone, central North Sea. *SEPM Journal of Sedimentary Research*, **57**, 2–7, https://doi.org/10.1306/212f8cd8-2b24-11d7-8648000102c1865d

Morton, A.C. and Hallsworth, C.R. 1999. Processes controlling the composition of heavy mineral assemblages in sandstones. *Sedimentary Geology*, **124**, 3–29, https://doi.org/10.1016/S0037-0738(98)00118-3

Morton, A.C. and Hallsworth, C. 2007. Stability of detrital heavy minerals during burial diagenesis. *Developments in Sedimentology*, **58**, 215–245, https://doi.org/10.1016/S0070-4571(07)58007-6

Morton, A., Hallsworth, C. and Chalton, B. 2004. Garnet compositions in Scottish and Norwegian basement terrains: a framework for interpretation of North Sea sandstone provenance. *Marine and Petroleum Geology*, **21**, 393–410, https://doi.org/10.1016/j.marpetgeo.2004.01.001

Morton, A.C., Borg, G., Hansley, P.L., Haughton, P.D.W., Krinsley, D.H. and Trusty, P. 1989. The origin of faceted garnets in sandstones: dissolution or overgrowth? *Sedimentology*, **36**, 927–942, https://doi.org/10.1111/j.1365-3091.1989.tb01754.x

Morton, A.C., Claoué-Long, J.C. and Berge, C. 1996. SHRIMP constraints on sediment provenance and transport history in the Mesozoic Statfjord Formation, North Sea. *Journal of the Geological Society*, **153**, 915–929, https://doi.org/10.1144/gsjgs.153.6.0915

Moser, A.C., Hacker, B.R., Gehrels, G.E., Seward, G.G.E., Kylander, A.R.C. and Garber, J.M. 2022. Linking titanite U–Pb dates to coupled deformation and dissolution–reprecipitation. *Contributions to Mineralogy and Petrology*, **177**, 1–27, https://doi.org/10.1007/s00410-022-01906-9

Motta, R.G., Fitzsimons, I.C.W., Moraes, R., Johnson, T.E., Schuindt, S. and Benetti, B.Y. 2021. Recovering *P–T–t* paths from ultra-high temperature (UHT) felsic orthogneiss: an example from the Southern Brasília Orogen, Brazil. *Precambrian Research*, **359**, 106222, https://doi.org/10.1016/j.precamres.2021.106222

Moyen, J.-F. and Martin, H. 2012. Forty years of TTG research. *Lithos*, **148**, 312–336.

Mulder, J.A. and Cawood, P.A. 2022. Evaluating preservation bias in the continental growth record against the monazite archive. *Geology*, **50**, 243–247, https://doi.org/10.1130/G49416.1

Murakami, T., Chakoumakos, B.C., Ewing, R., Lumpkin, G.R. and Weber, W.J. 1991. Alpha-decay event damage in zircon. *American Mineralogist*, **76**, 1510–1532.

Nagy, K.L. 1995. Dissolution and precipitation kinetics of sheet silicates. *Reviews in Mineralogy & Geochemistry*, **31**, 173–234, https://doi.org/10.1515/9781501509650-007

Najman, Y., Garzanti, E., Pringle, M., Bickle, M., Stix, J. and Khan, I. 2003. Early-Middle miocene paleodrainage and tectonics in the Pakistan Himalaya. *Bulletin of the Geological Society of America*, **115**, 1265–1277, https://doi.org/10.1130/B25165.1

Nebel, O., Rapp, R.P. and Yaxley, G.M. 2014. The role of detrital zircons in Hadean crustal research, *Lithos* 190–191, 313–327, https://doi.org/10.1016/j.lithos.2013.12.010

Nédélec, A. and Bouchez, J.-L. 2015. Precambrian granitic rocks. *In*: Nédélec, A., Bouchez, J.-L. and Bowden, P. (eds) *Granites: Petrology, Structure, Geological Setting, and Metallogeny*, Oxford, https://doi.org/10.1093/acprof:oso/9780198705611.003.0013

Nutman, A.P. 2007. Apatite recrystallisation during prograde metamorphism, Cooma, southeast Australia: implications for using an apatite-graphite association as a biotracer in ancient metasedimentary rocks. *Australian Journal of Earth Sciences*, **54**, 1023–1032, https://doi.org/10.1080/08120090701488321

Olierook, H.K.H., Taylor, R.J.M. *et al.* 2019. Unravelling complex geologic histories using U–Pb and trace element systematics of titanite. *Chemical Geology*, **504**, 105–122, https://doi.org/10.1016/j.chemgeo.2018.11.004

Oliver, G.J.H., Chen, F., Buchwaldt, R. and Hegner, E. 2000. Fast tectonometamorphism and exhumation in the type area of the Barrovian and Buchan zones. *Geology*, **28**, 459–462, https://doi.org/10.1130/0091-7613(2000)28<459:FTAEIT>2.0.CO;2

O'Sullivan, G.J., Chew, D.M. and Samson, S.D. 2016. Detecting magma-poor orogens in the detrital record. *Geology*, **44**, 871–874, https://doi.org/10.1130/G38245.1

O'Sullivan, G.J., Chew, D.M., Morton, A.C., Mark, C. and Henrichs, I.A. 2018. An integrated apatite geochronology and geochemistry tool for sedimentary provenance analysis. *Geochemistry, Geophysics, Geosystems*, **19**, 1309–1326, https://doi.org/10.1002/2017GC007343

O'Sullivan, G., Chew, D., Kenny, G., Henrichs, I. and Mulligan, D. 2020. The trace element composition of apatite and its application to detrital provenance studies. *Earth-Science Reviews*, **201**, 103044, https://doi.org/10.1016/j.earscirev.2019.103044

Papapavlou, K., Darling, J.R., Lightfoot, P.C., Lasalle, S., Gibson, L., Storey, C.D. and Moser, D. 2018*a*. Polyorogenic reworking of ore-controlling shear zones at the South Range of the Sudbury impact structure: a telltale story from in situ U–Pb titanite geochronology. *Terra Nova*, **30**, 254–261, https://doi.org/10.1111/ter.12332

Papapavlou, K., Darling, J.R. *et al.* 2018*b*. U–Pb isotopic dating of titanite microstructures: potential implications for the chronology and identification of large impact structures. *Contributions to Mineralogy and Petrology*, **173**, 1–15, https://doi.org/10.1007/s00410-018-1511-0

Papoulis, D., Tsolis-Katagas, P. and Katagas, C. 2004. Monazite alteration mechanisms and depletion measurements in kaolins. *Applied Clay Science*, **24**, 271–285, https://doi.org/10.1016/j.clay.2003.08.011

Park, C., Song, Y., Chung, D., Kang, I.M., Khulganakhuu, C. and Yi, K. 2016. Recrystallization and hydrothermal growth of high U–Th zircon in the Weondong deposit, Korea: record of post-magmatic alteration. *Lithos*, **260**, 268–285, https://doi.org/10.1016/j.lithos.2016.05.026

Parman, S.W. 2015. Time-lapse zirconography: imaging punctuated continental evolution. *Geochemical Perspectives Letters*, **1**, 43–52, https://doi.org/10.7185/geochemlet.1505

Parrish, R. 1990. U–Pb dating of monazite and its application to geological problems. *Canadian Journal of Earth* Sciences, 27, Number 11, https://doi.org/10.1139/e90-152.

Patchett, P.J., Kouvo, O., Hedge, C.E. and Tatsumoto, M. 1981. Evolution Of Continental-Crust And Mantle Heterogeneity - Evidence From Hf Isotopes. *Contributions to Mineralogy and Petrology*, **78**, 279–297.

Paul, A.N., Spikings, R.A., Chew, D. and Daly, J.S. 2019. The effect of intra-crystal uranium zonation on apatite U–Pb thermochronology: a combined ID–TIMS and LA–MC–ICP–MS study. *Geochimica et Cosmochimica Acta*, **251**, 15–35, https://doi.org/10.1016/j.gca.2019.02.013

Payne, J.L., McInerney, D.J., Barovich, K.M., Kirkland, C.L., Pearson, N.J. and Hand, M. 2016. Strengths and limitations of zircon Lu–Hf and O isotopes in modelling crustal growth. *Lithos*, **248–251**, 175–192, https://doi.org/10.1016/j.lithos.2015.12.015

Pereira, I. and Storey, C.D. 2023. Detrital rutile: records of the deep crust, ores and fluids. *Lithos*, **438–439**, 107010, https://doi.org/10.1016/j.lithos.2022.107010

Pereira, I., Storey, C., Darling, J., Lana, C. and Alkmim, A.R. 2019. Two billion years of evolution enclosed in hydrothermal rutile: recycling of the São Francisco Craton Crust and constraints on gold remobilisation processes. *Gondwana Research*, **68**, 69–92, https://doi.org/10.1016/j.gr.2018.11.008

Pereira, I., Storey, C.D., Strachan, R.A., Bento dos Santos, T. and Darling, J.R. 2020. Detrital rutile ages can deduce the tectonic setting of sedimentary basins. *Earth and Planetary Science Letters*, **537**, 116193, https://doi.org/10.1016/j.epsl.2020.116193

Pereira, I., Storey, C.D., Darling, J.R., Moreira, H., Strachan, R.A. and Cawood, P.A. 2021. Detrital rutile tracks the first appearance of subduction zone low *T*/*P* paired metamorphism in the Palaeoproterozoic. *Earth and Planetary Science Letters*, **570**, 117069, https://doi.org/10.1016/j.epsl.2021.117069

Petrus, J.A., Chew, D.M., Leybourne, M.I. and Kamber, B.S. 2017. A new approach to laser-ablation inductively-coupled-plasma mass-spectrometry (LA–ICP–MS) using the flexible map interrogation tool 'Monocle'. *Chemical Geology*, **463**, 76–93, https://doi.org/10.1016/j.chemgeo.2017.04.027

Pettijohn, F.J. 1957. *Sedimentary Rocks*, 2nd edn. *Geological Magazine*, **94**, 6, https://doi.org/10.1017/S0016756800070254

Pettingill, H.S. and Patchett, P.J. 1981. Lu–Hf total-rock age for the Amîtsoq gneisses, West Greenland. *Earth and Planetary Science Letters*, **55**, 150–156, https://doi.org/10.1016/0012-821X(81)90093-5

Pidgeon, R.T., O'Neil, J.R. and Silver, L.T. 1966. Uranium and lead isotopic stability in a metamict zircon under experimental hydrothermal conditions. *Science (New York)*, **154**, 1538–1540, https://doi.org/10.1126/science.154.3756.1538

Pidgeon, R.T., Nemchin, A.A. and Whitehouse, M.J. 2017. The effect of weathering on U–Th–Pb and oxygen isotope systems of ancient zircons from the Jack Hills, Western Australia. *Geochimica et Cosmochimica Acta*, **197**, 142–166, https://doi.org/10.1016/j.gca.2016.10.005

Pidgeon, R.T., Nemchin, A.A., Roberts, M.P., Whitehouse, M.J., Bellucci, J.J. and Hills, J. 2019. The accumulation of non-formula elements in zircons during weathering: ancient zircons from the Jack Hills, Western Australia. *Chemical Geology*, **530**, 119310, https://doi.org/10.1016/j.chemgeo.2019.119310

Pitra, P., Poujol, M., Van den Driessche, J., Bretagne, E., Lotout, C. and Cogné, N. 2022. Late Variscan (315 Ma) subduction or deceptive zircon REE patterns and U–Pb dates from migmatite-hosted eclogites? (Montagne Noire, France). *Journal of Meta*, **40**, 39–65, https://doi.org/10.1111/jmg.12609

Plavsa, D., Reddy, S.M., Agangi, A., Clark, C., Kylander-Clark, A. and Tiddy, C.J. 2018. Microstructural, trace element and geochronological characterization of TiO_2 polymorphs and implications for mineral exploration. *Chemical Geology*, **476**, 130–149, https://doi.org/10.1016/j.chemgeo.2017.11.011

Poldervaart, A. and Eckelmann, F.D. 1955. Growth Phenomena in zircon of autochthonous granites. *Bulletin of the Geological Society of America*, **66**, 947–948, https://doi.org/10.1130/0016-7606(1955)66[947:GPIZOA]2.0.CO;2

Porten, K.W., Walderhaug, O. and Torkildsen, G. 2015. Apatite overgrowth cement as a possible diagenetic temperature-history indicator. *Journal of Sedimentary Research*, **85**, 1478–1491, https://doi.org/10.2110/jsr.2015.99

Porter, J.K., McNaughton, N.J., Evans, N.J. and McDonald, B.J. 2020. Rutile as a pathfinder for metals exploration. *Ore Geology Reviews*, **120**, https://doi.org/10.1016/j.oregeorev.2020.103406

Pupin, J.P. 1980. Zircon and granite petrology. *Contributions to Mineralogy and Petrology*, **73**, 207–220, https://doi.org/10.1007/BF00381441

Qasim, M., Ding, L., Asif, M., Baral, U., Jadoon, I.A.K., Umar, M. and Imran, M. 2020. Provenance of the Hangu Formation, Lesser Himalaya, Pakistan: insight from the detrital zircon U–Pb dating and spinel geochemistry. *Palaeoworld*, **29**, 729–743, https://doi.org/10.1016/j.palwor.2019.12.003

Raimondo, T., Payne, J., Wade, B., Lanari, P., Clark, C. and Hand, M. 2017. Trace element mapping by LA–ICP–MS: assessing geochemical mobility in garnet. *Contributions to Mineralogy and Petrology*, **172**, 1–22, https://doi.org/10.1007/s00410-017-1339-z

Ranjan, S., Upadhyay, D., Lochan, K. and Nanda, J.K. 2020. Detrital zircon evidence for change in geodynamic regime of continental crust formation 3.7–3. 6 billion years ago. *Earth and Planetary Science Letters*, **538**, 116206, https://doi.org/10.1016/j.epsl.2020.116206

Rapp, J.F., Klemme, S., Butler, I.B. and Harley, S.L. 2010. Extremely high solubility of rutile in chloride and fluoride-bearing metamorphic fluids: an experimental investigation. *Geology*, **38**, 323–326, https://doi.org/10.1130/G30753.1

Rasmussen, B. and Muhling, J.R. 2007. Monazite begets monazite: evidence for dissolution of detrital monazite and reprecipitation of syntectonic monazite during low-grade regional metamorphism. *Contributions to Mineralogy and Petrology*, **154**, 675–689, https://doi.org/10.1007/s00410-007-0216-6

Rasmussen, B. and Muhling, J.R. 2009. Reactions destroying detrital monazite in greenschist-facies sandstones from the Witwatersrand basin, South Africa. *Chemical Geology*, **264**, 311–327, https://doi.org/10.1016/j.chemgeo.2009.03.017

Reimink, J.R., Davies, J.H.F.L. and Ielpi, A. 2021. Global zircon analysis records a gradual rise of continental crust throughout the Neoarchean. *Earth and Planetary Science Letters*, **554**, 116654, https://doi.org/10.1016/j.epsl.2020.116654

Richter, F., Lana, C., Stevens, G., Buick, I., Pedrosa-Soares, A.C., Alkmim, F.F. and Cutts, K. 2016. Sedimentation, metamorphism and granite generation in a back-arc region: records from the Ediacaran Nova Venécia Complex (Araçuaí Orogen, Southeastern Brazil). *Precambrian Research*, **272**, 78–100, https://doi.org/10.1016/j.precamres.2015.10.012

Roberts, N.M.W. and Spencer, C.J. 2015. The zircon archive of continent formation through time. *Geological Society, London, Special Publications*, **389**, 197–225, https://doi.org/10.1144/SP389.14

da Rocha, B.C., de Moraes, R., Möller, A., Cioffi, C.R. and Jercinovic, M.J. 2017. Timing of anatexis and melt crystallization in the Socorro–Guaxupé Nappe, SE Brazil: insights from trace element composition of zircon, monazite and garnet coupled to UPb geochronology. *Lithos*, **277**, 337–355, https://doi.org/10.1016/j.lithos.2016.05.020

Rollinson, H. 2017. There were no large volumes of felsic continental crust in the early Earth. *Geosphere*, **13**, 235–246, https://doi.org/10.1130/GES01437.1

Ross, G.M., Parrish, R.R. and Dudas, F.O. 1991. Provenance of the Bonner Formation (Belt Supergroup), Montana: insights from U–Pb and Sm–Nd analyses of detrital minerals. *Geology*, **19**, 340–343, https://doi.org/10.1130/0091-7613(1991)019<0340:POTBFB>2.3.CO;2

Roth, A.S., Bourdon, B., Mojzsis, S.J., Rudge, J.F., Guitreau, M. and Blichert-Toft, J. 2014. Combined ^{147,146}Sm–^{143,142}Nd constraints on the longevity and residence time of early terrestrial crust. *Geochemistry, Geophysics, Geosystems*, **15**, 2329–2345, https://doi.org/10.1002/2014GC005313

Rubatto, D. 2002. Zircon trace element geochemistry: partitioning with garnet and the link between U–Pb ages and metamorphism. *Chemical Geology*, **184**, 123–138, https://doi.org/10.1016/S0009-2541(01)00355-2

Rubatto, D. 2017. Zircon: the metamorphic mineral. *Reviews in Mineralogy and Geochemistry*, **83**, 261–295, https://doi.org/10.2138/rmg.2017.83.9

Rubatto, D. and Hermann, J. 2003. Zircon formation during fluid circulation in eclogites (Monviso, Western Alps): implications for Zr and Hf budget in subduction zones. *Geochimica et Cosmochimica Acta*, **67**, 2173–2187, https://doi.org/10.1016/S0016-7037(02)01321-2

Rubatto, D., Müntener, O., Barnhoorn, A. and Gregory, C.J. 2008. Dissolution–reprecipitation of zircon at low-temperature, high-pressure conditions (Lanzo Massif, Italy). *American Mineralogist*, **93**, 1519–1529, https://doi.org/10.2138/am.2008.2874

Schärer, U. and Allègre, C.J. 1982. Uranium–lead system in fragments of a single zircon grain. *Nature*, **295**, 585–587, https://doi.org/10.1038/295585a0

Schiller, D. and Finger, F. 2019. Application of Ti-in-zircon thermometry to granite studies: problems and possible solutions. *Contributions to Mineralogy and Petrology*, **174**, 1–16, https://doi.org/10.1007/s00410-019-1585-3

Schmidt, D., Schmidt, T.S., Mullis, J., Mählmann, R.F. and Frey, M. 1997. Very low grade metamorphism of the Taveyanne formation of western Switzeland. *Contributions to Mineralogy and Petrology*, **129**, 385–403, https://doi.org/10.1007/s004100050344

Schoene, B. 2014. U–Th–Pb Geochronology. *In*: Heinrich D. Holland and Karl K. Turekian (eds) *Treatise on Geochemistry* (2nd edn), Elsevier, 341–378, https://doi.org/10.1016/B978-0-08-095975-7.00310-7

Schönig, J., Meinhold, G., Von Eynatten, H. and Lünsdorf, N.K. 2018. Tracing ultrahigh-pressure metamorphism at the catchment scale. *Scientific Reports*, **8**, 1–11, https://doi.org/10.1038/s41598-018-21262-8

Schönig, J., von Eynatten, H., Meinhold, G. and Lünsdorf, N.K. 2019. Diamond and coesite inclusions in detrital garnet of the Saxonian Erzgebirge, Germany. *Geology*, **47**, 715–718, https://doi.org/10.1130/G46253.1

Schönig, J., von Eynatten, H., Meinhold, G., Lünsdorf, N.K., Willner, A.P. and Schulz, B. 2020. Deep subduction of felsic rocks hosting UHP lenses in the central Saxonian Erzgebirge: implications for UHP terrane exhumation. *Gondwana Research*, **87**, 320–329, https://doi.org/10.1016/j.gr.2020.06.020

Schönig, J., von Eynatten, H., Tolosana-Delgado, R. and Meinhold, G. 2021*a*. Garnet major-element composition as an indicator of host-rock type: a machine learning approach using the random forest classifier. *Contributions to Mineralogy and Petrology*, **176**, 1–21, https://doi.org/10.1007/s00410-021-01854-w

Schönig, J., Von Eynatten, H. and Meinhold, G. 2021*b*. Life-cycle analysis of coesite-bearing garnet. *Geological Magazine*, **158**, 1421–1440, https://doi.org/10.1017/S0016756821000017

Schönig, J., von Eynatten, H., Meinhold, G. and Lünsdorf, N.K. 2022. The sedimentary record of ultrahigh-pressure metamorphism: a perspective review. *Earth-Science Reviews*, **227**, https://doi.org/10.1016/j.earscirev.2022.103985

Schulz, B. 2021. Monazite microstructures and their interpretation in petrochronology. *Frontiers in Earth Science*, **9**, 1–22, https://doi.org/10.3389/feart.2021.668566

Schulze, D. 2003. A classification scheme for mantle-derived garnets in kimberlite: a tool for investigating the mantle and exploring for diamonds. *Lithos*, **71**, 195–213, https://doi.org/10.1016/s0024-4937(03)00113-0

Scibiorski, E.A. and Cawood, P.A. 2022. Titanite as a petrogenetic indicator. *Terra Nova*, **34**, 177–183, https://doi.org/10.1111/ter.12574

Scibiorski, E., Kirkland, C.L., Kemp, A.I.S., Tohver, E. and Evans, N.J. 2019. Trace elements in titanite: a potential tool to constrain polygenetic growth processes and timing. *Chemical Geology*, **509**, 1–19, https://doi.org/10.1016/j.chemgeo.2019.01.006

Seman, S., Stockli, D.F. and McLean, N.M. 2017. U-Pb geochronology of grossular-andradite garnet. *Chemical Geology*, **460**, 106–116, https://doi.org/10.1016/j.chemgeo.2017.04.020

Şengün, F. and Zack, T. 2016. Trace element composition of rutile and Zr-in-rutile thermometry in meta-ophiolitic rocks from the Kazdağ Massif, NW Turkey. *Mineralogy and Petrology*, **110**, 547–560, https://doi.org/10.1007/s00710-016-0433-7

Shields, G. and Veizer, J. 2002. Precambrian marine carbonate isotope database: version 1.1. *Geochemistry, Geophysics, Geosystems*, **3**, https://doi.org/10.1029/2001GC000266

Silver, L.T. and Deutsch, S. 1963. Uranium-lead isotopic variations in zircons: a case study. *The Journal of Geology*, **71**, 721–758, https://doi.org/10.1086/626951

Simpson, A., Gilbert, S. *et al.* 2021. In-situ Lu–Hf geochronology of garnet, apatite and xenotime by LA ICP MS/MS. *Chemical Geology*, **577**, 120299, https://doi.org/10.1016/j.chemgeo.2021.120299

Sláma, J. and Koler, J. 2012. Effects of sampling and mineral separation on accuracy of detrital zircon studies. *Geochemistry, Geophysics, Geosystems*, **13**, 1–17, https://doi.org/10.1029/2012GC004106

Smythe, D.J. and Brenan, J.M. 2016. Magmatic oxygen fugacity estimated using zircon-melt partitioning of cerium. *Earth and Planetary Science Letters*, **453**, 260–266, https://doi.org/10.1016/j.epsl.2016.08.013

Spandler, C., Hermann, J. and Rubatto, D. 2004. Exsolution of thortveitite, yttrialite, and xenotime during low-temperature recrystallization of zircon from New Caledonia, and their significance for trace element incorporation in zircon. *American Mineralogist*, **89**, 1795–1806, https://doi.org/10.2138/am-2004-11-1226

Spear, F.S. and Kohn, M.J. 1996. Trace element zoning in garnet as a monitor of crustal melting. *Geology*, **24**, 1099–1102, https://doi.org/10.1130/0091-7613(1996)024<1099:TEZIGA>2.3.CO;2

Spear, F.S. and Pyle, J.M. 2002. Apatite, monazite, and xenotime in metamorphic rocks. *Phosphates: Geochemical, Geobiological and Materials Importance*, **48**, 293–336, https://doi.org/10.2138/rmg.2002.48.7

Spencer, C.J. 2020. Continuous continental growth as constrained by the sedimentary record. *American Journal of Science*, **320**, 373–401, https://doi.org/10.2475/04.2020.02

Spencer, C.J., Kirkland, C.L., Roberts, N.M.W., Evans, N.J. and Liebmann, J. 2020. Strategies towards robust interpretations of in situ zircon Lu–Hf isotope analyses. *Geoscience Frontiers*, **11**, 843–853, https://doi.org/10.1016/j.gsf.2019.09.004

Steiger, R.H. and Wasserburg, G.J. 1966. Systematics in the Pb^{208}-Th^{232}, Pb^{207}-U^{235}, and Pb^{206}-U^{238} Systems. *Journal of Geophysical Research*, **71**, 24, 6065–6090, https://doi.org/10.1029/JZ071i024p06065

Stevenson, R.K. and Patchett, P.J. 1990. Implications for the evolution of continental crust from Hf isotope systematics of Archean detrital zircons. *Geochimica et Cosmochimica Acta*, **54**, 1683–1697, https://doi.org/10.1016/0016-7037(90)90400-F

Stöckhert, B., Duyster, J., Trepmann, C. and Massonne, H.J. 2001. Microdiamond daughter crystals precipitated from supercritical COH+ silicate fluids included in garnet, Erzgebirge, Germany. *Geology*, **29**, https://doi.org/10.1130/0091-7613(2001)029<0391:MDCPFS>2.0.CO;2

Storey, C.D., Smith, M.P. and Jeffries, T.E. 2007. In situ LA–ICP–MS U–Pb dating of metavolcanics of Norrbotten, Sweden: records of extended geological histories in complex titanite grains. *Chemical Geology*, **240**, 163–181, https://doi.org/10.1016/j.chemgeo.2007.02.004

Stutenbecker, L., Berger, A. and Schlunegger, F. 2017. The potential of detrital garnet as a provenance proxy in the Central Swiss Alps. *Sedimentary Geology*, **351**, 11–20, https://doi.org/10.1016/j.sedgeo.2017.02.002

Stutenbecker, L., Tollan, P.M.E., Madella, A. and Lanari, P. 2019. Miocene basement exhumation in the Central Alps recorded by detrital garnet geochemistry in foreland basin deposits. *Solid Earth*, **10**, 1581–1595, https://doi.org/10.5194/se-10-1581-2019

Suggate, S.M. and Hall, R. 2014. Using detrital garnet compositions to determine provenance: a new compositional database and procedure. *Geological Society, London, Special Publications*, **386**, 373–393, https://doi.org/10.1144/SP386.8

Sullivan, G.J.O. and Chew, D.M. 2020. The clastic record of a Wilson Cycle: evidence from detrital apatite petrochronology of the Grampian-Taconic fore-arc. *Earth and Planetary Science Letters*, **552**, 116588, https://doi.org/10.1016/j.epsl.2020.116588

Suzuki, K., Adachi, M. and Kajizuka, I. 1994. Electron microprobe observations of Pb diffusion in metamorphosed detrital monazites. *Earth and Planetary Science Letters*, **128**, 391–405, https://doi.org/10.1016/0012-821X(94)90158-9

Tang, M., Chen, K. and Rudnick, R.L. 2016. Archean upper crust transition from mafic to felsic marks the onset of plate tectonics. *Science (New York)*, **351**, 372–375, https://doi.org/10.1126/science.aad5513

Tang, M., Ji, W., Chu, X., Wu, A. and Chen, C. 2021. Reconstructing crustal thickness evolution from europium anomalies in detrital zircons. *Geology*, **49**, 76–80, https://doi.org/10.1130/G47745.1/5144012/g47745.pdf

Tatsumoto, M. and Patterson, C. 1964. Age studies of zircon and feldspar concentrates from the Franconia sandstone. *The Journal of Geology*, **72**, 232–242, https://doi.org/10.1086/626978

Taylor, S.R. and McLennan, S.M. 1985. The continental crust: its composition and evolution, Oxford.

Tedeschi, M., Lanari, P. *et al.* 2017. Reconstruction of multiple *PTt* stages from retrogressed mafic rocks: subduction v. collision in the Southern Brasília orogen (SE Brazil). *Lithos*, **294**, 283–303, https://doi.org/10.1016/j.lithos.2017.09.025

Tedeschi, M., Leonardo, P. *et al.* 2023. Unravelling the protracted U–Pb zircon geochronological record of high to ultrahigh temperature metamorphic rocks: implications for provenance investigations. *Geoscience Frontiers*, **14**, 101515, https://doi.org/10.1016/j.gsf.2022.101515

Tilton, G.R. 1960. Volume diffusion as a mechanism for discordant lead ages. *Journal of Geophysical Research*, **65**, 2933–2945, https://doi.org/10.1029/jz065i009p02933

Tilton, G.R., Patterson, C., Brown, H., Inghram, M., Hayden, R., Hess, D. and Larssen, L.J. 1955. Isotopic composition and distribution of lead, uranium and thorium in precambrian granite. *Bulletin of the Geological Society of America*, **86**, 1131–1148, https://doi.org/10.1130/0016-7606(1955)66[1131:ICADOL]2.0.CO;2

Tilton, G.R., Davis, G.L., Wetherill, G.W. and Aldrich, L.T. 1957. Isotopic ages of zircon from granites and pegmatites. *Eos, Transactions American Geophysical Union*, **38**, 360–371, https://doi.org/10.1029/TR038i003p00360

Tole, M.P. 1985. The kinetics of dissolution of zircon ($ZrSiO_4$). *Geochimica et Cosmochimica Acta*, **49**, 453–458, https://doi.org/10.1016/0016-7037(85)90036-5

Tolosana-Delgado, R., von Eynatten, H., Krippner, A. and Meinhold, G. 2018. A multivariate discrimination scheme of detrital garnet chemistry for use in sedimentary provenance analysis. *Sedimentary Geology*, **375**, 14–26, https://doi.org/10.1016/j.sedgeo.2017.11.003

Tomaschek, F., Kennedy, A.K., Villa, I.M., Lagos, M. and Ballhaus, C. 2003. Zircons from Syros, Cyclades, Greece – recrystallization and mobilization of zircon during high-pressure metamorphism. *Journal of Petrology*, **44**, 1977–2002, https://doi.org/10.1093/petrology/egg067

Tomkins, H.S., Powell, R. and Ellis, D.J. 2007. The pressure dependence of the zirconium-in-rutile thermometer. *Journal of Metamorphic Geology*, **25**, 703–713, https://doi.org/10.1111/j.1525-1314.2007.00724.x

Trail, D., Watson, E.B. and Tailby, N.D. 2012. Ce and Eu anomalies in zircon as proxies for the oxidation state of magmas. *Geochimica et Cosmochimica Acta*, **97**, 70–87, https://doi.org/10.1016/j.gca.2012.08.032

Triebold, S., von Eynatten, H., Luvizotto, G.L. and Zack, T. 2007. Deducing source rock lithology from detrital rutile geochemistry: an example from the Erzgebirge, Germany. *Chemical Geology*, **244**, 421–436, https://doi.org/10.1016/j.chemgeo.2007.06.033

Triebold, S., von Eynatten, H. and Zack, T. 2012. A recipe for the use of rutile in sedimentary provenance analysis. *Sedimentary Geology*, **282**, 268–275, https://doi.org/10.1016/j.sedgeo.2012.09.008

Turner, G. and Morton, A.C. 2007. The effects of burial diagenesis on detrital heavy mineral grain surface textures. *Developments in Sedimentology*, **58**, 393–412, https://doi.org/10.1016/S0070-4571(07)58014-3

Twenhofel, W.H. 1941. The frontiers of sedimentary mineralogy and petrology. *Journal of Sedimentary Research*, **11**, 53–63, https://doi.org/10.1306/D42690EE-2B26-11D7-8648000102C1865D

Tyler, S.A. 1931. The petrography of some bottom samples from the North Pacific Ocean. *Journal of Sedimentary Research*, **1**, 12–22.

Valley, J.W. 2003. Oxygen isotopes in zircon. *Reviews in Mineralogy & Geochemistry*, **53**, 343–385, https://doi.org/10.2113/0530343

Vavra, G. 1990. On the kinematics of zircon growth and its petrogenetic significance: a cathodoluminescence study. *Contributions to Mineralogy and Petrology*, **106**, 90–99, https://doi.org/10.1007/BF00306410

Vermeesch, P. 2004. How many grains are needed for a provenance study? *Earth and Planetary Science Letters*, **224**, 441–451, https://doi.org/10.1016/j.epsl.2004.05.037

Vermeesch, P. 2012. On the visualisation of detrital age distributions. *Chemical Geology*, **312–313**, 190–194, https://doi.org/10.1016/j.chemgeo.2012.04.021

Vermeesch, P., Rittner, M., Petrou, E., Omma, J., Mattinson, C. and Garzanti, E. 2017. High throughput petrochronology and sedimentary provenance analysis by automated phase mapping and LAICPMS. *Geochemistry, Geophysics, Geosystems*, **18**, 4096–4109, https://doi.org/10.1002/2017GC007109

Vervoort, J.D. and Kemp, A.I.S. 2016. Clarifying the zircon Hf isotope record of crust–mantle evolution. *Chemical Geology*, **425**, 65–75, https://doi.org/10.1016/j.chemgeo.2016.01.023

Voice, P.J., Kowalewski, M. and Eriksson, K.A. 2011. Quantifying the timing and rate of crustal evolution: global compilation of radiometrically dated detrital zircon grains. *Journal of Geology*, **119**, 109–126, https://doi.org/10.1086/658295

Volante, S., Collins, W.J. *et al.* 2022. Spatio–temporal evolution of Mesoproterozoic magmatism in NE Australia: a hybrid tectonic model for final Nuna assembly. *Precambrian Research*, **372**, 106602, https://doi.org/10.1016/J.PRECAMRES.2022.106602

Volante, S., Blereau, E., Guitreau, Tedeschi, M., van Schijndel, V. and Cutts, K. 2023. Current applications using key mineral phases in igneous and metamorphic geology: perspectives for the future. *Geological Society, London, Special Publications*, 537, https://doi.org/10.1144/SP537-2022-254

von Eynatten, H., Tolosana-Delgado, R., Karius, V., Bachmann, K. and Caracciolo, L. 2016. Sediment generation in humid Mediterranean setting: grain-size and source-rock control on sediment geochemistry and mineralogy (Sila Massif, Calabria). *Sedimentary Geology*, **336**, 68–80, https://doi.org/10.1016/j.sedgeo.2015.10.008

Wang, C., Lai, Y.J., Foley, S.F., Liu, Y., Belousova, E., Zong, K. and Hu, Z. 2020. Rutile records for the cooling history of the Trans-North China orogen from assembly to break-up of the Columbia supercontinent. *Precambrian Research*, **346**, 105763, https://doi.org/10.1016/j.precamres.2020.105763

Watson, E.B., Wark, D.A. and Thomas, J.B. 2006. Crystallization thermometers for zircon and rutile. *Contributions to Mineralogy and Petrology*, **151**, 413–433, https://doi.org/10.1007/s00410-006-0068-5

Wetherill, G.W. 1963. Discordant uranium-lead ages: 2. Disordant ages resulting from diffusion of lead and uranium. *Journal of Geophysical Research*, **68**, 2957–2965, https://doi.org/10.1029/jz068i010p02957

White, A.F. 1995. Chemical weathering rates of silicate minerals in soils. *Reviews in Mineralogy & Geochemistry*, **31**, 407–462, https://doi.org/10.1515/9781501509650-011

White, A.F., Blum, A.E., Bullen, T.D., Vivit, D.V., Schulz, M. and Fitzpatrick, J. 1999. The effect of temperature on experimental and natural chemical weathering rates of granitoid rocks. *Geochimica et Cosmochimica Acta*, **63**, 3277–3291, https://doi.org/10.1016/S0016-7037(99)00250-1

Wilde, S.A., Valley, J.W., Peck, W.H. and Graham, C.M. 2001. Evidence from detrital zircons for the existence of continental crust and oceans on the Earth 4.4 Gyr ago. *Nature*, **409**, 175–178, https://doi.org/10.1038/35051550

Xiong, Q., Zheng, J., Griffin, W.L., O'Reilly, S.Y. and Pearson, N.J. 2012. Decoupling of U–Pb and Lu–Hf isotopes and trace elements in zircon from the UHP North Qaidam orogen, NE Tibet (China): tracing the deep subduction of continental blocks. *Lithos*, **155**, 125–145, https://doi.org/10.1016/j.lithos.2012.08.022

Zack, T. and Kooijman, E. 2017. Petrology and Geochronology of Rutile. *Reviews in Mineralogy and Geochemistry*, **83**, 443–467, https://doi.org/10.2138/rmg.2017.83.14

Zack, T., Kronz, A., Foley, S.F. and Rivers, T. 2002. Trace element abundances in rutiles from eclogites and associated garnet mica schists. *Chemical Geology*, **184**, 97–122, https://doi.org/10.1016/S0009-2541(01)00357-6

Zack, T., von Eynatten, H. and Kronz, A. 2004. Rutile geochemistry and its potential use in quantitative provenance studies. *Sedimentary Geology*, **171**, 37–58, https://doi.org/10.1016/j.sedgeo.2004.05.009

Zeh, A., Gerdes, A., Klemd, R. and Barton, J.M. 2008. U–Pb and Lu–Hf isotope record of detrital zircon grains from the Limpopo Belt – evidence for crustal recycling at the Hadean to early-Archean transition. *Geochimica et Cosmochimica Acta*, **72**, 5304–5329, https://doi.org/10.1016/j.gca.2008.07.033

Zeh, A., Stern, R.A. and Gerdes, A. 2014. The oldest zircons of Africa – their U–Pb–Hf–O isotope and trace element systematics, and implications for Hadean to Archean crust – mantle evolution. *Precambrian Research*, **241**, 203–230, https://doi.org/10.1016/j.precamres.2013.11.006

Zeh, A., Cabral, A.R., Koglin, N. and Decker, M. 2018. Rutile alteration and authigenic growth in metasandstones of the Moeda Formation, Minas Gerais, Brazil – a result of Transamazonian fluid–rock interaction. *Chemical Geology*, **483**, 397–409, https://doi.org/10.1016/j.chemgeo.2018.03.007

Zhang, R., Lehmann, B., Seltmann, R., Sun, W. and Li, C. 2017. Cassiterite U–Pb geochronology constrains magmatic-hydrothermal evolution in complex evolved granite systems: the classic Erzgebirge tin province (Saxony and Bohemia). *Geology*, **45**, 1095–1098, https://doi.org/10.1130/G39634.1

Zhang, L., Wu, J.L., Tu, J.R., Wu, D., Li, N., Xia, X.P. and Ren, Z.Y. 2020. RMJG rutile: a new natural reference material for microbeam U–Pb dating and Hf isotopic analysis. *Geostandards and Geoanalytical Research*, **44**, 133–145, https://doi.org/10.1111/ggr.12304

Zhong, X., Andersen, N.H., Dabrowski, M. and Jamtveit, B. 2019. Zircon and quartz inclusions in garnet used for complementary Raman thermobarometry: application to the Holsnøy eclogite, Bergen Arcs, Western Norway. *Contributions to Mineralogy and Petrology*, **174**, 1–17, https://doi.org/10.1007/s00410-019-1584-4

Zhou, G., Fisher, C.M., Luo, Y., Pearson, D.G., Li, L., He, Y. and Wu, Y. 2020. A clearer view of crustal evolution: U–Pb, Sm–Nd, and Lu–Hf isotope systematics in five detrital minerals unravel the tectonothermal history of northern China. *GSA Bulletin*, **132**, 2367–2381, https://doi.org/10.1130/b35515.1

Zhou, T., Qiu, K., Wang, Y., Yu, H. and Hou, Z. 2022. Apatite Eu/Y–Ce discrimination diagram: a big data based approach for provenance classification. *Acta Petrologica Sinica*, **38**, 291–299, https://doi.org/10.18654/1000-0569/2022.01.19

Zhu, Z., Zhang, W. *et al*. 2023. Magmatic crystallization drives zircon Zr isotopic variations in a large granite batholith. *Geochimica et Cosmochimica Acta*, **342**, 15–30, https://doi.org/10.1016/j.gca.2022.12.003

Zoleikhaei, Y., Mulder, J.A. and Cawood, P.A. 2021. Integrated detrital rutile and zircon provenance reveals multiple sources for Cambrian sandstones in North Gondwana. *Earth-Science Reviews*, **213**, 103462, https://doi.org/10.1016/j.earscirev.2020.103462

Current applications using key mineral phases in igneous and metamorphic geology: perspectives for the future

Silvia Volante[1,2,3]*, Eleanore Blereau[4], Martin Guitreau[5], Mahyra Tedeschi[6,7], Valby van Schijndel[8] and Kathryn Cutts[9]

[1]Structural Geology and Tectonics Group, Geological Institute, Department of Earth Sciences, ETH Zürich

[2]Institute of Geology, Mineralogy and Geophysics, Ruhr-Universität Bochum, Universitätsstraße 150, 44801 Bochum, Germany

[3]ISOTOPIA Lab., School of Earth, Atmosphere and Environment, Monash University, Wellington Rd, Clayton, VIC 3800, Australia

[4]Institute of Geophysics and Tectonics, School of Earth and Environment, University of Leeds, Woodhouse, Leeds LS2 9JT, UK

[5]Université Clermont Auvergne, CNRS-UMR 6524, IRD-UMR 163, OPGC, Laboratoire Magmas et Volcans, F-63178 Aubière, France

[6]Programa de Pós-Graduação em Geologia, Universidade Federal de Minais Gerais, Centro de Pesquisas Manoel Teixeira da Costa, Instituto de Geociênciais, Av. Antônio Carlos, 6627, Belo Horizonte 31270-901, Brazil

[7]Institute of Geological Sciences, University of Bern, 3012 Bern, Switzerland

[8]Institute of Geosciences, University of Potsdam, 14476 Potsdam, Germany

[9]Geological Survey of Finland, P.O. Box 96, FI-02151 Espoo, Finland

SV, 0000-0001-8807-4087; EB, 0000-0001-8850-397X
*Correspondence: svolante@ethz.ch/silvia.volante89@gmail.com

Abstract: The study of magmatic and metamorphic processes is challenged by geological complexities like geochemical variations, geochronological uncertainties and the presence/absence of fluids and/or melts. However, by integrating petrographic and microstructural studies with geochronology, geochemistry and phase equilibrium diagram investigations of different key mineral phases, it is possible to reconstruct insightful pressure–temperature–deformation–time histories. Using multiple geochronometers in a rock can provide a detailed temporal account of its evolution, as these geological clocks have different closure temperatures. Given the continuous improvement of existing and new *in situ* analytical techniques, this contribution provides an overview of frequently utilized petrochronometers such as garnet, zircon, titanite, allanite, rutile, monazite/xenotime and apatite, by describing the geological record that each mineral can retain and explaining how to retrieve this information. These key minerals were chosen as they provide reliable age information in a variety of rock types and, when coupled with their trace element (TE) composition, form powerful tools to investigate crustal processes at different scales. This review recommends best applications for each petrochronometer, highlights limitations to be aware of and discusses future perspectives. Finally, this contribution underscores the importance of integrating information retrieved by multi-petrochronometer studies to gain an in-depth understanding of complex thermal and deformation crustal processes.

Unravelling the composite tectono-magmatic and metamorphic evolution of terrains can be challenging because their multistage histories may have erased past information. One way to interrogate and address these complexities is to investigate geological processes throughout Earth's history that are encapsulated in key minerals such as garnet, zircon, titanite, allanite, rutile, monazite/xenotime and apatite. The mineral phases chosen for this contribution are the most used in petrochronological studies as they are reliable chronometers, and their chemical variability has been used to investigate metamorphic and magmatic crustal processes. As petrochronometers, they can retain information about the petrogenesis of their protoliths, (multiple) pressures (P), temperatures (T) and timing (t) at which they (re)crystallized in magmatic, metamorphic and/or hydrothermal events. Investigations employing these minerals

From: van Schijndel, V., Cutts, K., Pereira, I., Guitreau, M., Volante, S. and Tedeschi, M. (eds) 2024. *Minor Minerals, Major Implications: Using Key Mineral Phases to Unravel the Formation and Evolution of Earth's Crust.* Geological Society, London, Special Publications, **537**, 57–121.
First published online May 31, 2023, https://doi.org/10.1144/SP537-2022-254

have provided new breakthroughs in Earth Sciences. Scientific findings through petrochronological investigations include (but are not limited to) the study of the oldest minerals on Earth (Valley *et al.* 2014), the understanding of trace element (TE) partitioning processes between different key mineral phases (e.g. Rubatto 2002) and the study of fluid and magma sources during crust evolution (e.g. Valley 2003; Bouvier *et al.* 2012; Mulder *et al.* 2021). Magmatic and/or metamorphic overprinting stages are heterogeneously recorded by these petrochronometers, which preserve snapshots of a rock evolution. The timing, conditions of formation and distinct closure temperatures (i.e. *Tc* as the temperature of a geochronological system at the time corresponding to its apparent age; Dodson 1973) that characterize minerals and isotope systems (Fig. 1) allow, for example, the U–Pb system in garnet (Mezger *et al.* 1989*a*) to retain information at higher temperatures than in rutile, which resets its internal clock as a function of *T*, grain size and cooling rate. Thus, carrying out multi-mineral studies (Fig. 1) is an incredibly powerful tool to investigate complex magmatic and metamorphic histories (e.g. Cutts *et al.* 2014; Stearns *et al.* 2015; Manzotti *et al.* 2018; Volante *et al.* 2020*c*; Fumes *et al.* 2022; Odlum *et al.* 2022).

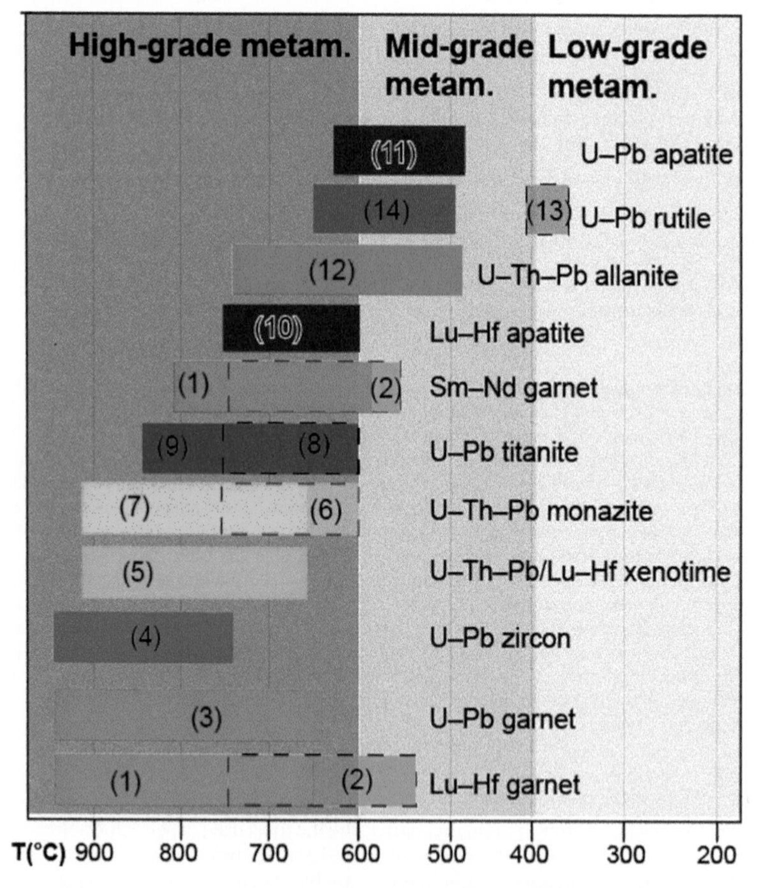

Fig. 1. Closure temperatures (Dodson 1973) for various commonly used geochronometers. References cited in this figure: (1) Smit *et al.* (2010, 2013); (2) Scherer *et al.* (2000); (3) Mezger *et al.* (1989*a*), Li *et al.* (2022); (4) Lee *et al.* (1997), Cherniak and Watson (2001); (5) Cherniak (2006); (6) Copeland *et al.* (1988), Kingsbury *et al.* (1993), Suzuki *et al.* (1994), Smith and Giletti (1997); (7) Spear and Parrish (1996), Vry *et al.* (1996), Braun *et al.* (1998), Kamber *et al.* (1998), Rubatto *et al.* (2001), Asami *et al.* (2002), Schmitz and Bowring (2003), Cherniak *et al.* (2004); (8) Mezger *et al.* (1991), Scott and St-Onge (1995), Cherniak (1995), Frost *et al.*(2001); (9) Schärer *et al.* (1994), Zhang and Schärer (1996), Kohn and Corrie (2011), Spencer *et al.* (2013), Stearns *et al.* (2015), Kohn (2017), Hartnady *et al.* (2019), Kirkland *et al.* (2020); (10) Cherniak (2000*a*), Barfod *et al.* (2005); (11) Cherniak *et al.* (1991), Krogstad and Walker (1994); (12) Oberli *et al.* (2004), Gregory *et al.* (2012); (13) Mezger *et al.* (1989*a*); (14) Cherniak (2000*a*), Vry and Baker (2006), Kooijman *et al.* (2010).

Over the past decades, our ability to interrogate the preserved rock record of crustal metamorphism and magmatism has significantly advanced thanks to continuously improving *in situ* analytical techniques, measurement methodologies and protocols, as well as availability of reference materials that have been calibrated through worldwide inter-laboratory collaborations. Most commonly used instruments include secondary ion mass spectrometry (SIMS) (e.g. Chamberlain *et al.* 2010; Ushikubo *et al.* 2014; Zhou *et al.* 2016; Chaussidon *et al.* 2017) and inductively coupled plasma mass spectrometers (single and multi-collector, sector-field, single and triple quadrupoles) associated with a laser ablation system (LA-ICP-MS, LA-MC-ICP-MS or LA-ICP-MS/MS) either used in tandem (split stream) or separately (e.g. Kylander-Clark *et al.* 2013; Hacker *et al.* 2015; Simpson *et al.* 2021). Additionally, improvements in microanalytical techniques, such as (but not limited to) electron backscatter diffraction (EBSD) mapping (e.g. Cavosie *et al.* 2015), TESCAN integrated mineral analyser (TIMA; e.g. Porter *et al.* 2020), field emission scanning electron microscope (FESEM; Tacchetto *et al.* 2022) for qualitative analysis acquisition and field emission gun (FEG) transmission electron microscope (TEM) for atomic lattice resolution imaging and nanoscale microanalysis (Reddy *et al.* 2020; Tacchetto *et al.* 2021), have significantly contributed in advancing the micro- to nanoscale petrographic, chemical, physical and structural characterization of sample material. The integration of continuously updated thermodynamic modelling studies of monazite (e.g. Janots *et al.* 2007; Kelsey *et al.* 2008; Spear and Pyle 2010; Hoschek 2016), allanite (Spear 2010), zircon (Kelsey *et al.* 2008; Kelsey and Powell 2011; Yakymchuk and Brown 2014) and apatite (Spear and Pyle 2010; Yakymchuk 2017) allows for the construction of a fundamental basis from which to overcome problems related to qualitative interpretations of accessory minerals within the *P–T* space, for which rutile (e.g. Fumes *et al.* 2022; Holtmann *et al.* 2022; Horton *et al.* 2022; Vanardois *et al.* 2022) and titanite (e.g. Kapp *et al.* 2009; Kirkland *et al.* 2016, 2020; Apen *et al.* 2020; Walters *et al.* 2022) already greatly contribute. The above-mentioned analytical advances combined with more traditional petro-structural (micro)analyses (Volante *et al.* 2020*b*) and continuously updated thermodynamic datasets used for modelling pressure (*P*), temperature (*T*) and composition (*X*) are continuously contributing to improve our understanding of crustal processes by allowing a more detailed characterization of mineral phases.

This review aims to showcase snapshots of recent analytical developments such as advancement in understanding the capabilities of LA-ICP-MS/MS and their applications, including new findings in the uptake of water in zircon crystal lattice and its influence on oxygen isotope composition, development and improvement in generating *in situ* quantitative elemental maps that can be directly linked with petrochronological and textural information using key mineral phases in igneous and metamorphic studies. In particular, we focus on the most employed minerals in the literature that can be used not only as chronometers in a great variety of rock lithologies but also as tracers of fluid–rock interactions at crustal conditions, and they include: garnet, zircon, titanite, rutile, allanite, monazite/xenotime and apatite (Fig. 2). In the following section, each mineral is described including its potential as chronometer, thermobarometer and chemical/isotopic tracer applied to crustal magmatic and metamorphic processes. Ultimately, the Discussion section provides examples of applications that use these petrochronometers in magmatic and metamorphic studies with a focus on the complementary set of information obtained when multi-mineral investigations are used to unravel different steps in the evolution of a volume of rock. Future perspectives on the use of these key mineral phases to investigate metamorphic and igneous processes are also included.

Mineral capsules

Garnet

Garnet (Fig. 2a-c; Fig. 3) is a major rock-forming mineral, occurring in metamorphic (felsic and mafic compositions) and felsic peraluminous magmatic rocks (Villaros *et al.* 2009; Bartoli *et al.* 2013; Liu *et al.* 2014; Volante *et al.* 2020*a*; R. Li *et al.* 2021). Garnet is a key metamorphic indicator mineral (Barrow 1893, 1912) and its composition directly relates to the *P–T* conditions it experienced during growth (the garnet stability field extends from greenschist-facies to high-pressure (*HP*) and high-temperature (*HT*) conditions, dependent on rock composition), making it extremely useful as a geothermobarometer. Garnet grains can shield mineral inclusions whose identity or composition can indicate earlier assemblages/conditions experienced by the rock (e.g. St-Onge 1987; Lü *et al.* 2008; Thomas and Davidson 2012; Pourteau *et al.* 2019; Schönig *et al.* 2019; Godet *et al.* 2022). These mineral inclusions can also preserve or trace earlier foliations, which can be used to infer older foliation trends (e.g. Sayab 2006; Aerden *et al.* 2013, 2021; Sayab *et al.* 2015, 2016; Volante *et al.* 2020*b*). Like other porphyroblasts, garnet can also infer kinematics via grain rotation (Fig. 3a; Passchier and Simpson 1986; Johnson 1999). Garnet can record a prolonged history, preserving a growth hiatus, where textural, chemical or chronological evidence shows that garnet grew during more than one

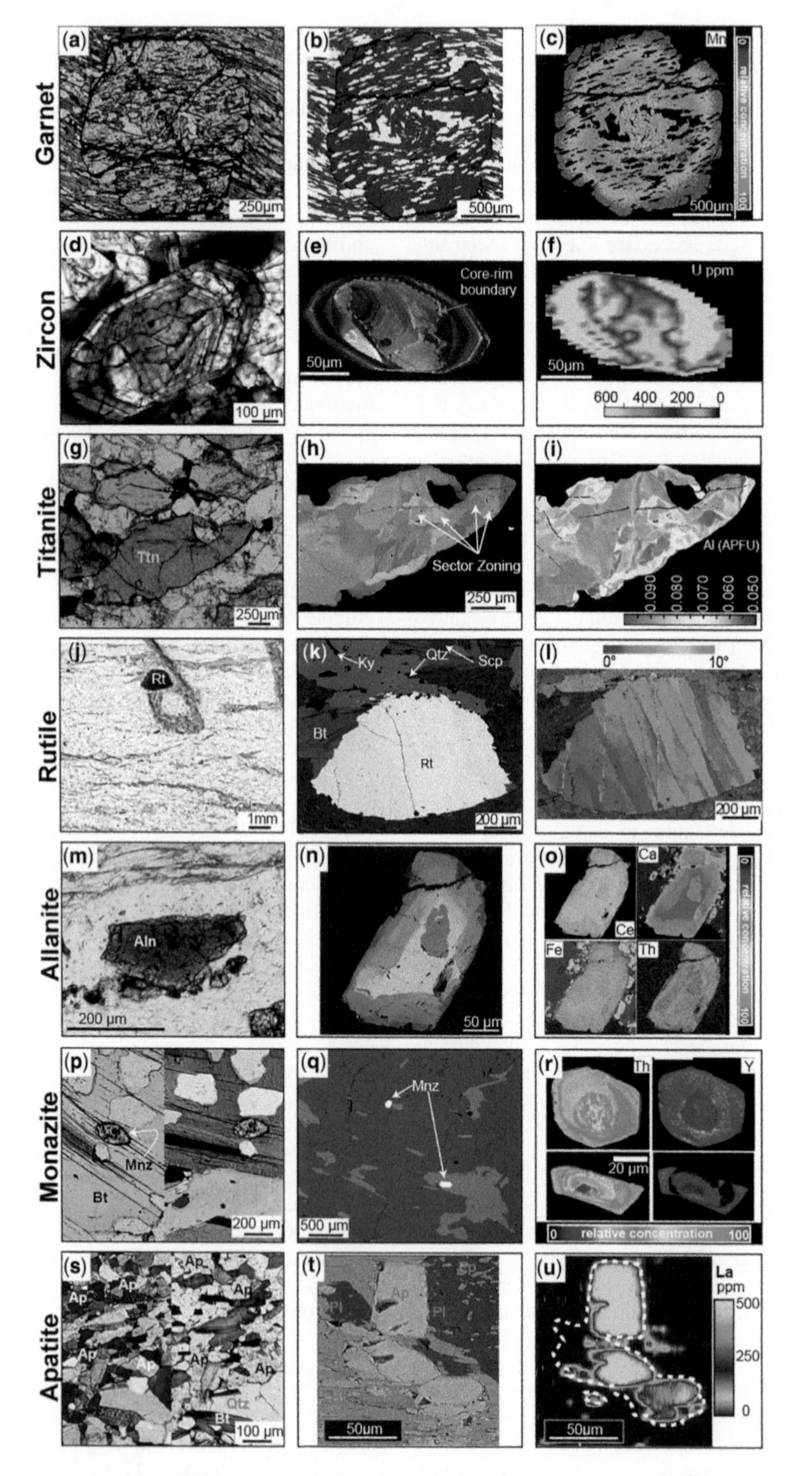

Fig. 2. Representative textures for the key mineral phases described in this contribution. (**a**) Plane polarized light. (**b**) Backscattered electron (BSE) and (**c**) electron microprobe Mn elemental map images showing the early and late stages of fabric development in core and rims, respectively, during garnet growth. (**d**) Plane polarized light,

metamorphic and/or magmatic event (Fig. 3b; e.g. Vance *et al.* 1998; Cutts *et al.* 2010; Ortolano *et al.* 2014; Kulhánek *et al.* 2021; Massonne and Li 2022).

Geochemical tools. Garnet is a powerful tool for linking accessory mineral geochronological and/or isotopic data to the evolution of the major mineral assemblage using trace elements. Linking these two data sources together is the cornerstone of petrochronology (Engi *et al.* 2017). Trace elements such as rare earth elements (REE) partition strongly into garnet (Bea *et al.* 1994; Hermann and Rubatto 2014), particularly the middle to heavy rare earth elements (M-HREE; Fig. 3c). Other minerals that also readily accommodate M-HREE, such as zircon and monazite, compete with garnet, resulting in modified trace element patterns when the accessory phase grows in the presence or absence of garnet. The use and effectiveness of these trace element partitioning relationships (D_{REE}) were first presented for zircon and garnet (Fig. 3d; Rubatto 2002). Zircon grown in the presence of garnet has a M-HREE slope of *c.* 1 effectively flat, and a slightly curved to near flat D_{REE} pattern (Whitehouse and Platt 2003; Hokada and Harley 2004; Kelly and Harley 2005; Harley and Kelly 2007; Wu *et al.* 2008*a*, *b*; Fornelli *et al.* 2014, 2018). Since these first studies, trace element partitioning has grown into a widely used tool in petrochronology (Whitehouse and Platt 2003; Hokada and Harley 2004; Baldwin and Brown 2008; Clark *et al.* 2009; Harley and Nandakumar 2014), and some studies conducted quantitative partitioning experiments to constrain temperature dependent patterns for D_{REE} (Zrc/Grt) (Rubatto and Hermann 2007; Taylor *et al.* 2015), building upon empirical studies (Rubatto 2002; Hermann and Rubatto 2003; Whitehouse and Platt 2003; Hokada and Harley 2004; Kelly and Harley 2005; Buick *et al.* 2006). To address the challenge of handling extensive datasets generated by LA-ICP-MS, Taylor *et al.* (2017) developed an array plot for trace element partitioning data that can also show additional trends compared to the standard D_{REE} plot (Fig. 3e). D_{REE} (Mnz/Grt) is also a useful partitioning system (see Monazite section for more details). Trace element mapping of garnet via LA-ICP-MS is an area that is currently undergoing significant advances (e.g. Chew *et al.* 2017, 2021; Raimondo *et al.* 2017; Rubatto *et al.* 2020). M-HREE, Y and Cr zoning in garnet provides additional information about the growth history and mineral relationships not apparent in major element zonation (Raimondo *et al.* 2017).

Geochronology. Since the late 1980s, U–Pb and Sm–Nd geochronology (Mearns 1986; Mezger *et al.* 1989*b*) used multiple mineral phases to determine the age of whole-rock. The application of these methods led to the discovery of a favourable spread in Sm–Nd ratios between garnet and whole rock (Humphries and Cliff 1982). U–Pb garnet dating was found to be affected by U-rich inclusions such as zircon, monazite and apatite. Thus, for the following 30 years, garnet geochronology has been conducted using Sm–Nd and, subsequently, once instrumentation advanced sufficiently to measure low Hf contents in garnet, the Lu–Hf system was used for garnet dating (Duchêne *et al.* 1997). Inclusions such as monazite and apatite for the Sm–Nd system were treated with a HCl leaching step method that would remove significant monazite Nd components (Scherer *et al.* 2000), whereas Lu–Hf garnet geochronology is greatly affected by inclusions such as zircon, with samples containing significant contents of zircon being avoided for garnet whole-rock dating (Scherer *et al.* 2000). The following studies have compared the two isotopic systems for garnet dating via dissolution (Vervoort 2013), and they have shown that the Lu–Hf system (*Tc c.* 650–900°C) has slightly higher *Tc* than Sm–Nd (*Tc c.* 600–800°C) (Fig. 1; Scherer *et al.* 2000; Smit *et al.* 2013; Johnson *et al.* 2018). Traditionally, both Sm–Nd and Lu–Hf systems via isotope dilution–thermal ionization mass spectrometry (ID-TIMS) or MC-ICP-MS dissolution methods require the separation and purification of garnet prior to dissolution, column separation and measurement that is commonly associated with loss of the textural context of the analysed grains (cf. Pollington and Baxter 2010). Prolonged (e.g. Baxter and Scherer 2013; Bollen *et al.* 2022) or incremental garnet growth makes a single age from garnet separate, potentially geologically meaningless (e.g. Pollington and Baxter 2010). To overcome these issues, garnet has often been separated based on colour (e.g. Vance *et al.* 1998) or by using a microdrill to target garnet growth rings in order to determine the growth rate of garnet crystals (Fig. 3f; see Pollington and Baxter

Fig. 2. *Continued.* (**e**) cathodoluminescence (CL) and (**f**) laser ablation ICP–MS U (ppm) elemental map images of zircon. (**g**) Plane polarized light, (**h**) BSE and (**i**) quantitative map of Al (a.p.f.u., atom per formula unit) images of titanite. (**j**) Plane polarized light, (**k**) BSE and (**l**) EBSD map images for rutile. (**m**) Plane and crossed polarized light, (**n**) BSE and (**o**) electron microprobe Th and Y elemental map images of monazite. (**p**) Plane and crossed polarized, (**q**) BSE and (**r**) LA–ICP–MS trace element map images of apatite. (**s**) Plane polarized light, (**t**) BSE and (**u**) X-ray elemental map images of allanite. Source: (c) Volante *et al.* (2020*c*); (f) Chew *et al.* (2017); (i) Walters *et al.* (2022); (l) Moore *et al.* (2020*a*, *b*); (n) Barrote *et al.* (2022*a*); (o) Volante *et al.* (2020*c*); (r) Henrichs *et al.* (2019); (u) Corti *et al.* (2020).

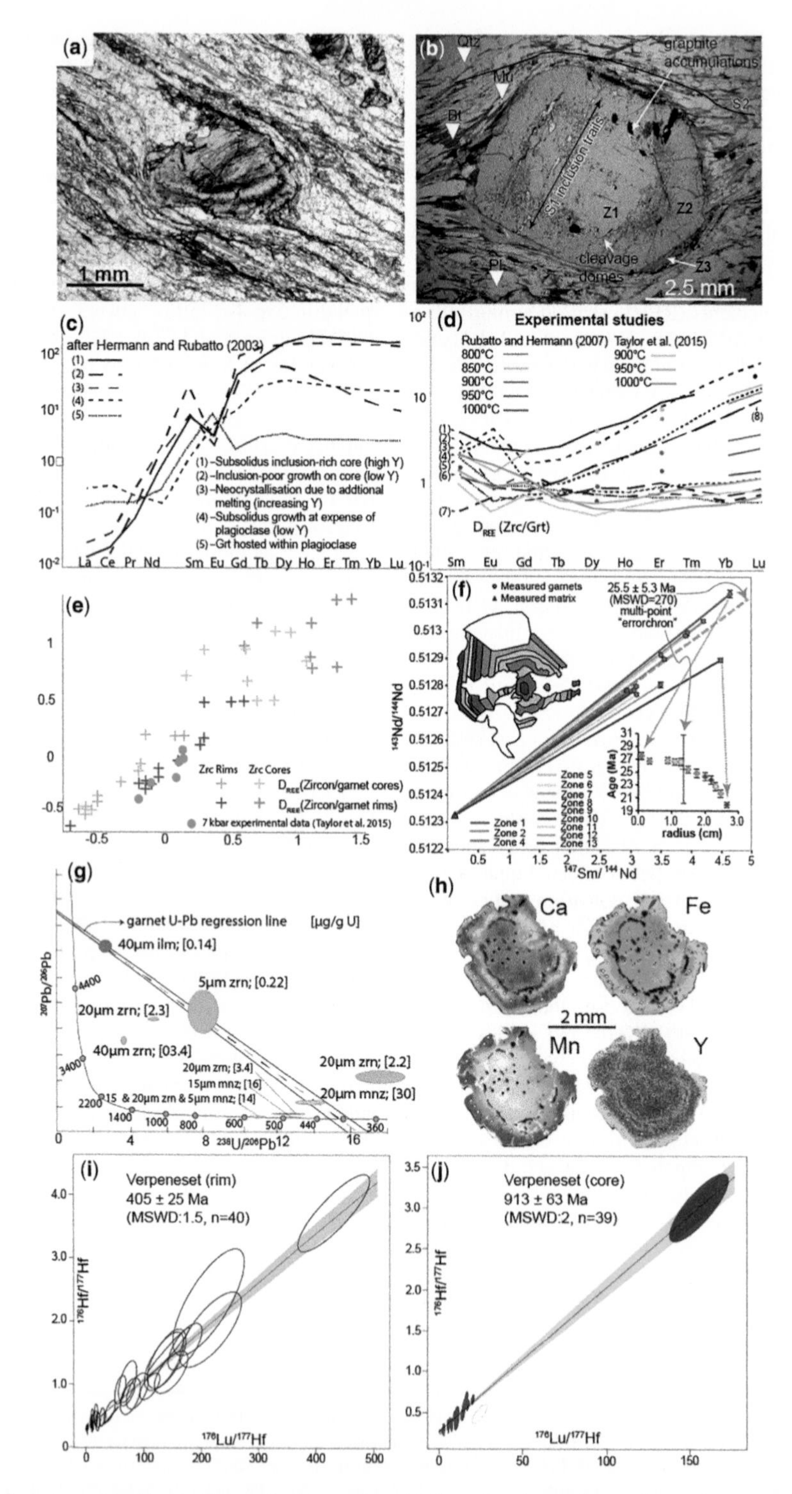

Fig. 3. (**a**) Garnet porphyroblast indicating the kinematics of deformation (arrows) as a sigma-clast and exhibiting inclusion trails of an internal foliation. (**b**) Polymetamorphic garnet grain from Polnish, Scotland (see Vance *et al.* 1998; Cutts *et al.* 2009). The core (Z1) and first rim (Z2) zones preserve the same early foliation (S1) but have

2010, 2011; Dragovic *et al.* 2012, 2015; Schmidt *et al.* 2015; Tual *et al.* 2022). Results of such studies have showcased how effectively garnet can be used for geochronology, the caveat being that these methods are expensive, time consuming (*c.* 2 months) and the presence of inclusions must be considered and evaluated.

The recent introduction of ICP-MS/MS, which eliminates isotopic interferences for Lu–Hf measurements (i.e. ^{176}Lu and ^{176}Yb on ^{176}Hf), coupled with an increase in machine sensitivity allowing to measure elements at very low concentrations (i.e. part per billion, ppb), has permitted significant advances in garnet geochronology. These analytical advances not only are at the foundation of the development of *in situ* Lu–Hf measurements in garnet (see Simpson *et al.* 2021; Brown *et al.* 2022; Ribeiro *et al.* 2022; Tamblyn *et al.* 2022; Simpson *et al.* 2023), but also resulted in revisiting of the U–Pb system in garnet with *in situ* analyses now possible using LA-ICP-MS or -SC (single collector) -MS (Seman *et al.* 2017; Gevedon *et al.* 2018; Millonig *et al.* 2020). Since garnet commonly has low U contents, the presence of U-rich inclusions remains an issue for *in situ* methods. Nonetheless, Uranium content in garnet can easily be tested via LA-ICP-MS spot analysis prior to U–Pb analysis and coupled with detailed textural imaging to identify potential U-rich inclusions. Grossular and andradite garnet tend to have enough U (>1 part per million, ppm), and many studies have used this method as a means of dating skarns related to mineralization (Seman *et al.* 2017; Gevedon *et al.* 2018; Wafforn *et al.* 2018; Duan *et al.* 2020). More recently, other works have also found almandine-pyrope-rich garnet with lower but enough U (>0.02 ppm) to produce reliable metamorphic ages (Millonig *et al.* 2020; Cerva-Alves *et al.* 2021; Schannor *et al.* 2021; O'Sullivan *et al.* 2023). Nonetheless, U-rich inclusions such as monazite, apatite and zircon shift data points on a Tera–Wasserburg (TW) diagram to the left, right or along the regression, reflecting older, younger or the same age as garnet, respectively (Fig. 3g; Millonig *et al.* 2020). Besides textural observations, to screen and filter out inclusion-dominated analyses it is crucial to monitor the masses on the ICP-MS such as Y and Zr for zircon, or light rare element (LREE) for monazite and allanite. The *in situ* U–Pb garnet dating method (*Tc c.* 650–900°C) has the advantage of minimal sample preparation, rapid analysis (2 hours) and the possibility of monitoring the signal. The Lu–Hf method is limited to garnet that contains sufficient Lu for measurement and relatively inclusion free, with zircon inclusions significantly affecting the age results (Simpson *et al.* 2021). This method has larger uncertainties than solution-based Lu–Hf analysis, but numerous advantages. Low Lu or high Hf (i.e. zircon) inclusions can be targeted to produce an isochron from a single garnet grain. The ablated spots (*c.* 40–120 µm) can target different chemical domains, including separate core and rim isochron ages (Fig. 3h–j; Tamblyn *et al.* 2022). The ability to produce an isochron from a single garnet grain means that this method can also be used to investigate detrital garnet (see Pereira *et al.* this volume, in press). An alternative to directly date garnet is targeting datable primary mineral inclusions such as monazite (e.g. Mahan *et al.* 2006; Cutts *et al.* 2010; Williams *et al.* 2017). However, possible preservation of detrital or older inclusions must be evaluated (Martin *et al.* 2007; Cutts *et al.* 2009, 2013; Peixoto *et al.* 2018).

Isotope geochemistry. Oxygen isotopes have been applied to whole-rock and mineral samples to trace the sources of metamorphic/metasomatic (e.g. Chamberlain and Conrad 1991; Crowe *et al.* 2001; Page *et al.* 2010; Martin *et al.* 2011, 2014; Vho *et al.* 2020) and magmatic/mantle fluids (e.g. King and Valley 2001; Lackey *et al.* 2006; Harris and Vogeli 2010). Initially, oxygen isotopes were applied to garnet to determine pressure estimates, based on the temperature dependent fractionation

Fig. 3. *Continued.* cleavage domes and graphite inclusions on their border. A thin, inclusion-rich rim (Z3) which has been poorly preserved is observed. (**c**) Representative chondrite-normalized REE patterns for different types of garnet (Blereau 2017). (**d**) Partitioning between zircon and garnet: (1) Buick *et al.* (2006); (2) Hokada and Harley (2004); (3) GL7 in Rubatto (2002); (4) GP7 in Rubatto (2002); (5) Harley *et al.* (2001); (6) Kelly and Harley (2005); (7) Whitehouse and Platt (2003); (8) Hermann and Rubatto (2003). (**e**) Partitioning array plot of zircon/garnet compared to the experimental data of Taylor *et al.* (2015). The axes of this plot are: x, $\log(D_{Yb})$; y, $\log(D_{slope}) = \log(D_{Yb/DGd})$ (Blereau 2017). (**f**) Sm–Nd dating of a large garnet porphyroblast by Pollington and Baxter (2010). The various garnet rims were obtained using a microdrill (the drill lines are marked in black on the garnet). Coloured zones correspond to the coloured isochrons and age points in the plots indicate that garnet grew over *c.* 8 myr, from 28 to 20 Ma. (**g**) TW diagram showing how the grain size and the amount of ablated U-rich inclusions affect garnet U–Pb dating. (**h**) Garnet compositional map of Ca, Fe, Mn and Y (from top left) of sample Verpeneset from Tamblyn *et al.* (2022). Black and white dots represent the position of core and rim analyses, respectively. (**i** and **j**) Lu/Hf isochrons obtained from garnet (i) core and (j) rim. Uncertainties consider overdispersion of the data and isochron ages are calculated with age offset (see Tamblyn *et al.* 2022 for full details). Source: (a) Photo courtesy of J. Pownall; (c–d) modified after Blereau (2017); (f) adapted from Pollington and Baxter (2010), Baxter *et al.* (2017); (g) modified after Millonig *et al.* (2020).

between co-existing mineral pairs (experimentally or empirically determined; Baxter *et al.* 2017 and references therein). This method utilizes slow diffusion of oxygen within garnet, thus peak temperature isotope fractionation is retained (Kohn and Valley 1998; Baxter *et al.* 2017). Improvements in analytical methods (analysis via SIMS) allowed for *in situ* analysis of oxygen isotopes in garnet to track the origin of metamorphic/metasomatic fluids (e.g. Raimondo *et al.* 2012, 2017; Russell *et al.* 2013). For example, garnet can preserve variation of fluid compositions as it grows, particularly in eclogites or skarns (D'Errico *et al.* 2012; Rubatto and Angiboust 2015). Oxygen in garnet is very similar to the bulk rock $\delta^{18}O$ value, but deviations can occur following temperature variations affecting equilibration factors between minerals (i.e. garnet–quartz), changes in modal proportions and mineral assemblage during metamorphism as well as the presence of externally sourced fluids. Nonetheless, experimentally determined oxygen diffusion rates in garnet show that original isotope compositions can be retained despite partial re-equilibration of major elements (Scicchitano *et al.* 2021). Other stable isotopes such as Fe or Li may be applied in garnet (Bebout *et al.* 2014, 2022; An *et al.* 2017; Gerrits *et al.* 2019; Penniston-Dorland *et al.* 2020; Hoover *et al.* 2021, 2022), with limitations of these largely depending on analytical ability.

Thermobarometry. Geothermometry in garnet (Ferry and Spear 1978; Baxter *et al.* 2017) was initially applied via Fe–Mg exchange between garnet and biotite (Ferry and Spear 1978; Perchuk and Lavrent'Eva 1983; Ganguly and Saxena 1984) and geobarometry via the 'GASP' (garnet–aluminosilicate–silicate–plagioclase) end-member reaction, which is pressure sensitive due to the large molar volume difference between reactants and products (e.g. Ghent 1976; Newton and Haselton 1981; Holdaway 2001; Caddick and Thompson 2008). Later, large thermodynamic datasets (e.g. Berman 1988; Holland and Powell 1990, 1998, 2011) allowed for *P–T* pseudosection calculation, which was combined with the use of garnet end-member compositions to constrain the growth path or max *P* and *T* conditions (e.g. Spear *et al.* 1984; Vance and Mahar 1998). A caveat in the use of this method is the isolation of grown garnet (i.e. core domain) from the whole-rock composition used for *P–T* pseudosection calculation (e.g. Marmo *et al.* 2002; Evans 2004; Tinkham and Ghent 2005; Cutts *et al.* 2009, 2010). This approach is particularly useful where garnet growth domains are the result of different orogenic events (e.g. Cutts *et al.* 2014). Software for modelling evolving bulk compositions during garnet growth along a *P–T* path was developed to produce more reliable *P–T* estimates (Theria_G, Gaidies *et al.* 2008; GRTMod, Lanari *et al.* 2017). For *P–T* modelling, garnet compositions (i) may record prograde growth, which is typically defined by bell-shaped Mn enrichment in the garnet core (e.g. Hollister 1966; Atherton 1968; Cygan and Lasaga 1985; Schwandt *et al.* 1996; Dziggel *et al.* 2009; Lanari and Engi 2017; Spear 2017; Dempster *et al.* 2020), or (ii) alternatively, at high temperatures, garnet chemistry may be used to determine the max *T* conditions experienced by the sample where the prograde growth major element zonation is partially or completely re-equilibrated (Caddick *et al.* 2010). Also, fluid and/or melt inclusions hosted in peritectic garnet from migmatites have been used to evaluate the starting composition of the anatectic melt and fluid regime during anatexis (e.g. Cesare *et al.* 2009, 2011; Ferrero *et al.* 2012, 2021; Carvalho *et al.* 2019; Borghini *et al.* 2020). Inclusions such as zircon and quartz in garnet can also be used to produce pressure estimates using Raman barometry (Zhong *et al.* 2019; Spear and Wolfe 2020). After being trapped in a host mineral during part of the cooling and exhumation history of a rock, inclusions may develop residual pressures due to the differences in thermal expansivity and compressibility between the host and the inclusions (Gonzalez *et al.* 2019; Zhong *et al.* 2019, 2020). Within both metamorphic and magmatic studies, however, garnet-based thermobarometry should always be applied with a thorough petrological and microstructural relationships characterization and an appropriate estimate of equilibrium volume of the rock sample. Results that are inconsistent with the observed mineral assemblage should be crosschecked with other comparable tools and should be interpreted with care. Additionally, taking a step back and carefully reanalysing field and textural context of the investigated rocks, as well as reassessing the documentation of sample preparation, analytical procedure and data processing, may also reveal missed information.

Zircon

The mineral zircon ($ZrSiO_4$) has traditionally been the most robust accessory phase and widely used geochronological and geochemical tool in petrochronology to investigate crustal formation and evolution, and to untangle complex rock histories that most minerals fail to retain (e.g. Froude *et al.* 1983; Scherer *et al.* 2007; Guitreau *et al.* 2019). For instance, in the Archean Lewisian Gneiss Complex, NW Scotland, zircon is the only accessory mineral that has recorded the several Archean magmatic phases (cores) and the multiple ultra-high temperature (UHT) to HT/medium (M)T overprinting events (rims) (e.g. Whitehouse and Kemp 2010; Taylor *et al.* 2020; Fischer *et al.* 2021), whereas others such as monazite and titanite record only parts of the metamorphic events (e.g. Zhu and O'Nions

1999; Goodenough *et al.* 2013). Its resistance to erosive processes makes zircon a prime choice when studying detrital crystals to unravel past formation of continental crust in deeply eroded areas, or simply ancient terranes (e.g. Iizuka *et al.* 2005; Dhuime *et al.* 2012; Næraa *et al.* 2012; Nordsvan *et al.* 2018; Gardiner *et al.* 2019; Kirkland *et al.* 2021; Tedeschi *et al.* 2023; Pereira *et al.* this volume, in press). Over the years, many techniques have been improved to study zircon and retrieve various types of information from it (Table 1).

When zircon crystallizes, it can develop various textures depending on the conditions at which it forms, such as sharp concentric growth zoning and/or marginal resorption if zircon grew in a magmatic environment or more irregular domains of homogeneous zoning cutting discordantly the growth domains in the post-magmatic environment (e.g. Corfu *et al.* 2003 and references therein). These textures are commonly revealed by routine imaging of cathodoluminescence (CL) or backscattered electrons (BSE) using a scanning electron microscope (SEM). Cathodoluminescence response is mostly due to interaction between the primary electron-beam and zircon lattice through a complex energy transfer reaction that ends with the emission of photons by tetravalent REE, especially Dy (e.g. Nasdala *et al.* 2003). Visible textures range from broad to fine oscillatory zoning in igneous zircon (Fig. 2d–f; Fig. 4a, b) to large homogeneous to patchy domains in metamorphic conditions (e.g. Corfu *et al.* 2003). Raman spectroscopy can be used to quantitatively map radiation damage zoning in single zircon crystals (i.e. metamictization; Balan *et al.* 2001) and, hence, evaluate the effects of radiation damage on the final U/Pb ratio (Anderson *et al.* 2020) or on any elemental and/or isotopic signature. Igneous zircon most commonly form when the

Table 1. *Synoptic table of most used tools for zircon investigations*

Tool*	Interest	Example of reference
Morphology	Nature of parental magma (alkaline, peraluminous or calc-alkaline)	Pupin (1980); Vavra (1993)
CL and BSE images	Pristineness of crystal lattice and crystallization conditions (igneous, metamorphic or weathered)	Vavra (1990); Hanchar and Miller (1993); Corfu *et al.* (2003); Guitreau *et al.* (2018)
Inclusions (mineral and melt)	Nature of parental magma and/or fluids (for primary inclusions either as melt or mineral)	Thomas *et al.* (2003); Hopkins *et al.* (2008); Bell *et al.* (2015)
EBSD	Orientation and deformation of crystals	Tolometti *et al.* (2022); Cox *et al.* (2022)
Trace element concentrations	Nature of parental magma, magmatic or metamorphic origin and pristineness	Grimes *et al.* (2007, 2015); Laurent *et al.* (2021); Guitreau *et al.* (2022)
(Ti)	Crystallization temperature	Watson *et al.* (2006); Fu *et al.* (2008)
(K) and (Ca)	Primitive v. altered zircon	Bouvier *et al.* (2012); McCubbin *et al.* (2016)
Th/U	Zircon crystallization conditions (e.g. magmatic v. metamorphic origin) and weathering	Vavra *et al.* (1999); Kirkland *et al.* (2015); Yakymchuk *et al.* (2018); Guitreau and Flahaut (2019)
Ce anomaly	Redox state of parental magma	Smythe and Brenan (2016); Trail *et al.* (2012)
U–Pb	Dating (absolute age determination)	Wetherill (1956, 1963); Davis *et al.* (2003 and reference therein); Schoene (2014)
Lu–Hf	Source tracing (e.g. mantle or crust)	Patchett (1983); Amelin *et al.* (2000); Gerdes and Zeh (2009); Guitreau *et al.* (2012)
O	Source tracing (e.g. involvement of clay-rich sediments)	Valley (2003); Iizuka *et al.* (2013)
Si isotopes	Source tracing, parental magma SiO_2 content and crystallization history	Trail *et al.* (2018); Guitreau *et al.* (2020, 2022)
Zr isotopes	Magma crystallization history	Ibañez-Mejia and Tissot (2019); Guo *et al.* (2020); Tompkins *et al.* (2020)

*BSE, Backscattered electron; CL, cathodoluminescence; EBSD, electron backscatter diffraction.

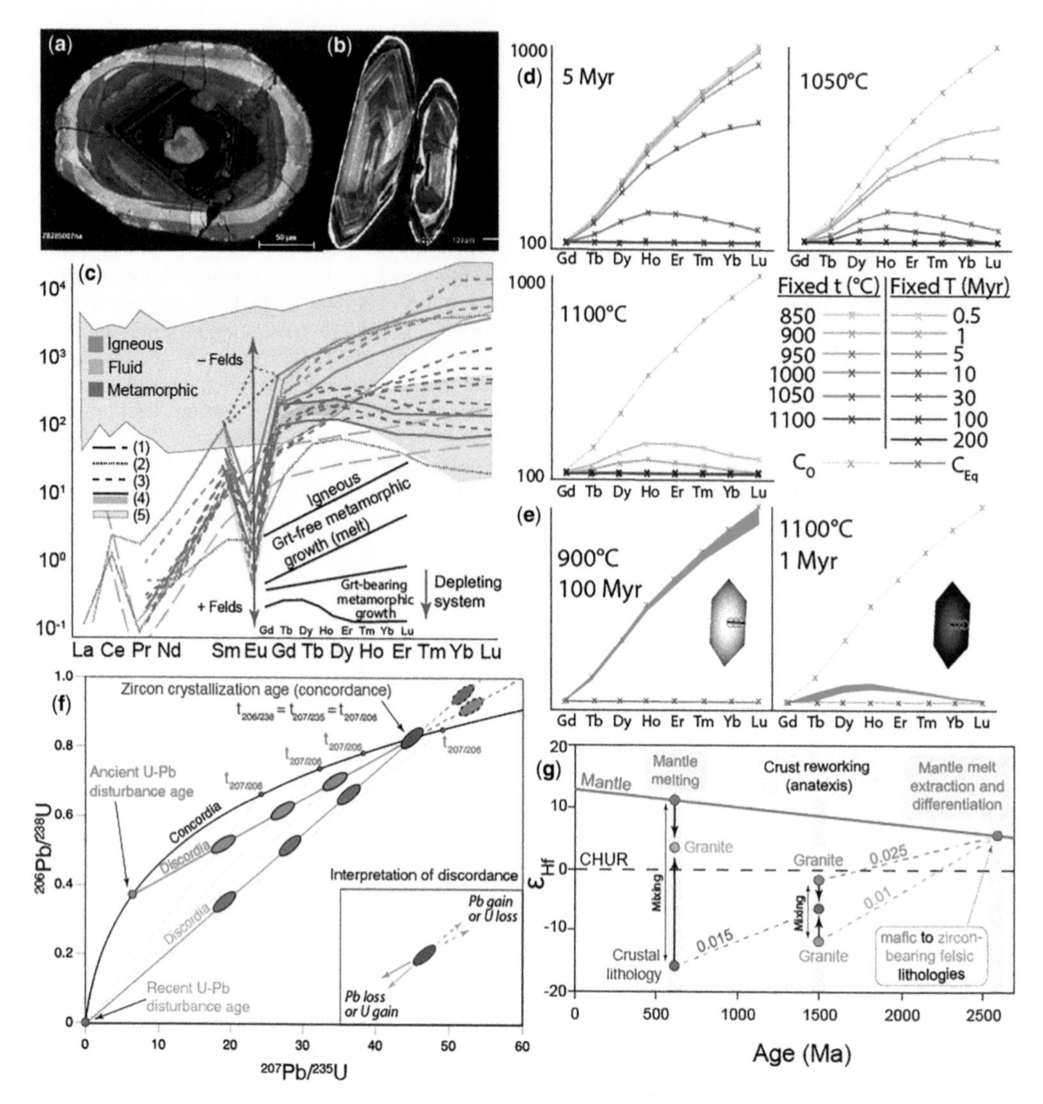

Fig. 4. (**a**) Cathodoluminescence image of a complexly zoned zircon crystal from the Napier Complex (Antarctica) showing a (recrystallized) bright metamorphic core surrounded by a fine-oscillatory zoned magmatic domain, itself resorbed and overgrown by three distinct metamorphic overgrowths. (**b**) Igneous zircon crystals exhibiting fine-oscillatory zoning from core to rim except for a thin metamorphic outer overgrowth. (**c**) Representative chondrite normalized REE patterns for different types of zircon. (1) Rubatto and Hermann (2007); (2) Hoskin and Schaltegger (2003); (3) Rubatto *et al.* (2013); (4) Taylor *et al.* (2014); (5) Li *et al.* (2018). (**d**) Modelled diffusion modified M-HREE compositions expected from *c.* 20 μm diameter SHRIMP spot analysis on the edge of a modelled theoretical zircon. C_0, original unmodified composition; C_{Eq}, equilibrated composition. (**e**) M-HREE spread generated by incomplete modification of a modelled zircon after different *T–t* conditions. (**f**) Concordia diagram illustrating the interpretation of zircon U–Pb data in terms of ages depending on whether the U–Pb system evolved in a closed or open system, and when disturbance occurred. (**g**) Graphical representation of Lu–Hf isotope evolution of distinct geological reservoirs. This graph provides a theoretical framework for the interpretation of zircon Hf isotope signatures. Source: (d) and (e) after Blereau *et al.* (2022).

activity of ZrO_2 and SiO_2 is optimal. This is the case in most magmas with SiO_2 concentrations over 57 wt%, although rare crystals in magmas with lower SiO_2 concentrations do exist (e.g. Aranovich *et al.* 2017; Fischer *et al.* 2021; Bea *et al.* 2022). Zircon precipitation from melts was investigated early on by Watson (1979) and Watson and Harrison (1984), and revisited by Boehnke *et al.* (2013).

The nature of mineral inclusions can provide a first order evidence of phases that co-precipitated with zircon and, thus, give information about the nature/composition of the melt in which it crystallized and the pressure and temperature conditions at which it grew (Delavault *et al.* 2016; Emo *et al.* 2018; Antoine *et al.* 2020). The same can be done with melt inclusions, but these can also be used to estimate partition coefficients for trace elements if they represent bulk-melt composition and their chemistry remained unmodified since entrapment (Thomas *et al.* 2003; Gudelius *et al.* 2020).

Geochemical tools. Zircon incorporates large quantities of trace elements from ppm to thousands of ppm level (e.g. Belousova *et al.* 2001; Hoskin and Schaltegger 2003), making it an easy target for analysis using electron microprobe (EPMA), ICP-MS, LA-(MC-)ICP-MS (Košler *et al.* 2005; Chew *et al.* 2017; Petrus *et al.* 2017) and secondary ionization mass spectrometers or sensitive high-resolution ion microprobe (SIMS/SHRIMP). The rapid technological improvements using LA-(MC-)ICP-MS allow now the acquisition and extraction of quantitative data from a mapped area, including chemical and isotopic information of accessory phases such as zircon (Petrus *et al.* 2017; Chew *et al.* 2021). Experiments and natural observations indicate very slow diffusion for most elements within the zircon lattice (see Lee *et al.* 1997; Cherniak and Watson 2003 for values). Pristine igneous zircon elemental variability is usually relatively small, making this mineral a poorly sensitive tracer of melt composition, except for sharp variations (i.e. strongly alkaline, felsic and mafic; e.g. Belousova *et al.* 2002; Grimes *et al.* 2007; Guitreau *et al.* 2022). However, when zircon undergoes significant radiation damage, it can be sensitive to thermal events that result in incorporation of measurable quantities of light REEs (e.g. Hoskin and Schaltegger 2003; Bouvier *et al.* 2012; Pidgeon *et al.* 2017). This modification of zircon composition is a good proxy for secondary alteration and/or weathering. Along the same line, non-formula elements, such as Ca, K and Al, can enter zircon easily once it becomes porous due to significant radiation damage accumulation (e.g. Holland and Gottfried 1955; Ewing *et al.* 2003; McCubbin *et al.* 2016; Pidgeon *et al.* 2019). Monitoring LREEs and non-formula elements allows for filtering of pristine zircon crystals or domains from altered and weathered ones. In contrast, rare earth element patterns in metamorphic zircon are more useful because equilibrium with other phases has a forcing effect on partition coefficients such that zircon cannot incorporate as much HREE as it would normally do (Fig. 4c). This is well illustrated when zircon forms in equilibrium with garnet (e.g. Kelly and Harley 2005; Blereau 2017), resulting in flattened to steepened heavy REE patterns (Fig. 4c). Whilst extremely sluggish under most conditions of metamorphism (Cherniak and Watson 2003), under extreme metamorphic conditions (>1000°C), the REE content in zircon may also be modified as a function of temperature and time (Watson 1996; Blereau *et al.* 2022), although others have suggested decoupling of REE from U–Pb above 850°C (Kunz *et al.* 2018; Jiao *et al.* 2020*b*; Durgalakshmi *et al.* 2021). REEs within zircon are impurities that are expelled at high-*T* resulting in the removal of internal zoning (Hoskin and Black 2000), with garnet acting as a sink for the released REEs (see Discussion). Diffusional modelling of REE-in-zircon shows that an initially igneous zircon (i.e. an inherited grain with steep M-HREE patterns) within a garnet-bearing metapelite is in disequilibrium with the garnet, and when exposed to metamorphic temperatures, the zircon attempts to reach equilibrium with garnet (i.e. a *c.* 1:1 D_{REE} pattern) (Blereau *et al.* 2022; Fig. 4d). Zircon with a 50 µm radius can be re-equilibrated with the host metamorphic assemblage during both short (1100°C for 1–5 Ma) and extended periods (1050°C for 10–30 Ma or 1000°C for 200 Ma) of UHT metamorphism (Blereau *et al.* 2022; Fig. 4e). Conversely, unless diffusion is enhanced by fluids or other processes, below 900°C, zircon will largely preserve its pre-metamorphic REE signature, even when metamorphism is prolonged (>100 Ma) (Blereau *et al.* 2022; Fig. 4e). The change of valence of Ce and Eu to 4+ and 2+, respectively, in igneous zircon can be used to estimate redox conditions through Ce and Eu anomalies (e.g. Trail *et al.* 2012; Smythe and Brenan 2016; W.T. Li *et al.* 2021). However, some authors suggest using these proxies with caution because of melt cooling and chemical evolution (e.g. Loader *et al.* 2022). Europium anomalies can also be used to track feldspar fractionation and to estimate whether zircon crystallized in equilibrium with feldspar in igneous and metamorphic zircon, respectively (Rubatto 2002).

The ratio of Th over U is often used in zircon because it provides information about co-precipitating phases (e.g. Kunz *et al.* 2018). It can help discriminate igneous (Th/U *c.* 0.2–0.8) from metamorphic (Th/U < 0.1 or >1) zircon domains/crystals, distinguish between zircon formed under amphibolite-facies and granulite-facies and can trace low-temperature weathering of radiation-damaged zircon (e.g. Vavra *et al.* 1999; Hoskin and Black 2000; Kirkland *et al.* 2015; Yakymchuk *et al.* 2018; Guitreau *et al.* 2019; Guitreau and Flahaut 2019; Barrote *et al.* 2020). However, Th/U in zircon is not always a faithful recorder of metamorphic processes (Möller *et al.* 2003; Harley and Kelly 2007), and it should be used with caution on a case-by-case basis. Typically, U–Pb LA-ICP-MS geochronology routines do not include appropriate internal standard elements, therefore, the U, Th and Pb concentrations are semi-quantitative only.

Geochronology. Zircon is most commonly dated using the U–Pb isotope system (e.g. Schoene 2014) as, for this systematics, it has several advantages compared to other minerals. Zircon U–Pb isotope measurements are done using a great variety of techniques from solution-based such as TIMS for the most precise ages and MC-ICP-MS/ICP-MS to *in situ* microbeam techniques (i.e. SIMS, SHRIMP, LA-ICP-MS, LA-MC-ICP-MS). One of the main advantages is that Pb is essentially excluded from the zircon lattice because of its size and valence, resulting in virtually all measurable Pb in zircon being radiogenic – produced by the radioactive decay of U and Th isotopes. This means that the parent/daughter ratio measured in pristine zircon combined with U decay constants can be directly converted into an age (Schoene 2014). However, zircon lattice can become damaged by radioactive decay over time, which results in metamictization. This process makes zircon porous to external agents that may incorporate Pb, and other non-formula elements, into the crystal lattice, thus compromising the determined age. Another major benefit of the U–Pb isotope system compared to others is the fact that it contains two isotope systems (i.e. ^{238}U–^{206}Pb and ^{235}U–^{207}Pb) with distinct decay constants (e.g. Le Roux and Glendenin 1963; Jaffey *et al.* 1971), allowing open- v. closed-system evolution to be assessed. For easy visualization, U–Pb data are commonly plotted in ^{238}U–^{206}Pb and ^{235}U–^{207}Pb Wetherill concordia diagram (Fig. 4f; Wetherill 1956). When both dates lie on the concordia curve, the date is called concordant and is geologically meaningful since it likely reflects a closed U–Pb system evolution. By contrast, when both dates are different, the measured date is discordant and evaluation on the geological meaning will vary case by case. A third date can also be derived directly from the $^{207}Pb/^{206}Pb$ ratio that inherently assumes closed-system evolution. This date is the oldest of the three in the case of 'normally' discordant data (i.e. datapoint is located below the concordia curve). Unless information regarding the validity of intercepts in concordia diagram is available (e.g. multiple analyses from the same crystal or growth zones), $^{207}Pb/^{206}Pb$ ages are commonly more precise for zircons >1.5 Ga, whereas ^{238}U–^{206}Pb ages are used for crystals <1.5 Ga (Spencer *et al.* 2016).

Depending on when the U–Pb isotope system disturbance occurred, the memory of primary crystallization may or may not be preserved. If U–Pb disturbance is recent, U–Pb data are discordant and distribute along a discordia line that passes through zero (blue points). Since Pb isotopes are not fractionated in detectable proportions in such disturbance, $^{207}Pb/^{206}Pb$ ratios still provide the primary crystallization age, which corresponds to the discordia upper-intercept (Fig. 4f) and is identical to the purple-filled ellipse which represents a concordant analysis from either the same zircon or the same population that remained unmodified. In contrast, for an old U–Pb disturbance, the datapoints would align along a discordia line that connects a lower-intercept corresponding to the age of U–Pb disturbance and an upper-intercept representing the actual crystallization age of zircon (red-filled ellipses in Fig. 4f). Most cases of discordance correspond to Pb-loss, which graphically corresponds to a migration of datapoints towards the lower end of a discordia line, and less common cases are associated with reverse discordance which reflects Pb accumulation (e.g. Williams *et al.* 1984; Kusiak *et al.* 2013). The case in which zircon contains measurable amounts of common-Pb (Pb_c) has not been presented here because, in most cases, it is a sign of zircon post-crystallization modification and/or advanced alteration. Consequently, data are generally discarded when ^{204}Pb is detected, since the Pb_c correction relies on knowledge of the isotopic composition of Pb when it entered the crystal, which is difficult to know in zircon.

Isotope geochemistry. The source of zircon parental magma can be assessed using multiple isotope systems, with the most common being Lu–Hf and O isotopes (e.g. Patchett 1983; Valley 2003). Lu–Hf isotopes in zircon are normally measured using MC-ICP-MS, either in solution or laser-ablation mode (e.g. Fisher *et al.* 2014*a*, *b*). The Lu–Hf isotope system is radiogenic and based on the decay of radioactive ^{176}Lu into ^{176}Hf with a half-life of *c.* 36 Ga (Scherer *et al.* 2001; Söderlund *et al.* 2004). The measured $^{176}Hf/^{177}Hf$ tracks the time-integrated fractionation of Lu from Hf in a reservoir (Fig. 4g). The principle of this technique is that during partial melting of most mantle and crustal lithologies, Hf and Lu are fractionated from each other, which results in magmas having Lu/Hf lower than that of the melting residue. Over time, melting residues (refractory mantle) develop very radiogenic (elevated) $^{176}Hf/^{177}Hf$ ratios (positive ε_{Hf}), and crustal lithologies (former magmas) comparatively low radiogenic ratios (negative ε_{Hf}) (Fig. 4g). For global interpretation, Hf isotope compositions are normalized to the Chondritic Uniform Reservoir (CHUR; Blichert-Toft and Albarède 1997; Bouvier *et al.* 2008; Iizuka *et al.* 2015) and transformed into epsilon Hf notation (ε_{Hf}), with CHUR approximating the bulk silicate Earth composition. For instance, Figure 4g illustrates a magma source produced by partial melting of the mantle at 2600 Ma (green spot in Fig. 4g) and that differentiates into mafic and felsic lithologies, which are both characterized by specific $^{176}Lu/^{177}Hf$ content. These lithologies evolve after crystallization until 1500 Ma, when they are reworked by partial melting forming different granites (blue and red spots in Fig. 4g). Some of the generated melts mix with each other forming hybrid granites (grey spot in Fig. 4g). At 600 Ma, magmas extracted from the mantle (green spot in Fig. 4g) mix with crustal melts forming new hybrid granites (orange spot in Fig. 4g). Note that all crustal lithologies contain zircon, which allows the evolution of these reservoirs to be followed through time. Newly formed zircon domains (grains or overgrowths) within a single rock sample mostly have higher initial $^{176}Hf/^{177}Hf$ than older domains due to incomplete dissolution of detrital or magmatic zircon grains (hosting most of the non-radiogenic Hf). Also, metamorphic zircon incorporates additional radiogenic ^{176}Hf formed by ^{176}Lu decay in the rock's matrix between successive zircon growth events (Gerdes and Zeh 2009).

Oxygen isotopes are stable isotopes and the measured $^{18}O/^{16}O$, and possibly $^{17}O/^{16}O$ when the triple-isotope system is used, tracks isotope fractionations due to magmatic processes and/or sources. These light isotopes do not fractionate much during magmatic processes (i.e. partial melting and fractional crystallization), but differences can be measured (e.g. Valley 2003; Bindeman 2008). Most O isotope studies in zircon use the source-tracing potential of O isotopes, taking advantage of large O isotope fractionation caused by low-temperature alteration of crustal lithologies inducing clay formation and resulting in enrichment of ^{18}O relative to ^{16}O. Much like the Lu–Hf isotope system, O isotope ratios are difficult to interpret as numerical values. Therefore, O isotope ratios are normalized relative to the international Vienna standard mean ocean water (VSMOW) and reported in the standard delta ($\delta^{18}O$) notation. Zircon formed from a reworked clay-rich crustal lithology may have $\delta^{18}O$ up to +12‰ (Valley *et al.* 2005; Kemp *et al.* 2007), and mantle zircons have consistent $\delta^{18}O$ of $+5.3 \pm 0.6$‰ (Valley 2003). Zircon $\delta^{18}O$ values below that of the mantle indicate that zircon parental melt interacted with meteoric fluids and/or was altered at high temperatures (Valley 2003 and references therein). Most studies using O isotopes in zircon interpret O isotope variations as evidence for mixtures between mantle- and crustal-derived melts and/or fluids or crustal contaminants (e.g. Kemp *et al.* 2007; Smithies *et al.* 2021). Recent studies have demonstrated that the uptake of water into the zircon crystal lattice can significantly modify its oxygen isotopic composition (Pidgeon *et al.* 2017; Liebmann *et al.* 2021). Therefore, monitoring of the $^{16}O^{1}H/^{16}O$ ratio during SIMS oxygen isotope measurements to assess secondary modification of O isotope composition by water addition is recommended (Liebmann *et al.* 2021).

Recently, new stable isotope systems such as Si and Zr have been applied to zircon. Different techniques such as SIMS, MC-ICP-MS and LA-MC-ICP-MS have been employed to measure Si isotopes in zircon, which have resulted in good precision and closely mimic the natural variability of high-temperature processes (Trail *et al.* 2018, 2019; Guitreau *et al.* 2020, 2022). Si isotopes can be used to trace the origin of zircon parental magma due to the various silicon isotope signatures found in different types of igneous rocks (e.g. I, A, S, tonalite–trondhjemite–granodiorite (TTG); Savage *et al.* 2014; Deng *et al.* 2019). Si isotope compositions, expressed as permil deviations from a quartz standard ($\delta^{29}Si$ and $\delta^{30}Si$), are sensitive to the SiO_2 content, which reflects the degree of polymerization and crystallization temperature (Qin *et al.* 2016; Trail *et al.* 2019; Guitreau *et al.* 2022). This allows for reconstructions of the magma evolution. Changes in Si isotope compositions can occur due to metamorphic alteration and/or zircon recrystallization, which are dependent on the metamorphic grade (Guitreau *et al.* 2022).

Zr isotopes can be measured using conventional MC-ICP-MS instruments in solution or laser mode and applied to track fractional crystallization processes (Ibañez-Mejia and Tissot 2019; Zhang *et al.* 2019; Guo *et al.* 2020; Tian *et al.* 2020). Recent studies attributed measurable Zr isotope variations to kinetic fractionation in response to chemical gradients rather than equilibrium processes (Chen *et al.* 2020; Méheut *et al.* 2021). Zirconium isotope compositions of metamorphic zircon compared to igneous zircon may also result from chemical gradient effects (Zhang *et al.* 2019). This technique can hence provide insights into magmatic crystallization dynamics.

Thermometry. Zircon is used as a mineral pair thermometer based on incorporation of Ti into zircon at *HT* coupled with Zr substitution in rutile (Zack *et al.* 2004*a*; Watson and Harrison 2005; Watson *et al.* 2006; Ferry and Watson 2007). This technique requires the presence of both zircon and rutile within the mineral paragenesis, otherwise all temperatures are minimum estimates (Watson and Harrison 2005). Magmatic and metamorphic temperatures can also be overestimated at low pressures (<5 kbar) (Rubatto 2017). Temperature variations within an investigated sample have been interpreted to reflect waves of magmatic pulses within a magma chamber generating thermal and compositional heterogeneities (e.g. Collins *et al.* 2016; Volante *et al.* 2020*a*). Low diffusivity of Ti within zircon was previously measured perpendicular to the crystallographic c-axis (Cherniak and Watson 2007), making this thermometry a widely used tool. In contrast, recent experiments conducted parallel to the c-axis found significant anisotropy in the diffusivity of Ti (Bloch *et al.* 2022). When extrapolated, the resulting diffusivities are *c.* 7.5–11 times faster at 950–650°C than the original experiments, indicating that this thermometer can be modified under elevated crustal temperatures, slow cooling and/or small grain sizes (Bloch *et al.* 2022). Moreover, since this thermometer is dependent upon alpha-quartz and t-TiO_2 activities, software such as Rhyolite-MELTS Gualda *et al.* (2012) have been proposed to improve temperature estimates' precision (e.g. Schiller and Finger 2019). The latter tool is useful and intuitive to use, however its caveats (Volante *et al.* 2020*a*) and its relatively low precision of about 50°C (Guitreau *et al.* (2022)) should be considered.

Titanite

Titanite ($Ca[Ti,Al,Fe^{3+}]SiO_4[O,OH,F,Cl]$) (Fig. 2g-i; Fig. 5) can occur as minor or accessory mineral and it is a particularly efficient chemical reactant with other major mineral phases that contain Ca and Ti, commonly leading to re-crystallization (Frost *et al.* 2001). Titanite commonly grows during magmatic (e.g. Ca-rich granitic rocks), metamorphic (e.g. calc-silicate, amphibolite) and hydrothermal events (Fig. 5a-c), with each different setting resulting in different titanite REE compositions that can be tracked by trace element fingerprinting (see reviews by Frost *et al.* 2001; Kohn 2017). Titanite commonly forms in mafic and calc-silicate rocks, and its growth can provide valuable time constraints in cases where other accessory minerals do not develop. Titanite crystallizes over a broad *P–T* range, including growth after breakdown of rutile, which is commonly stable at higher pressures (>1.4 GPa), ilmenite, Fe–Ti oxide stable at higher temperatures (>650/700°C), magnetite and/or clinozoisite (Frost *et al.* 2001; Kohn 2017). The combined acquisition of geochronological and geochemical data in texturally controlled titanite has robustly discriminated between distinct metamorphic, hydrothermal and magmatic events, and provided the opportunity to evaluate physical and chemical processes at the micro-scale in titanite-bearing rocks (e.g. Stearns *et al.* 2015; Garber *et al.* 2017; Olierook *et al.* 2019; Cavosie *et al.* 2022; Walters *et al.* 2022). However, petrological and chemical complexities can challenge U–Pb and *P–T* data interpretation. For instance, overgrowths of multiple metamorphic and hydrothermal titanite generations can significantly affect and modify the U–Pb system (e.g. Storey *et al.* 2006; Marsh and Smye 2017; Kirkland *et al.* 2018) and trace element compositions (e.g. Gordon *et al.* 2021), inducing decoupling of titanite U–Pb dates and trace element compositions (e.g. Romer and Rötzler 2001, 2011; Castelli and Rubatto 2002; Bonamici and Blum 2020; Walters *et al.* 2022).

Geochemical tools. Texturally, both magmatic and metamorphic (or hydrothermally altered) titanite can exhibit complex compositional zoning (Fig. 5a–c) and associated trace element (TE) patterns related to fluid–rock interaction (e.g. Smith *et al.* 2009; Garber *et al.* 2017; Olierook *et al.* 2019; Walter *et al.* 2021; Walters *et al.* 2022). Titanite preferentially incorporates minor and TE including U, high field-strength elements (HFSEs) such as Zr and REEs (e.g. Tiepolo *et al.* 2002; Lucassen *et al.* 2012), compared to other accessories (e.g. zircon, monazite). Trace element in titanite can provide information about crystallization pressure and temperature estimates (Hayden *et al.* 2008), oxygen fugacity (e.g. King *et al.* 2013; Cao *et al.* 2015) and fractionation processes (e.g. Piccoli *et al.* 2000; John *et al.* 2011). Geochemical and experimental studies on titanite suggest that it preferentially incorporates the medium rare earth element (MREE) in its crystal structure when in equilibrium with melt (e.g. Tiepolo *et al.* 2002; Prowatke and Klemme 2005; Olin and Wolff 2012), whereas recent petrochronological works show that melt-present and melt-absent metamorphic titanite have different REE patterns strongly depending on the presence and/or abundance of other phases in equilibrium with them (e.g. Garber *et al.* 2017; Walter *et al.* 2021; Walters *et al.* 2022). For instance, different TE uptake in titanite allowed Garber *et al.* (2017) to discriminate between (i) LREE-enriched (Precambrian) igneous cores, (ii) LREE-depleted and HREE-rich (Caledonian) recrystallized metamorphic rims likely associated with fluid or melt and (iii) (Caledonian) neocrystallized metamorphic titanite reflecting a negative or more positive REE slope based on whether titanite grew in equilibrium with hornblende and garnet or with allanite, apatite, plagioclase and biotite, respectively, during prograde and/or retrograde amphibolite-facies metamorphism (Fig. 5d, e; see also Cioffi *et al.* 2019; Walter *et al.* 2021).

Recent studies have highlighted the potential of TE in titanite as petrogenetic discriminator/indicator in magmatic, metamorphic and detrital studies (e.g. Ma *et al.* 2019; Olierook *et al.* 2019; Scibiorski and Cawood 2022). For example, Olierook *et al.* (2019) show that Fe concentrations and Th/U and Th/Pb ratios are systematically higher in magmatic than metamorphic titanite, reflecting useful petrogenetic discriminators when plotted against Al and Zr-in-titanite temperature (*T*°C), respectively (Fig. 5f–h). Also, normalized LREE/MREE, LREE/HREE and Eu anomalies reflect negative and/or positive REE slope correlations, which are useful to complement the characterization of titanite

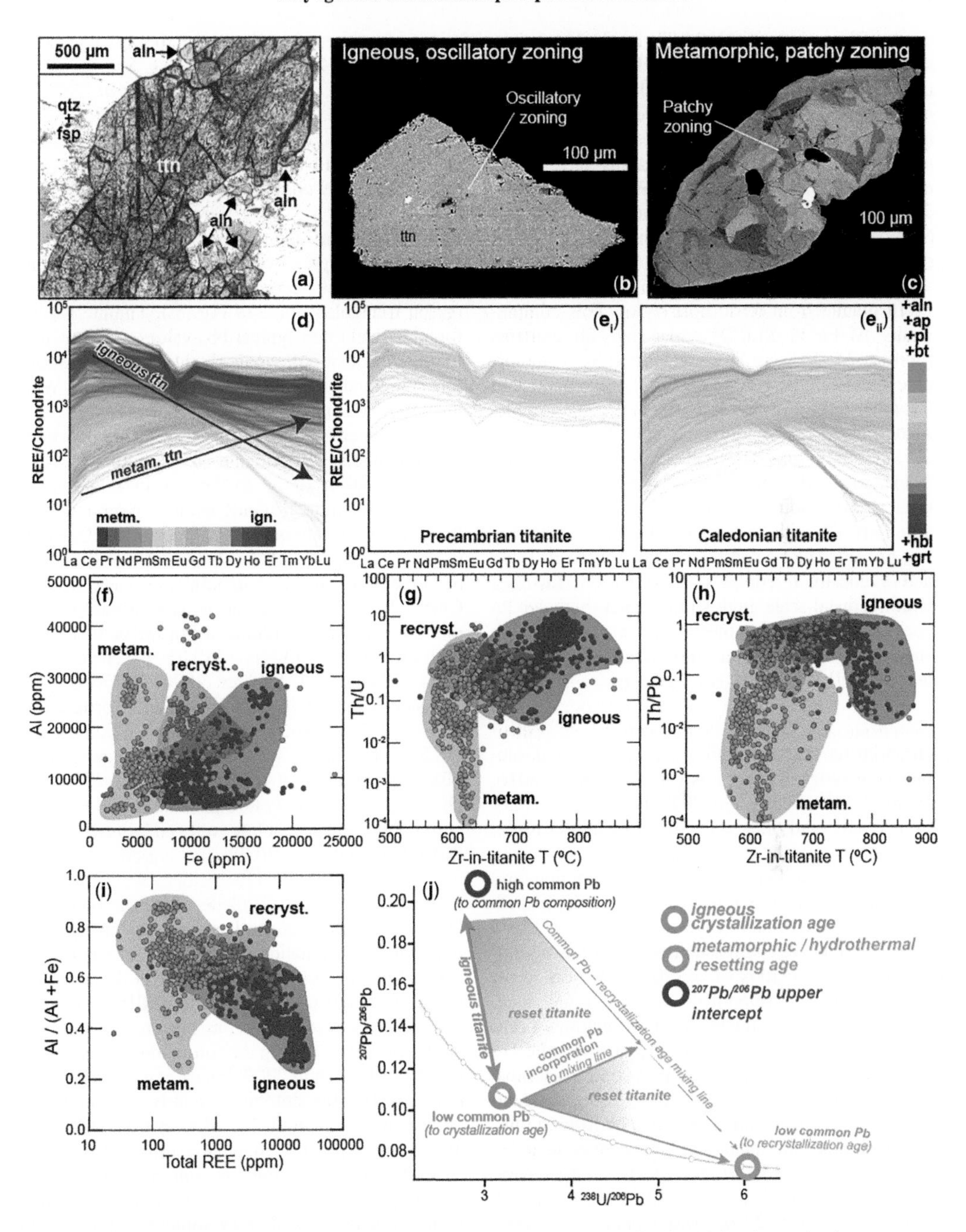

Fig. 5. Titanite. (**a**) titanite in orthogneisses. (**b**) Oscillatory zoning in igneous titanite. (**c**) Patchy zoning in metamorphic titanite. (**d**) REE patterns in magmatic and metamorphic titanite. (**e**) Slopes of REE patterns are affected by other mineral phases growing in equilibrium with (i) igneous Precambrian and (ii) metamorphic Caledonian titanite grains. (**f**) Al v. Fe. (**g**) Th/U v. Zr-in-titanite. (**h**) Th/Pb v. Zr-in-titanite. (**i**) Al/(Al + Fe) v. total REE discrimination diagrams to distinguish between igneous (in red), recrystallized (in purple) and metamorphic (in green) titanite. (**j**) Schematic Tera–Wasserburg concordia diagram for interpreting U–Pb titanite data. Source: (a), (c), (d) and (e) modified after Garber *et al.* (2017); (b), (i) and (j) modified after Olierook *et al.* (2019).

origin (Olierook *et al.* 2019). TE-based discrimination can be corroborated with detailed textural investigation of titanite grains, which are commonly characterized by oscillatory or sector zoning when magmatic in origin, with patchy and/or homogeneous zoning in metamorphic grains (e.g. Smith *et al.* 2009; Garber *et al.* 2017; Walters and Kohn 2017; Cioffi *et al.* 2019; Holder and Hacker 2019). Scibiorski and Cawood (2022) show that different titanite host-rock lithologies are reflected in the variation of TE chemistry, with low Zr/Y and high Fe in titanite from felsic host-rocks. This complements Al/Fe v. ΣLREE and U v. Th contents (ppm) used to discriminate magmatic v. metamorphic titanite (Olierook *et al.* 2019; Scibiorski and Cawood 2022; Fig. 5i) and between metamorphic, recrystallized and inherited igneous titanite (fig. 2 in Scibiorski and Cawood 2022), respectively. Also, systematic integration of titanite microstructural investigations by electron backscatter diffraction (EBSD) with *in situ* U–Pb petrochronology was proved to be a powerful tool not only to untangle and reconstruct complex deformation histories within crustal-scale high-strain zones but also to constrain deformation mechanism associated with shock and thermal metamorphism (Papapavlou *et al.* (2017); McGregor *et al.* 2021).

Geochronology. Titanite is a powerful U–Pb petrochronometer recording different primary and secondary geological processes (Spandler *et al.* 2016; Ma *et al.* 2019; Fisher *et al.* 2020; Barla 2021). However, titanite and other accessories including rutile, apatite and allanite tend to incorporate significant Pb_c (e.g. Kirkland *et al.* 2017, 2018; Bonamici and Blum 2020), which is reflected in discordant U–Pb ratios (e.g. Marsh and Smye 2017), making U–Pb dates interpretation of these accessory phases challenging (e.g. Olierook *et al.* 2019; Walters *et al.* 2022). Figure 5j (modified after Olierook *et al.* 2019) represents a schematic TW diagram (Tera and Wasserburg 1972) illustrating how titanite U–Pb data (and the other phases that commonly have Pb_c), including Pb_c (see Storey *et al.* 2006; Kirkland *et al.* 2017), crystallization and recrystallization mixing lines can be interpreted (e.g. Spencer *et al.* 2013; Bonamici *et al.* 2015; Garber *et al.* 2017; Papapavlou *et al.* 2017; Holder and Hacker 2019; Mottram *et al.* 2019; Timms *et al.* 2020; Gordon *et al.* 2021). As a result of Pb_c incorporation in the crystals, geochronological data typically form an intercept line (red line in Fig. 5j, 'igneous titanite ages') characterized by a $^{207}Pb/^{206}Pb$ upper intercept and lower concordia intercept (e.g. Spencer *et al.* 2013; Chew *et al.* 2014; Garber *et al.* 2017; Kirkland *et al.* 2017, 2018, 2020), with the igneous age being calculated using the lower one. However, when such intercept line is not statistically robust, then (i) a weighted mean of uncorrected dates can be calculated, if they are within 2SD uncertainty (e.g. Spencer *et al.* 2016; Olierook *et al.* 2019; Barrote *et al.* 2022*b*), or (ii) the upper intercept can be used to calculate concordant analyses with negligible Pb_c (Olierook *et al.* 2019). U–Pb dates of samples affected by overprinting events would fall in a space in the TW diagram defined by a $^{207}Pb/^{206}Pb$ upper intercept, a first lower concordia intercept recording the first magmatic/metamorphic event and a second lower concordia intercept recording the subsequent event (green triangular shape in Fig. 5j). Titanite is also found to yield non-typical Pb_c values, possibly due to inheritance of radiogenic Pb into the crystal structure (Kirkland *et al.* 2018; Mottram *et al.* 2019; Walters *et al.* 2022).

Additionally, an important factor to consider when dealing with titanite geochronology is *Tc* (Fig. 1; Dodson 1973). In the past two decades, it has been shown that titanite is much more retentive than previously envisaged (with *Tc* of *c.* 600–650° C, Mezger *et al.* 1991; 600–650°C, Scott and St-Onge 1995, Frost *et al.* 2001; 650 and 750°C, Cherniak 1993), increasing the temperature threshold for Pb and Zr volume diffusion in titanite as high as *c.* 750–840°C (*c.* 740°C, e.g. Schärer *et al.* 1994, Zhang and Schärer 1996; >750°C, e.g. Kylander-Clark *et al.* 2008, Kohn and Corrie 2011, Gao *et al.* 2012, Spencer *et al.* 2013, Stearns *et al.* 2015, Kohn 2017; >830°C, e.g. Hartnady *et al.* 2019, Holder *et al.* 2019, Kirkland *et al.* 2020) and challenging the assumption that titanite U–Pb dates commonly reflect cooling ages (Hartnady *et al.* 2019; Kirkland *et al.* 2020).

Isotope geochemistry. While less explored compared to other accessory minerals such as monazite, Sm–Nd isotope systematics in titanite (e.g. Yang *et al.* 2008; Amelin 2009; Fisher *et al.* 2011, 2020; Hammerli *et al.* 2014; Spandler *et al.* 2016; Ma *et al.* 2019; Zhang *et al.* 2021) has potential as a source tracer for understanding the formation and evolution of the crust (Amelin 2009; Fisher *et al.* 2020; Zhang *et al.* 2021). This isotopic system has been demonstrated to have high *Tc* (850–950°C; Cherniak 1995), surviving HT magmatic, metamorphic and hydrothermal conditions. *In situ* Sm–Nd isotopic systematics of titanite can be used to trace fluid, melt or crustal/juvenile rock sources (e.g. Lucassen *et al.* 2011; Hammerli *et al.* 2014; Spandler *et al.* 2016; Zhang *et al.* 2021).

Experimental studies have demonstrated that oxygen diffusion in accessory minerals is slower than in rock-forming minerals (e.g. Fortier and Giletti 1989), but oxygen diffusion in apatite is faster than in titanite, which is in turn faster than in zircon (Bruand *et al.* 2019). Oxygen isotopes in magmatic and metamorphic titanite have been used as a geochemical indicator to investigate magma petrogenesis as well as metamorphic and hydrothermal overprinting events (e.g. King *et al.* 2001; Bonamici *et al.* 2011, 2014, 2015; Bruand *et al.* 2019). For example, Bonamici *et al.* (2014) differentiated four generations of titanite with distinct $\delta^{18}O$ values and internal textural zoning. Also, consistent results between $\delta^{18}O$ in titanite and zircon indicate that titanite is a robust accessory mineral preserving the original magmatic $\delta^{18}O$ composition (Bruand *et al.* 2019).

Thermometry. A thermometer using the Zr content in titanite (Zr-in-titanite) was developed by Hayden *et al.* (2008), relying on the direct substitution of Zr^{4+} for Ti^{4+}, whereas pressures can be estimated from the net transfer reaction $_2Ca_2\ Al_3Si_3O_{12}(OH) + TiO_2 + SiO_2 = 3CaAl_2Si_2O_8 + CaTiSiO_5 + H_2O$ (referred to as TZARS; Kapp *et al.* 2009), using automated calculations in THERMOCALC (Holland and Powell 2011). The Zr-in-titanite thermometer covers a wide temperature range and is pressure dependent, resulting in large pressure uncertainties and rutile activity in rutile-absent rocks, limiting the reliability of temperature estimates. Therefore, modelling titanite within a mineral paragenesis using software such as THERMOCALC allows for pressure estimates with errors less than *c.* 0.1 GPa (Kohn *et al.* 2017).

Rutile

Rutile (Fig. 2j-l; Fig. 6) is the high-temperature TiO_2 polymorph (Dachille *et al.* 1968) and a common accessory mineral that occurs in metamorphic (U) high-pressure (UHP) mafic rocks (Fig. 6a; e.g. Zack *et al.* 2004*b*; Triebold *et al.* 2012; Zack and Kooijman 2017; Böhnke *et al.* 2019), in moderate-pressures and moderate-temepratures to *HP* and *HT* metapelitic rocks (Fig. 6b; e.g. Hart *et al.* 2018; Gonçalves *et al.* 2019) and in low-temperature/hydrothermal metapelitic rocks (Fig. 6c; e.g. Plavsa *et al.* 2018; Salama *et al.* 2018; Agangi *et al.* 2019; Porter *et al.* 2020; Schirra and Laurent 2021). In contrast, rutile occurrence in magmatic rocks is limited to *HT* and dry alkaline, kimberlite or pegmatitic rocks (e.g. Carruzzo *et al.* 2006; Cerny *et al.* 2007). At low-*T* (LT) and low-*P* (LP) conditions, rutile is no longer stable, and it is replaced by polymorphs anatase (tetragonal) and brookite (orthorhombic), respectively, whereas rutile *HP* polymorph is TiO_2(II) (Dachille *et al.* 1968; Jamieson and Olinger 1969). During prograde metamorphism, rutile growth is attributed to breakdown of ilmenite, titanite and biotite (Zack and Kooijman 2017). Rutile popularity as a petrochronometer has exponentially increased due to its multifaceted potential to investigate tectonic processes as single-mineral thermometer (e.g. Zack *et al.* 2004*a*; Watson *et al.* 2006; Tomkins *et al.* 2007; Luvizotto and Zack 2009), geochronometer (e.g. Mezger *et al.* 1989*a*), geospeedometer (Cruz-Uribe *et al.* 2014; Kohn *et al.* 2016), as a provenance indicator (e.g. Triebold *et al.* 2012; Pereira and Storey 2023) and as an isotopic tracer (e.g. Ewing *et al.* 2011). Additionally, the increasing sensitivity of analytical equipment has enabled dating of low-U phases such as rutile (Luvizotto *et al.* 2009; Axelsson *et al.* 2018; Verberne *et al.* 2019; Moore *et al.* 2020*b*).

Geochemical tools. Rutile primarily incorporates high field strength elements (HFSE; i.e. Nb, Ta, Zr, Hf and Cr; Rudnick *et al.* 2000; Zack *et al.* 2002; Schmidt *et al.* 2009). These can be used (i) to characterize micro-scale processes associated with growth of rutile crystals (e.g. Hart *et al.* 2018; Verberne *et al.* 2022*a*); (ii) as a pressure-proxy, such that trace element budget (i.e. Na and Ta) in the melt rutile crystallized in reflects the depth at which partial melting of the source rock occurred (e.g. Foley *et al.* 2002; Moyen and Stevens 2006; Meyer *et al.* 2011; Kooijman *et al.* 2012); or (iii) as geochemical pathfinders for mineralized rocks (Clark and Williams-Jones 2004; Smythe *et al.* 2008; Pochon *et al.* 2017; Plavsa *et al.* 2018; Agangi *et al.* 2019; Ballouard *et al.* 2020; Porter *et al.* 2020; Sciuba and Beaudoin 2021). While HFSE are compatible in rutile, other trace elements including Sr, Th and REEs are incompatible (Klemme *et al.* 2005; Meyer *et al.* 2011). Trace element concentrations between the three TiO_2 polymorphs (rutile, anatase and brookite) systematically differ, leading to erroneous results when applying the Zr-in-rutile thermometer or Cr and Nb discrimination diagrams (see below) to phases other than rutile (Triebold *et al.* 2012).

Lithology discrimination schemes from Triebold *et al.* (2012) and Meinhold *et al.* (2008) using Cr and Nb (ppm) content in rutile as well as Zr/Hf and Nb/Ta ratios are useful to identify the protolith composition (metapelitic v. metabasic). However, care must be taken when applying this technique for source discrimination in provenance studies due to documented unsystematic Nb/Cr ratios for rutile from amphibolite-facies rocks potentially reflecting TE disturbance during retrogression or prolonged HT metamorphism (Meyer *et al.* 2011; Kooijman *et al.* 2012). Additionally, the application of principal component analysis (PCA) on rutile from Precambrian UHT and Phanerozoic HP terranes (Hart *et al.* 2018), as well as on ore-bearing and barren metamorphic and magmatic rocks (Plavsa *et al.* 2018; Pereira *et al.* 2019; Porter *et al.* 2020) indicates that several TE can be used for lithological discrimination (van Schijndel *et al.* 2021) and to distinguish rutile from mineralized v. barren rocks. For example,

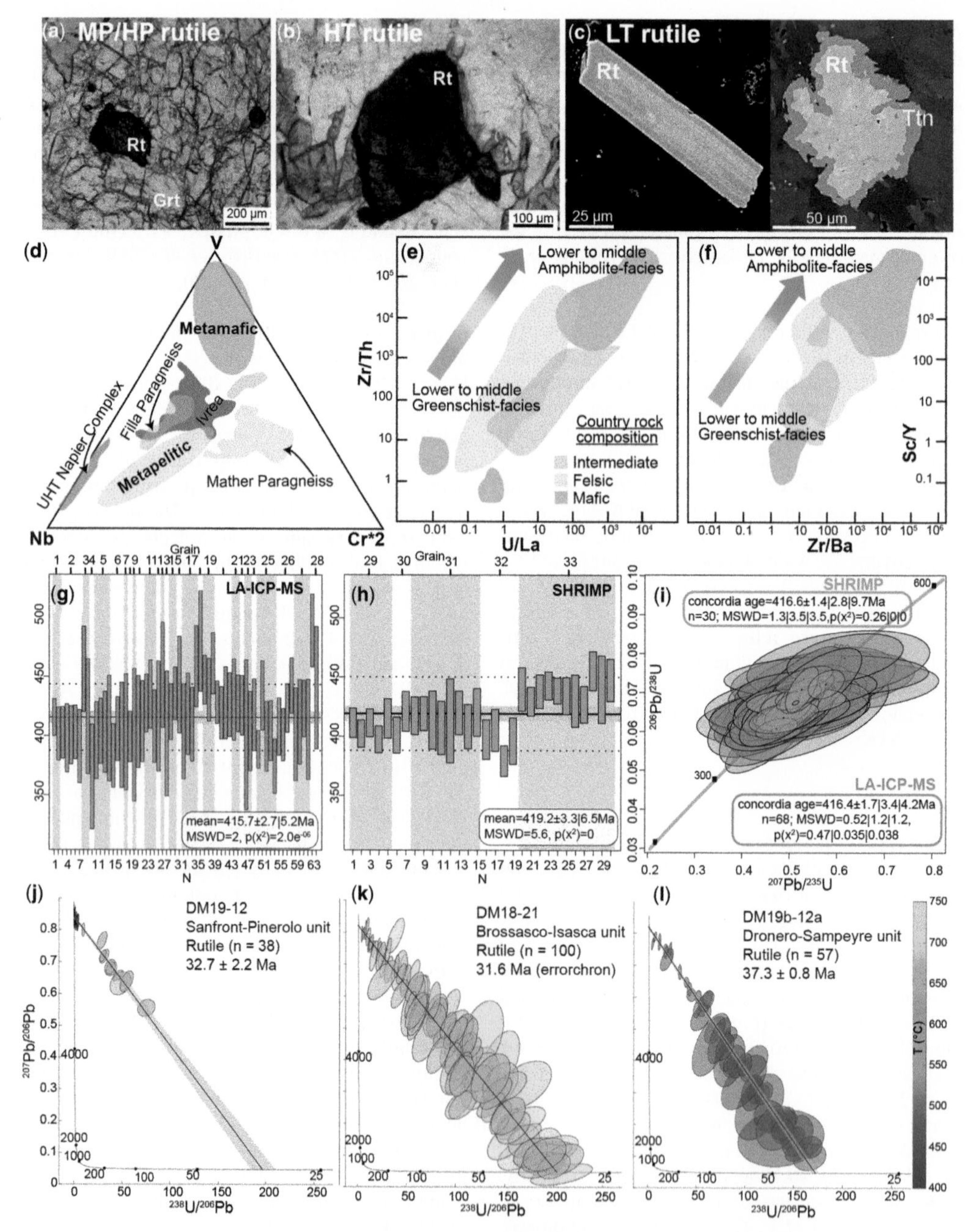

Fig. 6. Rutile textures in (**a**) rutile inclusion in garnet from a HP mafic eclogite from the Sanbagawa belt, SW Japan. (**b**) HT rutile from an orthopyroxene–cordierite granulite from Madagascar. (**c**) LT rutile. Left: a BSE image of tabular rutile with a high W core (W-rt) from the Speewah carbonatite, Australia. Right: rutile grain from the Boddington Au–Cu deposit with patchy W zonation and coronitic titanite. (**d**) Nb–V–Cr ternary discrimination diagram for rutile. Binary plots of rutile (**e**) U/La v. Zr/Th and (**f**) Zr/Ba v. Sc/Y trace element ratios from orogenic gold deposits; (**g–i**) U–Pb data for LA-ICP-MS and SHRIMP analyses for porphyroblastic rutile. Weighted mean age of individual analyses for (g) LA-ICP-MS and (h) SHRIMP data. (i) U–Pb concordia diagram, ellipses representing the 2σ uncertainty; (**j–l**) TW plots and isochrons/errorchrons combined with Zr-in-rutile thermometry. Source: (a and b) photo courtesy of Pereira; (c) modified after Porter *et al.* (2020); (d) modified after Hart *et al.* (2018); (e) and (f) modified after Sciuba and Beaudoin (2021); (g–i) modified after Moore *et al.* (2020*b*); (j–l) modified after Bonnet *et al.* (2022).

the Nb–V–Cr ternary diagram shows that rutile from HP metamafic rocks is commonly V-rich, whereas in metapelites is more Nb-rich (Fig. 6d; Hart *et al.* 2018). When associated with mineralized systems, rutile contains anomalous concentrations of V, Sn, Sb, W, Ni, Cu, Cr, Ta, Nb and Fe (e.g. Scott and Radford 2007; Plavsa *et al.* 2018; Porter *et al.* 2020), where W and Cr variability is used to discriminate its origin from mineralized or barren rocks (Porter *et al.* 2020). It is also possible to identify rutile deriving from pegmatitic rocks (high Nb, Ta, Sn) and Au-ore, which are enriched in Sb. This can also be done by combining multi-element clustering of PCA analysis (Porter *et al.* 2020; Sciuba and Beaudoin 2021). However, due to the systematic differences in trace element contents between the three TiO_2 polymorphs, a detailed characterization of the analysed grains is recommended using either EBSD or Raman spectroscopy (Plavsa *et al.* 2018; Porter *et al.* 2020; Sciuba and Beaudoin 2021). Multi-variant statistical analysis of rutile TE composition, particularly variations in Sc, REE, Y, Ca, Ba, Th, Zr, U and V, can reflect different metamorphic grades experienced by the country rock (Sciuba and Beaudoin 2021). For example, mafic and ultramafic, lower to middle greenschist-facies country rocks have lower Zr/Th and U/La ratios than intermediate and sedimentary greenschist facies one (Fig. 6e; Sciuba and Beaudoin 2021). A similar trend is observed for relative concentrations of Sc/Y and Zr/Ba, where lower and higher contents reflect lower and higher metamorphic grade, respectively (Fig. 6f; Sciuba and Beaudoin 2021). Also, the identification of localized TE enrichment along twin interface in rutile grains using atom probe tomography is interpreted to occur via volume diffusion during *HT* metamorphism (Verberne *et al.* 2022*a*).

Geochronology. Vry and Baker (2006) calculated the *Tc* of Pb diffusion in rutile to be between 500 and 540°C, based on natural samples (Fig. 1). For a spherical rutile of 200 µm, experimental data from Cherniak (2000*a*) predict whole-grain *Tc* of *c.* 600°C at an average cooling rate of 2–3°C/Ma, and above *c.* 640°C for Zr in rutile (2°C/Ma cooling rate; Cherniak *et al.* 2007; Dohmen *et al.* 2019). This moderate *Tc* for rutile implies that it is possible to find crystallization ages in cores of larger rutile crystals, especially if cooling is very fast. The preservation of a crystallization/growth age depends on several factors including max metamorphic temperatures, cooling rates and grain size (Kylander-Clark *et al.* 2008; Zack and Kooijman 2017; Moore *et al.* 2020*b*), and a multi-proxy approach may be needed to distinguish between different metamorphic and deformation stages. For example, Moore *et al.* (2020*b*) used rutile Zr thermometry, SHRIMP U–Pb age determination and electron backscatter diffraction (EBSD) microstructural analyses to identified two-stage rutile age populations which were not distinguishable using LA-ICP-MS data alone (Fig. 6g–i). Due to low-U content in rutile (<0.1 ppm; see Zack *et al.* 2011), it is important to first determine the U content to obtain meaningful metamorphic and/or magmatic ages (Zack *et al.* 2002, 2004*b*, 2011). Low-U content can lead to a high common v. radiogenic Pb ratio; therefore, monitoring the $^{206}Pb/^{208}Pb$ ratio and application of the ^{208}Pb correction method is recommended (Zack *et al.* 2011). An application of this method allows the investigation of age variations within a single rutile grain, where transects from core to rim give younger ages towards the rim as a result of Pb diffusion during cooling and provide a temperature–time trajectory (Kooijman *et al.* 2010). In contrast to rutile geochronology obtained with LASS-ICP-MS, recent atom probe investigations indicate evidence for heterogeneous Pb and trace element distribution at the nanoscale (Verberne *et al.* 2020). Nonetheless, at the microscale (>20 µm), TE variations in a rutile single grain are negligible and concordant U–Pb dating is obtained, indicating that nanoscale defects do not significantly impact the micro-scale analysis (Verberne *et al.* 2020).

Isotope geochemistry. In situ analysis of Hf isotopes in rutile (Sláma *et al.* 2007; Ewing *et al.* 2011, 2014) has been applied to trace metasomatic processes in the lithospheric mantle (Choukroun *et al.* 2005; Aulbach *et al.* 2008) and recycling of continental material in the mantle (Ewing and Müntener 2018). Despite the relatively low Hf content (<300 ppm Hf), the $^{176}Hf/^{177}Hf$ of rutile can be accurately measured *in situ* by LA-MC-ICP-MS, provided care is taken with the analytical protocol and data reduction process (Ewing *et al.* 2011; Ewing and Müntener 2018). Matrix matched standards and a ^{176}Hf signal intensity above 10 mV are necessary, requiring a large spot size of >160 µm (Yang *et al.* 2015; Ewing and Müntener 2018). This technique is particularly interesting for rocks that lack zircon but contain rutile, such as mafic lithologies, with low Hf concentrations (Ewing *et al.* 2011). In contrast, rutile from (U)HT felsic granulites can contain much higher Hf contents, ranging from 20 to 400 ppm (Ewing *et al.* 2013).

Thermometry. The solubility of ZrO_2 in rutile is strongly temperature-dependent, and Zr-in-rutile has been identified as a useful thermometer (ZiR) when the rutile coexists with the appropriate buffer assemblage, i.e. zircon + quartz (Zack *et al.* 2004*a*; Watson *et al.* 2006; Tomkins *et al.* 2007; Hofmann *et al.* 2013). Underestimation of the calculated temperatures occurs when the rutile grows in the absence of zircon and/or in partially reset mineral assemblages (Zack *et al.* 2004*a*; Harley 2008). Possible biases include micro-inclusions of zircon (Zack *et al.* 2004*a*), prograde relict grains or incomplete equilibration with quartz (high Zr-rutile) or zircon (low-Zr rutile) due to slow diffusion along grain boundaries (see discussions in Taylor-Jones and Powell 2015; Kohn *et al.* 2016; Kohn 2020). Decoupling between Ti-in-zircon and Zr-in-rutile thermometry during UHT metamorphism is recorded in rutile that occurs as inclusions in zircon (Lei *et al.* 2020). Diffusion of Zr within rutile and Zr loss are closely related to the distribution of Zr, duration of UHT metamorphism and rutile grain size (Dohmen *et al.* 2019; Lei *et al.* 2020). The Zr concentration in rutile (Zr-in-rutile) is temperature-sensitive over a large range of geologically significant temperatures (e.g. Zack *et al.* 2004*a*; Ewing *et al.* 2013; Wawrzenitz *et al.* 2015; Pape *et al.* 2016; Böhnke *et al.* 2019; Clark *et al.* 2019; Moore *et al.* 2020*b*; Adlakha and Hattori 2021; Campomenosi *et al.* 2021; Bonnet *et al.* 2022). This relationship has been experimentally calibrated by Watson *et al.* (2006) and Ferry and Watson (2007), and a significant pressure effect has been calibrated by Tomkins *et al.* (2007). The Zr-in-rutile thermometer was recently refined and now predicts temperatures up to 40°C lower for $T \leq 550°C$, and systematically higher temperatures for $T > 800°C$ (Kohn 2020). With the new calibrations, precisions of $\pm 5°C$ and accuracy of *c.* $\pm 15°C$ may be possible, although a variable rutile composition may lead to larger uncertainties (Kohn 2020). Taylor-Jones and Powell (2015) showed that Zr can leave rutile and move along grain boundaries towards existing distal zircon.

Zr-in-rutile temperatures can record HT events, whereas U–Pb in rutile records cooling ages due to low *Tc* (Fig. 1). Rutile U–Pb ages likely postdate Zr temperatures following high-grade metamorphism and subsequent simple cooling, although a more complex history of episodic cooling and reheating may lead to more significant decoupling between Zr temperatures and U–Pb ages (e.g. Ewing *et al.* 2015). Bonnet *et al.* (2022) show the combined use of Zr-in-rutile and U–Pb ages for rutile occurrence in subduction complexes that may be interpreted as crystallization ages for the units that experienced high-pressure, low-temperature metamorphism, but not for the high-grade units (Fig. 6j–l).

Allanite

The term 'allanite' has been used to designate minerals from the allanite subgroup of the epidote group (Armbruster *et al.* 2006; Mills *et al.* 2009), which occurs in magmatic (Fig. 2m-o; Fig. 7a), metamorphic (from greenschist- to granulite-facies) and hydrothermal rocks (Janots *et al.* 2006, 2009; Rubatto *et al.* 2011; Airaghi *et al.* 2019). The allanite subgroup comprises a series of REE minerals with an ideal structural formula of $Ca(LREE^{3+})(Al)_2$ $(Fe^{2+}, Fe^{3+})(SiO_4)(Si_2O_7)O(OH)$. The presence of LREEs as major constituents, the various element substitutions, the broad *P* and *T* stability field and the preservation of growth stages, make allanite one of the burgeoning protagonists in petrochronology (e.g. Rubatto *et al.* 2011; Manzotti *et al.* 2018; Airaghi *et al.* 2019). Compositionally, two main factors are crucial for its use in petrochronology: (i) REEs' sites can be occupied by Th^{4+} and U^{4+}, pivotal elements for geochronology (Gieré and Sorensen 2004), and (ii) REEs are exceptional tracers of geological processes (Hermann 2002; Engi 2017). Allanite often exhibits complex chemical zoning (Fig. 7a; Romer and Xiao 2005; Rubatto *et al.* 2011; Airaghi *et al.* 2019), requiring investigation via *in situ* methods (e.g. Burn 2016; Zhang *et al.* 2022).

Geochemical tools. Trace element composition of allanite depends on the interplay between the bulk rock composition and fractionation of these elements during mineral reactions. For instance, a decrease in the LREE in magmatic allanite can be observed in Ca-rich rocks → diorite and granodiorite → granite → syenite (Smye *et al.* 2014; Engi 2017). The Th/U ratios can be often used as a complementary tool (together with initial Pb_c lead values) to distinguish between magmatic (>100) and metamorphic (<50) grains or domains (Fig. 7b; Gregory *et al.* 2007, 2012; Di Rosa *et al.* 2020). REEs have also been used to discriminate distinct allanite growth stages in metamorphic and magmatic rocks (Fig. 7c; Manzotti *et al.* 2018; Corti *et al.* 2020). Corti *et al.* (2020) compared composition and internal structure of allanite crystals from metagranitoids recording different strain rates during HP and LT metamorphism, concluding that the matrix and allanite crystals accommodated plastic and brittle deformation, respectively. Allanite resistance to plastic deformation is a noteworthy characteristic, as evidenced by relicts preserved in the sheared eclogites from Monte Mucrone (Stünitz and Tullis 2001; Cenki-Tok *et al.* 2011). However, fracturing during brittle deformation can disturb its isotopic system (Burn 2016).

Despite the degree of deformation, the allanite grains exhibit the same sequence of chemical zoning pattern (evident for Ca and Ce), but with different textures and LREE contents, suggesting that deformation facilitates the release of LREEs (Fig. 7a, d and e; Corti *et al.* 2020). Gregory *et al.* (2012) used Th/U v. La/Sm and Eu/Eu* v. La/Sm discrimination diagrams to distinguish low- from high-temperature magmatic allanite, revealing useful information about the amount of melt present during allanite growth. However, no geochemical ratios

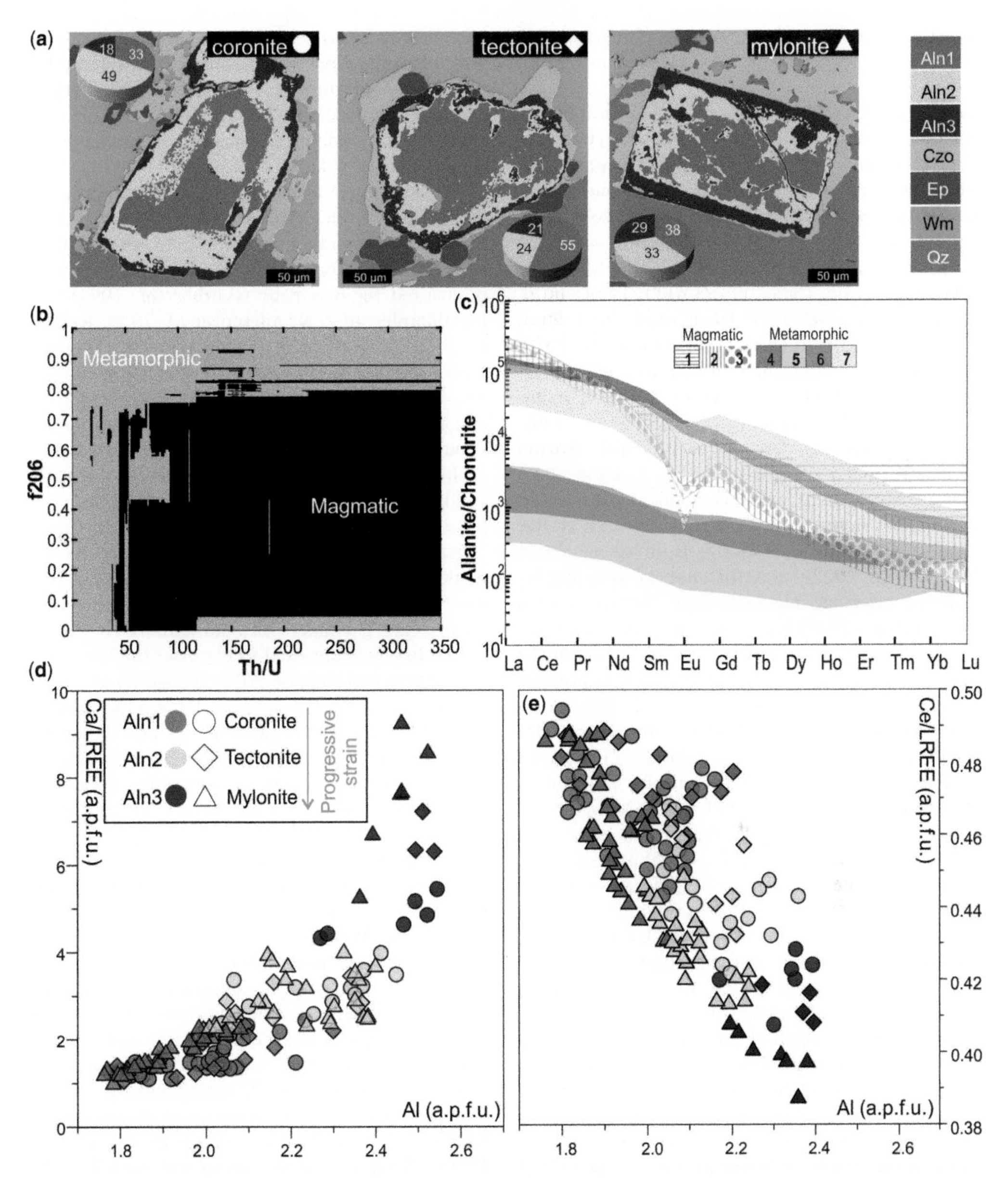

Fig. 7. (**a**) Quantitative X-ray Map Analyser images (Q–XRMA; Ortolano *et al.* 2018) used to distinguish generations of metaigneous allanite grains recording different strain rates. (**b**) Classification diagram to distinguish magmatic from metamorphic allanite based on their Th/U ratio and the fraction of initial ^{206}Pb (f206) values produced using machine learning (Random Forest algorithm). (**c**) Chondrite-normalized REE patterns of magmatic (pattern) v. metamorphic (filled) allanite: (1) Renna *et al.* (2007); (2) Gregory *et al.* (2012); (3) Zhang *et al.* (2022); (4) Regis *et al.* (2014); (5) Boston *et al.* (2017); (6) Vho *et al.* (2020); (7) Di Rosa *et al.* (2020). Use of geochemical data to distinguish magmatic and metamorphic allanite. Despite the degree of deformation of allanite crystals presented in (a) the same sequence of chemical zoning can be observed, with progressively deformed allanite crystals exhibiting variations of (**d**) Ca and (**e**) Ce contents in relation to LREE contents. Source: (a) from Corti *et al.* (2020); (b) modified after Di Rosa *et al.* (2020); (c) modified after Di Rosa *et al.* (2020).

have yet been found to systematically discriminate metamorphic from magmatic allanite (Di Rosa *et al.* 2020). In migmatitic rocks, allanite incorporates significant amounts of Th relative to melt (e.g. Hermann and Rubatto 2009), and it may exhibit intermediate Pb_c values, between those for magmatic and metamorphic allanite (Gregory *et al.* 2012).

Geochronology. The favourable composition of allanite allows dating using the $^{232}Th/^{208}Pb$, $^{238}U/^{206}Pb$ and $^{235}U/^{207}Pb$ systems. Allanite dating techniques range from single and multi-grain ID-TIMS (e.g. von Blackenburg 1992; Oberli *et al.* 2004; Smye *et al.* 2014; López-Moro *et al.* 2017) to *in situ* analysis using SHRIMP, SIMS or LA-ICP-MS (e.g. Catlos *et al.* 2000; Janots *et al.* 2009; Darling *et al.* 2012; Regis *et al.* 2014; Burn *et al.* 2017; Giuntoli *et al.* 2018; Liao *et al.* 2020; Vho *et al.* 2020). Challenges in the use of allanite reference material for LA-ICP-MS and SHRIMP methods (Burn 2016) include: (i) chemical and isotopic heterogeneities (e.g. Gregory *et al.* 2007; Boston *et al.* 2017; Giuntoli *et al.* 2018); (ii) excess ^{206}Pb in magmatic allanite, which is the most widely used reference material (e.g. BONA, CAP and TARA allanite; Gregory *et al.* 2007; Burn *et al.* 2017; Yang *et al.* 2022); and (iii) use of non-matrix-matched reference materials (e.g. NIST610 glass by McFarlane 2016; Plešovice zircon in Burn *et al.* 2017). Nonetheless, robust LA-ICP-MS results have been obtained also by using zircon as primary reference material (Darling *et al.* 2012; Burn *et al.* 2017; Vho *et al.* 2020), which has a similar structure to allanite. Rastering (Darling *et al.* 2012) or a spot analyses routine (Burn 2016) can also be used to minimize matrix sensitivity in LA-ICP-MS analysis. As other accessory minerals, allanite may also incorporate non-radiogenic (^{204}Pb) as well as radiogenic Pb (intermediate nuclei from ^{238}U decay) affecting dates and uncertainties (Romer and Siegesmund 2003; Darling *et al.* 2012; Engi 2017). Furthermore, the structure of allanite can complicate geochronological procedures, as matrix matching reference materials is needed and the decay from ^{232}Th, ^{235}U and ^{238}U causes structural damage to the crystal lattice (Burn 2016; McFarlane 2016). The destruction of the crystalline structure promotes Pb-loss and/or actinide remobilization with the formation of Th- and U-rich mineral phases (e.g. Barth *et al.* 1994; Smye *et al.* 2014). Additionally, deformation and interaction with fluids may open isotopic systems and play an important role in mineral re-equilibration, resorption and precipitation (Radulescu *et al.* 2009; Airaghi *et al.* 2019; Corti *et al.* 2020). Thus, syn-kinematic allanite has been used to date deformation processes at upper to middle crustal levels (Cenki-Tok *et al.* 2011). Investigations of magmatic allanite from an intensely deformed Mesoproterozoic granite in southern Norway indicate that higher mobility of Th than Pb during deformation processes results in the U–Pb system being more reliable than the Th–Pb system (Burn 2016). Nevertheless, allanite crystal structure may protractedly recover by annealing (Karioris *et al.* 1981), forming preserved (non-metamict) crystals that may yield younger dates (Catlos *et al.* 2000). More recently, allanite has been used as primary and secondary reference material for epidote dating (Peverelli *et al.* 2022).

When dating allanite, one of the three approaches discussed by Burn (2016) and Engi (2017) should be utilized to deal with initial or Pb_c correction. (i) Consider Pb_c evolution models. This approach is more often used for magmatic (Barth *et al.* 1994) than metamorphic (e.g. Radulescu *et al.* 2009; Rubatto *et al.* 2011) rocks due to the main issue of using global (silicate Earth or mantle) evolution models. Th, U and Pb contents differ from rock to rock, and their distribution within minerals is heterogeneous. These factors depend on several variables such as local effective bulk composition and fluids availability (Lanari and Engi 2017). (ii) Measuring Pb_c in phases that coexist with allanite (e.g. Cenki-Tok *et al.* 2014), which can be hampered by the difficulty in interpreting coexisting phases. Finally, (iii) the 'intercept approach', in which intercepts from uncorrected TW and $^{206}Pb_c$ normalized Th–Pb isochron diagrams are used to estimate the initial Pb_c (e.g. Janots and Rubatto 2014; Airaghi *et al.* 2019). Gregory *et al.* (2012) indicate that igneous allanite tends to have smaller amounts of non-radiogenic ^{208}Pb than high-grade metamorphic rocks, whereas allanite crystals formed at subsolidus conditions exhibit the highest non-radiogenic ^{208}Pb values.

Isotope geochemistry. Allanite enrichment in LREE and Sr allows for both *in situ* Nd and Sr isotopic-based petrogenetic information to be combined with U–Th–Pb dating, making allanite an important petrogenetic tool (Hoshino *et al.* 2007). Heterogeneous Pb and Sr isotopic concentrations in allanite are demonstrated to be inherited from precursor minerals involved in allanite-producing metamorphic reactions (Romer and Xiao 2005). Zhang *et al.* (2022) combined U–Th–Pb dating and Nd isotopes of allanite with U–Pb–Hf analyses of zircon, demonstrating good correlations between the two systems to investigate crustal formation and evolution. Nd isotopes in allanite analysis were used by Su *et al.* (2021) to unravel the hydrothermal history of an iron oxide copper–gold deposit from ore formation (multi-source) to the post-ore overprinting tectonothermal events.

Monazite/xenotime

Monazite (LREE,Y,Th,Ca,Si)PO_4 is a REE-rich phosphate mineral (Fig. 2p-r) that occurs in a variety of rock compositions and from diagenetic (Evans and Zalasiewicz 1996; Pereira *et al.* this volume, in press) to granulite-facies conditions (Black *et al.*

1984). Monazite is common in peraluminous granites (e.g. Montel 1993; Förster 1998), syenite, granitic and quartz veins (e.g. Piechocka *et al.* 2017) and carbonatitic plutons (e.g. Anenburg *et al.* 2021; Kamenetsky *et al.* 2021), and can exhibit various textures (from sector, firtree, oscillatory to lobate; Fig. 8a). However, under certain fluid conditions, in particular alkali-bearing fluids, monazite undergoes coupled dissolution–reprecipitation (CDR) (Vavra and Schaltegger 1999; Harlov and Hetherington 2010; Harlov *et al.* 2011; Kelly *et al.* 2012; Taylor *et al.* 2014, 2016; Bosse and Villa 2019; Weinberg *et al.* 2020; Salminen *et al.* 2022). CDR of monazite leaves very distinctive, lobate textures but can also modify U–Pb systematics (Fig. 8b; e.g. Vavra and Schaltegger 1999; Taylor *et al.* 2014; Blereau *et al.* 2016; Prent *et al.* 2020), cause U and Th loss (Williams *et al.* 2011) or Th gain (Harlov *et al.* 2011). Monazite is a common accessory mineral in metapelitic rocks experiencing low-amphibolite facies metamorphism (e.g. Rubatto 2002) to high-grade granulites, migmatites and charnockites (Fig. 8b; e.g. Laurent *et al.* 2018; Dev *et al.* 2021; Williams *et al.* 2022), whereas it is less common in *HP* rocks (Finger and Krenn 2007). Whilst geochronological disruption causes resetting of the oldest growth of the mineral, the reactivity of monazite is useful for tracing or determining the timing of fluid activity, reactivation of metamorphic processes and/or ore genesis. On the other hand, xenotime (Y, HREE)PO_4 is a HREE carrier, which also makes it a critical source of HREEs (Strzelecki *et al.* 2022). Xenotime occurs in metapelitic, granitic and carbonatitic rocks, and is less abundant in mafic and calc-silicate rocks (Spear and Pyle 2002). Recent phase equilibria modelling studies have demonstrated that xenotime usually occurs at pressures lower than 8 kbar and temperatures lower than 750°C (e.g. Shrestha *et al.* 2019). In contrast, monazite is found to be stable at higher pressures and temperatures, where water availability strongly controls the monazite *P–T* stability field and its preservation

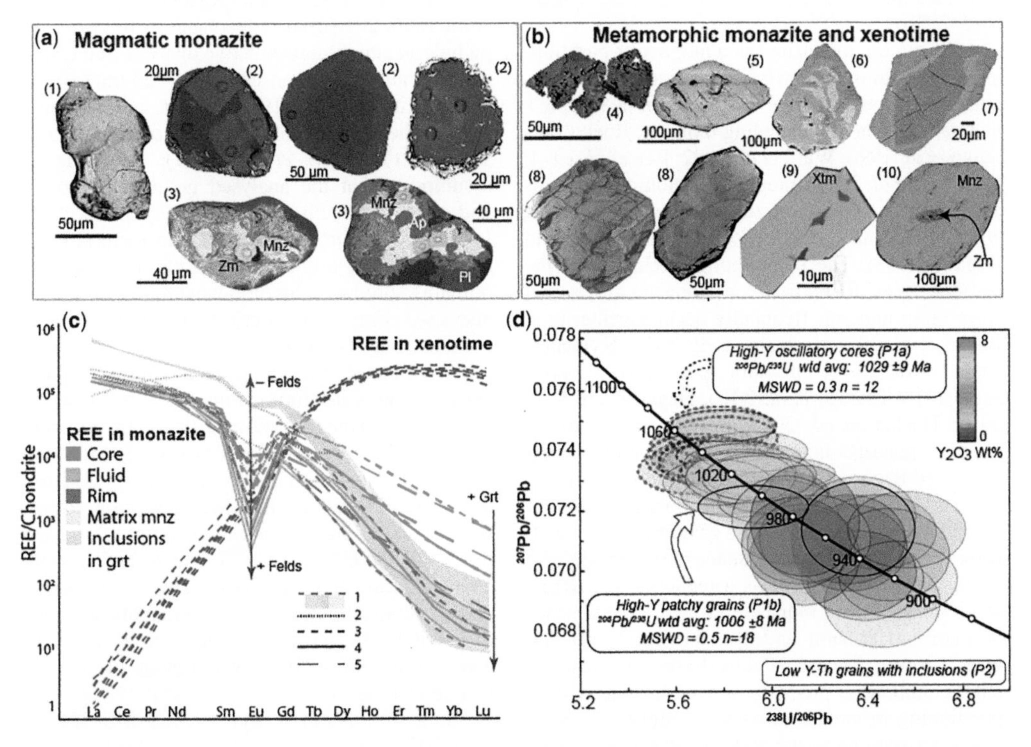

Fig. 8. (**a**) Magmatic textures of monazite crystals from (1) Barrote *et al.* (2020); (2) Volante *et al.* (2020*a*); (3) Piechocka *et al.* (2017). (**b**) Metamorphic textures of monazite and xenotime crystals: (4) Volante *et al.* (2020*c*); (5) Photo courtesy of Cutts; (6) Cutts *et al.* (2018); (7) Blereau *et al.* (2016); (8) Laurent *et al.* (2018); (9) Manzotti *et al.* (2018); (10) Barrote *et al.* (2020). (**c**) Representative chondrite normalized REE patterns for different types of monazite and xenotime: (1) Manzotti *et al.* (2018); (2) Rasmussen and Muhling (2007); (3) Rubatto *et al.* (2013); (4) Taylor *et al.* (2014); (5) Buick *et al.* (2010). (**d**) U–Pb data for monazite grains from sapphirine–cordierite UHT gneisses in Norway. Monazite ages are colour-coded based on Y_2O_3 content. Source: (d) after Laurent *et al.* (2018).

along the prograde path at much higher *P–T* conditions than previously envisaged (Larson *et al.* 2022).

Geochemical tools. Both monazite and xenotime are REE-rich mineral phases, containing a critical amount of REEs in addition to Y (Schulz 2021 and references therein). Monazite REE patterns are usually characterized by a negative slope, whereas xenotime commonly exhibits a positive REE pattern more like zircon (Fig. 8c). Sr-enrichment in monazite has been interpreted to reflect *HP* monazite growing in the absence of feldspars (Finger and Krenn 2007). Additionally, a REE signature like *HP* zircon, with low HREE and absence of a negative Eu anomaly, has been identified in monazite grains from the Kokchetav UHP rocks in Kazakhstan, and UHP rocks in Norway, suggesting this as a geochemical signature for HP monazite (Hermann and Rubatto 2014; Hacker *et al.* 2015). Furthermore, REEs in monazite have been utilized as geochemical discriminators with U–Pb geochronology to distinguish between magmatic and metamorphic monazite in complex deformed and metamorphosed terranes (e.g. Pe-Piper *et al.* 2014; Prent *et al.* 2019; Itano *et al.* 2020).

Like zircon, monazite has a number of partitioning relationships that can be useful in integrating various data sources to the evolution of a sample's mineral paragenesis, including monazite/melt (Yurimoto *et al.* 1990; Ward *et al.* 1992; Bea *et al.* 1994; Stepanov *et al.* 2012), monazite/xenotime (Andrehs and Heinrich 1998), monazite/K-feldspar (Villaseca *et al.* 2003) and monazite/garnet (Hermann and Rubatto 2003). The most applied technique is the partitioning of REEs between monazite and garnet since these minerals frequently occur together (e.g. Buick *et al.* 2006; Rubatto *et al.* 2006; Kylander-Clark *et al.* 2013; Mottram *et al.* 2014; Taylor *et al.* 2014; Blereau *et al.* 2016; Hagen-Peter *et al.* 2016; Hacker *et al.* 2019; Warren *et al.* 2019). Despite monazite being relatively poor in HREEs compared to zircon and garnet, the presence of garnet still impacts the relative concentration of HREEs in monazite. Monazite growing syn- to post-garnet or modified in the presence of garnet (see also Discussion) typically shows a reduction in HREEs and Y compared to monazite grown in the absence of garnet (Hermann and Rubatto 2003; Rubatto *et al.* 2006). Recent studies have demonstrated more complicated processes associated with HREE partitioning in monazite, where partitioning coefficients between monazite and garnet within investigated metapelitic rocks did not reproduce the expected values (e.g. Larson *et al.* 2019, 2022; Shrestha *et al.* 2019). Their temperature dependence is also found to be more relevant than initially envisaged (Hacker *et al.* 2019; Warren *et al.* 2019; Jiao *et al.* 2021). Also, recent work has highlighted by modelling Sm–Eu–Gd partitioning in suprasolidus systems that even though fractionation of Eu by feldspar growth can dominantly control the Eu budget, at equilibrium, other factors such as oxygen fugacity were shown to play an important role (Holder *et al.* 2020).

Geochronology. Monazite has proved a reliable geochronometer using the U–Pb system (Parrish 1990; Williams *et al.* 2007). Generally, monazite does not incorporate Pb into its crystal structure but has high Th (often several wt%) and U (several thousand ppm), meaning that it has low Pb_c and high radiogenic Pb. Monazite is also inferred to have a high *Tc* of up to 900°C (Fig. 1; Cherniak *et al.* 2004). However, monazite can be quite reactive (via CDR, see above), resulting in resetting of ages in fluid-dominated systems (Seydoux-Guillaume *et al.* 2002), whereas monazite in dry rocks (i.e. granulites) has been found to preserve detrital ages (e.g. Suzuki and Adachi 1994; Cutts *et al.* 2013; Guo *et al.* 2020). The high U and Th contents of monazite allow dating via chemical U–Th–Pb dating using EPMA (e.g. Suzuki and Adachi 1994; Montel *et al.* 1996, 2018). The advantages of this approach include *in situ*, non-destructive analysis and a small spot size allowing monazite grains <20 µm to be targeted (Ning *et al.* 2019; Williams and Jercinovic 2002). The main pitfall is that Pb_c correction is not possible (Williams *et al.* 2017), involving the assumption that the analysed grain is concordant and contains no Pb_c. The CHIME (chemical Th–U-total Pb isochron method) dating method is like chemical dating but targets multiple spots in an age domain to produce a 'pseudo-isochron'. This method also uses compositional criteria to determine if the monazite age is concordant, resulting in a more robust method than traditional chemical dating (Suzuki and Kato 2008). Improvements in EPMA sensitivity and the method make this a powerful technique moving forward (Konečný *et al.* 2018; Montel *et al.* 2018; Ning *et al.* 2019). Prior to EPMA dating, many studies utilized ID-TIMS geochronology of monazite (i.e. Smith and Barreiro 1990), and later the LA-ICP-MS approach (Machado and Gauthier 1996; Poitrasson *et al.* 2000). Due to its high U and Th contents, monazite is extremely amenable to LA-ICP-MS, where small spot sizes (8–12 µm) can be used. Similarly, geochronology of xenotime can be analysed *in situ* by both SIMS (Cross 2009; Fielding *et al.* 2017) and LA-ICP-MS (Lawley *et al.* 2015; Simpson *et al.* 2021) to collect texturally contextualized isotopic dates. Currently, no matrix-matched reference materials for xenotime are available as no homogeneous natural xenotime has been found. Therefore, first-order matrix corrections have been applied on xenotime reference materials such as z6413 (Stern and Rayner 2003) and MG-1

(Fletcher *et al.* 2004) for U–Pb geochronology, whereas glass NIST 610 is currently used as primary reference material for *in situ* Lu–Hf dating (Simpson *et al.* 2021). Due to the wealth of information provided by REE data, TEs are commonly collected either separately or simultaneously with geochronological data using one or two mass spectrometers (Holder *et al.* 2013; e.g. Kylander-Clark *et al.* 2013; Hacker *et al.* 2015; Volante *et al.* 2020*c*; Barrote *et al.* 2022*b*). Nanoscale geochronological analysis of monazite (Fougerouse *et al.* 2020, 2021) and xenotime (Joseph *et al.* 2021) via atom probe (ATP) is also possible. Analytical development has allowed to investigate diffusion and migration of atoms along monazite grain boundaries and/or crystal defects, and relates e.g. intracrystalline deformation with age resetting due to radiogenic Pb loss (Fougerouse *et al.* 2021). Great potential to determine ages by using the ATP was shown also on small xenotime crystals (Joseph *et al.* 2021). *In situ* studies allow monazite ages to be directly related to mineral textures present in the rock, so the age of deformation can be attributed to metamorphic events in complex terranes (e.g. Smith and Barreiro 1990; Foster *et al.* 2002; Štípská *et al.* 2015; Piechocka *et al.* 2017; Prent *et al.* 2019, 2020; Jiao *et al.* 2020*a*; Volante *et al.* 2020*c*).

Isotope geochemistry. In granitic rocks, high LREE minerals, like monazite, are great competitors for trace elements, including Sm–Nd and REE, in the host magmas (e.g. Fisher *et al.* 2017; Hammerli and Kemp 2021). The Sm–Nd isotope tracer system in monazite can complement the more commonly applied Lu–Hf system in zircon (e.g. Fisher *et al.* 2017; Martin *et al.* 2020; Mulder *et al.* 2021; Barrote *et al.* 2022*a*; Volante *et al.* 2022) and whole-rock Sm–Nd and/or Lu–Hf (e.g. Mark 2001; Hoffmann *et al.* 2011; Caxito *et al.* 2021), providing insights into the formation, evolution and differentiation of the continental crust (Hammerli and Kemp 2021). Significant contributions have been made to improve *in situ* Sm–Nd precision using LASS-MC-ICP-MS (e.g. Barrote *et al.* 2022*a*) to achieve a comparable level of precision to Lu–Hf in zircon (e.g. Fisher *et al.* 2011, 2020; Goudie *et al.* 2014; Spencer *et al.* 2020). Furthermore, monazite is less prone to weathering than other Sm–Nd-bearing major minerals (e.g. plagioclase), making it a valuable tool to investigate primary isotopic signatures (e.g. Barrote *et al.* 2022*a*).

Thermometry. To estimate minimum magmatic temperatures, monazite thermometry relies on whole-rock REE content in peraluminous and metaluminous granitic rocks (Montel 1993; Plank *et al.* 2009; Stepanov *et al.* 2012). While the Montel (1993) equation is calibrated only at low pressure, Stepanov *et al.* (2012) corrected it by extending experiments to higher pressures. Recent studies have presented the importance of applying variable H_2O contents for granitic rocks to determine monazite saturation temperatures, with this approach producing consistent zircon and monazite saturation temperatures across the same granitic samples (Volante *et al.* 2020*a*). The Y-HREE fractionation between monazite and xenotime results in an asymmetric miscibility gap in YPO_4-(REE)PO_4. This allows for the Y content of monazite to be used as a geothermometer, which is largely independent of pressure (e.g. Gratz and Heinrich 1997, 1998; Heinrich *et al.* 1997; Andrehs and Heinrich 1998). This thermometer can only be used when monazite has grown in the presence of xenotime. However, identifying and ascertaining that these two phases grew in equilibrium can be challenging (Pyle and Spear 1999). Like garnet, xenotime also preferentially incorporates HREE. Hence, xenotime is usually consumed during prograde metamorphism by garnet growth, and it commonly reappears during garnet breakdown at post-peak conditions or along the retrograde path (e.g. Hallett and Spear 2015). In this view, compositional variations in monazite (e.g. Regis *et al.* 2016; Manzotti *et al.* 2018) have been crucial to assess xenotime saturation within the rock system (e.g. Krenn and Finger 2010). For example, Y-rich monazite domains have been suggested to be plausible targets to use for thermometry (e.g. Viskupic and Hodges 2001; Krenn *et al.* 2012; Laurent *et al.* 2018). Y content can also be used to distinguish between different monazite age populations (i.e. Fig. 8d), and can be combined with thermometry to indicate the temperature–time evolution of the sample (e.g. Laurent *et al.* 2018). Some attempts were made to thermodynamically model monazite and xenotime (e.g. Kelsey *et al.* 2007, 2008; Spear and Pyle 2010; Shrestha *et al.* 2019). However, this approach only works with low-Ca bulk-rock compositions and requires the simplification of a complicated system.

Apatite

Apatite $Ca_5(PO_4)_3(F,Cl,OH)$ is a calcium phosphate (Fig. 2s-u; Fig. 9) that is often present as an accessory mineral in several rock types, including various igneous rocks (ultramafic–felsic) to metamorphic and sedimentary, and even in meteorites (e.g. Harlov 2015; McCubbin and Jones 2015; Webster and Piccoli 2015). Apatite can readily incorporate large amounts of TE (e.g. U, Th, REE, Y, Sr), which replace Ca in the crystal lattice with charge compensation mechanisms (Engi 2017), making apatite important for the trace element budget of a rock. In addition, apatite commonly contains significant amounts of water and/or halogens and volatile

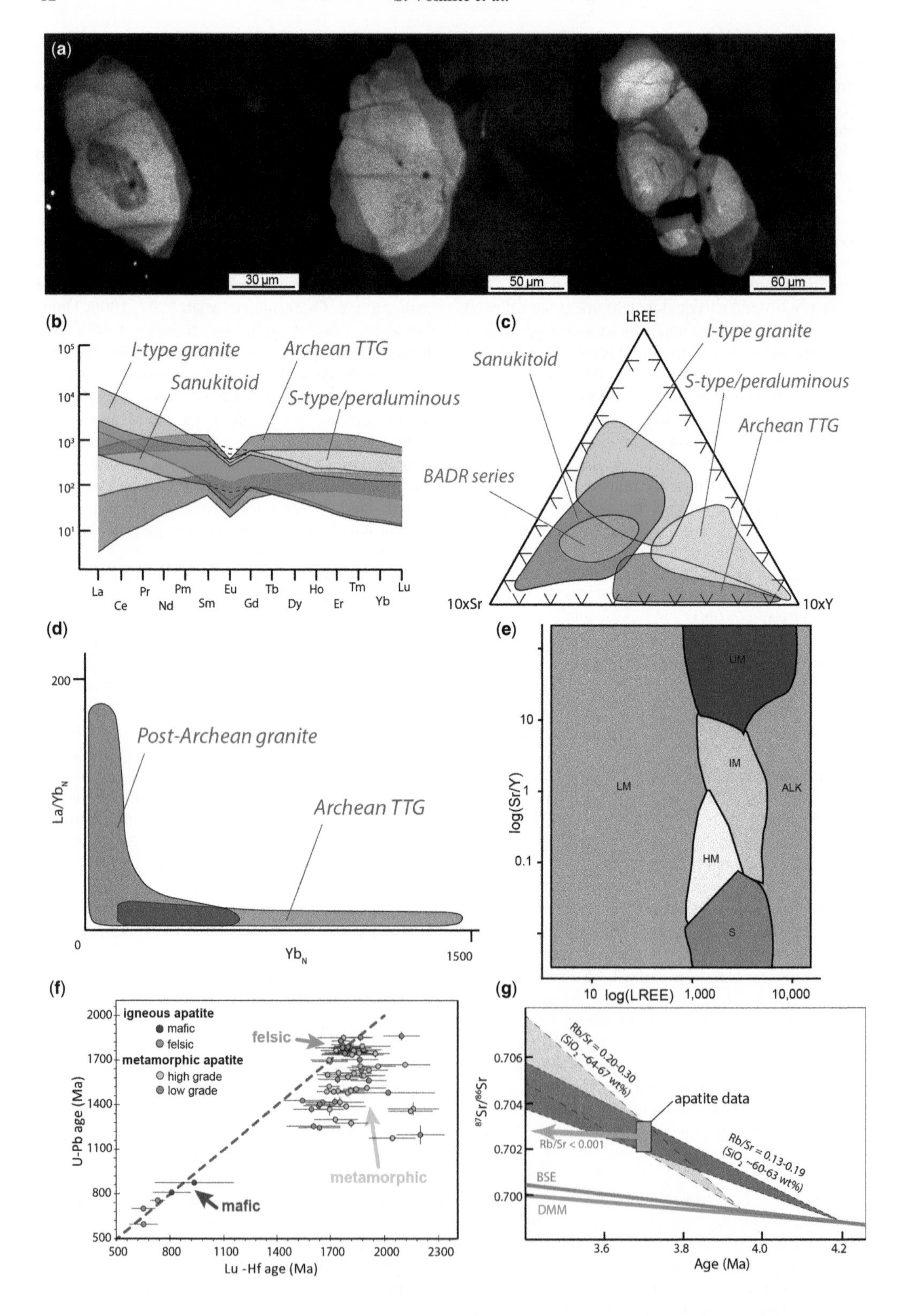
(a)
30 µm
50 µm
60 µm
(b)
I-type granite
Archean TTG
Sanukitoid
S-type/peraluminous
La Ce Pr Nd Pm Sm Eu Gd Tb Dy Ho Er Tm Yb Lu
(c)
LREE
Sanukitoid
I-type granite
S-type/peraluminous
Archean TTG
BADR series
10xSr
10xY
(d)
200
La/Yb$_N$
Post-Archean granite
Archean TTG
0
Yb$_N$
1500
(e)
UM
IM
LM
ALK
HM
S
log(Sr/Y)
log(LREE)
(f)
igneous apatite
mafic
felsic
metamorphic apatite
high grade
low grade
felsic
metamorphic
mafic
U-Pb age (Ma)
Lu -Hf age (Ma)
(g)
Rb/Sr = 0.20-0.30
(SiO$_2$ ~64-67 wt%)
apatite data
Rb/Sr < 0.001
Rb/Sr = 0.13-0.19
(SiO$_2$ ~60-63 wt%)
BSE
DMM
^{87}Sr/^{86}Sr
Age (Ma)

species (e.g. F and Cl; Harrison and Watson 1984; Piccoli and Candela 2002), and records the evolution of metasomatic and hydrothermal fluids (e.g. Harlov 2015; Lin *et al.* 2023). Combined, the geochemical characteristics of apatite make it suitable for many geochemical and petrological applications, with measurements being possible with a large variety of instruments (e.g. SIMS, (LA)-ICP-MS, (LA)-MC-ICP-MS, EPMA), imaging with BSE (Fig. 9a) and TE mapping with LA-ICP-MS (Fig. 2r). Depending on its crystallization conditions, apatite can exhibit various textures visible in CL or BSE images (e.g. Mühlberg *et al.* 2021) and geochemical signatures that make this mineral a very interesting proxy for rock formation and evolution as well as for chronological information (Chew *et al.* 2011; Hammerli *et al.* 2014; Kirkland *et al.* 2017; Antoine *et al.* 2020; Fisher *et al.* 2020; Prent *et al.* 2020; Paul *et al.* 2021).

Geochemical tools. Igneous apatite has reasonably similar total REE contents regardless of the protolith lithology (e.g. Chu *et al.* 2009), except for apatite in alkali-rich rocks, which are commonly significantly REE-rich (e.g. Zirner *et al.* 2015). However, relative REE concentrations, and Y and Sr, are variable within different rock-types and particularly effective for discriminating magma types (e.g. Bea 1996; Belousova *et al.* 2001; Bruand *et al.* 2020). Negative Eu anomalies are present in apatite from most igneous rocks except ultramafics (e.g. Chakhmouradian *et al.* 2017; Bruand *et al.* 2020), and the magnitude of Eu anomalies is typically positively correlated with SiO_2 content. In contrast, Sr content in apatite is negatively correlated with SiO_2 content, with the highest Sr contents being observed in ultramafic rocks (Ihlen *et al.* 2014) and the lowest Sr contents in felsic rocks (Sha and Chappell 1999). Also, Sr in apatite strongly correlates with Sr content in the corresponding whole-rock (Belousova *et al.* 2001; Jennings *et al.* 2011; Bruand *et al.* 2014). Variations in Sr content in TTGs are interpreted to directly correlate with the melting depth of the TTGs' source, where a deep source is in equilibrium with garnet and rutile, but no plagioclase (Moyen and Martin 2012). Moreover, apatite chemistry can be used as a tracer of different magmatic petrogenetic processes, reflecting geodynamic changes during crustal evolution (Bruand *et al.* 2020) such as $(La/Yb)_N$ v. Yb_N discrimination diagram for TTGs (Antoine *et al.* 2020; Fig. 9b–d).

Metamorphic apatite also exhibits contrasting signatures depending on its metamorphic grade and its textural context (e.g. O'Sullivan and Chew 2020; Prent *et al.* 2020). Total REE content increases with the metamorphic grade (e.g. El Korh *et al.* 2009), such that in high-grade rocks it is indistinguishable from that of igneous apatite (e.g. Bingen *et al.* 1996). In addition, negative Eu anomalies characterize apatite growing in low-grade metamorphic rocks (Henrichs *et al.* 2018). Also, metamorphic apatite contains higher Sr concentrations in greenschist- and blueschist-facies rocks (e.g. Nishizawa *et al.* 2005), and lower ones in migmatites and granulite-facies rocks (e.g. Nutman 2007). In felsic igneous and metamorphic rocks, apatite $(La/Lu)_N$ ratio is commonly ≤ 1, whereas in all other rocks $(La/Lu)_N$ ratio is >1, with the highest values recorded in ultramafic rocks (e.g. O'Reilly and Griffin 2000). A discrimination diagram using Sr/Y v. ΣLREE biplots from various igneous and metamorphic/metasomatic rocks (Fig. 9e) can be used to statistically categorize source lithologies (O'Sullivan *et al.* 2020). Additional discriminant diagrams based on the enrichment in LREEs and MREEs, Eu anomalies, the tetrad effect, Mn and Sr contents, total REEs and Y in apatite can be used to highlight the variability between different types of mineralization (Decrée *et al.* 2023). For example, bell-shaped REE patterns defined by MREE enrichment and positive Eu anomalies were found to be a unique indicator of hydrothermal apatite formed in a mineralized, reducing hydrothermal system (e.g. Krneta *et al.* 2018; Lin *et al.* 2023).

Geochronology. Apatite has long been the target for geochronological studies (e.g. Schoene and Bowring 2007; Chew *et al.* 2011; Glorie *et al.* 2022) either using the U–Th–Pb or Lu–Hf isotope system (e.g. Barfod *et al.* 2005; Chew *et al.* 2011; Simpson *et al.* 2021; Glorie *et al.* 2023), or Sm–Nd (Fisher *et al.* 2020). Apatite commonly incorporates

Fig. 9. (**a**) BSE images for magmatic apatite. (**b**) Distinct REE patterns for apatite from different types of granitoids. (**c**) Discrimination diagram for apatite from various granitoids. BADR, basalt–andesite–dacite–rhyolite series. (**d**) Chondrite-normalized La/Yb_N v. Yb_N diagram of apatite from post-Archean granites and Archean TTGs. (**e**) Lithological discrimination diagram. Abbreviations: ALK, alkali-rich igneous rocks; IM, mafic I-type granitoids and mafic igneous rocks; LM, low- and medium-grade metamorphic and metasomatic; HM, partial-melts/leucosomes/high-grade metamorphic; S, S-type granitoids and high aluminum saturation index (ASI) 'felsic' I-types; UM, ultramafic rocks including carbonatites, lherzolites and pyroxenites. (**f**) Binary diagram U–Pb v. Lu–Hf ages (Ma) in apatite. (**g**) $^{87}Sr/^{86}Sr$ v. age diagram showing how initial apatite Sr isotope signatures can be interpreted in terms of source lithology, derived from time-integrated Rb/Sr ratios, and resulting model age (intersection between grey fans and BSE evolution). Depleted MORB (Mid-ocean ridge basalt) mantle (DMM). Source: (b) and (c) modified after Bruand *et al.* (2020); (d) modified after Antoine *et al.* (2020); (e) modified after O'Sullivan *et al.* (2020); (f) rock lithologies following discrimination diagram from O'Sullivan *et al.* (2020); (g) modified after Emo *et al.* (2018).

significant amounts of Pb_c that vary from limited amounts in igneous apatite to larger amounts in hydrothermal apatite crystals (e.g. Kirkland *et al.* 2017). Low-grade metamorphic apatite typically provides poor precision for U–Pb dating due to low U concentrations (<5 ppm) and high initial Pb_c contents, while apatite from igneous rocks is typically U-rich (>20 ppm) and has lower Pb_c, leading to more precise U–Pb ages (e.g. Henrichs *et al.* 2018, 2019; O'Sullivan *et al.* 2018). Because of its Pb_c content, apatite can often produce U–Pb dates that are strongly discordant (also see Titanite section). This Pb_c presents a particular challenge for young samples that have had little time to accumulate substantial radiogenic Pb (Pb*), or for apatite grains with low concentrations of U. Following to the development of apatite U–Th–Pb age reference materials (e.g. Chew *et al.* 2011; Thomson *et al.* 2012; Apen *et al.* 2022; Lana *et al.* 2022) and data reduction schemes employing ^{208}Pb-, ^{207}Pb- or ^{204}Pb-based Pb_c corrections to age reference materials and unknowns, it is now possible to routinely date apatite both precisely and accurately (e.g. Andersson *et al.* 2008, 2022; Chew *et al.* 2011; Thomson *et al.* 2012; Antoine *et al.* 2020; Prent *et al.* 2020; Glorie *et al.* 2022). However, some complications in data interpretation can occur in apatite crystals that have undergone metamorphism due to diffusion effects (e.g. Paul *et al.* 2019). Apatite U–Pb dates may be reset by deformation and fluids, causing the age to represent the last major deformation event or latest dissolution–precipitation and/or chemical exchange (Odlum *et al.* 2022).

Recent advances including the development of LA-ICP-MS/MS technology allowed resolution of isobaric interferences for Lu–Hf *in situ* dating of apatite crystals (e.g. Barfod *et al.* 2003, 2005; Larsson and Söderlund 2005; Simpson *et al.* 2021; Gillespie *et al.* 2022; Glorie *et al.* 2022, this volume, in press) and Sm–Nd (e.g. Hammerli *et al.* 2014, 2019; Doucelance *et al.* 2020; Fisher *et al.* 2020), which is extremely useful to investigate post-metamorphic cooling history and/or compare crystallization ages. Apatite is commonly combined with other mineral phases that exhibit various parent/daughter ratios (i.e. Lu/Hf and Sm/Nd) to obtain a more robust isotope isochron (e.g. Hammerli *et al.* 2014; Laurent *et al.* 2017; Simpson *et al.* 2021). Advantages of the Lu–Hf isotope system in apatite over U–Pb include higher *Tc* of the former (Fig. 1), resulting in an age closer to the apatite crystallization age (e.g. Henrichs *et al.* 2019). Comparisons of the two systems can have significant advantages (Fig. 9f; Glorie *et al.* 2022).

Isotope geochemistry. The Sm–Nd isotope system in apatite can provide insights into timing and source of apatite host-rock (e.g. Fisher *et al.* 2020). Among stable isotopes, oxygen isotopes appear to be proxies for magma sources and to record fluid circulation and metamorphism processes (e.g. Bruand *et al.* 2019). Chlorine and hydrogen isotopes have also been used as proxies for volatilization/condensation processes (e.g. Potts *et al.* 2018; Wudarska *et al.* 2020). Isotope systematics such as ^{87}Rb–^{87}Sr can be measured using either (LA-)MC-ICP-MS or LA-ICP-MS/MS, TIMS or SIMS, and coupled to age information (Fig. 9g; Emo *et al.* 2018; Ravindran *et al.* 2020; Gillespie *et al.* 2021). The ^{87}Sr/^{86}Sr ratio of apatite may be used together with K-rich or Rb-bearing minerals such as micas for *in situ* Rb–Sr geochronology to create a combined isochron (Olierook *et al.* 2020).

Thermometry. There is no currently developed thermometer for apatite like those for zircon (Ti-in-zircon) or rutile (Zr-in-rutile), but the U–Pb *Tc* (350–570°C; Fig. 1) can be used as an indirect proxy for tracking metamorphic or magmatic cooling in combination with other chronometers in apatite (Fig. 1; e.g. Cherniak 2000*b*; Barfod *et al.* 2003, 2005; Cochrane *et al.* 2014; Chew and Spikings 2015; Kirkland *et al.* 2018; Ferreira *et al.* 2022).

Discussion

Applications of multi-mineral petrochronometers to metamorphic processes

Interplay between garnet–zircon–monazite–xenotime. Geochronological and geochemical data retrieved from metamorphic accessory minerals are commonly integrated with information from the major mineral paragenesis, *P–T* constraints from phase equilibrium diagram calculations and microstructures. Figure 10 reflects an example of two synoptic, theoretical *P–T–t* evolutions where petrochronological information of metamorphic accessory minerals such as monazite, zircon and garnet is linked to *P–T* information. The first *P–T* path (Fig. 10a) represents common clockwise granulite-facies metamorphism of a pelitic protolith hosting inherited zircon (Fig. 10a, i). During prograde metamorphism, garnet growth begins and inherited zircon grains start to recrystallize with increasing temperatures (Fig. 10a ii). This prograde part of the metamorphic evolution is often difficult to accurately constrain due to open-system processes, overprinting and/or exceeding closure temperatures of earlier phases. Zircon has been shown to grow along the prograde path connected to the movement of locally-derived melts (e.g. Harley and Nandakumar 2014; Harley 2016; Weinberg *et al.* 2020) as well as injected, externally-derived melts (Andersson *et al.* 2002; Flowerdew *et al.* 2006; Wu *et al.* 2007).

Once the sample crosses the solidus, partial melting begins. Inherited zircons may be consumed and or dissolved in the melt that becomes enriched in zirconium (Watson and Harrison 1984; Boehnke *et al.* 2013; Gervasoni *et al.* 2016), forming embayment and relict core textures in zircon. Monazite might also grow along the prograde and retrograde path (e.g. Rubatto 2002; Shrestha *et al.* 2019; Larson *et al.* 2022), but also during (Kelsey *et al.* 2008; Rubatto *et al.* 2009; Larson *et al.* 2022) and/or just after max T conditions are reached (Fig. 10a iii, v, vii; Clark *et al.* 2014; Shrestha *et al.* 2019), making it a more suitable time-capsule than zircon by preserving a greater *P–T–t* window at subsolidus conditions (e.g. Ambrose *et al.* 2015; Mottram *et al.* 2015; Hacker *et al.* 2019; Shrestha *et al.* 2020; Larson *et al.* 2022). Recent works modelled the *P–T* stability fields for monazite and xenotime, with monazite present during prograde growth of garnet, max T(/P) conditions and along the retrograde path during garnet breakdown at both sub- and suprasolidus conditions (Shrestha *et al.* 2019). In contrast, xenotime appears only at low *P–T* conditions during the early stages of the prograde path and/or along the retrograde path, when it would start to assimilate most of the Y and HREE, resulting in a decrease in monazite proportions (Shrestha *et al.* 2019). Prograde garnet sees further growth at suprasolidus conditions, combined with an additional generation of peritectic garnet (Fig. 10a iv) that could be distinguished based on different REE patterns and inclusions (Fig. 3f). Trace element partitioning is mostly utilized for max T(/P) to retrograde portion of the *P–T–t* history as this information is most likely preserved. For example, Figure 10a shows that garnet growth and metamorphic 'soccer ball' zircon (e.g. Vavra and Schaltegger 1999; Blereau *et al.* 2016; Taylor *et al.* 2016) grow in the same geochemical system and have a near 1:1 partitioning of M-HREE. How well these partitioning relationships are preserved depends on the conditions of metamorphism and how these affect the equilibration volume of the sample. Typically, this is largest at max T conditions, where rates of diffusion and potential fluids and/or melt increase. Approaching the solidus on the retrograde path, Zr-bearing phases may breakdown, e.g. garnet to cordierite (Fig. 10a vi) resulting in additional zircon growth (e.g. Fraser *et al.* 1997; Degeling *et al.* 2001; Wu *et al.* 2007; Kelsey *et al.* 2008), from rims to completely new grains (Fig. 10a vii). A subsequent fluid event, e.g. upon crystallization of local melt, can modify monazite and zircon textures, perturbing age and geochemical data (Poitrasson *et al.* 2000; Seydoux-Guillaume *et al.* 2012; Kröner *et al.* 2014; Taylor *et al.* 2014; Blereau *et al.* 2016; Prent *et al.* 2019). In more complex garnet textures, such as the unusual atoll garnet (e.g. Jonnalagadda *et al.* 2017; Kulhánek *et al.* 2021; Godet *et al.* 2022; Massonne and Li 2022), TE content recorded by subsequent growth of concentric garnet rings can also provide useful information about the interchange of TE within the system due to growth and breakdown of other accessory phases (Godet *et al.* 2022). For example, by using LA-ICP-MS trace element mapping, Godet *et al.* (2022) reported enrichment in V and Ti in the garnet inner rim compared to the core, which was attributed to rutile breakdown, whereas Cr, Y, LREE and MREE enrichment in the outer rim was interpreted to reflect allanite and monazite breakdown.

The second *P–T–t* evolution (Fig. 10b) represents a more residual scenario, where the protolith has already been partially melted. In this case, recrystallization and preservation of existing major and accessory minerals are promoted due to the lack and/or reduction in volume of partial melt being produced, which also limits the generation of new assemblages (Fig. 10b, i, ii, iii; Bea and Montero 1999; White and Powell 2002). Monazite and zircon contrast in their behaviour under these conditions. Monazite may grow in larger amounts than zircon, despite being in a melt poor environment, due to higher reactivity (e.g. Högdahl *et al.* 2012; Rubatto *et al.* 2013; Morrissey *et al.* 2016), potentially recording information completely missed by zircon. As in the first scenario (Fig. 10a), relict garnet breaks down during decompression to an intergrowth of plagioclase + orthopyroxene, another micro zircon permitting reaction (Fig. 10b). Local hydration or different amounts of melt loss (Morrissey *et al.* 2016; Larson *et al.* 2022) is reflected by a second *P–T* event (Fig. 10b, v, vi, vii) recorded only by more reactive high-strain microsites. The interpretation of geochronological data obtained from studies of high-grade metamorphic rocks can also be extremely challenging as most terranes record a spread of ages rather than statistical populations (e.g. Whitehouse and Kemp 2010; Farias *et al.* 2020; Taylor *et al.* 2020; Finch *et al.* 2021; Gutieva *et al.* 2021; Mulder and Cawood 2021; Salminen *et al.* 2022; Whitehouse *et al.* 2022). For example, in Figure 10c, the younger age population could represent either a single prolonged metamorphic event or potentially two events, with the second modifying the first and many other possible combinations. Determining the correct age interpretation depends on mineral textures, where the analyses and sample come together with complementary geochemical and isotopic information.

Interplay between allanite–monazite–xenotime–apatite–(titanite–rutile). Metamorphic reactions between REE-rich phases such as allanite–monazite–xenotime–apatite are illustrated with those of titanite–rutile on the same theoretical *P–T–t–(d)* path, though these sets of minerals may occur in

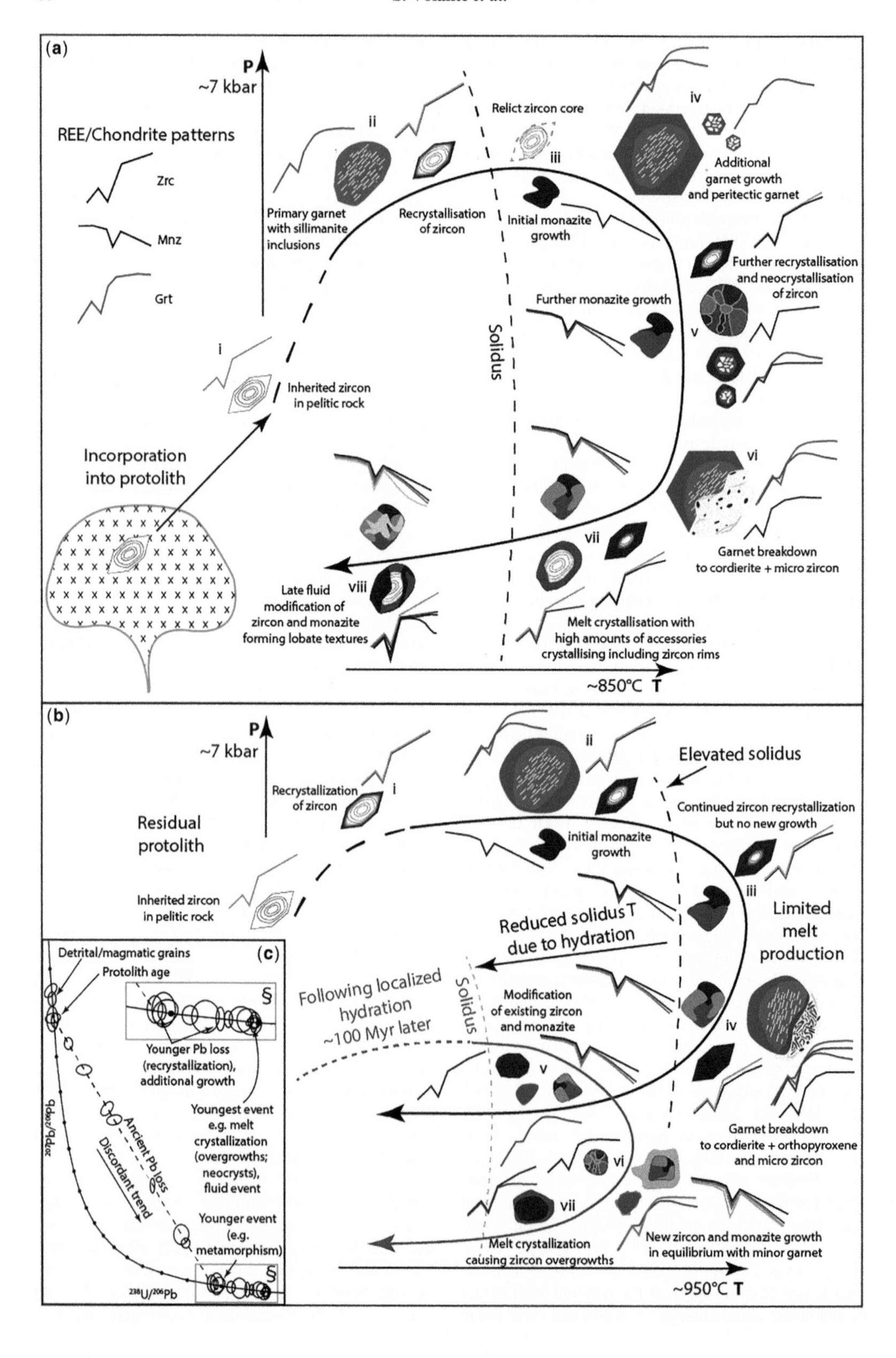

(a)
P
~7 kbar
REE/Chondrite patterns
Zrc
Mnz
Grt
ii
Primary garnet with sillimanite inclusions
Recrystallisation of zircon
Relict zircon core
iii
Initial monazite growth
iv
Additional garnet growth and peritectic garnet
Further recrystallisation and neocrystallisation of zircon
v
Further monazite growth
Solidus
i
Inherited zircon in pelitic rock
Incorporation into protolith
vi
Garnet breakdown to cordierite + micro zircon
vii
viii
Late fluid modification of zircon and monazite forming lobate textures
Melt crystallisation with high amounts of accessories crystallising including zircon rims
~850°C T
(b)
P
~7 kbar
Residual protolith
Recrystallization of zircon
i
ii
Elevated solidus
Continued zircon recrystallization but no new growth
initial monazite growth
iii
Inherited zircon in pelitic rock
Reduced solidus T due to hydration
Limited melt production
Following localized hydration ~100 Myr later
Solidus
Modification of existing zircon and monazite
iv
v
Garnet breakdown to cordierite + orthopyroxene and micro zircon
vi
vii
Melt crystallization causing zircon overgrowths
New zircon and monazite growth in equilibrium with minor garnet
~950°C T
(c)
Detrital/magmatic grains
Protolith age
§
Younger Pb loss (recrystallization), additional growth
Youngest event e.g. melt crystallization (overgrowths; neocrysts), fluid event
Ancient Pb loss
Discordant trend
Younger event (e.g. metamorphism)
§
207Pb/206Pb
238U/206Pb

different chemical systems (Fig. 11). The timing and sequence of metamorphic reactions (i.e. growth of accessory phases) depend not only on the variation of pressure and temperature conditions, including the residence time of a volume of rock at certain conditions, but also on the bulk-rock composition of the protolith and on strain rate, which strongly influence the reaction sequence during fluid–rock interaction at different crustal levels. Reconstructing the formation and breakdown of these minerals and growth of other major and minor phases in the rock usually requires a detailed petrographic investigation that includes X-ray compositional maps and analysis of Y and REE partitioning between the phases (e.g. Janots *et al.* 2008; Garber *et al.* 2017; Manzotti *et al.* 2018; Airaghi *et al.* 2019). Additionally, tools such as TE mapping by LA-ICP-MS (e.g. Raimondo *et al.* 2017; Chew *et al.* 2021; Sliwinski and Stoll 2021; Godet *et al.* 2022) allow for greater precision in determining chemical composition effects on growth or break-down of minerals associated with different *P–T* conditions and fluid-present or absent processes.

During the prograde evolution of a time capsule mineral such as garnet, key inclusions can be incorporated and provide insights regarding the *P* and *T* conditions a rock volume experienced during the early stages of the evolution path (Fig. 11). For example, identification of rutile inclusions in garnet core (Fig. 11) and prograde allanite can indicate that rutile grew during the prograde stages towards max *P* conditions (Boston *et al.* 2017). In this case, prograde growth of garnet sequestered most of the Mn and HREE, resulting in allanite cores with a relatively HREE-depleted pattern in relation to the rim. Manzotti *et al.* (2018) investigated rocks in which Y-rich garnet cores enclosed a first generation of rutile and allanite inclusions, whereas Y-poor garnet rims included monazite grains that crystallized aligned parallel to the matrix foliation (Fig. 11). At this stage, monazite may form after breakdown of allanite or other phosphates such as apatite and/or xenotime (Engi 2017; Manzotti *et al.* 2018), recording peak pressure conditions (e.g. Boston *et al.* 2017; Manzotti *et al.* 2018; Volante *et al.* 2020*c*; Barrote *et al.* 2022*b*; Fumes *et al.* 2022). Monazite REE patterns may exhibit low Y content together with a steeply negative HREE slope, suggesting that garnet was still stable during monazite growth, whereas pronounced negative Eu anomalies commonly suggest plagioclase stability or oxygen fugacity variations (Fig. 11; e.g. Boston *et al.* 2017; Holder *et al.* 2020). At low- to medium-metamorphic grade, allanite and monazite may replace each other, and this process depends on different factors including bulk-rock (e.g. Wing *et al.* 2003; Janots *et al.* 2008; Spear 2010) and fluid (Budzyń *et al.* 2010) compositions, as well as oxygen fugacity (Janots *et al.* 2011) and pressure and temperature conditions (Janots *et al.* 2007). Post-metamorphic pressure peak, along the lower-pressure, possibly higher-temperature path, a second generation of allanite and apatite can form at the expenses of monazite grains commonly exhibiting complex coronitic dissolution textures (Fig. 11; e.g. Manzotti *et al.* 2018). At this stage, in mafic and Ca-rich systems, titanite crystals may grow as a replacement product of rutile, whereas ilmenite would grow in a Ca-poor system and at lower pressures. During retrograde, low-grade hydrothermal metamorphism in pelitic systems, new xenotime and monazite may crystallize following allanite, apatite, residual monazite relicts and garnet breakdown, with xenotime associated with chlorite and minor biotite aggregates replacing garnet porphyroblasts (Fig. 11; Manzotti *et al.* 2018). Most of the Y released by garnet would be incorporated by xenotime and monazite in minor amounts. New rutile crystals may grow during the retrograde cooling stage at the expense of titanite and as rutile exsolution, growing parallel to biotite cleavage during its replacement by chlorite (Fig. 11). During this late hydrothermal stage, rutile may grow with apatite and record the cooling stages (e.g. Apen *et al.* 2020).

Applications of multi-mineral petrochronometers to magmatic processes

In this section, we emphasize that a combined multi-mineral approach is a powerful tool to provide

Fig. 10. Synoptic *P–T* evolutions of a metapelite in different hypothetical scenarios with associated types of zircon, monazite, xenotime, garnet growth/breakdown textures and potential REE patterns. (**a**) Singular clockwise *P–T* evolution: (i) inherited zircon; (ii) primary garnet and recrystallization of inherited zircon; (iii) relict zircon core, initial monazite growth; (iv) growth of garnet rim and peritectic garnet; (v) neocrystallized 'soccer ball' zircon, additional monazite and recrystallization of zircon; (vi) garnet breakdown; (vii) melt crystallization; (viii) late fluid modification. (**b**) A residual protolith along a clockwise polymetamorphic *P–T* path: (i) recrystallization of zircon; (ii) prograde monazite growth in the presence of relict garnet; (iii) at the elevated solidus, minor melting causes additional monazite but no new zircon growth; (iv) garnet breakdown, final monazite growth and complete 'ghost zoning' to recrystallized zircon. After a localized hydrous retrograde event, the hydrated areas follow a secondary *P–T* path; (v) melt modifies zircon and monazite to relict cores stable with relict garnet grains; (vi) new zircon and monazite; (vii) final melt crystallization. (**c**) Synthetic Tera–Wasserburg plot with interpretation based on the nature of analysed textures; § enlarged inset of younger ages. Source: (a–c) modified after Blereau (2017).

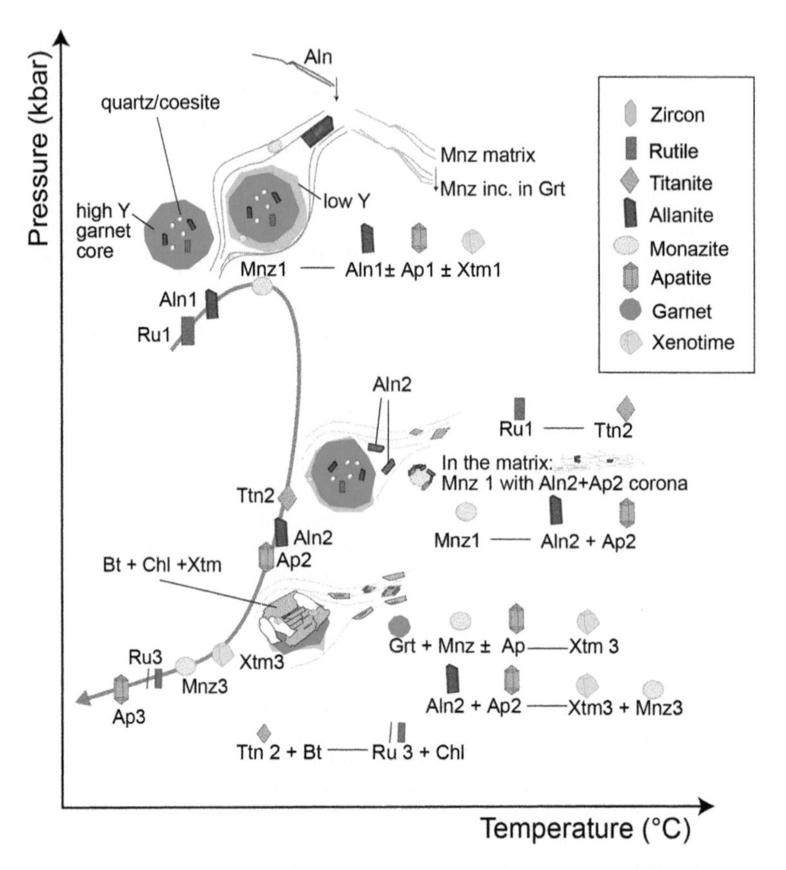

Fig. 11. Synoptic *P–T* evolution diagram including reactions of key petrochronometers such as garnet, allanite, monazite, rutile, apatite, xenotime and titanite. Source: modified after Manzotti *et al.* (2018) including key mineral reactions from Boston *et al.* (2017); Fumes *et al.* (2022); Manzotti *et al.* (2022).

constraints on magma sources (e.g. mineralogy, composition), differentiation processes (e.g. mixing, fractional crystallization) and timing of magma formation and crystallization. The first scenario corresponds to simple fractional crystallization of a single magma batch (Fig. 12). The second is a more complex scenario wherein magma is contaminated by a distinct lithology and/or mingles with a different magma. The third scenario refers to Precambrian gneisses, wherein the objective is to recover the initial magmatic signatures despite the occurrence of a metamorphic overprinting event.

To constrain the crystallization age of a magmatic rock, zircon is arguably the easiest mineral to date among those presented here but it is not necessarily observed in all rock types. Consequently, the timing of igneous rock formation can be addressed using other phases such as titanite, rutile, monazite, apatite or allanite, provided that they record a magmatic age or fast cooling. In a scenario where a magmatic intrusion does not experience post-crystallization disturbance and contains most mineral phases presented in this contribution, multi-mineral dating becomes useful as these phases likely record different ages reflecting the magma cooling history (Fig. 12; e.g. Schaltegger *et al.* 2009; Jayananda *et al.* 2015). A potential issue is the precision on determined ages, as granite batholiths can cool over relatively short timescales (e.g. Coleman *et al.* 2004; Schaltegger *et al.* 2009; Barboni *et al.* 2013). In the same scenario, Lu–Hf in zircon/apatite, Sm–Nd in monazite/titanite/apatite and Rb–Sr in apatite, combined, allow these source tracing radiogenic isotope systems to be directly compared, and coupled to geochronological data (i.e. U–Th–Pb and Lu–Hf in apatite, and Sm–Nd in monazite/titanite/apatite). Also, comparisons of these isotopic systems at the whole-rock and mineral scale can help constrain source composition or detect small discrepancies that may be important for understanding rock formation (e.g. Guitreau *et al.* 2012; Fisher *et al.* 2020; Barrote *et al.* 2022*a*; Volante *et al.* 2022; Zhang *et al.* 2022).

The second scenario includes a magmatic body that experienced mingling, mixing or contamination.

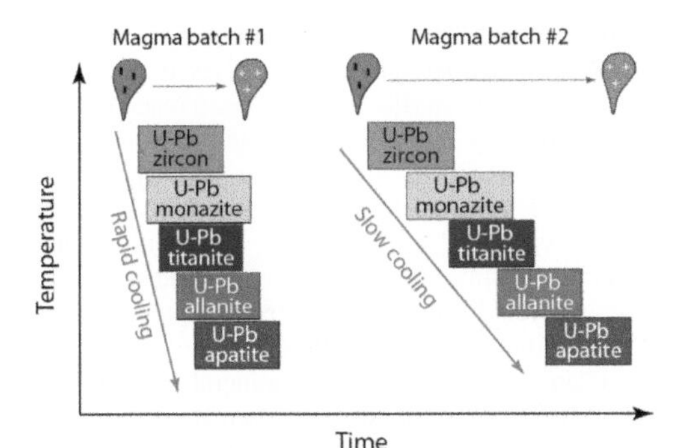

Fig. 12. Hypothetical temperature–time paths for two distinct intrusions with one having cooled rapidly and the other one slowly. Note that closure temperatures do not overlap between techniques and minerals for illustrative purposes only. Also, cooling paths are presented as straight lines which may not accurately depict actual cooling paths.

Radiogenic systems allow end-members to be identified due to specific sensitivity to contamination, source characteristics and respective elemental concentrations. For example, Laurent *et al.* (2017) used Rb–Sr and Sm–Nd in apatite and titanite, respectively, from hybrid granitoids to show that these minerals did not crystallize at the same time and that while apatite formed in both magmas before mingling and after mingling (zonation), titanite crystallized only in the mingled magma. To identify end-member isotopic compositions, another geochemical study investigated zircon, apatite and titanite, which revealed mingling of mantle and crust-derived magmas (Sun *et al.* 2010). If one of the end-members is crustal derived, zircon xenocrysts can provide further information about crustal diversity (e.g. Villaros *et al.* 2012; Bea *et al.* 2021). Also, monazite can inherit Nd isotope heterogeneities from the source while losing the original U–Pb crystallization age (Fisher *et al.* 2017). Trace element patterns and ratios can provide further insights into magma evolution after and/or before mixing/mingling (e.g. Sun *et al.* 2010; Laurent *et al.* 2017). Melt and/or mineral inclusions in accessory minerals can also provide information about the melt source composition (e.g. Bruand *et al.* 2017; Antoine *et al.* 2020; Ferrero *et al.* 2021).

The third scenario considers Precambrian rocks, which often experienced polyphase histories that involved at least one metamorphic episode potentially compromising one or more isotope systems (e.g. Black 1988; Hammerli *et al.* 2014; Emo *et al.* 2018; Guitreau *et al.* 2018; Antoine *et al.* 2020; Fisher *et al.* 2020; Hammerli and Kemp 2021). For example, *in situ* Sm–Nd investigation of apatite, allanite, titanite, xenotime and monazite in Precambrian metasedimentary rocks revealed re-equilibration of heterogeneous Nd isotope signature in apatite over 550°C, after which it retains its Nd isotope signature throughout anatexis (Hammerli *et al.* 2014). While titanite and allanite equilibrate at LT (350–400°C), REE-rich accessory phases exhibit homogeneous Nd isotopic signatures at HT (600°C) and behave as open systems during partial melting and anatexis (Hammerli *et al.* 2014).

Perspectives: unconventional analytical techniques. Where to?

From the conceptualization of petrochronology (as summarized in Engi *et al.* 2017; Kohn *et al.* 2017), significant analytical improvements have occurred in the field of geoscience and have been applied to all key mineral phases that are described in this review. New analytical methods have been developed (e.g. atom probe dating, Fougerouse *et al.* 2021; *in situ* Lu–Hf dating using LA-ICP-MS/MS, Simpson *et al.* 2021, 2022, 2023), previous analytical methods have been continuously updated and improved (e.g. *in situ* EPMA dating, Montel *et al.* 2018; *in situ* U–Pb dating, Millonig *et al.* 2020), new and updated numerical tools for quantitative petrology have been developed and enhanced, including imaging and modelling tools (e.g. XMapTools, Lanari *et al.* 2014, 2019; Q–XRMA, Ortolano *et al.* 2018, Zucali *et al.* 2021; Bingo-Antidote, Duesterhoeft and Lanari 2020) as well as the use of TE maps (e.g. Raimondo *et al.* 2017; George *et al.* 2018; Lanari and Piccoli 2020; Rubatto *et al.* 2020; Gaidies *et al.* 2021), and new boundaries have been pushed to investigate melting mechanisms (e.g. Ferrero *et al.* 2021). In this section, we briefly discuss the exciting and bright future in front of the discipline of petrochronology that has only started to push analytical advances to the limit of conventional methods, allowing for a better understanding of plate- to nanoscale crustal processes.

Garnet is the ultimate petrochronometer (Baxter *et al.* 2017), with the most exciting advance being the development of *in situ* Lu–Hf geochronology (Simpson *et al.* 2021). The use of inclusions as proxies for whole-rock isotopic composition also means that this method may be suitable for detrital garnet studies (Maneiro *et al.* 2019; Mark *et al.* 2022). Future work should focus on enhancing accuracy and precision of Lu–Hf and U–Pb measurement in secondary reference material e.g. Aysal *et al.* 2023 and finding primary reference material (i.e. isotopically homogeneous) for common Hf-bearing phases such as garnet (Simpson *et al.* 2021). Additionally, incorporation of Mn in the activity–composition relations for amphibole and clinopyroxene in the thermodynamic dataset for mafic systems (Green *et al.* 2016) would also greatly enhance accuracy for garnet modelling (White *et al.* 2014). Recent advances using *in situ* U–Pb dating of garnet have also demonstrated significant potential (e.g. Seman *et al.* 2017; Millonig *et al.* 2020; Schannor *et al.* 2021) and applicability to almandine compositions (e.g. Cerva-Alves *et al.* 2021; Schannor *et al.* 2021). Further advances aim to extend the range of metasomatic and metamorphic garnet composition that can be targeted using this technique (Millonig *et al.* 2022; O'Sullivan *et al.* 2023). Finally, the continuous development of quantitative elemental and mineral map tools (e.g. XMapTools) along with EPMA and LA-ICP-MS trace element maps (e.g. Chew *et al.* 2021) will prove an extremely powerful tool to enhance *in situ* microanalytical investigations and our understanding of complex micro-chemical processes. The Bingo-Antidote add-on to XMapTools performs thermodynamic calculations by comparing modelled and observed mineral assemblages, modes and compositions (Duesterhoeft and Lanari 2020).

Zircon can preserve a realm of information that can be retrieved from images (optical, SEM), mineral inclusions and elemental and isotope data (e.g. Th/U, REE, $\delta^{18}O$, ε_{Hf}, $\delta^{30}Si$). Although zircon has long been used in igneous and metamorphic studies, recent technical and methodological advances allow to further extend our use of this mineral. These include analytical improvements in $\delta^{30}Si$ and $\delta^{94}Zr$ data acquisition to enhance and extend the use of these isotopes in zircon studies, and the integration of EBSD data with petrochronological investigation to better understand TE diffusion in zircon crystals under various *P–T* conditions. Pioneer work on elastic geobarometry for a noncubic host-inclusion system indicates potential of anisotropic quartz-in-zircon elastic model for elastic thermobarometry, and its potential wide applicability to crustal rocks (Gonzalez *et al.* 2021).

Monazite is an increasingly useful accessory mineral as it can be used to trace the presence of certain fluids (e.g. Weinberg *et al.* 2020; Salminen *et al.* 2022), and can potentially grow throughout metamorphism (e.g. Larson *et al.* 2022). With improvement of analytical and spatial resolution, even if highly zoned, monazite will only become more applicable for unravelling complex and multifaceted processes. However, even with the collection of *in situ* data, it can still be challenging to tie monazite growth to deformation and fluid-related processes. Further work into monazite behaviour under deformation combined with the use of EBSD may prove a powerful tool. Additionally, the integration of EBSD, crystallographic vorticity axis analysis and petrochronology should be applied to monazite within deformational structures as it has been done for zircon (Brown *et al.* 2022). More work needs to be done on monazite TE partitioning to better understand temperature dependencies (e.g. Larson *et al.* 2022). Some attempts have been made to thermodynamically model monazite and xenotime (e.g. Kelsey *et al.* 2007, 2008; Spear and Pyle 2010; Shrestha *et al.* 2019), however, this approach only works with low-Ca bulk-rock compositions and requires the simplification of a complicated system. These current limitations represent an extensive platform for improvements in this field.

Titanite is a very reactive mineral compared to other accessory minerals; however, its tendency to incorporate initial Pb requires combining U–Pb dating with trace-element composition, zoning and microstructural information to accurately filter out any additional, unrelated radiogenic Pb (Walters *et al.* 2022). Analytical improvement in acquisition of Sm–Nd and O isotopic analysis in titanite may be extremely beneficial to trace magmatic sources (e.g. Bonamici *et al.* 2014; Bruand *et al.* 2019, 2020) and the origin of fluids playing a major role in crustal-scale shear zones (Gordon *et al.* 2021; Moser *et al.* 2022) and ore deposit (e.g. Marfin *et al.* 2020) genesis. The assessment of ilmenite U–Pb dating by LA-ICP-MS (Thompson *et al.* 2021) may be a valuable contribution for evaluating residual initial Pb affecting titanite dating. It may also be useful to investigate the timing of replacement of these Ti-rich phases. Advancing and developing the recent application of elastic geothermobarometry of host-inclusion systems (e.g. Mazzucchelli *et al.* 2021) to titanite grains may provide a complementary non-destructive method to estimate the *P* and *T* at which inclusions were trapped in the hosting titanite (Nestola 2021).

Rutile is a crucial mineral to unravel the early history of rock's evolution, including the prograde, UHP, burial stage (e.g. Rezvukhina *et al.* 2021; Manzotti *et al.* 2022), to trace melt sources in magmatic systems using Lu–Hf as isotopic tracer (e.g. Ewing *et al.* 2011) and hydrothermal, metasomatic processes associated with ore deposit genesis (e.g. Agangi *et al.* 2019; Ballouard *et al.* 2020; Porter *et al.* 2020; Sciuba and Beaudoin 2021). Therefore, analytical advances in microanalytical and petrochronological investigations of this mineral phase are essential to contribute to our understanding of orogenic systems, tectonic environments and exploration strategies. A better micro- to nanoscale understanding of trace element mobility in rutile (Kooijman *et al.* 2012; Ewing *et al.* 2013; Kohn *et al.* 2016; Pape *et al.* 2016; Penniston-Dorland *et al.* 2018; Smye *et al.* 2018) may improve U–Pb dating and associated trace element distribution and isotopic information (Verberne *et al.* 2022*a*, *b*), including the investigation of Zr transport through the rock's matrix (Ewing *et al.* 2013; Kohn *et al.* 2016). Also, advancing the use of integrated EBSD investigations of rutile grains within different rock-types with LA-ICP-MS trace element mapping may significantly enhance our understanding of trace element partitioning and concentration in this mineral.

Allanite has overcome various challenges as a petrochronometer. Indeed, the highly variable composition of natural allanite complicates petrological interpretations and applications. Thermobarometric, isotopic (e.g. *in situ* oxygen) and TE-based petrological investigations have to be carried out using microanalytical techniques such as EBSD and/or LA-ICP-MS TE maps to explore these compositional variations. The use of compositional maps to distinguish different generations of allanite has proven to be an essential step in geochronological and petrological studies (e.g. Burn 2016; Airaghi *et al.* 2019; Corti *et al.* 2020), and the use of TE maps obtained with LA-ICP-MS can improve classification and understanding of allanite growth history. The chemical distinction between magmatic and metamorphic allanite has been enhanced by machine learning (e.g. Di Rosa *et al.* 2020), but more data are required to improve this classification. Nd and Sr isotopes are valuable geochemical tools that have been utilized in previous studies. Recent analytical advancements and developments have further strengthened the importance of these isotopic systems as tracers in understanding magmatic and metamorphic processes, especially in investigations of hydrothermal processes. Also, a better comprehension of the stability conditions (pressure, temperature and fluid) of allanite and correlated accessory minerals (e.g. monazite) is required. This would involve experiments, improvements on the thermodynamic dataset and investigation of inclusions (mineral, fluid and melt) in allanite.

Apatite is a useful tool for studying crustal processes by means of a variety of petrochronology methods, continuously enhanced in their accuracy and precision. Glorie *et al.* (2022, this volume, in press) show the promising apatite Lu–Hf geochronology method using LA ICP-MS/MS, which is more likely to reveal primary apatite growth ages in reworked terranes, due to the higher *Tc*, than the U–Pb system (Fig. 1). This may be important since Odlum *et al.* (2022) show that apatite REEs and U–Pb behaviour are decoupled in high-grade gneiss samples, suggesting REEs record higher-temperature processes than U–Pb isotopic systems. LA-ICP-MS trace element mapping is a promising technique to monitor the behaviour of trace elements in accessory mineral phases, and it should be more routinely applied. In apatite, it can be used to distinguish between metamorphic and magmatic grains or identify growth zones (Henrichs *et al.* 2018, 2019; Chew *et al.* 2021). Recent techniques such as age-depth profiles and laser ablation split stream make it possible to identify individual apatite generations and can aid with interpreting data from apatite domains with complex thermal histories (Kirkland *et al.* 2018). New apatite reference material is being characterized for *in situ* apatite U–Pb petrochronology and Sr–Nd isotope geochemistry (e.g. Apen *et al.* 2022; Kennedy *et al.* 2022) to obtain more precise and robust data.

In addition to the key mineral phases discussed in this contribution, we encourage and see potential in integrating other mineral phases in magmatic and metamorphic studies. Multi-mineral studies are paramount to combine major and accessory mineral phases that record different steps of the prograde and retrograde history of a volume of rock with lower temperature petrochronometers, which may retain useful information on the youngest evolution of the system. Critical and crucial advancement in petrochronology in the near future includes *in situ* investigations, where microanalytical studies are linked to the chemical and chronological information of different minerals and associated fabrics in a rock volume. Extreme potential is foreseen in the development and acquisition of quantitative LA-ICP-MS TE (e.g. George *et al.* 2018; Muñoz-Montecinos, J. *et al.* 2023) and geochronological maps (e.g. Chew *et al.* 2021). In this scenario, by enabling near-simultaneous detection, the time of flight (TOF) detector is a key component of LA-ICP-TOF-MS, a promising technique that captures the complete elemental mass spectrum for each laser pulse (Chew *et al.* 2021). Generating quantitative chemical and geochronological maps of a whole mineral using LA-ICP-MS can facilitate exploring the relationships between internal textures and trace and major elemental variation over time. This analytical technique will allow to discriminate distinct textural domains based on the ages and major, minor and trace elements, possibly avoiding issues such as mixing domains to help the interpretation of complex mineral textures. The advances made in LA-ICP-MS imaging have made it a critical analytical technique in the geosciences, as it enables the acquisition of high-resolution geochemical data that can help to constrain the timing and nature of igneous, metamorphic, ore-forming, sedimentary and diagenetic processes.

Acknowledgements We thank Catherine Mottram and an anonymous reviewer for their constructive and insightful comments that greatly improved the manuscript. We also thank Inês Pereira for helpful comments and for her editorial handling. Martin Engi and Daniela Rubatto are thanked for insightful discussion on allanite. Vitor Barrote is thanked for long insightful discussion throughout the preparation of the manuscript about the pros and cons of using different analytical techniques in petrochronology applied to the mineral phases discussed in this manuscript.

M. Tedeschi was supported by the National Council for Scientific and Technological Development – CNPq through a Research Productivity Grant 308733/2021-5. This is contribution no. 591 of the ClerVolc program of the International Research Center for Disaster Sciences and Sustainable Development of the University of Clermont Auvergne.

Competing interests The authors declare that they have no known competing financial interests or personal relationships that could have appeared to influence the work reported in this paper.

Author contributions **SV**: conceptualization (lead), formal analysis (lead), project administration (lead), writing – original draft (lead), writing – review & editing (lead); **EB**: conceptualization (equal), formal analysis (equal), writing – original draft (equal), writing – review & editing (equal); **MG**: conceptualization (equal), formal analysis (equal), writing – original draft (equal), writing – review & editing (equal); **MT**: conceptualization (supporting), formal analysis (supporting), writing – original draft (supporting), writing – review & editing (supporting); **VVS**: conceptualization (supporting), data curation (supporting), writing – original draft (supporting), writing – review & editing (supporting); **KC**: conceptualization (supporting), data curation (supporting), writing – original draft (supporting), writing – review & editing (supporting).

Funding This research received no specific grant from any funding agency in the public, commercial or not-for-profit sectors.

Data availability Data sharing is not applicable to this article as no datasets were generated or analysed during the current study.

References

Adlakha, E. and Hattori, K. 2021. Thermotectonic events recorded by U–Pb geochronology and Zr-in-rutile thermometry of Ti oxides in basement rocks along the P2 fault, eastern Athabasca Basin, Saskatchewan, Canada. *GSA Bulletin*, **134**, 567–576, https://doi.org/10.1130/B35820.1

Aerden, D.G.A.M., Bell, T.H., Puga, E., Sayab, M., Lozano, J.A. and Diaz de Federico, A. 2013. Multi-stage mountain building v. relative plate motions in the Betic Cordillera deduced from integrated microstructural and petrological analysis of porphyroblast inclusion trails. *Tectonophysics*, **587**, 188–206, https://doi.org/10.1016/j.tecto.2012.11.025

Aerden, D.G.A.M., Ruiz-Fuentes, A., Sayab, M. and Forde, A. 2021. Kinematics of subduction in the Ibero-Armorican arc constrained by 3D microstructural analysis of garnet and pseudomorphed lawsonite porphyroblasts from Île de Groix (Variscan belt). *Solid Earth*, **12**, 971–992, https://doi.org/10.5194/se-12-971-2021

Agangi, A., Reddy, S.M., Plavsa, D., Fougerouse, D., Clark, C., Roberts, M. and Johnson, T.E. 2019. Antimony in rutile as a pathfinder for orogenic gold deposits. *Ore Geology Reviews*, **106**, 1–11, https://doi.org/10.1016/j.oregeorev.2019.01.018

Airaghi, L., Janots, E., Lanari, P., de Sigoyer, J. and Magnin, V. 2019. Allanite petrochronology in fresh and retrogressed garnet–biotite metapelites from the Longmen Shan (Eastern Tibet). *Journal of Petrology*, **60**, 151–176, https://doi.org/10.1093/petrology/egy109

Ambrose, T.K., Larson, K.P., Guilmette, C., Cottle, J.M., Buckingham, H. and Rai, S. 2015. Lateral extrusion, underplating, and out-of-sequence thrusting within the Himalayan metamorphic core, Kanchenjunga, Nepal. *Lithosphere*, **7**, 441–464, https://doi.org/10.1130/L437.1

Amelin, Y. 2009. Sm–Nd and U–Pb systematics of single titanite grains. *Chemical Geology*, **261**, 53–61, https://doi.org/10.1016/j.chemgeo.2009.01.014

Amelin, Y., Lee, D.-C. and Halliday, A.N. 2000. Early–middle Archaean crustal evolution deduced from Lu–Hf and U–Pb isotopic studies of single zircon grains. *Geochimica et Cosmochimica Acta*, **64**, 4205–4225, https://doi.org/10.1016/S0016-7037(00)00493-2

An, Y., Huang, J.-X., Griffin, W.L., Liu, C. and Huang, F. 2017. Isotopic composition of Mg and Fe in garnet peridotites from the Kaapvaal and Siberian cratons. *Geochimica et Cosmochimica Acta*, **200**, 167–185, https://doi.org/10.1016/j.gca.2016.11.041

Anderson, A.J., Hanchar, J.M., Hodges, K.V. and van Soest, M.C. 2020. Mapping radiation damage zoning in zircon using Raman spectroscopy: implications for zircon chronology. *Chemical Geology*, **538**, article 119494, https://doi.org/10.1016/j.chemgeo.2020.119494

Andersson, J., Möller, C. and Johansson, L. 2002. Zircon geochronology of migmatite gneisses along the Mylonite Zone (S Sweden): a major Sveconorwegian terrane boundary in the Baltic Shield. *Precambrian Research*, **114**, 121–147, https://doi.org/10.1016/S0301-9268(01)00220-0

Andersson, J., Bingen, B., Cornell, D., Johansson, L., Söderlund, U. and Möller, C. 2008. The Sveconorwegian orogen of southern Scandinavia: setting, petrology and geochronology of polymetamorphic high-grade terranes. *33rd International Geological Congress*, 6–14 August 2008, Oslo, Norway, Field Excursion Guide.

Andersson, J.B.H., Logan, L. *et al.* 2022. U–Pb zircon–titanite–apatite age constraints on basin development and basin inversion in the Kiruna mining district, Sweden. *Precambrian Research*, **372**, article 106613, https://doi.org/10.1016/j.precamres.2022.106613

Andrehs, G. and Heinrich, W. 1998. Experimental determination of REE distributions between monazite and xenotime: potential for temperature-calibrated geochronology. *Chemical Geology*, **149**, 83–96, https://doi.org/10.1016/S0009-2541(98)00039-4

Anenburg, M., Broom-Fendley, S. and Chen, W. 2021. Formation of rare earth deposits in carbonatites. *Elements*, **17**, 327–332. https://doi.org/10.2138/gselements.17.5.327

Antoine, C., Bruand, E., Guitreau, M. and Devidal, J. 2020. Understanding preservation of primary signatures in apatite by comparing matrix and zircon-hosted crystals from the Eoarchean Acasta Gneiss Complex (Canada). *Geochemistry, Geophysics, Geosystems*, **21**, article e2020GC008923, https://doi.org/10.1029/2020GC008923

Apen, F.E., Rudnick, R.L., Cottle, J.M., Kylander-Clark, A.R.C., Blondes, M.S., Piccoli, P.M. and Seward, G. 2020. Four-dimensional thermal evolution of the East African Orogen: accessory phase petrochronology of crustal profiles through the Tanzanian Craton and Mozambique Belt, northeastern Tanzania. *Contributions to Mineralogy and Petrology*, **175**, article 97, https://doi.org/10.1007/s00410-020-01737-6

Apen, F.E., Wall, C.J., Cottle, J.M., Schmitz, M.D., Kylander-Clark, A.R. and Seward, G.G. 2022. Apatites for destruction: reference apatites from Morocco and Brazil for U–Pb petrochronology and Nd and Sr isotope geochemistry. *Chemical Geology*, **590**, article 120689, https://doi.org/10.1016/j.chemgeo.2021.120689

Aranovich, L.Y., Bortnikov, N. *et al.* 2017. Morphology and impurity elements of zircon in the oceanic lithosphere at the Mid-Atlantic ridge axial zone (6–13 N): evidence of specifics of magmatic crystallization and postmagmatic transformations. *Petrology*, **25**, 339–364, https://doi.org/10.1134/S0869591117040026

Armbruster, T., Bonazzi, P. *et al.* 2006. Recommended nomenclature of epidote-group minerals. *European Journal of Mineralogy*, **18**, 551–567, https://doi.org/10.1127/0935-1221/2006/0018-0551

Asami, M., Suzuki, K. and Grew, E.S. 2002. Chemical Th–U–total Pb dating by electron microprobe analysis of monazite, xenotime and zircon from the Archean Napier Complex, East Antarctica: evidence for ultra-high-temperature metamorphism at 2400 Ma. *Precambrian Research*, **114**, 249–275, https://doi.org/10.1016/S0301-9268(01)00228-5

Atherton, M.P. 1968. The variation in garnet, biotite and chlorite composition in medium grade pelitic rocks from the Dalradian, Scotland, with particular reference to the zonation in garnet. *Contributions to Mineralogy and Petrology*, **18**, 347–371, https://doi.org/10.1007/BF00399696

Aulbach, S., O'Reilly, S.Y., Griffin, W. and Pearson, N.J. 2008. Subcontinental lithospheric mantle origin of high niobium/tantalum ratios in eclogites. *Nature Geoscience*, **1**, 468–472, https://doi.org/10.1038/ngeo226

Axelsson, E., Pape, J., Berndt, J., Corfu, F., Mezger, K. and Raith, M.M. 2018. Rutile R632 – a new natural reference material for U–Pb and Zr determination. *Geostandards and Geoanalytical Research*, **42**, 319–338, https://doi.org/10.1111/ggr.12213

Aysal, N., Guillong, M. *et al.* 2023. A New Natural Secondary Reference Material for Garnet U-Pb Dating by TIMS and LA-ICP-MS. *Geostandards and Geoanalytical Rsearch*, https://doi.org/10.1111/ggr.12493

Balan, E., Neuville, D.R., Trocellier, P., Fritsch, E., Muller, J.-P. and Calas, G. 2001. Metamictization and chemical durability of detrital zircon. *American Mineralogist*, **86**, 1025–1033, https://doi.org/10.2138/am-2001-8-909

Baldwin, J.A. and Brown, M. 2008. Age and duration of ultrahigh-temperature metamorphism in the Anápolis–Itauçu Complex, Southern Brasília Belt, central Brazil – constraints from U–Pb geochronology, mineral rare earth element chemistry and trace-element thermometry. *Journal of Metamorphic Geology*, **26**, 213–233, https://doi.org/10.1111/j.1525-1314.2007.00759.x

Ballouard, C., Massuyeau, M., Elburg, M.A., Tappe, S., Viljoen, F. and Brandenburg, J.-T. 2020. The magmatic and magmatic-hydrothermal evolution of felsic igneous rocks as seen through Nb–Ta geochemical fractionation, with implications for the origins of rare-metal mineralizations. *Earth-Science Reviews*, **203**, article 103115, https://doi.org/10.1016/j.earscirev.2020.103115

Barboni, M., Schoene, B., Ovtcharova, M., Bussy, F., Schaltegger, U. and Gerdes, A. 2013. Timing of incremental pluton construction and magmatic activity in a back-arc setting revealed by ID-TIMS U/Pb and Hf isotopes on complex zircon grains. *Chemical Geology*, **342**, 76–93, https://doi.org/10.1016/j.chemgeo.2012.12.011

Barfod, G.H., Otero, O. and Albarède, F. 2003. Phosphate Lu–Hf geochronology. *Chemical Geology*, **200**, 241–253, https://doi.org/10.1016/S0009-2541(03)00202-X

Barfod, G.H., Krogstad, E.J., Frei, R. and Albarède, F. 2005. Lu–Hf and PbSL geochronology of apatites from Proterozoic terranes: a first look at Lu–Hf isotopic closure in metamorphic apatite. *Geochimica et Cosmochimica Acta*, **69**, 1847–1859, https://doi.org/10.1016/j.gca.2004.09.014

Barla, A. 2021. Titanite petrochronology by LA-ICPMS: method and significance in deciphering igneous processes, Msc Thesis, University of Arizona, USA.

Barrote, V.R., McNaughton, N.J., Tessalina, S.G., Evans, N.J., Talavera, C., Zi, J.-W. and McDonald, B.J. 2020. The 4D evolution of the Teutonic Bore Camp VHMS deposits, Yilgarn Craton, Western Australia. *Ore Geology Reviews*, **120**, article 103448, https://doi.org/10.1016/j.oregeorev.2020.103448

Barrote, V.R., Nebel, O., Wainwright, A.N., Cawood, P.A., Hollis, S.P. and Raveggi, M. 2022*a*. Testing the advantages of simultaneous *in-situ* SmNd, UPb and elemental analysis of igneous monazite for petrochronological studies. An example from the late Archean, Penzance granite, Western Australia. *Chemical Geology*, **594**, article 120760, https://doi.org/10.1016/j.chemgeo.2022.120760

Barrote, V.R., Volante, S., Blereau, E.R., Rosière, C.A. and Spencer, C.J. 2022*b*. Implications of the dominant *LP–HT* deformation in the Guanhães Block for the Araçuaí West-Congo Orogen evolution. *Gondwana Research*, **107**, 154–175, https://doi.org/10.1016/j.gr.2022.03.012

Barrow, G. 1893. On an intrusion of muscovite–biotite gneiss in the South-eastern Highlands of Scotland, and its accompanying metamorphism. *Quarterly Journal of the Geological Society*, **49**, 330–358, https://doi.org/10.1144/GSL.JGS.1893.049.01-04.52

Barrow, G. 1912. On the geology of lower Dee-side and the southern Highland border. *Proceedings of the Geologists' Association*, **23**, 274–290, https://doi.org/10.1016/S0016-7878(12)80018-6

Barth, S., Oberli, F. and Meier, M. 1994. ThPb v. UPb isotope systematics in allanite from co-genetic rhyolite and granodiorite: implications for geochronology. *Earth and Planetary Science Letters*, **124**, 149–159, https://doi.org/10.1016/0012-821X(94)00073-5

Bartoli, O., Cesare, B., Poli, S., Bodnar, R.J., Acosta-Vigil, A., Frezzotti, M.L. and Meli, S. 2013. Recovering the composition of melt and the fluid regime at the onset of crustal anatexis and S-type granite formation. *Geology*, **41**, 115–118, https://doi.org/10.1130/G33455.1

Baxter, E.F. and Scherer, E.E. 2013. Garnet geochronology: timekeeper of tectonometamorphic processes. *Elements*, **9**, 433–438, https://doi.org/10.2113/gselements.9.6.433

Baxter, E.F., Caddick, M.J. and Dragovic, B. 2017. Garnet: a rock-forming mineral petrochronometer. *Reviews in Mineralogy and Geochemistry*, **83**, 469–533, https://doi.org/10.2138/rmg.2017.83.15

Bea, F. 1996. Residence of REE, Y, Th and U in granites and crustal protoliths; implications for the chemistry of crustal melts. *Journal of Petrology*, **37**, 521–552, https://doi.org/10.1093/petrology/37.3.521

Bea, F. and Montero, P. 1999. Behavior of accessory phases and redistribution of Zr, REE, Y, Th, and U during metamorphism and partial melting of metapelites in the lower crust: an example from the Kinzigite Formation of Ivrea–Verbano, NW Italy. *Geochimica et Cosmochimica Acta*, **63**, 1133–1153, https://doi.org/10.1016/S0016-7037(98)00292-0

Bea, F., Pereira, M.D. and Stroh, A. 1994. Mineral/leucosome trace-element partitioning in a peraluminous migmatite (a laser ablation-ICP-MS study). *Chemical Geology*, **117**, 291–312, https://doi.org/10.1016/0009-2541(94)90133-3

Bea, F., Morales, I., Molina, J.F., Montero, P. and Cambeses, A. 2021. Zircon stability grids in crustal partial melts: implications for zircon inheritance. *Contributions to Mineralogy and Petrology*, **176**, 1–13, https://doi.org/10.1007/s00410-020-01755-4

Bea, F., Bortnikov, N. *et al.* 2022. Zircon crystallization in low-Zr mafic magmas: possible or impossible? *Chemical Geology*, **602**, article 120898, https://doi.org/10.1016/j.chemgeo.2022.120898

Bebout, G.E., Tsujimori, T., Ota, T., Shimaki, Y., Kunihiro, T., Carlson, W.D. and Nakamura, E. 2014. Lithium behavior during growth of metasedimentary garnets from the Cignana UHP locality, Italy. *American Geophysical Union Fall Meeting*, 15–19 December 2014, San Francisco, CA, Abstracts, V31D-4783.

Bebout, G.E., Ota, T., Kunihiro, T., Carlson, W.D. and Nakamura, E. 2022. Lithium in garnet as a tracer of subduction zone metamorphic reactions: the record in ultrahigh-pressure metapelites at Lago di Cignana, Italy. *Geosphere*, **18**, 1020–1029, https://doi.org/10.1130/GES02473.1

Bell, E.A., Boehnke, P., Hopkins-Wielicki, M.D. and Harrison, T.M. 2015. Distinguishing primary and secondary inclusion assemblages in Jack Hills zircons. *Lithos*, **234**, 15–26, https://doi.org/10.1016/j.lithos.2015.07.014

Belousova, E.A., Walters, S., Griffin, W.L. and O'Reilly, S.Y. 2001. Trace-element signatures of apatites in granitoids from the Mt Isa Inlier, northwestern Queensland. *Australian Journal of Earth Sciences*, **48**, 603–619, https://doi.org/10.1046/j.1440-0952.2001.00879.x

Belousova, E., Griffin, W.L., O'Reilly, S.Y. and Fisher, N. 2002. Igneous zircon: trace element composition as an indicator of source rock type. *Contributions to Mineralogy and Petrology*, **143**, 602–622, https://doi.org/10.1007/s00410-002-0364-7

Berman, R.G. 1988. Internally-consistent thermodynamic data for minerals in the system Na_2O–K2O–CaO–MgO–FeO–Fe_2O_3–Al_2O_3–SiO_2–TiO_2–H_2O–CO_2. *Journal of Petrology*, **29**, 445–522, https://doi.org/10.1093/petrology/29.2.445

Bindeman, I. 2008. Oxygen isotopes in mantle and crustal magmas as revealed by single crystal analysis. *Reviews in Mineralogy and Geochemistry*, **69**, 445–478, https://doi.org/10.2138/rmg.2008.69.12

Bingen, B., Demaiffe, D. and Hertogen, J. 1996. Redistribution of rare earth elements, thorium, and uranium over accessory minerals in the course of amphibolite to granulite facies metamorphism: the role of apatite and monazite in orthogneisses from southwestern Norway. *Geochimica et Cosmochimica Acta*, **60**, 1341–1354, https://doi.org/10.1016/0016-7037(96)00006-3

Black, L.P. 1988. Isotopic resetting of U–Pb zircon and Rb–Sr and Sm–Nd whole-rock systems in Enderby Land, Antarctica: implications for the interpretation of isotopic data from polymetamorphic and multiply deformed terrains. *Precambrian Research*, **38**, 355–365, https://doi.org/10.1016/0301-9268(88)90033-2

Black, L.P., Fitzgerald, J.D. and Harley, S.L. 1984. Pb isotopic composition, colour, and microstructure of monazites from a polymetamorphic rock in Antarctica. *Contributions to Mineralogy and Petrology*, **85**, 141–148, https://doi.org/10.1007/BF00371704

Blereau, E., Clark, C., Taylor, R.J., Johnson, T., Fitzsimons, I. and Santosh, M. 2016. Constraints on the timing and conditions of high-grade metamorphism, charnockite formation and fluid–rock interaction in the Trivandrum Block, southern India. *Journal of Metamorphic Geology*, **34**, 527–549, https://doi.org/10.1111/jmg.12192

Blereau, E. 2017. A Petrochronological Investigation of Metamorphic, Melt and Fluid Related Processes in Lower Crustal Rocks from Southwestern Norway and Southern India. PhD Thesis, Curtin University, Western Australia. http://hdl.handle.net/20.500.11937/59704

Blereau, E., Clark, C., Kinny, P.D., Sansom, E., Taylor, R.J.M. and Hand, M. 2022. Probing the history of ultra-high temperature metamorphism through rare earth element diffusion in zircon. *Journal of Metamorphic Geology*, **40**, 329–357, https://doi.org/10.1111/jmg.12630

Blichert-Toft, J. and Albarède, F. 1997. The Lu–Hf isotope geochemistry of chondrites and the evolution of the mantle–crust system. *Earth and Planetary Science Letters*, **148**, 243–258, https://doi.org/10.1016/S0012-821X(97)00040-X

Bloch, E., Jollands, M. *et al.* 2022. Diffusion anisotropy of Ti in zircon and implications for Ti-in-zircon thermometry. *Earth and Planetary Science Letters*, **578**, article 117317, https://doi.org/10.1016/j.epsl.2021.117317

Boehnke, P., Watson, E.B., Trail, D., Harrison, T.M. and Schmitt, A.K. 2013. Zircon saturation re-revisited. *Chemical Geology*, **351**, 324–334, https://doi.org/10.1016/j.chemgeo.2013.05.028

Böhnke, M., Bröcker, M., Maulana, A., Klemd, R., Berndt, J. and Baier, H. 2019. Geochronology and Zr-in-rutile thermometry of high-pressure/low temperature metamorphic rocks from the Bantimala complex, SW

Sulawesi, Indonesia. *Lithos*, **324–325**, 340–355, https://doi.org/10.1016/j.lithos.2018.11.020

Bollen, E.M., Stowell, H.H., Aronoff, R.F., Stotter, S.V., Daniel, C.G., McFarlane, C.R.M. and Vervoort, J.D. 2022. Reconciling garnet Lu–Hf and Sm–Nd and monazite U–Pb ages for a prolonged metamorphic event, northern New Mexico. *Journal of Petrology*, **63**, article egac031, https://doi.org/10.1093/petrology/egac031

Bonamici, C.E. and Blum, T.B. 2020. Reconsidering initial Pb in titanite in the context of *in situ* dating. *American Mineralogist: Journal of Earth and Planetary Materials*, **105**, 1672–1685, https://doi.org/10.2138/am-2020-7274

Bonamici, C.E., Kozdon, R., Ushikubo, T. and Valley, J.W. 2011. High-resolution P–T–t paths from $\delta^{18}O$ zoning in titanite: a snapshot of late-orogenic collapse in the Grenville of New York. *Geology*, **39**, 959–962, https://doi.org/10.1130/G32130.1

Bonamici, C.E., Kozdon, R., Ushikubo, T. and Valley, J.W. 2014. Intragrain oxygen isotope zoning in titanite by SIMS: cooling rates and fluid infiltration along the Carthage-Colton Mylonite Zone, Adirondack Mountains, NY, USA. *Journal of Metamorphic Geology*, **32**, 71–92, https://doi.org/10.1111/jmg.12059

Bonamici, C.E., Fanning, C.M., Kozdon, R., Fournelle, J.H. and Valley, J.W. 2015. Combined oxygen-isotope and U–Pb zoning studies of titanite: new criteria for age preservation. *Chemical Geology*, **398**, 70–84, https://doi.org/10.1016/j.chemgeo.2015.02.002

Bonnet, G., Chopin, C., Locatelli, M., Kylander-Clark, A.R.C. and Hacker, B.R. 2022. Protracted subduction of the European hyperextended margin revealed by rutile U–Pb geochronology across the Dora-Maira Massif (Western Alps). *Tectonics*, **41**, article e2021 TC007170, https://doi.org/10.1029/2021TC007170

Borghini, A., Ferrero, S., O'Brien, P.J., Laurent, O., Günter, C. and Ziemann, M.A. 2020. Cryptic metasomatic agent measured *in situ* in Variscan mantle rocks: melt inclusions in garnet of eclogite, Granulitgebirge, Germany. *Journal of Metamorphic Geology*, **38**, 207–234, https://doi.org/10.1111/jmg.12519

Bosse, V. and Villa, I.M. 2019. Petrochronology and hygrochronology of tectono-metamorphic events. *Gondwana Research*, **71**, 76–90, https://doi.org/10.1016/j.gr.2018.12.014

Boston, K.R., Rubatto, D., Hermann, J., Engi, M. and Amelin, Y. 2017. Geochronology of accessory allanite and monazite in the Barrovian metamorphic sequence of the Central Alps, Switzerland. *Lithos*, **286**, 502–518, https://doi.org/10.1016/j.lithos.2017.06.025

Bouvier, A., Vervoort, J.D. and Patchett, P.J. 2008. The Lu–Hf and Sm–Nd isotopic composition of CHUR: constraints from unequilibrated chondrites and implications for the bulk composition of terrestrial planets. *Earth and Planetary Science Letters*, **273**, 48–57, https://doi.org/10.1016/j.epsl.2008.06.010

Bouvier, A.-S., Ushikubo, T., Kita, N.T., Cavosie, A.J., Kozdon, R. and Valley, J.W. 2012. Li isotopes and trace elements as a petrogenetic tracer in zircon: insights from Archean TTGs and sanukitoids. *Contributions to Mineralogy and Petrology*, **163**, 745–768, https://doi.org/10.1007/s00410-011-0697-1

Braun, I., Montel, J.-M. and Nicollet, C. 1998. Electron microprobe dating of monazites from high-grade gneisses and pegmatites of the Kerala Khondalite Belt, southern India. *Chemical Geology*, **146**, 65–85, https://doi.org/10.1016/S0009-2541(98)00005-9

Brown, D.A., Simpson, A., Hand, M., Morrissey, L.J., Gilbert, S., Tamblyn, R. and Glorie, S. 2022. Laser-ablation Lu–Hf dating reveals Laurentian garnet in subducted rocks from southern Australia. *Geology*, **50**, 837–842, https://doi.org/10.1130/G49784.1

Bruand, E., Storey, C. and Fowler, M. 2014. Accessory mineral chemistry of high Ba–Sr granites from northern Scotland: constraints on petrogenesis and records of whole-rock signature. *Journal of Petrology*, **55**, 1619–1651, https://doi.org/10.1093/petrology/egu037

Bruand, E., Fowler, M., Storey, C. and Darling, J. 2017. Apatite trace element and isotope applications to petrogenesis and provenance. *American Mineralogist*, **102**, 75–84, https://doi.org/10.2138/am-2017-5744

Bruand, E., Storey, C., Fowler, M. and Heilimo, E. 2019. Oxygen isotopes in titanite and apatite, and their potential for crustal evolution research. *Geochimica et Cosmochimica Acta*, **255**, 144–162, https://doi.org/10.1016/j.gca.2019.04.002

Bruand, E., Fowler, M. *et al.* 2020. Accessory mineral constraints on crustal evolution: elemental fingerprints for magma discrimination. *Geochemical Perspectives Letters*, **13**, 7–12, https://doi.org/10.7185/geochemlet.2006

Budzyń, B., Hetherington, C.J., Williams, M.L., Jercinovic, M.J. and Michalik, M. 2010. Fluid–mineral interactions and constraints on monazite alteration during metamorphism. *Mineralogical Magazine*, **74**, 659–681, https://doi.org/10.1180/minmag.2010.074.4.659

Buick, I.S., Hermann, J., Williams, I.S., Gibson, R.L. and Rubatto, D. 2006. A SHRIMP U–Pb and LA-ICP-MS trace element study of the petrogenesis of garnet–cordierite–orthoamphibole gneisses from the Central Zone of the Limpopo Belt, South Africa. *Lithos*, **88**, 150–172, https://doi.org/10.1016/j.lithos.2005.09.001

Buick, I.S., Clark, C., Rubatto, D., Hermann, J., Pandit, M. and Hand, M. 2010. Constraints on the Proterozoic evolution of the Aravalli–Delhi Orogenic Belt (NW India) from monazite geochronology and mineral trace element geochemistry. *Lithos*, **120**, 511–528, https://doi.org/10.1016/j.lithos.2010.09.011

Burn, M. 2016. LA-ICP-QMS Th–U/Pb allanite dating: methods and applications, Doctoral dissertation, Philosophisch-naturwissenschfatliche Fakultät der Universität Bern, Germany.

Burn, M., Lanari, P., Pettke, T. and Engi, M. 2017. Non-matrix-matched standardisation in LA-ICP-MS analysis: general approach, and application to allanite Th–U–Pb dating. *Journal of Analytical Atomic Spectrometry*, **32**, 1359–1377, https://doi.org/10.1039/C7JA00095B

Caddick, M.J. and Thompson, A.B. 2008. Quantifying the tectono-metamorphic evolution of pelitic rocks from a wide range of tectonic settings: mineral compositions in equilibrium. *Contributions to Mineralogy and Petrology*, **156**, 177–195, https://doi.org/10.1007/s00410-008-0280-6

Caddick, M.J., Konopásek, J. and Thompson, A.B. 2010. Preservation of garnet growth zoning and the duration of prograde metamorphism. *Journal of Petrology*, **51**, 2327–2347, https://doi.org/10.1093/petrology/egq059

Campomenosi, N., Scambelluri, M. *et al.* 2021. Using the elastic properties of zircon–garnet host-inclusion pairs for thermobarometry of the ultrahigh-pressure Dora-Maira whiteschists: problems and perspectives. *Contributions to Mineralogy and Petrology*, **176**, article 36, https://doi.org/10.1007/s00410-021-01793-6

Cao, M., Qin, K., Li, G., Evans, N.J. and Jin, L. 2015. *In situ* LA-(MC)-ICP-MS trace element and Nd isotopic compositions and genesis of polygenetic titanite from the Baogutu reduced porphyry Cu deposit, Western Junggar, NW China. *Ore Geology Reviews*, **65**, 940–954, https://doi.org/10.1016/j.oregeorev.2014.07.014

Carruzzo, S., Clarke, D.B., Pelrine, K.M. and MacDonald, M.A. 2006. Texture, composition, and origin of rutile in the South Mountain Batholith, Nova Scotia. *The Canadian Mineralogist*, **44**, 715–729, https://doi.org/10.2113/gscanmin.44.3.715

Carvalho, B.B., Bartoli, O. *et al.* 2019. Anatexis and fluid regime of the deep continental crust: new clues from melt and fluid inclusions in metapelitic migmatites from Ivrea Zone (NW Italy). *Journal of Metamorphic Geology*, **37**, 951–975, https://doi.org/10.1111/jmg.12463

Castelli, D. and Rubatto, D. 2002. Stability of Al- and F-rich titanite in metacarbonate: petrologic and isotopic constraints from a polymetamorphic eclogitic marble of the internal Sesia Zone (Western Alps). *Contributions to Mineralogy and Petrology*, **142**, 627–639, https://doi.org/10.1007/s00410-001-0317-6

Catlos, E., Sorensen, S.S. and Harrison, T.M. 2000. Th–Pb ion-microprobe dating of allanite. *American Mineralogist*, **85**, 633–648, https://doi.org/10.2138/am-2000-5-601

Cavosie, A.J., Erickson, T.M. *et al.* 2015. A terrestrial perspective on using *ex situ* shocked zircons to date lunar impacts. *Geology*, **43**, 999–1002, https://doi.org/10.1130/G37059.1

Cavosie, A.J., Spencer, C.J., Evans, N., Rankenburg, K., Thomas, R.J. and Macey, P.H. 2022. Granular titanite from the Roter Kamm crater in Namibia: product of regional metamorphism, not meteorite impact. *Geoscience Frontiers*, **13**, article 101350, https://doi.org/10.1016/j.gsf.2022.101350

Caxito, F.A., Basto, C.F. *et al.* 2021. Neoproterozoic magmatic arc volcanism in the Borborema Province, NE Brazil: possible flare-ups and lulls and implications for western Gondwana assembly. *Gondwana Research*, **92**, 1–25, https://doi.org/10.1016/j.gr.2020.11.015

Cenki-Tok, B., Oliot, E. *et al.* 2011. Preservation of Permian allanite within an Alpine eclogite facies shear zone at Mt Mucrone, Italy: mechanical and chemical behavior of allanite during mylonitization. *Lithos*, **125**, 40–50, https://doi.org/10.1016/j.lithos.2011.01.005

Cenki-Tok, B., Darling, J.R., Rolland, Y., Dhuime, B. and Storey, C.D. 2014. Direct dating of mid-crustal shear zones with synkinematic allanite: new *in situ* U–Th–Pb geochronological approaches applied to the Mont Blanc Massif. *Terra Nova*, **26**, 29–37, https://doi.org/10.1111/ter.12066

Cerny, P., Novak, M., Chapman, R. and Ferreira, K.J. 2007. Subsolidus behavior of niobian rutile from the Pisek region, Czech Republic: a model for exsolution in W- and $Fe^{2+} >> Fe^{3+}$ -rich phases. *Journal of Geosciences*, **52**, 143–159, https://doi.org/10.3190/jgeosci.008

Cerva-Alves, T., Hartmann, L.A., Queiroga, G.N., Lana, C., Castro, M.P., Maciel, L.A.C. and Remus, M.V.D. 2021. Metamorphic evolution of the juvenile Serrinha forearc basin in the southern Brasiliano Orogen. *Precambrian Research*, **365**, article 106394, https://doi.org/10.1016/j.precamres.2021.106394

Cesare, B., Ferrero, S., Salvioli-Mariani, E., Pedron, D. and Cavallo, A. 2009. 'Nanogranite' and glassy inclusions: the anatectic melt in migmatites and granulites. *Geology*, **37**, 627–630, https://doi.org/10.1130/G25759A.1

Cesare, B., Acosta-Vigil, A., Ferrero, S., Bartoli, O. and Forster, M. 2011. Melt inclusions in migmatites and granulites. *Journal of the Virtual Explorer*, **38**, article 2, https://doi.org/10.3809/jvirtex.2011.00268

Chakhmouradian, A.R., Reguir, E.P. *et al.* 2017. Apatite in carbonatitic rocks: compositional variation, zoning, element partitioning and petrogenetic significance. *Lithos*, **274**, 188–213, https://doi.org/10.1016/j.lithos.2016.12.037

Chamberlain, C.P. and Conrad, M.E. 1991. Oxygen isotope zoning in garnet. *Science*, **254**, 403–406, https://doi.org/10.1126/science.254.5030.403

Chamberlain, K.R., Schmitt, A.K. *et al.* 2010. *In situ* U–Pb SIMS (IN-SIMS) micro-baddeleyite dating of mafic rocks: method with examples. *Precambrian Research*, **183**, 379–387, https://doi.org/10.1016/j.precamres.2010.05.004

Chaussidon, M., Deng, Z., Villeneuve, J., Moureau, J., Watson, B., Richter, F. and Moynier, F. 2017. *In situ* analysis of non-traditional isotopes by SIMS and LA-MC-ICP-MS: key aspects and the example of Mg isotopes in olivines and silicate glasses. *Reviews in Mineralogy and Geochemistry*, **82**, 127–163, https://doi.org/10.2138/rmg.2017.82.5

Chen, X., Wang, W., Zhang, Z., Nie, N.X. and Dauphas, N. 2020. Evidence from Ab initio and transport modeling for diffusion-driven zirconium isotopic fractionation in igneous rocks. *ACS Earth and Space Chemistry*, **4**, 1572–1595, https://doi.org/10.1021/acsearthspacechem.0c00146

Cherniak, D.J. 1993. Lead diffusion in titanite and preliminary results on the effects of radiation damage on Pb transport. *Chemical Geology*, **110**, 177–194, https://doi.org/10.1016/0009-2541(93)90253-F

Cherniak, D.J. 1995. Sr and Nd diffusion in titanite. *Chemical Geology*, **125**, 219–232, https://doi.org/10.1016/0009-2541(95)00074-V

Cherniak, D.J. 2000*a*. Pb diffusion in rutile. *Contributions to Mineralogy and Petrology*, **139**, 198–207, https://doi.org/10.1007/PL00007671

Cherniak, D.J. 2000*b*. Rare earth element diffusion in apatite. *Geochimica et Cosmochimica Acta*, **64**, 3871– 3885, https://doi.org/10.1016/S0016-7037(00)00467-1

Cherniak, D.J. 2006. Pb and rare earth element diffusion in xenotime. *Lithos*, **88**, 1–14, https://doi.org/10.1016/j.lithos.2005.08.002

Cherniak, D.J. and Watson, E.B. 2001. Pb diffusion in zircon. *Chemical Geology*, **172**, 5–24, https://doi.org/10.1016/S0009-2541(00)00233-3

Cherniak, D.J. and Watson, E.B. 2003. Diffusion in zircon. *Reviews in Mineralogy and Geochemistry*, **53**, 113–143, https://doi.org/10.2113/0530113

Cherniak, D.J. and Watson, E. 2007. Ti diffusion in zircon. *Chemical Geology*, **242**, 470–483, https://doi.org/10.1016/j.chemgeo.2007.05.005

Cherniak, D.J., Lanford, W.A. and Ryerson, F.J. 1991. Lead diffusion in apatite and zircon using ion implantation and Rutherford backscattering techniques. *Geochimica et Cosmochimica Acta*, **55**, 1663–1673, https://doi.org/10.1016/0016-7037(91)90137-T

Cherniak, D.J., Watson, E.B., Grove, M. and Harrison, T.M. 2004. Pb diffusion in monazite: a combined RBS/SIMS study. *Geochimica et Cosmochimica Acta*, **68**, 829–840, https://doi.org/10.1016/j.gca.2003.07.012

Cherniak, D.J., Manchester, J. and Watson, E. 2007. Zr and Hf diffusion in rutile. *Earth and Planetary Science Letters*, **261**, 267–279, https://doi.org/10.1016/j.epsl.2007.06.027

Chew, D.M. and Spikings, R.A. 2015. Geochronology and thermochronology using apatite: time and temperature, lower crust to surface. *Elements*, **11**, 189–194, https://doi.org/10.2113/gselements.11.3.189

Chew, D.M., Sylvester, P.J. and Tubrett, M.N. 2011. U–Pb and Th–Pb dating of apatite by LA-ICPMS. *Chemical Geology*, **280**, 200–216, https://doi.org/10.1016/j.chemgeo.2010.11.010

Chew, D.M., Petrus, J.A. and Kamber, B.S. 2014. U–Pb LA-ICPMS dating using accessory mineral standards with variable common Pb. *Chemical Geology*, **363**, 185–199, https://doi.org/10.1016/j.chemgeo.2013.11.006

Chew, D.M., Petrus, J.A., Kenny, G.G. and McEvoy, N. 2017. Rapid high-resolution U–Pb LA-Q-ICPMS age mapping of zircon. *Journal of Analytical Atomic Spectrometry*, **32**, 262–276, https://doi.org/10.1039/C6JA00404K

Chew, D., Drost, K., Marsh, J.H. and Petrus, J.A. 2021. LA-ICP-MS imaging in the geosciences and its applications to geochronology. *Chemical Geology*, **559**, article 119917, https://doi.org/10.1016/j.chemgeo.2020.119917

Choukroun, M., O'Reilly, S.Y., Griffin, W.L., Pearson, N.J. and Dawson, J.B. 2005. Hf isotopes of MARID (mica–amphibole–rutile–ilmenite–diopside) rutile trace metasomatic processes in the lithospheric mantle. *Geology*, **33**, 45–48, https://doi.org/10.1130/G21084.1

Chu, M.-F., Wang, K.-L., Griffin, W.L., Chung, S.-L., O'Reilly, S.Y., Pearson, N.J. and Iizuka, Y. 2009. Apatite composition: tracing petrogenetic processes in Transhimalayan granitoids. *Journal of Petrology*, **50**, 1829–1855, https://doi.org/10.1093/petrology/egp054

Cioffi, C.R., da Campos Neto, M.C., Möller, A. and Rocha, B.C. 2019. Titanite petrochronology of the southern Brasília Orogen basement: effects of retrograde net-transfer reactions on titanite trace element compositions. *Lithos*, **344–345**, 393–408, https://doi.org/10.1016/j.lithos.2019.06.035

Clark, C., Collins, A.S., Santosh, M., Taylor, R. and Wade, B.P. 2009. The P–T–t architecture of a Gondwanan suture: REE, U–Pb and Ti-in-zircon thermometric constraints from the Palghat Cauvery shear system, South India. *Precambrian Research*, **174**, 129–144, https://doi.org/10.1016/J.PRECAMRES.2009.07.003

Clark, C., Kirkland, C.L., Spaggiari, C.V., Oorschot, C., Wingate, M.T.D. and Taylor, R.J. 2014. Proterozoic granulite formation driven by mafic magmatism: an example from the Fraser Range Metamorphics, Western Australia. *Precambrian Research*, **240**, 1–21, https://doi.org/10.1016/J.PRECAMRES.2013.07.024

Clark, C., Taylor, R.J.M., Johnson, T.E., Harley, S.L., Fitzsimons, I.C.W. and Oliver, L. 2019. Testing the fidelity of thermometers at ultrahigh temperatures. *Journal of Metamorphic Geology*, **37**, 917–934, https://doi.org/10.1111/jmg.12486

Clark, J.R. and Williams-Jones, A.E. 2004. *Rutile as a Potential Indicator Mineral for Metamorphosed Metallic Ore Deposits*. Rapport Final de DIVEX, Sous-projet SC2, Montréal, Canada.

Cochrane, R., Spikings, R.A. *et al.* 2014. High temperature (>350°C) thermochronology and mechanisms of Pb loss in apatite. *Geochimica et Cosmochimica Acta*, **127**, 39–56, https://doi.org/10.1016/j.gca.2013.11.028

Coleman, D.S., Gray, W. and Glazner, A.F. 2004. Rethinking the emplacement and evolution of zoned plutons: geochronologic evidence for incremental assembly of the Tuolumne Intrusive Suite, California. *Geology*, **32**, 433–436, https://doi.org/10.1130/G20220.1

Collins, W.J., Huang, H.-Q. and Jiang, X. 2016. Water-fluxed crustal melting produces Cordilleran batholiths. *Geology*, **44**, 143–146, https://doi.org/10.1130/G37398.1

Copeland, P., Parrish, R.R. and Harrison, T.M. 1988. Identification of inherited radiogenic Pb in monazite and its implications for U–Pb systematics. *Nature*, **333**, 760–763, https://doi.org/10.1038/333760a0

Corfu, F., Hanchar, J.M., Hoskin, P.W.O. and Kinny, P. 2003. Atlas of zircon textures. *Reviews in Mineralogy and Geochemistry*, **53**, 469–500, https://doi.org/10.2113/0530469

Corti, L., Zanoni, D., Gatta, G.D. and Zucali, M. 2020. Strain partitioning in host rock controls light rare earth element release from allanite-(Ce) in subduction zones. *Mineralogical Magazine*, **84**, 93–108, https://doi.org/10.1180/mgm.2020.4

Cox, M.A., Cavosie, A.J. *et al.* 2022. Impact and habitability scenarios for early Mars revisited based on a 4.45-Ga shocked zircon in regolith breccia. *Science Advances*, **8**, article eabl7497, https://doi.org/10.1126/sciadv.abl7497

Cross, A.J. 2009. *SHRIMP U–Pb Xenotime Geochronology and Its Application to Dating Mineralisation, Sediment Deposition and Metamorphism*. PhD thesis, Australian National University, https://doi.org/10.25911/5d778658a52a2

Crowe, D.E., Riciputi, L.R., Bezenek, S. and Ignatiev, A. 2001. Oxygen isotope and trace element zoning in hydrothermal garnets: windows into large-scale fluid-flow behavior. *Geology*, **29**, 479–482, https://doi.org/10.1130/0091-7613(2001)029<0479:OIATEZ>2.0.CO;2

Cruz-Uribe, A.M., Feineman, M.D., Zack, T. and Barth, M. 2014. Metamorphic reaction rates at ~650–800°C from diffusion of niobium in rutile. *Geochimica et*

Cosmochimica Acta, **130**, 63–77, https://doi.org/10.1016/j.gca.2013.12.015

Cutts, K.A., Hand, M., Kelsey, D.E., Wade, B., Strachan, R.A., Clark, C. and Netting, A. 2009. Evidence for 930 Ma metamorphism in the Shetland Islands, Scottish Caledonides: implications for Neoproterozoic tectonics in the Laurentia–Baltica sector of Rodinia. *Journal of the Geological Society, London*, **166**, 1033–1047, https://doi.org/10.1144/0016-76492009-006

Cutts, K.A., Kinny, P.D. *et al.* 2010. Three metamorphic events recorded in a single garnet: integrated phase modelling, *in situ* LA-ICPMS and SIMS geochronology from the Moine Supergroup, NW Scotland. *Journal of Metamorphic Geology*, **28**, 249–267, https://doi.org/10.1111/j.1525-1314.2009.00863.x

Cutts, K.A., Kelsey, D.E. and Hand, M. 2013. Evidence for late Paleoproterozoic (*ca* 1690–1665 Ma) high- to ultrahigh-temperature metamorphism in southern Australia: implications for Proterozoic supercontinent models. *Gondwana Research*, **23**, 617–640, https://doi.org/10.1016/j.gr.2012.04.009

Cutts, K.A., Stevens, G., Hoffmann, J.E., Buick, I.S., Frei, D. and Münker, C. 2014. Paleo- to Mesoarchean polymetamorphism in the Barberton Granite–Greenstone Belt, South Africa: constraints from U–Pb monazite and Lu–Hf garnet geochronology on the tectonic processes that shaped the belt. *GSA Bulletin*, **126**, 251–270, https://doi.org/10.1130/B30807.1

Cutts, K., Lana, C., Alkmim, F. and Peres, G.G. 2018. Metamorphic imprints on units of the southern Araçuaí belt, SE Brazil: the history of superimposed Transamazonian and Brasiliano orogenesis. *Gondwana Research*, **58**, 211–234, https://doi.org/10.1016/j.gr.2018.02.016

Cygan, R.T. and Lasaga, A.C. 1985. Self-diffusion of magnesium in garnet at 750 degrees to 900 degrees C. *American Journal of Science*, **285**, 328–350, https://doi.org/10.2475/ajs.285.4.328

Dachille, F., Simons, P.Y. and Roy, R. 1968. Pressure–temperature studies of anatase, brookite, rutile and TiO2-II. *American Mineralogist*, **53**, 1929–1939.

Darling, J.R., Storey, C.D. and Engi, M. 2012. Allanite U–Th–Pb geochronology by laser ablation ICPMS. *Chemical Geology*, **292**, 103–115, https://doi.org/10.1016/j.chemgeo.2011.11.012

Davis, D.W., Krogh, T.E. and Williams, I.S. 2003. Historical development of zircon geochronology. *Reviews in Mineralogy and Geochemistry*, **53**, 145–181, https://doi.org/10.2113/0530145

Decrée, S., Coint, N., Debaille, V., Hagen-Peter, G., Leduc, T. and Schiellerup, H. 2023. The potential for REEs in igneous-related apatite deposits in Europe. *Geological Society, London, Special Publications*, **526**, 2021-2175, https://doi.org/10.1144/SP526-2021-175

Degeling, H., Eggins, S. and Ellis, D. 2001. Zr budgets for metamorphic reactions, and the formation of zircon from garnet breakdown. *Mineralogical Magazine*, **65**, 749–758, https://doi.org/10.1180/002646101656 0006

Delavault, H., Dhuime, B., Hawkesworth, C.J., Cawood, P.A., Marschall, H. and the Edinburgh Ion Microprobe Facility 2016. Tectonic settings of continental crust formation: insights from Pb isotopes in feldspar inclusions in zircon. *Geology*, **44**, 819–822, https://doi.org/10.1130/G38117.1

Dempster, T.J., Coleman, S., Kennedy, R., Chung, P. and Brown, R.W. 2020. Growth zoning of garnet porphyroblasts: grain boundary and microtopographic controls. *Journal of Metamorphic Geology*, **38**, 1011–1027, https://doi.org/10.1111/jmg.12558

Deng, Z., Chaussidon, M., Guitreau, M., Puchtel, I.S., Dauphas, N. and Moynier, F. 2019. An oceanic subduction origin for Archaean granitoids revealed by silicon isotopes. *Nature Geoscience*, **12**, 774–778, https://doi.org/10.1038/s41561-019-0407-6

D'Errico, M., Lackey, J. *et al.* 2012. A detailed record of shallow hydrothermal fluid flow in the Sierra Nevada magmatic arc from low-$\delta^{18}O$ skarn garnets. *Geology*, **40**, 763–766, https://doi.org/10.1130/G33008.1

Dev, J.A., Tomson, J.K., Sorcar, N. and Nandakumar, V. 2021. Combined U–Pb/Hf isotopic studies and phase equilibrium modelling of HT–UHT metapelites from Kambam ultrahigh-temperature belt, South India: constraints on tectonothermal history of the terrane. *Lithos*, **406–407**, article 106531, https://doi.org/10.1016/j.lithos.2021.106531

Dhuime, B., Hawkesworth, C.J., Cawood, P.A. and Storey, C.D. 2012. A change in the geodynamics of continental growth 3 billion years ago. *Science*, **335**, 1334–1336, https://doi.org/10.1126/science.1216066

Di Rosa, M., Farina, F., Lanari, P. and Marroni, M. 2020. Pre-Alpine thermal history recorded in the continental crust from Alpine Corsica (France): evidence from zircon and allanite LA-ICP-MS dating. *Swiss Journal of Geosciences*, **113**, article 19, https://doi.org/10.1186/s00015-020-00374-2

Dodson, M.H. 1973. Closure temperature in cooling geochronological and petrological systems. *Contributions to Mineralogy and Petrology*, **40**, 259–274, https://doi.org/10.1007/BF00373790

Dohmen, R., Marschall, H.R., Ludwig, T. and Polednia, J. 2019. Diffusion of Zr, Hf, Nb and Ta in rutile: effects of temperature, oxygen fugacity, and doping level, and relation to rutile point defect chemistry. *Physics and Chemistry of Minerals*, **46**, 311–332, https://doi.org/10.1007/s00269-018-1005-7

Doucelance, R., Bruand, E., Matte, S., Bosq, C., Auclair, D. and Gannoun, A.-M. 2020. *In-situ* determination of Nd isotope ratios in apatite by LA-MC-ICPMS: challenges and limitations. *Chemical Geology*, **550**, article 119740, https://doi.org/10.1016/j.chemgeo.2020.119740

Dragovic, B., Samanta, L.M., Baxter, E.F. and Selverstone, J. 2012. Using garnet to constrain the duration and rate of water-releasing metamorphic reactions during subduction: an example from Sifnos, Greece. *Chemical Geology*, **314–317**, 9–22, https://doi.org/10.1016/j.chemgeo.2012.04.016

Dragovic, B., Baxter, E.F. and Caddick, M.J. 2015. Pulsed dehydration and garnet growth during subduction revealed by zoned garnet geochronology and thermodynamic modeling, Sifnos, Greece. *Earth and Planetary Science Letters*, **413**, 111–122, https://doi.org/10.1016/j.epsl.2014.12.024

Duan, Z., Gleeson, S.A., Gao, W.-S., Wang, F.-Y., Li, C.-J. and Li, J.-W. 2020. Garnet U–Pb dating of the Yinan Au–Cu skarn deposit, Luxi District, North China

Craton: implications for district-wide coeval Au–Cu and Fe skarn mineralization. *Ore Geology Reviews*, **118**, article 103310, https://doi.org/10.1016/j.oregeorev.2020.103310

Duchêne, S., Blichert-Toft, J., Luais, B., Télouk, P., Lardeaux, J.-M. and Albarède, F. 1997. The Lu–Hf dating of garnets and the ages of the Alpine high-pressure metamorphism. *Nature*, **387**, 586–589, https://doi.org/10.1038/42446

Duesterhoeft, E. and Lanari, P. 2020. Iterative thermodynamic modelling – part 1: a theoretical scoring technique and a computer program (Bingo-Antidote). *Journal of Metamorphic Geology*, **38**, 527–551, https://doi.org/10.1111/jmg.12538

Durgalakshmi, Sajeev, K. *et al.* 2021. The timing, duration and conditions of UHT metamorphism in remnants of the former eastern Gondwana. *Journal of Petrology*, **62**, article egab068, https://doi.org/10.1093/petrology/egab068

Dziggel, A., Wulff, K., Kolb, J., Meyer, F.M. and Lahaye, Y. 2009. Significance of oscillatory and bell-shaped growth zoning in hydrothermal garnet: evidence from the Navachab gold deposit, Namibia. *Chemical Geology*, **262**, 262–276, https://doi.org/10.1016/j.chemgeo.2009.01.027

El Korh, A., Schmidt, S.T., Ulianov, A. and Potel, S. 2009. Trace element partitioning in HP–LT metamorphic assemblages during subduction-related metamorphism, Ile de Groix, France: a detailed LA-ICPMS study. *Journal of Petrology*, **50**, 1107–1148, https://doi.org/10.1093/petrology/egp034

Emo, R.B., Smit, M.A. *et al.* 2018. Evidence for evolved Hadean crust from Sr isotopes in apatite within Eoarchean zircon from the Acasta Gneiss Complex. *Geochimica et Cosmochimica Acta*, **235**, 450–462, https://doi.org/10.1016/j.gca.2018.05.028

Engi, M. 2017. Petrochronology based on REE-minerals: monazite, allanite, xenotime, apatite. *Reviews in Mineralogy and Geochemistry*, **83**, 365–418, https://doi.org/10.2138/rmg.2017.83.12

Engi, M., Lanari, P. and Kohn, M.J. 2017. Significant ages – an introduction to petrochronology. *Reviews in Mineralogy and Geochemistry*, **83**, 1–12, https://doi.org/10.2138/rmg.2017.83.1

Evans, J. and Zalasiewicz, J. 1996. UPb, PbPb and SmNd dating of authigenic monazite: implications for the diagenetic evolution of the Welsh Basin. *Earth and Planetary Science Letters*, **144**, 421–433, https://doi.org/10.1016/S0012-821X(96)00177-X

Evans, T.P. 2004. A method for calculating effective bulk composition modification due to crystal fractionation in garnet-bearing schist: implications for isopleth thermobarometry. *Journal of Metamorphic Geology*, **22**, 547–557, https://doi.org/10.1111/j.1525-1314.2004.00532.x

Ewing, R.C., Meldrum, A., Wang, L., Weber, W.J. and Corrales, L.R. 2003. Radiation effects in zircon. *Reviews in Mineralogy and Geochemistry*, **53**, 387–425, https://doi.org/10.2113/0530387

Ewing, T.A. and Müntener, O. 2018. The mantle source of island arc magmatism during early subduction: evidence from Hf isotopes in rutile from the Jijal Complex (Kohistan arc, Pakistan). *Lithos*, **308–309**, 262–277, https://doi.org/10.1016/j.lithos.2018.03.005

Ewing, T.A., Rubatto, D., Eggins, S.M. and Hermann, J. 2011. *In situ* measurement of hafnium isotopes in rutile by LA-MC-ICPMS: protocol and applications. *Chemical Geology*, **281**, 72–82, https://doi.org/10.1016/j.chemgeo.2010.11.029

Ewing, T.A., Hermann, J. and Rubatto, D. 2013. The robustness of the Zr-in-rutile and Ti-in-zircon thermometers during high-temperature metamorphism (Ivrea–Verbano Zone, northern Italy). *Contributions to Mineralogy and Petrology*, **165**, 757–779, https://doi.org/10.1007/s00410-012-0834-5

Ewing, T.A., Rubatto, D. and Hermann, J. 2014. Hafnium isotopes and Zr/Hf of rutile and zircon from lower crustal metapelites (Ivrea–Verbano Zone, Italy): implications for chemical differentiation of the crust. *Earth and Planetary Science Letters*, **389**, 106–118, https://doi.org/10.1016/j.epsl.2013.12.029

Ewing, T.A., Rubatto, D., Beltrando, M. and Hermann, J. 2015. Constraints on the thermal evolution of the Adriatic margin during Jurassic continental break-up: U–Pb dating of rutile from the Ivrea–Verbano Zone, Italy. *Contributions to Mineralogy and Petrology*, **169**, 1–22, https://doi.org/10.1007/s00410-015-1135-6

Farias, P., Weinberg, R., Sola, A. and Becchio, R. 2020. From crustal thickening to orogen-parallel escape: the 120-myr-long HT–LP evolution recorded by titanite in the Paleozoic Famatinian Backarc, NW Argentina. *Tectonics*, **39**, article e2020TC006184, https://doi.org/10.1029/2020TC006184

Ferreira, J.A., Pereira, I., Bento dos Santos, T. and Mata, J. 2022. U–Pb age constraints on the protolith, cooling and exhumation of a Variscan middle crust migmatite complex from the Central Iberian Zone: insights into the Variscan metamorphic evolution and Ediacaran palaeogeographic implications. *Journal of the Geological Society*, **179**, jgs2021-072, https://doi.org/10.1144/jgs2021-072

Ferrero, S., Bartoli, O. *et al.* 2012. Microstructures of melt inclusions in anatectic metasedimentary rocks. *Journal of Metamorphic Geology*, **30**, 303–322, https://doi.org/10.1111/j.1525-1314.2011.00968.x

Ferrero, S., Wannhoff, I. *et al.* 2021. Embryos of TTGs in Gore Mountain garnet megacrysts from water-fluxed melting of the lower crust. *Earth and Planetary Science Letters*, **569**, article 117058, https://doi.org/10.1016/j.epsl.2021.117058

Ferry, J.M. and Spear, F. 1978. Experimental calibration of the partitioning of Fe and Mg between biotite and garnet. *Contributions to Mineralogy and Petrology*, **66**, 113–117, https://doi.org/10.1007/BF00372150

Ferry, J.M. and Watson, E.B. 2007. New thermodynamic models and revised calibrations for the Ti-in-zircon and Zr-in-rutile thermometers. *Contributions to Mineralogy and Petrology*, **154**, 429–437, https://doi.org/10.1007/s00410-007-0201-0

Fielding, I.O.H., Johnson, S.P. *et al.* 2017. Using *in situ* SHRIMP U–Pb monazite and xenotime geochronology to determine the age of orogenic gold mineralization: an example from the Paulsens mine, southern Pilbara Craton. *Economic Geology*, **112**, 1205–1230, https://doi.org/10.5382/econgeo.2017.4507

Finch, M., Weinberg, R., Barrote, V. and Cawood, P. 2021. Hf isotopic ratios in zircon reveal processes of anatexis and pluton construction. *Earth and Planetary Science

Letters, **576**, article 117215, https://doi.org/10.1016/j.epsl.2021.117215

Finger, F. and Krenn, E. 2007. Three metamorphic monazite generations in a high-pressure rock from the Bohemian Massif and the potentially important role of apatite in stimulating polyphase monazite growth along a PT loop. *Lithos*, **95**, 103–115, https://doi.org/10.1016/j.lithos.2006.06.003

Fischer, S., Prave, A.R., Johnson, T.E., Cawood, P.A., Hawkesworth, C.J., Horstwood, M.S.A. and EIMF 2021. Using zircon in mafic migmatites to disentangle complex high-grade gneiss terrains – terrane spotting in the Lewisian complex, NW Scotland. *Precambrian Research*, **355**, article 106074, https://doi.org/10.1016/j.precamres.2020.106074

Fisher, C.M., McFarlane, C.R.M., Hanchar, J.M., Schmitz, M.D., Sylvester, P.J., Lam, R. and Longerich, H.P. 2011. Sm–Nd isotope systematics by laser ablation-multicollector-inductively coupled plasma mass spectrometry: methods and potential natural and synthetic reference materials. *Chemical Geology*, **284**, 1–20, https://doi.org/10.1016/j.chemgeo.2011.01.012

Fisher, C.M., Vervoort, J.D. and DuFrane, S.A. 2014*a*. Accurate Hf isotope determinations of complex zircons using the 'laser ablation split stream' method. *Geochemistry, Geophysics, Geosystems*, **15**, 121–139, https://doi.org/10.1002/2013GC004962

Fisher, C.M., Vervoort, J.D. and Hanchar, J.M. 2014*b*. Guidelines for reporting zircon Hf isotopic data by LA-MC-ICPMS and potential pitfalls in the interpretation of these data. *Chemical Geology*, **363**, 125–133, https://doi.org/10.1016/j.chemgeo.2013.10.019

Fisher, C.M., Hanchar, J.M., Miller, C.F., Phillips, S., Vervoort, J.D. and Whitehouse, M.J. 2017. Combining Nd isotopes in monazite and Hf isotopes in zircon to understand complex open-system processes in granitic magmas. *Geology*, **45**, 267–270, https://doi.org/10.1130/G38458.1

Fisher, C.M., Bauer, A.M. *et al.* 2020. Laser ablation split-stream analysis of the Sm–Nd and U–Pb isotope compositions of monazite, titanite, and apatite – improvements, potential reference materials, and application to the Archean Saglek Block gneisses. *Chemical Geology*, **539**, article 119493, https://doi.org/10.1016/j.chemgeo.2020.119493

Fletcher, I.R., McNaughton, N.J., Aleinikoff, J.A., Rasmussen, B. and Kamo, S.L. 2004. Improved calibration procedures and new standards for U–Pb and Th–Pb dating of Phanerozoic xenotime by ion microprobe. *Chemical Geology*, **209**, 295–314, https://doi.org/10.1016/j.chemgeo.2004.06.015

Flowerdew, M., Millar, I.L., Vaughan, A., Horstwood, M. and Fanning, C. 2006. The source of granitic gneisses and migmatites in the Antarctic Peninsula: a combined U–Pb SHRIMP and laser ablation Hf isotope study of complex zircons. *Contributions to Mineralogy and Petrology*, **151**, 751–768, https://doi.org/10.1007/s00410-006-0091-6

Foley, S., Tiepolo, M. and Vannucci, R. 2002. Growth of early continental crust controlled by melting of amphibolite in subduction zones. *Nature*, **417**, 837–840, https://doi.org/10.1038/nature00799

Fornelli, A., Langone, A., Micheletti, F., Pascazio, A. and Piccarreta, G. 2014. The role of trace element partitioning between garnet, zircon and orthopyroxene on the interpretation of zircon U–Pb ages: an example from high-grade basement in Calabria (southern Italy). *International Journal of Earth Sciences*, **103**, 487–507, https://doi.org/10.1007/s00531-013-0971-8

Fornelli, A., Langone, A., Micheletti, F. and Piccarreta, G. 2018. REE partition among zircon, orthopyroxene, amphibole and garnet in a high-grade metabasic system. *Geological Magazine*, **155**, 1705–1726, https://doi.org/10.1017/S001675681700067X

Förster, H.-J. 1998. The chemical composition of REE–Y–Th–U-rich accessory minerals in peraluminous granites of the Erzgebirge–Fichtelgebirge region, Germany, part I: the monazite-(Ce)-brabantite solid solution series. *American Mineralogist*, **83**, 259–272, https://doi.org/10.2138/am-1998-3-409

Fortier, S.M. and Giletti, B.J. 1989. An empirical model for predicting diffusion coefficients in silicate minerals. *Science*, **245**, 1481–1484, https://doi.org/10.1126/science.245.4925.1481

Foster, G., Gibson, H.D., Parrish, R., Horstwood, M., Fraser, J. and Tindle, A. 2002. Textural, chemical and isotopic insights into the nature and behaviour of metamorphic monazite. *Chemical Geology*, **191**, 183–207, https://doi.org/10.1016/S0009-2541(02)00156-0

Fougerouse, D., Kirkland, C.L., Saxey, D.W., Seydoux-Guillaume, A.-M., Rowles, M.R., Rickard, W.D.A. and Reddy, S.M. 2020. Nanoscale isotopic dating of monazite. *Geostandards and Geoanalytical Research*, **44**, 637–652, https://doi.org/10.1111/ggr.12340

Fougerouse, D., Cavosie, A.J. *et al.* 2021. A new method for dating impact events – thermal dependency on nanoscale Pb mobility in monazite shock twins. *Geochimica et Cosmochimica Acta*, **314**, 381–396, https://doi.org/10.1016/j.gca.2021.08.025

Fraser, G., Ellis, D. and Eggins, S. 1997. Zirconium abundance in granulite-facies minerals, with implications for zircon geochronology in high-grade rocks. *Geology*, **25**, 607–610, https://doi.org/10.1130/0091-7613(1997)025<0607:ZAIGFM>2.3.CO;2

Frost, B.R., Chamberlain, K.R. and Schumacher, J.C. 2001. Sphene (titanite): phase relations and role as a geochronometer. *Chemical Geology*, **172**, 131–148, https://doi.org/10.1016/S0009-2541(00)00240-0

Froude, D.O., Ireland, T.R., Kinny, P.D., Williams, I.S., Compston, W., Williams, I.R. and Myers, J.S. 1983. Ion microprobe identification of 4,100–4,200 Myr-old terrestrial zircons. *Nature*, **304**, 616–618, https://doi.org/10.1038/304616a0

Fu, B., Page, F.Z. *et al.* 2008. Ti-in-zircon thermometry: applications and limitations. *Contributions to Mineralogy and Petrology*, **156**, 197–215, https://doi.org/10.1007/s00410-008-0281-5

Fumes, R.A., Luvizotto, G.L. *et al.* 2022. Petrochronology of high-pressure granulite facies rocks from Southern Brasília Orogen, SE Brazil: combining quantitative compositional mapping, single-element thermometry and geochronology. *Journal of Metamorphic Geology*, **40**, 517–552, https://doi.org/10.1111/jmg.12637

Gaidies, F., de Capitani, C. and Abart, R. 2008. THERIA_G: a software program to numerically model prograde garnet growth. *Contributions to Mineralogy and Petrology*, **155**, 657–671, https://doi.org/10.1007/s00410-007-0263-z

Gaidies, F., Morneau, Y.E., Petts, D.C., Jackson, S.E., Zagorevski, A. and Ryan, J.J. 2021. Major and trace element mapping of garnet: unravelling the conditions, timing and rates of metamorphism of the Snowcap assemblage, west–central Yukon. *Journal of Metamorphic Geology*, **39**, 133–164, https://doi.org/10.1111/jmg.12562

Ganguly, J. and Saxena, S.K. 1984. Mixing properties of aluminosilicate garnets: constraints from natural and experimental data, and applications to geothermobarometry. *American Mineralogist*, **69**, 88–97.

Gao, X.-Y., Zheng, Y.-F., Chen, Y.-X. and Guo, J. 2012. Geochemical and U–Pb age constraints on the occurrence of polygenetic titanites in UHP metagranite in the Dabie orogen. *Lithos*, **136–139**, 93–108, https://doi.org/10.1016/j.lithos.2011.03.020

Garber, J.M., Hacker, B.R., Kylander-Clark, A.R.C., Stearns, M. and Seward, G. 2017. Controls on trace element uptake in metamorphic titanite: implications for petrochronology. *Journal of Petrology*, **58**, 1031– 1057, https://doi.org/10.1093/petrology/egx046

Gardiner, N.J., Kirkland, C.L., Hollis, J., Szilas, K., Steenfelt, A., Yakymchuk, C. and Heide-Jørgensen, H. 2019. Building Mesoarchaean crust upon Eoarchaean roots: the Akia Terrane, West Greenland. *Contributions to Mineralogy and Petrology*, **174**, article 20, https://doi.org/10.1007/s00410-019-1554-x

George, F., Gaidies, F. and Boucher, B. 2018. Population-wide garnet growth zoning revealed by LA-ICP-MS mapping: implications for trace element equilibration and syn-kinematic deformation during crystallisation. *Contributions to Mineralogy and Petrology*, **173**, 1–22, https://doi.org/10.1007/s00410-018-1503-0

Gerdes, A. and Zeh, A. 2009. Zircon formation v. zircon alteration – new insights from combined U–Pb and Lu–Hf *in-situ* LA-ICP-MS analyses, and consequences for the interpretation of Archean zircon from the Central Zone of the Limpopo Belt. *Chemical Geology*, **261**, 230–243, https://doi.org/10.1016/j.chemgeo.2008.03.005

Gerrits, A.R., Inglis, E.C., Dragovic, B., Starr, P.G., Baxter, E.F. and Burton, K.W. 2019. Release of oxidizing fluids in subduction zones recorded by iron isotope zonation in garnet. *Nature Geoscience*, **12**, 1029–1033, https://doi.org/10.1038/s41561-019-0471-y

Gervasoni, F., Klemme, S., Rocha-Júnior, E.R.V. and Berndt, J. 2016. Zircon saturation in silicate melts: a new and improved model for aluminous and alkaline melts. *Contributions to Mineralogy and Petrology*, **171**, article 21, https://doi.org/10.1007/s00410-016-1227-y

Gevedon, M., Seman, S., Barnes, J.D., Lackey, J.S. and Stockli, D.F. 2018. Unraveling histories of hydrothermal systems via U–Pb laser ablation dating of skarn garnet. *Earth and Planetary Science Letters*, **498**, 237–246, https://doi.org/10.1016/j.epsl.2018.06.036

Ghent, E.D. 1976. Plagioclase–garnet–Al2SiO5–quartz; a potential geobarometer–geothermometer. *American Mineralogist*, **61**, 710–714.

Gieré, R. and Sorensen, S.S. 2004. Allanite and other REE-rich epidote-group minerals. *Reviews in Mineralogy and Geochemistry*, **56**, 431–493, https://doi.org/10.2138/gsrmg.56.1.431

Gillespie, J., Kinny, P.D., Kirkland, C.L., Martin, L., Nemchin, A.A., Cavosie, A.J. and Hasterok, D. 2021. Isotopic modelling of Archean crustal evolution from comagmatic zircon–apatite pairs. *Earth and Planetary Science Letters*, **575**, article 117194, https://doi.org/10.1016/j.epsl.2021.117194

Gillespie, J., Kirkland, C.L., Kinny, P.D., Simpson, A., Glorie, S. and Rankenburg, K. 2022. Lu–Hf, Sm–Nd, and U–Pb isotopic coupling and decoupling in apatite. *Geochimica et Cosmochimica Acta*, **338**, 121–135, https://doi.org/10.1016/j.gca.2022.09.038

Giuntoli, F., Lanari, P., Burn, M., Kunz, B.E. and Engi, M. 2018. Deeply subducted continental fragments – part 2: insight from petrochronology in the central Sesia Zone (western Italian Alps). *Solid Earth*, **9**, 191–222, https://doi.org/10.5194/se-9-191-2018

Glorie, S., Gillespie, J. *et al.* 2022. Detrital apatite Lu–Hf and U–Pb geochronology applied to the southwestern Siberian margin. *Terra Nova*, **34**, 201–209, https://doi.org/10.1111/ter.12580

Glorie, S. *et al.* 2023. Robust laser ablation Lu–Hf dating of apatite: an empirical evaluation. *Geological Society, London, Special Publications*, **537**, https://doi.org/10.1144/SP537-2022-205

Godet, A., Raimondo, T. and Guilmette, C. 2022. Atoll garnet: insights from LA-ICP-MS trace element mapping. *Contributions to Mineralogy and Petrology*, **177**, article 57, https://doi.org/10.1007/s00410-022-01924-7

Gonçalves, G.O., Lana, C., Buick, I.S., Alkmim, F.F., Scholz, R. and Queiroga, G. 2019. Twenty million years of post-orogenic fluid production and hydrothermal mineralization across the external Araçuaí orogen and adjacent São Francisco craton, SE Brazil. *Lithos*, **342–343**, 557–572, https://doi.org/10.1016/j.lithos.2019.04.022

Gonzalez, J.P., Thomas, J.B., Baldwin, S.L. and Alvaro, M. 2019. Quartz-in-garnet and Ti-in-quartz thermobarometry: methodology and first application to a quartzofeldspathic gneiss from eastern Papua New Guinea. *Journal of Metamorphic Geology*, **37**, 1193–1208, https://doi.org/10.1111/jmg.12508

Gonzalez, J.P., Mazzucchelli, M.L., Angel, R.J. and Alvaro, M. 2021. Elastic geobarometry for anisotropic inclusions in anisotropic host minerals: quartz-in-zircon. *Journal of Geophysical Research: Solid Earth*, **126**, article e2021JB022080, https://doi.org/10.1029/2021JB022080

Goodenough, K.M., Crowley, Q.G., Krabbendam, M. and Parry, S.F. 2013. New U–Pb age constraints for the Laxford Shear Zone, NW Scotland: evidence for tectono-magmatic processes associated with the formation of a Paleoproterozoic supercontinent. *Precambrian Research*, **233**, 1–19, https://doi.org/10.1016/j.precamres.2013.04.010

Gordon, S.M., Kirkland, C.L. *et al.* 2021. Deformation-enhanced recrystallization of titanite drives decoupling between U–Pb and trace elements. *Earth and Planetary Science Letters*, **560**, article 116810, https://doi.org/10.1016/j.epsl.2021.116810

Goudie, D.J., Fisher, C.M., Hanchar, J.M., Crowley, J.L. and Ayers, J.C. 2014. Simultaneous *in situ* determination of U–Pb and Sm–Nd isotopes in monazite by laser ablation ICP-MS. *Geochemistry, Geophysics, Geosystems*, **15**, 2575–2600, https://doi.org/10.1002/2014GC005431

Gratz, R. and Heinrich, W. 1997. Monazite–xenotime thermobarometry: experimental calibration of the miscibility gap in the binary system $CePO_4$–YPO_4. *American Mineralogist*, **82**, 772–780, https://doi.org/10.2138/am-1997-7-816

Gratz, R. and Heinrich, W. 1998. Monazite–xenotime thermometry. III. Experimental calibration of the partitioning of gadolinium between monazite and xenotime. *European Journal of Mineralogy*, **10**, 579–588, https://doi.org/10.1127/ejm/10/3/0579

Green, E.C.R., White, R.W., Diener, J.F.A., Powell, R., Holland, T.J.B. and Palin, R.M. 2016. Activity–composition relations for the calculation of partial melting equilibria in metabasic rocks. *Journal of Metamorphic Geology*, **34**, 845–869, https://doi.org/10.1111/jmg.12211

Gregory, C.J., Rubatto, D., Allen, C.M., Williams, I.S., Hermann, J. and Ireland, T. 2007. Allanite micro-geochronology: a LA-ICP-MS and SHRIMP U–Th–Pb study. *Chemical Geology*, **245**, 162–182, https://doi.org/10.1016/j.chemgeo.2007.07.029

Gregory, C.J., Rubatto, D., Hermann, J., Berger, A. and Engi, M. 2012. Allanite behaviour during incipient melting in the southern Central Alps. *Geochimica et Cosmochimica Acta*, **84**, 433–458, https://doi.org/10.1016/j.gca.2012.01.020

Grimes, C.B., John, B.E. *et al.* 2007. Trace element chemistry of zircons from oceanic crust: a method for distinguishing detrital zircon provenance. *Geology*, **35**, 643–646, https://doi.org/10.1130/G23603A.1

Grimes, C.B., Wooden, J.L., Cheadle, M.J. and John, B.E. 2015. 'Fingerprinting' tectono-magmatic provenance using trace elements in igneous zircon. *Contributions to Mineralogy and Petrology*, **170**, 1–26.

Gualda, G.A.R., Ghiorso, M.S., Lemons, R.V. and Carley, T.L. 2012. Rhyolite-MELTS: a modified calibration of MELTS optimized for silica-rich, fluid-bearing magmatic systems. *Journal of Petrology*, **53**, 875–890, https://doi.org/10.1093/petrology/egr080

Gudelius, D., Zeh, A., Almeev, R.R., Wilson, A.H., Fischer, L.A. and Schmitt, A.K. 2020. Zircon melt inclusions in mafic and felsic rocks of the Bushveld Complex – constraints for zircon crystallization temperatures and partition coefficients. *Geochimica et Cosmochimica Acta*, **289**, 158–181, https://doi.org/10.1016/j.gca.2020.08.027

Guitreau, M. and Flahaut, J. 2019. Record of low-temperature aqueous alteration of Martian zircon during the late Amazonian. *Nature Communications*, **10**, 1–9, https://doi.org/10.1038/s41467-019-10382-y

Guitreau, M., Blichert-Toft, J., Martin, H., Mojzsis, S.J. and Albarède, F. 2012. Hafnium isotope evidence from Archean granitic rocks for deep-mantle origin of continental crust. *Earth and Planetary Science Letters*, **337**, 211–223, https://doi.org/10.1016/j.epsl.2012.05.029

Guitreau, M., Mora, N. and Paquette, J. 2018. Crystallization and disturbance histories of single zircon crystals from Hadean–Eoarchean Acasta gneisses examined by LA-ICP-MS U–Pb traverses. *Geochemistry, Geophysics, Geosystems*, **19**, 272–291, https://doi.org/10.1002/2017GC007310

Guitreau, M., Boyet, M. *et al.* 2019. Hadean protocrust reworking at the origin of the Archean Napier Complex (Antarctica). *Geochemical Perspectives Letters*, **12**, 7–11, https://doi.org/10.7185/geochemlet.1927

Guitreau, M., Gannoun, A., Deng, Z., Marin-Carbonne, J., Chaussidon, M. and Moynier, F. 2020. Silicon isotope measurement in zircon by laser ablation multiple collector inductively coupled plasma mass spectrometry. *Journal of Analytical Atomic Spectrometry*, **35**, 1597–1606, https://doi.org/10.1039/D0JA00214C

Guitreau, M., Gannoun, A., Deng, Z., Chaussidon, M., Moynier, F., Barbarin, B. and Marin-Carbonne, J. 2022. Stable isotope geochemistry of silicon in granitoid zircon. *Geochimica et Cosmochimica Acta*, **316**, 273–294, https://doi.org/10.1016/j.gca.2021.09.029

Guo, J.-L., Wang, Z., Zhang, W., Moynier, F., Cui, D., Hu, Z. and Ducea, M.N. 2020. Significant Zr isotope variations in single zircon grains recording magma evolution history. *Proceedings of the National Academy of Sciences of the USA*, **117**, 21125–21131, https://doi.org/10.1073/pnas.2002053117

Gutieva, L., Dziggel, A., Volante, S. and Johnson, T. 2021. Zircon U–Pb and Lu–Hf record from the Archean Lewisian Gneiss Complex, NW Scotland. *Goldschmidt Conference 2021*, 4–9 July 2021, virtual event, https://doi.org/10.7185/gold2021.4082

Hacker, B.R., Kylander-Clark, A.R., Holder, R., Andersen, T.B., Peterman, E.M., Walsh, E.O. and Munnikhuis, J.K. 2015. Monazite response to ultrahigh-pressure subduction from U–Pb dating by laser ablation split stream. *Chemical Geology*, **409**, 28–41, https://doi.org/10.1016/j.chemgeo.2015.05.008

Hacker, B., Kylander-Clark, A. and Holder, R. 2019. REE partitioning between monazite and garnet: implications for petrochronology. *Journal of Metamorphic Geology*, **37**, 227–237, https://doi.org/10.1111/jmg.12458

Hagen-Peter, G., Cottle, J.M., Smit, M. and Cooper, A.F. 2016. Coupled garnet Lu–Hf and monazite U–Pb geochronology constrain early convergent margin dynamics in the Ross orogen, Antarctica. *Journal of Metamorphic Geology*, **34**, 293–319, https://doi.org/10.1111/jmg.12182

Hallett, B.W. and Spear, F.S. 2015. Monazite, zircon, and garnet growth in migmatitic pelites as a record of metamorphism and partial melting in the East Humboldt Range, Nevada. *American Mineralogist*, **100**, 951–972, https://doi.org/10.2138/am-2015-4839

Hammerli, J. and Kemp, T.I.S. 2021. Combined Hf and Nd isotope microanalysis of co-existing zircon and REE-rich accessory minerals: high resolution insights into crustal processes. *Chemical Geology*, **581**, article 120393, https://doi.org/10.1016/j.chemgeo.2021.120393

Hammerli, J., Kemp, A.I.S. and Spandler, C. 2014. Neodymium isotope equilibration during crustal metamorphism revealed by *in situ* microanalysis of REE-rich accessory minerals. *Earth and Planetary Science Letters*, **392**, 133–142, https://doi.org/10.1016/j.epsl.2014.02.018

Hammerli, J., Kemp, A.I.S. and Whitehouse, M.J. 2019. *In situ* trace element and Sm–Nd isotope analysis of accessory minerals in an Eoarchean tonalitic gneiss from Greenland: implications for Hf and Nd isotope decoupling in Earth's ancient rocks. *Chemical Geology*,

524, 394–405, https://doi.org/10.1016/j.chemgeo.2019.06.025

Hanchar, J.M. and Miller, C.F. 1993. Zircon zonation patterns as revealed by cathodoluminescence and backscattered electron images: implications for interpretation of complex crustal histories. *Chemical Geology*, **110**, 1–13, https://doi.org/10.1016/0009-2541(93)90244-D

Harley, S.L. 2008. Refining the P–T records of UHT crustal metamorphism. *Journal of Metamorphic Geology*, **26**, 125–154, https://doi.org/10.1111/j.1525-1314.2008.00765.x

Harley, S.L. 2016. A matter of time: the importance of the duration of UHT metamorphism. *Journal of Mineralogical and Petrological Sciences*, **111**, 50–72, https://doi.org/10.2465/jmps.160128

Harley, S.L. and Kelly, N.M. 2007. The impact of zircon–garnet REE distribution data on the interpretation of zircon U–Pb ages in complex high-grade terrains: an example from the Rauer Islands, East Antarctica. *Chemical Geology*, **241**, 62–87, https://doi.org/10.1016/j.chemgeo.2007.02.011

Harley, S.L. and Nandakumar, V. 2014. Accessory mineral behaviour in granulite migmatites: a case study from the Kerala Khondalite Belt, India. *Journal of Petrology*, **55**, 1965–2002, https://doi.org/10.1093/petrology/egu047

Harley, S.L., Kinny, P., Snape, I. and Black, L.P. 2001. Zircon chemistry and the definition of events in Archaean granulite terrains. *Extended Abstracts of the 4th International Archaean Symposium*, 24–28 September, Perth, Australia. AGSO-Geoscience Australia, 511–513.

Harlov, D.E. 2015. Apatite: a fingerprint for metasomatic processes. *Elements*, **11**, 171–176, https://doi.org/10.2113/gselements.11.3.171

Harlov, D.E. and Hetherington, C.J. 2010. Partial high-grade alteration of monazite using alkali-bearing fluids: experiment and nature. *American Mineralogist*, **95**, 1105–1108, https://doi.org/10.2138/am.2010.3525

Harlov, D.E., Wirth, R. and Hetherington, C.J. 2011. Fluid-mediated partial alteration in monazite: the role of coupled dissolution–reprecipitation in element redistribution and mass transfer. *Contributions to Mineralogy and Petrology*, **162**, 329–348, https://doi.org/10.1007/s00410-010-0599-7

Harris, C. and Vogeli, J. 2010. Oxygen isotope composition of garnet in the Peninsula Granite, Cape Granite Suite, South Africa: constraints on melting and emplacement mechanisms. *South African Journal of Geology*, **113**, 401–412, https://doi.org/10.2113/gssajg.113.4.401

Harrison, T.M. and Watson, E.B. 1984. The behavior of apatite during crustal anatexis: equilibrium and kinetic considerations. *Geochimica et Cosmochimica Acta*, **48**, 1467–1477, https://doi.org/10.1016/0016-7037(84)90403-4

Hart, E., Storey, C., Harley, S.L. and Fowler, M. 2018. A window into the lower crust: trace element systematics and the occurrence of inclusions/intergrowths in granulite-facies rutile. *Gondwana Research*, **59**, 76–86, https://doi.org/10.1016/j.gr.2018.02.021

Hartnady, M.I.H., Kirkland, C.L., Clark, C., Spaggiari, C.V., Smithies, R.H., Evans, N.J. and McDonald, B.J. 2019. Titanite dates crystallization: slow Pb diffusion during super-solidus re-equilibration. *Journal of Metamorphic Geology*, **37**, 823–838, https://doi.org/10.1111/jmg.12489

Hayden, L.A., Watson, E.B. and Wark, D.A. 2008. A thermobarometer for sphene (titanite). *Contributions to Mineralogy and Petrology*, **155**, 529–540, https://doi.org/10.1007/s00410-007-0256-y

Heinrich, W., Rehs, G. and Franz, G. 1997. Monazite–xenotime miscibility gap thermometry. I. An empirical calibration. *Journal of Metamorphic Geology*, **15**, 3–16, https://doi.org/10.1111/j.1525-1314.1997.t01-1-00052.x

Henrichs, I.A., O'Sullivan, G., Chew, D.M., Mark, C., Babechuk, M.G., McKenna, C. and Emo, R. 2018. The trace element and U–Pb systematics of metamorphic apatite. *Chemical Geology*, **483**, 218–238, https://doi.org/10.1016/j.chemgeo.2017.12.031

Henrichs, I.A., Chew, D.M., O'Sullivan, G.J., Mark, C., McKenna, C. and Guyett, P. 2019. Trace element (Mn–Sr–Y–Th–REE) and U–Pb isotope systematics of metapelitic apatite during progressive greenschist- to amphibolite-facies Barrovian metamorphism. *Geochemistry, Geophysics, Geosystems*, **20**, 4103–4129, https://doi.org/10.1029/2019GC008359

Hermann, J. 2002. Allanite: thorium and light rare earth element carrier in subducted crust. *Chemical Geology*, **192**, 289–306, https://doi.org/10.1016/S0009-2541(02)00222-X

Hermann, J. and Rubatto, D. 2003. Relating zircon and monazite domains to garnet growth zones: age and duration of granulite facies metamorphism in the Val Malenco lower crust. *Journal of Metamorphic Geology*, **21**, 833–852, https://doi.org/10.1046/j.1525-1314.2003.00484.x

Hermann, J. and Rubatto, D. 2009. Accessory phase control on the trace element signature of sediment melts in subduction zones. *Chemical Geology*, **265**, 512–526, https://doi.org/10.1016/j.chemgeo.2009.05.018

Hermann, J. and Rubatto, D. 2014. Subduction of continental crust to mantle depth: geochemistry of ultrahigh-pressure rocks. *In*: Holland, H.D. and Turekian, K.K. (eds) *Treatise on Geochemistry*. 2nd edn, 309–340, https://doi.org/10.1016/B978-0-08-095975-7.00309-0.

Hoffmann, J.E., Münker, C., Polat, A., Rosing, M.T. and Schulz, T. 2011. The origin of decoupled Hf–Nd isotope compositions in Eoarchean rocks from southern West Greenland. *Geochimica et Cosmochimica Acta*, **75**, 6610–6628, https://doi.org/10.1016/j.gca.2011.08.018

Hofmann, A.E., Baker, M.B. and Eiler, J.M. 2013. An experimental study of Ti and Zr partitioning among zircon, rutile, and granitic melt. *Contributions to Mineralogy and Petrology*, **166**, 235–253, https://doi.org/10.1007/s00410-013-0873-6

Högdahl, K., Majka, J., Sjöström, H., Nilsson, K.P., Claesson, S. and Konečný, P. 2012. Reactive monazite and robust zircon growth in diatexites and leucogranites from a hot, slowly cooled orogen: implications for the Palaeoproterozoic tectonic evolution of the central Fennoscandian Shield, Sweden. *Contributions to Mineralogy and Petrology*, **163**, 167–188, https://doi.org/10.1007/s00410-011-0664-x

Hokada, T. and Harley, S.L. 2004. Zircon growth in UHT leucosome: constraints from zircon-garnet rare earth elements (REE) relations in Napier Complex, East Antarctica. *Journal of Mineralogical and Petrological*

Sciences, **99**, 180–190, https://doi.org/10.2465/jmps.99.180

Holdaway, M. 2001. Recalibration of the GASP geobarometer in light of recent garnet and plagioclase activity models and versions of the garnet-biotite geothermometer. *American Mineralogist*, **86**, 1117–1129, https://doi.org/10.2138/am-2001-1001

Holder, R.M. and Hacker, B.R. 2019. Fluid-driven resetting of titanite following ultrahigh-temperature metamorphism in southern Madagascar. *Chemical Geology*, **504**, 38–52, https://doi.org/10.1016/j.chemgeo.2018.11.017

Holder, R.M., Hacker, B.R. and Kylander-Clark, A.R. 2013. Monazite petrochronology from the UHP Western Gneiss region, Norway. *American Geophysical Union Fall Meeting*, 9–13 December 2013, San Francisco, CA, Abstracts, *2013AGUFM. V23A2759H.*

Holder, R.M., Hacker, B.R., Seward, G.G.E. and Kylander-Clark, A.R.C. 2019. Interpreting titanite U–Pb dates and Zr thermobarometry in high-grade rocks: empirical constraints on elemental diffusivities of Pb, Al, Fe, Zr, Nb, and Ce. *Contributions to Mineralogy and Petrology*, **174**, article 42, https://doi.org/10.1007/s00410-019-1578-2

Holder, R.M., Yakymchuk, C. and Viete, D.R. 2020. Accessory mineral Eu anomalies in suprasolidus rocks: beyond feldspar. *Geochemistry, Geophysics, Geosystems*, **21**, article e2020GC009052, https://doi.org/10.1029/2020GC009052

Holland, H.D. and Gottfried, D. 1955. The effect of nuclear radiation on the structure of zircon. *Acta Crystallographica*, **8**, 291–300, https://doi.org/10.1107/S0365110X55000947

Holland, T.J.B. and Powell, R. 1990. An enlarged and updated internally consistent thermodynamic dataset with uncertainties and correlations: the system K_2O–Na_2O–CaO–MgO–MnO–FeO–Fe_2O_3–Al_2O_3–TiO_2–SiO_2–C–H_2–O_2. *Journal of Metamorphic Geology*, **8**, 89–124, https://doi.org/10.1111/j.1525-1314.1990.tb00458.x

Holland, T.J.B. and Powell, R. 1998. An internally consistent thermodynamic data set for phases of petrological interest. *Journal of Metamorphic Geology*, **16**, 309–344, https://doi.org/10.1111/j.1525-1314.1998.00140.x

Holland, T.J.B. and Powell, R. 2011. An improved and extended internally consistent thermodynamic dataset for phases of petrological interest, involving a new equation of state for solids. *Journal of Metamorphic Geology*, **29**, 333–383, https://doi.org/10.1111/j.1525-1314.2010.00923.x

Hollister, L.S. 1966. Garnet zoning: an interpretation based on the Rayleigh fractionation model. *Science*, **154**, 1647–1651, https://doi.org/10.1126/science.154.3757.1647

Holtmann, R., Muñoz-Montecinos, J. *et al.* 2022. Cretaceous thermal evolution of the closing Neo-Tethyan realm revealed by multi-method petrochronology. *Lithos*, **422–423**, 106731, https://doi.org/10.1016/j.lithos.2022.106731

Hoover, W.F., Penniston-Dorland, S.C., Baumgartner, L.P., Bouvier, A.-S., Baker, D., Dragovic, B. and Gion, A. 2021. A method for secondary ion mass spectrometry measurement of lithium isotopes in garnet: the utility of glass reference materials. *Geostandards and Geoanalytical Research*, **45**, 477–499, https://doi.org/10.1111/ggr.12383

Hoover, W.F., Penniston-Dorland, S. *et al.* 2022. Episodic fluid flow in an eclogite-facies shear zone: insights from Li isotope zoning in garnet. *Geology*, **50**, 746–750, https://doi.org/10.1130/G49737.1

Hopkins, M., Harrison, T.M. and Manning, C.E. 2008. Low heat flow inferred from >4 Gyr zircons suggests Hadean plate boundary interactions. *Nature*, **456**, 493–496, https://doi.org/10.1038/nature07465

Horton, F., Holder, R.M. and Swindle, C.R. 2022. An extensive record of orogenesis recorded in a Madagascar granulite. *Journal of Metamorphic Geology*, **40**, 287–305, https://doi.org/10.1111/jmg.12628.

Hoschek, G. 2016. Phase relations of the REE minerals florencite, allanite and monazite in quartzitic garnet–kyanite schist of the Eclogite Zone, Tauern Window, Austria. *European Journal of Mineralogy*, **28**, 735–750.

Hoshino, M., Kimata, M., Arakawa, Y., Shimizu, M., Nishida, N. and Nakai, S. 2007. Allanite-(Ce) as an indicator of the origin of granitic rocks in Japan: importance of Sr–Nd isotopic and chemical composition. *The Canadian Mineralogist*, **45**, 1329–1336, https://doi.org/10.3749/canmin.45.6.1329

Hoskin, P.W.O. and Black, L.P. 2000. Metamorphic zircon formation by solid-state recrystallization of protolith igneous zircon. *Journal of Metamorphic Geology*, **18**, 423–439, https://doi.org/10.1046/j.1525-1314.2000.00266.x

Hoskin, P.W.O. and Schaltegger, U. 2003. The composition of zircon and igneous and metamorphic petrogenesis. *Reviews in Mineralogy and Geochemistry*, **53**, 27–62, https://doi.org/10.2113/0530027

Humphries, F.J. and Cliff, R.A. 1982. Sm–Nd dating and cooling history of Scourian granulites, Sutherland. *Nature*, **295**, 515–517, https://doi.org/10.1038/295515a0

Ibañez-Mejia, M. and Tissot, F.L. 2019. Extreme Zr stable isotope fractionation during magmatic fractional crystallization. *Science Advances*, **5**, article eaax8648, https://doi.org/10.1126/sciadv.aax8648

Ihlen, P.M., Schiellerup, H., Gautneb, H. and Skår, Ø. 2014. Characterization of apatite resources in Norway and their REE potential – a review. *Ore Geology Reviews*, **58**, 126–147, https://doi.org/10.1016/j.oregeorev.2013.11.003

Iizuka, T., Hirata, T., Komiya, T., Rino, S., Katayama, I., Motoki, A. and Maruyama, S. 2005. U–Pb and Lu–Hf isotope systematics of zircons from the Mississippi River sand: implications for reworking and growth of continental crust. *Geology*, **33**, 485–488, https://doi.org/10.1130/G21427.1

Iizuka, T., Campbell, I.H., Allen, C.M., Gill, J.B., Maruyama, S. and Makoka, F. 2013. Evolution of the African continental crust as recorded by U–Pb, Lu–Hf and O isotopes in detrital zircons from modern rivers. *Geochimica et Cosmochimica Acta*, **107**, 96–120, https://doi.org/10.1016/j.gca.2012.12.028

Iizuka, T., Yamaguchi, T., Hibiya, Y. and Amelin, Y. 2015. Meteorite zircon constraints on the bulk Lu−Hf isotope composition and early differentiation of the Earth. *Proceedings of the National Academy of Sciences of the*

USA, **112**, 5331–5336, https://doi.org/10.1073/pnas.1501658112

Itano, K., Ueki, K., Iizuka, T. and Kuwatani, T. 2020. Geochemical discrimination of monazite source rock based on machine learning techniques and multinomial logistic regression analysis. *Geosciences*, **10**, article 63, https://doi.org/10.3390/geosciences10020063

Jaffey, A., Flynn, K., Glendenin, L., Bentley, W.T. and Essling, A. 1971. Precision measurement of half-lives and specific activities of ^{235}U and ^{238}U. *Physical Review C*, **4**, article 1889, https://doi.org/10.1103/PhysRevC.4.1889

Jamieson, J.C. and Olinger, B. 1969. Pressure–temperature studies of anatase, brookite rutile, and Ti02(II): a discussion. *American Mineralogist*, **54**, 1477–1481.

Janots, E. and Rubatto, D. 2014. U–Th–Pb dating of collision in the external Alpine domains (Urseren zone, Switzerland) using low temperature allanite and monazite. *Lithos*, **184–187**, 155–166, https://doi.org/10.1016/j.lithos.2013.10.036

Janots, E., Negro, F., Brunet, F., Goffé, B., Engi, M. and Bouybaouène, M.L. 2006. Evolution of the REE mineralogy in HP–LT metapelites of the Sebtide Complex, Rif, Morocco: monazite stability and geochronology. *Lithos*, **87**, 214–234, https://doi.org/10.1016/j.lithos.2005.06.008

Janots, E., Brunet, F., Goffé, B., Poinssot, C., Burchard, M. and Cemič, L. 2007. Thermochemistry of monazite-(La) and dissakisite-(La): implications for monazite and allanite stability in metapelites. *Contributions to Mineralogy and Petrology*, **154**, 1–14, https://doi.org/10.1007/s00410-006-0176-2

Janots, E., Engi, M., Berger, A., Allaz, J., Schwarz, J. and Spandler, C. 2008. Prograde metamorphic sequence of REE minerals in pelitic rocks of the Central Alps: implications for allanite–monazite–xenotime phase relations from 250 to 610°C. *Journal of Metamorphic Geology*, **26**, 509–526, https://doi.org/10.1111/j.1525-1314.2008.00774.x

Janots, E., Engi, M., Rubatto, D., Berger, A., Gregory, C. and Rahn, M. 2009. Metamorphic rates in collisional orogeny from *in situ* allanite and monazite dating. *Geology*, **37**, 11–14, https://doi.org/10.1130/G25192A.1

Janots, E., Berger, A. and Engi, M. 2011. Physico-chemical control on the REE minerals in chloritoid-grade metasediments from a single outcrop (Central Alps, Switzerland). *Lithos*, **121**, 1–11, https://doi.org/10.1016/j.lithos.2010.08.023

Jayananda, M., Chardon, D., Peucat, J.-J. and Fanning, C.M. 2015. Paleo- to Mesoarchean TTG accretion and continental growth in the western Dharwar craton, southern India: constraints from SHRIMP U–Pb zircon geochronology, whole-rock geochemistry and Nd–Sr isotopes. *Precambrian Research*, **268**, 295–322, https://doi.org/10.1016/j.precamres.2015.07.015

Jennings, E.S., Marschall, H., Hawkesworth, C. and Storey, C. 2011. Characterization of magma from inclusions in zircon: apatite and biotite work well, feldspar less so. *Geology*, **39**, 863–866, https://doi.org/10.1130/G32037.1

Jiao, S., Fitzsimons, I.C.W., Zi, J.-W., Evans, N.J., Mcdonald, B.J. and Guo, J. 2020*a*. Texturally controlled U–Th–Pb monazite geochronology reveals Paleoproterozoic UHT metamorphic evolution in the Khondalite Belt, North China Craton. *Journal of Petrology*, **61**, article egaa023, https://doi.org/10.1093/petrology/egaa023

Jiao, S., Guo, J., Evans, N.J., Mcdonald, B.J., Liu, P., Ouyang, D. and Fitzsimons, I.C.W. 2020*b*. The timing and duration of high-temperature to ultrahigh-temperature metamorphism constrained by zircon U–Pb–Hf and trace element signatures in the Khondalite Belt, North China Craton. *Contributions to Mineralogy and Petrology*, **175**, article 66, https://doi.org/10.1007/s00410-020-01706-z

Jiao, S., Evans, N.J., Mitchell, R.N., Fitzsimons, I.C.W. and Guo, J. 2021. Heavy rare-earth element and Y partitioning between monazite and garnet in aluminous granulites. *Contributions to Mineralogy and Petrology*, **176**, article 50, https://doi.org/10.1007/s00410-021-01808-2

John, T., Klemd, R., Klemme, S., Pfänder, J.A., Elis Hoffmann, J. and Gao, J. 2011. Nb–Ta fractionation by partial melting at the titanite–rutile transition. *Contributions to Mineralogy and Petrology*, **161**, 35–45, https://doi.org/10.1007/s00410-010-0520-4

Johnson, S.E. 1999. Porphyroblast microstructures: a review of current and future trends. *American Mineralogist*, **84**, 1711–1726, https://doi.org/10.2138/am-1999-11-1202

Johnson, T.A., Vervoort, J.D., Ramsey, M.J., Aleinikoff, J.N. and Southworth, S. 2018. Constraints on the timing and duration of orogenic events by combined Lu–Hf and Sm–Nd geochronology: an example from the Grenville orogeny. *Earth and Planetary Science Letters*, **501**, 152–164, https://doi.org/10.1016/j.epsl.2018.08.030

Jonnalagadda, M.K., Karmalkar, N.R., Duraiswami, R.A., Harshe, S., Gain, S. and Griffin, W.L. 2017. Formation of atoll garnets in the UHP eclogites of the Tso Morari Complex, Ladakh, Himalaya. *Journal of Earth System Science*, **126**, article 107, https://doi.org/10.1007/s12040-017-0887-y

Joseph, C., Fougerouse, D., Saxey, D.W., Verberne, R., Reddy, S.M. and Rickard, W.D.A. 2021. Xenotime at the nanoscale: U–Pb geochronology and optimisation of analyses by atom probe tomography. *Geostandards and Geoanalytical Research*, **45**, 443–456, https://doi.org/10.1111/ggr.12398

Kamber, B.S., Frei, R. and Gibb, A.J. 1998. Pitfalls and new approaches in granulite chronometry: an example from the Limpopo Belt, Zimbabwe. *Precambrian Research*, **91**, 269–285, https://doi.org/10.1016/S0301-9268(98)00053-9

Kamenetsky, V.S., Doroshkevich, A.G., Elliott, H.A.L. and Zaitsev, A.N. 2021. Carbonatites: contrasting, complex, and controversial. *Elements*, **17**, 307–314, https://doi.org/10.2138/gselements.17.5.307

Kapp, P., Manning, C.E. and Tropper, P. 2009. Phase-equilibrium constraints on titanite and rutile activities in mafic epidote amphibolites and geobarometry using titanite–rutile equilibria. *Journal of Metamorphic Geology*, **27**, 509–521, https://doi.org/10.1111/j.1525-1314.2009.00836.x

Karioris, F., Gowda, K.A. and Cartz, L. 1981. Heavy ion bombardment of monoclinic $ThSiO_4$, ThO_2 and monazite. *Radiation Effects*, **58**, 1–3, https://doi.org/10.1080/01422448108226520

Kelly, N.M. and Harley, S.L. 2005. An integrated microtextural and chemical approach to zircon geochronology: refining the Archaean history of the Napier Complex, east Antarctica. *Contributions to Mineralogy and Petrology*, **149**, 57–84, https://doi.org/10.1007/s00410-004-0635-6

Kelly, N.M., Harley, S.L. and Möller, A. 2012. Complexity in the behavior and recrystallization of monazite during high-T metamorphism and fluid infiltration. *Chemical Geology*, **322**, 192–208, https://doi.org/10.1016/j.chemgeo.2012.07.001

Kelsey, D.E. and Powell, R. 2011. Progress in linking accessory mineral growth and breakdown to major mineral evolution in metamorphic rocks: A thermodynamic approach in the Na_2O-CaO-K_2O-FeO-MgO-Al_2O_3-SiO_2-H_2O-TiO_2-ZrO_2 system. *Journal of Metamorphic Geology*, **29**, 151–166.

Kelsey, D.E., Hand, M., Clark, C. and Wilson, C.J.L. 2007. On the application of *in situ* monazite chemical geochronology to constraining P–T–t histories in high-temperature (>850°C) polymetamorphic granulites from Prydz Bay, East Antarctica. *Journal of the Geological Society, London*, **164**, 667–683, https://doi.org/10.1144/0016-76492006-013

Kelsey, D.E., Clark, C. and Hand, M. 2008. Thermobarometric modelling of zircon and monazite growth in melt-bearing systems: examples using model metapelitic and metapsammitic granulites. *Journal of Metamorphic Geology*, **26**, 199–212, https://doi.org/10.1111/j.1525-1314.2007.00757.x

Kemp, A., Hawkesworth, C. *et al.* 2007. Magmatic and crustal differentiation history of granitic rocks from Hf–O isotopes in zircon. *Science*, **315**, 980–983, https://doi.org/10.1126/science.1136154

Kennedy, A.K., Wotzlaw, J.-F. *et al.* 2022. Apatite reference materials for SIMS microanalysis of isotopes and trace elements. *Geostandards and Geoanalytical Research*, **47**, 215–467, https://doi.org/10.1111/ggr.12477

King, E.M. and Valley, J.W. 2001. The source, magmatic contamination, and alteration of the Idaho wbatholith. *Contributions to Mineralogy and Petrology*, **142**, 72–88, https://doi.org/10.1007/s004100100278

King, E.M., Valley, J.W., Davis, D.W. and Kowallis, B.J. 2001. Empirical determination of oxygen isotope fractionation factors for titanite with respect to zircon and quartz. *Geochimica et Cosmochimica Acta*, **65**, 3165–3175, https://doi.org/10.1016/S0016-7037(01)00639-1

King, P.L., Sham, T.-K., Gordon, R.A. and Dyar, M.D. 2013. Microbeam X-ray analysis of Ce^{3+}/Ce^{4+} in Ti-rich minerals: a case study with titanite (sphene) with implications for multivalent trace element substitution in minerals. *American Mineralogist*, **98**, 110–119, https://doi.org/10.2138/am.2013.3959

Kingsbury, J.A., Miller, C.F., Wooden, J.L. and Harrison, T.M. 1993. Monazite paragenesis and U–Pb systematics in rocks of the eastern Mojave Desert, California, USA: implications for thermochronometry. *Chemical Geology*, **110**, 147–167, https://doi.org/10.1016/0009-2541(93)90251-D

Kirkland, C.L., Smithies, R.H., Taylor, R.J.M., Evans, N. and McDonald, B. 2015. Zircon Th/U ratios in magmatic environs. *Lithos*, **212**, 397–414, https://doi.org/10.1016/j.lithos.2014.11.021

Kirkland, C.L., Spaggiari, C.V. *et al.* 2016. Grain size matters: Implications for element and isotopic mobility in titanite. *Precambrian Research*, **278**, 283–302, https://doi.org/10.1016/j.precamres.2016.03.002.

Kirkland, C.L., Hollis, J., Danisik, M., Petersen, J., Evans, N. and Mcdonald, B. 2017. Apatite and titanite from the Karrat Group, Greenland; implications for charting the thermal evolution of crust from the U–Pb geochronology of common Pb bearing phases. *Precambrian Research*, **300**, 107–120, https://doi.org/10.1016/j.precamres.2017.07.033

Kirkland, C.L., Fougerouse, D., Reddy, S.M., Hollis, J. and Saxey, D.W. 2018. Assessing the mechanisms of common Pb incorporation into titanite. *Chemical Geology*, **483**, 558–566, https://doi.org/10.1016/j.chemgeo.2018.03.026

Kirkland, C.L., Yakymchuk, C., Gardiner, N.J., Szilas, K., Hollis, J., Olierook, H. and Steenfelt, A. 2020. Titanite petrochronology linked to phase equilibrium modelling constrains tectono-thermal events in the Akia Terrane, West Greenland. *Chemical Geology*, **536**, article, 119467, https://doi.org/10.1016/j.chemgeo.2020.119467

Kirkland, C.L., Hartnady, M.I.H., Barham, M., Olierook, H.K.H., Steenfelt, A. and Hollis, J.A. 2021. Widespread reworking of Hadean-to-Eoarchean continents during Earth's thermal peak. *Nature Communications*, **12**, article 331, https://doi.org/10.1038/s41467-020-20514-4

Klemme, S., Prowatke, S., Hametner, K. and Günther, D. 2005. Partitioning of trace elements between rutile and silicate melts: implications for subduction zones. *Geochimica et Cosmochimica Acta*, **69**, 2361–2371, https://doi.org/10.1016/j.gca.2004.11.015

Kohn, M.J. 2017. Titanite petrochronology. *Reviews in Mineralogy and Geochemistry*, **83**, 419–441, https://doi.org/10.2138/rmg.2017.83.13

Kohn, M.J. 2020. A refined zirconium-in-rutile thermometer. *American Mineralogist: Journal of Earth and Planetary Materials*, **105**, 963–971, https://doi.org/10.2138/am-2020-7091

Kohn, M.J. and Corrie, S.L. 2011. Preserved Zr-temperatures and U–Pb ages in high-grade metamorphic titanite: evidence for a static hot channel in the Himalayan orogen. *Earth and Planetary Science Letters*, **311**, 136–143, https://doi.org/10.1016/j.epsl.2011.09.008

Kohn, M.J. and Valley, J.W. 1998. Effects of cation substitutions in garnet and pyroxene on equilibrium oxygen isotope fractionations. *Journal of Metamorphic Geology*, **16**, 625–639, https://doi.org/10.1111/j.1525-1314.1998.00162.x

Kohn, M.J., Penniston-Dorland, S.C. and Ferreira, J.C.S. 2016. Implications of near-rim compositional zoning in rutile for geothermometry, geospeedometry, and trace element equilibration. *Contributions to Mineralogy and Petrology*, **171**, article 78, https://doi.org/10.1007/s00410-016-1285-1

Kohn, M.J., Engi, M. and Lanari, P. 2017. Petrochronology: methods and applications. *Reviews in Mineralogy and Geochemistry*, **83**, https://doi.org/10.1515/9783110561890

Konečný, P., Kusiak, M.A. and Dunkley, D.J. 2018. Improving U–Th–Pb electron microprobe dating using monazite age references. *Chemical Geology*, **484**, 22– 35, https://doi.org/10.1016/j.chemgeo.2018.02.014

Kooijman, E., Mezger, K. and Berndt, J. 2010. Constraints on the U–Pb systematics of metamorphic rutile from *in situ* LA-ICP-MS analysis. *Earth and Planetary Science Letters*, **293**, 321–330, https://doi.org/10.1016/j.epsl.2010.02.047

Kooijman, E., Smit, M.A., Mezger, K. and Berndt, J. 2012. Trace element systematics in granulite facies rutile: implications for Zr geothermometry and provenance studies. *Journal of Metamorphic Geology*, **30**, 397–412, https://doi.org/10.1111/j.1525-1314.2012.00972.x

Košler, J., Wiedenbeck, M., Wirth, R., Hovorka, J., Sylvester, P. and Míková, J. 2005. Chemical and phase composition of particles produced by laser ablation of silicate glass and zircon–implications for elemental fractionation during ICP-MS analysis. *Journal of Analytical Atomic Spectrometry*, **20**, 402–409, https://doi.org/10.1039/B416269B

Krenn, E. and Finger, F. 2010. Unusually Y-rich monazite-(Ce) with 6–14 wt.% Y_2O_3 in a granulite from the Bohemian Massif: implications for high-temperature monazite growth from the monazite–xenotime miscibility gap thermometry. *Mineralogical Magazine*, **74**, 217–225, https://doi.org/10.1180/minmag.2010.073.2.217

Krenn, E., Schulz, B. and Finger, F. 2012. Three generations of monazite in Austroalpine basement rocks to the south of the Tauern Window: evidence for Variscan, Permian and Eo-Alpine metamorphic events. *Swiss Journal of Geosciences*, **105**, 343–360, https://doi.org/10.1007/s00015-012-0104-6

Krneta, S., Ciobanu, C.L., Cook, N.J. and Ehrig, K.J. 2018. Numerical modeling of REE fractionation patterns in fluorapatite from the Olympic Dam Deposit (South Australia). *Minerals*, **8**, article 342, https://doi.org/10.3390/min8080342

Krogstad, E.J. and Walker, R.J. 1994. High closure temperatures of the U–Pb system in large apatites from the Tin Mountain pegmatite, Black Hills, South Dakota, USA. *Geochimica et Cosmochimica Acta*, **58**, 3845–3853, https://doi.org/10.1016/0016-7037(94)90367-0

Kröner, A., Wan, Y., Liu, X. and Liu, D. 2014. Dating of zircon from high-grade rocks: which is the most reliable method? *Geoscience Frontiers*, **5**, 515–523, https://doi.org/10.1016/j.gsf.2014.03.012

Kulhánek, J., Faryad, S.W., Jedlicka, R. and Svojtka, M. 2021. Dissolution and reprecipitation of garnet during eclogite-facies metamorphism; major and trace element transfer during atoll garnet formation. *Journal of Petrology*, **62**, article egab077, https://doi.org/10.1093/petrology/egab077

Kunz, B.E., Regis, D. and Engi, M. 2018. Zircon ages in granulite facies rocks: decoupling from geochemistry above 850°C? *Contributions to Mineralogy and Petrology*, **173**, article 26, https://doi.org/10.1007/s00410-018-1454-5

Kusiak, M.A., Whitehouse, M.J., Wilde, S.A., Nemchin, A.A. and Clark, C. 2013. Mobilization of radiogenic Pb in zircon revealed by ion imaging: implications for early Earth geochronology. *Geology*, **41**, 291–294, https://doi.org/10.1130/G33920.1

Kylander-Clark, A.R.C., Hacker, B.R. and Mattinson, J.M. 2008. Slow exhumation of UHP terranes: titanite and rutile ages of the Western Gneiss Region, Norway. *Earth and Planetary Science Letters*, **272**, 531–540, https://doi.org/10.1016/j.epsl.2008.05.019

Kylander-Clark, A.R.C., Hacker, B.R. and Cottle, J.M. 2013. Laser-ablation split-stream ICP petrochronology. *Chemical Geology*, **345**, 99–112, https://doi.org/10.1016/J.CHEMGEO.2013.02.019

Lackey, J.S., Valley, J.W. and Hinke, H.J. 2006. Deciphering the source and contamination history of peraluminous magmas using $\delta^{18}O$ of accessory minerals: examples from garnet-bearing plutons of the Sierra Nevada batholith. *Contributions to Mineralogy and Petrology*, **151**, 20–44, https://doi.org/10.1007/s00410-005-0043-6

Lana, C., Gonçalves, G.O. *et al.* 2022. Assessing the U–Pb, Sm–Nd and Sr–Sr isotopic compositions of the Sumé apatite as a reference material for LA-ICP-MS analysis. *Geostandards and Geoanalytical Research*, **46**, 71–95, https://doi.org/10.1111/ggr.12413

Lanari, P. and Engi, M. 2017. Local bulk composition effects on metamorphic mineral assemblages. *Reviews in Mineralogy and Geochemistry*, **83**, 55–102, https://doi.org/10.2138/rmg.2017.83.3

Lanari, P and Piccoli, F 2020. *IOP Conference Series: Materials Science and Engineering*, **891**, 012016, https://doi.org/10.1088/1757-899X/891/1/012016

Lanari, P., Vidal, O., De Andrade, V., Dubacq, B., Lewin, E., Grosch, E.G. and Schwartz, S. 2014. XMapTools: a MATLAB©-based program for electron microprobe X-ray image processing and geothermobarometry. *Computers & Geosciences*, **62**, 227–240, https://doi.org/10.1016/j.cageo.2013.08.010

Lanari, P., Giuntoli, F., Loury, C., Burn, M. and Engi, M. 2017. An inverse modeling approach to obtain P–T conditions of metamorphic stages involving garnet growth and resorption. *European Journal of Mineralogy*, **29**, 181–199, https://doi.org/10.1127/ejm/2017/0029-2597

Lanari, P., Ferrero, S., Goncalves, P. and Grosch, E.G. 2019. Metamorphic geology: progress and perspectives. *Geological Society, London, Special Publications*, **478**, 1–12, https://doi.org/10.1144/SP478-2018-186

Larson, K.P., Ali, A., Shrestha, S., Soret, M., Cottle, J.M. and Ahmad, R. 2019. Timing of metamorphism and deformation in the Swat Valley, northern Pakistan: insight into garnet-monazite HREE partitioning. *Geoscience Frontiers*, **10**, 849–861, https://doi.org/10.1016/j.gsf.2018.02.008

Larson, K.P., Shrestha, S., Cottle, J.M., Guilmette, C., Johnson, T.A., Gibson, H.D. and Gervais, F. 2022. Re-evaluating monazite as a record of metamorphic reactions. *Geoscience Frontiers*, **13**, article 101340, https://doi.org/10.1016/j.gsf.2021.101340

Larsson, D. and Söderlund, U. 2005. Lu–Hf apatite geochronology of mafic cumulates: an example from a Fe–Ti mineralization at Smålands Taberg, southern Sweden. *Chemical Geology*, **224**, 201–211, https://doi.org/10.1016/j.chemgeo.2005.07.007

Laurent, A.T., Duchene, S., Bingen, B., Bosse, V. and Seydoux-Guillaume, A.-M. 2018. Two successive phases of ultrahigh temperature metamorphism in Rogaland, S. Norway: evidence from Y-in-monazite thermometry. *Journal of Metamorphic Geology*, **36**, 1009–1037, https://doi.org/10.1111/jmg.12425

Laurent, O., Zeh, A., Gerdes, A., Villaros, A., Gros, K. and Słaby, E. 2017. How do granitoid magmas mix with each other? Insights from textures, trace element and Sr–Nd isotopic composition of apatite and titanite from the Matok pluton (South Africa). *Contributions to Mineralogy and Petrology*, **172**, article 80, https://doi.org/10.1007/s00410-017-1398-1

Laurent, O., Moyen, J.-F., Wotzlaw, J.-F., Björnsen, J. and Bachmann, O. 2021. Early Earth zircons formed in residual granitic melts produced by tonalite differentiation. *Geology*, **50**, 437–441, https://doi.org/10.1130/G49232.1

Lawley, C.J.M., Creaser, R.A. *et al.* 2015. Unraveling the western Churchill Province Paleoproterozoic gold metallotect: constraints from Re–Os arsenopyrite and U–Pb xenotime geochronology and LA-ICP-MS arsenopyrite trace element chemistry at the BIF-hosted Meliadine gold district, Nunavut, Canada. *Economic Geology*, **110**, 1425–1454, https://doi.org/10.2113/econgeo.110.6.1425

Lee, J.K.W., Williams, I.S. and Ellis, D.J. 1997. Pb, U and Th diffusion in natural zircon. *Nature*, **390**, 159–162, https://doi.org/10.1038/36554

Lei, H., Xu, H. and Liu, P. 2020. Decoupling between Ti-in-zircon and Zr-in-rutile thermometry during ultrahigh temperature metamorphism of the Dabie Orogen, China. *Geological Journal*, **55**, 6442–6449, https://doi.org/10.1002/gj.3819

Le Roux, L. and Glendenin, L. 1963. Half-life of ^{232}Th. In: *Proceedings of the National Meeting on Nuclear Energy*, Pretoria, South Africa. **83**, 94.

Li, B., Ge, J. and Zhang, B. 2018. Diffusion in garnet: a review. *Acta Geochimica*, **37**, 19–31, https://doi.org/10.1007/s11631-017-0187-x

Li, D., Fu, Y. *et al.* 2022. PL57 garnet as a new natural reference material for *in situ* U–Pb isotope analysis and its perspective for geological applications. *Contributions to Mineralogy and Petrology*, **177**, article 19, https://doi.org/10.1007/s00410-021-01884-4

Li, R., Collins, W.J., Yang, J.-H., Blereau, E. and Wang, H. 2021. Two-stage hybrid origin of Lachlan S-type magmas: a re-appraisal using isotopic microanalysis of lithic inclusion minerals. *Lithos*, **402–403**, article 106378, https://doi.org/10.1016/j.lithos.2021.106378

Li, W.-T., Jiang, S.-Y., Fu, B., Liu, D.-L. and Xiong, S.-F. 2021. Zircon Hf–O isotope and magma oxidation state evidence for the origin of Early Cretaceous granitoids and porphyry Mo mineralization in the Tongbai–Hong'an–Dabie orogens, eastern China. *Lithos*, **398**, article 106281, https://doi.org/10.1016/j.lithos.2021.106281

Liao, X., Li, Q., Whitehouse, M.J., Yang, Y. and Liu, Y. 2020. Allanite U–Th–Pb geochronology by ion microprobe. *Journal of Analytical Atomic Spectrometry*, **35**, 489–497, https://doi.org/10.1039/C9JA00426B

Liebmann, J., Spencer, C.J., Kirkland, C.L., Xia, X.-P. and Bourdet, J. 2021. Effect of water on δ^{18}O in zircon. *Chemical Geology*, **574**, article 120243, https://doi.org/10.1016/j.chemgeo.2021.120243

Lin, S., Hu, K., Cao, J., Liu, Y., Liu, S. and Zhang, B. 2023. Geochemistry and origin of hydrothermal apatite in Carlin-type Au deposits, southwestern China (Gaolong deposit). *Ore Geology Reviews*, **154**, article 105312, https://doi.org/10.1016/j.oregeorev.2023.105312

Liu, Z.-C., Wu, F.-Y., Ji, W.-Q., Wang, J.-G. and Liu, C.-Z. 2014. Petrogenesis of the Ramba leucogranite in the Tethyan Himalaya and constraints on the channel flow model. *Lithos*, **208–209**, 118–136, https://doi.org/10.1016/j.lithos.2014.08.022

Loader, M.A., Nathwani, C.L., Wilkinson, J.J. and Armstrong, R.N. 2022. Controls on the magnitude of Ce anomalies in zircon. *Geochimica et Cosmochimica Acta*, **328**, 242–257, https://doi.org/10.1016/j.gca.2022.03.024

López-Moro, F.J., Romer, R., López-Plaza, M. and Sanchez, M.G. 2017. Zircon and allanite U–Pb ID-TIMS ages of vaugnerites from the Calzadilla pluton, Salamanca (Spain): dating mantle-derived magmatism and post-magmatic subsolidus overprint. *Geologica Acta*, **15**, 395–408, https://doi.org/10.1344/GeologicaActa2017.15.4.9

Lü, Z., Zhang, L., Du, J. and Bucher, K. 2008. Coesite inclusions in garnet from eclogitic rocks in western Tianshan, northwest China: convincing proof of UHP metamorphism. *American Mineralogist*, **93**, 1845–1850, https://doi.org/10.2138/am.2008.2800

Lucassen, F., Franz, G., Dulski, P., Romer, R.L. and Rhede, D. 2011. Element and Sr isotope signatures of titanite as indicator of variable fluid composition in hydrated eclogite. *Lithos*, **121**, 12–24, https://doi.org/10.1016/j.lithos.2010.09.018

Lucassen, F., Franz, G. and Rhede, D. 2012. Small-scale transport of trace elements Nb and Cr during growth of titanite: an experimental study at 600°C, 0.4 GPa. *Contributions to Mineralogy and Petrology*, **164**, 987–997, https://doi.org/10.1007/s00410-012-0784-y

Luvizotto, G.L. and Zack, T. 2009. Nb and Zr behavior in rutile during high-grade metamorphism and retrogression: an example from the Ivrea–Verbano Zone. *Chemical Geology*, **261**, 303–317, https://doi.org/10.1016/j.chemgeo.2008.07.023

Luvizotto, G.L., Zack, T. *et al.* 2009. Rutile crystals as potential trace element and isotope mineral standards for microanalysis. *Chemical Geology*, **261**, 346–369, https://doi.org/10.1016/j.chemgeo.2008.04.012

Ma, Q., Evans, N.J., Ling, X.-X., Yang, J.-H., Wu, F.-Y., Zhao, Z.-D. and Yang, Y.-H. 2019. Natural titanite reference materials for *in situ* U–Pb and Sm–Nd isotopic measurements by LA-(MC)-ICP-MS. *Geostandards and Geoanalytical Research*, **43**, 355–384, https://doi.org/10.1111/ggr.12264

Machado, N. and Gauthier, G. 1996. Determination of ^{207}Pb/^{206}Pb ages on zircon and monazite by laser-ablation ICPMS and application to a study of sedimentary provenance and metamorphism in southeastern Brazil. *Geochimica et Cosmochimica Acta*, **60**, 5063–5073, https://doi.org/10.1016/S0016-7037(96)00287-6

Mahan, K.H., Goncalves, P., Williams, M.L. and Jercinovic, M.J. 2006. Dating metamorphic reactions and fluid flow: application to exhumation of high-P granulites in a crustal-scale shear zone, western Canadian Shield. *Journal of Metamorphic Geology*, **24**, 193–217, https://doi.org/10.1111/j.1525-1314.2006.00633.x

Maneiro, K.A., Baxter, E.F., Samson, S.D., Marschall, H.R. and Hietpas, J. 2019. Detrital garnet geochronology: application in tributaries of the French Broad River, southern Appalachian Mountains, USA. *Geology*, **47**, 1189–1192, https://doi.org/10.1130/G46840.1

Manzotti, P., Bosse, V., Pitra, P., Robyr, M., Schiavi, F. and Ballèvre, M. 2018. Exhumation rates in the Gran Paradiso Massif (Western Alps) constrained by *in situ* U–Th–Pb dating of accessory phases (monazite, allanite and xenotime). *Contributions to Mineralogy and Petrology*, **173**, article 24, https://doi.org/10.1007/s00410-018-1452-7

Manzotti, P., Schiavi, F., Nosenzo, F., Pitra, P. and Ballèvre, M. 2022. A journey towards the forbidden zone: a new, cold, UHP unit in the Dora-Maira Massif (Western Alps). *Contributions to Mineralogy and Petrology*, **177**, 1–22, https://doi.org/10.1007/s00410-022-01923-8

Marfin, A.E., Ivanov, A.V., Kamenetsky, V.S., Abersteiner, A., Yakich, T.Y. and Dudkin, T.V. 2020. Contact metamorphic and metasomatic processes at the Kharaelakh intrusion, Oktyabrsk deposit, Norilsk–Talnakh ore district: application of LA-ICP-MS dating of perovskite, apatite, garnet, and titanite. *Economic Geology*, **115**, 1213–1226, https://doi.org/10.5382/econgeo.4744

Mark, C., Stutenbecker, L. *et al.* 2022. Detrital garnet Lu–Hf and U–Pb geochronometry coupled with compositional analysis: possibilities and limitations as a sediment provenance indicator. *EGU General Assembly, Conference Abstracts*, 23–27 May 2022, Vienna, Austria, EGU22-6405, https://doi.org/10.5194/egusphere-egu22-6405.

Mark, G. 2001. Nd isotope and petrogenetic constraints for the origin of the Mount Angelay igneous complex: implications for the origin of intrusions in the Cloncurry district, NE Australia. *Precambrian Research*, **105**, 17–35, https://doi.org/10.1016/S0301-9268(00)00101-7

Marmo, B., Clarke, G. and Powell, R. 2002. Fractionation of bulk rock composition due to porphyroblast growth: effects on eclogite facies mineral equilibria, Pam Peninsula, New Caledonia. *Journal of Metamorphic Geology*, **20**, 151–165, https://doi.org/10.1046/j.0263-4929.2001.00346.x

Marsh, J.H. and Smye, A.J. 2017. U–Pb systematics and trace element characteristics in titanite from a high-pressure mafic granulite. *Chemical Geology*, **466**, 403–416, https://doi.org/10.1016/j.chemgeo.2017.06.029

Martin, A.J., Gehrels, G.E. and DeCelles, P.G. 2007. The tectonic significance of (U, Th)/Pb ages of monazite inclusions in garnet from the Himalaya of central Nepal. *Chemical Geology*, **244**, 1–24, https://doi.org/10.1016/j.chemgeo.2007.05.003

Martin, E.L., Collins, W.J. and Spencer, C.J. 2020. Laurentian origin of the Cuyania suspect terrane, western Argentina, confirmed by Hf isotopes in zircon. *GSA Bulletin*, **132**, 273–290, https://doi.org/10.1130/B35150.1

Martin, L.A.J., Ballèvre, M., Boulvais, P., Halfpenny, A., Vanderhaeghe, O., Duchêne, S. and Deloule, E. 2011. Garnet re-equilibration by coupled dissolution–reprecipitation: evidence from textural, major element and oxygen isotope zoning of 'cloudy' garnet. *Journal of Metamorphic Geology*, **29**, 213–231, https://doi.org/10.1111/j.1525-1314.2010.00912.x

Martin, L.A.J., Rubatto, D., Crépisson, C., Hermann, J., Putlitz, B. and Vitale-Brovarone, A. 2014. Garnet oxygen analysis by SHRIMP-SI: matrix corrections and application to high-pressure metasomatic rocks from Alpine Corsica. *Chemical Geology*, **374**, 25–36, https://doi.org/10.1016/j.chemgeo.2014.02.010

Massonne, H.-J. and Li, B. 2022. Eclogite with unusual atoll garnet from the southern Armorican Massif, France: pressure–temperature path and geodynamic implications. *Tectonophysics*, **823**, article 229183, https://doi.org/10.1016/j.tecto.2021.229183

Mazzucchelli, M.L., Angel, R.J. and Alvaro, M. 2021. EntraPT: an online platform for elastic geothermobarometry. *American Mineralogist*, **106**, 830–837, https://doi.org/10.2138/am-2021-7693CCBYNCND

McCubbin, F.M. and Jones, R.H. 2015. Extraterrestrial apatite: planetary geochemistry to astrobiology. *Elements*, **11**, 183–188, https://doi.org/10.2113/gselements.11.3.183

McCubbin, F.M., Boyce, J.W. *et al.* 2016. Geologic history of Martian regolith breccia Northwest Africa 7034: evidence for hydrothermal activity and lithologic diversity in the Martian crust. *Journal of Geophysical Research: Planets*, **121**, 2120–2149, https://doi.org/10.1002/2016JE005143

McFarlane, C.R. 2016. Allanite U–Pb geochronology by 193 nm LA ICP-MS using NIST610 glass for external calibration. *Chemical Geology*, **438**, 91–102, https://doi.org/10.1016/j.chemgeo.2016.05.026

McGregor, M., Erickson, T.M. *et al.* 2021. High-resolution EBSD and SIMS U–Pb geochronology of zircon, titanite, and apatite: insights from the Lac La Moinerie impact structure, Canada. *Contributions to Mineralogy and Petrology*, **176**, article 76, https://doi.org/10.1007/s00410-021-01828-y

Mearns, E.W. 1986. Sm–Nd ages for Norwegian garnet peridotite. *Lithos*, **19**, 269–278, https://doi.org/10.1016/0024-4937(86)90027-7

Méheut, M., Ibañez-Mejia, M. and Tissot, F.L. 2021. Drivers of zirconium isotope fractionation in Zr-bearing phases and melts: the roles of vibrational, nuclear field shift and diffusive effects. *Geochimica et Cosmochimica Acta*, **292**, 217–234, https://doi.org/10.1016/j.gca.2020.09.028

Meinhold, G., Anders, B., Kostopoulos, D. and Reischmann, T. 2008. Rutile chemistry and thermometry as provenance indicator: an example from Chios Island, Greece. *Sedimentary Geology*, **203**, 98–111, https://doi.org/10.1016/j.sedgeo.2007.11.004

Meyer, M., John, T., Brandt, S. and Klemd, R. 2011. Trace element composition of rutile and the application of Zr-in-rutile thermometry to UHT metamorphism (Epupa Complex, NW Namibia). *Lithos*, **126**, 388–401, https://doi.org/10.1016/j.lithos.2011.07.013

Mezger, K., Hanson, G.N. and Bohlen, S.R. 1989*a*. High-precision U–Pb ages of metamorphic rutile: application to the cooling history of high-grade terranes. *Earth and Planetary Science Letters*, **96**, 106–118, https://doi.org/10.1016/0012-821X(89)90126-X

Mezger, K., Hanson, G.N. and Bohlen, S.R. 1989*b*. U–Pb systematics of garnet: dating the growth of garnet in the Late Archean Pikwitonei granulite domain at Cauchon and Natawahunan lakes, Manitoba, Canada. *Contributions to Mineralogy and Petrology*, **101**, 136–148, https://doi.org/10.1007/BF00375301

Mezger, K., Rawnsley, C.M., Bohlen, S.R. and Hanson, G.N. 1991. U–Pb garnet, sphene, monazite, and rutile ages: implications for the duration of high-grade metamorphism and cooling histories, Adirondack Mts, New York. *The Journal of Geology*, **99**, 415–428, https://doi.org/10.1086/629503

Millonig, L.J., Albert, R., Gerdes, A., Avigad, D. and Dietsch, C. 2020. Exploring laser ablation U–Pb dating of regional metamorphic garnet – The Straits Schist, Connecticut, USA. *Earth and Planetary Science Letters*, **552**, article 116589, https://doi.org/10.1016/j.epsl.2020.116589

Millonig, L.J., Beranoaguirre, A., Albert, R., Marschall, H., Baxter, E. and Gerdes, A. 2022. Garnet U–Pb dating by LA-ICPMS: opportunities, limitations, and applications. *EGU General Assembly Abstracts*, Vienna, Austria, EUG22-7077, https://doi.org/10.5194/egusphere-egu22-7077

Mills, S.J., Hatert, F., Nickel, E.H. and Ferraris, G. 2009. The standardisation of mineral group hierarchies: application to recent nomenclature proposals. *European Journal of Mineralogy*, **21**, 1073–1080, https://doi.org/10.1127/0935-1221/2009/0021-1994

Möller, A., O'Brien, P.J., Kennedy, A. and Kröner, A. 2003. Linking growth episodes of zircon and metamorphic textures to zircon chemistry: an example from the ultrahigh-temperature granulites of Rogaland (SW Norway). *Geological Society, London, Special Publications*, **220**, 65–81, https://doi.org/10.1144/GSL.SP.2003.220.01.04

Montel, J.-M. 1993. A model for monazite/melt equilibrium and application to the generation of granitic magmas. *Chemical Geology*, **110**, 127–146, https://doi.org/10.1016/0009-2541(93)90250-M

Montel, J.-M., Foret, S., Veschambre, M., Nicollet, C. and Provost, A. 1996. Electron microprobe dating of monazite. *Chemical Geology*, **131**, 37–53, https://doi.org/10.1016/0009-2541(96)00024-1

Montel, J.-M., Kato, T., Enami, M., Cocherie, A., Finger, F., Williams, M. and Jercinovic, M. 2018. Electron-microprobe dating of monazite: the story. *Chemical Geology*, **484**, 4–15, https://doi.org/10.1016/j.chemgeo.2017.11.001

Moore, J., Beinlich, A., Piazolo, S., Austrheim, H. and Putnis, A. 2020*a*. Metamorphic differentiation via enhanced dissolution along high permeability zones. *Journal of Petrology*, **61**, article egaa096, https://doi.org/10.1093/petrology/egaa096

Moore, J., Beinlich, A. *et al.* 2020*b*. Microstructurally controlled trace element (Zr, U–Pb) concentrations in metamorphic rutile: an example from the amphibolites of the Bergen Arcs. *Journal of Metamorphic Geology*, **38**, 103–127, https://doi.org/10.1111/jmg.12514

Morrissey, L.J., Hand, M., Lane, K., Kelsey, D.E. and Dutch, R.A. 2016. Upgrading iron–ore deposits by melt loss during granulite facies metamorphism. *Ore Geology Reviews*, **74**, 101–121, https://doi.org/10.1016/j.oregeorev.2015.11.012

Moser, A.C., Hacker, B.R., Gehrels, G.E., Seward, G.G.E., Kylander-Clark, A.R.C. and Garber, J.M. 2022. Linking titanite U–Pb dates to coupled deformation and dissolution–reprecipitation. *Contributions to Mineralogy and Petrology*, **177**, article 42, https://doi.org/10.1007/s00410-022-01906-9

Mottram, C.M., Warren, C.J., Regis, D., Roberts, N.M.W., Harris, N.B.W., Argles, T.W. and Parrish, R.R. 2014. Developing an inverted Barrovian sequence; insights from monazite petrochronology. *Earth and Planetary Science Letters*, **403**, 418–431, https://doi.org/10.1016/j.epsl.2014.07.006

Mottram, C.M., Parrish, R.R., Regis, D., Warren, C.J., Argles, T.W., Harris, N.B.W. and Roberts, N.M.W. 2015. Using U–Th–Pb petrochronology to determine rates of ductile thrusting: time windows into the Main Central Thrust, Sikkim Himalaya. *Tectonics*, **34**, 1355–1374, https://doi.org/10.1002/2014TC003743

Mottram, C.M., Cottle, J.M. and Kylander-Clark, A.R.C. 2019. Campaign-style U–Pb titanite petrochronology: along-strike variations in timing of metamorphism in the Himalayan metamorphic core. *Geoscience Frontiers*, **10**, 827–847, https://doi.org/10.1016/j.gsf.2018.09.007

Moyen, J.F. and Martin, H. 2012. Forty years of TTG research. *Lithos*, **148**, 312–336.

Moyen, J.F. and Stevens, G. 2006. Experimental constraints on TTG petrogenesis: implications for Archean geodynamics. *AGU Geophysical Monograph Series*, **164**, 149–175, http://doi.org/10.1029/164GM11

Mühlberg, M., Stevens, G., Moyen, J.-F., Kisters, A.F. and Lana, C. 2021. Thermal evolution of the Stolzburg Block, Barberton granitoid–greenstone terrain, South Africa: implications for Paleoarchean tectonic processes. *Precambrian Research*, **359**, article 106082, https://doi.org/10.1016/j.precamres.2020.106082

Mulder, J.A. and Cawood, P.A. 2021. Evaluating preservation bias in the continental growth record against the monazite archive. *Geology*, **50**, 243–247, https://doi.org/10.1130/G49416.1

Mulder, J.A., Nebel, O., Gardiner, N.J., Cawood, P.A., Wainwright, A.N. and Ivanic, T.J. 2021. Crustal rejuvenation stabilised Earth's first cratons. *Nature Communications*, **12**, 1–7, https://doi.org/10.1038/s41467-021-23805-6

Muñoz-Montecinos, J., Angiboust, S., Garcia-Casco, A. and Raimondo, T. 2023. Shattered veins elucidate brittle creep processes in the deep slow slip and tremor region. *Tectonics*, **42**, e2022TC007605, https://doi.org/10.1029/2022TC007605

Næraa, T., Scherstén, A., Rosing, M.T., Kemp, A., Hoffmann, J., Kokfelt, T. and Whitehouse, M. 2012. Hafnium isotope evidence for a transition in the dynamics of continental growth 3.2 Gyr ago. *Nature*, **485**, 627–630, https://doi.org/10.1038/nature11140

Nasdala, L., Zhang, M., Kempe, U., Panczer, G., Gaft, M., Andrut, M. and Plötze, M. 2003. Spectroscopic methods applied to zircon. *Reviews in Mineralogy and Geochemistry*, **53**, 427–467, https://doi.org/10.2113/0530427

Nestola, F. 2021. How to apply elastic geobarometry in geology. *American Mineralogist*, **106**, 669–671, https://doi.org/10.2138/am-2021-7845

Newton, R. and Haselton, H. 1981. Thermodynamics of the garnet–plagioclase–Al_2SiO_5–quartz geobarometer. *In*: Newton, R.C., Navrotsky, A. and Wood, B.J. (eds) *Thermodynamics of Minerals and Melts*. Springer, 131–147.

Ning, W., Wang, J., Xiao, D., Li, F., Huang, B. and Fu, D. 2019. Electron probe microanalysis of monazite and its applications to U–Th–Pb dating of geological samples. *Journal of Earth Science*, **30**, 952–963, https://doi.org/10.1007/s12583-019-1020-8

Nishizawa, M., Takahata, N., Terada, K., Komiya, T., Ueno, Y. and Sano, Y. 2005. Rare-earth element, lead, carbon, and nitrogen geochemistry of apatite-bearing metasediments from the ~3.8 Ga Isua Supracrustal Belt, West Greenland. *International Geology Review*, **47**, 952–970, https://doi.org/10.2747/0020-6814.47.9.952

Nordsvan, A.R., Collins, W.J. *et al.* 2018. Laurentian crust in northeast Australia: implications for the assembly of the supercontinent Nuna. *Geology*, **46**, 251–254, https://doi.org/10.1130/G39980.1

Nutman, A.P. 2007. Apatite recrystallisation during prograde metamorphism, Cooma, southeast Australia: implications for using an apatite–graphite association as a biotracer in ancient metasedimentary rocks. *Australian Journal of Earth Sciences*, **54**, 1023–1032, https://doi.org/10.1080/08120090701488321

Oberli, F., Meier, M., Berger, A., Rosenberg, C.L. and Gieré, R. 2004. U–Th–Pb and $^{230}Th/^{238}U$ disequilibrium isotope systematics: precise accessory mineral chronology and melt evolution tracing in the Alpine Bergell intrusion. *Geochimica et Cosmochimica Acta*, **68**, 2543–2560, https://doi.org/10.1016/j.gca.2003.10.017

Odlum, M.L., Levy, D.A., Stockli, D.F., Stockli, L.D. and DesOrmeau, J.W. 2022. Deformation and metasomatism recorded by single-grain apatite petrochronology. *Geology*, **50**, 697–703, https://doi.org/10.1130/G49809.1

Olierook, H.K.H., Taylor, R.J.M. *et al.* 2019. Unravelling complex geologic histories using U–Pb and trace element systematics of titanite. *Chemical Geology*, **504**, 105–122, https://doi.org/10.1016/j.chemgeo.2018.11.004

Olierook, H.K.H., Rankenburg, K. *et al.* 2020. Resolving multiple geological events using *in situ* Rb–Sr geochronology: implications for metallogenesis at Tropicana, Western Australia. *Geochronology*, **2**, 283–303, https://doi.org/10.5194/gchron-2-283-2020

Olin, P.H. and Wolff, J.A. 2012. Partitioning of rare earth and high field strength elements between titanite and phonolitic liquid. *Lithos*, **128–131**, 46–54, https://doi.org/10.1016/j.lithos.2011.10.007

O'Reilly, S.Y. and Griffin, W. 2000. Apatite in the mantle: implications for metasomatic processes and high heat production in Phanerozoic mantle. *Lithos*, **53**, 217– 232, https://doi.org/10.1016/S0024-4937(00)00026-8

Ortolano, G., Visalli, R., Cirrincione, R. and Rebay, G. 2014. PT-path reconstruction via unraveling of peculiar zoning pattern in atoll shaped garnets via image assisted analysis: an example from the Santa Lucia del Mela garnet micaschists (northeastern Sicily-Italy). *Periodico di Mineralogia*, **83**, 257–297.

Ortolano, G., Visalli, R., Godard, G. and Cirrincione, R. 2018. Quantitative X-ray map analyser (Q-XRMA): a new GIS-based statistical approach to mineral image analysis. *Computers & Geosciences*, **115**, 56–65, https://doi.org/10.1016/j.cageo.2018.03.001

O'Sullivan, G.J. and Chew, D.M. 2020. The clastic record of a Wilson Cycle: evidence from detrital apatite petrochronology of the Grampian–Taconic fore-arc. *Earth and Planetary Science Letters*, **552**, article 116588, https://doi.org/10.1016/j.epsl.2020.116588

O'Sullivan, G.J., Chew, D., Morton, A., Mark, C. and Henrichs, I. 2018. An integrated apatite geochronology and geochemistry tool for sedimentary provenance analysis. *Geochemistry, Geophysics, Geosystems*, **19**, 1309–1326, https://doi.org/10.1002/2017GC007343

O'Sullivan, G., Chew, D., Kenny, G., Henrichs, I. and Mulligan, D. 2020. The trace element composition of apatite and its application to detrital provenance studies. *Earth-Science Reviews*, **201**, article 103044, https://doi.org/10.1016/j.earscirev.2019.103044

O'Sullivan, G., Hoare, B., Mark, C., Drakou, F. and Tomlinson, E. 2023. Uranium–lead geochronology applied to pyrope garnet with very low concentrations of uranium. *Geological Magazine*, **160**, 1010–1019, https://doi.org/10.1017/S0016756823000122

Page, F.Z., Kita, N.T. and Valley, J.W. 2010. Ion microprobe analysis of oxygen isotopes in garnets of complex chemistry. *Chemical Geology*, **270**, 9–19, https://doi.org/10.1016/j.chemgeo.2009.11.001

Papapavlou, K., Darling, J.R., Storey, C.D., Lightfoot, P.C., Moser, D.E. and Lasalle, S. 2017. Dating shear zones with plastically deformed titanite: new insights into the orogenic evolution of the Sudbury impact structure (Ontario, Canada). *Precambrian Research*, **291**, 220–235, https://doi.org/10.1016/j.precamres.2017.01.007

Pape, J., Mezger, K. and Robyr, M. 2016. A systematic evaluation of the Zr-in-rutile thermometer in ultra-high temperature (UHT) rocks. *Contributions to Mineralogy and Petrology*, **171**, article 44, https://doi.org/10.1007/s00410-016-1254-8

Parrish, R.R. 1990. U–Pb dating of monazite and its application to geological problems. *Canadian Journal of Earth Sciences*, **27**, 1431–1450, https://doi.org/10.1139/e90-152

Passchier, C.W. and Simpson, C. 1986. Porphyroclast systems as kinematic indicators. *Journal of Structural Geology*, **8**, 831–843, https://doi.org/10.1016/0191-8141(86)90029-5

Patchett, P.J. 1983. Importance of the Lu–Hf isotopic system in studies of planetary chronology and chemical evolution. *Geochimica et Cosmochimica Acta*, **47**, 81–91, https://doi.org/10.1016/0016-7037(83)90092-3

Paul, A.N., Spikings, R.A., Chew, D. and Daly, J.S. 2019. The effect of intra-crystal uranium zonation on apatite U–Pb thermochronology: a combined ID-TIMS and LA-MC-ICP-MS study. *Geochimica et Cosmochimica Acta*, **251**, 15–35, https://doi.org/10.1016/j.gca.2019.02.013

Paul, A.N., Spikings, R.A. and Gaynor, S.P. 2021. U–Pb ID-TIMS reference ages and initial Pb isotope compositions for Durango and Wilberforce apatites. *Chemical Geology*, **586**, article 120604, https://doi.org/10.1016/j.chemgeo.2021.120604

Peixoto, E., Alkmim, F.F., Pedrosa-Soares, A., Lana, C. and Chaves, A.O. 2018. Metamorphic record of collision and collapse in the Ediacaran–Cambrian Araçuaí orogen, SE-Brazil: insights from P–T pseudosections and monazite dating. *Journal of Metamorphic Geology*, **36**, 147–172, https://doi.org/10.1111/JMG.12287

Penniston-Dorland, S.C., Kohn, M.J. and Piccoli, P.M. 2018. A mélange of subduction temperatures: evidence from Zr-in-rutile thermometry for strengthening of the subduction interface. *Earth and Planetary Science Letters*, **482**, 525–535, https://doi.org/10.1016/j.epsl.2017.11.005

Penniston-Dorland, S.C., Baumgartner, L.P., Dragovic, B. and Bouvier, A.-S. 2020. Li isotope zoning in garnet from Franciscan eclogite and amphibolite: the role of subduction-related fluids. *Geochimica et Cosmochimica Acta*, **286**, 198–213, https://doi.org/10.1016/j.gca.2020.07.025

Pe-Piper, G., Piper, D.J. and Triantafyllidis, S. 2014. Detrital monazite geochronology, Upper Jurassic–Lower Cretaceous of the Scotian Basin: significance for tracking first-cycle sources. *Geological Society, London, Special Publications*, **386**, 293–311, https://doi.org/10.1144/SP386.13

Perchuk, L. and Lavrent'Eva, I. 1983. Experimental investigation of exchange equilibria in the system cordierite–garnet–biotite. *Advances in Physical Geochemistry*, **3**, 199–239.

Pereira, I. and Storey, C.D. 2023. Detrital rutile: records of the deep crust, ores and fluids. *Lithos*, 438–439, article 107010, https://doi.org/10.1016/j.lithos.2022.107010

Pereira, I., Storey, C., Darling, J., Lana, C. and Alkmim, A.R. 2019. Two billion years of evolution enclosed in hydrothermal rutile: recycling of the São Francisco Craton crust and constraints on gold remobilisation processes. *Gondwana Research*, **68**, 69–92, https://doi.org/10.1016/j.gr.2018.11.008

Pereira, I. *et al.* In press. A review of detrital heavy mineral contributions to furthering our understanding of continental crust formation and evolution. *Geological Society, London, Special Publications*, https://doi.org/SP537-2022-250

Petrus, J.A., Chew, D.M., Leybourne, M.I. and Kamber, B.S. 2017. A new approach to laser-ablation inductively-coupled-plasma mass-spectrometry (LA-ICP-MS) using the flexible map interrogation tool 'Monocle'. *Chemical Geology*, **463**, 76–93, https://doi.org/10.1016/j.chemgeo.2017.04.027

Peverelli, V., Berger, A., Mulch, A., Pettke, T., Piccoli, F. and Herwegh, M. 2022. Epidote U–Pb geochronology and H isotope geochemistry trace pre-orogenic hydration of midcrustal granitoids. *Geology*, **50**, 1073–1077, https://doi.org/10.1130/G50028.1

Piccoli, P.M. and Candela, P.A. 2002. Apatite in igneous systems. *Reviews in Mineralogy and Geochemistry*, **48**, 255–292, https://doi.org/10.2138/rmg.2002.48.6

Piccoli, P., Candela, P. and Rivers, M. 2000. Interpreting magmatic processes from accessory phases: titanite – a small-scale recorder of large-scale processes. *Earth and Environmental Science Transactions of the Royal Society of Edinburgh*, **91**, 257–267, https://doi.org/10.1017/S0263593300007422

Pidgeon, R.T., Chapman, P.G., Danišík, M. and Nemchin, A.A. 2017. Dry annealing of metamict zircon: a differential scanning calorimetry study. *American Mineralogist*, **102**, 1066–1072, https://doi.org/10.2138/am-2017-5901

Pidgeon, R.T., Nemchin, A., Roberts, M., Whitehouse, M.J. and Bellucci, J. 2019. The accumulation of non-formula elements in zircons during weathering: ancient zircons from the Jack Hills, Western Australia. *Chemical Geology*, **530**, article 119310, https://doi.org/10.1016/j.chemgeo.2019.119310

Piechocka, A.M., Gregory, C.J., Zi, J.-W., Sheppard, S., Wingate, M.T.D. and Rasmussen, B. 2017. Monazite trumps zircon: applying SHRIMP U–Pb geochronology to systematically evaluate emplacement ages of leucocratic, low-temperature granites in a complex Precambrian orogen. *Contributions to Mineralogy and Petrology*, **172**, article 63, https://doi.org/10.1007/s00410-017-1386-5

Plank, T., Cooper, L.B. and Manning, C.E. 2009. Emerging geothermometers for estimating slab surface temperatures. *Nature Geoscience*, **2**, 611–615, https://doi.org/10.1038/ngeo614

Plavsa, D., Reddy, S., Clark, C. and Agangi, A. 2018. *Capricorn Orogen Rutile Study: a Combined Electrons Backscatter Diffraction (EBSD) and Laser Ablation Split Stream (LASS) Analytical Approach.* Government of Western Australia.

Pochon, A., Beaudoin, G., Branquet, Y., Boulvais, P., Gloaguen, E. and Gapais, D. 2017. Metal mobility during hydrothermal breakdown of Fe–Ti oxides: insights from Sb–Au mineralizing event (Variscan Armorican Massif, France). *Ore Geology Reviews*, **91**, 66–99, https://doi.org/10.1016/j.oregeorev.2017.10.021

Poitrasson, F., Chenery, S. and Shepherd, T.J. 2000. Electron microprobe and LA-ICP-MS study of monazite hydrothermal alteration: implications for U–Th–Pb geochronology and nuclear ceramics. *Geochimica et Cosmochimica Acta*, **64**, 3283–3297, https://doi.org/10.1016/S0016-7037(00)00433-6

Pollington, A.D. and Baxter, E.F. 2010. High resolution Sm–Nd garnet geochronology reveals the uneven pace of tectonometamorphic processes. *Earth and Planetary Science Letters*, **293**, 63–71, https://doi.org/10.1016/j.epsl.2010.02.019

Pollington, A.D. and Baxter, E.F. 2011. High precision microsampling and preparation of zoned garnet porphyroblasts for Sm–Nd geochronology. *Chemical Geology*, **281**, 270–282, https://doi.org/10.1016/j.chemgeo.2010.12.014

Porter, J.K., McNaughton, N.J., Evans, N.J. and McDonald, B.J. 2020. Rutile as a pathfinder for metals exploration. *Ore Geology Reviews*, **120**, article 103406, https://doi.org/10.1016/j.oregeorev.2020.103406

Potts, N.J., Barnes, J.J., Tartèse, R., Franchi, I.A. and Anand, M. 2018. Chlorine isotopic compositions of apatite in Apollo 14 rocks: evidence for widespread vapor-phase metasomatism on the lunar nearside *c.* 4 billion years ago. *Geochimica et Cosmochimica Acta*, **230**, 46–59, https://doi.org/10.1016/j.gca.2018.03.022

Pourteau, A., Scherer, E.E., Schorn, S., Bast, R., Schmidt, A. and Ebert, L. 2019. Thermal evolution of an ancient subduction interface revealed by Lu–Hf garnet geochronology, Halilbaği Complex (Anatolia). *Geoscience Frontiers*, **10**, 127–148, https://doi.org/10.1016/j.gsf.2018.03.004

Prent, A.M., Beinlich, A., Morrissey, L.J., Raimondo, T., Clark, C. and Putnis, A. 2019. Monazite as a monitor for melt–rock interaction during cooling and exhumation. *Journal of Metamorphic Geology*, **37**, 415–438, https://doi.org/10.1111/jmg.12471

Prent, A.M., Beinlich, A., Raimondo, T., Kirkland, C.L., Evans, N.J. and Putnis, A. 2020. Apatite and monazite: an effective duo to unravel superimposed fluid-flow and deformation events in reactivated shear zones. *Lithos*, **376–377**, article 105752, https://doi.org/10.1016/j.lithos.2020.105752

Prowatke, S. and Klemme, S. 2005. Effect of melt composition on the partitioning of trace elements between titanite and silicate melt. *Geochimica et Cosmochimica Acta*, **69**, 695–709, https://doi.org/10.1016/j.gca.2004.06.037

Pupin, J.P. 1980. Zircon and granite petrology. *Contributions to Mineralogy and Petrology*, **73**, 207–220, https://doi.org/10.1007/BF00381441

Pyle, J.M. and Spear, F.S. 1999. Yttrium zoning in garnet: coupling of major and accessory phases during metamorphic reactions. *Geological Materials Research*, **1**, 1–49.

Qin, T., Wu, F., Wu, Z. and Huang, F. 2016. First-principles calculations of equilibrium fractionation of O and Si isotopes in quartz, albite, anorthite, and zircon. *Contributions to Mineralogy and Petrology*, **171**, 1–14, https://doi.org/10.1007/s00410-015-1217-5

Radulescu, I.G., Rubatto, D., Gregory, C. and Compagnoni, R. 2009. The age of HP metamorphism in the Gran Paradiso Massif, Western Alps: a petrological and geochronological study of 'silvery micaschists'. *Lithos*, **110**, 95–108, https://doi.org/10.1016/j.lithos.2008.12.008

Raimondo, T., Clark, C., Hand, M., Cliff, J. and Harris, C. 2012. High-resolution geochemical record of fluid–rock interaction in a mid-crustal shear zone: a comparative study of major element and oxygen isotope transport in garnet. *Journal of Metamorphic Geology*, **30**, 255–280, https://doi.org/10.1111/j.1525-1314.2011.00966.x

Raimondo, T., Payne, J., Wade, B., Lanari, P., Clark, C. and Hand, M. 2017. Trace element mapping by LA-ICP-MS: assessing geochemical mobility in garnet. *Contributions to Mineralogy and Petrology*, **172**, article 17, https://doi.org/10.1007/s00410-017-1339-z

Rasmussen, B. and Muhling, J.R. 2007. Monazite begets monazite: evidence for dissolution of detrital monazite and reprecipitation of syntectonic monazite during low-grade regional metamorphism. *Contributions to Mineralogy and Petrology*, **154**, 675–689, https://doi.org/10.1007/s00410-007-0216-6

Ravindran, A., Mezger, K., Balakrishnan, S., Kooijman, E., Schmitt, M. and Berndt, J. 2020. Initial $^{87}Sr/^{86}Sr$ as a sensitive tracer of Archaean crust–mantle evolution: constraints from igneous and sedimentary rocks in the western Dharwar Craton, India. *Precambrian Research*, **337**, article 105523, https://doi.org/10.1016/j.precamres.2019.105523

Reddy, S.M., Saxey, D.W., Rickard, W.D.A., Fougerouse, D., Montalvo, S.D., Verberne, R. and van Riessen, A. 2020. Atom probe tomography: development and application to the geosciences. *Geostandards and Geoanalytical Research*, **44**, 5–50, https://doi.org/10.1111/ggr.12313

Regis, D., Rubatto, D., Darling, J., Cenki-Tok, B., Zucali, M. and Engi, M. 2014. Multiple metamorphic stages within an eclogite-facies terrane (Sesia Zone, Western Alps) revealed by Th–U–Pb petrochronology. *Journal of Petrology*, **55**, 1429–1456, https://doi.org/10.1093/petrology/egu029

Regis, D., Warren, C.J., Mottram, C.M. and Roberts, N.M.W. 2016. Using monazite and zircon petrochronology to constrain the P–T–t evolution of the middle crust in the Bhutan Himalaya. *Journal of Metamorphic Geology*, **34**, 617–639, https://doi.org/10.1111/jmg.12196

Renna, M.R., Tribuzio, R. and Tiepolo, M. 2007. Origin and timing of the post-Variscan gabbro–granite complex of Porto (Western Corsica). *Contributions to Mineralogy and Petrology*, **154**, 493–517, https://doi.org/10.1007/s00410-007-0205-9

Rezvukhina, O.V., Skublov, S.G., Rezvukhin, D.I. and Korsakov, A.V. 2021. Rutile in diamondiferous metamorphic rocks: new insights from trace-element composition, mineral/fluid inclusions, and U–Pb ID-TIMS dating. *Lithos*, **394–395**, article 106172, https://doi.org/10.1016/j.lithos.2021.106172

Ribeiro, B.V., Finch, M.A. *et al.* 2022. From microanalysis to supercontinents: insights from the Rio Apa Terrane into the Mesoproterozoic SW Amazonian Craton evolution during Rodinia assembly. *Journal of Metamorphic Geology*, **40**, 631–663, https://doi.org/10.1111/jmg.12641

Romer, R.L. and Rötzler, J. 2001. P–T–t evolution of ultrahigh-temperature granulites from the Saxon Granulite Massif, Germany. Part II: geochronology. *Journal of Petrology*, **42**, 2015–2032, https://doi.org/10.1093/petrology/42.11.2015

Romer, R.L. and Rötzler, J. 2011. The role of element distribution for the isotopic dating of metamorphic minerals. *European Journal of Mineralogy*, **23**, 17–33, https://doi.org/10.1127/0935-1221/2011/0023-2081

Romer, R.L. and Siegesmund, S. 2003. Why allanite may swindle about its true age. *Contributions to Mineralogy and Petrology*, **146**, 297–307, https://doi.org/10.1007/s00410-003-0494-6

Romer, R.L. and Xiao, Y. 2005. Initial Pb–Sr (–Nd) isotopic heterogeneity in a single allanite–epidote crystal: implications of reaction history for the dating of minerals with low parent-to-daughter ratios. *Contributions to Mineralogy and Petrology*, **148**, 662–674, https://doi.org/10.1007/s00410-004-0630-y

Rubatto, D. 2002. Zircon trace element geochemistry: partitioning with garnet and the link between U–Pb ages and metamorphism. *Chemical Geology*, **184**, 123–138, https://doi.org/10.1016/S0009-2541(01)00355-2

Rubatto, D. 2017. Zircon: the metamorphic mineral. *Reviews in Mineralogy and Geochemistry*, **83**, 261–295, https://doi.org/10.2138/rmg.2017.83.9

Rubatto, D. and Angiboust, S. 2015. Oxygen isotope record of oceanic and high-pressure metasomatism: a P–T–time–fluid path for the Monviso eclogites (Italy).

Contributions to Mineralogy and Petrology, **170**, article 44, https://doi.org/10.1007/s00410-015-1198-4
Rubatto, D. and Hermann, J. 2007. Experimental zircon/melt and zircon/garnet trace element partitioning and implications for the geochronology of crustal rocks. *Chemical Geology*, **241**, 38–61, https://doi.org/10.1016/j.chemgeo.2007.01.027
Rubatto, D., Williams, I.S. and Buick, I.S. 2001. Zircon and monazite response to prograde metamorphism in the Reynolds Range, central Australia. *Contributions to Mineralogy and Petrology*, **140**, 458–468, https://doi.org/10.1007/PL00007673
Rubatto, D., Hermann, J. and Buick, I.S. 2006. Temperature and bulk composition control on the growth of monazite and zircon during low-pressure anatexis (Mount Stafford, central Australia). *Journal of Petrology*, **47**, 1973–1996, https://doi.org/10.1093/petrology/egl033
Rubatto, D., Hermann, J., Berger, A. and Engi, M. 2009. Protracted fluid-induced melting during Barrovian metamorphism in the Central Alps. *Contributions to Mineralogy and Petrology*, **158**, 703–722, https://doi.org/10.1007/s00410-009-0406-5
Rubatto, D., Regis, D., Hermann, J., Boston, K., Engi, M., Beltrando, M. and McAlpine, S.R. 2011. Yo-yo subduction recorded by accessory minerals in the Italian Western Alps. *Nature Geoscience*, **4**, 338–342, https://doi.org/10.1038/ngeo1124
Rubatto, D., Chakraborty, S. and Dasgupta, S. 2013. Timescales of crustal melting in the Higher Himalayan Crystallines (Sikkim, Eastern Himalaya) inferred from trace element-constrained monazite and zircon chronology. *Contributions to Mineralogy and Petrology*, **165**, 349–372, https://doi.org/10.1007/s00410-012-0812-y
Rubatto, D., Burger, M. *et al.* 2020. Identification of growth mechanisms in metamorphic garnet by high-resolution trace element mapping with LA-ICP-TOFMS. *Contributions to Mineralogy and Petrology*, **175**, article 61, https://doi.org/10.1007/s00410-020-01700-5
Rudnick, R.L., Barth, M., Horn, I. and McDonough, W.F. 2000. Rutile-bearing refractory eclogites: missing link between continents and depleted mantle. *Science*, **287**, 278–281, https://doi.org/10.1126/science.287.5451.278
Russell, A.K., Kitajima, K., Strickland, A., Medaris, L.G., Schulze, D.J. and Valley, J.W. 2013. Eclogite-facies fluid infiltration: constraints from $\delta^{18}O$ zoning in garnet. *Contributions to Mineralogy and Petrology*, **165**, 103–116, https://doi.org/10.1007/s00410-012-0794-9
Salama, W., Anand, R. and Roberts, M. 2018. Cassiterite and rutile as indicator minerals for exploring the VMS system. *ASEG Extended Abstracts*, **2018**, 1–4, https://doi.org/10.1071/aseg2018abt7_2d
Salminen, P.E., Hölttä, P., Lahtinen, R. and Sayab, M. 2022. Monazite record for the Paleoproterozoic Svecofennian orogeny, SE Finland: an over 150-Ma spread of monazite dates. *Lithos*, **416–417**, article 106654, https://doi.org/10.1016/j.lithos.2022.106654
Savage, P.S., Armytage, R.M., Georg, R.B. and Halliday, A.N. 2014. High temperature silicon isotope geochemistry. *Lithos*, **190**, 500–519, https://doi.org/10.1016/j.lithos.2014.01.003
Sayab, M. 2006. Decompression through clockwise P–T path: implications for early N–S shortening orogenesis in the Mesoproterozoic Mt Isa Inlier (NE Australia). *Journal of Metamorphic Geology*, **24**, 89–105, https://doi.org/10.1111/j.1525-1314.2005.00626.x
Sayab, M., Suuronen, J.-P., Hölttä, P., Aerden, D., Lahtinen, R. and Kallonen, A.P. 2015. High-resolution X-ray computed microtomography: a holistic approach to metamorphic fabric analyses. *Geology*, **43**, 55–58, https://doi.org/10.1130/G36250.1
Sayab, M., Suuronen, J.-P. *et al.* 2016. Three-dimensional textural and quantitative analyses of orogenic gold at the nanoscale. *Geology*, **44**, 739–742, https://doi.org/10.1130/G38074.1
Schaltegger, U., Brack, P. *et al.* 2009. Zircon and titanite recording 1.5 million years of magma accretion, crystallization and initial cooling in a composite pluton (southern Adamello batholith, northern Italy). *Earth and Planetary Science Letters*, **286**, 208–218, https://doi.org/10.1016/j.epsl.2009.06.028
Schannor, M., Lana, C., Nicoli, G., Cutts, K., Buick, I., Gerdes, A. and Hecht, L. 2021. Reconstructing the metamorphic evolution of the Araçuaí orogen (SE Brazil) using *in situ* U–Pb garnet dating and P–T modelling. *Journal of Metamorphic Geology*, **39**, 1145–1171, https://doi.org/10.1111/jmg.12605
Schärer, U., Lian-Sheng, Z. and Tapponnier, P. 1994. Duration of strike-slip movements in large shear zones: the Red River belt, China. *Earth and Planetary Science Letters*, **126**, 379–397, https://doi.org/10.1016/0012-821X(94)90119-8
Scherer, E.E., Cameron, K.L. and Blichert-Toft, J. 2000. Lu–Hf garnet geochronology: closure temperature relative to the Sm–Nd system and the effects of trace mineral inclusions. *Geochimica et Cosmochimica Acta*, **64**, 3413–3432, https://doi.org/10.1016/S0016-7037(00)00440-3
Scherer, E., Münker, C. and Mezger, K. 2001. Calibration of the lutetium–hafnium clock. *Science*, **293**, 683–687, https://doi.org/10.1126/science.1061372
Scherer, E.E., Whitehouse, M.J. and Munker, C. 2007. Zircon as a monitor of crustal growth. *Elements*, **3**, 19–24, https://doi.org/10.2113/gselements.3.1.19
Schiller, D. and Finger, F. 2019. Application of Ti-in-zircon thermometry to granite studies: problems and possible solutions. *Contributions to Mineralogy and Petrology*, **174**, article 51, https://doi.org/10.1007/s00410-019-1585-3
Schirra, M. and Laurent, O. 2021. Petrochronology of hydrothermal rutile in mineralized porphyry Cu systems. *Chemical Geology*, **581**, article 120407, https://doi.org/10.1016/j.chemgeo.2021.120407
Schmidt, A., Weyer, S., John, T. and Brey, G.P. 2009. HFSE systematics of rutile-bearing eclogites: new insights into subduction zone processes and implications for the Earth's HFSE budget. *Geochimica et Cosmochimica Acta*, **73**, 455–468, https://doi.org/10.1016/j.gca.2008.10.028
Schmidt, A., Pourteau, A., Candan, O. and Oberhänsli, R. 2015. Lu–Hf geochronology on cm-sized garnets using microsampling: new constraints on garnet growth rates and duration of metamorphism during continental collision (Menderes Massif, Turkey). *Earth and*

Planetary Science Letters, **432**, 24–35, https://doi.org/10.1016/j.epsl.2015.09.015

Schmitz, M.D. and Bowring, S.A. 2003. Ultrahigh-temperature metamorphism in the lower crust during Neoarchean Ventersdorp rifting and magmatism, Kaapvaal Craton, southern Africa. *GSA Bulletin*, **115**, 533–548, https://doi.org/10.1130/0016-7606(2003)115<0533:UMITLC>2.0.CO;2

Schoene, B. 2014. U–Th–Pb geochronology. *In*: Holland, H.D. and Turekian, K.K. (eds) *Treatise on Geochemistry*, 2nd edn, 341–378, https://doi.org/10.1016/B978-0-08-095975-7.00310-7

Schoene, B. and Bowring, S.A. 2007. Determining accurate temperature–time paths from U–Pb thermochronology: an example from the Kaapvaal Craton, southern Africa. *Geochimica et Cosmochimica Acta*, **71**, 165–185, https://doi.org/10.1016/j.gca.2006.08.029

Schönig, J., von Eynatten, H., Meinhold, G. and Lünsdorf, N.K. 2019. Diamond and coesite inclusions in detrital garnet of the Saxonian Erzgebirge, Germany. *Geology*, **47**, 715–718, https://doi.org/10.1130/G46253.1

Schulz, B. 2021. Monazite microstructures and their interpretation in petrochronology. *Frontiers in Earth Science*, **9**, https://doi.org/10.3389/feart.2021.668566

Schwandt, C.S., Papike, J.J. and Shearer, C.K. 1996. Trace element zoning in pelitic garnet of the Black Hills, South Dakota. *American Mineralogist*, **81**, 1195–1207, https://doi.org/10.2138/am-1996-9-1018

Scibiorski, E.A. and Cawood, P.A. 2022. Titanite as a petrogenetic indicator. *Terra Nova*, **34**, 177–183, https://doi.org/10.1111/ter.12574

Scicchitano, M.R., Spicuzza, M.J., Ellison, E.T., Tuschel, D., Templeton, A.S. and Valley, J.W. 2021. *In situ* oxygen isotope determination in serpentine minerals by SIMS: addressing matrix effects and providing new insights on serpentinisation at Hole BA1B (Samail ophiolite, Oman). *Geostandards and Geoanalytical Research*, **45**, 161–187, https://doi.org/10.1111/ggr.12359

Sciuba, M. and Beaudoin, G. 2021. Texture and trace element composition of rutile in orogenic gold deposits. *Economic Geology*, **116**, 1865–1892, https://doi.org/10.5382/econgeo.4857

Scott, D.J. and St-Onge, M.R. 1995. Constraints on Pb closure temperature in titanite based on rocks from the Ungava orogen, Canada: implications for U–Pb geochronology and P–T–t path determinations. *Geology*, **23**, 1123–1126, https://doi.org/10.1130/0091-7613(1995)023<1123:COPCTI>2.3.CO;2

Scott, K.M. and Radford, N.W. 2007. Rutile compositions at the Big Bell Au deposit as a guide for exploration. *Geochemistry: Exploration, Environment, Analysis*, **7**, 353–361, https://doi.org/10.1144/1467-7873/07-135

Seman, S., Stockli, D.F. and McLean, N.M. 2017. U–Pb geochronology of grossular-andradite garnet. *Chemical Geology*, **460**, 106–116, https://doi.org/10.1016/j.chemgeo.2017.04.020

Seydoux-Guillaume, A.-M., Paquette, J.-L., Wiedenbeck, M., Montel, J.-M. and Heinrich, W. 2002. Experimental resetting of the U–Th–Pb systems in monazite. *Chemical Geology*, **191**, 165–181, https://doi.org/10.1016/S0009-2541(02)00155-9

Seydoux-Guillaume, A.-M., Montel, J.-M. *et al.* 2012. Low-temperature alteration of monazite: fluid mediated coupled dissolution–precipitation, irradiation damage, and disturbance of the U–Pb and Th–Pb chronometers. *Chemical Geology*, **330–331**, 140–158, https://doi.org/10.1016/j.chemgeo.2012.07.031

Sha, L.-K. and Chappell, B.W. 1999. Apatite chemical composition, determined by electron microprobe and laser-ablation inductively coupled plasma mass spectrometry, as a probe into granite petrogenesis. *Geochimica et Cosmochimica Acta*, **63**, 3861–3881, https://doi.org/10.1016/S0016-7037(99)00210-0

Shrestha, S., Larson, K.P., Duesterhoeft, E., Soret, M. and Cottle, J.M. 2019. Thermodynamic modelling of phosphate minerals and its implications for the development of P–T–t histories: a case study in garnet-monazite bearing metapelites. *Lithos*, **334–335**, 141–160, https://doi.org/10.1016/j.lithos.2019.03.021

Shrestha, S., Larson, K.P., Martin, A.J., Guilmette, C., Smit, M.A. and Cottle, J.M. 2020. The Greater Himalayan thrust belt: insight into the assembly of the exhumed Himalayan metamorphic core, Modi Khola Valley, Central Nepal. *Tectonics*, **39**, article e2020TC006252, https://doi.org/10.1029/2020TC006252

Simpson, A., Gilbert, S. *et al.* 2021. *In-situ* Lu–Hf geochronology of garnet, apatite and xenotime by LA ICP MS/MS. *Chemical Geology*, **577**, article 120299, https://doi.org/10.1016/j.chemgeo.2021.120299

Simpson, A., Glorie, S., Hand, M., Spandler, C., Gilbert, S. and Cave, B. 2022. *In-situ* Lu–Hf geochronology of calcite. *GChron*, **4**, 353–372, https://doi.org/10.5194/gchron-4-353-2022

Simpson, A., Glorie, S., Hand, M., Spandler, C. and Gilbert, S. 2023. Garnet Lu-Hf speed dating: A novel method to rapidly resolve polymetamorphic histories. *Gondwana Research*, https://doi.org/10.1016/j.gr.2023.04.011

Sláma, J., Košler, J. and Pedersen, R. 2007. Behaviour of zircon in high-grade metamorphic rocks: evidence from Hf isotopes, trace elements and textural studies. *Contributions to Mineralogy and Petrology*, **154**, 335–356, https://doi.org/10.1007/s00410-007-0196-6

Sliwinski, J.T. and Stoll, H.M. 2021. Combined fluorescence imaging and LA-ICP-MS trace element mapping of stalagmites: microfabric identification and interpretation. *Chemical Geology*, **581**, article 120397, https://doi.org/10.1016/j.chemgeo.2021.120397

Smit, M.A., Scherer, E.E., Bröcker, M. and van Roermund, H.L.M. 2010. Timing of eclogite facies metamorphism in the southernmost Scandinavian Caledonides by Lu–Hf and Sm–Nd geochronology. *Contributions to Mineralogy and Petrology*, **159**, 521–539, https://doi.org/10.1007/s00410-009-0440-3

Smit, M.A., Scherer, E.E. and Mezger, K. 2013. Lu–Hf and Sm–Nd garnet geochronology: chronometric closure and implications for dating petrological processes. *Earth and Planetary Science Letters*, **381**, 222–233, https://doi.org/10.1016/J.EPSL.2013.08.046

Smith, H.A. and Barreiro, B. 1990. Monazite U–Pb dating of staurolite grade metamorphism in pelitic schists. *Contributions to Mineralogy and Petrology*, **105**, 602–615, https://doi.org/10.1007/BF00302498

Smith, H.A. and Giletti, B.J. 1997. Lead diffusion in monazite. *Geochimica et Cosmochimica Acta*, **61**, 1047–1055, https://doi.org/10.1016/S0016-7037(96)00396-1

Smith, M., Storey, C., Jeffries, T. and Ryan, C. 2009. *In situ* U–Pb and trace element analysis of accessory minerals in the Kiruna district, Norrbotten, Sweden: new constraints on the timing and origin of mineralization. *Journal of Petrology*, **50**, 2063–2094, https://doi.org/10.1093/petrology/egp069

Smithies, R.H., Lu, Y. *et al.* 2021. Oxygen isotopes trace the origins of Earth's earliest continental crust. *Nature*, **592**, 70–75, https://doi.org/10.1038/s41586-021-03337-1

Smye, A.J., Roberts, N.M., Condon, D.J., Horstwood, M.S. and Parrish, R.R. 2014. Characterising the U–Th–Pb systematics of allanite by ID and LA-ICPMS: implications for geochronology. *Geochimica et Cosmochimica Acta*, **135**, 1–28, https://doi.org/10.1016/j.gca.2014.03.021

Smye, A.J., Marsh, J., Vermeesch, P., Garber, J. and Stockli, D. 2018. Applications and limitations of U–Pb thermochronology to middle and lower crustal thermal histories. *Chemical Geology*, **494**, 1–18, https://doi.org/10.1016/j.chemgeo.2018.07.003

Smythe, D.J. and Brenan, J.M. 2016. Magmatic oxygen fugacity estimated using zircon–melt partitioning of cerium. *Earth and Planetary Science Letters*, **453**, 260–266, https://doi.org/10.1016/j.epsl.2016.08.013

Smythe, D., Schulze, D. and Brenan, J. 2008. Rutile as a kimberlite indicator mineral: minor and trace element geochemistry. *International Kimberlite Conference: Extended Abstracts*, Frankfurt, Germany, **9**. https://doi.org/10.29173/ikc3434

Söderlund, U., Patchett, P.J., Vervoort, J.D. and Isachsen, C.E. 2004. The ^{176}Lu decay constant determined by Lu–Hf and U–Pb isotope systematics of Precambrian mafic intrusions. *Earth and Planetary Science Letters*, **219**, 311–324, https://doi.org/10.1016/S0012-821X(04)00012-3

Spandler, C., Hammerli, J., Sha, P., Hilbert-Wolf, H., Hu, Y., Roberts, E. and Schmitz, M. 2016. MKED1: a new titanite standard for *in situ* analysis of Sm–Nd isotopes and U–Pb geochronology. *Chemical Geology*, **425**, 110–126, https://doi.org/10.1016/j.chemgeo.2016.01.002

Spear, F.S. 2010. Monazite–allanite phase relations in metapelites. *Chemical Geology*, **279**, 55–62, https://doi.org/10.1016/j.chemgeo.2010.10.004

Spear, F.S. 2017. Garnet growth after overstepping. *Chemical Geology*, **466**, 491–499, https://doi.org/10.1016/j.chemgeo.2017.06.038

Spear, F.S. and Parrish, R.R. 1996. Petrology and cooling rates of the Valhalla complex, British Columbia, Canada. *Journal of Petrology*, **37**, 733–765, https://doi.org/10.1093/petrology/37.4.733

Spear, F.S. and Pyle, J.M. 2002. Apatite, monazite, and xenotime in metamorphic rocks. *Reviews in Mineralogy and Geochemistry*, **48**, 293–335, https://doi.org/10.2138/rmg.2002.48.7

Spear, F.S. and Pyle, J.M. 2010. Theoretical modeling of monazite growth in a low-Ca metapelite. *Chemical Geology*, **273**, 111–119, https://doi.org/10.1016/j.chemgeo.2010.02.016

Spear, F.S. and Wolfe, O.M. 2020. Revaluation of 'equilibrium' P–T paths from zoned garnet in light of quartz inclusion in garnet (QuiG) barometry. *Lithos*, **372–373**, article 105650, https://doi.org/10.1016/j.lithos.2020.105650

Spear, F.S., Selverstone, J., Hickmott, D., Crowley, P. and Hodges, K.V. 1984. P–T paths from garnet zoning: a new technique for deciphering tectonic processes in crystalline terranes. *Geology*, **12**, 87–90, https://doi.org/10.1130/0091-7613(1984)12<87:PPFGZA>2.0.CO;2

Spencer, C.J., Kirkland, C.L. and Taylor, R.J.M. 2016. Strategies towards statistically robust interpretations of *in situ* U–Pb zircon geochronology. *Geoscience Frontiers*, **7**, 581–589, https://doi.org/10.1016/j.gsf.2015.11.006

Spencer, C.J., Kirkland, C., Roberts, N., Evans, N. and Liebmann, J. 2020. Strategies towards robust interpretations of *in situ* zircon Lu–Hf isotope analyses. *Geoscience Frontiers*, **11**, 843–853, https://doi.org/10.1016/j.gsf.2019.09.004

Spencer, K.J., Hacker, B.R. *et al.* 2013. Campaign-style titanite U–Pb dating by laser-ablation ICP: implications for crustal flow, phase transformations and titanite closure. *Chemical Geology*, **341**, 84–101, https://doi.org/10.1016/j.chemgeo.2012.11.012

Stearns, M.A., Hacker, B.R., Ratschbacher, L., Rutte, D. and Kylander-Clark, A.R.C. 2015. Titanite petrochronology of the Pamir gneiss domes: implications for middle to deep crust exhumation and titanite closure to Pb and Zr diffusion. *Tectonics*, **34**, 784–802, https://doi.org/10.1002/2014TC003774

Stepanov, A.S., Hermann, J., Rubatto, D. and Rapp, R.P. 2012. Experimental study of monazite/melt partitioning with implications for the REE, Th and U geochemistry of crustal rocks. *Chemical Geology*, **300**, 200–220, https://doi.org/10.1016/j.chemgeo.2012.01.007

Stern, R.A. and Rayner, N.M. 2003. *Ages of Several Xenotime Megacrysts by ID-TIMS. Potential Reference Materials for Ion Microprobe U–Pb Geochronology*. Geological Survey of Canada.

Štípská, P., Hacker, B.R., Racek, M., Holder, R., Kylander-Clark, A.R.C., Schulmann, K. and Hasalová, P. 2015. Monazite dating of prograde and retrograde P–T–d paths in the Barrovian terrane of the Thaya window, Bohemian Massif. *Journal of Petrology*, **56**, 1007–1035, https://doi.org/10.1093/petrology/egv026

St-Onge, M.R. 1987. Zoned poikiloblastic garnets: P–T paths and syn-metamorphic uplift through 30 km of structural depth, Wopmay Orogen, Canada. *Journal of Petrology*, **28**, 1–21, https://doi.org/10.1093/petrology/28.1.1

Storey, C.D., Jeffries, T.E. and Smith, M. 2006. Common lead-corrected laser ablation ICP-MS U–Pb systematics and geochronology of titanite. *Chemical Geology*, **227**, 37–52, https://doi.org/10.1016/j.chemgeo.2005.09.003

Strzelecki, A.C., Reece, M. *et al.* 2022. Crystal chemistry and thermodynamics of HREE (Er, Yb) mixing in a xenotime solid solution. *ACS Earth and Space Chemistry*, **6**, 1375–1389, https://doi.org/10.1021/acsearthspacechem.2c00052

Stünitz, H. and Tullis, J. 2001. Weakening and strain localization produced by syn-deformational reaction of plagioclase. *International Journal of Earth Sciences*, **90**, 136–148, https://doi.org/10.1007/s005310000148

Su, J.-H., Zhao, X.-F., Li, X.-C., Su, Z.-K., Liu, R., Qin, Z.-J. and Chen, M. 2021. Fingerprinting REE mineralization and hydrothermal remobilization history of the carbonatite-alkaline complexes, Central China: constraints from *in situ* elemental and isotopic analyses of phosphate minerals. *American Mineralogist: Journal of Earth and Planetary Materials*, **106**, 1545–1558, https://doi.org/10.2138/am-2021-7746

Sun, J.-F., Yang, J.-H., Wu, F.-Y., Li, X.-H., Yang, Y.-H., Xie, L.-W. and Wilde, S.A. 2010. Magma mixing controlling the origin of the Early Cretaceous Fangshan granitic pluton, North China Craton: *in situ* U–Pb age and Sr-, Nd-, Hf- and O-isotope evidence. *Lithos*, **120**, 421–438, https://doi.org/10.1016/j.lithos.2010.09.002

Suzuki, K. and Adachi, M. 1994. Middle Precambrian detrital monazite and zircon from the Hida gneiss on Oki-Dogo Island, Japan: their origin and implications for the correlation of basement gneiss of southwest Japan and Korea. *Tectonophysics*, **235**, 277–292, https://doi.org/10.1016/0040-1951(94)90198-8

Suzuki, K. and Kato, T. 2008. CHIME dating of monazite, xenotime, zircon and polycrase: protocol, pitfalls and chemical criterion of possibly discordant age data. *Gondwana Research*, **14**, 569–586, https://doi.org/10.1016/j.gr.2008.01.005

Suzuki, K., Adachi, M. and Kajizuka, I. 1994. Electron microprobe observations of Pb diffusion in metamorphosed detrital monazites. *Earth and Planetary Science Letters*, **128**, 391–405, https://doi.org/10.1016/0012-821X(94)90158-9

Tacchetto, T., Reddy, S.M., Saxey, D.W., Fougerouse, D., Rickard, W.D.A. and Clark, C. 2021. Disorientation control on trace element segregation in fluid-affected low-angle boundaries in olivine. *Contributions to Mineralogy and Petrology*, **176**, article 59, https://doi.org/10.1007/s00410-021-01815-3

Tacchetto, T., Clark, C., Erickson, T., Reddy, S.M., Bhowany, K. and Hand, M. 2022. Weakening the lower crust: conditions, reactions and deformation. *Lithos*, **422–423**, article 106738, https://doi.org/10.1016/j.lithos.2022.106738

Tamblyn, R., Hand, M., Simpson, A., Gilbert, S., Wade, B. and Glorie, S. 2022. *In situ* laser ablation Lu–Hf geochronology of garnet across the Western Gneiss Region: campaign-style dating of metamorphism. *Journal of the Geological Society, London*, **179**, article jgs2021-094, https://doi.org/10.1144/jgs2021-094

Taylor, R.J.M., Clark, C., Fitzsimons, I.C.W., Santosh, M., Hand, M., Evans, N. and McDonald, B. 2014. Post-peak, fluid-mediated modification of granulite facies zircon and monazite in the Trivandrum Block, southern India. *Contributions to Mineralogy and Petrology*, **168**, article 1044, https://doi.org/10.1007/s00410-014-1044-0

Taylor, R.J.M., Harley, S.L., Hinton, R.W., Elphick, S., Clark, C. and Kelly, N.M. 2015. Experimental determination of REE partition coefficients between zircon, garnet and melt: a key to understanding high-T crustal processes. *Journal of Metamorphic Geology*, **33**, 231–248, https://doi.org/10.1111/jmg.12118

Taylor, R.J.M., Kirkland, C.L. and Clark, C. 2016. Accessories after the facts: constraining the timing, duration and conditions of high-temperature metamorphic processes. *Lithos*, **264**, 239–257, https://doi.org/10.1016/j.lithos.2016.09.004

Taylor, R.J.M., Clark, C., Harley, S.L., Kylander-Clark, A.R.C., Hacker, B.R. and Kinny, P.D. 2017. Interpreting granulite facies events through rare earth element partitioning arrays. *Journal of Metamorphic Geology*, **35**, 759–775, https://doi.org/10.1111/jmg.12254

Taylor, R.J.M., Johnson, T.E., Clark, C. and Harrison, R.J. 2020. Persistence of melt-bearing Archean lower crust for >200 my – an example from the Lewisian Complex, northwest Scotland. *Geology*, **48**, 221–225, https://doi.org/10.1130/G46834.1

Taylor-Jones, K. and Powell, R. 2015. Interpreting zirconium-in-rutile thermometric results. *Journal of Metamorphic Geology*, **33**, 115–122, https://doi.org/10.1111/jmg.12109

Tedeschi, M., Rossi Vieira, P.L. *et al.* 2023. Unravelling the protracted U–Pb zircon geochronological record of high to ultrahigh temperature metamorphic rocks: implications for provenance investigations. *Geoscience Frontiers*, **14**, article 101515, https://doi.org/10.1016/j.gsf.2022.101515

Tera, F. and Wasserburg, G.J. 1972. U–Th–Pb systematics in lunar highland samples from the Luna 20 and Apollo 16 missions. *Earth and Planetary Science Letters*, **17**, 36–51, https://doi.org/10.1016/0012-821X(72)90257-9

Thomas, J.B., Bodnar, R.J., Shimizu, N. and Chesner, C.A. 2003. Melt inclusions in zircon. *Reviews in Mineralogy and Geochemistry*, **53**, 63–87, https://doi.org/10.2113/0530063

Thomas, R. and Davidson, P. 2012. Water in granite and pegmatite-forming melts. *Ore Geology Reviews*, **46**, 32–46, https://doi.org/10.1016/j.oregeorev.2012.02.006

Thompson, J.M., Goemann, K., Belousov, I., Jenkins, K., Kobussen, A., Powell, W. and Danyushevsky, L. 2021. Assessment of the mineral ilmenite for U–Pb dating by LA-ICP-MS. *Journal of Analytical Atomic Spectrometry*, **36**, 1244–1260, https://doi.org/10.1039/D1JA00069A

Thomson, S.N., Gehrels, G.E., Ruiz, J. and Buchwaldt, R. 2012. Routine low-damage apatite U–Pb dating using laser ablation–multicollector–ICPMS. *Geochemistry, Geophysics, Geosystems*, **13**, https://doi.org/10.1029/2011GC003928

Tian, S., Inglis, E.C. *et al.* 2020. The zirconium stable isotope compositions of 22 geological reference materials, 4 zircons and 3 standard solutions. *Chemical Geology*, **555**, article 119791, https://doi.org/10.1016/j.chemgeo.2020.119791

Tiepolo, M., Oberti, R. and Vannucci, R. 2002. Trace-element incorporation in titanite: constraints from experimentally determined solid/liquid partition coefficients. *Chemical Geology*, **191**, 105–119, https://doi.org/10.1016/S0009-2541(02)00151-1

Timms, N.E., Kirkland, C.L. *et al.* 2020. Shocked titanite records Chicxulub hydrothermal alteration and impact age. *Geochimica et Cosmochimica Acta*, **281**, 12–30, https://doi.org/10.1016/j.gca.2020.04.031

Tinkham, D.K. and Ghent, E.D. 2005. Estimating PT conditions of garnet growth with isochemical phase-diagram sections and the problem of effective bulk-composition. *The Canadian Mineralogist*, **43**, 35–50, https://doi.org/10.2113/gscanmin.43.1.35

Tolometti, G.D., Erickson, T.M., Osinski, G.R., Cayron, C. and Neish, C.D. 2022. Hot rocks: constraining the thermal conditions of the Mistastin Lake impact melt deposits using zircon grain microstructures. *Earth and Planetary Science Letters*, **584**, article 117523, https://doi.org/10.1016/j.epsl.2022.117523

Tomkins, H.S., Powell, R. and Ellis, D.J. 2007. The pressure dependence of the zirconium-in-rutile thermometer. *Journal of Metamorphic Geology*, **25**, 703–713, https://doi.org/10.1111/j.1525-1314.2007.00724.x

Tompkins, H.G., Zieman, L.J., Ibanez-Mejia, M. and Tissot, F.L. 2020. Zirconium stable isotope analysis of zircon by MC-ICP-MS: methods and application to evaluating intra-crystalline zonation in a zircon megacryst. *Journal of Analytical Atomic Spectrometry*, **35**, 1167–1186, https://doi.org/10.1039/C9JA00315K

Trail, D., Watson, E.B. and Tailby, N.D. 2012. Ce and Eu anomalies in zircon as proxies for the oxidation state of magmas. *Geochimica et Cosmochimica Acta*, **97**, 70–87, https://doi.org/10.1016/j.gca.2012.08.032

Trail, D., Boehnke, P., Savage, P.S., Liu, M.-C., Miller, M.L. and Bindeman, I. 2018. Origin and significance of Si and O isotope heterogeneities in Phanerozoic, Archean, and Hadean zircon. *Proceedings of the National Academy of Sciences of the USA*, **115**, 10287–10292, https://doi.org/10.1073/pnas.1808335115

Trail, D., Savage, P.S. and Moynier, F. 2019. Experimentally determined Si isotope fractionation between zircon and quartz. *Geochimica et Cosmochimica Acta*, **260**, 257–274, https://doi.org/10.1016/j.gca.2019.06.035

Triebold, S., von Eynatten, H. and Zack, T. 2012. A recipe for the use of rutile in sedimentary provenance analysis. *Sedimentary Geology*, **282**, 268–275, https://doi.org/10.1016/j.sedgeo.2012.09.008

Tual, L., Smit, M.A., Cutts, J., Kooijman, E., Kielman-Schmitt, M., Majka, J. and Foulds, I. 2022. Rapid, paced metamorphism of blueschists (Syros, Greece) from laser-based zoned Lu–Hf garnet chronology and LA-ICPMS trace element mapping. *Chemical Geology*, **607**, article 121003, https://doi.org/10.1016/j.chemgeo.2022.121003

Ushikubo, T., Williford, K.H., Farquhar, J., Johnston, D.T., van Kranendonk, M.J. and Valley, J.W. 2014. Development of *in situ* sulfur four-isotope analysis with multiple Faraday cup detectors by SIMS and application to pyrite grains in a Paleoproterozoic glaciogenic sandstone. *Chemical Geology*, **383**, 86–99, https://doi.org/10.1016/j.chemgeo.2014.06.006

Valley, J.W. 2003. Oxygen isotopes in zircon. *Reviews in Mineralogy and Geochemistry*, **53**, 343–385, https://doi.org/10.2113/0530343

Valley, J.W., Lackey, J.S. *et al.* 2005. 4.4 billion years of crustal maturation: oxygen isotope ratios of magmatic zircon. *Contributions to Mineralogy and Petrology*, **150**, 561–580, https://doi.org/10.1007/s00410-005-0025-8

Valley, J.W., Cavosie, A.J. *et al.* 2014. Hadean age for a post-magma-ocean zircon confirmed by atom-probe tomography. *Nature Geoscience*, **7**, 219–223, https://doi.org/10.1038/ngeo2075

Vanardois, J., Roger, F. *et al.* 2022. Exhumation of deep continental crust in a transpressive regime: The example of Variscan eclogites from the Aiguilles-Rouges massif (Western Alps). *Journal of Metamorphic Geology*, **40**, 1087–1120, https://doi.org/10.1111/jmg.12659.

Vance, D. and Mahar, E. 1998. Pressure–temperature paths from P–T pseudosections and zoned garnets: potential, limitations and examples from the Zanskar Himalaya, NW India. *Contributions to Mineralogy and Petrology*, **132**, 225–245, https://doi.org/10.1007/s004100050419

Vance, D., Strachan, R.A. and Jones, K.A. 1998. Extensional v. compressional settings for metamorphism: garnet chronometry and pressure–temperature–time histories in the Moine Supergroup, northwest Scotland. *Geology*, **26**, 927–930, https://doi.org/10.1130/0091-7613(1998)026<0927:EVCSFM>2.3.CO;2

van Schijndel, V., Stevens, G., Lana, C., Zack, T. and Frei, D. 2021. De Kraalen and Witrivier greenstone belts, Kaapvaal Craton, South Africa: characterisation of the Palaeo–Mesoarchaean evolution by rutile and zircon U–Pb geochronology combined with Hf isotopes. *South African Journal of Geology*, **124**, 17–36, https://doi.org/10.25131/sajg.124.0011

Vavra, G. 1990. On the kinematics of zircon growth and its petrogenetic significance: a cathodoluminescence study. *Contributions to Mineralogy and Petrology*, **106**, 90–99, https://doi.org/10.1007/BF00306410

Vavra, G. 1993. A guide to quantitative morphology of accessory zircon. *Chemical Geology*, **110**, 15–28, https://doi.org/10.1016/0009-2541(93)90245-E

Vavra, G. and Schaltegger, U. 1999. Post-granulite facies monazite growth and rejuvenation during Permian to Lower Jurassic thermal and fluid events in the Ivrea Zone (Southern Alps). *Contributions to Mineralogy and Petrology*, **134**, 405–414, https://doi.org/10.1007/s004100050493

Vavra, G., Schmid, R. and Gebauer, D. 1999. Internal morphology, habit and U–Th–Pb microanalysis of amphibolite-to-granulite facies zircons: geochronology of the Ivrea Zone (Southern Alps). *Contributions to Mineralogy and Petrology*, **134**, 380–404, https://doi.org/10.1007/s004100050492

Verberne, R., Saxey, D.W., Reddy, S.M., Rickard, W.D.A., Fougerouse, D. and Clark, C. 2019. Analysis of natural rutile (TiO_2) by laser-assisted atom probe tomography. *Microscopy and Microanalysis*, **25**, 539–546, https://doi.org/10.1017/S1431927618015477

Verberne, R., Reddy, S.M. *et al.* 2020. The geochemical and geochronological implications of nanoscale trace-element clusters in rutile. *Geology*, **48**, 1126–1130, https://doi.org/10.1130/G48017.1

Verberne, R., Reddy, S.M. *et al.* 2022*a*. Dislocations in minerals: fast-diffusion pathways or trace-element traps? *Earth and Planetary Science Letters*, **584**, article 117517, https://doi.org/10.1016/j.epsl.2022.117517

Verberne, R., van Schrojenstein Lantman, H.W. *et al.* 2022*b*. Trace-element heterogeneity in rutile linked to dislocation structures: implications for Zr-in-rutile geothermometry. *Journal of Metamorphic Geology*, **41**, 3–24, https://doi.org/10.1111/jmg.12686

Vervoort, J. 2013. Lu–Hf dating: the Lu–Hf isotope system. *In*: Rink, W.J. and Thompson, J. (eds) *Encyclopedia of Scientific Dating Methods*. Springer, 1–20.

Vho, A., Rubatto, D., Putlitz, B. and Bouvier, A.-S. 2020. New reference materials and assessment of matrix effects for SIMS measurements of oxygen isotopes in garnet. *Geostandards and Geoanalytical Research*, **44**, 459–471, https://doi.org/10.1111/ggr.12324

Villaros, A., Stevens, G. and Buick, I.S. 2009. Tracking S-type granite from source to emplacement: clues from garnet in the Cape Granite Suite. *Lithos*, **112**, 217–235, https://doi.org/10.1016/j.lithos.2009.02.011

Villaros, A., Buick, I. and Stevens, G. 2012. Isotopic variations in S-type granites: an inheritance from a heterogeneous source? *Contributions to Mineralogy and Petrology*, **163**, 243–257, https://doi.org/10.1007/s00410-011-0673-9

Villaseca, C., Romera, C.M., De la Rosa, J. and Barbero, L. 2003. Residence and redistribution of REE, Y, Zr, Th and U during granulite-facies metamorphism: behaviour of accessory and major phases in peraluminous granulites of central Spain. *Chemical Geology*, **200**, 293–323, https://doi.org/10.1016/S0009-2541(03)00200-6

Viskupic, K. and Hodges, K.V. 2001. Monazite–xenotime thermochronometry: methodology and an example from the Nepalese Himalaya. *Contributions to Mineralogy and Petrology*, **141**, 233–247, https://doi.org/10.1007/s004100100239

Volante, S., Collins, W.J. *et al.* 2020*a*. Reassessing zircon–monazite thermometry with thermodynamic modelling: insights from the Georgetown igneous complex, NE Australia. *Contributions to Mineralogy and Petrology*, **175**, article 110, https://doi.org/10.1007/s00410-020-01752-7

Volante, S., Collins, W.J., Pourteau, A., Li, Z.X., Li, J. and Nordsvan, A.R. 2020*b*. Structural evolution of a 1.6 Ga orogeny related to the final assembly of the supercontinent Nuna: coupling of episodic and progressive deformation. *Tectonics*, **39**, article e2020TC006162, https://doi.org/10.1029/2020TC006162

Volante, S., Pourteau, A. *et al.* 2020*c*. Multiple P–T–d–t paths reveal the evolution of the final Nuna assembly in northeast Australia. *Journal of Metamorphic Geology*, **38**, 593–627, https://doi.org/10.1111/jmg.12532

Volante, S., Collins, W.J. *et al.* 2022. Spatio-temporal evolution of Mesoproterozoic magmatism in NE Australia: a hybrid tectonic model for final Nuna assembly. *Precambrian Research*, **372**, article 106602, https://doi.org/10.1016/j.precamres.2022.106602

von Blackenburg, F. 1992. Combined high-precision chronometry and geochemical tracing using accessory minerals: applied to the Central-Alpine Bergell intrusion (central Europe). *Chemical Geology*, **100**, 19–40, https://doi.org/10.1016/0009-2541(92)90100-J

Vry, J.K. and Baker, J.A. 2006. LA-MC-ICPMS Pb–Pb dating of rutile from slowly cooled granulites: confirmation of the high closure temperature for Pb diffusion in rutile. *Geochimica et Cosmochimica Acta*, **70**, 1807–1820, https://doi.org/10.1016/j.gca.2005.12.006

Vry, J., Compston, W. and Cartwright, I. 1996. SHRIMP II dating of zircons and monazites: reassessing the timing of high-grade metamorphism and fluid flow in the Reynolds Range, northern Arunta Block, Australia. *Journal of Metamorphic Geology*, **14**, 335–350, https://doi.org/10.1111/j.1525-1314.1996.00335.x

Wafforn, S., Seman, S., Kyle, J.R., Stockli, D., Leys, C., Sonbait, D. and Cloos, M. 2018. Andradite garnet U–Pb geochronology of the Big Gossan skarn, Ertsberg–Grasberg mining district, Indonesia. *Economic Geology*, **113**, 769–778, https://doi.org/10.5382/econgeo.2018.4569

Walter, B.F., Giebel, R.J., Steele-MacInnis, M., Marks, M.A., Kolb, J. and Markl, G. 2021. Fluids associated with carbonatitic magmatism: a critical review and implications for carbonatite magma ascent. *Earth-Science Reviews*, **215**, article 103509, https://doi.org/10.1016/j.earscirev.2021.103509

Walters, J.B. and Kohn, M.J. 2017. Protracted thrusting followed by late rapid cooling of the Greater Himalayan Sequence, Annapurna Himalaya, Central Nepal: insights from titanite petrochronology. *Journal of Metamorphic Geology*, **35**, 897–917, https://doi.org/10.1111/jmg.12260

Walters, J.B., Cruz-Uribe, A.M., Song, W.J., Gerbi, C. and Biela, K. 2022. Strengths and limitations of *in situ* U–Pb titanite petrochronology in polymetamorphic rocks: an example from western Maine, USA. *Journal of Metamorphic Geology*, **40**, 1043–1066, https://doi.org/10.1111/jmg.12657

Ward, C., McArthur, J. and Walsh, J. 1992. Rare earth element behaviour during evolution and alteration of the Dartmoor granite, SW England. *Journal of Petrology*, **33**, 785–815, https://doi.org/10.1093/petrology/33.4.785

Warren, C.J., Greenwood, L.V., Argles, T.W., Roberts, N.M.W., Parrish, R.R. and Harris, N.B.W. 2019. Garnet-monazite rare earth element relationships in sub-solidus metapelites: a case study from Bhutan. *Geological Society, London, Special Publications*, **478**, 145–166, https://doi.org/10.1144/SP478.1

Watson, E.B. 1979. Zircon saturation in felsic liquids: experimental results and applications to trace element geochemistry. *Contributions to Mineralogy and Petrology*, **70**, 407–419, https://doi.org/10.1007/BF00371047

Watson, E.B. 1996. Dissolution, growth and survival of zircons during crustal fusion: kinetic principals, geological models and implications for isotopic inheritance. *Earth and Environmental Science Transactions of the Royal Society of Edinburgh*, **87**, 43–56, https://doi.org/10.1017/S0263593300006465

Watson, E.B. and Harrison, T.M. 1984. Accessory minerals and the geochemical evolution of crustal magmatic systems: a summary and prospectus of experimental approaches. *Physics of the Earth and Planetary Interiors*, **35**, 19–30, https://doi.org/10.1016/0031-9201(84)90031-1

Watson, E.B. and Harrison, T.M. 2005. Zircon thermometer reveals minimum melting conditions on earliest Earth. *Science (New York, NY)*, **308**, 841–844, https://doi.org/10.1126/science.1110873

Watson, E.B., Wark, D.A. and Thomas, J.B. 2006. Crystallization thermometers for zircon and rutile. *Contributions to Mineralogy and Petrology*, **151**, 413–433, https://doi.org/10.1007/s00410-006-0068-5

Wawrzenitz, N., Krohe, A., Baziotis, I., Mposkos, E., Kylander-Clark, A.R.C. and Romer, R.L. 2015. LASS

U–Th–Pb monazite and rutile geochronology of felsic high-pressure granulites (Rhodope, N Greece): effects of fluid, deformation and metamorphic reactions in local subsystems. *Lithos*, **232**, 266–285, https://doi.org/10.1016/j.lithos.2015.06.029

Webster, J.D. and Piccoli, P.M. 2015. Magmatic apatite: a powerful, yet deceptive, mineral. *Elements*, **11**, 177–182, https://doi.org/10.2113/gselements.11.3.177

Weinberg, R.F., Wolfram, L.C., Nebel, O., Hasalová, P., Závada, P., Kylander-Clark, A.R.C. and Becchio, R. 2020. Decoupled U–Pb date and chemical zonation of monazite in migmatites: the case for disturbance of isotopic systematics by coupled dissolution–reprecipitation. *Geochimica et Cosmochimica Acta*, **269**, 398–412, https://doi.org/10.1016/j.gca.2019.10.024

Wetherill, G.W. 1956. Discordant uranium–lead ages, I. *Eos, Transactions American Geophysical Union*, **37**, 320–326, https://doi.org/10.1029/TR037i003p00320

Wetherill, G.W. 1963. Discordant uranium–lead ages: 2. Discordant ages resulting from diffusion of lead and uranium. *Journal of Geophysical Research (1896–1977)*, **68**, 2957–2965, https://doi.org/10.1029/JZ068i010p02957

White, R.W. and Powell, R. 2002. Melt loss and the preservation of granulite facies mineral assemblages. *Journal of Metamorphic Geology*, **20**, 621–632, https://doi.org/10.1046/j.1525-1314.2002.00206_20_7.x

White, R.W., Powell, R. and Johnson, T.E. 2014. The effect of Mn on mineral stability in metapelites revisited: new a–x relations for manganese-bearing minerals. *Journal of Metamorphic Geology*, **32**, 809–828, https://doi.org/10.1111/jmg.12095

Whitehouse, M.J. and Kemp, A.I.S. 2010. On the difficulty of assigning crustal residence, magmatic protolith and metamorphic ages to Lewisian granulites: constraints from combined *in situ* U–Pb and Lu–Hf isotopes. *Geological Society, London, Special Publications*, **335**, 81–101, https://doi.org/10.1144/SP335.5

Whitehouse, M.J. and Platt, J.P. 2003. Dating high-grade metamorphism – constraints from rare-earth elements in zircon and garnet. *Contributions to Mineralogy and Petrology*, **145**, 61–74, https://doi.org/10.1007/s00410-002-0432-z

Whitehouse, M.J., Kemp, A.I.S. and Petersson, A. 2022. Persistent mildly supra-chondritic initial Hf in the Lewisian Complex, NW Scotland: implications for Neoarchean crust–mantle differentiation. *Chemical Geology*, **606**, article 121001, https://doi.org/10.1016/j.chemgeo.2022.121001

Williams, I.S., Compston, W., Black, L., Ireland, T. and Foster, J. 1984. Unsupported radiogenic Pb in zircon: a cause of anomalously high Pb–Pb, U–Pb and Th–Pb ages. *Contributions to Mineralogy and Petrology*, **88**, 322–327, https://doi.org/10.1007/BF00376756

Williams, M.A., Kelsey, D.E. and Rubatto, D. 2022. Thorium zoning in monazite: a case study from the Ivrea–Verbano Zone, NW Italy. *Journal of Metamorphic Geology*, **40**, 1015–1042, https://doi.org/10.1111/jmg.12656

Williams, M.L. and Jercinovic, M.J. 2002. Microprobe monazite geochronology: putting absolute time into microstructural analysis. *Journal of Structural Geology*, **24**, 1013–1028, https://doi.org/10.1016/S0191-8141(01)00088-8

Williams, M.L., Jercinovic, M.J. and Hetherington, C.J. 2007. Microprobe monazite geochronology: understanding geologic processes by integrating composition and chronology. *Annual Review of Earth and Planetary Sciences*, **35**, 137–175, https://doi.org/10.1146/annurev.earth.35.031306.140228

Williams, M.L., Jercinovic, M., Harlov, D., Budzyń, B. and Hetherington, C. 2011. Resetting monazite ages during fluid-related alteration. *Chemical Geology*, **283**, 218–225, https://doi.org/10.1016/j.chemgeo.2011.01.019

Williams, M.L., Jercinovic, M.J., Mahan, K.H. and Dumond, G. 2017. Electron microprobe petrochronology. *Reviews in Mineralogy and Geochemistry*, **83**, 153–182, https://doi.org/10.2138/rmg.2017.83.5

Wing, B.A., Ferry, J.M. and Harrison, T.M. 2003. Prograde destruction and formation of monazite and allanite during contact and regional metamorphism of pelites: petrology and geochronology. *Contributions to Mineralogy and Petrology*, **145**, 228–250, https://doi.org/10.1007/s00410-003-0446-1

Wu, Y.-B., Zheng, Y., Zhang, S., Zhao, Z., Wu, F. and Liu, X. 2007. Zircon U–Pb ages and Hf isotope compositions of migmatite from the North Dabie terrane in China: constraints on partial melting. *Journal of Metamorphic Geology*, **25**, 991–1009, https://doi.org/10.1111/j.1525-1314.2007.00738.x

Wu, Y.-B., Gao, S., Zhang, H.-F., Yang, S.-H., Jiao, W.-F., Liu, Y.-S. and Yuan, H.-L. 2008*a*. Timing of UHP metamorphism in the Hong'an area, western Dabie Mountains, China: evidence from zircon U–Pb age, trace element and Hf isotope composition. *Contributions to Mineralogy and Petrology*, **155**, 123–133, https://doi.org/10.1007/s00410-007-0231-7

Wu, Y.-B., Zheng, Y.-F., Gao, S., Jiao, W.-F. and Liu, Y.-S. 2008*b*. Zircon U–Pb age and trace element evidence for Paleoproterozoic granulite-facies metamorphism and Archean crustal rocks in the Dabie Orogen. *Lithos*, **101**, 308–322, https://doi.org/10.1016/j.lithos.2007.07.008

Wudarska, A., Wiedenbeck, M. *et al.* 2020. SIMS- and IRMS-based study of apatite reference materials reveals new analytical challenges for oxygen isotope analysis. *22nd EGU General Assembly Conference,* 4–8 May 2020, online event, Abstracts, 18841.

Yakymchuk, C. 2017. Behaviour of apatite during partial melting of metapelites and consequences for prograde suprasolidus monazite growth. *Lithos*, **274**, 412–426.

Yakymchuk, C. and Brown, M. 2014. Behaviour of zircon and monazite during crustal melting. *Journal of the Geological Society*, **171**, 465–479.

Yakymchuk, C., Kirkland, C.L. and Clark, C. 2018. Th/U ratios in metamorphic zircon. *Journal of Metamorphic Geology*, **36**, 715–737, https://doi.org/10.1111/jmg.12307

Yang, M., Yang, Y.-H. *et al*. 2022. Natural allanite reference materials for *in situ* U–Th–Pb and Sm–Nd isotopic measurements by LA-(MC)-ICP-MS. *Geostandards and Geoanalytical Research*, **46**, 169–203, https://doi.org/10.1111/ggr.12417

Yang, Y., Sun, J., Xie, L., Fan, H. and Wu, F. 2008. *In situ* Nd isotopic measurement of natural geological materials by LA-MC-ICPMS. *Chinese Science Bulletin*, **53**, 1062–1070, https://doi.org/10.1007/s11434-008-0166-z

Yang, Z., Wang, B., Cui, H., An, H., Pan, Y. and Zhai, J. 2015. Synthesis of crystal-controlled TiO_2 nanorods by a hydrothermal method: rutile and brookite as highly active photocatalysts. *The Journal of Physical Chemistry C*, **119**, 16905–16912, https://doi.org/10.1021/acs.jpcc.5b02485

Yurimoto, H., Duke, E., Papike, J. and Shearer, C. 1990. Are discontinuous chondrite-normalized REE patterns in pegmatitic granite systems the results of monazite fractionation? *Geochimica et Cosmochimica Acta*, **54**, 2141–2145, https://doi.org/10.1016/0016-7037(90)90277-R

Zack, T. and Kooijman, E. 2017. Petrology and geochronology of rutile. *Reviews in Mineralogy and Geochemistry*, **83**, 443–467, https://doi.org/10.2138/rmg.2017.83.14

Zack, T., Kronz, A., Foley, S.F. and Rivers, T. 2002. Trace element abundances in rutiles from eclogites and associated garnet mica schists. *Chemical Geology*, **184**, 97–122, https://doi.org/10.1016/S0009-2541(01)00357-6

Zack, T., Moraes, R. and Kronz, A. 2004*a*. Temperature dependence of Zr in rutile: empirical calibration of a rutile thermometer. *Contributions to Mineralogy and Petrology*, **148**, 471–488, https://doi.org/10.1007/s00410-004-0617-8

Zack, T., von Eynatten, H. and Kronz, A. 2004*b*. Rutile geochemistry and its potential use in quantitative provenance studies. *Sedimentary Geology*, **171**, 37–58, https://doi.org/10.1016/j.sedgeo.2004.05.009

Zack, T., Stockli, D.F., Luvizotto, G.L., Barth, M.G., Belousova, E., Wolfe, M.R. and Hinton, R.W. 2011. *In situ* U–Pb rutile dating by LA-ICP-MS: ^{208}Pb correction and prospects for geological applications. *Contributions to Mineralogy and Petrology*, **162**, 515–530, https://doi.org/10.1007/s00410-011-0609-4

Zhang, H.-X., Jiang, S.-Y., Yuan, F. and Liu, S.-Q. 2022. LA-(MC)-ICP-MS U–Th–Pb dating and Nd isotopes of allanite in NYF pegmatite from lesser Qingling orogenic belt, central China. *Ore Geology Reviews*, **145**, article 104893, https://doi.org/10.1016/j.oregeorev.2022.104893

Zhang, L., Wu, J.-L., Zhang, Y.-Q., Yang, Y.-N., He, P.-L., Xia, X.-P. and Ren, Z.-Y. 2021. Simultaneous determination of Sm–Nd isotopes, trace-element compositions and U–Pb ages of titanite using a laser-ablation split-stream technique with the addition of water vapor. *Journal of Analytical Atomic Spectrometry*, **36**, 2312–2321, https://doi.org/10.1039/D1JA00246E

Zhang, L.-S. and Schärer, U. 1996. Inherited Pb components in magmatic titanite and their consequence for the interpretation of U–Pb ages. *Earth and Planetary Science Letters*, **138**, 57–65, https://doi.org/10.1016/0012-821X(95)00237-7

Zhang, W., Wang, Z. *et al.* 2019. Determination of Zr isotopic ratios in zircons using laser-ablation multiple-collector inductively coupled-plasma mass-spectrometry. *Journal of Analytical Atomic Spectrometry*, **34**, 1800–1809, https://doi.org/10.1039/C9JA00192A

Zhong, X., Andersen, N.H., Dabrowski, M. and Jamtveit, B. 2019. Zircon and quartz inclusions in garnet used for complementary Raman thermobarometry: application to the Holsnøy eclogite, Bergen Arcs, western Norway. *Contributions to Mineralogy and Petrology*, **174**, article 50, https://doi.org/10.1007/s00410-019-1584-4

Zhong, X., Moulas, E. and Tajčmanová, L. 2020. Post-entrapment modification of residual inclusion pressure and its implications for Raman elastic thermobarometry. *Solid Earth*, **11**, 223–240, https://doi.org/10.5194/se-11-223-2020

Zhou, Y., Yao, J. *et al.* 2016. Improving the molecular ion signal intensity for *in situ* liquid SIMS analysis. *Journal of the American Society for Mass Spectrometry*, **27**, 2006–2013, https://doi.org/10.1007/s13361-016-1478-x

Zhu, X.K. and O'Nions, R.K. 1999. Monazite chemical composition: some implications for monazite geochronology. *Contributions to Mineralogy and Petrology*, **137**, 351–363, https://doi.org/10.1007/s004100050555

Zirner, A.L., Marks, M.A., Wenzel, T., Jacob, D.E. and Markl, G. 2015. Rare earth elements in apatite as a monitor of magmatic and metasomatic processes: the Ilímaussaq complex, South Greenland. *Lithos*, **228**, 12–22, https://doi.org/10.1016/j.lithos.2015.04.013

Zucali, M., Corti, L., Roda, M., Ortolano, G., Visalli, R. and Zanoni, D. 2021. Quantitative X-ray maps analysis of composition and microstructure of Permian high-temperature relicts in acidic rocks from the Sesia–Lanzo Zone eclogitic continental crust, Western Alps. *Minerals*, **11**, article 1421, https://doi.org/10.3390/min11121421

Al and H incorporation and Al-diffusion in natural rutile and its high-pressure polymorph TiO_2 (II)

Bastian Joachim-Mrosko[1]*, Jürgen Konzett[1], Thomas Ludwig[2], Thomas Griffiths[3], Gerlinde Habler[3], Eugen Libowitzky[4] and Roland Stalder[1]

[1]University of Innsbruck, Faculty of Geo- and Atmospheric Sciences, Institute of Mineralogy and Petrography, Innrain 52, 6020 Innsbruck, Austria

[2]Institute of Earth Sciences, Heidelberg University, Im Neuenheimer Feld 234-236, 69120 Heidelberg, Germany

[3]Department of Lithospheric Research, University of Vienna, Josef-Holaubek-Platz 2, 1090 Vienna, Austria

[4]Department of Mineralogy and Crystallography, University of Vienna, Josef-Holaubek-Platz 2, 1090 Vienna, Austria

BJ-M, 0000-0003-2134-004X
*Correspondence: bastian.joachim@uibk.ac.at

Abstract: Rutile is an important accessory mineral in metamorphic rocks and is used as a geothermobarometer or geochronometer. This study aims to bridge the gap between diffusion studies in simplified and complex natural systems by investigating the incorporation and mobility of Al in natural rutile and its high-pressure polymorph TiO_2 (II).

Experiments were performed at 0.1 MPa to 7 GPa, 1223–1373 K, at buffered $\mu(Al_2O_3)$ and with fO_2 constrained to $\leq$CCO, which is the equilibrium between graphite and a CO-CO_2 gas phase. Based on electron probe microanalysis, secondary ion mass spectrometry and Fourier transform infrared analyses, we suggest a complex combination of mechanisms to explain the incorporation of Al and H in natural rutile and TiO_2 (II). This includes: (1) the incorporation of Al^{3+} on octahedral Ti-sites charge balanced by the formation of oxygen vacancies; and (2) the incorporation of oxygen in interstitial positions charge balanced by hydrogen interstitials.

Determined Al-diffusivities in natural TiO_2 are approximately eight to nine orders of magnitude faster compared to previously published data. A possible explanation includes a significantly enhanced rate of ionic diffusion through the combined effect of hydrolytic weakening, enhanced Al-diffusion through extended defects and to a minor extent oxygen fugacity variations. Consequently, results of this study question that the inferred high closure temperatures for the Al-in-rutile geothermobarometer can be applied to all natural systems.

Supplementary material: Detailed information on (1) EPMA and FTIR analyses of the starting materials, (2) reference materials for the H_2O determination in rutile and TiO_2 (II), (3) the calibration for H_2O analyses using SIMS, (4) EPMA and SIMS data of the concentration profiles in rutile and TiO_2 (II) and (5) FSD and EBSD analyses are available at https://doi.org/10.6084/m9.figshare.c.6829367

Rutile (TiO_2) is a common accessory mineral in many medium- to ultrahigh-grade metamorphic rocks. Because of its comparatively high density and mechanical and chemical stability, it is also a widespread detrital phase in sedimentary rocks. Rutile is frequently used as a tool to deduce a rock's *P–T–t* history as a geothermometer (e.g. Zack *et al.* 2004; Watson and Harrison 2005; Watson *et al.* 2006; Tomkins *et al.* 2007; Kooijman *et al.* 2015), geobarometer (Hwang *et al.* 2000), geochronometer (e.g. Mezger *et al.* 1989; Vry and Baker 2006; Kooijman *et al.* 2010; Li *et al.* 2011) or geospeedometer (e.g. Blackburn *et al.* 2012; Cruz-Uribe *et al.* 2014; Kohn 2020). Rutile shows a great affinity towards the incorporation of hydrogen and has been described as one of the most 'hydrous' nominally anhydrous minerals (NAMs) (e.g. Vlassopoulos *et al.* 1993; Bromiley *et al.* 2004). This is of particular importance for the transport of water into the deep Earth's interior through subduction of oceanic crust, because hydrogen incorporation in NAMs such as rutile and its high-pressure polymorphs may represent an important mechanism for storage and release of water at pressure and temperature conditions beyond the stability of hydrous minerals, i.e. beyond the depths of arc magmatism and even into the lower mantle (Bell and Rossmann 1992; Bromiley *et al.* 2004).

Rutile thermometry, geochronology and geospeedometry, as well as rutile's ability to record

From: van Schijndel, V., Cutts, K., Pereira, I., Guitreau, M., Volante, S. and Tedeschi, M. (eds) 2024. *Minor Minerals, Major Implications: Using Key Mineral Phases to Unravel the Formation and Evolution of Earth's Crust.* Geological Society, London, Special Publications, **537**, 123–147.
First published online October 18, 2023, https://doi.org/10.1144/SP537-2022-187

elemental and isotopic signatures and to store and transport hydrogen all depend on solid state diffusion of trace elements through its crystal lattice (Dohmen *et al.* 2019). A very good knowledge of trace element incorporation mechanisms, mobilities and solubilities in rutile are therefore a prerequisite for these applications. Trace elements may be incorporated into rutile through homovalent $Ti^{4+} \leftrightarrow M^{4+}$ (Zr^{4+}, Mo^{4+}, Sn^{4+}, Hf^{4+}, U^{4+}) exchange or as coupled substitution mechanisms that can include hexavalent (W^{6+}, U^{6+}), pentavalent (Nb^{5+}, Sb^{5+}, Ta^{5+}), trivalent (Al^{3+}, Sc^{3+}, V^{3+}, Cr^{3+}, Fe^{3+}, Y^{3+}) and divalent (Fe^{2+}, and to a lesser degree Mn^{2+}, Zn^{2+}) cations as well as oxygen defects (Meinhold 2010). The key mechanisms for Al and H incorporation in rutile and its high-pressure polymorphs that are relevant for this particular study are listed in Table 1. Numerous studies have already investigated the effect of extrinsic parameters such as pressure, temperature or oxygen fugacity on solubility and/or trace element diffusion processes in rutile and its high-pressure polymorphs. Investigated trace elements include for example in chronological order H (Johnson *et al.* 1975), Sc, Zr, Cr, Mn, Fe, Co and Ni (Sasaki *et al.* 1985), H (Swope and Smyth 1995), Pb (Cherniak 2000), H and ferric iron (Bromiley *et al.* 2004), H, Al, Cr, Ga, Mg and Ca (Bromiley and Hilaret 2005), Zr and Hf (Cherniak *et al.* 2007), He (Cherniak and Watson 2011), Al (Escudero *et al.* 2011), V and Nb (Liu *et al.* 2014), Zr, Hf, Nb and Ta (Dohmen *et al.* 2019), Al and Si (Cherniak and Watson 2019) and Al (Hoff and Watson 2021).

The concentration of Zr in rutile coexisting with quartz and zircon is based on the homovalent $Zr^{4+} \leftrightarrow Ti^{4+}$ exchange and represents an important example of a geothermometer that may preserve post-crystallization equilibration temperatures (Sasaki *et al.* 1985; Degeling 2003; Zack *et al.* 2004; Watson *et al.* 2006; Cherniak *et al.* 2007; Ferry and Watson 2007; Tomkins *et al.* 2007; Dohmen *et al.* 2019). However, comparatively fast Zr diffusivities in rutile imply intracrystalline redistribution operating at rather low temperatures, so that the closure temperature for this geothermometer is in the range of 750°C (Dohmen *et al.* 2019). Recent results of Cherniak and Watson (2019) indicated that the diffusivity of Si is approximately two to four and the diffusivity of Al approximately five to seven orders of magnitude slower than Zr diffusion in rutile, implying that the Al-in-rutile geothermobarometer devised by Hoff and Watson (2018) has the potential to be a much more robust indicator of post-magmatic or metamorphic crystallization/equilibration pressures and temperatures.

The application of the Al-in-rutile geothermometer, and indeed of all other trace element concentrations in rutile for geochronology or geospeedometry, requires two assumptions: firstly that element mobilities during geological events subsequent to initial equilibration are well understood; and secondly that closure temperatures are well calibrated. However, the presence of dislocation microstructures such as twin boundaries or low-angle to high-angle grain boundaries have been proposed to affect the distribution and mobility of trace elements in natural rutile (Zack and Kooijman 2017; Moore *et al.* 2020; Verberne *et al.* 2022). If this was true, a direct application of simplified laboratory experiments, which usually use synthetic or gem-quality rutile starting materials, to natural systems could be problematic. The present study aims to test if the available diffusion data for Al in rutile and its high-pressure polymorph TiO_2 (II) are generally applicable to all natural systems by experimentally investigating the diffusivity of Al in natural rutile embedded in a matrix of natural metapelitic gneiss from the Rhodope Massif in Greece, at pressures ranging from 0.1 MPa to 7 GPa, temperatures ranging from 1223 to 1373 K, at buffered $\mu(Al_2O_3)$, and with oxygen fugacity constrained to values $\leq$CCO, which is the equilibrium between graphite and a $CO\text{-}CO_2$ gas phase.

Table 1. Overview of important incorporation mechanisms for Al, Fe and H in rutile and its high-pressure polymorph TiO_2 (II) that are relevant for this particular study

Incorporation mechanism	References
$Ti^X_{Ti} \leftrightarrow Fe'_{Ti} + H^{\cdot}_{i}$	Swope and Smyth (1995)
$Ti^X_{Ti} \leftrightarrow Ti'_{Ti} + H^{\cdot}_{i}$	Khomenko *et al.* (1998)
$Ti^X_{Ti} \leftrightarrow Al'_{Ti} + H^{\cdot}_{i}$	Johnson *et al.* (1973)
IR-band at 3279 cm^{-1}: H potentially unassociated with compositional impurities (charge balance not solved)	Khomenko *et al.* (1998)
$Ti^X_{Ti} + 2O^X_O \leftrightarrow V''''_{Ti} + Ti^{\cdot\cdot\cdot}_{i} + 2V^{\cdot\cdot}_{O} + 3e'$	Kofstad (1972)
$2Ti^X_{Ti} \leftrightarrow 2Al'_{Ti} + V^{\cdot\cdot}_{O}$	Sayle *et al.* (1995); Gesenhues and Rentschler (1999)
$3Ti^X_{Ti} \leftrightarrow 3Al'_{Ti} + Al^{\cdot\cdot\cdot}_{i}$ (only in TiO_2(II))	Escudero *et al.* (2012)
$2Ti^X_{Ti} \leftrightarrow Al'_{Ti} + M^{\cdot}_{Ti}$ ($M = Nb^{5+}$, Ta^{5+}etc.)	Johnson *et al.* (1973); Vlassopoulos *et al.* (1993)

Note that incorporation mechanisms are given in simplified form using Kröger-Vink notation (Kröger and Vink 1956).

Table 2. Average compositions of starting materials: a garnet–kyanite–muscovite gneiss from the Rhodope Massif in northern Greece (RG) was used as matrix

	SiO_2	Al_2O_3	Fe_2O_3	MnO	MgO	CaO	Na_2O	K_2O	TiO_2	P_2O_5	Cr_2O_3	V_2O_3	LOI	Total
RG	60.58	17.29	7.94	0.293	3.97	0.57	1.3	4.12	1.06	0.06			1.84	99.02
Rutile		0.03 (2)	0.27 (4)						99.67 (63)		0.13 (2)	0.45 (5)		100.5 (7)

Natural rutile grains were separated from an eclogite from the Tauern Window in the Eastern Alps. The average RG bulk rock composition was analysed by Activation Laboratories Ltd (Report No. A03-2940); the average rutile composition is based on 44 EPMA analyses (Supplementary material 1). Uncertainties are given as 1σ. LOI, loss on ignition.

Experimental details

Starting materials

Each experimental charge contained three to five natural rutile grains with a size of *c.* 200–500 µm in diameter that were separated from an eclogite of the Tauern Window located in the Eastern Alps. These grains were embedded in a fine-grained powder of a natural garnet–kyanite–muscovite gneiss originating from the Rhodope Massif in northern Greece (Mpokos *et al.* 2009). Averages of both starting material compositions are presented in Table 2. A detailed characterization of the rutile grains used as starting material is presented in Supplementary material 1 proving that all Al and H concentration profiles presented in this study cannot be inherited from the starting material. In one experiment (BJ-IBK-ru4), the natural gneiss matrix was replaced by pure synthetic Al_2O_3 powder. Starting materials were placed into 2.0 (outer)/1.8 mm (inner) diameter Pt100 capsules fitted with an inner graphite liner and welded shut. The graphite liner was used to keep the experimental charge separated from the Pt and thus to prevent loss of iron to the platinum during the course of the experiment. In addition, the graphite constrained the oxygen fugacity to values $\leq$CCO.

Experiments

Before the experiments, we made sure that the rutile grains used as starting material showed homogeneous trace element distributions so that we could exclude that the analysed concentration profiles were inherited from the starting material (Table 2; Supplementary material 1). Experimental run conditions are presented in Table 3. Experiments at ambient pressures were performed in a conventional box furnace. High-pressure experiments were performed in a piston-cylinder apparatus (PC) at $P \leq 3.5$ GPa or in a multi-anvil press (MA) at $P > 3.5$ GPa.

Box furnace. For all three experiments performed at 0.1 MPa, the capsules were placed in a platinum crucible and transferred to a conventional box furnace that was pre-heated to the experimental target temperature of 1373 K. After 2 hours and 35 minutes, the experiments were quenched instantly by dropping the capsule into cold water.

Table 3. Experimental conditions and observed phase assemblages with minor phases in parentheses

Sample	Pressure (GPa)	Temperature (K)	Duration (h:min)	Phase assemblage
BJ-IBK-ru3	0.0001	1373	2:35	Rt + Grt + Qz + melt (+ Mnz + Zrn + Ky)
BJ-IBK-ru4	0.0001	1373	2:35	Rt + Crd
JKI_172	2	1373	80:00	Rt + Grt + Qz + melt (Zrn + Ky + Mnz)
BJ-IBK-ru1	2.5	1323	24:42	Rt + Grt + Qz + melt (+ Ky + Opx + Zrn + Mnz)
JKI_175	3	1373	80:00	Rt + Grt + Qz (Zrn + Ky + Mnz)
BJ-IBK-ru2	3.5	1373	24:35	Rt + Grt + Qz (Zrn + Ky + Mnz)
MA31	4	1273	187:22	Rt + Grt + Phn + Coe + melt (+ Cpx + Ky)
MA83	4	1273	2:03	Rt + Grt + Phn + Coe (+ Cpx + Ky)
MA81	4	1273	23:59	Rt + Grt + Phn + Coe (+ Cpx + Ky)
MA22	5	1273	145:30	Rt + Grt + Phn + Coe (+ Cpx + Ky)
MA1	6	1223	239:55	Rt + Grt + Phn + Coe (+ Ky)
MA27	7	1273	169:15	TiO_2 (II) + Grt + Phn + Coe (+ Cpx + Ky + Mnz)

In experiment BJ-IBK-ru4, pure synthetic Al_2O_3 powder was used as matrix.
Coe, coesite; Crd, corundum; Cpx, clinopyroxene; Grt, garnet; Ky, kyanite; Mnz, monazite; Opx, orthopyroxene; Phn, phengite; Qz, quartz; Rt, rutile; Zrn, zircon.

Piston-cylinder apparatus. PC experiments were performed with a Boyd-and-England-type piston-cylinder apparatus (Boyd and England 1960). The pressure assembly consisted of a 1/2′ talc-Pyrex tube and a straight-walled graphite heater with the capsule (length 4–5 mm) positioned in a container made of sintered MgO powder near the hotspot of the graphite furnace. Sample pressures were calibrated using the CsCl melting reaction (McDade *et al.* 2002) based on the CsCl melting curve established by Clark (1959) with an overall uncertainty in pressure determination of ± 0.1 GPa. Temperatures were determined with an uncertainty of $\pm 10°$ C by using a $Pt_{90}R_{10}$–Pt (S-type) thermocouple that was positioned at the bottom of the sample capsule. Runs were terminated by cutting off the transformer power. Sample capsules were then removed from the assembly, mounted longitudinally in epoxy and ground to expose the centre of the experimental charge. This procedure ensures that at least two to three TiO_2 grains embedded in the recrystallized gneiss matrix became available for examination and that the original phase distribution in the sample capsule is preserved. In a final step, the surface of the charges was polished to a 0.25 µm finish.

Multi-anvil press. MA experiments were performed using a 1000 t Walker-type device with 18/11 assemblies consisting of pre-fabricated MgO–Cr_2O_3 octahedra with graphite furnaces and 32 mm WC cubes. The preparation of sample capsules for PC and MA runs was almost identical with the only difference being that the total capsule length for MA experiments was 3.0–3.5 mm. Run temperatures were measured using D-type W–Re thermocouples placed at the bottom of the respective sample capsule. Experimental and calibration procedures are similar to those described by Keppler and Frost (2005) and Konzett and Frost (2009). After termination of the runs and removal from their assemblies, the capsules were prepared for analysis by analogy with procedures used for PC runs.

Analyses

Electron microprobe. Major and minor element compositions of TiO_2 grains were analysed with a JEOL 8100 Superprobe electron microprobe (EPMA) using a 15 kV acceleration voltage, 50 nA beam current and the following counting times on peaks and backgrounds of the respective $K\alpha$ X-ray lines: Ti: 20 s on peak and 10 s on background (20/10); Al, Cr and V: 60/30; Fe and Nb: 100/50. These analytical conditions lead to the following typical instrumental relative 2σ errors for element concentrations of an individual EPMA analysis: 4.8% for 0.1 wt% Al_2O_3 and 1.4% for 1.0 wt% Al_2O_3; 13% for 0.15 wt% Fe_2O_3 and 3% for 0.5 wt% Fe_2O. The following reference materials were used: Ti and Al: pure synthetic TiO_2 and Al_2O_3; Fe, V and Cr: natural almandine, V-grossular and chromite, respectively; Nb: pure synthetic $Nb_2O_{4.98}$. Raw counts were converted to element concentrations with the φ–ρ–Z correction procedure. Starting points for analytical profiles on the rims of the TiO_2 grains after the experiments were located adjacent to a silica polymorph (quartz or coesite) grain or epoxy resin (in case parts of the recrystallized gneiss matrix were lost during sample preparation) to avoid beam overlap with phases high in Al and Fe. Special care was taken that these profiles were analysed at positions that were not affected by any dissolution or recrystallization (see 'Results' section for details). The experimental sample size and geometry did not allow orienting the rutile grains before EPMA and secondary ion mass spectrometry (SIMS) analyses. The analytical profiles were analysed perpendicular to the respective TiO_2 grain boundary with a total length of 52–130 µm and a minimum increment of 3 µm between two individual analyses.

The spatial distribution of the oxygen concentration in the outermost portion of one TiO_2 grain from sample MA27 was determined using the relative intensity of the O-$K\alpha$ line analysed with an LDE1 crystal in a 60×20 µm^2 area.

Forescatter detector imaging and electron backscatter diffraction analyses. Following microprobe analysis, samples JKI_175, MA22, BJ-IBK-ru1 and BJ-IBK-ru2 were polished for 3 hours with colloidal silica to achieve the surface quality necessary for forescatter detector (FSD) imaging and electron backscatter diffraction (EBSD) analyses, which were carried out using an FEI Quanta 3D FEG field-emission gun scanning electron microscope (FEG-SEM) at the laboratory for field-emission scanning electron microscopy and focused ion beam applications at the Faculty of Geosciences, Geography and Astronomy, University of Vienna. The SEM is equipped with an FSD mounted on the EBSD tube of an EDAX Apollo XV system (elevation angle 5°), which comprises a Digiview 5 EBSD camera. All analyses were acquired with the sample tilted to 70°, at a working distance of 14 to 16 mm, using a 15 kV accelerating voltage, analytical mode (1 mm SEM aperture) and spot size 1 (*c.* 4 nA probe current). FSD images were acquired on all samples with the detector retracted by *c.* 10 mm from the EBSD position to maximize orientation contrast (OC). Detailed EBSD scans of sample JKI_175 used a camera binning of 8×8 pixels, a Hough θ step size of 1°, binned pattern size of 140 pixels and a 9×9 convolution mask, and were acquired on square grids with step sizes ranging from 80 to 180 nm. The orientations of rutile grains in the remaining three samples were acquired using maps with step sizes >1 µm, using a 2 × 2 pixel camera binning, but with Hough parameters unchanged. EBSD data were processed and analysed using the Matlab toolbox MTEX (Bachmann *et al.* 2010), version 5.6.1. EBSD data with confidence index values below 0.1 were reclassified as not indexed. Pixels belonging to grains (15° misorientation angle threshold) less than 2 pixels in size were removed before recalculating grains with a 2° threshold for final analysis.

Secondary ion mass spectrometry. H concentration profiles were analysed by SIMS using a Cameca ims3f ion probe at Heidelberg University. Primary $^{16}O^-$ ions with an energy of 14.5 keV, a beam current of *c.* 1 nA and a spot diameter of *c.* 5 µm were used. Secondary ions were accelerated to 4.5 keV, and we applied energy filtering with an energy window of 75 ± 20 eV. The mass resolving power (@ 10% of peak intensity) was *c.* 400 and the imaged field was limited to a diameter of 6 µm. The H background in the sample chamber was reduced by cooling a plate in the sample chamber with liquid nitrogen. H, ^{27}Al, ^{50}Ti and ^{56}Fe were analysed with integration times of 8 s/cycle for H and 2 s/cycle for the other isotopes. Each spot was pre-sputtered for 300 s followed by 16 acquisition cycles. Al and Fe concentrations were calculated using relative ion yields (RIY). RIYs for Al and Fe were determined on NIST SRM610 glass (note that these semiquantitative Al and Fe concentrations were only used to compare the SIMS profiles with the EPMA profiles).

H concentrations were quantified using a working curve with a fixed intercept of zero determined on two reference rutile samples (12 grains of sample Sb-Rt and 14 grains of sample CM15/01, see Supplementary material 2) using Fourier transform infrared spectroscopy (FTIR) H concentration data. The low H RIY (which is the slope of the working curve if the intercept is zero) of $5.1 \times 10^{-3} \pm 3 \times 10^{-4}$ (1σ standard error) is most likely a consequence of H_2O loss during the SIMS analyses (this applies to both the reference rutiles and the unknown samples) and its large standard deviation is caused by large inter- and intra-sample variability (Supplementary materials 2 and 3). Intra-sample variations (variation between grains) can be attributed to the small volume sampled by SIMS (<500 μm^3) compared with the volume sampled by FTIR spectroscopy (>1.5×10^5 μm^3). The accuracy of absolute H_2O concentrations is therefore not the best while the reproducibility of analyses on a single crystal is typically much better (1σ <5%). H concentrations were not corrected for the H background of the SIMS analyses. H_2O concentration uncertainties based on error propagation of the working curve fit for all analyses are presented in Supplementary material 4. These uncertainties are large (+23%, −45% 1σ) for the lowest H_2O concentrations measured (*c.* 260 µg g^{-1}) and less than ±20% (1σ) for the higher concentrations.

Fourier transform infrared spectroscopy. Infrared (IR) spectra and images were acquired using a Bruker Vertex 70 FTIR spectrometer combined with a Hyperion 3000 microscope equipped with a Globar light source, a mercury–cadmium–telluride (MCT) detector and a focal plane array (FPA) detector. Measurements were performed using unpolarized light in transmission mode in the 3000–4000 cm^{-1} wavenumber range, which is the characteristic interval for vibrations of hydrous species. Thirty-two scans per spectra were averaged with a spectral resolution of 2 cm^{-1}. The aperture size was 20 × 100 μm^2. Backgrounds were subtracted by using a third-order polynomial line function in the 3000–3600 cm^{-1} region.

Detailed images were recorded with the FPA detector displaying the OH-defect distribution in a TiO_2 grain. The FPA detector consists of 64 × 64 MCT detectors generating an image consisting of 4096 pixels that are focused on a 170 × 170 μm^2 area of the sample. Images were recorded with a spectral resolution of 4 cm^{-1} in a 2 × 2 binning mode, leading to a spatial pixel resolution of 5.4 × 5.4 μm^2, which is close to the theoretical resolution of 3.75 × 3.75 μm^2 given by the numerical aperture (0.4) and the wavelength of the IR radiation (e.g. 3 µm at 3300 cm^{-1}). For the presentation of integrated absorbances, a linear background between 3100 and 3450 cm^{-1} was subtracted.

Results

Phase assemblage

A detailed list of the observed major and minor phases in each experiment is provided in Table 3. In the temperature range of this study (1223–1373 K), temperature variations have no effect on the observed phase assemblage. Experiments performed at pressures above 3.5 GPa show the following major phase assemblage: rutile below 7 GPa or α-PbO_2-type TiO_2 (II) at 7 GPa + garnet + phengite + coesite (Fig. 1a). At pressures between 0.1 MPa and 3.5 GPa, experimental run products show the major phase assemblage rutile + garnet + quartz + melt (Fig. 1b). It is important to note that kyanite was observed as a minor phase and quartz or coesite as a major phase in all experiments that used natural Rhodope gneiss as matrix and were widely distributed throughout the whole matrix with no particular textural relationship (Fig. 1c), so that it can be safely assumed that the presence of kyanite + quartz or coesite buffers $\mu(Al_2O_3)$. Garnet zonation

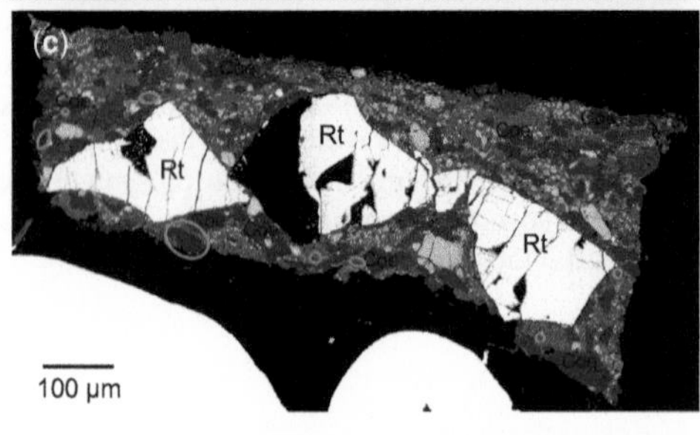

Fig. 1. (**a**, **b**) Exemplary backscattered electron (BSE) images of phase assemblages obtained after (a) a high-pressure, MA27 (7 GPa, 1273 K, 169, 25 h) and (b) a comparably lower-pressure, JKI_172 (2.5 GPa, 1373 K, 80 h) experiment. Both experiments show a large rutile grain in the centre of the capsule. (**a**) The experiment performed at 7 GPa shows α-PbO-type TiO_2 (II) and as additional major phases garnet (Grt), coesite (Coe) and phengite (Phn). (**b**) In the experiment performed at 2 GPa, rutile and as additional major phases quartz (Qz), garnet (Grt) and melt were identified. Minor phases that are visible in this particular BSE image are monazite (Mnz) and kyanite (Ky). White arrows denote irregular crystal shapes that indicate rutile (Rt) dissolution or (re-)crystallization on the micrometre-scale during the experiment. Mineral abbreviations after Warr (2021). (**c**) Overview of the exposed sample surface after experiment MA31. Kyanite (red circles) and coesite are distributed all over the sample with no specific textural relationship indicating that $\mu(Al_2O_3)$ is buffered throughout the experiment.

that can be observed in all samples (Fig. 1a–c), is not dependent on *P–T* conditions or experimental run durations and is thus inherited from the starting material. Experiment BJ-IBK-ru4 is an exception, because pure synthetic Al_2O_3 powder was used as a matrix in this particular sample, so that the observed phase assemblage consists of only rutile and corundum with no additional phases (Table 3). After polishing, all experiments show a section of at least two large TiO_2 grains with sizes of 0.2–0.5 mm exposed at the sample surface. Some TiO_2 grains show a few isolated short boundary segments (white arrows in Fig. 1b) that indicate dissolution or recrystallization.

Al-diffusion profiles in rutile and TiO_2 (II) and equilibrium concentrations with the surrounding matrix

Al, Fe and H are homogeneously distributed in rutile grains used as starting material with no concentration profiles towards the grain boundary, so it can be excluded that the observed concentration profiles are inherited from the starting material (Table 2; Supplementary material 1).

Microprobe and SIMS line scans reveal concentration profiles of Al_2O_3 with a length of 20 μm to more than 140 μm, with most profiles extending between 30 and 80 μm until the respective background level concentrations are reached. The Al-diffusion profiles presented in this study are those that decrease to the lowest background concentration in each sample, which ensures that the effect of the 3D-TiO_2 grain geometry below the surface is minimized (Fig. 2a, b; Supplementary material 4). An example of a set of concentration profiles of sample MA27 with concentrations normalized to one at the TiO_2 grain boundary is presented in Figure 2a. Good agreement of the two displayed Al_2O_3 and FeO profiles in sample MA27 (Fig. 2a), which were analysed with two independent methods, EPMA and SIMS, at an identical position, give confidence that the analysed concentration profiles are correct. A time series for Al-profiles in rutile is presented in Figure 2b. Data for this time series plotted as concentration v. X^2/t fall all within error on the same line (Fig. 2c), which shows that Al-diffusivities are not affected by the experiment duration. Only μ (Al_2O_3) was buffered throughout the experiment through the presence of kyanite + quartz or coesite buffers (Fig. 1c), while the activities of H and Fe are unbuffered, so only Al-diffusivities were determined. Al-diffusion profiles were fitted to the solution for a one-dimensional semi-infinite diffusion setting with a constant concentration boundary condition (fixed boundary condition C_i) at the margin and initial concentration C_0 according to Crank

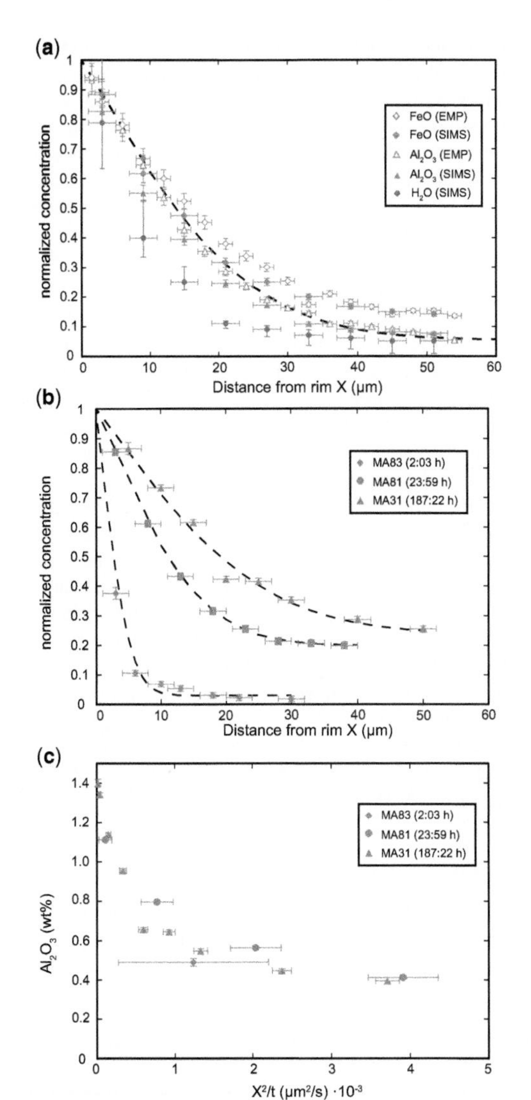

Fig. 2. (**a**, **b**) Elemental diffusion profiles in TiO_2analysed perpendicular to the orientation of the grain boundary. The respective concentration at the grain boundary is normalized to 1. Uncertainties are given as 1σ-values and do not include the uncertainty of the respective oxide concentrations in standard materials (see text for details). (**a**) Exemplary diffusion profiles of FeO_{tot}, Al_2O_3 and H_2O in α-PbO-type TiO_2 (II) in experiment MA27. Al_2O_3 (open triangles) and FeO_{tot} (open diamonds) concentrations were analysed using the electron microprobe (EMP). The accuracy of both methods was cross-checked by analysing Al_2O_3 (filled triangles) and FeO_{tot} (filled diamonds) concentrations on the identical rutile grain in sample MA27 with secondary ion mass spectrometry (SIMS). H_2O abundances (filled circles) were determined using SIMS. Diffusion profiles for FeO_{tot} and Al_2O_3 are based on data points displayed as open symbols. The mobility of hydrogen in TiO_2 (II) is significantly slower compared to that of FeO_{tot} and Al_2O_3. (**b**) Time series showing the Al-diffusion into rutile at 4 GPa and 1273 K for different run durations. The dotted line for sample MA83 represents not a fit but the calculated diffusion profile for $\log D_{Al}^{Rt}$ ($m^2\ s^{-1}$) $= -15$, which agrees well with diffusivities determined for samples MA81 and MA31 (Table 3). (**c**) Data for the identical time series plotted as Al_2O_3 (wt%) v. X^2/t. Good agreement of the individual datapoints of all three concentration profiles indicates that the Al_2O_3 diffusivity is not affected by the experimental duration.

(1975), which is $\frac{C-C_0}{C_i-C_0} = \mathrm{erf}\left(\frac{X}{\sqrt{4Dt}}\right)$ (Table 4). Note that this analytical approach comes with limitations that are mainly related to the experimental approach, in which irregular-shaped grains were imbedded in a powder matrix. These are discussed in detail below. The obtained diffusivities D for Al show a strong temperature dependence while the effect of pressure seems to be small, if present at all (Fig. 3). However, the experiments in this study do not allow us to provide an Arrhenius relationship for Al-diffusivities as the effect of P and fO_2 cannot be excluded (see 'Experimental or analytical artefacts' section for details). It is therefore important to note that the Al-diffusivities are only correct for the P–T–$\mu(Al_2O_3)$–fO_2 conditions chosen in this particular study. Furthermore, the chosen conditions reflect typical natural conditions and thus our findings have potentially wide implications for the interpretation of natural samples (see 'Discussion' section for details).

The intercept of a calculated diffusion profile with the y-axis represents the equilibrium concentration of the respective element in TiO_2 with the surrounding matrix if boundary conditions were constant throughout the experiment (Table 4). This is the case for Al as $\mu(Al_2O_3)$ was buffered through the presence of kyanite + quartz. The Al_2O_3 equilibrium concentration in TiO_2 shows a significant increase from 0.3 (1) to 1.0 (1) wt% between 0.1 MPa and 2 GPa at T varying between 1323 and 1373 K (Table 4) before reaching a plateau value between 2.0 and 3.5 GPa. Between 3.5 GPa/1373 K and 4 GPa/1273 K, a stepwise increase can be observed (1.0 (1) to 1.5 (1) wt%). A further stepwise increase (1.5 (1) to 2.2 (1) wt%) can be observed between 6 GPa/1223 K and 7 GPa/1273 K (Table 4; Fig. 4).

FeO concentration profiles appear to have the same lengths compared to Al_2O_3 concentration profiles, while hydrogen transport into TiO_2 seems to be significantly slower compared to Al and Fe between 3.5 and 7 GPa and at temperatures below 1373 K (Fig. 2a; Supplementary material 4). In contrast, Cr_2O_3 abundances in TiO_2 show no concentration variations at least between 3.5 and 7 GPa (Supplementary material 4). However, Fe, H and Cr mobilities are not discussed further in this study because

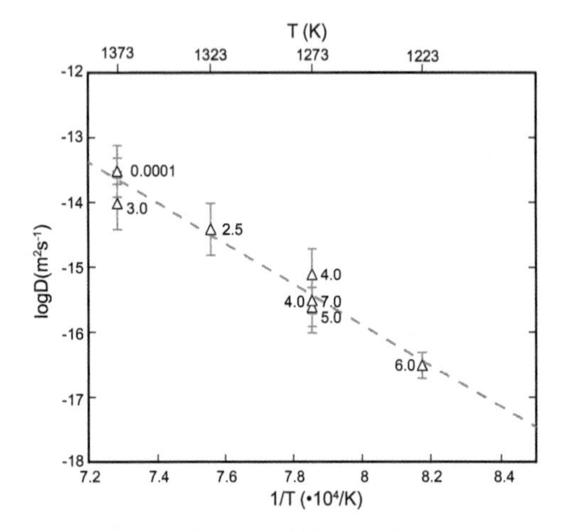

Fig. 3. Diagram of inverse aluminium diffusivities v. temperature in TiO_2. Uncertainties are given as 1σ-values. Labels at each datapoint represent pressures in GPa. An activation energy is not given, as a potential minor effect of pressure or oxygen fugacity on the Al-diffusivity cannot be excluded. A strong temperature dependence on Al-diffusivities is indicated by the dashed line.

the activities of these elements at the TiO_2–matrix interface were not buffered throughout the experiments.

Oxygen defect distribution

A map and a ten-point moving average profile line of the oxygen distribution in TiO_2 in sample MA27 are presented in Figure 5. A complex oxygen concentration profile is observed. The map and profile show a slight decrease in oxygen counts from the TiO_2 grain boundary towards the centre of the grain, with the minimum abundance being located at a distance of *c.* 25 μm from the TiO_2 grain boundary. This is followed by an increase within 5 μm before reaching background level abundance.

OH-defect distribution

IR spectra analysed at the rim of the TiO_2 grains show in the OH stretching region an overlap of three absorption bands located at 3278, 3290 and 3324 cm^{-1} (Fig. 6a). IR spectra also reveal a significant change in absorbance from the rim towards the centre of the grain (Fig. 6b), leading to only two overlapping absorption bands at 3278 and 3290 cm^{-1} (Fig. 6c) in the centre. In addition, while the absorbance of the band at 3278 cm^{-1} increases by approximately a factor of four, the band located at 3290 cm^{-1} is reduced by approximately a factor of two relative to IR spectra analysed at the rim of the TiO_2 grain (Fig. 6b). The band at 3324 cm^{-1} appears towards the rim of the TiO_2 grain (Fig. 6a, b), which can be clearly visualized through FTIR imaging (Fig. 7a). As a consequence, the total water concentration in the respective TiO_2 increases accordingly (Fig. 7b).

Rutile microstructure

The crystallographic orientation of grains in four different samples was determined to investigate if anisotropy effects can have an effect on the

Table 4. Calculated Al-diffusivities and equilibrium concentrations of Al_2O_3 in $TiO_2$with the surrounding matrix at fixed $\mu(Al_2O_3)$

Sample	$\log D_{Al}^{Rt}$ ($m^2\ s^{-1}$)	Al_2O_3 (wt%)	Angle of profile vector to *a* (°)	Angle of profile vector to *c* (°)	*uvw* of EPMA profile vector[†]
BJ-IBK-ru3	−13.5 (2)	0.3 (1)			
BJ-IBK-ru4*	n.d.	0.45 (9)			
JKI_172	n.d.	0.94 (9)			
BJ-IBK-ru1	−14.4 (4)	0.8 (1)	57.8	33.4	−1 −4 10
JKI_175	−14.0 (4)	1.08 (7)	34.0	57.6	5 1 5
BJ-IBK-ru2	−13.5 (4)	1.0 (1)	55.8	51.7	4 −4 7
MA31	−15.5 (4)	1.6 (1)			
MA81	−15.1 (4)	1.3 (1)			
MA22	−15.6 (4)	1.5 (1)	47.4	55.7	5 −7 9
MA1	−16.5 (2)	1.5 (1)			
MA27	−15.5 (2)	2.2 (2)			

Uncertainties are given as 2σ values. All analysed Al_2O_3 concentration profiles in sample BJ-IBK-ru4 and JKI_172 show a plateau (Supplementary material 4), so that no diffusion coefficient could be determined. For the four samples where EBSD data are available, the orientation of the compositional profile is expressed relative to the *a*- and *c*-axes of rutile, and as a vector in crystal coordinates.
[*]Al-saturated conditions.
[†]Rounded to an individual [*uvw*] maximum value of ten.

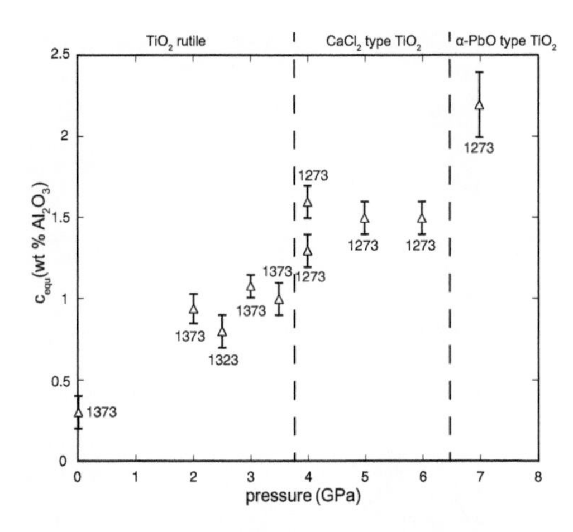

Fig. 4. Calculated equilibrium concentration of Al_2O_3 in rutile and TiO_2 (II) with the surrounding matrix at fixed $\mu(Al_2O_3)$ v. pressure. Only experiments that used gneiss as matrix are presented in this study. Labels at each datapoint represent experimental temperatures in K. The Al_2O_3 equilibrium concentration shows an increase between 3.5 and 4 GPa attributed to the transition of tetragonal rutile to an orthorhombic $CaCl_2$-type TiO_2 structure during quenching, which is associated with the incorporation of aluminium on octahedral interstices in the rutile lattice (Escudero *et al.* 2012). It also shows a stepwise increase between 6 and 7 GPa, which can be related to a phase transition to α-PbO_2-type TiO_2 (II) at high pressures (Bendeliany *et al.* 1966). The apparent increase of the Al_2O_3 equilibrium concentration in rutile between 1 bar and 2 GPa is not related to a phase transition.

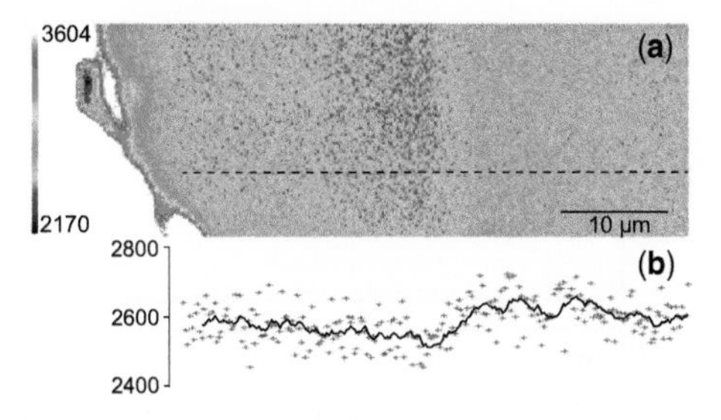

Fig. 5. Relative distribution of the oxygen abundance in a TiO_2 grain in sample MA27 analysed using the electron microprobe. Numbers on the scale bars refer to counts on the detector. (**a**) Mapping of the relative oxygen distribution. The mapping parameters are as follows: 15 kV acceleration voltage and 50 nA beam current; mapping area 300 × 100 pixels with 0.2 μm pixel size and O-Kα counting time of 1200 ms/pixel. The dotted line marks the position of the profile displayed in (**b**) that shows all individual datapoints analysed along the profile (light grey crosses) and a ten-point moving average (black line). The starting point of the ten-point moving average profile line is set to a distance of 3 μm to the TiO_2 grain boundary to make sure that the observed relative changes in oxygen abundances are solely related to the actual oxygen abundance variations in the TiO_2 grain and not affected by the matrix. Both the map and the profile reveal an oxygen minimum abundance at a distance of *c.* 25 μm from the rim of the TiO_2 grain.

determined diffusivities. Note that all samples were indexed as tetragonal rutile with very good fit values, because the $CaCl_2$-type polymorph (sample MA22) has only a very minor deviation from a tetragonal structure (Escudero *et al.* 2012). EBSD results show that all diffusion profiles were measured in different, but general, orientations. The angles of the profile direction relative to the rutile *a*- and *c*-axes are provided in Table 4, along with the [*uvw*] indices of the rutile direction parallel to the profile vector.

The microstructure of two rutile grains in sample JKI_175 was studied in detail using OC (FSD) imaging and EBSD. Note that FSD images of TiO_2 grains in samples MA22, BJ-IBK-ru1 and BJ-IBK-ru2 show identical features, so the results reported for sample JKI_175 are highly likely representative of all starting material in the experiments. Multiple sets of lamellae with different orientations are visible in OC images, some completely cross-cutting rutile grains and others tapering to end at sharp points within grains (Fig. 8a, b; Supplementary material Figs 5_1a, b and 5_2a, b). EBSD measurements reveal that all lamellae are {101} rutile twins (Fig. 8f; Supplementary material Figs 5_1e, 5_2c). Some twins terminate against differently oriented lamellae with tapered points. In other cases, one twin set is cross-cut and offset by another, in some cases with a small subgrain located at the site of intersection (Supplementary material Fig. 5_2). Finally, some wider lamellae contain thin twins of a second lamella orientation, which do not extend beyond them (Supplementary material Fig. 5_1a, b). While these features are present in both rutile grains in sample JKI_175, the frequency of twins varies considerably between the two grains.

FSD images also reveal gradual orientation changes within the large rutile grains (including within twin lamellae) and abrupt curved boundaries between differently oriented domains (Fig. 8a; Supplementary material Fig. 5_1a). Orientation variations are larger and abrupt transitions more frequent towards the rims of the rutile grains. EBSD crystal orientation mapping confirms and allows quantification of these observations. Profiles and maps of misorientation angle (relative to reference points far from grain rims) show total lattice rotations of up to *c.* 10° towards grain edges (Fig. 8c, d; Supplementary material Fig. 5_1c and d). Most of this lattice rotation occurs within the outermost *c.* 30 μm of the large grains, where lattice

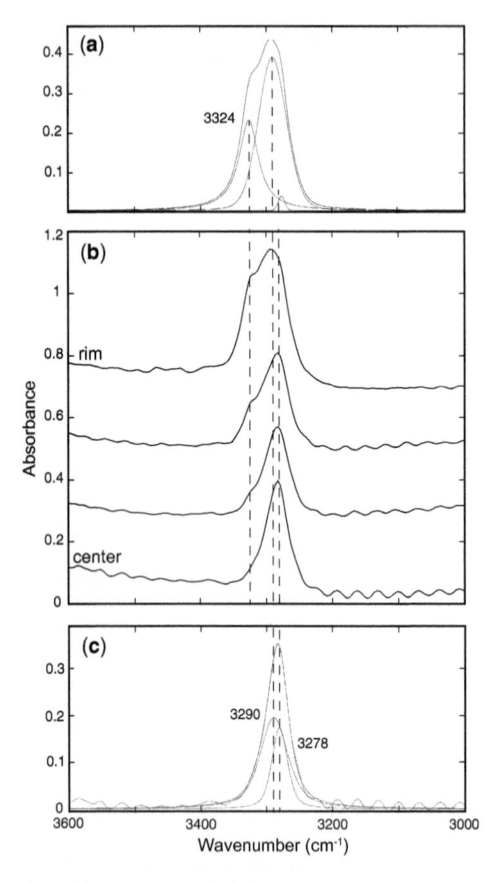

Fig. 6. Example of thickness-normalized IR spectra of a TiO_2 grain in experiment MA27 displayed between 3000 and 3600 cm^{-1}. (**a**) An IR spectrum at the rim of the grain shows a strong band at 3290 cm^{-1}, a small band at 3276 cm^{-1} and a shoulder that can be assigned to a band at 3324 cm^{-1}. A detailed discussion of the potential assignment of these bands to defects in the TiO_2 structure is given in the text. (**b**) Spectral evolution from the rim towards the centre of a TiO_2 grain that clearly shows increasing absorbance of a shoulder located at 3324 cm^{-1} towards the rim. (**c**) An IR spectrum at the centre of the TiO_2 grain shows decreasing absorbance of the band at 3290 cm^{-1} and increasing absorbance of the band at 3278 cm^{-1}, if compared to IR spectra analysed at the rim of the grain. The band located at 3324 cm^{-1} is not visible in this spectrum.

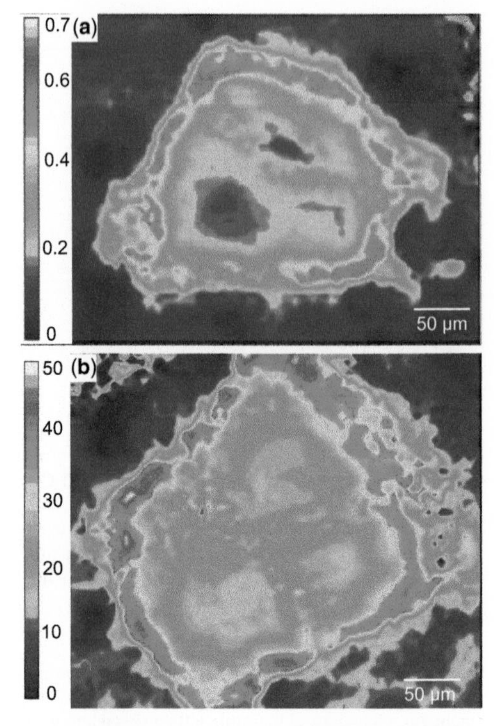

Fig. 7. (**a**) Fourier transform infrared (FTIR) image visualizing the linear absorbance ratio of the shoulder at 3324 cm^{-1} v. the main band at 3290 cm^{-1} (Fig. 6) of the TiO_2 grain in sample MA27. The systematic increase towards the TiO_2 grain boundary illustrates that the increase of this ratio is not a local effect. (**b**) FTIR image of another grain in sample MA27 visualizing relative differences in the total absorbance between 3150 and 3400 cm^{-1}. Again, the systematic increase towards the TiO_2 grain boundary illustrates that the increase of the water abundance towards the rim is not a local effect. Absence of these features in rutiles used as starting material (Supplementary material 1) proves that these features are not inherited but a result of the experiment.

curvatures of up to 1°/μm are maintained over lengths of several micrometres and orientation changes of ≥0.2°/μm are sustained over distances of at least 20 μm (Fig. 8c; Supplementary material Fig. 5_1c, d). Long-range orientation variations in inner regions of rutile grains are much smaller (Fig. 8d), with misorientation angles varying by less than 1° over distances of >50 μm (corresponding to ≤0.02°/μm). Although misorientation increases overall towards rims, regions with both positive and negative gradients are visible in misorientation angle profiles. Abrupt contrast changes in FSD images mostly correspond to misorientation changes of <2°. EBSD maps reveal that abrupt orientation changes occur over multiple EBSD pixels. As these small orientation changes still take place over distances of <1 μm, they are nonetheless referred to as low-angle grain boundaries in this paper.

Rarely, grains of rutile 10–20 μm in size are observed at the rim of the large rutile crystals (Fig. 8a; Supplementary material Fig. 5_1a, b). These grains contain no internal orientation variations or twin lamellae, and FSD images show in many cases a topographic contrast at their boundary

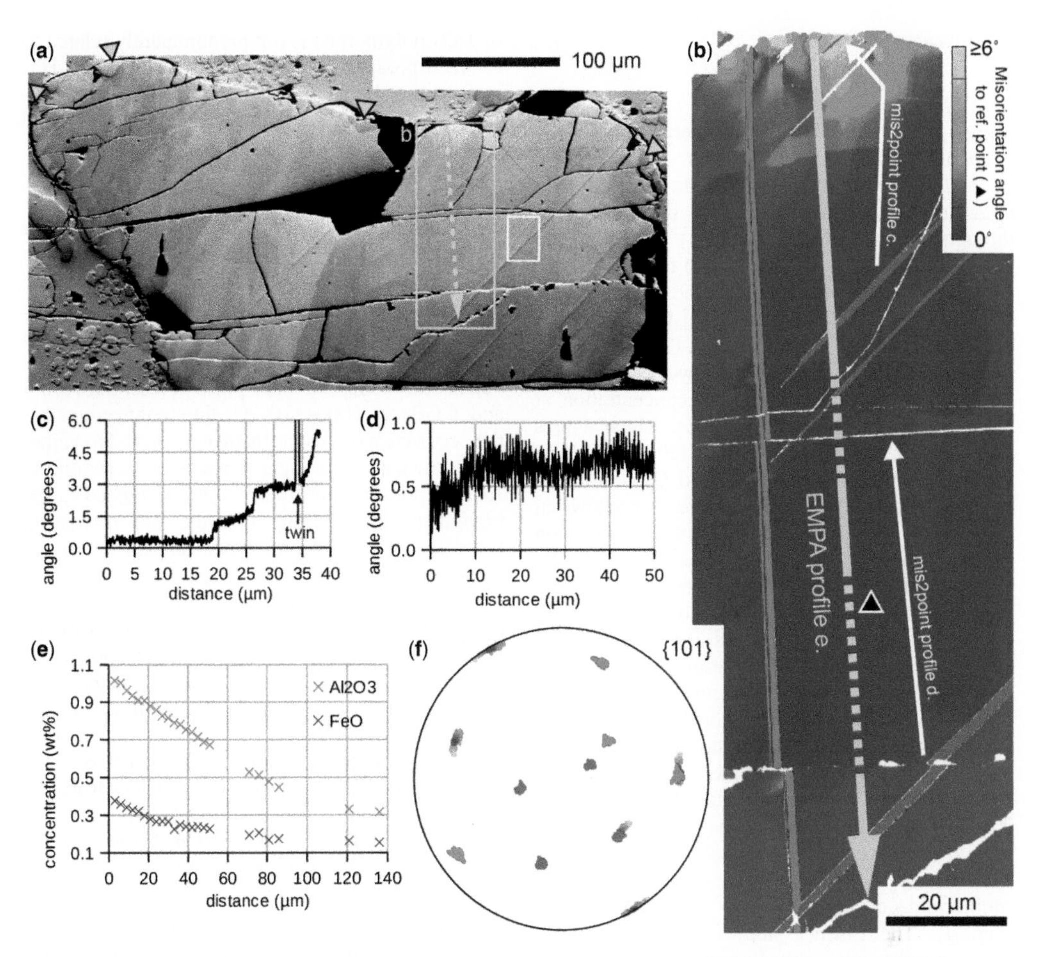

Fig. 8. Internal microstructures in a rutile grain from experiment JKI_175 and their relation to a compositional profile. (**a**) Forescatter detector (FSD) orientation contrast image of part of the grain. Locations of small grains with homogeneous orientation at the rim of the large distorted grain are indicated by yellow triangles. The location of the electron backscatter diffraction (EBSD) map in (b) is marked by the labelled yellow box, with the approximate trace of the electron microprobe (EPMA) profile in (e) within it. The white box indicates the location of the EBSD map in Supplementary material Fig. 5_2. (**b**) EBSD map (step size 80 nm) of the location marked in (a). The colour coding from blue to yellow indicates the misorientation angle of EBSD pixels in the main grain to the reference point (black triangle), from 0 to 6°. Twin lamellae of different orientations are highlighted in red and purple, but no information on exact pixel orientation is shown for twin lamellae. Decompression cracks are visible as horizontal traces offsetting twins. White arrows mark the position and direction of misorientation angle to (first) point profiles in (c) and (d). The acquisition direction and approximate trace of the EPMA profile in (e) is marked by the yellow arrow, with dashed sections indicating longer gaps in the profile. Note that the position of the profile is not exact, due to polishing in between EPMA and EBSD analysis. (**c, d**) Profiles of misorientation angle relative to the initial point of each profile, for the profile lines marked in (b). (**e**) Compositional profile for Al_2O_3 and FeO concentrations (both in wt%) that are based on EPMA from the rim of the crystal inwards, as indicated in (b). (**f**) Stereographic projection, upper hemisphere pole figure of rutile {101} plane normal orientations for the map shown in (b). Colour coding corresponds exactly to colour coding used in (b). Owing to the twinning relationship, some {101} maxima of the main orientation are partially obscured by points from the twin lamellae.

with the large grains. The outer boundary of these grains follows the outline of the large rutile crystal, while the inner boundary projects into the distorted crystal with a convex shape. EBSD analysis confirms that these grains are free from lattice distortion and reveals that they have no specific orientation relationship to the neighbouring large grain (Supplementary material Fig. 5_1e).

Discussion

TiO_2 phase transitions

The Al_2O_3 concentration at the outermost boundary of the rutile and TiO_2 (II) grains represents the Al_2O_3 concentration that is in equilibrium with the surrounding matrix. With $\mu(Al_2O_3)$ being buffered through the presence of kyanite + quartz, it can safely be assumed that variations in the equilibrium concentration and variations in the solubility are caused by similar processes. The Al_2O_3 equilibrium concentration at 10^5 Pa and 1373 K (0.3 (1) wt%) is approximately a factor of three lower compared to the respective Al_2O_3 equilibrium concentration at 2 GPa and 1373 K (0.94 (1) wt%), which agrees with findings from Escudero *et al.* (2012) and Hoff and Watson (2021) that describe Al_2O_3 solubilities in TiO_2. Data from this study further show an increase in the Al_2O_3 equilibrium concentration between 3.5 and 4 GPa, and between 6 and 7 GPa (Fig. 4). The first increase between 3.5 and 4 GPa may be attributed to the incorporation of aluminium into 0 ½ 0 octahedral interstices in the rutile lattice (Escudero *et al.* 2012). As a result, rutile remains tetragonal at experimental temperatures but is transformed into a $CaCl_2$-type polymorph during quenching due to an orthorhombic distortion of the rutile structure. This implies that the $CaCl_2$-type polymorph is solely a quench product (Escudero *et al.* 2012). Nevertheless, the additional incorporation of Al into interstices at pressures above 3.5 GPa provides an explanation for the increase of the Al_2O_3 equilibrium concentration in tetragonal rutile (Fig. 4). The distinct stepwise increase in the Al_2O_3 equilibrium concentration between 6 and 7 GPa is related to the transition to an orthorhombic α-PbO_2-type TiO_2 (II) polymorph, which agrees well with *P–T* conditions reported for the transition of pure TiO_2 (Akaogi *et al.* 1992; Withers *et al.* 2003; Escudero *et al.* 2012). Other studies have shown that the solubility of Al_2O_3 in TiO_2 increases not only with increasing pressure but also with increasing temperature (Hoff and Watson 2021), particularly at high pressures ≥4 GPa (Stebbins 2007; Escudero *et al.* 2012). Experimental data obtained from this study were not designed to test if this is also the case for the equilibrium concentration, as temperatures in experiments performed at pressures ≥4 GPa differ by only 50 K (Table 3).

Al and H incorporation mechanisms in TiO_2

The absence of any clearly observable absorption band at 3308 cm^{-1} (Fig. 6), which would be related to the formation of an Al^{3+} defect on a Ti-site charge balanced by the incorporation of H^+ (Bromiley and Hilaret 2005), indicates that hydrogen incorporation in TiO_2 polymorphs is not predominantly related to the incorporation of Al^{3+} charge balanced by H^+. In fact, the dominant Al_2O_3 incorporation mechanism into rutile is the substitution of Ti^{4+} by Al^{3+} either charge balanced by the formation of oxygen vacancies (Sayle *et al.* 1995; Gesenhues and Rentschler 1999) or by incorporation of Al^{3+} on interstitial sites (Escudero *et al.* 2012). The latter mechanism, which is responsible for a change in symmetry to a $CaCl_2$-type TiO_2 polymorph during quenching, has been observed above *c.* 2 GPa at 1300°C (Escudero *et al.* 2012). Results of this study imply that the incorporation of Al into 0 ½ 0 octahedral interstices takes place at least between 4 and 6 GPa at 1000–1100°C, thus confirming that the occurrence of this mechanism is shifted to higher pressures with decreasing temperature. In addition, Al incorporation into rutile might be related to the high dislocation density in rutile (see discussion below), as such microstructures can host significant amounts of impurities.

The absorption band at 3278 cm^{-1} may be assigned to H not associated with any compositional defect (Khomenko *et al.* 1998; Bromiley and Hilaret 2005), which is in accordance with spectroscopic investigations of natural rutile (Johnson *et al.* 1968; Vlassopoulos *et al.* 1993). If correct, the question of how charge balance is accomplished remains unsolved for this particular case. The absorption band at 3290 cm^{-1} may be attributed to a coupled substitution of Ti^{4+} by Fe^{3+} and H^+. However, the possibility that this band is related to the association of H^+ with Ti^{3+} cannot be ruled out completely (Bromiley and Hilaret 2005). In fact, the experimental conditions of this study, i.e. high temperatures (1223–1373 K) and a low oxygen fugacity (≤CCO), even favour this mechanism (Colasanti *et al.* 2011). Another potential mechanism that might explain the hydrogen incorporation is the coupling of H^+ with Cr^{3+} resulting in Cr'_{Ti}-$H_i^{\bullet}$ defects (using Kröger and Vink 1956 notation). However, this defect should result in an absorption band at 3257 cm^{-1} (Bromiley and Hilaret 2005), which cannot be observed in the respective IR spectra (Fig. 6).

Figure 9 depicts the hydrogen v. trivalent cation abundance in TiO_2 in sample MA27 in atoms per formula unit (apfu) at varying distances to the grain boundary (data are listed in Table 5). The maximum hydrogen concentration is above the 1:1 line for [Fe] and [Fe + Cr], which indicates that the sum of [Fe^{3+} + Cr^{3+}] on a Ti-site charge balanced by H^+ on an interstitial position cannot fully explain the incorporation of H^+ into the α-PbO-type TiO_2 structure. This agrees with findings of other studies (Bromiley and Hilaret 2005; Escudero *et al.* 2012). On the other hand, [Al^{3+} + Fe^{3+} + Cr^{3+}] overcompensate proton defects, except at the grain boundary. This implies that other substitution mechanisms for metal

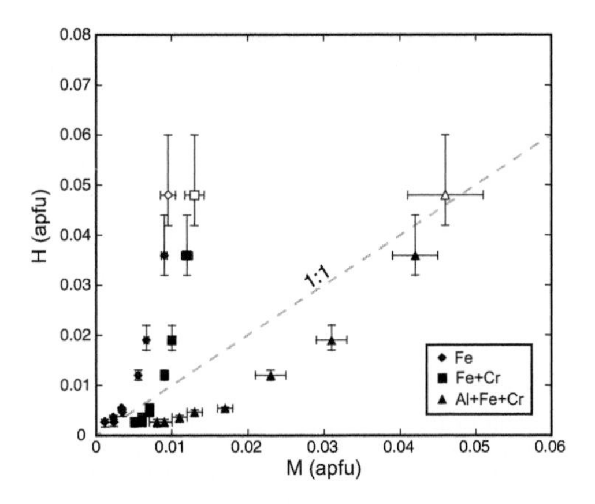

Fig. 9. Plot of hydrogen abundance v. Fe (diamond), Fe + Cr (square) and Al + Fe + Cr (triangle) cations in TiO_2 in atoms per formula unit (apfu). Open symbols represent calculated maximum concentrations at the boundary of the TiO_2 grain. Uncertainties are given as 1σ-values. Plotted data do not correlate with a 1:1 (dotted line) trend indicating that the incorporation of (Fe + Cr + Al) into the TiO_2 structure cannot be fully explained by the formation of $M'_{Ti}-H_i^{\cdot}$ defects alone except close to the grain boundary. In addition, the incorporation of H cannot be fully explained by charge balance of (Fe + Cr) and needs an additional substitution mechanism.

cations are necessary. This leaves the possibility that at least some of the protons are incorporated by additional mechanisms, which is supported by the fact that hydrogen concentration profiles are shorter compared to Al and Fe concentration profiles at least in experiments BJ-IBK-ru2, MA22 and MA27 (Fig. 2a; Supplementary material 4).

The complex oxygen profile (Fig. 5) can be used to evaluate additional potential substitution mechanisms that involve oxygen defects. A minimum oxygen concentration is observed at a distance of *c.* 25 µm from the grain boundary followed by a distinct increase. Note that that the formation of Magnéli phases, which form in oxygen deficient TiO_2 structures, cannot be excluded (Bursill and Hyde 1972). Formation of stable oxygen defects ($V_O^{\cdot\cdot}$) can be related to the incorporation of Al^{3+} on a Ti^{4+} position (Al'_{Ti}) charge balanced by oxygen defect formation (Sayle *et al.* 1995; Gesenhues and Rentschler 1999).

$$\frac{1}{2}Al_2O_3 + Ti^x_{Ti} + \frac{1}{2}O^x_O = Al'_{Ti} + \frac{1}{2}V_O^{\cdot\cdot} + TiO_2 \quad (1)$$

This mechanism would imply that the rate of oxygen defect formation can be correlated with the diffusivity of Al in TiO_2. Literature data on oxygen diffusivities in rutile only partially support this assumption. While the order of magnitude of those literature data determined at 1273 K and dry conditions parallel to *c* fit quite well to our determined Al-diffusivities, which is in particular the case for rutile that is doped with Cr-defects (Arita *et al.* 1979), other data determined at hydrothermal conditions are approximately two orders of magnitude slower (Moore *et al.* 1998). However, all these studies used synthetic rutile crystals, and a much wider scatter in diffusion coefficients may be expected in natural rutile samples (van Orman and Crispin 2010), which can explain the deviation from the results of this study. Further studies are therefore required that investigate the diffusivity of oxygen in natural TiO_2. If we assume that this mechanism is correct, this would imply that the minimum oxygen abundance should be located at the rim of the TiO_2 grain followed by an increase in oxygen concentration towards the grain's interior. Thus, formation of oxygen defects can explain the increase in

Table 5. Cations in atoms per formula unit in TiO_2 in sample MA27 with increasing distance to the rim of the grain boundary

Distance (µm)	Al	Fe	Cr	Al + Fe + Cr	H	Ti
0	*0.033(3)*	*0.0095(10)*	*0.0034(1)*	*0.046(5)*	*0.048(+12/−6)*	*0.903(91)*
3	0.030(2)	0.0090(5)	0.0035(1)	0.042(3)	0.036(+8/−4)	0.918(28)
9	0.021(1)	0.0066(3)	0.0035(1)	0.031(2)	0.019(+3/−2)	0.948(28)
15	0.014(1)	0.0055(3)	0.0036(1)	0.023(2)	0.012(+1/−1)	0.964(29)
21	0.010(5)	0.0033(2)	0.0036(1)	0.017(1)	0.0054(+2/−4)	0.978(29)
27	0.006(3)	0.0034(2)	0.0036(1)	0.013(1)	0.0047(+4/−9)	0.982(29)
33	0.005(2)	0.0022(1)	0.0036(1)	0.011(1)	0.0036(+4/−9)	0.986(30)
39	0.003(2)	0.0023(1)	0.0036(1)	0.009(1)	0.0027(+5/−9)	0.988(30)
45	0.003(2)	0.0011(1)	0.0036(1)	0.008(1)	0.0027(+5/−10)	0.989(30)
51	0.003(2)	0.0011(1)	0.0036(1)	0.008(1)	0.0027(+5/−10)	0.989(30)

Values in italics were calculated based on the estimated respective maximum concentration at the grain boundary (Supplementary material 4). Errors were calculated based on the uncertainties of the individual analyses. Error for H is the propagated error of the working curve fit (1 standard error).

oxygen concentration at a distance of *c.* 25–30 μm from the TiO_2 grain boundary but not the decrease in the oxygen abundance that is observable on the first 25 μm of the profile (Fig. 5).

Consequently, a second mechanism is required to explain the decrease in oxygen abundance, which must involve incorporation of oxygen into the TiO_2 structure at a slower rate compared to the formation of oxygen defects. A potential candidate for this mechanism is the incorporation of oxygen in interstitial positions charge balanced by hydrogen interstitials. In fact, it has been reported that oxygen can be present as oxygen interstitials in anatase and rutile TiO_2 (Kamisaki and Yamashita 2011).

$$H_2O_g = 2H_i^{\cdot} + O_i'' \qquad (2)$$

Several arguments support the assumption that this mechanism might be responsible for the combined incorporation of hydrogen and oxygen into the TiO_2 structure:

(1) The sum of metal impurity-related defects ($Al^{3+} + Fe^{3+} + Cr^{3+}$) cannot balance the incorporation of H^+ into the α-PbO-type TiO_2 (II) structure at the rim of the grain (Fig. 9), if we assume incorporation as an $M'_{Ti}-H_i^{\cdot}$ defect and consider that the incorporation of Al is at least partly charge balanced through the formation of oxygen defects (equation 1). This implies that hydrogen incorporation must be at least partially related to Ti^{3+} or oxygen defects.

(2) The position of the minimum oxygen abundance at a distance of 25 μm from the TiO_2 grain boundary (Fig. 5) coincides with the length of the H_2O concentration profile (Fig. 2), thus suggesting a correlation between the incorporation of hydrogen into the TiO_2 structure and the distribution of oxygen defects.

(3) Self-diffusion of hydrogen in rutile is at least four orders of magnitude faster than Al, Fe or oxygen diffusion due to its small size (van Orman and Crispin 2010), while the hydrogen concentration profiles in this study are shorter compared to those of Al and Fe (Fig. 2a; Supplementary material 4). This implies that hydrogen incorporation into TiO_2 must be related to the incorporation of another slow-diffusing component into the rutile structure. Incorporation of oxygen in interstitial positions charge balanced by hydrogen interstitials would fit to the observed slow diffusivities.

(4) IR spectra show increasing absorbance of a band located at 3324 cm^{-1} towards the rim of the TiO_2 grains (Figs 6 & 7) and it has been suggested that this band might be interpreted as $Ti'_{Ti}-H_i^{\cdot}$ defects (Khomenko *et al.* 1998). Formation of this defect would require only mobility of hydrogen and an electron and cannot explain the observed slow hydrogen diffusivities. However, the fact that this absorbance band has not been interpreted as indicating metal impurity-related defects fits well to the proposed mechanism (2), with incorporation of hydrogen interstitials being related to the formation of oxygen interstitials in TiO_2.

Al-diffusivities and the applicability of rutile as a geothermobarometer

Al and Si have recently been found to be among the slowest diffusing species in rutile studied so far (Cherniak and Watson 2019), justifying rutile as a potential robust indicator of crystallization/equilibration *P–T* conditions of rutile-bearing rocks. However, Al-diffusivities in TiO_2 determined in this study are approximately eight to nine orders of magnitude faster compared to those determined by Cherniak and Watson (2019) (Fig. 10). This discrepancy casts serious doubts on the general applicability of the Al-in-rutile geothermobarometer in all natural samples. The potential reasons for the vastly differing Al-diffusivities will be evaluated in the following section. It will first be shown that experimental and analytical artefacts (even in the most unfavourable combination) by far cannot explain the observed discrepancy between Al-diffusivities determined in this study and that determined by Cherniak and Watson (2019). It will then be shown that the combined effect of differences in the rutile and TiO_2 (II) microstructure, hydrolytic weakening and the effect of oxygen fugacity might serve as a possible explanation for the mismatch.

Experimental or analytical artefacts. Analytical results show that the analysed Al, Fe or H concentration profiles were not inherited from the starting material (Supplementary material 1). For Al incorporation into the TiO_2 grains, we can further safely assume a fixed boundary condition at the grain–matrix interface, with kyanite (Fig. 1c) and quartz or coesite being widely distributed in the matrix of each sample with no specific textural relationship, so that $\mu(Al_2O_3)$ is buffered. A time series performed at 4 GPa and 1273 K gives (within uncertainty) identical Al-diffusivities for experimental durations of 24 and 187 hours, which also fit to the concentration profile after an experimental duration of 2 hours (Fig. 2b, c). This shows that the experimental duration has no effect on the determined Al-diffusivities at a constant pressure of 4 GPa and a constant temperature of 1273 K. With these three prerequisites being fulfilled, Al-diffusion profiles were fitted to the solution for a one-dimensional semi-infinite diffusion setting with a constant concentration boundary condition (fixed boundary condition C_i) at the margin and initial concentration C_0 according to Crank (1975).

In this study, rutile grains were transferred from an eclogite to a garnet–kyanite–muscovite gneiss, which raises the question of whether dissolution or (re-)crystallization of rutile during the course of the experiments may have led to the obtained differences in Al-diffusivity. Indeed, some rutile grains show disequilibrium features indicating dissolution or overgrowth (white arrows in Fig. 1b). Therefore, positions of the analytical profiles were carefully selected to make sure that the analysed concentration profiles are not affected by any dissolution or (re-) crystallization features at least within the resolution of the backscattered electron (BSE) images. This implies that any dissolution or (re-)crystallization feature at the beginning of a profile must have a size of well below 0.1 µm, if present at all, which is equal to the uncertainty for the determination of the exact TiO_2-grain–matrix interface and very short compared to the length of the diffusion profiles that have a length of 20–140 µm, with the average length of Al-diffusion profiles being between 30 and 80 µm (Fig. 2; Supplementary material 4). Furthermore, if (1) the concentration profiles were strongly influenced by (re-)crystallization processes, and (2) Al-diffusion was extremely slow, then the obtained Al-concentration profiles would be controlled by Al-partitioning into rutile and we would expect a concentration profile with a plateau value for the Al-concentration in the area affected by rutile (re)crystallization, which is not the case. Nevertheless, an error of 0.1 log units for the determination of all Al-diffusion coefficients (Table 4; Fig. 3) has been added as the presence of such features cannot entirely be excluded.

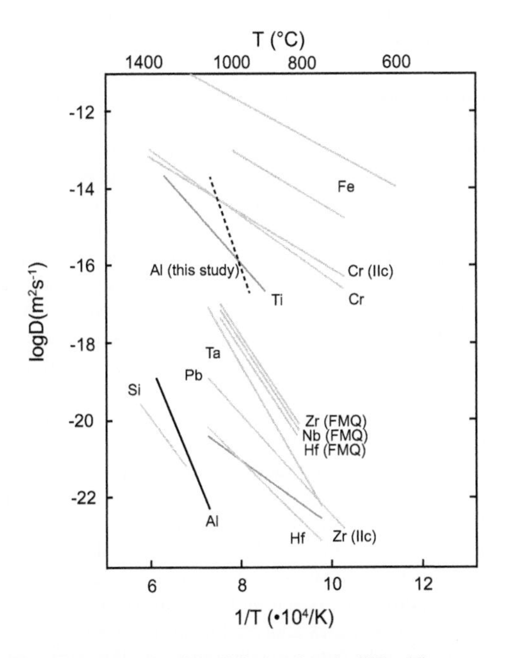

Fig. 10. Selected cation diffusion data in TiO_2. The bold dotted line highlights the Al-diffusivities in TiO_2 determined in this study (line is identical to the line presented in Fig. 3), which are approximately eight to nine orders of magnitude faster compared to Al-diffusivities determined by Cherniak and Watson (2019; bold line) in the identical temperature range. Source: data for Ti from Akse and Whitehurst (1978); Cr (perpendicular to *c*-axis), Cr (IIc, parallel to *c*-axis) and Fe from Sasaki *et al.* (1985); Pb from Cherniak (2000); Hf and Zr from Cherniak *et al.* (2007); Zr (fayalite–magnetite–quartz, FMQ), Nb (FMQ) and Hf (FMQ) from Dohmen *et al.* (2019); Ta from Marschall *et al.* (2013); Si and Al from Cherniak and Watson (2019); Al from this study.

The use of a 1D-model to estimate diffusion coefficients is a simplification, which is necessary as the experimental setup that was chosen to simulate natural conditions does not allow the reconstruction of the 3D-structure of the rutile grains above and below the exposed sample surface. In every sample, two to four Al-diffusion profiles were analysed at different positions. The Al-diffusion profiles presented in this study (Supplementary material 4) are those that decrease to the lowest background concentration in each sample, which ensures that the effect of the 3D-TiO_2 grain geometry below the surface is minimized. However, even these concentration profiles, particularly those of long experiments, show indeed that the Al core composition is slightly elevated compared to the initial composition (Table 2; Supplementary material 4). This affects the quality of the fit, as the background concentration C_i and all datapoints in the profile were slightly affected by 3D-diffusion related to the grain geometry. However, even in these cases, the 1D-model still provides diffusion data that are in very good agreement with the short diffusion profiles unaffected by this limitation (Figs 2 & 3; Table 4). This indicates that a 3D fit would only have a very minor effect on the determined diffusivities. Nevertheless, to account for this limitation, a factor of two has been added to the uncertainty of all Al-diffusivity estimates for diffusion profiles that show core concentrations slightly above the initial concentration (Fig. 3; Table 4).

With temperatures varying between 1273 and 1373 K, pressures varying between 10^5 Pa and 7 GPa, and with oxygen fugacity being constrained to values $\leq$CCO but not fixed, the individual effect of each of these parameters on Al-diffusivities can only be estimated (see details below). However, it cannot be excluded that each of these parameters has a (minor) effect on the estimated Al-diffusivities. This implies that each of the determined

Al-diffusivities (Table 4) are only valid for the *P*–*T*–fO_2 conditions of the individual experiment, while no activation energy or preexponential factor can be determined.

In summary, an experimental setup that aims to reproduce natural conditions comes with limitations. These are, however, not able to provide an explanation for the observed differences in Al-diffusivities between the study of Cherniak and Watson (2019) and this study.

Oxygen fugacity. Sample materials in this study were surrounded by an inner graphite capsule, which constrained fO_2 to values ≤CCO. At 1 bar and 1373 K, this is approximately five orders of magnitude below the oxygen fugacity in most of the experiments performed by Cherniak and Watson (2019), which were largely buffered at nickel–nickel oxide (NNO) by using a gas mixing furnace. At higher pressures, this difference is reduced to ≤2 log units between 2 and 7 GPa with the CCO buffer being more reducing between 2 and 4 GPa and more oxidizing between 4 and 7 GPa. The diffusivity of Al in TiO_2 can at least partly be correlated to the formation of oxygen defects (equation 1). This implies that the diffusivity of Al can be enhanced at more reducing conditions, if oxygen defect formation is the rate limiting factor for Al-diffusion. However, it has been shown in a recent study that a decrease in fO_2 by 5 log units will increase trace element diffusivities in rutile by only *c.* 2 log units (Dohmen *et al.* 2019). This is in line with further experimental results presented by Cherniak and Watson (2019), who performed three experiments buffered at iron-wuestite (IW). These show slightly enhanced Al-diffusivities in rutile of 0.4–0.7 log units compared to the NNO-buffered experiments. An experiment in an unbuffered ampoule, which likely resulted in more oxidizing conditions, slightly decreased the Al-diffusivity by *c.* 0.1 log units compared to experiments performed at NNO (Cherniak and Watson 2019). The relatively minor effect of fO_2 on Al-diffusivities in rutile is further supported by experimental results showing that the self-diffusion coefficient of oxygen in rutile at 1373 K increases by only one order of magnitude, if fO_2 decreases by six orders of magnitude (Millot and Picard 1988). This implies that the difference in oxygen fugacity can to a minor extent but not fully explain the observed difference in Al-diffusivities between the two studies.

Microstructures: twinning and extended defects. Another major difference between the two studies is that Cherniak and Watson (2019) used synthetic rutile as starting material to determine Al-diffusivities while the rutile used in this study is natural. The question arises of whether synthetic and natural rutiles differ in their defect structure strongly enough to cause significantly different Al-diffusion behaviour. This is of particular importance as the presence of dislocation microstructures such as twin boundaries or low-angle to high-angle grain boundaries have been proposed to affect the distribution and mobility of trace elements in natural rutile (Zack and Kooijman 2017; Moore *et al.* 2020; Verberne *et al.* 2022), because diffusion of elements along planar and line defects is orders of magnitude faster than diffusion through the lattice (Dohmen and Milke 2010) due to the local departure from the ideal crystal structure (Ruoff and Balluffi 1963). Both dislocations and planar defects have also been assumed to enhance diffusion rates in other geological materials, including feldspar (Yund *et al.* 1981), zircon (Timms *et al.* 2011; Piazolo *et al.* 2016), garnet (Griffiths *et al.* 2014; Hawemann *et al.* 2019), pyroxene (Chapman *et al.* 2019) and pyrite (Dubosq *et al.* 2019).

FSD images and EBSD maps of rutile grains in the experimental products reveal continuous lattice rotation, low-angle boundaries and multiple sets of twin lamellae (Fig. 8; Supplementary material 5). Lattice rotation and low-angle grain boundaries are clear evidence of crystal plastic deformation. Furthermore, the tapered lamella terminations and the existence of offsets and cross-cutting relationships between different orientations of twin lamellae indicate the sequential formation of twins and mechanical interaction between them, implying that these are deformation twins. Lattice rotation also occurs within twin lamellae (Fig. 8f; Supplementary material Fig. 5_1b, e), suggesting that crystal plastic deformation was simultaneous with or postdated deformation twinning. Overall, the microstructures observed are very similar to those observed in EBSD studies of rutile grains deformed under amphibolite-facies conditions in a shear zone from the Bergen Arc (Moore *et al.* 2020). The deformation microstructures in rutile are thus inferred to originate during metamorphism of their eclogite host rock, and are assumed to be present in all rutile crystals used in our experiments, based on their confirmed presence in the four samples analysed using FSD/EBSD.

The deformation-free grains at the rims of deformed crystals cross-cut all deformation microstructures and exhibit high-angle boundaries with the distorted crystal (Supplementary material Fig. 5_1), implying an origin via static recrystallization of the more strongly deformed lattice at crystal rims (Boneh *et al.* 2021). It is not possible to determine via EBSD whether static recrystallization occurred during the metamorphic history of the grains or during diffusion experiments. However, the statically recrystallized grains are rare, which is in line with observations on BSE images (Fig. 2a). Concentration profiles were not analysed at positions where these features occur; any significant effect of static recrystallization on the determined Al-diffusivities can therefore be excluded.

The EBSD data reveal the presence of several types of extended lattice defect in the rutile crystals: (1) continuous lattice rotation indicates distributed dislocations (line defects); (2) low-angle boundaries imply planar networks of dislocations; while (3) high-angle boundaries indicate continuous planar regions where atoms are arranged differently to the bulk lattice (Sutton and Balluffi 1995; Wheeler *et al.* 2009). Twin boundaries are a special case of high-angle boundary, where the twin orientation relationship leads to lower deviation from the regular lattice structure (Lee *et al.* 1993). The undeformed synthetic rutile crystals used by Cherniak and Watson (2019) are assumed to be free of extended defects. To assess whether the high content of extended defects observed in the natural rutile used as starting material in this study can serve as an explanation for the much higher bulk Al-diffusivity, it is important to understand how fast diffusion pathways may contribute to the respective bulk diffusivity.

The Al-concentration profiles indicate diffusion-controlled transport of Al into rutile or TiO_2 (II) and may represent one of two different end-member fast diffusion regimes (in the following type A and type C; Harrison 1961). Type A behaviour occurs at the limit of long timescales, where the average diffusion distance of an atom in the lattice over the duration of the experiment is much greater than the separation between fast pathways. In the type A regime, the measured bulk diffusion coefficient lies between that of the lattice and the fast pathways (Lee 1995). Type C behaviour occurs at the limit of short times and/or very large differences between lattice and fast pathway diffusion coefficients and/or very strong partitioning of an element into fast pathways. As a first approximation, diffusion occurs only along fast pathways in the type C regime, and the measured bulk diffusion coefficient is that of the fast pathways. Type B represents an intermediate regime between these two end-members.

Enhanced diffusion along the infrequent low-angle and general (non-twin) high-angle boundaries in the studied rutiles cannot explain the high Al bulk diffusion coefficients obtained in this study. These boundaries are too far apart to contribute to enhancement of the bulk diffusion coefficient in the type A regime (Al-diffusion profiles of Cherniak and Watson (2019) are only hundreds of nanometres long for multi-day experiments, compared to boundary separations of tens of micrometres; Fig. 8a, b; Supplementary material Fig. 5_2a, b). Furthermore, type B or C diffusion exclusively along these rare boundaries would lead to increased Al concentrations only where these boundaries intersect measurement profiles, which is not observed. In contrast, dislocations and twin boundaries are widespread in the deformed rutile, and thus the possibility that they are responsible for the discrepancy between the Al-diffusion coefficient of Cherniak and Watson (2019) and the current results merits further discussion. Note that although perfect twin boundaries should not represent efficient fast pathways, the ubiquitous lattice rotation in the deformed rutile leads to deviations from the perfect twin relationship of up to 14° (Supplementary material Fig. 5_2b). This requires the presence of dislocations at the twin boundaries, making them equivalent to low-angle boundaries for the purposes of enhancing diffusion.

In order to determine whether type A diffusion is a possible explanation, it is first necessary to determine the spacings between extended defects. The minimum spacing of twin lamellae visible in OC images is *c.* 500 nm (Supplementary material Fig. 5_2a). The separation between individual dislocations contributing to orientation gradients cannot be directly assessed. However, a rough estimate of dislocation density for a given misorientation angle and EBSD pixel size, and thus a given misorientation angle gradient, can be obtained by assuming all dislocations have the same Burgers vector and are of edge character, and then approximating the dislocations required to generate the pixel-to-pixel misorientation with an ordered array with the same net Burgers vector (Tucker *et al.* 2000). From this calculated density, the expected mean nearest neighbour distance between edge dislocations can be obtained, assuming the parallel edge dislocations are in fact randomly distributed (Jerram *et al.* 1996). By using the smallest known Burgers vector in rutile for the calculation, maximum dislocation densities and minimum mean separation distances for representative long-range orientation gradients of 1°, 0.2° and 0.02°/µm are calculated (Table 6).

In order to evaluate whether twin boundaries and dislocations are close enough in the deformed rutile to allow fast-pathway-enhanced bulk Al-diffusion in the type A regime, the separation distance d is compared with the average diffusion distance of an atom in the lattice over the duration of the experiment, which is on the order of $\sqrt{D_{\mathrm{lattice}} t}$. For type A behaviour $D < 2\sqrt{Dt}$ has been suggested as the minimum requirement (Ruoff and Balluffi 1963). Other authors advocate more restrictive criteria of $D < \sqrt{Dt}$ (Harrison 1961) or $D < 0.2\sqrt{Dt}$ (Joesten 1991). The $D < 2\sqrt{Dt}$ limit is used here to indicate the lower limit of dislocation density compatible with type A behaviour. The quantity $2\sqrt{Dt}$ calculated with the Al-diffusion coefficient of Cherniak and Watson (2019) for the conditions of experiment JKI_175 (1373 K, 80 hours) is 6.6 nm, ten times shorter than the minimum estimated dislocation spacing for even the highest orientation gradients

Table 6. Rutile [001] edge dislocation densities calculated using the method of Tucker *et al.* (2000) for representative misorientation angle gradients, and the mean nearest neighbour separation calculated for each density assuming a random distribution (Jerram *et al.* 1996)

Misorientation angle gradient	Calculated dislocation density in m^{-2} ([001] edge dislocations, b = 0.296 nm)	Corresponding mean nearest neighbour separation in nm (assuming random distribution)
1°/μm	5.9×10^{13}	65.1
0.2°/μm	1.2×10^{13}	146
0.02°/μm	1.2×10^{12}	460

observed, and far below the twin spacing. For the short-duration atmospheric pressure experiments BJ-IBK-ru3 and BJ-IBK-ru4, which also show high Al-diffusion coefficients, $2\sqrt{Dt}$ falls even lower, to 1.2 nm. Therefore, at first glance, lattice diffusion at the rate determined by Cherniak and Watson (2019) is strongly incompatible with type A diffusion behaviour of Al in experiments in this study. However, three possibilities must be considered before discarding type A diffusion entirely: (1) estimates of *d* might be too high due to ‘invisible’ extended defects; (2) mobile dislocations might remove the requirement that $D < 2\sqrt{Dt}$; or (3) the Al lattice diffusion coefficient of Cherniak and Watson (2019) might not apply in the current samples. Points (1) and (2) will be addressed immediately, while point (3) will be returned to after considering the possibility of type C diffusion behaviour.

EBSD only detects the excess dislocations of one sign that are responsible for net lattice rotation, referred to as geometrically necessary dislocations (GNDs). Total dislocation density in the rutile crystals may thus be higher than inferred from lattice rotation alone. However, to reach even a 7 nm average spacing for randomly distributed dislocations, a total dislocation density of $>5 \times 10^{15}$ m^{-2} would be required throughout the crystal. This value is extremely high, comparable to rutile from shock deformation experiments (Casey *et al.* 1988). Such high dislocation densities would also lead to deterioration in Kikuchi pattern quality, which was not observed, even when viewing higher resolution Kikuchi patterns. Twin lamellae with widths below the pixel size of the FSD images (*c.* 60 nm) would also not show up in EBSD maps, so that the abundance of twin boundaries could also be higher than observed. However, alternating twin lamellae with a spacing of ≤7 nm (smaller than the interaction volume of an EBSD point analysis) would lead to mixed and thus low-contrast Kikuchi patterns, which, again, were not observed. In summary, while dislocation and twin boundary densities could easily be higher than those directly measured, it is highly unlikely that they reach the abundance necessary to allow type A diffusion behaviour if the Al lattice diffusion coefficient under the experimental conditions is that of Cherniak and Watson (2019).

If dislocations were mobile during the diffusion experiment, this could allow diffusing atoms to encounter multiple defects despite short diffusion length scales. An effect due to plastic deformation during high-pressure runs can be excluded, as Al-diffusivity is equally high in the atmospheric pressure runs. Alternatively, annealing pure rutile above 1020°C is known to allow migration of dislocations into mobile subgrain boundaries, leaving behind recovered domains of low dislocation density (Bell *et al.* 1972; Muhammad *et al.* 2021). However, the current samples show mostly continuous orientation gradients, suggesting incomplete recovery, consistent with the fact that Al suppresses recovery in rutile (Hatta *et al.* 1996). Therefore, type A behaviour due to mobile dislocations is unlikely.

As long as the spacing between Al-bearing defects is considerably smaller than the interaction volume of compositional analysis spots, type C diffusion can also be the cause of the analysed Al-diffusion profiles (Fig. 2a, b). In this scenario, instead of being homogeneously distributed in the bulk, the measured Al is contained (and travels) exclusively within extended defects, which must therefore form an interconnected network. While this interpretation avoids the difficulties associated with assuming type A behaviour, the maximum amount of Al incorporated (e.g. *c.* 1.6 at% in the rim of JKI_175) is problematic. Even assuming 100% replacement of Ti by Al in extended defects, this would require extended defects to take up 1.6% of the entire lattice volume. If Al is assumed to lie only in dislocation cores of 1 nm diameter, this is impossible for even the most extreme total dislocation densities discussed above. Increasing the allowed diameter of Al-enrichment around dislocations is only a partial solution. Even if an extreme upper estimate for the cross-sectional area of the Al-enriched zone where enhanced diffusion can occur is used (60×60 nm^2, based on the size of the charged area around rutile dislocations estimated by Muhammad *et al.* 2021), the concentration of Al in dislocation-adjacent regions would still need to exceed 7 at% for the highest GND density from Table 6. Thus, the incorporation of the observed amount of Al at extended defects alone appears to require formation of a different, Al-rich phase at those defects. Such a phenomenon is not completely unknown: for grain boundaries in Al-rich rutile samples, observed segregation required up to eight monolayers of Al that cover the boundary surface (Ikeda *et al.* 1993). Diaspore and corundum clusters were observed at (101) twin boundaries in natural rutiles that developed during progressive crystallization from a precursor phase (Daneu *et al.* 2014). However, unless and until such a phenomenon is proven to occur for dislocations and Al can be shown to segregate to regions much larger than dislocation cores, the total amount of Al incorporated in rutile rims is a significant obstacle to the hypothesis of type C diffusion in these experiments if using the Cherniak and Watson (2019) lattice diffusion coefficient. Note that invoking a mixed (type B) diffusion regime does not solve this issue, as using the slow lattice diffusion coefficient, the average diffusion distance into the lattice adjacent to fast pathways is <10 nm over experimental timescales (see above). Another obstacle for type C diffusion is that this regime requires significant Al transport along twin planes to enhance the Al bulk diffusivity in rutile by orders of magnitude. Twin boundaries in rutile are oriented in distinct directions with the most common twin planes being (011), (101) and (301). This should result in anisotropic Al transport properties, which were not observed (Table 3).

However, there are several theoretical reasons to expect that extended defects, in particular dislocations, should represent efficient pathways for Al-diffusion in rutile in particular. *Ab initio* modelling studies of the structure of rutile [001] edge dislocations (Sun *et al.* 2015; Maras *et al.* 2019) reveal not only distortion of bond lengths (the general reason to expect enhanced diffusion at all dislocations), but also mixed 3^+ and 4^+ valency of Ti in dislocation cores. The presence of Ti^{3+} could potentially allow direct substitution of Al for Ti on both lattice and interstitial sites, with no need for movement of associated oxygen vacancies. Additionally, [001] dislocations are oxygen deficient at reducing conditions (Maras *et al.* 2019), which would promote Al-diffusion even in the absence of Ti^{3+}. <101> dislocations, the other common Burgers vector in rutile, may be positively (and not negatively) charged (Muhammad *et al.* 2021), but unfortunately, no *ab initio* studies comparable to those for [001] dislocations exist. While this rules out Ti^{3+} on lattice sites in <101> dislocation cores, it is consistent with higher concentrations of oxygen vacancies, and thus again, potentially enhanced Al incorporation and diffusion. This is also in line with our earlier discussion saying that the formation of stable oxygen defects ($V_O^{\bullet\bullet}$) can be related to the incorporation of Al^{3+} on a Ti^{4+} position (Al'_{Ti}) charge balanced by oxygen defect formation (equation 1). These crystal-chemical considerations may explain why Al is known to strongly partition into rutile dislocations (Verberne *et al.* 2022) and subgrain boundaries (Ikeda *et al.* 1993). The consequences of increasing partitioning of an element to fast pathways is to increase the size of the effect on the bulk diffusion coefficient for a given volume fraction of fast pathways, with the end-member of extremely strong partitioning resulting in diffusion only along fast pathways (Lee 1995).

The composition of the matrix, experimental temperatures and pressures and diffusional anisotropy. All these parameters cannot explain the mismatch in Al-diffusivities between the two studies.

To test the effect of a synthetic pure Al_2O_3 v. a natural matrix as the source of Al, an additional experiment was performed that was identical to experiment BJ-IBK-ru3 with the exception that solely pure synthetic Al_2O_3 powder was used as a matrix (BJ-IBK-ru4; Tables 3 and 4). Results show a constant Al_2O_3 concentration of 0.45 (9) wt% throughout the whole rutile grain, which is approximately one order of magnitude above the Al_2O_3 concentration in the original rutile that was used as starting material (0.03 wt%; Table 2; Supplementary material 1). This implies that we can safely assume that rutile grains are Al-saturated in this particular experiment. Consequently, the nature of the matrix material (synthetic v. natural) cannot explain a substantially slower Al-diffusivity.

Figure 3 clearly shows that Al-diffusivities in rutile and its high-pressure polymorph TiO_2 (II) increase with increasing temperature. However, this does not provide an explanation for the difference in Al-diffusivities between this study and the study of Cherniak and Watson (2019), as Al-diffusivities in both studies show a roughly similar temperature dependence and the investigated temperature ranges overlap (Fig. 10). The determined Al-diffusivity at 1373 K and 0.1 MPa in this study ($\log D_{Al}^{TiO2} = -13.5\ m^2\ s^{-1}$) is almost nine orders of magnitude higher compared to the respective value determined by Cherniak and Watson (2019; $\log D_{Al}^{TiO2} = -22.42\ m^2\ s^{-1}$). Furthermore, Al-diffusivities in this study do not show any observable effect of pressure at a constant temperature of 1273 K between 4 and 7 GPa, and at 1373 K between 1 bar and 3.5 GPa (Tables 3 and 4; Fig. 3). This implies that pressure has at most a very minor effect on the Al-diffusivity, and pressure and temperature can be safely ruled out as a reason for the observed difference in Al-diffusivity.

One parameter known to affect diffusivities is a potential anisotropy. For rutile (TiO_2), this does not apply to tetravalent and pentavalent ions such as Zr, Ta, Nb or Hf, which show almost no anisotropy in diffusion (Dohmen *et al.* 2019). However, divalent ions show a strong anisotropy, with diffusion rates parallel to the *c*-axis being orders of magnitude faster compared to those perpendicular to *c*. In this context it is important to note that elemental concentration profiles in this study were obtained from unoriented samples. EBSD analyses of four selected samples confirm that all profiles were obtained in general orientations, but with the angle between the respective concentration profiles and the rutile *c*-axis varying between 33.4 and 57.8° (Table 3). This implies that any strong anisotropy would be visible in the determined Al-diffusivities, which indicates that a major diffusional anisotropy can be excluded. This fits well with findings of Sasaki *et al.* (1985), who argued that impurity diffusion of trivalent ions in rutile can show minor anisotropy of approximately a factor of two with faster diffusivities along *c*, if diffusion is related to an interstitialcy mechanism. It is also in good agreement with recent results of Cherniak and Watson (2019), which indicated little diffusional anisotropy for Si and Al in rutile.

In conclusion, if the Cherniak and Watson (2019) Al-diffusion coefficients are assumed to represent lattice diffusion in the current samples, the observed abundance of extended defects could be sufficient to explain the extremely high Al bulk diffusion coefficient obtained, in the case that Al-diffusion occurred largely along extended defects (type B or C regimes). Based on bulk Al concentrations, this would require an Al-rich phase replacing TiO_2 around such defects and, if twin boundaries are involved, an explanation for the fact that anisotropic Al transport properties were not determined. If bulk diffusion instead occurred in the type A regime, this requires that although bulk diffusion is indeed enhanced by the defect content of the studied rutile crystals, the lattice diffusion coefficient must also be higher than that measured by Cherniak and Watson (2019). A potential reason for this is to a minor extent the lower oxygen fugacity in the experiments in this study, which might enhance the formation of oxygen defects and in succession the Al-diffusivity. Another reason is based on the initial water concentration (*c.* 0.03 wt %; Supplementary materials 1 and 3) in the natural rutile starting material used in this study. It is well known that the strengths of NAMs decrease systematically with increasing hydrogen concentration (e.g. Kronenberg and Tullis 1984; Post *et al.* 1996; Karato and Jung 2003) through hydrolytic weakening (Griggs 1967) or protonic weakening (Mackwell and Kohlstedt 1990), which might cause significantly enhanced rates of ionic diffusion (Kohlstedt 2006). This implies that the geological history of a particular sample may have a strong effect on the observed bulk diffusivities and must be considered before diffusivities from simplified diffusion experiments using pure synthetic starting materials are extrapolated to complex natural systems.

Implications for Al contents in rutile in natural samples

The highest Al contents in rutiles known so far have been reported from upper mantle settings (e.g. Kaminsky *et al.* 2000; Sobolev and Yefimova 2000; Afanasyev *et al.* 2009; Meinhold 2010). These include rutiles with 1.1 and 2.6 wt% Al_2O_3 from kyanite-bearing eclogites sampled by diamondiferous kimberlites (Spetsius and Safronov 1986) and rutiles with 1.4–2.3 wt% Al_2O_3 occurring as diamond inclusions (Mvuemba Ntanda *et al.* 1982). This indicates that high pressure is a major factor favouring high Al concentrations in rutile and its high-pressure polymorph TiO_2 (II) and confirms the findings of this and of previous high-*P–T* experimental studies (Konzett 1997; Escudero *et al.* 2012; Hoff and Watson 2021).

The wide overall range in Al_2O_3 contents of rutiles reported from the upper mantle with values <0.1 wt% (Spetsius and Safronov 1986; Sobolev *et al.* 1997; Kaminsky *et al.* 2000; Sobolev and Yefimova 2000; Afanasyev *et al.* 2009) indicates, however, that additional factors affect Al abundances in natural rutile. At least one of these is $\mu(Al_2O_3)$ during rutile crystallization/equilibration, which can be buffered by specific phase combinations known to occur in the upper mantle. At any given *P* and *T*, Al content in rutile is expected to decrease from highest possible values buffered by corundum ($\mu(Al_2O_3) = \mu(\text{corundum})$) to values buffered by kyanite + coesite ($\mu(Al_2O_3) = \mu(\text{kyanite}) - \mu(\text{coesite})$) and to those buffered by garnet + orthopyroxene ($\mu(Al_2O_3) = \mu(\text{pyrope}) - 1.5\mu(\text{enstatite})$). This effect of the phase assemblage superimposed on the pressure effect is indeed consistent with the range in Al contents of rutiles determined in mantle xenoliths stemming from Siberian diamondiferous kimberlites (Spetsius and Safronov 1986). Kyanite-bearing eclogites tend to contain rutiles with higher Al_2O_3 contents than those from kyanite-absent eclogites or websterites. Thus, the high Al rutile inclusions in diamonds are most likely derived from kyanite- and/or corundum-bearing eclogites or grospydites.

In addition, rutiles with > *c.* 1 wt% Al_2O_3 have only been reported so far from mantle xenolith and diamond inclusion samples that were exposed to an extreme rate of *P–T* decrease during transport to the surface via an ascending kimberlite melt (Mvuemba Ntanda *et al.* 1982; Spetsius and Safronov 1986). Rutiles from ultrahigh-pressure rocks including diamond-bearing ones brought to the surface by tectonic processes show, however, very low Al_2O_3 contents not exceeding *c.* 0.2 wt% (Smith and Pinet 1985; Hammer and Beran 1991). Some reported Al_2O_3 contents are even orders of magnitude lower. For example, a natural kyanite–quartz granulite (1.45 GPa, 930°C) shows a consistent abundance of 0.04 wt% Al_2O_3 in rutile (Pauly *et al.* 2016). This indicates a tendency of Al to be returned to coexisting silicates through diffusion in the wake of retrogressive re-equilibration and agrees with the comparatively fast Al-diffusivities found in this study.

Key parameters to preserve high Al contents in rutile thus seem to be: (1) high pressures corresponding to the diamond stability field; (2) a rapid decrease in *P* and *T* facilitated by xenolith transport in an ultra-fast ascending kimberlite magma; and (3) in the case of diamond inclusions, a separation of the rutiles from their matrix prohibiting Al-loss through re-equilibration.

Conclusions

- Results of this study suggest that water is mainly incorporated into the TiO_2 structure through incorporation of oxygen in interstitial positions charge balanced by hydrogen interstitials.
- The Al equilibrium concentration in rutile and TiO_2 (II) with the surrounding matrix at fixed μ (Al_2O_3) shows an increase between 3.5 and 4 GPa that is related to the incorporation of Al into 0 ½ 0 octahedral interstices, and an increase between 6 and 7 GPa, which is related to the transition of rutile-TiO_2 to α-PbO-type TiO_2 (II).
- Al-diffusivities in TiO_2 determined in this study are eight to nine orders of magnitude faster compared to those determined by Cherniak and Watson (2019). This can potentially be explained by a combination of:
  - Oxygen fugacity variations, which might affect the formation rate of oxygen defects. This will affect Al-diffusivities, as Al incorporation in TiO_2 is at least partly charge balanced by oxygen defects (equation 1). However, as discussed above, it has been shown that oxygen fugacity variation has only a relatively minor effect on the Al-diffusivity in rutile.
  - The presence or absence of small amounts of water as defects in the starting material, because hydrolytic or protonic weakening might cause significantly enhanced rates of ionic diffusion.
  - The enhancement of Al-diffusion through extended defects, which are present in natural rutiles and were likely introduced into the mineral structure during high-grade metamorphism. This defect structure is presumably absent in synthetic samples, which implies that the geological history of a sample may have a direct effect on the Al-mobility in natural rutile. Even if this aspect mostly relates to natural mechanically deformed grains, it deserves special attention, since geothermobarometers are applied to natural samples.
- If the internal defect microstructure and the water content does indeed have a major effect on bulk diffusivities, then great caution is required when diffusivities from diffusion experiments are extrapolated to complex natural systems. Further studies are required to investigate whether this holds true.

Acknowledgements We thank Horst Marschall and two anonymous reviewers as well as Elias Bloch, who revised an earlier version of this manuscript, for detailed reviews that significantly improved the quality of this paper. We also thank Inês Pereira for careful editorial handling. We further thank M. Tribus for her support during microprobe analyses and A. Proyer for providing the garnet–kyanite–muscovite gneiss used as the starting material.

Competing interests The authors declare that they have no known competing financial interests or personal relationships that could have appeared to influence the work reported in this paper.

Author contributions **BJ-M**: conceptualization (lead), investigation (lead), methodology (equal), project administration (lead), visualization (lead), writing – original draft (lead), writing – review & editing (equal); **JK**: conceptualization (supporting), investigation (supporting), methodology (equal), writing – review & editing (equal); **TL**: methodology (equal), writing – review & editing (equal); **TG**: investigation (supporting), methodology (equal), writing – original draft (supporting), writing – review & editing (equal); **GH**: methodology (supporting), writing – review & editing (equal); **EL**: methodology (supporting), writing – review & editing (equal); **RS**: conceptualization (supporting), investigation (supporting), methodology (equal), writing – review & editing (equal).

Funding The study was financially supported by the young researcher support 2017 fund and the Institute of Mineralogy and Petrography at the University of Innsbruck, which is gratefully acknowledged.

Data availability All data generated or analysed during this study are included in this published article and its supplementary information files.

References

Afanasyev, V.P., Agashev, A.M. and Orihashi, Y. 2009. Paleozoic U-Pb age of rutile inclusions in diamonds of the V–VII variety from placers of the northeast Siberian platform. *Doklady Earth Sciences*, **428**, 1151–1155, https://doi.org/10.1134/S1028334X09070253

Akaogi, M., Kusaba, K. *et al.* 1992. High-pressure high-temperature stability of α-PbO_2-type TiO_2 and $MgSiO_3$ majorite: calorimetric and *in situ* X-ray diffraction studies. *American Geophysical Union, Geophysical Monograph*, **67**, 447–455, https://doi.org/10.1029/GM067p0447

Akse, J.R. and Whitehurst, H.B. 1978. Diffusion of titanium in slightly reduced rutile. *Journal of Physics and Chemistry of Solids*, **39**, 457–465, https://doi.org/10.1016/0022-3697(78)90022-7

Arita, M., Hosoya, M., Kobayashi, M. and Someno, M. 1979. Depth profile measurement by secondary ion mass spectrometry for determining tracer diffusivity of oxygen in rutile. *Journal of the American Ceramic Society*, **62**, 443–446, https://doi.org/10.1111/j.1151-2916.1979.tb19101.x

Bachmann, F., Hielscher, R. and Schaeben, H. 2010. Texture analysis with MTEX-Free and open source software toolbox. *Solid State Phenomena*, **160**, 63–68, https://doi.org/10.4028/http://www.scientific.net/SSP.160.63

Bell, D.R. and Rossmann, G.R. 1992. Water in the Earth's mantle: the role of nominally anhydrous minerals. *Science (New York, NY)*, **255**, 1391–1397, https://doi.org/10.1126/science.255.5050.1391

Bell, H., Krishnamachari, V. and Jones, J.T. 1972. Recovery of high-temperature creep-resistant substructure in rutile. *Journal of the American Ceramic Society*, **55**, 6–10, https://doi.org/10.1111/j.1151-2916.1972.tb13407.x

Bendeliany, N.A., Popova, S.V. and Vereschagin, L.F. 1966. A new modification of titanium dioxide stable at high pressures. *Geokhimiya*, **5**, 499–502.

Blackburn, T., Shimizu, N., Bowring, S.A., Schoene, B. and Mahan, K.H. 2012. Zirconium in rutile speedometry: new constraints on lower crustal cooling rates and residence temperatures. *Earth and Planetary Science Letters*, **317–318**, 231–240, https://doi.org/10.1016/j.epsl.2011.11.012

Boneh, Y., Chin, E.J., Chilson-Parks, B.H., Saal, A.E., Hauri, E., Carter Hearn, B. and Hirth, G. 2021. Microstructural shift due to post-deformation annealing in the upper mantle. *Geochemistry, Geophysics, Geosystems*, **22**, e09377, https://doi.org/10.1029/2020GC009377

Boyd, F.R. and England, J.L. 1960. Apparatus for phase-equilibrium measurements at pressures up to 50 kilobars and temperatures up to 1750°C. *Journal of Geophysical Research*, **65**, 741–748, https://doi.org/10.1029/JZ065i002p00741

Bromiley, G.D. and Hilaret, N. 2005. Hydrogen and minor element incorporation in synthetic rutile. *Mineralogical Magazine*, **69**, 345–358, https://doi.org/10.1180/0026461056930256

Bromiley, G., Hilaret, N. and McCammon, C. 2004. Solubility of hydrogen and ferric iron in rutile and TiO_2 (II): implications for phase assemblages during ultrahigh-pressure metamorphism and for the stability of silica polymorphs in the lower mantle. *Geophysical Research Letters*, **31**, L04610, https://doi.org/10.1029/2004GL019430

Bursill, L.A. and Hyde, B.G. 1972. Crystallographic shear in the higher titanium oxides: structure, texture, mechanisms and thermodynamics. *Progress in Solid State Chemistry*, **7**, 177–253, https://doi.org/10.1016/0079-6786(72)90008-8

Casey, W.H., Carr, M.J. and Graham, R.A. 1988. Crystal defects and the dissolution kinetics of rutile. *Geochimica et Cosmochimica Acta*, **52**, 1545–1556, https://doi.org/10.1016/0016-7037(88)90224-4

Chapman, T., Clarke, G.L., Piazolo, S., Robbins, V.A. and Trimby, P.W. 2019. Grain-scale dependency of metamorphic reaction on crystal plastic strain. *Journal of Metamorphic Geology*, **37**, 1021–1036, https://doi.org/10.1111/jmg.12473

Cherniak, D.J. 2000. Pb diffusion in rutile. *Contributions to Mineralogy and Petrology*, **139**, 198–207, https://doi.org/10.1007/PL00007671

Cherniak, D.J. and Watson, E.B. 2011. Helium diffusion in rutile and titanite, and consideration of the origin and implications of diffusional anisotropy. *Chemical Geology*, **288**, 149–161, https://doi.org/10.1016/j.chemgeo.2011.07.015

Cherniak, D.J. and Watson, E.B. 2019. Al and Si diffusion in rutile. *American Mineralogist*, **104**, 1638–1649, https://doi.org/10.2138/am-2019-7030

Cherniak, D.J., Manchester, J. and Watson, E.B. 2007. Zr and Hf diffusion in rutile. *Earth and Planetary Science Letters*, **261**, 267–279, https://doi.org/10.1016/j.epsl.2007.06.027

Clark, S.P. 1959. Effect of pressure on the melting point of eight alkali halides. *Journal of Chemical Physics*, **31**, 1526–1531, https://doi.org/10.1063/1.1730648

Colasanti, V.C., Johnson, E.A. and Manning, C.E. 2011. An experimental study of OH solubility in rutile at 500–900°C, 0.5–2 GPa, and a range of oxygen fugacities. *American Mineralogist*, **96**, 1291–1299, https://doi.org/10.2138/am.2011.3708

Crank, J. 1975. *The Mathematics of Diffusion*, 2nd edn. Oxford University Press, London.

Cruz-Uribe, A.M., Feineman, M.D., Zack, T. and Barth, M. 2014. Metamorphic reaction rates at ~650–800°C from diffusion of niobium in rutile. *Geochimica et Cosmochimica Acta*, **130**, 63–77, https://doi.org/10.1016/j.gca.2013.12.015

Daneu, N., Recnik, A. and Mader, W. 2014. Atomic structure and formation mechanism of (101) rutile twins from Diamantina (Brazil). *American Mineralogist*, **99**, 612–624, https://doi.org/10.2138/am.2014.4672

Degeling, H.S. 2003. *Zr Equilibria in Metamorphic Rocks*. PhD thesis, Australian National University, Canberra.

Dohmen, R. and Milke, R. 2010. Diffusion in polycrystalline materials: grain boundaries, mathematical models, and experimental data. *Reviews in Mineralogy and Geochemistry*, **72**, 921–970, https://doi.org/10.2138/rmg.2010.72.21

Dohmen, R., Marschall, H.R., Ludwig, T. and Polednia, J. 2019. Diffusion of Zr, Hf, Nb and Ta in rutile: effects of temperature, oxygen fugacity, and doping level, and relation to rutile point defect chemistry. *Physics and Chemistry of Minerals*, **46**, 311–332, https://doi.org/10.1007/s00269-018-1005-7

Dubosq, R., Rogowitz, A., Schweinar, K., Gault, B. and Schneider, D.A. 2019. A 2D and 3D nanostructural study of naturally deformed pyrite: assessing the links between trace element mobility and defect structures. *Contributions to Mineralogy and Petrology*, **174**, 72, https://doi.org/10.1007/s00410-019-1611-5

Escudero, A., Delevoye, L. and Langenhorst, F. 2011. Aluminum incorporation in TiO_2 rutile at high pressures: an XRD and high-resolution ^{27}Al NMR study. *Journal of Physics and Chemistry*, **115**, 12196–12201, https://doi.org/10.1021/jp202930r

Escudero, A., Langenhorst, F. and Müller, W.F. 2012. Aluminum solubility in rutile at high pressure and experimental evidence for a $CaCl_2$-structured polymorph. *American Mineralogist*, **97**, 1075–1082, https://doi.org/10.2138/am.2012.4049

Ferry, J.M. and Watson, E.B. 2007. New thermodynamic models and revised calibrations for the Ti-in-zircon and Zr-in-rutile thermometers. *Contributions to Mineralogy and Petrology*, **154**, 429–437, https://doi.org/10.1007/s00410-007-0201-0

Gesenhues, U. and Rentschler, T. 1999. Crystal growth and defect structure of Al^{3+}-doped rutile. *Journal of Solid State Chemistry*, **143**, 210–218, https://doi.org/10.1006/jssc.1998.8088

Griffiths, T.A., Habler, G., Rhede, D., Wirth, R., Ram, F. and Abart, R. 2014. Localization of submicron inclusion re-equilibration at healed fractures in host garnet. *Contributions to Mineralogy and Petrology*, **168**, 1–21, https://doi.org/10.1007/s00410-014-1077-4

Griggs, D.T. 1967. Hydrolitic weakening of quartz and other silicates. *Geophysical Journal of the Royal Astronomical Society*, **14**, 19–31, https://doi.org/10.1111/j.1365-246X.1967.tb06218.x

Hammer, M.F.V. and Beran, A. 1991. Variations in the OH concentration of rutiles from different geological environments. *Mineralogy and Petrology*, **45**, 1–9, https://doi.org/10.1007/BF01164498

Harrison, L.G. 1961. Influence of dislocations on diffusion kinetics in solids with particular reference to the alkali halides. *Transactions of the Faraday Society*, **57**, 1191–1199, https://doi.org/10.1039/TF9615701191

Hatta, K., Higuchi, M., Takahashi, J. and Kodaira, K. 1996. Floating zone growth and characterization of aluminum-doped rutile single crystals. *Journal of Crystal Growth*, **163**, 279–284, https://doi.org/10.1016/0022-0248(95)00972-8

Hawemann, F., Mancktelow, N., Wex, S., Pennacchioni, G. and Camacho, A. 2019. Fracturing and crystal plastic behaviour of garnet under seismic stress in the dry lower continental crust (Musgrave Ranges, Central Australia). *Solid Earth*, **10**, 1635–1649, https://doi.org/10.5194/se-10-1635-2019

Hoff, C.M. and Watson, E.B. 2018. Aluminum in rutile as a recorder of temperature and pressure. *Goldschmidt Abstracts*, **2018**, 1035.

Hoff, C.M. and Watson, E.B. 2021. Aluminum solubility in rutile (TiO_2). *Physics and Chemistry of Minerals*, **48**, 45, https://doi.org/10.1007/s00269-021-01169-z

Hwang, S.L., Shen, P., Chu, H.T. and Yui, T.F. 2000. Nanometer-size-α-PbO_2-type TiO_2 in garnet: a thermobarometer for ultrahigh-pressure metamorphism. *Science (New York, NY)*, **288**, 321–324, https://doi.org/10.1126/science.288.5464.321

Ikeda, J.A.S., Chiang, Y.-M., Garratt-Reed, A.J. and Sande, J.B.V. 1993. Space charge segregation at grain boundaries in titanium dioxide: II, model experiments. *Journal of the American Ceramic Society*, **76**, 2447–2459, https://doi.org/10.1111/j.1151-2916.1993.tb03965.x

Jerram, D.A., Cheadle, M.J., Hunter, R.H. and Elliott, M.T. 1996. The spatial distribution of grains and crystals in rocks. *Contributions to Mineralogy and Petrology*, **125**, 60–74, https://doi.org/10.1007/s004100050206

Joesten, R. 1991. Grain-boundary diffusion kinetics in silicate and oxide minerals. *In*: Ganguly, J. (ed.) *Diffusion, Atomic Ordering, and Mass Transport*. Springer US, 345–395.

Johnson, O., Ohlsen, W. and Kingsbury, P.I. , , Jr 1968. Defects in rutile (iii). Optical and electrical properties of impurities and charge carriers. *Physical Review*, **175**, 1102–1108, https://doi.org/10.1103/PhysRev.175.1102

Johnson, O.W., DeFord, J. and Shaner, J.W. 1973. Experimental technique for the precise determination of H and D concentration in rutile (TiO_2). *Journal of Applied Physics*, **44**, 3008, https://doi.org/10.1063/1.1662697

Johnson, O.W., Paek, S.-H. and DeFord, J.W. 1975. Diffusion of H and D in TiO_2. Suppression of internal fields by isotope exchange. *Journal of Applied Physics*, **46**, 1026–1033, https://doi.org/10.1063/1.322206

Karato, S.I. and Jung, H. 2003. Effects of pressure on high-temperature dislocation creep in olivine. *Philosophical Magazine*, **83**, 401–414, https://doi.org/10.1080/0141861021000025829

Kaminsky, F.V., Zakharchenko, O.D., Griffin, W.L., Channer, D.M.D.R. and Khachatryan-Blinova, G.K. 2000. Diamond from the Guaniamo area, Venezuela. *The Canadian Mineralogist*, **38**, 1347–1370, https://doi.org/10.2113/gscanmin.38.6.1347

Kamisaki, H. and Yamashita, K. 2011. Theoretical study of the interstitial oxygen atom in anatase and rutile TiO_2: electron trapping and the elongation of the r(O-O) bond. *The Journal of Physical Chemistry*, **115**, 8265–8273, https://doi.org/10.1021/jp110648q

Keppler, H. and Frost, D.J. 2005. Introduction to minerals under extreme conditions. *European Mineralogical Union Notes in Mineralogy*, **7**, 1–30, https://doi.org/10.1180/EMU-notes.7.1

Khomenko, V.M., Langer, K., Rager, H. and Fett, A. 1998. Electronic absorption by Ti^{3+} ions and electron delocalization in synthetic blue rutile. *Physics and Chemistry of Minerals*, **25**, 338–346, https://doi.org/10.1007/s002690050124

Kofstad, P. 1972. *Non-Stochiometry, Diffusion and Electrical Conductivity in Binary Oxides*. Wiley-Interscience, New York.

Kohlstedt, D.L. 2006. The role of water in high-temperature rock deformation. *Reviews in Mineralogy and Geochemistry*, **62**, 377–396, https://doi.org/10.2138/rmg.2006.62.16

Kohn, M.J. 2020. A refined zirconium-in-rutile thermometer. *American Mineralogist*, **105**, 963–971, https://doi.org/10.2138/am-2020-7091

Konzett, J. 1997. Phase relations and chemistry of Ti-rich K-richterite-bearing mantle assemblages: an experimental study to 8.0 GPa in a Ti-KNCMASH system. *Contributions to Mineralogy and Petrology*, **128**, 385–404, https://doi.org/10.1007/s004100050316

Konzett, J. and Frost, D.J. 2009. The high *P–T* stability of hydroxyl-apatite in natural and simplified MORB – an experimental study to 15 GPa with implications for transport and storage of phosphorus in subduction zones. *Journal of Petrology*, **50**, 2043–2062, https://doi.org/10.1093/petrology/egp068

Kooijman, E., Mezger, K. and Berndt, J. 2010. Constraints on the U-Pb systematics of metamorphic rutile from in situ LA-ICP-MS analysis. *Earth and Planetary Science Letters*, **293**, 321–330, https://doi.org/10.1016/j.epsl.2010.02.047

Kooijman, E., Hacker, B.R., Smit, M.A. and Kylander-Clark, A.R.C. 2015. Rutile thermochronology constrains time-resolved cooling histories in orogenic belts. *Goldschmidt Abstracts*, **2015**, 1657.

Kröger, F.A. and Vink, H.J. 1956. Relations between the concentrations of imperfections in crystalline solids. *Solid State Physics*, **3**, 307–435.

Kronenberg, A.K. and Tullis, J. 1984. Flow strengths of quartz aggregates: grain size and pressure effects due to hydrolytic weakening. *Journal of Geophysical Research*, **89**, 4281–4297, https://doi.org/10.1029/JB089iB06p04281

Lee, J.K.W. 1995. Multipath diffusion in geochronology. *Contributions to Mineralogy and Petrology*, **120**, 60–82, https://doi.org/10.1007/BF00311008

Lee, W.-Y., Bristowe, P.D., Gao, Y. and Merkle, K.L. 1993. The atomic structure of twin boundaries in rutile. *Philosophical Magazine Letters*, **68**, 309–314, https://doi.org/10.1080/09500839308242908

Li, Q., Lin, W., Su, W., Li, X., Shi, Y., Liu, Y. and Tang, G. 2011. SIMS U-Pb rutile age of low-temperature eclogites from southwestern Chinese Tianshan, NW China. *Lithos*, **122**, 76–86, https://doi.org/10.1016/j.lithos.2010.11.007

Liu, L., Xiao, Y., Aulbach, S., Li, D. and Hou, Z. 2014. Vanadium and niobium behavior in rutile as a function of oxygen fugacity: evidence from natural samples. *Contributions to Mineralogy and Petrology*, **167**, 1026, https://doi.org/10.1007/s00410-014-1026-2

Mackwell, S.J. and Kohlstedt, D.L. 1990. Diffusion of hydrogen in olivine: implications for water in the mantle. *Journal of Geophysical Research*, **90**, 11319–11333, https://doi.org/10.1029/JB095iB04p05079

Maras, E., Saito, M., Inoue, K., Jónsson, H., Ikuhara, Y. and McKenna, K.P. 2019. Determination of the structure and properties of an edge dislocation in rutile TiO_2. *Acta Materialia*, **163**, 199–207, https://doi.org/10.1016/j.actamat.2018.10.015

Marschall, H. R., Dohmen, R. and Ludwig, T. 2013. Diffusion-induced fractionation of niobium and tantalum during continental crust formation. *Earth and Planetary Science Letters*, **375**, 361–371, https://doi.org/10.1016/j.epsl.2013.05.055

McDade, P., Wood, B.J. *et al.* 2002. Pressure corrections for a selection of piston-cylinder cell assemblies. *Mineralogical Magazine*, **66**, 1021–1028, https://doi.org/10.1180/0026461026660074

Meinhold, G. 2010. Rutile and its applications in earth sciences. *Earth-Science Reviews*, **102**, 1–28, https://doi.org/10.1016/j.earscirev.2010.06.001

Mezger, K., Hanson, G.N. and Bohlen, S.R. 1989. High-precision U-Pb ages of metamorphic rutile: application to the cooling history of high-grade terranes. *Earth and Planetary Science Letters*, **96**, 106–118, https://doi.org/10.1016/0012-821X(89)90126-X

Millot, F. and Picard, C. 1988. Oxygen self-diffusion in non-stoichiometric rutile in TiO_2-x at high temperature. *Solid State Ionics*, **28**, 1344–1348, https://doi.org/10.1016/0167-2738(88)90384-0

Moore, D.K., Cherniak, D.J. and Watson, E.B. 1998. Oxygen diffusion in rutile from 750 to 1000°C and 0.1 to 1000 MPa. *American Mineralogist*, **83**, 700–711, https://doi.org/10.2138/am-1998-7-803

Moore, J., Beinlich, A. *et al.* 2020. Microstructurally controlled trace element (Zr, U–Pb) concentrations in metamorphic rutile: an example from the amphibolites of the Bergen Arcs. *Journal of Metamorphic Geology*, **38**, 103–127, https://doi.org/10.1111/jmg.12514

Mpokos, E., Perraki, M. and Palikari, S. 2009. Single and multiphase inclusions in metapelitic garnets of the Rhodope Metamorphic Province, NE Greece. *Spectrochimica Acta Part A*, **73**, 477–483, https://doi.org/10.1016/j.saa.2008.12.035

Muhammad, Q.K., Porz, L. *et al.* 2021. Donor and acceptor-like self-doping by mechanically induced dislocations in bulk TiO_2. *Nano Energy*, **85**, 105944, https://doi.org/10.1016/j.nanoen.2021.105944

Mvuemba Ntanda, F., Moreau, J. and Meyer, H.O.A. 1982. Particularites des inclusions cristallines primaires des diamants du Kasai, Zaire. *The Canadian Mineralogist*, **20**, 59–64.

Pauly, J., Marschall, H.R., Meyer, H.P., Chatterjee, N. and Monteleone, B.D. 2016. Prolonged Ediacaran-Cambrian metamorphic history and short-lived high-pressure granulite-facies metamorphism in the H.U. Sverdrupfjella, Droning Maud Land (East Antarctica): evidence for continental collision during Gondwana assembly. *Journal of Petrology*, **57**, 185–228, https://doi.org/10.1093/petrology/egw005

Piazolo, S., La Fontaine, A., Trimby, P., Harley, S., Yang, L., Armstrong, R. and Cairney, J.M. 2016. Deformation-induced trace element redistribution in zircon revealed using atom probe tomography. *Nature Communications*, **7**, 10490, https://doi.org/10.1038/ncomms10490

Post, A.D., Tullis, J. and Yund, R.A. 1996. Effects of chemical environment on dislocation creep of quartzite. *Journal of Geophysical Research*, **101**, 22143–22155, https://doi.org/10.1029/96JB01926

Ruoff, A.L. and Balluffi, R.W. 1963. Strain-enhanced diffusion in metals. II. Dislocation and grain-boundary short-circuiting models. *Journal of Applied Physics*, **34**, 1848–1853, https://doi.org/10.1063/1.1729698

Sasaki, J., Peterson, N.L. and Hoshino, K. 1985. Tracer impurity diffusion in single-crystal rutile (TiO_{2-x}). *Journal of Physics and Chemistry of Solids*, **46**, 1267–1283, https://doi.org/10.1016/0022-3697(85)90129-5

Sayle, D.C., Catlow, C.R.A., Perrin, M.A. and Nortier, P. 1995. Computer simulation study of the defect chemistry of rutile TiO_2. *Journal of Physics and Chemistry of Solids*, **56**, 799–805, https://doi.org/10.1016/0022-3697(94)00270-3

Smith, D.C. and Pinet, M. 1985. Petrochemistry of opaque minerals in eclogites from the Western Gneiss Region, Norway: 2. Chemistry of the ilmenite mineral group.

Chemical Geology, **50**, 251–266, https://doi.org/10.1016/0009-2541(85)90123-8

Sobolev, N.V. and Yefimova, E.S. 2000. Composition and petrogenesis of Ti-oxides associated with diamonds. *International Geology Review*, **42**, 758–767, https://doi.org/10.1080/00206810009465110

Sobolev, N.V., Kaminsky, F.V., Griffin, W.L., Yefimova, E.S., Win, T.T., Ryan, C.G. and Botkunov, A.I. 1997. Mineral inclusions in diamonds from the Sputnik kimberlite pipe, Yakutia. *Lithos*, **39**, 135–157, https://doi.org/10.1016/S0024-4937(96)00022-9

Spetsius, V. and Safronov, A.F. 1986. Some compositional characteristics of rutile from eclogitic associations and in paragenesis with diamond. *Zap Vseross Mineral O-va*, **115**, 699–705 [in Russian].

Stebbins, J.F. 2007. Aluminum substitution in rutile titanium dioxide: new constraints from high-resolution ^{27}Al NMR. *Chemistry of Materials*, **19**, 1862–1869, https://doi.org/10.1021/cm0629053

Sun, R., Wang, Z., Shibata, N. and Ikuhara, Y. 2015. A dislocation core in titanium dioxide and its electronic structure. *RSC Advances*, **5**, 18506–18510, https://doi.org/10.1039/C4RA15278F

Sutton, A.P. and Balluffi, R.W. 1995. *Interfaces in Crystalline Materials*. Clarendon Press, Oxford.

Swope, J.R. and Smyth, J.R. 1995. H in rutile type compounds: I. Single-crystal neutron and X-ray diffraction study of H in rutile. *American Mineralogist*, **80**, 448–453, https://doi.org/10.2138/am-1995-5-604

Timms, N.E., Kinny, P.D., Reddy, S.M., Evans, K., Clark, C. and Healy, D. 2011. Relationship among titanium, rare earth elements, U–Pb ages and deformation microstructures in zircon: implications for Ti-in-zircon thermometry. *Chemical Geology*, **280**, 33–46, https://doi.org/10.1016/j.chemgeo.2010.10.005

Tomkins, H.S., Powell, R. and Ellis, D.J. 2007. The pressure dependence of the zirconium-in-rutile thermometer. *Journal of Metamorphic Geology*, **25**, 703–713, https://doi.org/10.1111/j.1525-1314.2007.00724.x

Tucker, A.T., Wilkinson, A.J., Henderson, M.B., Ubhi, H.S. and Martin, J.W. 2000. Measurement of fatigue crack plastic zones in fine grained materials using electron backscattered diffraction. *Materials Science and Technology*, **16**, 457–462, https://doi.org/10.1179/026708300101507910

Van Orman, J.A. and Crispin, K.L. 2010. Diffusion in oxides. *Reviews in Mineralogy and Geochemistry*, **17**, 757–826, https://doi.org/10.1515/9781501508394-018

Verberne, R., Reddy, S.M. *et al.* 2022. Dislocations in minerals: fast-diffusion pathways or trace-element traps? *Earth and Planetary Science Letters*, **584**, 117517, https://doi.org/10.1016/j.epsl.2022.117517

Vlassopoulos, D., Rossman, G.R. and Haggerty, S. 1993. Coupled substitution of H and minor elements in rutile and the implications of high OH contents in Nb- and Cr-rich rutile from the upper mantle. *American Mineralogist*, **78**, 1181–1191.

Vry, J.K. and Baker, J.A. 2006. LA-MC-ICPMS Pb-Pb dating of rutile from slowly cooled granulites: confirmation of the high closure temperature for Pb diffusion in rutile. *Geochimica et Cosmochimica Acta*, **70**, 1807–1820, https://doi.org/10.1016/j.gca.2005.12.006

Warr, L.N. 2021. IMA-CNMNC approved mineral symbols. *Mineralogical Magazine*, **85**, 291–320, https://doi.org/10.1180/mgm.2021.43

Watson, E.B. and Harrison, T.M. 2005. Zircon thermometer reveals minimum melting conditions on earliest Earth. *Science (New York, NY)*, **308**, 841–844, https://doi.org/10.1126/science.1110873

Watson, E.B., Wark, D.A. and Thomas, J.B. 2006. Crystallization thermometers for zircon and rutile. *Contributions to Mineralogy and Petrology*, **151**, 413–433, https://doi.org/10.1007/s00410-006-0068-5

Wheeler, J., Mariani, E., Piazolo, S., Prior, D.J., Trimby, P. and Drury, M.R. 2009. The weighted Burgers vector: a new quantity for constraining dislocation densities and types using electron backscatter diffraction on 2D sections through crystalline materials. *Journal of Microscopy*, **233**, 482–494, https://doi.org/10.1111/j.1365-2818.2009.03136.x

Withers, A.C., Essene, E.J. and Zhang, Y. 2003. Rutile/TiO_2II phase equilibria. *Contributions to Mineralogy and Petrology*, **145**, 199–204, https://doi.org/10.1007/s00410-003-0445-2

Yund, R.A., Smith, B.M. and Tullis, J. 1981. Dislocation-assisted diffusion of oxygen in albite. *Physics and Chemistry of Minerals*, **7**, 185–189, https://doi.org/10.1007/BF00307264

Zack, T. and Kooijman, E. 2017. Petrology and geochronology of rutile. *Reviews in Mineralogy and Geochemistry*, **83**, 443–467, https://doi.org/10.2138/rmg.2017.83.14

Zack, T., Moraes, R. and Kronz, A. 2004. Temperature dependence of Zr in rutile: empirical calibration of a rutile thermometer. *Contributions to Mineralogy and Petrology*, **148**, 471–488, https://doi.org/10.1007/s00410-004-0617-8

Experimental alteration of allanite at 200°C: the role of pH and aqueous ligands

Axel Denys[1], Anne-Line Auzende[1], Emilie Janots[1]*, German Montes-Hernandez[1], Nathaniel Findling[1], Pierre Lanari[2] and Valérie Magnin[1]

[1]Univ. Grenoble Alpes, Univ. Savoie Mont Blanc, CNRS, IRD, IFSTTAR, ISTerre, 38000 Grenoble, France

[2]Institute of Geological Sciences, University of Bern, Baltzerstrasse 1+3, CH-3012 Bern, Switzerland

EJ, 0000-0002-0658-3644
*Correspondence: emilie.janots@univ-grenoble-alpes.fr

Abstract: Allanite is a major host of rare earth elements (REEs) in the continental crust. In this study, reaction mechanisms behind allanite alteration are investigated through batch experiment runs on natural allanite grains in carbonate-bearing hydrothermal fluids at 200°C, with initial acidic (pH = 4) or alkaline (pH = 8) conditions and with different aqueous ligands (120 mmol kg^{-1} of F, Cl, P or S). Time-series experiment runs in F-doped systems at different durations between 15 and 180 days reached a steady state at 120 days. The pH efficiently controls the allanite alteration process, with initial high pH, alkaline conditions being more reactive (75% alteration compared with 25% under acidic conditions). The ligand also significantly influences the alteration process under initial acidic conditions with the P-doped system (70%) almost non-reactive for the Cl- and S-doped systems (<5%). In the alteration rim, REEs are mainly redistributed in REE-bearing phases either as carbonates (F-doped) or phosphates (P-doped). The relatively flat REE-normalized patterns of the recovered experimental fluids suggest a fractionation of light rare earth elements (LREEs) over heavy rare earth elements (HREEs) during the course of the alteration reactions. It is proposed that secondary REE mineral precipitation at the reaction front creates a local disequilibrium in the solution and a steep chemical gradient promoting allanite dissolution and thus its alterability.

Supplementary material: Tables S1, S2, S3 corresponding to EMP data and analytical conditions, Figure S4 showing BSE images of the starting allanite material and Figure S5 presenting EMP compositions (REE versus Ca) of the Calcite and BGM experimental products are available at https://doi.org/10.6084/m9.figshare.c.6699992

Allanite, a mineral of the epidote group with the ideal formula $CaREEFe^{2+}(Al, Fe^{3+})_2(Si_2O_7)(SiO_4)O(OH)$, is a major REE carrier mineral in the continental crust, with preferential incorporation of the light REEs (LREEs: La to Gd) over the heavy REEs (HREEs: Tb to Lu + Y). Primary allanite occurs as an accessory phase in magmatic and metamorphic rocks (Gieré and Sorensen 2004 and references therein). It is a good petrological proxy and geochronometer for metamorphic processes (e.g. Engi 2017), magma sourcing (e.g. Anenburg *et al.* 2015) or mineralization under hydrothermal conditions (e.g. Pal *et al.* 2011). Allanite has also been described as the main primary REE and U source in supergene and hydrothermal systems (Caruso and Simmons 1985; Berger *et al.* 2008; Ichimura *et al.* 2020), reaching economic levels (Chabiron and Cuney 2001; Corriveau *et al.* 2007). Hydrothermal alteration of allanite is common (Poitrasson 2002) and often occurs as partial replacement of primary allanite by secondary REE minerals. These include fluorocarbonates (e.g. Middleton *et al.* 2013), phosphates (Berger *et al.* 2008) and silicates (Smith *et al.* 2002). Frequently, secondary Th-minerals are also described in association with allanite replacement (Middleton *et al.* 2013). As a main REE host, these alteration reactions are thus important for understanding REE mass transfer, with their economic implications as strategic metals. Furthermore, understanding associated actinide mobility in REE-rich hydrothermal systems is also crucial, because it can be decisive for REE mining (as by-products or nuclear waste).

It is widely accepted that the greater sensitivity of allanite to alteration, compared with that of epidote (Price *et al.* 2005), is partly due to its metamict state, which is caused by α-particle bombardment damaging the structure, even for a low Th and U content (Ewing *et al.* 1987; Ercit 2002). The role of other inherent factors, such as the crystal chemistry of the allanite and the physicochemical properties of the fluid, remains poorly understood. While numerous examples of natural allanite alteration have been reported, its experimental reactivity in the presence

From: van Schijndel, V., Cutts, K., Pereira, I., Guitreau, M., Volante, S. and Tedeschi, M. (eds) 2024. *Minor Minerals, Major Implications: Using Key Mineral Phases to Unravel the Formation and Evolution of Earth's Crust.*
Geological Society, London, Special Publications, **537**, 149–164.
First published online September 21, 2023, https://doi.org/10.1144/SP537-2023-21

of hydrothermal fluids has received little attention, only at high pressure and high temperature conditions (Krenn *et al.* 2012) or as a product of monazite and xenotime alteration (Budzyń *et al.* 2011, 2017). In order to fill this gap, allanite alteration experiments have been conducted on natural homogeneous crystalline grains in the presence of carbonate-bearing hydrothermal fluids doped with various ligands (120 mM of F, Cl, P, S) under initial acidic and high-pH alkaline conditions at 200°C and P_{sat} for durations of 15 to 180 days. The role of added ligands was investigated for initial acidic conditions (pH around 4 at room temperature) common under hydrothermal conditions (Seward *et al.* 2014). The investigated ligands are elements of importance for REE mobility in hydrothermal systems due to strong aqueous complexation at 200°C (Gammons *et al.* 1996). The kinetics of the alteration reactions were investigated through time-dependent experiments using both acidic and high-pH alkaline fluids (pH = 8 at room temperature) in the presence of F. In this study, the experimental alteration of allanite is strongly controlled by the fluid composition and the precipitation of secondary REE minerals in the form of fluorocarbonates and phosphates.

Analytical methods and experimental procedure

Starting material

All experiments were performed using fragments from a monocrystal of allanite-Ce (henceforth allanite) from the Frontenac Formation in the Central Metasedimentary Belt of the Greenville Orogen (Ontario, Canada). Crystallization ages, based on associated titanite U–Pb geochronology, are around 1157–1178 Ma (Mezger *et al.* 1993). The composition, determined by electron probe microanalyser (EPMA), is homogeneous and corresponds to an intermediate composition between ferriallanite and allanite (general formula: $Ca_{1-1.2}REE_{0.6-0.8}Al_{1.5-1.7}Fe^{3+}_{0.1-0.5}Fe^{2+}_{0.8-1.1}Mg_{0.1}Si_{3.1-3.3}O_{12}(OH)$) (Table S1, supp. mat.). The content of radionuclides (such as 0.36–0.79 wt% ThO_2 and UO_2) ranges mostly below the detection limit (DL).

The crystal was crushed manually and then ground mechanically using a planetary micromill Fritsch Pulverisette 7. To ensure maximized kinetics and yet a suitable grain size for post-experimental characterization, we selected an initial grain size of 20–50 μm after sieving. Traces of REE-carbonates in microcracks (observed by scanning electron microscope but not detected on the X-ray diffraction pattern) were removed by soaking the allanite powder in a 1 M acetic acid solution in an ultrasonic bath for 10 min. After centrifugation (3500 rpm, 10 min), the solid residue was filtered through a 2.7 μm glass fibre filter and dried overnight at 50° C. The material was stored in spectroscopic plastic tubes in the dark at ambient temperature.

Experimental procedure

Experiments were conducted by reacting 150 mg of powdered allanite (Fig. S4, supp. mat.) with 1.5 ml of the aqueous solution (fluid/solid ratio = 10) in 3 ml Teflon cell reactors sealed into a steel autoclave without agitation ('static batch reactor') and placed in a multi-oven at 200°C (P_{sat} ≈ 16 bar or 16 bar + pCO_2, when CO_2 was added). However, pCO_2 varies during alteration and these variations were not quantified. Acidic solutions with an initial pH = 4 were obtained by adding to ultrapure water 99.9% certified pure carbonic ice (around 30 mg). For alkaline systems, ultrapure water was replaced by 1.5 ml of a 1 M $NaHCO_3$ solution (pH = 8.7, Lafay *et al.* 2014). These solutions were mixed with 120 mM (equivalent to the REE molar content in allanite) of F (introduced as solid NaF), P (as Na_3PO_4, $12H_2O$), S (as Na_2SO_4) or Cl (as NaCl). An initial time series of experiments using NaF as ligand in acidic and alkaline systems was performed (15, 30, 60, 120 and 180 days) to investigate the kinetics of the reaction processes (Table 1). Experiments with other ligands were run for 120 days. At the end of each experiment, the sealed reactor was rapidly quenched in cold water. Recovered solutions were carefully collected with a syringe, filtered to remove solid residue (0.2 μm), diluted 5 times in ultrapure water slightly acidified with nitric acid and immediately stored at 4°C in an ion-free tube for further characterization. The residual solid was collected, dried at 60°C overnight, weighted and stored at ambient temperature. A fraction of each solid run product was mounted in epoxy resin and finely polished (mirror surface) for microscopic and electron microprobe analyses.

Solid characterization methods

X-ray diffraction and Rietveld refinement. Mineral modal abundances of initial and post-experimental solids were characterized by X-ray diffraction (XRD) at ISTerre (Grenoble, France). Samples were ground in ethanol using a McCrone micronizing mill, oven-dried overnight and prepared as a randomly oriented mount. The XRD patterns were recorded with a Bruker D8 powder diffractometer equipped with a SolXE Si(Li) solid-state detector from Baltic Scientific Instruments using CuKα_1 + 2 radiation. Intensities were recorded at 0.026° 2θ step intervals from 5 to 90° (10 s counting time per step). Eva Bruker software associated with the International Centre for Diffraction Data (ICDD)

Table 1. *Experimental conditions and solid products*

Set	Exp.	Carbonate source	Initial pH	Ligands (120 mmol kg^{-1})	Duration (days)	Aln (%)	Secondary solid products
B1	B1015	Carbonic ice[1]	4	NaF	15	93	Bsn (1.2%); Syn (1.5%); Flr (2.0%); Hem (2.1%)
	B1030	Carbonic ice[1]	4	NaF	30	92	Bsn (2.2%); Syn (1.0%); Flr (2.5%); Hem (2.0%)
	B1060	Carbonic ice[1]	4	NaF	60	95	Bsn (3.2%); Flr (3.6%); Hem (2.4%); Ana (4.9%)
	B1120	Carbonic ice[1]	4	NaF	120	76	Bsn (5.1%); Syn (<1%); Flr (4.3%); Hem (2.2%); Ana (12%)
	B1180	Carbonic ice[1]	4	NaF	180	77	Bsn (5.6%); Flr (5.8%); Hem (4.7%); Ana (7.3%)
B2	B2015	$NaHCO_3^-$ 1M 1.5 ml	8.7	NaF	15	36	Bsn (2.0%); Syn (1.0%); Pst (6.6%); BGM (8.9%); Cal (4.5%); Hem (8.9%); Ana (22%); Sme (6.7%); Nsd (3.1%)
	B2030	$NaHCO_3^-$ 1M 1.5 ml	8.7	NaF	30	29	Bsn (3.1%); Syn (1.0%); Pst (8.9%); BGM (5.7%); Cal (4.5%); Hem (8.4%); Ana (26%); Sme (7.7%); Nsd (4.1%)
	B2060	$NaHCO_3^-$ 1M 1.5 ml	8.7	NaF	60	31	Bsn (3.1%); Syn (1.7%); Pst (8.9%); BGM (9.4%); Cal (8.4%); Hem (9.1%); Ana (24%); Sme (7.6%); Nsd (1.8%)
	B2120	$NaHCO_3^-$ 1M 1.5 ml	8.7	NaF	120	23	Bsn (5.2%); Syn (<1%); Pst (3.3%); BGM (5.9%); Cal (9.5%); Hem (9.9%); Ana (31%); Sme (7.2%); Nsd (3.8%)
B3	B3P120	Carbonic ice[1]	4	$Na_3PO_4.12H_2O$	120	27	Mnz (15%); Hap (13%); Hem (6.8%); Ana (31%); Sme (6.3%)
	B3S120	Carbonic ice[1]	4	Na_2SO_4	120	97	Anh (1.6%); Ana (1.6%)
	B3Cl120	Carbonic ice[1]	4	NaCl	120	98	Hl (2.0%)

[1]Carbonic ice is certified 100% pure CO_2 – around 30 mg (after the epoxy reactor closure). Estimated standard deviation is <2% for values >10% and does not exceed 10% for lower quantification. Aln, allanite; Ana, analcime; Anh, anhydrite; BGM, burbankite-group mineral; Bsn, bastnäsite; Cal, calcite; Chl, chlorite; Flr, fluorite; Hap, hydroxyapatite; Hem, hematite; Hl, halite; Mnz, monazite; Nsd, nordstrandite; Pst, parisite; Sme, smectite; Syn, synchysite. For a better reading comprehension with Aln (allanite), Ana designate analcime instead of the common abbreviation Anl. Source: abbreviations from Warr L.N. (2021).

Powder Diffraction File (PDF) database was used to determine the modal composition of the powder on a significant part of the recovered solid product for all alteration experiments. This enables distinguishing between the different REE minerals in the solid product. Rietveld refinement with Profex/BGMN software was then performed to precisely quantify mineral abundances. The quality of the Rietveld refinement is assessed by the χ^2 factor, which lies between 2.5 and 4 for all experiments.

Scanning electron microscopy. The mineral distribution and microstructure of the experimental solids were investigated by using a ZEISS Gemini 500 scanning electron microscope (SEM). The measurements were carried out in high-vacuum mode using a high tension (EHT) of 3 kV for a working distance (WD) between 3.4 and 4.8 mm. Samples were sputtered with a 15 nm thick cover of Au-Pd. Complementary investigations were performed using the conventional environmental SEM Tescan Vega 3. The measurements were carried out in high-vacuum mode (9.9×10^{-9} bar), using an accelerating voltage of 16 kV, with 90 nm spot size and 15 mm WD. Samples were sputtered with 20 nm of carbon. The microscope was equipped with a 30 mm^2 Energy Dispersive X-Ray Spectrometer (EDS) manufactured by Rayspec with SamX's electronic system and software.

Electron probe microanalyser. Quantitative chemical analyses of initial and post-experimental allanite and secondary products (when the grain size was suitable) were carried out using a JEOL JXA-8230 EPMA equipped with five wavelength-dispersive spectrometers (WDS) at ISTerre (Grenoble, France). Analytical conditions were 15 kV acceleration voltage, 12 nA beam current and 1 to 5 μm beam size (details in Table S3, supp. mat.). The ZAF (atomic number, absorption and fluorescence) correction procedure was applied using the JEOL software for quantitative analysis. The DLs range between 0.01 and 0. 04 wt% using the 2σ criterion (Batanova *et al.* 2018).

Trace element analysis. Trace element analysis of allanite was performed at the Institute of Geological Sciences (University of Bern) using laser ablation inductively-coupled plasma mass spectrometry (LA-ICP-MS), which consisted of a Geolas Pro 193 nm ArF excimer laser coupled to an Elan DRC-e quadrupole ICP-MS. A He–H_2 gas mixture (1 and 0.008 L/min, respectively) was used as the aerosol transport gas. Allanite trace element analyses were performed with laser beam diameters of 16, 24 and 32 μm, frequencies of 9 and 7 Hz and energy densities on the sample of 5.0 J/cm^2. Sample analyses were calibrated using GSD-1Gg and accuracy was monitored using a reference glass NIST SRM 612 (Jochum *et al.* 2005, 2011). Data reduction was performed using the SILLS software package (Guillong *et al.* 2008) and LOD values obtained with the method of Pettke *et al.* (2011).

Fluid characterization methods

Inductively-coupled plasma spectrometry. All recovered solutions were stored using metal-free tubes (from VWR). For an accurate quantification of trace elements, measurements were performed by ICP-MS using a Thermo Scientific XSERIES 2 spectrometer. Recovered solutions were diluted 3 times with 2% HNO_3 solution to a volume of 6 ml. Finally, 0.5 ml of an In solution was systematically added as an internal standard to correct for the drift of the ICP-MS. Collision cell technology (CCT) was used for some elements (Ca, Fe and Mn) in order to reduce polyatomic interferences with 5% H_2 in He gas. Measurement quality was evaluated by duplicating the measurement of standards that were analysed 5 times on the ICP-MS. Calculations to extract concentrations from the integration of peak signals were performed off-line. Reproducibility depends on the nature of the analysed element. It ranges from 1% to 19% for the REEs and from 3% to 30% for other trace elements. The DLs are defined as 3 times the average of the blank measurements. All data below the DL were excluded. Because the torch for ICP-MS was sheathed in quartz, the concentration of Si was then determined by atomic emission spectrometry (ICP-AES) using a Perkin Elmer Optima 3000 DV ICP-AES. Solutions were diluted 5 times using a 2% HNO_3 solution providing the minimum analysable volume. The same HNO_3 solution was used to prepare standards and blanks. The analytical error for Si is 8%.

Capillary electrophoresis. The anionic content (Cl^-, SO_4^{2-}, F^-, PO_4^{3-}, HCO_3^-) in the recovered solutions was quantified using a capillary electrophoresis (CE) system by WATERS®. The CE apparatus was equipped with a fuse capillary (75 μm i.d. × 60 cm total length) and a diode detector. The CE was operated at 20°C and at a voltage of 20 kV. Electrophoregrams were recorded with indirect mode detection at 254 nm using an Hg lamp. The background electrolyte (BGE) was composed of 4.6 mmol/LNa_2CrO_4 solution, 0.5 mmol/L OFMOH™ from WATERS™ and an H_3BO_4 solution (pH = 8.0). Prior to each measurement series, the capillary was conditioned by flushing with 1 mol/L NaOH and 0.1 mol/L NaOH (5 min each) followed by a 10 min flush with deionized water and a BGE solution (15 min flush). The capillary was preconditioned prior to each measurement by flushing the BGE for 1 min. All samples were measured in duplicate using hydrostatic injection mode.

Results

Allanite alteration as a function of the initial pH

The role of initial pH was investigated by time-series experiments in order to evaluate, together with the final alteration extent, the kinetics of the reaction. The experiments were performed for an F-doped system at a duration of between 15 and 180 days under initial acidic and high-pH alkaline conditions (Table 1).

The recovered experimental solids, characterized by XRD, displayed alteration evidence with secondary phases in the run products (Fig. 1). The extent of alteration was estimated on the basis of remaining allanite in the recovered samples. The constancy of the allanite composition between the final and initial materials (Table S1 – Supp. Mat) attested to no secondary allanite/epidote precipitation. The reaction progress was estimated from the allanite abundance. The run products' modal composition showed that the kinetics and extent of alteration greatly varied with the initial pH (Fig. 1). The alteration rate was much higher in the high-pH alkaline system, with 65% alteration reached within the first 15 days and a rapid stabilization at around 70–75% alteration from 30 to 120 days. In contrast, the extent of alteration in the acidic experiments was scarce after 15 days and only achieved 25% alteration for the longest durations (120 and 180 days). However, under both the acidic and high-pH alkaline conditions, with an F-doped solution, the same mineral phases grew at the expense of allanite (Fig. 1): analcime (Na-Al silicate), hematite (Fe_2O_3), and REE-bearing carbonates (Table 2). The REE-bearing carbonates, determined from XRD analyses, changed depending on the initial pH of the solution. They consisted of: (1) bastnäsite (general formula: $LREECO_3F$) and synchysite (general formula: $CaLREE(CO_3)_2F$) under acidic conditions; and (2) a burbankite-group mineral (BGM, with general formula: $(Na,Ca)_3(Sr,Ba,Ce)_3(CO_3)_5$) and REE-fluorocarbonates with parisite (general formula: $CaLREE_2(CO_3)_2F_2$) associated with bastnäsite and minor synchysite in the high-pH alkaline system. Besides these phases, fluorite (CaF_2) appeared in the acidic experiments, while calcite ($CaCO_3$), smectite and minor nordstrandite (general formula: $Al(OH)_3$) crystallized in the high-pH alkaline system. High REE contents up to 2 wt% and 8 wt% $(REE + Y)_2O_3$ were measured in fluorite and calcite, respectively (Table 2; Table S2 supp. mat.).

The microstructures of the recovered solids differed depending on the initial pH, which can be attributed to the reaction progress. In acidic systems, due to low alteration rate (15 days), the initial shape of the allanite grains, characterized by typical conchoidal edges, was mostly preserved while fluorite and analcime crystals grew around allanite from the bulk fluid (Fig. 2a). The allanite surface was pitted and covered by smectite. At this stage, REE-(fluoro)carbonates were restricted to inherited fractures in the allanite. A higher reaction extent (25%) in the acidic system (120 days) resulted in alteration rims surrounding some of the allanite grains. Their surfaces were characterized by a

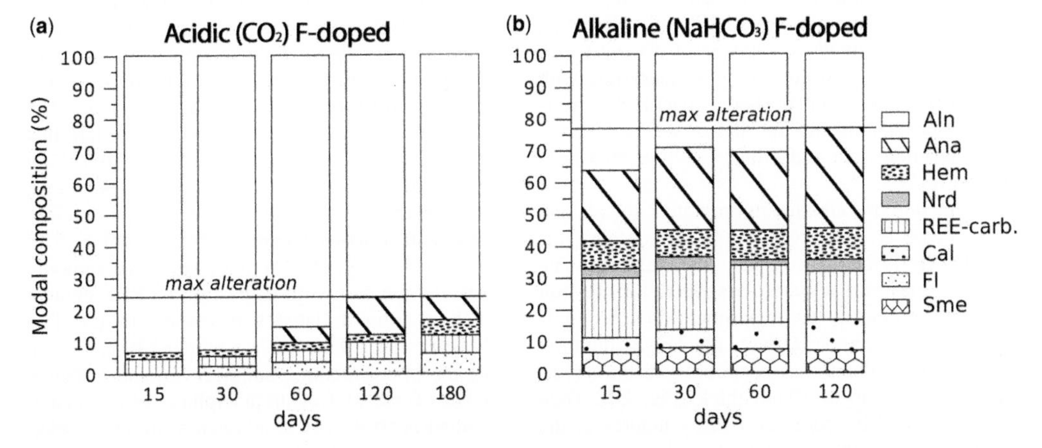

Fig. 1. Evolution of the proportion of the starting material and run products, utilizing Rietveld refinement (in %) for (**a**) initial acidic and (**b**) high-pH alkaline F-doped systems. The group of REE-carb (REE-carbonate minerals) represents bastnäsite + synchysite in acidic system and parisite + bastnäsite + synchysite + the burbankite-group mineral in the high-pH alkaline system. These REE-carbonate minerals were identified by X-ray diffraction (XRD) analyses. Full lines represent the maximum allanite replacement in a state close to equilibrium. Numerical values are presented in Table 1. Aln, allanite; Ana, analcime; Cal, calcite; Fl, fluorite; Hem, hematite; Nrd, nordstrandite; Sme, smectite.

Table 2. *Microprobe selected analyses of major run products (wt%)*

	Analcime		Calcite		BGM		Fluorite		Anhydrite
System	$HCO_3^- + F^-$	$CO_2 + PO_4^{2-}$	$HCO_3^- + F^-$		$HCO_3^- + F^-$		$CO_2 + F^-$		$CO_2 + SO_4^{2-}$
Days	120	120	120	120	120	120	120	180	120
SiO_2	49.1	49.0							
Al_2O_3	24.5	24.1					0.22	0.32	
FeO	0.09	0.13	0.23	0.32	0.54		0.09	0.21	0.10
CaO			51.2	53.0	21.08	6.07	66.35	67.05	41.75
Na_2O	15.8	15.2	0.62	0.39	0.37	1.66	0.5	0.51	0.05
P_2O_5		0.15	0.12				0.14	0.18	
SO_3									47.47
F			0.30		0.18	0.57	48.02	47.75	
La_2O_3			1.90	0.90	14.69	15.17	0.68	0.35	0.17
Ce_2O_3			4.13	2.30	23.51	30.52	1.20	0.69	0.51
Pr_2O_3			0.46	0.24	1.68	2.57	0.19	–	0.14
Nd_2O_3			1.16	0.60	3.56	6.35	0.23	0.12	0.16
Sm_2O_3									
Gd_2O_3		0.13	0.21						
Dy_2O_3					0.23				
Y_2O_3			0.12	0.11			0.12		
SrO			0.25	0.26	1.02	0.68	1.62	1.59	0.61
ThO_2			0.09	0.15	0.84	0.90	0.27		
PbO						0.13			
Total*	89.50	88.66	60.63	58.27	67.63	64.37	99.41	98.68	90.94
∑(REE)†		0.13	7.99	4.16	43.68	54.60	2.42	1.17	0.97
La/Y	n.d.	n.d.	17.4	8.79	n.d.	n.d.	6.02	n.d.	n.d.
Ce/Ce*‡	n.d.	n.d.	1.07	1.19	1.14	1.18	0.8	2.84	0.82

Notes: Values in italic (%) column are mean relative errors, and 2σ is the standard deviation; n.d. not determined.
*Total is corrected for $-O = F_2$ values.
†∑REE refers to the sum of $(REE + Y)_2O_3$.
‡Ce/Ce* $CeN/(LaN*PrN)^{1/2}$. Empty cells are concentrations below detection; n.d. not determined.

pervasive sawtooth-shaped reaction front, highlighting more extensive dissolution (Fig. 2b). Reaction rims were sequentially composed of discontinuous layers of hematite followed by nanocrystals with the granular and acicular shape of REE-fluorocarbonates penetrating through the dissolving allanite at the reaction front and filling newly-opened fractures (Fig. 2c).

In the high-pH alkaline run products, the allanite grain shape is preserved on the microscale, surrounded by layers of complex microtextures. The morphologies and textures described for the run product from the 15-day experiment do not significantly change compared with the longer duration experiments. Allanite has penetrative reaction rims that can reach up to 10 µm thick (Fig. 2d). They were delimited by complex microstructures at the reaction front, such as nanoscale etch pits or sawtooth surfaces (Fig. 2e). Close to the reaction front, REE-fluorocarbonates also precipitated at the surface, within the etch pits or in inherited microfractures, with nanogranular, acicular or prismatic shapes (Fig. 2e). The submicron size of these phases prevented quantitative chemical analyses by EMPA. The allanite surface was overlain with a thin layer of hematite crystals of around 10 to 500 nm in size. Smectites were also ubiquitous and clearly identifiable by their fibrous (honeycomb) morphology and platelet growth oriented towards the fluid. The BGMs occur as microscale euhedral crystals that randomly precipitated from the reactive bulk fluid (decoupled from the allanite replacement products). They are mainly prismatic and more or less elongated with a size generally varying from *c.* 5 to 30 µm in size (Fig. 2f). The BGM crystals commonly display a zonation with respect to the REE content which is anti-correlated with respect to Ca (Fig. S5, supp. mat.). They have a higher LREE content than the initial allanite, but with Sm and Y below the DL (Table 2). Calcite precipitates as aggregates of euhedral crystals 10 to 20 microns in size or intergrown with relic allanite (Fig. 2f).

Allanite alteration as a function of ligands

To investigate the effects of ligands on the alteration of allanite, experiments were run under the initial acidic conditions with P-doped, S-doped and

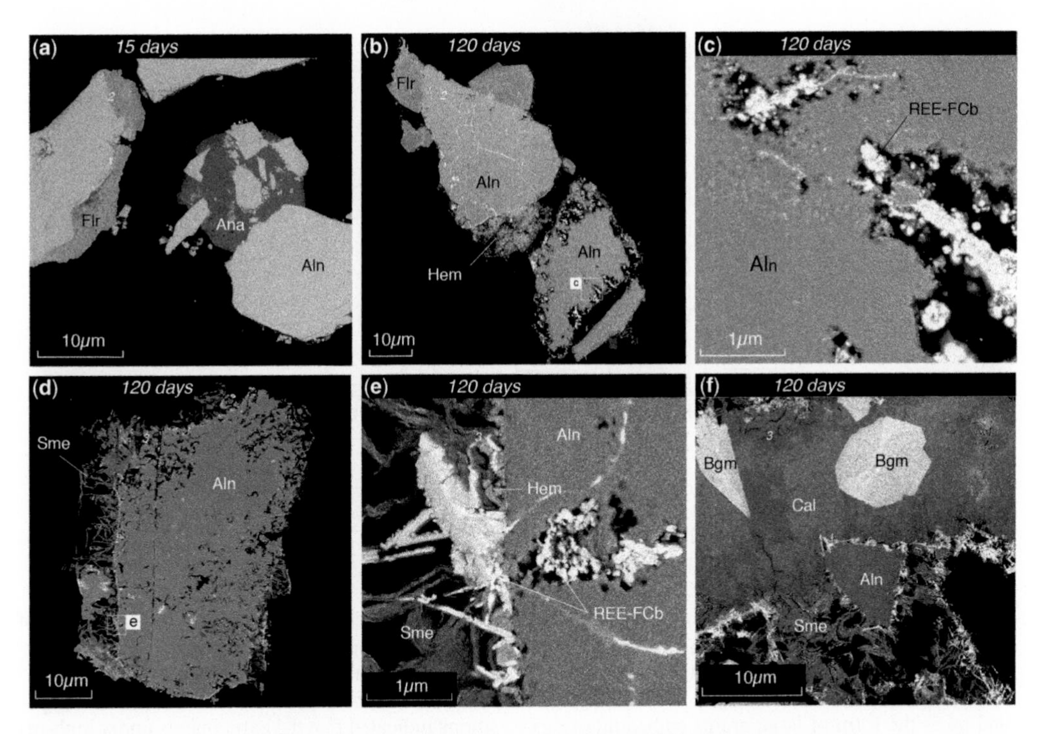

Fig. 2. Scanning electron microscopy images using backscattered electron (BSE) imaging of allanite and secondary products from acidic (a, b and c) and high-pH alkaline (d, e and f) F-doped runs for different times. (**a**) Relatively unaltered allanite with fluorite and analcime growing along the allanite grain rims. (**b**) Relatively unaltered allanite (top left) with internal fractures filled by REE phases along with more reacted allanite (bottom right) with sawtooth-shaped grain rims after 120 days. (**c**) Magnification of (b) showing reaction interface with allanite composed of granular nanometric REE-fluorocarbonates replacing allanite. (**d**) Allanite grain displaying edge pitting with a porosity that progresses anisotropically to the grain centre. Edges are rimmed by saponite whiskers, which probably formed during quenching. (**e**) Detail of an allanite edge showing a sharp eroded surface rimmed by REE-fluorocarbonates with a granular, prismatic and needle-like shape, along with hematite and saponite. Porous cavities are filled with REE-fluorocarbonates. (**f**) Cluster of burbankite-group minerals with a large crystal of calcite growing in the interstitial space between the minerals. Aln, allanite; Ana, analcime; Bgm, burbankite-group mineral; Cal, calcite; Flr, fluorite; Hem, hematite; REE-FCb, REE-fluorocarbonates (bastnäsite, parisite, synchysite); Sap, saponite; Sme, smectite.

Cl-doped solutions for 120 days (Table 1). Similar to that observed for the pH, the reaction progress was also significantly affected by the ligands (Fig. 3). The most reactive system was the P-doped one, which achieved 73% allanite alteration. This reaction extent was similar to that seen in the F-doped system under high-pH alkaline conditions (75% of reaction), but much higher than that obtained under similar acidic conditions (23% of reaction) over the same duration (120 days). The S- and Cl-bearing systems were less reactive compared with the others, with less than 5% secondary minerals.

In the reactive P-doped experiments, analcime and hematite were present in major proportions in the recovered solid, as for the F-doped system. Smectite was also an alteration product of allanite, as in the high-pH alkaline system. The main difference between the P- and F-doped systems was the nature of the mineral phases that accommodated REEs and Ca, such as monazite (general formula: $LREEPO_4$, 21% of solid product) and hydroxyapatite (general formula: $Ca_5(PO_4)_3OH$, 18% of solid product). The alteration microstructures were similar to those previously described in the F-bearing system. Allanite was largely affected by dissolution, as illustrated by the numerous etch pits scattered on the surface (Fig. 4a). Allanite alteration resulted in thick reaction rims made up of a nanomixture of monazite-hematite and hydroxylapatite with an apparent microscale spatial distribution from the reaction front towards the reactive fluid (Fig. 4b). Submicron monazite crystals precipitated directly at the interface with the allanite (Fig. 4c). Similar to the F-doped systems, hematite occurred as a thin, quasi-continuous corona around the allanite grains. Finally, euhedral micrometric-sized grains

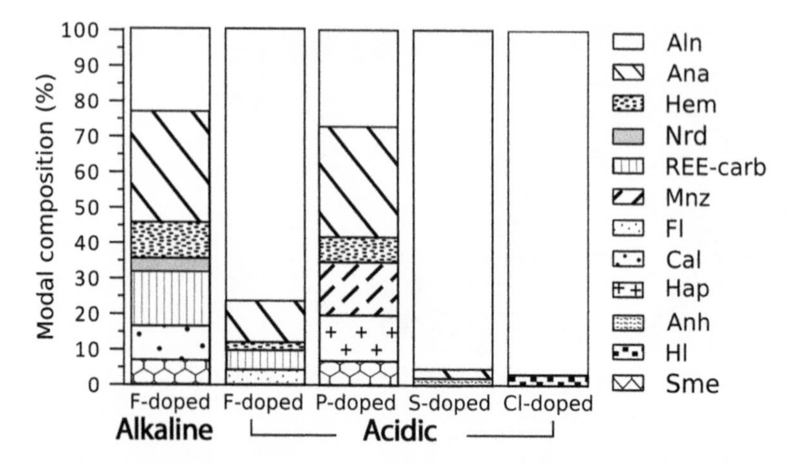

Fig. 3. Comparison of mineral modal compositions after 120 days for the F-doped high-pH alkaline systems and the F-, P-, S- and Cl-doped systems (respectively shown in columns), which were identified by X-ray diffraction (XRD) and Rietveld refinement (in %). Aln, allanite; Ana, analcime; Anh, anhydrite; Cal, calcite; Fl, fluorite; Hl, halite; Hap, hydroxylapatite; Hem, hematite; Mnz, monazite; Nrd, nordstrandite; REE-Carb, REE-fluorocarbonates; Sme, smectite.

of hydroxylapatite, mixed with smectite filaments which probably formed during the quench, are seen along the outer edge of the alteration rim (Fig. 4b, c). Analcime remains the major alteration phase, and takes the form of large grains embedding relict allanite along the allanite reaction rims (Fig. 4a).

Recovered fluid chemistry

Beside solid product characterization, the fluid composition was also analysed for each experiment (Table 3; Fig. 5). While the fluid compositions can be modified by internal and external factors through the course of the reaction (water consumption by alteration products, permeability limits of the Teflon reactors, quenching effects), the reproducibility of the results supports the general qualitative significance of the fluid chemistry dataset.

In time-series experiments, the final fluid compositions indicated that the experiments under high-pH alkaline conditions were already in a steady-state (approaching constant concentrations of all measured elements with time) after 15 days (Fig. 5b), which is in agreement with the mineralogical results. In the initially acidic system, elemental concentrations of Ca, REE, Th and U evolved until reaching a near plateau only after 120 days (Fig. 5a). At that stage, Si, Al and Ca reached similar concentrations in the F-doped acidic and high-pH alkaline systems, whereas REEs, U, and Th were lower in the acidic

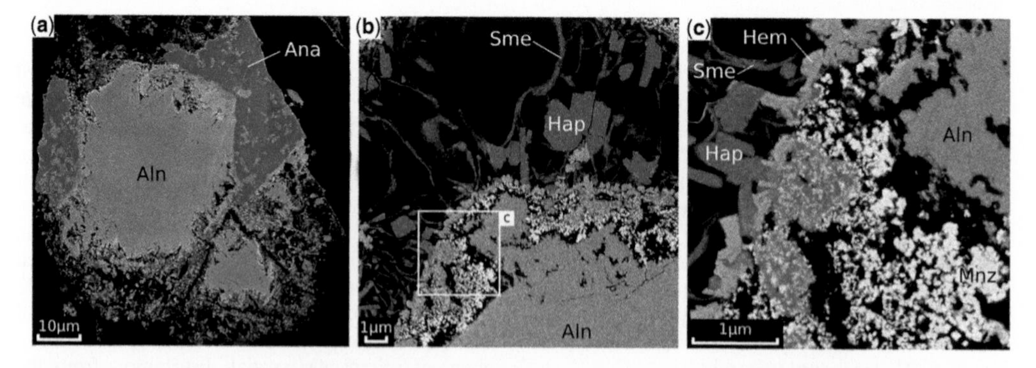

Fig. 4. Secondary electron (SE) and backscattered electron (BSE) images of allanite alteration in an acidic P-doped system. (**a**) Typical allanite grain with eroded grain boundaries and with large crystals of analcime partially embedding the other secondary minerals. (**b**) Continuous reaction front composed of monazite replacing allanite and a thin (<200 nm) rim of hematite outlining the original shape of the allanite. Outwards from the reaction front is rimmed by euhedral hydroxyapatite (Hap) that precipitated in the interstitial space between filaments of saponite (Sap). (**c**) Detail from (b) showing that the abundance of the nanosize monazite increases in the vicinity of the eroded allanite. Aln, allanite; Ana, analcime; Cal, calcite; Hem, hematite Mnz, monazite; Sme, smectite.

Table 3. *Composition of fluids after allanite batch experiments*

System	Experimental set: $CO_2 + F^-$					$NaHCO_3 + F^-$				$CO_2 + PO_4^{2-}$	$CO_2 + SO_4^{2-}$	$CO_2 + Cl^-$
Experiment	B1015	B1030	B1060	B1120	B1180	B2015	B2030	B2060	B2120	B3P120	B3S120	B3Cl120
Days	15	30	60	120	180	15	30	60	120	120	120	120
mol/kg × 10^{-4}												
[1] Si	38.0	15.8	13.1	16.8	16.0	19.4	22.3	25.4	43.0	62.1	14.9	27.4
Al	1.20	1.35	0.51	5.88	5.76	2.40	2.06	1.50	0.50	2.48	0.35	0.11
Fe						0.21	0.20	0.16	0.12	0.045		
Ca				0.11	0.069	0.14	0.14	0.16	0.17	0.082		0.077
Respective ligands	950	976	869	668	598	1781	1661	1412	1211		1477	2877
HCO_3^-						10 980	10 791	7132	6604	2489		
mol/kg × 10^{-8}												
Y				4.34	0.20	140	175	227	297	4.85	0.26	1.27
La		0.25	0.40	4.49	1.51	30.3	24.8	41.6	52.9	15	4.90	2.88
Ce	0.07	0.25	0.35	6.48	5.28	81.4	73.5	94.2	75.1	8.04	1.92	0.93
Pr			0.033	0.48	0.23	9.50	11.3	12.5	14.8	1.14	0.27	0.25
Nd				1.97	0.52	37.2	45.6	49.1	55.6	5.23	0.94	4.26
Sm				0.24	0.037	9.26	11.4	11.8	12.7	0.44	0.04	0.19
Eu				0.21		1.69	2.10	2.22	2.59	0.36	0.02	0.20
Gd				0.45	0.077	9.80	12.1	13.8	15.3	0.76	0.06	0.25
Tb				0.0031		1.70	2.15	2.54	2.85	0.013	0.0022	0.0035
Dy				1.13		10.9	14.0	16.7	19.6	1.21	0.045	0.38
Ho						2.45	3.15	3.84	4.55			
Er				2.25		8.89	11.7	13.5	16.6	2.15	0.073	0.67
Tm						1.64	2.12	2.33	2.59			
Yb				3.26		13.6	17.6	17.6	20.2	2.67	0.094	1.05
Lu						2.69	3.47	3.31	3.59			
Th		0.0077	0.033	0.24	0.24	349	216	304	193	0.30	0.025	
U	16.0	18.1	4.89	8.76	12.4	2795	2972	2570	3120	6.54		

Respective ligands refer to the anion initially used for the experiments (measured as F^-, PO_4^{3-}, SO_4^{2-}, Cl^-). Maximum analytical error is <4% for Si, La, Ce, Pr, Ho, Tm; <8% for Al, Fe, Eu, Lu, Y, Th; 9% for U; <15% for Nd, Sm, Yb; <19% for Gd, Eu; 34% for Ca. Empty cells are concentrations below detection.

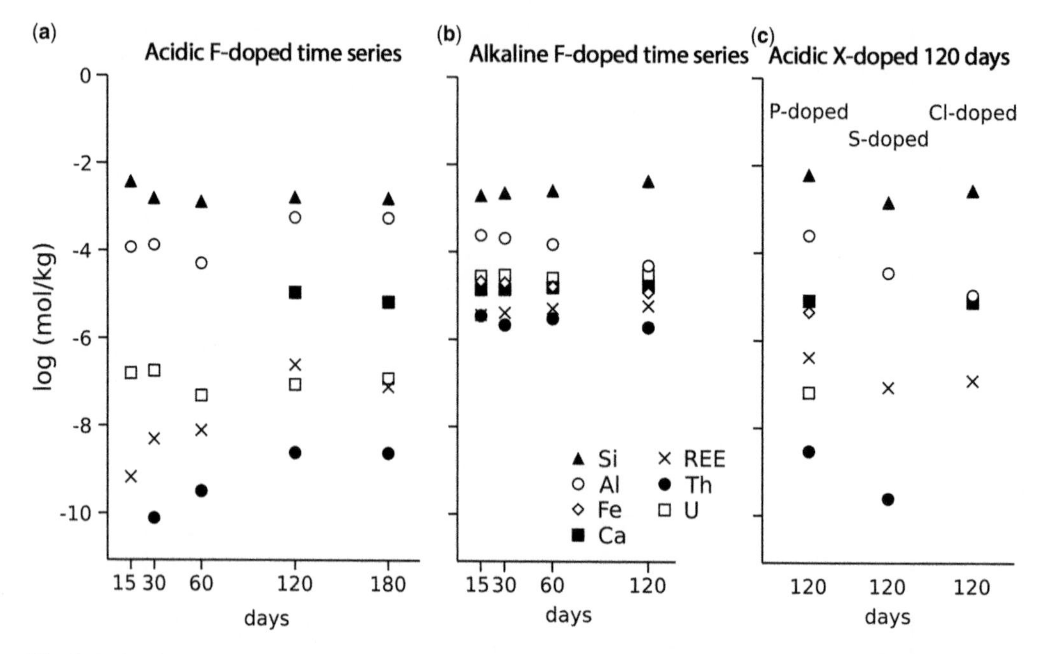

Fig. 5. Major elements, REE and actinide concentrations (log) in experimental fluids for the time-series F-doped experiments in the acidic system (**a**), high-pH alkaline system (**b**), and in the 120 day experiments for the P-, S- and Cl-doped systems (**c**).

system compared with the high-pH alkaline system by 2 to 4 orders of magnitude.

In the P-doped system, which was the most reactive system under acidic conditions, elemental concentrations are similar to the concentrations of the F-doped in the acidic system at 120 days (Fig. 5c). In the unreactive Cl- and S-doped systems, Si and REE concentrations are comparable with those measured in the F- and P-doped systems under acidic conditions. The other elements were generally at lower concentrations.

In terms of REEs, the chondrite-normalized patterns plot relatively flat for the high-pH alkaline systems (Fig. 6). In acidic fluids, patterns plot also relatively flat but with a slight depletion in Sm, Gd and Dy, with no dependence on the ligand.

Discussion

Allanite alteration mechanisms

In the batch experiment runs (Table 1), the alteration of allanite ranges from a limited (<5%) up to an extensive (77%) degree, depending on the fluid chemistry after 120 days. The pH has the first effect on the alteration of allanite, as the kinetics for the high-pH alkaline system are fastest (65%) after 15 days, and the more advanced (70–75%) after 120 days in the two time-series experiment runs conducted in an F-doped system. Under acidic conditions, the nature of the ligand significantly affects the extent of alteration. The presence of P enhances the allanite alterability, reaching 73% of the reaction rate after 120 days, while allanite reactivity is minor in the Cl- or S-doped systems (<5%). F-doped systems display moderate alteration at the same duration (25%).

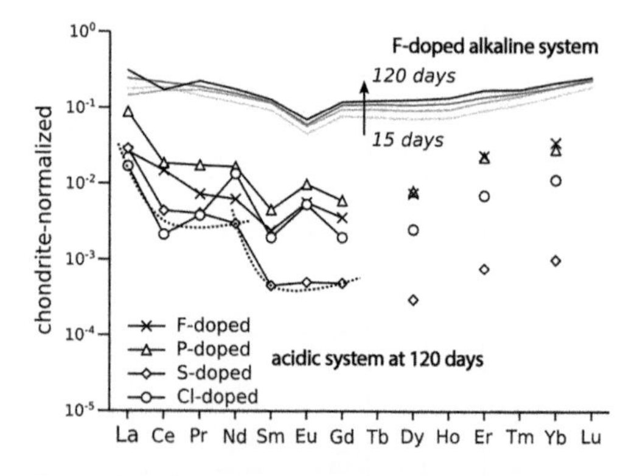

Fig. 6. Chondrite-normalized REE spectra of fluids after 120 days from the slightly acidic, F-doped, P-doped, S- doped, Cl-doped and time-series, high-pH alkaline, F-doped experiments. The HREEs with odd numbers are below the detection limits (DLs) or have been removed from the diagram because of artificial anomalies due to being close to the DLs (Table 1). The lanthanide tetrad effect is discernible within the LREEs for the slightly acid experiments (dashed lines).

Higher allanite reactivity in a high-pH alkaline fluid shows that high-pH fluids efficiently promote silicate dissolution rates, while dissolution is more limited in near neutral fluids (Hellmann 1994). This effect is also demonstrated for epidote group minerals (Rose 1991). Phosphorus seems to have a similar effect on allanite, though with a lower extent of alteration.

In the most reactive systems (high-pH alkaline and P-doped conditions), the alteration of allanite is promoted by increasing dissolution coupled with the precipitation of other minerals. On the one hand, dissolution can be promoted due to a solubility change for the dissolving elements in the bulk solution, by modifying element complexation, the concentration in the solution and the chemical affinity per the dissolution reaction between allanite and the fluid. On the other hand, the mineral microstructures evidenced here also point to the crucial role of secondary precipitation on the alteration rate. Alteration microstructures from highly altered experiment products show well-developed dissolution features (etch pits, fractures, porosity) with a penetrative replacement by an alteration rim made up of secondary minerals with a complex mineralogical zonation. General preservation of the initial pristine shape of allanite suggests a mechanism of replacement by interfacially-coupled dissolution-precipitation (Putnis 2002; Putnis and Putnis 2007; Harlov *et al.* 2011; Hellmann *et al.* 2012), which indicate disequilibrium between the solid and the fluid (Putnis 2009; Ruiz-Agudo *et al.* 2014). Such alteration processes can lead to an apparent incongruent dissolution due to a preferential precipitation of low solubility phases (with different composition than the altered phase) at the alteration interface (Ruiz-Agudo *et al.* 2012). Such apparent incongruent dissolution has already been demonstrated for epidote dissolution (Kalinowski *et al.* 1998), and seems also to apply here to allanite alteration as seen by the mineralogical gradation from the reaction front to the bulk solution. In the reaction rim, the precipitation of submicron, low-solubility secondary phases takes in elements from the solution and changes their concentration at the reaction interface. This is the case for hematite, which nucleates as a thin rim at the interface with the allanite. This is particularly true for REE mineral phases (REE-fluorocarbonates or monazite, depending on the ligand), which also occur as a discontinuous rim of nanoscale crystallites propagating anisotropically inwards into the pristine grain and along fractures in the allanite. The growth of other main phases with a higher solubility in the solution, e.g. analcime, fluorite and calcite, is spatially decoupled from the alteration interface with precipitation from the bulk solution as larger euhedral crystals. Such precipitation from the bulk fluid away from the rim of the dissolving mineral have been described in other alkaline systems (Lafay *et al.* 2014, 2018). In the batch experiment runs, the preferential precipitation of REE phases at the reaction front is proposed to efficiently maintain significant dissolution rates by producing steep concentration gradients in the fluids close to the reactive surface, which act to renew the solutions (Frugier *et al.* 2008; Ruiz-Agudo *et al.* 2016).

Coupled with the chemical gradient at the interface, the precipitation of REE mineral phases will further strongly modify the geometry of the reaction front. The complex microstructures at the reaction interface, with etch pits, indentations and secondary fractures, are the result of reaction-induced fracturing due to molar volume change and the force of crystallization during the replacement of allanite by secondary phases (e.g. Jamtveit *et al.* 2009; Lafay *et al.* 2018). This increase of the reactive surface also enhances allanite dissolution.

In the two non-reactive systems (Cl- and S-doped), there was limited precipitation of secondary phases (<5%). In batch experiment runs, this drop in the dissolution rates can occur when element concentrations progressively approach saturation in the fluid or when precipitation of an inert passivation layer isolates the reacting mineral from the reactive fluid (Montes-Hernandez *et al.* 2012). In unreactive systems, secondary precipitation observed at the grain surface is sufficiently low such that allanite remains accessible to the fluid throughout the experiment runs. In contrast, concentrations in fluids similar to those of reactive systems indicate that they reach conditions approaching saturation. Since precipitation of analcime and hematite is not chemically restricted, the only limiting factor here appears to concern the stability of the REE phases.

In the investigated reactive systems, the precipitation of secondary REE phases is thus proposed to be the main driving force behind allanite alteration by lowering the activities of REEs in the interfacial fluid. In the absence of efficient REE mineral precipitation (Cl- and S-doped) at the allanite interface, 'steady state' concentrations measured in the bulk fluid are assumed to be more readily reached, thus decreasing reaction rates. Therefore, allanite alteration remains low. These results are in good agreement with natural observations. The secondary, experimental REE mineral phases, i.e. REE-fluorocarbonates and/or REE-phosphates, are typical of low-temperature alteration products (e.g. Berger *et al.* 2008; Ondrejka *et al.* 2018).

REE redistribution during allanite alteration

In all reactive systems, comparison between a simple mass balance calculation from the low REE concentrations in the recovered fluid and the allanite composition and alteration rates indicate that the REEs released during alteration are mostly in secondary phases. An allanite alteration of 70% would provide 100 µmol of the REE released in high-pH alkaline experiment runs. However, the REE content in the final fluids are 4 orders of magnitude below. The main REE minerals (REE-carbonates or REE-phosphates identified by XRD) occur as submicronic crystals in the alteration rim, preventing accurate determination of their REE content. Based on the theoretical compositions of REE-fluorocarbonates and monazite, along with their XRD modal abundance, rough mass balance calculations confirm that they are a major sink for the REEs released by allanite. In the P-doped system, the hydroxylapatite grains are also too small to determine their REE content, though it could be up to a few wt% (Budzyń *et al.* 2017).

While the composition of secondary phases in the altered rim cannot be analysed precisely for their REE content, minerals precipitating from the bulk fluid are large enough for evaluating their REE content by EPMA. In F-doped systems, fluorite represents 25% of the secondary products and can incorporate up to 1–2 wt% REE_2O_3. The REE content in fluorite has been extensively studied in hydrothermal systems (Möller *et al.* 1998; Schwinn and Markl 2005; Schönenberger *et al.* 2008; Gob *et al.* 2011), in economical REE deposits, such as the Bayan Obo Complex (Xu *et al.* 2012) or by thermodynamic modelling (Kolonin and Shironosova 2007). It shows that REEs in fluorite, while extremely variable, can reach up to >10 wt% in yttrofluorite (Pekov *et al.* 2009). Although a coupled substitution involving Na is often considered preponderant for incorporating the REEs in fluorite, i.e. $REE^{3+} + Na^{+} \leftrightarrow 2\ Ca^{2+}$ (Möller *et al.* 1998), there is no real correlation between the REE and Na contents in the fluorite from these experiments, despite the high Na concentrations. The BGM (identified from XRD) precipitating from the bulk fluid also accommodates significant REEs, but with Na concentrations that are significantly lower compared with burbankite *sensu stricto* (Belovitskaya and Pekov 2004). The BGM grains are zoned with a typical hourglass sector zoning, suggesting crystallographic control on REE incorporation (Fig. 2f). Integration of the REEs is directly correlated to the size and geometry of the crystallographic sites, which favours the LREEs in calcic minerals such as tourmaline (van Hinsberg *et al.* 2010). Burbankite is a hydrothermal mineral encountered in alkaline pegmatites and associated carbonatites (Zaitsev *et al.* 2002). In experiments, the precipitation of BGMs is probably favoured by the Na concentration in the fluid. Finally, the REE concentrations in the calcite are considerably higher than those normally encountered in nature (Stipp *et al.* 2006) but are thermodynamically stable (Rimstidt *et al.* 1998), as has been experimentally demonstrated (Toyama and Terakado 2014; Gabitov *et al.* 2017). In calcite, two coupled substitution mechanisms are proposed (Perry and Gysi 2018): $REE^{3+} + Na^{+} \leftrightarrow 2Ca^{2+}$ and $2REE^{3+} + \square \leftrightarrow 3Ca^{2+}$ (square represents site vacancies). The composition of the calcite produced in these experiments indicates that both mechanisms occur under the experimental conditions of this study (Fig. S5, supp. mat.).

REE fractionation between fluid and solid

Though the REEs are mainly stored in secondary phases, minor REE concentrations have been recovered in the fluids. Though precise quantitative fluid concentrations are limited by the batch experimental setup, our qualitative results clearly indicate a significant difference in REE fractionation between the solid and the fluid, whatever the pH and the complexing ligands. Experimental fluids display relatively flat chondrite-normalized REE spectra, indicating that the experimental alteration of allanite ultimately produces a fluid enriched in HREEs relative to the initial LREE-rich allanite composition. This implies in turn the preferential fractionation of LREEs over HREEs in the secondary mineral precipitates relative to the fluid. This is in good agreement with the limited incorporation of HREEs in fluorocarbonates and monazite, as demonstrated for $T <$ 450°C (Heinrich *et al.* 1997; Poitrasson *et al.* 2000; Janots *et al.* 2008; Budzyń *et al.* 2010, 2017; Grand'Homme *et al.* 2018). Also, secondary minerals that precipitate from the bulk fluid (calcite, fluorite, BGM) are enriched in LREEs over HREEs but with lower La/Y compared with allanite, again supporting the fractionation of the LREEs over the HREEs in the bulk fluid compared with fluid at the reaction front. In these secondary mineral phases, the Y values are typically at the same level as in allanite, which suggests that the HREEs are more mobile compared with the LREEs, as seen in numerous natural environments, e.g. during monazite alteration (Hentschel *et al.* 2020).

The fluid compositions measured in this study have numerous implications for REE deposits. Here the flat or gently incurved REE-normalized pattern indicates that REEs are not released congruently, but that speciation in the fluids or precipitation of secondary products favours HREE fractionation over LREEs in the fluid compared with the initial allanite composition.

Th and U behaviour during allanite alteration

Actinides seem to mostly partition into the fluid as opposed to secondary minerals. Simplified qualitative calculations show that virtually all the U released by allanite accumulates in the fluid under these conditions. Actinide concentrations are higher in the high-pH alkaline system (with higher carbonate activities) than in the acidic system. This agrees well with studies that show that the solubility of actinides increases with the concentration of the aqueous carbonate or phosphate ligands (Rai *et al.* 1994; Sandino and Bruno 1998). Recent studies also show that actinides can be highly mobile in the presence of ligands such as S-, Cl- or F-complexes for temperatures close to 200°C (Nisbet *et al.* 2018, 2019; Migdisov *et al.* 2019). In the experimental runs under initial acidic conditions, U release is at least 1 to 2 orders of magnitude lower than that under high-pH alkaline conditions. Regardless of the chemical system, Th is systematically lower in the fluid compared with U, while it is higher in the starting allanite, indicating U/Th fractionation during allanite alteration. According to Rai *et al.* (1994), ThO_2 solubility is higher than that of UO_2, suggesting that tetravalent U is likely oxidized in its hexavalent state during the allanite alteration reaction. Preferential incorporation of tetravalent Th in secondary REE mineral phases may in turn enhance Th/U fractionation between the fluid and secondary products, as observed in natural monazite and allanite precipitated from hydrothermal systems (Janots *et al.* 2012).

Conclusions

Allanite has a complex composition and its experimental alteration under low temperature conditions results in a high diversity of mineralogical assemblages and microstructures. Allanite can be highly reactive in certain fluids, reaching more than 75% of alteration at 200°C and $P_{sa} \approx 16$ bar, after only 15 days. The pH and the nature of the complexing ligand added to the fluid will strongly affect the alteration rate of the allanite, with the high-pH alkaline system being the most reactive. In carbonate-bearing fluids, F and P will promote allanite alteration, while allanite shows negligible alteration in the presence of Cl and S. The main driving force behind the alteration of allanite resides in the precipitation at a reactive front of secondary REE minerals, whose chemistry depends on the complexing ligands. These precipitated minerals maintain a local disequilibrium close to the reaction interface between the fluid and the solid, thus sustaining allanite dissolution. Though REEs are mostly stored in the secondary mineral phases, there is a preferential fractionation of the LREEs over the HREEs into the solid compared with the fluid, while U is strongly partitioned into the fluid.

Acknowledgments Electron microscopy was performed with the kind help of Rachel Martin at the CMTC characterization platform of Grenoble INP supported by the Centre of Excellence of Multifunctional Architectured Materials 'CEMAM' n°AN-10-LABX-44–01 funded by the Investments for the Future Program. We are very grateful to S. Campillo, S. Bureau and M. Lanson for fluid analyses and following discussions. This paper benefited from insightful comments of C. Cordier, F. Brunet and A. Fernandez-Martinez. This PhD work was further supported by the TelluS Program of CNRS/INSU and local BQR funding. The work was mainly done at ISTerre (Université Grenoble Alpes), which is part of Labex OSUG@2020 (ANR10 LABX56). The authors are extremely thnankful to D. Harlow, B. Budzyń and F. Poitrasson for their successful evaluations of the manuscript.

Competing interests The authors declare that they have no known competing financial interests or personal relationships that could have appeared to influence the work reported in this paper.

Author contributions **AD**: data curation (lead), formal analysis (lead), investigation (lead), writing – original draft (lead); **A-LA**: supervision (equal), writing – review and editing (equal); **EJ**: supervision (equal), writing – review and editing (equal); **GM-H**: conceptualization (supporting); **NF**: data curation (supporting); **PL**: data curation (equal); **VM**: data curation (supporting).

Funding This research was funded by Labex OSUG@2020 (ANR10LABX56), award ID0EORAG4608 to Axel Denys.

Data availability All data generated or analysed during this study are included in this published article (and, if present, its supplementary information files).

References

Anenburg, M., Katzir, Y., Rhede, D., Jöns, N. and Bach, W. 2015. Rare earth element evolution and migration in plagiogranites: a record preserved in epidote and allanite of the Troodos ophiolite. *Contributions to Mineralogy and Petrology*, **169**, 1–19, https://doi.org/10.1007/s00410-015-1114-y

Batanova, V.G., Sobolev, A.V. and Magnin, V. 2018. Trace element analysis by EPMA in geosciences: detection limit, precision and accuracy. *IOP Conference Series: Materials Science and Engineering*, **304**, 012001, https://doi.org/10.1088/1757-899X/304/1/012001

Belovitskaya, Y.V. and Pekov, I.V. 2004. Genetic mineralogy of the Burbankite group. *New Data on Minerals*, **39**, 15.

Berger, A., Gnos, E., Janots, E., Fernandez, A. and Giese, J. 2008. Formation and composition of rhabdophane, bastnäsite and hydrated thorium minerals during alteration: implications for geochronology and low-temperature processes. *Chemical Geology*, **254**, 238–248, https://doi.org/10.1016/j.chemgeo.2008.03.006

Budzyń, B., Hetherington, C.J., Williams, M.L., Jercinovic, M.J. and Michalik, M. 2010. Fluid-mineral interactions and constraints on monazite alteration during metamorphism. *Mineralogical Magazine*, **74**, 659–681, https://doi.org/10.1180/minmag.2010.074.4.659

Budzyń, B., Harlov, D.E., Williams, M.L. and Jercinovic, M.J. 2011. Experimental determination of stability relations between monazite, fluorapatite, allanite, and REE-epidote as a function of pressure, temperature, and fluid composition. *American Mineralogist*, **96**, 1547–1567, https://doi.org/10.2138/am.2011.3741

Budzyń, B., Harlov, D.E., Kozub-Budzyń, G.A. and Majka, J. 2017. Experimental constraints on the relative stabilities of the two systems monazite-(Ce) – allanite-(Ce) – fluorapatite and xenotime-(Y) – (Y, HREE)-rich epidote – (Y,HREE)-rich fluorapatite, in high Ca and Na-Ca environments under P-T conditions of 200–1000 MPa and 450–750°C. *Mineralogy and Petrology*, **111**, 183–217, https://doi.org/10.1007/s00710-016-0464-0

Caruso, L. and Simmons, G. 1985. Uranium and microcracks in a 1,000-meter core, Redstone, New Hampshire. *Contributions to Mineralogy and Petrology*, **90**, 1–17, https://doi.org/10.1007/BF00373036

Chabiron, A. and Cuney, M. 2001. Altération de l'allanite dans les granites sous la caldeira de Streltsovka (Transbaïkalie, Russie). Une source possible d'uranium pour les gisement. *C R Acad Sci Paris, Sciences de la Terre et des planètes / Earth and Planetary Sciences*, **332**, 99–105, https://doi.org/10.1016/S1251-8050(00)01509-3

Corriveau, L., Ootes, L. *et al.* 2007. Alteration vectoring to IOCG(U) deposits in frontier volcano-plutonic terrains, Canada. *Proceedings of Exploration 07: Fifth Decennial International Conference on Mineral Exploration. B.*, Milkereit, 1171–1177.

Engi, M. 2017. Petrochronology based on REE-minerals: Monazite, allanite, xenotime, apatite. *Reviews in Mineralogy and Geochemistry*, **83**, 365–418, https://doi.org/10.2138/rmg.2017.83.12

Ercit, T.S. 2002. The mess that is 'Allanite.' *The Canadian Mineralogist*, **40**, 1411–1419, https://doi.org/10.2113/gscanmin.40.5.1411

Ewing, R.C., Chakoumakos, B.C., Lumpkin, G.R. and Murakami, T. 1987. The metamict state. *MRS Bulletin*, **12**, 58–66, https://doi.org/10.1557/S0883769400067865

Frugier, P., Gin, S. *et al.* 2008. SON68 nuclear glass dissolution kinetics: current state of knowledge and basis of the new GRAAL model. *Journal of Nuclear Materials*, **380**, 8–21, https://doi.org/10.1016/j.jnucmat.2008.06.044

Gabitov, R., Sadekov, A. and Migdisov, A. 2017. REE incorporation into calcite individual crystals as one time spike addition. *Minerals*, **7**, 204, https://doi.org/10.3390/min7110204

Gammons, C.H., Wood, S.A. and Williams-Jones, A.E. 1996. The aqueous geochemistry of the rare earth elements and yttrium. VI: Stability of neodymium chloride complexes from 25 to 300°C. *Geochimica et Cosmochimica Acta*, **60**, 4615–4630, https://doi.org/10.1016/S0016-7037(96)00262-1

Gieré, R. and Sorensen, S.S. 2004. Allanite and other REE-rich epidote-group minerals. *Reviews in Mineralogy and Geochemistry*, **56**, 431–493, https://doi.org/10.2138/gsrmg.56.1.431

Gob, S., Wenzel, T., Bau, M., Jacob, D.E., Loges, A. and Markl, G. 2011. The redistribution of Rare-Earth Elements in secondary minerals of hydrothermal veins, Schwarzwald, Southwestern Germany. *The Canadian Mineralogist*, **49**, 1305–1333, https://doi.org/10.3749/canmin.49.5.1305

Grand'Homme, A., Janots, E. *et al.* 2018. Mass transport and fractionation during monazite alteration by anisotropic replacement. *Chemical Geology*, **484**, 51–68, https://doi.org/10.1016/j.chemgeo.2017.10.008

Guillong, M., Meier, D.L., Allan, M.M., Heinrich, C.A. and Yardley, B.W.D. 2008. SILLS: a Matlab-based program for the reduction of laser ablation ICP-MS data of homogeneous materials and inclusions. Mineralogical Association of Canada, Vancouver, BC. Appendix A6, 328–333.

Harlov, D.E., Wirth, R. and Hetherington, C.J. 2011. Fluid-mediated partial alteration in monazite: the role of coupled dissolution–reprecipitation in element redistribution and mass transfer. *Contributions to Mineralogy and Petrology*, **162**, 329–348, https://doi.org/10.1007/s00410-010-0599-7

Heinrich, W., Rehs, G. and Franz, G. 1997. Monazite–xenotime miscibility gap thermometry. I: An empirical calibration. *Journal of Metamorphic Geology*, **15**, 3–16, https://doi.org/10.1111/j.1525-1314.1997.t01-1-00052.x

Hellmann, R. 1994. The albite-water system. Part I: The kinetics of dissolution as a function of pH at 100, 200, and 300°C. *Geochemica et Cosmochemica Acta*, **58**, 595–611, https://doi.org/10.1016/0016-7037(94)90491-X

Hellmann, R., Wirth, R. *et al.* 2012. Unifying natural and laboratory chemical weathering with interfacial dissolution–reprecipitation: a study based on the nanometer-scale chemistry of fluid–silicate interfaces. *Chemical Geology*, **294–295**, 203–216, https://doi.org/10.1016/j.chemgeo.2011.12.002

Hentschel, F., Janots, E., Trepmann, C.A., Magnin, V. and Lanari, P. 2020. Corona formation around monazite and xenotime during greenschist-facies metamorphism and deformation. *European Journal of Mineralogy*, **32**, 521–544, https://doi.org/10.5194/ejm-32-521-2020

Ichimura, K., Sanematsu, K., Kon, Y., Takagi, T. and Murakami, T. 2020. REE redistributions during granite weathering: implications for Ce anomaly as a proxy for paleoredox states. *American Mineralogist*, **105**, 848–859, https://doi.org/10.2138/am-2020-7148

Jamtveit, B., Putnis, C.V. and Malthe-Sørenssen, A. 2009. Reaction induced fracturing during replacement processes. *Contributions to Mineralogy and Petrology*, **157**, 127–133, https://doi.org/10.1007/s00410-008-0324-y

Janots, E., Engi, M., Berger, A., Allaz, J., Schwarz, J.-O. and Spandler, C. 2008. Prograde metamorphic sequence of REE minerals in pelitic rocks of the Central Alps: implications for allanite–monazite–xenotime phase relations from 250 to 610°C. *Journal of Metamorphic Geology*, **26**, 509–526, https://doi.org/10.1111/j.1525-1314.2008.00774.x

Janots, E., Berger, A., Gnos, E., Whitehouse, M., Lewin, E. and Pettke, T. 2012. Constraints on fluid evolution during metamorphism from U–Th–Pb systematics in Alpine hydrothermal monazite. *Chemical Geology*, **326–327**, 61–71, https://doi.org/10.1016/j.chemgeo.2012.07.014

Jochum, K.P., Willbold, M., Raczek, I., Stoll, B. and Herwig, K. 2005. Chemical characterisation of the USGS reference glasses GSA-1G, GSC-1G, GSD-1G,

GSE-1G, BCR-2G, BHVO-2G and BIR-1G using EPMA, ID-TIMS, ID-ICP-MS and LA-ICP-MS. *Geostandards and Geoanalytical Research*, **29**, 285–302, https://doi.org/10.1111/j.1751-908X.2005.tb00901.x

Jochum, K.P., Weis, U. *et al.* 2011. Determination of reference values for NIST SRM 610-617 glasses following ISO Guidelines. *Geostandards and Geoanalytical Research*, **35**, 397–429, https://doi.org/10.1111/j.1751-908X.2011.00120.x

Kalinowski, B.E., Faith-Ell, C. and Schweda, P. 1998. Dissolution kinetics and alteration of epidote in acidic solutions at 25°C. *Chemical Geology*, **151**, 181–197, https://doi.org/10.1016/S0009-2541(98)00079-5

Kolonin, G.R. and Shironosova, G.P. 2007. REE distribution between fluorite and ore-forming fluid based on results of thermodynamic modeling. *Doklady Earth Sciences*, **414**, 661–665, https://doi.org/10.1134/S1028334X0704037X

Krenn, E., Harlov, D.E., Finger, F. and Wunder, B. 2012. LREE-redistribution among fluorapatite, monazite, and allanite at high pressures and temperatures. *American Mineralogist*, **97**, 1881–1890, https://doi.org/10.2138/am.2012.4005

Lafay, R., Montes-Hernandez, G., Janots, E., Chiriac, R., Findling, N. and Toche, F. 2014. Simultaneous precipitation of magnesite and lizardite from hydrothermal alteration of olivine under high-carbonate alkalinity. *Chemical Geology*, **368**, 63–75, https://doi.org/10.1016/j.chemgeo.2014.01.008

Lafay, R., Montes-Hernandez, G., Renard, F. and Vonlanthen, P. 2018. Intracrystalline reaction-induced cracking in olivine evidenced by hydration and carbonation experiments. *Minerals*, **8**, 412, https://doi.org/10.3390/min8090412

Mezger, K., Essene, E.J., van der Pluijm, B.A. and Halliday, A.N. 1993. U-Pb geochronology of the Grenville Orogen of Ontario and New York: constraints on ancient crustal tectonics. *Contributions to Mineralogy and Petrology*, **114**, 13–26, https://doi.org/10.1007/BF00307862

Middleton, A.W., Förster, H.-J., Uysal, I.T., Golding, S.D. and Rhede, D. 2013. Accessory phases from the Soultz monzogranite, Soultz-sous-Forêts, France: implications for titanite destabilisation and differential REE, Y and Th mobility in hydrothermal systems. *Chemical Geology*, **335**, 105–117, https://doi.org/10.1016/j.chemgeo.2012.10.047

Migdisov, A., Guo, X., Nisbet, H., Xu, H. and Williams-Jones, A.E. 2019. Fractionation of REE, U, and Th in natural ore-forming hydrothermal systems: thermodynamic modeling. *The Journal of Chemical Thermodynamics*, **128**, 305–319, https://doi.org/10.1016/j.jct.2018.08.032

Möller, P., Bau, M., Dulski, P. and Lüders, V. 1998. REE and yttrium fractionation in fluorite and their bearing on fluorite formation. *Proceedings of the Ninth Quadrennial IAGOD Symposium*, Schweizerbart, Stuttgart, 575–592.

Montes-Hernandez, G., Chiriac, R., Toche, F. and Renard, F. 2012. Gas–solid carbonation of $Ca(OH)_2$ and CaO particles under non-isothermal and isothermal conditions by using a thermogravimetric analyzer: implications for CO_2 capture. *International Journal of Greenhouse Gas Control*, **11**, 172–180, https://doi.org/10.1016/j.ijggc.2012.08.009

Nisbet, H., Migdisov, A. *et al.* 2018. An experimental study of the solubility and speciation of thorium in chloride-bearing aqueous solutions at temperatures up to 250° C. *Geochimica et Cosmochimica Acta*, **239**, 363–373, https://doi.org/10.1016/j.gca.2018.08.001

Nisbet, H., Migdisov, A.A., Williams-Jones, A.E., Xu, H., van Hinsberg, V.J. and Roback, R. 2019. Challenging the thorium-immobility paradigm. *Scientific Reports*, **9**, 17035, https://doi.org/10.1038/s41598-019-53571-x

Ondrejka, M., Bačík, P. *et al.* 2018. Minerals of the rhabdophane group and the alunite supergroup in microgranite: products of low-temperature alteration in a highly acidic environment from the Velence Hills, Hungary. *Mineralogical Magazine*, **82**, 1277–1300, https://doi.org/10.1180/mgm.2018.137

Pal, D.C., Chaudhuri, T., McFarlane, C., Mukherjee, A. and Sarangi, A.K. 2011. Mineral chemistry and *in situ* dating of allanite, and geochemistry of its host rocks in the Bagjata uranium mine, Singhbhum Shear Zone, India: implications for the chemical evolution of REE mineralization and mobilization. *Economic Geology*, **106**, 1155–1171, https://doi.org/10.2113/econgeo.106.7.1155

Pekov, I.V., Krivovichev, S.V., Zolotarev, A.A., Yakovenchuk, V.N., Armbruster, T. and Pakhomovsky, Y.A. 2009. Crystal chemistry and nomenclature of the lovozerite group. *European Journal of Mineralogy*, **21**, 1061–1071, https://doi.org/10.1127/0935-1221/2009/0021-1957

Perry, E.P. and Gysi, A.P. 2018. Rare Earth Elements in mineral deposits: speciation in hydrothermal fluids and partitioning in calcite. *Geofluids*, **2018**, 1–19, https://doi.org/10.1155/2018/5382480

Pettke, T., Oberli, F., Audétat, A., Wiechert, U., Harris, C.R. and Heinrich, C.A. 2011. Quantification of transient signals in multiple collector inductively coupled plasma mass spectrometry: accurate lead isotope ratio determination by laser ablation of individual fluid inclusions. *Journal of Analytical Atomic Spectrometry*, **26**, 475–492, https://doi.org/10.1039/C0JA00140F

Poitrasson, F. 2002. *In situ* investigations of allanite hydrothermal alteration: examples from calc-alkaline and anorogenic granites of Corsica (southeast France). *Contributions to Mineralogy and Petrology*, **142**, 485–500, https://doi.org/10.1007/s004100100303

Poitrasson, F., Chenery, S. and Shepherd, T.J. 2000. Electron microprobe and LA-ICP-MS study of monazite hydrothermal alteration: implications for the U-Th-Pb geochronology and nuclear ceramics. *Geochimica et Cosmochimica Acta*, **64**, 3283–3297, https://doi.org/10.1016/S0016-7037(00)00433-6

Price, J.R., Velbel, M.A. and Patino, L.C. 2005. Allanite and epidote weathering at the Coweeta Hydrologic Laboratory, western North Carolina, USA. *American Mineralogist*, **90**, 101–114, https://doi.org/10.2138/am.2005.1444

Putnis, A. 2009. Mineral replacement reactions. *Reviews in Mineralogy and Geochemistry*, **70**, 87–124, https://doi.org/10.2138/rmg.2009.70.3

Putnis, A. 2002. Mineral replacement reactions: from macroscopic observations to microscopic mechanisms. *Mineralogical Magazine*, **66**, 689–708, https://doi.org/10.1180/0026461026650056

Putnis, A. and Putnis, C.V. 2007. The mechanism of re-equilibration of solids in the presence of a fluid phase. *Journal of Solid State Chemistry*, **180**, 1783–1786, https://doi.org/10.1016/j.jssc.2007.03.023

Rai, D., Felmy, A.R., Moore, D.A. and Mason, M.J. 1994. The solubility of Th(IV) and U(IV) hydrous oxides in concentrated $NAHCO_3$ and Na_2Co_3 solutions. *MRS Proceedings*, **353**, 1143, https://doi.org/10.1557/PROC-353-1143

Rimstidt, J.D., Balog, A. and Webb, J. 1998. Distribution of trace elements between carbonate minerals and aqueous solutions. *Geochimica et Cosmochimica Acta*, **62**, 1851–1863, https://doi.org/10.1016/S0016-7037(98)00125-2

Rose, N.M. 1991. Dissolution rates of prehnite, epidote, and albite. *Geochimica et Cosmochimica Acta*, **55**, 3273–3286, https://doi.org/10.1016/0016-7037(91)90488-Q

Ruiz-Agudo, E., Putnis, C.V., Rodriguez-Navarro, C. and Putnis, A. 2012. Mechanism of leached layer formation during chemical weathering of silicate minerals. *Geology*, **40**, 947–950, https://doi.org/10.1130/G33339.1

Ruiz-Agudo, E., Putnis, C.V. and Putnis, A. 2014. Coupled dissolution and precipitation at mineral–fluid interfaces. *Chemical Geology*, **383**, 132–146, https://doi.org/10.1016/j.chemgeo.2014.06.007

Ruiz-Agudo, E., King, H.E., Patiño-López, L.D., Putnis, C.V., Geisler, T., Rodriguez-Navarro, C. and Putnis, A. 2016. Control of silicate weathering by interface-coupled dissolution-precipitation processes at the mineral–solution interface. *Geology*, **44**, 567–570, https://doi.org/10.1130/G37856.1

Sandino, A. and Bruno, J. 1998. The solubility of $(UO_2)_3$-$(PO_4)_2.4H_20_{(s)}$ and the formation of U(VI) phosphate complexes: their influence in uranium speciation in natural waters. *Geochemica et Cosmochemica Acta*, **56**, 11.

Schönenberger, J., Köhler, J. and Markl, G. 2008. REE systematics of fluorides, calcite and siderite in peralkaline plutonic rocks from the Gardar Province, South Greenland. *Chemical Geology*, **247**, 16–35, https://doi.org/10.1016/j.chemgeo.2007.10.002

Schwinn, G. and Markl, G. 2005. REE systematics in hydrothermal fluorite. *Chemical Geology*, **216**, 225–248, https://doi.org/10.1016/j.chemgeo.2004.11.012

Seward, T.M., Williams-Jones, A.E. and Migdisov, A.A. 2014. The chemistry of metal transport and deposition by ore-forming hydrothermal fluids. *In*: *Treatise on Geochemistry*. Elsevier, 29–57.

Smith, M.P., Henderson, P. and Jeffries, T. 2002. The formation and alteration of allanite in skarn from the Beinn an Dubhaich granite aureole, Skye. *European Journal of Mineralogy*, **14**, 471–486, https://doi.org/10.1127/0935-1221/2002/0014-0471

Stipp, S.L.S., Christensen, J.T., Lakshtanov, L.Z., Baker, J.A. and Waight, T.E. 2006. Rare earth element (REE) incorporation in natural calcite: upper limits for actinide uptake in a secondary phase. *Radiochimica Acta*, **94**, https://doi.org/10.1524/ract.2006.94.9-11.523

Toyama, K. and Terakado, Y. 2014. Experimental study of rare earth element partitioning between calcite and sodium chloride solution at room temperature and pressure. *Geochemical Journal*, **48**, 463–477, https://doi.org/10.2343/geochemj.2.0322

van Hinsberg, V.J., Migdisov, A.A. and Williams-Jones, A.E. 2010. Reading the mineral record of fluid composition from element partitioning. *Geology*, **38**, 847–850, https://doi.org/10.1130/G31112.1

Warr, L.N. 2021. IMA–CNMNC approved mineral symbols. *Mineralogical Magazine*, **85**, 291–320.

Xu, C., Taylor, R.N., Li, W., Kynicky, J., Chakhmouradian, A.R. and Song, W. 2012. Comparison of fluorite geochemistry from REE deposits in the Panxi region and Bayan Obo, China. *Journal of Asian Earth Sciences*, **57**, 76–89, https://doi.org/10.1016/j.jseaes.2012.06.007

Zaitsev, A.N., Demény, A., Sindern, S. and Wall, F. 2002. Burbankite group minerals and their alteration in rare earth carbonatites: source of elements and fluids (evidence from C–O and Sr–Nd isotopic data). *Lithos*, **62**, 15–33, https://doi.org/10.1016/S0024-4937(02)00084-1

Robust laser ablation Lu–Hf dating of apatite: an empirical evaluation

Stijn Glorie[1]*, Martin Hand[1], Jacob Mulder[1], Alexander Simpson[1], Robert B. Emo[2], Balz Kamber[2], Nicholas Fernie[1], Angus Nixon[1] and Sarah Gilbert[3]

[1]Department of Earth Sciences, University of Adelaide, Adelaide, SA 5005, Australia

[2]School of Earth and Atmospheric Sciences, Queensland University of Technology, QLD 4000, Australia

[3]Adelaide Microscopy, University of Adelaide, Adelaide, SA 5005, Australia

SG, 0000-0002-3107-9028
*Correspondence: stijn.glorie@adelaide.edu.au

Abstract: Recent developments in laser-ablation Lu–Hf dating have opened a new opportunity to rapidly obtain apatite ages that are potentially more robust to isotopic resetting compared to traditional U–Pb dating. However, the robustness of the apatite Lu–Hf system has not been systematically examined. To address this knowledge gap, we conducted four case studies to determine the resistivity of the apatite Lu–Hf system compared to the zircon and apatite U–Pb system. In all cases, the apatite U–Pb system records a secondary (metamorphic or metasomatic) overprint. The apatite Lu–Hf system, however, preserves primary crystallization ages in unfoliated granitoids at temperatures of at least *c.* 660°C. Above *c.* 730°C, the Lu–Hf system records isotopic resetting by volume diffusion. Hence, in our observations for apatites of 'typical' grain sizes in granitoids (*c.* 0.01–0.03 mm^2), the closure temperature of the Lu–Hf system is between *c.* 660 and *c.* 730°C, consistent with theoretical calculations. In foliated granites, the Lu–Hf system records the timing of recrystallization, while the apatite U–Pb system tends to record younger cooling ages. We also present apatite Lu–Hf dates for lower crustal xenoliths erupted with young alkali basalts, demonstrating that the Lu–Hf system can retain a memory of primary ages when exposed to magmatic temperatures for a relatively short duration. Hence, the apatite Lu–Hf system is a new insightful addition to traditional zircon (or monazite) U–Pb dating, particularly when zircons/monazites are absent or difficult to interpret due to inheritance or when U and Pb isotopes display open system behaviour. The laser-ablation-based Lu–Hf method allows campaign-style studies to be conducted at a similar rate to U–Pb studies, opening new opportunities for magmatic and metamorphic studies.

Supplementary material: Data spreadsheets and supporting documentation are available at https://doi.org/10.6084/m9.figshare.c.6365962

The mainstay of geochronology is the U–Pb systems of the mineral zircon ($ZrSiO_4$), which is extremely robust to isotopic resetting (e.g. Gehrels 2014; Schaltegger *et al.* 2015). More recently, other geochronometers have been developed to complement zircon geochronology with the specific aims to: (1) expand the range of dateable lithologies to rocks that do not crystallize zircons, such as (ultra)mafic rocks (e.g. O'Sullivan *et al.* 2016; Pochon *et al.* 2016; Ackerman *et al.* 2020); (2) date low-temperature metamorphic or metasomatic processes that are usually not recorded by the zircon U–Pb systems (e.g. Glorie *et al.* 2019; Odlum and Stockli 2020); or (3) overcome issues with radiation damage, metamictization, and associated fluid-assisted Pb-loss (e.g. Herrmann *et al.* 2021), especially when zircon is very old and/or very rich in U and Th (e.g. Oosthuyzen and Burger 1973). Among these complementary mineral chronometers, apatite ($Ca_5(PO_4)_3(F,Cl,OH)$) stands out for being common and typically abundant in most rock types, including mafic and even ultramafic rocks. Apart from igneous crystallization, it can also grow under a variety of metamorphic and metasomatic conditions (Belousova *et al.* 2002; Nutman 2007; Harlov 2015; Gillespie *et al.* 2018; O'Sullivan *et al.* 2020). However, the apatite U–Pb system has a much lower nominal closure temperature (*c.* 350–570°C; e.g. Chew and Spikings 2021) compared to zircon (*c.* 900°C; e.g. Lee *et al.* 1997). Consequently, in regionally metamorphosed terranes, apatite U–Pb ages often record the timing of secondary isotopic disturbances, either by volume diffusion or recrystallization (e.g. Kirkland *et al.* 2018; Glorie *et al.* 2019; Henrichs *et al.* 2019). While the comparatively low closure temperature in apatite and its higher susceptibility

From: van Schijndel, V., Cutts, K., Pereira, I., Guitreau, M., Volante, S. and Tedeschi, M. (eds) 2024. *Minor Minerals, Major Implications: Using Key Mineral Phases to Unravel the Formation and Evolution of Earth's Crust.*
Geological Society, London, Special Publications, **537**, 165–184.
First published online March 10, 2023, https://doi.org/10.1144/SP537-2022-205

to secondary processes allows the post-crystallization history of a study area to be investigated, it can also pose an obstacle to resolve primary crystallization ages in rocks where apatite is the only phase amenable to dating. A further complication is that apatite generally incorporates a significant concentration of initial Pb (Chew *et al.* 2014), which can induce large uncertainties to apatite U–Pb dates or even make U–Pb dating effectively untenable.

The recently developed *in situ* Lu–Hf dating method (Simpson *et al.* 2021) can overcome a number of these challenges. The concentration of initial Hf is generally low in apatite, especially in felsic rocks, and where it is more significant, it is generally sufficiently variable between individual grains to allow calculation of robust ages (Glorie *et al.* 2022). More significantly, the closure temperature of the Lu–Hf system in apatite is much higher than that of the U–Pb system, theoretically on the order of *c.* 675–750°C (Barfod *et al.* 2005; Chew and Spikings 2015), making it less susceptible to resetting. To date, this closure temperature estimate is yet to be systematically interrogated with rocks from different environments and over geological timescales. To address this gap in knowledge we here use samples from a range of settings to compare the robustness of the *in situ* apatite Lu–Hf and U–Pb systems. Specifically, the case studies investigate: (1) the extent to which the Lu–Hf system can endure metasomatism and deformation; (2) the temperatures to which the apatite Lu–Hf system retains primary age information; and (3) if apatite grains contained in crustal xenoliths can preserve a memory of their crystallization history prior to magmatic entrainment.

Sample descriptions and geological background

Taratap Granodiorite

The Taratap Granodiorite in the Cambro-Ordovician Delamerian Orogenic belt in South Australia was selected because (1) it has a tightly constrained ID-TIMS zircon U–Pb reference age of 497.11 ± 0.56 Ma ($^{206}Pb/^{238}U$ weighted mean age, 95% confidence interval uncertainty, MSWD = 1.8; Curtis *et al.* 2022) and (2) there is some evidence for a protracted history of tectonic strain (Burtt and Abbot 1998), which might influence the isotopic systems in apatite. The Taratap Granodiorite is largely buried beneath Cenozoic cover and is exposed in a small outcrop, just north of Kingston SE (Fig. 1). The

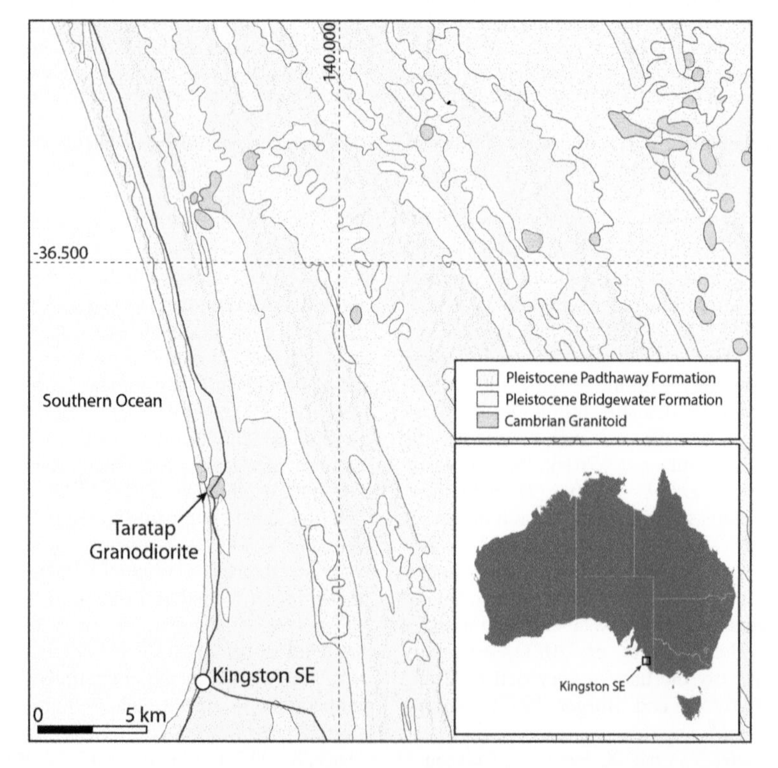

Fig. 1. Location map of the Taratap Granodiorite within the Delamerian Orogenic Belt in South Australia.

intrusion truncates folded metasediments, suggesting it was likely emplaced during the late stages of the Delamerian Orogeny (*c.* 514–490 Ma; Foden *et al.* 2006). The granodiorite is classified as S-type, calc-alkaline with a composition dominated by microcline megacrysts (*c.* 3–4 cm in length), which define a NNE-trending magmatic fabric in a coarse-grained groundmass of plagioclase, quartz, K-feldspar and biotite, with accessory zircon, apatite, and monazite. Low-temperature alteration is evident in thin section by the presence of chlorite–muscovite–titanite and minor allanite. Comagmatic enclaves of doleritic, dioritic and lesser microgranitic composition are present, as well as layered, sulfide-rich metasedimentary xenoliths (Burtt and Abbot 1998).

The granodiorite locally contains crystal-rich zones of densely packed imbricated megacrysts, produced by filter pressing of the magma during shearing. Evidence for solid-state strain is preserved in feldspars and quartz crystals, including patchy/undulose extinction, deformation twinning in plagioclase, and interlobate grain boundaries between quartz and biotite. Some quartz grains have irregular boundaries, providing evidence for mild grain boundary migration recrystallization. Hence, it is interpreted that the regional tectonic strain persisted after crystallization of the pluton (Curtis et al. 2022).

Reynolds–Anmatjira Range granites and gneisses

A series of granite and gneiss samples were selected from the Reynolds–Anmatjira Range in the Aileron Province in the southern North Australian Craton (Fig. 2), with the aim of investigating the behaviour of the apatite U–Pb and Lu–Hf systems during progressive high-temperature, low-pressure (HT–LP) metamorphism in a slowly cooled terrane. The Reynolds-Anmatjira Range provides a unique opportunity for this purpose as it continuously exhumes a sequence of metapelites and granite-gneisses from greenschist to granulite facies conditions (Dirks *et al.* 1991; Hand and Buick 2001). The thermal and temporal evolution of the Reynolds–Anmatjira Range has been extensively studied and is well understood. The oldest metasedimentary rocks of the *Lander Rock Formation* were deposited between *c.* 1.84 and 1.81 Ga and the granitic intrusions occurred in two stages at *c.* 1.81–1.79 Ga and *c.* 1.78–1.77 Ga (Hand and Buick 2001). HT–LP metamorphism (peaking >850°C) affected the area between *c.* 1680–1550 Ma, indicative of a long-lived, slowly cooled metamorphic system (Rubatto *et al.* 2001; Anderson *et al.* 2013; Morrissey *et al.* 2014; Alessio *et al.* 2020). Subsequently, pegmatites that overprint the regional gneissic fabric were emplaced at *c.* 1.52–1.51 Ga (Morrissey *et al.* 2014).

Five samples were taken from (meta)granitic rocks transected by metamorphic isograds between sub-greenschist facies and upper amphibolite facies metamorphism (Fig. 2). The lowest grade sample (RR08) was taken from a granite at sub-greenschist facies conditions (<350°C). Surrounding sediments have stable muscovite–chlorite, demonstrating the sample is below the biotite isograd. Samples RR02 and RR05 were taken from foliated granite deformed at upper greenschist facies conditions (incipient new biotite growth). The highest-grade samples (RR06 and RR01) were sourced from a granite and gneiss, respectively, with a metamorphic grade consistent with incipient migmatization (*c.* 650°C), evidenced by the presence of thin felsic veins within shear bands that transect the foliation. Sample RR01 was taken less than a kilometre away from metapelites with coarse sillimanite and K-feldspar, further demonstrating upper amphibolite facies conditions. Therefore, the sample suite contains both foliated (RR02, RR05, RR01) and unfoliated samples (RR08, RR06) from different metamorphic grades. This allows evaluation of the robustness of the apatite U–Pb and Lu–Hf systems to both volume diffusion (temperature increase) and recrystallization (foliation development).

Point Sir Isaac Granite

The Point Sir Isaac Granite was sampled to assess the response of the apatite Lu–Hf system to high temperature metamorphism without recrystallization. The peraluminous Point Sir Isaac Granite forms part of the Dutton Suite upper crustal batholith and outcrops along the northern coast of the Coffin Bay Peninsula in South Australia (Fig. 3; Dutch *et al.* 2008; Dutch and Hand 2010). The sampled granite is unfoliated and contains garnet and cordierite in addition to K-feldspar, quartz and plagioclase with minor biotite, zircon, monazite, and apatite. The granite crystallization age is constrained to *c.* 2.41 Ga based on monazite U–Pb and garnet Sm–Nd dates (Dutch and Hand 2010). The granite underwent high-grade reworking (*c.* 730°C and 1 GPa) expressed by the formation of mylonitic and migmatitic high-strain zones spaced 50–200 m apart (Dutch *et al.* 2008). SHRIMP U–Pb titanite dating in mafic rocks, and monazite U–Pb in granitic gneiss high-strain zones gave ages between *c.* 1.74 Ga and 1.69 Ga (Dutch *et al.* 2008). The sampled unfoliated granite was taken 50 m from the nearest *c.* 1.74–1.69 Ga high strain zone and shows no signs of incipient alteration or deformation. The apatites are *c.* 0.01–0.03 mm^2 in size and occur as inclusions within large

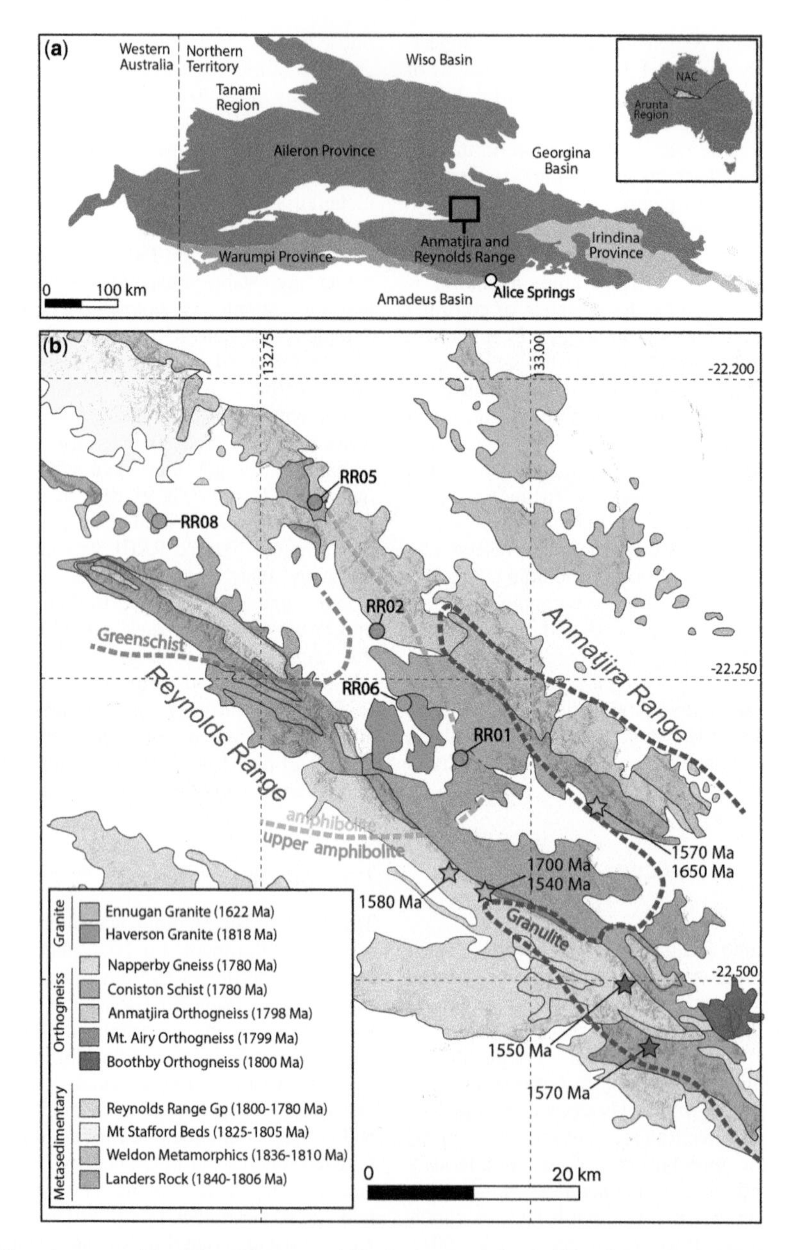

Fig. 2. (**a**) Overview map of the Aileron Provence within the Arunta region in Central Australia. (**b**) Geological map of the Anmatjira and Reynolds Ranges within the Aileron Province. Isograds are indicated by dashed lines. The sub-greenshist zone is defined by muscovite–chlorite; the greenshist zone is defined by muscovite–biotite; the lower amphibolite zone is defined by biotite–andalusite–muscovite ± cordierite; the upper amphibolite zone is defined by biotite–sillimanite–cordierite ± garnet and the granulite zone is defined by cordierite–spinel ± garnet–orthopyroxene (± biotite or sillimanite). Circle symbols (purple = orthogneiss, blue = granite) define the sample locations for this study. Star symbols show the sample locations for previous monazite geochronology studies in the upper amphibolite–granulite facies zones of the study area (from Rubatto *et al.* (2001), Anderson *et al.* (2013) and Alessio *et al.* (2020)). Source: maps are modified from Alessio *et al.* (2020) and the isograds were defined based on Hand and Buick (2001).

homogenous and euhedral plagioclase crystals. Optically, there is no evidence that the apatites have been recrystallized under metamorphic conditions, and therefore, this case study provides an opportunity to put a high temperature constraint on Hf diffusion in apatite.

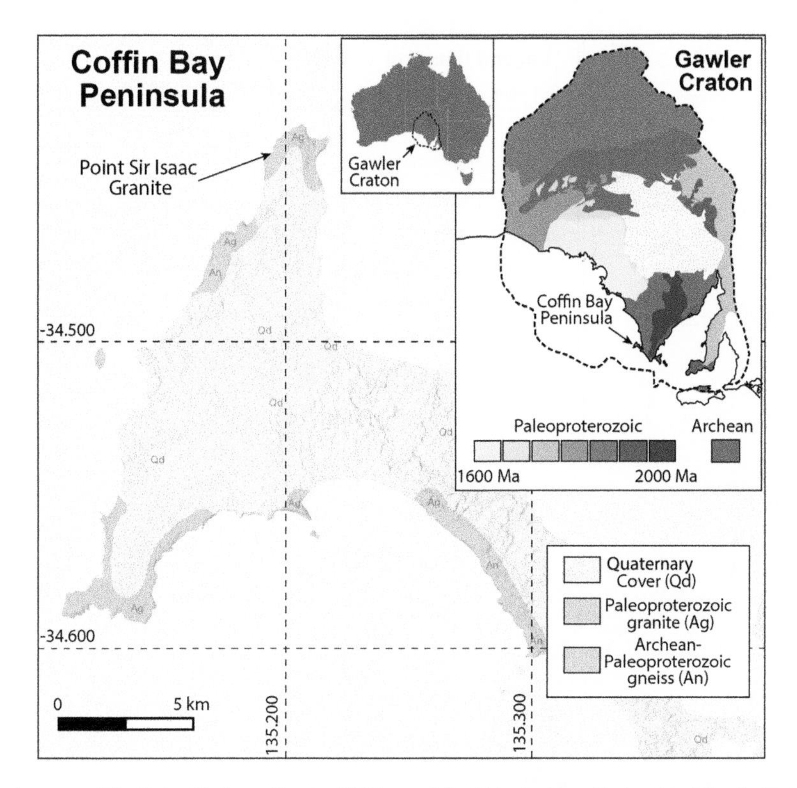

Fig. 3. Location map of the Point Sir Isaac Granite (PSI3 sample) within the Eyre Peninsula of South Australia.

Mt St Martin lower crustal xenoliths

In order to evaluate the robustness of the Lu–Hf system in apatite to short-lived, high-temperature events, mafic granulite xenoliths were sampled from young (*c.* 3 Ma; Sutherland *et al.* 1977; Griffin *et al.* 1987), pyroclastic alkali basalts of Mt St Martin, central Queensland, Australia (Fig. 4). The xenoliths were derived from lower continental crust underlying the Permian–Triassic Bowen Basin and granitoids of the New England Orogen (Rosenbaum 2018; Siegel *et al.* 2020). Textural evidence of the xenoliths and the co-occurrence of pristine mantle xenoliths indicate rapid dislodgement, entrainment and eruption for this locality. The xenolith host–magma grain boundaries are sharp and there is no geochemical evidence in the xenolith for extensive host–xenolith interaction/mingling (Emo and Kamber 2021). Two xenoliths were selected for this study as they contain abundant, large apatite grains (≥100 µm) amenable for *in situ* U–Pb and Lu–Hf dating. They have a simple plagioclase, clinopyroxene, and orthopyroxene mineralogy with accessory apatite and ilmenite (Emo and Kamber 2021). The aim of this case study is to investigate if apatite grains in xenoliths entrained in basaltic melts can retain a memory of their crystallization history in the lower crust.

Analytical methods

Zircon, monazite, titanite, and apatite U–Pb geochronology

Zircon and apatite grains were liberated using conventional crushing and magnetic and heavy liquid separation methods. With the exception of the central Queensland samples (samples 19RBE48 and 19RBE54), zircon and apatite U–Pb and trace element analysis was conducted using a RESOlution-LR 193 nm excimer laser ablation system, with a 30 µm beam size, coupled to an Agilent 7900 ICP-MS, using identical analytical parameters as in Gillespie *et al.* (2018) and Glorie *et al.* (2019). Isotope ratios were calculated in LADR (Norris and Danyushevsky 2018). For the zircon samples, GJ-1 (ID-TIMS U–Pb age of 600.7 ± 1.1 Ma; Jackson *et al.* 2004) was used as primary standard and Plešovice zircon (U–Pb age of 337.13 ± 0.37 Ma; Sláma *et al.* 2008) as secondary standard for accuracy verification purposes. Our obtained age for Plešovice zircon (U–Pb Concordia age of 337.8 ± 1.0 Ma) is in excellent agreement with the published age (Supplementary File 1). For the apatite samples, MAD (ID-TIMS U–Pb age 473.5 ± 0.7 Ma; Thomson *et al.* 2012; Chew *et al.* 2014) was used as

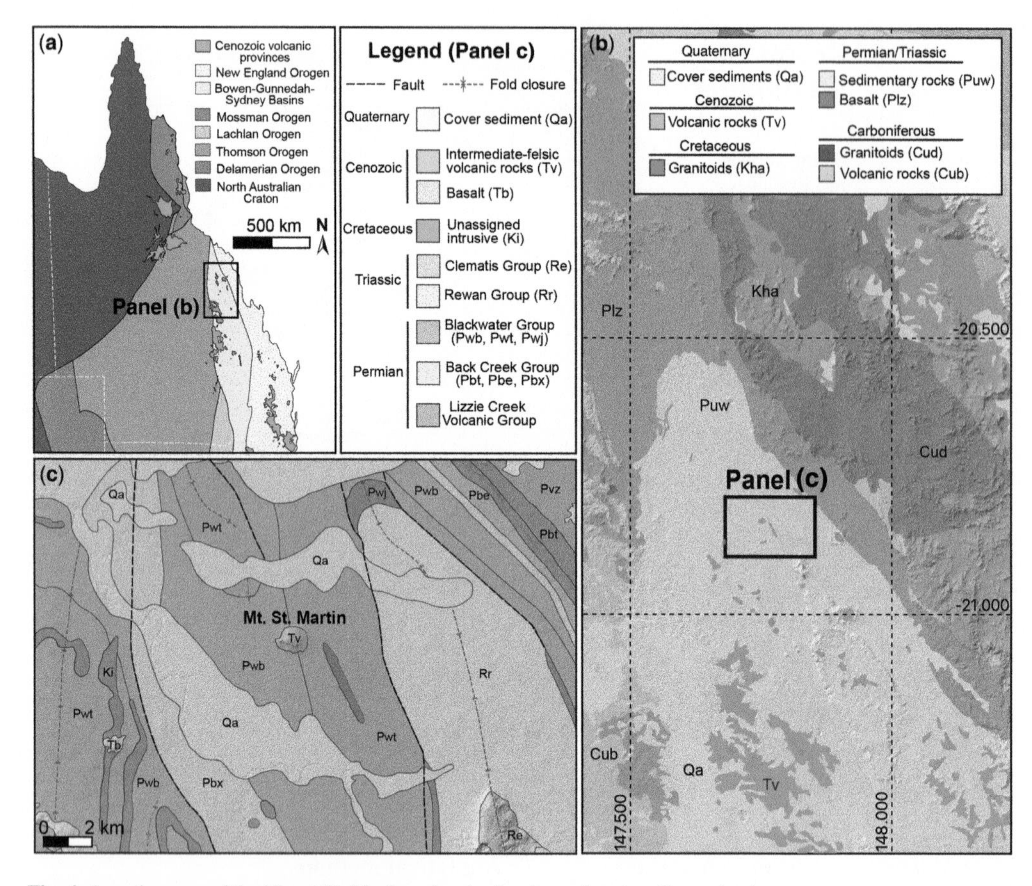

Fig. 4. Location map of the Mount St. Martin volcanic plug in northeastern Queensland.

primary reference material and McClure apatite (ID-TIMS U–Pb age 523.51 ± 1.47 Ma; Schoene and Bowring 2006) as secondary standard. Our obtained weighted mean $^{206}Pb/^{238}U$ ages for McClure apatite (527.6 ± 15.2 Ma; 527.4 ± 15. 05 Ma; 526.6 ± 8.6 Ma) are internally consistent over different analytical sessions and in good agreement with the published age (Supplementary File 1). Resulting zircon and apatite U–Pb ages were calculated in IsoplotR (Vermeesch 2018). Apatite trace element concentrations were obtained simultaneously and processed in LADR using ^{43}Ca as the internal standard element. Rare earth element (REE) spidergraphs were calculated and plotted in GCDkit 3.6.0 (JanouŠek *et al.* 2006) using REE primitive mantle normalization (McDonough and Sun 1995).

For the central Queensland samples, the U–Pb and trace element data were collected in thin section using a split stream (SS) LA-ICP-MS system with two Agilent 7900 ICP-MS instruments connected to a Photon Machines Excite + laser ablation system at the Central Analytical Research Facility, Queensland University of Technology. Spot analyses (35–80 µm) were used with a 10 Hz repetition rate, *c.* 2.5 J cm^{-2} laser energy, a 30 s dwell time and a 25–30 s washout time. MAD apatite was used as an external calibrant for apatite U–Pb analyses (Thomson *et al.* 2012), and Durango (apatite weighted mean $^{206}Pb/^{238}U$ ID-TIMS age of 32.29 ± 0.12 Ma; Paul *et al.* 2021) as a secondary standard. The $^{207}Pb/^{206}Pb$-anchored discordia age of 31.6 ± 1.0 Ma for Durango apatite is in good agreement with published data (Supplementary File 1). The apatite U–Pb data were reduced using Iolite v4 with the VisualAge_UComPbine data reduction scheme (Petrus and Kamber 2012).

Titanite and monazite from the Taratap Granodiorite were analysed within thin section at Adelaide Microscopy using a RESOlution-LR 193 nm excimer laser ablation system, coupled to an Agilent 8900 ICP-MS/MS. Laser beam diameters of 43 µm and 13 µm were used for titanite and monazite, respectively. Isotope ratios were calculated in LADR (Norris and Danyushevsky 2018). For monazite, MAdel monazite was used as primary standard and 222 monazite as secondary standard for accuracy verification (Payne *et al.* 2008). The obtained weighted mean $^{206}Pb/^{238}U$ age of 453.5 ± 2.1 Ma for the 222 monazite reference material (Supplementary File 1) is in agreement with the published age of 449.7 ± 6.8 Ma by Richter *et al.* (2019). For titanite, MKED1 was used as a primary standard (Spandler *et al.* 2016) and an in-house titanite from Mount Painter as a secondary standard. The obtained weighted mean $^{206}Pb/^{238}U$ age of 443.7 ± 2.5 Ma for the Mount Painter titanite reference material (Supplementary File 1) agrees within uncertainty with the published value of 442.6 ± 1.8 Ma (Elburg *et al.* 2003).

Apatite Lu–Hf geochronology

Subsequent to U–Pb dating, apatite crystals that were sufficiently large to allow a second ablation spot were analysed for Lu–Hf isotopes. Analyses were conducted in four analytical sessions using a RESOlution-LR 193 nm excimer laser ablation system, with a variable beam size between 67 μm and 170 μm, coupled to an Agilent 8900 ICP-MS/MS. See Supplementary File 2 for analytical conditions. The laser-based Lu–Hf method uses NH_3 gas in the reaction-cell of the mass spectrometer, which allows high-order reaction products of ^{176}Hf and ^{178}Hf to be measured free from isobaric interferences at masses 258 and 260 amu, respectively. ^{177}Hf is subsequently calculated from ^{178}Hf, assuming natural abundances. ^{175}Lu is measured on mass as a proxy for ^{176}Lu (see details in Simpson *et al.* 2021). Isotope ratios were calculated in LADR (Norris and Danyushevsky 2018) using NIST 610 as a primary standard (Nebel *et al.* 2009), and corrected for matrix-induced fractionation (cf. Roberts *et al.* 2017) using OD-306 apatite (1597 ± 7 Ma; Thompson *et al.* 2016). Lu–Hf ages were calculated as inverse isochrons using IsoplotR (Vermeesch 2018; Li and Vermeesch 2021). For samples with highly radiogenic $^{177}Hf/^{176}Hf$ ratios (< *c.* 0.1), it is difficult to calculate a robust free isochron as the initial $^{177}Hf/^{176}Hf$ ratio is poorly defined. In this case, the isochron was anchored to an initial $^{177}Hf/^{176}Hf$ composition of 3.55 ± 0.05, which spans the entire range of initial $^{177}Hf/^{176}Hf$ ratios of the terrestrial reservoir (e.g. Spencer *et al.* 2020). Additionally, an alternative age calculation has been applied for data with highly radiogenic Lu–Hf ratios ($^{177}Hf/^{176}Hf$ ratios <0.1), by using a common-Hf correction on the ^{176}Lu–^{176}Hf ratios. The approach is detailed in Simpson *et al.* (2022). Subsequently, a weighted mean Lu–Hf age can be calculated from the common-Hf corrected ^{176}Lu–^{176}Hf ratios. The in-house reference apatites Bamble-1 (Bamble Sector, SE Norway; corrected Lu–Hf age: 1097 ± 5 Ma) and HR-1 (Harts Range, NT Australia; corrected Lu–Hf age: 343 ± 2 Ma) were monitored for accuracy checks and are in excellent agreement with previously published data (Simpson *et al.* 2021; Glorie *et al.* 2022). Table 1 provides a summary of the calculated ages for the reference materials. Associated isochron and weighted mean calculations are presented in Supplementary File 3.

TitaniQ titanium-in-quartz geothermometry

TitaniQ titanium-in-quartz geothermometry was used to estimate temperatures in assemblages containing the dated apatites. This approach was adopted because the granitic bulk compositions do not contain mineral assemblages amenable to conventional geothermometry. Quartz crystals were ablated in polished rock blocks using a NewWave NWR213 laser ablation system, with a beam size of 30 μm, coupled to an Agilent 7900 ICP-MS. Analytical conditions are summarized in Supplementary

Table 1. *Apatite Lu–Hf results for the reference materials used in this study*

Session	OD-306 (uncorrected)			Bamble-1 (corrected)			HR-1 (corrected)		
	Age (Ma)	2σ (Ma)	MSWD (*n*)	Age (Ma)	2σ (Ma)	MSWD (*n*)	Age (Ma)	2σ (Ma)	MSWD (*n*)
Tararap	1666	14	0.7 (19)	–	–	–	344.0	5.4	0.7 (19)
Reynolds	1652	19	0.7 (22)	–	–	–	344.5	6.9	0.7 (17)
PSI3	1658	8	1.2 (30)	1098	7	0.7 (32)	345.2	3.3	1.0 (35)
QLD	1659	11	1.4 (19)	1102	10	0.3 (20)	345.0	5.2	1.1 (19)
Expected*	1597†	7	0.2	1097‡	5	0.8 (70)	344.0‡	2	0.7 (96)

The session names refer to the four study areas.
*The ratio between the measured and expected age for OD-306 is used as a session-dependent calibration factor for the Lu–Hf ratios for each sample and secondary reference material.
†The expected age for the matrix-matched age correction standard (OD-306) is the isotope dilution U–Pb age from Thompson *et al.* (2016).
‡The expected ages for the Bamble-1 and Hr-1 secondary standards (corrected against OD-306) are long-term weighted mean apatite Lu–Hf ages from Glorie *et al.* (2022). (*n*) = number of analyses.

File 2. NIST-610 was used a primary standard and NIST-612 as an accuracy check for the obtained Ti concentrations. ^{29}Si was used as the internal standard element. The obtained weighted mean ^{47}Ti concentration for NIST-612, over both analytical sessions, is 37.6 ± 0.2 ppm (MSWD = 0.36). This value is *c.* 5.2% lower than the expected value of 39.6 ± 0.3 ppm, obtained by LA-ICP-MS, using the same wavelength laser (Jochum *et al.* 2011). The *c.* 5.2% offset on NIST-612 has been propagated to the analytical uncertainty of the Ti concentrations in the analysed samples. Quartz (re)crystallization temperatures were subsequently calculated using the TitaniQ titanium-in-quartz geothermometer (Wark and Watson 2006). All dated samples where TitaniQ titanium-in-quartz geothermometry was applied are granitic in composition and contain high-Ti biotite and ilmenite. None of the dated samples contain rutile, meaning the titanium content in quartz is unbuffered. The TiO_2 activity ($aTiO_2$) for such samples is generally estimated to be between 0.55 and 1 (Wark and Watson 2006) and an arbitrary $aTiO_2$ value of 0.75 was adopted, similar to Ehrlich *et al.* (2012). For the range of measured Ti contents, the effects of varying $aTiO_2$ by 0.2 results in 30° of variation in the calculated temperatures.

Results

The geochronology results for the zircon U–Pb, apatite U–Pb, and apatite Lu–Hf methods are summarized in Table 2 and described below. Corresponding data tables can be found in Supplementary File 4 (zircon U–Pb data), Supplementary File 5 (apatite, monazite and titanite U–Pb and trace element data) and Supplementary File 6 (apatite Lu–Hf data). The results for the Ti-in-quartz geothermometer are tabulated in Supplementary File 7.

Taratap Granodiorite

The apatite U–Pb data for the Taratap Granodiorite plot along a linear array with little isotopic dispersion (MSWD = 0.64), from which a $^{238}U–^{206}Pb$ age of 439 ± 34 Ma is calculated (Fig. 5). The Lu–Hf data are mostly highly-radiogenic (35/38 analyses with $^{177}Hf/^{176}Hf$ ratios <0.1) and define an anchored Lu–Hf isochron age of 497.1 ± 5.5 Ma (MSWD = 1.1; Fig. 5). Alternatively, calculating the weighted mean common-Hf corrected Lu–Hf age (for apatites with $^{177}Hf/^{176}Hf$ ratios <0.1) returns an identical age (within uncertainty) of 498.1 ± 6.9 Ma (MSWD = 0.88; Fig. 5). The apatite REE spidergraphs reveal that the apatites are significantly depleted in light REEs (LREEs; La/Sm ratio of 0.64 ± 0.10) with flat heavy REE (HREE) profiles (Supplementary File 8). This suggests apatite growth or recrystallization while competing with another mineral for LREEs (e.g. Glorie *et al.* 2019), which is supported by observations in thin section of cogenetic allanite with apatite, as well as monazite overgrowths on apatite (Supplementary File 9). The monazite U–Pb data plot close to concordia with a $^{206}Pb/^{238}U$ weighted mean age of 499.6 ± 6.7 Ma (MSWD = 6; Fig. 6). For titanite, occurring within chlorite alteration zones within biotite, an overdispersed dataset was obtained. Based on the concentration of the LREE (La + Ce + Pr + Nd), two titanite populations were separated. Analyses with low LREE concentrations (<120 ppm) plot along an isochron with a lower intercept U–Pb age of 507 ± 10 Ma (MSWD = 1.1). Analyses with higher LREE concentrations define a younger U–Pb isochron of 445 ± 12 Ma (MSWD = 1.2; Fig. 6). The only textural difference between the old and young analyses is that the latter tend to cluster near the centre of the elongated titanite grains (Supplementary File 9).

Ti concentrations in quartz vary significantly between different ablated quartz grains, ranging between 156 and 2 ppm. For most analysed grains, the intra-grain variability of the Ti concentration is relatively low (*c.* 12–16% standard deviation of the mean), while the inter-grain variability is >50%. Resulting quartz crystallization temperatures vary between 843 ± 17°C and 433 ± 10°C (Supplementary File 7).

Reynolds–Anmatjira Range granites and gneisses

Five Reynolds–Anmatjira Range samples were analysed for zircon U–Pb, apatite U–Pb and apatite Lu–Hf isotopes. The zircons produce highly scattered data in concordia plots, consistent with multiple age populations and significant isotopic open system behaviour (i.e. Pb-loss) (Supplementary File 10). For each sample, concordia ages were calculated for the youngest concordant (discordance <5%) data cluster, and these were taken as the best estimate of the magmatic crystallization ages of the granites or the protoliths of the granitic gneisses. Gneiss samples RR01, RR02 and RR05 yield consistent concordant zircon U–Pb ages of 1808 ± 6 Ma (MSWD = 0.8), 1814 ± 7 Ma (MSWD = 0.3) and 1808 ± 5 Ma (MSWD = 0.9), respectively (Fig. 7). Granite samples RR06 and RR08 produced slightly younger zircon U–Pb crystallization ages of 1781 ± 6 Ma (MSWD = 1.2) and 1783 ± 5 Ma (MSWD = 0.9), respectively. These zircon ages are in excellent agreement with two groups of previously reported granite ages of *c.* 1.81–1.79 Ga and *c.* 1.78–1.77 Ga (Hand and Buick 2001; Howlett *et al.* 2015).

The apatite U–Pb ages from the same samples are significantly younger than the zircon ages. For

Table 2. *Summary of apatite U–Pb and Lu–Hf ages, compared to zircon U–Pb and monazite (italic) reference ages for the crystallization age of each analysed sample*

Sample	Zr and/or *Mz* U–Pb			Ap U–Pb			Ap Lu–Hf					
	Age (Ma)	2σ (Ma)	MSWD (*n*)	Iso age (Ma)	2σ (Ma)	MSWD (*n*)	Iso age (Ma)	2σ (Ma)	MSWD (*n*)	WM age (Ma)	2σ (Ma)	MSWD (*n*)
Taratap (gr)	497.1*	0.6	1.8 (6)	439	34	0.7 (24)	497.1	5.5	1.1 (38)	498.1	6.9	0.88 (35)
RR01 (gns)	1808	6	0.8 (5)	1527[†]	8	1.1 (56)	1642	28	0.6 (18)	1636	35	0.4 (16)
RR02 (gns)	1814	7	0.3 (4)	1511[†]	14	1.4 (35)	1604	45	1.2 (8)	1591	44	1.4 (8)
RR05 (gns)	1808	5	0.9 (9)	1453[‡]	55	1.9 (26)	1617	23	1.5 (19)	1608	26	1.1 (19)
RR06 (gr)	1781	6	1.2 (5)	1570[§]	17	6.8 (25)	1754	24	1.1 (16)	1751	31	0.7 (13)
RR08 (gr)	1783	5	0.9 (9)	1590[§]	7	2.5 (23)	1754	45	1.0 (10)	1755	49	0.7 (10)
PSI3 (gr)	*2414*[‖]	*6*	*0.95*	1704[¶]	83	6.4 (14)	1745	10	1.3 (35)	–	–	–
19BRE48	–	–	–	14	43	1.1	35.7	9.1	1.6 (30)	–	–	–
19BRE54	–	–	–	14	73	0.4	182	10	2.3 (41)	–	–	–

gr, granitic; gns, foliated (orthogneiss). The apatite Lu–Hf ages are corrected against OD-306. Both isochron (Iso age) and weighted mean (WM) apatite Lu–Hf ages are presented. The WM age is a common-Hf corrected weighted mean $^{176}Hf/^{176}Lu$ age, filtered for $^{176}Hf/^{177}Hf$ ratios > 1 (see text for further details). For the apatite U–Pb system, isochron ages are presented. Zircon U–Pb ages for the Reynolds Range samples are presented in Supplementary file 10. Apatites with low Nd concentrations (<400 ppm) were rejected from the apatite U–Pb isochron regression for sample PSI3. (*n*) = number of analyses.

*The Taratap zircon U–Pb age is an ID-TIMS age from Curtis *et al.* (2022).

†The apatite U–Pb ages for RR01 and RR02 correspond with monazite dates in late pegmatites (Morrissey *et al.* 2014).

‡The apatite U–Pb dates for RR05 are a highly dispersed array. The dataset was filtered arbitrary until an MSWD <2 was obtained. This age can be regarded as an interpreted minimum age for the sample.

§The apatite U–Pb results for RR06 and RR08 are scattered data arrays. Data with low LREE (<1500 ppm) were rejected from the isochron regression. The resulting ages correspond with zircon and monazite dates in Rubatto *et al.* (2001) and Morrissey *et al.* (2014).

‖The monazite U–Pb age for the Point Sir Isaac sample is from Dutch and Hand (2010).

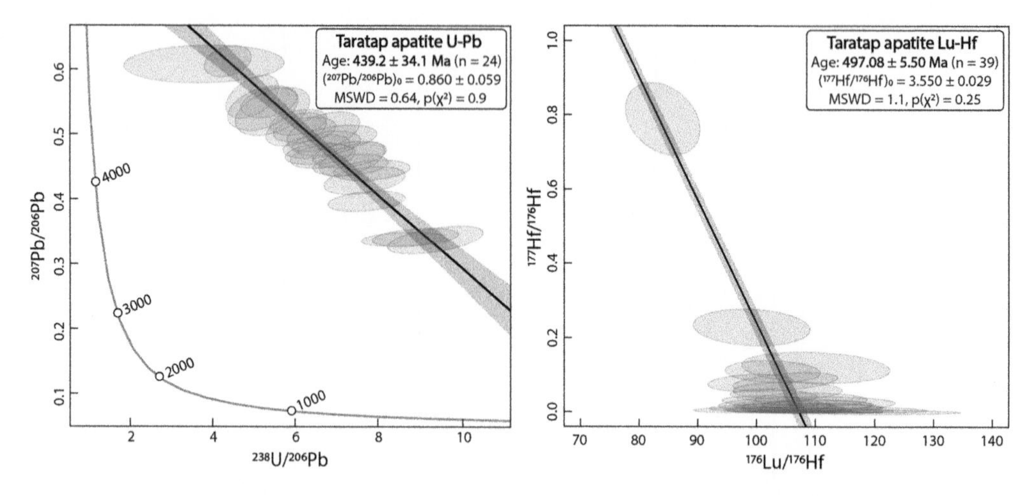

Fig. 5. Apatite U–Pb (left) and Lu–Hf (right) isochron ages for the Taratap Granodiorite (calculated in IsoplotR; Vermeesch 2018). MSWD = mean squared weighted deviation and P(χ^2) = Chi-squared probability for a single data population.

RR01, the apatite U–Pb data plot along a single linear array with little isotopic dispersion (MSWD = 1.1), producing an isochron age of 1527 ± 8 Ma (Fig. 8). The apatite U–Pb data for sample RR02 are more scattered. A minimum isochron age (MSWD = 1.4) of 1510 ± 14 Ma can be calculated based on 35 out of 44 analyses (Fig. 8). This age overlaps within uncertainty with the apatite U–Pb age for RR01. The grains excluded from the isochron could be related to isotopic disturbance (U-loss) or (partial) inheritance from an older event. Similar to RR02, the U–Pb data for sample RR05 are highly scattered, which can likely be attributed to isotopic open system behaviour (such as U-loss). A relatively poorly constrained minimum apatite U–Pb age of 1453 ± 55 Ma (MSWD = 1.9) was calculated (Fig. 8), which overlaps within uncertainty with the apatite U–Pb age obtained for RR02, but not with RR01. The apatite U–Pb ages obtained from granite samples RR06 and RR08 are significantly older

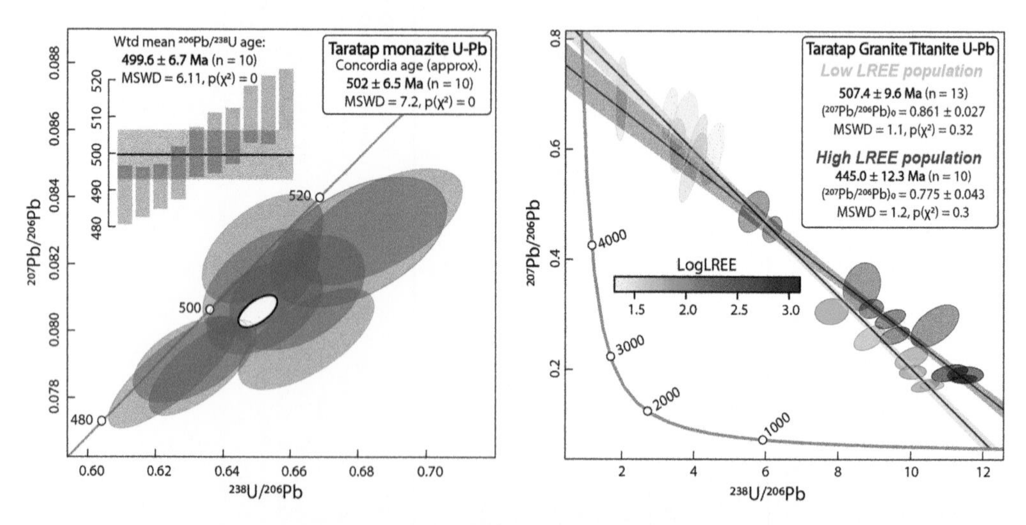

Fig. 6. Monazite (left) and titanite (right) U–Pb data for the Taratap Granodiorite (calculated in IsoplotR; Vermeesch 2018). MSWD = mean squared weighted deviation and P(χ^2) = Chi-squared probability for a single data population. For the monazite data, both a concordia age and weighted mean ^{206}Pb/^{238}U age are presented. For the titanite data, two free isochron regressions were calculated by subdividing the population based on LREE (=sum of La, Ce, Pr, Nd) concentrations. The low LREE population (arbitrary set as <120 ppm) constrains an older titanite U–Pb age compared to the higher LREE population (see text for discussion).

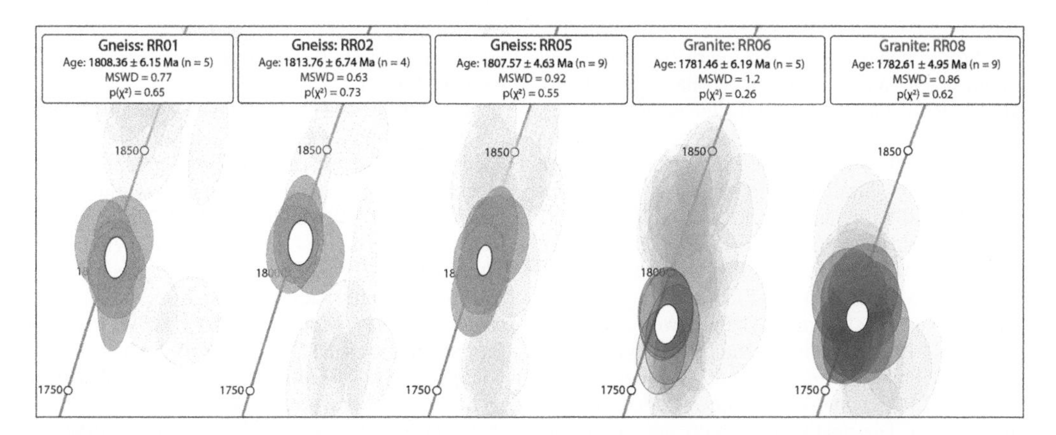

Fig. 7. Zircon U–Pb concordia plots for the Reynolds Range granite (blue) and orthogneiss (purple) samples, zoomed in on the youngest concordant data clusters (calculated in IsoplotR; Vermeesch 2018). Full zircon U–Pb plots are presented in Supplementary File 10. MSWD = mean squared weighted deviation and $P(\chi^2)$ = Chi-squared probability for a single data population.

compared to those from the gneiss samples (RR01, RR02 and RR05). Sample RR06 yields highly dispersed U–Pb data that can be subdivided into two groups, based on LREE compositions. Apatites with higher LREE concentrations (arbitrarily filtered at >1500 ppm) plot along a linear array that defines

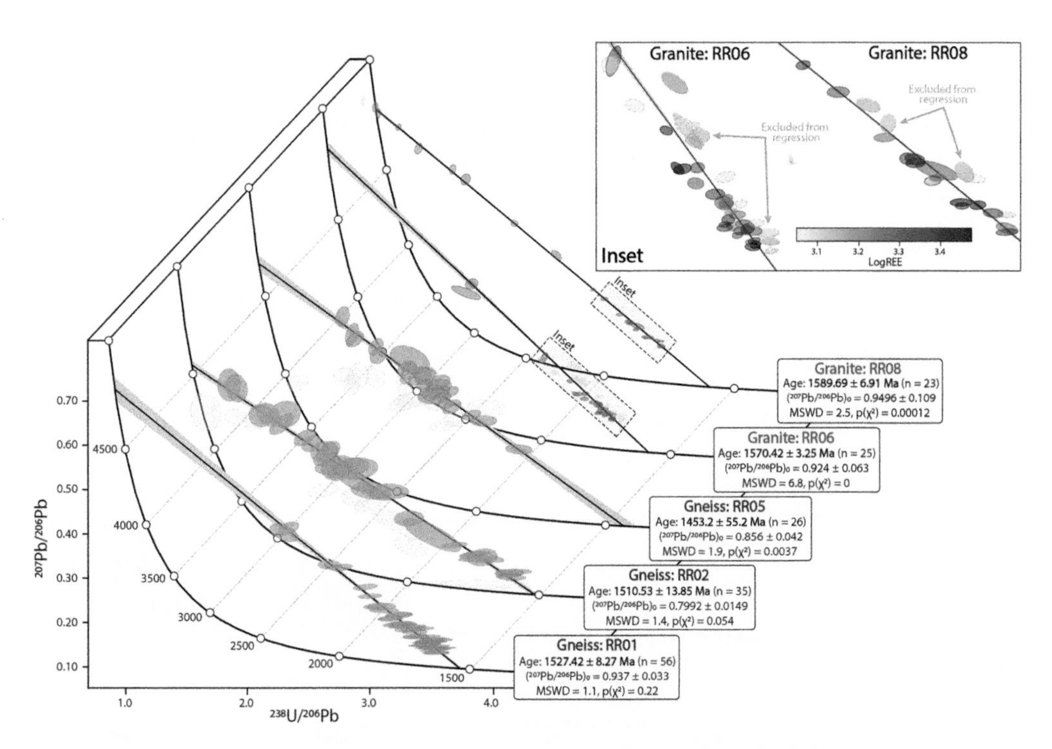

Fig. 8. Stacked apatite U–Pb isochron plots for the Reynolds granite (blue) and orthogneiss (purple) samples (calculated in IsoplotR; Vermeesch 2018). MSWD = mean squared weighted deviation and $P(\chi^2)$ = Chi-squared probability for a single data population. The inset magnifies the higher radiogenic section of the isochrons from granite samples RR06 and RR08, where the data are colour-coded following the LREE composition of analysed apatites. Low LREE apatites (<1500 ppm) are thought to record a secondary low-temperature reset and were excluded from the isochron calculations.

an isochron with a U–Pb age of 1570 ± 18 Ma. The isochron remains statistically relatively poorly constrained (MSWD = 6.8) but can be regarded as a maximum apatite U–Pb age for the sample (Fig. 8). The younger (LREE < 1500 ppm) population appears highly scattered, precluding a useful age determination. Similarly, the apatite U–Pb data for sample RR08 appears overdispersed (MSWD = 6.3). Applying the same filter as for RR06 (LREE >1500 ppm), produces a more robust isochron of 1590 ± 7 Ma (MSWD = 2.5). In summary, the apatite U–Pb data for the gneiss samples were calculated as *c.* 1.50–1.53 Ga, whereas the granites yield older apatite U–Pb ages of *c.* 1.57–1.59 Ga when apatites with low LREE compositions are excluded from the regressions. The apatite REE plots are reasonably consistent between the analysed samples, with moderately to strongly depleted LREE profiles (Supplementary File 8).

The apatite Lu–Hf ages for the Reynolds–Anmatjira Range samples are consistently older than the apatite U–Pb dates. For the gneiss samples, the inverse isochron ages are consistent within uncertainty. The Lu–Hf data for RR01 define an inverse isochron age of 1642 ± 28 Ma (MSWD = 0.6). The common-Hf corrected weighted mean Lu–Hf age for the highly radiogenic analyses (16/18) was calculated as 1636 ± 35 Ma (MSWD = 0.43; Fig. 9). Samples RR02 and RR05 both yield one younger analysis, leading to slightly more dispersed and younger isochron ages of 1604 ± 45 Ma (MSWD = 1.2) and 1617 ± 23 Ma (MSWD =

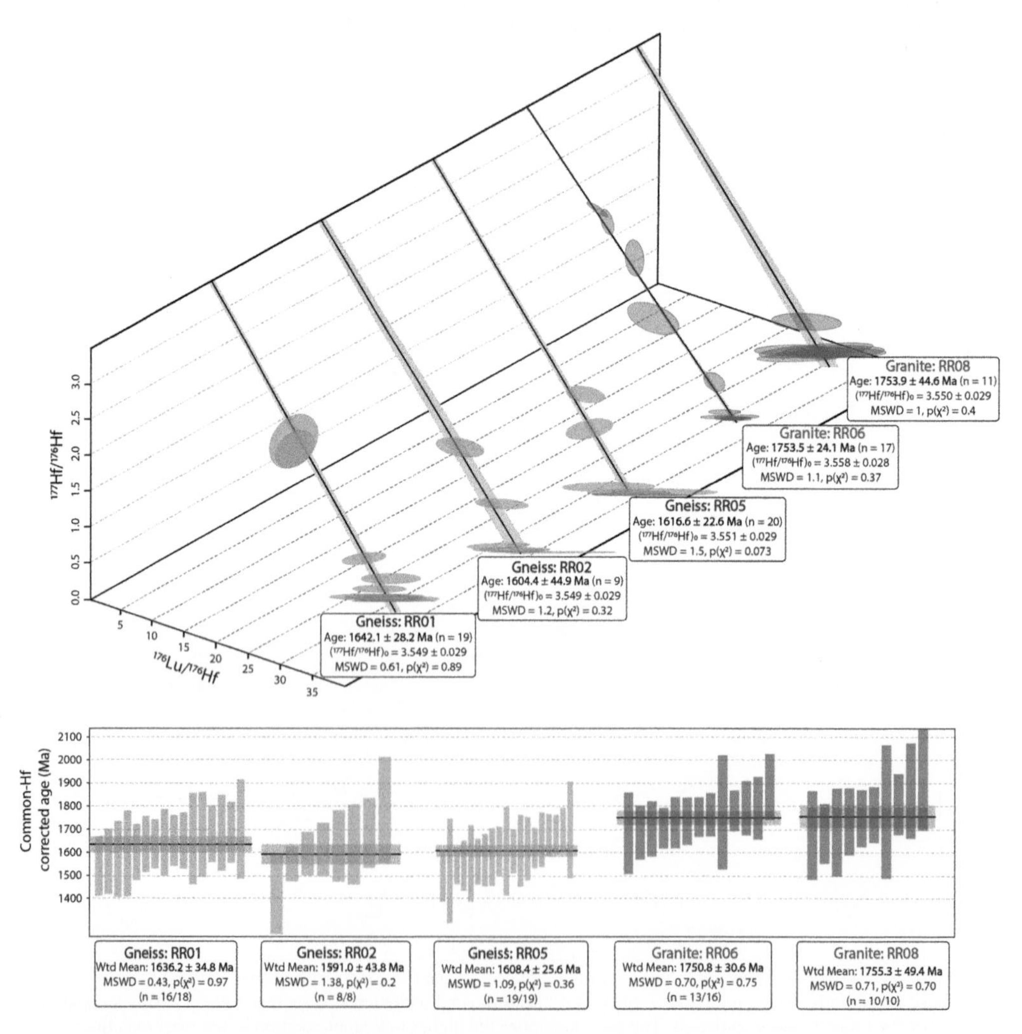

Fig. 9. Stacked apatite Lu–Hf isochron and weighted mean plots for the Reynolds granite (blue) and orthogneiss (purple) samples (calculated in IsoplotR; Vermeesch 2018). MSWD = mean squared weighted deviation and P(χ^2) = Chi-squared probability for a single data population.

1.5), respectively. The common-Hf corrected weighted mean Lu–Hf ages are 1591 ± 44 Ma (MSWD = 1.4) and 1608 ± 26 Ma (MSWD = 1.1), respectively (Fig. 9). In sharp contrast, the Lu–Hf ages for the granite samples (RR06 and RR08) are significantly older. The Lu–Hf data for sample RR08 are all highly radiogenic. The anchored inverse isochron defines a Lu–Hf age of 1754 ± 45 Ma (MSWD = 1.0) and the common-Hf corrected Lu–Hf age was calculated as 1755 ± 49 Ma (MSWD = 0.7). Sample RR06 has more variable $^{177}Hf/^{176}Hf$ ratios, defining an inverse isochron of 1754 ± 24 Ma (MSWD = 1.1), while the common-Hf corrected weighted mean Lu–Hf age is 1751 ± 31 Ma (MSWD = 0.7; Fig. 9). Hence, the apatite Lu–Hf ages for the granites are consistent (*c.* 1.75 Ga) and agree within uncertainty with the zircon U–Pb ages, whereas the gneiss samples record significantly younger apatite Lu–Hf ages of *c.* 1.64–1.59 Ga.

The Ti in Quartz concentrations in the granite samples (RR06 and RR08) show significant variability (34% and 87% standard deviation of the mean, respectively). Excluding three outliers, the calculated mean temperature for sample RR06 is 820 ± 7°C, while the lowest temperature for an additional analysis was calculated at 657 ± 14°C (Supplementary File 7). For sample RR08, the calculated temperatures can be subdivided into two populations (both based on multiple analyses in three different quartz grains) with mean temperatures of 872 ± 7°C and 475 ± 8°C. For the gneiss samples (in order of metamorphic grade: RR05, RR02, RR01), the intra-sample variation in Ti concentrations is relatively low, resulting in calculated mean temperatures of 621 ± 3°C, 646 ± 3°C, and 656 ± 2°C, respectively (Supplementary File 7). Reported uncertainties are standard errors of the mean.

Point Sir Isaac Granite

Apatites from the Point Sir Isaac Granite can be grouped into two populations based on REE profiles. One population is characterized by depleted LREE profiles with negative Eu anomalies, whereas the other population has flat REE profiles with positive Eu anomalies, and consistently low total REE concentrations (Supplementary File 8). The difference in Nd concentrations effectively distinguishes the two apatite populations. The U–Pb data are highly dispersed on the Tera-Wasserburg concordia with the low-Nd population defining a broadly horizontal trend to lower $^{238}U/^{206}Pb$ values, away from the main linear data array. Filtering to Nd concentrations >400 ppm, a linear regression defines an overdispersed U–Pb isochron age of 1704 ± 83 Ma (MSWD = 6.4; Fig. 10). The Lu–Hf data yield sufficient variability in $^{177}Hf/^{176}Hf$ ratios to calculate a robust inverse isochron regression, resulting in a Lu–Hf age of 1745 ± 10 Ma (MSWD = 1. 3; Fig. 10). Hence, both apatite chronometers are in agreement within uncertainty. Given the temperature of high-grade metamorphism is well constrained to 730°C (Dutch *et al.* 2008), Ti-in-quartz geothermometry was not conducted.

Mt St Martin lower crustal xenoliths

The apatites from xenolith samples 19RBE48 and 19RBE54 provide poorly constrained U–Pb

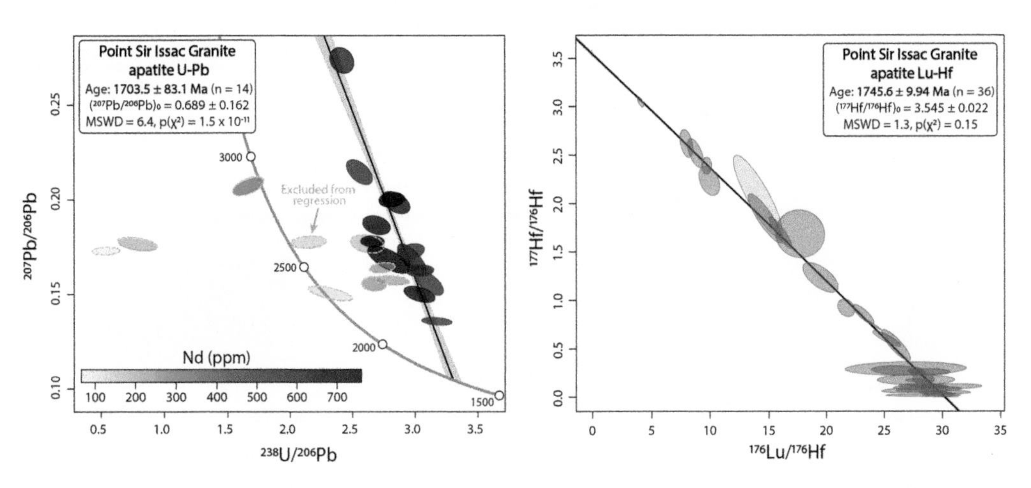

Fig. 10. Apatite U–Pb (left) and Lu–Hf (right) isochron ages for the Point Sir Isaac (PSI3) Granite (calculated in IsoplotR; Vermeesch 2018). MSWD = mean squared weighted deviation and $P(\chi^2)$ = Chi-squared probability for a single data population. The apatite U–Pb data are highly dispersed. Apatites with low (< 400 ppm) Nd concentrations are interpreted to record isotopic disturbance and were excluded from the isochron regression.

discordia ages of 14 ± 44 Ma and 14 ± 73 Ma, respectively (Fig. 11). The large uncertainties are due to the low U concentrations of the apatite grains (U = 1.2–5.5 ppm) relative to initial Pb, reflecting the mafic and refractory nature of the xenoliths, and the young age of the carrier magma that brought the xenoliths to the surface (*c.* 3 Ma) (Sutherland *et al.* 1977; Griffin *et al.* 1987). However, the U–Pb data are consistent with Cenozoic (re) crystallization of the apatite grains, most likely during entrapment, transport and emplacement of the host magma. Apatite REE profiles (Supplementary File 8) are typical for mafic igneous apatites with a consistently negative slope and modest negative Eu anomaly (e.g. Glorie *et al.* 2020; O'Sullivan *et al.* 2020). The Lu–Hf data for the apatites are mostly low in radiogenic Hf ($^{177}Hf/^{176}Hf$ ratios >1), defining poorly constrained isochron ages of 36 ± 9 Ma (19RBE48) and 182 ± 10 Ma (19RBE54). Both samples fail the χ^2 test, suggesting the data do not constitute a single age population (Fig. 11). In an attempt to statistically pull apart the dispersed Lu–Hf data for both samples, single-grain ages were calculated by applying a common-Hf correction (Supplementary File 6). In addition, grain size dimensions were obtained by optical microscopy to evaluate if single-grain Lu–Hf ages correlate with grain size variations. Figure 12 displays the results of this exercise (combined for both samples), with fragmented grains plotted separately from those that preserve the majority of the original crystals. The analysis shows that single-grain Lu–Hf ages decrease with decreasing grain size. Relatively consistent Lu–Hf ages of *c.* 220–150 Ma were obtained for >0.08 mm^2 grains. For smaller grains, the Lu–Hf ages decrease systematically until as low as 11 ± 8 Ma for *c.* 0.04–0.06 mm^2 grains. Hence, grain size seems to control the retentivity of the Lu–Hf age in apatite for the xenolith samples.

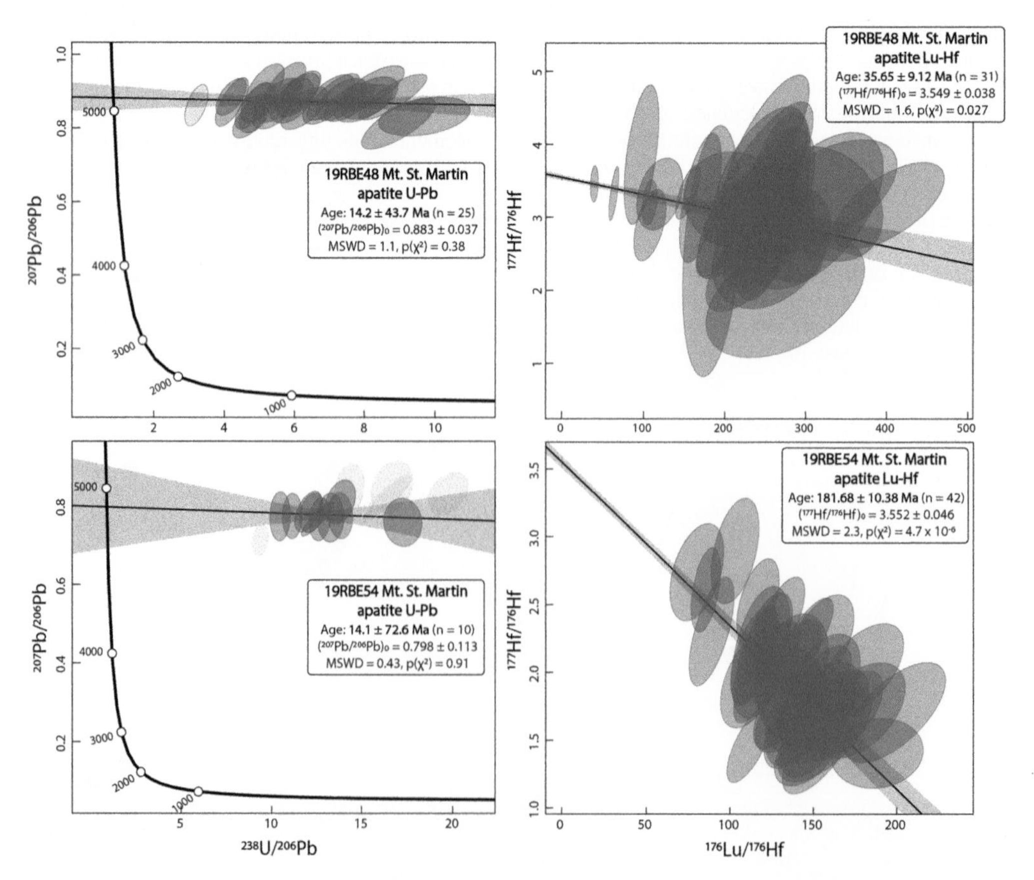

Fig. 11. Apatite U–Pb (left) and Lu–Hf (right) isochron dates for the Mount St Martin samples from northeastern Queensland (calculated in IsoplotR; Vermeesch 2018). MSWD = mean squared weighted deviation and P(χ^2) = Chi-squared probability for a single data population. The apatite U–Pb data produce a meaningless age with a large uncertainty. The Lu–Hf data are relatively low radiogenic and dispersed.

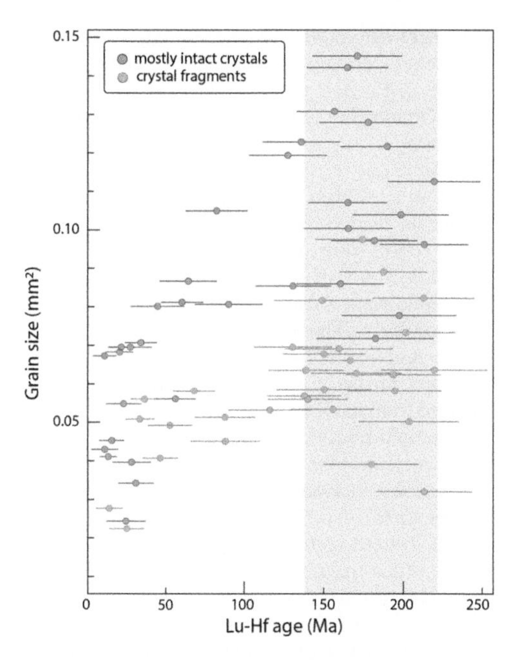

Fig. 12. Apatite Lu–Hf age v. grain size biplot for the Mount St Martin Lu–Hf data. Grain sizes were measured by optical microscopy and classified in two groups: mostly intact grains (blue symbols) and crystal fragments (purple symbols). The blue window highlights apatite grains that have preserved a memory of the crystallization age of the lower crust, while others have undergone varying degrees of Lu–Hf isotopic disturbance.

Discussion

Combined, the four case studies enable an empirical evaluation of the robustness of the Lu–Hf system in apatite, relative to the zircon U–Pb and apatite U–Pb systems. The Taratap Granodiorite records evidence for accommodation of regional post-crystallization solid-state strain, as demonstrated by patchy/undulose extinction in quartz, deformation twinning in plagioclase and minor grain boundary migration in quartz. Additionally, the presence of titanite, allanite cogenetic with apatite, and monazite overgrowing apatite, can explain the depleted LREE trends in the apatite trace element chemistry (e.g. Harlov 2015). Thus, although there is little evidence for alteration or metasomatism in hand-specimen, microscopic observations suggest a secondary event affected the sampled rock, which is likely the source for the high variability of calculated quartz crystallization temperatures. While the majority of the Ti-in-quartz analyses constrain the temperature of magmatic crystallization, the lowest calculated temperature of 433 ± 10°C can be interpreted as a maximum estimate of the temperature associated with a secondary deformation/alteration event. The apatite U–Pb system records a U–Pb age of *c.* 440 Ma, which is significantly younger than the precise zircon crystallization age of 497 ± 1 Ma (Curtis *et al.* 2022). The apatite Lu–Hf age of 497 ± 6 Ma is, however, identical to the zircon U–Pb age, as well as to the monazite U–Pb age of 499.6 ± 6.7 Ma. Hence, while a secondary process has affected the apatite U–Pb system, the Lu–Hf system remained undisturbed, recording the timing of apatite crystallization during magmatic emplacement. The oldest titanite U–Pb population (507 ± 10 Ma) is also consistent with the zircon and monazite U–Pb and apatite Lu–Hf age. However, the younger population, with relatively enriched LREE concentrations, is interpreted to be isotopically disturbed. Although more data are required to confirm the timing of Pb-loss in apatite and titanite, the calculated titanite U–Pb isochron age of 445 ± 12 Ma is consistent with the apatite U–Pb age, possibly suggesting that low-temperature (<440°C) alteration or recrystallization occurred at *c.* 445–440 Ma. Importantly, this case study illustrates that apatite Lu–Hf dating can resolve the primary crystallization age of altered S-type granites, where the apatite (and titanite) U–Pb system records isotopic disturbance that may or may not be geological meaningful in the context of the subsequent cooling history.

The Reynolds–Anmatjira Range samples were sampled from different metamorphic grades within the same study area. The gneiss samples (RR01, RR02, RR05) were taken from a granitic protolith with a zircon U–Pb crystallization age of *c.* 1.81–1.80 Ga, while the unfoliated samples (RR06, RR08) were emplaced at *c.* 1.78 Ga. Although tectonic fabrics are absent in samples RR06 and RR08, there is evidence for minor quartz recrystallization as illustrated by the variability in calculated Ti-in-quartz temperatures. While the calculated *c.* 870–820°C temperatures are interpreted as magmatic crystallization temperatures, the secondary populations of *c.* 475°C (RR08) and *c.* 660°C (RR06) are interpreted as maximum estimates for metamorphic temperatures, with the latter being consistent with the mean quartz crystallization temperature of gneiss sample RR01 (Fig. 2, Supplementary File 7).

The apatite U–Pb ages are reset with respect to the magmatic crystallization ages for all Reynolds–Anmatjira Range samples, regardless of metamorphic grade. The granite samples record apatite U–Pb ages of *c.* 1.59–1.57 Ga, which agree with zircon and monazite U–Pb dates from nearby higher-grade (granulite facies) rocks (Rubatto *et al.* 2001; Morrissey *et al.* 2014). In these granulites, peak metamorphic conditions exceeded 850°C (Anderson *et al.* 2013; Morrissey *et al.* 2014). However, the rocks sampled in this study record maximum metamorphic temperatures ranging from *c.* 475°C (RR08) to *c.* 660°C (RR06 and RR01), as recorded by Ti-in-quartz thermochronometry. The apatite Lu–Hf ages of *c.* 1.75 Ga are within uncertainty of the zircon U–Pb crystallization ages, suggesting the Lu–Hf system preserves primary crystallization ages in non-foliated granites up to *c.* 660°C. The magmatic crystallization age was difficult to constrain due to (1) the large amount of inherited zircons, as is common in S-type granites, as well as (2) extensive Pb-loss (Fig. 7; Supplementary File 10). Hence, the actual emplacement age might be closer to *c.* 1.75 Ga as constrained by the apatite Lu–Hf age, illustrating the power of the apatite Lu–Hf method to date S-type granites.

The gneiss samples record the youngest apatite U–Pb ages for the study area of *c.* 1.53–1.51 Ga (the younger age obtained for RR05 is excluded here, given its large uncertainty). These apatite U–Pb dates are consistent with crystallization ages for pegmatites that overprint the gneissic fabric (*c.* 1.52–15.1 Ga) (Morrissey *et al.* 2014). The apatite Lu–Hf dates for the same sample are significantly older (*c.* 1.60–1.64 Ga) but much younger than the zircon U–Pb crystallization ages. Given that the Lu–Hf system was not reset in undeformed granites of similar metamorphic grade (Lu–Hf age for RR06 is *c.* 1.75 Ga), the 1.60–1.64 Ga dates (from apatites with similar dimensions across all samples from the Reynolds–Anmatjira Range) cannot be associated with volume diffusion resetting. Instead, we interpret these dates as the timing of apatite recrystallization in the gneissic rocks and thus date the development of the gneissic fabric. Our Ti-in-quartz geothermometry results suggest recrystallization temperatures reached *c.* 620–660°C. Comparable monazite ages (*c.* 1.65 Ga) were obtained in higher-grade (upper amphibolite- to granulite-grade) rocks (Anderson *et al.* 2013; Alessio *et al.* 2020), suggesting the timing of apatite recrystallization was coeval with monazite growth in the metasediments.

In summary, the Lu–Hf system appears to preserve primary crystallization ages in granites in greenschist to mid-amphibolite facies (*c.* 400–660°C) metamorphic terranes, but records the timing of recrystallization in foliated rocks under the same metamorphic conditions. The apatite U–Pb system (with a lower nominal closure temperature) may record subsequent age resetting in response to volume diffusion during the thermal waning stages of regional metamorphism.

The Point Sir Isaac Granite provides an opportunity to explore the maximum limit of the robustness of the apatite Lu–Hf system to volume diffusion. In contrast to the Reynold–Anmatjira Range samples, the Point Sir Isaac Granite experienced granulite facies temperatures (730°C, 1 GPa) during a regional-scale tectonic event without recrystallizing (Dutch *et al.* 2008). Given the comparatively low closure temperature of the apatite U–Pb system, the apatite U–Pb age of *c.* 1.7 Ga for the Point Sir Isaac Granite can unsurprisingly be correlated with the timing of metamorphism (*c.* 1.71 Ga SIMS U–Pb titanite age; *c.* 1.74 and 1.69 Ga monazite ages; Dutch *et al.* 2008). The crystallization age of the granite is well defined at *c.* 2.41 Ga (Dutch and Hand 2010). Thus, the obtained apatite Lu–Hf age of *c.* 1.745 ± 0.010 Ga demonstrates that the apatites underwent (nearly complete) Lu–Hf isotopic resetting by volume diffusion at peak metamorphic temperatures of *c.* 730°C. The apatite grains from the Point Sir Isaac granite have similar dimensions to those from the Reynolds Range samples, suggesting isotopic resetting by volume diffusion in long-lived metamorphic systems generally occurs at temperatures between *c.* 660°C and 730°C. Once temperatures of c. 730°C are reached, the apatite Lu–Hf system records complete resetting and the timing of the metamorphic event.

In contrast, for a system in which apatite-bearing rocks experienced shorter-lived thermal perturbations, such as volcanic eruptions, our data from Mt St Martin in central Queensland demonstrate that the Lu–Hf system can retain a memory of primary ages at temperatures in excess of 730°C. Whereas the apatite U–Pb system was completely reset during, or just before, volcanic eruption, recording broadly Cenozoic U–Pb ages, the Lu–Hf system records variable resistance to isotopic resetting. Significantly, there is a correlation between extent of resetting and apatite grain size. Lu–Hf ages of *c.* 220–150 Ma were obtained from >0.08 mm^2 grains, which can be correlated with the waning stages of the New England Orogeny (e.g. Rosenbaum 2018). Hence, the lower crust under the Bowen Basin was either (1) hotter than the apatite U–Pb closure temperature but colder than the Lu–Hf closure temperature at the time of xenolith entrainment in the host basaltic magma; or (2) the apatite may have resided at a temperature below the U–Pb closure temperature in the lower crust and was completely (U–Pb) isotopically reset during its residence time in the host basaltic magma. In either scenario, neither the ambient temperature of the lower crust, nor the residence time in the basaltic magma were sufficient to reset the Lu–Hf system in the larger apatite grains, suggesting the apatite Lu–Hf geochronometer can potentially survive short-lived entrainment in a tholeiite melt (in excess of 1200°C).

Conclusions

The presented case studies advance understanding of the robustness of the apatite Lu–Hf system relative to the more resistive zircon (and monazite) U–Pb and less resistive apatite U–Pb systems. From our observations, a number of conclusions can be drawn.

(1) In unfoliated granitoids, the apatite Lu–Hf system preserves primary crystallization ages at temperatures of at least 660°C. Hence, in cases of extensive zircon inheritance and/or zircon U–Pb loss, typical of (old) S-type granites, the apatite Lu–Hf system can potentially better resolve magmatic crystallization ages compared to the zircon U–Pb system.
(2) In foliated granitoids, apatite tends to recrystallize relatively easily. In this scenario, the apatite Lu–Hf system records the timing of recrystallization, whereas, the apatite U–Pb system often records cooling ages that can be either geologically meaningless (mixing ages) or related to the waning stages of orogenic events.
(3) In this study, at temperatures in excess of 730° C, the Lu–Hf system was completely reset by volume diffusion. Hence, for typical grain sizes (*c.* 0.01–0.03 mm^2), the closure temperature for Hf volume diffusion in apatite is between *c.* 660°C and 730°C, consistent with theoretical calculations (Barfod *et al.* 2005; Chew and Spikings 2015).
(4) For larger apatites (>0.08 mm^2) in granulite xenoliths that were entrained in mafic magma and extruded in young volcanic eruptions, the Lu–Hf system can preserve a memory of the lower-crustal crystallization age. This opens up new opportunities to determine the age of xenoliths that do not contain zircon (such as most mantle and many lower crustal xenoliths).

Acknowledgements Stacey Curtis is thanked for sharing the zircon CA-TIMS U–Pb age and thin section for the Taratap sample. Laura Morrissey is thanked for assistance with finding dateable titanite and tiny monazite in thin section. Chris Mark and Gary ÒSullivan are thanked for their constructive reviews.

Competing interest The authors declare that they have no known competing financial interests or personal relationships that could have appeared to influence the work reported in this paper.

Author contributions **SG**: conceptualization (lead), formal analysis (lead), funding acquisition (lead), investigation (lead), methodology (equal), supervision (lead), writing – original draft (lead), writing – review & editing (lead); **MH**: conceptualization (equal), investigation (supporting), supervision (supporting), validation (equal), writing – review & editing (supporting); **JM**: investigation (supporting), validation (supporting), visualization (lead), writing – review & editing (supporting); **AS**: investigation (supporting), methodology (equal), validation (supporting), writing – review & editing (supporting); **RBE**: formal analysis (supporting), investigation (supporting), validation (supporting), writing – review & editing (supporting); **BK**: conceptualization (supporting), funding acquisition (supporting), investigation (supporting), supervision (supporting), writing – review & editing (supporting); **NF**: formal analysis (supporting), investigation (supporting); **AN**: formal analysis (supporting), investigation (supporting), writing – review & editing (supporting); **SG**: methodology (equal), validation (equal), writing – review & editing (supporting).

Funding This paper was supported by research grant DP200101881 from the Australian Research Council (ARC). SG was also supported by an ARC Future Fellowship (FT210100906).

Data availability All data generated or analysed during this study are included in this published article (and its supplementary information files).

References

Ackerman, L., Kotková, J., Čopjaková, R., Sláma, J., Trubač, J. and Dillingerová, V. 2020. Petrogenesis and Lu–Hf dating of (ultra)mafic rocks from the Kutná Hora Crystalline Complex: implications for the Devonian evolution of the Bohemian Massif. *Journal of Petrology*, **61**, https://doi.org/10.1093/petrology/egaa075

Alessio, K.L., Hand, M., Hasterok, D., Morrissey, L.J., Kelsey, D.E. and Raimondo, T. 2020. Thermal modelling of very long-lived (>140 Myr) high thermal gradient metamorphism as a result of radiogenic heating in the Reynolds Range, central Australia. *Lithos*, **352–353**, 105280, https://doi.org/10.1016/j.lithos.2019.105280

Anderson, J.R., Kelsey, D.E., Hand, M. and Collins, W.J. 2013. Conductively driven, high-thermal gradient metamorphism in the Anmatjira Range, Arunta region, central Australia. *Journal of Metamorphic Geology*, **31**, 1003–1026, https://doi.org/10.1111/jmg.12054

Barfod, G.H., Krogstad, E.J., Frei, R. and Albarède, F. 2005. Lu–Hf and PbSL geochronology of apatites from Proterozoic terranes: A first look at Lu–Hf isotopic closure in metamorphic apatite. *Geochimica et Cosmochimica Acta*, **69**, 1847–1859, https://doi.org/10.1016/j.gca.2004.09.014

Belousova, E.A., Griffin, W.L., O'Reilly, S.Y. and Fisher, N.I. 2002. Apatite as an indicator mineral for mineral exploration: trace-element compositions and their relationship to host rock type. *Journal of Geochemical Exploration*, **76**, 45–69, https://doi.org/10.1016/S0375-6742(02)00204-2

Burtt, A.C. and Abbot, P.J. 1998. The Taratap Granodioritite, South-East South Australia. *MESA Journal*, **10**, 35–39.

Chew, D.M. and Spikings, R.A. 2015. Geochronology and thermochronology using apatite: time and temperature, lower crust to surface. *Elements*, **11**, 189–194, https://doi.org/10.2113/gselements.11.3.189

Chew, D.M. and Spikings, R.A. 2021. Apatite U–Pb thermochronology: a review. *Minerals*, **11**, 1095, https://doi.org/10.3390/min11101095

Chew, D.M., Petrus, J.A. and Kamber, B.S. 2014. U–Pb LA-ICPMS dating using accessory mineral standards with variable common Pb. *Chemical Geology*, **363**, 185–199, https://doi.org/10.1016/j.chemgeo.2013.11.006

Curtis, S., Payne, J., Jagodzinski, E., Preiss, W., Pawley, M., McGee, L. and Morrissey, L.J. 2022. Magmatic evolution during rapid back-arc opening and inversion in a continental subduction margin: an example from the Delamerian Orogeny. *Chemical Geology*, in review.

Dirks, P.H.G.M., Hand, M. and Powell, R. 1991. The P–T–deformation path for a mid-Proterozoic, low-pressure terrane: the Reynolds Range, central Australia. *Journal of Metamorphic Geology*, **9**, 641–661, https://doi.org/10.1111/j.1525-1314.1991.tb00553.x

Dutch, R. and Hand, M. 2010. Retention of Sm–Nd isotopic ages in garnets subjected to high-grade thermal reworking: implications for diffusion rates of major and rare earth elements and the Sm–Nd closure temperature in garnet. *Contributions to Mineralogy and Petrology*, **159**, 93–112, https://doi.org/10.1007/s00410-009-0418-1

Dutch, R., Hand, M. and Kinny, P.D. 2008. High-grade Paleoproterozoic reworking in the southeastern Gawler Craton, South Australia. *Australian Journal of Earth Sciences*, **55**, 1063–1081, https://doi.org/10.1080/08120090802266550

Ehrlich, K., Vers, E., Kirs, J. and Soesoo, A. 2012. Using a titanium-in-quartz geothermometer for crystallization temperature estimation of the Palaeoproterozoic Suursaari quartz porphyry. *Estonian Journal of Earth Sciences*, **61**, 195–204, https://doi.org/10.3176/earth.2012.4.01

Elburg, M.A., Bons, P.D., Foden, J. and Brugger, J. 2003. A newly defined Late Ordovician magmatic-thermal event in the Mt Painter Province, Northern Flinders Ranges, South Australia. *Australian Journal of Earth Sciences*, **50**, 611–631, https://doi.org/10.1046/j.1440-0952.2003.01016.x

Emo, R.B. and Kamber, B.S. 2021. Evidence for highly refractory, heat producing element-depleted lower continental crust: Some implications for the formation and evolution of the continents. *Chemical Geology*, **580**, 120389, https://doi.org/10.1016/j.chemgeo.2021.120389

Foden, J., Elburg, M.A., Dougherty-Page, J. and Burtt, A. 2006. The timing and duration of the Delamerian orogeny: Correlation with the Ross Orogen and implications for Gondwana assembly. *Journal of Geology*, **114**, 189–210, https://doi.org/10.1086/499570

Gehrels, G. 2014. Detrital Zircon U–Pb Geochronology Applied to Tectonics. *Annual Review of Earth and Planetary Sciences*, **42**, 127–149, https://doi.org/10.1146/annurev-earth-050212-124012

Gillespie, J., Glorie, S., Khudoley, A. and Collins, A.S. 2018. Detrital apatite U–Pb and trace element analysis as a provenance tool: Insights from the Yenisey Ridge (Siberia). *Lithos*, **314–315**, 140–155, https://doi.org/10.1016/j.lithos.2018.05.026

Glorie, S., Jepson, G. *et al.* 2019. Thermochronological and geochemical footprints of post-orogenic fluid alteration recorded in apatite: implications for mineralisation in the Uzbek Tian Shan. *Gondwana Research*, **71**, 1–15, https://doi.org/10.1016/j.gr.2019.01.011

Glorie, S., March, S. *et al.* 2020. Apatite U–Pb dating and geochemistry of the Kyrgyz South Tian Shan (Central Asia): Establishing an apatite fingerprint for provenance studies. *Geoscience Frontiers*, **11**, 2003–2015, https://doi.org/10.1016/j.gsf.2020.06.003

Glorie, S., Gillespie, J. *et al.* 2022. Detrital apatite Lu–Hf and U–Pb geochronology applied to the southwestern Siberian margin. *Terra Nova*, **34**, 201–209, https://doi.org/10.1111/ter.12580

Griffin, W.L., Sutherland, F.L. and Hollis, J.D. 1987. Geothermal profile and crust mantle transition beneath east-central Queensland – volcanology, xenolith petrology and seismic data. *Journal of Volcanology and Geothermal Research*, **31**, 177–203, https://doi.org/10.1016/0377-0273(87)90067-9

Hand, M. and Buick, I.S. 2001. Tectonic evolution of the Reynolds-Anmatjira Ranges: A case study in terrain reworking from the Arunta Inlier, central Australia. *Geological Society, London, Special Publications*, **184**, 237–260, https://doi.org/10.1144/GSL.SP.2001.184.01.12

Harlov, D.E. 2015. Apatite: a fingerprint for metasomatic processes. *Elements*, **11**, 171–176, https://doi.org/10.2113/gselements.11.3.171

Henrichs, I.A., Cheve, D.M., O'Sullivan, G.J., Mark, C., McKenna, C. and Guyett, P. 2019. Trace element (Mn–Sr–Y–Th–REE) and U–Pb isotope systematics of metapelitic apatite during progressive greenschist-to amphibolite-facies Barrovian metamorphism. *Geochemistry Geophysics Geosystems*, **20**, 4103–4129, https://doi.org/10.1029/2019gc008359

Herrmann, M., Söderlund, U., Scherstén, A., Næraa, T., Holm-Alwmark, S. and Alwmark, C. 2021. The effect of low-temperature annealing on discordance of U–Pb zircon ages. *Scientific Reports*, **11**, 7079, https://doi.org/10.1038/s41598-021-86449-y

Howlett, D., Raimondo, T. and Hand, M. 2015. Evidence for 1808–1770 Ma bimodal magmatism, sedimentation, high-temperature deformation and metamorphism in the Aileron Province, central Australia. *Australian Journal of Earth Sciences*, **62**, 831–852, https://doi.org/10.1080/08120099.2015.1108364

Jackson, S.E., Pearson, N.J., Griffin, W.L. and Belousova, E.A. 2004. The application of laser ablation-inductively coupled plasma-mass spectrometry to in situ U–Pb zircon geochronology. *Chemical Geology*, **211**, 47–69, https://doi.org/10.1016/j.chemgeo.2004.06.017

JanouŠek, V., Farrow, C.M. and Erban, V. 2006. Interpretation of whole-rock geochemical data in igneous geochemistry: introducing Geochemical Data Toolkit (GCDkit). *Journal of Petrology*, **47**, 1255–1259, https://doi.org/10.1093/petrology/egl013

Jochum, K.P., Weis, U. *et al.* 2011. Determination of reference values for NIST SRM 610–617 glasses following ISO guidelines. *Geostandards and Geoanalytical Research*, **35**, 397–429, https://doi.org/10.1111/j.1751-908X.2011.00120.x

Kirkland, C.L., Yakymchuk, C., Szilas, K., Evans, N., Hollis, J., McDonald, B. and Gardiner, N.J. 2018. Apatite: a U–Pb thermochronometer or geochronometer? *Lithos*,

318, 143–157, https://doi.org/10.1016/j.lithos.2018.08.007

Lee, J.K.W., Williams, I.S. and Ellis, D.J. 1997. Pb, U and Th diffusion in natural zircon. *Nature*, **390**, 159–162, https://doi.org/10.1038/36554

Li, Y. and Vermeesch, P. 2021. Short communication: Inverse isochron regression for Re–Os, K–Ca and other chronometers. *Geochronology*, **3**, 415–420, https://doi.org/10.5194/gchron-3-415-2021

McDonough, W.F. and Sun, S.S. 1995. The composition of the Earth. *Chemical Geology*, **120**, 223–253, https://doi.org/10.1016/0009-2541(94)00140-4

Morrissey, L.J., Hand, M., Raimondo, T. and Kelsey, D.E. 2014. Long-lived high-T, low-P granulite facies metamorphism in the Arunta Region, central Australia. *Journal of Metamorphic Geology*, **32**, 25–47, https://doi.org/10.1111/jmg.12056

Nebel, O., Morel, M.L.A. and Vroon, P.Z. 2009. Isotope dilution determinations of Lu, Hf, Zr, Ta and W, and Hf isotope compositions of NIST SRM 610 and 612 glass wafers. *Geostandards and Geoanalytical Research*, **33**, 487–499, https://doi.org/10.1111/j.1751-908X.2009.00032.x

Norris, A. and Danyushevsky, L. 2018. *Towards Estimating the Complete Uncertainty Budget of Quantified Results Measured By LA-ICP-MS*. Goldschmidt, Boston, USA.

Nutman, A.P. 2007. Apatite recrystallisation during prograde metamorphism, Cooma, southeast Australia: implications for using an apatite–graphite association as a biotracer in ancient metasedimentary rocks. *Australian Journal of Earth Sciences*, **54**, 1023–1032, https://doi.org/10.1080/08120090701488321

Odlum, M.L. and Stockli, D.F. 2020. Geochronologic constraints on deformation and metasomatism along an exhumed mylonitic shear zone using apatite U–Pb, geochemistry, and microtextural analysis. *Earth and Planetary Science Letters*, **538**, 116177, https://doi.org/10.1016/j.epsl.2020.116177

Oosthuyzen, E.J. and Burger, A.J. 1973. The suitability of apatite as an age indicator by the uranium-lead isotope method. *Earth and Planetary Science Letters*, **18**, 29–36, https://doi.org/10.1016/0012-821X(73)90030-7

O'Sullivan, G.J., Chew, D.M. and Samson, S.D. 2016. Detecting magma-poor orogens in the detrital record. *Geology*, **44**, 871–874, https://doi.org/10.1130/G38245.1

O'Sullivan, G., Chew, D., Kenny, G., Henrichs, I. and Mulligan, D. 2020. The trace element composition of apatite and its application to detrital provenance studies. *Earth-Science Reviews*, **201**, 103044, https://doi.org/10.1016/j.earscirev.2019.103044

Paul, A.N., Spikings, R.A. and Gaynor, S.P. 2021. U–Pb ID-TIMS reference ages and initial Pb isotope compositions for Durango and Wilberforce apatites. *Chemical Geology*, **586**, 120604, https://doi.org/10.1016/j.chemgeo.2021.120604

Payne, J.L., Hand, M., Barovich, K.M. and Wade, B.P. 2008. Temporal constraints on the timing of high-grade metamorphism in the northern Gawler Craton: implications for assembly of the Australian Proterozoic. *Australian Journal of Earth Sciences*, **55**, 623–640, https://doi.org/10.1080/08120090801982595

Petrus, J.A. and Kamber, B.S. 2012. VizualAge: a novel approach to laser ablation ICP-MS U–Pb geochronology data reduction. *Geostandards and Geoanalytical Research*, **36**, 247–270, https://doi.org/10.1111/j.1751-908X.2012.00158.x

Pochon, A., Poujol, M., Gloaguen, E., Branquet, Y., Cagnard, F., Gumiaux, C. and Gapais, D. 2016. U–Pb LA-ICP-MS dating of apatite in mafic rocks: Evidence for a major magmatic event at the Devonian-Carboniferous boundary in the Armorican Massif (France). *American Mineralogist*, **101**, 2430–2442, https://doi.org/10.2138/am-2016-5736

Richter, M., Nebel-Jacobsen, Y., Nebel, O., Zack, T., Mertz-Kraus, R., Raveggi, M. and Rösel, D. 2019. Assessment of five monazite reference materials for U–Th/Pb dating using laser-ablation ICP-MS. *Geosciences*, **9**, https://doi.org/10.3390/geosciences9090391

Roberts, N.M.W., Rasbury, E.T., Parrish, R.R., Smith, C.J., Horstwood, M.S.A. and Condon, D.J. 2017. A calcite reference material for LA-ICP-MS U–Pb geochronology. *Geochemistry, Geophysics, Geosystems*, **18**, 2807–2814, https://doi.org/10.1002/2016GC006784

Rosenbaum, G. 2018. The Tasmanides: Phanerozoic tectonic evolution of Eastern Australia. *Annual Review of Earth and Planetary Sciences*, **46**, 291–325, https://doi.org/10.1146/annurev-earth-082517-010146

Rubatto, D., Williams, I.S. and Buick, I.S. 2001. Zircon and monazite response to prograde metamorphism in the Reynolds Range, central Australia. *Contributions to Mineralogy and Petrology*, **140**, 458–468, https://doi.org/10.1007/PI00007673

Schaltegger, U., Schmitt, A.K. and Horstwood, M.S.A. 2015. U–Th–Pb zircon geochronology by ID-TIMS, SIMS, and laser ablation ICP-MS: Recipes, interpretations, and opportunities. *Chemical Geology*, **402**, 89–110, https://doi.org/10.1016/j.chemgeo.2015.02.028

Schoene, B. and Bowring, S.A. 2006. U–Pb systematics of the McClure Mountain syenite: thermochronological constraints on the age of the 40Ar/39Ar standard MMhb. *Contributions to Mineralogy and Petrology*, **151**, 615, https://doi.org/10.1007/s00410-006-0077-4

Siegel, C., Bryan, S.E., Allen, C.M., Gust, D.A. and Purdy, D.J. 2020. Crustal evolution in the New England Orogen, Australia: repeated igneous activity and scale of magmatism govern the composition and isotopic character of the continental crust. *Journal of Petrology*, **61**, https://doi.org/10.1093/petrology/egaa078

Simpson, A., Gilbert, S. *et al.* 2021. In-situ LuHf geochronology of garnet, apatite and xenotime by LA ICP MS/MS. *Chemical Geology*, **577**, 120299, https://doi.org/10.1016/j.chemgeo.2021.120299

Simpson, A., Glorie, S., Hand, M., Spandler, C., Gilbert, S. and Cave, B. 2022. In situ Lu–Hf geochronology of calcite. *Geochronology*, **4**, 353–372, https://doi.org/10.5194/gchron-4-353-2022

Sláma, J., Košler, J. *et al.* 2008. Plešovice zircon – A new natural reference material for U–Pb and Hf isotopic microanalysis. *Chemical Geology*, **249**, 1–35, https://doi.org/10.1016/j.chemgeo.2007.11.005

Spandler, C., Hammerli, J., Sha, P., Hilbert-Wolf, H., Hu, Y., Roberts, E. and Schmitz, M. 2016. MKED1: A new titanite standard for in situ analysis of Sm–Nd isotopes and U–Pb geochronology. *Chemical Geology*,

425, 110–126, https://doi.org/10.1016/j.chemgeo.2016.01.002

Spencer, C.J., Kirkland, C.L., Roberts, N.M.W., Evans, N.J. and Liebmann, J. 2020. Strategies towards robust interpretations of in situ zircon Lu–Hf isotope analyses. *Geoscience Frontiers*, **11**, 843–853, https://doi.org/10.1016/j.gsf.2019.09.004

Sutherland, F.L., Stubbs, D. and Green, D.C. 1977. K–Ar ages of Cainozoic volcanic suites, Bowen-St Lawrence Hinterland, North Queensland (with some implications for petrologic models). *Journal of the Geological Society of Australia*, **24**, 447–460, https://doi.org/10.1080/00167617708729004

Thompson, J., Meffre, S. *et al.* 2016. Matrix effects in Pb/U measurements during LA-ICP-MS analysis of the mineral apatite. *Journal of Analytical Atomic Spectrometry*, **31**, 1206–1215, https://doi.org/10.1039/C6JA00048G

Thomson, S.N., Gehrels, G.E., Ruiz, J. and Buchwaldt, R. 2012. Routine low-damage apatite U–Pb dating using laser ablation–multicollector–ICPMS. *Geochemistry, Geophysics, Geosystems*, **13**, https://doi.org/10.1029/2011GC003928

Vermeesch, P. 2018. IsoplotR: A free and open toolbox for geochronology. *Geoscience Frontiers*, **9**, 1479–1493, https://doi.org/10.1016/j.gsf.2018.04.001

Wark, D.A. and Watson, E.B. 2006. TitaniQ: a titanium-in-quartz geothermometer. *Contributions to Mineralogy and Petrology*, **152**, 743–754, https://doi.org/10.1007/s00410-006-0132-1

Monazite–xenotime thermometry: a review of best practices and an example from the Caledonides of northern Scotland

Calvin A. Mako[1,2]*, Mark J. Caddick[1], Richard D. Law[1] and J. Ryan Thigpen[3]

[1]Department of Geosciences, Virginia Tech, Blacksburg, VA 24061, USA

[2]Arizona Geological Survey, Tucson, AZ 85721, USA

[3]Department of Earth and Environmental Sciences, University of Kentucky, Lexington, KY 40506, USA

CAM, 0000-0002-0969-3080; MJC, 0000-0001-8795-8438;
RDL, 0000-0001-6256-1944; JRT, 0000-0002-3075-5178
*Correspondence: cmako@arizona.edu

Abstract: Monazite–xenotime thermometry is a potentially powerful technique for understanding the evolution of Earth systems. While a rich set of experimental and empirical datasets are available for monazite–xenotime equilibria, five different thermometric calibrations yield significantly different results, making this technique difficult to apply in practice. To clarify best practices for monazite–xenotime thermometry, a compilation of published compositional data for monazite and xenotime with independently determined pressure–temperature conditions is evaluated. For each existing thermometer, we examine how closely estimated temperatures match independent empirical temperatures and consider how best to calculate monazite end-members for each thermometer. Monazite–xenotime thermometry is applied to samples from the Northern Highlands Terrane of northern Scotland, which experienced amphibolite–upper greenschist facies metamorphism and penetrative deformation during the Scandian orogeny. Thermometry data in conjunction with U–Pb dating define relatively slow regional cooling across the Scandian thrust nappes. Thermometry data closely match quartz *c*-axis fabric-based deformation thermometry across the structurally lower nappes, suggesting that monazite and xenotime record the timing and temperature of penetrative deformation and shearing. The data suggest that ductile deformation in the hinterland nappes of the Scandian orogen in Scotland occurred as late as 415–410 Ma.

Supplementary material: Compilations of published compositional and pressure-temperature data for monazite and xenotime, and U-Pb-trace element data from this study are available at https://doi.org/10.6084/m9.figshare.c.6724396

Monazite is commonly used to unravel the metamorphic, igneous and even sedimentary record of the Earth's evolution. This rare earth phosphate mineral ([REE]PO_4) participates in metamorphic reactions (Kohn and Malloy 2004; Dumond *et al.* 2015), is sensitive to fluid alteration (Hetherington *et al.* 2010; Harlov *et al.* 2011; Grand'Homme *et al.* 2016; Mako *et al.* 2021), crystallizes from melts (Montel 1993; Kelsey *et al.* 2008) and frequently endures erosion and transport as detrital grains (Hietpas *et al.* 2010). In the material sciences, monazite is of interest for storing radioactive waste and for other applications (McCarthy *et al.* 1978; Ewing and Wang 2002; Mogilevsky 2007; Dacheux *et al.* 2013). Monazite's compositional and textural characteristics can provide a detailed record of various geological processes and, perhaps most importantly, allow absolute time constraints to be added to metamorphic pressure–temperature paths (Parrish 1990; Spear and Parrish 1996; Foster *et al.* 2004; Mottram *et al.* 2014, 2015; Laurent *et al.* 2018). Linking time constraints to monazite dates typically relies on semiquantitative textural relationships and observations of monazite's rare earth element (REE) contents in relation to coexisting metamorphic phases such as garnet. However, quantitative estimates of temperature can be made using the composition of monazite in equilibrium with xenotime, another rare earth phosphate ([Y,REE]PO_4) (Hetherington *et al.* 2008). Indeed, multiple monazite–xenotime thermometers have been calibrated using both empirical and experimental data (Gratz and Heinrich 1997, 1998; Heinrich *et al.* 1997; Pyle *et al.* 2001; Seydoux-Guillaume *et al.* 2002).

This contribution aims to clarify the usage of monazite–xenotime thermometry and recommend

From: van Schijndel, V., Cutts, K., Pereira, I., Guitreau, M., Volante, S. and Tedeschi, M. (eds) 2024. *Minor Minerals, Major Implications: Using Key Mineral Phases to Unravel the Formation and Evolution of Earth's Crust.*
Geological Society, London, Special Publications, **537**, 185–208.
First published online August 23, 2023, https://doi.org/10.1144/SP537-2022-246

best practices for successful analysis. The basis for quantifying monazite–xenotime equilibria is straightforward and well established (Franz *et al.* 1996; Gratz and Heinrich 1997, 1998; Heinrich *et al.* 1997; Seydoux-Guillaume *et al.* 2002), but the technique is under-utilized compared with how frequently monazite petrochronology is applied to metamorphic systems. This is likely due to various factors, including that: (1) there are five calibrated monazite–xenotime thermometers with differing thermodynamic foundations that can give greatly conflicting results, and (2) it is often unclear what inputs are most appropriate for each of these thermometers. Despite this, monazite–xenotime thermometry provides the potential to link temperature conditions to geochronological constraints quantitatively. Here, we provide a detailed review of each thermometer and evaluate their effectiveness based on natural and empirical datasets for which independently determined pressures and temperatures are also available. Our analysis should allow monazite-based thermometry to be used more confidently and clarify where future work is necessary to improve its predictive power further. To highlight the utility of monazite–xenotime thermometry, we apply it to a series of samples from the Northern Highlands Terrane (NHT) in Scotland, where our results address the timing and rates of metamorphism, penetrative deformation and exhumation.

Background

Monazite–xenotime equilibria

Monazite and xenotime frequently coexist in metamorphic rocks. At equilibrium, there is a miscibility gap in Y + REE phosphate composition space (Fig. 1), with the monoclinic monazite end-member occupying the light rare earth element (LREE)-rich limb and tetragonal xenotime occupying the yttrium + heavy rare earth element (Y + HREE)-rich limb (Heinrich *et al.* 1997). The monazite and xenotime components are thus defined as $[LREE]PO_4$ and $[Y, HREE]PO_4$, respectively (Heinrich *et al.* 1997; Pyle *et al.* 2001). Note that elements La–Eu are generally regarded as light rare earths and Gd–Lu as heavy rare earths. While Y is not an REE, its behaviour is very similar to an HREE, so it is often included with that group. In simpler experimental systems, the monazite component has been approximated as $CePO_4$ and xenotime as YPO_4 (Gratz and Heinrich 1997, 1998; Heinrich *et al.* 1997). The monazite–xenotime miscibility gap narrows with increasing temperature, with an increased xenotime component in monazite, and vice versa. Additionally, both monazite and xenotime exhibit solid solutions with cheralite ($Ca[Th,U](PO_4)_2$) and thorite ($[Th,U]SiO_4$) (e.g. Engi 2017). These U- and Th-bearing components are typically less than 5–20% in most natural monazites and xenotimes but can substantially affect monazite–xenotime equilibrium compositions at higher amounts (Seydoux-Guillaume *et al.* 2002).

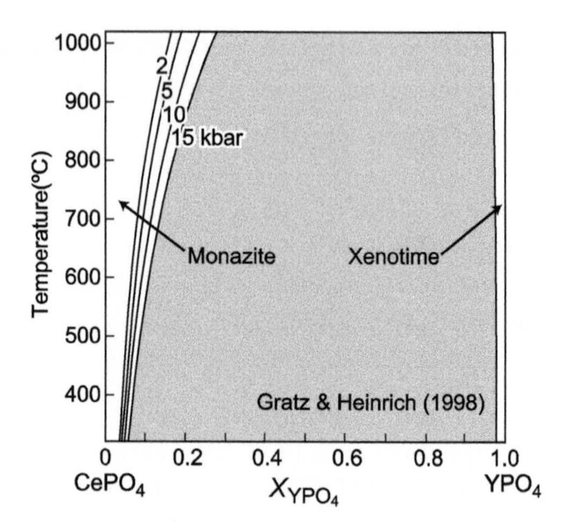

Fig. 1. Temperature dependence of the monazite–xenotime miscibility gap in the $CePO_4$–YPO_4 system modified after Gratz and Heinrich (1997). Source: reproduced by permission of E. Schweizerbart Science Publishers, http://www.schweizerbart.de/journals/ejm.

Monazite thermometry carries similar potential benefits as other systems allowing direct temperature–time constraints from single laser ablation or microprobe analysis on a mineral that is also dateable using U–Th–Pb systematics, with examples including Zr-in-titanite (Hayden *et al.* 2008), Ti-in-zircon (Watson *et al.* 2006) and Zr-in-rutile (Watson *et al.* 2006). Such trace element or 'single mineral' thermometers rely on the presence of a buffering phase to saturate the target mineral in the element of interest, usually at tens to hundreds of ppm. In contrast, the xenotime content of monazite is much higher than the trace element level, generally greater than 4–5 wt% in metamorphic rocks. Indeed, the equilibrium between monazite and xenotime yields a solvus thermometer (Philpotts and Ague 2022) that geometrically resembles muscovite–paragonite thermometry (Guidotti *et al.* 1994) or calcite–dolomite thermometry (Anovitz and Essene 1987). Thus, 'monazite–xenotime thermometry' might be considered a more fitting term than 'Y-in-monazite thermometry', which has occasionally been used in the literature.

Previous experimental and empirical studies

A wealth of experimental and natural data provides the underpinnings for monazite–xenotime thermometry (Fig. 2; Table 1). Franz *et al.* (1996) quantified monazite and xenotime compositions in pelitic

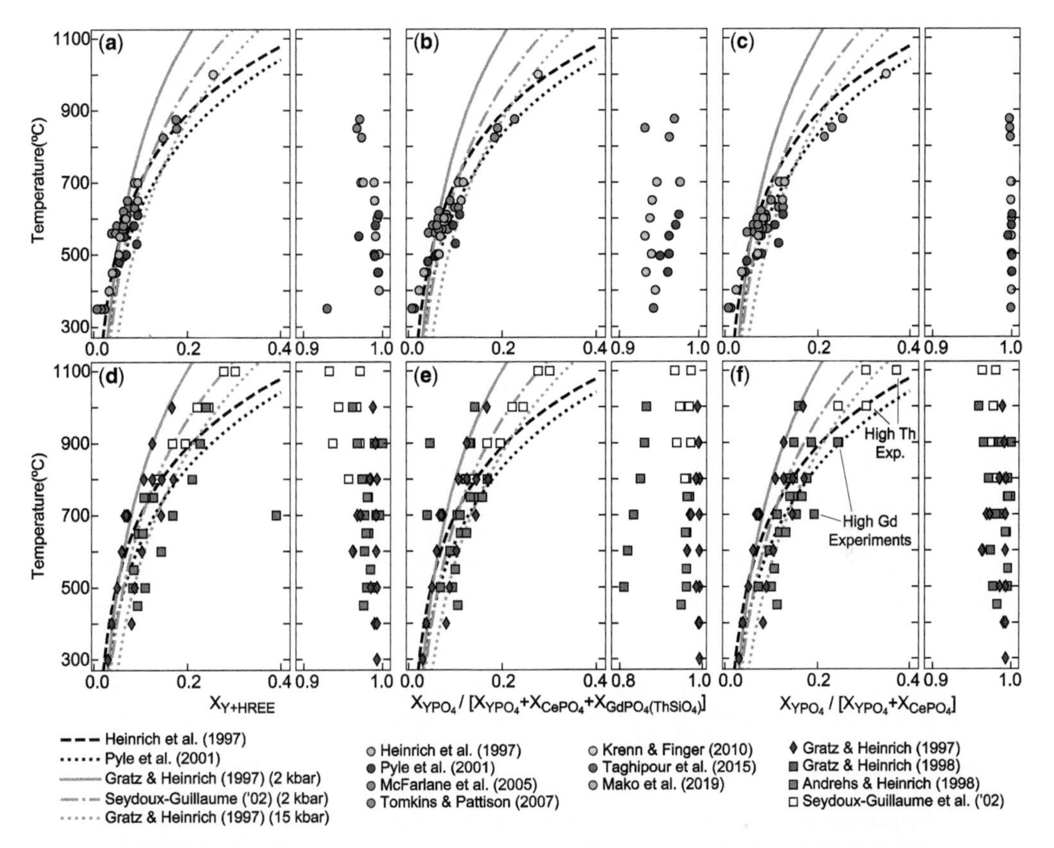

Fig. 2. Compilation of available monazite and xenotime compositional data from the literature and calibrated thermometers. (**a–c**) show natural monazite and xenotime data with the mole fraction of xenotime calculated by different methods. (**d–f**) show experimental monazite and xenotime data with the mole fraction of xenotime calculated by different methods (see text for further explanation). Note that each method of calculating the mole fraction of xenotime leads to markedly different calculated temperatures for each thermometer.

rocks ranging from greenschist to granulite facies, providing the basis for quantitative analysis of the miscibility gap and element partitioning (Heinrich *et al.* 1997). Heinrich *et al.* (1997) did not provide an equation expressing the temperature dependence of this partitioning. Subsequent experimental studies examined monazite and xenotime compositions in the simpler $CePO_4$–YPO_4 (Gratz and Heinrich 1997) and $CePO_4$–YPO_4–$GdPO_4$ (Gratz and Heinrich 1998) systems, as well as a more comprehensive REE-PO_4 system (Andrehs and Heinrich 1998). These studies quantified the crystallographic parameters of monazite and xenotime and the miscibility gap's temperature–pressure dependence, thus providing quantitative thermobarometric relationships. Additional experimental work by Seydoux-Guillaume *et al.* (2002) examined the effects of adding $ThSiO_4$ to $CePO_4$–YPO_4, which produced three-phase monazite–xenotime–thorite assemblages in high $ThSiO_4$ bulk compositions and altered the monazite–xenotime partitioning of Y and Ce. Seydoux-Guillaume *et al.* (2002) calibrated a thermometer that accounts for high $ThSiO_4$ contents in monazite. Finally, Pyle *et al.* (2001) examined a large dataset of natural monazite and xenotime compositions, for which pressure and temperature were independently constrained. They derived two monazite–xenotime thermometric relationships, one based on the data of Heinrich *et al.* (1997) – hereafter, 'the Heinrich *et al.* (1997) thermometer' – and one based on their own data – hereafter, 'the Pyle *et al.* (2001) thermometer'. Each of these existing thermometers is calibrated based on expressions for the composition of monazite rather than xenotime because there is far less variability in the measured xenotime composition. Still, it is important to clarify that they each implicitly rely on the buffering presence of xenotime in equilibrium with the monazite.

It has previously been noted that each of these thermometer calibrations, compared in Figure 2,

Table 1. *Sources of published monazite and xenotime data that are used in this study*

Publication	Compositional system	Independent temperature	Temperature uncertainty	Thermometer calculated
Experimental data				
Gratz and Heinrich (1997)	Ce, Y	Experimental	3°C	Yes
Gratz and Heinrich (1998)	Ce, Gd, Y	Experimental	3°C	Yes
Andrehs and Heinrich (1998)	Y, La to Lu [-Eu, Ho, Tm]	Experimental	3°C	
Seydoux-Guillaume *et al.* (2002)	Ce, Th, Y	Experimental	3°C	Yes
Empirical data				
Heinrich *et al.* (1997)	Natural	Metamorphic field gradient	50°C	
Pyle *et al.* (2001)	Natural	Garnet–biotite thermometry	25°C	Yes
McFarlane *et al.* (2005)	Natural	Garnet–orthopyroxene thermometry	25°C	
Tomkins and Pattison (2007)	Natural	Silicate phase equilibria	25°C	
Krenn and Finger (2010)	Natural	Ternary fsp thermometry, garnet–cpx thermometry	50°C	
Taghipour *et al.* (2015)	Natural	Fluid inclusion thermometry	50°C	
Mako *et al.* (2019)	Natural	Qtz–Fsp microstructures	50°C	

can yield slightly to substantially conflicting temperature estimates (Pyle *et al.* 2001; Daniel and Pyle 2006; Tomkins and Pattison 2007; Mako *et al.* 2019). Indeed, the Heinrich *et al.* (1997) and Pyle *et al.* (2001) thermometers yield an *c.* 80°C difference in temperature estimate for a given monazite composition. A further challenge is discerning how the input monazite end-member composition should be calculated when using a thermometer derived from a simple system ($CePO_4$–YPO_4) to evaluate a complex natural sample ([Ca,Th,U,Y,REE][P,Si]O_4). In this contribution, we attempt to clarify this question and the relationship between simple and complex monazite–xenotime systems.

The second goal of this work is to verify in a practical way how well each thermometer can reproduce estimates of independently determined metamorphic temperatures. To accomplish this, we use data from several studies that provide the compositions of both natural monazite and xenotime (Heinrich *et al.* 1997; Pyle *et al.* 2001; McFarlane *et al.* 2005; Taghipour *et al.* 2015; Mako *et al.* 2019), as well as studies that provide only monazite compositional data but note the presence of xenotime in the equilibrium metamorphic assemblage (Tomkins and Pattison 2007; Krenn and Finger 2010) (Table 1). Critically, each of these studies also provides independent temperature constraints that do not utilize monazite composition. Several previous studies have compared and commented on the effectiveness of each available monazite–xenotime thermometer, with some concluding that the Heinrich *et al.* (1997) calibration produces the 'best' or most appropriate results (McFarlane *et al.* 2005; Tomkins and Pattison 2007; Mako *et al.* 2019, 2021), while other studies favour the Pyle *et al.* (2001) calibration (Daniel and Pyle 2006; Krenn and Finger 2010; Betkowski *et al.* 2017) or that of Seydoux-Guillaume *et al.* (2002) (Laurent *et al.* 2018). The independent temperatures used for calibration of the Heinrich *et al.* (1997) thermometer were not based on quantitative petrology, but rather are assumed based on the metamorphic field gradient in their study area. Therefore, we provide additional supporting analysis of the accuracy of this thermometer.

Geological background: northern Scotland

Following a comparison of the various monazite–xenotime thermometers, we show how their application can aid understanding of the metamorphic history of the northern part of the NHT of Scotland, which experienced multiple phases of Neoproterozoic and Caledonian (Ordovician–Silurian) metamorphism and deformation, culminating in the dominantly preserved record of Scandian orogenesis at 435–415 Ma (Dallmeyer *et al.* 2001; Strachan *et al.* 2002, 2020; Cutts *et al.* 2010; Bird *et al.* 2013; Thigpen *et al.* 2013; Ashley *et al.* 2015; Mako *et al.* 2019) (Fig. 3). We focus on samples collected to the north and NE of the Assynt area, where a series of ductile thrust nappes are exposed. From structurally highest to lowest, these includes the Skinsdale, Naver, Ben Hope and Moine nappes. A Scandian metamorphic field gradient is preserved across this nappe stack, ranging from upper

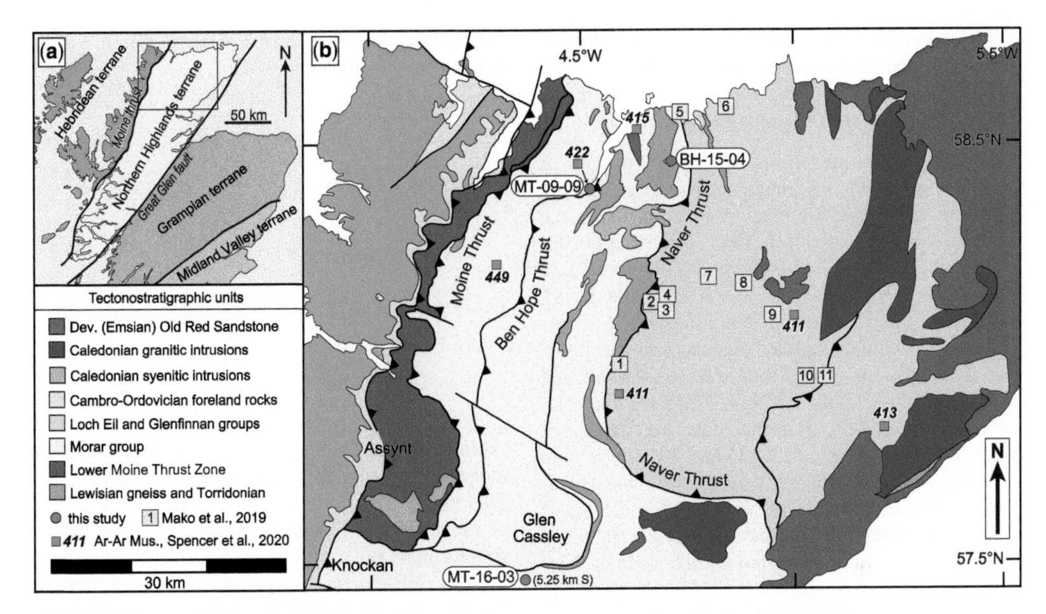

Fig. 3. Geological maps of northern Scotland. (**a**) Terrane map of Scotland modified after Cawood *et al.* (2015). (**b**) Geological map of the northern part of the NHT. Sample locations from this study, Mako (2019), Mako *et al.* (2019) and Spencer *et al.* (2020) are shown. Note that sample MT-16-03 was collected 5.25 km to the south of position shown on the map. Ar–Ar muscovite ages for the samples of Spencer *et al.* (2020) are indicated.

amphibolite facies in the hinterland-positioned Naver nappe to greenschist facies in the foreland-positioned western edge of the Moine nappe (Winchester 1974; Barr *et al.* 1986; Thigpen *et al.* 2013; Ashley *et al.* 2015; Mako *et al.* 2019). The Moine Thrust Zone developed under sub-greenschist facies conditions and grain-scale deformation is dominantly brittle (Knipe 1990; Goodenough *et al.* 2011), with local mylonitization at structurally higher levels (Law *et al.* 1986; see review in Law and Johnson 2010). Latest peak metamorphism in the Naver nappe occurred at 425 Ma, reaching temperatures of *c.* 700°C (Mako *et al.* 2019), but this was likely preceded by earlier high-temperature metamorphism at *c.* 445 Ma and *c.* 470 Ma (Kinny *et al.* 1999; Bird *et al.* 2013). Scandian orogenesis was followed by rapid cooling and orogenic collapse (Spencer *et al.* 2020, 2021). By 407–403 Ma, Emsian age sediments were deposited directly on high-grade Scandian metamorphic rocks (Wellman 2015). The orogenic wedge exposed in the NHT provides an excellent natural laboratory for investigating the relationship between metamorphism and deformation in orogenic systems. Diffusion speedometry studies in the Moine nappe yield extremely fast rates of metamorphic heating (Ashley *et al.* 2015), which are potentially related to the rapid emplacement of hot ductile thrust nappes (Thigpen *et al.* 2017, 2021). Here, we aim to directly constrain temperature–time points that can be used to understand and model the tectono-metamorphic evolution of the orogenic wedge.

Methods

Compilation of existing monazite–xenotime data

To evaluate the effectiveness of various monazite–xenotime thermometers, compositional data were compiled from original published sources. End-member compositions were recalculated according to the methods outlined below, with complete sets of monazite composition, xenotime composition and pressure–temperature data presented in Table 1 in the supplementary material.

Electron beam methods

Scanning electron microscopy (SEM) and electron probe microanalysis (EPMA) were used to identify and locate monazite, xenotime and other metamorphic mineral phases in the NHT samples. X-ray maps of monazite and xenotime were made using EPMA to identify compositional zoning and guide laser ablation analyses. A 100–300 nA current and 15 kV accelerating voltage were used for X-ray imaging of monazite and xenotime. A simultaneous image processing strategy was employed to standardize X-ray intensities to colour in grain images.

EPMA analyses were conducted at both the University of Massachusetts and Virginia Tech.

Laser ablation split stream analysis

Trace element compositions and U–Pb isotopes of monazite and xenotime were measured simultaneously on the same volumes of material using the laser ablation split stream (LASS) system at the University of California, Santa Barbara (Kylander-Clark *et al.* 2013; McKinney *et al.* 2015). The LASS employs a Nu Instruments Nu Plasma HR multicollector inductively coupled plasma mass spectrometer (ICP-MS) for U–Pb isotope measurements and an Agilent 7700S quadrupole ICP-MS for trace element measurements. The raw data were analysed using Iolite (Paton *et al.* 2011) and Isoplot version 4.1 (Ludwig 2008). Internal uncertainty was calculated as the additional error required for the set of standard analyses to be considered a single statistical population and was propagated to each analysis. Natural monazite standard 44069 (Aleinikoff *et al.* 2006) was used as the primary U–Pb standard to correct for instrument drift, and natural monazite Bananeira (Kylander-Clark *et al.* 2013) was used as the primary trace element standard for monazite and xenotime. An additional 1% external uncertainty was added to final age calculations to account for the reproducibility of the secondary isotope standard (Bananeira), which was typically within 1% of the accepted value. A xenotime reference material was not used because sufficiently accurate results can be obtained using a monazite reference material, as demonstrated by the successful use by McKinney *et al.* (2015) of 44069 monazite as a standard for xenotime. Xenotime likely does not have significantly different laser ablation matrix effects than monazite. All data quoted below include the full external uncertainty. U–Pb trace element data used in this study are available in Table 2 in the supplementary material.

Results

Calculating monazite and xenotime end-member compositions

Careful consideration of input parameters is required before the appropriate application of all geothermometers. The Pyle *et al.* (2001) and Heinrich *et al.* (1997) monazite–xenotime thermometer calibrations each have the form $T\,(°C) = A\ln[X] + B$, where B is a fitted constant and X refers to the mole fraction of Y + HREE in monazite (X_{Y+HREE}), calculated using the following expressions (Pyle *et al.* 2001).

$$X_{A_iPO_4} = A_i/D \quad (1)$$

where A_i = Y, La, Ce, Pr, Nd, Sm, Gd, Dy, Ho, Er, Yb in moles. The mole fractions of the cheralite (formerly 'brabantite') and thorite (formerly 'huttonite') components are calculated as

$$X_{cher} = 2Ca\ /\ D \quad (2)$$

$$X_{thor} = (Th + U + Pb - Ca)\ /\ D. \quad (3)$$

In each of the above equations, D is

$$D = [Y + La + Ce + Pr + Nd + Sm + Gd + Dy + Ho + Er + Yb + (2Ca) + (Th + U + Pb - Ca)]. \quad (4)$$

Note that some HREE (Tb, Tm and Lu) are excluded here, likely because their abundances in monazite are typically below detection by EPMA. This makes a negligible difference in the calculation of monazite end-member fractions but does have a small impact on xenotime end-member calculations because HREEs are more abundant in xenotime. Finally, the monazite and xenotime components are formulated as follows:

$$X_{LREE} = X_{LaPO_4} + X_{CePO_4} + X_{PrPO_4} + X_{NdPO_4} + X_{SmPO_4} \quad (5)$$

$$X_{Y+HREE} = X_{YPO_4} + X_{GdPO_4} + X_{DyPO_4} + X_{HoPO_4} + X_{ErPO_4} + X_{YbPO_4}. \quad (6)$$

Any usage of the Pyle *et al.* (2001) or Heinrich *et al.* (1997) thermometers should closely follow this set of expressions to avoid introducing additional error to the calculated temperature. The mole fraction, X_{Y+HREE}, as given in equation (6), is the direct input for both the Pyle *et al.* (2001) and Heinrich *et al.* (1997) thermometers (see fig. 8b of Pyle *et al.* 2001).

Unlike the calibrations based on natural sample datasets, monazite–xenotime thermometer calibrations based on experimental work generally utilize simpler compositional systems. When applying experimentally derived thermometers, it might be expected that the calculation of the monazite and xenotime mole fractions should reflect the compositional range of the experimental system. For example, Gratz and Heinrich (1997) used an experimental system that included only $CePO_4$ and YPO_4, so the mole fraction of xenotime in monazite should be calculated as $YPO_4/CePO_4 + YPO_4$ for application in this thermometer. It is currently unclear, however, whether it is more appropriate to use $YPO_4/CePO_4 + YPO_4$ or X_{Y+HREE}, as formulated in equations (1)–(6), to calculate the temperature of natural monazite crystals. Both formulations represent the xenotime component in monazite, and previous studies have been inconsistent in which approach is used. The importance of correctly calculating the xenotime content of monazite is illustrated in Figure 2, where the monazite component is expressed differently for the same empirical and experimental datasets. Widely varying temperature estimates would result for each case.

Monazite–xenotime element partitioning

A compilation of five natural and four experimental datasets suggests that the partitioning of all REEs between monazite and xenotime is at least weakly temperature-dependent (Fig. 4), as expected (Andrehs and Heinrich 1998). For the natural datasets, the partitioning of Y and Dy are tightly correlated with temperature, while there is significant scatter for most other elements. LREE contents are negatively correlated with temperature, while Y and HREE are positively correlated with temperature. The partitioning behaviour of experimental monazite and xenotime (Fig. 4e, f) is broadly similar to natural

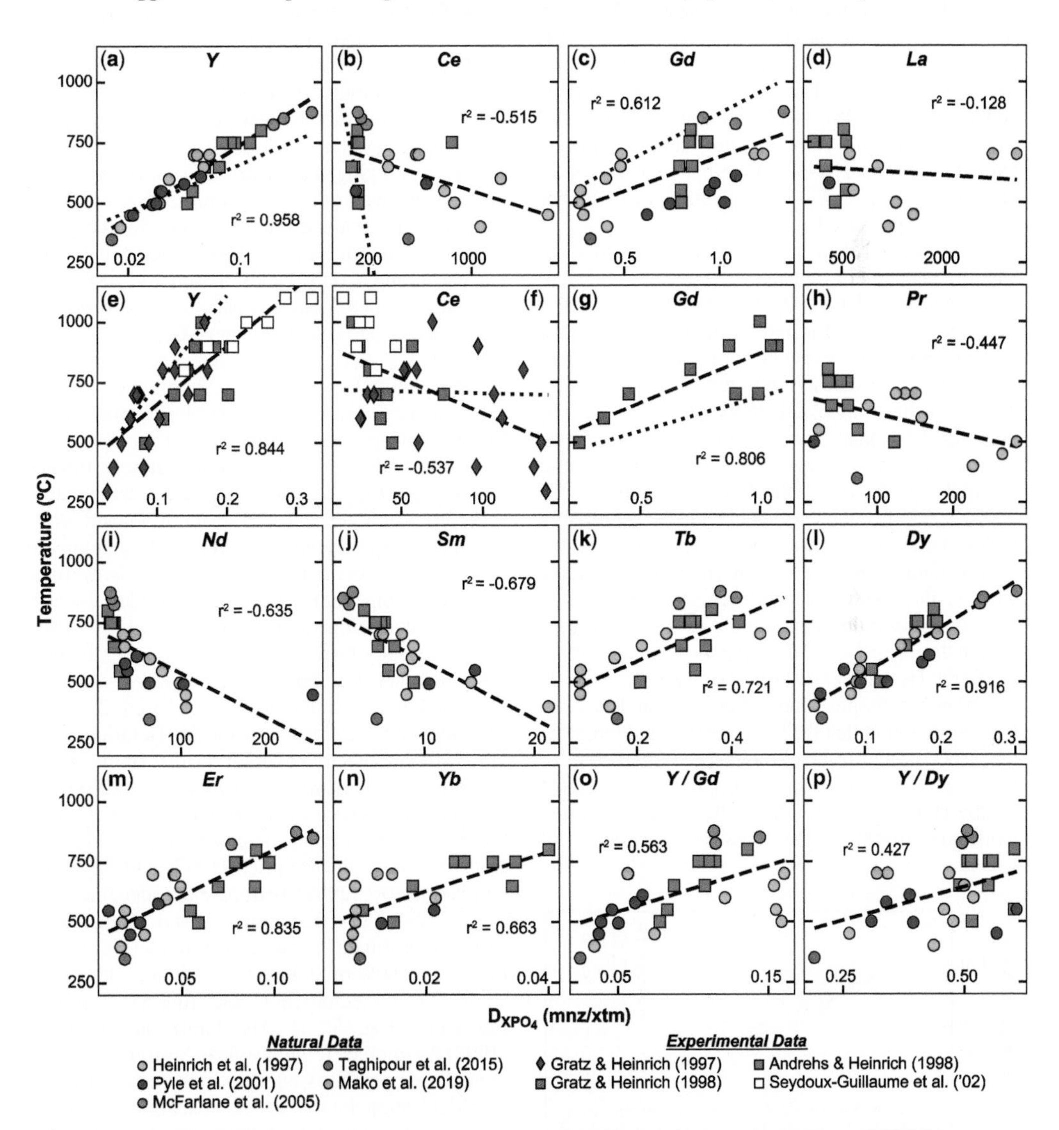

Fig. 4. Temperature dependence of element partitioning between monazite and xenotime. XPO_4 is calculated according to equation (1) for each element. Experimental and natural data are treated separately, except for the data from Andrehs and Heinrich (1998), which are included with the natural data. The best-fitting lines through each dataset are simple linear regressions and do not account for uncertainties. In (**a–c**), the dashed line is a regression of the data shown and the dotted line is a regression of the experimental data for the same elements shown in (**e–g**). In (**e–g**), the dotted line is a regression of the natural data shown in (**a–c**).

samples, with the tightest correlation for Y. Data for compositionally simple systems (Y–Ce ± Gd ± Th) show that adding a third cation can significantly affect Y and Ce partitioning (Gratz and Heinrich 1998; Seydoux-Guillaume *et al.* 2002). Notably, the experimental data of Andrehs and Heinrich (1998), which utilized a more complete range of REEs rather than a simple system, are very similar to natural monazite–xenotime data. Therefore, we include the Andrehs and Heinrich (1998) data in our analysis of natural monazite datasets. Figure 5 shows the distribution coefficients plotted as REE profiles, coloured by temperature, for all natural monazite–xenotime datasets and that of Andrehs and Heinrich (1998). While substantial scatter and overlap exists between various samples, lower-temperature monazite–xenotime pairs (blue) generally have steeper K_D-REE profiles than higher temperature pairs (red).

Experimental data have clearly shown that the xenotime content of monazite is pressure-dependent (Gratz and Heinrich 1997) (Fig. 2). Although this is expected to be true for natural monazite as well, it is challenging to evaluate pressure dependence using the natural datasets in Figures 2 and 4 because nearly all the samples are reported to have equilibrated at 2.5–6 kbar. Based on the published experimental data, which cover 2–15 kbar (Gratz and Heinrich 1997, 1998; Seydoux-Guillaume *et al.* 2002), a significant pressure dependence is probably undetectable over the narrow pressure range of the natural data. Thus, we cannot rigorously evaluate whether the pressure dependence of the monazite–xenotime miscibility gap is more or less significant than in the experimental data. It may be inappropriate to use the Heinrich *et al.* (1997) or Pyle *et al.* (2001) thermometers for pressures <2 or >7 kbar; however, it is worth noting that the high pressure and temperature (1000°C and 16 kbar) monazite compositions of Krenn and Finger (2010) are close to those predicted by the Heinrich *et al.* (1997) thermometer. Future work that examines the pressure dependence of compositionally complex monazite and xenotime would likely be valuable.

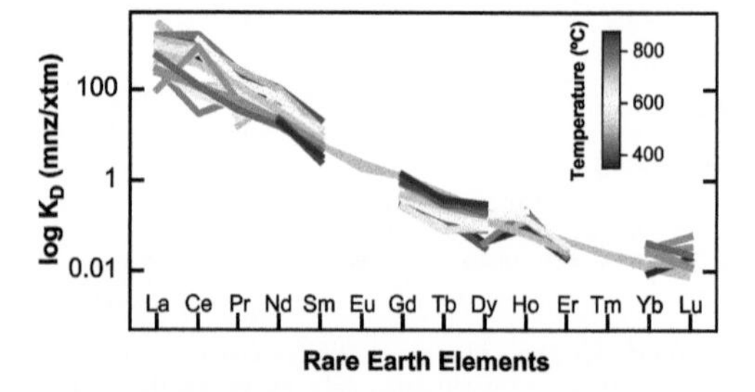

Fig. 5. Monazite–xenotime distribution coefficients for REEs in natural samples compiled from the literature. Profiles are coloured according to the independently determined temperature for each sample.

Evaluation of existing thermometers

We use our compilation of monazite and xenotime compositions with independently determined temperature constraints (Table 1) to evaluate the five published monazite–xenotime thermometers (Fig. 6). X_{Y+HREE} (equation 6) is used to calculate temperatures for the Heinrich *et al.* (1997), Pyle *et al.* (2001), Gratz and Heinrich (1997) and Seydoux-Guillaume *et al.* (2002) thermometers. Although the Gratz and Heinrich (1997) and Seydoux-Guillaume *et al.* (2002) thermometers were calibrated in binary or ternary systems ($X_{CePO4}/X_{CePO4} + X_{YPO4} \pm X_{ThSiO4}$), utilizing an identical formulation yields poor results, with only a few samples plotting within ±50°C of the empirical temperature. Similarly, the Gratz and Heinrich (1998) thermometer, based on Gd partitioning, only yields reasonable results for our dataset when X_{GdPO4} is calculated according to equations (1)–(5), rather than $X_{GdPO4}/X_{CePO4} + X_{YPO4} + X_{GdPO4}$ as in the original experiments.

The Heinrich *et al.* (1997) thermometer yields temperature estimates that most closely correspond to the independently determined temperatures (Fig. 6a), with 35 out of 48 temperature estimates within ±50°C of the empirical temperature (Fig. 6g). The Pyle *et al.* (2001) thermometer, while reasonably reproducing the same study's empirical temperatures, seems to significantly underestimate empirical temperatures from other studies. Indeed, the estimated temperatures from Pyle *et al.* (2001) are consistently lower for a given X_{Y+HREE} than the other datasets (Fig. 6a–f). If the Pyle *et al.* (2001) data used in their calibration are excluded, 33 out of 40 temperature estimates using the Heinrich *et al.* (1997) thermometer fall within ±50°C of the empirical temperature. However, the Pyle *et al.* (2001) data do not appear as such an outlier with respect to Y partitioning between monazite and xenotime (Fig. 4a). Both the Gratz and Heinrich (1997) and Seydoux-Guillaume *et al.* (2002) thermometers appear to overestimate empirical temperatures at high temperatures and underestimate at low temperatures (Fig. 6c, d). Fewer than 50% of temperature estimates are within ± 50°C of the empirical temperature for both thermometers (Fig. 6i, j). The Gratz and Heinrich (1998) thermometer, while able to reproduce experimental temperatures, greatly overestimates the empirical temperatures for most natural monazites (Fig. 6e). Finally, temperatures are also calculated using a slightly different input value for the Heinrich *et al.* (1997) thermometer (Fig. 6f). Using $X_{Y+HREE}/X_{Y+HREE} + X_{LREE}$ effectively recalculates the ternary monazite, xenotime, cheralite + thorite system (Seydoux-Guillaume *et al.* 2002) as a pseudo-binary

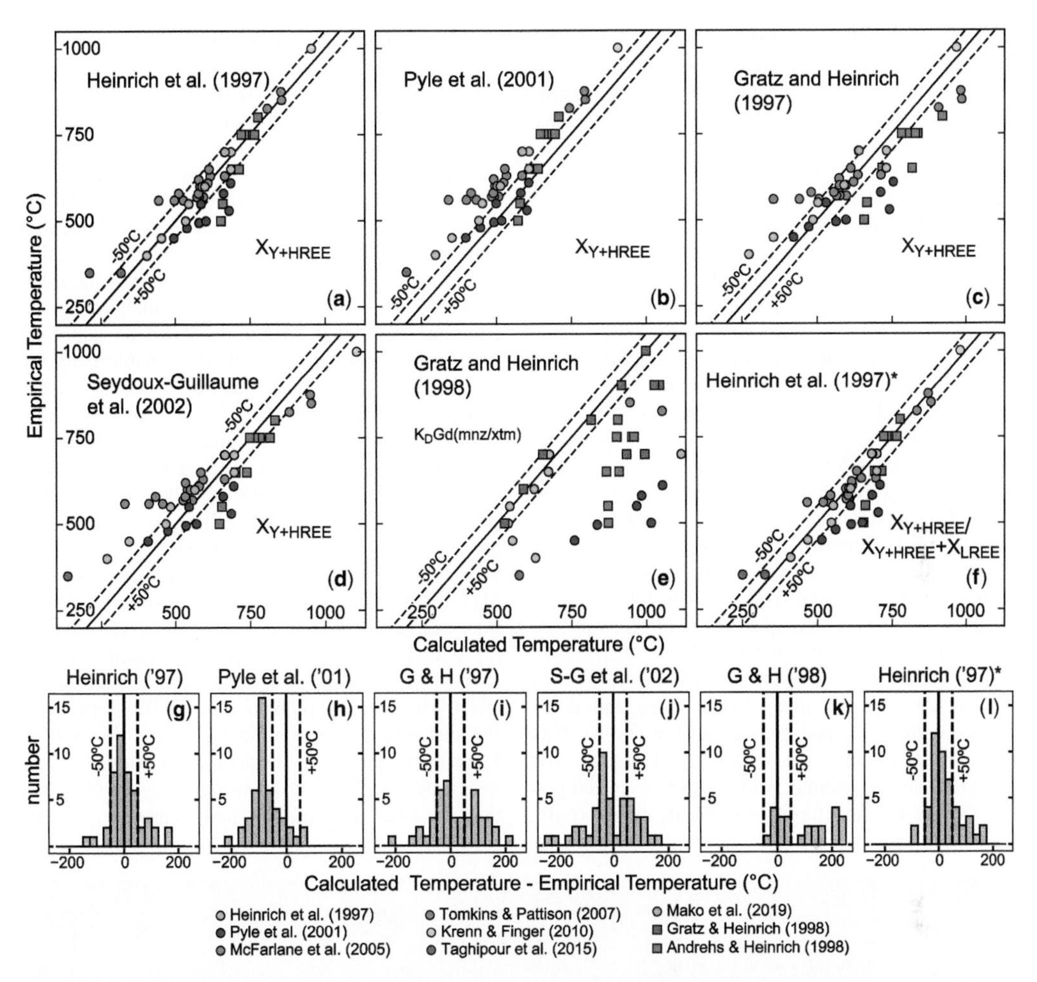

Fig. 6. Comparison of calculated and empirical (independently determined) monazite temperatures for each available thermometer. In each panel, the method for calculating the xenotime content of monazite is indicated with X_{Y+HREE} calculated according to equations (1)–(6). (**a–f**) Direct comparison of calculated and empirical temperatures with the solid line indicating a perfect match and dashed lines indicating a ± 50°C difference. (**g–l**) Histograms of the difference between calculated and empirical temperatures, corresponding to graphs in (**a–f**), respectively.

system, ignoring the actinide-bearing end-members. This approach has little effect on most of the data, but for high Th monazites (Krenn and Finger 2010, yellow circle at the top right of panels in Fig. 6; Taghipour *et al.* 2015, grey circles at the lower left of panels in Fig. 6), the monazite–xenotime temperature estimates are closer to the empirical temperature estimates. This may be an effective means of estimating temperatures for actinide-rich monazites.

Northern Highlands sample petrology and monazite–xenotime petrochronology

We have analysed monazite and xenotime in a suite of three samples from the Moine and Ben Hope nappes in the NHT (Fig. 3). Further details on the phase equilibria, petrography and geological setting of these samples are given in (Mako 2019), and a complementary dataset for the Naver nappe is also presented in Mako *et al.* (2019). Here we summarize the petrology, U–Pb ages and monazite–xenotime thermometry from the Moine and Ben Hope nappe samples. Given the above evaluation of the various monazite–xenotime thermometers, the temperature estimates we report here are calculated using the Heinrich *et al.* (1997) thermometer. This analysis aims to establish temperature–time constraints for the Moine and Ben Hope nappes to understand better the timing of metamorphism and deformation in northern Scotland. For monazite–xenotime temperature estimates, we have used a generalized ± 25°C uncertainty based on adding ±10% to the monazite composition. For trace element compositions determined by LASS, ±10% is a conservative compositional uncertainty. No previous studies have included any statistical analysis that would allow more meaningful uncertainties to be assigned to each thermometric relationship.

Sample MT-09-09 (UK grid reference: NC 55375 53025) is a pelitic schist from the immediate footwall of the Ben Hope thrust at the southern tip of the Kyle of Tongue (Fig. 3). It is composed of quartz, muscovite, biotite, garnet, plagioclase, staurolite and chlorite with accessory ilmenite, rutile, apatite, tourmaline, monazite and xenotime. This sample has been analysed extensively in Thigpen *et al.* (2013) and Ashley *et al.* (2015), with peak temperatures estimated at 590–600°C using various geothermometers and pseudosection-based phase equilibria modelling (Ashley *et al.* 2015). Staurolite in this sample forms 'pressure caps' (e.g. Passchier and Trouw 2005), often grown on the edges of garnet grains oriented parallel to the matrix fabric, rather than on edges perpendicular to foliation. Monazite and xenotime often occur near one another and in similar microstructural settings. Monazite is most commonly found sharing crystal faces and intergrown with biotite and muscovite (Fig. 7a, b). This textural relationship may suggest: (1) either that monazite and xenotime grain shapes were defined by the pre-existing biotite–muscovite fabric (post-tectonic mimetic growth, implying ages post-date deformation); (2) or that monazite and xenotime growth was coeval with deformation (syn-tectonic grown, implying ages are synchronous with penetrative deformation). Additionally, several monazite and xenotime grains are included in staurolite, but most of these grains were too small to date. In one case, a xenotime grain is apparently included in staurolite, with monazite grown in cracks leading from the xenotime to the edge of the staurolite (Fig. 7c). This observation likely evidences significant fluid-mediated REE mobility. This xenotime grain gave ages that are indistinguishable from the matrix population. Backscattered electron images and X-ray maps of Y, Th, Ca and Nd reveal little compositional variation in monazite or xenotime.

U–Pb isotopic and trace element data for monazite and xenotime in MT-09-09 were collected in two thin sections cut from the same billet from this sample during LASS sessions in January 2015 and June 2017. The ages calculated from each of these runs (412 ±4 v. 415 ±4 Ma; see text below) are within uncertainty of one another, and illustrate the inherent reproducibility of unknown analyses using this technique. We refer to one of these thin sections as MT-09-09 and the other as MT-09-09_2, or when the distinction is insignificant, simply MT-09-09. Concordia ages were calculated for monazite and xenotime in MT-09-09_2 using analyses with less than ±10% discordance and rejecting obvious

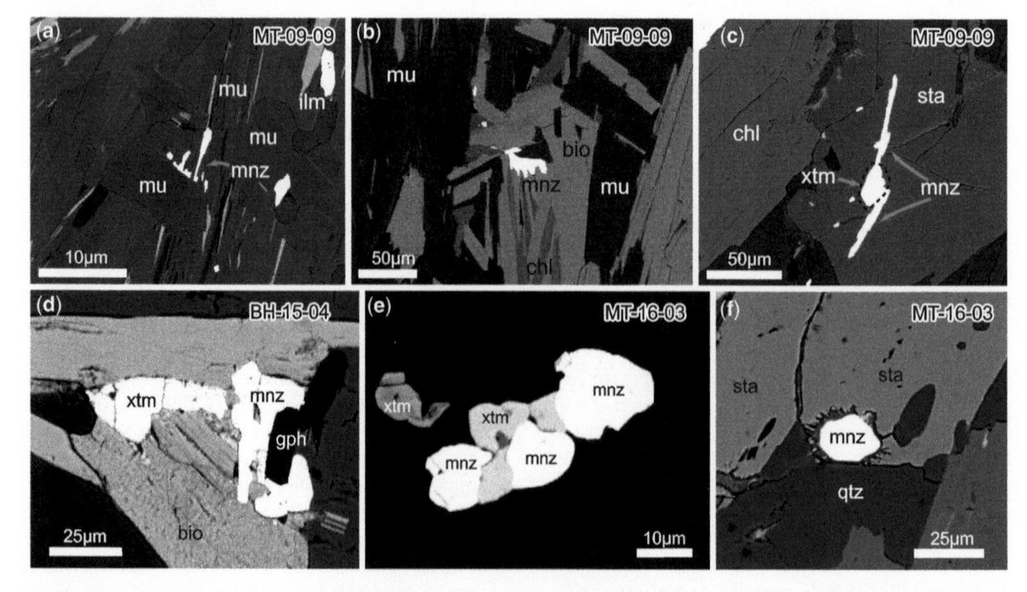

Fig. 7. Backscatter electron images of monazite and xenotime for samples from this study. (**a**, **b**) Monazite growing along muscovite and biotite grain edges in MT-09-09. (**c**) Monazite and xenotime in contact, included within staurolite in MT-09-09. (**d**) Monazite and xenotime grains nearly in contact in BH-15-04. (**e**) Aggregate of fine-grained monazite and xenotime in MT-16-03. Image was collected at high contrast and low brightness to emphasize distinct monazite and xenotime grains. (**f**) Monazite included in the outer edge of staurolite in MT-16-03.

outliers. Concordia ages are 412 ± 4 Ma (MSWD = 1.6, $n = 28$) and 413 ± 5 Ma (MSWD = 1.2, $n = 27$) for monazite and xenotime, respectively (Fig. 8a, b). REE profiles are typical of monazite and xenotime (Fig. 8c). The average X_{Y+HREE} content of monazite in MT-09-09_2 is 0.0589, corresponding to a temperature of 557 ± 25°C (Fig. 8d). For MT-09-09, ages were calculated using analyses with less than ± 10% discordance and rejecting obvious outliers. Concordia age calculations for this thin section yielded discordant results, so the $^{206}Pb/^{238}U$ age of an x–y weighted average (Ludwig 2008; Isoplot 4.1) was used. The use of the $^{206}Pb/^{238}U$ age is consistent with the established laboratory procedure at University of California, Santa Barbara. This yields ages of 415 ± 4 Ma (MSWD = 1.6, $n = 42$) and 420 ± 6 Ma (MSWD = 2.2, $n = 7$) for monazite and xenotime, respectively (Fig. 8e, f). REE profiles are typical of monazite and xenotime (Fig. 8g). The average X_{Y+HREE}

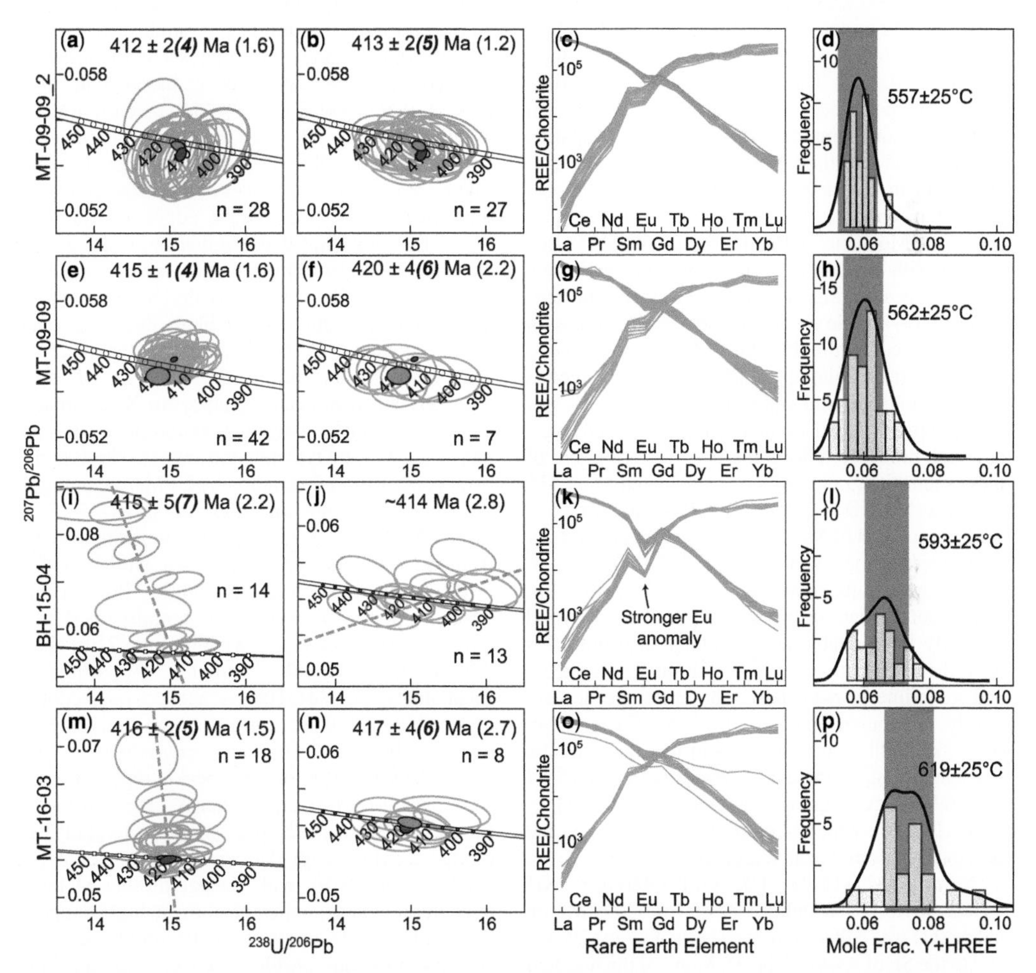

Fig. 8. Monazite and xenotime geochronology, trace elements and monazite X_{Y+HREE} contents for samples (**a–d**) MT-09-09_2, (**e–h**) MT-09-09, (**i–l**) BH-15-04 and (**m–p**) MT-16-03. In (**a**, **e**, **i**, **m**) U–Pb analyses of monazite are shown by the light blue ellipses, with the filled dark blue and red ellipses representing the concordia age ellipses for monazite and xenotime, respectively, for each sample. In (**b**, **f**, **j**, **n**) U–Pb analyses of xenotime are shown by the light red ellipses, with the filled dark blue and red ellipses representing the concordia age ellipses for monazite and xenotime, respectively, for each sample. The dashed grey lines in (**i**, **j**, **m**) are isochrons used to calculated intercept ages. Concordia ages or lower intercept ages are shown for each sample. All ellipses are plotted at 2σ uncertainty. The numerical uncertainties in each plot include the internal uncertainty related just to the analysis and weighted average calculation, and the full uncertainty including propagated external uncertainty in parenthesis. MSWD is given in the second set of parentheses. In (**c**, **g**, **k**, **o**) REE/chondrite values are plotted for monazite (blue) and xenotime (red). Relative probability diagrams and histograms are shown in (**d**, **h**, **l**, **p**) for monazite analyses in each sample.

content of monazite in MT-09-09 is 0.0599, corresponding to a temperature of 562 ± 25°C (Fig. 8h). Note that this is 30–40°C less than the previously inferred peak temperature (Thigpen *et al.* 2013; Ashley *et al.* 2015). The REE/chondrite profiles for MT-09-09 and MT-09-09_2 are indistinguishable, with a tight distribution and minimal Eu anomaly (Fig. 8c, g).

Sample BH-15-04 (UK grid reference NC 66743 58028) is a pelitic schist from the immediate footwall of the Naver thrust in the upper part of the Ben Hope nappe (Fig. 3), where peak metamorphic temperatures have been constrained by garnet–biotite thermometry at 675 ± 25°C (Ashley *et al.* 2015, their sample MT-09-12). The BH-15-04 sample comprises quartz, plagioclase, biotite, muscovite and garnet with accessory graphite, ilmenite, rutile, apatite, monazite and xenotime. Monazite and xenotime usually occur in the same microstructural domains and are often almost in contact with one another (Fig. 7d). Monazite and xenotime grains also often contact with basal planes of biotite and muscovite, texturally post-dating these micas. Monazite and xenotime in this sample are discordant along a regression away from concordia (Fig. 8i, j). This may be due to the inadvertent ablation of Pb-containing minerals (apatite, feldspar) and cracks during the analysis of these relatively small monazite grains. A lower intercept age probably reflects the crystallization age of these monazite grains, which is calculated to be 415 ± 7 Ma (MSWD = 2.2, $n = 14$) (Fig. 8i). An intercept age of *c.* 414 Ma ($n = 13$) was calculated for xenotime, but the calculated $^{207}Pb/^{206}Pb$ is anomalously low and its significance is unclear (Fig. 8j). The xenotime data in this sample thus do not yield a readily interpretable age, but their distribution is nevertheless permissive of synchronous monazite and xenotime growth. REE/chondrite profiles are tightly distributed and have a strong Eu anomaly (Fig. 8k), unlike other samples in this dataset. The average X_{Y+HREE} content of monazite in BH-15-04 is 0.0670, corresponding to a temperature of 593 ± 25°C (Fig. 8l), which is more than 80°C below the previously inferred peak temperature.

Sample MT-16-03 (UK grid reference: NH 45183 96154) is a pelitic schist from the Moine nappe that was collected to the south of Glen Cassley and to the SE of Assynt (Fig. 3). Many of the regional scale structures bend to the east around the Assynt – Glen Cassley Culmination (Fig. 3), including the Moine, Ben Hope, Achness and Naver thrusts (e.g. Leslie *et al.* 2010). Traced to the south, both the Ben Hope and Achness thrusts are mapped as merging along the thrust transport-parallel Oykel Transverse Zone (Leslie *et al.* 2010), which in turn links to the west with the Moine thrust (Fig. 3). In this structural scheme, the Ben Hope and Achness thrusts are viewed as subsidiary thrusts with the regionally larger-scale Moine nappe located to the north of the Oykel Transverse Zone (Leslie *et al.* 2010; Krabbendam *et al.* 2011). Sample MT-16-03 is located *c.* 5 km to the south of the Oykel Transverse Zone and, although still located within the regional scale Moine nappe, is located further towards the hinterland than samples MT-09-09 and BH-15-04 (measured in map view parallel to the WNW transport direction).

Sample MT-16-03 is composed of quartz, garnet, staurolite, muscovite and biotite with accessory ilmenite, rutile, kyanite, chlorite, apatite, monazite and xenotime. Preliminary pseudosection-based thermodynamic modelling suggests that this sample reached temperatures of *c.* 660°C, based on the presence of kyanite (Mako 2019). Monazite often forms as clusters of small ameboid grains in mica-rich domains, often with intergrown xenotime (Fig. 7e). There are several occurrences of monazite and xenotime inclusions in the outer edges of staurolite (Fig. 7f), suggesting that original monazite growth and late-stage staurolite growth were broadly contemporaneous. There are many discordant monazite analyses in this sample, again likely due to the inadvertent ablation of Pb-bearing minerals (apatite, feldspar) and cracks during the analysis of these relatively small monazite grains. No X-ray images were collected for monazite or xenotime in this sample, as grains were typically too small to show significant compositional variation by EPMA mapping. A lower intercept age for monazite is 416 ± 5 Ma (MSWD = 1.5, $n = 18$), which is indistinguishable from a concordia age of analyses with less than ±10% discordance (Fig. 8m). The $^{206}Pb/^{238}U$ age of an *x–y* weighted average for xenotime is 416 ± 6 Ma (MSWD = 2.7, $n = 8$) (Fig. 8n). REE/chondrite profiles are tightly distributed, with a minimal Eu anomaly (Fig. 8o). The average X_{Y+HREE} content of monazite is 0.0738, corresponding to a temperature of 619 ± 25°C (Fig. 8p), approximately 40°C below the previously inferred peak temperature.

A clear pattern emerges when K_D-REE profiles of Moine and Ben Hope Nappe samples (this study) are plotted together with data from the Naver nappe (Mako *et al.* 2019), with steeper profiles at lower temperatures and gentler profiles at high temperatures (Fig. 9). In contrast to the highly scattered and discontinuous profiles from the literature data (Fig. 5), the northern Scotland data are smooth and 'rotate' about Gd. Monazite–xenotime thermometry on samples across the Moine and Ben Hope nappes indicates a progressive increase in temperature toward the hinterland (Fig. 10a). Our temperature estimates are in excellent agreement with quartz fabric opening angle thermometry across the same structural positions (Fig. 10a). Deformation temperature data (Fig. 10a) in this area were initially presented in Thigpen *et al.* (2013) and Law (2014), calculated

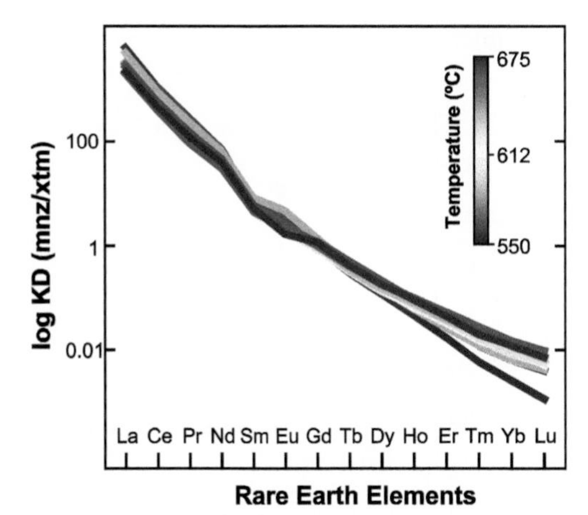

Fig. 9. Monazite–xenotime distribution coefficients for REEs in samples from this study and Mako *et al.* (2019). Profiles are coloured according to the temperature calculated using the Heinrich *et al.* (1997) monazite–xenotime thermometer. Each profile represents simple average compositions from a population of monazite and xenotime within a given sample.

based on the Kruhl (1998) calibration of quartz *c*-axis fabric opening angles. In the present study, they have been recalculated according to the Faleiros *et al.* (2016) linear calibration, which reduces the estimated temperatures by a few tens of degrees. Monazite and xenotime ages in the Moine and Ben Hope nappes are within uncertainty of one another, though slightly younger toward the foreland, (Fig. 10b), and are younger than ages from the Naver nappe. Taken together with Ar–Ar muscovite ages from the thrust nappes (Fig. 10b), these data indicate relatively slow regional cooling at $12 \pm 3°C\ Myr^{-1}$ from 425 to 415 Ma, followed by rapid cooling until final exhumation (e.g. Spencer *et al.* 2020).

Two samples from the Naver nappe (map locations shown in Fig. 3b) were included with the compilation of the literature data that we used above (light blue ellipses in Figs 2, 4, 6). The monazite and xenotime compositional data for these samples were previously published by Mako *et al.* (2019). Here, we briefly review the petrographic observations that allow us to estimate an independent temperature that can be compared with estimates from various thermometers. NT-08 is composed of quartz, plagioclase, orthoclase and minimal biotite, in addition to monazite and xenotime. Quartz exhibits blocky extinction with tilt walls apparently parallel to basal and prism planes, possibly indicating chessboard extinction and minimum deformation temperatures of *c.* 630°C (Fazio *et al.* 2020). Extreme undulose extinction is found in some orthoclase grains, and highly sinuous boundaries between quartz and feldspars are common, likely indicating grain boundary migration (GBM) recrystallization. NT-06 is composed of biotite, white mica, plagioclase, orthoclase and quartz, with accessory monazite and xenotime. Again, highly sinuous boundaries between quartz and feldspars are common, indicating GBM recrystallization. Some plagioclase grains exhibit deformation twinning and undulose extinction, and a few quartz grains show weakly developed blocky, chessboard-like, extinction. GBM recrystallization in quartz and particularly feldspar is generally indicative of relatively high-temperature deformation, likely exceeding 700°C (Passchier and Trouw 2005; Fazio *et al.* 2020). Previous work, including quantitative petrology, has indicated that peak temperatures across the Naver nappe reached about 700°C, with estimates ranging from 600 to 750°C (see Mako *et al.* 2019 for in-depth discussion). It is unlikely that temperatures substantially exceeded 700°C given that evidence of pervasive melting is not generally observed. Based on deformation microstructures and regional petrological constraints, a temperature of $700 \pm 50°C$ is used for NT-06 and NT-08.

Discussion

A compilation of coexisting monazite and xenotime compositions from the literature has been used to evaluate published calibrations of the monazite–xenotime thermometer. Our analysis shows that the Heinrich *et al.* (1997) thermometer yields the 'best' results, giving temperature estimates that most closely match independently determined empirical temperatures (Fig. 6). Calculating the xenotime content of monazite as a pseudo-binary between X_{Y+HREE} and X_{LREE} further improves temperature estimates for high Th monazites. This finding agrees with previous studies (McFarlane *et al.* 2005; Tomkins and Pattison 2007), and the agreement between quartz fabric opening angle based deformation temperatures and monazite–xenotime thermometry in our NHT samples (Fig. 10) further validates the Heinrich *et al.* (1997) calibration. This conclusion is perhaps surprising because the Heinrich *et al.* (1997) calibration is based on only six samples with no quantitative petrology, while the Pyle *et al.* (2001) calibration is based on more than a dozen samples with detailed quantitative geothermometry. However, the Pyle *et al.* (2001) calibration appears to underestimate independently constrained, empirically derived temperatures consistently. Though the reason for this is unclear, it may result from the temperature estimates initially used to underpin the Pyle *et al.* (2001) calibration. These utilized the Hodges and Spear (1982) garnet–biotite thermometer, and it is possible that alternative garnet–biotite calibrations would yield quite different temperatures (e.g. Wu

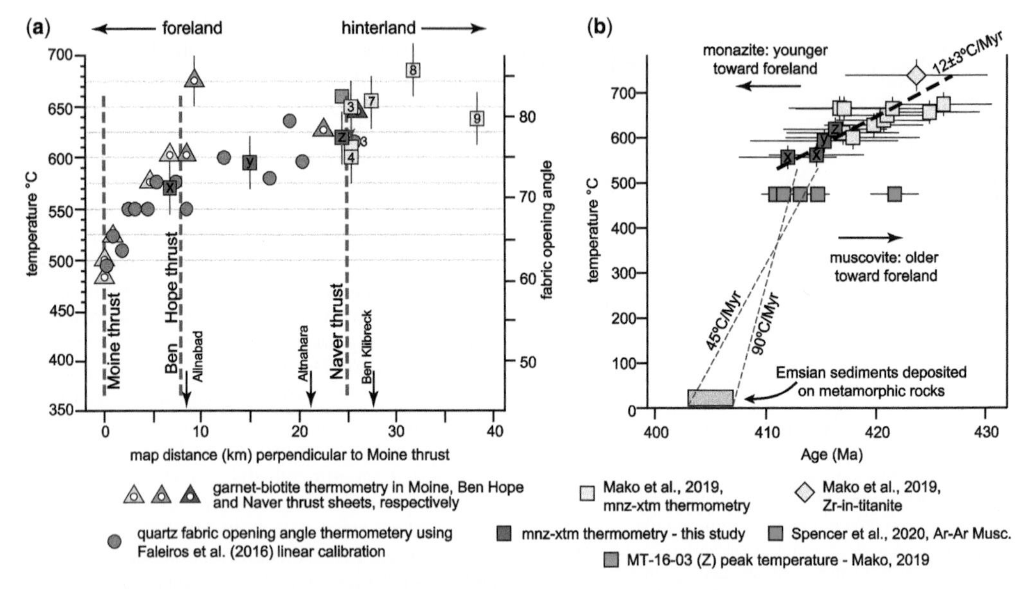

Fig. 10. Compilation of thermometry and geochronology data for the northern part of the NHT. (**a**) Quartz *c*-axis fabric-based deformation thermometry (Thigpen *et al.* 2013; Law 2014; Faleiros *et al.* 2016), garnet–biotite thermometry (Thigpen *et al.* 2013; Ashley *et al.* 2015), pseudosection-based temperature constraints (Mako 2019) and monazite–xenotime thermometry (this study; Mako *et al.* 2019) at varying distances from the Moine thrust. (**b**) Temperature–time graph showing all monazite–xenotime thermometry from this study and Mako *et al.* (2019), with ages from U–Pb geochronology. Data points from this study are labelled as 'x' (MT-09-09), 'y' (BH-15-04) and 'z' (MT-16-03). Muscovite Ar–Ar geochronology data from Spencer *et al.* (2020) are also shown.

and Cheng 2006). Also, recent experimental work has shown that oxygen fugacity can substantially affect monazite–xenotime equilibria (Chowdhury and Trail 2021). Differences in the redox state may explain the systematic difference in the Pyle *et al.* (2001) samples, but this appears unlikely given the range of locations from which the samples were sourced. While it is difficult to explain these discrepancies, our results suggest that the Heinrich *et al.* (1997) calibration most accurately recovers equilibration temperature for a wider range of samples. Each thermometer based on a simpler experimental system (Fig. 6; Gratz and Heinrich 1997, 1998; Seydoux-Guillaume *et al.* 2002) can reliably reproduce empirical temperatures only over a restricted temperature range and may not produce meaningful temperature estimates for natural samples.

Appropriate accounting for X-ray peak interferences among multiple elements is an essential part of analysing REEs and trace elements in monazite and xenotime by EPMA (Pyle *et al.* 2002; Jercinovic and Williams 2005). Some of the analytical choices related to peak interference corrections may account for variations in composition and thus the differences in Y + HREE v. temperature relationship from different datasets. Both Franz *et al.* (1996) and Andrehs and Heinrich (1998) used peak interference correction factors based on Amli and Griffin (1975), while Pyle *et al.* (2001) developed their own correction factors. McFarlane *et al.* (2005) used a compilation of correction factors in Pyle *et al.* (2002), and Tomkins and Pattison (2007) appear not to have corrected for peak interferences. At face value, there does not appear to be any systematic relationship between the choice of correction factors and calculated Y + HREE content. Indeed, there is no great difference among the various correction factors from different sources (see compilation in Pyle *et al.* 2002). The most significant difference is the correction factor for Gd used by Pyle *et al.* (2001), and we experimented with changing the Gd correction factor for the Pyle *et al.* (2001) data to that of Amli and Griffin (1975). While this resulted in much better agreement in calculated Gd partitioning between the Pyle *et al.* (2001) data and other datasets (Fig. 4c), it had a negligible effect on the total X_{Y+HREE} content of the Pyle *et al.* (2001) data. Although the choice of correction factor may explain some of the scatter in the data (Figs 2, 4), it does not appear sufficient to explain the discrepancy between the thermometers.

Assessing monazite–xenotime equilibrium

A constant challenge of applying techniques such as monazite–xenotime thermometry is the evaluation of whether coexisting phases represent a thermodynamic equilibrium (Pyle *et al.* 2001; Spear and Pyle 2002). Several compositional criteria have been proposed that may provide evidence that monazite and xenotime in a given sample were in equilibrium. Pyle *et al.* (2001) suggested that the partitioning of Y/Gd and Y/Dy between monazite and xenotime are relatively constant, meaning that a certain range of values would be evidence for equilibrium. However, Figures 4(o) and 4(p) show that both Y/Gd and Y/Dy partitioning apparently have a weak temperature dependence, with much scatter among the empirical datasets. This implies, for example, that xenotime that grew at 450°C and monazite that grew at 600°C in a given sample could be associated with Y/Gd and Y/Dy partitioning values that would still be in the 'accepted' range of equilibrium values (i.e. would suggest equilibration). Similarly, Andrehs and Heinrich (1998) proposed that K_DSm/K_DNd values between 0.2 and 0.3 indicated that monazite–xenotime equilibrium was attained. This ratio ranges from 0.1 to 0.5 for the data compiled in this contribution, with a high degree of scatter. Again, it would be easy to pair monazite and xenotime compositions that were clearly not in equilibrium and still obtain partitioning values that fall in the generally accepted ranges. As such, it is difficult to use such ratios as evidence of equilibration between monazite and xenotime.

A range of additional observations may provide evidence supporting or refuting equilibrium between monazite and xenotime. Textural observations that are, to varying degrees, consistent with equilibrium are thoroughly reviewed by Pyle *et al.* (2001) and Spear and Pyle (2002). Age is another observation that may be useful in assessing equilibrium. Cases where the age of monazite and xenotime overlap within uncertainty (Fig. 8) should generally be regarded as permissive, though not diagnostic, of equilibrium. Conversely, significantly different ages of monazite and xenotime likely demonstrate that equilibrium was not attained. Especially with laser ablation inductively coupled plasma mass spectrometry (LA-ICPMS) geochronology, it is often now relatively easy to date coexisting monazite and xenotime in a sample. Thus, it is recommended that both monazite and xenotime be identified and analysed in petrochronological studies, wherever possible. Similarly, comparing REE profiles in monazite and xenotime can provide valuable information. Our data from the NHT show, by inspection, that monazite and xenotime have similar Eu anomalies when equilibrium is attained (Figs 8c, g, k, o). It is unlikely that this would occur if, for example, xenotime crystallized much earlier during prograde metamorphism than monazite and if these phases then remained out of equilibrium. Thus, similar Eu anomalies might be considered evidence that both monazite and xenotime equilibrated with the same Eu-sequestering assemblage. The NHT data also show a consistent pattern in the slope of K_D v. REE profiles across metamorphic grade, with profiles steepening at lower metamorphic grade (Fig. 9). This might also be viewed as consistent with monazite–xenotime equilibration, and temperatures obtained from samples that are anomalous in such a plot might be viewed with suspicion. In cases where monazite and xenotime are demonstrably out of equilibrium, monazite compositions reflect the effective bulk Y + REE composition and the minimum temperature of the system.

A monazite–xenotime temperature may reflect solely the initial growth conditions or continued re-equilibration with changing temperature. Data from the NHT seem to indicate that monazite and xenotime continued to equilibrate during progressive retrograde metamorphism and deformation. This is indicated by the agreement between quartz *c*-axis fabric opening angle based deformation temperatures and monazite–xenotime temperatures (Fig. 10a). If monazite–xenotime temperatures were related to an initial monazite-in reaction, we would expect the temperature to be the same across all structural positions. While monazite ages are often interpreted to record peak metamorphic conditions, previous constraints on peak metamorphism in the Moine and Ben Hope nappes indicate temperatures of 600°C and 675°C (Ashley *et al.* 2015), respectively, 40–90°C above the results of our monazite–xenotime thermometry for each sample from this study. Similarly, the observed monazite inclusions in staurolite (a typical prograde phase) may be interpreted to indicate that monazite and xenotime formed on the prograde path, although retrograde staurolite is possible (e.g. Kohn *et al.* 1997). However, few of the monazite and xenotime inclusions in staurolite were actually dated in this study, so it is difficult to evaluate whether any staurolite inclusions are older than the dominant matrix populations. Given the 'pressure cap' texture of staurolite in MT-09-09 and regional geochronological constraints (see section 'Thermal and structural evolution of the Scandian nappes' below), it is possible, though not conclusive, that staurolite in our samples is a retrograde phase. It is well known that monazite and xenotime are highly susceptible to fluid-mediated recrystallization (Harlov *et al.* 2007; Hetherington *et al.* 2010; Williams *et al.* 2011; Grand'Homme *et al.* 2016, 2018) and that monazite–xenotime equilibrium can be maintained during fluid alteration (Mako *et al.* 2021). The continued or episodic presence of fluids and deformation likely caused monazite and xenotime to progressively recrystallize throughout the retrograde metamorphic path. However, a pressure effect could also explain the apparent trend of increasing temperatures toward the hinterland. Progressively higher peak pressures are recorded toward the hinterland in the Scandian orogenic wedge (Ashley *et al.* 2015), and experimental data (Gratz and Heinrich 1997) have demonstrated that the xenotime component in monazite increases with increasing pressure (Fig. 1). Given that the Heinrich *et al.* (1997) thermometer does not incorporate pressure effects, the apparently higher temperatures toward the hinterland could be solely due to increasing pressure at monazite growth. The current lack of data on the effects of pressure for natural monazite–xenotime pairs limits a full consideration of this point and highlights the need for further work.

Non-ideality and thermodynamic components

Experimental data clearly demonstrate that the partitioning of Y and Ce in monazite and xenotime is strongly controlled by the addition of Th and Gd (Gratz and Heinrich 1998; Seydoux-Guillaume *et al.* 2002). Increased $ThSiO_4$ and $GdPO_4$ contents increase the YPO_4 (xenotime) content of monazite at a given temperature and pressure (Fig. 2f). Indeed, Seydoux-Guillaume *et al.* (2002) calculated a monazite–xenotime thermometer for high Th monazite (Fig. 2). These observations indicate thermodynamically non-ideal mixing between Y, Ce and Gd monazites, as well as monazite and thorite components. Using thermometers calculated in the Y–Ce phosphate system can thus give erroneous results when monazite contains significant amounts of other elements, as is common in natural monazites. Furthermore, Mogilevsky (2007) presented a detailed theoretical study of monazite–xenotime miscibility gaps and showed that the shape of miscibility gaps in various binary systems are markedly different. For example, the monazite–xenotime miscibility gap in the YPO_4–$CePO_4$ system is not identical to that of the $LaPO_4$–$LuPO_4$ system, and multi-element natural systems are likely also different. Thus, for natural monazites that incorporate all the REEs, one should not necessarily expect meaningful results from thermometers based on simplified experimental systems. This reduced predictive capacity of experimentally determined Y–Ce partitioning (Fig. 6) is likely due to non-ideality in natural multicomponent monazite and xenotime. Conversely, the experimental data of Andrehs and Heinrich (1998), which used all the REEs except Eu, Ho and Tm, generally fit well with the natural data. Although simpler REE-phosphate experimental systems have provided important mineralogical and crystallographic data, they may not be ideal for developing thermometers for natural, multi-component systems.

The Heinrich *et al.* (1997) and Pyle *et al.* (2001) thermometers both calibrate temperatures using the xenotime component in monazite (X_{Y+HREE}), and this appears to be the most effective way of dealing with natural monazite. This approach essentially treats monazite ($[La–Eu]PO_4$) and xenotime ($[Y, Gd–Lu]PO_4$) as thermodynamic components even though multiple elements reside in the same cation site. While this is an atypical approach, it is consistent with how thermodynamic components are defined. Individual HREEs or LREEs are insufficiently independent to require treatment as distinct thermodynamic components in metamorphic systems. Y + HREE, for example, all tend to behave similarly during specific metamorphic reactions. REEs typically represent major constituents only of monazite and xenotime in pelitic samples, forming minor or trace components of most other phases. The observation that this approach is practical for monazite and xenotime indicates that an important simplification could be made in thermodynamic modelling of REE-bearing metamorphic systems. It may not be necessary to model each REE as an individual component, and it is likely inadequate to simplify REE-bearing systems to simple binaries, such as YPO_4–$CePO_4$. For example, although thermodynamic datasets that use data for just YPO_4 and $CePO_4$ (e.g. Spear and Pyle 2010) probably do not capture the full behaviour of natural monazite, it may not be necessary to develop datasets and activity models that incorporate all REE components individually. Grouping HREEs and LREEs as thermodynamic components may be sufficient.

Thermal and structural evolution of the Scandian nappes

Monazite and xenotime in MT-09-09, BH-15-04 and MT-16-03 crystallized at *c.* 420–412 Ma, and temperatures constrained from monazite–xenotime thermometry are 560–620°C (Fig. 10). Despite many ages and temperatures from these samples being within uncertainty of one another, there is a trend of younger age and lower temperature toward the foreland (compare especially samples from the Naver and Moine nappes in Fig. 10). Our data show that relatively high temperatures of at least 550°C persisted until at least *c.* 415 Ma in the Moine nappe and likely across all the Scandian nappes exposed in northernmost mainland Scotland. At various structural positions, monazite–xenotime temperatures are very similar to quartz fabric opening angle based deformation temperatures (Fig. 10a). This potentially indicates that monazite and xenotime record the time at which penetrative deformation ceased in the structurally lower Moine and Ben Hope nappes. Quartz is likely to continue to recrystallize and presumably reset the crystallographic fabric until plastic deformation ceases, and if monazite–xenotime temperatures and quartz fabric-based temperatures are essentially the same, we would expect that monazite dates record the cessation of deformation. Using this line of reasoning, the monazite dates appear to record the timing of younger and waning stages of shearing within the thrust nappes. We note that it is highly unlikely that monazite records prograde ages in the Moine and Ben Hope nappes given that this would significantly conflict with regional geochronology data (Mako *et al.* 2019; Spencer *et al.* 2020; Strachan *et al.* 2020). For example, if monazite records the timing and temperature of the prograde path, peak metamorphism must postdate 412 $\pm$ 4 Ma (monazite), pre-date *c.* 415–410 Ma Ar–Ar muscovite ages, and be exhumed by 407–403 Ma (Wellman 2015). This would be an extraordinarily fast metamorphic cycle. While the textural relationship between monazite and fabric-forming minerals is somewhat ambiguous, indicating either syn- or post-tectonic growth, the association of monazite thermometry with quartz fabric-based deformation thermometry more clearly indicates syn-deformation monazite and xenotime recrystallization and resetting. Importantly, monazite and xenotime in our samples may be much younger than the grain shape fabric forming minerals (muscovite and biotite). White mica K–Ar, Ar–Ar and Rb–Sr dates that are significantly older (450–430 Ma) further south (Kelley 1988; Freeman *et al.* 1998) therefore do not necessarily conflict with the young monazite dates presented here. However, our ages are in broad agreement with Ar–Ar and Rb–Sr geochronology by Dallmeyer *et al.* (2001). It is possible that ductile deformation may have lasted longer in the present study area, which exposes a deeper structural level of the Scandian orogen than the area to the south of the Cassley Culmination (Law and Johnson 2010; Strachan *et al.* 2020; Law *et al.* 2021), and that although the regional penetrative fabric is continuous and correlative, the timing of final deformation may be diachronous. Despite extensive sampling, monazite- and xenotime-bearing rocks have not been identified in the Moine nappe south of Glen Cassley, preventing a similar study from being conducted in more southerly exposures of the orogen.

The observation that monazite from this study records the timing of deformation, with younger ages toward the foreland, is consistent with progressive localization of deformation toward the Moine thrust during the later stages of Scandian orogenesis. Indeed, Lusk and Platt (2020) used the same model in a rheological study across the Moine and Ben Hope nappes. Our geochronology data appear to confirm the validity of that approach, though the deformation temperatures determined by Lusk and Platt (2020) using Ti-in-quartz thermometry are of the order of 150°C less than our monazite–xenotime thermometry and quartz fabric-based thermometry at similar structural positions (Fig. 10). This could indicate that deformation persisted long after the time recorded in our samples, but it is also possible that the Ti-in-quartz thermometry used by Lusk and Platt (2020) significantly underestimated deformation temperatures. Regardless, it is clear from our data that relatively high-temperature ductile deformation persisted in this part of Scandian orogen until *c.* 415 Ma and possibly as late as 410 Ma.

Monazite–xenotime temperatures in each of our samples are lower than peak temperatures or garnet–biotite temperatures determined for these samples or other samples at similar structural positions (Thigpen *et al.* 2013; Ashley *et al.* 2015). Taken with textural considerations, this implies monazite–xenotime equilibration during retrogression. The temperature–time data from the Moine and Ben Hope nappes fall along a clear linear trend with data from the Naver nappe and allow for tighter constraints to be placed on the regional rates of cooling in the Scandian orogenic wedge of northernmost mainland Scotland (Fig. 10). The fact that this cooling path is apparently shared by the Moine, Ben Hope and Naver nappes may suggest a common exhumation and cooling mechanism across the whole orogen following peak temperatures in the Naver nappe at *c.* 425 Ma (Mako *et al.* 2019; Thigpen *et al.* 2021). A regression of all monazite–xenotime data from the Moine, Ben Hope and Naver nappes that are interpreted to fall on the retrograde path (Fig. 10b) yields a cooling rate of 12 $\pm$ 3°C Myr^{-1}. The younger ages from this study accentuate the apparent increase in cooling rate that must occur at 415–405 Ma to exhume the thrust nappes to the surface by the Early Emsian (407–403 Ma) deposition of the overlying Old Red Sandstone age conglomerates (Wellman 2015) across northern Scotland (Fig. 10b). Spencer *et al.* (2020) constrained final cooling rates in northern Scotland at 45–90°C Myr^{-1}, based on $^{40}Ar/^{39}Ar$ geochronology of muscovite, which ranged in age from 410.9 $\pm$ 1.3 to 414.7 $\pm$ 1.1 Ma for the Ben Hope, Naver and Skinsdale nappes (see also Dallmeyer *et al.* 2001). At such rapid cooling rates, the muscovite closure temperature (Dodson 1973) is calculated to be 475–500°C (assuming a grain size of 250–500 µm and Ar diffusivities from Harrison *et al.* 2009). These data are consistent with monazite–xenotime petrochronology of this study and together provide strong evidence for very rapid cooling and exhumation.

It is increasingly clear from geochronology data across the hinterland Scandian nappes and in the Moine Thrust Zone that the northern part of the NHT records a complex sequence of alternating or contemporaneous foreland and hinterland deformation (Strachan *et al.* 2020). Early deformation and metamorphism occurred during the Caledonian orogeny's 470–460 Ma Grampian phase (Kinny *et al.* 1999; Friend *et al.* 2000; Cutts *et al.* 2010; Bird *et al.* 2013; Dunk *et al.* 2019). Garnet geochronology constrains some deformation and folding in the Moine nappe during what is termed 'Grampian II orogenesis' at 450–445 Ma (Bird *et al.* 2013). In the Moine Thrust Zone, intrusive igneous rocks that cross-cut brittle and brittle–ductile thrusts are dated at 431–429 Ma (Goodenough *et al.* 2011). These precise dates in the foreland were thought to constrain the end of Scandian orogenesis in Scotland. However, renewed dating of syn-tectonic hinterland granites in the more hinterland-positioned Naver and Ben Hope nappes has demonstrated that high-temperature deformation at high structural levels occurred at 432–426 Ma (Strachan *et al.* 2020). Peak Scandian metamorphism in northern Scotland occurred at *c.* 425 Ma following rapid heating (*c.* 50°C Myr^{-1}) up to temperatures of 700°C (Mako *et al.* 2019).

The present study demonstrates that relatively slow cooling following peak Scandian metamorphism across the orogen accompanied deformation and progressive strain localization to lower structural levels. This progressive shear zone narrowing model (e.g. Lusk and Platt 2020) is potentially problematic given the presence of inliers of Lewisian basement gneiss across the nappes that delineate the presence of discrete thrust planes (e.g. Ben Hope and Naver thrusts) and basement cored isoclinal folds (Strachan and Holdsworth 1988). Such structures indicate deformation related to discrete thrust emplacement rather than a broad zone of homogeneous simple shear deformation. However, the thrust planes may in fact be much older and part of an earlier 'in sequence' deformation history, potentially formed during Grampian I, Grampian II or earlier Scandian orogenesis (Bird *et al.* 2013; Strachan *et al.* 2020). Prograde metamorphism and thrust-related heating in the Moine and Ben Hope nappes (Ashley *et al.* 2015; Thigpen *et al.* 2017, 2021) might be associated with this part of the deformation sequence. The emplacement of significant volumes of mantle-derived granitic magmas, related to the Newer Granites, at 430–425 Ma (Oliver *et al.* 2008; Strachan *et al.* 2020; Archibald *et al.* 2022), may have thermally weakened the hinterland of the orogen (Mako *et al.* 2019) causing renewed pervasive deformation that overprinted the earlier basement-involved structures in the same orientation. At higher temperatures, distributed deformation would be favoured over deformation on discrete thrust planes. Thus, the model of progressive strain localization across the Ben Hope and Moine nappes does not necessarily contradict the observation of discretely defined thrust planes.

Initial slow cooling transitioned to rapid cooling and exhumation at 415–410 Ma during the final stages of orogenic collapse (Spencer *et al.* 2020, 2021). A change in the mechanism of exhumation likely occurred at this time, leading to the cessation of deformation in the ductile nappes. Erosion and exhumation rates may have been enhanced by late Scandian age normal faulting or lower crustal flow (Spencer *et al.* 2020, 2021). The Great Glen fault, a major strike-slip boundary located to the SE of the study area, was likely active at this time (Fig. 3) through at least 399 Ma, after the complete exhumation of the hinterland nappes (Stewart *et al.* 2001; Dewey and Strachan 2003; Mendum and Noble 2010; Holdsworth *et al.* 2015; Strachan *et al.* 2020). Additionally, some faults within the broader Moine Thrust Zone may have been active as late as *c.* 408 Ma (Freeman *et al.* 1998; recalculated at 415 Ma using the revised ^{87}Rb decay constant of Villa *et al.* 2015), potentially forming in an orogen-scale transpressive environment at the onset of rapid exhumation (Holdsworth *et al.* 2015; Strachan *et al.* 2020). Indeed, transpressive deformation is associated with rapid exhumation (Thompson *et al.* 1997; Fossen and Tikoff 1998; Cochran *et al.* 2017), and transpression was likely important during the final stages of Scandian orogenesis in northern Scotland (Dewey and Strachan 2003).

Conclusion and recommendations

Monazite–xenotime thermometry is a powerful but under-utilized technique for quantifying the temperature–time evolution of metamorphic rocks. The analysis in this contribution clarifies some of the challenges associated with applying monazite–xenotime thermometry. The calculation of monazite end-members is a critical part of meaningfully using any monazite–xenotime thermometer and is best done employing the equations presented here (originally from Pyle *et al.* 2001). Based on an evaluation of published data, it appears that the Pyle *et al.* (2001) calibration of the Heinrich *et al.* (1997) data provides the most meaningful temperature estimates. For high Th monazite, it is best to calculate the xenotime content of monazite as a binary, ignoring the thorite and cheralite components (Fig. 6f). While it is possible to estimate temperature adequately, it is not possible with current data and calibrations to evaluate pressure and oxygen fugacity effects, which are both known to be important (Gratz and Heinrich 1997; Chowdhury and Trail 2021). Future experimental and empirical work will likely focus on this problem. The current experimentally derived calibrations can yield good results over a narrow temperature range, but ultimately the use of thermometers based on simple experimental systems is inappropriate in cases where significant non-ideality is not captured in the simplified system. As such, the binary monazite–xenotime systems likely do not adequately reflect the thermodynamics of multi-element natural monazite and xenotime.

Monazite and xenotime must be in equilibrium to apply monazite–xenotime thermometry, so it is essential to evaluate whether equilibrium was attained. To that end, it is advisable to always analyse both monazite and xenotime major and trace element compositions (and ages) when possible. They can be analysed with the same techniques, so this is often trivial, and the result yields higher confidence temperature–time points. In addition to textural considerations (Pyle *et al.* 2001), similarities in age and Eu anomaly between monazite and xenotime and trends in K_D-REE plots can help to evaluate the likely attainment of equilibrium. While obtaining temperature–time points from single spot analyses is desirable, in practice it is necessary to analyse a population of both monazite and xenotime, in conjunction with thorough textural observations and X-ray imaging, to confidently determine temperature–time points for a sample.

Detailed accessory phase geochronology has been critical for unravelling the dynamics and tectonics of the NHT in Scotland. Monazite–xenotime thermometry is useful in understanding the higher-temperature parts of the heating and cooling history of this terrane, as well as the timing of deformation. The similarity between our monazite–xenotime temperatures and quartz fabric opening angle based deformation temperatures across the hinterland nappes indicate that monazite is likely recording the timing and temperature of waning penetrative deformation. High-temperature ductile deformation occurred as late as 415–410 Ma in the hinterland Scandian nappes. We speculate that our finding that monazite and xenotime record the timing of deformation may be generalized. If monazite and xenotime grains are unzoned, fine-grained and hosted within an actively deforming, fluid-rich zone, it is likely they will continue to recrystallize and equilibrate to reflect the conditions of deformation. The younger and cooler temperatures recorded toward the foreland of the NHT are consistent with progressive strain localization and shear zone narrowing across the nappes during the late stages of Scandian orogenesis (e.g. Lusk and Platt 2020). However, in conjunction with previous geochronology data (Goodenough *et al.* 2011; Bird *et al.* 2013; Mako *et al.* 2019; Strachan *et al.* 2020), our data are consistent with deformation alternating between foreland and hinterland positions, or simultaneous deformation across the orogen. Following high-temperature ductile deformation, rapid exhumation and cooling were possibly related to normal faulting at higher structural levels (Spencer *et al.* 2020) and/or the onset of a transpressive tectonic regime (Dewey and Strachan 2003).

Acknowledgements The authors gratefully acknowledge analytical assistance from Bob Tracy and Luca Fedele at Virginia Tech. We thank two anonymous reviewers and Craig Storey for their comments, which helped improve the quality of this manuscript substantially. We also thank Kathryn Cutts for her editorial handling and comments.

Competing interests The authors declare that they have no known competing financial interests or personal relationships that could have appeared to influence the work reported in this paper.

Author contributions **CAM**: conceptualization (lead), data curation (equal), formal analysis (lead), investigation (lead), methodology (lead), visualization (lead), writing – original draft (lead); **MJC**: conceptualization (supporting), formal analysis (supporting), investigation (supporting), methodology (supporting), resources (equal), supervision (equal), validation (supporting), writing – review and editing (equal); **RDL**: conceptualization (supporting), data curation (equal), formal analysis (supporting), funding acquisition (lead), investigation (supporting), supervision (equal), validation (equal), visualization (supporting), writing – review and editing (supporting); **JRT**: conceptualization (supporting), formal analysis (supporting), investigation (supporting), validation (supporting), writing – review and editing (supporting).

Funding The LA-ICPMS analyses at University of California Santa Barbara were funded by National Science Foundation grant EAR 1220138 to RDL. The EPMA and SEM analyses were funded by grants to CAM from the Geological Society of America, Sigma Xi and the Department of Geosciences at Virginia Tech.

Data availability The datasets generated and analysed during the current study are available in the supplementary information files.

References

Aleinikoff, J.N., Schenck, W.S., Plank, M.O., Srogi, L.A., Fanning, C.M., Kamo, S.L. and Bosbyshell, H. 2006. Deciphering igneous and metamorphic events in high-grade rocks of the Wilmington complex, Delaware: morphology, cathodoluminescence and backscattered electron zoning, and SHRIMP U-Pb geochronology of zircon and monazite. *Bulletin of the Geological Society of America*, **118**, 39–64, https://doi.org/10.1130/B25659.1

Amli, R. and Griffin, W.L. 1975. Microprobe analysis of REE minerals using empirical correction factors. *American Mineralogist*, **60**, 599–606.

Andrehs, G. and Heinrich, W. 1998. Experimental determination of REE distributions between monazite and xenotime: potential for temperature-calibrated geochronology. *Chemical Geology*, **149**, 83–96, https://doi.org/10.1016/S0009-2541(98)00039-4

Anovitz, L.M. and Essene, E.J. 1987. Phase equilibria in the system CaCO3-MgCO3-FeCO3. *Journal of Petrology*, **28**, 389–415, https://doi.org/10.1093/petrology/28.2.389

Archibald, D.B., Murphy, J.B., Fowler, M., Strachan, R.A. and Hildebrand, R.S. 2022. Testing petrogenetic models for contemporaneous mafic and felsic to intermediate magmatism within the 'Newer Granite' suite of the Scottish and Irish Caledonides. *In*: Kuiper, Y.D., Murphy, J.B., Nance, R.D., Strachan, R.A. and Thompson, M.D. (eds) *New Developments in the Appalachian-Caledonian-Variscan Orogen. Geological Society of America Special Paper*, **554**, 375–399, https://doi.org/10.1130/2021.2554(15)

Ashley, K.T., Thigpen, J.R. and Law, R.D. 2015. Prograde evolution of the Scottish Caledonides and tectonic implications. *Lithos*, **224–225**, 160–178, https://doi.org/10.1016/j.lithos.2015.03.011

Barr, D., Holdsworth, R.E. and Roberts, A.M. 1986. Caledonian ductile thrusting in a Precambrian metamorphic complex: the Moine of northwestern Scotland. *Geological Society of America Bulletin*, **97**, 754–764, https://doi.org/10.1130/0016-7606(1986)97<754:CDTIAP>2.0.CO;2

Betkowski, W.B., Rakovan, J. and Harlov, D.E. 2017. Geochemical and textural characterization of phosphate accessory phases in the vein assemblage and metasomatically altered Llallagua tin porphyry. *Mineralogy and Petrology*, **111**, 547–568, https://doi.org/10.1007/s00710-017-0510-6

Bird, A.F., Thirlwall, M.F., Strachan, R.A. and Manning, C.J. 2013. Lu–Hf and Sm–Nd dating of metamorphic garnet: evidence for multiple accretion events during the Caledonian orogeny in Scotland. *Journal of the Geological Society, London*, **170**, 301–317, https://doi.org/10.1144/jgs2012-083

Cawood, P.A., Strachan, R.A. *et al.* 2015. Neoproterozoic to early Paleozoic extensional and compressional history of East Laurentian margin sequences: the Moine Supergroup, Scottish Caledonides. *Bulletin of the Geological Society of America*, **127**, 349–371, https://doi.org/10.1130/B31068.1

Chowdhury, W. and Trail, D. 2021. Monazite-xenotime Y + REE partitioning as a marker of ancient metamorphic conditions. *Goldschmidt Meeting Abstracts*, Virtual, https://conf.goldschmidt.info/goldschmidt/2021/goldschmidt/2021/meetingapp.cgi/Paper/3899 (accessed July 2022)

Cochran, W.J., Spotila, J.A., Prince, P.S. and McAleer, R.J. 2017. Rapid exhumation of Cretaceous arc-rocks along the Blue Mountains restraining bend of the Enriquillo-Plantain Garden fault, Jamaica, using thermochronometry from multiple closure systems. *Tectonophysics*, **721**, 292–309, https://doi.org/10.1016/j.tecto.2017.09.021

Cutts, K.A., Kinny, P.D. *et al.* 2010. Three metamorphic events recorded in a single garnet: integrated phase modelling, in situ LA-ICPMS and SIMS geochronology from the Moine Supergroup, NW Scotland. *Journal of Metamorphic Geology*, **28**, 249–267, https://doi.org/10.1111/j.1525-1314.2009.00863.x

Dacheux, N., Clavier, N. and Podor, R. 2013. Monazite as a promising long-term radioactive waste matrix: Benefits of high-structural flexibility and chemical durability. *American Mineralogist*, **98**, 833–847, https://doi.org/10.2138/am.2013.4307

Dallmeyer, R.D., Strachan, R.A., Rogers, G., Watt, G.R. and Friend, C.R.L. 2001. Dating deformation and cooling in the Caledonian thrust nappes of north Sutherland, Scotland: insights from 40Ar/39Ar and Rb–Sr chronology. *Journal of the Geological Society*, **158**, 501–512, https://doi.org/10.1144/jgs.158.3.501

Daniel, C.G. and Pyle, J.M. 2006. Monazite-xenotime thermochronometry and Al2SiO5 reaction textures in the Picuris range, Northern New Mexico, USA: new evidence for a 1450–1400 Ma orogenic event. *Journal of Petrology*, **47**, 97–118, https://doi.org/10.1093/petrology/egi069

Dewey, J.F. and Strachan, R.A. 2003. Changing Silurian-Devonian relative plate motion in the Caledonides: sinistral transpression to sinistral transtension. *Journal of the Geological Society, London*, **160**, 219–229, https://doi.org/10.1144/0016-764902-085

Dodson, M.H. 1973. Closure temperature in cooling geochronological and petrological systems. *Contributions to Mineralogy and Petrology*, **40**, 259–274, https://doi.org/10.1007/BF00373790

Dumond, G., Goncalves, P., Williams, M.L. and Jercinovic, M.J. 2015. Monazite as a monitor of melting, garnet growth and feldspar recrystallization in continental lower crust. *Journal of Metamorphic Geology*, **33**, 735–762, https://doi.org/10.1111/jmg.12150

Dunk, M., Strachan, R.A. *et al.* 2019. Evidence for a late Cambrian juvenile arc and a buried suture within the Laurentian Caledonides of Scotland: Comparisons with hyperextended Iapetan margins in the Appalachian Mountains (North America) and Norway. *Geology*, **47**, 734–738, https://doi.org/10.1130/G46180.1

Engi, M. 2017. Petrochronology based on REE-minerals: monazite, allanite, xenotime, apatite. *Reviews in Mineralogy & Geochemistry*, **83**, 365–418, https://doi.org/10.2138/rmg.2017.83.12

Ewing, R.C. and Wang, L.M. 2002. Phosphates as nuclear waste forms. *Reviews in Mineralogy and Geochemistry*, **48**, 673–700, https://doi.org/10.2138/rmg.2002.48.18

Faleiros, F.M., Moraes, R., Pavan, M. and Campanha, G.A.C. 2016. A new empirical calibration of the quartz c-axis fabric opening-angle deformation thermometer. *Tectonophysics*, **671**, 173–182, https://doi.org/10.1016/j.tecto.2016.01.014

Fazio, E., Fiannacca, P., Russo, D. and Cirrincione, R. 2020. Submagmatic to solid-state deformation microstructures recorded in cooling granitoids during exhumation of Late-Variscan crust in north-eastern Sicily. *Geosciences*, **10**, 1–29, https://doi.org/10.3390/geosciences10080311

Fossen, H. and Tikoff, B. 1998. Extended models of transpression and transtension, and application to tectonic settings. *Geological Society, London, Special Publications*, **135**, 15–33, https://doi.org/10.1144/GSL.SP.1998.135.01.02

Foster, G., Parrish, R.R., Horstwood, M.S.A., Chenery, S., Pyle, J. and Gibson, H.D. 2004. The generation of prograde P-T-t points and paths; a textural, compositional, and chronological study of metamorphic monazite. *Earth and Planetary Science Letters*, **228**, 125–142, https://doi.org/10.1016/j.epsl.2004.09.024

Franz, G., Andrehs, G. and Rhede, D. 1996. Crystal chemistry of monazite and xenotime from Saxothuringian-Moldanubian metapelites, NE Bavaria, Germany. *European Journal of Mineralogy*, **8**, 1097–1118, https://doi.org/10.1127/ejm/8/5/1097

Freeman, S.R., Butler, R.W.H., Cliff, R.A. and Rex, D.C. 1998. Direct dating of mylonite evolution: a multidisciplinary geochronological study from the Moine Thrust Zone, NW Scotland. *Journal of the Geological Society, London*, **155**, 745–758, https://doi.org/10.1144/gsjgs.155.5.0745

Friend, C.R.L., Jones, K.A. and Burns, I.M. 2000. New high-pressure granulite event in the Moine Supergroup, northern Scotland: implications for Taconic (early Caledonian) crustal evolution. *Geology* , **28**, 543–546, https://doi.org/10.1130/0091-7613(2000)28<543:NHGEIT>2.0.CO;2

Goodenough, K.M., Millar, I., Strachan, R.A., Krabbendam, M. and Evans, J.A. 2011. Timing of regional deformation and development of the Moine Thrust Zone in the Scottish Caledonides: constraints from the U-Pb geochronology of alkaline intrusions. *Journal of the Geological Society, London*, **168**, 99–114, https://doi.org/10.1144/0016-76492010-020

Grand'Homme, A., Janots, E., Guillaume, D., Bosse, V. and Magnin, V. 2016. Partial resetting of the U-Th-Pb systems in experimentally altered monazite: nanoscale evidence of incomplete replacement. *Geology*, **44**, 431–434, https://doi.org/10.1130/G37770.1

Grand'Homme, A., Janots, E. *et al.* 2018. Mass transport and fractionation during monazite alteration by anisotropic replacement. *Chemical Geology*, **484**, 51–68, https://doi.org/10.1016/j.chemgeo.2017.10.008

Gratz, R. and Heinrich, W. 1997. Monazite-xenotime thermobarometry: experimental calibration of the miscibility gap in the binary system CePO4-YPO4. *American Mineralogist*, **82**, 772–780, https://doi.org/10.2138/am-1997-7-816

Gratz, R. and Heinrich, W. 1998. Monazite-xenotime thermometry. III. Experimental calibration of the partitioning of gadolinium between monazite and xenotime. *European Journal of Mineralogy*, **10**, 579–588, https://doi.org/10.1127/ejm/10/3/0579

Guidotti, C.V., Sassi, F.P., Blencoe, J.G. and Selverstone, J. 1994. The paragonite-muscovite solvus: I. P-T-X limits derived from the Na-K compositions of natural, quasibinary paragonite-muscovite pairs. *Geochimica et Cosmochimica Acta*, **58**, 2269–2275, https://doi.org/10.1016/0016-7037(94)90009-4

Harlov, D.E., Wirth, R. and Hetherington, C.J. 2007. The relative stability of monazite and huttonite at 300–900°C and 200–1000 MPa: metasomatism and the propagation of metastable mineral phases. *American Mineralogist*, **92**, 1652–1664, https://doi.org/10.2138/am.2007.2459

Harlov, D.E., Wirth, R. and Hetherington, C.J. 2011. Fluid-mediated partial alteration in monazite: the role of coupled dissolution-reprecipitation in element redistribution and mass transfer. *Contributions to Mineralogy and Petrology*, **162**, 329–348, https://doi.org/10.1007/s00410-010-0599-7

Harrison, T.M., Célérier, J., Aikman, A.B., Hermann, J. and Heizler, M.T. 2009. Diffusion of 40Ar in muscovite. *Geochimica et Cosmochimica Acta*, **73**, 1039–1051, https://doi.org/10.1016/j.gca.2008.09.038

Hayden, L.A., Watson, E.B. and Wark, D.A. 2008. A thermobarometer for sphene (titanite). *Contributions to Mineralogy and Petrology*, **155**, 529–540, https://doi.org/10.1007/s00410-007-0256-y

Heinrich, W., Andrehs, G. and Franz, G. 1997. Monazite–xenotime miscibility gap thermometry. I. An empirical calibration. *Journal of Metamorphic Geology*, **15**, 3–16, https://doi.org/10.1111/j.1525-1314.1997.t01-1-00052.x

Hetherington, C.J., Jercinovic, M.J., Williams, M.L. and Mahan, K. 2008. Understanding geologic processes with xenotime: composition, chronology, and a protocol for electron probe microanalysis. *Chemical Geology*, **254**, 133–147, https://doi.org/10.1016/j.chemgeo.2008.05.020

Hetherington, C.J., Harlov, D.E. and Budzyn, B. 2010. Experimental metasomatism of monazite and xenotime: mineral stability, REE mobility and fluid composition. *Contributions to Mineralogy and Petrology*, **99**, 165–184, https://doi.org/10.1007/s00710-010-0110-1

Hietpas, J., Samson, S., Moecher, D. and Schmitt, A.K. 2010. Recovering tectonic events from the sedimentary record: detrital monazite plays in high fidelity. *Geology*, **38**, 167–170, https://doi.org/10.1130/G30265.1

Hodges, K.V. and Spear, F.S. 1982. Geothermometry, geobarometry and the Al_2SiO_5 triple point at Mt. Moosilauke, New Hampshire. *American Mineralogist*, **67**, 1118–1134.

Holdsworth, R.E., Dempsey, E. *et al.* 2015. Silurian–Devonian magmatism, mineralization, regional exhumation and brittle strike-slip deformation along the Loch Shin Line, NW Scotland. *Journal of the Geological Society, London*, **172**, 748–762, https://doi.org/10.1144/jgs2015-058

Jercinovic, M.J. and Williams, M.L. 2005. Analytical perils (and progress) in electron microprobe trace element analysis applied to geochronology: background acquisition, interferences, and beam irradiation effects. *American Mineralogist*, **90**, 526–546, https://doi.org/10.2138/am.2005.1422

Kelley, S. 1988. The relationship between K-Ar mineral ages, mica grainsizes and movement on the Moine Thrust Zone, NW Highlands, Scotland. *Journal of the Geological Society, London*, **145**, 1–10, https://doi.org/10.1144/gsjgs.145.1.0001

Kelsey, D.E., Clark, C. and Hand, M. 2008. Thermobarometric modelling of zircon and monazite growth in melt-bearing systems: examples using model metapelitic and metapsammitic granulites. *Journal of Metamorphic Geology*, **26**, 199–212, https://doi.org/10.1111/j.1525-1314.2007.00757.x

Kinny, P.D., Friend, C.R.L., Strachan, R.A., Watt, G.R. and Burns, I.M. 1999. U-Pb geochronology of regional migmatites in East Sutherland, Scotland: evidence for crustal melting during the Caledonian orogeny. *Journal of the Geological Society, London*, **156**, 1143–1152, https://doi.org/10.1144/gsjgs.156.6.1143

Knipe, R.J. 1990. Microstructural analysis and tectonic evolution in thrust systems: examples from the Assynt region of the Moine Thrust Zone, Scotland. *In*: Barber, D.J. and Meredith, P.G. (eds) *Deformation Processes in Minerals, Ceramics and Rocks*. Unwin Hyman, London, 228–261.

Kohn, M.J. and Malloy, M.A. 2004. Formation of monazite via prograde metamorphic reactions among common silicates: implications for age determinations. *Geochimica et Cosmochimica Acta*, **68**, 101–113, https://doi.org/10.1016/S0016-7037(03)00258-8

Kohn, M.J., Spear, F.S. and Valley, J.W. 1997. Dehydration-melting and fluid recycling during metamorphism: Rangeley Formation, New Hampshire, USA. *Journal of Petrology*, **38**, 1255–1277, https://doi.org/10.1093/petroj/38.9.1255

Krabbendam, M., Strachan, R.A., Leslie, A.G., Goodenough, K.M. and Bonsor, H.C. 2011. The internal structure of the Moine Nappe Complex and the stratigraphy of the Morar Group in the Fannichs–Beinn Dearg area, NW Highlands. *Scottish Journal of Geology*, **47**, 1–20, https://doi.org/10.1144/0036-9276/01-419

Krenn, E. and Finger, F. 2010. Unusually Y-rich monazite-(Ce) with 6–14 wt.% Y_2O_3 in a granulite from the Bohemian Massif: implications for high-temperature monazite growth from the monazite-xenotime miscibility gap thermometry. *Mineralogical Magazine*, **74**, 217–225, https://doi.org/10.1180/minmag.2010.073.2.217

Kruhl, J.H. 1998. Reply: Prism- and basal-plane parallel subgrain boundaries in quartz: a microstructural geothermobarometer. *Journal of Metamorphic Geology*, **16**, 142–146, https://doi.org/10.1046/j.1525-1314.1996.00413.x

Kylander-Clark, A.R.C., Hacker, B.R. and Cottle, J.M. 2013. Laser-ablation split-stream ICP petrochronology. *Chemical Geology*, **345**, 99–112, https://doi.org/10.1016/j.chemgeo.2013.02.019

Laurent, A.T., Duchene, S., Bingen, B., Bosse, V. and Seydoux-Guillaume, A.M. 2018. Two successive phases of ultrahigh temperature metamorphism in Rogaland, S. Norway: evidence from Y-in-monazite thermometry. *Journal of Metamorphic Geology*, **36**, 1009–1037, https://doi.org/10.1111/jmg.12425

Law, R.D. 2014. Deformation thermometry based on quartz c-axis fabrics and recrystallization microstructures: a review. *Journal of Structural Geology*, **66**, 129–161, https://doi.org/10.1016/j.jsg.2014.05.023

Law, R.D. and Johnson, M.R.W. 2010. Microstructures and crystal fabrics of the Moine Thrust zone and Moine nappe: history of research and changing tectonic interpretations. *Geological Society, London, Special Publications*, **335**, 443–503, https://doi.org/10.1144/SP335.21

Law, R.D., Casey, M. and Knipe, R.J. 1986. Kinematic and tectonic significance of microstructures and crystallographic fabrics within quartz mylonites from the Assynt and Eriboll regions of the Moine thrust zone, NW Scotland. *Transactions of the Royal Society of Edinburgh: Earth Sciences*, **77**, 99–125, https://doi.org/10.1017/S0263593300010774

Law, R.D., Thigpen, J.R. *et al.* 2021. Tectonic transport directions, shear senses and deformation temperatures indicated by quartz c-axis fabrics and microstructures in a NW-SE transect across the Moine and Sgurr Beag thrust sheets, Caledonian orogen of northern Scotland. *Geosciences*, **11**, 411, https://doi.org/10.3390/geosciences11100411

Leslie, A.G., Krabbendam, M., Kimbell, G.S. and Strachan, R.S. 2010. Regional-scale lateral variation and linkage in ductile thrust architecture: the Oykel Transverse Zone, and 137 mullions, in the Moine Nappe, NW Scotland. *Geological Society, London, Special Publications*, **335**, 359–381, https://doi.org/10.1144/SP335.17

Ludwig, K.R. 2008. *Manual for Isoplot 3.7*. Special Publication 4. Berkeley Geochronology Center, Berkeley CA.

Lusk, A.D.J. and Platt, J.P. 2020. The deep structure and rheology of a plate boundary-scale shear zone: constraints from an exhumed Caledonian shear zone, NW Scotland. *Lithosphere*, **2020**, 1–33, https://doi.org/10.2113/2020/8824736

Mako, C.A. 2019. *Thermal and metamorphic evolution of the Northern Highlands Terrane, Scotland*. PhD thesis, Virginia Tech, Blacksburg VA.

Mako, C.A., Law, R.D., Caddick, M.J., Thigpen, J.R., Ashley, K.T., Cottle, J. and Kylander-Clark, A. 2019. Thermal evolution of the Scandian hinterland, Naver nappe, northern Scotland. *Journal of the Geological Society, London*, **176**, 669–688, https://doi.org/10.1144/jgs2018-224

Mako, C.A., Law, R.D. *et al.* 2021. Growth and fluid-assisted alteration of accessory phases before, during and after Rodinia breakup: U-Pb geochronology from the Moine Supergroup rocks of northern Scotland. *Precambrian Research*, **355**, 106089, https://doi.org/10.1016/j.precamres.2020.106089

McCarthy, G.J., White, W.B. and Pfoertsch, D.E. 1978. Synthesis of nuclear waste monazite, ideal actinide hosts of geologic disposal. *Materials Research Bulletin*, **13**, 1239–1245, https://doi.org/10.1016/0025-5408(78)90215-5

McFarlane, C.R.M., Connelly, J.N. and Carlson, W.D. 2005. Monazite and xenotime petrogenesis in the contact aureole of the Makhavinekh Lake Pluton, northern Labrador. *Contributions to Mineralogy and Petrology*, **148**, 524–541, https://doi.org/10.1007/s00410-004-0618-7

McKinney, S.T., Cottle, J.M. and Lederer, G.W. 2015. Evaluating rare earth element (REE) mineralization mechanisms in Proterozoic Gneiss, Music Valley, California. *Bulletin of the Geological Society of America*, **127**, 1135–1152, https://doi.org/10.1130/B31165.1

Mendum, J.R. and Noble, S.R. 2010. Mid-Devonian sinistral transpressional movements on the Great Glen Fault: the rise of the Rosemarkie Inlier and the Acadian Event in Scotland. *Geological Society, London, Special Publications*, **335**, 161–187, https://doi.org/10.1144/SP335.8

Mogilevsky, P. 2007. On the miscibility gap in monazite-xenotime systems. *Physics and Chemistry of Minerals*, **34**, 201–214, https://doi.org/10.1007/s00269-006-0139-1

Montel, J.M. 1993. A model for monazite/melt equilibrium and application to the generation of granitic magmas. *Chemical Geology*, **110**, 127–146, https://doi.org/10.1016/0009-2541(93)90250-M

Mottram, C.M., Warren, C.J., Regis, D., Roberts, N.M.W., Harris, N.B.W., Argles, T.W. and Parrish, R.R. 2014. Developing an inverted Barrovian sequence; insights from monazite petrochronology. *Earth and Planetary Science Letters*, **403**, 418–431, https://doi.org/10.1016/j.epsl.2014.07.006

Mottram, C.M., Warren, C.J., Halton, A.M., Kelley, S.P. and Harris, N.B.W. 2015. Argon behaviour in an inverted Barrovian sequence, Sikkim Himalaya: the consequences of temperature and timescale on40Ar/39Ar mica geochronology. *Lithos*, **238**, 37–51, https://doi.org/10.1016/j.lithos.2015.08.018

Oliver, G.J.H., Wilde, S.A. and Wan, Y. 2008. Geochronology and geodynamics of Scottish granitoids from the late Neoproterozoic break-up of Rodinia to Palaeozoic collision. *Journal of the Geological Society, London*, **165**, 661–674, https://doi.org/10.1144/0016-76492007-105

Parrish, R.R. 1990. U–Pb dating of monazite and its application to geological problems. *Canadian Journal of Earth Sciences*, **27**, 1431–1450, https://doi.org/10.1139/e90-152

Passchier, C.W. and Trouw, R.A.J. 2005. *Microtectonics*, 2nd edn. Springer, New York.

Paton, C., Hellstrom, J., Paul, B., Woodhead, J. and Hergt, J. 2011. Iolite: freeware for the visualisation and processing of mass spectrometric data. *Journal of Analytical Atomic Spectroscopy*, **26**, 2508–2518, https://doi.org/10.1039/C1JA10172B

Philpotts, A.R. and Ague, J.J. 2022. *Principles of Igneous and Metamorphic Petrology*, 3rd edn. Cambridge University Press.

Pyle, J.M., Spear, F.S., Rudnick, R.L. and Mcdonough, W.F. 2001. Monazite–xenotime–garnet equilibrium in metapelites and a new monazite–garnet thermometer. *Journal of Petrology*, **42**, 2083–2107, https://doi.org/10.1093/petrology/42.11.2083

Pyle, J.M., Spear, F.S. and Wark, D.A. 2002. Electron microprobe analysis of REE in apatite, monazite and xenotime: protocols and pitfalls. *Reviews in Mineralogy and Geochemistry*, **48**, 337–362, https://doi.org/10.2138/rmg.2002.48.8

Seydoux-Guillaume, A.-M., Wirth, R., Heinrich, W. and Montel, J.-M. 2002. Experimental determination of Thorium partitioning between monazite and xenotime using analytical electron microscopy and X-ray diffraction Rietveld analysis. *European Journal of Mineralogy*, **14**, 869–878, https://doi.org/10.1127/0935-1221/2002/0014-0869

Spear, F.S. and Parrish, R.R. 1996. Petrology and Cooling Rates of the Valhalla Complex, British Columbia, Canada. *Journal of Petrology*, **37**, 733–763, https://doi.org/10.1093/petrology/37.4.733

Spear, F.S. and Pyle, J.M. 2002. Apatite, monazite, and xenotime in metamorphic rocks. *Reviews in Mineralogy and Geochemistry*, **48**, 293–335, https://doi.org/10.2138/rmg.2002.48.7

Spear, F.S. and Pyle, J.M. 2010. Theoretical modeling of monazite growth in a low-Ca metapelite. *Chemical Geology*, **273**, 111–119, https://doi.org/10.1016/j.chemgeo.2010.02.016

Spencer, B.M., Thigpen, J.R., Law, R.D., Mako, C.A., McDonald, C.S., Hodges, K.V. and Ashley, K.T. 2020. Rapid cooling during late-stage orogenesis and implications for the collapse of the Scandian retrowedge, northern Scotland. *Journal of the Geological Society, London*, **178**, https://doi.org/10.1144/jgs2020-022

Spencer, B.M., Thigpen, J.R., Gallen, S.F., Dortch, J.M., Hodges, K.V., Law, R.D. and Mako, C.A. 2021. An evaluation of erosional-geodynamic thresholds for rapid orogenic denudation. *Journal of Geophysical Research: Solid Earth*, **126**, https://doi.org/10.1029/2021JB022353

Stewart, M., Strachan, R.A., Martin, M.W. and Holdsworth, R.E. 2001. Constraints on early sinistral displacements along the Great Glen Fault Zone, Scotland: structural setting, U-Pb geochronology and emplacement of the syn-tectonic Clunes tonalite. *Journal of the Geological Society, London*, **158**, 821–830, https://doi.org/10.1144/jgs.158.5.821

Strachan, R.A. and Holdsworth, R.E. 1988. Basement-cover relationships and structure within the Moine rocks and southeast Sutherland. *Journal of the Geological Society, London*, **145**, 23–36, https://doi.org/10.1144/gsjgs.145.1.0023

Strachan, R.A., Harris, A.L., Fettes, D.J. and Smith, M. 2002. The Highland and Grampian terranes. *In*: Trewin, N.H. (ed.) *The Geology of Scotland*, 4th edn. The Geological Society, London, 81–148.

Strachan, R.A., Alsop, G.I., Ramezani, J., Frazer, R.E., Burns, I.M. and Holdsworth, R.E. 2020. Patterns of Silurian deformation and magmatism during sinistral oblique convergence, northern Scottish Caledonides. *Journal of the Geological Society, London*, **177**, 893–910, https://doi.org/10.1144/jgs2020-039

Taghipour, S., Kananian, A., Harlov, D. and Oberhänsli, R. 2015. Kiruna-type iron oxide-apatite deposits, Bafq District, Central Iran: Fluid-aided genesis of fluorapatite-monazite-xenotime assemblages. *Canadian Mineralogist*, **53**, 479–496, https://doi.org/10.3749/canmin.4344

Thigpen, J.R., Law, R.D. *et al.* 2013. Thermal structure and tectonic evolution of the Scandian orogenic wedge, Scottish Caledonides: integrating geothermometry, deformation temperatures and conceptual kinematic-thermal models. *Journal of Metamorphic Geology*, **31**, 813–842, https://doi.org/10.1111/jmg.12046

Thigpen, J.R., Ashley, K.T. and Law, R.D. 2017. Evaluating kinematic displacement rate effects on transient thermal processes in thrust belts using coupled thermomechanical finite-element models. *In*: Law, R.D., Thigpen, J.R., Merschat, A.J. and Stowell, H.H. (eds) *Linkages and Feedbacks in Orogenic Systems. Geological Society of America Memoir*, **213**, 1–23, https://doi.org/10.1130/2017.1213(01)

Thigpen, J.R., Ashley, K.T., Mako, C., Law, R.D. and Spencer, B. 2021. Interplay between crustal-scale thrusting, high metamorphic heating rates, and the development of inverted thermal-metamorphic gradients: numerical models and examples from the Caledonides of Northern Scotland. *Tectonics*, **40**, https://doi.org/10.1029/2021TC006716

Thompson, A.B., Schulmann, K. and Jezek, J. 1997. Thermal evolution and exhumation in obliquely convergent (transpressive) orogens. *Tectonophysics*, **280**, 171–184, https://doi.org/10.1016/S0040-1951(97)00144-3

Tomkins, H.S. and Pattison, D.R.M. 2007. Accessory phase petrogenesis in relation to major phase assemblages in pelites from the Nelson contact aureole, southern British Columbia. *Journal of Metamorphic Geology*, **25**, 401–421, https://doi.org/10.1111/j.1525-1314.2007.00702.x

Villa, I.M., De Bièvre, P., Holden, N.E. and Renne, P.R. 2015. IUPAC-IUGS recommendation on the half life of 87Rb. *Geochimica et Cosmochimica Acta*, **164**, 382–385, https://doi.org/10.1016/j.gca.2015.05.025

Watson, E.B., Wark, D.A. and Thomas, J.B. 2006. Crystallization thermometers for zircon and rutile. *Contributions to Mineralogy and Petrology*, **151**, 413–433, https://doi.org/10.1007/s00410-006-0068-5

Wellman, C.H. 2015. Spore assemblages from the Lower Devonian 'Lower Old Red Sandstone' deposits of the

Northern Highlands of Scotland: the Berriedale Outlier. *Earth and Environmental Science Transactions of the Royal Society of Edinburgh*, **105**, 227–238, https://doi.org/10.1017/S1755691015000055

Williams, M.L., Jercinovic, M.J., Harlov, D.E., Budzyń, B. and Hetherington, C.J. 2011. Resetting monazite ages during fluid-related alteration. *Chemical Geology*, **283**, 218–225, https://doi.org/10.1016/j.chemgeo.2011.01.019

Winchester, J.A. 1974. The zonal pattern of regional metamorphism in the Scottish Caledonides. *Journal of the Geological Society, London*, **130**, 509–524, https://doi.org/10.1144/gsjgs.130.6.0509

Wu, C-M. and Cheng, B-H. 2006. Valid garnet–biotite (GB) geothermometry and garnet–aluminum silicate–plagioclase–quartz (GASP) geobarometry in metapelitic rocks. *Lithos*, **89**, 1–23, https://doi.org/10.1016/j.lithos.2005.09.002

In situ Pb–Pb garnet geochronology as a tool for investigating polymetamorphism: a case for Paleoarchean lateral tectonic thickening

K. A. Cutts[1,2*], C. Lana[1], G. Stevens[3] and I. S. Buick[1,3]

[1]Departamento de Geologia, Escola de Minas, Universidade Federal de Ouro Preto, Morro do Cruzeiro, 35400-000 Ouro Preto, MG, Brazil

[2]Geological Survey of Finland, P.O. Box 96, FI-02151 Espoo, Finland

[3]Centre for Crustal Petrology, Department of Earth Sciences, Stellenbosch University, Private Bag X1, Matieland 7602, South Africa

KAC, 0000-0002-7190-1944
*Correspondence: kathryn.cutts@gmail.com; kathryn.cutts@gtk.fi

Abstract: The Barberton Granite–Greenstone Belt remains a key location in the debate concerning the nature of Archean tectonic processes. Much work has focused on deciphering the tectonic significance of the *c.* 3.23 Ga metamorphism, as this has been correlated with lower geothermal gradient conditions potentially indicating Archean subduction. However, several studies also found evidence of an earlier, 3.45 Ga metamorphic episode, overprinted by the 3.23 Ga event. Here we apply *in situ* Pb–Pb dating and *P–T* modelling to a large (3 cm diameter) garnet crystal, allowing for the direct dating of the metamorphic conditions obtained from the garnet. The garnet core produced an isochron age of 3435 ± 45 Ma, corresponding to an increase in *P* and *T* evolution reaching peak conditions of at least 7 kbar and 700°C. Analyses obtained from the garnet rim give an isochron age of 3245 ± 41 Ma, corresponding to *P–T* conditions reaching 8–9 kbar and 700°C. The preservation of two moderate- to high-pressure events occurring 200 million years apart is consistent with lateral tectonic processes producing crustal thickening at 3.2 Ga and may also be a viable process for the earlier event.

Supplementary material: Description of CT imaging, analytical settings for Pb-Pb age dating and supplemental P-T diagrams, garnet mineral chemistry, whole rock XRF data, garnet trace element data and garnet and standard Pb-Pb age data are available at https://doi.org/10.6084/m9.figshare.c.6724419

The Barberton Granite–Greenstone Belt (BGGB) is the oldest and southeasternmost in a series of NE–SW-trending greenstone belts in the northeastern Kaapvaal Craton (Fig. 1a). It constitutes a key locality in the debate concerning the onset of plate tectonics (Anhaeusser *et al.* 1983; Kisters *et al.* 2003; Moyen *et al.* 2006; Van Kranendonk *et al.* 2009, 2014; Cutts *et al.* 2014, 2015; Wang *et al.* 2019). This debate has largely concerned the nature of the *c.* 3.23 Ga event in the BGGB with numerous studies favouring a vertical tectonic process producing this event (Van Kranendonk *et al.* 2009, 2014; Van Kranendonk 2011; Wang *et al.* 2019) and a similarly large number of studies advocating for a horizontal tectonic process (De Wit *et al.* 1992; De Ronde and De Wit 1994; Kisters *et al.* 2003; Dziggel *et al.* 2006; Moyen *et al.* 2006; Cutts *et al.* 2014, 2015; Diener and Dziggel 2021). Several of these studies have revealed the presence of an earlier event at *c.* 3.45 Ga that has previously been interpreted as a contact or regional metamorphic event related to intrusion of the early trondhjemite–tonalite–granodiorite rocks (TTGs) (Cutts *et al.* 2014), duplication of rocks in the southern part of the BGGB (Kisters *et al.* 2010) or a major tectonic uplift and accretion event at 3.43 Ga (Grosch *et al.* 2011). In the southern BGGB, the high grade of metamorphism related to the subsequent *c.* 3.23 Ga event has largely obscured the nature of the earlier event (Armstrong *et al.* 1990; Dziggel *et al.* 2002; Van Kranendonk *et al.* 2009; Cutts *et al.* 2014; Wang *et al.* 2019). The two most recent studies identifying the *c.* 3.45 Ga event have attempted to discern *P–T* conditions. Cutts *et al.* (2014) investigated two samples from the same locality, one producing a monazite U–Pb age of 3436 ± 18 Ma, and another with garnet cores containing a relict foliation that indicated peak *P–T* conditions of 4–5 kbar and 550°C based on *P–T* pseudosection modelling (Fig. 1b, point 4). Wang *et al.* (2019) found metamorphic zircon ages of 3443 ± 13 Ma and 3419 ± 8 Ma from samples that give *P–T* conditions of 550–625°C and 8.5–9 kbar, and 650–725°C and 6–8 kbar, respectively, obtained using conventional

From: van Schijndel, V., Cutts, K., Pereira, I., Guitreau, M., Volante, S. and Tedeschi, M. (eds) 2024. *Minor Minerals, Major Implications: Using Key Mineral Phases to Unravel the Formation and Evolution of Earth's Crust.* Geological Society, London, Special Publications, **537**, 209–229.
First published online September 8, 2023, https://doi.org/10.1144/SP537-2022-339

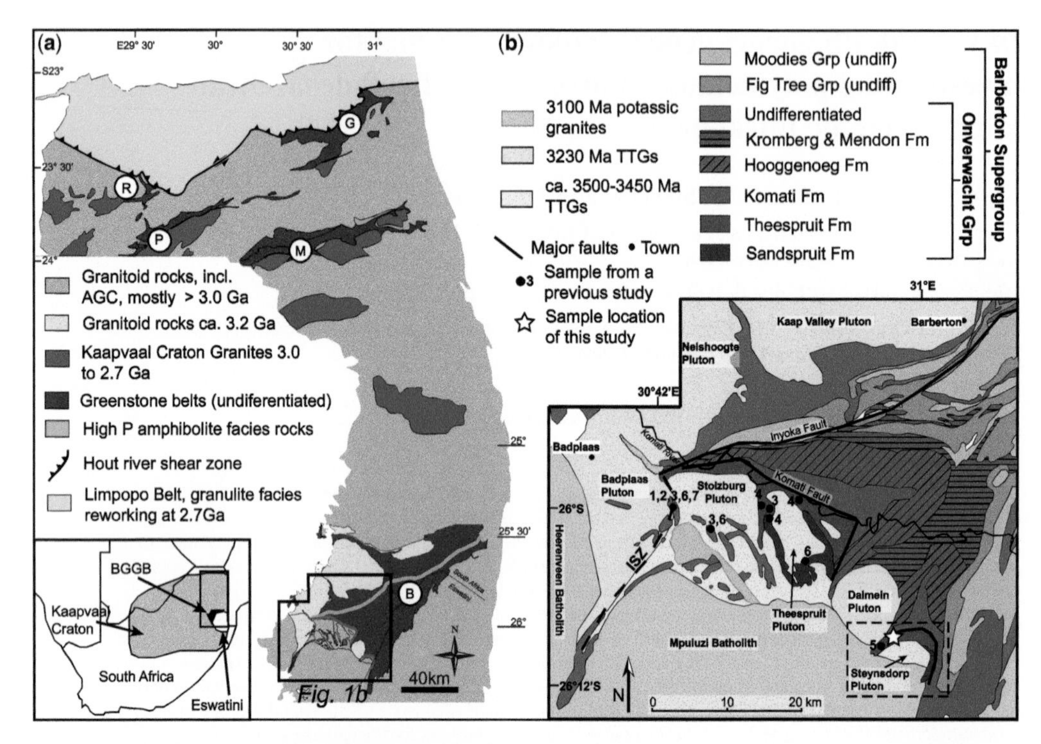

Fig. 1. (**a**) Simplified geological map of the northeastern Kaapvaal Craton showing the major, approximately NE–SW-oriented greenstone belts: P, Pietersburg (<2.9 Ga); R, Renosterkopies; G, Gyiani (3.2–2.8 Ga); M, Murchison (3.1–2.9 Ga); B, Barberton (3.45–3.2 Ga); AGC, Ancient Gneiss Complex (3.66-3.4 Ga). The location of this figure is indicated by the box on the inset map. The green line indicates the location of the Inyoka Fault and the purple line is the Komati Fault. The location of Figure 1b is also indicated. (**b**) Detailed geological map of the southwestern Barberton Granite–Greenstone Belt (BGGB) with the location of the sample used in this study indicated by the white star. The black dashed box indicates the location of the Steynsdorp area. The numbered black spots 1 to 7 are the locations of samples used in previous studies. ISZ, Inyoni Shear Zone; TTGs, trondhjemite–tonalite–granodiorite rocks. Source: (a) P: De Wit *et al.* (1992); G and M: Block *et al.* (2013); (b) 1 – Moyen *et al.* (2006); 2 – Nédélec *et al.* (2012); 3 – Diener and Dziggel (2021); 4 – Cutts *et al.* (2014); 5 – Lana *et al.* (2010); 6 – Wang *et al.* (2019); 7 – Cutts *et al.* (2021).

thermobarometry. Cutts *et al.* (2014) interpreted this early event as potentially a result of thermal metamorphism due to intrusion of the nearby *c.* 3.44 Ga Theespruit Pluton, although the garnet cores that they interpret to be associated with this event contain preserved foliations indicating they are syn- or post-tectonic. Wang *et al.* (2019) interpret both the 3.4 Ga and the later 3.2 Ga event to be associated with magmatism and result from partial convective overturn. Both previous studies produced *P–T* estimates using samples from directly adjacent to the Theespruit and Stolzburg plutons, which are also dated at *c.* 3.45 Ga (Fig. 1b; Kamo and Davis 1994; Cutts *et al.* 2014; Wang *et al.* 2019). This study targets a sample obtained from the Theespruit Formation occurring adjacent to the *c.* 3.51 Ga Steynsdorp Pluton (Fig. 1b). If the older metamorphism is related to intrusion of the TTGs then a sample from next to the Steynsdorp Pluton should produce an older metamorphic age.

The objective of this study is to resolve the nature of this early event by directly dating the garnet using the Pb–Pb system and using the same garnet grain to determine *P–T* conditions for both the 3.4 and 3.2 Ga events. There is presently a resurgence of interest in garnet dating with the advent of increasing sensitivity and sophistication of instrumentation (i.e. Seman *et al.* 2017; Millonig *et al.* 2020; Schannor *et al.* 2021; Simpson *et al.* 2021, 2023; Tamblyn *et al.* 2022). The utility of garnet ages is clear. As the ultimate petrochronometer (Baxter *et al.* 2017), it is possible to infer precise metamorphic conditions from the same garnet crystal that it is now possible to date with a variety of methods (U–Pb and Lu–Hf; Seman *et al.* 2017; Simpson *et al.* 2021). This study adds to the growing number that utilize garnets as a geochronometer and highlights the utility of obtaining age and *P–T* conditions from one crystal in order to unravel Archean polymetamorphism and gain insight into Archean tectonism.

Geological setting

The Barberton Granite–Greenstone Belt

The Paleo–Mesoarchean BGGB is situated in the Mpumalanga Province of South Africa and consists of a sequence of 3.55–3.22 Ga volcano-sedimentary successions known as the Barberton Supergroup (Fig. 1; De Wit *et al.* 1992; De Ronde and de Wit 1994; Kamo and Davis 1994; Dziggel *et al.* 2002, 2006; Kisters *et al.* 2003, 2010; Drabon *et al.* 2019). The BGGB is oriented NE–SW and is one in a sequence of NE–SW-oriented greenstone belts that developed on the northeastern edge of the Kaapvaal Craton (Fig. 1a). The Barberton Supergroup is broadly divided into three groups: (1) the Onverwacht Group (ultramafic, mafic and felsic volcanic rocks; *c.* 3.55–3.3 Ga); (2) the Fig Tree Group (clastic to volcaniclastic sediments, *c.* 3.26–3.225 Ga); and (3) the Moodies Group (sandstones and conglomerates, *c.* 3.225–3.215 Ga; Anhaeusser 1976; Sanchez-Garrido *et al.* 2011). The structurally lowermost units of the Onverwacht Group are the Sandspruit and Theespruit formations. These two units occur in the Stolzburg Domain in the southern BGGB (Fig. 1b) and are intruded by trondhjemite–tonalite–granodiorite (TTG) plutons: Steynsdorp (*c.* 3.51 Ga), Theespruit (*c.* 3.45 Ga) and Stolzburg (*c.* 3.45 Ga). The Stolzburg Domain is located in the region south of the Komati Fault and east of the Inyoni Shear Zone (Fig. 1b). Rocks in the Stolzburg Domain reached amphibolite facies conditions (Dziggel *et al.* 2002; Stevens *et al.* 2002; Kisters *et al.* 2003; Diener *et al.* 2005). A study by Cutts *et al.* (2014) utilized phase diagram modelling to indicate peak conditions of *c.* 8.5 kbar and 640°C during the *c.* 3.2 Ga event. Moyen *et al.* (2006) investigated rocks from the Inyoni Shear Zone using the average *P–T* method of THERMOCALC and indicated peak conditions of 12–15 kbar and 600–650°C, which they attributed to the 3.23 Ga event. Later *P–T* phase diagram modelling by Cutts *et al.* (2021) suggested more modest conditions of *c.* 8–10 kbar and 650°C, combined with garnet Sm–Nd ages of 3202–3200 Ma, indicating that the metamorphism in the Inyoni Shear Zone may be slightly younger than the regional metamorphism. Diener and Dziggel (2021) used *P–T* modelling on samples from the Inyoni Shear Zone, the central Stolzburg Domain and the Tjakastad Schist Belt, which occurs between the Stolzburg and Theespruit plutons (Fig. 1b). Their results indicated that the whole of the Stolzburg Domain experienced similar peak *P–T* conditions of *c.* 10 kbar and 650–700°C for the *c.* 3.2 Ga event. The upper units of the Onverwacht Group (Komati, Hooggenoeg, Kromberg and Mendon formations) north of the Komati Fault experienced only greenschist facies metamorphism (Kisters *et al.* 2003). The Komati Fault is a 1 km-wide ductile–brittle extensional detachment (Dziggel *et al.* 2002, 2006; Kisters *et al.* 2003).

The Steynsdorp area

The Steynsdorp area occurs in the southernmost part of the BGGB (dashed box in Fig. 1b) and is composed of 3511–3502 Ma TTG gneisses (Kröner *et al.* 1996) and the overlying 3540–3530 Ma Theespruit Formation (Kamo and Davis 1994; Kröner *et al.* 1996). In the south of the Steynsdorp area, the gneisses are intruded by the 3.1 Ga Mpuluzi potassic granite batholith (Fig. 1b; Lana *et al.* 2010). Lana *et al.* (2010) present a detailed study of the Steynsdorp area, indicating that the supracrustal rocks reached peak metamorphic conditions of 10–13 kbar and 640–660°C, which they interpret to have occurred during the regional metamorphic peak at 3.23 Ga. These are similar to the conditions obtained from the Inyoni Shear Zone by Moyen *et al.* (2006). The Steynsdorp Pluton and Theespruit Formation contain a unidirectional northeastward-directed lineation (see Lana *et al.* 2010, Fig. 2). This, together with consistent granitoid-up–greenstone-down kinematics, point to extrusion of the TTG gneisses along an extensional detachment. The contact between the amphibolite facies Theespruit Formation and the greenschist facies Komati Formation is a 20–50 m-wide zone where altered pillow lavas of the Komati Formation become elongated and highly sheared rods approach the contact with the Theespruit Formation (Lana *et al.* 2010). This boundary represents a metamorphic break with *P–T* estimates of 2.5 kbar and 350°C for the Komati Formation (Cloete 1999), indicating 6–8 kbar of difference between these units. Lana *et al.* (2010) presented no metamorphic ages but indicate the presence of large garnet porphyroblasts with Mn zonation indicative of two metamorphic cycles. They interpret that the cores of these garnets record a pre-3.2 Ga metamorphic event.

Sample description

Sample BKC-3 was sampled from the Theespruit Formation in a river exposure of the Steynsdorp area (26° 09′ 18.5″ S, 030° 57′ 07.5″ E; WGS84; Fig. 1b). The outcrop consists of a strongly foliated grey gneiss with fine (mm–cm scale) dark and light banding, as well as thin quartzose layers (Fig. 2a). The largely fine-grained matrix contains large garnet porphyroblasts (Fig. 2b). The matrix consists of biotite, chlorite, epidote, quartz, plagioclase and amphibole with minor ilmenite, apatite and titanite. In the

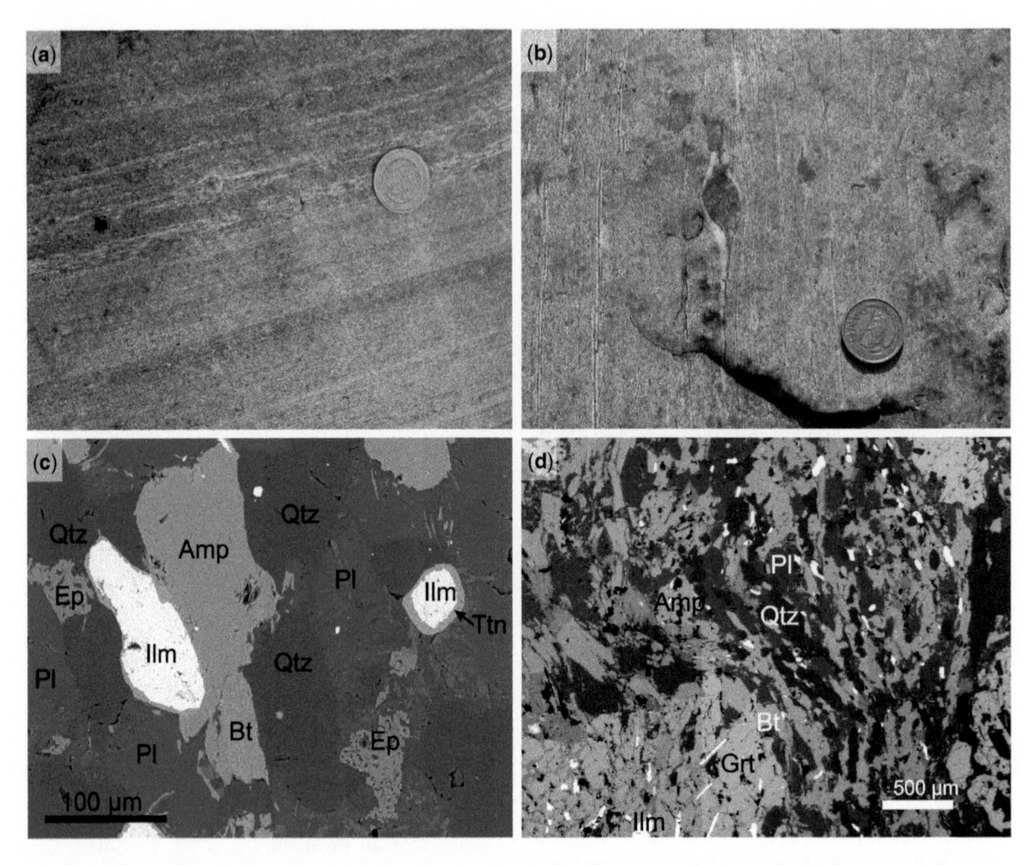

Fig. 2. (**a**) Outcrop image from the sample locality showing the banding present in the gneisses, the quartzose layers and the strong fabric. Coin is 2.5 cm in diameter. (**b**) Outcrop image showing one of the large garnet porphyroblasts wrapped by a quartz rich band. Several melanocratic layers are also apparent in this image, generally next to the quartz bands. Coin is 2.5 cm in diameter. (**c**) Backscattered electron (BSE) image of the matrix of the sample. The elongate amphibole (Amp), biotite (Bt) and ilmenite (Ilm) are aligned parallel to the foliation of the sample. Fine-grained epidote (Ep) grows adjacent to or inside plagioclase (Pl). Titanite (Ttn) occurs exclusive as rims on ilmenite. (**d**) BSE image of the matrix adjacent to a garnet grain. Here it is apparent that elongate quartz (Qtz) and plagioclase (Pl) are also parallel to the foliation and the foliation wraps around the garnet (Grt).

matrix of the sample, ilmenite is rimmed by titanite, there are also symplectites of quartz and epidote in the matrix, and plagioclase has albitic compositions on the rims where these symplectites occur (Fig. 2c). The foliation defining minerals are amphibole, biotite and ilmenite (Fig. 2d). Biotite in the matrix occurs as small, elongate, rounded grains with amphibole oriented parallel to the matrix foliation (Fig. 2c), but can also occur as large euhedral grains on garnet rims or in fractures in garnet grains. Where quartz and plagioclase are elongate, they are aligned parallel to the foliation (Fig. 2d). Some garnet grains are surrounded by a thin film of quartz, which increases in size in the pressure shadows of the garnet and displays a dextral shear sense (Fig. 2b). The thin quartzose layers are associated with thin dark bands consisting of biotite and amphibole, which can also wrap garnet grains (Fig. 2b).

Garnet grains are large (up to 3 cm in diameter; Fig. 2b), commonly fractured and contain numerous inclusions. The larger inclusions occur dominantly as a discrete band, which is a focus of large fractures in the grain (Fig. 3a–c). This band occurs between the largely inclusion-free core and rim (Fig. 3a–c). Minerals included in garnet are quartz, plagioclase, ilmenite, apatite and biotite. Titanite, epidote, allanite and chlorite occur exclusively on the large fractures in the garnet. The garnet core has fine inclusions (<50 µm, usually much less) of ilmenite, apatite, biotite and quartz. There does not seem to be a preferred orientation for these, although in some places the ilmenite appears to occur in clumps. Generally, the matrix foliation wraps the garnet grains; however, in some places it is possibly continuous with ilmenite grains in the garnet that seem to have a preferred orientation.

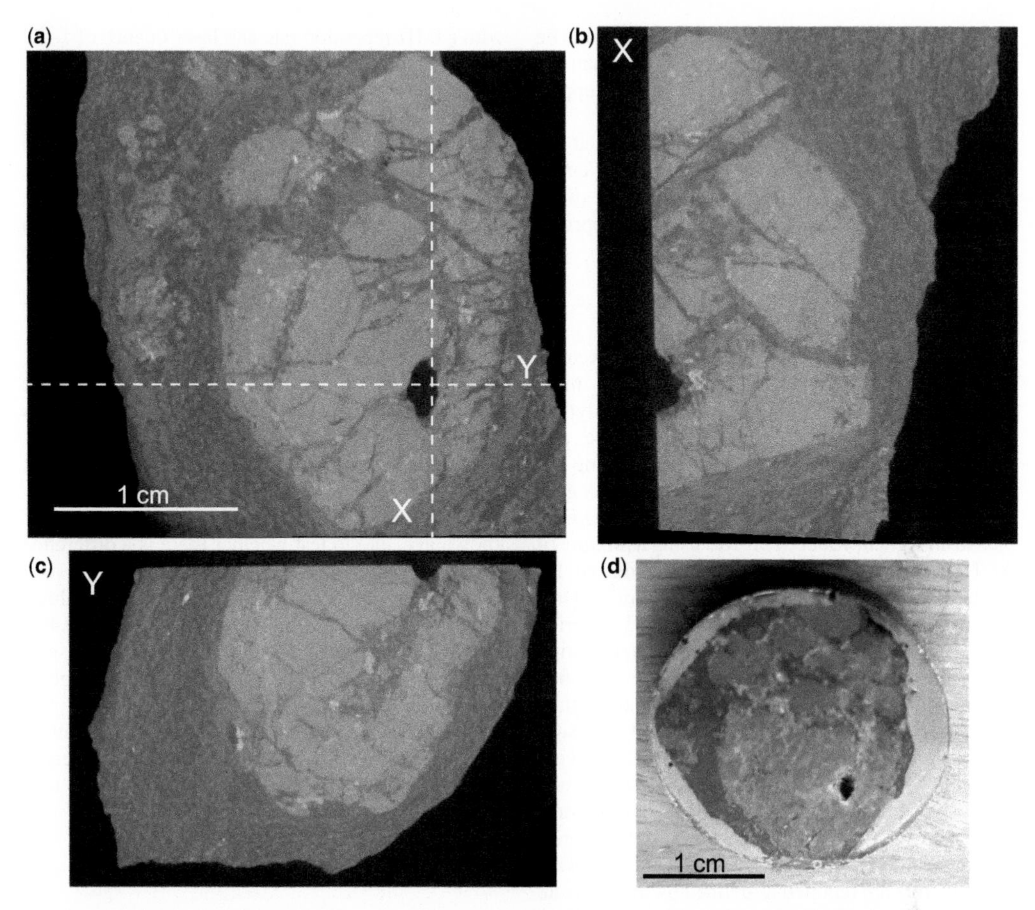

Fig. 3. (**a–c**) CT images of garnet from sample BKC-3 (see Supplementary Material for the analytical methods of this technique). These images were collected in order to locate the ideal garnet grain for this study, and they display density contrasts. They also allow for the imaging of the garnet grain in three dimensions with the investigated surface displayed in (a) (compare with Fig. 3d) and cross-sections X and Y shown in (b, c). Large fractures in the garnet grain are apparent surrounding the largely inclusion free core. These fractures are focused in the inclusion-rich zone (bright inclusions are ilmenite). This garnet was later cut out of the sample and mounted in epoxy (**d**).

Analytical methods

Mineral chemistry

Following CT imaging (see Supplementary Material for details), a single large garnet was cut out of sample BKC-3 and the weathered surface ground off before being mounted in epoxy and polished (Fig. 3d). Major element X-ray maps were conducted for Fe, Mg, Mn and Ca using a JEOL SuperProbe JXA-8200 microprobe hosted at the Steinmann-Institut, Bonn. X-ray maps were collected in wavelength dispersive mode with an accelerating voltage of 15 kV, a 50 nA beam current and a dwell time of 150 ms. Mineral major element compositions were analysed using a Zeiss EVO MA15 Scanning Electron Microscope in the Central Analytical Facility at Stellenbosch University. Textures were studied in backscattered electron (BSE) mode and mineral compositions quantified by EDX (Energy Dispersive X-ray) analysis using an X-Max 20 mm^2 ED X-ray detector and Oxford INCA software. Beam conditions were 20 kV accelerating voltage and 1.5 nA probe current, with a working distance of 8.5 mm and a specimen beam current of −19.0 to −20 nA. X-ray counts were typically *c.* 7000 cps and the counting time was 10 s live-time. Analyses were quantified using natural mineral standards. Comparisons between measured and accepted compositions of control standards analysed within this laboratory have been published by Diener *et al.* (2005) and Moyen *et al.* (2006) as a reflection of the accuracy of the analytical technique. The amount of Fe_2O_3 was calculated for garnet and ilmenite using the method of Droop (1987), although it was found to be low in both minerals (<1 wt%, usually less than 0.5 wt%). All garnet analyses are included in Supplementary Table 1.

The sample bulk composition was obtained via whole-rock XRF analysis carried out at the Central Analytical Facility at Stellenbosch University (Supplementary Table 2). The sample size was approximately 15 cm × 5 cm × 3 cm.

Garnet trace element analysis

Garnet trace elements analyses were undertaken by laser-ablation inductively-coupled-plasma mass spectrometry (LA-ICP-MS) at the Central Analytical Facility, Stellenbosch University. The trace and major element traverses were carried out along the same line. A 40 µm-diameter ablation spot was generated by a New Wave 213 nm Nd-YAG Laser coupled to an Agilent 7500ce ICP-MS using a mixture of Ar–He as the carrier gas. Operating conditions for the laser were 5 Hz frequency and 6 J cm^{-2}. Data were acquired in time-resolved mode (Longerich *et al.* 1996), which allowed potential contamination from mineral inclusions or fractures to be identified and excluded from the analysis. NIST-610 glass was used as a primary reference standard and SiO_2 (39 wt%) was used for internal normalization. Accuracy and reproducibility of multiple analyses was established from the analysis of NIST 612. Results were better than 5% relative for most elements. Data were processed using Glitter (v 4.4.2) software and 1σ errors are reported as defined by the software (Supplementary Table 3). Chondrite-normalized trace element values were corrected using the normalization values of McDonough and Sun (1995).

Garnet Pb–Pb dating

Garnet Pb–Pb dating was conducted on the same garnet grain mounted in epoxy resin. Analyses were conducted at the Universidade Federal de Ouro Preto using a Thermo-Finnigan Neptune multi-collector ICP-MS coupled with a Photon-Machines 193 nm excimer laser system (LA-MC-ICP-MS). For Pb–Pb data acquisition, the magnet was settled on a virtual mass (*c.* 223.2) in the centre of the array and all relevant masses were measured simultaneously (^{202}Hg, ^{204}($^{204}Pb + ^{204}Hg$), ^{206}Pb, ^{207}Pb, ^{208}Pb, ^{232}Th and ^{238}U). Data collection consisted of 550 cycles (or sweeps of *c.* 0.1 s each) of all the relevant masses and ratios (the method of Lana *et al.* (2017) was followed), and a table detailing all analytical conditions for the Pb–Pb dating is included in the Supplementary Material). A 15 s background was collected prior to 40 s of sample acquisition, with a typical analysis time of 55 s. A spot size of 130 µm was used for all garnet analyses, with a 6 Hz repetition rate and laser fluence of 1–2 J cm^{-2}. Soda-lime glass SRM-NIST 614 was used as the primary reference material (with a spot size of 85 µm), and BHVO (Woodhead and Hergt 2007) and a garnet grain of known age (internal standard Zeek) were analysed in the same analytical session. Raw data were corrected offline using Saturn software (Silva *et al.* 2022). The $^{207}Pb/^{206}Pb$ ratio was corrected for mass bias (0.08%) and the $^{206}Pb/^{238}U$ ratio for inter-element fractionation (*c.* 25%), including drift over the 2 h of sequence time, using SRM-NIST 614 ($n = 10$).

Plots and ages were calculated using IsoplotR (Vermeesch 2018). All uncertainties are reported at the 2σ level. Analyses of unknowns and standards are included in Supplementary Table 4.

A garnet from a skarn in the Bushveld contact aureole known as Zeek was analysed in order to test the setup using a matrix matched material of known age, and to test whether there is a difference between Pb–Pb and U–Pb ages in garnet (i.e. as a result of Pb loss). Zeek produced a Pb–Pb isochron age of 2031 ± 127 Ma ($n = 19$, MSWD = 0.48). This garnet is usually used as a U–Pb internal standard, previously producing a U–Pb intercept age of 2058 ± 3.5 Ma ($n = 9$, MSWD = 1.09; Marques *et al.* 2023). The Bushveld intrusion age is 2059 ± 1 Ma (Buick *et al.* 2001), although more recent work indicates an emplacement interval of *c.* 5 myr from 2060 to 2055 Ma (Scoates *et al.* 2021).

Results

Garnet major and trace element chemistry

The garnet preserves complicated major element zonation patterns (Fig. 4a, b). Cores are rich in X_{Sps} ($= Mn/(Mn + Ca + Mg + Fe^{2+})$) with a maximum value of 0.16, but this drops to 0.03 moving towards the inclusion-rich fracture zone between the core and rim (Fig. 4a, b). The core of the garnet is defined as the region with the highest X_{Sps} content and extends towards the first low in X_{Sps} (0.03, seen on the left-hand side of the garnet traverse at 2.5 mm on the *x*-axis of Fig. 4a). In the rim, X_{Sps} rises abruptly to 0.1 before gradually dropping towards the grain edge (0.02; Fig. 4a). X_{Pyr} ($= Mg/(Mn + Ca + Mg + Fe^{2+})$) rises steadily from core until the inclusion-rich fracture zone (0.04 to 0.07). In the rim, X_{Pyr} abruptly drops to 0.05 before rising again towards the grain edge to 0.07. In the core, X_{Grs} ($= Ca/(Mn + Ca + Mg + Fe^{2+})$) gradually decreases from 0.15 to 0.11 before rising sharply within the inclusion-rich fracture zone from 0.19 to 0.22. After the inclusion-rich fracture zone, at the start of the rim, X_{Grs} abruptly drops before rising steadily in the rim (0.15 to 0.25). X_{Alm} ($= Fe^{2+}/(Mn + Ca + Mg + Fe^{2+})$) rises gradually towards

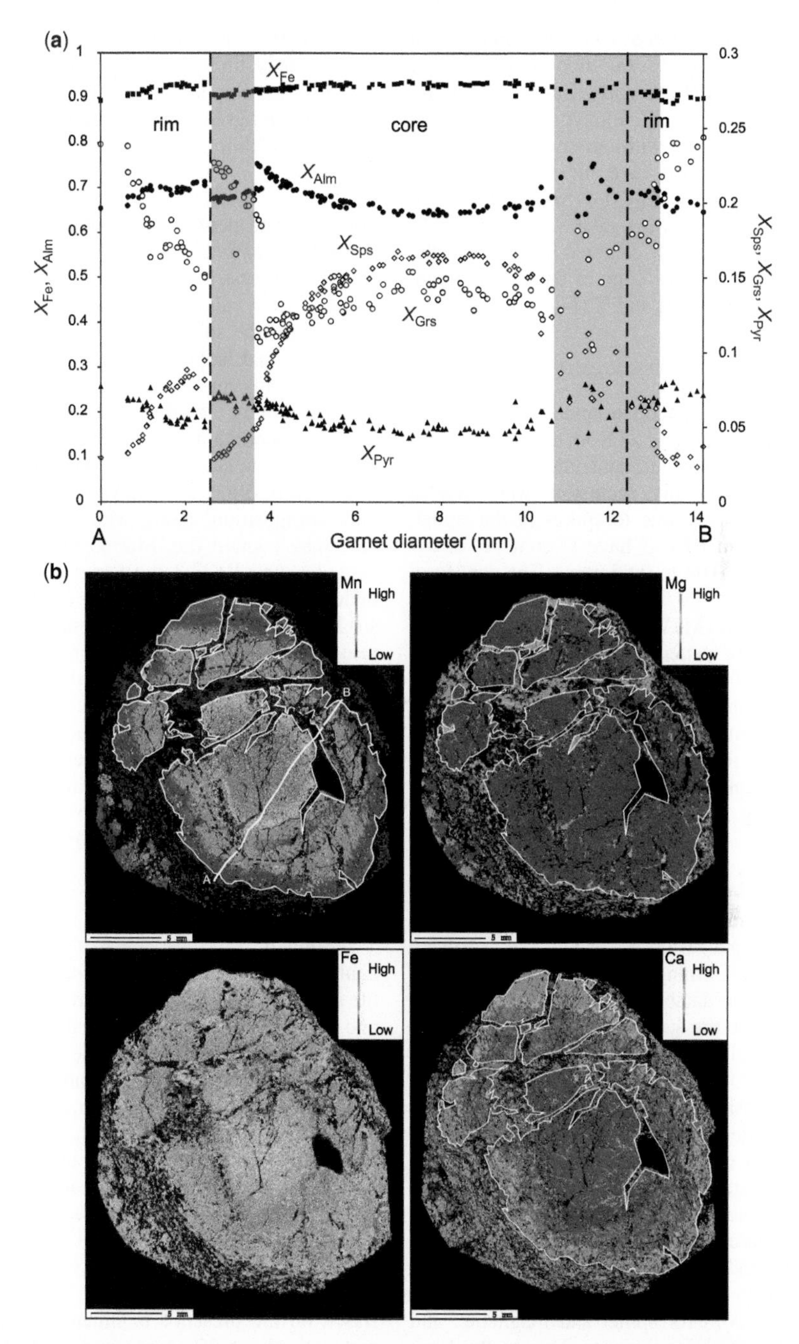

Fig. 4. (**a**) Major element traverse of garnet from sample BKC-3. The red fields indicate the zones where a large fracture and numerous inclusions interrupt the traverse. The traverse location is indicated on the Mn map in part (b). $X_{Alm} = (Fe^{2+}/(Mn + Ca + Mg + Fe^{2+}))$; $X_{Fe} = (Fe^{2+}/(Fe^{2+} + Mg))$; $X_{Grs} = (Ca/(Mn + Ca + Mg + Fe^{2+}))$; $X_{Pyr} = (Mg/(Mn + Ca + Mg + Fe^{2+}))$; $X_{Sps} = (Mn/(Mn + Ca + Mg + Fe^{2+}))$. (**b**) Elemental maps of Mn, Mg, Fe and Ca. The core–rim boundary is also marked on the Mn map with a dashed black line.

the inclusion-rich domain from 0.64 to 0.73. The composition changes abruptly at the start of the inclusion-rich domain to 0.7 and then gradually drops to 0.65. On the core–rim boundary, X_{Alm} abruptly increases to 0.72, then gradually decreases towards the grain edge (0.65).

The garnet trace element traverse indicates that garnet cores are slightly enriched in heavy rare earth elements (HREEs; $Gd_N/Lu_N = 0.05–0.48$, with an average of 0.17) relative to garnet rims ($Gd_N/Lu_N = 0.07–4.7$, with an average of 1.21; Fig. 5a). The cores also have slightly elevated Eu anomalies ($Eu/Eu^* = Eu_N/(\sqrt{(Sm_N^*Gd_N)})$) of 1.18 to 3.61, whereas rims give values of 0.8 to 1.8. Zoning profiles show elevated Ti and V in the cores (Fig. 5b, c). On the edge of the core domain, there is an increase in Lu and Y (Fig. 5b).

Garnet Pb–Pb geochronology

In sample BKC-3, 20 uncontaminated analyses were obtained from the core and 28 from the rim (Fig. 6a). All analyses that have significant amounts of U or Th were removed. Generally, inclusions were easy to detect during analysis due to spikes in the signal. Garnet cores from BKC-3 have U contents below 1 ppm (generally 0.04 to 0.85 ppm). The data form a $^{206}Pb/^{204}Pb$ v. $^{207}Pb/^{204}Pb$ isochron with an age of 3435 ± 45 Ma (MSWD: 1.3; $n = 20$; Fig. 6b). Garnet rims have generally overlapping to higher U contents of 0.02 to 2.19 ppm. Two analyses plotted off the isochron, potentially due to contamination, and are removed from age calculations. The rest of the data form a $^{206}Pb/^{204}Pb$ v. $^{207}Pb/^{204}Pb$ isochron with an age of 3245 ± 41 Ma (MSWD: 1.1; $n = 26$; Fig. 6c).

P–T modelling

Pressure–temperature pseudosections were calculated for sample BKC-3 using the software package Theriak-Domino (De Capitani and Petrakakis 2010) and the database of Holland and Powell (2011; ds62) for the geologically realistic system MnNCKFMASHTO ($MnO–Na_2O–CaO–K_2O–FeO–MgO–Al_2O_3–SiO_2–H_2O–TiO_2–Fe_2O_3$).

The 'metabasite set' of models from Green *et al.* (2016), converted to Theriak-Domino format by Doug Tinkham (see Jorgensen *et al.* 2019), were applied. These are: White *et al.* (2014) for orthopyroxene, garnet, biotite, muscovite and chlorite; Green *et al.* (2016) for clinoamphibole, augite and metabasite melt; Holland and Powell (2011) for olivine and epidote; Holland and Powell (2003) for plagioclase; White *et al.* (2002) for spinel and magnetite; and White *et al.* (2000) for ilmenite. Due to the large amount of Mn present in the garnet, MnO was included in the system despite a lack of Mn end-members in clinoamphibole, augite and ilmenite. The effect of this will be a larger stability field for garnet and potentially the presence of higher MnO contents in the modelled garnet than what appears in the sample. A diagram without MnO was also calculated using the same Fe^{3+} content as the final diagram (90% Fe^{2+}) and garnet only occurred in an extremely limited region of *P–T* space (high *P* and *T* – see Supplementary Figure 1). Given the size of garnet grains in this sample, including MnO in the modelling seems to reflect reality more accurately.

As the mineral assemblages and field observations indicate peak metamorphic conditions to be sub-solidus or not significantly melt-bearing, H_2O was set in excess for the sub-solidus parts of the diagram. A H_2O value allowing for a wet solidus was selected for the supra-solidus parts of the diagram. Compositional isopleths for garnet were calculated to aid with interpretation of the *P–T* path.

Determining the effective bulk composition seen by samples containing large porphyroblasts can be problematic because as the porphyroblasts grow, their composition is removed from the effective bulk composition, along with any inclusions that they may contain (i.e. Marmo *et al.* 2002). In the case of sample BKC-3, the garnet contains numerous inclusions of apatite and ilmenite (with 5 wt% MnO), so along with determining the optimum Fe^{3+} content to use, we also explored how changes in the MnO and CaO contents would affect the topology of diagrams and calculated chemical compositions.

Initially, a temperature-composition ($T–M_{Fe_2O_3}$) diagram was calculated in order to determine the most suitable Fe^{3+} value to set in order to represent the mineralogy that we observe in the sample (Fig. 7a). It was found that if Fe^{3+} content is too high, magnetite becomes stable. Since no magnetite is observed in the samples, an Fe^{3+} content of 10% was used (Fig. 7a).

Following this, the effect of reducing the CaO content was investigated. Based on an estimated apatite content of 2–3%, CaO was reduced by 1.5 wt% and a $T–M_{CaO}$ diagram was calculated (Fig. 7b). The reduction of CaO reduces the stability of epidote and increases the stability of white mica, neither of which are interpreted as peak minerals. This diagram was also contoured for the X_{Grs} content of garnet. The left-hand side of the diagram has X_{Grs} values of 0.28–0.32, whereas garnet in the sample has X_{Grs} ranging from 0.1 to 0.25. In order to better match the X_{Grs} values observed in the sample, the CaO content was reduced to 0.8 of this diagram (a reduction of approximately 1.2 wt%; Fig. 7b). Based on the presence of ilmenite in garnet (estimated at about 2% mode and containing 5 wt% MnO), the amount of MnO was reduced by 0.1 wt%.

As indicated above, the growth of large porphyroblasts can result in fractionation of their composition from the bulk rock composition (i.e. Marmo *et al.* 2002). Thus, when modelling the second metamorphic event (garnet rim growth), the composition of the garnet core was removed from the whole rock composition. Since the garnet changes composition

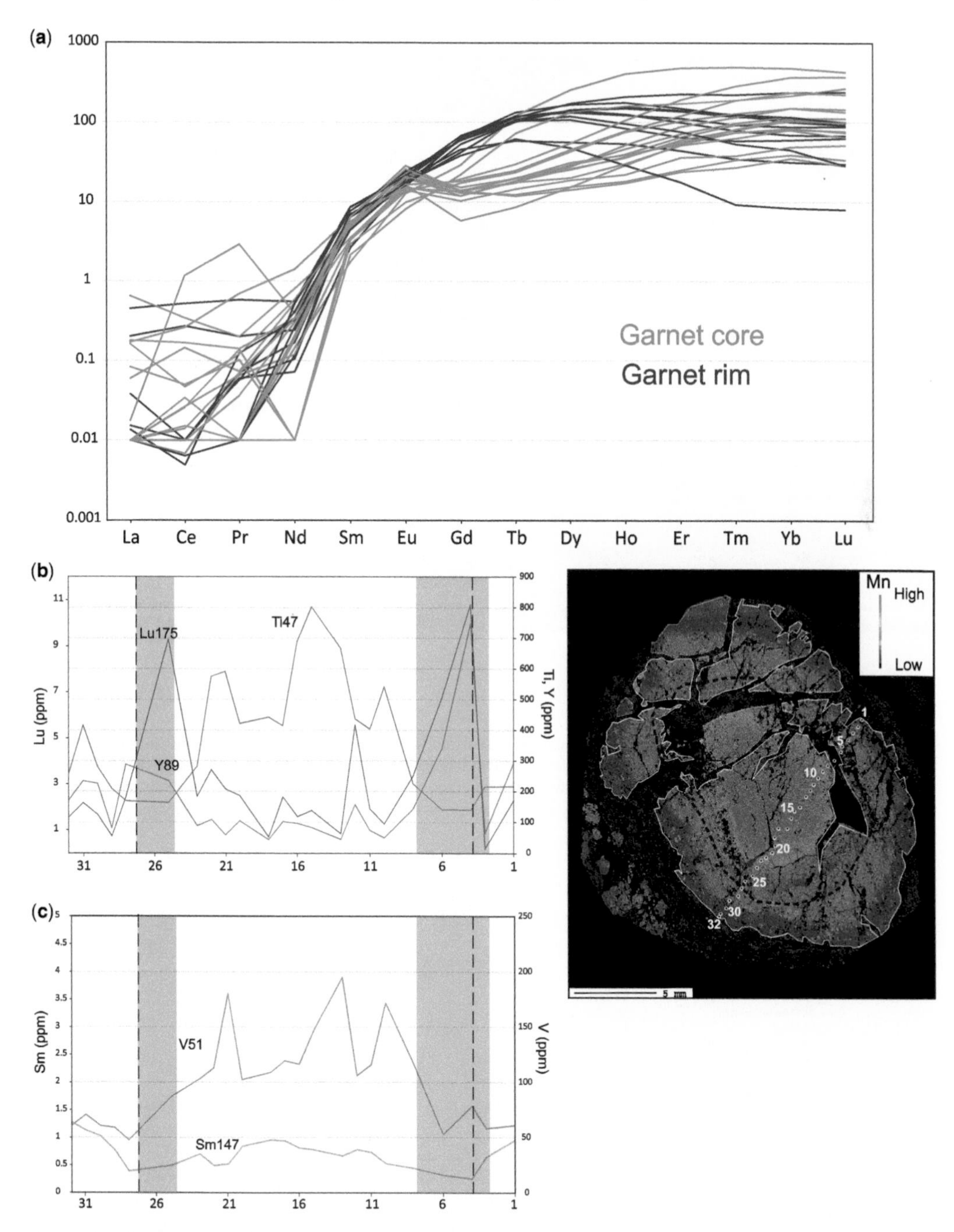

Fig. 5. (**a**) Chondrite-normalized trace element plot with analyses shaded depending on the location in the grain. Core analyses are red, rim analyses are blue. (**b**) Garnet compositional traverse showing Ti, Lu and Y variation. The red field indicates the zone where a large fracture and numerous inclusions interrupt the traverse. The *x*-axis indicates analysis numbers, which correspond to the numbered analysis locations given on the Mn elemental map of the garnet. Spot sizes have been exaggerated to be visible on the figure. The length of the profile is 1.4 cm (as in Fig. 4a). (**c**) Garnet compositional traverse showing Sm and V. Source: (a) normalized values from McDonough and Sun (1995).

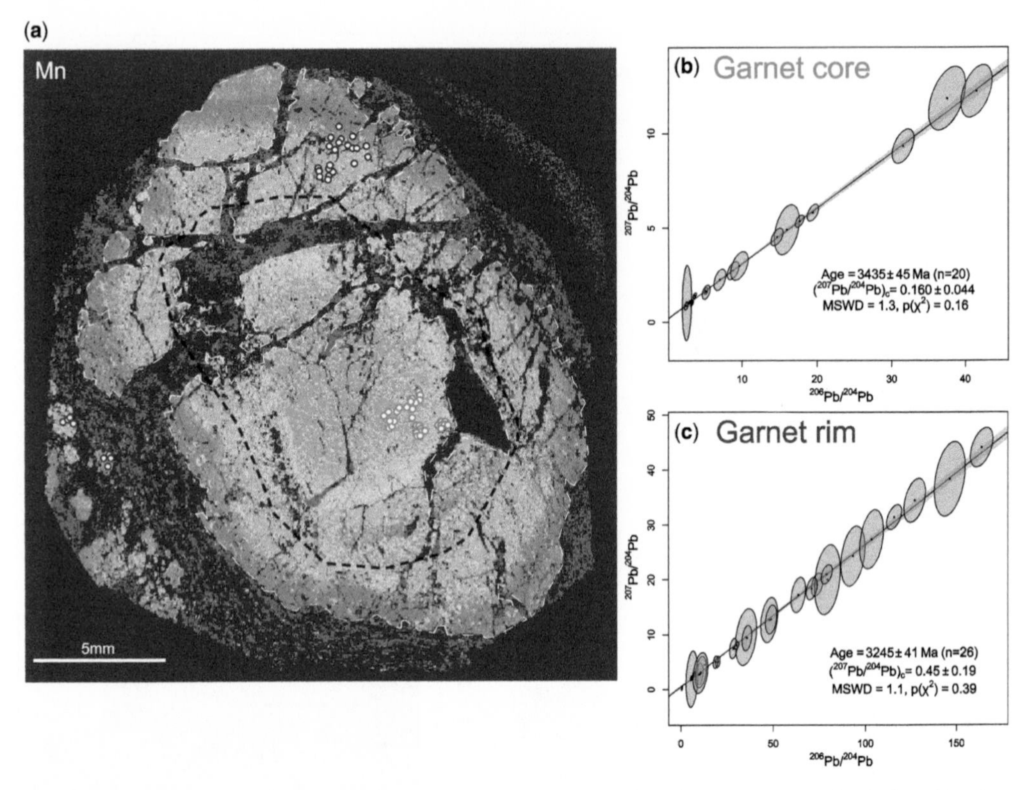

Fig. 6. (**a**) Garnet Mn elemental map with the location of analysed spots from the garnet core (white circles with red rims) and rim (white circles with blue rims). Spots are shown larger than actual size, so they are visible on the figure. The dashed black line indicates the boundary between the core and the rim. (**b**) Pb–Pb isochron for analyses obtained from the garnet core. (**c**) Pb–Pb isochron of analyses obtained from the garnet rim. MSWD, mean square weighted deviation.

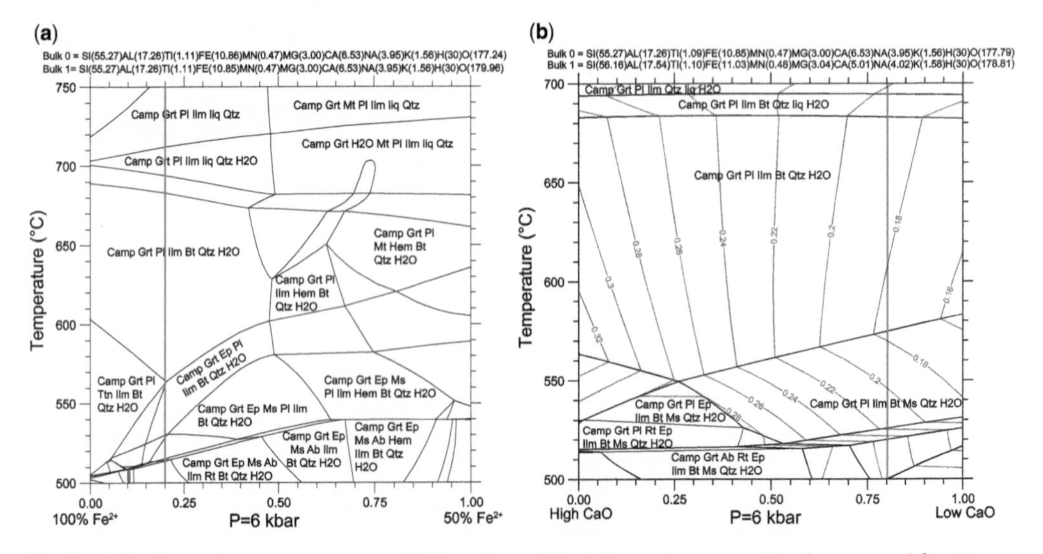

Fig. 7. (**a**) $T-M_{Fe_2O_3}$ diagram at a pressure of 6 kbar. The red line indicates the composition that was used for part (b) and for $P–T$ pseudosection calculations. (**b**) $T–M_{CaO}$ diagram at a pressure of 6 kbar. The grey lines indicate the composition of X_{Grs} in this figure. The red line indicates the amount of CaO used for $P–T$ pseudosection calculations.

from core to rim, the removed composition was determined using a stepwise approach. Each step represented a small domain of similar composition. In each step the composition was averaged and then weighted by volume and removed from the whole rock composition. An estimated 2% (of the rock volume) of the inner garnet core composition was removed, followed by an additional 3% of the garnet outer core composition. The garnet mode was estimated based on field and hand sample observations (while garnet porphyroblasts are large, they are sparse, making up less than 10% of the rock).

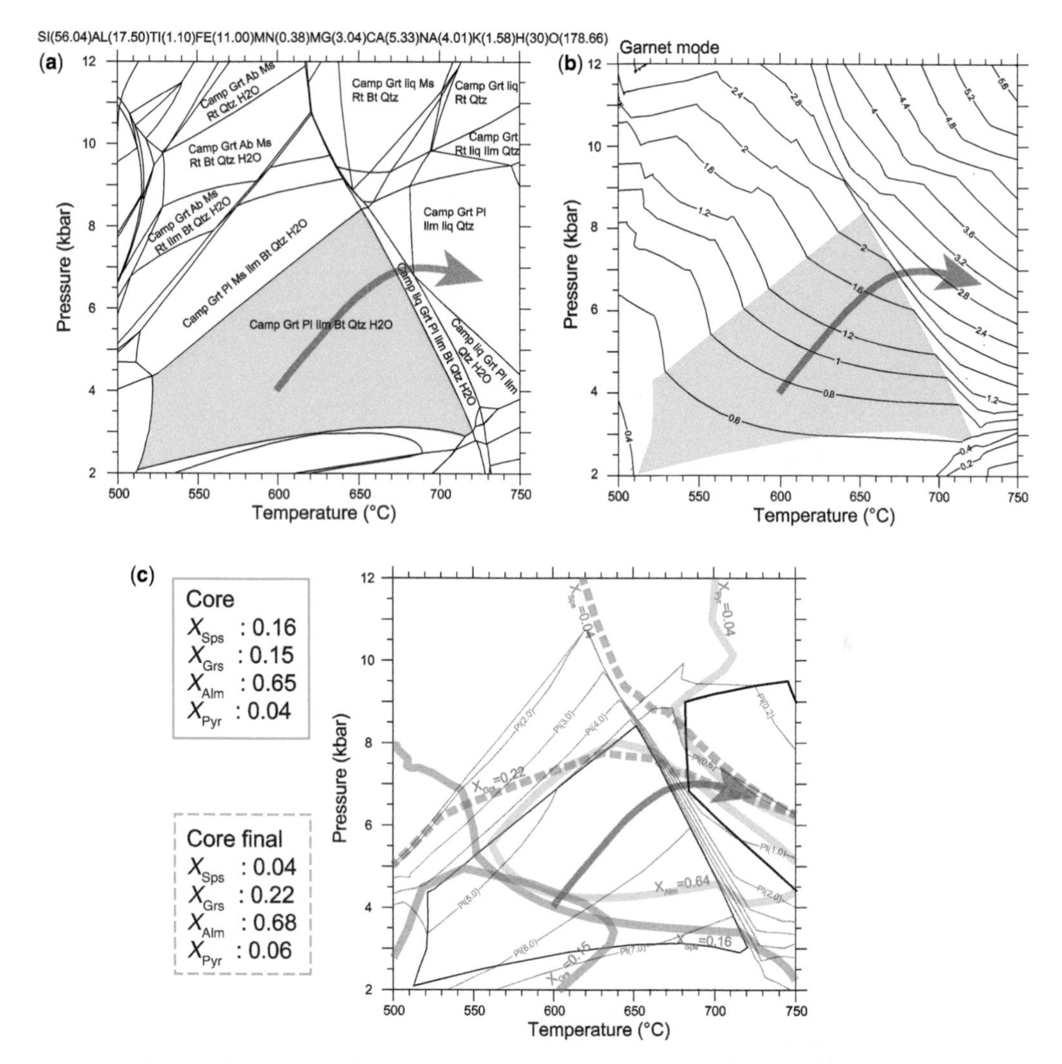

Fig. 8. (**a**) Calculated *P–T* pseudosections for the whole rock composition of the sample. The grey arrow represents the interpreted *P–T* trajectory for growth of the garnet core and transition zone based on the change in mineral assemblage, garnet mode, plagioclase mode and garnet composition. The red shaded field indicates the mineral assemblage field during garnet growth based on the inclusion assemblage. Ilm, ilmenite; Qtz, quartz; Grt, garnet; Camp, amphibole; Pl, plagioclase; Bt, biotite; Ms, muscovite; Ab, albite; Rt, rutile; liq, melt. (**b**) Diagram illustrating variation in garnet mode, with the interpreted *P–T* path represented by the grey arrow and showing an increase in garnet mode during the *P–T* evolution. The red shaded field indicates the mineral assemblage field during garnet growth based on the inclusion assemblage. (**c**) Summary *P–T* diagram showing the X_{Sps}, X_{Grs}, X_{Alm} and X_{Pyr} compositions for the core (solid lines) and final core (dashed lines). The X_{Alm} and X_{Pyr} compositions are presented as fainter because these were not used for the *P–T* path interpretation. The red boxes on the left indicate the values used for core (solid lines) and final core (dashed lines). The prograde field is outlined with a thin black line. The interpreted peak field is outlined in bold. The grey labelled lines refer to plagioclase modes. For definitions of X_{Alm}, X_{Grs}, X_{Pyr} and X_{Sps}, see Figure 4.

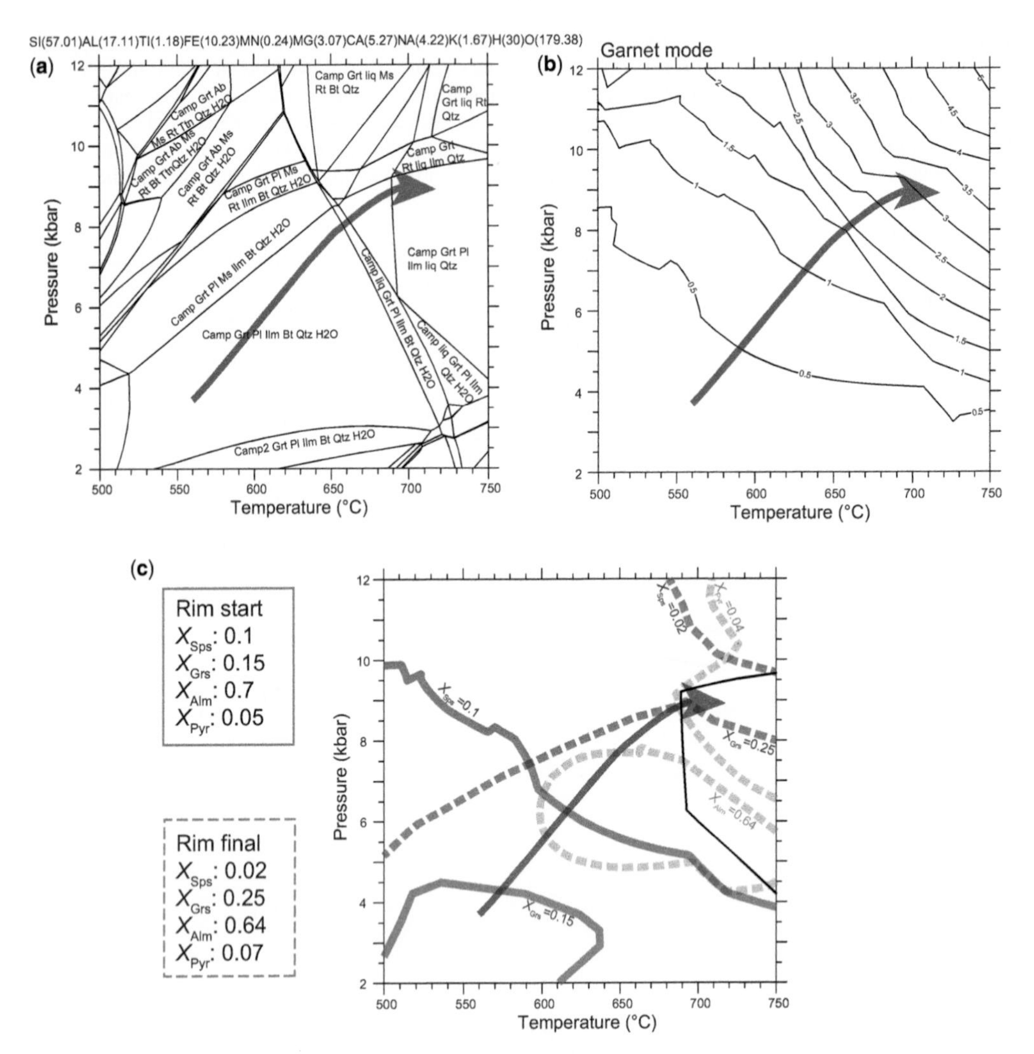

Fig. 9. (**a**) Calculated *P–T* pseudosections for the sample with the core zone of the garnet removed from the whole rock composition. The grey arrow represents the interpreted *P–T* trajectory for growth of the garnet rim based on the change in mineral assemblage, garnet mode and garnet composition. Bt, biotite; Ilm, ilmenite; Pl, plagioclase; Qtz, quartz; Grt, garnet; Camp, amphibole; Camp2, a second amphibole; Ms, muscovite; Ab, albite; Rt, rutile; Ttn, titanite; liq, melt. (**b**) Diagram illustrating variation in garnet mode, with the interpreted *P–T* path represented by the grey arrow and showing an increase in garnet mode during the *P–T* evolution. (**c**) Summary *P–T* diagram showing the X_{Sps}, X_{Grs}, X_{Alm} and X_{Pyr} compositions for the core (solid lines) and rim (dashed lines). The X_{Alm} and X_{Pyr} compositions are presented as fainter because these were not used for the *P–T* path interpretation. The blue boxes indicate the values used for rim start composition (solid lines) and rim final composition (dashed lines). The interpreted peak field is outlined in bold. For definitions of X_{Alm}, X_{Grs}, X_{Pyr} and X_{Sps}, see Figure 4.

The removal of the garnet core from the starting composition resulted in the calculation of two *P–T* diagrams (Figs 8 & 9). The first utilizes the composition of the whole rock with the appropriate Fe^{3+} content and a slightly reduced CaO and MnO. Based on the inclusion assemblage of garnet, the mineral assemblage stable during initial garnet growth is garnet + plagioclase + biotite + ilmenite + amphibole + quartz. The mineral assemblage occurs in a large field at 2–8 kbar and 500–700°C (Fig. 8a). The prograde *P–T* evolution is defined by an increase in garnet mode (Fig. 8b). The major element composition of the garnet core was used to attempt to constrain the *P–T* conditions of garnet core growth with the X_{Grs}, X_{Sps} and X_{Alm} values of the core (0.15, 0.16 and 0.64, respectively) overlapping at *P–T* conditions of 4–5 kbar and 560–630°C (Fig. 8c; full isopleth figures are included as

Supplementary Figure 2). The interpreted prograde *P–T* path is consistent with a decrease in plagioclase mode (from 5.5 to 0.8%; Fig. 8c). This may account for the positive Eu anomaly of the garnet core (Fig. 8c). The peak conditions for the garnet core zone are interpreted to occur in the garnet–amphibole–plagioclase–quartz–ilmenite–melt field. It is impossible to know for sure whether melt was present during the growth of the garnet core. However, the final compositions obtainable from the garnet core just before the edge of the rim are consistent with peak conditions in this field. This *P–T* evolution is also consistent with the change in garnet composition during core growth (increase in X_{Grs}, X_{Alm} and X_{Pyr} and decrease in X_{Sps}).

A subsequent diagram was made with the garnet core removed from the composition modelled in Figure 8 (see Figs 4 & 9a). Since there is not a large amount of garnet present in the sample, and the core is volumetrically inferior, this did not result in a significantly different diagram (compare Figs 8a & 9a). The first rim composition (highest X_{Sps} values for the rim) is used to define the start of the *P–T* evolution for the rim (X_{Sps} of 0.10, X_{Grs} of 0.15). The final rim composition indicates peak conditions, and the interpreted *P–T* path runs between these points and is based on an increase in garnet mode (Fig. 9b, c). Maximum pressure conditions are constrained by the absence of rutile (which occurs on the *P–T* pseudosection above 9 kbar; Fig. 9a). The peak field for the rim is the garnet + amphibole + plagioclase + ilmenite + quartz + melt field with peak conditions of 690–750°C and 8–9.5 kbar (Fig. 9a). The rocks have been strongly deformed; however, there are clear quartz-rich bands that increase in size in the pressure shadows of garnet (Fig. 2b), which are commonly bordered by thin mafic bands.

Discussion and conclusions

Assumptions involved in determining the P–T evolution of the garnet

The method used here for determining the *P–T* evolution of these rocks by removing garnet from the whole-rock composition to model subsequent garnet growth events relies on several interpretations which will be outlined in detail here. The main interpretations that we have made are the selection of the core–rim location, whether the profile sees the centre of the garnet and whether the outer core was preserved (resorption prior to rim growth).

The core–rim boundary was selected based on the discontinuity of X_{Sps} zoning (see Fig. 4); this is most clearly seen on the left-hand side of the profile, and can also be seen in the Mn elemental map (Fig. 4b), where a dashed line shows this as the boundary between the blue and green zones. The core zone has a continuous bell-shaped X_{Sps} profile, interpreted as a result of fractional crystallization during continuous garnet growth (i.e. Hollister 1966; Ikeda 1993). However, there is another compositional break in the garnet profile at around 4 mm (see Fig. 4a), this time with an abrupt increase in X_{Grs} and decrease in X_{Alm}. This also corresponds to the zone with many inclusions occurring in the final part of the core zone. It is plausible that this represents a separate growth zone which has continuous X_{Sps} with the garnet core (i.e. Argles *et al.* 1999). It was not possible to get an age for this zone due to the abundance of large inclusions that were present. Since it is not possible to date this zone, we have chosen to interpret the core and this inclusion-rich zone to be continuous with the break in X_{Grs} and X_{Alm} resulting from changes to garnet growth speed (and major element supply). Vielzeuf *et al.* (2021) also see enrichment in Ca content in garnet rims as well as inclusion-rich rims, which they associate with melting or influx of melt into the system. Further study is required to ascertain whether this is the case for sample BKC-3.

An additional issue is whether the investigated compositions are from the actual core of the garnet. The CT images (Fig. 3a–c) show that the BKC-3 garnet is not spherical, but rather an ellipsoid, making it difficult to discern the location of the actual core. If we did not hit the actual core then it is possible that our maximum X_{Sps} values are not the highest values present in garnet in sample BKC3. Generally, in trace element profiles for growth zoning, garnet cores are enriched in Lu (i.e. Raimondo *et al.* 2017; Rubatto *et al.* 2020). This is not the case for BKC-3 (see Fig. 5), which has Lu enrichment only in the outer core and middle rare earth element (MREE) depletion in the core. The implication for not sampling the true core would be that the amount of Mn in the core is underestimated, which means that too little Mn may have been removed from the whole rock composition for the modelling of the rim growth. This would put the low X_{Sps} contents we observe in the garnet rim at higher *P–T* conditions in the modelled *P–T* pseudosection, resulting in an overestimate of the peak *P–T* conditions. The mismatch between the observed peak assemblage (with biotite) and the modelled peak assemblage (without biotite) already suggests that the modelled X_{Sps} contents for garnet are too high. We suggest that peak conditions of 700°C and 9 kbar are more suitable for the garnet rim.

The final interpretation relates to whether the outer core was preserved prior to rim growth. The increase in Lu and Y in the final part of the core may be a result of resorption of the outer core followed by later garnet growth (i.e. Skora *et al.*

2006; Cruz-Uribe *et al.* 2015) or breakdown of Y and HREE accessory minerals such as allanite (i.e. Pyle and Spear 1999; Gieré *et al.* 2011).

Breakdown of the garnet core prior to rim growth (or garnet resorption following the first metamorphic event) would also result in Mn resorption into garnet (i.e. Carlson 2002), which has not been observed in our samples. In BKC-3, the X_{Sps} profile has a distinct break between core and rim with no indication of an increase in X_{Sps} on the outer part of the core profile (Fig. 4a). This break is also observed in the Mn elemental map (Fig. 4b). This suggests that for the retrograde evolution of the garnet core, garnet was not resorbed. This may be due to a lack of fluid in the rock preventing the alteration of garnet to chlorite (Guiraud *et al.* 2001).

Despite the lack of evidence for retrogression of the sample following the growth of the garnet core, we think it is likely that the sample was exhumed prior to the growth of the garnet rim. The peak temperatures of the core are 700°C and 7 kbar. If the garnet remained at these conditions for 200 million years, even with slow diffusion rates and large grain sizes, some diffusion of the core zoning in the garnet would be observed (Carlson 2006; Caddick *et al.* 2010).

Also, in order to grow a garnet rim, the rock must experience an increase in garnet mode. Even with the garnet core removed, the bulk rock composition for the rim growth would likely increase garnet mode starting from low to moderate *P–T* (compare Figs 8b & 9b). This suggests that the hiatus in garnet growth between the garnet core and rim occurred due to exhumation.

Significance and implications of garnet ages

In this study we have applied Pb–Pb dating due to the age of the sample. With an age >3.0 Ga, a garnet with initially low U does not have much U remaining (generally much less than 1 ppm (see Supplementary Table 4), making the U–Pb method less reliable. The Pb–Pb isochron method can produce a statistically robust age for the garnet core of BKC-3, producing an isochron age of 3436 ± 45 Ma (Fig. 6b). This age is similar to that of the Theespruit Pluton (*c.* 3443 ± 4 Ma; Lana *et al.* 2010) but slightly younger than the directly adjacent Steynsdorp Pluton (*c.* 3510 Ma; Lana *et al.* 2010). Plausibly this difference in age between sample BKC-3 and the Steynsdorp Pluton could be a result of Pb loss. The U–Pb method allows for the evaluation of lead loss, but it cannot be detected when using the Pb–Pb method alone. During the analytical session, an internal standard from the Bushveld metamorphic aureole was analysed and produced a Pb–Pb isochron age of 2031 ± 127 Ma ($n = 19$, MSWD = 0.48). This is within error of the LA-MC-ICP-MS U–Pb age produced for the same standard material (2058 ± 3.5 Ma; Marques *et al.* 2023) and the Bushveld intrusion age (2060–2055 Ma; Scoates *et al.* 2021). Both U and Pb diffusion rates in garnet are proposed to be extremely slow to negligible, with estimates of U–Pb closure temperatures of >800°C for garnet grains larger than 1 mm in diameter (Mezger *et al.* 1989, 1991; Burton *et al.* 1995; Zhu *et al.* 1997). Considering that the garnet grain investigated in this study has a grain diameter of 3 cm (with a core of 1 cm) and the *c.* 3.2 Ga metamorphic event is estimated to have reached peak temperature conditions of 600–650°C, it is unlikely that the garnet core age is a cooling age or has been affected by diffusional re-equilibration.

Figure 10a presents a compilation of all ages and *P–T* conditions obtained for the *c.* 3.45 Ga event. A U–Pb monazite age of *c.* 3436 Ma was obtained from the Onverwacht Group in the Stolzburg Domain (point 4, Fig. 1b; Cutts *et al.* 2014), interpreted to correspond to *P–T* conditions of 4–5 kbar and up to 600°C. Wang *et al.* (2019) found metamorphic zircon ages of 3443 ± 13 Ma and 3419 ± 8 Ma from samples that give *P–T* conditions of 550–625°C and 8–9 kbar and 650–725°C and 6–8 kbar, respectively (point 6, Fig. 1b). Even with a large error (3436 ± 45 Ma), the age obtained from BKC-3 is not within error of the emplacement age of the Steynsdorp Pluton (3511–3502 Ma; Kröner *et al.* 1996). This would suggest that the event resulting in garnet growth and metamorphism reaching conditions of 6–7 kbar and 700°C as determined in this study was in fact not related to the TTG emplacement. Additionally, the samples of Wang *et al.* (2019) and Cutts *et al.* (2014) were obtained from the Stolzburg Domain to the NW of the Steynsdorp area. Wang *et al.* (2019) presented only peak *P–T* estimates, but Cutts *et al.* (2014) suggested a clockwise *P–T* evolution for the *c.* 3.45 Ga event. This resulted in lower peak *P–T* conditions (*c.* 4.5 kbar and 550°C) but occurred on a similar geothermal gradient (Fig. 10a). One of the samples investigated by Wang *et al.* (2019) from the Stolzburg Domain (A2 in Fig. 10a) produced similar peak *P–T* conditions to sample BKC-3. Thus, the present data suggest that this event produced similar peak *P–T* conditions and *P–T* evolutions over the whole southern BGGB (Fig. 10a).

The garnet rim age of 3245 ± 41 Ma is within error of the age of the major metamorphic event in Barberton at 3.23 Ga (Dziggel *et al.* 2002; Stevens *et al.* 2002; Kisters *et al.* 2003; Diener *et al.* 2005; Cutts *et al.* 2014; Wang *et al.* 2019; Fig. 10b). Wang *et al.* (2019) obtained peak *P–T* conditions of 600–700°C and 9–10 kbar at 3222 ± 8 Ma (D1, Fig. 10b) and 8.4–12.7 kbar and 630–700°C at 3249 ± 6 Ma (D2, Fig. 10b) using conventional thermobarometry. Cutts *et al.* (2014) presented

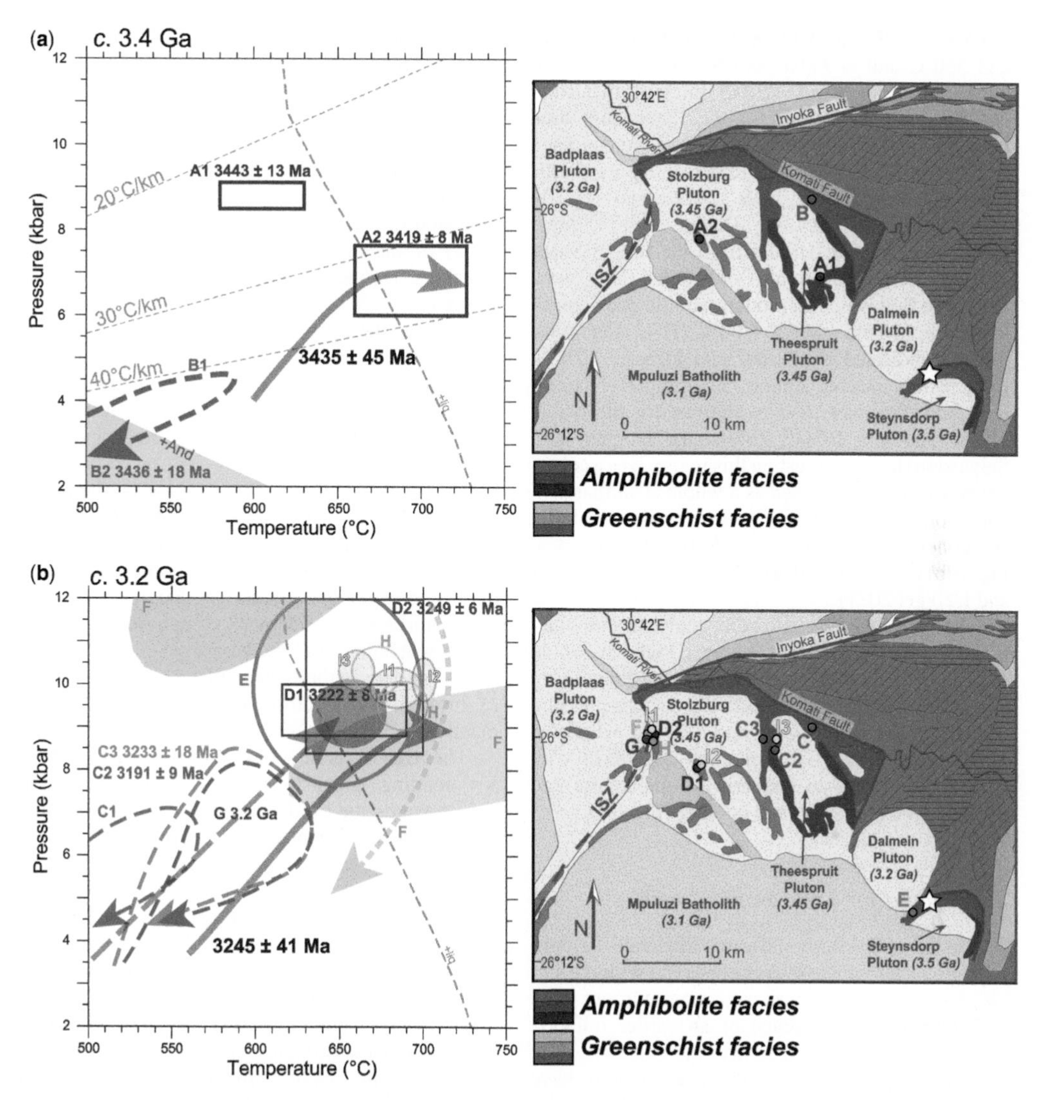

Fig. 10. (**a**) A compilation of the *P–T* results obtained from the Stolzburg Domain, which are interpreted to represent the 3.45 Ga event. Samples A1 and A2 correspond to samples 14SA04 and 15SA36, respectively. Samples B1 and B2 correspond to samples BKC-10 and BKC-8, respectively. The black age and dark grey arrow indicate the *P–T* path and age obtained in this study. The location of samples is indicated on the included map, with the star representing the sample location of this study. (**b**) A compilation of *P–T* results for the 3.2 Ga event. Ages given were obtained from the same samples as the *P–T* conditions. Samples with no ages given are presumed to be metamorphosed at *c.* 3.2 Ga. Samples C1, C2 and C3 correspond to samples BKC10, BKC-16 and BKC-23, respectively. Samples D1 and D2 are samples 15SA36 and 14SA35, respectively. E represents average *P–T* calculations using a garnet rim and matrix assemblage. F represents average *P–T* calculations from the Inyoni Shear Zone using garnet rim compositions and their interpreted *P–T* evolution. G is a summary of interpreted prograde evolution from Inyoni Shear Zone samples with estimated peak conditions given by the purple circle. H indicates the *P–T* conditions of two samples, with the higher-pressure circle representing their sample INY134 and the lower-pressure circle indicating B34A2. Samples I1, I2 and I3 are samples representing estimated peak conditions from the Inyoni Shear Zone, the Central Stolzburg Domain and the Tjakastad Schist Belt (see map for point locations). Source: (a) samples A1 and A2 from Wang *et al.* 2019; samples B1 and B2 from Cutts *et al.* 2014; (b) samples C1, C2 and C3 from Cutts *et al.* 2014; samples D1 and D2 from Wang *et al.* 2019; E: from Lana *et al.* 2010; F: from Moyen *et al.* 2006; G: from Cutts *et al.* 2014; H: modelled by Nédélec *et al.* 2012; samples I1, I2 and I3 from Diener and Dziggel 2021.

clockwise *P–T* paths with peak conditions of 7 kbar and 560°C and 8–9 kbar and 600–650°C (C1-C3, Fig. 10b). Lana *et al.* (2010) inferred that their obtained *P–T* conditions for the Steynsdorp area relate to the 3.23 Ga event (they used the rim zone and matrix minerals to calculate average *P–T* estimates) and indicated peak conditions of 10–13 kbar and 640–660°C. Diener and Dziggel (2021) investigated samples from the Inyoni Shear Zone, the central Stolzburg Domain and the Tjakastad Schist Belt which all produce similar peak conditions of *c.* 10 kbar and 650–700°C (I1-I3, Fig. 10b). As also indicated by Diener and Dziggel (2021), these results indicate similar maximum *P–T* conditions and clockwise *P–T* evolutions for the whole of the southern BGGB (Stolzburg Domain and Steynsdorp area), suggesting this region was a coherent block, metamorphosed as a whole at similar *P–T* conditions. Observed *P–T* paths usually involve a component of heating at depth (i.e. C2 and C3, Fig. 10b) consistent with a collisional setting (Diener and Dziggel 2021).

Therefore, the results of this study, in combination with previous work, suggest that the whole southern BGGB (Stolzburg Domain and Steynsdorp area) was affected by two moderate- to high-pressure metamorphic events at 3.45 and 3.23 Ga, with both having clockwise *P–T* evolutions consistent with crustal thickening, an increase in temperature and subsequent exhumation and cooling (Fig. 10).

Implications for Archean geodynamics

The results of this study show two moderate- to high-pressure metamorphic events preserved within a single garnet crystal. The peak pressure conditions of up to 7 kbar and the presence of an earlier foliation (Cutts *et al.* 2014) together with the clockwise *P–T* evolution suggest the *c.* 3.45 Ga event may have been the result of a collision between early continental fragments, perhaps followed by orogenic collapse that caused basement doming. The paucity of evidence for the *c.* 3.45 Ga event makes assigning a tectonic setting difficult. It is possible that the *c.* 3.45 Ga event results from subsidence of dense greenstones into hot TTG crust, as suggested by Wang *et al.* (2019) for both the 3.45 and 3.23 Ga events. However, while models of this process have shown multiple *P–T* evolution shapes (Francois *et al.* 2014), the mechanism for exhuming the greenstone rocks is unclear. Moreover, if BKC-3 was at high *P–T* conditions for 200 million years, the major element zoning in the garnet would not be so well preserved. The rocks of the Steynsdorp area must have been exhumed prior to burial in the 3.23 Ga event. Also, the partial convective overturn model needs to be reconciled with the older Steynsdorp Pluton, meaning some delay between TTG intrusion at 3.5 Ga and peak metamorphism at 3.45 Ga.

At 3.23 Ga, the main collisional event was preserved in the BGGB, producing higher *P–T* conditions and elevated temperatures at depth over the whole southern BGGB (Diener and Dziggel 2021). This was followed by orogenic collapse with lateral extrusion (directed to the SW, Fig. 1; Kisters *et al.* 2003) of the amphibolite facies lower crust (the Stolzburg Domain and Steynsdorp area; Fig. 10). Wang *et al.* (2019) also indicate partial convective overturn as the cause for the 3.23 Ga metamorphic event. This model is inconsistent with the similarity of all investigated samples in the southern BGGB, with all presenting clockwise *P–T* evolution and peak conditions ranging from 7–11 kbar and 550–700°C (Fig. 10b).

P–T evidence alone can only indicate the possibility of Archean plate tectonics. We also require supporting tectonic and geochemical evidence from multiple locations. Besides what is presented here, the evidence for horizontal plate tectonic processes in the northeastern Kaapvaal Craton before 3.0 Ga is overwhelming, and increasingly studies utilizing novel isotope methods are indicating the presence of surface derived isotopic signatures in the Archean mantle (i.e. Lewis *et al.* 2023).

The BGGB is the oldest of a series of NE–SW-trending greenstone belts, with greenstone belts towards the north becoming younger towards the edge of the Kaapvaal Craton (Fig. 1a). Such a series of belts could be produced by accretion of microplates.

Isotopic studies investigating the Bushveld Complex (*c.* 2.05 Ga) have found sulfur isotope signatures indicating the presence of recycled Archean surface material (Magalhães *et al.* 2019). Surface sulfur isotope signatures have also been found in the Itsaq Gneiss Complex in southern West Greenland, indicating mantle domains that were fertilized by surface-derived material (Lewis *et al.* 2023).

Zircon Hf isotopes from the Bushveld Complex, which indicate an enriched mantle source, are suggested to be the result of prolonged subduction on the northern margin of the Kaapvaal Craton prior to collision with the Zimbabwe Craton from 3.2 to 2.6 Ga (Zeh *et al.* 2009; Zirakparvar *et al.* 2014).

Smart *et al.* (2016) show that diamonds from the Kaapvaal Craton have nitrogen and oxygen isotopic compositions consistent with the inclusion of isotopically light surface material into the mantle source. All of the isotopic investigations outlined above indicate recycling of crustal material into the mantle during the Archean by some process that also results in collision and medium- to high-pressure metamorphism.

Isotopic studies can also present contrasting points of view, with Rollinson (2021) presenting trace element, Nd and Hf data to suggest that mixed isotopic signatures are the result of mixing between the lower and upper mantle facilitated by a mantle plume or heat pump. Johnson *et al.* (2017) presented a *P–T* modelling and trace element study of basalts from the Pilbara Terrane, Western Australia to suggest that Archean TTG formation was a protracted, multi-stage process that likely occurred near the base of thick, plateau-like basaltic crust. This would indicate that production of early Archean TTGs did not require subduction.

Lateral plate motion does not necessarily require modern-style subduction, with several contributions suggesting alternate mechanisms for lateral plate motion which are not driven by subduction. Chowdhury *et al.* (2017, 2020) suggested that convergence may have happened in the Archean as a result of lithospheric peeling. This process involves delamination or peeling off of the subcontinental lithospheric mantle and the lower continental crust. It is proposed to occur due to higher mantle temperatures; it allows for crustal recycling into the mantle and can account of the appearance of paired metamorphism, MP-HP TTGs and K-granites (Chowdhury *et al.* 2017, 2020).

Alternatively, Strong *et al.* (2023) used an isotopic study to present an accordion tectonic model for the Superior Province involving cratonic growth interspersed with periods of disaggregation with formation of rifts and volcanosedimentary successions with later collisions reassembling the fragments.

For the Isua area of West Greenland, surface isotope signatures for oxygen (Gauthiez-Putallaz *et al.* 2020) are recorded in metasedimentary garnet with a multistage metamorphic history. They propose that the source of these rocks was exposed to surficial alteration then buried to mid-crustal levels to form high $\delta^{18}O$ garnet within a period of 10–50 myr and suggest that the tectonic setting for this was flat subduction followed by collisional orogeny at 3.69–3.66 Ga (Gauthiez-Putallaz *et al.* 2020).

Thus, if we place our *P–T* results for Barberton in this broader context, prograde metamorphism achieving *P–T* conditions of 7 kbar at 3.45 Ga and then 9–10 kbar at 3.23 Ga could feasibly be the result of lateral collision of microcontinents and tectonic thickening.

Acknowledgements Kathryn Cutts acknowledges CNPq for her Science without Borders, Jovem Talento Scholarship. Thorsten Nagel is thanked for assistance in collecting the garnet elemental maps. We thank reviewers Gautier Nicoli, Annika Dziggel and Craig Storey for their comments, which significantly improved the manuscript. We also thank Valby van Schijndel for the editorial handling.

Competing interests The authors declare that they have no known competing financial interests or personal relationships that could have appeared to influence the work reported in this paper.

Author contributions **KAC**: conceptualization (lead), formal analysis (lead), investigation (lead), methodology (equal), writing – original draft (lead), writing – review and editing (lead); **CL**: formal analysis (supporting), investigation (supporting), methodology (equal), writing – original draft (supporting), writing – review and editing (supporting); **GS**: conceptualization (supporting), methodology (supporting), writing – original draft (supporting), writing – review and editing (supporting); **ISB**: methodology (supporting), writing – review and editing (supporting).

Funding Cristiano Lana acknowledges funding from CNPQ_PQ Processo 307353/2019-2.

Data availability All data generated or analysed during this study are included in this published article (and, if present, its supplementary information files).

References

Anhaeusser, C.R. 1976. The geology of the Sheba Hills area of the Barberton Mountain Land, South Africa, with particular reference to the Eureka syncline. *Geological Society of South Africa Transactions*, **79**, 253–280.

Anhaeusser, C.R., Robb, L.J. and Viljoen, M.J. 1983. Notes on the provisional geological map of the Barberton greenstone belt and surrounding granitic terrane, eastern Transvaal and Swaziland (1:250,000 colour map). Spec. Publ. Geol. Soc. S. Afr., 9 (1983), pp. 221–223.

Argles, T.W., Prince, C.I., Foster, G.L. and Vance, D. 1999. New garnets for old? Cautionary tales from young mountain belts. *Earth and Planetary Science Letters*, **172**, 301–309, https://doi.org/10.1016/S0012-821X(99)00209-5

Armstrong, R.A., Compston, W., De Wit, M.J. and Williams, I.S. 1990. The stratigraphy of the 3.5–3.2 Ga Barberton greenstone belt revisited: a single zircon ion microprobe study. *Earth and Planetary Science Letters*, **101**, 90–106, https://doi.org/10.1016/0012-821X(90)90127-J

Baxter, E.F., Caddick, M.J. and Dragovic, B. 2017. Garnet: a rock-forming mineral petrochronometer. *Reviews in Mineralogy and Geochemistry*, **83**, 469–533, https://doi.org/10.2138/rmg.2017.83.15

Block, S., Moyen, J.-F., Zeh, A., Poujol, M., Jaguin, J. and Paquette, J.-L. 2013. The Murchison greenstone belt, South Africa: accreted slivers with contrasting metamorphic conditions. *Precambrian Research*, **227**, 77–98, https://doi.org/10.1016/j.precamres.2012.03.005

Buick, I.S., Maas, R. and Gibson, R. 2001. Precise U-Pb titanite age constraints on the emplacement of the Bushveld Complex, South Africa. *Journal of the Geological Society*, **158**, 3–6, https://doi.org/10.1144/jgs.158.1.3

Burton, K.W., Kohn, M.J., Cohen, A.S. and O'Nions, R.K. 1995. The relative diffusion of Pb, Nd, Sr and O in garnet. *Earth and Planetary Science Letters*, **133**,

199–211, https://doi.org/10.1016/0012-821X(95)00067-M

Caddick, M.J., Konopásek, J. and Thompson, A.B. 2010. Preservation of garnet growth zoning and the duration of prograde metamorphism. *Journal of Petrology*, **51**, 2327–2347, https://doi.org/10.1093/petrology/egq059

Carlson, W.D. 2002. Scales of disequilibrium and rates of equilibration during metamorphism. *American Mineralogist*, **87**, 185–204, https://doi.org/10.2138/am-2002-2-301

Carlson, W.D. 2006. Dana Lecture. Rates of Fe, Mg, Mn and Ca diffusion in garnet. *American Mineralogist*, **91**, 1–11, https://doi.org/10.2138/am.2006.2043

Chowdhury, P., Gerya, T. and Chakraborty, S. 2017. Emergence of silicic continents as the lower crust peels off on a hot plate-tectonic Earth. *Nature Geoscience*, **10**, 698–703, https://doi.org/10.1038/ngeo3010

Chowdhury, P., Chakraborty, S., Gerya, T., Cawood, P. and Capitanio, F. 2020. Peel-back controlled lithospheric convergence explains the secular transitions in Archean metamorphism and magmatism. *Earth and Planetary Science Letters*, **538**, 116224, https://doi.org/10.1016/j.epsl.2020.116224

Cloete, M. 1999. Aspects of volcanism and metamorphism of the Onverwacht Group lavas in the southwestern portion of the Barberton Greenstone Belt. *Memoir of the Geological Survey of South Africa*, **84**.

Cruz-Uribe, A.M., Hoisch, T.D., Wells, M.L., Vervoort, J.D. and Mazdab, F.K. 2015. Linking thermodynamic modelling, Lu-Hf geochronology and trace elements in garnet: new P-T-t paths from the Sevier hinterland. *Journal of Metamorphic Geology*, **33**, 763–781, https://doi.org/10.1111/jmg.12151

Cutts, K.A., Stevens, G., Hoffmann, J.E., Buick, I.S., Frei, D. and Münker, C. 2014. Paleo-mesoarchean polymetamorphism in the Barberton granite-greenstone belt, South Africa: constraints from U-Pb monazite and Lu-Hf garnet geochronology on the tectonic processes that shaped the belt. *Geological Society of America Bulletin*, **126**, 251–270, https://doi.org/10.1130/B30807.1

Cutts, K.A., Stevens, G. and Kisters, A. 2015. Reply to 'Paleo- to Mesoarchean polymetamorphism in the Barberton granite-greenstone belt, South Africa: Constraints from U-Pb monazite and Lu-Hf garnet geochronology on the tectonic processes that shaped the belt: Discussion' by M. Brown. *Geological Society of America Bulletin*, **127**, 1558–1563, https://doi.org/10.1130/B31304.1

Cutts, K.A., Maneiro, K.A., Stevens, G. and Baxter, E. 2021. Metamorphic evolution for the Inyoni Shear Zone: investigating the geodynamic evolution of a 3.20 Ga terrane boundary in the Barberton granitoid greenstone terrane, South Africa. *South African Journal of Geology*, **124**, 163–180, https://doi.org/10.25131/sajg.124.0009

De Capitani, C. and Petrakakis, K. 2010. The computation of equilibrium assemblage diagrams with Theriak/Domino software. *American Mineralogist*, **95**, 1006–1016, https://doi.org/10.2138/am.2010.3354

De Ronde, C.E.J. and De Wit, M. 1994. The tectonic history of the Barberton greenstone belt, South Africa: 490 million years of Archean crustal evolution. *Tectonics*, **13**, 983–1005, https://doi.org/10.1029/94TC00353

De Wit, M.J., Roering, C. *et al.* 1992. Formation of an Archean continent. *Nature*, **357**, 553–562, https://doi.org/10.1038/357553a0

Diener, J.F.A. and Dziggel, A. 2021. Can mineral equilibrium modelling provide additional details on metamorphism of the Barberton garnet amphibolites? *South African Journal of Geology*, **124**, 211–224, https://doi.org/10.25131/sajg.124.0003

Diener, J.F.A., Stevens, G., Kisters, A.F.M. and Poujol, M. 2005. Metamorphism and exhumation of the basal parts of the Barberton greenstone belt, South Africa: constraining the rates of Mesoarchean tectonism. *Precambrian Research*, **143**, 87–112, https://doi.org/10.1016/j.precamres.2005.10.001

Drabon, N., Galić, A., Mason, P.R.D. and Lowe, D.R. 2019. Provenance and tectonic implications of the 3.28–3.23 Ga Fig Tree Group, central Barberton greenstone belt, South Africa. *Precambrian Research*, **325**, 1–19, https://doi.org/10.1016/j.precamres.2019.02.010

Droop, G.T.R. 1987. A general equation for estimating Fe^{3+} concentrations in ferromagnesian silicates and oxides from microprobe analyses, using stoichiometric criteria. *Mineralogical Magazine*, **51**, 431–435, https://doi.org/10.1180/minmag.1987.051.361.10

Dziggel, A., Stevens, G., Poujol, M., Anhaeusser, C.R. and Armstrong, R.A. 2002. Metamorphism of the granite-greenstone terrane south of the Barberton greenstone belt, South Africa: an insight into the tectono-thermal evolution of the 'lower' portions of the Onverwacht Group. *Precambrian Research*, **114**, 221–247, https://doi.org/10.1016/S0301-9268(01)00225-X

Dziggel, A., Knipfer, S., Kisters, A.F.M. and Meyer, F.M. 2006. P-T and structural evolution during exhumation of high-T, medium-P basement rocks in the Barberton Mountain Land, South Africa. *Journal of Metamorphic Geology*, **24**, 535–551, https://doi.org/10.1111/j.1525-1314.2006.00653.x

Francois, C., Philippot, P., Rey, P. and Rubatto, D. 2014. Burial and exhumation during Archean sagduction in the East Pilbara Granite-Greenstone Terrane. *Earth and Planetary Science Letters*, **396**, 235–251, https://doi.org/10.1016/j.epsl.2014.04.025

Gauthiez-Putallaz, L., Nutman, A., Bennett, V. and Rubatto, D. 2020. Origins of high $\delta^{18}O$ in 3.7–3.6 Ga crust: a zircon and garnet record in Isua clastic metasedimentary rocks. *Chemical Geology*, **537**, 119474, https://doi.org/10.1016/j.chemgeo.2020.119474

Gieré, R., Rumble, D., Günther, D., Connolly, J. and Caddick, M.J. 2011. Correlation of growth and breakdown of major and accessory minerals in metapelites from Campolungo, Central Alps. *Journal of Petrology*, **52**, 2293–2334, https://doi.org/10.1093/petrology/egr043

Green, E.C.R., White, R.W., Diener, J.F.A., Powell, R., Holland, T.J.B. and Palin, R.M. 2016. Activity-composition relations for the calculation of partial melting equilibria in metabasic rocks. *Journal of Metamorphic Geology*, **34**, 845–869, https://doi.org/10.1111/jmg.12211

Grosch, E.G., Kosler, J., McLoughlin, N., Drost, K., Slama, J. and Pedersen, R.B. 2011. Paleoarchean detrital zircon ages from the earliest tectonic basin in the Barberton

Greenstone Belt, Kaapvaal Craton, South Africa. *Precambrian Research*, **191**, 85–99, https://doi.org/10.1016/j.precamres.2011.09.003

Guiraud, M., Powell, R. and Rebay, G. 2001. H_2O in metamorphism and unexpected behaviour in the preservation of metamorphic mineral assemblages. *Journal of Metamorphic Geology*, **19**, 445–454, https://doi.org/10.1046/j.0263-4929.2001.00320.x

Holland, T.J.B. and Powell, R. 2003. Activity-composition relations for phases in petrological calculations: an asymmetric multicomponent formulation. *Contribution to Mineralogy and Petrology*, **145**, 492–501, https://doi.org/10.1007/s00410-003-0464-z

Holland, T.J.B. and Powell, R. 2011. An improved and extended internally consistent thermodynamic dataset for phases of petrological interest, involving a new equation of state for solids. *Journal of Metamorphic Geology*, **29**, 333–383, https://doi.org/10.1111/j.1525-1314.2010.00923.x

Hollister, L. 1966. Garnet zoning: an interpretation based on the Rayleigh fractionation model. *Science (New York)*, **154**, 1647–1651, https://doi.org/10.1126/science.154.3757.1647

Ikeda, T. 1993. Compositional zoning patterns of garnet during prograde metamorphism from the Yanai district, Ryoke metamorphic belt, southwest Japan. *Lithos*, **30**, 109–121, https://doi.org/10.1016/0024-4937(93)90010-A

Johnson, T.E., Brown, M., Gardiner, N., Kirkland, C. and Smithies, H. 2017. Earth's first stable continents did not form by subduction. *Nature*, **543**, 239–242, https://doi.org/10.1038/nature21383

Jorgensen, T.R.C., Tinkham, D.K. and Lecher, C.M. 2019. Low-P and high-T metamorphism of basalts: insights from the Sudbury impact melt sheet aureole and thermodynamic modelling. *Journal of Metamorphic Geology*, **37**, 271–313, https://doi.org/10.1111/jmg.12460

Kamo, S.L. and Davis, D.W. 1994. Reassessment of Archean crustal development in the Barberton Mountain Land, South Africa based on U-Pb dating. *Tectonics*, **13**, 167–192, https://doi.org/10.1029/93TC02254

Kisters, A.F.M., Stevens, G., Dziggel, A. and Armstrong, R.A. 2003. Extensional detachment faulting and core-complex formation in the southern Barberton granite-greenstone terrain, South Africa: evidence for a 3.2 Ga orogenic collapse. *Precambrian Research*, **127**, 355–378, https://doi.org/10.1016/j.precamres.2003.08.002

Kisters, A.F.M., Belcher, R.W., Poujol, M. and Dziggel, A. 2010. Continental growth and convergence-related arc plutonism in the Mesoarchaean: evidence from the Barberton granitoid-greenstone terrain, South Africa. *Precambrian Research*, **178**, 15–26, https://doi.org/10.1016/j.precamres.2010.01.002

Kröner, A., Hegner, E., Wendt, J.I. and Byerly, G.R. 1996. The oldest part of the Barberton granitoid-greenstone terrain, South Africa: evidence for crust formation between 3.5 and 3.7 Ga. *Precambrian Research*, **78**, 105–124, https://doi.org/10.1016/0301-9268(95)00072-0

Lana, C., Kisters, A.F.M. and Stevens, G. 2010. Exhumation of Mesoarchean TTG gneisses from the middle crust: insights from the Steynsdorp core complex, Barberton granitoid-greenstone terrain, South Africa. *Geological Society of America Bulletin*, **122**, 183–197, https://doi.org/10.1130/B26580.1

Lana, C., Farina, F., Gerdes, A., Alkmim, A., Gonçalves, G.O. and Jardim, A.C. 2017. Characterization of zircon reference materials via high precision U-Pb LA-MC-ICP-MS. *Journal of Analytical Atomic Spectrometry*, **32**, 2011–2023, https://doi.org/10.1039/c7ja00167c

Lewis, J.A., Hoffmann, J.E., Schwarzenbach, E.M., Strauss, H., Li, C., Münker, C. and Rosing, M.T. 2023. Sulfur isotope evidence from peridotite enclaves in southern West Greenland for recycling of surface material into Eoarchean depleted mantle domains. *Chemical Geology*, **633**, 121568, https://doi.org/10.1016/j.chemgeo.2023.121568

Longerich, H.P., Günther, D. and Jackson, S.E. 1996. Elemental fractionation in laser ablation inductively coupled plasma mass spectrometry, *Fresenius. Journal of Analytical Chemistry*, **355**, 538–542, https://doi.org/10.1007/s0021663550538

Magalhães, N., Farquhar, J., Bybee, G., Penniston-Dorland, S., Rumble, D., Kinnaird, J. and McCreesh, M. 2019. Multiple sulfur isotopes reveal a possible non-crustal source of sulfur for the Bushveld Province, southern Africa. *Geology*, **47**, 982–986, https://doi.org/10.1130/G46282.1

Marmo, B.A., Clarke, G.L. and Powell, R. 2002. Fractionation of bulk rock composition due to porphyroblast growth: effects on eclogite facies mineral equilibria, Pam Peninsula, New Caledonia. *Journal of Metamorphic Geology*, **20**, 151–165, https://doi.org/10.1046/j.0263-4929.2001.00346.x

Marques, C., Cutts, K.A., Cabral, A.R., Lana, C., Rios, F.J. and Buick, I. 2023. Dating hydrothermal processes related to the formation of the Lagoa Real uranium deposit using *in situ* U-Pb dating of andradite and titanite. *Journal of South American Earth Sciences*, **123**, 104239, https://doi.org/10.1016/j.jsames.2023.104239

McDonough, W.F. and Sun, S.S. 1995. The composition of the Earth. *Chemical Geology*, **120**, 223–253, https://doi.org/10.1016/0009-2541(94)00140-4

Mezger, K., Hanson, G.N. and Bohlen, S.R. 1989. U-Pb systematics of garnet: dating the growth of garnet in the Late Archean Pikwitonei granulite domain at Cauchon and Natawahunan lakes, Manitoba, Canada. *Contributions to Mineralogy and Petrology*, **101**, 136–148, https://doi.org/10.1007/BF00375301

Mezger, K., Rawnsley, C.M., Bohlen, S.R. and Hanson, G.N. 1991. U-Pb garnet, sphene, monazite and rutile ages: implications for the duration of high-grade metamorphism and cooling histories, Adirondack Mts, New York. *The Journal of Geology*, **99**, 415–456, https://doi.org/10.1086/629503

Millonig, L.J., Albert, R., Gerdes, A., Avigad, D. and Dietsch, C. 2020. Exploring laser ablation U-Pb dating of regional metamorphic garnet: The Straits Schist, Connecticut, USA. *Earth and Planetary Science Letters*, **552**, 116589, https://doi.org/10.1016/j.epsl.2020.116589

Moyen, J.F., Stevens, G. and Kisters, A. 2006. Record of mid-Archean subduction from metamorphism in the Barberton terrain, South Africa. *Nature*, **442**, 559–562, https://doi.org/10.1038/nature04972

Nédélec, A., Chevrel, M.O., Moyen, J.F., Ganne, J. and Fabre, S. 2012. TTGs in the making: natural evidence from Inyoni Shear Zone (Barberton, South Africa). *Lithos*, **153**, 25–38, https://doi.org/10.1016/j.lithos.2012.05.029

Pyle, J.M. and Spear, F.S. 1999. Yttrium zoning in garnet: coupling of major and accessory phases during metamorphic reactions. *Geological Materials Research*, **1**, 1–49.

Raimondo, T., Payne, J., Wade, B., Lanari, P., Clark, C. and Hand, M. 2017. Trace element mapping by LA-ICP-MS: assessing geochemical mobility in garnet. *Contributions to Mineralogy and Petrology*, **172**, 17, https://doi.org/10.1007/s00410-017-1339-z

Rollinson, H. 2021. No plate tectonics necessary to explain Eoarchean rocks at Isua (Greenland). *Geology*, **50**, 147–151, https://doi.org/10.1130/G49278.1

Rubatto, D., Burger, M. *et al.* 2020. Identification of growth mechanisms in metamorphic garnet by high-resolution trace element mapping with LA-ICP-TOFMS. *Contributions to Mineralogy and Petrology*, **175**, 61, https://doi.org/10.1007/s00410-020-01700-5

Sanchez-Garrido, C.J., Stevens, G., Armstrong, R.A., Moyen, J.F., Martin, H. and Doucelance, R. 2011. Diversity in Earth's early felsic crust: Paleoarchean peraluminous granites of the Barberton Greenstone Belt. *Geology*, **39**, 963–966, https://doi.org/10.1130/G32193.1

Schannor, M., Lana, C., Nicoli, G., Cutts, K., Buick, I., Gerdes, A. and Hecht, L. 2021. Reconstructing the metamorphic evolution of the Araçuaí orogen (SE Brazil) using *in situ* U-Pb garnet dating and P-T modelling. *Journal of Metamorphic Geology*, **39**, 1145–1171, https://doi.org/10.1111/jmg.12605

Scoates, J.S., Wall, C.J., Friedman, R.M., Weis, D., Mathez, E.A. and VanTongeren, J.A. 2021. Dating the Bushveld Complex: time of crystallization, duration of magmatism, and cooling of the world's largest layered intrusion and related rocks. *Journal of Petrology*, **62**, https://doi.org/10.1093/petrology/egaa107

Seman, S., Stockli, D.F. and McLean, N.M. 2017. U-Pb geochronology of grossular-andradite garnet. *Chemical Geology*, **460**, 106–116, https://doi.org/10.1016/j.chemgeo.2017.04.020

Silva, J.P.A., Lana, C., Mazoz, A., Buick, I. and Scholz, R. 2022. U-Pb Saturn: a new U-Pb/Pb-Pb data reduction software for LA-ICP-MS. *Geostandards and Geoanalytical Research*, **47**, 49–66, https://doi.org/10.1111/ggr.12474

Simpson, A., Gilbert, S. *et al.* 2021. *In-situ* Lu-Hf geochronology of garnet, apatite and xenotime by LA-ICP MS/MS. *Chemical Geology*, **577**, 120299, https://doi.org/10.1016/j.chemgeo.2021.120299

Simpson, A., Glorie, S., Hand, M., Spandler, C. and Gilbert, S. 2023. Garnet Lu-Hf speed dating: a novel method to rapidly resolve polymetamorphic histories. *Geoscience Frontiers*, **121**, 215–234, https://doi.org/10.1016/j.gr.2023.04.011

Skora, S., Baumgartner, L., Mahlen, N., Johnson, C.M., Pilet, S. and Hellebrand, E. 2006. Diffusion-limited REE uptake by eclogite garnets and its consequences for Lu-Hf and Sm-Nd geochronology. *Contributions to Mineralogy and Petrology*, **152**, 703–720, https://doi.org/10.1007/s00410-006-0128-x

Smart, K., Tappe, S., Stern, R.A., Webb, S.J. and Ashwal, L.D. 2016. Early Archaean tectonics and mantle redox recorded in Witwatersrand diamonds. *Nature Geoscience*, **9**, 255–259, https://doi.org/10.1038/ngeo2628

Stevens, G., Droop, G.T.R., Armstrong, R.A. and Anhaeusser, C.R. 2002. Amphibolite facies metamorphism in the Schapenburg schist belt: a record of the mid-crustal response to *c.* 3.23 Ga terrane accretion in the Barberton greenstone belt. *South African Journal of Geology*, **105**, 273–286, https://doi.org/10.2113/1050271

Strong, J.W.D., Mulder, J.A., Cawood, P.A., Cruden, A.R. and Nebel, O. 2023. Isotope evidence for Archean accordion-tectonics in the Superior Province. *Precambrian Research*, **393**, 107096, https://doi.org/10.1016/j.precamres.2023.107096

Tamblyn, R., Hand, M., Simpson, A., Gilbert, S., Wade, B. and Glorie, S. 2022. *In situ* laser ablation Lu–Hf geochronology of garnet across the Western Gneiss Region: campaign-style dating of metamorphism. *Journal of the Geological Society*, **179**, https://doi.org/10.1144/jgs2021-094

Van Kranendonk, M.J. 2011. Cool greenstone drips and the role of partial convective overturn in Barberton Greenstone Belt evolution. *Journal of African Earth Sciences*, **60**, 346–352, https://doi.org/10.1016/j.jafrearsci.2011.03.012

Van Kranendonk, M.J., Kröner, A., Hegner, E. and Connelly, J. 2009. Age, lithology and structural evolution of the c. 3.53 Ga Theespruit Formation in the Tjakastad area, southwestern Barberton Greenstone Belt, South Africa, with implications for Archean tectonics. *Chemical Geology*, **261**, 115–139, https://doi.org/10.1016/j.chemgeo.2008.11.006

Van Kranendonk, M.J., Kröner, A., Hoffman, E.J., Nagel, T. and Anhaeusser, C.R. 2014. Just another drip: reanalysis of a proposed Mesoarchean suture from the Barberton Mountain Land, South Africa. *Precambrian Research*, **254**, 19–35, https://doi.org/10.1016/j.precamres.2014.07.022

Vermeesch, P. 2018. IsoplotR: a free and open toolbox for geochronology. *Geoscience Frontiers*, **9**, 1479–1493, https://doi.org/10.1016/j.gsf.2018.04.001

Vielzeuf, D., Paquette, J.-L., Clemens, J.D., Stevens, G., Gannoun, A., Suchorski, K. and Saúl, A. 2021. Age, duration and mineral markers of magma interactions in the deep crust: an example from the Pyrenees. *Contributions to Mineralogy and Petrology*, **176**, 39, https://doi.org/10.1007/s00410-021-01789-2

Wang, H., Yang, J.-H., Kröner, A., Zhu, Y.-S. and Li, R. 2019. Non-subduction origin for 3.2 Ga high-pressure metamorphic rocks in the Barberton granitoids-greenstone terrane, South Africa. *Terra Nova*, **31**, 373–380, https://doi.org/10.1111/ter.12397

White, R.W., Powell, R., Holland, T.J.B. and Worley, B.A. 2000. The effect of TiO_2 and Fe_2O_3 on metapelitic assemblages at greenschist and amphibolite facies conditions: mineral equilibria calculations in the system K_2O-FeO-MgO-Al_2O_3-SiO_2-H_2O-TiO_2-Fe_2O_3. *Journal of Metamorphic Geology*, **18**, 497–511, https://doi.org/10.1046/j.1525-1314.2000.00269.x

White, R.W., Powell, R. and Clarke, G.L. 2002. The interpretation of reaction textures in Fe-rich metapelitic granulites of the Musgrave Block, central Australia:

constraints from mineral equilibria calculations in the system K_2O-FeO-MgO-Al_2O_3-SiO_2-H_2O-TiO_2-Fe_2O_3. *Journal of Metamorphic Geology*, **20**, 41–55, https://doi.org/10.1046/J.0263-4929.2001.00349.X

White, R.W., Powell, R. and Johnson, T.E. 2014. The effect of Mn on mineral stability in metapelites revisited: new a-x relations for manganese-bearing minerals. *Journal of Metamorphic Geology*, **32**, 809–828, https://doi.org/10.1111/jmg.12095

Woodhead, J.D. and Hergt, J.M. 2007. Pb-isotope analyses of USGS reference materials. *Geostandards Newsletter*, **24**, 33–38, https://doi.org/10.1111/j.1751-908X.2000.tb00584.x

Zeh, A., Gerdes, A. and Barton, J.M.J. 2009. Archaean accretion and crustal evolution of the Kalahari Craton: the zircon age and Hf isotope record of granitic rocks from Barberton/Swaziland to the Francistown arc. *Journal of Petrology*, **50**, 933–966, https://doi.org/10.1093/petrology/egp027

Zhu, Z.K., O'Nions, R.K., Belshaw, N.S. and Gibb, A.J. 1997. Lewisian crustal history from *in situ* SIMS mineral chronometry and related metamorphic textures. *Chemical Geology*, **136**, 205–218, https://doi.org/10.1016/S0009-2541(96)00143-X

Zirakparvar, N.A., Mathez, E.A., Scoates, J.S. and Wall, C.J. 2014. Zircon Hf isotope evidence for an enriched mantle source for the Bushveld Igneous Complex. *Contributions to Mineralogy and Petrology*, **168**, 1050, https://doi.org/10.1007/s00410-014-1050-2

Trace element changes in rutile from quartzite through increasing *P–T* from lower amphibolite to eclogite facies conditions

Regiane A. Fumes[1]*, George L. Luvizotto[1], Inês Pereira[2,3] and Renato Moraes[4]

[1]Department of Geology, São Paulo State University, Av. 24A, 1515, 13506-900 Rio Claro, Brazil

[2]Université Clermont Auvergne, CNRS, IRD, OPGC, Laboratoire Magmas et Volcans, F-63000 Clermont-Ferrand, France

[3]University of Coimbra, Geosciences Center, Earth Sciences Department, Rua Sílvio Lima, 3030-790 Coimbra, Portugal

[4]Department of Mineralogy and Geotectonics, University of São Paulo, Rua do Lago, 562, 05508-080 São Paulo, Brazil

RAF, 0000-0003-4055-7906; GLL, 0000-0002-6150-8292; IP, 0000-0001-9028-2483; RM, 0000-0001-6917-3696
*Correspondence: regiane.fumes@unesp.br

Abstract: Low concentrations of Na, Ca, K, Fe, Mg and Al in quartzite commonly prevent the crystallization of index metamorphic minerals, inhibiting the obtainability of thermobarometric calculations. Quartzite typically contains quartz, zircon and rutile; therefore, single-element thermometers, such as Zr-in-rutile, may be applied. We investigate changes in trace-element composition of rutile from quartzite through increasing metamorphic conditions. Studied samples derive from a quartzite package (Luminárias Nappe, Minas Gerais, Brazil) where previous thermobarometric constraints on metapelites showed an increasing metamorphic grade southwards, from high-pressure lower amphibolite facies (580°C; 0.9 GPa) to eclogite facies (630°C; 1.4 GPa). Rutile from the lower-grade facies samples show a large spread in Zr concentrations, with the highest values corresponding to temperature estimates higher than metamorphic conditions affecting those units, and thus interpreted as inherited detrital signatures. A narrower spread in Zr concentration is observed in rutile grains from the higher-grade facies, and estimated Zr-in-rutile temperatures agree with previous thermobarometric constraints. Therefore, we show that at 630°C, Zr contents in detrital rutile from quartzites re-equilibrate. The comparison between the quartzite- and metapelite-hosting rutile grains from the same area shows that the resetting of the geothermometer in the latter seems to occur at slightly lower temperatures (∼50°C lower).

Supplementary material: A full dataset of analized trace elements in rutile is available at https://doi.org/10.6084/m9.figshare.c.6793887

Rutile has attracted significant attention in recent years, since it can be used in metamorphic studies to calculate crystallization temperatures (Zack *et al.* 2004*a*; Watson *et al.* 2006; Tomkins *et al.* 2007; Ewing *et al.* 2013), in provenance studies to differentiate between mafic and pelitic sources (Zack *et al.* 2004*b*; Triebold *et al.* 2007, 2012), to deduce the tectonic setting of sedimentary basins (Pereira *et al.* 2020) and in geochronological studies to determine ages of metamorphic events (Mezger *et al.* 1989; Zack *et al.* 2011; Bracciali *et al.* 2013; Ewing *et al.* 2015; Verberne *et al.* 2020). Furthermore, rutile is a common accessory mineral in metamorphic and sedimentary rocks. It may also be present in mafic and ultramafic igneous rocks, although the high Fe contents may favour crystallization of ilmenite. Rutile is also known to mirror the host rock's high field strength elements (HFSE) composition (Zack *et al.* 2002), and thus is an important mineral to understand the chemical evolution of the continental crust (Schmidt *et al.* 2009; Marschall *et al.* 2013).

Quartzite is a very common rock type in metasedimentary sequences. Although its protolith may be of different sources (e.g. chert, hydrothermally altered granite and detrital sediments), it usually represents the metamorphic product of mature detrital sedimentary rocks where, besides quartz, only weathering-resistant minerals such as rutile, tourmaline, zircon, apatite, ilmenite and iron oxides are present. Quartzite is seldom used in metamorphic

From: van Schijndel, V., Cutts, K., Pereira, I., Guitreau, M., Volante, S. and Tedeschi, M. (eds) 2024. *Minor Minerals, Major Implications: Using Key Mineral Phases to Unravel the Formation and Evolution of Earth's Crust.* Geological Society, London, Special Publications, **537**, 231–248.
First published online October 3, 2023, https://doi.org/10.1144/SP537-2022-207

studies since the lack or very low concentration of elements such as Na, Ca, K, Fe, Mg and Al prevents the crystallization of metamorphic index minerals. Some exceptions are sapphirine quartzite (Harley and Motoyoshi 2000) and kyanite quartzite (Zhang *et al.* 2002), which may be used as proxies for ultra-high temperature and high pressure/ultra-high pressure rocks, respectively.

Previous studies have discussed the application of Zr-in-rutile thermometry to quartzite and have shown that the Zr content in inherited rutile in quartzite may reset at higher temperatures in comparison to crystallization temperatures of rutile in metapelitic rocks, i.e. above ~600°C (Triebold *et al.* 2007; Luvizotto and Zack 2009). The lack of Fe–Ca-bearing phases, which makes quartzite less reactive than other rock types, leads to a diminished replacement of rutile by titanite or ilmenite (Triebold *et al.* 2007; Luvizotto and Zack 2009). However, the application of Zr-in-rutile thermometry to rocks equilibrated at temperatures <600°C requires caution, due to slow diffusion of HFSE elements in rutile (e.g. Cruz-Uribe *et al.* 2018). While this is true, studies using rutile in blueschist-facies rocks have also shown that even at temperatures below 600°C, Zr-in-rutile temperatures are in agreement with independent *P–T* estimates for those rocks (Spear *et al.* 2006; Hart *et al.* 2016). It is equally important to take into consideration the whole geological evolution of the studied area and not only the *P–T* conditions, since fluid pressure, deformation strains and the timescale of processes may play a significant role in the resetting of these geothermometers. For example, studies on Ti-in-quartz thermometry have been carried out in quartzite (e.g. Kidder *et al.* 2013), where the authors documented Ti resetting in quartz during deformation, within low temperature (~360°C) fabric realignment (bulging recrystallization).

The extraction of meaningful pressure and temperature conditions from quartzite is particularly useful in areas with rocks that lack metamorphic index minerals. Furthermore, if trace-element compositions of inherited rutile are preserved in medium-temperature facies quartzite, rutile trace-element geochemistry may be used for provenance, to assess rock types (mafic or pelitic), crystallization or metamorphic temperature of source rocks (Triebold *et al.* 2012) and be coupled to metamorphic ages through U–Pb dating (Luvizotto *et al.* 2009). Together, these tools have been applied to detrital rutile to provide more constraints on the evolution of the continental crust (Krabbendam *et al.* 2017; Barber *et al.* 2019; Zhou *et al.* 2020; Pereira *et al.* 2021).

In the present study, we investigate trace element changes in rutile in quartzites through increasing *P–T* conditions, from lower amphibolite to eclogite facies. The main aim is to evaluate under which *P–T* conditions resetting of trace elements in detrital/inherited rutile takes place. Particular attention is given to the resetting of Zr contents and the application of the Zr-in-rutile geothermometer. To tackle this question, we investigated a continuous layer of rutile- and zircon-bearing quartzite from the Luminárias Nappe, Minas Gerais, Brazil, along which, the regional metamorphic conditions increased from greenschist/lower-amphibolite facies to amphibolite/eclogite facies (Trouw *et al.* 1980; Ribeiro and Heilbron 1982; Silva 2010; Fumes *et al.* 2019). These are the metamorphic conditions where resetting of Zr in rutile is expected to take place (Stendal *et al.* 2006; Cherniak *et al.* 2007*a*; Triebold *et al.* 2007; Kohn *et al.* 2016; Cruz-Uribe *et al.* 2018), and thus constitute an ideal case study. Possible causes involved in this resetting are investigated and discussed.

Geologic setting and sample description

The Neoproterozoic Southern Brasília Orogen is the result of an Ediacaran to Cambrian collision between the passive margin of the São Francisco Plate and the active continental margin of the Paranapanema Plate (Brito Neves *et al.* 1999; Campos Neto 2000; Trouw *et al.* 2000). In the southernmost portion of the Brasília Orogen, this collision is characterized by a syn-metamorphic stack of nappes represented, from top to bottom, by (Campos Neto 2000): (i) the Socorro-Guaxupé Nappe derived from the lower crust of a magmatic arc root and consisting of granulite–migmatite–granite association (Campos Neto and Caby 1999; Campos Neto 2000); (ii) the Andrelândia Nappe System (Trouw *et al.* 1983, 1984; Campos Neto and Caby 1999; Campos Neto and Caby 2000; Campos Neto *et al.* 2007), which is composed of highly deformed metasedimentary rocks associated with forearc and accretionary–prism segments; (iii) the Carrancas Nappe System (Campos Neto *et al.* 2004), which is composed of metasedimentary rocks interpreted as the passive continental margin of the Sanfranciscan Plate (Paciullo *et al.* 2000, 2003; Trouw *et al.* 2000; Westin *et al.* 2019).

The Carrancas Nappe System includes the Luminárias Nappe (Fig. 1; Trouw *et al.* 2000; Campos Neto *et al.* 2004). Rocks of the Carrancas Nappe System belong to the Carrancas Group (Trouw *et al.* 1980), divided into two units: the basal São Tomé das Letras Unit, which is composed of quartzite and muscovite quartzite, and it is the unit from where the studied samples derive; and the upper Campestre Unit, which is composed of schist or phyllite with frequent lenses of quartzite. In the Carrancas Nappe System, the maximum depositional

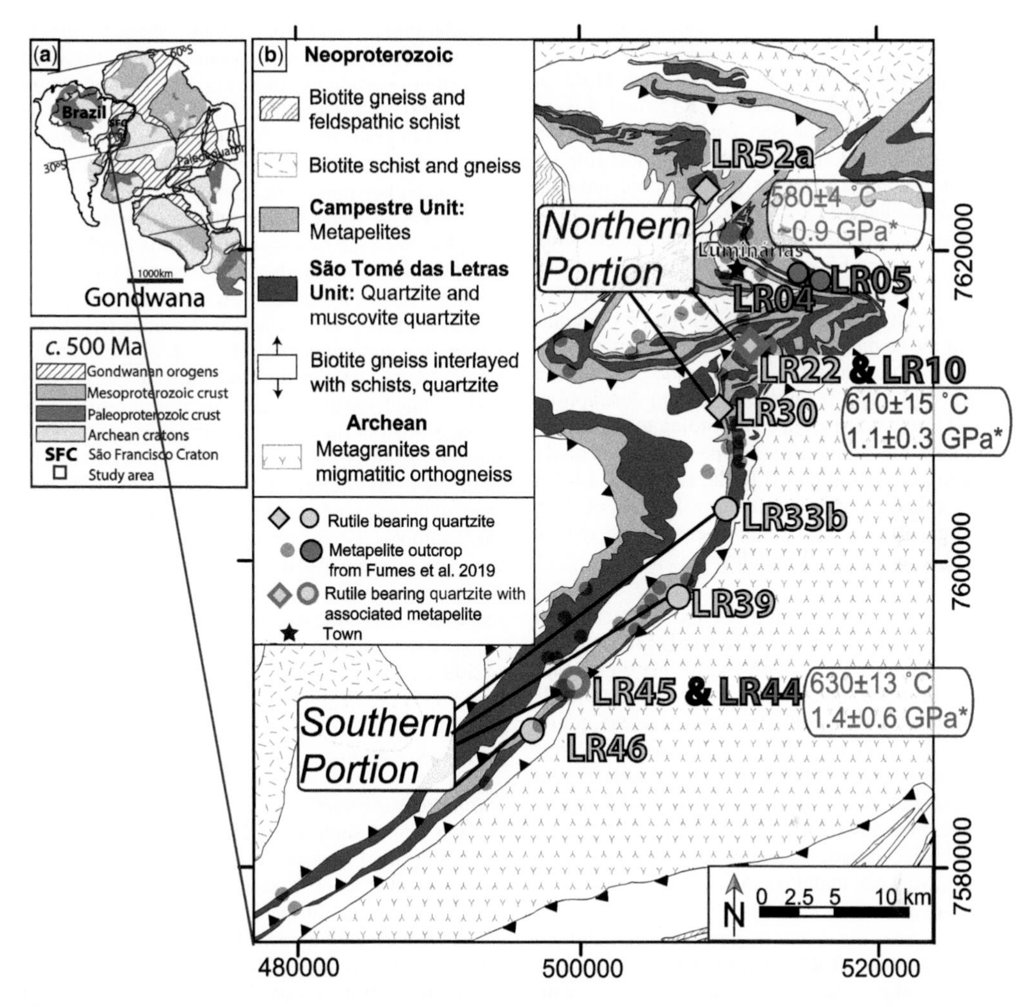

Fig. 1. (**a**) Gondwana map (situation ~500 myr ago) showing the location of the study area (grey rectangle). (**b**) Simplified geological map of Luminárias Nappe showing sample location (UTM, WGS84, Zone 23 K). Ms, muscovite; Qtz, quartz. Source: (a) Extracted from Spencer *et al.* (2013); (b) Modified after Quéméneur *et al.* (2002); Trouw *et al.* (2002); Paciullo *et al.* (2003); Nunes *et al.* (2008). *P–T conditions presented in the figure are those of Fumes *et al.* (2019).

age is *c.* 920 Ma (Valeriano *et al.* 2004; Westin and Campos Neto 2013; Westin *et al.* 2019; Kuster *et al.* 2020; Marimon *et al.* 2020).

A regional metamorphic gradient that extends throughout the Carrancas Group, with metamorphic conditions increasing southward from greenschist to amphibolite, has long been described in the literature (Trouw *et al.* 1980, 2000, 2013; Ribeiro and Heilbron 1982; Reno *et al.* 2012; Coelho *et al.* 2017). However, refined *P–T* conditions reflecting high pressures inside or close to the eclogite facies were determined more recently (Silva 2010; Fumes *et al.* 2019). The Luminárias Nappe records a section of the regional metamorphic gradient (Fig. 1) where the metamorphic conditions increase along the geological beds. According to Fumes *et al.* (2019), three diagnostic mineral peak assemblages are present in metapelitic rocks from the Luminárias Nappe: (a) the Chl + Ky + St + Ms + Qz + Rt assemblage (mineral abbreviations are after Whitney and Evans 2010) occurs in the northern portion at 580 ± 4°C and ~0.9 GPa, and indicate high-pressure lower amphibolite facies; (b) the St + Bt + Grt + Ms + Qz + Rt assemblage occurs in the central portion at 600 ± 15°C and 1.1 ± 0.3 GPa, and indicates high-pressure amphibolite facies; and (c) the St + Ky + Grt + Ms + Qz + Rt assemblage occurs in the southern portion at 630 ± 13°C and 1.4 ± 0.6 GPa, and indicates eclogite facies.

Sample description

Seven quartzite samples of ~20 kg each (LR52a, LR22, LR30, LR33b, LR39, LR45 and LR46) were collected from the São Tomé das Letras Unit covering a 40 km long north–south transect (Fig. 1; Table 1), along which, the metamorphic gradient increases southward.

In order to assure an independent characterization of peak *P–T* conditions, samples derive from sites where nearby metapelitic schists have been collected and studied by Fumes *et al.* (2019). The *P–T* conditions of the studied samples, as well as their correlation with those studied by Fumes *et al.* (2019), are presented in Table 1. For the following description, studied samples are grouped according to their location along the Luminárias Nappe (Fig. 1), namely northern portion samples (LR52a, LR22 and LR30) and southern portion samples (LR33b, LR39, LR45 and LR46). The studies were carried out on mineral separates in all samples.

Quartzite from the northern portion. The northern portion quartzite (LR52a, LR22 and LR30 samples) shows granoblastic texture with inequigranular, polygonal shaped quartz grains. Accessory minerals comprise less than 1 vol% (muscovite, zircon, rutile, magnetite, tourmaline, apatite and ilmenite). Average grain sizes are 200 μm for quartz and 150 μm for muscovite (Fig. 2a). Rock fabric is anisotropic, with oriented muscovite and elongated quartz grains, with tectonic foliation parallel to bedding. Quartz grains have weak undulose extinction.

Quartzite from the southern portion. The southern portion quartzite (LR33, LR39, LR45 and LR46 samples) also has granoblastic texture, with inequigranular and polygonal shaped quartz (Fig. 2c). The rock is coarser grained when compared to northern portion samples (Fig. 2c). Average grain size is 300 μm for quartz and 350 μm for muscovite (maximum grain size of 500 and 600 μm for quartz and muscovite, respectively). Samples have a slightly higher muscovite content (average of less than 3 vol%) when compared to northern portion samples. We note that sample LR46 is banded, and in muscovite-rich bands the concentration of this mineral reaches up to 15%. Accessory minerals are rutile, ilmenite, zircon, apatite, monazite, pyrite, tourmaline and magnetite. The rock displays anisotropic fabric, defined by elongated quartz crystals and oriented muscovite crystals; the anisotropy is stronger than in the northern portion quartzite. Tectonic foliation is parallel to bedding. Reduction of grain size by deformation, mainly by dynamic recrystallization of quartz, is locally observed in these samples. Quartz crystals have interlobate contacts and show intense undulose extinction. Irregular grain boundaries are interpreted to have formed in response to dynamic recrystallization of quartz associated with subgrain rotation and high-temperature (≥500°C) grain boundary migration, as described by Stipp *et al.* (2002).

Rutile description

Rutile from the northern portion. Most of the rutile crystals from samples LR52a, LR22 and LR30 have subround shapes (Fig. 3a–c). Few crystals have irregular shapes, including elongated ones in the LR22 sample. In the LR52a sample, separated rutile crystals vary from 80 to 180 μm and zircon inclusions are frequent. In the LR22 sample, separated rutile crystals vary from 80 to 120 μm and

Table 1. Location and metamorphic conditions of the studied quartzite samples

Quartzite	UTM E	UTM S	Metapelite sample	Nappe portion	*T* (°C)	*P* (GPa)	Min. assemblage metapelite*
LR-52a	511460	7625910	LR50†	Northern	580 ± 4	*c.* 0.9	Ky + Chl + Ctd + Ms
LR-22	513310	7615351	LR10†	Northern	600 ± 15	1.1 ± 0.3	St + Grt + Bt + Ms
LR30	510406	7611675	LR32	Northern			Grt + St + Ms
LR33b	511089	7604960	LR33	Southern			Grt + St + Ms
LR-39	506012	7598967	LR43	Southern			Ky + Grt + St + Ms
LR-45	501289	7593927	LR44†	Southern	630 ± 13	1.4 ± 0.6	Ky + Grt + St + Ms
LR-46	498605	7590930	LR47	Southern			Ky + Grt + Ms

Corresponding metapelite samples studied by Fumes *et al.* (2019) and their peak mineral assemblage are also presented. Coordinates are in Universal Transverse Mercator (UTM) (Zone 23 K, WGS84).
*Peak mineral assemblage according to Fumes *et al.* (2017).
†Metamorphic modelling according to Fumes *et al.* (2019).

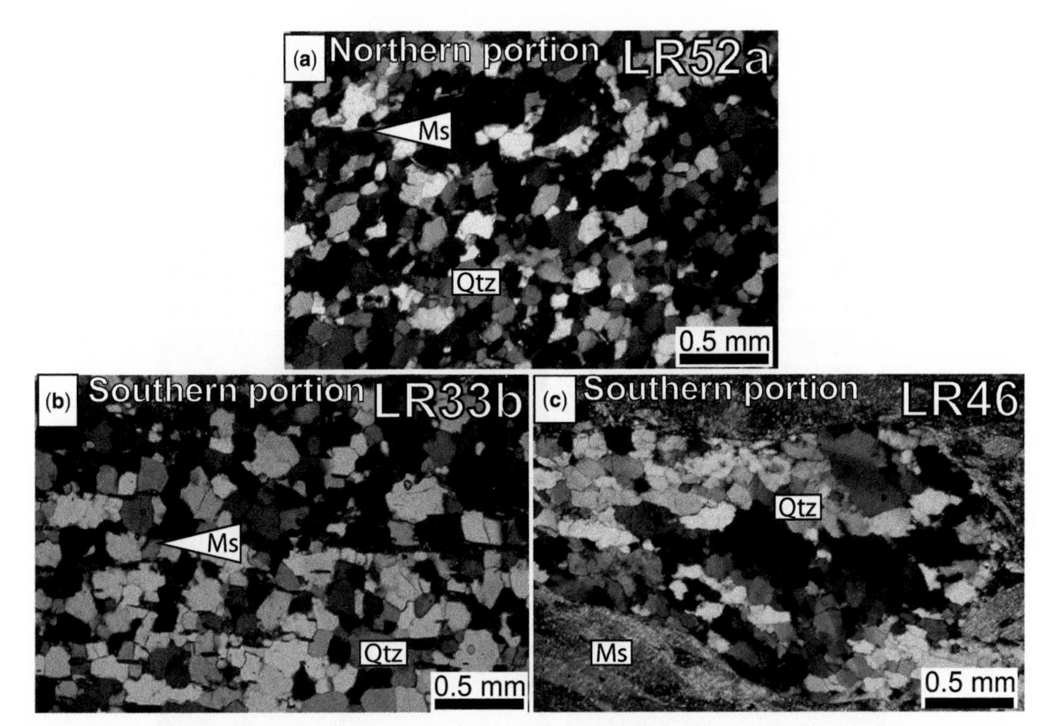

Fig. 2. Representative transmitted light photomicrographs (cross polarized light) of studied quartzite. (**a**) Photomicrograph from northern portion quartzite showing polygonal inequigranular texture, medium-grained (sample LR52a). (**b**) Photomicrograph from the southern portion quartzite showing granular texture (sample LR33b). (**c**) Photomicrograph from the southern portion quartzite showing inter-lobate texture, medium- to coarse-grained and strong undulose extinction in quartz (sample LR46). Qtz, quartz; Ms, muscovite.

the grains in many cases have inclusions of zircon and monazite. In the LR30 sample, separated rutile crystals vary from 90 to 250 μm (longest axis) and zircon inclusions and ilmenite lamellae occur. Rarely a zoning in rutile is observed in backscattered electron (BSE) images (Fig. 3b); some dissolution textures are visible (Fig. 3b), with porosity closer to the grain edges and fractures. The inclusions in the rutile from these samples have irregular shapes (Fig. 3a, b).

Fine-grained rutile crystals are observed only in thin sections from sample LR52a (average of 15 μm, longest axis), with needle-shaped crystals. Due to the small size, rutile crystals were not recovered during the heavy mineral separation process.

Rutile from the southern portion. In LR33 and LR39 samples, most of the rutile crystals are subrounded (Fig. 3e), on the other hand, some grains have angular, euhedral shapes (Fig. 3d). In LR45 and LR46 samples, most of the rutile crystals have elongated and prismatic euhedral shapes (Fig. 3f, g), and few rutile crystals have rounded shapes. Grain size varies from 50 to 120 μm in LR33 sample, 70 to 190 μm in LR39 sample, 100 to 340 μm at the longest axis in LR45 sample and 70 to 280 μm at the longest axis in LR46 sample. Ilmenite lamellae and zircon inclusion may occur in rutile grains from all samples (Fig. 3d). However, zircon inclusions in rutile grains are less frequent than in those from the northern portion. Zonation observed in BSE images and dissolution textures are rare (Fig. 3e).

Smaller rutile grains observed in thin section from samples from the southern portion are in many cases in contact with muscovite and have subhedral shape, occupying interstitial spaces between rock forming minerals. These rutile grains are oriented following the schistosity, indicating that they crystallized during peak metamorphism, and also the main deformation stage.

Large and rounded rutile crystals (70 to 340 μm, longest axis) are present most frequently in the northern portion (Fig. 3a–c), the size and shape pointing to preservation of the detrital shape. The smaller rutile grains, as well as subhedral and needle-shaped crystals, fill the interstitial spaces between rock forming minerals and, together with the larger euhedral crystals (Fig. 3d–g), are interpreted to have crystallized during metamorphism.

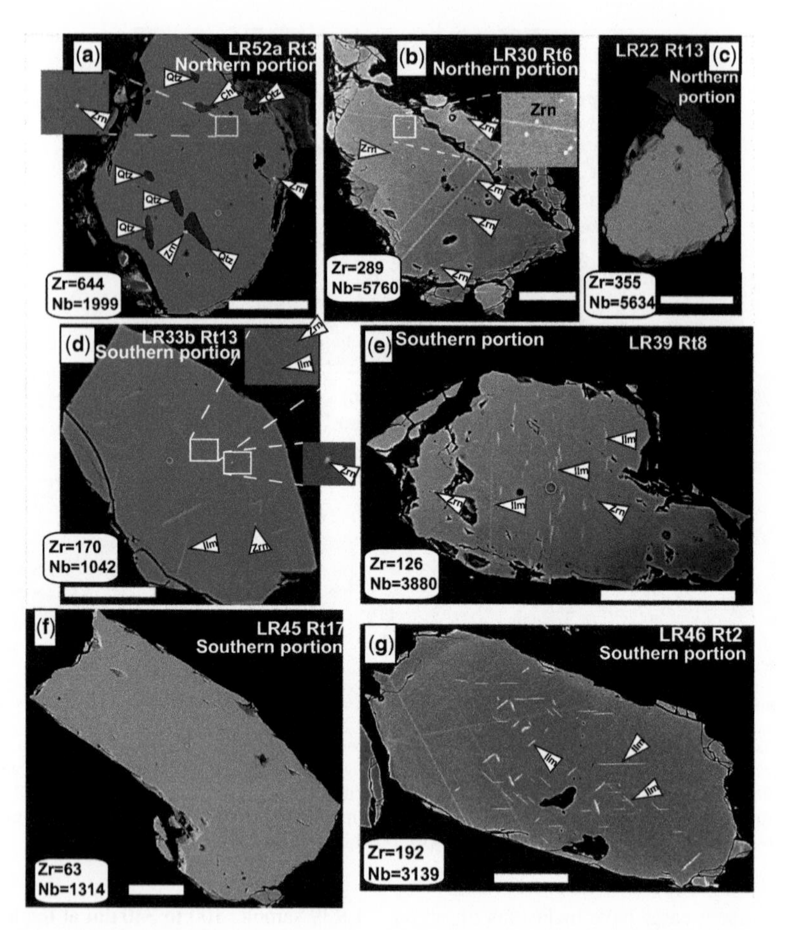

Fig. 3. Backscattered electron (BSE) images of representative rutile grains from crushed samples with concentrations of Zr and Nb in µg/g. (**a**) Rutile (Rt3) from LR52a sample with rounded detrital shape and zircon inclusion. (**b**) Rutile (Rt6) from LR30 sample with detrital shape and zircon inclusion. (**c**) Rutile (Rt13) from LR22 sample with detrital shape. (**d**) Rutile (Rt13) from LR33b sample with zircon inclusion and ilmenite lamella. (**e**) Rutile (Rt8) from LR39 sample with ilmenite lamella and subhedral shape. (**f**) Rutile (Rt17) from LR45 sample with elongated, euhedral shape. (**g**) Rutile (Rt4) from LR46 sample with ilmenite lamella. Chl, chlorite; Ilm, ilmenite; Qtz, quartz; Zrn, zircon; Scale bar measures 50 µm in all images. Spots' positions are indicated by the circles. Grains are labelled according to Appendix A in the Supplementary Material.

Methods

Sample preparation and imaging

Mineral separation followed the procedures of Triebold *et al.* (2007) and involved crushing and grain separation via sieving (63–200 µm fraction), heavy-liquid and magnetic separation. Mineral separates from the heavy, magnetic and non-magnetic fractions were hand-picked under stereoscopic microscope. Grains were embedded in epoxy discs that were subsequently polished and coated with carbon. We note that a rutile with the average diameter of 100 µm relates via hydrodynamic equivalence to a typical fine- to medium-grained sand (quartz with grain size of *c.* 200 µm).

Grain mounts were observed under scanning electron microscope to identify sub-microscopic inclusions in rutile. BSE images were collected using a JEOL JSM 6010 scanning electron microscope, using 15 to 20 kV. With these settings, the smallest imageable particle is *c.* 1 µm.

Electron microprobe analysis

Separated rutile grains from all samples (LR52a, LR22, LR30, LR33b, LR39, LR45 and LR46) were analysed by electron microprobe microanalyser (EPMA).

Electron microprobe analyses of rutile were carried out at the Department of Geology of São

Paulo State University in Rio Claro, Brazil with a JEOL JXA-8230 equipped with five wavelength-dispersive spectroscopy (WDS) detectors. Acceleration voltage was set to 20 kV and sample current to 80 nA. The following elements were analysed: Si, Al, Cr, Sb, Sn, W, Ta, Fe, Ti, Hf, Nb and Zr. A summary of the operational conditions is presented in Table 2. The Sy and R10 rutile mineral standards were routinely analysed to evaluate the quality of the analyses (Luvizotto *et al.* 2009). Furthermore, repeated analyses on the Sy rutile were used to assure 'true' zero concentration count rates on the peak, since all trace-element concentrations in this reference material are far below the EPMA detection limit (Luvizotto *et al.* 2009). Si contents in rutile were used as a quality control to detect and avoid contamination associated with submicroscopic zircon inclusions (following the method outlined by Zack *et al.* 2004*a*). Measurements with Si concentrations higher than 300 μg/g were excluded from the dataset. BSE images were used to select grain locations to be analysed (Fig. 3).

LA-ICP-MS analyses

Rutile crystals were also analysed using the laser ablation inductively coupled plasma mass spectrometry (LA-ICP-MS) instrument at the Department of Geology of São Paulo State University in Rio Claro, Brazil.

Rutile analyses were carried out using a Photon Machines Excimer Laser 193 nm coupled to a sector field inductively coupled plasma mass spectrometer (LA-SF-ICP-MS; Thermo Scientific Element 2). Trace elements were analysed with a 40 or 50 μm diameter laser beam, with a surface energy of 5.9 J/cm^2 and a repetition rate of 10 Hz. The following isotopes were analysed: ^{49}Ti, ^{51}V, ^{90}Zr, ^{93}Nb, ^{95}Mo, ^{118}Sn, ^{121}Sb, ^{178}Hf and ^{181}Ta. Each analysis consisted of 15 s of background, 46 s for the analysis signal and 4 s for system wash out time. A He–Ar mixture was used as the carrier gas. A sample bracketing method was used, where a block of 10 unknowns was interspersed by 2 to 3 analyses of the following reference materials: R10 (Luvizotto *et al.* 2009), NIST SRM 610 (Jochum *et al.* 2011), KL2-G (Jochum *et al.* 2000) and ML3B-G (Jochum *et al.* 2000). TiO_2 was used as an internal reference element using 59.95 wt% as the concentration of Ti. Trace-element data were processed using iolite 4 (Paton *et al.* 2011). Trace-element data and the uncertainty of reference values are listed in Appendix A in the supplementary material.

Results

Rutile trace-element geochemistry

As an average, 25 representative rutile crystals were analysed for each of the seven samples. Trace-element contents in rutile are presented in Figure 4, and the complete dataset is available in Appendix A (including all the analysed elements). In 27 grains, multiple analyses were performed to evaluate trace element intra-grain homogeneity. For grains with more than one analysis, only one representative result is presented (Fig. 4). Four out of 27 rutile grains show intra-grain variation of W, Ta and Nb, which is above the overall sample variation for the corresponding element. Intra-grain variation of Zr is smaller than the overall sample variation for this element in all analysed rutile grains. For Zr, the range of intra-grain variation (percentage difference from the average for each grain) is 9.4–121% for the northern portion samples, and 1.9–49.5% for the southern portion samples.

For Nb, a larger spread in concentrations (60–11 000 μg/g) is displayed by rutile grains from the northern portion samples (Fig. 4). A similar pattern is observed for Ta (LR22 and LR30 samples, 98–1700 μg/g). In southern portion samples, the average Nb concentration is 2840 μg/g, and Ta concentration is 365 μg/g. Two rutile crystals from the LR46 thin section record Nb content higher (8586 and 9482 μg/g) than those from the mineral separate (maximum concentration of 7997 μg/g). The average Nb/Ta ratio in the northern portion is higher

Table 2. Electron microprobe conditions applied for the rutile trace element analysis. TAP, thallium acid pthalate; PETJ, pentaerythritol; LIFL, lithium fluoride.

20 kV/80 nA	Si	Al	Cr	Sb	Sn	W	Ta	Fe	Ti	Nb	Zr
Crystal	TAP	TAP	PETJ	PETJ	PETJ	LIFL	LIFL	LIFL	PETJ	PETJ	PETJ
Line	Kα	Kα	Kα	Kα	Lα	Lα	Lα	Kα	Kβ	Lα	Lα
Peak sec*	300	300	150	150	150	150	150	150	30	300	300
Bkg. sec†	150	150	50	50	50	50	50	50	15	150	150
DL‡	25	20	50	80	75	95	85	40	55	40	45

*Count time on peak position in seconds.
†Count time on background position in seconds.
‡2σ detection limit, based on repeated measurement of variation on background, values in ppm.

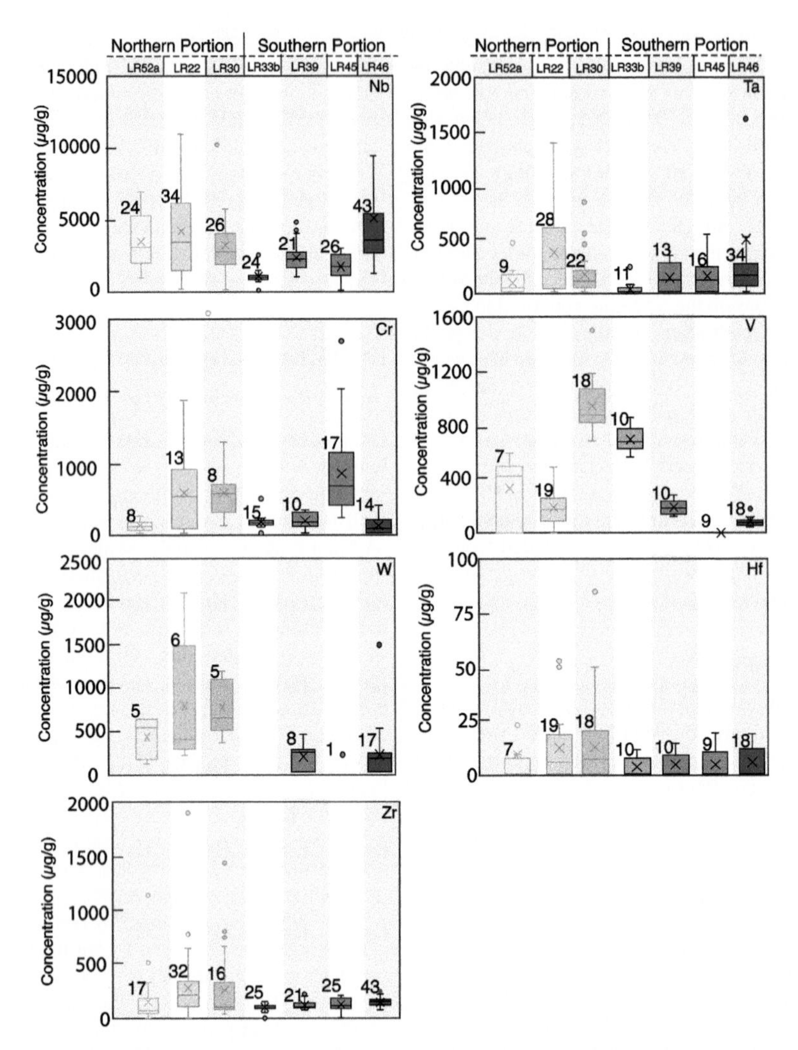

Fig. 4. Box and whisker plots showing concentration (in μg/g) of trace elements (Nb, Ta, Cr, V, W, Hf and Zr) in rutile crystals from studied samples. Whiskers represent the 5th and 95th percentiles, and boxes represent the second (bottom 25%) and third quartiles (top 75%). For rutile grains with more than one spot, only one representative analysis is plotted. The minimum and maximum values are plotted as small horizontal lines that limit the vertical lines of the box plots; the 'x' represents the median value and the lines represent the mean value. The numbers on top of each box represent the number of analyses for that sample. For samples in which less than five analyses are above detection limits, the values of each analysis are plotted as circles (Krzywinski and Altman 2014). Only analyses above the minimum detection limit are presented.

(average 21; range from 2 to 57) than in the southern portion (average 17; range from 3 to 67).

Rutile from LR22, LR30 and LR45 samples show a higher variation in concentration of Cr (60–2700 μg/g) compared to the other samples (53–495 μg/g) (Fig. 4).

Rutile crystals from LR52a, LR22 and LR30 samples (northern portion) and LR46 sample (southern portion) record the highest contents of W (102–1533 μg/g; Fig. 4). In the LR39 sample, the average concentration is within the 103–440 μg/g range.

Hafnium concentrations in all samples vary from 3 to 101 μg/g. LR22 and LR30 samples have a higher spread in Hf content (4–85 μg/g). The V concentration varies from 43 to 1509 μg/g; LR30 sample has the highest V values (680–1509 μg/g) and LR45 sample has the lowest values (43–210 μg/g).

Zirconium data show a distinct pattern (Fig. 4). Northern portion samples (LR52a, LR22 and LR30) show a large spread in Zr content. Several of the results are above 400 μg/g. Samples from the southern portion display a much narrower spread in Zr concentration, with maximum concentration not exceeding 309 µg/g. Zr contents in rutile crystals from thin sections and mineral separates are the same within the error.

The Zr/Hf ratio is similar for all samples with an average value of 16, which varies from 6 to 34.

EPMA elemental intensity profiling was carried out along rutile grains from mineral separates and thin sections to investigate possible elemental zoning within grains. Profiles in rutile from thin sections are particularly useful to evaluate the influence of neighbouring minerals on rutile geochemistry. Representative profiles of Nb and Zr are presented in Figure 5. We chose Zr and Nb because the former is a relevant element for its temperature dependence in rutile (Zack *et al.* 2004*b*), and the latter for its high compatibility in rutile under most conditions (Brenan *et al.* 1994; Ayers *et al.* 1997; Klemme *et al.* 2005), which usually translates into high concentrations in rutile (thousands of µg/g). In rutile grains from the northern portion samples, Zr distribution is homogeneous in some grains (Fig. 5;

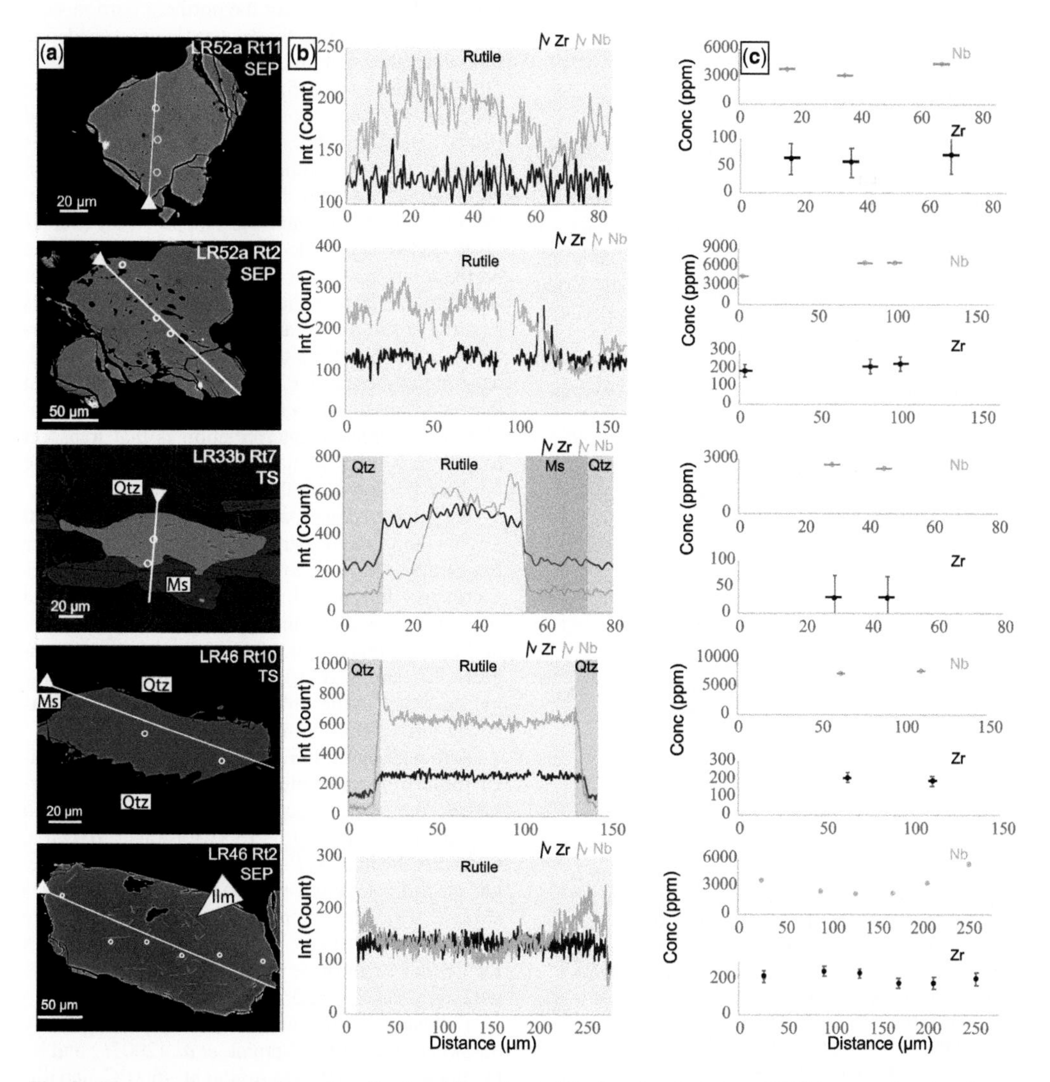

Fig. 5. Representative trace element profiles in rutile grains. (**a**) Backscattered electron (BSE) images of analysed rutile grains; white line indicates where the profiles were measured, and white circles indicate the location of quantitative analyses. (**b**) Data from line profiles of Zr and Nb (raw intensities on peak position for the elements). (**c**) Quantitative analyses along the profile; horizontal bars represent the spot size of analyses, and vertical bars are error bars indicating the 2σ uncertainty of Zr and Nb compositions. Value in µg/g. Conc, concentration; Ilm, Ilmenite; Int, intensity; Ms, muscovite; Qtz, quartz; SEP, separated rutile grains; TS, rutile in thin section.

LR52a–Rt11 and Rt2) and heterogeneous in others (Appendix A; LR52a–Rt2 and Rt3, LR22–Rt10), with no systematic distribution. Distribution of Nb along all analysed grains is heterogeneous, although the variation does not define a systematic core to rim pattern. Rutile grains from the southern portion samples exhibit flat Zr profiles, while showing heterogeneous Nb distributions (Fig. 5; LR33b and LR46). Higher intensities of Nb are observed in some regions close to rutile rims. In many cases, these Nb peaks occur where muscovite is in contact or near the rutile (Fig. 5; LR33b–Rt7 and LR46–Rt10).

Discussion

Application of Zr-in-rutile thermometry

In Figure 6, we present the calculated temperatures of all analysed rutile grains. In order to compare these results to those obtained by Fumes *et al.* (2019), the same calibration used by those authors is employed here, the calibration by Tomkins *et al.* (2007). The more recent calibration of Kohn (2020) would result in lower temperature estimates, with an offset of ~35°C (Appendix A).

The Zr content in rutile grains from the northern portion (LR52a, LR22 and LR30 samples) shows a large spread. Based on the metamorphic peak conditions obtained from neighbouring metapelites by Fumes *et al.* (2019), we used a pressure of 1.0 GPa for the northern portion samples and of 1.4 GPa for the southern portion samples to calculate Zr-in-rutile temperatures. Although the pressure effect on the Zr-in-rutile thermometer is fairly small, a potential uncertainty of ~40°C is added to the temperature uncertainty estimated for the northern portion samples, if a P range of 0.5 to 1.5 GPa is considered. Zr-in-rutile temperatures for the southern portion samples show a much narrower spread in comparison to results obtained for the northern portion samples (Fig. 6). If the third quartile is considered, following the recommendation of Tomkins *et al.* (2007), values of 615, 613, 633 and 638°C are obtained for samples LR33b, LR39, LR45 and LR46, respectively. Zr-in-rutile temperatures obtained for the southern portion samples are in good agreement with the temperature presented in the literature for Luminárias Nappe rocks (Fumes *et al.* 2019), unlike temperatures obtained for the northern portion (lower-grade conditions). One possible interpretation is that these results point towards a partial or full preservation of detrital rutile temperatures in samples from the northern portion, and that resetting of these relict Zr signatures was attained at the conditions that affected the southern portion samples. An alternative interpretation is that transport limitation restricted inter-grain equilibration in the northern portion more than in the southern portion.

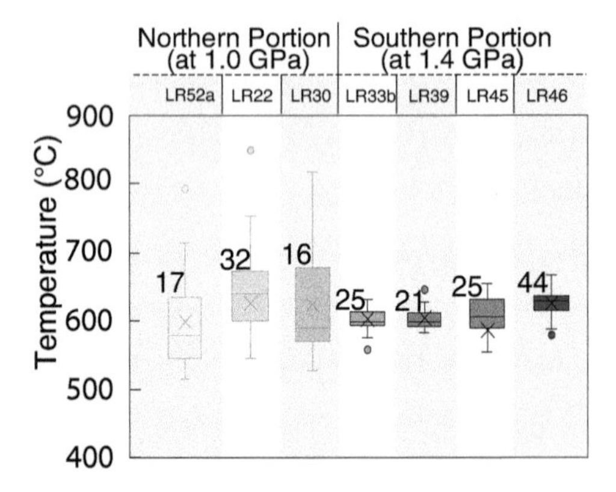

Fig. 6. Box and whisker plots showing Zr-in-rutile temperatures calculated for the studied rutile grains (after calibration of Tomkins *et al.* 2007). Whiskers represent the 5th and 95th percentiles, and boxes represent the second (bottom 25%) and third quartiles (top 75%). For rutile grains with more than one spot, only one representative analysis is plotted. The minimum and maximum values are plotted as small horizontal lines that limit the vertical lines of the box plots; the 'x' represents the median value, and the lines represent the mean value. The numbers on top of each box represent the number of analyses for that sample. For samples in which less than five analyses are above detection limits, the values of each analysis are plotted as circles.

Our preferred interpretation of an original detrital origin is supported by grain morphologies, which are frequently rounded in the lower-grade samples (northern), interpreted as preserving their detrital shape (Fig. 3a–c), against the typically more euhedral crystals that are found in the higher-grade samples (Fig. 3d, f & g). The euhedral shape may indicate that rutile in the southern portion was newly formed during metamorphism.

Although minimum Zr re-equilibration T in detrital rutile from quartzite cannot be precisely determined, our data indicate that Zr re-equilibration was attained at *c.* 630 ± 13°C for 1.4 ± 0.6 GPa, which are the highest $P–T$ conditions for the metapelite from the southern portion, according to Fumes *et al.* (2019). These conditions are within the range of those that have been suggested in previous studies for detrital Zr-in-rutile resetting (550–600°C, Triebold *et al.* 2007; ~620°C, Stendal *et al.* 2006).

Diffusion coefficients of Zr in rutile obtained by Sasaki *et al.* (1985), Cherniak *et al.* (2007*b*) and by Dohmen *et al.* (2019) are similar at ~600°C, and diffusion is slower at lower temperatures. According to Dohmen *et al.* (2019), at least some diffusion of Zr may take place at 630°C. The Zr closure temperature calculated in Dohmen *et al.* (2019) for a rutile of *c.* 100 μm at a cooling rate of 10°C/Ma is about 600°

C, similar to what is observed in our study. However, the Dodson (1973) closure temperature model is not realistic for Zr in rutile, because the model was developed for Ar loss, where transport limitation in the rock matrix is much less of a factor. This implies that rutile grains from the southern portion of the Luminárias Nappe very likely reset their Zr contents due to diffusion mechanisms.

Trace element systematics in rutile from the Luminárias Nappe quartzites

It has been shown that the Nb and Cr contents in rutile can be used to discriminate rutile derived from metamafic and metafelsic protoliths, where a higher Nb/Cr ratio (>1) is typical of metafelsic sources (Zack *et al.* 2004*a*; Triebold *et al.* 2012). Results obtained for the studied samples show large intra- and inter-sample variation in concentrations of Nb and Cr (Fig. 4), something that is quite common to find in metamorphic rocks and terranes, especially for Nb (Kooijman *et al.* 2012; Pereira and Storey 2023). Chromium concentrations between rutile grains within the samples of the northern portion vary more strongly (from 0.2 to 1888 µg/g) than in most samples from the southern portion (varying <400 µg/g) (Fig. 4). This is also true for Nb (Fig. 4). The Nb/Cr discrimination diagram (following Triebold *et al.* 2012) shows that rutile grains from all samples have, with very few exceptions (*n* = 16), a metapelitic signature (Fig. 7b). Since these rutile grains are hosted in quartzites, this discrimination does not provide clues about their detrital or recrystallized nature. The only five grains with metamafic affinity may indeed represent detrital grains, but due to some uncertainty (*c.* 10%) in the fidelity of the Nb/Cr discrimination (Meinhold *et al.* 2008; Pereira *et al.* 2021), interpretations cannot rely solely on Nb/Cr.

Considering the sample distribution along the Luminárias Nappe and the fact that metamorphic conditions increase southwards (Fumes *et al.* 2019), a few correlations can be observed between *P–T* conditions and W and Zr contents in the rutile (Fig. 4).

Due to variable diffusivities of different elements in rutile (see review in Pereira and Storey 2023), certain trace elements may become more diffusive to increasing metamorphic temperatures. For example, experimental data have shown that Fe and Cr are more diffusive than Nb, Ta or Hf (Sasaki *et al.* 1985; Marschall *et al.* 2013) at given temperatures. Indeed, Fe contents within our samples show a very similar range of concentrations among different samples (*c.* 2000 up to 7000 µg/g), with only one sample (LR45), in the southern domain, exhibiting lower average concentrations and a smaller variation (428–2391 µg/g) (Fig. 4). As metamorphic reactions take place, and certain minerals present in the sandstone break down, they may release given elements that have a strong preference for rutile. These elements may be taken up by rutile through diffusion or by (re)crystallization. One such element is V, which may be introduced via ilmenite or magnetite breakdown. This may explain why V contents in rutile grains from the high-temperature facies quartzites are more variable and reach higher values than in rutile grains from the low-temperature facies quartzites, which show more restricted intra-sample variation, and lower V contents (Fig. 7c). Additionally, as we compare rutile temperatures (estimated from Zr) with W (Fig. 7d) and Ta (Fig. 7e) contents, a similar pattern is observable: intra-sample variation of W and Ta in samples from the southern portion (high-temperature facies) is small, yielding low contents (<500 µg/g W and <500 µg/g Ta), whilst intra-sample variation of those same elements in samples from the northern portion (low-temperature facies) is larger (up to 2000 µg/g for both elements). Metamorphic rutile commonly yields lower W and Ta contents (Luvizotto *et al.* 2009; Agangi *et al.* 2020) when compared to rutile from leucogranites or fluid-altered rocks (1000 s of µg/g; Carruzzo *et al.* 2006; Agangi *et al.* 2020; Carocci *et al.* 2021). We interpret the higher variation in rutile grains from the low-temperature facies samples (northern portion) as reflecting their detrital nature, sourcing rocks with variable rutile compositions. Decreasing intra-sample variation coupled to decreasing average concentrations take place due to increasing metamorphic facies and re-homogenization of W and Ta during metamorphism, either by recrystallization or complete dissolution–reprecipitation.

Comparison between trace element in rutile from quartzite and metapelite rocks from the Luminárias Nappe

As described above, in the northern portion of the Luminárias Nappe, Zr concentrations in rutile grains from quartzite show a large spread. However, rutile grains from the neighbouring metapelitic rocks, and thus under the same metamorphic conditions, show a much smaller intra-sample Zr variation (Fumes *et al.* 2019). On the other hand, in the southern portion, where metamorphic conditions are slightly higher (by 50°C; Fumes *et al.* 2019), the concentrations of Zr in rutile from both quartzite and metapelite are similar and with small variation (Fig. 8). In the northern portion, these data are further accompanied by differences in rutile grain textures; detrital and metamorphic rutile grains are observed in quartzites, while strictly metamorphic rutile occurs in metapelite samples (see fig. 5 in Fumes *et al.* 2019). Variable *T* equilibration of Zr contents

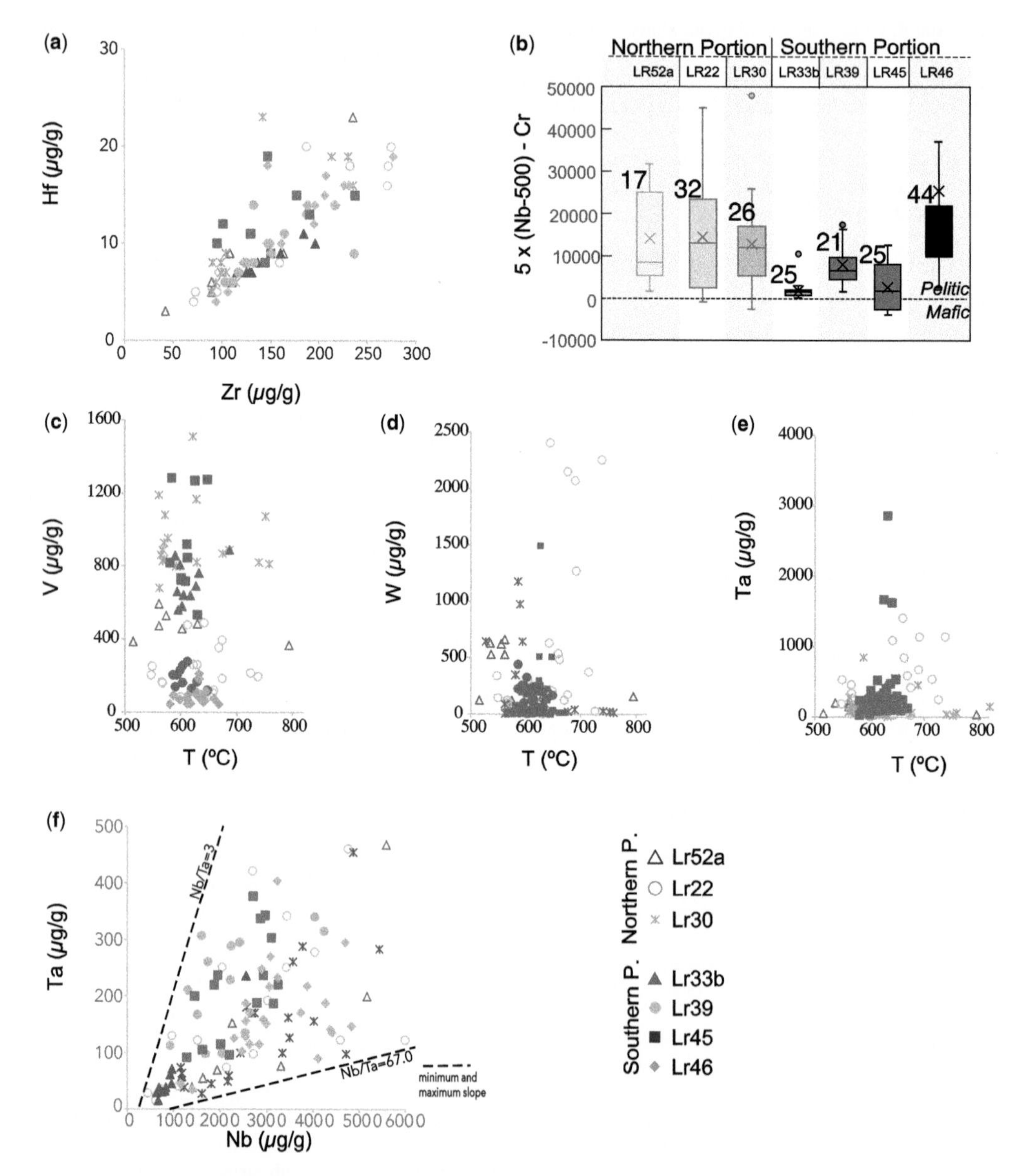

Fig. 7. (**a**) Zr v. Hf diagram for rutiles of different portions. (**b**) Box and whisker plots showing Pelitic/mafic discrimination line. Calculations are based on Nb and Cr concentrations and follow the method outlined by Triebold *et al.* (2012). Positive values indicate pelitic source, whereas negative values indicate mafic source for the rutile grains. Whiskers represent the 5th and 95th percentiles, and boxes represent the second (bottom 25%) and third quartiles (top 75%). For rutile grains with more than one spot, only one representative analysis is plotted. The minimum and maximum values are plotted as small horizontal lines that limit the vertical lines of the box plots; the 'x' represents the median value, and the lines represent the mean value. The numbers on top of each box represent the number of analyses for that sample. For samples in which less than five analyses are above detection limits, the values of each analysis are plotted as circles. Only analyses above the minimum detection limit are presented. (**c**) V v. temperature diagram for rutiles of different portions. (**d**) W v. temperature diagram for rutiles of different portions. (**e**) Ta v. temperature diagram for rutiles of different portions. (**f**) Nb v. Ta diagram for rutiles.

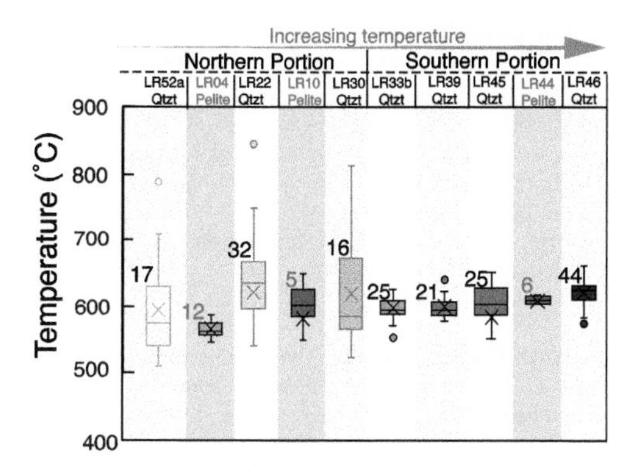

Fig. 8. Box and whisker plots showing Zr-in-rutile temperatures from quartzite and metapelite samples. The metapelite content is shown in green, and the quartzite is in grey and black. Whiskers represent the 5th and 95th percentiles, and boxes represent the second (bottom 25%) and third quartiles (top 75%). For rutile grains with more than one spot, only one representative analysis is plotted. The minimum and maximum values are plotted as small horizontal lines that limit the vertical lines of the box plots; the 'x' represents the median value, and the lines represent the mean value. The numbers on top of each box represent the number of analyses for that sample. For samples in which less than five analyses are above detection limits, the values of each analysis are plotted as circles. Source: The metapelite sample data are extracted from Fumes *et al.* (2019).

in rutile grains during prograde metamorphism between quartzite and metapelites had already been reported by Triebold *et al.* (2007) and Luvizotto and Zack (2009).

Differences in the texture and in the resetting temperature of Zr in rutile between quartzite and metapelite can be explained by the difference in the chemical system: quartzite has very low concentration of Na, Ca, K, Fe, Mg and Al, when compared to metapelite. The more complex chemical system of the metapelite favours metamorphic reactions involving rutile. Our study confirms these observations, and places a temperature constraint of 580° C, above which, rutile in quartzite may still preserve its source information, but where it already starts to re-equilibrate in metapelites. At about 630°C, Zr in all detrital rutile re-equilibrates with the mineral assemblage to reflect their metamorphic conditions.

Trace element redistribution in rutile during metamorphism

Line profiles carried out in rutile grains show an intra-grain enrichment of Nb where rutile is in contact or near muscovite (Fig. 5). It seems likely that mineral contact relationships control Nb concentration in some of these studied rutile grains, since high Nb contents occur in rutile rims that are in contact or near muscovite. Muscovite modes are higher in the southern portion (~3 vol%) than in the northern portion (~1 vol%). Whereas Nb contents change along some rutile grains, Zr contents are homogenous in all grains. Since Nb is highly compatible in rutile, the increase in Nb relates to the fact that some Ti is taken up by the mica, whereas Nb is not. As a result, the Nb back-diffusion in rutile leads to higher concentrations in the rutile rim. Higher contents of Nb in rutile rims have also been observed in previous studies (Lucassen *et al.* 2010; Cruz-Uribe *et al.* 2014; Kohn *et al.* 2016). According to Lucassen *et al.* (2010), rutile rimmed by titanite yields higher HFSE, including Nb, that are redistributed by back-diffusion into rutile. Experimental data (Dohmen *et al.* 2019) show that the diffusion coefficient D_{Zr} is approximately equal to D_{Nb} in rutile. The different behaviour shown by Nb and Zr can be explained by the fact that Zr is mainly hosted in zircon, and its incorporation in rutile is strongly temperature dependent, with a logarithmic increase of Zr with increasing temperature. In contrast, Nb is almost exclusively hosted in rutile, and its concentration in rutile is a direct function of the host-rock Nb/Ti ratio in rocks that contain no other Ti minerals. In such scenario, as soon as other Ti minerals form (e.g. ilmenite, magnetite, amphibole or mica), some of the Ti – but no Nb – is taken from the rutile, such that the Nb content of the remaining rutile increases. It is important to notice that Nb is highly compatible in rutile, and rutile grains may show growth zoning governed by Rayleigh fractionation or by back-diffusion at the grain margins during partial resorption. This can later be modified by intra-grain solid-state diffusion and by dissolution–reprecipitation. Considering the Nb diffusion data of Dohmen *et al.* (2019), volume diffusion will be too slow to reset internal zoning in rutile at greenschist facies conditions (350–400°C). Yet, internal zonation in 100–200 μm large rutile grains would be obliterated at around 750°C (upper amphibolite facies).

The Zr/Hf ratio is constant in all studied samples (Fig. 7a), with Hf tightly coupled to Zr. In a few samples, Zr/Hf yields a nicely defined positive slope (LR33b, LR39), suggesting that Hf may also be temperature dependent. This has been previously suggested by Luvizotto and Zack (2009) and Ewing *et al.* (2014).

The Nb/Ta ratio in the rutile varies from 2 to 67 (Fig. 7f). Decoupling of Nb and Ta could be assigned to variations between their diffusion coefficients, which are much higher for Nb than Ta (1.6–18 times) (Marschall *et al.* 2013; Dohmen *et al.* 2019). Alternatively, the decoupling can also be attributed to local dissolution and reprecipitation processes at relatively low temperature (e.g. Pereira *et al.* 2019).

Concluding remarks

Our data show that Zr contents in detrital rutile may re-equilibrate in quartzite, but commonly at higher temperatures than for rutile in associated metapelites. For the studied quartzites, minimum conditions for inter-grain Zr exchange are *c.* 630 ± 30°C for 1.4 ± 0.6 GPa. Although these conditions cannot be taken as a threshold temperature above which Zr-in-rutile temperatures in quartzite can be generally considered as recording metamorphic temperatures, our results indicate that rutile from high-temperature facies quartzite may indeed be used to calculate peak metamorphic conditions. Re-equilibration of rutile grains in quartzites takes place at temperatures that are approximately 50°C higher than in metapelites.

- The different behaviour of Zr in rutile in comparison to other trace elements is accredited to equilibrium of the geothermometer, where Zr and Si are buffered by the presence of zircon and quartz in the rock.
- Niobium is highly compatible in rutile, and grains will show Nb growth zoning due to Rayleigh fractionation or by back-diffusion at the grain margins during partial resorption (e.g. Marschall *et al.* 2013). This zoning can be (partially) homogenized by intra-grain solid-state diffusion and by dissolution–reprecipitation.
- The obtained results also indicate that re-equilibration of W, Ta, Cr and Nb in rutile from quartzite does not take place below *c.* 580°C at 1.0 GPa. Since rutile is the only mineral hosting Nb, Ta and W in the studied rocks, it would be possible to evaluate the rock budget for these elements and interpret the analysed contents in terms of inter-grain exchange and intra-grain homogenization. At higher temperatures during metamorphism, W, Ta and Nb re-homogenized either by recrystallization or complete dissolution–reprecipitation.
- The re-equilibration of Zr in rutile grains from neighbouring metapelites from the Luminárias Nappe seems to occur at slightly lower *P–T* conditions than for those in the quartzite. This implies that the detrital rutile signature should be better preserved in quartzitic samples. Thus, studies targeting detrital Zr-in-rutile thermometry should target quartzites instead of metapelites, even under lower-amphibolite conditions, provided that post-depositional fluid circulation was minimal.

Acknowledgements RAF and GLL are affiliated with the Instituto GeoAtlântico, a National Institute of Science and Technology, CNPq-Brazil (process no. 405653/2022-0). The authors would like to acknowledge Horst Marschall and an anonymous reviewer for their comments and suggestions, and Valby van Schijndel for her editorial work. We are grateful to Daniela Rubatto, Alicia M. Cruz-Uribe and an anonymous reviewer for their detailed and insightful comments on a previous version of this manuscript. The authors would also like to thank Mônica Heilbron, Ticiano José Saraiva dos Santos, Maurício Pavan Silva, Cauê Rodrigues Cioffi, Juliana Okubo and Mauly Bottene for their help with improving the manuscript.

Competing interests The authors declare that they have no known competing financial interests or personal relationships that could have appeared to influence the work reported in this paper.

Author contributions **RAF**: conceptualization (equal), data curation (equal), formal analysis (equal), investigation (equal), methodology (equal), writing – original draft (lead), writing – review & editing (lead); **GLL**: conceptualization (equal), data curation (supporting), formal analysis (equal), funding acquisition (lead), investigation (equal), methodology (equal), writing – original draft (supporting), writing – review & editing (supporting); **IP**: data curation (supporting), investigation (supporting), writing – original draft (supporting), writing – review & editing (supporting); **RM**: conceptualization (equal), data curation (supporting), writing – original draft (supporting), writing – review & editing (supporting).

Funding This work was supported by the São Paulo Research Foundation (FAPESP) through grants 2015/07750-0 and 2015/05230-0, RAF and GLL, respectively. GLL and RM were supported by the National Council for Scientific and Technological Development (CNPq, Brazil) (grant numbers 311606/2019-9 and 305720/2020-1, respectively). RAF was supported by the National Council for Scientific and Technological Development-CNPq (PhD scholarship 141604/2018-2) and by the Brazilian Federal Agency for Support and Evaluation of Graduate Education (CAPES). IP was supported by the Portuguese National Science Foundation (FCT) through a fellowship contract under the Individual Call to Scientific Employment Stimulus (2021.01616.CEECIND) and grants UIDB/00073/2020 and UIDP/00073/2020 awarded to Centro de Geociências.

Data availability All data generated or analysed during this study are included in this published article (and its supplementary information files).

References

Agangi, A., Plavsa, D., Reddy, S.M., Olierook, H. and Kylander-Clark, A. 2020. Compositional modification and trace element decoupling in rutile: insight from the Capricorn Orogen, Western Australia. *Precambrian Research*, **345**, article 105772, https://doi.org/10.1016/j.precamres.2020.105772

Ayers, J.C., Dittmer, S.K. and Layne, G.D. 1997. Partitioning of elements between peridotite and H2O at 2.0–3.0

GPa and 900–1100 C, and application to models of subduction zone processes. *Earth and Planetary Science Letters*, **150**, 381–398.

Barber, D.E., Stockli, D.F. and Galster, F. 2019. The Proto-Zagros foreland basin in Lorestan, western Iran: insights from multimineral detrital geothermochronometric and trace elemental provenance analysis. *Geochemistry, Geophysics, Geosystems*, **20**, 2657–2680, https://doi.org/10.1029/2019GC008185

Bracciali, L., Parrish, R.R., Horstwood, M.S., Condon, D.J. and Najman, Y. 2013. U–Pb LA-(MC)-ICP-MS dating of rutile: new reference materials and applications to sedimentary provenance. *Chemical Geology*, **347**, 82–101, https://doi.org/10.1016/j.chemgeo.2013.03.013

Brenan, J.M., Shaw, H.F., Phinney, D.L. and Ryerson, F.J. 1994. Rutile–aqueous fluid partitioning of Nb, Ta, Hf, Zr, U and Th: implications for high field strength element depletions in island-arc basalts. *Earth and Planetary Science Letters*, **128**, 327–339, https://doi.org/10.1016/0012-821X(94)90154-6

Brito Neves, B.B., Campos Neto, M.D.C. and Fuck, R.A. 1999. From Rodinia to Western Gondwana: an approach to the Brasiliano–Pan African Cycle and orogenic collage. *Episodes: Journal of International Geoscience*, **22**, 155–166.

Campos Neto, M.D.C. 2000. Orogenic systems from southwestern Gondwana: an approach to Brasiliano–Pan African Cycle and orogenic collage in southeastern Brazil. *In*: Cordani, U.G. *et al.* (eds) *Tectonic Evolution of South America*. Companhia de pesquisa de recursos minerais, Rio de Janerio, 335–365.

Campos Neto, M.C. and Caby, R. 1999. Neoproterozoic high-pressure metamorphism and tectonic constraint from the nappe system south of the São Francisco Craton, southeast Brazil. *Precambrian Research*, **97**, 3–26, https://doi.org/10.1016/S0301-9268(99)00010-8

Campos Neto, M.D.C. and Caby, R. 2000. Lower crust extrusion and terrane accretion in the Neoproterozoic nappes of southeast Brazil. *Tectonics*, **19**, 669–687.

Campos Neto, M.C., Basei, M.A.S., Vlach, S.R.F., Caby, R., Szabó, G.A.J. and Vasconcelos, P. 2004. Migração de orógenos e superposição de orogêneses: um esboço da colagem brasiliana no Sul do Cráton do São Francisco, SE-Brasil. *Geologia USP Série Científica*, **4**, 13–40, https://doi.org/10.5327/S1519-874x2004000100002

Campos Neto, M.C., Janasi, V.A., Basei, M.A.S. and Siga, O., Jr 2007. Sistema de nappes Andrelândia, setor oriental: litoestratigrafia e posição estratigráfica. *Revista Brasileira de Geociências*, **37**, 47–60, https://doi.org/10.25249/0375-7536.200737s44760

Carocci, E., Marignac, C., Cathelineau, M., Truche, L., Poujol, M., Boiron, M.-C. and Pinto, F. 2021. Incipient wolframite deposition at Panasqueira (Portugal): W-rich rutile and tourmaline compositions as proxies for the early fluid composition. *Economic Geology*, **116**, 123–146, https://doi.org/10.5382/econgeo.4783

Carruzzo, S., Clarke, D.B., Pelrine, K.M. and MacDonald, M.A. 2006. Texture, composition, and origin of rutile in the South Mountain Batholith, Nova Scotia. *Canadian Mineralogist*, **44**, 715–729, https://doi.org/10.2113/gscanmin.44.3.715

Cherniak, D.J., Manchester, J. and Watson, E.B. 2007*a*. Zr and Hf diffusion in rutile. *Earth and Planetary Science Letters*, **261**, 267–279, https://doi.org/10.1016/j.epsl.2007.06.027

Cherniak, D.J., Watson, E.B. and Wark, D.A. 2007*b*. Ti diffusion in quartz. *Chemical Geology*, **236**, 65–74, https://doi.org/10.1016/j.chemgeo.2006.09.001

Coelho, M.B., Trouw, R.A.J., Ganade, C.E., Vinagre, R., Mendes, J.C. and Sato, K. 2017. Constraining timing and P–T conditions of continental collision and late overprinting in the Southern Brasília Orogen (SE-Brazil): U–Pb zircon ages and geothermobarometry of the Andrelândia Nappe System. *Precambrian Research*, **292**, 194–215, https://doi.org/10.1016/j.precamres.2017.02.001

Cruz-Uribe, A.M., Feineman, M.D., Zack, T. and Barth, M. 2014. Metamorphic reaction rates at ~650–800°C from diffusion of niobium in rutile. *Geochimica et Cosmochimica Acta*, **130**, 63–77, https://doi.org/10.1016/j.gca.2013.12.015

Cruz-Uribe, A.M., Feineman, M.D., Zack, T. and Jacob, D.E. 2018. Assessing trace element (dis)equilibrium and the application of single element thermometers in metamorphic rocks. *Lithos*, **314–315**, 1–15, https://doi.org/10.1016/j.lithos.2018.05.007

Dodson, M.H. 1973. Closure temperature in cooling geochronological and petrological systems. *Contributions to Mineralogy and Petrology*, **40**, 259–274, https://doi.org/10.1007/BF00373790

Dohmen, R., Marschall, H.R., Ludwig, T. and Polednia, J. 2019. Diffusion of Zr, Hf, Nb and Ta in rutile: effects of temperature, oxygen fugacity, and doping level, and relation to rutile point defect chemistry. *Physics and Chemistry of Minerals*, **46**, 311–332, https://doi.org/10.1007/s00269-018-1005-7

Ewing, T.A., Hermann, J. and Rubatto, D. 2013. The robustness of the Zr-in-rutile and Ti-in-zircon thermometers during high-temperature metamorphism (Ivrea–Verbano Zone, northern Italy). *Contributions to Mineralogy and Petrology*, **165**, 757–779, https://doi.org/10.1007/s00410-012-0834-5

Ewing, T.A., Rubatto, D. and Hermann, J. 2014. Hafnium isotopes and Zr/Hf of rutile and zircon from lower crustal metapelites (Ivrea–Verbano Zone, Italy): implications for chemical differentiation of the crust. *Earth and Planetary Science Letters*, **389**, 106–118, https://doi.org/10.1016/j.epsl.2013.12.029

Ewing, T.A., Rubatto, D., Beltrando, M. and Hermann, J. 2015. Constraints on the thermal evolution of the Adriatic margin during Jurassic continental break-up: U–Pb dating of rutile from the Ivrea–Verbano Zone, Italy. *Contributions to Mineralogy and Petrology*, **169**, 1–22, https://doi.org/10.1007/s00410-015-1135-6

Fumes, R.A., Luvizotto, G.L., Moraes, R. and Ferraz, E.R.M. 2017. Petrografia, química mineral e geotermobarometria de metapelito do Grupo Carrancas na Nappe de Luminárias (MG). *Geociências*, **36**, 639–654, http://doi.org/10.5016/geociencias.v36i4.12150

Fumes, R.A., Luvizotto, G.L., Moraes, R., Heilbron, M. and Vlach, S.R.F. 2019. Metamorphic modeling and petrochronology of metapelitic rocks from the Luminárias Nappe, southern Brasília belt (SE Brazil).

Brazilian Journal of Geology, **49**, https://doi.org/10.1590/2317-4889201920180114

Harley, S.L. and Motoyoshi, Y. 2000. Al zoning in orthopyroxene in a sapphirine quartzite: evidence for >1120°C UHT metamorphism in the Napier Complex, Antarctica, and implications for the entropy of sapphirine. *Contributions to Mineralogy and Petrology*, **138**, 293–307, https://doi.org/10.1007/s004100050564

Hart, E., Storey, C., Bruand, E., Schertl, H.P. and Alexander, B.D. 2016. Mineral inclusions in rutile: a novel recorder of HP–UHP metamorphism. *Earth and Planetary Science Letters*, **446**, 137–148, https://doi.org/10.1016/j.epsl.2016.04.035

Jochum, K.P., Dingwell, D.B. *et al.* 2000. The preparation and preliminary characterisation of eight geological MPI-DING reference glasses for *in-situ* microanalysis. *Geostandards Newsletter*, **24**, 87–133, https://doi.org/10.1111/j.1751-908X.2000.tb00590.x

Jochum, K.P., Weis, U. *et al.* 2011. Determination of reference values for NIST SRM 610–617 glasses following ISO guidelines. *Geostandards and Geoanalytical Research*, **35**, 397–429, https://doi.org/10.1111/j.1751-908X.2011.00120.x

Kidder, S., Avouac, J.P. and Chan, Y.C. 2013. Application of titanium-in-quartz thermobarometry to greenschist facies veins and recrystallized quartzites in the Hsüehshan range, *Taiwan. Solid Earth*, **4**, 1–21, https://doi.org/10.5194/se-4-1-2013

Klemme, S., Prowatke, S., Hametner, K. and Günther, D. 2005. Partitioning of trace elements between rutile and silicate melts: implications for subduction zones. *Geochimica et Cosmochimica Acta*, **69**, 2361–2371, https://doi.org/10.1016/j.gca.2004.11.015

Kohn, M.J. 2020. A refined zirconium-in-rutile thermometer. *American Mineralogist*, **105**, 963–971, https://doi.org/10.2138/am-2020-7091

Kohn, M.J., Penniston-Dorland, S.C. and Ferreira, J.C.S. 2016. Implications of near-rim compositional zoning in rutile for geothermometry, geospeedometry, and trace element equilibration. *Contributions to Mineralogy and Petrology*, **171**, 1–15, https://doi.org/10.1007/s00410-016-1285-1

Kooijman, E., Smit, M.A., Mezger, K. and Berndt, J. 2012. Trace element systematics in granulite facies rutile: implications for Zr geothermometry and provenance studies. *Journal of Metamorphic Geology*, **30**, 397–412, https://doi.org/10.1111/j.1525-1314.2012.00972.x

Krabbendam, M., Bonsor, H., Horstwood, M.S.A. and Rivers, T. 2017. Tracking the evolution of the Grenvillian foreland basin: constraints from sedimentology and detrital zircon and rutile in the Sleat and Torridon groups, Scotland. *Precambrian Research*, **295**, 67–89, https://doi.org/10.1016/j.precamres.2017.04.027

Krzywinski, M. and Altman, N. 2014. Visualizing samples with box plots. *Nature Methods*, **11**, 119–120, https://doi.org/10.1038/nmeth.2813

Kuster, K., Ribeiro, A., Trouw, R.A.J., Dussin, I. and Marimon, R.S. 2020. The Neoproterozoic Andrelândia Group: evolution from an intraplate continental margin to an early collisional basin south of the São Francisco Craton, Brazil. *Journal of South American Earth Sciences*, **102**, article 102666, https://doi.org/10.1016/j.jsames.2020.102666

Lucassen, F., Dulski, P., Abart, R., Franz, G., Rhede, D. and Romer, R.L. 2010. Redistribution of HFSE elements during rutile replacement by titanite. *Contributions to Mineralogy and Petrology*, **160**, 279–295, https://doi.org/10.1007/s00410-009-0477-3

Luvizotto, G.L. and Zack, T. 2009. Nb and Zr behavior in rutile during high-grade metamorphism and retrogression: an example from the Ivrea–Verbano Zone. *Chemical Geology*, **261**, 303–317, https://doi.org/10.1016/j.chemgeo.2008.07.023

Luvizotto, G.L., Zack, T. *et al.* 2009. Rutile crystals as potential trace element and isotope mineral standards for microanalysis. *Chemical Geology*, **261**, 346–369, https://doi.org/10.1016/j.chemgeo.2008.04.012

Marimon, R.S., Johannes Trouw, R.A., Dantas, E.L. and Ribeiro, A. 2020. U–Pb and Lu–Hf isotope systematics on detrital zircon from the southern São Francisco Craton's Neoproterozoic passive margin: tectonic implications. *Journal of South American Earth Sciences*, **100**, article 102539, https://doi.org/10.1016/j.jsames.2020.102539

Marschall, H.R., Dohmen, R. and Ludwig, T. 2013. Diffusion-induced fractionation of niobium and tantalum during continental crust formation. *Earth and Planetary Science Letters*, **375**, 361–371, https://doi.org/10.1016/j.epsl.2013.05.055

Meinhold, G., Anders, B., Kostopoulos, D. and Reischmann, T. 2008. Rutile chemistry and thermometry as provenance indicator: an example from Chios Island, Greece. *Sedimentary Geology*, **203**, 98–111.

Mezger, K., Hanson, G.N. and Bohlen, S.R. 1989. High-precision U–Pb ages of metamorphic rutile: application to the cooling history of high-grade terranes. *Earth and Planetary Science Letters*, **96**, 106–118, https://doi.org/10.1016/0012-821X(89)90126-X

Nunes, R.P.M., Trouw, R.A.J. and Castro, E.O. 2008. *Mapa Geológico – Varginha.* 449000, Serviço Geológico do Brasil, https://www.sgb.gov.br/publique/media/geologia_basica/pgb/varginha.pdf

Paciullo, F.V.P., Ribeiro, A., Andreis, R.R. and Trouw, R.A.J. 2000. The Andrelândia Basin, a Neoproterozoic intraplate continental margin, southern Brasília belt, Brazil. *Revista Brasileira de Geociências*, **30**, 200–202, https://doi.org/10.25249/0375-7536.2000301200202

Paciullo, F.V.P., Ribeiro, A. and Trouw, R.A.J. 2003. *Geologia da Folha Andrelândia 1:100.000. Geologia e Recursos Minerais do Sudeste Mineiro Projeto Sul de Minas – Etapa I*, Companhia Mineradora de Minas Gerais, Belo Horizonte, 84–119.

Paton, C., Hellstrom, J., Paul, B., Woodhead, J. and Hergt, J. 2011. Iolite: freeware for the visualisation and processing of mass spectrometric data. *Journal of Analytical Atomic Spectrometry*, **26**, 2508–2518, https://doi.org/10.1039/c1ja10172b

Pereira, I. and Storey, C.D. 2023. Detrital rutile: records of the deep crust, ores and fluids. *Lithos*, 438–439, article 107010, https://doi.org/10.1016/j.lithos.2022.107010

Pereira, I., Storey, C., Darling, J., Lana, C. and Alkmim, A.R. 2019. Two billion years of evolution enclosed in hydrothermal rutile: recycling of the São Francisco

Craton crust and constraints on gold remobilisation processes. *Gondwana Research*, **68**, 69–92, https://doi.org/10.1016/j.gr.2018.11.008

Pereira, I., Storey, C.D., Strachan, R.A., Bento dos Santos, T. and Darling, J.R. 2020. Detrital rutile ages can deduce the tectonic setting of sedimentary basins. *Earth and Planetary Science Letters*, **537**, article 116193, https://doi.org/10.1016/j.epsl.2020.116193

Pereira, I., Storey, C.D., Darling, J.R., Moreira, H., Strachan, R.A. and Cawood, P.A. 2021. Detrital rutile tracks the first appearance of subduction zone low T/P paired metamorphism in the Palaeoproterozoic. *Earth and Planetary Science Letters*, **570**, article 117069, https://doi.org/10.1016/j.epsl.2021.117069

Quéméneur, J.J.G., Ribeiro, A., Trouw, R.A.J., Paciullo, F.V.P. and Heilbron, M. 2002. *Projeto Sul de Minas – Geologia da Folha Lavras*. Companhia Mineradora de Minas Gerais, Belo Horizonte, 259–319.

Reno, B.L., Piccoli, P.M., Brown, M. and Trouw, R.A.J. 2012. *In situ* monazite (U–Th)–Pb ages from the Southern Brasília Belt, Brazil: constraints on the high-temperature retrograde evolution of HP granulites. *Journal of Metamorphic Geology*, **30**, 81–112, https://doi.org/10.1111/j.1525-1314.2011.00957.x

Ribeiro, A. and Heilbron, M. 1982. Estratigrafia e metamorfismo dos Grupos Carrancas e Adrelândia, Sul de Minas Gerais. *Anais do XXXII Congresso Brasileiro de Geologia*. Slavador/BA Brazil, Sociedade Brasileira de Geologia, 177–186.

Sasaki, J., Peterson, N.L. and Hoshino, K. 1985. Tracer impurity diffusion in single-crystal rutile (TiO_{2-x}). *Journal of Physics and Chemistry of Solids*, **46**, 1267–1283, https://doi.org/10.1016/0022-3697(85)90129-5

Schmidt, A., Weyer, S., John, T. and Brey, G.P. 2009. HFSE systematics of rutile-bearing eclogites: new insights into subduction zone processes and implications for the Earth's HFSE budget. *Geochimica et Cosmochimica Acta*, **73**, 455–468, https://doi.org/10.1016/j.gca.2008.10.028

Silva, M.P. 2010. *Modelamento metamórfico de rochas das fácies xisto-verde e anfibolito com o uso de pseudosseções: exemplo das rochas da Klippe Carrancas, sul de Minas Gerais*. USP.

Spear, F.S., Wark, D.A., Cheney, J.T., Schumacher, J.C. and Watson, E.B. 2006. Zr-in-rutile thermometry in blueschists from Sifnos, Greece. *Contributions to Mineralogy and Petrology*, **152**, 375–385, https://doi.org/10.1007/s00410-006-0113-4

Spencer, C.J., Hawkesworth, C., Cawood, P.A. and Dhuime, B. 2013. Not all supercontinents are created equal: Gondwana–Rodinia case study. *Geology*, **41**, 795–798, https://doi.org/10.1130/G34520.1

Stendal, H., Toteu, S.F. *et al.* 2006. Derivation of detrital rutile in the Yaoundé region from the Neoproterozoic Pan-African belt in southern Cameroon (Central Africa). *Journal of African Earth Sciences*, **44**, 443–458, https://doi.org/10.1016/j.jafrearsci.2005.11.012

Stipp, M., Stünitz, H., Heilbronner, R. and Schmid, S.M. 2002. The eastern Tonale fault zone: a 'natural laboratory' for crystal plastic deformation of quartz over a temperature range from 250 to 700°C. *Journal of Structural Geology*, **24**, 1861–1884, https://doi.org/10.1016/S0191-8141(02)00035-4

Tomkins, H.S., Powell, R. and Ellis, D.J. 2007. The pressure dependence of the zirconium-in-rutile thermometer. *Journal of Metamorphic Geology*, **25**, 703–713, https://doi.org/10.1111/j.1525-1314.2007.00724.x

Triebold, S., von Eynatten, H., Luvizotto, G.L. and Zack, T. 2007. Deducing source rock lithology from detrital rutile geochemistry: an example from the Erzgebirge, Germany. *Chemical Geology*, **244**, 421–436, https://doi.org/10.1016/j.chemgeo.2007.06.033

Triebold, S., von Eynatten, H. and Zack, T. 2012. A recipe for the use of rutile in sedimentary provenance analysis. *Sedimentary Geology*, **282**, 268–275, https://doi.org/10.1016/j.sedgeo.2012.09.008

Trouw, R.A.J., Ribeiro, A. and Paciullo, F.V.P. 1980. Evolução estrutural e metamórfica de uma área a SE de Lavras – Minas Gerias. *Anais do XXXI Congresso Brasieliro de Geologia*. Camboriú/SC Brazil, October, 1980, Sociedade Brasileira de Geologia, 2273–2284.

Trouw, R.A.J., Ribeiro, A. and Paciullo, F.V.P. 1983. Geologia estrutural do Grupos São João del Rei, Carrancas e Andrelândia, sul de Minas Gerais. *Anais da Academia Brasileira de Ciências*, **55**, 71–87.

Trouw, R.A.J., Paciullo, F.V.P. and Heilbron, M. 1984. Os Grupos São João Del Rei, Carrancas e Andrelândia interpretados como a continuição dos Grupos Araxá e Canastra. *Anais do XXXIII Congresso Brasileiro de Geologia*. Rio de Janeiro/RJ Brazil, October 1984, Sociedade Brasileira de Geologia, 177–178.

Trouw, R.A.J., Heilbron, M. *et al.* 2000. The central segment of the Ribeira Belt. *In*: Cordani, U.G., Milani, E.J., Thomaz Filho, A. and Campos, D.A. (eds) *Tectonic Evolution of South America*. 1st edn. Companhia De Pesquisa De Recursos Minerais, 287–310.

Trouw, R.A.J., Ribeiro, A. and Paciullo, F.V.P. 2002. *Geologia da Folha Caxambu*. Companhia Mineradora de Minas Gerais, Belo Horizonte, 120–152.

Trouw, R.A.J., Peternel, R. *et al.* 2013. A new interpretation for the interference zone between the southern Brasília belt and the central Ribeira belt, SE Brazil. *Journal of South American Earth Sciences*, **48**, 43–57, https://doi.org/10.1016/j.jsames.2013.07.012

Valeriano, C.M., Machado, N., Simonetti, A., Valladares, C.S., Seer, H.J. and Simões, L.S.A. 2004. U–Pb geochronology of the southern Brasília belt (SE-Brazil): sedimentary provenance, Neoproterozoic orogeny and assembly of West Gondwana. *Precambrian Research*, **130**, 27–55, https://doi.org/10.1016/j.precamres.2003.10.014

Verberne, R., Reddy, S.M. *et al.* 2020. The geochemical and geochronological implications of nanoscale trace-element clusters in rutile. *Geology*, **48**, 1126–1130, https://doi.org/10.1130/G48017.1

Watson, E.B., Wark, D.A. and Thomas, J.B. 2006. Crystallization thermometers for zircon and rutile. *Contributions to Mineralogy and Petrology*, **151**, 413–433, https://doi.org/10.1007/s00410-006-0068-5

Westin, A. and Campos Neto, M.D.C. 2013. Provenance and tectonic setting of the external nappe of the Southern Brasília Orogen. *Journal of South American Earth*

Sciences, **48**, 220–239, https://doi.org/10.1016/j.jsames.2013.08.006

Westin, A., Campos Neto, M.C., Cawood, P.A., Hawkesworth, C.J., Dhuime, B. and Delavault, H. 2019. The Neoproterozoic southern passive margin of the São Francisco craton: insights on the pre-amalgamation of West Gondwana from U–Pb and Hf–Nd isotopes. *Precambrian Research*, **320**, 454–471, https://doi.org/10.1016/j.precamres.2018.11.018

Whitney, D.L. and Evans, B.W. 2010. Abbreviations for names of rock-forming minerals. *American Mineralogist*, **95**, 185–187.

Zack, T., Kronz, A., Foley, S.F. and Rivers, T. 2002. Trace element abundances in rutiles from eclogites and associated garnet mica schists. *Chemical Geology*, **184**, 97–122, https://doi.org/10.1016/S0009-2541(01)00357-6

Zack, T., Moraes, R. and Kronz, A. 2004*a*. Temperature dependence of Zr in rutile: empirical calibration of a rutile thermometer. *Contributions to Mineralogy and Petrology*, **148**, 471–488, https://doi.org/10.1007/s00410-004-0617-8

Zack, T., von Eynatten, H. and Kronz, A. 2004*b*. Rutile geochemistry and its potential use in quantitative provenance studies. *Sedimentary Geology*, **171**, 37–58, https://doi.org/10.1016/j.sedgeo.2004.05.009

Zack, T., Stockli, D.F., Luvizotto, G.L., Barth, M.G., Belousova, E., Wolfe, M.R. and Hinton, R.W. 2011. *In situ* U–Pb rutile dating by LA-ICP-MS: ^{208}Pb correction and prospects for geological applications. *Contributions to Mineralogy and Petrology*, **162**, 515–530, https://doi.org/10.1007/s00410-011-0609-4

Zhang, R.Y., Liou, J.G. and Shu, J.F. 2002. Hydroxyl-rich topaz in high-pressure and ultrahigh-pressure kyanite quartzites, with retrograde woodhouseite, from the Sulu terrane, eastern China. *American Mineralogist*, **87**, 445–453, https://doi.org/10.2138/am-2002-0408

Zhou, G., Fisher, C.M., Luo, Y., Pearson, D.G., Li, L., He, Y. and Wu, Y. 2020. A clearer view of crustal evolution: U–Pb, Sm–Nd, and Lu–Hf isotope systematics in five detrital minerals unravel the tectonothermal history of northern China. *Bulletin of the Geological Society of America*, **132**, 2367–2381, https://doi.org/10.1130/B35515.1

Electron probe petrochronology of monazite- and garnet-bearing metamorphic rocks in the Saxothuringian allochthonous domains (Erzgebirge, Granulite and Münchberg massifs)

Bernhard Schulz[1]* and Joachim Krause[2]

[1]Institute of Mineralogy, TU Bergakademie Freiberg, Brennhausgasse 14, D-09599 Freiberg/Saxony, Germany

[2]Helmholtz-Zentrum Dresden-Rossendorf, Helmholtz Institute Freiberg for Resource Technology, Chemnitzer Str. 40, D-09596 Freiberg/Saxony, Germany

BS, 0000-0001-5003-3431; JK, 0000-0001-8552-3318
*Correspondence: bernhard.schulz@mineral.tu-freiberg.de

Abstract: In the Saxothuringian Zone, a unique assemblage of high- to ultra-high-pressure and ultra-high-temperature metamorphic units is associated with medium- to low-pressure and temperature rocks. The units were studied in a campaign with garnet and monazite petrochronology of gneisses, micaschists and phyllites, and monazite dating in granites. P–T path segments of garnet crystallization were reconstructed by geothermobarometry and interpreted in terms of the monazite stability field, EPMA Th–U–Pb monazite ages and garnet Y + HREE zonations. One can recognize (1) Cambrian plutonism (512–503 Ma) with contact metamorphism in the Münchberg Massif. Subordinate monazite populations may indicate a (2) widespread but weak Silurian (444–418 Ma) thermal event. A (3) Devonian (389–360 Ma) high-pressure metamorphism prevails in the Münchberg and Frankenberg massifs. In the ultra-high-pressure and high-pressure units of the Erzgebirge the predominant (4) Carboniferous (336–327 Ma) monazites crystallized at the decompression paths. In the Saxonian Granulite Massif, prograde–retrograde P–T paths of cordierite-garnet gneisses can be related to monazite ages from 339 to 317 Ma. A (5) local hydrothermal overprint at 313–302 Ma coincides partly with post-tectonic (345–307 Ma) granite intrusions. Such diverse monazite age pattern and P–T time paths characterize the tectono-metamorphic evolution of each crustal segment involved in the Variscan Orogeny.

Supplementary material: Sampling locations and ages (S1), microscopy of polycrystalline quartz (S2), garnet composition maps (S3), silicates EPMA data and garnet LA-ICP-MS data (S4), monazite ThO_2*–PbO diagrams (S5) and monazite EPMA data (S6) are available at https://doi.org/10.6084/m9.figshare.c.6793959

Among the minor minerals, the phosphate monazite (LREE, Y, Th, U, Si, Ca)PO_4 – first described by Breithaupt (1829) at Freiberg/Saxony – is an accessory phase with unique chemical and crystal-physical properties (Spear and Pyle 2002; and references therein). These characteristics and *in situ* analytical methods applicable in its preserved textural context make it a perfect target for geochronology (e.g. Harrison *et al.* 2002). Among these methods, cost-effective and rapid *in situ* Th–U–Pb dating by electron probe microanalyser (EPMA) is based on: common Pb in monazite being negligible compared to radiogenic Pb resulting from decay of abundant Th and minor U (Parrish 1990; Montel *et al.* 1996); extremely low diffusion rates for radiogenic Pb at high temperatures (Cherniak *et al.* 2004); and its self-annealing capacities preventing accumulation of radiation damage (Meldrum *et al.* 1998; Nasdala *et al.* 2018).

Monazite occurs as an accessory phase in peraluminous granites, and in Ca-poor and Al-rich metapsammopelitic rocks at metamorphic grades above the upper greenschist facies (Spear and Pyle 2002). In contrast to properties of zircon in which U–Pb isotope systematics are mostly resilient at metamorphic temperatures, monazite can undergo dissolution–reprecipitation and recrystallization during metamorphism. It therefore appears as an incremental time-recorder and may even preserve multiple episodes of metamorphism (Schulz 2021*a*, *b*; Oriolo *et al.* 2022). This brings the almandine garnet into focus. Garnet is the most prominent recorder of changes in pressure and temperature during metamorphism. These are observed in its chemical zonations, as analysed by EPMA and laser ablation inductively coupled plasma mass spectrometry (LA-ICP-MS). These zonations reflect continuous metamorphic reactions in assemblages with biotite, muscovite,

From: van Schijndel, V., Cutts, K., Pereira, I., Guitreau, M., Volante, S. and Tedeschi, M. (eds) 2024. *Minor Minerals, Major Implications: Using Key Mineral Phases to Unravel the Formation and Evolution of Earth's Crust.* Geological Society, London, Special Publications, **537**, 249–284.
First published online September 25, 2023, https://doi.org/10.1144/SP537-2022-195

plagioclase, cordierite, aluminosilicates and quartz (Spear 1993).

The Saxothuringian Zone in the northern part of the Bohemian Massif is a prominent and well-studied part of the European Variscan orogenic belt (Kossmat 1927; Franke 2000, 2006; Kroner *et al.* 2007, 2020; Kroner and Romer 2010; Linnemann and Romer 2010; Franke *et al.* 2017; Martínez-Catalán *et al.* 2020, 2021; Machek *et al.* 2021). This zone comprises autochthonous and allochthonous domains, and preserves an exceptional variety of metamorphic rocks, ranging from low pressure–low temperature (LP–LT) over medium pressure–medium temperature (MP–MT) to high and ultra-high pressure (HP, UHP) and high to ultra-high temperature (HT, UHT) phyllites, micaschists, garnet gneisses and granulites (Pälchen 2011).

The UHT, UHP and HP rocks have been the target of numerous petrological studies (e.g. Klemd *et al.* 1991; Massonne 2001, 2011; Rötzler *et al.* 2004, 2008; Massonne *et al.* 2007*a*; Rötzler and Romer 2010; Collett *et al.* 2016, 2020, 2022; Sagawe *et al.* 2016; Schmädicke *et al.* 2018; Pohlner *et al.* 2021; Rötzler and Timmerman 2021; Závada *et al.* 2021; Gose and Schmädicke 2022; and references therein). Related low to high P and T rocks have received comparatively less attention (e.g. Reinhardt and Kleemann 1994; Mingram and Rötzler 1999; Waizenhöfer and Massonne 2017; Rahimi and Massonne 2018, 2020). There exists a wealth of U–Pb zircon age data from the meta-igneous rocks and the granites in the Saxothuringian Zone (e.g. Kröner *et al.* 1995; Tichomirowa *et al.* 2001, 2019*a*; Linnemann and Romer 2010; Koglin *et al.* 2018; Rötzler and Timmerman 2021; and references therein). The magmatic protolith ages range from Neoproterozoic (~550 Ma) over Cambrian and Ordovician to Carboniferous and Permian. Metamorphic zircon from the UHT, UHP and HT rocks produced mainly Carboniferous (345–330 Ma) U–Pb ages (e.g. Romer and Rötzler 2001; Massonne *et al.* 2007*a*; Collett *et al.* 2020; Rötzler and Timmerman 2021). In contrast, Devonian (390–360 Ma) and Carboniferous (358–310 Ma) ages resulted from K–Ar and Ar–Ar hornblende and mica dating (e.g. Kreuzer *et al.* 1989; Werner and Reich 1997; Werner and Lippolt 2000; Hallas *et al.* 2021).

The EPMA Th–U–Pb monazite dating of metapsammopelitic rocks has been scarcely applied in the Saxothuringian Zone when compared to K–Ar and Ar–Ar dating methods. In this contribution the geochronological information from extensive EPMA Th–U–Pb monazite dating is combined with P–T data calculated by geothermobarometry of garnet porphyroblasts and their related mineral assemblages in the main units of the Saxothuringian Zone. The potential of EPMA Th–U–Pb monazite dating to analyse grains less than 10 μm and to resolve complex metamorphic histories (e.g. Schulz and von Raumer 2011; Oriolo *et al.* 2022) results from an EPMA petrochronology campaign type of study where multiple samples were targeted, as single samples may not have recorded all the successive events (Schulz 2017). Furthermore, from the impressive variety of metamorphic units with different P–T paths in the Saxothuringian Zone an appropriate diversity of monazite microstructures can be expected. This dating method also allows comparative study of rocks with different metamorphic grade (e.g. UHP and MP–MT) within the same terrane. *In situ* geochronology and mineral chemistry (Y and REE) results show that even when they occur in the same sample and thin section, the metamorphic crystallization of garnet and monazite are not unequivocally related. Therefore, discussion arises about which stage or segment of a P–T evolution recorded by garnet-bearing assemblages is represented by a monazite age. Such P–T–time paths from various lithotectonic units are necessary for reconstruction and modelling of a subduction–accretion environment followed by continental collision. Our study from the Saxothuringian Zone furthermore appends to comparable data and monazite age distribution patterns from the Odenwald and Spessart, Armorican Massif, French Massif Central and the basement of the Alps (Schulz 2009, 2013, 2014, 2021*b*; Will *et al.* 2017). In this frame, the data contribute to a more detailed view on the multistage temporal and spatial evolution of an orogen, as exemplified by compiling studies and models of the Variscan (e.g. Kroner *et al.* 2020; Martínez-Catalán *et al.* 2021), or of other orogens (e.g. Volante *et al.* 2020; Cutts and Dyck 2023).

Subunits in the Saxothuringian Zone and sampling sites

The Saxothuringian Zone (ST) of the Variscan orogenic belt was defined by Kossmat (1927). It extends from eastern parts of the Bohemian Massif to the northern segment of the Armorican Massif (Fig. 1), and continues in the Ossa Morena Zone in the Iberian Massif (Kroner *et al.* 2020; Martínez-Catalán *et al.* 2020, 2021). The present study concerns the part of the ST extending from the Franconian Line to the SW towards the Elbe Zone to the NE (Fig. 2). To the NW the ST adjoins the Mid-German Crystalline Zone, as exposed in the Spessart and Odenwald (Will *et al.* 2017). Its southeastern border is the Tertiary Eger Rift Valley (Fig. 2).

According to Kroner *et al.* (2010), the ST can be subdivided into three sectors. These are: (1) a northern autochthonous domain; (2) a SE wrench and thrust domain; and (3) a medium- to high-grade metamorphic southern allochthonous domain

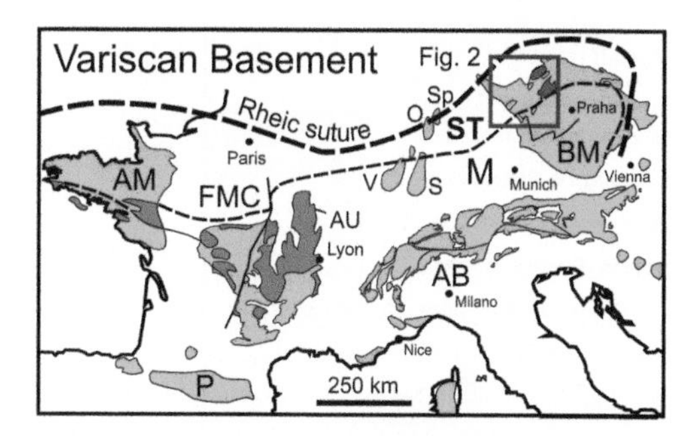

Fig. 1. Pre-Mesozoic basement areas of the Central and Western European Variscan Orogen to the south of the suture of the former Rheic Ocean. Position of study area (Fig. 2) within the Saxothuringian Zone (ST) in the northern Bohemian Massif (BM). AB, basement of the Alps; AM, Armorican Massif; FMC, French Massif Central; M, Moldanubian Zone; O, Odenwald; P, Pyrenees; S, Schwarzwald; Sp, Spessart; V, Vosges. Upper Allochthonous Units (AU) in AM and FMC. In the ST, the Upper Allochthon is shaded green, the Saxonian Granulite Massif and the Erzgebirge units are dark brown. For more details within the Bohemian Massif, see Martínez-Catalán *et al.* (2021).

which includes the Saxonian Granulite Massif with its Schist Cover, the Erzgebirge units and the Upper Allochthon with the Münchberg, Frankenberg and Wildenfels massifs. Sampling was focused on the allochthonous domain (3) (Supplementary material 1A).

The Saxonian Granulite Massif (SGM) occurs as a 15 × 44 km antiformal outcrop area surrounded by a Schist Cover (SC) with medium- to low-grade metamorphic rocks and a dissected Metagabbro–Serpentinite Unit (MSU), which may be considered as an ophiolite (Franke and Stein 2000; Rötzler and Timmerman 2021). Three samples (AJ201, GRO1, TRO2) from the Granulite Massif core delivered monazite of suitable size for analysis. The abundant felsic granulites bear only very small and few monazite grains, always attached to apatite. The most promising rocks containing monazite are sillimanite-bearing cordierite-garnet gneisses ('kinzigites') in the Cordierite Gneiss Unit (CGU) as outlined by Reinhardt and Kleemann (1994) and Rötzler and Timmerman (2021). This unit crops out in central parts of the SGM at Mohsdorf, and along the Chemnitz river (samples AB5A, AB5B, E3-11, MOH5, MOH6), but also in smaller lenses along the northwestern and southeastern margins next to the SC, so at Auerswalde (samples AB9A, AB9B). Garnet turned out to be rare in the SC, whereas monazite is abundant in numerous micaschist samples such as SW2AB and SW2AD near Auerswalde and BOH2 and BOH3 along the Striegis valley next to Böhrigen. Large bodies of anatectic granite such as those of Mittweida, but also numerous smaller metre- to decimetre-scale granite dykes, occur within the SGM core and rarely in the SC. These were sampled at the Chemnitz river (FS30), at Troischaufelsen (TRO3) and along the Mulde river (SGM4) at the eastern margin of the SGM (Fig. 2; Supplementary material 1A).

A garnet micaschist (MIL16) belonging to the Nossen-Wilsdruff Schists (NWS) was sampled at Munzig to the east of the SC and to the north of the Erzgebirge (Fig. 2). The structural assignment of the sample is not yet clear and it may belong to the SE wrench and thrust domain or to the southern allochthonous domain (Kroner *et al.* 2007). Lithotectonic units of the Erzgebirge are exposed in a SW–NE-striking antiformal large-scale structure which is subdivided into the Reitzenhain-Catherine dome to the SW and the Freiberg Gneiss Dome to the NE (Fig. 2). The gneiss domes experienced MP–MT metamorphism (Kröner *et al.* 1995; Rötzler 1995; Willner *et al.* 1997; Rötzler *et al.* 1998) and are overlain by several lithotectonic units that comprise partly HP and UHP metamorphism at various structural levels, alternating with MP rocks. The lithological units (Fig. 2) do not always coincide with the borders of the domains with HP and UHP metamorphism. This leads to inconsistencies in simplified geological maps which have been presented by various authors (Lorenz and Hoth 1990; Schmädicke 1994; Willner *et al.* 1997; Mingram *et al.* 2004, Tichomirowa *et al.* 2005, 2018; Sebastian 2013; Hallas *et al.* 2021). According to Sebastian (2013), the Freiberg Gneiss Dome comprises granitic orthogneisses, paragneisses (sample FG15) and a single garnet micaschist layer near Brand-Erbisdorf (sample BE2-1), and is the lowermost unit. An orthogneiss (SEI5) and paragneiss (SEI3, SEI4) samples come from the Seidewitz valley belonging to the Eastern Erzgebirge (Fig. 2). The Freiberg Gneiss Dome and the adjacent Eastern Erzgebirge next to the Elbe Zone have been assigned to a MP–MT Gneiss Amphibolite Unit (GAU) by Kröner *et al.* (1995), Mingram *et al.* (2004), Rötzler and Plessen (2010) and Tichomirowa and Köhler (2013).

In the hanging wall of the MP–MT the GAU follows the UHP Gneiss Eclogite Unit (GEU), also labelled as HP unit 1. Monazite has been found in granulite facies (Willner *et al.* 1997) kyanite garnet gneisses ('saidenbachites') which bear microdiamonds (Massonne *et al.* 2007*a*; Stöckhert *et al.* 2009) and polycrystalline quartz inclusions in garnet (Supplementary material 2). Such inclusions in garnet have been shown to be relicts and pseudomorphs of coesite (Lenze and Stöckhert 2008; Bidgood *et al.* 2020). These inclusions were also reported from detrital garnet in river sediments in the area (Schönig *et al.* 2020). Coesite and polycrystalline quartz inclusions as relicts of it have also been described from

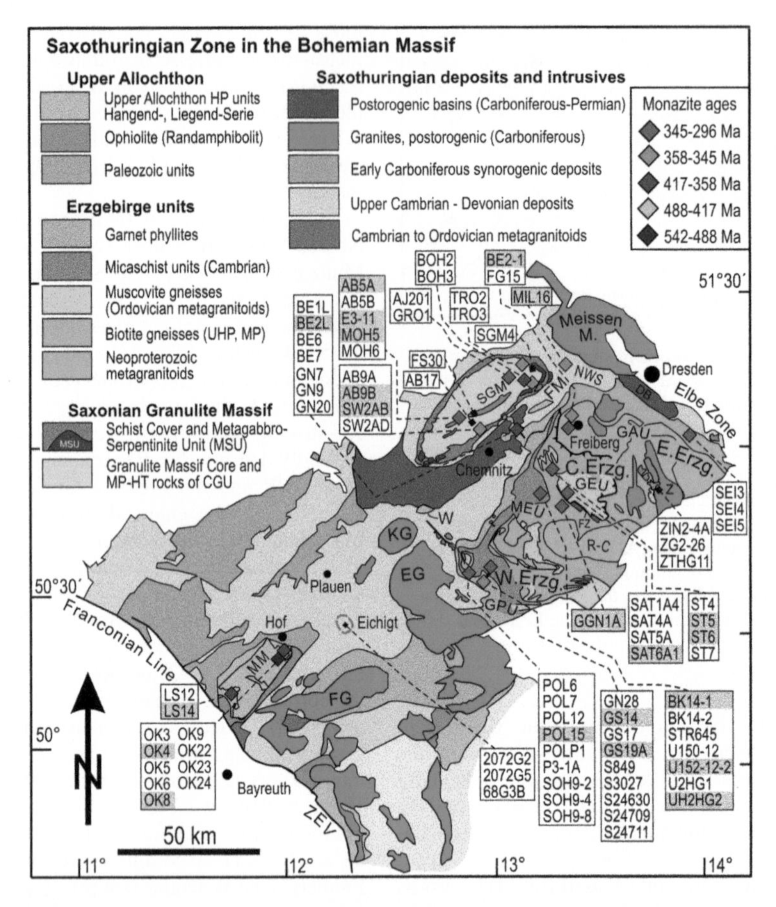

Fig. 2. Sketch map of the Saxothuringian Zone between the Franconian Line and the Elbe Zone, compiled from various sources (Kroner *et al.* 2007; Sebastian 2013; Koglin *et al.* 2018; Schmädicke *et al.* 2018; Tichomirowa *et al.* 2018; Martínez-Catalán *et al.* 2020, 2021; Rötzler and Timmerman 2021). Samples with monazite ages and garnet assemblage geothermobarometric data are marked. DB, Döhlen Basin; C. Erzg., Central Erzgebirge; CGU, Cordierite Gneiss Unit (part of Saxonian Granulite Massif); E. Erzg., Eastern Erzgebirge; FZ, Flöha Zone; GAU, Gneiss Amphibolite Unit with Freiberg Gneiss dome; GEU, Gneiss Eclogite Unit (UHP with thrust); GPU, Garnet Phyllite Unit; EG, Eibenstock Granite; FG, Fichtelgebirge granites; FM, Frankenberg Massif (klippe); KG, Kirchberg Granite; MM, Münchberg Massif; MSU, Metagabbro Serpentinite Unit (part of Schist Cover); MEU, Micaschist Eclogite Unit; NWS, Nossen-Wilsdruff Schists; R-C, Reitzenhain-Catherine dome; SGM, Saxonian Granulite Massif; W, Wildenfels (klippe); W. Erzg., Western Erzgebirge; Z, Zinnwald Greisen Granite; ZEV, Zone of Erbendorf-Vohenstrauß.

adjacent eclogites (Schmädicke 1991; O'Brien and Ziemann 2008). Garnet gneiss samples SAT1A4, SAT4A, SAT5A and SAT6A1 are from various locations around the Saidenbach Reservoir, as presented by Massonne (2001). Garnet gneiss GGN1A is from Zöblitz, as reported in Tichomirowa *et al.* (2018). A further set of monazite-bearing garnet gneisses where garnet shows no polycrystalline quartz inclusions replacing coesite has been sampled around an eclogite (Schmädicke 1994) to the south of Eppendorf (samples ST4, ST5, ST6, ST7). The eclogite of Eppendorf is interpreted to have maximal pressures of 16 kbar, thus is not UHP (Massonne and O'Brien 2003; Massonne 2011).

In the hanging wall of the GEU, the Micaschist Eclogite Unit (MEU) follows, also labelled as HP unit 2. In the Central Erzgebirge, this unit is exposed to the west of the UHP rocks. In the mine of Pockau-Lengefeld, the monazite-bearing garnet micaschists (samples BK14-1, BK14-2, U150-12, U152-12-2, U2HG1, UH2HG2) have been taken above and below the marble horizons. To the SW and in the hanging wall of the MEU, the Garnet Phyllite Unit (GPU) follows, labelled as HP unit 3, which also bears eclogites (Schmädicke *et al.* 1992). Garnet phyllites and micaschists of this unit were sampled along a SW-directed profile (samples GN28, GS14, GS17, GS19A, S849, S3027, S24630, S24709,

S24711). A further sampling site for the GPU are outcrops in the Pöhla-Hämmerlein visitor mine and its surrounding area. Within the visitor mine, a garnet micaschist (POL15) has been sampled among several other micaschists which are strongly mineralized with sphalerite and cassiterite, labelled as micaschist ore (samples P3-1A, SOH9-2, SOH9-4, SOH9-8). Further micaschists and paragneisses (POL6, POL7) were sampled around the mine. The garnet phyllites can be traced further to the west into the Elstergebirge (Faryad and Kachlík 2013), where metamorphic monazite between 370 and 325 Ma was reported by Rahimi and Massonne (2020). Also, Eibenstock Granite was sampled in an outcrop next to the mine (POL12). Further granite samples (2072G2, 68G3B, 2072G5) are from a drill core from Eichigt, where a two-mica granite was explored which is not exposed at the surface (Gottesmann *et al.* 2017). The Zinnwald Greisen Granite in Eastern Erzgebirge, with former mining of cassiterite and actual exploration on lithium-bearing mica has been sampled (samples ZIN2-4A, ZG2-26, ZTHG11) in a visitor mine and from drillcores (Fig. 2; Supplementary material 1A).

Münchberg, Wildenfels and Frankenberg massifs represent an Upper Allochthon in the Saxothuringian Zone (Klemd 2010; Koglin *et al.* 2018). They appear as klippes. The Münchberg Massif is composed of four nappes with an overall inverted metamorphic stratigraphy (Fig. 2). From top to bottom these are: Hangendserie ('hanging-wall series') with eclogites; Liegendserie ('footwall series'); Randamphibolit ('marginal amphibolite') and a greenschist facies Prasinit-Phyllit-Serie ('greenschist-phyllite series'). They overlie low-grade Paleozoic metasedimentary sequences (Stettner 1960; Höhn *et al.* 2017). The Hangendserie is mainly composed of banded hornblende gneisses and some felsic orthogneisses with eclogite lenses at the base. The magmatic protoliths belonged to an evolved oceanic to continental magmatic arc setting at about 480–450 Ma (Koglin *et al.* 2018). Protolith ages of felsic intrusions are at 550 Ma (Koglin *et al.* 2018). Eclogites lenses with N-MORB-type geochemical character and a 480 $\pm$ 23 Ma protolith age have been the subject of several isotopic and metamorphic studies, indicating a high-pressure metamorphic overprint at minimal pressures of 15 kbar and temperatures of 600° C, between 405 $\pm$ 7 Ma and 384 $\pm$ 2 Ma (Franz *et al.* 1986; Klemd 1989; Stosch and Lugmair 1990; Klemd *et al.* 1991; Okrusch *et al.* 1991; Klemd and Schmädicke 1994; Koglin *et al.* 2018; Pohlner *et al.* 2021). The Liegendserie consists of muscovite-biotite paragneisses, garnet micaschists and orthogneisses that were former peraluminous granitoids with two groups of protolith ages at 505–500 Ma and 485–481 Ma (Okrusch *et al.* 1990; Koglin *et al.* 2018). A similar range of monazite ages was reported by Waizenhöfer and Massonne (2017) from a metapsammopelite of the Liegendserie and interpreted as detrital monazites that were not affected by later metamorphism. The Liegendserie underwent amphibolite facies metamorphism with K–Ar cooling ages at ~380 Ma (Kreuzer *et al.* 1989; Okrusch *et al.* 1991). The Randamphibolit is a complex of metabasites with tholeiitic MORB-type compositions, with some interlayered thin marbles and calcsilicate rocks (Stettner 1960; Okrusch *et al.* 1989). The greenschist facies Prasinit-Phyllit-Serie at the bottom of the nappe pile shows transitions to volcaniclastic and siliciclastic rocks (Schüssler *et al.* 1986; Okrusch *et al.* 1989). The mafic-dominated and Ca-rich lithologies of the Hangendserie, Randamphibolit and Prasinit-Phyllit-Serie are not prospective for monazite occurrences; therefore, sampling was concentrated on the Liegendserie. At the NE rim of the Münchberg Massif numerous samples of garnet-bearing micaschists and paragneisses (OK3 to OK24) were analysed. Some samples of this sequence are also reported in the zircon–U–Pb study by Koglin *et al.* (2018). A geological cross-section from the Prasinit-Phyllit-Serie over the Randamphibolit to Liegendserie is exposed in the western part of the massif (Fig. 2). There, an ortho-augen-flasergneiss (LS12) and a garnet paragneiss (LS14) bear monazite.

The klippe of the Wildenfels Massif is poorly exposed and metabasite lithologies prevail; it was not sampled. The Frankenberg Massif within a synform at the SE border of the SGM is exposed in five isolated areas. An amphibolite to greenschist facies upper complex is there overlying very low-grade volcanosedimentary rocks. The upper metamorphic complex is subdivided into a Gneiss Unit with amphibolite, orthogneisses, paragneisses and phyllites, and a Prasinite Unit (Rötzler *et al.* 1999). Granitoid orthogneisses occur at Frankenberg (BE6, GN9) and at Sachsenburg (GN7). Garnet-bearing phyllites were sampled at various locations (BE1L, BE2L, BE7, GN20) along the valley at Langenstriegis (Supplementary material 1A).

Analytical methods

SEM-based automated mineralogy

Automated mineralogical methods (Schulz *et al.* 2020), based on a scanning electron microscope (SEM-AM) at the Geometallurgy Laboratory at TU Freiberg, were applied to a set of ~250 thin sections for selection of samples suitable for further analytical study by EPMA and LA-ICP-MS. A scanning electron microscope (SEM) Quanta 650-FEG-MLA by FEI Company, equipped with Bruker Dual X-Flash energy dispersive spectrometers for energy-dispersive spectrometry (EDS) analyses, was applied to study complete thin sections of garnet-bearing micaschists and gneisses by automated measurement routines. Electron beam conditions were set at 25 kV acceleration voltage and a 10 nA beam spot. A software package for mineral liberation analysis (MLA version 2.9.0.7 by FEI Company) was used for automated steerage of the electron beam for EDS identification of mineral grains and collection of numerous EDS spectra. Monazite and xenotime grains and their microstructural relationships within a distance of ~100 µm from target grains were detected by a sparse phase search routine (SPL). Data were used to select monazite grains for Th–U–Pb geochronology and mineral-chemical analysis by EPMA (Supplementary material 1B). The EDS spectral mapping routine (GXMAP) produces a narrow grid of ~1600 EDS spectra per mm^2 from garnet and biotite as phases of interest. For classification of minerals and compositions in SEM-AM measurements, identified reference EDS spectra were collected from matrix phases, and from defined parts of several garnet porphyroblasts (core–mid–rim). Reference spectra from garnet were generically labelled with corresponding garnet Fe, Mg, Mn and Ca compositions. When labelled spectra are arranged on a colour scale, semi-quantitative garnet zoning maps are established which can be compared to wavelength-dispersive spectrometry (WDS) element maps (Schulz 2017; Schulz and Krause 2021). The GXMAP measurements of complete thin sections allowed for selection of a few typical garnets out of dozens of porphyroblasts for quantitative WDS analysis and WDS element mapping with EPMA (Supplementary material 3).

Microstructurally controlled geothermobarometry by EPMA

Mineral-chemical analyses of silicate minerals in 36 garnet-bearing samples (~4000 analytical points) were performed with EPMA instruments at TU Freiberg and the Helmholtz Institute Freiberg for Resource Technology (JXA-8900RL, JXA-8230, JXA-8530F), using beam conditions of 15 kV, 20 nA and 2 µm, and ZAF correction procedures supplied by JEOL. Garnet zonation was analysed in detailed traverses. Compositions of associated feldspar and mica are documented by several single point analyses. The WDS element distribution maps were acquired with a JEOL JXA-8530F at an acceleration voltage of 15 kV, a beam current of 38.5 nA, beam diameters between 3 and 6 µm and a dwell time of 100 ms. Magnesium, Si, Fe, Ca and Mn were measured using WDS spectrometers, whereas Al, K and Ti were measured with EDS (Supplementary material 3 and 4A–D).

When local thermodynamic equilibrium is assumed, two principal methods for estimating P–T conditions recorded in metamorphic rocks can be applied. Both methods have the basic prerequisites and uncertainties of thermodynamic data in common. Classical thermobarometry employs mineral-chemical compositions of phases involved in continuous cation-exchange and net-transfer solid–solid reactions (Spear 1993; Powell and Holland 2008). Alternatively, pseudosection calculations involve an assumption that concerns the effective local chemical bulk composition in a rock domain for the determination of phase assemblage stability fields, metamorphic reactions and mineral compositions, especially garnet, that potentially occurred during the metamorphic evolution of a given rock sample (Holland and Powell 2011). In the Saxothuringian metapsammopelites compositional layering, strong variations in modal mineralogy along foliation planes and microlithons, and highly variable distribution of zoned garnet porphyroblasts imply significant compositional heterogeneities. Unknown sizes of the involved rock domain(s) impose uncertainty upon effective local reactant bulk composition(s) for pseudosection calculations. This especially concerns metamorphic P–T conditions for assemblages without garnet. Metamorphic P–T conditions during garnet crystallization in reference to the stability field of monazite crystallization are important for an interpretation of monazite ages in terms of a metamorphic evolution. Therefore, classical thermobarometry has been chosen here, in a similar way as it is implemented to support and refine results of pseudosection modelling of garnet-bearing assemblages. However, geothermobarometric calculations require combinations of mineral-chemical analyses from garnet, mica and plagioclase in textural equilibrium. Microstructural observations and the chemical evolution of the mineral phases provided criteria for a definition of local equilibria (Schulz 1993, 2017; Schulz *et al.* 2001). Despite controls by EDS spectral maps, traverses across garnet porphyroblasts may not have passed the entire core region and porphyroblasts may display zonation gaps. Some garnets may show only part of the complete garnet chemical evolution in a sample. Also, resorption of garnet may have occurred. Such complications are not necessarily evident by zonation profiles alone but can be checked in ternary diagrams with grossular–pyrope–spessartine mol% coordinates. When garnet is the main Mn fractionating phase, its Mn content during crystallization is controlled by Rayleigh fractionation (Spear 1993). This allows the recognition of a relative temporal chemical growth evolution of porphyroblasts. For each sample, a characteristic garnet mineral-chemical evolution trend can be derived from a compilation of the single core-rim zonation profiles in ternary grossular–pyrope–spessartine mol% coordinates (not shown here) and subsequently in *X*Mg–*X*Ca coordinates. Selected analyses out of garnet *X*Mg–*X*Ca chemical evolution trends then can be used for thermobarometry. The main foliation (S_2) in metapelites by phyllosilicates and long prismatic minerals, such as kyanite or fibrolitic sillimanite, is related to a later deformation stage as it surrounds millimetre-scale microlithons containing garnet, feldspar, mica, staurolite, aluminosilicates and quartz. Curved and planar internal foliations S_1i in the garnet porphyroblasts represent an early stage of deformation. External and internal structures of syn- and intertectonic garnet porphyroblasts are interpreted as successive incremental steps of progressive deformation (Passchier and Trouw 2005; Oriolo *et al.* 2022). The porphyroblast-S_1i to matrix-S_2 relationships allow one to establish the relative time of crystallization. This enables the mineral-chemical data from mica and plagioclase to be related to the garnet mineral-chemical evolution trends (Schulz 1993, 2017; Schulz *et al.* 2001; Passchier and Trouw 2005). For geothermobarometric calculations, garnet analyses from core, intermediate zones and rims were related to mica and unzoned plagioclase along the foliation and in the matrix. Temperatures were estimated using the garnet–biotite thermometer in combination with pressures calculated with garnet–aluminosilicate–plagioclase–quartz (GASP) and garnet–biotite–muscovite–plagioclase (GBMP) geobarometers, as provided in empirical re-calibrations by Wu *et al.* (2004), Wu and Zhao (2006, 2007) and Wu (2015, 2017, 2019). Geothermobarometric estimates include a minimum error of ± 50°C and ± 1 kbar. Uncertainties about P–T paths calculated from zoned garnets arise from quantitative systematic error in microprobe analyses and thermodynamic data. Shapes, or relative $\Delta P/\Delta T$ trends of P–T paths appear to be mainly preserved when uncertainties about corresponding plagioclase and mica are considered (Spear 1993). This concept of using the chemical evolution of minerals during progressive metamorphism for a reconstruction of the P–T evolution has been successfully applied to metapelites which have undergone single and polyphase metamorphic cycles (Schulz and von Raumer 2011; Schulz 2014, 2017; Schulz *et al.* 2019).

Laser ablation inductively coupled plasma mass spectrometry

A total of 906 measurements on garnet by transects of porphyroblasts from a selected set of 22 samples, were performed at the LA-ICP-MS laboratory at Mineralogisches Institut Westfälische Wilhelms-Universität Münster, Germany. Analyses were carried out with a Teledyne Photon Machines Analyte G2 193 nm Excimer LASER ablation system working at 193 nm wavelength coupled to a Thermo Scientific Element XR double focusing sector field mass spectrometer. The system was tuned on a NIST 612 glass (Jochum *et al.* 2011). Selected spots along garnet profiles previously analysed by EPMA were measured. The EPMA data were used for comparison and quantification. Blocks of 30–67 analyses were bracketed by the analysis of reference materials NIST-612, GSD-1G and GSE-1G in order to estimate the external precision. The overall time of a single analysis was 65 s (20 s for background, 30 s for peak after switching laser on, 15 s wash-out). A total of 34 isotopes (^{7}Li, ^{29}Si, ^{43}Ca, ^{47}Ti, ^{51}V, ^{59}Co, ^{62}Ni, ^{69}Ga, ^{73}Ge, ^{85}Rb, ^{88}Sr, ^{89}Y, ^{90}Zr, ^{93}Nb, ^{137}Ba, ^{139}La, ^{140}Ce, ^{141}Pr, ^{146}Nd, ^{147}Sm, ^{153}Eu, ^{157}Gd, ^{159}Tb, ^{163}Dy, ^{165}Ho, ^{166}Er, ^{169}Tm, ^{173}Yb, ^{175}Lu, ^{178}Hf, ^{181}Ta, ^{208}Pb, ^{232}Th and ^{238}U) were measured. The ^{29}Si was used as an internal standard element. Typical relative standard errors of measured concentrations are between 5 and 10% for most elements. Isotopes ^{51}V, ^{59}Co, ^{88}Sr and ^{89}Y had errors between 1 and 5%. The gas blank was subtracted from the signal. All signals were normalized to the internal standard isotope. A relative sensitivity factor (RSF) for each analysis is calculated using the NIST 612 reference glass. We applied the procedure given in Jochum *et al.* (2007) for RSF drift correction. Absolute concentrations of each mass and the limits of quantification are calculated by relating the intensities of each mass to the internal standard element and its known concentration in the NIST 612 reference glass (concentrations taken from Jochum *et al.* 2011), and the RSF (Supplementary material 4E).

Chemical Th–U–Pb monazite dating by EPMA

Electron microprobe Th–U–Pb dating of monazite is based on the assumption and the observation that common Pb in this mineral (LREE, Y, Si, Ca, Th, U)PO_4 is negligible when compared to radiogenic Pb resulting from decay of Th and U (Montel *et al.* 1996). Electron microprobe analysis of bulk Th, U and Pb concentrations in monazite, at a constant $^{238}U/^{235}U$, allows for the calculation of a chemical model age (CHIME) with a considerable error, provided that no further modification of Th/U/Pb ratios has occurred except by radioactive decay (Montel *et al.* 1996; Pyle *et al.* 2005*a*; Jercinovic *et al.* 2008; Suzuki and Kato 2008; Spear *et al.* 2009). EPMA measurements of monazite that led to isochrone ages were performed in thin sections of 71 samples. From 21 samples, both monazite ages and geothermobarometric data are presented. An individually adopted protocol for the monazite analysis with a JEOL JXA-8530F electron microprobe hosted at the Helmholtz Institute Freiberg for Resource Technology has been developed (Schulz *et al.* 2019). The electron beam was set at 20 kV acceleration voltage, 100 nA beam current and 3–5 µm beam diameter, dependent on the size of monazite grains. For calibration, the REE-ultra phosphates from ASTIMEX Ltd with Pb contents below the lower limit of detection, and the orthophosphates of the Smithsonian Institution, which partly contain residual Pb (Jarosewich and Boatner 1991; Donovan *et al.* 2003), were used as reference materials. Calibration of Pb was carried out on a natural crocoite. The U and Th contents were calibrated on metal reference materials. The $M\alpha1$ lines of Th and Pb and the $M\beta1$ lines for U of PETH crystal in a spectrometer with a capsuled Xe–Ar proportional counter were chosen. Counting times of 320 s (Pb), 80 s (U) and 40 s (Th) on peak were utilized. Interference of YLg on the PbMa line was corrected by linear extrapolation of correction factors gained from analysis of various Y-bearing standards (Montel *et al.* 1996). Interference of the ThMg line on the UMb line was also corrected. Potential problems of data comparability were avoided by using the same reference monazite, labelled *Madmon*, with validated special ThO_2*–PbO characteristics (Schulz and Schüssler 2013), for offline re-calibration of ThO_2 and for data control by repeated measurements during the analytical sessions.

The number of single analyses varies with the grain size of the monazites, e.g. 1–2 analyses in grains of <40 µm and up to ten analyses in grains of 100 µm in diameter. Monazite chemical ages were first calculated using the methods of Montel *et al.* (1996). A 1σ error deduced from the counting statistics (JEOL error) and an error $\varepsilon_{Pb} = \sqrt{}(\text{Cts}/s_{PEAK} + \text{Cts}/s_{BKG})/(\text{Cts}/s_{PEAK} - \text{Cts}/s_{BKG})$ was propagated to a 2σ error in Pb element%. For Pb the 2σ error in element% is ~0.004 (recalculated from the JEOL error) or ~0.001 (recalculated from ε_{Pb}) for the reference monazite *Madmon* with ~0.25 wt% Pb (Schulz *et al.* 2019). We applied an error in Pb element% of 0.004 to all single analyses, which propagates for the reference monazite *Madmon* with ~506 Ma typically to ±12 Ma (2σ), and for Carboniferous monazites (~0.085 wt% Pb) to 20–30 Ma (2σ). Ages were further determined using the ThO_2*–PbO isochrone method (CHIME) of Suzuki *et al.* (1994) and Montel *et al.* (1996) where ThO_2* is the sum of measured ThO_2 plus ThO_2 equivalent to measured UO_2. This is based on the slope of a regression line in ThO_2* v. PbO coordinates forced through zero. As the calculation of the regression line provides an underestimation of error, weighted average ages for monazite populations were calculated from the single analyses defining the regression line using Isoplot 3.0 (Ludwig 2001). Weighted average ages were also calculated when a large monazite allowed several analyses, and from clusters composed of many small monazites. Mostly the sizes of monazite grains were below 50 µm. In consequence it was intended to perform a narrow grid of full quantitative analyses from such grains. In all analysed samples, the model ages given by isochrone and weighted average methods coincided within the error. The age data are interpreted as the time of closure for the Th–U–Pb system of monazite during growth or recrystallization in the course of metamorphism and, in the case of granites, as the time of igneous crystallization (Supplementary material 5 and 6).

Garnet zonations and related geothermobarometry

Saxonian Granulite Massif and Schist Cover

In garnet gneiss samples (AB5A, E3-11, MOH5, AB9B) of the CGU in the SGM, round to elliptical garnets up to 5 mm in diameter are embedded in a foliated matrix defined by biotite, quartz, plagioclase, cordierite interspersed with sillimanite needles and lens-like aggregates of fibrolitic sillimanite. Hercynite occurs in aligned clusters of numerous small grains of <50 µm. Garnet porphyroblasts enclose biotite, plagioclase, quartz and numerous decussate sillimanite needles which are sometimes concentrated in distinct zones. Some porphyroblasts are mantled by thin rims of cordierite. The modes of garnet are always high and range between 20 and 40% (area%, recalculated from the EDS spectral maps). Modes of biotite (10–30%), plagioclase (5–10%), quartz (5–30%), K-feldspar (2–10%), sillimanite (20–30%), cordierite (3–20%) and hercynite (1–10% are highly variable in the thin sections. In garnet gneiss sample AB9B, beneath the SC the cordierite is rare, but one observes similar petrographic features as in the CGU samples from the centre of the SGM (Fig. 2). This provides an argument that the CGU is not restricted to the Granulite Massif core, but is also part of the SC at some locations, as outlined in geological maps (Rötzler and Timmerman 2021). Utilizing the EDS spectral and WDS element mapping allows one to distinguish two mineral-chemical populations of garnet porphyroblasts according to their zonations in Mn. An older garnet generation 1 has cores with low (10 mol%) pyrope (Mg), and high grossular (Ca) and spessartine (Mn) contents. Rims of these porphyroblasts have high pyrope contents (up to 30 mol%), and low grossular and spessartine contents. Garnet generation 2 has the same core compositions as Mg-rich garnet 1 rims. Towards the rims of garnet 2, and also in garnet 1 outer rims, the pyrope component strongly decreases (Fig. 3a–c, e–i; SM3; Figs 1a–o & 2a–c).

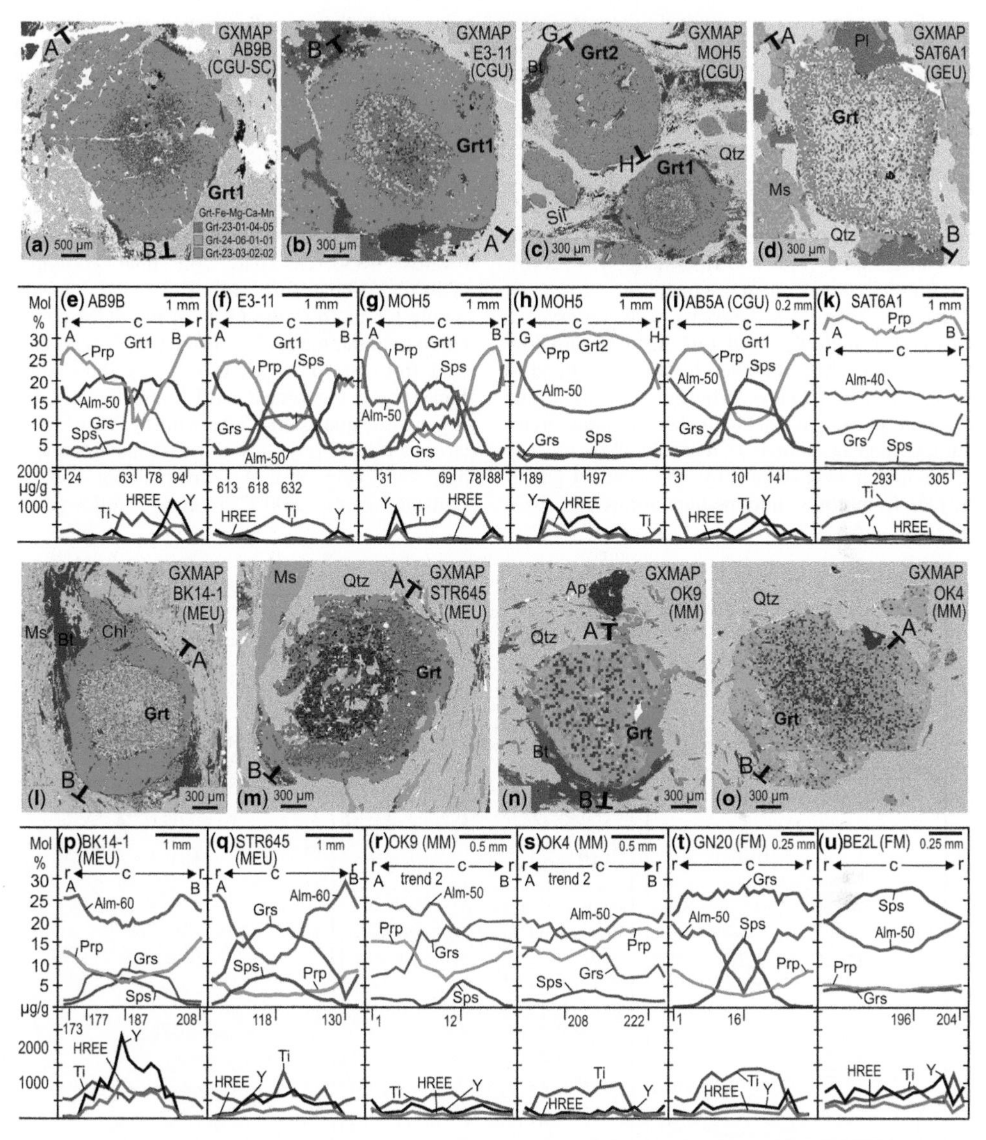

Fig. 3. Garnet mineral chemistry in energy dispersive spectral maps (GXMAP) produced by SEM. EDS spectra for garnet porphyroblasts with cores (c) and rims (r) are labelled in a generic way by Fe, Mg, Mn and Ca contents in normalized element wt% and by colours (see text). Locations of analytical profiles by EPMA are marked. Garnet zonations in almandine (Alm-50%, due to scale), pyrope (Prp), grossular (Grs) and spessartine (Sps) components (in mol%, calculated from mole fraction × 100). Garnet trace element zonations are reported by 10–20 LA-ICP-MS single spot analyses along profiles. HREE is the sum of Gd to Lu. Characteristic garnet analyses out of the zonation profiles, representing zonation trends and used for geothermobarometric calculations are marked by numbers along the baseline. See Figures 4a–c, 5a, b, 6a and b for XMg–XCa of the selected garnet analyses. (**a–c**) Zoned garnets in cordierite-garnet gneiss from the Saxonian Granulite Massif, Cordierite Gneiss Unit (CGU), partly below Schist Cover (SC), with Ca-rich garnet 1 core (Grt1) and Mg-rich rim, next to garnet 2 (Grt2) porphyroblast rich in Mg. (**d**) Poorly zoned garnet in UHP garnet gneiss from the Gneiss Eclogite Unit (GEU) in Central Erzgebirge. (**e–i**) Garnet zonations in cordierite-garnet gneisses of the CGU in the Saxonian Granulite Massif. (**k**) Garnet zonation in UHP garnet gneiss from the GEU. (**l**) and (**m**) Strongly zoned garnet in micaschists from the Micaschist Eclogite Unit (MEU) in Central Erzgebirge. (**n**) and (**o**) Poorly zoned garnet in garnet micaschists from the Liegendserie, Münchberg Massif (MM). (**p**) and (**q**) Garnet zonations in micaschists from the MEU. (**r**) and (**s**) Garnet zonations with trend 2 in micaschists from the Liegendserie in the MM. (**t**) and (**u**) Garnet zonations in garnet micaschist and phyllite from the Frankenberg Massif (FM). See also Supplementary material 3.

Accordingly garnet 1 cores mark the start of garnet crystallization, and garnet 1 rims and garnet 2 represent later stages of growth. In general, zonations of Y match the sum of HREE, while zonations of Ti are independent of Y and HREE zonations. Maximal Y contents in garnet 1 cores are at 600 $\mu g\ g^{-1}$ (Fig. 3i). In rims of garnet 1 and in cores of garnet 2 maximal Y contents are at 2000 $\mu g\ g^{-1}$. The Y-zonations display distinct peaks in the transition between garnet 1 cores and rims (Fig. 3e–h; SM3; Fig. 2a–c). Garnet main element zonations are resolved in *X*Mg–*X*Ca diagrams. One observes a first trend at high *X*Ca and low *X*Mg toward low *X*Ca and high *X*Mg, followed by a second trend at low *X*Ca from high to low *X*Mg (Fig. 4a–c). The garnet 1 core in sample E3-11 displays a unique trend with intermediate to high *X*Ca at low *X*Mg, preceding the higher *X*Ca stage (Fig. 4b). Distinct garnet analyses from cores, intermediate zones and rims, representing the overall zonation trend are shown in Figures 3e–i and 4a–c. These garnet analyses were combined with analyses of biotite and plagioclase enclosed in the garnet, and/or in the matrix next to the garnet porphyroblasts for P–T estimates. For sample AB9B, which belongs to the CGU situated below the SC, 550°C/7.5 kbar were calculated from Mg-poor garnet cores, then maximal pressures of ~800°C/8 kbar, followed by a decompression towards ~850°C/5 kbar, and final cooling to ~700°C/3.5 kbar (Fig. 4d). P–T conditions from sample E3-11 from the CGU in the SGM core are within a similar range (Fig. 4e). The P–T paths calculated from the other CGU samples show similar shapes within the same P and T limits (Fig. 4f). A felsic granulite gneiss AB17 from the SGM core shows garnet with strongly decreasing *X*Ca at comparably high *X*Mg (Fig. 4a). A corresponding P–T path segment from 800°C/15 kbar, followed by

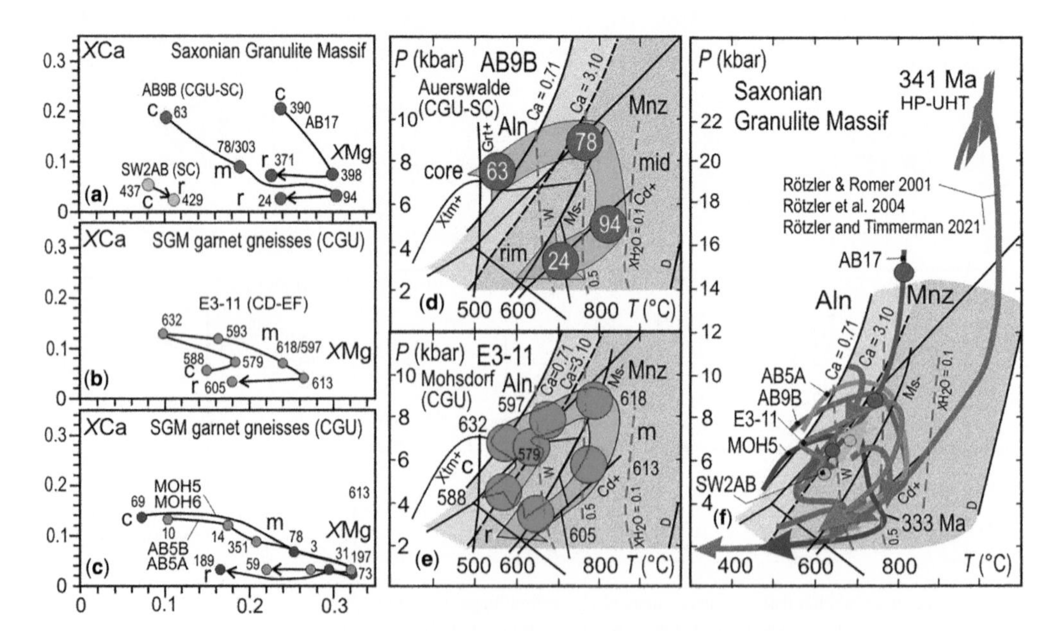

Fig. 4. Garnet mineral chemistry and geothermobarometry in the Saxonian Granulite Massif (SGM) and Schist Cover (SC). (**a–c**) Garnet zonations in *X*Mg–*X*Ca (mole fractions, based on cation formula with 12 O). Arrows between numbers (see Fig. 3) indicate core-to-rim (c, r) zonation trends from single garnet profiles. Numbers are selected analyses for geothermobarometry. (**d**) and (**e**) P–T estimates and core-to-rim P–T trends for garnet mineral assemblages in the Cordierite Gneiss Unit (CGU). Sample AB9B from the CGU below SC shows similar P–T path as sample E3-11 from the GCU in the Granulite Massif core. Numbers refer to garnet analyses in (a–c) and Figure 3, round symbols mark error of ±50°C/1.0 kbar on P–T estimates. The aluminosilicates (Ky, Sill), muscovite-out (Ms-), staurolite-in (Sta+) and cordierite-in (Cd+) univariant lines are after Spear (1993). (**f**) Summary of P–T data and paths from SGM and SC. Dark grey arrows and boxes are P–T data and ages in Ma as reported by Rötzler and Romer (2001), Rötzler *et al.* (2004), Rötzler and Timmerman (2021, and references therein), including data for the Metagabbro Serpentinite Unit and the SC. Data from the monazite-bearing samples are shown by coloured arrows (this study). Stability fields of monazite (Mnz) and allanite (Aln) at different bulk rock contents as a function of Ca wt %, and with xenotime (Xtm) stability field (Janots *et al.* 2007; Spear 2010). Red stippled lines are solidus curves in the granitic system Qtz–Ab–Or–H_2O at XH_2O of 1.0 (curve W, wet solidus), at XH_2O of 0.5 and 0.1 after Johannes and Holtz (1996).

decompression, then cooling to 650°C/6 kbar was calculated (Fig. 4f). Geothermobarometry of such garnet-bearing felsic granulites (sample AB17) appears problematic as the biotite in the foliated matrix may not be in equilibrium with garnet. The garnet–biotite thermometer provided maximal temperatures of 800°C at 15 kbar, which are significantly lower than the alternative temperature estimates by feldspar thermometry of other Granulite Massif core rocks reported by Rötzler and Timmerman (2021) and further references (Fig. 4f). Garnet is rare and often absent in the micaschists from the SC. An exception is a micaschist (sample SW2AB) with two isolated garnet porphyroblasts of 2 mm in diameter and microlithons with predominant quartz and minor plagioclase which are surrounded by a slightly anastomosing foliation composed of biotite and fibrolitic sillimanite. Garnet is zoned with slightly increasing *X*Mg at low *X*Ca, which allowed estimation of a short prograde P–T path segment from 600°C/5 kbar to 700°C/6.5 kbar (Fig. 4a, f).

Erzgebirge metamorphic units

Garnet-bearing metapsammopelites are rare in the Freiberg Gneiss Dome and in the Eastern Erzgebirge, which are both assigned to the GAU as the base of the Erzgebirge nappe pile. The micaschist sample BE2-1 exhibits small (<0.5 mm) garnet which encloses biotite and quartz. The porphyroblasts occur in microlithons with large plagioclase and quartz. Small garnet grains are sometimes enclosed in plagioclase. The matrix is an anastomosing foliation defined by biotite and muscovite. Garnet cores show high *X*Ca at low *X*Mg and a prograde zonation trend toward intermediate *X*Ca and *X*Mg in the rims (Fig. 5a; Supplementary material 3; Fig. 2d). These garnet porphyroblasts recorded a prograde P–T evolution from 500°C/9 kbar to 680°C/11 kbar (Fig. 5c).

Garnet in the UHP gneisses of the overlying GEU (samples SAT1A4, SAT4A, SAT5A, SAT6A1 and GGN1A) is 2–4 mm in diameter and appears in modes from 5 to 13%. The porphyroblasts are poorly round to elliptical and have sometimes sutured grain boundaries. They display numerous cracks and occasionally inclusions of polycrystalline quartz as pseudomorphs after coesite (Supplementary material 2; Fig. 1), monocrystalline quartz, mica, kyanite and zircon. An external foliation is defined by predominant muscovite (mode 5–36%) accompanied by small biotite (modes from 1 to 7%) and appears likewise with strictly or poorly preferentially aligned mica blades. Plagioclase (modes 12–30%) and quartz (28–35%) appear with equivalent grain sizes in the matrix. In samples without coesite relicts (ST5, ST6) garnet is slightly smaller with maximal 2 mm diameter. Biotite in these gneisses has higher modes (5–10%) compared to the UHP gneisses. Garnet is slightly zoned with high *X*Mg in the cores (Fig. 3d, k; Supplementary material 3; Figs 1p, 2e & f). In samples where garnet has polycrystalline quartz inclusions as relicts of coesite, as in GGN1A, SAT6A1, SAT5A1, a further slight increase of *X*Mg at constant *X*Ca is observed toward the garnet rims (Figs 3d, k & 5a). The garnet Y and HREE (up to 35 $\mu g\ g^{-1}$) are very low. Apparently, the Y and HREE are low in garnet when a high mode of garnet occurs. The Ti is up to 1000 $\mu g\ g^{-1}$ in the cores and decreases towards 500 $\mu g\ g^{-1}$ in the rims (SAT6A1, Fig. 3k; Supplementary material 3; Fig. 2e). In sample GGN1A the Ti is unzoned at very low concentrations (Supplementary material 3; Fig. 2f). Garnet in samples ST5 and ST6, which bear no polycrystalline quartz inclusions, displays a contrasting zonation trend with increasing *X*Ca and decreasing *X*Mg from cores to rims (Supplementary material 3; Fig. 2g). Albitic plagioclase, large muscovite and small biotite allowed P–T estimates from garnets in both gneiss sample sorts at around 800°C/20–16 kbar (Fig. 5c).

In the MEU above the GEU, micaschists with strongly zoned garnet are abundant. In samples STR45, BK14-1, U152-12-2, UH2HG2 a straight to anastomosing foliation S_2 defined by compact layers of biotite and muscovite surrounds lenticular microlithons. In the microlithons garnet up to 5 mm in diameter with modes from 5 to 10% is accompanied by quartz, plagioclase, opaque phases and isolated decussate mica blades. In some samples (BK14-1), crenulation folds by polygonal mica blades are developed between the S_2 planes. Garnet porphyroblasts have straight to curved inclusion trails of small quartz, plagioclase and opaque minerals which define an internal foliation S_1i discordant to the S_2 foliation. The porphyroblasts display cores with low pyrope and high spessartine and grossular contents. In rims there are higher pyrope contents (up to 15 mol%) at low grossular and spessartine (Fig. 3l, m, p & q). In *X*Mg – *X*Ca coordinates one observes coherent zonation trends with decreasing *X*Ca at increasing *X*Mg from garnet cores to rims (Fig. 5a). Garnet trace element zonations in the micaschists markedly differ from those in the UHP garnet gneisses. Mostly there are Y-rich cores with up to 2500 $\mu g\ g^{-1}$ and correspondingly high total HREE. The Ti contents are poorly zoned at intermediate levels around 500–1000 $\mu g\ g^{-1}$ (Fig. 3p, q). Coexisting plagioclase in the garnet-bearing assemblages is mostly oligoclase and in some cases albite. P–T trends for garnet core-to-rim growth are prograde in pressure and temperature and similar for all presented samples. They range from 400°C/2.5 kbar to 500°C/3.5 kbar (sample STR645), from 450°C/3.5 kbar to 600°C/7 kbar (sample U152-12-2) and from 500°C/5.5 kbar to

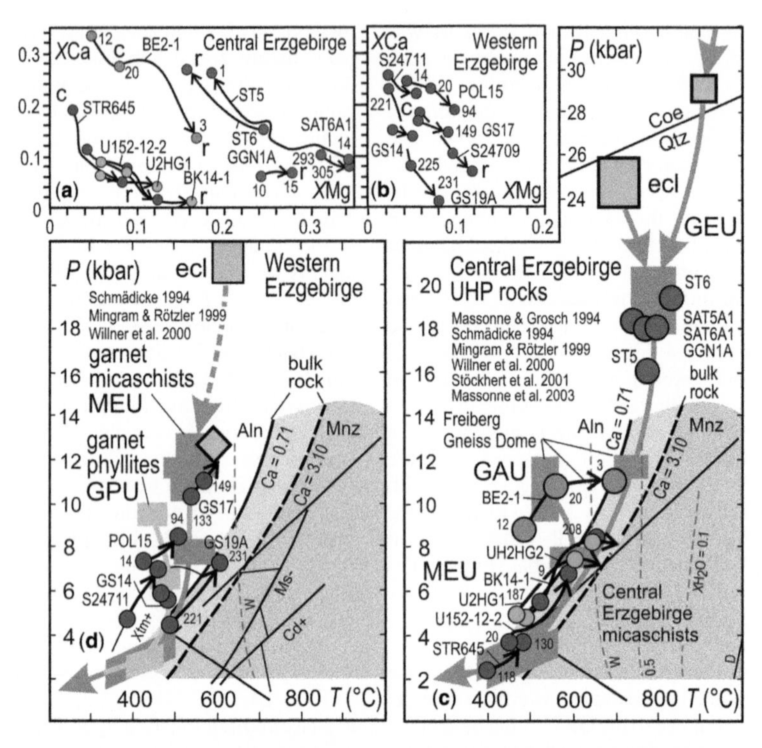

Fig. 5. Garnet mineral chemistry and geothermobarometry in the Erzgebirge units. (**a**) and (**b**) Garnet zonations in *X*Mg–*X*Ca (mole fractions, based on cation formula with 12 O). Arrows between numbers (see Fig. 3) indicate core-to-rim (c, r) zonation trends from single profiles. Numbers are selected analyses for geothermobarometry. (**c**) P–T estimates and core-to-rim P–T trends for garnet mineral assemblages in coloured symbols. Data from the UHP Gneiss-Eclogite Unit (GEU), the Micaschist-Eclogite Unit (MEU) in Central Erzgebirge and the Freiberg Gneiss Dome of the Gneiss-Amphibolite Unit (GAU). P–T data and P–T paths in dark grey shading and green symbols (ecl, eclogites) refer to reports in Massonne and Grosch (1994), Schmädicke (1994), Mingram and Rötzler (1999), Willner *et al.* (2000), Stöckhert *et al.* (2001) and Massonne and Nasdala (2003). (**d**) P–T estimates and core-to-rim P–T trends for garnet mineral assemblages in coloured symbols. Data from MEU in Western Erzgebirge and the Garnet Phyllite Unit (GPU). Numbers refer to garnet analyses in (a) and (b) and Figure 3; error of ± 50°C/1.0 kbar on P–T estimates. The aluminosilicates (Ky, Sill), muscovite-out (Ms-) and cordierite-in (Cd+) univariant lines are after Spear (1993). Stability fields of monazite (Mnz) and allanite (Aln) at different bulk rock contents as a function of Ca wt%, and with xenotime (Xtm) stability field (Janots *et al.* 2007; Spear 2010). Red stippled lines are solidus curves in the granitic system Qtz–Ab–Or–H_2O at XH_2O of 1.0 (curve W, wet solidus), at XH_2O of 0.5 and 0.1 after Johannes and Holtz (1996).

650°C/7–8 kbar (samples U2HG1, UH2HG2, BK14-1), as outlined in Figure 5c.

In garnet micaschists of the MEU and garnet phyllites in the GPU in the Western Erzgebirge, round to euhedral garnet up to 10 mm in diameter occurs in lens-like microlithons together with quartz and plagioclase. The porphyroblasts with modes from 5 to 8% sometimes display an internal foliation S_1i defined by quartz and elongated chloritoid discordant to the external S_2 foliation. Microlithons and layers by quartz are separated by a slightly anastomosing foliation composed of predominant muscovite, subordinate biotite and sometimes preferentially oriented chloritoid (samples GS17, GS19A). Garnet porphyroblasts display mostly marked zonations with core-to-rim decrease of Mn and *X*Ca, while *X*Mg increases (Fig. 5b; Supplementary material 3; Figs 1r, 2i & k). Also, garnet trace element zonations are similar as in the MEU in the Central Erzgebirge, with high Y and HREE contents in the cores which decrease toward the rims (Supplementary material 3; Fig. 2h–k). The P–T estimates with coexisting mica and oligoclase revealed prograde P–T path segments for all presented samples (Fig. 5d), ranging from 400°C/5 kbar to 500°C/7 kbar (S24711), 450°C/7 kbar to 500°C/8.5 kbar (POL15), 500°C/4.5 kbar to 600°C/7.5 kbar (GS19A) to 550°C/10 kbar to 600°C/11 kbar

(GS17). The HP conditions reported from the eclogite lenses (Schmädicke 1994) considerably exceed the maximal pressures which can be calculated from the garnet-bearing mineral assemblages in micaschists and phyllites (Fig. 5d).

Upper Allochthon units of Münchberg and Frankenberg massifs

Garnet in the micaschists of the Münchberg Massif Liegendserie has round to elliptical shapes and sometimes sutured grain boundaries. Sizes are maximal 4 mm in diameter and the modes range from 5 to 12%. The porphyroblasts have numerous inclusions of small quartz, biotite, muscovite and occasionally plagioclase. The inclusions are mostly randomly distributed. In rare cases the inclusions line up to a straight internal foliation S_1i which is at a high angle discordant to the external S_2. Microlithons composed of garnet, abundant plagioclase (mode 10–20%), quartz and some decussate mica do not have the elongated lens-like shape due to the strongly anastomosing external foliation S_2 defined by muscovite and biotite. Garnet porphyroblasts are poorly zoned. Garnet cores have high spessartine and high grossular content at low pyrope contents. Zonation trends towards the garnet rims display increasing pyrope (up to 18 mol%) at constant or decreasing grossular contents (Fig. 3n, o, r & s; Supplementary material 3; Fig. 2l, m). In the *X*Mg–*X*Ca plots garnet zonation trends appear slightly variable: trend 1 (samples OK8, OK22, LS14) starts at high or low *X*Ca and involves increasing *X*Mg at an almost constant level of *X*Ca. Trend 2 (samples OK4, OK9) starts at high *X*Ca and low *X*Mg and evolves toward lower *X*Ca and higher *X*Mg (Fig. 6a). The trace element zonations are similar in trend 1 and trend 2 garnet porphyroblasts. Elevated Ti contents at up to 1000 $\mu g\ g^{-1}$ in the broad

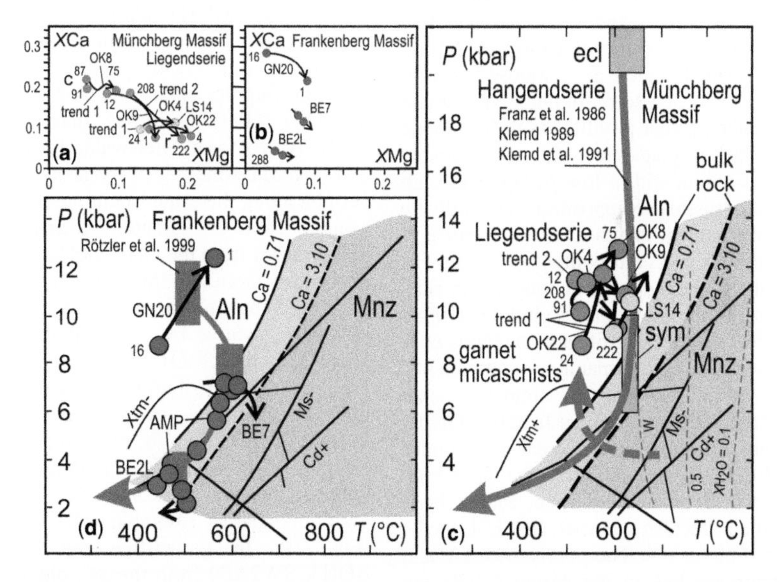

Fig. 6. Garnet mineral chemistry and geothermobarometry in the Upper Allochthon units with the Münchberg (MM) and Frankenberg massifs (FM). (**a**) and (**b**) Garnet zonations in *X*Mg–*X*Ca coordinates (mole fractions, based on cation formula with 12 O). Arrows between numbers (see Fig. 3) indicate core-to-rim (c, r) zonation trends from single profiles. Numbers are selected analyses for geothermobarometry. (**c**) P–T estimates and core-to-rim P–T trends for garnet mineral assemblages from MM in coloured symbols. P–T data and P–T paths in grey shading and green symbols (ecl, eclogites; sym, symplectites) for the Hangendserie refer to reports in Franz *et al.* (1986), Klemd (1989) and Klemd *et al.* (1991). Stippled dark grey arrow indicates P–T trend from Cambrian contact metamorphism to Devonian HP metamorphism in the Liegendserie. (**d**) P–T estimates and core-to-rim P–T trends from garnet mineral assemblages in the FM in coloured symbols. P–T data and P–T paths in grey symbols refer to Rötzler *et al.* (1999). Green symbols are amphibole analyses (AMP) in Rötzler *et al.* (1999) recalculated using the geothermobarometer of Zenk and Schulz (2004). Numbers refer to garnet analyses in (a) and (b) and Figure 3; error of $\pm 50°C/1.0$ kbar on P–T estimates. The aluminosilicates (Ky, Sill), muscovite-out (Ms-) and cordierite-in (Cd+) univariant lines are after Spear (1993). Stability fields of monazite (Mnz) and allanite (Aln) at different bulk rock contents as a function of Ca wt%, and with xenotime (Xtm) stability field (Janots *et al.* 2007; Spear 2010). Red stippled lines are solidus curves in the granitic system Qtz–Ab–Or–H_2O at XH_2O of 1.0 (curve W, wet solidus), at XH_2O of 0.5 and 0.1 after Johannes and Holtz (1996).

cores decrease toward the narrow rims. The Y and HREE are very poorly zoned and apparently have lower contents in the trend 1 garnet cores (Fig. 3r, s; Supplementary material 3; Fig. 2l, m). P–T estimates for the core-to-rim garnet zonations with coexisting plagioclase and mica resulted in two different P–T path segments: garnet of zonation trend 1 crystallized at increasing temperatures and pressures, in detail from 550–600°C/9–10 kbar to 600–630°C/ 10–13 kbar (Fig. 6c, blue symbols). In contrast, the garnet porphyroblasts with zonation trend 2 crystallized at increasing temperatures and decreasing pressures, from 550°C/11.5 kbar to 620°C/10.5–9. 5 kbar (Fig. 6c, golden symbols). As the P–T path segments from trend 1 and trend 2 garnet partly overlap and are all within a similar limited range at 550–630°C/9–13 kbar it is likely that they belong to the same metamorphic event and represent distinct parts of a single metamorphic cycle.

In the Frankenberg Massif, a garnet phyllite (sample GN20) contains garnet up to 2 mm in diameter with a mode of 8%. The porphyroblasts occur with abundant oligoclase and quartz in lens-like microlithons which are surrounded by a slightly anastomosing main foliation S_2 composed of predominant muscovite and some biotite. The sample contrasts the other garnet phyllites with less mode of garnet and lack of plagioclase. Garnet in sample GN20 displays a core with a low pyrope and high spessartine and grossular composition. Towards the rim, the pyrope content increases up to 8 mol% at quite constant high grossular contents of 25 mol%. Grossular only slightly decreases toward the rims (Figs 3t & 6b). In the other garnet phyllites (BE2L BE7), the garnet displays significant zonations only for the spessartine and almandine components, whereas pyrope and grossular are almost constant at a low level from core to rim (Figs 3u & 6b). Garnet trace element zonations are different in samples GN20 and BE2L. In sample GN20 the Ti is high (1500 $\mu g\ g^{-1}$) in a broad core zone, as has been observed in the Münchberg Massif. In BE2L the Ti is lower at <1000 $\mu g\ g^{-1}$. The poorly zoned Y and HREE contents are low in the garnet GN20 and higher in the BE2L cores (Fig. 3t, u). In sample GN20 the P–T segment starts at 450°C/9 kbar for garnet cores and ends at 580°C/12.5 kbar for the garnet rims (Fig. 6d). Lacking oligoclase in sample BE7 led to pressure estimates by the garnet–biotite geothermobarometer calibration by Wu (2019). Accordingly, garnet crystallized at increasing temperatures from 590 to 615°C at 7 kbar (Fig. 6d). A further garnet phyllite sample (BE2L) without oligoclase and with poorly zoned garnet delivered 500–512°C at decreasing low pressures of 2.4–2.1 kbar (Fig. 6d). Mineral-chemical data from metabasites which are intercalated to the metapsammopelites were presented by Rötzler *et al.* (1999). These given amphibole compositions within suitable assemblages allowed application of the amphibole geothermobarometer by Zenk and Schulz (2004). This yielded a retrograde P–T evolution from 600° C/6 kbar to 450°C/3 kbar, matching and combining the P–T data from samples BE7 and BE2L (Fig. 6d, green symbols).

Monazite EPMA Th–U–Pb ages in the Saxothuringian Zone

Saxonian Granulite Massif and Schist Cover

In the cordierite-garnet gneisses ('kinzigites') from the CGU, monazite in the matrix occurs in poorly zoned grains up to 150 µm large in Bulk Silicate Earth (BSE) images. Monazite grain boundaries sometimes appear as sutured due to adjacent and enclosed fibrolitic sillimanite (Fig. 7a–d). Also, many monazite grains are enclosed by garnet porphyroblasts. Matrix monazite in the cordierite-garnet gneiss MOH5 defines an isochrone at 326 ± 3 Ma; the isochrone from monazite enclosed in garnet is significantly older, at 339 ± 6 Ma (Fig. 8a; Supplementary material 5; Fig. 1a–c). In the ThO_2*–PbO isochrone diagrams the garnet gneiss AB9B from Auerswalde provides an age of 335 ± 6 Ma. Analyses of monazite enclosed in garnet plot along the same isochrone (Fig. 8b). Further samples from the Granulite Massif core (AJ201, TRO2) apart from the CGU yielded Lower Carboniferous isochrones at 356 ± 6 Ma and 348 ± 3 Ma (Fig. 8c; Supplementary material 3; Fig. d), but also at 338 ± 7 Ma as from an orthopyroxene-bearing felsic granulite GRO1 (Supplementary material 5; Fig. 1e). Age distributions within large grains are quite homogeneous, even when zonations appear in BSE images (Fig. 7d, e). Garnet-free micaschist samples of the SC contain aggregates of monazite with grain sizes of 10–15 µm. The straight monazite grain boundaries resemble static recrystallization (Fig. 7f). In garnet-free micaschists (BOH2, BOH3, SW2AD) from the SC, older isochrones at 372 ± 5 Ma, 358 ± 5 Ma and 358 ± 4 Ma are observed (Supplementary material 5; Fig. 1f–h). This is confirmed by a garnet-bearing micaschist (SW2AB) from the SC near Auerswalde with monazite along the mica-dominated foliation at 360 ± 5 Ma (Fig. 8d). In the SC a non-negligible number of single Silurian and Ordovician monazite analyses appears (Supplementary material 1B, Supplementary material 5; Fig. 1f–h).

Erzgebirge metamorphic units

The Freiberg Gneiss Dome is assigned to the GAU and underlies the UHP GEU in the Eastern

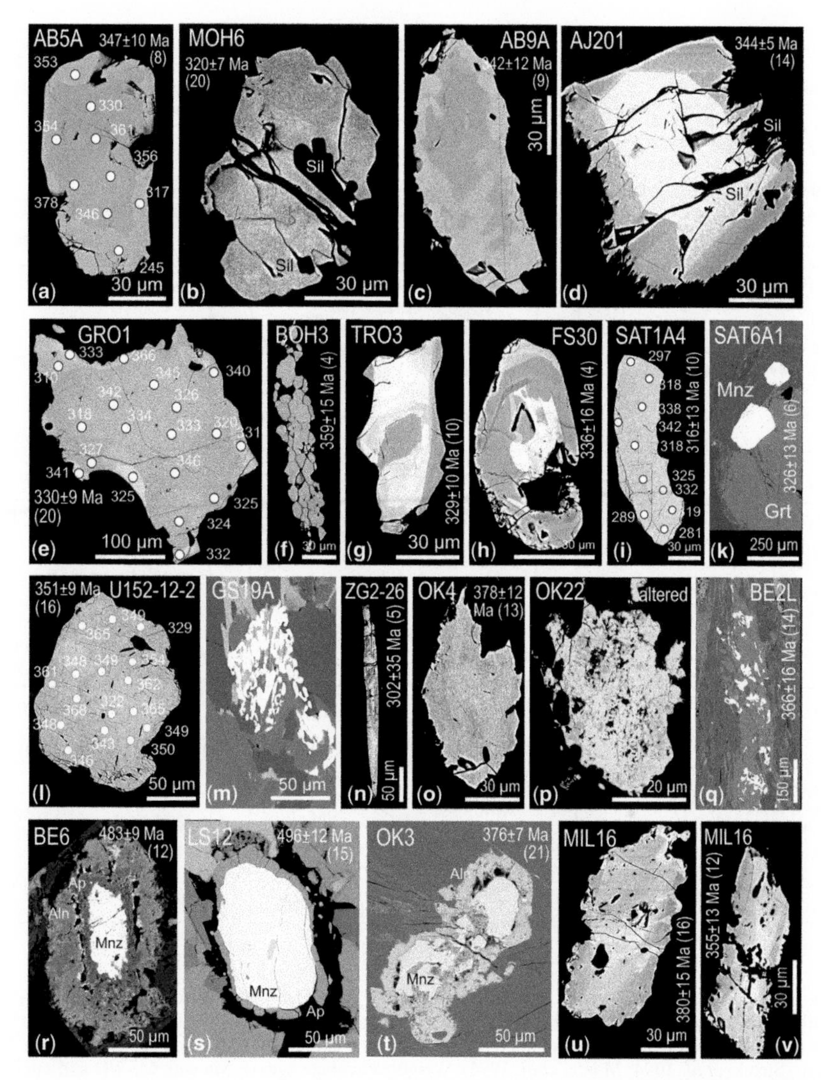

Fig. 7. Microstructures of monazite (Mnz) from the Saxothuringian Zone in backscattered electron images (BSE). Numbers are single Th–U–Pb ages in Ma. Weighted average ages with 2σ error (Ludwig 2001) are calculated from several analyses within a grain (number in brackets). (**a**) Weakly zoned and partly euhedral monazite in cordierite-garnet gneiss in the Cordierite Gneiss Unit (CGU) of the Saxonian Granulite Massif (SGM). (**b**) Large monazite with numerous inclusions of decussate prismatic sillimanite (Sil) in cordierite-garnet gneiss. (**c**) Large zoned monazite in garnet gneiss of the CGU below the Schist Cover. (**d**) Zoned monazite with sutured grain boundary in felsic granulite from SGM core. (**e**) Homogeneous monazite in felsic orthopyroxene garnet granulite, SGM core. (**f**) Polycrystalline aggregate composed of numerous small monazite with static grain boundaries in garnet-free micaschist, Schist Cover of SGM. (**g**) Zoned xenomorphic monazite in anatectic granite (SGM). (**h**) Euhedral zoned monazite in anatectic granite (SGM). (**i**) Homogeneous monazite in UHP garnet gneiss from the Gneiss Eclogite Unit (GEU) of Central Erzgebirge. (**k**) Monazite grain partly enclosed in garnet (Grt) with polycrystalline quartz inclusions (GEU). (**l**) Poorly zoned monazite in garnet micaschist of the Micaschist Eclogite Unit (MEU), Central Erzgebirge. (**m**) Monazite-allanite symplectites in garnet micaschist from the MEU, Western Erzgebirge. (**n**) Needle-shaped euhedral monazite from the Zinnwald Greisen Granite, Eastern Erzgebirge. (**o**) Poorly zoned subhedral Devonian monazite in garnet micaschist from the Liegendserie, Münchberg Massif (MM). (**p**) Altered sponge-like Cambrian monazite in garnet micaschist (MM). (**q**) Cluster structure of small monazite grains in phyllitic micaschist, Frankenberg Massif (FM). (**r**) Monazite with double corona texture of inner apatite (Ap) and outer allanite (Aln) in orthogneiss (FM). (**s**) Corona texture of monazite with apatite (Ap) in orthogneiss (MM). (**t**) Corona texture of allanite (Aln) around monazite in garnet micaschist (MM). (**u**) and (**v**) Zoned and sector-zoned monazite in andalusite garnet micaschist of the Nossen-Wilsdruff Schists (NWS).

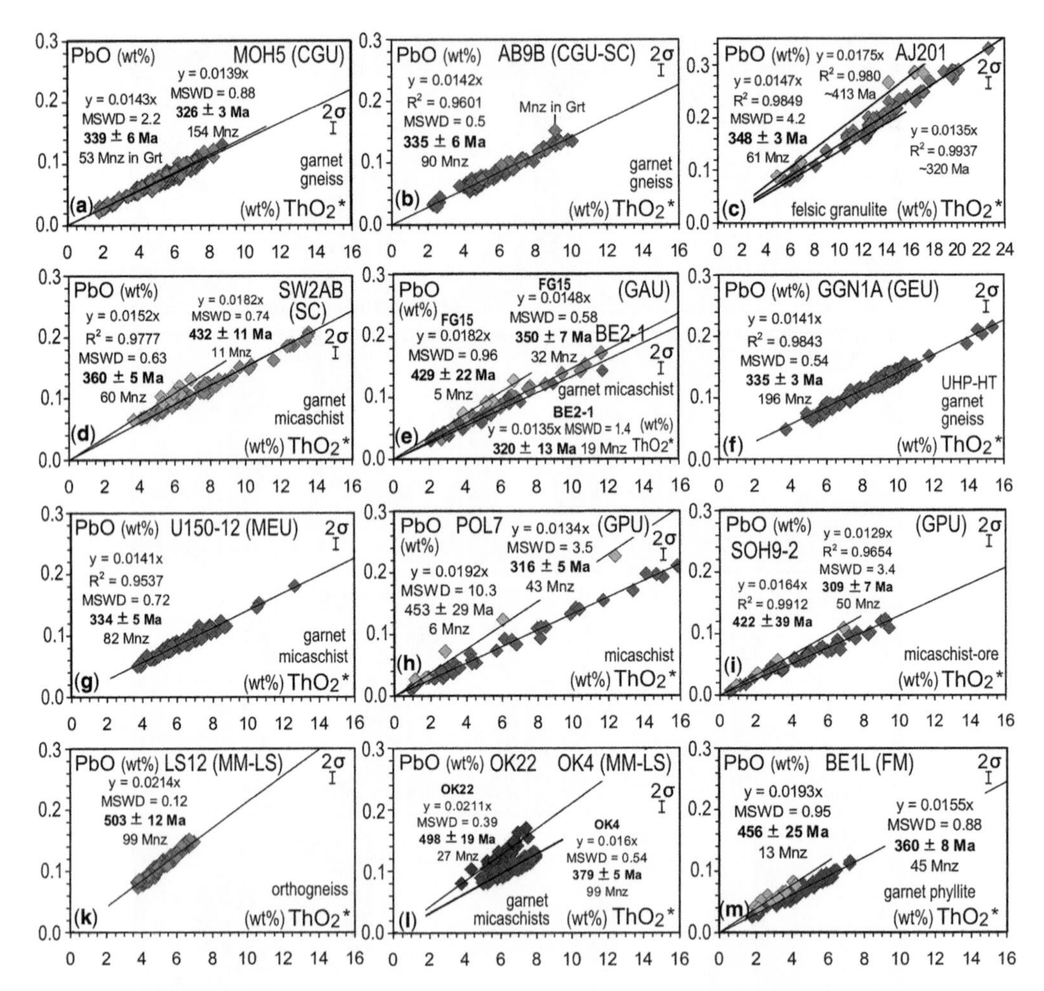

Fig. 8. Th–U–Pb chemical model ages of monazite. Total ThO_2* v. PbO (wt%) isochrone diagrams. ThO_2* is ThO_2 + UO_2 equivalents expressed as ThO_2. General minimal 2σ error on monazite PbO analysis is shown by a bar. Regression lines of monazite populations marked by colours give a coefficient of determination R^2 and are forced through zero (Suzuki *et al.* 1994; Montel *et al.* 1996). Weighted average ages in Ma with MSWD and minimal error of 2σ are calculated from the single analyses belonging to an isochrone according to Ludwig (2001). (**a–d**) Data from monazite in the Cordierite Gneiss Unit (CGU) in the Saxonian Granulite Massif core, in the CGU below the Schist Cover (CGU-SC) and a felsic granulite. (**e**) Monazite ages in 2 samples from the Freiberg Gneiss Dome in the Gneiss Amphibolite Unit (GAU). (**f**) Data from UHP garnet gneiss of the Gneiss Eclogite Unit (GEU) as reported in Tichomirowa *et al.* (2018). (**g**) Garnet micaschist of the Micaschist Eclogite Unit (MEU) in Central Erzgebirge. (**h**) and (**i**) Monazite data from garnet micaschists of the Garnet Phyllite Unit (GPU) in Western Erzgebirge. (**k**) Cambrian monazite in granitic orthogneiss in the Liegendserie of Münchberg Massif (MM). (**l**) Cambrian monazite ages in a garnet micaschist OK22 match the ages in the orthogneisses and contrast Devonian ages (OK4). (**m**) Mainly Devonian monazite ages in garnet phyllite of the Frankenberg Massif (FM). The coloured symbols mark analyses belonging to monazite age populations and defining isochrones, falling into Upper Carboniferous (brown), Lower Carboniferous (green), Devonian (blue), Silurian and Ordovician (light blue), and Cambrian (dark brown) ranges of ages. See Supplementary material 5 for further diagrams.

Erzgebirge (Fig. 2). A garnet micaschist from Brand-Erbisdorf at the margin of the Freiberg Gneiss Dome (BE2-1) yielded a Carboniferous isochrone at 320 ± 13 Ma, whereas a paragneiss representing the core of the Gneiss Dome from Freiberg town (FG15), yielded 350 ± 7 Ma, from monazite associated with mica in the foliation planes and with plagioclase in the microlithons. Also, some Silurian monazite ages occur (Fig. 8e, both samples shown in one diagram). An orthogneiss sample SEI5 from the Seidewitz valley provides a poorly defined isochrone at 544 ± 31 Ma, interpreted to date the protolith intrusion. Paragneiss samples (SEI3, SEI4) from the Seidewitz valley yielded a monazite isochrone at 336 ± 8 Ma and 333 ± 11 Ma with some Silurian ages (Supplementary material 1A, Supplementary material 5; Fig. 2a, b).

Monazite ages from UHP rocks from the GEU are of special interest for comparison with earlier studies (Massonne *et al.* 2007*a*; Závada *et al.* 2021). Garnet gneisses from this unit bear large monazite, mostly associated with muscovite, quartz and feldspar, with weak zonations in BSE images (Fig. 7i). They can be partly enclosed in garnet rims (Fig. 7k). Monazite isochrones from samples at the Saidenbach Reservoir vary from 336 ± 8 Ma to 327 ± 12 Ma and agree with 335 ± 3 Ma in a garnet gneiss sample GGN1A from Zöblitz (Fig. 8f; Supplementary material 5; Fig. 2c, d; Tichomirowa *et al.* 2018). Quite well-defined isochrones from gneiss samples without polycrystalline quartz inclusions as coesite relicts in garnet range from 339 ± 6 Ma (ST7) to 328 ± 10 Ma (ST4) to 321 ± 9 Ma (ST6), from samples taken at Eppendorf (Fig. 2; Supplementary material 5; Fig. 2e). The isochrones in garnet micaschist samples from the MEU in part match these Carboniferous ages. This is true for sample U150-12 with a 334 ± 5 Ma monazite isochrone (Fig. 8g). The other samples have older isochrones up to 344 ± 5 Ma (Supplementary material 5; Fig. 2f, g), as in U152-12-2, where a single large monazite grain has 351 ± 9 Ma as a weighted average from 16 analyses (Fig. 7l). Variations of isochrones of about ±10 Ma are observed from samples which are situated nearby, as shown in Figure 2.

In the MEU and GPU of the Western Erzgebirge, which are also assigned as HP units due to the occurrence of eclogite lenses (Schmädicke 1994), a broad variation in Carboniferous isochrone ages is observed. These range from 356 ± 7 Ma (S3027) and 353 ± 4 Ma (S849), through 343 ± 4 Ma (S24630) to 316 ± 5 Ma (POL7, Fig. 8h; Supplementary material 5; Fig. 2h, i). Monazite is mostly related to muscovite, quartz and garnet. In sample GS19A an isochrone at 336 ± 4 Ma is obtained from monazite grains forming symplectitic textures with allanite (Fig. 7m). At the visitor mine at Pöhla and the surrounding lithologies of the GPU, several garnet micaschists contain monazite grains associated with biotite, muscovite and plagioclase, which yield isochrones at around 328 ± 8 Ma (POL15, POLP1, P3-1A). Furthermore, in the Pöhla visitor mine sphalerite and cassiterite occur in veins and dispersed in micaschists. These rocks are referred to as micaschist ore. Monazite isochrones from the micaschist ore samples are markedly younger at 310 ± 9 Ma, 309 ± 7 Ma (sample SOH9-2, Fig. 8i) and 304 ± 7 Ma. In strongly mineralized micaschist ore, there is also monazite with ThO_2 <0.5 wt% which cannot be dated by the EPMA Th–U–Pb method (not shown).

Upper Allochthon units of Münchberg and Frankenberg massifs

In the Liegendserie of the Münchberg Massif, and in the Frankenberg Massif, coarse-grained granitoid orthogneisses are interlayered with paragneisses, micaschists and phyllites. In the metapsammopelites small monazite displays poor internal zonations and sutured grain boundaries (Fig. 7o). Sponge-like altered monazite (Fig. 7p), and monazite in cluster structures are also observed (Fig. 7q). In the granitoid orthogneisses some relict monazite grains are mantled by characteristic corona textures by apatite and allanite (Fig. 7r, s). Such corona textures are also observed in some micaschists (Fig. 7t). They are interpreted as an indicator of monazite decomposition during a retrograde metamorphic overprint (Finger *et al.* 1998; Janots *et al.* 2008; Manzotti *et al.* 2018; Schulz 2021*a*). A monazite isochrone in the orthogneiss with flaser and augen structure from the Münchberg Massif (LS12) indicates 512 ± 12 Ma (Fig. 8k). Similar ages at 512 ± 11 Ma (GN7), 507 ± 17 Ma (BE6) and 502 ± 14 Ma (GN9) occur in the Frankenberg Massif orthogneisses (Supplementary material 5; Fig. 3a–c). A Cambrian age of 498 ± 19 Ma is obvious in a garnet micaschist sample (OK22) in the Münchberg Massif Liegendserie (Fig. 8l). It coincides within the error with the Cambrian monazite ages of the granitoid orthogneisses which are protolith crystallization ages according to the U–Pb–zircon data (Koglin *et al.* 2018). This Cambrian age is an exception and contrasts with the metamorphic monazite ages in numerous other Münchberg Massif garnet micaschist samples of which single isochrones are 389 ± 9 Ma (LS14), 388 ± 6 Ma (OK8), 379 ± 5 Ma (OK4, Fig. 8l) and 376 ± 5 Ma (OK3). There are several samples that show two distinct isochrones, such as OK6 with a Devonian (366 ± 5 Ma) and a Carboniferous (302 ± 9 Ma) age population. There are also samples (OK5, OK23) where monazite data arrange in three isochrones, minor and poorly defined Silurian, dominant Devonian (379 ± 6 Ma, 375 ± 6 Ma) and also poorly defined Carboniferous isochrones at 313 ± 14 Ma and 308 ± 15 Ma (Supplementary material 5; Fig. 3f–h). In the garnet phyllites of the Frankenberg Massif, the Devonian monazite age population prevails, with isochrones at 362 ± 9 Ma (BE2L) and 360 ± 8 Ma (BE1L). A second subordinate population of Ordovician monazite ages is also observed (Fig. 8m; Supplementary material 5; Fig. 3i).

Th–U–Pb monazite ages in terms of metamorphism and magmatism

Monazite mineral chemistry and P–T stability field

A profound interpretation of the EPMA Th–U–Pb monazite ages in terms of magmatism, metamorphism, P–T evolution and fluid-triggered events requires consideration of the monazite mineral chemistry. Monazite mineral compositions are almost fully captured by the applied EPMA analytical protocol (Schulz *et al.* 2019). Nominally La, Ce and Nd together dominate with 2.8 to 3.2 cations per formula unit (p.f.u.) per 4 oxygens (Spear and Pyle 2002). The other LREE, such as Pr, Sm, Eu and Gd, occur in minor proportions (0.3–0.4 cations p.f.u.). The Th occurs with up to 0.25 cations (p.f.u.). Most monazites contain Ca, Si and HREE. Contrasting the very low HREE contents, the Y contents in monazite (~0.1 cations p.f.u.) are remarkable (see below).

Whether monazite or allanite occurs in a metamorphic rock is determined by the whole-rock composition and the metamorphic grade (Janots *et al.* 2007, 2008; Spear 2010). Accordingly, monazite is commonly stable in metapsammopelites under amphibolite facies conditions. The monazite stability field is shifted towards lower temperatures with decreasing bulk rock Ca. Also, the stability field of monazite is extended to lower temperature with increasing Al (e.g. Shrestha *et al.* 2019). As a consequence, monazite can be expected to crystallize at upper greenschist facies conditions in high-Al and low-Ca metapelites (Spear 2010; Spear and Pyle 2010; Larson *et al.* 2022). Monazite can react to allanite plus apatite and vice versa (Finger *et al.* 1998; Wing *et al.* 2003; Harlov *et al.* 2011; Skrzypek *et al.* 2018). A higher Ca whole-rock content expands the allanite stability field to higher temperatures and the monazite stability field then retreats to granulite facies conditions (Bingen *et al.* 1996). The extent of the monazite stability field at pressure above 14 kbar is yet poorly constrained by experiments and thermodynamic modelling (Spear 2010; Shrestha *et al.* 2019). However, the isopleths of the Ca bulk rock composition limiting the monazite stability field are mainly temperature-dependent, but also inclined with increasing pressure (Figs 4–6). Also, it is important to highlight that Al and Ca in metapsammopelites may have a very heterogeneous distribution due to layering and subdivision into foliation planes with mica (Al rich, Ca poor) and microlithons with quartz, feldspar and garnet. As a consequence, during a prograde metamorphic evolution, some monazite may crystallize early and at low temperatures in distinct Al-rich, Ca-poor microstructural domains, followed by a pervasive crystallization at higher temperatures and in Ca-richer microstructural domains (Salminen *et al.* 2022).

Numerous studies have outlined that Y or $XYPO_4$ in monazite increases with metamorphic grade and temperature, and Y in monazite as well as garnet–monazite–Y geothermometers have been proposed and applied (e.g. Franz *et al.* 1996; Gratz and Heinrich 1997; Heinrich *et al.* 1997; Förster 1998; Pyle *et al.* 2001; Spear and Pyle 2002; Wing *et al.* 2003; Cutts and Dyck 2023). These observations may be tested in the various metamorphic units of the Saxothuringian Zone.

P–T–time evolution of Saxonian Granulite Massif and Schist Cover

Monazite in garnet gneisses of the CGU with temperatures at around 800°C have, in general, low Y_2O_3 at <2 wt%, contrasting the higher Y_2O_3 (2–3 wt%) in the garnet-free metapelites in the SC with lower temperatures (700°C). Thus, increasing Y in monazite with temperature cannot be confirmed in this study. There is a general tendency of increasing Y_2O_3 with monazite age (Fig. 9a). Monazite enclosed in the garnet yielded slightly older isochrones than matrix monazite (Fig. 8a; Supplementary material 5; Fig. 1a, b). However, no significant differences in Y_2O_3 contents or $XGdPO_4$–$XYPO_4$ coordinates (Pyle *et al.* 2001) of matrix and garnet inclusion monazite can be stated (Fig. 9a, b). Most monazite grains follow a cheralite substitution trend in the Th + U v. Ca diagram and only the monazite in the felsic granulite sample AJ201 displays an individual trend at constant and low Ca (Fig. 9c). The monazite age isochrones in the CGU samples from the Granulite Massif core match those from samples AB9A–AB9B which can be assigned to the part of the CGU below the SC. Also, the garnet porphyroblasts in these samples display similar zonation trends (Figs 3e–i & 4a–c), which result in P–T paths with comparable shapes and maximal temperatures at 800°C and maximal pressures at 8 kbar (Fig. 4d–f). This provides arguments that the CGU exposed in the Granulite Massif core and the border rim next to the SC underwent a common metamorphic evolution. The prograde and retrograde P–T path segments of the CGU evolve almost entirely within the monazite stability field (Fig. 4f). Decussate sillimanite needles enclosed in the matrix monazite suggest that it crystallized contemporaneously or subsequently to sillimanite. This textural relationship indicates that monazite crystallized when the P–T path passed the sillimanite stability field or subsequently (Fig. 4f). The observation of the youngest matrix monazite isochrone at 317 ± 6 Ma (MOH6) and the oldest isochrone from monazite enclosed in

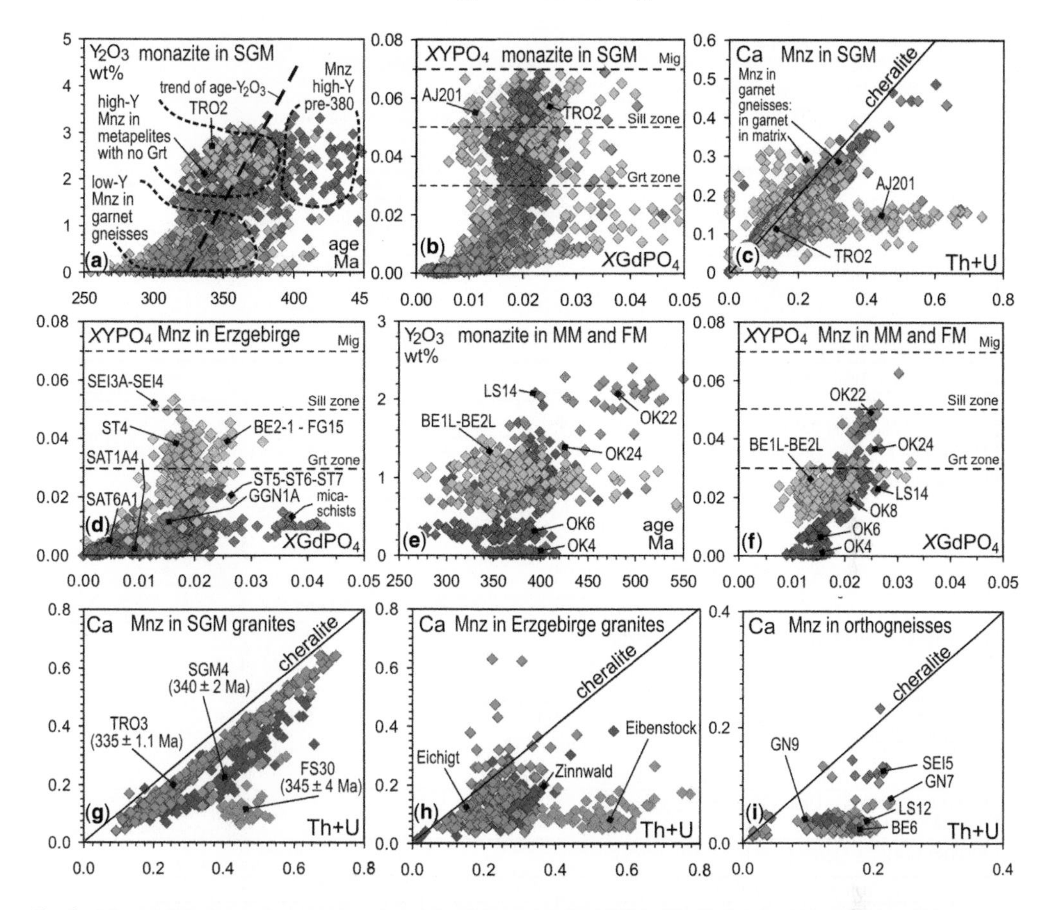

Fig. 9. Mineral chemistry of monazite and distributions of monazite (Mnz) Th–U–Pb chemical ages. (**a**) Monazite Y_2O_3 v. age in Ma in the Saxonian Granulite Massif (SGM) and Schist Cover. Younger low-Y monazite in garnet gneisses of the Cordierite Gneiss Unit (CGU) contrast older high-Y monazite in garnet-free metapelites. Note trend of age–Y_2O_3 and high-Y pre-380 Ma monazites. Garnet gneisses of CGU with monazite enclosed in garnet are in red, matrix monazite in garnet gneisses are in pink symbols. (**b**) Monazite XGdPO$_4$ v. XYPO$_4$ in the SGM with compositions in reference to the garnet (Grt), sillimanite (Sil) and migmatite (Mig) metamorphic mineral zones as defined by Pyle *et al.* (2001); same symbols as in (a). (**c**) Monazite compositions in Th + U v. Ca cations per formula unit (4 oxygens) in reference to the cheralite substitution trend. Note distinct trend in sample AJ201. Same symbols as in (a). (**d**) Monazite XGdPO$_4$ v. XYPO$_4$ in the Erzgebirge samples. Note low-Y and low-Gd monazite in UHP garnet gneisses. (**e**) Monazite Y_2O_3 v. age in Ma in the Münchberg and Frankenberg Massif (MM, FM). High-Y old (~500 Ma) metamorphic monazite in garnet micaschist OK22 contrasts Devonian metamorphic monazite in other samples. (**f**) Monazite XGdPO$_4$ v. XYPO$_4$ in the Münchberg and Frankenberg massifs (MM, FM). (**g–i**) Monazite in anatectic granites from the Saxonian Granulite Massif (SGM), and in Erzgebirge (EG) granites, and with considerably lower Th + U and Ca in granitoid orthogneisses, in Th + U v. Ca cations per formula unit (4 oxygens), in reference to the cheralite substitution trend.

garnet at 339 ± 6 Ma (MOH5) indicate that the corresponding P–T path segment within the sillimanite stability field may be bracketed between ~340 and ~320 Ma. In this case the Y and HREE zonations in garnet and the low Y in monazite can provide no further constraints on the crystallization of monazite in reference to garnet, as has been outlined in other studies (Pyle *et al.* 2002, 2005*b*; Rubatto *et al.* 2006; Larson *et al.* 2019, 2022; Shrestha *et al.* 2019). Monazite should have high Y when it has crystallized previous to the garnet with low Y, or when garnet underwent significant resorption during monazite growth (Hermann and Rubatto 2003). The overall low Y contents (Fig. 9a, b) merely suggest a contemporaneous crystallization of monazite with garnet, followed by further post-garnet growth, while Y is fractionated into the abundant garnet. Monazite is very rare in garnet-bearing felsic granulites. The garnet-free samples TRO2 and AJ201 provided Lower Carboniferous isochrones at 356

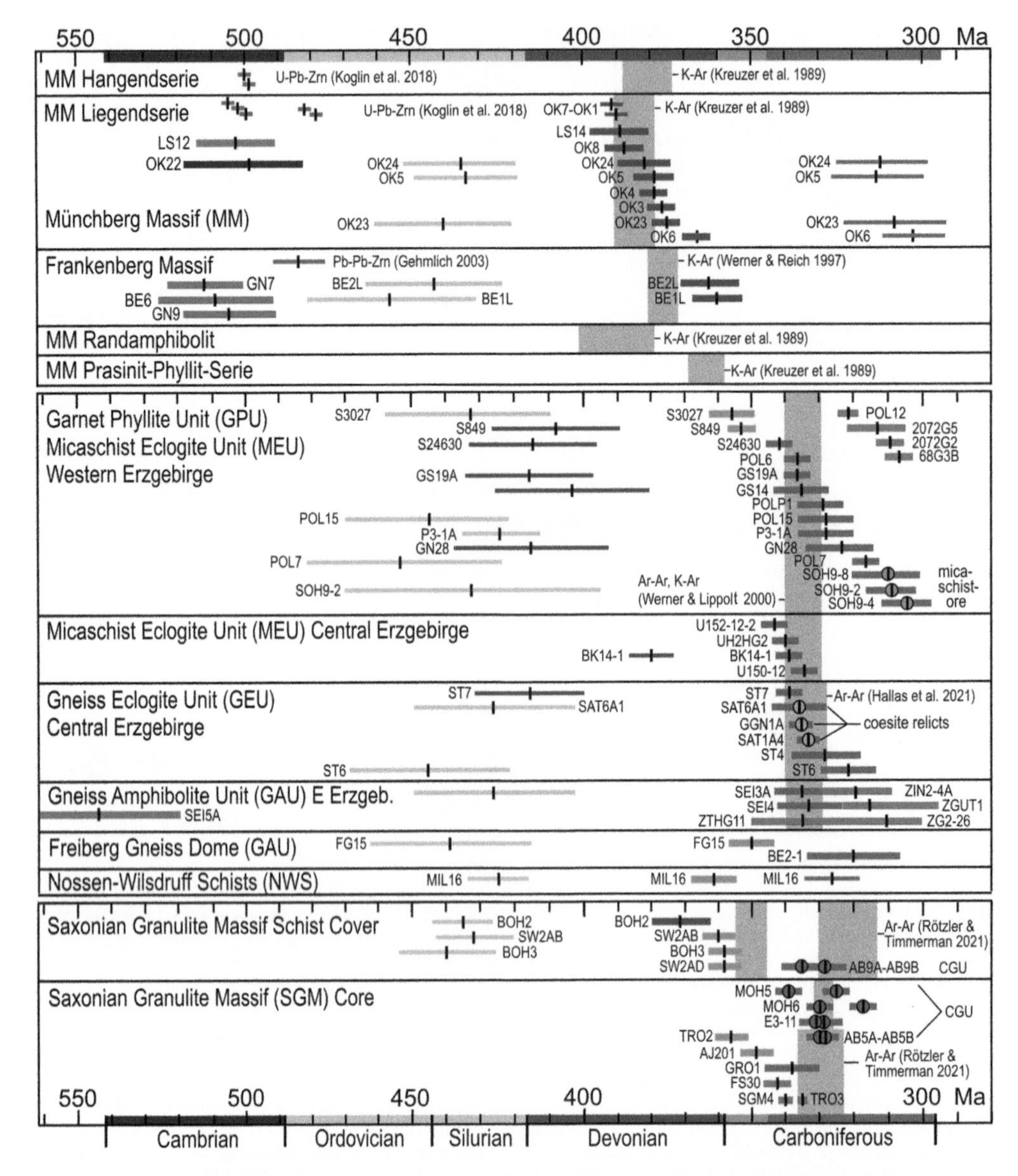

Fig. 10. Summary of EPMA Th–U–Pb monazite ages in various parts of the Saxothuringian Zone, sorted from upper to lower allochthonous units. K–Ar and Ar–Ar mica and hornblende data (metamorphic cooling ages) by other authors and U–Pb zircon data in Koglin *et al.* (2018) for Münchberg Massif units are shown for comparison. Ages of EPMA Th–U–Pb monazite isochrones are marked by black lines, errors are marked by coloured bars. The main isochrone in a sample is marked by a thick bar, subordinate isochrones are marked by thin bars. Data from granites and orthogneisses are marked by red bars. Data from micaschist-ore in the Pöhla visitor mine are marked by blue points; data from UHP garnet gneisses from the Gneiss Eclogite Unit are marked in green points, and from cordierite-garnet gneisses in the Cordierite Gneiss Unit (CGU) in purple points. Note non-negligible but poorly defined isochrones of Silurian metamorphic monazite.

$\pm$ 6 Ma and 348 $\pm$ 3 Ma (Figs 8c & 10). As there are indications that the granulite prograde compression path as well as the decompression path passed the monazite stability field, this extends the initiation of the metamorphism in the SGM to the Devonian, confirming the observations in Rötzler and Timmerman (2021). Monazite isochrones for the SC samples range from the Upper Devonian at 372 $\pm$ 9 Ma to the Lower Carboniferous at 358 $\pm$ 5 Ma (Fig. 8d). Rare garnet in sample SW2AB recorded a prograde P–T segment within the monazite stability field. This implies that the initiation of the metamorphism in the SC is in the Upper Devonian, also confirming and extending earlier reports (Kröner *et al.* 1998; Rötzler and Timmerman 2021; and references therein).

P–T–time evolution in the Erzgebirge metamorphic units

The andalusite-bearing micaschist MIL16 from the NWS (Fig. 2) bears significantly zoned garnet with a prograde core-to-rim zonation trend. Corresponding P–T estimates, involving coexisting oligoclase and mica yielded a P–T path section with increasing temperature of 570–600°C at decreasing pressure of 5 to 2 kbar, evolving from the sillimanite to the andalusite stability field and resembling overprint by contact metamorphism. The euhedral monazite is sector-zoned (Fig. 7u, v). The monazite ages can be arranged in subordinate Silurian (424 $\pm$ 8 Ma), a predominant Devonian (362 $\pm$ 6 Ma), and subordinate Carboniferous (326 $\pm$ 7 Ma) isochrones (Supplementary material 5; Fig. 1i). This sample also provided evidence of pre-Carboniferous metamorphism.

Geothermobarometry in the Freiberg Gneiss Dome of the GAU confirms previous results by Mingram and Rötzler (1999) and Willner *et al.* (2000). Maximum pressures at 11 kbar were attained at 680°C, which is at the lower temperature limits of the monazite stability field (Fig. 5c). Correspondingly, the monazite isochrone of 320 $\pm$ 13 Ma may refer to the subsequent decompression path, which entered the monazite stability field. An older isochrone at 350 $\pm$ 7 Ma (Fig. 8e) from a garnet-free sample (FG15) may refer to an early stage or the initiation of the metamorphism.

The P–T estimates from the UHP garnet gneisses in the GEU confirm previous data (Willner *et al.* 2000) and mark a high-pressure granulite facies episode or stopover at 800°C/20–16 kbar during the decompression of the coesite and microdiamond-bearing rocks (Massonne and Grosch 1994; Schmädicke 1994; Schmädicke *et al.* 1995; Massonne and Nasdala 2003). The decompression path of the GEU rocks passes the monazite stability field merely for Ca-poor bulk compositions (Fig. 5c). As a consequence, corresponding monazite ages should refer to this decompression path. When garnet rims partly enclose a 326 $\pm$ 13 Ma monazite, as in sample SAT6A1 (Fig. 7k), this implies that monazite should be of the same age or younger, but not older, than garnet. The age span of monazite isochrones in GEU gneisses ranges from 339 $\pm$ 6 Ma to 321 $\pm$ 9 Ma (Figs 4f & 10). Samples with garnet enclosing polycrystalline quartz as coesite relicts display a slightly narrower age span from 336 $\pm$ 3 Ma to 327 $\pm$ 12 Ma (Fig. 10). This coincides within the errors with the EPMA monazite ages from the Saidenbach locality by Massonne *et al.* (2007*a*). Závada *et al.* (2021) reported also older (355–340 Ma) monazite ages from metagranitoids in the Eger Complex.

Garnet in the micaschists of the MEU in the Central Erzgebirge crystallized during the prograde P–T evolution toward maximum conditions of 650°C/8 kbar (Fig. 5c), which is a new observation adding to the decompression paths as far as these were reported by Mingram and Rötzler (1999), Kröner and Willner (1998), Willner *et al.* (2000), and compiled by Sebastian (2013). These prograde P–T paths would allow monazite crystallization only for Ca-poor bulk rock compositions (Fig. 5c). The span of monazite isochrones from the MEU Central Erzgebirge micaschists ranges from 339 $\pm$ 3 Ma to 334 $\pm$ 5 Ma and it is similar to those from the GEU. In the GPU micaschist samples from the Western Erzgebirge, the garnet also recorded short segments of prograde P–T paths with maximum P–T conditions of 550°C/8 kbar (Fig. 5d). Distinct micaschist samples (GS17, GS19A), giving somewhat higher pressures and temperatures (580°C/11 kbar, 600°C/7 kbar), may belong to the underlying MEU, as the P–T data match the observations by Mingram and Rötzler (1999), Willner *et al.* (2000) and the compilation of P–T data by Sebastian (2013). The prograde P–T paths evolve outside the monazite stability field. The subsequent decompression paths pass the monazite stability field at $<$5 kbar (Fig. 5d). For the micaschist sample GS19A with a 336 $\pm$ 4 Ma monazite isochrone and symplectitic monazite–allanite intergrowths (Fig. 7m), the prograde P–T path is entirely in the monazite stability field for Ca-poor bulk rock compositions, as observed from the MEU in the Central Erzgebirge (Fig. 5c, d). The micaschist samples of the GPU from the Western Erzgebirge cover a much wider span of isochrones, from 356 $\pm$ 7 Ma (S3027) to 316 $\pm$ 5 Ma (POL7).

In the UHP and granulite facies, metamorphic garnet gneisses of the GEU in Central Erzgebirge, monazite with low XGdPO$_4$ and low XYPO$_4$ dominate (Fig. 9d). The Y and HREE contents in abundant garnet in the gneisses are also low. In contrast, garnet cores in the micaschist samples display high Y and HREE contents which decrease toward the rims. However, the mode of garnet in the micaschists is significantly lower as in the garnet gneisses of the GEU. In general, monazite has low Y in these micaschist samples, regardless of its age (Fig. 9d). The XYPO$_4$ in the micaschists from the MEU and the GPU do not exceed 0.05, which is the reference line for the sillimanite zone given by Pyle *et al.* (2001). As a consequence, the monazite Y contents in garnet-bearing metapsammopelites show no increase with metamorphic grade. There is no general tendency of increasing $XGdPO_4$ and $XYPO_4$ with increasing monazite age in the Erzgebirge units, in contrast to the observations in the SGM. To sum up, the Y distribution pattern provides no further indication for monazite crystallization in reference to the garnet. The Lower Carboniferous (356 ± 7 Ma, 353 ± 4 Ma) monazite in some of the micaschist samples may date the initiation of the metamorphism within microstructural sites with Ca-poor and Al-rich compositions, The bulk of monazite ages between 335 and 316 Ma may refer to the decompression and cooling paths. This interpretation is in general supported by the observation that EPMA monazite ages coincide with the 340–330 Ma K–Ar and $^{40}Ar/^{39}Ar$ ages, which are interpreted to date differential cooling of the various Erzgebirge nappe units (Werner and Lippolt 2000; Sebastian 2013; Hallas *et al.* 2021).

P–T–time evolution in the Upper Allochthon units of the Münchberg and Frankenberg massifs

Monazite in the granitoid orthogneisses in the Münchberg and Frankenberg massifs displays uniformly Cambrian isochrones, matching the U–Pb–zircon data by Gehmlich (2003) and Koglin *et al.* (2018). The Th–U–Pb monazite and the U–Pb zircon data in the orthogneisses are both interpreted as protolith ages, indicating a pervasive Cambrian (512–502 Ma) magmatic event. This magmatic activity supports the interpretation of monazite ages from some metapsammopelitic rocks (sample OK22, 498 ± 19 Ma) to be the result of contact metamorphism. Apparently, subsequent Devonian regional high-pressure metamorphism led to formation of apatite and allanite corona textures in Cambrian igneous monazite (Fig. 7r, s). Such corona structures are interpreted to indicate monazite decomposition during retrogression (Finger *et al.* 1998; Schulz 2021*a*; and references therein). This is supported by the P–T estimates from the garnet-bearing assemblages, which indicate a prograde Devonian metamorphism at conditions outside the monazite stability field (Fig. 6c).

In the Liegendserie of the Münchberg Massif and in the Frankenberg Massif the metamorphic monazites with Devonian isochrone ages are prevalent (Figs 10 & 11). Isochrones range from 389 to 362 Ma, and mostly coincide with K–Ar and Ar–Ar mica and amphibole data reported by Kreuzer *et al.* (1989), Werner and Reich (1997) and Rötzler *et al.* (1999). Geothermobarometric data from the garnet-bearing micaschists in the Liegendserie, and also the Frankenberg Massif (sample GN20), indicate a metamorphic overprint at 600°C/13 kbar and 650°C/10 kbar which is outside the monazite stability field for Ca-poor and Al-rich bulk rock compositions (Fig. 6c, d). P–T paths for decompression and decompression-cooling, as outlined by Franz *et al.* (1986), Klemd (1989) and Klemd *et al.* (1991), pass the monazite stability field only for Ca-poor bulk rock compositions, thus indicating a Devonian (389–360 Ma) age for these final P–T path segments (Fig. 6c). One micaschist sample (OK22) in which garnet follows the same prograde zonation trend 1 and recorded a similar P–T path segment as some other samples (Fig. 6a, c), has a significantly older Cambrian monazite age around a 498 ± 10 Ma isochrone (Fig. 8l), which differs from Devonian (389–360 Ma) monazite ages in the other garnet micaschists. In age–Y_2O_3 and also in $XGdPO_4$–$XYPO_4$ coordinates, monazite in sample OK22 displays considerably higher Y contents (~2 wt% Y_2O_3) as observed from the other samples, including OK4 (Fig. 9e, f). When monazite crystallizes at P–T conditions that allow no garnet crystallization at a given bulk composition, the available Y and HREE will fractionate into monazite and/or xenotime (Pyle *et al.* 2001). As a consequence, the Y-rich Cambrian monazite in sample OK22 should have crystallized previous to the high-pressure–low-temperature metamorphic Devonian garnet which is Y-poor throughout the Liegendserie micaschists (Fig. 3r, s; Supplementary material 3; Fig. 2l). At a given bulk rock composition, garnet will not crystallize at low pressures (Spear 1993). This suggests a low-pressure and/or contact metamorphism for the Cambrian (498 ± 10 Ma) monazite crystallization in micaschist sample OK22. More Ca-rich bulk compositions such as that of sample OK22 and the corresponding reduced monazite stability field which was then not passed during the decompression path, may have prevented crystallization of further Devonian (389–360 Ma) monazite in this sample, as observed in the other micaschists (Fig. 6c).

Sporadic Silurian (444–417 Ma) monazite ages

In metapsammopelites of the Upper Allochthon units, in the Erzgebirge units and in the autochthonous Nossen-Wildruff Schists, in SGM SC, but not in the core of the SGM, a subordinate but non-negligible number of monazite analyses with Silurian ages has been recorded (Figs 10 & 11). These single analyses occur mostly isolated in monazite grains where the other analyses display well-defined Devonian and Carboniferous ages. When compiled, these Silurian-age analyses are arranged in isochrones with low statistical definition and broad error (Figs 8e, h, i, m & 10; Supplementary material 5;

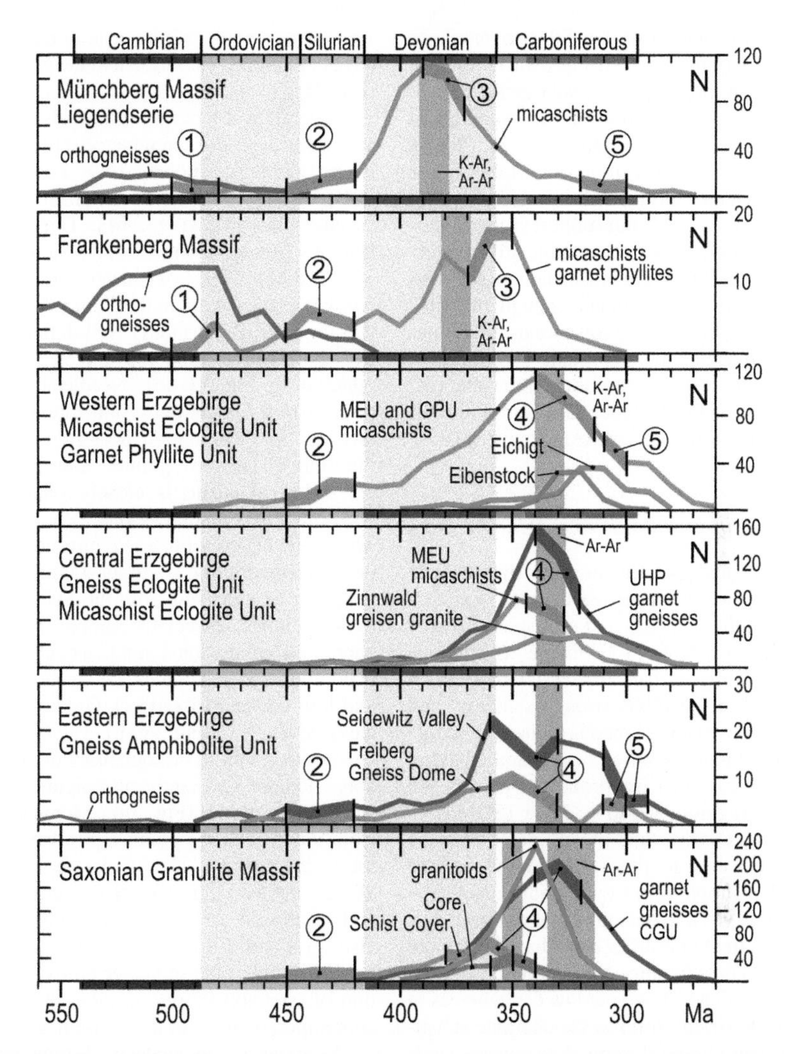

Fig. 11. Histogram view of EPMA Th–U–Pb monazite age pattern in the various units of the Saxothuringian Zone, with indications of metamorphic and magmatic events. Devonian and Ordovician era is marked in grey. Age scale is colour-coded: 345–296 Ma light brown (Upper Carboniferous); 358–345 Ma green (Lower Carboniferous); 417–358 Ma blue (Devonian); 488–417 Ma light blue (Ordovician and Silurian); 540–488 Ma dark brown (Cambrian). Thick lines between bars mark the spans of isochrones from single samples, from Figure 10. Dark grey domains mark age spans of K–Ar and Ar–Ar cooling ages by other authors (see Fig. 10). 1 – Cambrian contact or regional metamorphism; 2 – Silurian ages interpreted as a weak metamorphism; 3 – Devonian regional HP–LT metamorphism; 4 – Carboniferous UHP, HP, MP–MT and UHT metamorphism; 5 – Upper Carboniferous local hydrothermal event.

Figs 1g–i, 2a, d–f, i & 3g, i). Therefore, an interpretation of the geodynamic significance of these monazite ages is difficult. Early Ordovician plutonism and volcanism in the Saxothuringian Zone ceased at ~470 Ma. Prominent Late Ordovician or Silurian magmatism is not yet documented from the region, however, but corresponding U–Pb zircon ages were occasionally reported (Tichomirowa *et al.* 2001; Linnemann *et al.* 2010; Sagawe *et al.* 2016; Collett *et al.* 2020). During the Late Ordovician–Early Silurian the Saxothuringian Zone was a part of the northern peri-Gondwanan shelf along the southern margin of the Rheic Ocean (Kroner and Romer 2013; Žák *et al.* 2013; Stephan *et al.* 2018, 2019). In such a frame, and lacking other evidence of significant magmatic and tectonic events in the crust, the corresponding monazite ages may be explained by a widespread and diffuse thermal metamorphism at low pressures and without marked temporal or local peaks of temperature, as can be expected in a fading magmatic arc. Another explanation of the Silurian monazite ages could be mixing between ingrown radiogenic Pb of Ordovician monazite with radiogenic Pb produced during recrystallization at the subsequent metamorphic events (e.g. Barnes *et al.* 2021). However, similar Silurian monazite ages in the units with dominant Devonian and also Carboniferous isochrones (Fig. 10) do not support such an interpretation. Although it is not possible to uniquely identify inherited Pb by chemical dating, previous studies have reported significant enough inherited Pb in monazite to produce discordance (Barnes *et al.* 2021). If at all, possibly inherited common Pb seems to be restricted to local intra-granular domains, as the other Devonian and Carboniferous isochrones from these monazite grains give no indication for it.

Monazite ages in granitoids and orthogneisses

The SGM core and SC are intruded by anatectic granites (Fig. 2) which cut across the main foliation (S_2) of the granulite gneisses and micaschists. These granites range in size from kilometre- (Mittweida granite) to decimetre-scale dykes. Monazite in these granites partly exhibits strong oscillatory and sector zoning, as typical of igneous monazite (Schulz 2021*a*), and is found as xenomorphic and euhedral grains (Fig. 7g, h). Their isochrones vary from 345 $\pm$ 4 Ma (FS30) through 340 $\pm$ 2 Ma (SGM4) to 335 $\pm$ 1.1 Ma (TRO3), thus mostly older than the monazite isochrones in the surrounding gneisses and the CGU (Supplementary material 6; Fig. 1k–m). The Th + U v. Ca in igneous monazite from various granites in the SGM differ considerably: sample FS30 with the oldest isochrone has numerous monazite with intermediate Th + U at low Ca, whereas in TRO3 monazite strictly follows the cheralite substitution trend. This chemical trend plots towards cheralite substitution with progressively younger ages in the granitoid monazites (Fig. 9g). It appears that the monazite ages from the granite intrusions crosscutting the metamorphic structures are significantly older than those in the foliated host rocks, especially in the case of the CGU (Fig. 11). This could be explained as follows: (1) some of the monazites in the granites are significantly older (>350 Ma) than the main population that defines the isochrone (345 $\pm$ 4 Ma, sample FS30). This suggests the presence of inherited monazite grains. (2) Apparently, granitic melts were generated by a partial anatexis of granulitic crust. The granitic melts should have intruded and crystallized during the decompression paths of CGU and Granulite Massif core rocks at variable XH_2O, e.g. for nearly dry melts with XH_2O at 0.1–0.5 at around 800°C (Johannes and Holtz 1996). As the post-800°C host rock cooling paths evolved in the monazite stability field at low pressure, a significant amount of monazite in host rocks should have crystallized subsequent to the granite intrusions (Fig. 4f).

Several post-tectonic granitoids that are partly associated with greisen mineralizations are observed in the Erzgebirge (Förster 1998; Förster and Romer 2010). A monazite isochrone from the Eibenstock Granite plots at 321 $\pm$ 2 Ma. Three core samples from the Eichigt Granite drill hole to the SW yielded 313 $\pm$ 8 Ma, 309 $\pm$ 4 Ma and 307 $\pm$ 4 Ma, with overlap of the errors. In the Zinnwald Greisen Granite sample ZG2-26, a fairly well-defined isochrone at 311 $\pm$ 10 Ma can be recognized (Supplementary material 6; Fig. 2k–m). Needle-shaped monazite (Fig. 7n) may relate to crystallization during hydrothermal fluid circulation post-granite intrusion. In the other greisen granite samples (ZIN2-4A, ZTHG11) monazite is closely related to fluorite and sometimes displays excess Pb. Furthermore, in Th + U v. Ca coordinates, Zinnwald Greisen Granite monazite grains display substantial different compositions as observed from the SGM granites. The SGM granites, especially sample TRO3, follow a cheralite compositional trend, in contrast to the Zinnwald and Eichigt granites which show low Th + U and low Ca, and Eibenstock Granite (POL12) with elevated Th + U (Fig. 9g, h). The Cambrian monazite in the granitoid orthogneisses displays considerably lower Ca and no significant cheralite substitution trend (Fig. 9i). Monazite isochrones from the Erzgebirge granites postdate isochrones in their metamorphic host rocks, in contrast to the SGM granites (Fig. 10). However, in the Eichigt and Zinnwald granites some monazite analyses also give pre-Carboniferous ages. This may be caused by inherited monazite and/or the incorporation of common Pb during monazite crystallization contemporaneously to the greisenization process.

As expected in anatectic S-type peraluminous granites (Barbarin 1999), monazite is abundant in the Saxothuringian Carboniferous granite plutons. Monazite from Carboniferous granitoids is occasionally used as reference material for controls of EPMA Th–U–Pb monazite dating (Montel *et al.* 1996; Cocherie *et al.* 1998, 2005). One reason for that is U–Pb dating of the commonly associated zircon by various isotope methods to provide independent controls on the monazite age, assuming that no inherited zircon or parts of such have been analysed (Crowley *et al.* 2008). Also, monazite from such granitoids may be dated by U–Pb isotope analysis and evaluated for their appropriate potential to serve as reference material (Förster *et al.* 1999; Kempe *et al.* 2004; Förster and Romer 2010; Tichomirowa *et al.* 2019*a*, *b*). Despite well-defined isochrones, these granites may contain inherited or partly recrystallized monazite grains in some cases (Supplementary material 5; Figs 1k–m, 2k–m & 3a–c). The statistical definition of the isochrones may be low or hampered by excess or loss of common Pb (Schulz and Schüssler 2013). For controls of EPMA Th–U–Pb monazite dating, it should be possible to check the isochrone of the reference monazite by only a few single analyses during a single session. This requires that the ThO_2*–PbO analyses of the reference monazite grains should strictly and closely define an isochrone. This is given only in the cases of the TRO3 and Eibenstock (POL12) granites (Supplementary material 5; Figs 1m & 2l). However, due to their highly variable PbO and ThO_2 along the isochrone, such granitoid monazites are not eligible for the indispensable contemporaneous control of ThO_2, as it is given by the *Madmon* reference monazite used in this study (Schulz and Schüssler 2013).

Late Carboniferous (310–304 Ma) hydrothermal event

It is well known from experiments and natural observations that the crystallization, recrystallization and dissolution–reprecipitation of monazite is driven by metasomatic alkali-bearing, metamorphic and hydrothermal fluids (e.g. Poitrasson *et al.* 1996; Spear and Pyle 2002; Hetherington *et al.* 2010; Harlov *et al.* 2011; Williams *et al.* 2011). Fluid flow in shear zones may induce crystallization of monazite in distinct microstructures (Oriolo *et al.* 2018, 2022). Also, the intrusion of pegmatites into crustal domains can be accompanied by pervasive monazite crystallization in metamorphic host rocks (Schulz and Krause 2021). In the Erzgebirge GPU the micaschist ore samples from the Pöhla visitor mine show markedly younger monazite isochrones at 310–304 Ma (Figs 8h, i, 10 & 11). These ages fall within the range of previously reported 308–295 Ma U–Pb ages from type-C garnet-skarns (Burisch *et al.* 2019). Also, low-ThO_2 monazite occurs in some of the micaschist ores, typical for hydrothermal monazite (Poitrasson *et al.* 1996; Schandl and Gorton 2004). These observations indicate a distinct Late Carboniferous (310–304 Ma) hydrothermal event. This stage overlaps in time with monazite crystallization in the Zinnwald and Eichigt granites (Fig. 10; Supplementary material 5; Fig. 2k, m). In some micaschist samples from the Münchberg Massif a Late Carboniferous (313–302 Ma) isochrone is also observed. In contrast to the micaschist ore samples, the young monazite grains are scarce and dispersed within the older Devonian (389–366 Ma) crystals and do not define a single population (Supplementary material 5; Fig. 3f–h). Also, distinct hydrothermal ore minerals are lacking there. As a consequence, the young monazite ages in the Münchberg Massif may have documented a more disperse hydrothermal infiltration or a thrusting event with enhanced fluid activity.

Conclusions to the geodynamic evolution in the Saxothuringian Zone

A combined approach of EPMA Th–U–Pb monazite dating and microstructurally controlled geothermobarometry of garnet-bearing assemblages in metapsammopelites allowed us to reveal (1) Cambrian metamorphism (498 $\pm$ 19 Ma), (2) Silurian ages with an uncertain interpretation as a weak metamorphism, a dominant (3) Devonian to (4) Carboniferous metamorphism at various grades up to UHP and UHT conditions, and a local (5) Late Carboniferous hydrothermal overprint (Figs 10 & 11). Combining metamorphic P–T path segments of garnet crystallization in reference to the monazite stability field, the monazite microstructures and its mineral chemistry, allowed at least a partial temporal assignment of various stages of the metamorphic evolution for the Saxothuringian Zone during the Variscan Orogeny. These P–T time paths are arranged in a schematic NW–SE crustal section of the Saxothuringian Zone to the north of the Tertiary Eger Graben, following Martínez-Catalán *et al.* (2021). In this crustal section, the SGM, the SC, Upper Cambrian to Devonian deposits and the MP-MT GAU appear as the lowermost units. The Erzgebirge units with the UHP GEU at the base overlie these units. They all underwent a metamorphism in Carboniferous times. The Upper Allochthon with Münchberg and Frankenberg massifs constitutes the top of the pile and represents an early and Devonian stage of the evolution (Figs 11 & 12).

In the SGM, the CGU underwent clockwise prograde–retrograde P–T paths with maximum pressures at 700°C/8 kbar, followed by maximum temperatures at 750–800°C/4 kbar, and cooling at pressures below 4 kbar. The ThO_2*–PbO isochrones of monazite enclosed in garnets range from 339 to 332 Ma and are slightly older than those from the matrix monazites with 335–317 Ma (Figs 10, 11 & 12a). The monazite isochrones from felsic granulites in the massif core range from 356 to 338 Ma and signal an initiation of the metamorphism in the Lower Carboniferous. In the SC the monazite isochrones range from 372 to 335 Ma and thus indicate an even earlier initiation of MP metamorphism. Monazite isochrones in post-tectonic granitoid intrusions are apparently older (345–335 Ma) than those in many host rocks (Figs 10, 11 & 12e). This may be explained by (1) inherited older grains in the granites and (2) by continuing monazite crystallization in the host rocks during the post-intrusion cooling history.

The overall allochthonous character of the Erzgebirge UHP rocks in the GEU and its overlying HP units MEU and GPU is established by the geothermobarometry in the underlying Freiberg Gneiss Dome of the GAU. A prograde P–T path reached

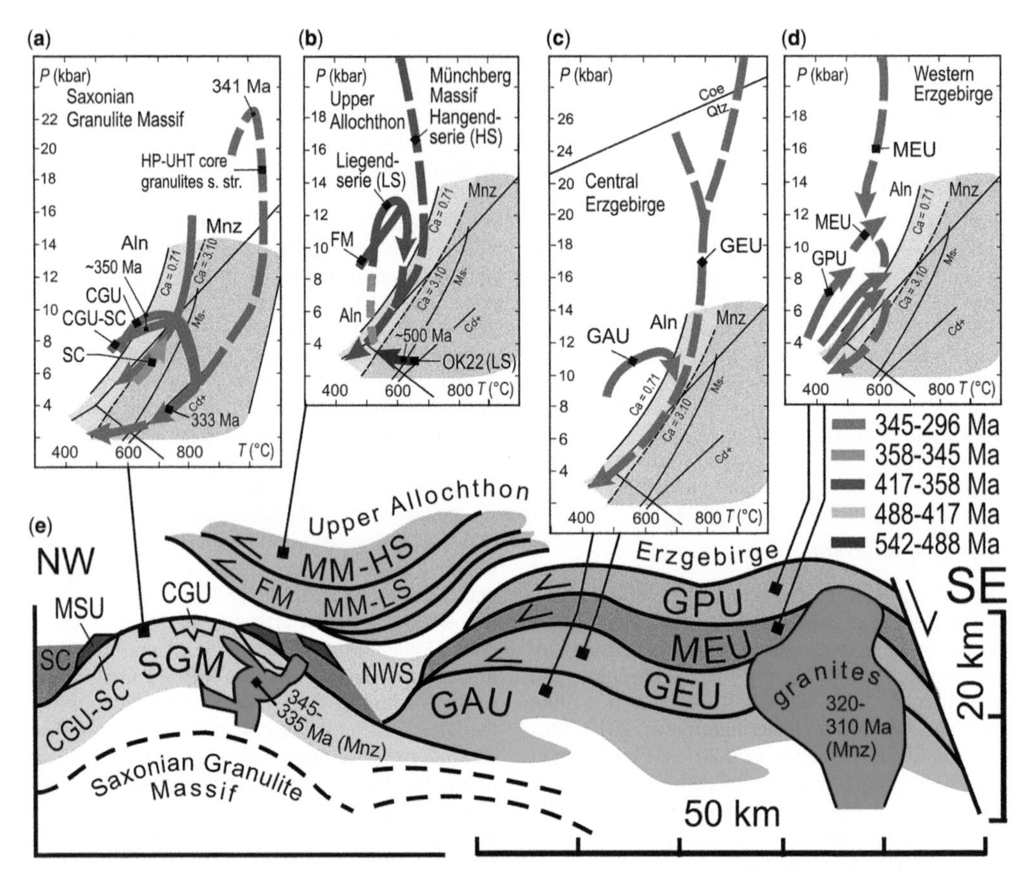

Fig. 12. Summary of pressure–temperature–time paths of Saxothuringian units in a schematic NW–SE cross-section of the crustal structure. P–T path sections from garnet assemblage geothermobarometry (this study) in full lines; P–T path sections given by other authors are shown with hatched lines. Time scale is given in colours: light brown – Carboniferous (345–296 Ma); green – Lower Carboniferous (358–345 Ma); blue – Devonian(417–358 Ma); light blue – Ordovician–Silurian (488–417 Ma); dark brown – Cambrian (540–488 Ma). The aluminosilicates, muscovite-out (Ms-) and cordierite-in (Cd+) univariant lines are after Spear (1993). Stability fields of monazite (Mnz) and allanite (Aln) at different bulk rock contents as a function of Ca wt% (Janots *et al.* 2007; Spear 2010). (**a**) P–T paths in the Saxonian Granulite Massif with core granulites (hatched), Cordierite Gneiss-Unit (CGU) and Schist Cover (SC); for details see Figure 4f. (**b**) Upper Allochthon with Münchberg Massif Hangendserie (MM-HS) and Liegendserie (MM-LS) and Frankenberg Massif (FM), with Devonian ages. Eclogite decompression path (hatched). Sample OK22 displays a Cambrian monazite population. For details see Figure 6c and d. (**c**) Central Erzgebirge with Carboniferous UHP Gneiss-Eclogite Unit (GEU) upon the MP–MT Gneiss-Amphibolite Unit (GAU). For details see Figure 5c. (**d**) Western Erzgebirge with prograde Carboniferous P–T paths reconstructed from metapsammopelites and decompression paths of eclogites in the Micaschist Eclogite Unit (MEU) and Garnet Phyllite Unit (GPU); for details see Figure 5d. (**e**) Crustal structure of the Saxothuringian Zone in a schematic cross-section, including the ophiolitic Metagabbro Serpentinite Unit (MSU) in the Schist Cover, as presented by Martínez-Catalán *et al.* (2021, and references therein), with addition of post-tectonic granite intrusions. Colours of units correspond to legend in Figure 2.

there maximal 680°C/11 kbar and a monazite isochrone at 320 Ma refers to subsequent decompression and cooling (Figs 11 & 12c). Garnet in the UHP GEU of the Central Erzgebirge encloses polycrystalline quartz aggregates as relicts of coesite (O'Brien and Ziemann 2008). Garnet-bearing assemblages crystallized at granulite facies conditions of 800°C/20–16 kbar, interpreted as a stopover event during the decompression path of the rocks from the earlier ultra-high-pressure conditions. Monazite crystallization between 336 and 327 Ma was related to the post-granulitic decompression path within the monazite stability field for low-Ca and high-Al bulk rock compositions (Figs

11 & 12c). Metapelite garnet in the MEU of Central Erzgebirge crystallized along prograde P–T paths towards 650°C/8 kbar. Prograde as well as decompression paths are in the monazite stability field only for Ca-poor bulk rock compositions. Correspondingly, the older monazite isochrones between 351 and 334 Ma refer to the initiation and prograde metamorphism and the younger isochrones belong to the decompression and cooling path (Figs 11 & 12d). In the Westerzgebirge with the GPU in the hanging wall of the MEU, garnet also recorded prograde P–T paths towards 520°C/8 kbar, which is outside the monazite stability field. Monazite isochrones range from 356 to 316 Ma. As in the Central Erzgebirge, the younger isochrones should refer to the decompression and cooling paths. A local hydrothermal overprint led to monazite recrystallization at 310–304 Ma in micaschist ore. Monazite mineral chemistry in the post-tectonic Erzgebirge granites differs from those in the SGM, and the monazite isochrones are markedly younger at 320–310 Ma (Fig. 12e).

In the upper allochthonous units with the Münchberg Massif Liegendserie (MM-LS) and the Frankenberg Massif (FM), monazite isochrones at 512–503 Ma refer to the protolith ages of Cambrian S-type granitoids, now orthogneisses (Fig. 11). Garnet crystallization in the Liegendserie micaschists occurred during a prograde metamorphism at 500–630°C and 9–12.5 kbar. Similar maximal P–T conditions can be reported from the FM. Monazite crystallization between 389 and 366 Ma (MM-LS) and slightly younger in the FM (362–360 Ma) refers to the decompression paths in these massifs. An apparent conflict of both Cambrian and Devonian monazite isochrones in similar garnet-bearing micaschists could be explained by a local Cambrian contact metamorphism in the vicinity of the granitoids (Figs 11 & 12b).

In samples with a high mode of garnet, as in the high-grade gneisses of the CGU in the SGM, and in the Erzgebirge GEU, mostly low Y and HREE with poor zonations are observed. In contrast, micaschist samples from the Erzgebirge MEU and GPU with lower mode of garnet display strongly zoned garnet with high Y and HREE in the cores. There are also micaschists with poorly zoned and low Y and HREE garnet, as in the Münchberg Massif. Apparently, the Y in monazite is not controlled by metamorphic temperature in garnet-bearing metapelites, but by the abundance and resorption of garnet (e.g. Larson *et al.* 2022). This is exemplified in the SGM where monazite in garnet-free micaschists of the SC have higher Y than monazite in the CGU garnet gneisses. In the micaschists of the Münchberg Massif it was possible to distinguish Y-rich Cambrian and Y-poor Devonian monazite populations, which are separated by a period of low Y garnet crystallization at high pressures (Figs 11 & 12b). However, the *in situ* Lu–Hf and U–Pb dating of metapelite garnet crystallization (e.g. Millonig *et al.* 2020; Simpson *et al.* 2021) may allow better resolution of garnet and monazite relationships in the future.

In many cases, the EPMA Th–U–Pb isochrones from the dominant monazite populations are similar or within the range of K–Ar and Ar–Ar hornblende and mica cooling ages (Figs 10 & 11). In the FM, the monazite age distribution maximum (370–360 Ma) slightly postdates the cooling ages (380–370 Ma; Werner and Reich 1997). In contrast, monazite age maxima in the SC (370–360 Ma) and SGM core gneisses (360–345 Ma) predate the cooling ages (~335–315 Ma). As reported by Rötzler and Timmerman (2021), there also exists an older group (353–345 Ma) of cooling ages in the SC. This indicates that the crystallization of monazite was approximately related to the decompression or uplift and cooling section of the P–T paths. However, this coincidence does not imply that the concept of cooling age and closure temperature as for hornblende and mica can be applied to the interpretation of the monazite ages. This coincidence of ages is merely a consequence that the uplift and cooling paths enter and pass the related stability field of a monazite crystallization. During uplift and cooling, fluids potentially enter the crust, leading to retrogression of peak metamorphic mineral assemblages. These fluids may also have enhanced the dissolution–reprecipitation and crystallization of monazite. This leads to a variation and time span of the monazite ages which is obvious in the ThO_2* v. PbO diagrams.

When interpreted in terms of garnet crystallization along P–T path sections, monazite stability fields at different bulk-rock Ca and Al compositions, mineral chemistry and isochrones, the EPMA Th–U–Pb monazite ages considerably enlarge the methodological spectrum of metamorphic age dating studies. This is evident when comparing the cooling ages to the complete monazite age distribution pattern (Figs 10 & 11). Apart from the major metamorphic events, the monazite data allow local features to be detected, like the Cambrian metamorphism and the Late Carboniferous hydrothermal event. However, there may also appear monazite ages like the widespread but subordinate Silurian population of which interpretation is doubtful and which possibly indicate a weak metamorphism. Several studies have shown that metamorphic units, in addition to their P–T paths, are characterized by a distinct EPMA Th–U–Pb monazite age distribution pattern in their metapelites (Schulz 2013, 2017, 2021*b*; Will *et al.* 2017). The recognition of such an age pattern requires a campaign-style investigation of numerous samples but will considerably enlarge the limited ranges of the K–Ar and Ar–Ar cooling age distributions.

Since the first discovery of the coesite and microdiamond-bearing UHP rocks by Massonne (2001, and references therein), the Saxothuringian Zone can be considered as a test site for models of UHP terrane exhumation. In an early approach Willner *et al.* (2002) stated that the Erzgebirge is composed of a complicated pile of tectonometamorphic units of very different size which were juxtaposed at very different depths. Accordingly, the UHP rocks were transported during the Carboniferous from a deep-sited crustal root toward the surface in an exhumation channel immediately after the delamination of the lithospheric mantle in a subduction–collision scenario (Willner *et al.* 2002; Hallas *et al.* 2021). Such a Carboniferous crustal root at depth would require the occurrence of a crustal thickening process already during the Devonian, as can be seen in the Upper Allochthon. Considerably more sophisticated general models were developed by Gerya and Stöckhert (2006) and Massonne *et al.* (2007*b*). These considered not only the UHP domain itself, but also the very different prograde and retrograde P–T–time paths in the neighbouring units in their distinctive tectonic positions within a subduction scenario. The tectonic positions of the Saxothuringian units after multistage subduction and continental collision can be seen exemplified in Figure 12. Models of UHP terrane exhumation are fundamental for constraining crustal evolution, as stated by Hacker and Gerya (2013). However, in their list of UHP terrane exhumation mechanisms, a regional example for channel flow is unknown and not given. Also, the coupled MP, HP, UHP and UHT Saxothuringian units have not been mentioned as a natural example (Hacker and Gerya 2013). To be meaningful, future UHP terrane exhumation models must reproduce the coupled P–T–time–deformation–space paths that have been reconstructed from the various domains, as in the Saxothuringian Zone.

Acknowledgements Colleagues U. Schüssler (University of Würzburg/Germany) and M. Lapp (Landesamt für Landwirtschaft, Umwelt und Geologie Sachsen) contributed with rock samples from collections. Numerous rock thin sections were prepared by R. Würkert and M. Stoll at the Helmholtz Institute Freiberg for Resource Technology. Technical support during electron probe microanalyses at the Institute of Materials Science, TU Bergakademie Freiberg/Saxony (INST-FUGG 267/156-1) were provided by D. Heger and A. Treichel. At the LA-ICP-MS laboratory of the Mineralogisches Institut at Westfälische Wilhelms-Universität Münster, Germany, J. Berndt-Gerdes supported the garnet trace element analyses. The SEM-based Automated Mineralogy studies in the Geometallurgy Laboratory at TU Bergakademie Freiberg were assisted by S. Gilbricht. Detailed and extensive comments by K. Cutts, J. Walters and several anonymous reviewers, and the profound editorial efforts by S. Volante are gratefully acknowledged and led to improvement and finalization of the manuscript.

Competing interests The authors declare that they have no known competing financial interests or personal relationships that could have appeared to influence the work reported in this paper.

Author contributions **BS**: conceptualization (lead), data curation (supporting), formal analysis (equal), funding acquisition (lead), investigation (lead), methodology (lead), project administration (lead), writing – original draft (lead), writing – review & editing (lead); **JK**: conceptualization (supporting), data curation (lead), formal analysis (equal), funding acquisition (supporting), resources (equal), validation (equal), writing – original draft (supporting), writing – review & editing (supporting).

Funding The sampling in the Saxothuringian Zone and subsequent analytical studies of B. Schulz and J. Krause in the Saxothuringian Zone were financed by grants from the Deutsche Forschungsgemeinschaft DFG (SCHU 676/21; KR-4549/2).

Data availability All data generated or analysed during this study are included in this published article (and, if present, its supplementary information files).

References

Barbarin, B. 1999. A review of the relationships of granitoid types, their origins and their geodynamic environments. *Lithos*, **46**, 605–626, https://doi.org/10.1016/s0024-4937(98)00085-1

Barnes, C.J., Majka, J. *et al.* 2021. Using Th–U–Pb geochronology to extract crystallization ages of Paleozoic metamorphic monazite contaminated by initial Pb. *Chemical Geology*, **582**, 120450, https://doi.org/10.1016/j.chemgeo.2021.120450

Bidgood, A., Parsons, A.J., Lloyd, G.E., Waters, D. and Goddard, R.M. 2020. EBSD-based criteria for coesite–quartz transformation. *Journal of Metamorphic Geology*, **39**, 165–180, https://doi.org/10.1111/jmg.12566

Bingen, B., Demaiffe, D. and Hertogen, J. 1996. Redistribution of rare earth elements, thorium, and uranium over accessory minerals in the course of amphibolite to granulite facies metamorphism; the role of apatite and monazite in orthogneisses from south western Norway. *Geochimica Cosmochimica Acta*, **60**, 1341–1354, https://doi.org/10.1016/0016-7037(96)00006-3

Breithaupt, A. 1829. Über den Monazit, eine neue Specie des Mineral-Reichs. *Journal für Chemie und Physik*, **55**, 301–303.

Burisch, M., Gerdes, A, Meinert, L.D., Albert, R., Seifert, T. and Gutzmer, J. 2019. The essence of time – fertile skarn formation in the Variscan Orogenic Belt. *Earth and Planetary Science Letters*, **519**, 165–170, https://doi.org/10.1016/j.epsl.2019.05.015

Cherniak, D.J., Watson, E.B., Grove, M. and Harrison, T.M. 2004. Pb diffusion in monazite: a combined RBS/SIMS study. *Geochimica Cosmochimica Acta*, **68**, 829–840, https://doi.org/10.1016/j.gca.2003.07.012

Cocherie, A., Legendre, O., Peucat, J.J. and Koumelan, A.N. 1998. Geochronology of polygenetic monazites constrained by in-situ microprobe Th–U–total lead determination: implications for lead behaviour in monazite. *Geochimica et Cosmochimica Acta*, **62**, 2475–2497, https://doi.org/10.1016/S0016-7037(98)00171-9

Cocherie, A., Be Mézème, E., Legendre, O., Fanning, C.M., Faure, M. and Rossi, P. 2005. Electron microprobe dating as a tool for determining the closure of Th–U–Pb systems in migmatitic monazites. *American Mineralogist*, **90**, 607–618, https://doi.org/10.2138/am.2005.1303

Collett, S., Štípská, P., Kusbach, V., Schulmann, K. and Marciniak, G. 2016. Dynamics of Saxothuringian subduction channel/wedge constrained by phase-equilibria modelling and micro-fabric analysis. *Journal of Metamorphic Geology*, **35**, 253–280, https://doi.org/10.1111/jmg.12226

Collett, S., Schulmann, K., Štípská, P. and Míková, J. 2020. Chronological and geochemical constraints on the pre-Variscan tectonic history of the Erzgebirge, Saxothuringian Zone. *Gondwana Research*, **79**, 27–48, https://doi.org/10.1016/j.gr.2019.09.009

Collett, S., Schulmann, K. *et al.* 2022. Reconstruction of the mid-Devonian HP–HT metamorphic event in the Bohemian Massif (European Variscan belt). *Geoscience Frontiers*, **13**, 101374, https://doi.org/10.1016/j.gsf.2022.101374

Crowley, J.L., Brown, R.L., Gervais, F. and Gibson, H.D. 2008. Assessing inheritance of zircon and monazite in granitic rocks from the Monashee complex, Canadian Cordillera. *Journal of Petrology*, **49**, 1915–1929, https://doi.org/10.1093/petrology/egn047

Cutts, J. and Dyck, B. 2023. Incipient collision of the Rae and Slave cratons at ca. 1.95 Ga. *Geological Society of America Bulletin*, **135**, 903–914, https://doi.org/10.1130/B36393.1

Donovan, J.J., Hanchar, J.M., Picolli, P.M., Schrier, M.D., Boatner, L.A. and Jarosewich, E. 2003. A re-examination of the rare-earth-element orthophosphate standards in use for electron-microprobe analysis. *Canadian Mineralogist*, **41**, 221–232, https://doi.org/10.2113/gscanmin.41.1.221

Faryad, S.W. and Kachlík, V. 2013. New evidence of blueschist facies rocks and their geotectonic implications for Variscan suture(s) in the Bohemian Massif. *Journal of Metamorphic Geology*, **31**, 63–82, https://doi.org/10.1111/jmg.12009

Finger, F., Broska, I., Roberts, M. and Schermaier, A. 1998. Replacement of primary monazite by allanite-epidote coronas in an amphibolite-facies granite gneiss from the eastern Alps. *American Mineralogist*, **83**, 248–258, https://doi.org/10.2138/am-1998-3-408

Förster, H.-J. 1998. The chemical composition of REE–Y–Th–U-rich accessory minerals in peraluminous granites of the Erzgebirge-Fichtelgebirge region, Germany, Part I: the monazite-(Ce)–brabantite solid solution series. *American Mineralogist*, **83**, 259–272, https://doi.org/10.2138/am-1998-3-409

Förster, H.J. and Romer, R.L. 2010. Carboniferous magmatism. *In*: Linnemann, U., Kroner, U. and Romer, R.L. (eds) *Pre-Mesozoic Geology of Saxo-Thuringia – From the Cadomian Active Margin to the Variscan Orogen*. Schweizerbart, Stuttgart, 287–310.

Förster, H.-J., Tischendorf, G., Trumbull, R.B. and Gottesmann, B. 1999. Late-Collisional granites in the Variscan Erzgebirge, Germany. *Journal of Petrology*, **40**, 1613–1645, https://doi.org/10.1093/petroj/40.11.1613

Franke, W. 2000. The mid-European segment of the Variscides: tectonostratigraphic units, terrane boundaries and plate tectonic evolution. *Geological Society, London, Special Publications*, **179**, 35–61, https://doi.org/10.1144/GSL.SP.2000.179.01.05

Franke, W. 2006. The Variscan orogen in Central Europe: construction and collapse. *Geological Society, London, Memoirs*, **32**, 333–343, https://doi.org/10.1144/GSL.MEM.2006.032.01.20

Franke, W. and Stein, E. 2000. Exhumation of high-grade rocks in the Saxo-Thuringian Belt; geological constraints and geodynamic concepts. *Geological Society, London, Special Publications*, **179**, 337–354, https://doi.org/10.1144/GSL.SP.2000.179.01.20

Franke, W., Cocks, L.R.M. and Torsvik, T.H. 2017. The Palaeozoic Variscan oceans revisited. *Gondwana Research*, **48**, 257–284, https://doi.org/10.1016/j.gr.2017.03.005

Franz, G., Thomas, S. and Smith, D.C. 1986. High-pressure decomposition in the Weissenstein eclogite, Münchberg Gneiss Massif, Germany. *Contributions to Mineralogy and Petrology*, **92**, 71–85, https://doi.org/10.1007/BF00373964

Franz, G., Andrehs, G. and Rhede, D. 1996. Crystal chemistry of monazite and xenotime from Saxothuringian–Moldanubian metapelites, NE Bavaria, Germany. *European Journal of Mineralogy*, **8**, 1097–1118, https://doi.org/10.1127/ejm/8/5/1097

Gehmlich, M. 2003. Die Cadomiden und Varisziden des Saxothuringischen Terranes – Geochronologie magmatischer Ereignisse. *Freiberger Forschungshefte Reihe C*, **500**, 1–129.

Gerya, T.V. and Stöckhert, B. 2006. Two-dimensional numerical modeling of tectonic and metamorphic histories at active continental margins. *International Journal Earth Sciences*, **95**, 250–274, https://doi.org/10.1007/s00531-005-0035-9

Gose, J. and Schmädicke, E. 2022. H_2O in omphacite of quartz and coesite eclogite from Erzgebirge and Fichtelgebirge, Germany. *Journal of Metamorphic Geology*, **40**, 665–686, https://doi.org/10.1111/jmg.12642

Gottesmann, B., Förster, H.-J., Müller, A.B. and Kämpf, H. 2017. The concealed granite massif of Eichigt–Schönbrunn (Vogtland, Germany): petrography, mineralogy, geochemistry and age of the Eichigt apical intrusion. *Freiberg Online Geoscience*, **49**, 1–46.

Gratz, R. and Heinrich, W. 1997. Monazite–xenotime thermobarometry: experimental calibration of the miscibility gap in the binary system $CePO_4$-YPO_4. *American Mineralogist*, **82**, 772–780, https://doi.org/10.2138/am-1997-7-816

Hacker, B.R. and Gerya, T.V. 2013. Paradigms, new and old, for ultrahigh-pressure tectonism. *Tectonophysics*, **603**, 79–88, https://doi.org/10.1016/j.tecto.2013.05.026

Hallas, P., Pfänder, J.A., Kroner, U. and Sperner, B. 2021. Microtectonic control of $^{40}Ar/^{39}Ar$ white mica age distributions in metamorphic rocks (Erzgebirge, N-

Bohemian Massif): constraints from combined step heating and multiple single grain total fusion experiments. *Geochimica et Cosmochimica Acta*, **314**, 178–208, https://doi.org/10.1016/j.gca.2021.08.043

Harlov, D.E., Wirth, R. and Hetherington, C.J. 2011. Fluid-mediated partial alteration in monazite: the role of coupled dissolution–reprecipitation in element redistribution and mass transfer. *Contributions to Mineralogy and Petrology*, **162**, 329–348, https://doi.org/10.1007/s00410-010-0599-7

Harrison, T.M., Catlos, E.J. and Montel, J.-M. 2002. U–Th–Pb dating of phosphate minerals. *Reviews in Mineralogy and Geochemistry*, **48**, 524–558, https://doi.org/10.2138/rmg.2002.48.14

Heinrich, W., Andrehs, G. and Franz, G. 1997. Monazite–xenotime miscibility gap thermometry. I. An empirical calibration. *Journal of Metamorphic Geology*, **15**, 3–16, https://doi.org/10.1111/j.1525-1314.1997.t01-1-00052.x

Hermann, J. and Rubatto, D. 2003. Relating zircon and monazite domains to garnet growth zones: age and duration of granulite facies metamorphism in the Val Malenco lower crust. *Journal of Metamorphic Geology*, **21**, 833–852, https://doi.org/10.1046/j.1525-1314.2003.00484.x

Hetherington, C.J., Harlov, D.E. and Budzyń, B. 2010. Experimental initiation of dissolution-reprecipitation reactions in monazite and xenotime: the role of fluid composition. *Mineralogy and Petrology*, **99**, 165–184, https://doi.org/10.1007/s00710-010-0110-1

Höhn, S., Koglin, N. *et al.* 2017. Geochronology, stratigraphy and geochemistry of Cambro-Ordovician, Silurian and Devonian volcanic rocks of the Saxothuringian Zone in NE Bavaria (Germany) – new constraints for Gondwana break up and Rheic ocean island magmatism. *International Journal Earth Sciences*, **107**, 359–377, https://doi.org/10.1007/s00531-017-1497-2

Holland, T.J.B. and Powell, R. 2011. An improved and extended internally consistent thermodynamic dataset for phases of petrological interest, involving a new equation of state for solids. *Journal of Metamorphic Geology*, **29**, 333–383, https://doi.org/10.1111/j.1525-1314.2010.00923.x

Janots, E., Brunet, F., Goffe, B., Poinssot, C., Burchard, M. and Cemic, L. 2007. Thermochemistry of monazite-(La) and dissakisite (La): implications for monazite and allanite stability in metapelites. *Contributions to Mineralogy and Petrology*, **154**, 1–14, https://doi.org/10.1007/s00410-006-0176-2

Janots, E., Engi, M., Berger, A., Allaz, J., Schwarz, J.-O. and Spandler, C. 2008. Prograde metamorphic sequence of REE minerals in pelitic rocks of the Central Alps: implications for allanite–monazite–xenotime phase relations from 250 to 610°C. *Journal of Metamorphic Geology*, **26**, 509–526, https://doi.org/10.1111/j.1525-1314.2008.00774.x

Jarosewich, E. and Boatner, L.A. 1991. Rare-earth element reference samples for electron microprobe analysis. *Geostandards Newsletter*, **15**, 397–399, https://doi.org/10.1111/j.1751-908X.1991.tb00115.x

Jercinovic, M.J., Williams, M.L. and Lane, E.D. 2008. In-situ trace element analysis of monazite and other fine-grained accessory minerals by EPMA. *Chemical Geology*, **254**, 197–215, https://doi.org/10.1016/j.chemgeo.2008.05.016

Jochum, K.P., Stoll, B., Herwig, K. and Willbold, M. 2007. Validation of LA-ICP-MS trace element analysis of geological glasses using a new solid-state 193 nm Nd: YAG laser and matrix-matched calibration. *Journal of Analytical Atomic Spectrometry*, **22**, 112–121, https://doi.org/10.1039/b609547j

Jochum, K., Weis, U., Stoll, B., Kuzmin, D., Yang, Q., Raczek, I. and Enzweiler, J. 2011. Determination of reference values for NIST SRM 610-617 glasses following ISO guidelines. *Geostandards and Geoanalytical Research*, **35/4**, 397–429, https://doi.org/10.1111/j.1751-908X.2011.00120.x

Johannes, W. and Holtz, F. 1996. *Petrogenesis and experimental petrology of granitic rocks*. Minerals and Rocks Series. Springer-Verlag, Berlin, **22**.

Kempe, U., Bombach, K., Matukov, D., Schlothauer, T., Hutschenreuter, J., Wolf, D. and Sergeev, S. 2004. Pb/Pb and U/Pb zircon dating of subvolcanic rhyolite as a time marker for Hercynian granite magmatism and Sn mineralisation in the Eibenstock granite, Erzgebirge, Germany: considering effects of zircon alteration. *Mineralium Deposita*, **39**, 646–669, https://doi.org/10.1007/s00126-004-0435-y

Klemd, R 1989. P–T evolution and fluid inclusion characteristics of retrograded eclogites, Münchberg gneiss complex, Germany. *Contributions to Mineralogy and Petrology*, **102**, 221–229, https://doi.org/10.1007/BF00375342

Klemd, R. 2010. Early Variscan allochthonous domains: the Münchberg Complex, Frankenberg, Wildenfels, and Góry Sowie. *In*: Linnemann, U. and Romer, R.L. (eds) *Pre-Mesozoic Geology of Saxo-Thuringia – From the Cadomian Active Margin to the Variscan Orogen*. Schweizerbart, Stuttgart, 221–232.

Klemd, R. and Schmädicke, E. 1994. High-pressure metamorphism in the Münchberg Gneiss Complex and the Erzgebirge Crystalline Complex: the influence of fluid and reaction kinetics. *Chemie der Erde – Geochemistry*, **54**, 241–261.

Klemd, R., Matthes, S. and Okrusch, M. 1991. High-pressure relics in meta-sediments intercalated with the Weissenstein eclogite, Münchberg gneiss complex, Bavaria. *Contributions to Mineralogy and Petrology*, **107**, 328–342, https://doi.org/10.1007/BF00325102

Koglin, N., Zeh, A., Franz, G., Schüssler, U., Glodny, J., Gerdes, A. and Brätz, H. 2018. From Cadomian magmatic arc to Rheic ocean closure: the geochronological-geochemical record of nappe protoliths of the Münchberg Massif, NE Bavaria (Germany). *Gondwana Research*, **55**, 135–152, https://doi.org/10.1016/j.gr.2017.11.001

Kossmat, F. 1927. Gliederung des varistischen Gebirgsbaues. *Abhandlungen des Sächsischen Geologischen Landesamts*, **1**, 1–39.

Kreuzer, H., Seidel, E., Schüssler, U., Okrusch, M., Lenz, K.-L. and Raschka, H. 1989. K-Ar geochronology of different tectonic units at the northwestern margin of the Bohemian Massif. *Tectonophysics*, **157**, 149–178, https://doi.org/10.1016/0040-1951(89)90348-X

Kröner, A. and Willner, A.P. 1998. Time of formation and peak of Variscan HP–HT metamorphism of quartz-feldspar rocks in the central Erzgebirge, Saxony,

Germany. *Contributions to Mineralogy and Petrology*, **132**, 1–20, https://doi.org/10.1007/s004100050401

Kröner, A., Willner, A.P., Hegner, E., Frischbutter, A., Hofmann, J. and Bergner, R. 1995. Latest Precambrian (Cadomian) zircon ages, Nd isotopic systematics and P–T evolution of granitoid orthogneisses of the Erzgebirge, Saxony and Czech Republic. *Geologische Rundschau*, **84**, 437–456, https://doi.org/10.1007/s005310050016

Kröner, A., Jaeckel, P., Reischmann, T. and Kroner, U. 1998. Further evidence for an early Carboniferous (~340 Ma) age of high-grade metamorphism in the Saxonian granulite complex. *International Journal of Earth Sciences*, **86**, 751–766, https://doi.org/10.1007/PL00009939

Kroner, U. and Romer, R.L. 2010. The Saxo-Thuringian Zone – tip of the Armorican Spur and part of the Gondwana plate. *In*: Linnemann, U. and Romer, R.L. (eds) *Pre-Mesozoic Geology of Saxo-Thuringia – From the Cadomian Active Margin to the Variscan Orogen*. Schweizerbart, Stuttgart, 371–394.

Kroner, U. and Romer, R.L. 2013. Two plates – many subduction zones: the Variscan orogeny reconsidered. *Gondwana Research*, **24**, 298–329, https://doi.org/10.1016/j.gr.2013.03.001

Kroner, U., Hahn, T., Romer, R.L. and Linnemann, U. 2007. The Variscan orogeny in the Saxo-Thuringian Zone – Heterogenous overprint of Cadomian/Palaeozoic Peri-Gondwana crust. *Geological Society of America Special Paper*, **423**, 153–172, https://doi.org/10.1130/2007.2423(06)

Kroner, U., Romer, R.L. and Linnemann, U. 2010. The Saxo-Thuringian Zone of the Variscan Orogen as part of Pangea. *In*: Linnemann, U. and Romer, R.L. (eds) *Pre-Mesozoic Geology of Saxo-Thuringia – From the Cadomian Active Margin to the Variscan Orogen*. Schweizerbart, Stuttgart, 3–16.

Kroner, U., Stephan, T., Romer, R.L. and Roscher, M. 2020. Paleozoic plate kinematics during the Pannotia–Pangaea supercontinent cycle. *Geological Society, London, Special Publications*, **503**, 83–104, https://doi.org/10.1144/SP503-2020-15

Larson, K.P., Ali, A., Shrestha, S., Soret, M., Cottle, J.M. and Ahmad, R. 2019. Timing of metamorphism and deformation in the Swat valley, northern Pakistan: insight into garnet-monazite HREE partitioning. *Geoscience Frontiers*, **10**, 849–861, https://doi.org/10.1016/j.gsf.2018.02.008

Larson, K.P., Shrestha, S., Cottle, J.M., Guilmette, C., Johnson, T.A., Gibson, H.D. and Gervais, F. 2022. Re-evaluating monazite as a record of metamorphic reactions. *Geoscience Frontiers*, **13**, 101340, https://doi.org/10.1016/j.gsf.2021.101340

Lenze, A. and Stöckhert, B. 2008. Microfabrics of quartz formed from coesite (Dora-Maira Massif, Western Alps). *European Journal of Mineralogy*, **20**, 811–826, https://doi.org/10.1127/0935-1221/2008/0020-1848

Linnemann, U. and Romer, R.L. 2010. *Pre-Mesozoic Geology of Saxo-Thuringia. From the Cadomian active margin to the Variscan Orogen*. Schweizerbart, Stuttgart.

Linnemann, U., Hofmann, M., Romer, R.L. and Gerdes, A. 2010. Transitional stages between the Cadomian and Variscan orogenies: basin development and tectono-magmatic evolution of the southern margin of the Rheic Ocean in the Saxo-Thuringian Zone (North Gondwana shelf). *In*: Linnemann, U. and Romer, R.L. (eds) *Pre-Mesozoic Geology of Saxo-Thuringia – From the Cadomian Active Margin to the Variscan Orogen*. Schweizerbart, Stuttgart, 59–98.

Lorenz, W. and Hoth, K. 1990. Lithostratigraphie im Erzgebirge – Konzeption, Entwicklung, Probleme und Perspektiven. *Abhandlungen des Staatlichen Museums für Mineralogie und Geologie Dresden*, **37**, 7–35.

Ludwig, K.R. 2001. *Users Manual for Isoplot/Ex (rev. 2.49): A Geochronological Toolkit for Microsoft Excel*. Berkeley Geochronology Center, Special Publication, Los Angeles, No. 1a.

Machek, M., Soejono, I., Sláma, J. and Žáčková, E. 2021. Timing and kinematics of the Variscan orogenic cycle at the Moldanubian periphery of the central Bohemian Massif. *Journal of the Geological Society, London*, **179**, jgs2021-096, https://doi.org/10.1144/jgs2021-096

Manzotti, P., Bosse, V., Pitra, P., Robyr, M., Schiavi, F. and Ballèvre, M. 2018. Exhumation rates in the Gran Paradiso Massif (Western Alps) constrained by in situ U–Th–Pb dating of accessory phases (monazite, allanite and xenotime). *Contributions to Mineralogy and Petrology*, **173**, 24, https://doi.org/10.1007/s00410-018-1452-7

Martínez-Catalán, J.R., Collett, S., Schulmann, K., Aleksandrowski, P. and Mazur, S. 2020. Correlation of allochthonous terranes and major tectonostratigraphic domains between NW Iberia and the Bohemian Massif, European Variscan belt. *International Journal Earth Sciences*, **109**, 1105–1131, https://doi.org/10.1007/s00531-019-01800-z

Martínez-Catalán, J., Schulmann, K. and Ghienne, J.-F. 2021. The Mid-Variscan Allochthon: keys from correlation, partial retrodeformation and plate-tectonic reconstruction to unlock the geometry of a non-cylindrical belt. *Earth-Science Reviews*, **220**, 103700, https://doi.org/10.1016/j.earscirev.2021.103700

Massonne, H.-J. 2001. First find of coesite in the UHP metamorphic area of the central Erzgebirge, Germany. *European Journal of Mineralogy*, **13**, 565–570, https://doi.org/10.1127/0935-1221/2001/0013-0565

Massonne, H.-J. 2011. Pre-conference field trip of 9. International Eclogite Conference: Erzgebirge (Ore Mountains), Germany and Czech Republic; German part of the Saxonian Erzgebirge. *Geolines*, **23**, 29–59.

Massonne, H.-J. and Grosch, U. 1994. P–T evolution of Paleozoic garnet peridotites from the Saxonian Erzgebirge, Germany, and their metamorphic evolution. *Proceedings of the 6th International Kimberlite Conference*, Novosibirsk, 353–355.

Massonne, H.-J. and Nasdala, L. 2003. Characterization of an early metamorphic stage through inclusions in zircon of a diamondiferous quartzofeldspathic rock from the Erzgebirge, Germany. *American Mineralogist*, **88**, 883–889, https://doi.org/10.2138/am-2003-5-618

Massonne, H.-J. and O'Brien, P.J. 2003. The Bohemian Massif and the NW Himalaya. *In*: Carswell, D.A. and Compagnoni, R. (eds) *Ultrahigh Pressure Metamorphism*. EMU Notes in Mineralogy, **5**, 145–187.

Massonne, H.J., Kennedy, A., Nasdala, L. and Theye, Z. 2007*a*. Dating of zircon and monazite from diamondiferous quartzofeldspathic rocks of the Saxonian

Erzgebirge – hints at burial and exhumation velocities. *Mineralogical Magazine*, **71**, 371–389, https://doi.org/10.1180/minmag.2007.071.4.407

Massonne, H.J., Willner, A.P. and Gerya, T.V. 2007*b*. Densities of metapelitic rocks at high to ultrahigh pressure conditions: what are the geodynamic consequences? *Earth and Planetary Science Letters*, **256**, 12–27, https://doi.org/10.1016/j.epsl.2007.01.013

Meldrum, A., Boatner, L.A., Weber, W.J. and Ewing, R.C. 1998. Radiation damage in zircon and monazite. *Geochimica et Cosmochimica Acta*, **62**, 2509–2520, https://doi.org/10.1016/s0016-7037(98)00174-4

Millonig, L.J., Albert, R., Gerdes, A., Avigad, D. and Dietsch, C. 2020. Exploring laser ablation U–Pb dating of regional metamorphic garnet – the Straits Schist, Connecticut, USA. *Earth and Planetary Science Letters*, **552**, 116589, https://doi.org/10.1016/j.epsl.2020.116589

Mingram, B. and Rötzler, K. 1999. Geochemische, petrologische und geochronologische Untersuchungen im Erzgebirgskristallin – Rekonstruktion des Krustenstapels. *Schriftenreihe fur Geologische Wissenschaften*, **9**, 1–80.

Mingram, B., Kröner, A., Hegner, E. and Krentz, O. 2004. Zircon ages, geochemistry, and Nd isotopic systematics of pre-Variscan orthogneisses from the Erzgebirge, Saxony (Germany), and geodynamic interpretation. *International Journal Earth Sciences*, **93**, 706–727, https://doi.org/10.1007/s00531-004-0414-7

Montel, J.-M., Foret, S., Veschambre, M., Nicollet, C. and Provost, A. 1996. Electron microprobe dating of monazite. *Chemical Geology*, **131**, 37–51, https://doi.org/10.1016/0009-2541(96)00024-1

Nasdala, L., Akhmadaliev, S., Artac, A., Chanmuang, N.C., Habler, G. and Lenz, C. 2018. Irradiation effects in monazite–(Ce) and zircon: raman and photoluminescence study of Au-irradiated FIB foils. *Physics and Chemistry of Minerals*, **45**, 855–871, https://doi.org/10.1007/s00269-018-0975-9

O'Brien, P.J. and Ziemann, M.A. 2008. Preservation of coesite in exhumed eclogite: insights from Raman mapping. *European Journal of Mineralogy*, **20**, 827–834, https://doi.org/10.1127/0935-1221/2008/0020-1883

Okrusch, M., Seidel, E., Schüssler, U. and Richter, P. 1989. Geochemical characteristics of metabasites in different tectonic units of the northeast Bavarian crystalline basement. *In*: Emmermann, R. and Wohlenberg, J. (eds) *The German Continental Deep Drilling Program (KTB), Site-Selection Studies in the Oberpfalz and Schwarzwald*. Springer, Berlin, 67–79.

Okrusch, M., Matthes, S. and Schmidt, K. 1990. Eklogite der Münchberger Gneismasse. *Beihefte zum European Journal of Mineralogy*, **2**, 55–84, https://doi.org/10.1127/ejm/2/1/0055

Okrusch, M., Matthes, S., Klemd, R., O'Brien, P.J. and Schmidt, K. 1991. Eclogites at the northwestern margin of the Bohemian Massif: a review. *European Journal of Mineralogy*, **3**, 707–730, https://doi.org/10.1127/ejm/3/4/0707

Oriolo, S., Wemmer, K., Oyhantçabal, P., Fossen, H., Schulz, B. and Siegesmund, S. 2018. Geochronology of shear zones – a review. *Earth-Science Reviews*, **185**, 665–683, https://doi.org/10.1016/j.earscirev.2018.07.007

Oriolo, S., Schulz, B. *et al.* 2022. The petrologic and petrochronological record of progressive vs polyphase deformation: opening the analytical toolbox. *Earth-Science Reviews*, **234**, 104235, https://doi.org/10.1016/j.earscirev.2022.104235

Pälchen, W. 2011. *Geologie von Sachsen, Band I*, 2nd edn. Schweizerbart, Stuttgart.

Parrish, R.R. 1990. U–Pb dating of monazite and its application to geological problems. *Canadian Journal Earth Sciences*, **27**, 1431–1450, https://doi.org/10.1139/e90-152

Passchier, C.W. and Trouw, R.A.J. 2005. *Microtectonics*, 2nd edn. Springer, Heidelberg.

Pohlner, J.E., El Korh, A., Klemd, R., Grobéty, B., Pettke, T. and Chiaradia, M. 2021. Trace element and oxygen isotope study of eclogites and associated rocks from the Münchberg Massif (Germany) with implications on the protolith origin and fluid-rock interactions. *Chemical Geology*, **579**, 120352, https://doi.org/10.1016/j.chemgeo.2021.120352

Poitrasson, F., Chenery, S. and Bland, D.J. 1996. Contrasted monazite hydrothermal alteration mechanisms and their geochemical implications. *Earth and Planetary Science Letters*, **145**, 79–96, https://doi.org/10.1016/S0012-821X(96)00193-8

Powell, R. and Holland, T.J.B. 2008. On thermobarometry. *Journal of Metamorphic Geology*, **26**, 155–179, https://doi.org/10.1111/j.1525-1314.2007.00756.x

Pyle, J.M., Spear, F.S., Rudnick, R.L. and McDonough, W.F. 2001. Monazite–xenotime–garnet equilibrium in metapelites and a new monazite-garnet thermometer. *Journal of Petrology*, **42**, 2083–2107, https://doi.org/10.1093/petrology/42.11.2083

Pyle, J.M., Spear, F.S. and Wark, D.A. 2002. Electron microprobe analysis of REE in apatite, monazite, and xenotime: protocols and pitfalls. *Reviews in Mineralogy and Geochemistry*, **48**, 337–362, https://doi.org/10.2138/rmg.2002.48.8

Pyle, J.M., Spear, F.S., Cheney, J.T. and Layne, G. 2005*a*. Monazite ages in the Chesham Pond Nappe, SW New Hampshire, USA: implications for assembly of central New England thrust sheets. *American Mineralogist*, **90**, 592–606, https://doi.org/10.2138/am.2005.1341

Pyle, J.M., Spear, F.S., Wark, D.A., Daniel, C.G. and Storm, L.C. 2005*b*. Contributions to precision and accuracy of monazite microprobe ages. *American Mineralogist*, **90**, 547–577, https://doi.org/10.2138/am.2005.1340

Rahimi, G. and Massonne, H.-J. 2018. Pressure–temperature–time evolution of a Variscan garnet-bearing micaschist from the northeastern Fichtelgebirge, NW Bohemian Massif in central Europe. *Lithos*, **316–317**, 366–384, https://doi.org/10.1016/j.lithos.2018.07.023

Rahimi, G. and Massonne, H.-J. 2020. Metamorphic evolution of chloritoid-bearing micaschist from the Variscan Elstergebirge: evidences for stacking of high-pressure rocks in the Saxothuringian Zone of Central Europe. *Journal of Earth Science*, **31**, 425–446, https://doi.org/10.1007/s12583-020-1300-3

Reinhardt, J. and Kleemann, U. 1994. Extensional unroofing of granulitic lower crust and related low pressure, high-temperature metamorphism in the Saxonian Granulite Massif, Germany. *Tectonophysics*, **238**, 71–94, https://doi.org/10.1016/0040-1951(94)90050-7

Romer, R.L. and Rötzler, J. 2001. P–T–t evolution of ultrahigh-temperature granulites from the Saxon Granulite Massif, Germany. Part II: geochronology. *Journal of Petrology*, **42**, 2015–2032, https://doi.org/10.1093/petrology/42.11.2015

Rötzler, J. and Romer, R.L. 2001. P-T-t evolution of ultrahigh-temperature granulites from the Saxon Granulite Massif, Germany. Part I: petrology. *Journal of Petrology*, **42**, 1995–2013, https://doi.org/10.1093/petrology/42.11.1995

Rötzler, J. and Romer, R.L. 2010. The Saxonian Granulite Massif: a key area for the geodynamic evolution of Variscan central Europe. *In*: Linnemann, U. and Romer, R.L. (eds) *Pre-Mesozoic Geology of Saxo-Thuringia – From the Cadomian Active Margin to the Variscan Orogen*. Schweizerbart, Stuttgart, 233–252.

Rötzler, J. and Timmerman, M.J. 2021. Geochronological and petrological constraints from the evolution in the Saxon Granulite Massif, Germany, on the Variscan continental collision orogeny. *Journal of Metamorphic Geology*, **39**, 3–38, https://doi.org/10.1111/jmg.12559

Rötzler, J., Carswell, D.A., Gerstenberger, H. and Haase, G. 1999. Transitional blueschist-epidote amphibolite facies metamorphism in the Frankenberg massif, Germany, and geotectonic implications. *Journal of Metamorphic Geology*, **17**, 109–125, https://doi.org/10.1046/j.1525-1314.1999.00183.x

Rötzler, J., Romer, R.L., Budzinski, H. and Oberhänsli, R. 2004. Ultrahigh-temperature high-pressure granulites from Tirschheim, Saxon Granulite Massif, Germany: P–T–t path and geotectonic implications. *European Journal of Mineralogy*, **16**, 917–937, https://doi.org/10.1127/0935-1221/2004/0016-0917

Rötzler, J., Hagen, B. and Hoernes, S. 2008. Geothermometry of the ultrahigh-temperature Saxon granulites revisited. Part I: new evidence from key mineral assemblages and reaction textures. *European Journal of Mineralogy*, **20**, 1097–1115, https://doi.org/10.1127/0935-1221/2008/0020-1857

Rötzler, K. 1995. *Die P-T-Entwicklung der Metamorphite des Mittel- und Westerzgebirges*. Scientific Technical Reports, **STR95/14**. GeoForschungsZentrum Potsdam, Germany.

Rötzler, K. and Plessen, B. 2010. The Erzgebirge: A pile of ultrahigh- to low-pressure nappes of Early Paleozoic rocks and their Cadomian basement. *In*: Linnemann, U. and Romer, R.L. (eds) *Pre-Mesozoic Geology of Saxothuringia*. Schweizerbart, Stuttgart, 253–270.

Rötzler, K., Schumacher, R., Maresch, W.V. and Willner, A.P. 1998. Characterization and geodynamic implications of contrasting metamorphic evolution in juxtaposed high-pressure units of the Western Erzgebirge (Saxony, Germany). *European Journal of Mineralogy*, **10**, 261–280, https://doi.org/10.1127/ejm/10/2/0261

Rubatto, D., Hermann, J. and Buick, I.S. 2006. Temperature and bulk composition control on the growth of monazite and zircon during low-pressure anatexis (Mount Stafford, Central Australia). *Journal of Petrology*, **47**, 1973–1996, https://doi.org/10.1093/petrology/egl033

Sagawe, A., Gärtner, A., Linnemann, U., Hofmann, M. and Gerdes, A. 2016. Exotic crustal components at the northern margin of the Bohemian Massif – implications from U–Th–Pb and Hf isotopes of zircon from the Saxonian Granulite Massif. *Tectonophysics*, **681**, 234–249, https://doi.org/10.1016/j.tecto.2016.04.013

Salminen, P.E., Hölttä, P., Lahtinen, R. and Sayab, M. 2022. Monazite record for the Paleoproterozoic Svecofennian orogeny, SE Finland: an over 150-Ma spread of monazite dates. *Lithos*, **416–417**, 106654, https://doi.org/10.1016/j.lithos.2022.106654

Schandl, E.S. and Gorton, M.P. 2004. A textural and geochemical guide to the identification of hydrothermal monazite: criteria for selection of samples for dating epigenetic hydrothermal ore deposits. *Economic Geology*, **99**, 1027–1035, https://doi.org/10.2113/gsecongeo.99.5.1027

Schmädicke, E. 1991. Quartz pseudomorphs after coesite in eclogites from the Saxonian Erzgebirge. *European Journal of Mineralogy*, **3**, 231–238, https://doi.org/10.1127/ejm/3/2/0231

Schmädicke, E. 1994. Die Eklogite des Erzgebirges. *Freiberger Forschungshefte Reihe C*, **456**, 1–338.

Schmädicke, E., Okrusch, M. and Schmidt, W. 1992. Eclogite-facies rocks in the Saxonian Erzgebirge, Germany: high pressure metamorphism under contrasting P–T conditions. *Contributions to Mineralogy and Petrology*, **110**, 226–241, https://doi.org/10.1007/BF00310740

Schmädicke, E., Mezger, K., Cosca, M.A. and Okrusch, M. 1995. Variscan Sm–Nd and Ar–Ar ages of eclogite facies rocks from the Erzgebirge, Bohemian Massif. *Journal of Metamorphic Geology*, **13**, 537–552, https://doi.org/10.1111/j.1525-1314.1995.tb00241.x

Schmädicke, E., Will, T.M., Ling, X., Li, X.-H. and Li, Q.-L. 2018. Rare peak and ubiquitous post-peak zircon in eclogite: constraints for the timing of UHP and HP metamorphism in Erzgebirge, Germany. *Lithos*, **322**, 250–267, https://doi.org/10.1016/j.lithos.2018.10.017

Schönig, J., von Eynatten, H., Meinhold, G., Lünsdorf, N.K., Willner, A.P. and Schulz, B. 2020. Deep subduction of felsic rocks hosting UHP lenses in the central Saxonian Erzgebirge: implications for UHP terrane exhumation. *Gondwana Research*, **87**, 320–329, https://doi.org/10.1016/j.gr.2020.06.020

Schulz, B. 1993. P–T–deformation paths of Variscan metamorphism in the Austroalpine basement: controls on geothermobarometry from microstructures in progressively deformed metapelites. *Swiss Bulletin of Mineralogy and Petrology*, **73**, 257–274.

Schulz, B. 2009. EMP-monazite age controls on P–T paths of garnet metapelites in the Variscan inverted metamorphic sequence of La Sioule, French Massif Central. *Bulletin de la Société Géologique de la France*, **180**, 171–182, https://doi.org/10.2113/gssgfbull.180.3.271

Schulz, B. 2013. Monazite EMP–Th–U–Pb age pattern in Variscan metamorphic units in the Armorican Massif (Brittany, France). *Zeitschrift der Deutschen Gesellschaft für Geowissenschaften (German Journal of Geosciences)*, **164**, 313–335, https://doi.org/10.1127/1860-1804/2013/0008

Schulz, B. 2014. Early Carboniferous P–T path from the Upper Gneiss Unit of Haut-Allier (French Massif

Central) – reconstructed by geothermobarometry and EMP–Th–U–Pb monazite dating. *Journal of Geosciences*, **59**, 327–349, https://doi.org/10.3190/jgeosci.178

Schulz, B. 2017. Polymetamorphism in garnet micaschists of the Saualpe Eclogite Unit (Eastern Alps, Austria), resolved by automated SEM methods and EMP–Th–U–Pb monazite dating. *Journal of Metamorphic Geology*, **35**, 141–163, https://doi.org/10.1111/jmg.12224

Schulz, B. 2021*a*. Monazite microstructures and their interpretation in petrochronology. *Geoscience Frontiers*, **9**, https://www.frontiersin.org/articles/10.3389/feart.2021.668566/full

Schulz, B. 2021*b*. Petrochronology of monazite-bearing garnet micaschists as a tool to decipher the metamorphic evolution of the Alpine basement. *Minerals*, **11**, 981, https://doi.org/10.3390/min11090981

Schulz, B. and Krause, J. 2021. Electron probe petrochronology of polymetamorphic garnet micaschists in the lower nappe units of the Austroalpine Saualpe basement (Carinthia, Austria). *Zeitschrift der Deutschen Gesellschaft für Geowissenschaften (German Journal of Geosciences)*, **172**, 19–46, https://doi.org/10.1127/zdgg/2021/0247

Schulz, B. and Schüssler, U. 2013. Electron-microprobe Th–U–Pb monazite dating in Early-Palaeozoic high-grade gneisses as a completion of U–Pb isotopic ages (Wilson Terrane, Antarctica). *Lithos*, **175–176**, 178–192, https://doi.org/10.1016/j.lithos.2013.05.008

Schulz, B. and von Raumer, J.F. 2011. Discovery of Ordovician–Silurian metamorphic monazite in garnet metapelites of the Alpine External Aiguilles Rouges Massif. *Swiss Journal of Geosciences*, **104**, 67–79, https://doi.org/10.1007/s00015-010-0048-7

Schulz, B., Triboulet, C., Audren, C. and Feybesse, J.-L. 2001. P–T paths from metapelite garnet zonations, and crustal stacking in the Variscan inverted metamorphic sequence of La Sioule, French Massif Central. *Zeitschrift der Deutschen Geologischen Gesellschaft*, **152**, 1–25, https://doi.org/10.1127/zdgg/152/2001/1

Schulz, B., Krause, J. and Zimmermann, R. 2019. Electron microprobe petrochronology of monazite-bearing garnet micaschists in the Oetztal-Stubai Complex (Alpeiner Valley, Stubai). *Swiss Journal of Geosciences*, **112**, 597–617, https://doi.org/10.1007/s00015-019-00351-4

Schulz, B., Sandmann, D. and Gilbricht, S. 2020. SEM-based automated mineralogy and its application in geo- and material sciences. *Minerals*, **10**, 1004, https://doi.org/10.3390/min10111004

Schüssler, U., Oppermann, U., Kreuzer, H., Seidel, E., Okrusch, M., Lenz, K.-L. and Raschka, H. 1986. Zur Alterstellung des ostbayerischen Kristallins. Ergebnisse neuer K-Ar-Datierungen. *Geologica Bavarica*, **89**, 21–47.

Sebastian, U. 2013. *Die Geologie des Erzgebirges*. Springer-Verlag, Berlin.

Shrestha, S., Larson, K.P., Duesterhoeft, E., Soret, M. and Cottle, J.M. 2019. Thermodynamic modelling of phosphate minerals and its implications for the development of P–T–t histories: a case study in garnet – monazite bearing metapelites. *Lithos*, **334–335**, 141–160, https://doi.org/10.1016/j.lithos.2019.03.021

Simpson, A., Gilbert, S. *et al.* 2021. In-situ Lu–Hf geochronology of garnet, apatite and xenotime by LA ICP MS/MS. *Chemical Geology*, **577**, 120299, https://doi.org/10.1016/j.chemgeo.2021.120299

Skrzypek, E., Kato, T., Kawakami, T., Sakata, S., Hattori, K., Hirata, T. and Ikeda, T. 2018. Monazite behaviour and time-scale of metamorphic processes along a low-pressure/high-temperature field gradient (Ryoke Belt, SW Japan). *Journal of Petrology*, **59/6**, 1109–1144, https://doi.org/10.1093/petrology/egy056

Spear, F.S. 1993. *Metamorphic Phase Equilibria and Pressure–Temperature–Time Paths*. Mineralogical Society of America Monograph Series, Washington.

Spear, F.S. 2010. Monazite–allanite phase relations in metapelites. *Chemical Geology*, **279**, 55–62, https://doi.org/10.1016/j.chemgeo.2010.10.004

Spear, F.S. and Pyle, J.M. 2002. Apatite, monazite, and xenotime in metamorphic rocks. *Reviews in Mineralogy and Geochemistry*, **48**, 293–335, https://doi.org/10.2138/rmg.2002.48.7

Spear, F.S. and Pyle, J.M. 2010. Theoretical modeling of monazite growth in a low-Ca metapelite. *Chemical Geology*, **273**, 111–119, https://doi.org/10.1016/j.chemgeo.2010.02.016

Spear, F.S., Pyle, J.M. and Cherniak, D. 2009. Limitations of chemical dating of monazite. *Chemical Geology*, **266**, 218–230, https://doi.org/10.1016/j.chemgeo.2009.06.007

Stephan, T., Kroner, U. and Romer, R.L. 2018. The pre-orogenic detrital zircon record of the Peri-Gondwanan crust. *Geological Magazine*, **156**, 281–307, https://doi.org/10.1017/s0016756818000031

Stephan, T., Kroner, U., Romer, R.L. and Rösel, D. 2019. From a bipartite Gondwanan shelf to an arcuate Variscan belt: the early Paleozoic evolution of northern Peri-Gondwana. *Earth Science Reviews*, **192**, 491–512, https://doi.org/10.1016/j.earscirev.2019.03.012

Stettner, G. 1960. Über Bau und Entwicklung der Münchberger Gneismasse. *Geologische Rundschau*, **49**, 350–375, https://doi.org/10.1007/BF01983033

Stöckhert, B., Duyster, J., Trepmann, C. and Massonne, H.-J. 2001. Microdiamond daughter crystals precipitated from supercritical COH silicate fluids included in garnet, Erzgebirge, Germany. *Geology*, **29**, 391–394, https://doi.org/10.1130/0091-7613(2001)029<0391:MDCPFS>2.0.CO;2

Stöckhert, B., Trepmann, C. and Massonne, H.-J. 2009. Decrepitated UHP fluid inclusions: about diverse phase assemblages and extreme decompression rates (Erzgebirge, Germany). *Journal of Metamorphic Geology*, **27**, 673–684, https://doi.org/10.1111/j.1525-1314

Stosch, H.-G. and Lugmair, G.W. 1990. Geochemistry and evolution of MORB-type eclogites from the Münchberg Massif, southern Germany. *Earth and Planetary Science Letters*, **99**, 230–249, https://doi.org/10.1016/0012-821X(90)90113-C

Suzuki, K. and Kato, T. 2008. CHIME dating of monazite, xenotime, zircon and polycrase: protocol, pitfalls and chemical criterion of possible discordant age data. *Gondwana Research*, **14**, 569–586, https://doi.org/10.1016/j.gr.2008.01.005

Suzuki, K., Adachi, M. and Kajizuka, I. 1994. Electron microprobe observations of Pb diffusion in

metamorphosed detrital monazites. *Earth and Planetary Science Letters*, **128**, 391–405, https://doi.org/10.1016/0012-821X(94)90158-9
Tichomirowa, M. and Köhler, R. 2013. Discrimination of protolithic v. metamorphic zircon ages in eclogites: constraints from the Erzgebirge metamorphic core complex (Germany). *Lithos*, **177**, 436–450, https://doi.org/10.1016/j.lithos.2013.07.013
Tichomirowa, M., Berger, H.J. *et al.* 2001. Zircon ages of high-grade gneisses in the Eastern Erzgebirge (Central European Variscides) – constraints on origin of the rocks and Precambrian to Ordovician magmatic events in the Variscan foldbelt. *Lithos*, **56**, 303–332, https://doi.org/10.1016/S0024-4937(00)00066-9
Tichomirowa, M., Whitehouse, M.J. and Nasdala, L. 2005. Resorption, growth, solid state recrystallisation, and annealing of granulite facies zircon – a case study from the Central Erzgebirge, Bohemian Massif. *Lithos*, **82**, 25–50, https://doi.org/10.1016/j.lithos.2004.12.005
Tichomirowa, M., Whitehouse, M., Gerdes, A. and Schulz, B. 2018. Zircon (Hf, O isotopes) as melt indicator: Melt infiltration and abundant new zircon growth within melt rich layers of granulite-facies lenses v. solid state recrystallization in hosting amphibolite-facies gneisses (central Erzgebirge, Bohemian Massif). *Lithos*, **302–303**, 65–68, https://doi.org/10.1016/j.lithos.2017.12.020
Tichomirowa, M., Käßner, A. *et al.* 2019*a*. Dating multiply overprinted granites: the effect of protracted magmatism and fluid flow on dating systems (zircon U–Pb: SHRIMP/SIMS, LA-ICP-MS, CAID-TIMS, and Rb–Sr, Ar–Ar) – granites from the Western Erzgebirge (Bohemian Massif, Germany). *Chemical Geology*, **519**, 11–38, https://doi.org/10.1016/j.chemgeo.2019.04.024
Tichomirowa, M., Gerdes, A., Lapp, M., Leonhardt, D. and Whitehouse, M. 2019*b*. The chemical evolution from older (323–318 Ma) towards younger highly evolved tin granites (315–314 Ma) – sources and metal enrichment in Variscan granites of the Western Erzgebirge (Central European Variscides, Germany). *Minerals*, **9**, 769, https://doi.org/10.3390/min9120769
Volante, S., Pourteau, A. *et al.* 2020. Multiple P–T–d–t paths reveal the evolution of the final Nuna assembly in northeast Australia. *Journal of Metamorphic Geology*, **38**, 593–627, https://doi.org/10.1111/jmg.12532
Waizenhöfer, F. and Massonne, H.-J. 2017. Monazite in a Variscan mylonitic paragneiss from the Münchberg Metamorphic Complex (NE Bavaria) records Cadomian protolith ages. *Journal of Metamorphic Geology*, **35**, 453–469, https://doi.org/10.1111/jmg.12240
Werner, O. and Lippolt, H.J. 2000. White mica $^{40}Ar/^{39}Ar$ ages of the Erzgebirge metamorphic rocks: simulating the chronological results by a model of Variscan crustal imbrication. *Geological Society, London, Special Publications*, **179**, 323–336, https://doi.org/10.1144/GSL.SP.2000.179.01.19
Werner, O. and Reich, S. 1997. $^{40}Ar/^{39}Ar$-Abkühlalter von Gesteinen mit unterschiedlicher P–T-Entwicklung aus dem Schiefermantel des Sächsischen Granulitgebirges. *Terra Nostra*, **97/5**, 196–198.
Will, T.M., Schulz, B. and Schmädicke, E. 2017. The timing of metamorphism in the Odenwald-Spessart basement, Mid-German Crystalline Zone. *International Journal Earth Sciences*, **106**, 1631–1649, https://doi.org/10.1007/s00531-016-1375-3
Williams, M.L., Jercinovic, M.J., Harlov, D.E., Budzyn, B. and Hetherington, C.J. 2011. Resetting monazite ages during fluid-related alteration. *Chemical Geology*, **283**, 218–225, https://doi.org/10.1016/j.chemgeo.2011.01.019
Willner, A.P., Rötzler, K. and Maresch, W.V. 1997. Pressure–temperature and fluid evolution of quartzo-feldspathic metamorphic rocks with a relic high-pressure, granulite-facies history from the Central Erzgebirge (Saxony, Germany). *Journal of Petrology*, **38**, 307–336, https://doi.org/10.1093/petroj/38.3.307
Willner, A.P., Krohe, A. and Maresch, W.V. 2000. Interrelated P–T–t–d paths in the Variscan Erzgebirge dome (Saxony, Germany): constraints on the rapid exhumation of high-pressure rocks from the root zone of a collisional orogen. *International Geology Review*, **42**, 64–85, https://doi.org/10.1080/00206810009465070
Willner, A., Sebazungu, E., Gerya, T., Maresch, W. and Krohe, A. 2002. Numerical modelling of PT-paths related to rapid exhumation of high-pressure rocks from the crustal root in the Variscan Erzgebirge Dome (Saxony/Germany). *Journal of Geodynamics*, **33**, 281–314, https://doi.org/10.1016/S0264-3707(01)00071-0
Wing, B.A., Ferry, J.M. and Harrison, T.M. 2003. Prograde destruction and formation of monazite and allanite during contact and regional metamorphism of pelites: petrology and geochronology. *Contributions to Mineralogy and Petrology*, **145**, 228–250, https://doi.org/10.1007/s00410-003-0446-1
Wu, C.M. 2015. Revised empirical garnet–biotite–muscovite–plagioclase (GBMP) geobarometer in metapelites. *Journal of Metamorphic Geology*, **33**, 167–176, https://doi.org/10.1111/jmg.12115
Wu, C.M. 2017. Calibration of the garnet–biotite–Al_2SiO_5–quartz geobarometer for metapelites. *Journal of Metamorphic Geology*, **35**, 983–998, https://doi.org/10.1111/jmg.12264
Wu, C.M. 2019. Original calibration of a garnet geobarometer in metapelite. *Minerals*, **9**, 540, https://doi.org/10.3390/min9090540
Wu, C.M. and Zhao, G.C. 2006. Recalibration of the garnet–muscovite (GM) geothermometer and the garnet–muscovite–plagioclase–quartz (GMPQ) geobarometer for metapelitic assemblages. *Journal of Petrology*, **47**, 2357–2368, https://doi.org/10.1093/petrology/egl047
Wu, C.M. and Zhao, G.C. 2007. The metapelitic garnet–biotite–muscovite–aluminosilicate–quartz (GBMAQ) geobarometer. *Lithos*, **97**, 365–372, https://doi.org/10.1016/j.lithos.2007.01.003
Wu, C.M., Zhang, J. and Ren, L.D. 2004. Empirical garnet–biotite–plagioclase–quartz (GBPQ) geobarometry in medium- to high-grade metapelites. *Journal of Petrology*, **45**, 1907–1921, https://doi.org/10.1093/petrology/egh038
Žák, J., Kraft, P. and Hajná, J. 2013. Timing, styles and kinematics of Cambro-Ordovician extension in the Teplá-Barrandian Unit, Bohemian Massif, and its

bearing on the opening of the Rheic Ocean. *International Journal Earth Sciences*, **102**, 415–433, https://doi.org/10.1007/s00531-012-0811-2

Závada, P., Štípská, P. *et al.* 2021. Monazite geochronology in melt-percolated UHP meta-granitoids: an example from the Erzgebirge continental subduction wedge, Bohemian Massif. *Chemical Geology*, **559**, 119919, https://doi.org/10.1016/j.chemgeo.2020.119919

Zenk, M. and Schulz, B. 2004. Zoned Ca-amphiboles and related P–T evolution in metabasites from the classical Barrovian metamorphic zones in Scotland. *Mineralogical Magazine*, **68**, 769–786, https://doi.org/10.1180/0026461046850218

Zircon trace-element and isotopes (U–Pb, Lu–Hf, $\delta^{18}O$) response to fluid-deficient metamorphism of a subducted continental terrane (North Muya, Eastern Siberia)

Sergei Skuzovatov[1]*, Kuo-Lung Wang[2,3], Xian-Hua Li[4], Yoshiyuki Iizuka[2] and Vladislav Shatsky[1,5]

[1]Isotope Geochemistry Laboratory, Vinogradov Institute of Geochemistry, Russian Academy of Sciences, Siberian Branch, 1A Favorskogo str., Irkutsk 664033, Russia

[2]Institute of Earth Sciences, Academia Sinica, 128, Sec. 2, Academia Road, Nangang, Taipei 11529, Taiwan

[3]Department of Geosciences, National Taiwan University, No. 1, Sec. 4, Roosevelt Rd., Taipei 10617, Taiwan

[4]State Key Laboratory of Lithospheric Evolution, Institute of Geology and Geophysics, Chinese Academy of Sciences, Beijing 100027, China

[5]Laboratory of Crystal Growth and Mineralogy, Sobolev Institute of Geology and Mineralogy, Russian Academy of Sciences, Siberian Branch, 3 Koptyuga ave., Novosibirsk 630090, Russia

SS, 0000-0002-2253-6020
*Correspondence: skuzovatov@igc.irk.ru

Abstract: The orogenic continental crust of accretionary and collisional belts worldwide is dominated by felsic and metasedimentary rocks, which show variable responses to high-grade metamorphism. Transformation of felsic rocks is commonly limited as compared to that of the enclosed mafic rocks (including eclogites *sensu stricto*), which is widely attributed to availability of H_2O–CO_2 fluids, kinetically controlled growth of high-grade assemblages, and their preferential preservation in metabasites more competent to rehydration. We report on the results of studies of the geochemical behaviour of zircon (trace-element, U–Pb, Lu–Hf and $\delta^{18}O$) in three felsic samples (two metagranitoids and one paragneiss), which are spatially (geographically and at the outcrop-scale) juxtaposed with mafic eclogites within the North Muya block (Neoproterozoic Baikalides, northern Central Asian Orogenic Belt). The data imply that metagranitoids and metasediments within the buried continental lithosphere might follow a single subduction-related *P–T–t* trend, whereas contrasting degrees of mineralogical and zircon transformation were governed by mineral buffer reactions in the absence of external fluids. The latter was significant only in an H_2O-enriched protolith of metasediments. The formation of ^{18}O-depleted zircon recrystallization rims together with Mn enrichment of garnet rims indicate a distinct metamorphic stage without or with minor localized fluid infiltration, most likely, related to peak temperature conditions during collision.

Supplementary material: The data on the mineral composition, trace-element and U–Pb–Lu–Hf–O isotopic composition of zircon are available at https://doi.org/10.6084/m9.figshare.c.6794043

Blueschist- and eclogite-bearing metamorphic complexes exhumed from high-pressure (HP) and ultrahigh-pressure (UHP) depths and enclosed mainly in the Late Neoproterozoic and Phanerozoic subduction and accretionary complexes are vital indicators of the former convergent-margin processes (e.g. Liou and Tsujimori 2013). In the Alpine-type orogens, which are normally characterized by thicker lithosphere and hotter subduction geotherm, minor eclogites co-exist with metasedimentary and felsic rocks of continental origin (Ernst 2010). Because these lithologies are volumetrically dominant in subducted crustal units, their metamorphic modification plays a crucial role in controlling the patterns and dynamics of subduction and exhumation of HP rocks (Hacker *et al.* 2003; Agard *et al.* 2009). Nonetheless, resolving the issue of (in-)coherence of metamorphic histories between mafic eclogites and their felsic hosts is not always simple. This is mainly due to the kinetically limited reactive growth of HP assemblages, or their intensive recrystallization during exhumation, or even tectonic juxtaposition low-grade felsic lithologies with higher-grade mafic rocks during orogeny termination (Proyer 2003; Massonne 2009; Young and Kylander-Clark 2015; Schorn 2022). Thus, the assessment of subduction–

From: van Schijndel, V., Cutts, K., Pereira, I., Guitreau, M., Volante, S. and Tedeschi, M. (eds) 2024. *Minor Minerals, Major Implications: Using Key Mineral Phases to Unravel the Formation and Evolution of Earth's Crust.*
Geological Society, London, Special Publications, **537**, 285–311.
First published online September 29, 2023, https://doi.org/10.1144/SP537-2022-309

exhumation events in such orogens requires reliable petrological tracers of metamorphic transformation as well as sensitive geochronometers showing a systematic response to high-grade metamorphism.

Zircon isotope (U–Pb, $\varepsilon_{Hf}(t)$, $\delta^{18}O$) and trace-element composition are conventionally used as recorders and geochemical tracers of metamorphic modification of crustal lithologies (Rubatto and Hermann 2003; Martin *et al.* 2006, 2008; Chen and Zheng 2017). Nonetheless, the mineralogical response (i.e. the occurrence of HP assemblages) on a whole-rock level as well as the degree of the achieved chemical re-equilibration within single zircon crystals largely depend on the fluid regime on the course of prograde-to-peak evolution. In the case of low- to medium-temperature metamorphic felsic rocks (<600°C) and fluid-deficient subsolidus zircon transformation, incomplete zircon re-equilibration should be expected due to predominant dissolution–reprecipitation (e.g. Martin *et al.* 2008; McElhinney *et al.* 2022). Here, we examined the behaviour of zircon (textural, trace-element, U–Pb, Lu–Hf and $\delta^{18}O$ isotopes) in garnet-free and garnet-bearing felsic (metagranitoid and metasedimentary) rocks, which are tectonically juxtaposed with mafic eclogites in the Neoproterozoic North Muya HP complex (Siberia) (mainly up to ~650–670°C and 2.5–2.7 GPa) (Shatsky *et al.* 2012; Shatskii *et al.* 2014; Skuzovatov *et al.* 2019*b*; Skuzovatov 2022). Combining these zircon data with the results of petrographic observations and phase modelling results, we highlight contrasting responses of felsic and metasedimentary rocks to high-grade metamorphism within this HP unit and their lower-pressure evolution as compared with the eclogites. Variable internal/external fluid availability, which was likely a major driving force of mineralogical and zircon modification of different lithologies, might be governed by their position relative to the original subduction interface and thus their accessibility to pervasive fluid percolation.

Geological, thermobarometric and geochronological background

The Baikal–Muya Fold Belt (BMFB), located in the northeastern segment of the Central Asian Orogenic Belt (CAOB), is a terrane puzzle formed primarily during Early to Late Neoproterozoic convergence and the corresponding tectonomagmatic activity in the southern periphery of Siberia (Fig. 1). The geology and the major stages of the geodynamic evolution of the eastern BMFB have been recently summarized and reported by Skuzovatov *et al.* (2016*b*, 2017, 2019*a*, b). For decades, the Muya terrane, including the North Muya high-grade metamorphic block (NMB), has been considered an Archean cratonic unit (Salop 1964; Bulgatov 1983; Bulgatov and Gordienko 1999). The block comprises the metasedimentary sequences of the Kindikan (lower unit) and the Dzhaltuk series (upper unit), intruded by anatectic granitoids of the Ileir complex (Rytsk *et al.* 2001), and overlain by weakly metamorphosed volcanics, volcano-sedimentary and terrigenous rocks (Mitrofanov 1978; Rytsk *et al.* 2007). Recent studies argued that the NMB is Neoproterozoic and suggested its origin is related to the reworking of ancient crust (Shatsky *et al.* 2015; Skuzovatov *et al.* 2019*b*).

The eclogites of the North Muya complex (Eastern Siberia) are located within the Early Neoproterozoic metasedimentary and felsic rocks of the BMFB, and occur mostly as metre-scale to tens of metres-scale boudins and lenses among felsic country rocks within a 10 km wide belt from the Ileir–Samokut pass (marked as a star in Fig. 1b), trending NE for 30 km. The eclogites show subduction-related affinity, with large-ion lithophile (LILE) and light rare-earth element (LREE) enrichment and high field-strength element (HFSE) depletion signatures, similar to the exposed plutonic and volcanic rocks of the Early Neoproterozoic (Early Baikalian) subduction setting in the BMFB. Their radiogenic isotope signatures ($\varepsilon_{Nd}(t)$ of +6 to −1.4, $^{87}Sr/^{86}Sr$ ratios of 0.70498–0.71054) and mineral $\delta^{18}O$ systematics (+3.9–11.5‰ for garnet) indicate a continental or continental-arc affinity of eclogites and significant role of crustal recycling or contamination (up to 5–10%) from the Early Precambrian continental rocks. The estimated metamorphic conditions for both the burial and exhumation of rocks indicate a continental subduction setting with a relatively cold geotherm (~20–25°C $kbar^{-1}$). The Sm–Nd ages of both eclogite and its host rock (Shatsky *et al.* 2012), as well as the zircon U–Pb ages of eclogites and their felsic hosts (Shatsky *et al.* 2015; Skuzovatov *et al.* 2019*b*) suggest that the Neoproterozoic metamorphism occurred at approximately ~630 Ma. A detailed mineralogical study performed by Shatsky *et al.* (2012) indicated that eclogites equilibrated under a wide range of pressure and temperature conditions (500–750°C at minimum pressures of 1.4–1.8 GPa). More recent results of phase modelling assessment of the selected eclogites suggest that the Ediacarian HP metamorphic event for different rocks led to differential subduction and exhumation of metabasites, leading to the formation of porphyroblastic eclogites (~560–620°C at 2.5–2.7 GPa; Skuzovatov *et al.* 2019*b*) and medium-grained eclogites (~600–730°C at 1.7–2.1 GPa; Skuzovatov 2022), with the highest-grade condition recorded by a rare kyanite eclogite (up to ~750°C at 2.5–2.7 GPa).

Although a coherent tectonic evolution is assumed from the geochronological data for the whole

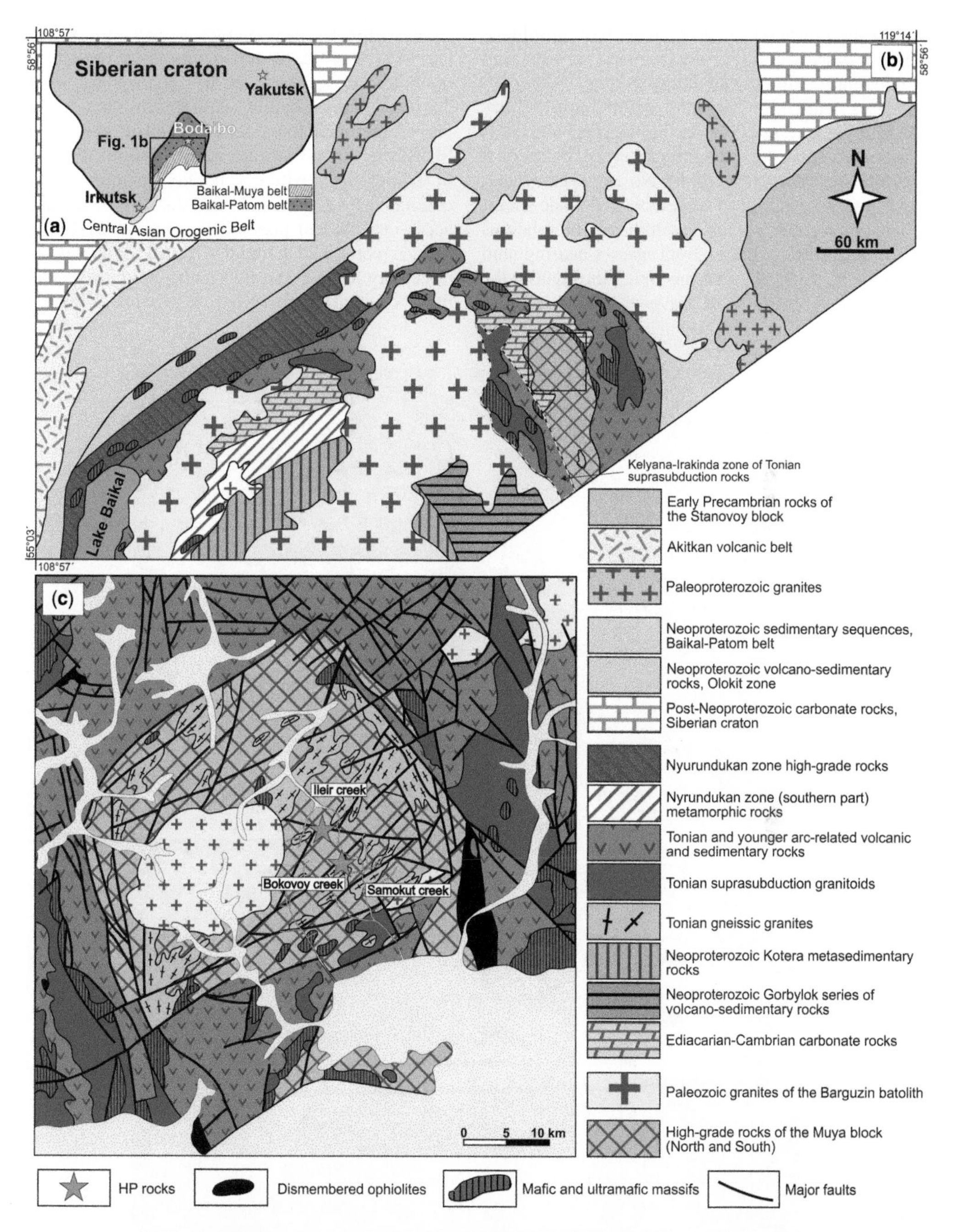

Fig. 1. Schematic map of the Baikal–Muya Fold Belt (BMFB) relative to the Siberian craton (**a**), general structure of the BMFB with key Neoproterozoic complexes (**b**), and detailed geological structure of the North Muya eclogite–gneiss complex (**c**). Cambrian and younger terrigenous and carbonate rocks are not shown. Source: panel (c) modified after Skuzovatov *et al.* (2016*b*).

NMB unit, a particular *P–T–t* history of their volumetrically dominant felsic hosts remains rather uncertain. Detailed sampling revealed most of the Neoproterozoic NMB structures to be composed of low- to medium-grade, simply foliated, fine-grained feldspar–quartz–biotite–clinozoisite and feldspar–

quartz–two-mica schists (Fig. 2a, b), rarer porphyroblastic garnet–biotite–feldspar–quartz schists and garnet–muscovite–biotite–feldspar–quartz–epidote gneisses with more complex deformation history (Fig. 2c–e), and two-feldspar biotite metagranitoids (orthogneisses) and paragneisses with occasional epidote and garnet (Fig. 2f–h). No evidence of anatexis and migmatization is observed in felsic rocks, either on an outcrop-scale or within mineral relations on a thin-section scale. Both garnet-bearing and garnet-free metasediments are found throughout the whole block in an S–N traverse, whereas Early Neoproterozoic metagranitoids lack garnet and almost lack any signs of metamorphism or deformations in the southern NMB (Skuzovatov *et al.* 2016*b*). Garnet–biotite equilibrium temperature ranges retrieved for a series of felsic gneisses by Shatsky *et al.* (2012) are mainly within the range of 620–640°C, whereas neither quantitative nor qualitative (e.g. based on the stability of micas or accessory phases) pressure constraints were available. Zircon U–Pb dating performed for one metasedimentary sample and two metagranitoids revealed their original Early Neoproterozoic (Tonian to

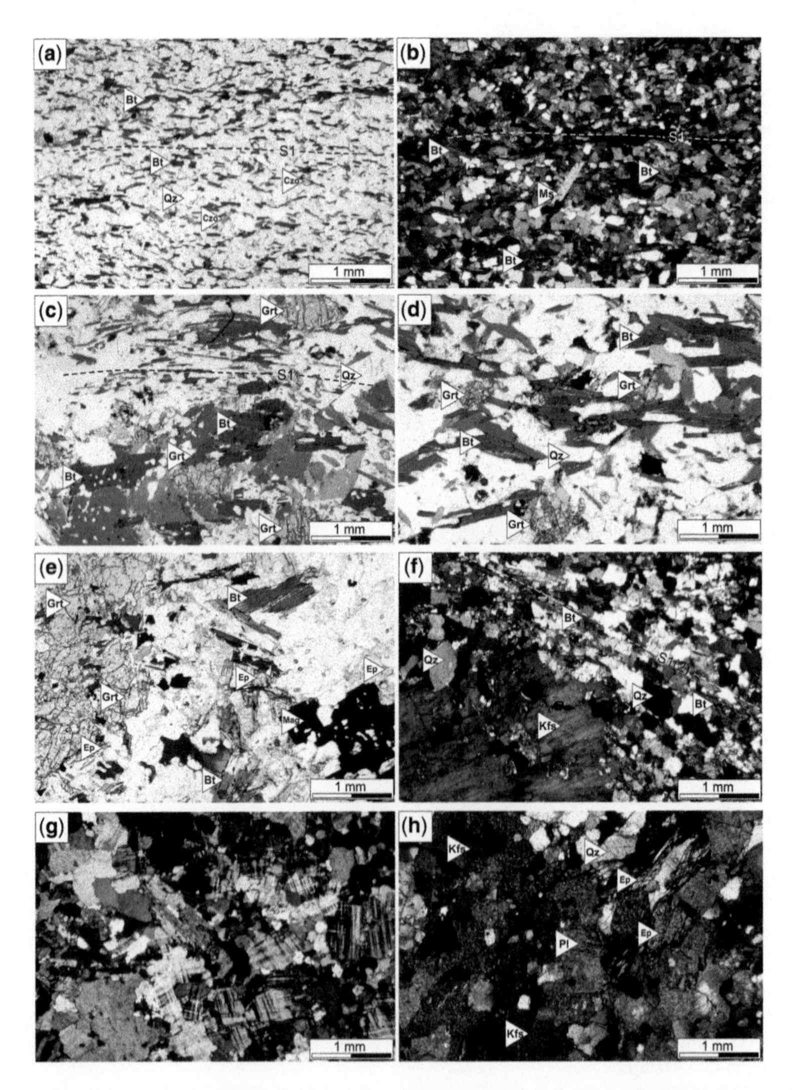

Fig. 2. Transmitted-light photomicrographs of thin sections representing major types of the NMB felsic rocks: low- to medium-grade, simply foliated, fine-grained feldspar–quartz–biotite–clinozoisite and feldspar–quartz–two-mica schists (**a**) and (**b**), rarer porphyroblastic garnet–biotite–feldspar–quartz schists and garnet–muscovite–biotite–feldspar–quartz–epidote gneisses with more complex deformation history (**c–e**), and two-feldspar biotite metagranitoids (orthogneisses) and paragneisses with occasional epidote and garnet (**f–h**).

Cryogenian) affinities, which is, for metagranitoids, expressed by crystallization ages of igneous zircons (764 ± 5 Ma and 763 ± 14 Ma, respectively) and – in case of a metasediment – the dominance of *c.* 941–793 Ma population among detrital grains. Although Shatsky *et al.* (2015) observed limited textural and microchemical evidence of re-equilibration, exemplified, for instance, by low-Th/U overgrowths or recrystallization rims and Pb loss trends, predominantly high preservation degrees of primary zircon and a low amount of analysed grains did not allow the use of zircon as a universal tracer of HP modification.

Samples and analytical methods

To check the bulk-rock mineralogical and response to metamorphic modification, we used three typical samples of felsic (metagranitoid and metasedimentary) rocks, which compose the northern, eclogite-bearing area of the North Muya block (Shatsky *et al.* 2012; Skuzovatov *et al.* 2019*b*). Electron-microprobe analyses were performed only for the two thick sections of garnet-bearing gneiss samples for the sake of thermobarometric assessment. The data on mineral composition of rock-forming and accessory phases for felsic gneisses and metasediments were acquired using a JEOL Superprobe JXA8200 electron-microprobe analyser equipped with five wavelength-dispersive (WDX) spectrometers, using a beam sized of 2 µm, a beam current of 15 nA and an accelerating voltage of 20 kV. The set of natural and synthetic standards used for calibration included albite (Na), pyrope for Al, K-feldspar (K), diopside (Si, Ca, Mg), olivine (Si, Mg, Fe), garnet (Si, Al, Fe, Mn), rutile (Ti) and chromite (Cr, Fe). On-peak and background counting times for each element were 10 seconds each, and analytical errors were typically within 0.01 (for minor elements with the content near corresponding determination levels) to 0.1–0.2 wt%. The calculation of mineral structural formulae was undertaken using the CALCMIN software (Brandelik 2009). Selected garnet grains were mapped using EPMA techniques to check their homogeneity for phase modelling assessment. Measurements were performed at the Centre of Isotopic and Geochemical Research (IGC, Irkutsk, Russia; Skuzovatov *et al.* 2022).

Zircons studied here were separated from 0.16–0.25 and 0.25–0.50 mm heavy mineral fractions, mounted in epoxy, and polished down to about half of their thickness for trace-element and isotopic analysis. Prior to analytical work, polished surfaces were examined for zoning and inclusions using a JEOL FE-SEM JSM-7100F field emission scanning electron microscope equipped with a Centaurus Scintillator cathodoluminescence (CL) detector (Institute of Earth Sciences, Academia Sinica), with CL images acquired at 15 kV and 75 µA. Observed CL images together with transmitted light microphotographs were used to choose areas free from abundant mineral inclusions for laser ablation analyses. The composition of mineral inclusions in zircon was analysed at the SEM/EPMA Lab of the Institute of Earth Sciences (Academia Sinica, Taipei) using an FE-EPMA JEOL JXA8500F microprobe analyser. Measurements were carried out with five spectrometers at 14 kV, 6 nA at 2 µm probe size.

Then, 137 zircons were further analysed in the three samples for U–Pb isotopic ages, of which only 97 were analysed by traces, 48 for oxygen isotopes and 81 for Lu–Hf, given the spatial limitations within each grain. Around one-half of the larger grains were examined using laser ablation inductively coupled plasma mass spectrometry (LA-ICP-MS) for U–Pb ages of key domains, followed by trace-element and Hf isotope measurements, where enough grain space was available. After clarification of major compositional trends, the second batch of smaller grains was independently prepared to oxygen isotope analyses (as the least destructive ones), followed by slight re-polishing and further *in situ* (multicollection, MC)-LA-ICP-MS measurements (U–Pb, trace-element, Lu–Hf isotope), where grain sizes allowed combining the analyses.

In situ oxygen isotope composition of selected zircon grains was analysed by secondary-ion mass spectrometry (SIMS) using a Cameca IMS-1280 mass spectrometer stored at the State Key Laboratory of Lithospheric Evolution of the Institute of Geology and Geophysics (Beijing, China). Detailed analytical parameters are provided by Li *et al.* (2010). The Cs^+ primary ion beam accelerated at 10 kV with an intensity of *c.* 2 nA was used on the 20 µm diameter spots. Oxygen isotopes were measured using the MC mode. Measured $^{18}O/^{16}O$ ratios were normalized using Vienna Standard Mean Ocean Water compositions (VSMOW; $^{18}O/^{16}O = 0.0020052$). The instrumental mass-fractionation correction was further employed by repeated measurements of the 91500 zircon reference material (9.9‰; Wiedenbeck *et al.* 2004), which yielded reproducibility of 0.10‰ (1 SD). The accuracy was controlled by multiple analyses of the Penglai (Li *et al.* 2010) and Qinghu (Li *et al.* 2013) zircon reference materials, which yielded the average values of 5.25 ± 0.09 (1 SD, $N = 8$) and 5.49 ± 0.14‰ (1 SD, $N = 6$), respectively. The results are reported in the conventional $\delta^{18}O$ notation with reference to VSMOW in per mil.

Zircon U–Pb dating, trace-element analyses and Lu–Hf isotope analyses were performed at the Institute of Earth Sciences (Academia Sinica, Taipei). The U–Pb dating analyses used a quadrupole Agilent

7900 Q-ICP-MS coupled with a Photon Machines Analyte G2 laser ablation system (Photon Machines, Inc. Redmond, USA) that utilizes a 193 nm ArF Excimer laser with a 5 ns pulse duration. The analyses were conducted using a spot diameter of 35 μm and produced laser pits ~30–35 μm deep. Calibration was performed using the GJ-1 zircon standard (Jackson *et al.* 2004*a*). Reference zircons 91500 (Wiedenbeck *et al.* 2004; with a weighted average $^{207}Pb/^{206}Pb$ age of 1066 ± 11 Ma, $N = 17$, at a recommended value of 1065 Ma) and Plešovice (Sláma *et al.* 2008; with a weighted average $^{206}Pb/^{238}U$ age of 336 ± 3 Ma, $N = 18$, at a recommended value of 337 Ma) were used as independent controls on the data reproducibility and accuracy. The ^{238}U decay constant of 1.55125×10^{-10} a^{-1} was used for recalculations. The GLITTER (http://www.es.mq.edu.au/gemoc/glitter) off-line software package was employed for the data reduction (including correction for mass fractionation and instrumental drift), age calculations and uncertainty propagation. The data were further processed using Isoplot software version 4.15 (Ludwig 2003).

For trace element analysis, the same instrumental setup was applied for spots 35 to 40 μm in diameter. The procedure involved bracketing 10–12 unknown samples with duplicated measurements of SRM NIST612 for the instrumental drift correction and single BCR-2 glass analyses to control the accuracy and reproducibility. The silicon content in zircon measured by EPMA was used as an internal standard and was within 32.6–33.6 wt% of SiO_2 for the analysed grains. The reproducibility of BCR-2 analyses was within 2–5% (2 SD, $N = 20$) for Ti, Y, Nb, most LREEs and middle rare-earth elements (MREEs) (La–Sm, Gd, Dy, Er) and Pb, and slightly worse (5–7%) for Eu, Tb, Ho, Tm–Lu, Ta, Th and U. The accuracy was evaluated on the same repeated BCR-2 analyses relative to the recommended values (Jochum *et al.* 2016), and the average values generally overlapped with or were close to the working values within the estimated uncertainties. Based on the acquired data, we may consider a potentially increased uncertainty for La and Pr measurements, as their contents approach the corresponding detection limits. For all isotopes measured, the resulting uncertainty was within a 2σ interval of the recommended values for BCR-2. To further ensure the accuracy and reproducibility for matrix-matching standards, especially those for low-content isotopes, we also obtained trace-element data for 91500 (Wiedenbeck *et al.* 2004) and Mud Tank (Black and Gulson 1978) reference materials. The studies for both zircon reference materials show good internal precisions well within 5% and high within-grain reproducibility for all isotopes of interest, excluding La and Pr. The obtained average element mass fractions closely reproduce the reference values for 91500 zircon for most elements within given uncertainties (Wiedenbeck *et al.* 2004), but are higher or near the upper limit of the previously reported recommended ranges for Mud Tank zircon (see the recent compilation by Gain *et al.* 2019), which is most likely due to originally heterogeneous (both on a within-grain and on an inter-grain scale) nature of Mud Tank carbonatitic zircon shown previously (Gain *et al.* 2019).

In situ Hf isotope determinations were performed as close as possible to the same location measured for the U–Pb analysis for each zircon grain using the same LA system coupled to a Nu Plasma MC-ICP-MS instrument. The analyses were conducted using a spot diameter of 50 μm and produced laser pits ~40–60 μm deep. MC-ICP-MS instrumental conditions and data acquisition procedures followed those reported by Griffin *et al.* (2000). For this study, 172, 175, 176, 177, 178, 179 and 180 masses were simultaneously analysed, and all analyses were carried out in a static collection mode. The yielded total Hf beam was within 3.4–11.9 V for the analysed zircon standards and unknown samples. The data were normalized to $^{179}Hf/^{177}Hf = 0.7325$, using an exponential correction for mass bias. The initial set-up of the instrument was done using a 50 ppb solution of AMES Hf metal. Isobaric interferences of ^{176}Lu and ^{176}Yb on ^{176}Hf were corrected by measuring the intensities of the interference-free ^{175}Lu and ^{172}Yb isotopes and using appropriate $^{176}Lu/^{175}Lu$ and $^{176}Yb/^{172}Yb$ ratios to calculate $^{176}Lu/^{177}Hf$ and $^{176}Yb/^{177}Hf$ ratios. In the method of Griffin *et al.* (2000), it is assumed that $f\text{Hf} = f\text{Yb} = f\text{Lu}$ (f is the mass fractionation coefficient) and the mass bias obtained for Hf is also applied to the Yb and Lu mass bias correction. With this approach, usually only one Yb isotope is measured (^{172}Yb in this study). Griffin *et al.* (2000) tested this correction method by analysing solutions of JMC475 spiked with Yb and JMC475 spiked with Lu. The 'true' values for $^{172}Yb/^{176}Yb$ and $^{175}Lu/^{176}Lu$ were adjusted to give the 'true' $^{176}Hf/^{177}Hf$ of JMC475. The Yb and Lu isotopic compositions derived from the solution analyses were then used to correct the laser analyses. The recommended $^{176}Lu/^{175}Lu$ and $^{176}Yb/^{172}Yb$ ratios of 0.02669 and 0.5865 were used for data reproduction. A Mud Tank zircon standard (Black and Gulson 1978) was used to address reproducibility and accuracy of the results and yielded an average $^{176}Hf/^{177}Hf$ value of 0.282490 ± 0.000032 (2 SD, $N = 18$) over the course of the analytical session, with a long-term value of 0.282495 ± 0.000030 (at recommended values within 0.282504–0.282523; Woodhead and Hergt 2005; Gain *et al.* 2019). As Mud Tank zircon has relatively low contribution from Yb ($^{176}Yb/^{177}Hf$ ratios an order of magnitude lower than those of the unknowns), the data were carefully checked to avoid a potential bias from the used correction approach. The $\varepsilon_{Hf}(t)$ values were calculated using chondritic ratios of $^{176}Hf/^{177}Hf$ (0.282785) and $^{176}Lu/^{177}Hf$ (0.0336) as derived by Bouvier *et al.* (2008). The ^{176}Lu decay constant of 1.865×10^{-11} a^{-1} reported by Scherer *et al.* (2001) was used in the calculation. The single-stage model ages (T_{DM}) were calculated relative to the depleted mantle with a present-day $^{176}Hf/^{177}Hf = 0.28325$ (Nowell *et al.* 1998) and $^{176}Lu/^{177}Hf = 0.0384$ (Griffin *et al.* 2000).

Results

Sample petrography and mineral composition

Metagranitoids. The exposed NMB metagranitoids are macroscopically nearly undeformed to moderately deformed coarse-grained and medium-grained rocks (Figs 2 and 3) with perthitic feldspar porphyroclasts (Kfs) up to 5 mm (40%), sodic plagioclase (Pl) (30%), quartz (Qz) (10–15%), biotite (Bt) (10–15%), occasional epidote (Ep) (<1–2%) and garnet (Grt) (<1%), as well as accessory apatite (Ap), titanite (Ttn) and zircon (Zrn). The rocks display different degrees of dynamic recrystallization, reflected by the occurrence of undulose extinction, deformation twinning, subgrain formation and sutured boundaries in quartz. At a thin-section scale, the Mu-93-75 metagranitoid sample is weakly deformed and has newly formed metamorphic mineral phases including rare resorbed and cracked Grt and Ep in the Qz–Pl matrix (Fig. 2h), with no evidence of white mica, rutile or clinopyroxene as HP phases. It is worth noting that the Mu-93-75 metagranitoid was sampled in the same area as the only occurrence of a kyanite eclogite (Shatsky *et al.* 2012). The opposite case is the Mu-93-43 sample, which is a moderately deformed metagranitoid with large Kfs porphyroclasts and a clear Bt foliation in the partially recrystallized Kfs–Pl–Qz matrix (Fig. 2f), and has accessory Ttn, Ap and Zrn. Unlike the above weakly

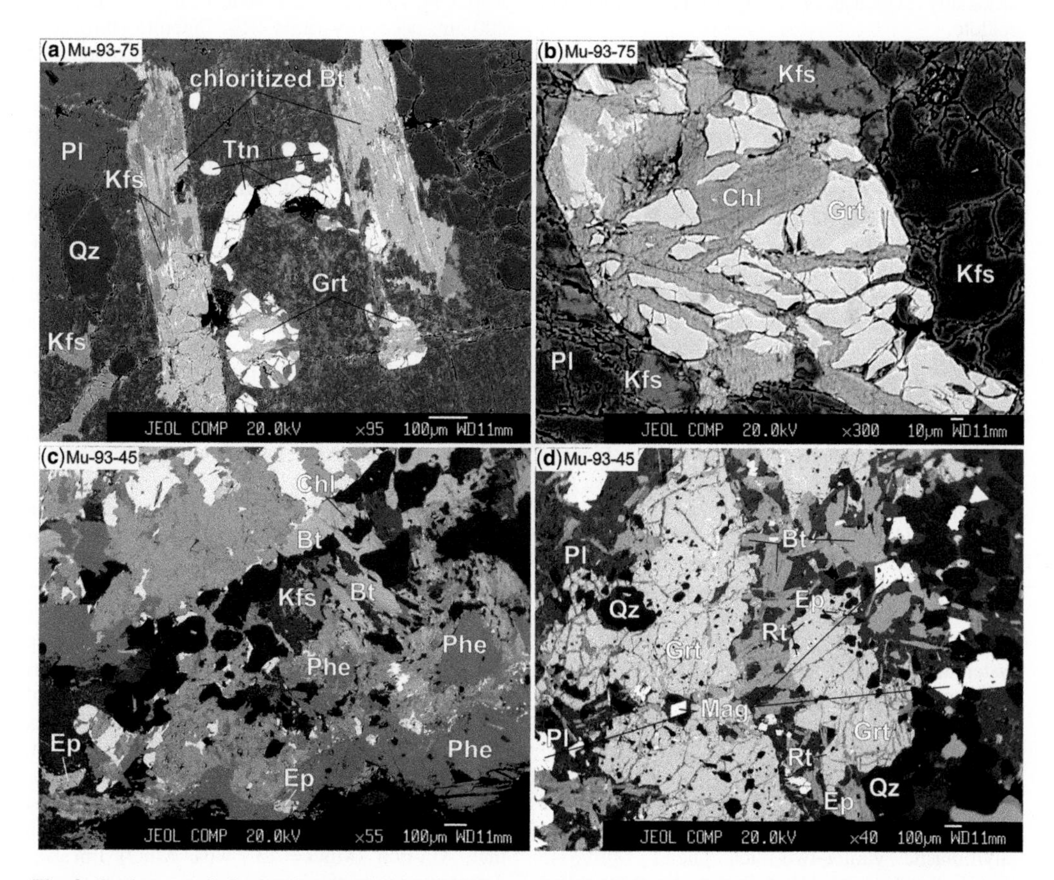

Fig. 3. Backscattered-electron images of garnet-bearing metagranitoid Mu-93-75 (**a**, **b**) and garnet paragneiss Mu-93-45 (**c**, **d**) studied in detail. (**a**, **b**) Preferential localization of garnet in the interstitial areas with partially to completely chloritized biotite, plagioclase and titanite. (**c**, **d**) Heterogeneous distribution of major newly grown metamorphic phases (garnet, epidote, rutile and phengite) in a paragneiss.

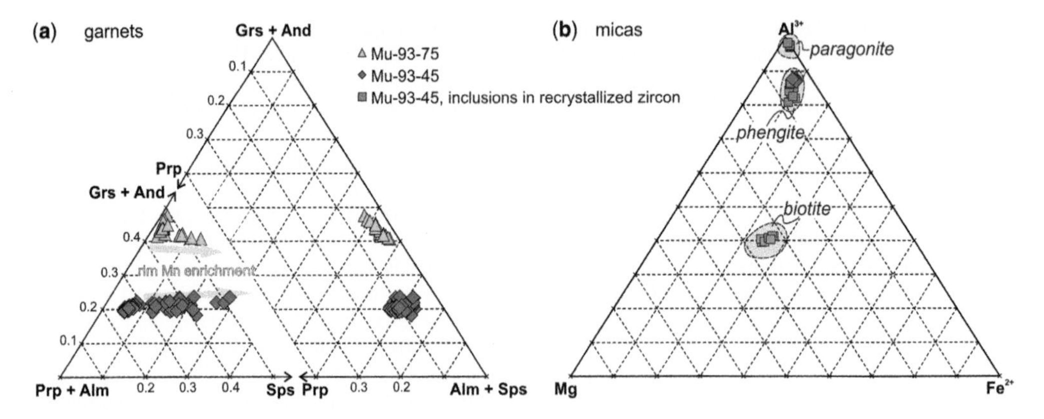

Fig. 4. Garnet (**a**) and mica (**b**) composition in the studied rocks. Garnets in both types of garnet-bearing rocks show relatively homogeneous cores and apparent MnO enrichment in the rims. For a paragneiss, the compositions of phengite and paragonite co-existing in zircon inclusions are shown.

deformed Mu-93-75 metagranitoid, this sample lacks any evidence of newly formed garnet and epidote.

Garnet in the Mu-93-75 metagranitoid is represented as separate isometric grains up to 200 μm (Fig. 3a, b), observed mainly together with Bt and partially modified Pl, which is replaced by sericite and clinozoisite–epidote. Garnet is extensively chloritized through cracks, which reflect the retrograde overprinting. Its composition is Mg-poor and varies within the $Alm_{44–52}Grs_{38–44}Prp_{2–5}Sps_{1–13}$ range (Fig. 4a), with most of the grain volume being weakly zoned (the cores slightly enriched in CaO) or rather homogeneous in distribution of major components. Nonetheless, a few separate grains or their relics were found to be enriched in modal spessartine proportion (MnO up to 3.43–5.68 wt% at <1 wt% for garnet cores). Clinozoisite/epidote has nearly constant X_{Fe3+} (formula ratio of $Fe^{3+}/(Fe^{3+} + Al)$) of ~0.15. Kfs has a limited contribution of Ab (2 mol. %), and Pl is almost a pure albite (1–2 mol. % of K-felsdpar component). Biotite is strongly chloritized, which precluded its precise chemical analyses in this sample. Titanite forms elongated, mainly euhedral grains up to 200–250 μm and has 3.1–4.1 wt% of Al_2O_3 and a systematic admixture of F (0.8–1.1 wt%). No measurements were made on the Mu-93-43 metagranitoid, as it was inapplicable for further thermobarometric assessment.

Metasedimentary rocks. The representative sample of typical NMB metasedimentary rocks is a porphyroblastic garnet–two-mica gneiss Mu-93-45 (Fig. 2e) with an unclear foliation at a thin-section level. The gneiss is composed of unevenly distributed, large and anhedral Grt porphyroblasts up to ~5–6 mm in size; smaller (up to 1 mm) and variably oriented flakes of chloritized Bt and white mica, matrix Pl, Kfs and Qz; anhedral Ep grains up 100–200 μm; accessory rutile (Rt), ilmenite (Ilm), Ttn, Zrn and varisized, commonly poikillitic magnetite (Mag) up to 1–1.5 mm.

Garnet has corroded and curvilinear outlines, bears inclusions of Qz observed throughout the Grt grains, which associate with ilmenite in the garnet cores and rutile – only within the outer zones of Grt porphyroblasts. Inclusions of Ep and Bt tend to be enclosed in the outer zones of Grt cores. The composition varies within the $Alm_{43–67}Grs_{13–21}Prp_{5–13}Sps_{5–28}$ range. Porphyroblasts are homogeneous in terms of MgO and especially CaO distribution (Figs 4a, 5e, 5f), but exhibit slight FeO (Fig. 5a, b) and significant MnO enrichment in the thin outermost porphyroblast rims (up to 4.5–12.0 wt% of MnO v. <3.0% in the core parts) (Figs 4a, 5c, 5d). Biotite is moderately Fe-enriched, with variable #Fe (calculated ratio of $Fe^{2+}/(Fe^{2+} + Mg)$) within 0.40–0.45. White mica of phengite composition commonly occurs within the aggregate of Bt or replaces it (Fig. 3c), has limited $Mg–Fe^{2+}$ contribution at #Fe = 0.63–0.71 and calculated Si values within 3.13–3.21 per formulae unit (Fig. 4b). Matrix Ep, including mass-deficient REE-bearing Ep (referred to as allanite by Shatsky *et al.* 2012), has an X_{Fe3+} of 0.20–0.22, whereas X_{Fe3+} values of Ep inclusions are more variable (0.15–0.24). Matrix Kfs has 23–26% of Ca (anorthite) component, whereas Pl is close to albite. Isometric ilmenite inclusions were detected in the core domains of Grt, most of which are Mn-poor (<0.5 wt% MnO). Along with that, Mn-rich (MnO = 13.7 wt% MnO) Ilm was detected once in the homogeneous Grt core. Small isometric and elongated Rt grains up to 20 μm were found enclosed in the outermost zones of Grt cores and were not detected in the sample matrix.

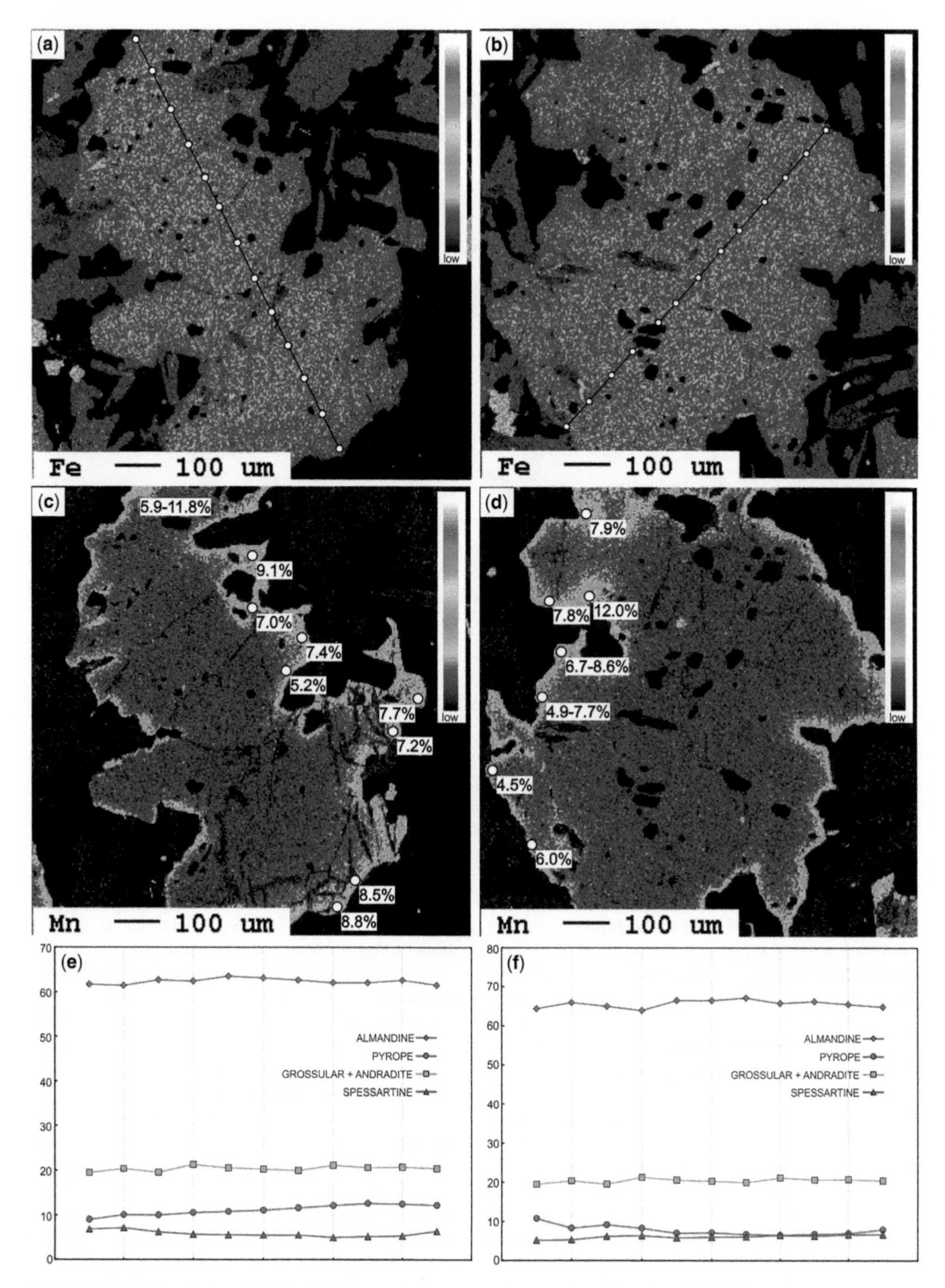

Fig. 5. X-ray elemental (Fe, Mn) maps (**a–d**) and cation distribution profiles (**e**, **f**) of the selected garnet porphyroblasts from the Mu-93-45 paragneiss. The data from detailed rim analysis of porphyroblasts are provided to show the variability of MnO enrichment in a single grain.

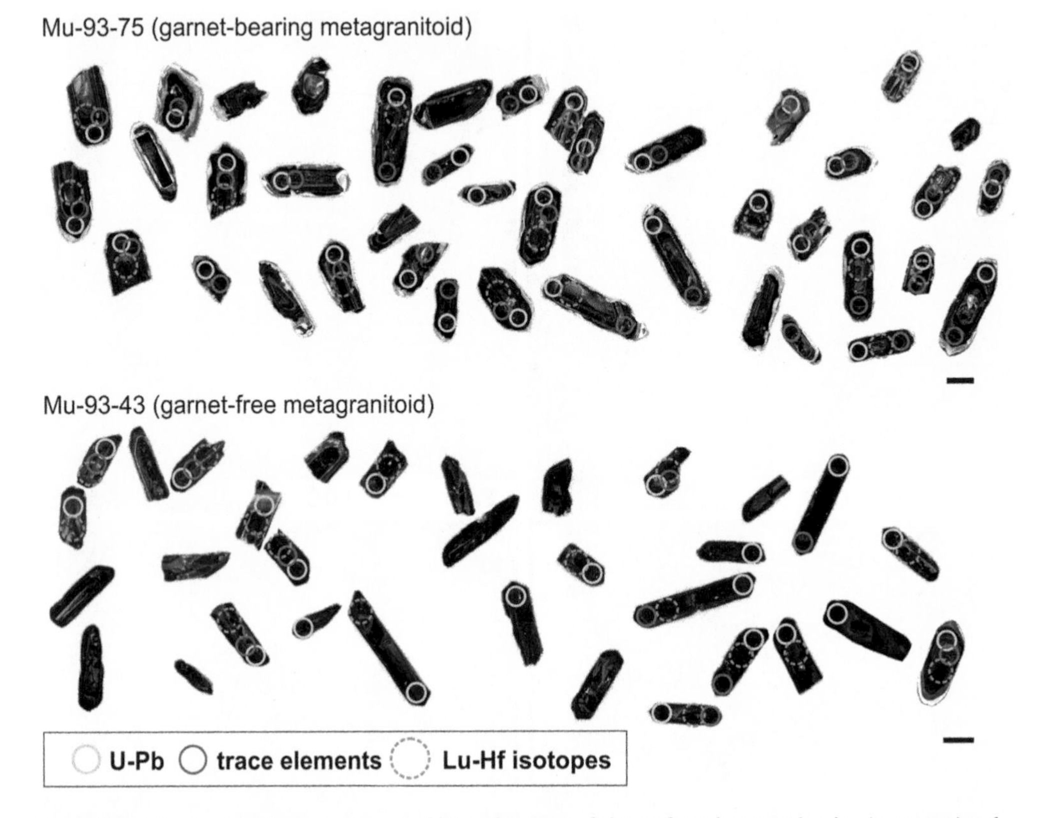

Fig. 6. Selected cathodoluminescence images of internal textures of zircons from the garnet-bearing (upper part) and garnet-free (lower part) metagranitoids. Scale bars are equal to 50 μm. The spot locations for *in situ* LA-ICP-MS (U–Pb, trace-element, Lu–Hf) are shown in grains, where multiple analyses were available.

Zircon morphology and textures

Zircons from the weakly deformed Mu-93-75 metagranitoid are light pinkish, transparent, prismatic grains up to 200 μm in size, with a 1 : 2.5 to 1 : 4 aspect ratio, and have well-preserved euhedral morphology and oscillatory zoning of primary magmatic zircon (Fig. 6, upper panel). Most grains exhibit thin up to 20 μm) and uneven recrystallization rims with bright CL, in contrast to dark CL of primary cores. Preliminary optical and Raman spectroscopic investigations of mineral inclusions revealed the predominance apatite, which typically forms long prismatic crystals no more than 10–15 μm in size.

Zircons from the moderately deformed, garnet-free Mu-93-43 metagranitoid are also represented by long prismatic pinkish and transparent grains, which are mainly within 180 μm in size and 1 : 4–1 : 6 aspect ratios, and are close to euhedral in shape. Although most zircons exhibit dark CL similar to that of zircons from the Mu-93-75 metagranitoid, zircon grains here commonly have the curvilinear shape of oscillatory zoning and patchy textures with recrystallization fronts, whereas recrystallization rims with bright CL are lacking (Fig. 6, lower panel). Prismatic apatite was identified as a major mineral phase among zircon inclusions.

Detrital zircons from the metasedimentary garnet gneiss Mu-93-45 are mainly short prismatic (aspect ratios within 1 : 1.5 to 1 : 3) grains with partially rounded, resorbed outlines (Fig. 7). Detailed assessment revealed two types of textures in the two analysed batches of zircons. First picked, generally larger grains, exhibit the systematic presence of apparent varisized cores, which show either bright or dark CL and partially preserved oscillatory zoning (mainly within ~15–30% vol%), followed by strongly heterogeneous recrystallization mantles and occasionally preserved thin recrystallization rims with bright CL (Fig. 8). Smaller zircons of the second batch, except for a few grains, lack the mentioned recrystallization mantles, but instead exhibit patchy or homogeneous cores and bright recrystallization rims. Unlike zircons from the two metagranitoids, the voluminous recrystallization mantle in detrital zircon from this sample bears multiple

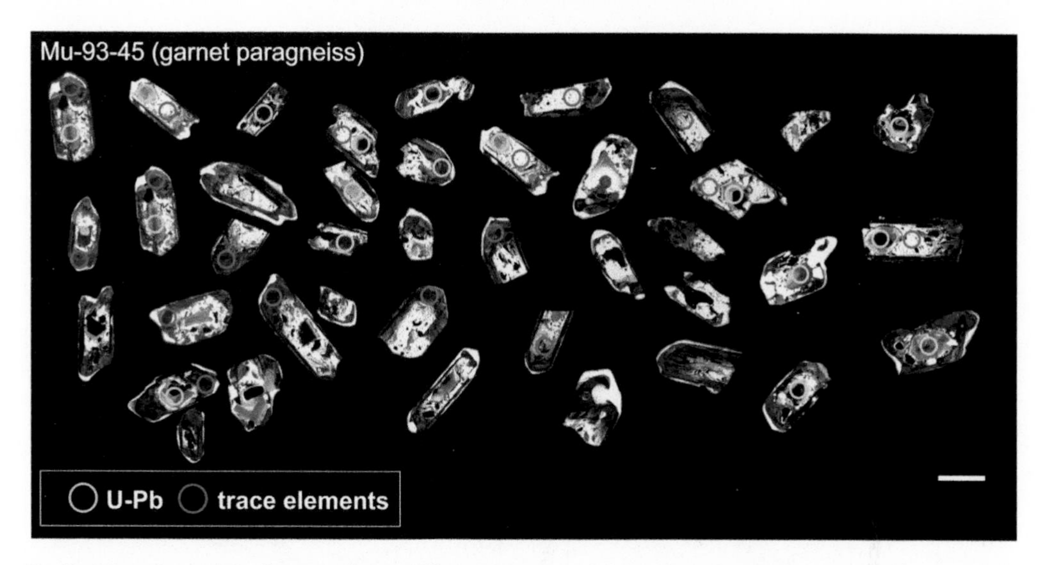

Fig. 7. Selected cathodoluminescence images of internal textures of zircons from the garnet paragneiss. The scale bar is equal to 100 μm. The spot locations for *in situ* LA-ICP-MS (U–Pb, trace-element) are shown in grains, where multiple analyses were available.

inclusions of metamorphic minerals, including white mica, Ep, Qz, rarer Ap, Ttn, albite, Kfs, as well as Rt and Grt. White mica is represented by Si-poor phengite similar in composition to matrix phengite (#Fe = 0.53–0.65; 3.08–3.19 Si p.f.u.; see Fig. 4b), along with paragonite (#Fe = 0.82–0.92; K/Na 0.09–0.13), which is absent in the matrix. Epidote shows a slightly lower X_{Fe3+} of 0.16–0.18 as compared to matrix Ep. Titanite has 2.8–3.4 wt% of Al_2O_3 and systematic admixture of F (0.1–1.3%). The inclusions of Rt and Grt are evidently rare and were detected only once.

Zircon U–Pb geochronology and trace-element composition

Overall, 143 U–Pb analyses on 137 grains were analysed via LA-ICP-MS techniques. The data have been compiled and are provided in the supplementary file.

Zircons from the Mu-93-75 metagranitoid yield very uniform U–Pb dates with a major cluster of age estimates ($N = 30$) corresponding to the 799 ± 4 Ma age of its igneous protolith (Fig. 9a).

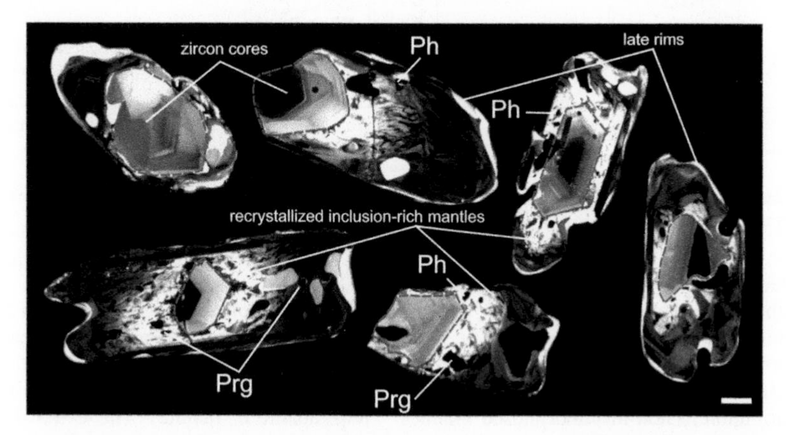

Fig. 8. Detailed cathodoluminescence (CL) images of selected zircons from the garnet paragneiss showing the complex, at least two-stage metamorphic recrystallization patterns. The latter include preserved oscillatory cores of igneous origin (outlined by purple dashed lines), heterogeneous recrystallization mantles with mineral inclusions (Ph denotes phengite, Prg denotes paragonite), and occasional outermost bright-CL recrystallization rims.

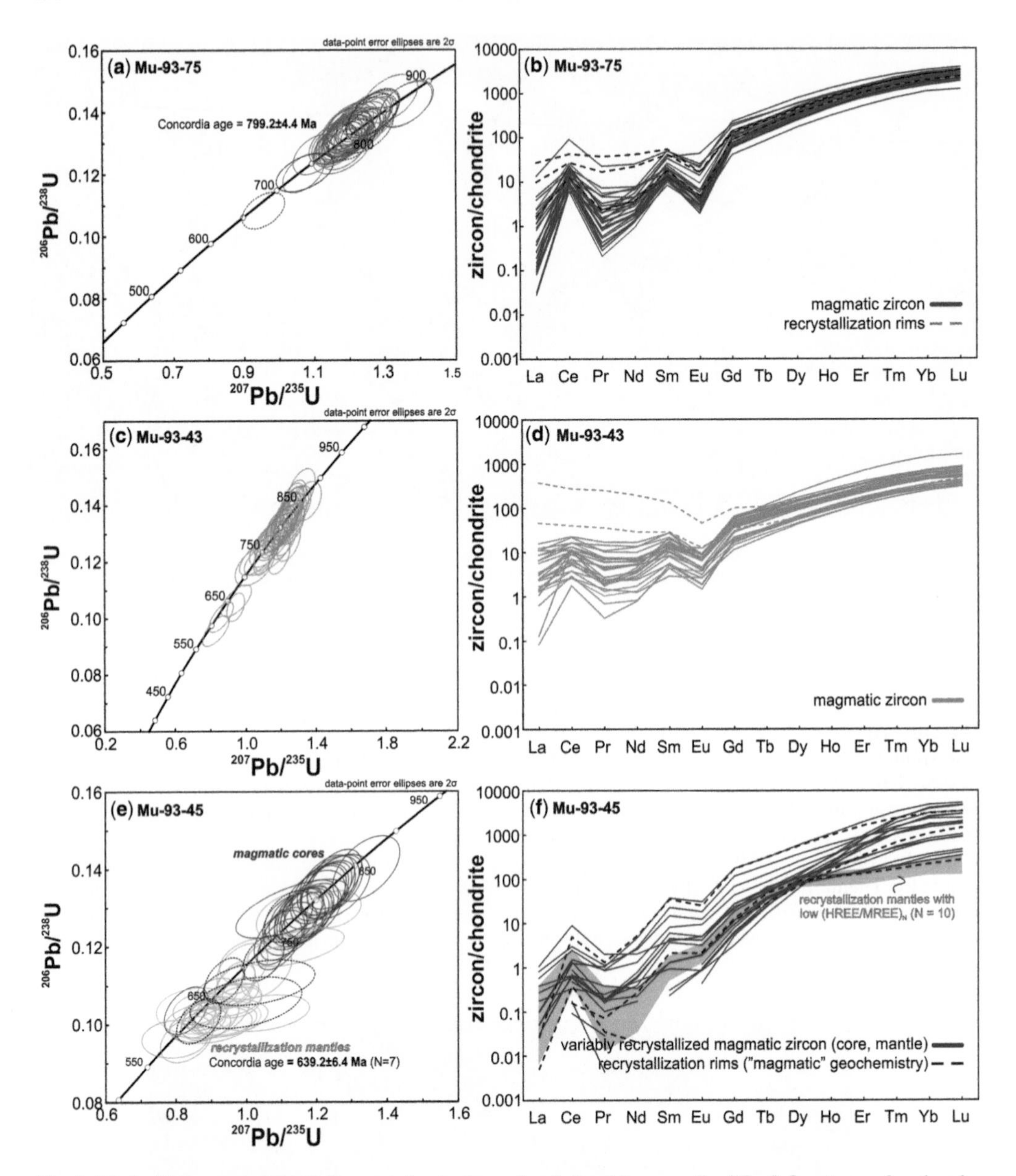

Fig. 9. Wetherill diagrams of the U–Pb age estimates (**a**, **c**, **e**) and chondrite-normalized (**b**, **d**, **f**) patterns of analysed zircons from three felsic samples.

Several older single-grain values, including four grains with concordant age estimates (826 ± 13 Ma), are either related to antecrysts or were trapped from more ancient granite intrusions in the NMB lithosphere, which are hinted at from the U–Pb age distribution in non-metamorphosed granites (Skuzovatov *et al.* 2016*b*). The analysed zircons exhibit uniform distribution patterns of REEs, with low $(LREE/HREE)_N$ ($(Sm/Yb)_N = 0.01–0.02$, $(Gd/Yb)_N = 0.03–0.07$), clear Eu/Eu*-minima (0.06–0.50) and Ce/Ce*-maxima (1.34–80.1), which are typical for igneous zircon from granitic rocks (Hoskin and Ireland 2000; see Fig. 9b). A few grains show LREE enrichment, which more probably refers to inclusions of LREE-rich phases (e.g. apatite or titanite). The U–Pb dates for these spots show normal or reverse discordance and thus were excluded from further consideration (three spots). A few available U–Pb analyses within thin recrystallization rims showed disturbance of the U–Pb system, which is stressed by either abundant reversely discordant age estimates, or normally discordant estimates variably shifted from concordia (^{206}Pb–^{238}U ages within 811–663 Ma). However, no systematic effect of recrystallization is observed in the trace-element systematics of these rims.

Much more notable variations were revealed for zircons from the deformed and garnet-free Mu-93-43 metagranitoid. The scatter of concordant single-grain U–Pb ages (846–758 Ma; Fig. 9c) precludes a precise concordia calculation of a potential igneous protolith age. Nonetheless, a clear age distribution peak for zircons with <5% discordance ($N =$ 34) is observed at *c.* 810 Ma. A few grains exhibit Late Neoproterozoic discordant (within a discordance exceeding 8%) estimates within ~670–600 Ma, which indicate progressive Pb loss by the primary zircon. While the U–Pb system was mobile and partially reset, REE systematics of the same zircon grains preserve the original magmatic affinities and show high $(HREE/LREE)_N$ ($(Sm/Yb)_N =$ 0.01–0.15, $(Gd/Yb)_N = 0.02$–0.07), Eu/Eu* of 0.12–0.51 and Ce/Ce* of 1.24–37.7 (Fig. 9d).

The metasedimentary gneiss sample Mu-93-45 presents distinct U–Pb and trace-element systematics in zircons. The analysis revealed that relic igneous cores with oscillatory zoning have their U–Pb isotope system well preserved and unmodified; the corresponding concordant (<5% discordance) ^{206}Pb–^{238}U age estimates lie within 831–745 Ma (Fig. 9e), whereas trace-element measurements were unavailable given limited grain space in the same domains. Trace-element signatures of recrystallization mantles, including their homogeneous counterparts, vary from typical 'igneous' (with high $(HREE/LREE)_N$ ratios, Eu/Eu* minima and Ce/Ce* maxima), to those with a nearly flat MREE–HREE distribution and the absence of Ce/Ce* and Eu/Eu* anomalies (Fig. 9f). More specifically, REE patterns of typical 'igneous' zircon exhibit two distinct types, the first exhibiting a clear Eu/Eu* (0.32–0.86) and moderately positive slope of HREE ($(Sm/Yb)_N = <0.02$, $(Gd/Yb)_N =$ 0.01–0.07), and another with a steeper HREE profile ($(Sm/Yb)_N = <0.01$, $(Gd/Yb)_N <0.01$) and almost absent Eu/Eu* anomalies. The concordant ^{238}Pb–^{206}U age estimates of zircons, whose REE patterns show flat MREE–HREE profiles (Eu/Eu* = 0.61–0.92, $(Sm/Yb)_N = <0.01$, $(Gd/Yb)_N =$ 0.04–0.14), yield a concordia intercept at 629 ± 6 Ma ($N = 16$). Importantly, limited analyses ($N =$ 7) of thin recrystallization rims revealed mainly discordant age estimates (Fig. 9e, dashed ellipses), which spread over concordia within ~776–619 Ma (i.e. from the suggested magmatic ages towards a metamorphic modification stage). Same rims display either magmatic trace-element signatures from either magmatic or metamorphic (high-grade) zircon described above for recrystallization mantles (Fig. 9f).

Zircon O isotopes

We carried out 60 analyses on 48 zircon grains from the three studied samples, and both zircon cores/mantles and thin recrystallization rims were analysed in grains, where enough space was available. The results are provided in the Supplementary File S3 and Figures 10 and 11(b). Zircon from the Mu-93-75 metagranitoid (Fig. 10a, b) displays rather limited variations of measured δ^{18}O in the core (magmatic) domains with dark CL, which range within +8.3 to +8.0‰ and thus are moderately elevated relative to a mantle value. The values tend to shift towards lighter, ^{18}O-depleted compositions in zircon grains, where recrystallization rims are formed, and yield a range from 'transitional' (+7.5 to +7.1‰) to terminal ones (+6.9‰ to +6.8‰), which are predominant in the rims. Seven grains with both core and rim measured have $\Delta\delta^{18}$O up to 0.6–1.6‰ difference between the ^{18}O-enriched core and ^{18}O-depleted rim. Importantly, two of three zircon cores with slightly depleted δ^{18}O (+6. 7 and +7.5) do not show any significant changes in U–Pb dates other than a slight reverse discordance, whereas one core (93-75_53d) was evidently affected by Pb loss. In turn, as mentioned above, U–Pb estimates in recrystallized rims variably shift from concordia, from nearly concordant to apparently discordant (mainly normally discordant) estimates with younger ^{206}Pb–^{238}U ages down to 663 Ma.

Zircon from the deformed Mu-93-43 metagranitoid, where zircon grains mostly lack recrystallization rims (Fig. 10a, c), overall have more scattered δ^{18}O values. However, 15 out of 20 analyses yielded a very narrow range of moderately elevated δ^{18}O values within +9.2 to +8.5‰, whereas only five out of 20 grains show ^{18}O-depletion (+8.4 to +7.7‰). The most remarkable are the δ^{18}O signatures of zircon from the metasedimentary sample Mu-93-45, where zircon is unusually high δ^{18}O (+14.6 to +13.8‰). Only one exception was found in a preserved magmatic core with δ^{18}O of +8.2‰ (Fig. 10a, d) with high δ^{18}O (14.3‰) in a recrystallization rim. Three grains, where both core/mantle and rim were measured, exhibit zero to ~0.8‰ shift towards ^{18}O-depletion in the rim.

Zircon Lu–Hf isotopes

We perfomed 81 Lu–Hf isotope analyses on 79 zircon grains from the three samples. The data are provided in the Supplementary File S4. Zircon from the Mu-93-75 metagranitoid displays moderate variations of juvenile Hf isotope signatures, with $\varepsilon_{Hf}(t)$

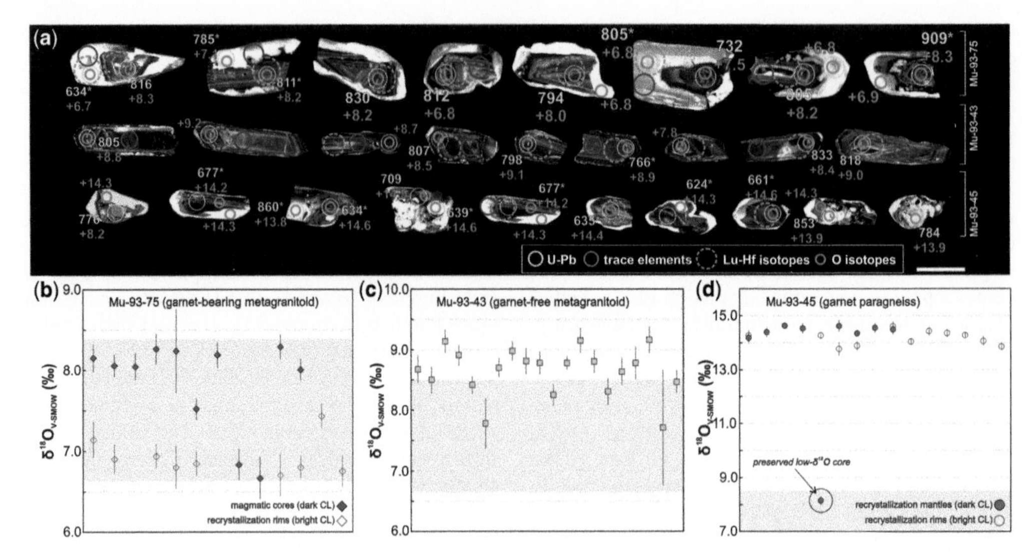

Fig. 10. Cathodoluminescence images of selected zircons analysed for U–Pb age and oxygen isotope composition (**a**) and $\delta^{18}O$ values of zircons from the three felsic samples (**b–d**). Yellow numbers are ^{206}Pb–^{238}U age estimates (Ma); those with an asterisk mark variably discordant age estimates. Red numbers are $\delta^{18}O$ values. The grey field marks the $\delta^{18}O$ range for zircon from the Mu-93-75 metagranitoid with a narrow range of core values and a systematic ^{18}O depletion of recrystallization rims. This range is used in (**c**) and (**d**) as a reference to highlight the scale of variations in a partially recrystallized zircon from the deformed metagranitoid Mu-93-43 and a uniquely high $\delta^{18}O$ value of zircon from the Mu-93-43 paragneiss.

values recalculated to the measured ^{238}Pb–^{206}U ages within +3.4 to +8.2 and Mesoproterozoic crustal model ages (T^{C}_{DM}) within 1.51–1.15 Ga (Fig. 11a). These values closely resemble $\varepsilon_{Hf}(t)$ variations for both metagranitoid and non-metamorphosed granite reported by Shatsky *et al.* (2015) and Skuzovatov *et al.* (2016*b*), and evidently reproduce the initial Hf isotope heterogeneity in granitic melts. In contrast, zircon from the deformed metagranitoid Mu-93-43 displays mainly negative, 'crustal' $\varepsilon_{Hf}(t)$ values from to −7.6 to +1.5 and T^{C}_{DM} within 1.58–2.10 Ga. These features are not common for the NMB felsic rocks but were reported as typical for the South Muya block granites and metasediments (Skuzovatov *et al.* 2019*a*). In contrast, zircon from the metasedimentary sample Mu-93-45 exhibits the most juvenile Hf isotope signatures, with $\varepsilon_{Hf}(t)$ ranging within +7.9 to +11.7 and T^{C}_{DM} within 0.93–1.10 Ga. Zircon from both the metasedimentary sample and deformed metagranitoid displays similar trends of $^{176}Hf/^{177}Hf$ evolution with Lu/Hf ratios below 0.015, which may result, to some extent, from recrystallization or Pb loss by primary igneous/detrital zircon. However, because apparent variations of $^{176}Hf/^{177}Hf$ (i.e. vertical arrays) are present in each sample due to a real scatter of values and instrumental uncertainties, the contribution of zircon recrystallization to Lu–Hf systematics cannot be unambiguously justified.

Phase equilibria modelling

We used the phase diagram analysis approach to assess the possible prograde-to-peak evolution of the chosen sample using the composition of Grt, Bt and Ph and the presence of critical Ti-bearing phases in the relevant mineral assemblages. As shown below, Grt in felsic rocks likely tends to have zonation in MnO and FeO that allows them to be used as robust markers of *P–T* conditions recorded during progressive burial and heating of the NMB. Thermodynamic modelling of *P–T* conditions for the two representative garnet-bearing samples (metagranitoid and metapsammite) was performed in the Na_2O–CaO–K_2O–FeO–MgO–MnO–Al_2O_3–SiO_2–TiO_2–H_2O (MnNCKFMASTOH) system using the Perple_X software (version 6.7.6) (Connolly 2005). Major-oxide bulk-rock composition determined using XRF by Shatsky *et al.* (2012) was used as an effective bulk composition (EBC) for the Mu-93-45 as it displays uniform mineral distributions, modal amounts and mineral composition in each thin section checked. The thermodynamic database of Holland and Powell (1998) (version 1998, updated to 2002) was applied for the pressure interval of 0.5–1.5 GPa and temperature interval of 500–700°C, assumed to be relevant for the observed mineral assemblages: the a–x solid-solution models of Ms–Ph, Grt and Chl after Holland and Powell

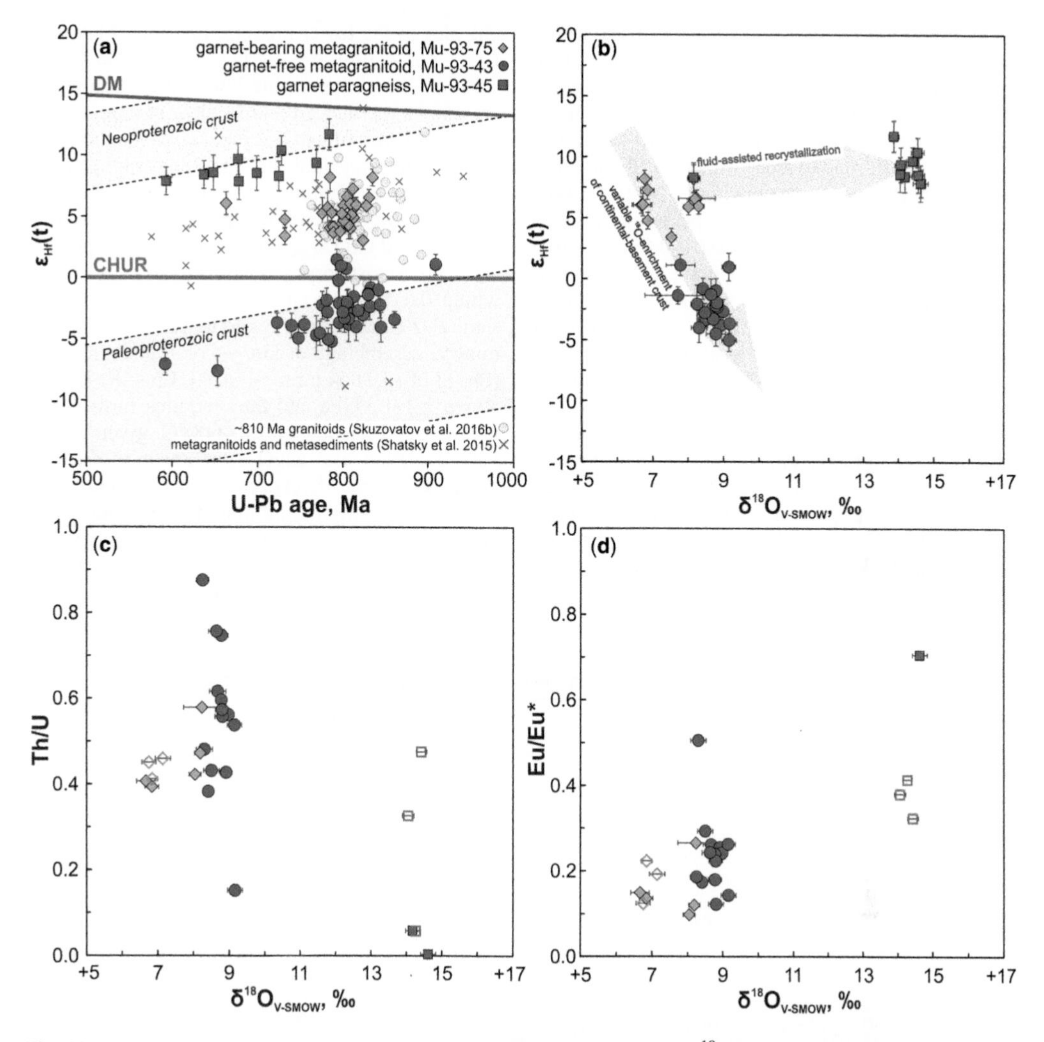

Fig. 11. $\varepsilon_{Hf}(t)$ v. U–Pb diagram (**a**) and selected trace-element ratios plotted v. $\delta^{18}O$ values (**b**, **c**, **d**), which demonstrate the effect of zircon modification during metamorphism. Please note the dispersion of values along the average crustal evolution (Lu/Hf = 0.015) trends, which derives from progressive recrystallization and Pb loss. The vertical spread of $\varepsilon_{Hf}(t)$ values in both metagranitoids reflects the potential heterogeneity of their melting sources.

(1998); Amp after Dale *et al.* (2005); ternary feldspar after Benisek *et al.* (2010); and the CORK model of Holland and Powell (1991).

Based on petrographic evidence, the observed metamorphic assemblage in the metasedimentary sample Mu-93-45 includes Grt, Ep, Bt and Ph. The prograde Ti-bearing phase during Grt nucleation corresponding to Grt cores was likely Ilm, which was then replaced in the assemblage by Ttn and Rt (from inclusions in recrystallized zircon and Grt rims). The abundance of Ep and Mag/Ilm in both matrix and Grt porphyroblasts requires a non-zero $Fe^{3+}/\Sigma Fe$ ratio used for modelling. Most of the Fe in metagranitoids was transferred from ilmenite + biotite to the newly formed Ep, Grt, Mag and minor Ph, which make Ep and Mag major carriers of Fe^{3+}. A few preliminary *P–T* pseudo-sections were plotted for $Fe^{3+}/\Sigma Fe$ values within 0–0.15 to track the effect of Fe^{3+} incorporation, which was found to be negligible. We thus further use an average $Fe^{3+}/\Sigma Fe$ value of 0.10 based on both Fe^{3+} content in these phases and their modal proportions, and the maximum H_2O was thus set at the level of measured loss on ignitions percentage. Additionally, we examined the effect of the increasing fluid (H_2O) amount, which was found to be completely buffered by the amount of hydrous Fe–Mg phases (i.e. biotite, epidote and white mica) in low-temperature fields (<550°C).

The assemblage of Qz + Bt + Grt + Ep + Ilm, which represents the prograde metamorphic stage based on the Grt porphyroblasts and their inclusions, is stable over a significant pressure range mainly at the low-temperature (Chl stable) to medium-temperature (Ep stable) area, limited by ilmenite-out and riebeckite-in reactions at ~1.0–1.1 GPa (Fig. 12a). Garnet isopleths for spessartine (X_{Sps} = 0.05–0.08), almandine (X_{Alm} = 0.61–0.67) and grossular (X_{Grs} = 0.13–0.20) components, corresponding to Grt cores, intersect best in the LT–LP field of ~540–560°C and ~0.8–1.0 GPa in the Kfs + Qz + Bt + Grt + Ep + Chl + Prg + Ilm stability field (Fig. 12a). Neither HP white mica (phengite) nor titanite/rutile are observed as stable phases, which is consistent with petrographic observations and relatively medium-temperature, medium-pressure mineral assemblage of the Mu-93-45 gneiss. Some uncertainties are present, though, between Fe^{2+} (X_{Alm}) – Mn (X_{Sps}) and Ca (X_{Grs}) isopleths. The isopleths with the lowest X_{Grs} values (down to 0.13) at relevant X_{Sps} (0.05–0.08) values of porphyroblastic Grt plot in the lower-pressure field of Chl + Qz + Kfs + Grt + Bt + Ilm + Mag (~540–560°C and 0.6–0.8 GPa), whereas the lowest X_{Alm} below 0.65 are poorly reproduced and localize at unreasonably high pressures (up to 1.2 GPa at similar temperatures). This may indicate late-stage, higher-T and higher-P re-equilibration and Fe^{2+}–Mg diffusion. For the same bulk rock composition, the growth of Mn-enriched Grt (X_{Grs} = 0.11–0.28), corresponding to the outermost porphyroblast rims, is possible only within 0.6–0.8 GPa and mainly below 600°C. The same P–T calculations suggest that the newly formed mineral assemblage of Grt + Prg + phengitic mica (Ph) (3.08–3.21 Si p.f.u.) + Ab + Ep + Rt is stable above 1.2–1.3 GPa and thus requires further compression and heating up to ~600°C, given a slight FeO depletion and MgO enrichment of Grt rimwards. To test the effect of prograde Grt volume fractionation on to Mn partitioning, for the modelling of late Grt growth, a new EBC was adjusted for the outer Grt cores and rims using the subtraction of 11% Grt with the average composition of core

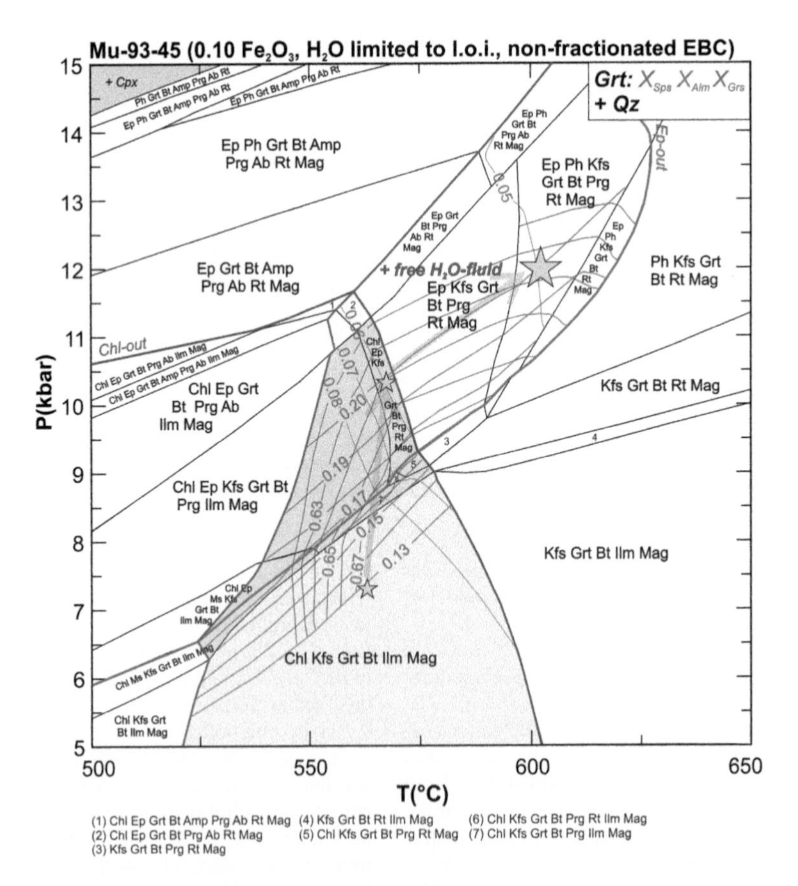

Fig. 12. Pressure–temperature pseudo-section diagram for the Mu-93-45 garnet paragneiss. Yellow stars mark major metamorphic stages based on the calculations. See the text for a detailed description of the calculation parameters.

domains. The effect was found to be negligible, as a phengite + rutile assemblage turns out to be stable at even lower pressures (0.8–1.0 GPa).

Discussion

Mineralogical indicators and P–T conditions of HP metamorphism: petrological context for zircon modification

The association of eclogites with evidently lower-pressure (<1.5 GPa) metagranitoids and paragneisses, which record less significant pressures, is common worldwide (e.g. Massonne 2015; Li and Massonne 2016). Such co-existence has been so far interpreted to result from tectonic juxtaposition of deeply subducted mafic rocks with shallower and more buoyant continent-derived felsic rocks at the final collisional stage (Li and Massonne 2016), or contrasting degrees of HP assemblage preservation in mafic and felsic rocks due to fluid/melt overprinting and dynamic recrystallization (Proyer 2003). Three studied gneiss samples, which associate geographically and at the outcrop-scale with eclogites (including the only kyanite eclogite found), exhibit distinct degrees of metamorphic modification and responses to strain partitioning. The two metagranitoid samples (Mu-93-43 and Mu-93-75) reveal, in the case of the former, no indicators of high-grade metamorphic phase nucleation (i.e. epidote, garnet or rutile) at a moderate degree of ductile deformation, and, in the latter case, the formation of garnet–epidote assemblage in the presence of titanite. Although no *P–T* estimations could be acquired in this study for metagranitoids, the maximum pressure, in the case of Mu-93-75, may be constrained by rutile–titanite transition by maximum of 1.1–1.3 GPa within ~500–650°C (Angiboust and Harlov 2017). Assuming a mainly CaO-poor (<3.0% CaO) composition of NMB granitic gneisses, a partial titanite–rutile transformation should be expected at ~0.7–1.2 GPa, which is not evident in the studied sample, and thus may further limit possible peak pressure for this sample below ~1 GPa. The absence of garnet in Mu-93-43 and its presence as a low-volume phase in Mu-93-75 implies either lower pressures (<0.6–0.7 GPa), or a significant impediment of garnet growth reactions in a dry felsic protolith (Zeh and Holness 2003; Schorn 2022).

For the Mu-93-45 metasedimentary sample, a more specific *P–T* history may be retrieved by combining the porphyroblastic garnet composition with mineral inclusions in both garnet and zircon. Based on the garnet composition, the prograde assemblage evolved from garnet nucleation in the epidote-absent conditions, likely within ~0.7 GPa and ~560°C, through the epidote- and ilmenite-present assemblage at ~0.8–1.0 GPa at similar temperatures, to the rutile-bearing assemblage above ~1.0 GPa and 570°C (second star at the mentioned *P–T* conditions in Fig. 12). This prograde pressure level only slightly underestimates the experimentally derived ilmenite to rutile transition curve lying at ~1.3 GPa (Angiboust and Harlov 2017). The almost isothermal *P–T* trajectory produces only limited variations of X_{Alm} and X_{Grs}, whereas even a more strict range of X_{Sps} compositions is generated, and thus no late-stage diffusional re-equilibration is necessary to account for garnet zoning. Importantly, paragonite is stable in the reconstructed prograde assemblage throughout this prograde path, which is consistent with its presence in recrystallized zircon (see the earlier subsection 'Zircon morphology and textures'). The calculations revealed the composition of the outermost garnet cores and phengite presence in the matrix reproduced at least above ~1.2 GPa and ~600°C (Fig. 12) as a result of further compression and only slight heating. The latter is consistent with the conventional garnet–biotite thermobarometry data, which record the average temperature within 590 ± 15 to 640 ± 26°C by different garnet–biotite thermometer calibrations (Shatsky *et al.* 2012). Overall, these peak conditions are consistent with ilmenite- and rutile-bearing assemblages reproduced below ~1.4 GPa and Ab = Jd + Q reaction curve, as no jadeite was identified in the recrystallized matrix assemblage (Fig. 12b). The presence of all major matrix phases, including white micas, epidote, quartz, feldspars, with rare garnet and rutile, as inclusions in zircon, assume that prograde zircon recrystallization proceeded continuously within 0.7–1.3 GPa and ~560–600°C. Although some suprasolidus reactions throughout the metamorphic evolution of rocks at above ~600°C cannot be excluded, there is neither field-based, nor petrographic (e.g. peritectic mineral growth or dehydration breakdown textures) evidence of anatexis and migmatization.

Although the reconstructed *P–T* evolution allows the occurrence of phengite partially overprinting a previously stable assemblage, the occurrence of Ttn in recrystallized Zrn and garnet porphyroblasts, the stability of phengite as a key HP phase, and MnO enrichment in garnet are rather uncertain. The reconstructed *P–T* evolution excludes both the prograde stability of titanite and prograde formation of titanite over ilmenite (Fig. 12). The occurrence of titanite in some recrystallized zircon grains (this study) and some garnet porphyroblasts (Shatsky *et al.* 2012) may thus be controlled by the local CaO activity in a metasedimentary protolith, or titanite might be a detrital phase in a metasedimentary protolith. According to the titanite–rutile barometry of Massonne and Schreyer (1987), phengite with Si p.f.u. within 3.08–3.19 in the equilibrium with biotite should be stable below ~0.5–0.6 GPa. However, for the given bulk composition, the highest Si-in-phengite values (up to 3.21) are reproduced at ~1.0–1.2 GPa and above 600°C, which is in agreement with the peak *P–T* constraints for the NMB gneisses outlined above.

The peak or post-peak nucleation of Mn-rich garnet may result from either re-heating of subducted rocks, or external post-peak fluid/melt infiltration (e.g. García-Casco *et al.* 2002; Nyström and Kriegsman 2003; Tan *et al.* 2020; Skuzovatov 2021). The *P–T* phase modelling performed for both bulk and fractionated (garnet-free) EBCs indicates similar trends of gradual prograde MnO depletion of a growing garnet, whereas Mn-rich garnet is reproduced only at <550°. Therefore, re-heating following peak pressure conditions was unlikely to be a driving force for enhanced MnO incorporation into garnet. Along with garnet, ilmenite may be a key Mn-bearing phase, which controls Mn budget in the medium-grade metasediments (Evans and Guidotti 1966; Itaya and Banno 1980; Pattison and Tinkham 2009; McCarron *et al.* 2019). The finding of Mn-rich ilmenite enclosed in porphyroblastic garnet supports the abundance of such ilmenite as a possible (but not necessary) source of MnO in metasediments. As prograde evolution of NMB metasediments implies ilmenite–rutile transition and favours MnO liberation into a metamorphic fluid (Fig. 12), MnO-enriched fluids may be responsible for Mn redistribution in gneisses from prograde ilmenite breakdown to back-reaction of low-volume fluids with the host metasedimentary rocks. Because Mn-rich garnet was identified also in a garnet-bearing metagranitoid Mu-93-75, where ilmenite is lacking as an accessory phase, Mn-rich fluids might have an external sedimentary source for both types of NMB felsic rocks. Thus, low-Si phengites in the Mu-93-45 paragneiss and the presence of Mn-rich garnet in both paragneiss and metagranitoid may be produced via post-peak modification of gneisses by low-volume hydrous Mn-rich fluids.

To summarize, limited abundance of high-grade phase assemblages (especially garnet + rutile + phengite in a paragneiss) and direct calculations from the composition of both matrix minerals and inclusions in the recrystallized zircon, suggest that the NMB felsic rocks might be involved in the same high-grade metamorphic episode. However, there is so far no petrological evidence that the studied felsic rocks reached the peak depths of eclogitic metamorphism (at least above ~1.2 GPa). The inconsistency between metamorphic evolution of deeply buried eclogites and their likely shallower hosts could result from a limited burial and metamorphic transformation of a continental margin. In this case, the exhumation and juxtaposition of eclogites with lower-grade felsic rocks might be tectonically controlled, for instance, by thrusting and nappe stacking during a final collision stage.

Zircon transformation during metamorphism: textural and chemical evidence of elemental and isotopic mobility

Zircon is well known as an accessory phase with a strong ability to retain the original U–Pb isotope age characteristics as it has the high closure temperature of Pb diffusion well above the range of most crustal processes (Cherniak and Watson 2003). Along with that, zircon may recrystallize in a solid state or dissolve-reprecipitate in the presence of fluids under a wide range of metamorphic conditions (Rubatto 2002*a*; Rubatto and Hermann 2007; Liu and Liou 2011; Kohn *et al.* 2015; Chen and Zheng 2017). Thus, a crucial point in the geochronological characterization of subduction-related high-grade metamorphic rocks is to properly distinguish primary igneous/detrital zircon from newly formed or recrystallized grains or growth zones, and the correspondence between zircon growth and metamorphic *P–T–t* conditions.

On the course of subduction-collision metamorphism, the NMB complexes evolved in subsolidus conditions (Shatsky *et al.* 2012; Skuzovatov *et al.* 2019*b*), which, for most of the felsic rocks, occurred below ~600–650°C (Shatsky *et al.* 2012; this study) and below the U–Pb closure temperature (Cherniak and Watson 2003). This favours the preferential preservation of initial zircon U–Pb systematics, which is only the case of the Mu-93-75 undeformed garnet-bearing metagranitoid, and is consistent with observations of well-preserved internal textures of igneous zircon (Fig. 6). In this case, only thin recrystallization rims form as a likely response to late-stage modification potential related to fluid infiltration. Evidently the coupled and limited development of HP metamorphic assemblages, which include only occasional garnet and epidote, and the unmodified state of the zircon U–Pb system, imply both a limited amount of H_2O preserved in mineral igneous phases (biotite) and the absence of external fluid during prograde transformation (Peterman *et al.* 2009; Young and Kylander-Clark 2015; Schorn 2022).

In the deformed, biotite-rich metagranitoid Mu-93-43, the observed convoluted zoning, patchy textures and a corresponding spread of U–Pb isotope dates over concordia (^{206}Pb–^{238}U ages of 846–758 Ma, with a number of younger discordant estimates) (Fig. 9c) indicate partial recrystallization of zircon. Grain textures and the absence of metamorphic minerals as inclusion phases indicate the limited scale of this recrystallization, which was enough to produce local intragrain mobility of Pb and its variable loss in a metamorphic and/or deformation episode. It is generally accepted that deformation-related zircon transformation is typical of high-temperature shear zones (Wayne and Sinha 1988), whereas examples of low- to medium-temperature U–Pb system disturbance are rare or unlikely (Gebauer and Grünenfelder 1976), unless dissolution–reprecipitation or annealing and partial recrystallization of high-U–Th zircon are involved. The absence of apparent metamorphic assemblages (i.e. garnet, epidote, phengite and rutile) suggests lower-temperature crustal storage of this metagranitoid far below a required *P–T* level of eclogitization, biotite dehydration and plagioclase breakdown (within 1.3–1.4 GPa for 600°C; Schorn 2022). Thus, mobile behaviour of Pb in this case may be related to post-metamorphic fluid-assisted deformations.

Zircon from both metagranitoids lacks any evidence of REE redistribution and perfectly preserves what is here interpreted to represent the original magmatic Eu/Eu* and Ce/Ce* signatures (Fig. 9b, d), similarly narrow ranges of HREE/MREE ratios (Fig. 13a, b). However, unlike zircon from the undeformed metagranitoid Mu-93-75, which displays typically narrow ranges of Th/U mainly within 0.3–0.5, the same ratios for zircon from the Mu-93-43 deformed gneiss scatter within 0.1–0.8 (Fig. 13c) and highlight the U mobility at predicted metamorphic temperatures (~570–600° C). The potential effect of partial recrystallization is stressed by Ti contents and calculated Ti-in-zircon temperatures, which are within a narrow range of 550–620°C (calculated here, for instance, after Watson *et al.* 2006 for $a_{TiO2} = 0.5$) for Mu-93-75, but spread over ~620–900°C for Mu-93-43 (Fig. 13d). This partial recrystallization did not affect Lu–Hf isotope systematics of primary zircon, as it exhibits similar ranges of $\varepsilon_{Hf}(t)$ variations for both metagranitoids (Fig. 11a), which relate to the initial Hf isotope heterogeneity and melt sources. Unlike Lu–Hf isotope signatures, $\delta^{18}O$ values of the deformed metagranitoid Mu-93-43 were more sensitive to this alteration (Fig. 14) and display a wider range of compositions (+7.7‰ to +9.2), as compared to that of zircon cores in the undeformed granitoid. Whether this recrystallization is related to subduction metamorphism or late-stage pervasive fluid-assisted deformations, which could obliterate the peak garnet-bearing assemblage, remains uncertain. The well-preserved crystal morphology of zircon from the two metagranitoids indicates no or very minor dissolution, which should have been expected during anatectic melting or prograde fluid overprint.

Contrasting patterns are exhibited by zircon from the Mu-93-45 metasedimentary sample, which experienced an apparent prograde-to-peak transformation in a subduction-collision episode. The obtained U–Pb ages of preserved zircon cores (831–745 Ma; Fig. 9e) closely resemble that of the widespread Early Neoproterozoic (Cryogenian) non-metamorphosed granitoids (*c.* 810 Ma; Skuzovatov *et al.* 2016*a*), their synmetamorphic (gneissic) counterparts of NMB (*c.* 764–763 Ma; Shatsky *et al.* 2015) and the age ranges of detrital zircons throughout the Muya block (941–793 Ma for NMB, Shatsky *et al.* 2015; ~940–780 Ma for the South Muya block, Skuzovatov *et al.* 2019*b*). Except for a few cores with the youngest ages, the concordance of age dates indicates the robustness of U–Pb records in zircon cores and incomplete remobilization of U–Pb isotopes within the crystal volume (Fig. 14). Although no $\varepsilon_{Hf}(t)$ data were recovered from these detrital cores, $\varepsilon_{Hf}(t)$ values of progressively recrystallized cores and mantles from the second batch of zircons from Mu-93-45 are exceptionally uniform and positive (+7.9 to +11.7). Such composition highlights the derivation of zircons and their parental granites from an unusual of juvenile mantle-derived Neoproterozoic substrate (e.g. underplated lower crust or arc crust in NMB). Zircon recrystallization and Pb loss commonly produce trends with low Lu/Hf ratios in $\varepsilon_{Hf}(t)$ v. U–Pb age diagrams. Garnet co-crystallization stemming from the abundance of recrystallized zircon with flat MREE–HREE patterns (Fig. 9f) typically yields a medium locally depleted in Lu and hence lower Lu/Hf ratios of zircon overgrowths or newly precipitated metamorphic grains with less radiogenic $\varepsilon_{Hf}(t)$ signatures. Thus, the observed dispersion along a trend with the Lu/Hf ratio close to the average crustal value (0.015) (Fig. 11a) either requires the contribution from another source with radiogenic Hf and/or a high Lu/Hf ratio, produced, for instance, via breakdown of prograde ferromagnesian phases under garnet-present conditions.

Zircon mantles and recrystallized cores exhibit textures of intensive partial recrystallization, which was accompanied by entrapment of metamorphic mineral inclusions at a recrystallization front (Fig. 14). The weighted average age retrieved from low-Th/U and high-Eu/Eu* recrystallization mantles (629 ± 6 Ma) is coeval with that of HP metamorphism (630 Ma by the Sm–Nd method on garnet-omphacite assemblage and the U–Pb method on zircon; Shatsky *et al.* 2012; Skuzovatov *et al.* 2019*b*). This emphasizes an effective Pb removal and U–Pb mobility during fluid-assisted zircon recrystallization at relevant temperatures of ~560–600°C, stressed by strongly variable Th/U ratios (Fig. 13a, c). However, both mineral inclusion assemblage coupled with *P–T* phase modelling and variable trace-element signatures of recrystallize zircon do not allow us to unambiguously link the zircon modification process to the HP metamorphic episode. Although garnet growth should have accompanied zircon recrystallization throughout

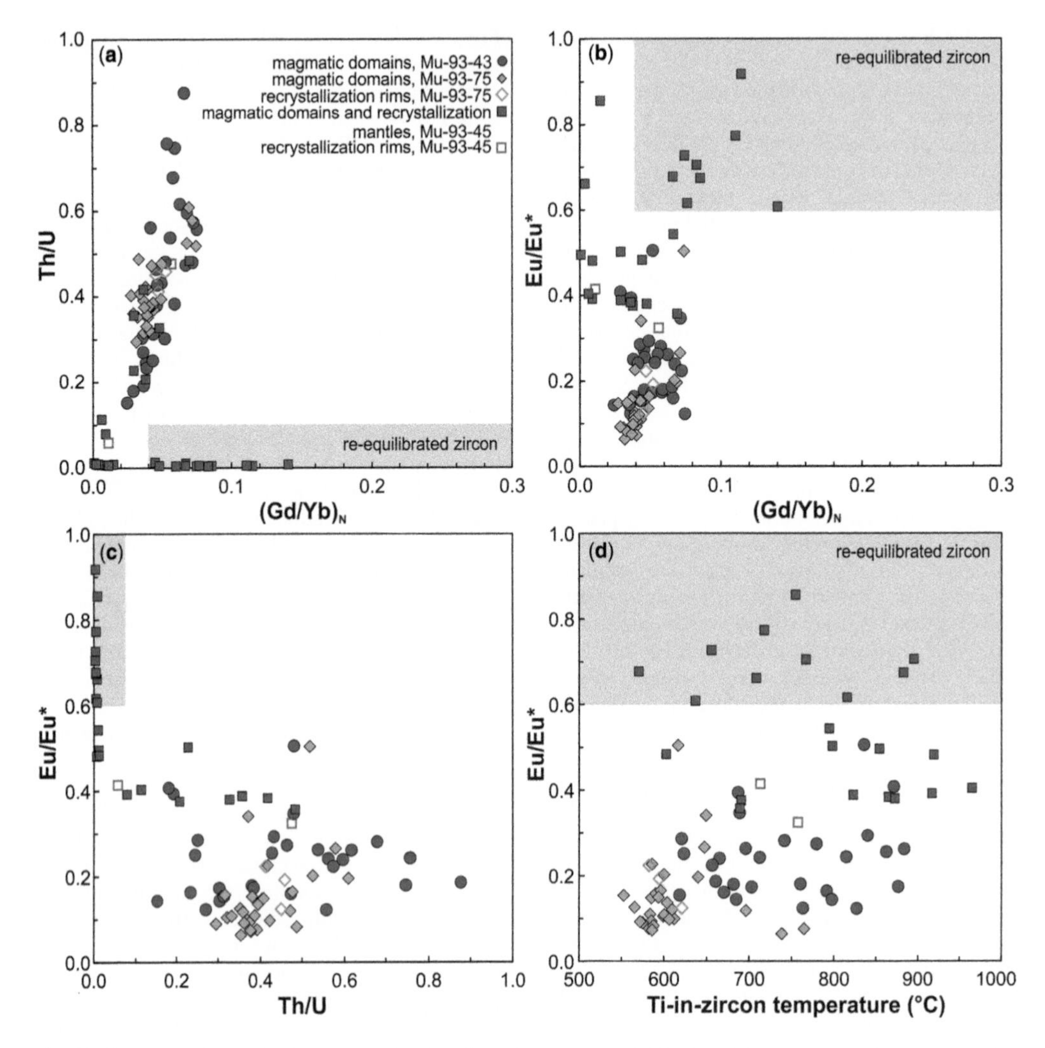

Fig. 13. Key trace-element indicators showing the effect of zircon recrystallization during metamorphic evolution of the NMB felsic rocks. The purple shading outlines the composition of zircon mantles showing a flat MREE–HREE distribution assumed as an indicator of high-grade dissolution-reprecipitation.

the prograde-to-peak path within ~0.7–1.2 GPa based on *P–T* reconstructions (Fig. 12), only a limited amount of recrystallized grains exhibit HREE-depleted, 'high-grade' REE patterns (Rubatto 2002*b*; Rubatto and Hermann 2007). Instead, some of recrystallized zircon displays MREE depletion at the absent or negligible Eu/Eu* anomaly, which may be related to preferential incorporation of MREE into epidote-group minerals at pressure levels of a prograde stage (above 0.8 GPa; Spandler *et al.* 2003; Xiao *et al.* 2016). If so, such zircon should be regarded as produced in response to high-grade recrystallization, but at the somewhat earlier, prograde burial stage. Alternatively, the observed inconsistency of the garnet co-growth effect on to zircon MREE–HREE patterns and the almost complete absence of garnet among mineral inclusions may relate to lithological control of garnet nucleation and zircon recrystallization (McElhinney *et al.* 2022), or reflect differences in zircon recrystallization mechanisms (Chen *et al.* 2010). The range of recalculated $\varepsilon_{Hf}(t)$ values of zircon from this sample is quite narrow (Fig. 11a), which means the effect of garnet with high Lu/Hf on to REE and Hf isotope systematics is in fact limited. Assuming the precision of the U–Pb age estimate for recrystallization mantles (629 $\pm$ 6 Ma), these observations may also imply a short-lived burial episode, which does not allow temporal resolution between prograde and peak metamorphic modification.

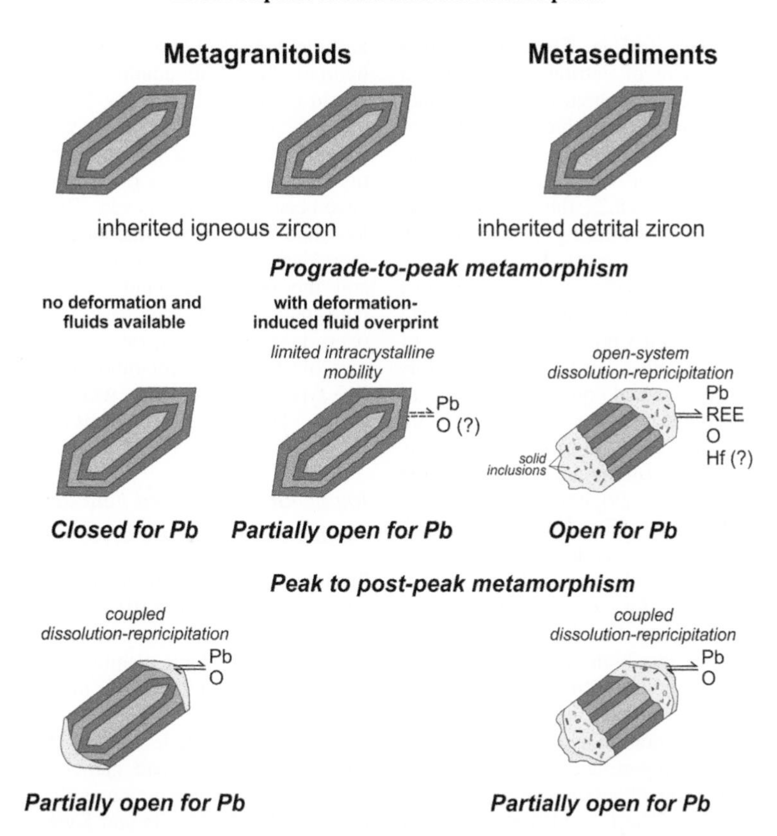

Fig. 14. Schematic illustration of zircon behaviour in the felsic rocks variably affected by high-pressure metamorphism. Preferential preservation of REE and U–Pb–Hf–O isotopic signatures is expected in the case of metagranitoids with limited fluid availability during high-grade metamorphism (left column), with only minor Pb and O mobility during late-stage alteration. Limited Pb and O isotope redistribution via intracrystalline diffusion (centre column) may be recorded in felsic rocks, which do not display metamorphic assemblages (garnet, epidote, rutile), but might have been subjected to partial deformation-induced recrystallization. Enhanced redistribution of REE and U–Pb–O is expected during zircon dissolution–reprecipitation in metasedimentary rocks (right column), where particular trace-element patterns and isotopic signatures are governed by co-crystallizing mineral assemblages (e.g. the presence of garnet) and *in situ* produced, metamorphic fluid.

Oxygen isotope variations, the role of hydrous fluids and nature of high $\delta^{18}O$ zircon

The studied NMB metagranitoids exhibit relatively limited (<1‰) variations of zircon $\delta^{18}O$ values in the magmatic core domains. Both metagranitoids resemble $\delta^{18}O$ values (+7.7‰ to +8.3‰) well above the mantle values (5.5–5.9‰; Bindeman 2008) and close to the average continental crust (above +8.0‰; Spencer *et al.* 2014). This implies granitic melt formation with a contribution from the altered mafic crust or sediments, which experienced interaction with hydrous fluids. Although the background $\delta^{18}O$ data for the NMB complexes are almost lacking, the present data and $\delta^{18}O$ variations exemplified by NMB eclogites (mainly within +4 to +8‰; Skuzovatov *et al.* 2019*b*) may potentially constrain the input of such fluid-altered crust into melt generation as minor to moderate. However, two different processes and/or metamorphic stages might have affected the primary $\delta^{18}O$ values of zircon in the studied rocks. Notable initial ^{18}O-enrichment is observed in all but one zircon in Mu-93-45 (+14.6 to +13.8‰) up to values typical for modern bulk sediments (~15‰; Spencer *et al.* 2014) and many metamorphic zircons (Peck *et al.* 2003; Martin *et al.* 2006, 2008; Cavosie *et al.* 2011). Because diffusion-limited ^{18}O enrichment is known to be slow under both dry and wet conditions (Watson and Cherniak 1997; Peck *et al.* 2003; Bindeman *et al.* 2018), only zircon reprecipitation with hydrous fluid involved may result in prominent $\delta^{18}O$ deviation from the original igneous/detrital $\delta^{18}O$ levels (Martin *et al.* 2006, 2008; Chen *et al.* 2016).

Two possible scenarios are therefore possible to explain such uniquely high $\delta^{18}O$ values of zircons in a metasedimentary sample. First, this feature could be representative of detrital zircons and a unique high-$\delta^{18}O$ melting source of their parental granite originated, for instance, through melting of supracrustal sediments (e.g. muds; Spencer *et al.* 2017). Although a core in one grain was found to have much lower $\delta^{18}O$ (+8.2), this may be interpreted as trapped xenocryst during melting or melt passage through the upper crust. However, there were no reports of purely sediment-derived granites in the NMB, which could resemble such high $\delta^{18}O$ values, or at least exhibit whole-rock signatures of S-type granites. Secondly, detrital zircons in this sample might once have had compositions close to those of the studied metagranitoids (i.e. $\delta^{18}O$ within ~7–9‰) and the only low-$\delta^{18}O$ core (+8.2). These values are much more common elsewhere in the NMB including mafic eclogites (mineral $\delta^{18}O$ mainly within +4 and +8‰; Skuzovatov *et al.* 2019*a*). The homogeneous textures of some zircon cores reflect their possible recrystallization that could provide pathways for oxygen diffusion from a fluid of 'metasedimentary' composition. It would therefore be possible that high zircon $\delta^{18}O$ values in HP metasediments resulted from pervasive prograde zircon dissolution–reprecipitation aided by internally sourced dehydration-derived fluid. Nonetheless, the only fact supporting this hypothesis is a preserved magmatic core of zircon with $\delta^{18}O$ typical of metagranitoids (+8.2‰; Fig. 11b) at high $\delta^{18}O$ (14.3‰) in a recrystallization rim.

In both garnet-bearing rocks, some zircons exhibit outermost bright-CL recrystallization rims with preservation (within the analytical uncertainties) of the initial $\delta^{18}O$ or ^{18}O-depletion (up to 0.8‰ for a garnet paragneiss and up to 1.6‰ for a garnet-bearing metagranitoid), and infer variable involvement of a single process, which affected the two rock types modified by prograde transformation. In this process, the $\delta^{18}O$ values tend to shift towards lighter, ^{18}O-depleted compositions typical for discordant recrystallized zircon (Booth *et al.* 2005). Chen *et al.* (2016) interpreted such ^{18}O-depletion to result from hydrous overprint by a low-$\delta^{18}O$ fluid. Given the metamorphic temperatures of ~560–600°C, in a closed system, $\delta^{18}O$ levels of a fluid calculated using zircon–water fractionation of ~1‰ (Zheng 1993) should have had contrasting origins and are expected to be at least 5.8–6.1‰ for Mu-93-75 and 12.9–13.3‰ for Mu-93-45. Furthermore, whereas this depletion is a typical feature in the Mu-93-45 metagranitoid (Fig. 10b), only a few zircons from metasedimentary samples resemble this trend (Fig. 10d). Importantly, garnet from both rock types revealed domains or porphyroblast rims drastically enriched in MnO. If the processes of low-volume garnet regrowth and zircon recrystallization were coupled and linked to late-stage infiltration of a single, low-$\delta^{18}O$ (at least below <5.8‰) Mn-rich hydrous fluid, the equilibrium ^{18}O fractionation would ideally lead to similar $\delta^{18}O$ of zircon rims in the two samples. The absence of such patterns favours the idea that the equilibrium oxygen fractionation during zircon rim recrystallization might have been far from achieved. This might be due to low fluid/rock ratios, which are expected for metamorphism of continental-crust rocks (Zheng 2009) and the occasional fluid involvement in zircon recrystallization. In this case, only the local fluid overprint event at the peak or retrograde stage was possible with different fluid compositions for the two types of felsic rocks, which recorded variable zircon ^{18}O depletion in the rims (Booth *et al.* 2005).

Implications to limited fluid activity in the subducted continental lithosphere

Felsic rocks of the continental terranes are known to variably transform into HP assemblages, which significantly limits the understanding of metamorphic evolution and tectonic coupling of different lithologies. For metagranitoids, this may be due to purely kinetical reasons as dry quartz–feldspathic experience early prograde dehydration and fluid-consuming epidote- and garnet-forming reactions at relatively moderate pressures (Proyer 2003; Peterman *et al.* 2009; Young and Kylander-Clark 2015; Schorn 2022). This is especially the case for garnet-bearing and garnet-free metagranitoids of the continental basement, which may experience early metamorphism conversion and reside in the deep crust at relatively cool conditions (~500–600°C; Mareschal and Jaupart 2013) and pressures up to ~1.0–1.1 GPa (Williams *et al.* 2014). Such polycyclic metamorphic conditions are common for HP continental slivers within the orogenic belts (Lardeaux and Spalla 1991; Shatsky *et al.* 2021) and require an overprint by hydrous fluids for partial or complete eclogitization (e.g. Austrheim and Griffin 1985; Jackson *et al.* 2004*b*; Jolivet *et al.* 2005).

For the eastern BMFB, a possible pre-HP, medium- to high-temperature metamorphic stage was previously highlighted by geochronology of metasediments (Skuzovatov *et al.* 2019*a*), but is also hinted at by CO_2-dominated fluid inclusions in some of the eclogite-associated felsic rocks in the NMB (Shatsky *et al.* 2012). Thus, it is possible that at least some of the NMB felsic rocks might be subjected to early metamorphism and dehydration prior to the onset of the Ediacarian HP subduction stage. Petrological reconstructions have revealed that the NMB metasedimentary rocks evolved nearly isothermally towards pressures of ~1.0 GPa, which implies a nearly constant geotherm and the original position of metasediments in the middle crust

(Simon *et al.* 2022 and references therein). However, relatively cool conditions of crustal storage allowed preservation of H_2O (e.g. in chlorite as a major hydrous phase) and further HP transformation of metasediments during the Ediacarian subduction. Although H_2O-rich metasedimentary rocks are transformed into HP assemblages more easily than metagranitoids, dehydration of NMB metasediments led to only limited and local-scale fluid migration as pinpointed by zircon $\delta^{18}O$ systematics. Limited externally sourced H_2O was also available at the relevant level of the continental crust during crustal exhumation based on the eclogites studies (Skuzovatov 2022).

Conclusions

We report on the results of petrographic, mineralogical and detailed zircon (trace-element, U–Pb, Lu–Hf, $\delta^{18}O$) studies of Neoproterozoic felsic samples (metagranitoids and paragneiss), which are spatially juxtaposed with mafic eclogites within the North Muya block in the Neoproterozoic complexes of the northern CAOB. The felsic rocks, which host eclogites, exhibit contrasting degrees of whole-rock and zircon high-grade transformation, which, based on the study, mostly depend on the amount of internally sourced H_2O preserved in hydrous mineral phases. The prograde transformation of metagranitoids, which are generally poor in hydrous Mg–Fe-bearing phases, leads to only local nucleation of high-grade minerals (garnet + epidote) and preferentially preserves the original trace element, U–Pb age and Lu–Hf–O systematics of igneous zircon. Conversely, abundant discordance due to partial Pb loss and wider $\delta^{18}O$ variations is expected in similar metagranitoids affected by deformations at the revealed metamorphic temperatures (~570–600°C), while their peak metamorphic assemblages may be completely obliterated. Progressive metamorphic transformation is typical for metasedimentary rocks, with garnet + epidote + rutile + phengite as markers of the HP event. In this process, zircon experiences dissolution–reprecipitation, which leads to entrapment of metamorphic minerals as inclusions, partial to complete resetting of trace-elements ('eclogitic' REE patterns), U–Pb system and $\delta^{18}O$ signatures. We interpret this to reflect variable discharge of metamorphic fluids, which, at a HP stage, were low-volume and internally sourced, and significant only in an H_2O-enriched protolith of a garnet gneiss. The evidence of ^{18}O-depleted zircon recrystallization rims after both 'magmatic' and 'eclogitic' zircon, together with Mn enrichment of garnet rims, potentially indicates a distinct metamorphic stage, which could result from heating and/or fluid infiltration linked to peak crustal stacking and thickening. The data imply that both metagranitoids and metasediments might follow a similar metamorphic evolution to that of eclogites during a prograde burial of the continental lithosphere, but either had much more limited burial depths, or successfully back-reacted into medium- to low-pressure assemblages during retrograde hydration.

Acknowledgements We acknowledge the effort from two anonymous referees and the editorial handling of Martin Guitreau, whose thorough consideration and comments helped to significantly improve the presentation. The authors are grateful to Olga Belozerova (IGC) for assisting with the EPMA studies of thin sections as well as to Fu-Long Lin (IES) for assisting with the zircon studies.

Competing interests The authors declare that they have no known competing financial interests or personal relationships that could have appeared to influence the work reported in this paper.

Author contributions **SS**: conceptualization (lead), data curation (lead), formal analysis (lead), funding acquisition (equal), investigation (equal), methodology (equal), project administration (lead), validation (lead), visualization (lead), writing – original draft (lead), writing – review & editing (lead); **K-LW**: funding acquisition (equal), investigation (equal), methodology (equal), resources (equal), writing – review & editing (equal); **X-HL**: data curation (equal), investigation (equal), methodology (equal), resources (equal), writing – review & editing (equal); **YI**: investigation (supporting), methodology (supporting), resources (supporting); **VS**: conceptualization (supporting), writing – review & editing (supporting).

Funding This study was supported by the Russian Science Foundation (Grant 21-77-10038).

Data availability All data generated or analysed during this study are included in this published article (and, if present, its supplementary information files).

References

Agard, P., Yamato, P., Jolivet, L. and Burov, E. 2009. Exhumation of oceanic blueschists and eclogite in subduction zones: timing and mechanisms. *Earth-Science Review*, **92**, 53–79, https://doi.org/10.1016/j.earscirev.2008.11.002

Angiboust, S. and Harlov, D. 2017. Ilmenite breakdown and rutile-titanite stability in metagranitoids: natural observations and experimental results. *American Mineralogist*, **102**, 1696–1708, https://doi.org/10.2138/am-2017-6064

Austrheim, H. and Griffin, W.L. 1985. Shear deformation and eclogite formation within granulite-facies anorthosites of the Bergen Arcs, western Norway. *Chemical*

Geology, **50**, 267–281, https://doi.org/10.1016/0009-2541(85)90124-X

Benisek, A., Dachs, E. and Kroll, H. 2010. A ternary feldspar-mixing model based on calorimetric data: development and application. *Contributions to Mineralogy and Petrology*, **160**, 327–337, https://doi.org/10.1007/s00410-009-0480-8

Bindeman, I. 2008. Oxygen isotopes in mantle and crustal magmas as revealed by single crystal analysis. *Reviews in Mineralogy and Geochemistry*, **69**, 445–478, https://doi.org/10.2138/rmg.2008.69.12

Bindeman, I.N., Schmitt, A.K., Lundstrom, C.C. and Hervig, R.L. 2018. Stability of zircon and its isotopic ratios in high-temperature fluids: long-term (4 months) isotope exchange experiment at 850°C and 50 MPa. *Frontiers in Earth Science*, **6**, 59, https://doi.org/10.3389/feart.2018.00059

Black, L.P. and Gulson, B.L. 1978. The age of the mud tank carbonatite, strangways range, northern territory. *BMR Journal of Australian Geology and Geophysics*, **3**, 227–232

Booth, A., Kolodny, Y., Chamberlain, P., McWilliams, M., Schmitt, A.K. and Wooden, J. 2005. Oxygen isotopic composition and U-Pb discordance in zircon. *Geochimica et Cosmochimica Acta*, **69**, 4895–4905, https://doi.org/10.1016/j.gca.2005.05.013

Bouvier, A., Vervoort, J.D. and Patchett, P.J. 2008. The Lu-Hf and Sm-Nd isotopic composition of CHUR: constraints from unequilibrated chondrites and implications for the bulk composition of terrestrial planets. *Earth and Planetary Science Letters*, **273**, 48–57, https://doi.org/10.1016/j.epsl.2008.06.010

Brandelik, A. 2009. CALCMIN – an EXCEL™ Visual Basic application for calculating mineral structural formulae from electron microprobe analyses. *Computers & Geosciences*, **35**, 1540–1551, https://doi.org/10.1016/j.cageo.2008.09.011

Bulgatov, A.N. 1983. *Tectonotype of Baikalids*. Nauka, Novosibirsk (in Russian).

Bulgatov, A.N. and Gordienko, I.V. 1999. Terranes of the Baikal mountainous region and location of gold deposits. *Geology of Ore Deposits*, **41**, 230–240 (in Russian).

Cavosie, A.J., Valley, J.W., Kita, N., Spicuzza, M.J., Ushikubo, T. and Wilde, S.A. 2011. The origin of high $\delta^{18}O$ zircons: marbles, megacrysts, and metamorphism. *Contributions to Mineralogy and Petrology*, **162**, 961–974, https://doi.org/10.1007/s00410-011-0634-3

Chen, R.-X. and Zheng, Y. 2017. Metamorphic zirconology of continental subduction zones. *Journal of Asian Earth Sciences*, **145**, 149–176, https://doi.org/10.1016/j.jseaes.2017.04.029

Chen, R.-X., Zheng, Y.-F. and Xie, L. 2010. Metamorphic growth and recrystallization of zircon: distinction by simultaneous in-situ analyses of trace elements, U–Th–Pb and Lu–Hf isotopes in zircons from eclogite-facies rocks in the Sulu orogen. *Lithos*, **114**, 132–154, https://doi.org/10.1016/j.lithos.2009.08.006

Chen, Y.-X., Schertl, H.-P., Zheng, Y.-F., Huang, F., Zhou, K. and Gong, Y.-Z. 2016. Mg–O isotopes trace the origin of Mg-rich fluids in the deeply subducted continental crust of Western Alps. *Earth and Planetary Science Letters*, **456**, 157–167, https://doi.org/10.1016/j.epsl.2016.09.010

Cherniak, D.J. and Watson, E.B. 2003. Diffusion in zircon. *Reviews in Mineralogy and Geochemistry*, **53**, 113–143, https://doi.org/10.2113/0530113

Connolly, J.A.D. 2005. Computation of phase equilibria by linear programming: a tool for geodynamic modeling and its application to subduction zone decarbonation. *Earth and Planetary Science Letters*, **236**, 524–541, https://doi.org/10.1016/j.epsl.2005.04.033

Dale, J., Powell, R., White, R.W. and Elmer, F.L. 2005. A thermodynamic model for Ca–Na clinoamphiboles in Na_2O–CaO–FeO–MgO–Al_2O_3–SiO_2–H_2O–O for petrological calculations. *Journal of Metamorphic Geology*, **23**, 771–791, https://doi.org/10.1111/j.1525-1314.2005.00609.x

Ernst, W.G. 2010. Subduction-zone metamorphism, calc-alkaline magmatism, and convergent-margin crustal evolution. *Gondwana Research*, **18**, 8–16, https://doi.org/10.1016/j.gr.2009.05.010

Evans, B.W. and Guidotti, C.V. 1966. The sillimanite-potash feldspar isograd in western Maine, USA. *Contributions to Mineralogy and Petrology*, **12**, 25–62, https://doi.org/10.1007/BF02651127

Gain, S.E.M., Gréau, Y., Henry, H., Belousova, E., Dainis, I., Griffin, W.L. and O'Reilly, S.Y. 2019. Mud Tank zircon: long-term evaluation of a reference material for U–Pb dating, Hf-isotope analysis and trace element analysis. *Geostandards and Geoanalytical Research*, **43**, 339–354, https://doi.org/10.1111/ggr.12265

García-Casco, A., Torres-Roldán, R.L., Millán, G., Monié, P. and Schneider, J. 2002. Oscillatory zoning in eclogitic garnet and amphibole, Northern Serpentinite Melange, Cuba: a record of tectonic instability during subduction? *Journal of Metamorphic Geology*, **20**, 581–598, https://doi.org/10.1046/j.1525-1314.2002.00390.x

Gebauer, D. and Grünenfelder, M. 1976. U–Pb zircon and Rb–Sr whole rock dating of low-grade metasediments, example: Montagne Noire (Southern France). *Contributions to Mineralogy and Petrology*, **59**, 13–32, https://doi.org/10.1007/BF00375108

Griffin, W.L., Pearson, N.J., Belousova, E., Jackson, S.E., van Achterberg, E., O'Reilly, S.Y. and Shee, S.R. 2000. The Hf isotope composition of cratonic mantle: LAM-MC-ICPMS analysis of zircon megacrysts in kimberlites. *Geochimica et Cosmochimica Acta*, **64**, 133–147, https://doi.org/10.1016/S0016-7037(99)00343-9

Hacker, B.R., Abers, G.A. and Peacock, S.M. 2003. Subduction factory 1. Theoretical mineralogy, densities, seismic wave speeds, and H_2O contents. *Journal of Geophysical Research*, **108**, 2029, https://doi.org/10.1029/2001JB001127

Holland, T. and Powell, R. 1991. A Compensated-Redlich-Kwong (CORK) equation for volumes and fugacities of CO_2 and H_2O in the range 1 bar to 50 kbar and 100–1600°C. *Contributions to Mineralogy and Petrology*, **109**, 265–273, https://doi.org/10.1007/BF00306484

Holland, T.J.B. and Powell, R. 1998. An internally-consistent thermodynamic dataset for phases of petrological interest. *Journal of Metamorphic Geology*, **16**, 309–344, https://doi.org/10.1111/j.1525-1314.1998.00140.x

Hoskin, P.W.O. and Ireland, T.R. 2000. Rare earth element chemistry of zircon and its use as a provenance indicator. *Geology*, **28**, 627–630, https://doi.org/10.1130/0091-7613(2000)28<627:REECOZ>2.0.CO;2

Itaya, T. and Banno, S. 1980. Paragenesis of titanium-bearing accessories in pelitic schists of the Sanbagawa metamorphic belt, central Shikoku, Japan. *Contributions to Mineralogy and Petrology*, **73**, 267–276, https://doi.org/10.1007/BF00381445

Jackson, J.A., Austrheim, H., McKenzie, D. and Priestley, K. 2004*a*. Metastability, mechanical strength, and the support of mountain belts. *Geology*, **32**, 625–628, https://doi.org/10.1130/G20397.1

Jackson, S.E., Pearson, N.J., Griffin, W.L. and Belousova, E.A. 2004*b*. The application of laser ablation-inductively coupled plasma-mass spectrometry to in situ U-Pb zircon geochronology. *Chemical Geology*, **211**, 47–69, https://doi.org/10.1016/j.chemgeo.2004.06.017

Jochum, K.P., Weis, U., Schwager, B., Stoll, B., Wilson, S.A., Haug, G.H., Andreae, M.O. and Enzweiler, J. 2016. Reference values following ISO guidelines for frequently requested rock reference materials. *Geostandards and Geoanalytical Research*, **40**, 333–350, https://doi.org/10.1111/j.1751-908X.2015.00392.x

Jolivet, L., Raimbourg, H., Labrousse, L., Avigad, D., Leroy, Y., Austrheim, H. and Andersen, T.B. 2005. Softening triggered by eclogitization, the first step toward exhumation during continental subduction. *Earth and Planetary Science Letters*, **237**, 532–547, https://doi.org/10.1016/j.epsl.2005.06.047

Kohn, M.J., Corrie, S.L. and Markley, C. 2015. The fall and rise of metamorphic zircon. *American Mineralogist*, **100**, 897–908, https://doi.org/10.2138/am-2015-5064

Lardeaux, J.M. and Spalla, M.I. 1991. From granulites to eclogites in the Sesia zone (Italian Western Alps): a record of the opening and closure of the Piedmont ocean. *Journal of Metamorphic Geology*, **9**, 35–59, https://doi.org/10.1111/j.1525-1314.1991.tb00503.x

Li, B.T. and Massonne, H.-J. 2016. Early Variscan P–T evolution of an eclogite body and adjacent orthogneiss from the Northern Malpica–Tuy shear-zone in NW Spain. *European Journal of Mineralogy*, **28**, 1131–1154, https://doi.org/10.1127/ejm/2016/0028-2569

Li, X.H., Long, W.-G. *et al.* 2010. Penglai zircon megacryst: a potential new working reference for microbeam analysis of Hf-O isotopes and U-Pb age. *Geostandards and Geoanalytical Research*, **34**, 117–134, https://doi.org/10.1111/j.1751-908X.2010.00036.x

Li, X.-H., Tang, G.-Q. *et al.* 2013. Qinghu zircon: a working reference for microbeam analysis of U-Pb age and Hf and O isotopes. *Chinese Science Bulletin*, **58**, 4647–4654, https://doi.org/10.1007/s11434-013-5932-x

Liu, F.L. and Liou, J.G. 2011. Zircon as the best mineral for P-T-time history of UHP metamorphism: a review on mineral inclusions and U-Pb SHRIMP ages of zircons from the Dabie-Sulu UHP rocks. *Journal of Asian Earth Sciences*, **40**, 1–39, https://doi.org/10.1016/j.jseaes.2010.08.007

Liou, J.G. and Tsujimori, T. 2013. The fate of subducted continental crust: evidence from recycled UHP-UHT minerals. *Elements*, **9**, 248–250, https://doi.org/10.2113/gselements.9.4.248

Ludwig, K.R. 2003. *ISOPLOT 3.0-a Geochronological Toolkit for Microsoft Excel*. Berkeley Geochronology Center Special Publication.

Mareschal, J.-C. and Jaupart, C. 2013. Radiogenic heat production, thermal regime and evolution of continental crust. *Tectonophysics*, **609**, 524–534, https://doi.org/10.1016/j.tecto.2012.12.001

Martin, L., Duchêne, S., Deloule, E. and Vanderhaeghe, O. 2006. The isotopic composition of zircon and garnet: a record of the metamorphic history of Naxos, Greece. *Lithos*, **87**, 174–192, https://doi.org/10.1016/j.lithos.2005.06.016

Martin, L.A.J., Duchêne, S., Deloule, E. and Vanderhaeghe, O. 2008. Mobility of trace elements and oxygen in zircon during metamorphism: consequences for geochemical tracing. *Earth and Planetary Science Letters*, **267**, 161–174, https://doi.org/10.1016/j.epsl.2007.11.029

Massonne, H.-J. 2009. Hydration, dehydration, and melting of metamorphosed granitic and dioritic rocks at high- and ultrahigh-pressure conditions. *Earth and Planetary Science Letters*, **288**, 244–254, https://doi.org/10.1016/j.epsl.2009.09.028

Massonne, H.-J. 2015. Derivation of P–T paths from high-pressure metagranites – examples from the Gran Paradiso Massif, western Alps. *Lithos*, **226**, 265–279, https://doi.org/10.1016/j.lithos.2014.12.024

Massonne, H.J. and Schreyer, W. 1987. Phengite geobarometry based on the limiting assemblage with K-feldspar, phlogopite, and quartz. *Contributions to Mineralogy and Petrology*, **96**, 212–224, https://doi.org/10.1007/BF00375235

McCarron, T., McFarlane, C.R.M. and Gaidies, F. 2019. The significance of Mn-rich ilmenite and the determination of P–T paths from zoned garnet in metasedimentary rocks from the western Cape Breton Highlands, Nova Scotia. *Journal of Metamorphic Geology*, **37**, 1171–1192, https://doi.org/10.1111/jmg.12507

McElhinney, T.R., Dempster, T.J. and Chung, P. 2022. The influence of microscale lithological layering and fluid availability on the metamorphic development of garnet and zircon: insights into dissolution–reprecipitation processes. *Mineralogical Magazine*, **86**, 9–26, https://doi.org/10.1180/mgm.2021.97

Mitrofanov, G.L. 1978. Evolution of tectonic structures and stages of the continental crust formation of North-Western Transbaikalia. *In*: *Tectonics and Metallogeny of Eastern Siberia*. Irkutsk State University, Irkutsk, 38–56 (in Russian).

Nowell, G.M., Kempton, P.D., Noble, S.R., Fitton, J.G., Saunders, A.D., Mahoney, J.J. and Taylor, R.N. 1998. High precision Hf isotope measurements of MORB and OIB by thermal ionisation mass spectrometry: insights into the depleted mantle. *Chemical Geology*, **149**, 211–233.

Nyström, A. and Kriegsman, L.M. 2003. Prograde and retrograde reactions, garnet zoning patterns, and accessory phase behaviour in SW Finland migmatites, with implications for geochronology. *Geological Society, London, Special Publications*, **220**, 213–230, https://doi.org/10.1144/GSL.SP.2003.220.01.13

Pattison, D.R.M. and Tinkham, D.K. 2009. Interplay between equilibrium and kinetics in prograde metamorphism of pelites: an example from the Nelson aureole, British Columbia. *Journal of Metamorphic Geology*, **27**, 249–279, https://doi.org/10.1111/j.1525-1314.2009.00816.x

Peck, W.H., Valley, J.W. and Graham, C.M. 2003. Slow oxygen diffusion rates in igneous zircons from metamorphic rocks. *American Mineralogist*, **88**, 1003–1014, https://doi.org/10.2138/am-2003-0708

Peterman, E.M., Hacker, B.R. and Baxter, E.F. 2009. Phase transformations of continental crust during subduction and exhumation: Western Gneiss Region, Norway. *European Journal of Mineralogy*, **21**, 1097–1118, https://doi.org/10.1127/0935-1221/2009/0021-1988

Proyer, A. 2003. The preservation of high-pressure rocks during exhumation: metagranites and metapelites. *Lithos*, **70**, 183–194, https://doi.org/10.1016/S0024-4937(03)00098-7

Rubatto, D. 2002*a*. Zircon trace element geochemistry: distribution coefficients and the link between U-Pb ages and metamorphism. *Chemical Geology*, **184**, 123–138, https://doi.org/10.1016/S0009-2541(01)00355-2

Rubatto, D. 2002*b*. Zircon trace element geochemistry: partitioning with garnet and the link between U-Pb ages and metamorphism. *Chemical Geology*, **184**, 123–138, https://doi.org/10.1016/S0009-2541(01)00355-2

Rubatto, D. and Hermann, J. 2003. Zircon formation during fluid circulation in eclogites (Monviso, Western Alps): implications for Zr and Hf budget in subduction zones. *Geochimica et Cosmochimica Acta*, **67**, 2173–2187, https://doi.org/10.1016/S0016-7037(02)01321-2

Rubatto, D. and Hermann, J. 2007. Zircon behavior in deeply subducted rocks. *Elements*, **3**, 31–35, https://doi.org/10.2113/gselements.3.1.31

Rytsk, E.Y., Amelin, Y.V. *et al.* 2001. Age of rocks in the Baikal-Muya Foldbelt. *Strarigraphy and Geological Correlation*, **9**, 3–15.

Rytsk, E.Y., Kovach, V.P., Yarmolyuk, V.V. and Kovalenko, V.I. 2007. Structure and evolution of the continental crust in the Baikal Fold Region. *Geotectonics*, **41**, 440–464, https://doi.org/10.1134/S0016852107060027

Salop, L.I. 1964. *Geology of Baikal Mountain Area*. Nedra, Moscow (in Russian).

Scherer, E., Münker, C. and Mezger, K. 2001. Calibration of the Lutetium-Hafnium Clock. *Science*, **293**, 683–687.

Schorn, S. 2022. Self-induced incipient 'eclogitization' of metagranitoids at closed-system conditions. *Journal of Metamorphic Geology*, **40**, 1271–1290, https://doi.org/10.1111/jmg.12665

Shatskii, V.S., Skuzovatov, S.Y., Ragozin, A.L. and Dril, S.I. 2014. Evidence of Neoproterosoic continental subduction in the Baikal-Muya fold belt. *Doklady Earth Sciences*, **495**, 1442–1445, https://doi.org/10.1134/S1028334X14110166

Shatsky, V.S., Sitnikova, E.S., Tomilenko, A.A., Ragozin, A.L., Koz'menko, O.A. and Jagoutz, E. 2012. Eclogite-gneiss complex of the Muya block (East Siberia): age, mineralogy, geochemistry, and petrology. *Russian Geology and Geophysics*, **53**, 501–521, https://doi.org/10.1016/j.rgg.2012.04.001

Shatsky, V.S., Malkovets, V.G., Belousova, E.A. and Skuzovatov, S.Y. 2015. Evolution history of the Neoproterozoic eclogite-bearing complex of the Muya dome (Central Asian Orogenic Belt): constraints from zircon U-Pb age, Hf and whole-rock Nd isotopes. *Precambrian Research*, **261**, 1–11, https://doi.org/10.1016/j.precamres.2015.01.013

Shatsky, V.S., Ragozin, A.L., Skuzovatov, S.Y., Kozmenko, O.A. and Yagoutz, E. 2021. Isotope-geochemical evidence of the nature of protoliths of diamond-bearing rocks of the Kokchetav subduction-collision zone (Northern Kazakhstan). *Russian Geology and Geophysics*, **62**, 547–556, https://doi.org/10.2113/RGG20204278

Simon, M., Pitra, P., Yamato, P. and Poujol, M. 2022. Isothermal compression of an eclogite from the Western Gneiss Region (Norway). *Journal of Metamorphic Geology*, **41**, 181–203, https://doi.org/10.1111/jmg.12692

Skuzovatov, S.Y. 2021. Nature and (in-)coherent metamorphic evolution of subducted continental crust in the Neoproterozoic accretionary collage of SW Mongolia. *Geoscience Frontiers*, **12**, 101097, https://doi.org/10.1016/j.gsf.2020.10.004

Skuzovatov, S.Y. 2022. Differential fluid activity in a single exhumed continental subduction unit from local P-T-M(H_2O) records of zoned amphiboles (North Muya, Eastern Siberia). *Minerals*, **12**, 217, https://doi.org/10.3390/min12020217

Skuzovatov, S.Y., Sklyarov, E.V., Shatsky, V.S., Wang, K.-L., Kulikova, K.V. and Zarubina, O.V. 2016*a*. Granulites of the South-Muya block (Baikal-Muya Foldbelt): age of metamorphism and nature of protolith. *Russian Geology and Geophysics*, **57**, 451–463, https://doi.org/10.1016/j.rgg.2016.03.007

Skuzovatov, S.Y., Wang, K.-L., Shatsky, V.S. and Buslov, M.M. 2016*b*. Geochemistry, zircon U-Pb age and Hf isotopes of the North Muya block granitoids (Central Asian Orogenic belt): constraints on petrogenesis and geodynamic significance of felsic magmatism. *Precambrian Research*, **280**, 14–30, https://doi.org/10.1016/j.precamres.2016.04.015

Skuzovatov, S.Y., Shatsky, V.S. and Dril, S.I. 2017. High-pressure mafic granulites of the South Muya block (Central Asian Orogenic Belt). *Doklady Earth Sciences*, **473**, 423–426, https://doi.org/10.1134/S1028334X17040067

Skuzovatov, S.Y., Wang, K.-L., Dril, S.I., Iizuka, Y. and Lee, H.-Y. 2019*a*. Geochemistry, zircon U-Pb and Lu-Hf systematics of high-grade metasedimentary sequences from the South Muya block (northeastern Central Asian Orogenic Belt): reconnaissance of polymetamorphism and accretion of Neoproterozoic exotic blocks in southern Siberia. *Precambrian Research*, **321**, 34–53, https://doi.org/10.1016/j.precamres.2018.11.022

Skuzovatov, S.Y., Shatsky, V.S. and Wang, K.-L. 2019*b*. Continental subduction during arc-microcontinent collision in the southern Siberian craton: constraints on protoliths and metamorphic evolution of the North Muya complex eclogites (Eastern Siberia). *Lithos*, **342–343**, 76–96, https://doi.org/10.1016/j.lithos.2019.05.022

Skuzovatov, S.Y., Belozerova, O.Y. *et al.* 2022. Centre of Isotopic and Geochemical Research (IGC SB RAS):

current state of micro- and macroanalysis. *Geodynamics & Tectonophysics*, **13**, 0585, https://doi.org/10.5800/GT-2022-13-2-0585

Sláma, J., Košler, J. *et al.* 2008. Plesovice zircon – a new natural reference material for U-Pb and Hf isotopic microanalysis. *Chemical Geology*, **249**, 1–35, https://doi.org/10.1016/j.chemgeo.2007.11.005

Spandler, C., Hermann, J., Arculus, R. and Mavrogenes, J. 2003. Redistribution of trace elements during prograde metamorphism from lawsonite blueschist to eclogite facies; implications for deep subduction-zone processes. *Contributions to Mineralogy and Petrology*, **146**, 205–222, https://doi.org/10.1007/s00410-003-0495-5

Spencer, C.J., Cawood, P.A., Hawkesworth, C.J., Raub, T.D., Prave, A.R. and Roberts, N.M.W. 2014. Proterozoic onset of crustal reworking and collisional tectonics: reappraisal of the zircon oxygen isotope record. *Geology*, **42**, 451–454, https://doi.org/10.1130/G35363.1

Spencer, C.J., Cavosie, A.J. *et al.* and EIMF 2017. Evidence for melting mud in Earth's mantle from extreme oxygen isotope signatures in zircon. *Geology*, **45**, 975–978, https://doi.org/10.1130/G39402.1

Tan, Z., Agard, P., Gao, J., Hong, T. and Wan, B. 2020. Concordant pulse in Mn, Y and HREEs concentrations during UHP eclogitic garnet growth: transient rock dynamics along a cold subduction plate interface. *Earth and Planetary Science Letters*, **530**, 115908, https://doi.org/10.1016/j.epsl.2019.115908

Watson, E.B. and Cherniak, D.J. 1997. Oxygen diffusion in zircon. *Earth and Planetary Science Letters*, **148**, 527–544, https://doi.org/10.1016/S0012-821X(97)00057-5

Watson, E.B., Wark, D.A. and Thomas, J.B. 2006. Crystallization thermometers for zircon and rutile. *Contributions to Mineralogy and Petrology*, **151**, 413–433, https://doi.org/10.1007/s00410-006-0068-5

Wayne, D.M. and Sinha, A.K. 1988. Physical and chemical response of zircons to deformation. *Contributions to Mineralogy and Petrology*, **98**, 109–121, https://doi.org/10.1007/BF00371915

Wiedenbeck, M., Hanchar, J.M. *et al.* 2004. Further characterisation of the 91500 zircon crystal. *Geostandards and Geoanalytical Research*, **28**, 9–39, https://doi.org/10.1111/j.1751-908X.2004.tb01041.x

Williams, M.L., Dumond, G., Mahan, K., Regan, S. and Holland, M. 2014. Garnet-forming reactions in felsic orthogneiss: implications for densification and strengthening of the lower continental crust. *Earth and Planetary Science Letters*, **405**, 207–219, https://doi.org/10.1016/j.epsl.2014.08.030

Woodhead, J.D. and Hergt, J.M. 2005. A preliminary appraisal of seven natural zircon reference materials for in situ Hf isotope determination. *Geostandards and Geoanalaytical Research*, **29**, 183–195.

Xiao, Y., Niu, Y., Wang, K.-L., Lee, D.-C. and Iizuka, Y. 2016. Geochemical behaviours of chemical elements during subduction-zone metamorphism and geodynamic significance. *International Geology Review*, **58**, 1253–1277, https://doi.org/10.1080/00206814.2016.1147987

Young, D.J. and Kylander-Clark, A.R.C. 2015. Does continental crust transform during eclogite facies metamorphism? *Journal of Metamorphic Geology*, **33**, 331–357, https://doi.org/10.1111/jmg.12123

Zeh, A. and Holness, M. 2003. The effect of reaction overstep on garnet microtextures in metapelitic rocks of the Ilesha Schist Belt, SW Nigeria. *Journal of Petrology*, **44**, 967–994, https://doi.org/10.1093/petrology/44.6.967

Zheng, Y.-F. 1993. Calculation of oxygen isotope fractionation in anhydrous silicate minerals. *Geochimica et Cosmochimica Acta*, **57**, 1079–1091, https://doi.org/10.1016/0016-7037(93)90042-U

Zheng, Y.-F. 2009. Fluid regime in continental subduction zones: petrological insights from ultrahigh-pressure metamorphic rocks. *Journal of the Geological Society*, **166**, 763–782, https://doi.org/10.1144/0016-76492008-016R

Re-evaluating metamorphism in the southern Natal Province, South Africa

Eleanore Blereau[1,2]* and Christopher Spencer[1,3]

[1]School of Earth and Planetary Sciences, Institute for Geoscience Research (TIGeR), Curtin University, GPO Box U1987, Perth, WA 6845, Australia

[2]John de Laeter Centre, Curtin University, GPO Box U1987, Perth, WA 6845, Australia

[3]Department of Geological Sciences and Geological Engineering, Queen's University, Kingston, Ontario, Canada K7L 3N6

EB, 0000-0001-8850-397X; CS, 0000-0003-4264-3701

Present addresses: EB, School of Earth and Environment, University of Leeds, Leeds LS2 9JT, UK

*Correspondence: earebl@leeds.ac.uk

Abstract: The metamorphic conditions of the Natal Metamorphic Province (NMP) have been the focus of previous studies to assist with Rodinia reconstructions but there are limited constraints on the age of metamorphism. We use a combination of modern techniques to provide new constraints on the conditions and timing of metamorphism in the two southernmost terranes: the Mzumbe and Margate. Metamorphism reached granulite facies, 780–834°C at 3.9–7.8 kbar in the Mzumbe Terrane and 850–892°C at 5.7–6.1 kbar in the Margate Terrane. The new pressure and temperature constraints are supportive of isobaric cooling in the Margate Terrane as previously proposed. Peak metamorphism of the two terranes is shown to have occurred *c.* 40 myr apart, which contrasts strongly with previous assumptions of coeval metamorphism. While the age of peak metamorphism of the Margate Terrane (1032.7 ± 4.7 Ma) coincides with the tectonism and magmatism associated with the emplacement of the Oribi Gorge Suite (*c.* 1050–1030 Ma), the age of metamorphism of the Mzumbe Terrane (987.4 ± 8.1 Ma) occurs *c.* 30–40 myr after tectonism is previously thought to have finished. We propose that models of advective cooling during transcurrent shearing can explain the metamorphic conditions and timing of the NMP.

Supplementary material: T–XH2O diagrams used for the generation of final P–T pseudosections and complete monazite U-Pb geochronology and garnet REE datasets are available at https://doi.org/10.6084/m9.figshare.c.6488823

Metamorphic terranes, in particular high-grade terranes, present a complex challenge when constraining and differentiating between interconnected geological processes (e.g. partial melting and fluid flux: Taylor *et al.* 2014; Blereau *et al.* 2016; Carvalho *et al.* 2019; Wang *et al.* 2021) and overprinting from subsequent events (e.g. polymetamorphism: Blereau *et al.* 2017, 2019; Laurent *et al.* 2018*a*, *b*). This amalgamation of processes and events often makes it difficult to determine clearly the duration and conditions of discrete metamorphic episodes (i.e. a discrete pressure–temperature–time (*P–T–t*) path). The petrochronological approach (Kylander-Clark *et al.* 2013; Kohn 2016; Engi *et al.* 2017), a multidisciplinary investigation of processes that connect major silicate mineral evolutions (petrology and *P–T* constraints) to a range of analytical data from major and accessory minerals (e.g. geochemistry and geochronology), has been applied to an array of geological systems. Petrochronology has allowed us to refine our understanding metamorphic processes (e.g. REE systematics in zircon during high-grade metamorphism: Whitehouse 2003; Holder *et al.* 2015; Taylor *et al.* 2016; Rubatto 2017; Taylor *et al.* 2017; Blereau *et al.* 2022) and constrain *P–T–t* paths for short (e.g. Viete and Lister 2017) and prolonged metamorphic events (e.g. Clark *et al.* 2018). Petrochronology is a powerful field of research as all data used in this approach require the retainment of the broader geological context on the microscopic level and potentially up to the field level.

The Mzumbe and Margate terranes of the Natal Metamorphic Province (NMP) have historically been interpreted to have been metamorphosed together during the emplacement of the Oribi Gorge Suite in an early stage of metamorphism

From: van Schijndel, V., Cutts, K., Pereira, I., Guitreau, M., Volante, S. and Tedeschi, M. (eds) 2024. *Minor Minerals, Major Implications: Using Key Mineral Phases to Unravel the Formation and Evolution of Earth's Crust.* Geological Society, London, Special Publications, **537**, 313–331.
First published online May 18, 2023, https://doi.org/10.1144/SP537-2022-222

(M_1) (McCourt *et al.* 2006; Eglington *et al.* 2010; Grantham *et al.* 2012); however, metamorphism in this area is poorly dated (the only direct metamorphic ages are from Spencer *et al.* 2015) and needs the development of a metamorphic and deformational geochronological framework. This geochronology has additional importance as this region provides information on the early history of Rodinia, as the NMP comprises several terranes that accreted onto the Kaapvaal Craton in the late Mesoproterozoic during to the assembly of Rodinia (Jacobs and Thomas 1994). We revisit the metamorphic evolution of the NMP in order to better understand the relationship between the Mzumbe and Margate terranes during metamorphism, and the re-evaluate the peak metamorphic conditions experienced.

Regional geology

The NMP is part of the Rodinia-age (*c.* 1.1 Ga) Namaqua–Natal Belt and is subdivided into three terranes: the Tugela Terrane, the Mzumbe Terrane and

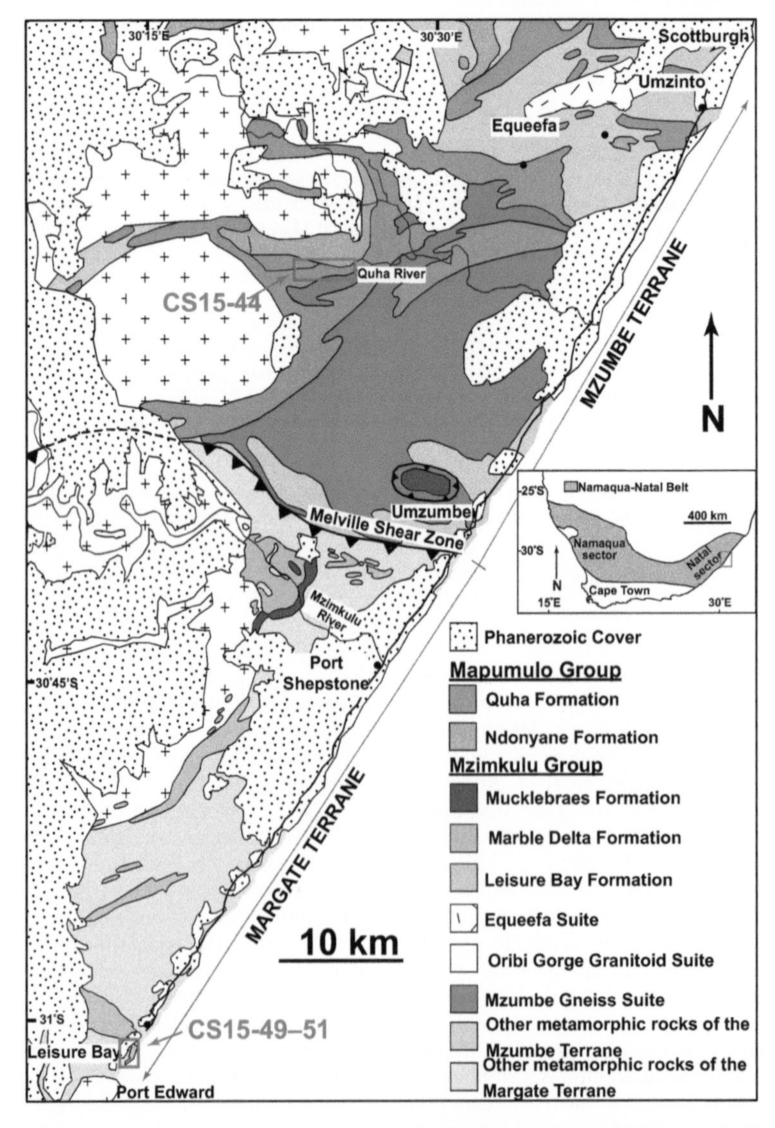

Fig. 1. Simplified geological map of the Margate and Mzumbe terranes within the Natal Metamorphic Province showing the locations of samples analysed in this study. Modified after Thomas (1990), Grantham *et al.* (1991), Thomas *et al.* (1991*a*, *b*) and Thomas (1992*a*, *b*). Inset map of the Namaqua–Natal Belt showing the location of the field area, modified after Spencer *et al.* (2015).

Table 1. *Summary of existing constraints on the metamorphic evolution of the Mzumbe and Margate terranes*

Temperature (*T*) (°C)	Pressure (*P*) (kbar)	Metamorphic event	Method	Reference
Mzumbe Terrane				
No constraints	No constraints	Earlier event?	Not applicable	Evans (1984); Evans *et al.* (1987)
750–800	6–8	Dominant event	Compilation of experimental stabilities of different mineral assemblages	Evans *et al.* (1987)
c. 650	*c.* 6	Retrograde event	Epidote growth	Evans *et al.* (1987)
850/1100		Dominant event	Two pyroxene thermometry for Cpx and Ca-rich Cpx, respectively	Thomas *et al.* (1992*b*)
	5–8	Dominant event	Geobarometry	Grantham (1983); Evans *et al.* (1987); Thomas *et al.* (1992*a*)*
830–1030		Dominant event	Thermo-Calc (version from Powell and Holland 1988)	Grantham *et al.* (1993)
Margate Terrane				
434–1110		Dominant event	Garnet pair thermometer	Mendonidis and Grantham (2003)
>850	*c.* 4	Dominant event	Mineral stability in *P–T* grids and Thermo-Calc version 2.7	Mendonidis and Grantham (2003)
c. 800	7.5–9	Secondary event	GAES barometer	Mendonidis and Grantham (2003)

*Pressure conditions, except for those in Evans *et al.* (1987), are all from samples from the neighbouring Margate Terrane, not from the Mzumbe Terrane, but are used in the Mzumbe Terrane.

the Margate Terrane (Thomas 1989). This study focuses on metamorphic rocks within the Mzumbe and Margate terranes, which are separated by the Melville Shear Zone (Fig. 1). The metamorphic history of the Mzumbe and Margate terranes (Table 1) is historically assumed to have occurred coevally at *c.* 1090–1040 Ma, despite there being no direct metamorphic ages available for the Mzumbe, and to be similar in tectonic style (Eglington *et al.* 2003; Mendonidis *et al.* 2015; Spencer *et al.* 2015). Previous attempts to establish ages of metamorphism were achieved through dating cross-cutting igneous rocks. The Mzumbe and Margate terranes (along with the Tugela Terrane north of the Mzumbe Terrane) were assembled during a multistage accretion event during the Natal Orogeny (Spencer *et al.* 2015; Mendonidis and Thomas 2019). The Mzumbe Terrane is interpreted to record protracted oceanic arc magmatism from *c.* 1200–1160 Ma (represented by the Mzumbe plutonic suite) followed by accretion onto the southern margin of the Kaapvaal Craton at *c.* 1150 Ma (Spencer *et al.* 2015 and references therein). The Margate Terrane largely records a similar magmatic history but has been interpreted to represent a separate, coeval, oceanic arc (Mendonidis *et al.* 2015). Following accretion and metamorphism, transcurrent deformation and syn- to post-orogenic plutons (Oribi and Sezela plutonic suites) are proposed to have dominated the region from *c.* 1080 to *c.* 1030 Ma (Eglington *et al.* 2003; Mendonidis *et al.* 2015; Spencer *et al.* 2015). In the following subsections we summarize the existing constraints on each terrane from the original literature, as more recent literature often misquotes these constraints.

Mzumbe Terrane

The Mzumbe Terrane comprises the intermediate-mafic and psammitic Quha Formation, acid Ndonyane Formation (both part of the Mapumulo Group), and a number of magmatic suites including the Equeefa Suite, the intensely deformed Mzumbe Gneiss Suite (or Mzumbe Granitoid Suite) and a suite of S-type granites (Thomas *et al.* 1991*a*; Thomas 1992*a*). The Quha Formation is the oldest supracrustal gneiss sequence in the Mzumbe Terrane with a minimum formation age of *c.* 1235 Ma (Thomas and Eglington 1990; Thomas *et al.* 1999), and is a layered sequence of semi-pelitic, pelitic, calc-silicate and magnesian gneisses, with minor amounts of marble and amphibolite. The Mzumbe Terrane is interpreted to have a polymetamorphic history with an earlier episode that has no constrained *P–T* conditions (M_1) and a more dominant metamorphic event (M_2). M_1 is represented by small quartzo-feldspathic veins intruding the Quha Formation (i.e. Banded Gneiss Formation in Evans *et al.* 1987) and pre-M_2 garnets, with an assemblage of biotite, hornblende, cordierite and fibrolite tentatively interpreted as the assemblage of this event (Evans 1984; Evans *et al.* 1987). Evans *et al.* (1987) compiled *P–T* constraints on the dominant metamorphic event (M_2) based on a number of lithological assemblages around the Umizinto region (north in Fig. 1). The synthesis of the experimental stabilities for different mineral assemblage and reactions yielded granulite-facies conditions of 750–800°C at 6–8 kbar (Evans *et al.* 1987). Following peak-M_2 metamorphism, cooling caused the development of Fe-rich epidote, with *P–T* conditions estimated to have been below *c.* 6 kbar and 650°C.

Also within the Mzumbe Terrane, the mafic Equeefa Suite was used in an attempt to constrain M_2 metamorphism. Two-pyroxene thermometry (after Lindsley 1983) yielded magmatic conditions of *c.* 1100°C from clinopyroxene and an interpreted metamorphic temperature of *c.* 850°C from more calcic clinopyroxene (Thomas *et al.* 1992*b*). Using an early version of the Thermo-Calc software (Powell and Holland 1988), Grantham *et al.* (1993) modelled reaction curves relating to two corona reactions (between olivine–plagioclase and phlogopite–plagioclase) and the reaction between olivine and phlogopite within the olivine melanorite within the Equeefa Suite. The pressure conditions used in the aforementioned phase equilibria models were based on previous geobarometry estimates by Grantham (1983), Evans *et al.* (1987) and Thomas *et al.* (1992*a*) (5–8 kbar). Although, aside from Evans *et al.* (1987), the pressure constraints used within Grantham (1983) and Thomas *et al.* (1992*a*) are all from granulite-facies rocks within the neighbouring Margate Terrane not from the Mzumbe Terrane. The stability of the Ol–Pl corona reaction in the olivine melanorite (olivine next to plagioclase is overgrown by orthopyroxene then clinoamphibole) was shown to be affected by water activity (a_{H_2O}), with a change in a_{H_2O} from 1.0 to 0.1 reducing the reaction temperature from *c.* 1030 to 830°C at 7 kbar. The Phl–Pl coronas (pargasite at the Phl–Pl interface) were difficult to model due to extensive solid solutions in all phases, making it difficult to define end members as a result of the applied models containing no solid solutions. Solid solutions were not introduced into Thermo-Calc until 1990 (Guiraud *et al.* 1990) and activity compositions appropriate for modelling melt-bearing mafic compositions only became available in 2016 (Green *et al.* 2016). The Ol–Phl reaction was also shown to be sensitive to a_{H_2O} but had to assume the presence of earlier K-feldspar, although no K-feldspar remains within the olivine melanorite.

Margate Terrane

The Margate Terrane contains three main formations: the Leisure Bay, Marble Delta and Mucklebraes formations, which form the Mzimkulu Group. Similar to the Quha Formation, the Leisure Bay Formation is the oldest supracrustal gneiss sequence within the Margate Terrane, and is a layered sequence of pelitic, semi-pelitic and calcic paragneisses, with minor kinzigite and metabasic gneisses (Thomas *et al.* 1991*c*; Thomas 1992*b*). *P–T* estimates for this region are based on a number of garnet mineral pair thermometers (i.e. Grt–Bt, Grt–Opx, Grt–Crd and Grt–Ilm), yielding a wide range of possible temperatures from as low as *c.* 434°C to as high as *c.* 1110°C, which is likely to be due to re-equilibration upon cooling (Mendonidis and Grantham 2003). Using experimental *P–T* grids and an early versions of the Thermo-Calc software (version 2.7, *c.* 1998), *P–T* estimates for M_1 based on the stability of spinel, cordierite and hypersthene yielded temperatures >850°C and pressures of *c.* 4 kbar, followed by isobaric cooling (Mendonidis and Grantham 2003). The Leisure Bay rocks utilized by Mendonidis and Grantham (2003) are intercalated with the Munster Suite that intruded at *c.* 1090 Ma, which was interpreted to have resulted in M_1, but the Leisure Bay rocks have not been directly dated. M_2 was restricted to fertile rocks and is less evident in the Leisure Bay rocks due to previous partial melting during M_1, with the breakdown of biotite in fertile lithologies leading to garnet growth and partial melt formation (Mendonidis and Grantham 2003). M_2 occurred at a higher pressure (7.5–9 kbar: GAES barometer) in order for only garnet to form from incongruent melting instead of garnet + cordierite + hypersthene as seen in M_1, with anatexis occurring at *c.* 800°C (Mendonidis and Grantham 2003). M_1 metamorphism in the Margate Terrane and M_2 in the Mzumbe Terrane are both interpreted to follow *P–T* paths with limited pressure variations (Grantham *et al.* 1994; Mendonidis and Grantham 2003). The *P–T* path for M_2 in the Margate Terrane is interpreted to be clockwise but at higher pressure (Mendonidis and Grantham 2003). There is a lot of variability within these pre-existing constraints and, with the lack of a firm geochronological framework for the metamorphic history of this region, a re-examination and collection of *in situ* data and more modern techniques could clarify existing data.

Thanks to updated activity models we now have the opportunity to refine and re-evaluate the existing *P–T* constraints to determine whether the style and conditions between the Margate and Mzumbe terranes are truly similar. Using monazite U–Pb geochronology we investigated the age of peak metamorphism to provide some much-needed new data. The samples collected for this study will not only aid in evaluating the metamorphic conditions and timing but will further refine the tectonic history between the Mzumbe Terrane and the Margate Terrane.

Methods

Electron microprobe analyser

Electron microprobe analyses and X-ray compositional maps were made using a Cameca SX-50 electron microprobe at the Department of Geological Sciences, Brigham Young University, Utah, USA. Backscatter electron images and element maps of Fe, Mg, Ca, Mn were made of the phases selected for probe analyses. X-ray maps were used to determine appropriate locations for analyses, and were collected with an acceleration voltage of 15 kV, a current of 40 nA and a time per pixel of 20 ms. Point analyses and/or transects were conducted across garnet, biotite, muscovite and plagioclase to further characterize the compositional zoning and to find appropriate areas for thermobarometric calculations. The analytical conditions used for quantitative analyses of silicates were 15 kV acceleration voltage, 20 s count time and 10–20 nA current. Natural minerals were used as standards to calibrate the compositions of unknown minerals.

Laser ablation inductively coupled plasma mass spectrometer (LA-ICP-MS)

Garnet trace element analyses. Rare earth element (REE) and other trace element compositions of garnet were measured by LA-ICP-MS using an ASI RESOlution M-50A-LR laser ablation system, using a Compex 193 nm Ar-F excimer laser and an Agilent 7700 inductively coupled plasma mass spectrometer at Curtin University. Garnet was analysed in polished thin sections using a 50 µm spot size and 30 s analyses at a repetition rate of 7 Hz. NIST glass 610 (Pearce *et al.* 1997) was used as the primary trace element reference materials, with NIST 612 as the secondary standard. Stoichiometric Si (18 wt%) was assumed for calibration of garnet trace elements. Time-resolved data were processed using Iolite software (version 3.1: Paton *et al.* 2010, 2011). Trace elements were normalized relative to chondrite based on the values of Anders and Grevesse (1989).

Monazite U–Pb geochronology. Individual monazite grains (mounted and polished in 1 inch epoxy rounds) were ablated using a Resonetics RESOlution M-50A-LR laser ablation system, incorporating a Compex 102 excimer laser. Following a 15–20 s period of background analysis, samples were spot ablated for 30 s at a 7 Hz repetition rate using a 23 µm beam and laser energy of 1.7 J cm^{-2} at the sample surface. The sample cell was flushed by ultra-high purity He (0.68 l min^{-1}) and N_2 (2.8 ml min^{-1}). Isotopic intensities were measured using an Agilent 7700s quadrupole ICP-MS with high-purity Ar as the plasma gas (flow rate 0.98 l min^{-1}). The dwell time for ^{204}Pb, ^{206}Pb, ^{207}Pb and ^{208}Pb was 0.03 s, and 0.0125 s for ^{232}Th and ^{238}U.

The primary reference material used for U–Pb dating in this study was 44069 ($^{206}Pb/^{238}U$ age 424.9 ± 0.4 Ma: Aleinikoff *et al.* 2006), with Moacyr ($^{206}Pb/^{238}U$ age 515.7 ± 0.7 Ma: Horstwood *et al.* 2016), Stern ($^{206}Pb/^{238}U$ age 512.4 ± 0.3 Ma: Horstwood *et al.* 2016) and Trebilcock ($^{206}Pb/^{238}U$ age 272 ± 2 Ma: Tomascak *et al.* 1996) used as secondary age reference materials. During the analytical session, 44069 yielded a $^{206}Pb/^{238}U$ weighted average age of 423.5 ± 2.7 Ma (MSWD = 0.6, $n = 18$; self-normalized). Moacyr yielded a $^{206}Pb/^{238}U$ weighted average age of 508.1 ± 4.6 Ma (MSWD = 1.7, $n = 9$), Stern a $^{206}Pb/^{238}U$ weighted average age of 503.7 ± 3.0 Ma (MSWD = 0.3, $n = 18$) and Trebilcock a $^{206}Pb/^{238}U$ weighted average age of 270.6 ± 1.9 Ma (MSWD = 0.3, $n = 13$). $^{206}Pb/^{238}U$ ages calculated for the secondary reference materials, treated as unknowns, were found to be within 2% of the accepted value and therefore no addition of excess variance to the systematic uncertainty was warranted. The time-resolved mass spectra were reduced using the U_Pb_Geochronology3 data reduction scheme in Iolite version 3.1 (Paton *et al.* 2011 and references therein). U–Pb data (including weighted mean and MSWD interpretation) were evaluated using the methodology of Spencer *et al.* (2016) and plotted using a Java-based computer application, KDX (Spencer *et al.* 2017).

Phase equilibrium modelling

Metamorphic *P–T* conditions were constrained using pseudosections modelled in the Na_2O–CaO–K_2O–FeO–MgO–Al_2O_3–SiO_2–H_2O–TiO_2–O (NCKFMASHTO) system. Thermo-Calc version 3.40i and the internally consistent dataset of Holland and Powell (2011) (tc-ds62 generated on 6 February 2014) was used with the activity composition models from White *et al.* (2014*a*). Mn-bearing solution models are available (White *et al.* 2014*b*); however, the effect of Mn at high temperatures is negligible (Johnson *et al.* 2015) and was not considered for this study. Calculations considered the phases: garnet, silicate melt, plagioclase, K-feldspar, epidote, biotite, orthopyroxene, cordierite, spinel-magnetite, ilmenite, rutile, sillimanite, kyanite, quartz, muscovite, sphene and chlorite.

Bulk rock compositions of the metapelites were determined by X-ray fluorescence (XRF) analysis on a Siemens SRS 303 X-ray fluorescence spectrometer at the Department of Geological Sciences at Brigham Young University, USA. The XRF bulk compositions are given in Figure 9. The material used for the bulk composition for CS15-44 contained visible garnet that was not able to be captured in thin section but was assumed to be part of the peak assemblage. The Thermo-Calc-normalized bulk compositions (expressed in mol% oxides) used in the creation of the *P–T* pseudosections are given in Figure 9. The ferric iron content was assumed to equal 20% of the ferrous iron for all samples, which replicated the observed assemblages and appropriate oxides. Modelled H_2O contents in the metapelites were constrained using *T–X* pseudosections ranging from a quantity assuming all analysed loss on ignition as H_2O (loss on ignition (LOI): $X = 0$) to lower values (0.1 mol% at $X = 1$). The H_2O content chosen for the *P–T* modelling of the metapelites was such that the solidus was close to the field containing the peak assemblage whilst also avoiding the stabilization of sillimanite, which is absent in all samples, to higher temperatures (see Supplementary material Figures S1–S4).

TESCAN Integrated Mineral Analysis

Mineral phase maps based on energy dispersive X-ray spectroscopy (EDX) and backscattered electron (BSE) responses were created using the TESCAN Integrated Mineral Analysis (TIMA) multidetector scanning electron microscope and Oxford Instrument Aztec software located at the John de Laeter Centre at Curtin University. These phase maps were used to quantify the modality of minerals as a direct output from the Aztec software and confirm the identity of all mineral phases.

Sample descriptions

Quha Formation of the Mzumbe Terrane

The single gneissic sample (CS15-44) from the Quha Formation was collected from Stratotype B from Thomas *et al.* (1991*b*) along the Mkomazi River (Figs 1 & 2a). The stratotype comprises semi-pelitic gneiss, pelitic gneiss, amphibolite, calc-silicate gneiss, marble, psammite and quartzite.

CS15-44 (−30.13440° S, 30.55254° E). CS15-44 is a quartzofeldspathic semi-pelitic gneiss (Figs 2b & 3). The sample is mainly composed of plagioclase (*c.* 34.5%), K-feldspar (*c.* 31%) and quartz (*c.* 31%) (Fig. 3), with rare inclusion-rich garnet porphyroblasts (Fig. 2b, inset) and minor ilmenite and magnetite. No garnets were present within our thin section but they were present within localized patches of incipient melt within the melanosome in outcrop (Fig. 2b). Biotite (*c.* 2.5%) is present in minor amounts with all matrix phases. Even smaller amounts of muscovite are interpreted as retrograde. Evidence of another unknown, likely mafic, mineral is seen in rare completely chloritized pseudomorphs. Minor myrmekite is also present near plagioclase. Monazite is located along the foliation bounded by quartz, plagioclase and K-feldspar.

Leisure Bay Formation of the Margate Terrane

The three metapelitic samples (CS15-49–CS15-51: Figs 4 & 5) from the Leisure Bay Formation were collected at Port Edward from the key outcrop between the Port Edward and Nicholson's Point granites (Figs 1 & 6a). This is the type locality for this formation. The samples fall in a transect between the two plutons, with sample CS15-49 being the closest and CS15-51 being the most distal to the Nicholson's Point granite, and the inverse in relation to the Port Edward granite. Sample CS15-49 is within *c.* 50 m of the contact of the Nicholson's Point granite and sample CS15-51 is within *c.* 1 m of the Port Edward granite. All samples show evidence of partial melting at the outcrop scale with centimetre-scale leucosomes (Fig. 6b). Thin sections were made predominantly from the melanosome of the migmatite with only small-scale leucosomes.

CS15-49 (−31.02408° S, 30.24483° E). CS15-49 is a pelitic gneiss with large anhedral garnet porphyroblasts (*c.* 0.5–2.5 cm; *c.* 31%) (Figs 4a & 5a, b). The porphyroblasts contain large sporadic inclusions of biotite and quartz, and smaller pyrrhotite inclusions. The matrix is largely composed of K-feldspar (*c.* 24%) followed by lesser amounts of cordierite (*c.* 18%), quartz (*c.* 13%), plagioclase (*c.* 7%), biotite (*c.* 5%), and trace amounts of ilmenite and pyrrhotite. Quartz fills embayments within the garnet. Cordierite is partly replaced by muscovite (sericite), and ilmenite is seen breaking down to an intergrowth of rutile, quartz and pyrrhotite. Monazite occurs within the matrix as well as at the margins of or within garnet. Trace amounts of apatite occur within the matrix.

CS15-50 (−31.02599° S, 30.24484° E). CS15-50 is a pelitic gneiss with two distinct compositional layers (Figs 4b & 5c–d). Anhedral–subhedral garnet porphyroblasts (*c.* 19%; 0.5–4 mm) are mainly found within K-feldspar-rich layers (*c.* 9%) but also as smaller grains in lesser amounts (0.5–1.5 mm) within plagioclase-rich layers (*c.* 46%). Garnet porphyroblasts preserve biotite, ilmenite, pyrrhotite and quartz inclusions across both layers. The garnet and K-feldspar-rich layers record a larger amount of late randomly orientated biotite (*c.* 10%) than the plagioclase-rich layers, with biotite abutting

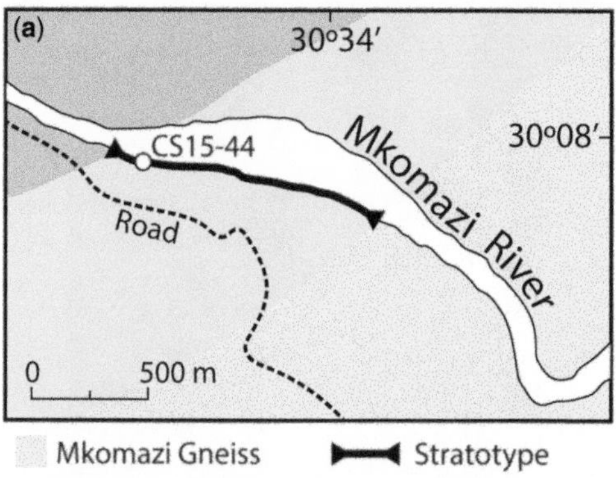

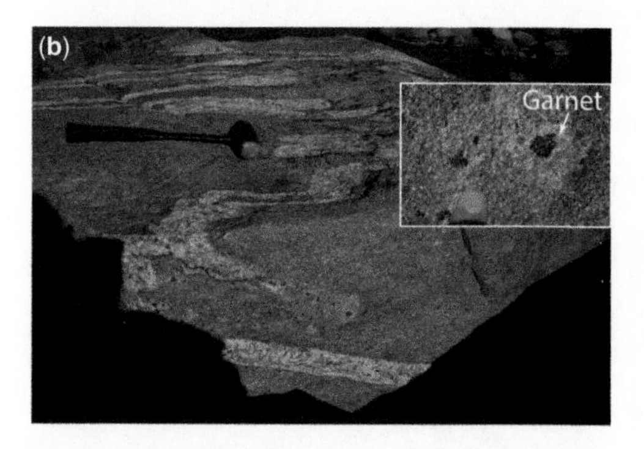

Fig. 2. (**a**) Simplified geological map of the Quha Formation stratotype B showing the sample locations. (**b**) Field photograph of the Quha quartzofeldspathic semi-pelitic gneiss with folded leucosomes. Rare garnet porphyroblasts rich in inclusions are visible, with a closer view shown in the inset. The geological hammer is for scale.

and growing around garnet porphyroblasts and with feldspar. The K-feldspar-rich layers also contain minor amounts of cordierite (*c.* 2%) and quartz (*c.* 5%). The plagioclase-rich layers contain a larger amount of quartz as well as orthopyroxene (*c.* 6%), which varies in grain size from similar sizes as the smaller garnet porphyroblasts to fine grained material (<0.5–2 mm). Quartz melt films can be seen at the edges of garnet and orthopyroxene within the plagioclase-rich layer, as well as rare intergrowths between biotite and quartz. Rare muscovite, trace pyrrhotite and ilmenite are seen in the matrix of both layers. Monazite occurs in the matrix, as well as grains at the margins of garnet or as inclusions. Trace amounts of apatite occur within the matrix.

CS15-51 (−31.02662° S, 30.24458° E). CS15-51 is also a pelitic gneiss and is the most homogeneous sample of the three collected from the Leisure Bay Formation (Figs 4c & 5e–f). Garnet porphyroblasts (*c.* 13%; <0.5–1 mm) are anhedral and embayed with large quartz inclusions, as well as with minor ilmenite and trace pyrrhotite inclusions. Some grains of garnet show fine wormy inclusions/intergrowth of quartz. Fine- to coarse-grained orthopyroxene (*c.* 3.5%) appears to have grown coevally with garnet, based on the lack of orthopyroxene inclusions and corona structures, and also shows irregular grain shapes such as for garnet. Similar to garnet, orthopyroxene has been disaggregated from *c.* 2 mm grains to grains that are 1 mm or smaller. The matrix is predominantly composed of quartz (*c.* 42%) and plagioclase (*c.* 34.5%), with minor K-feldspar (*c.* 1%), cordierite (*c.* 1.4%), ilmenite, muscovite and trace pyrrhotite. Biotite (*c.* 3%) appears as a late phase, reacting near orthopyroxene and garnet, crystallizing when H_2O is released as the partial melt crystallizes. Cordierite grains are rimmed by plagioclase, with garnet encapsulated or rimmed by quartz, both of which are likely to be melt films. Monazite occurs mainly in the matrix, with some grains at the margins of garnet. Trace amounts of apatite are present within the matrix.

Results

EPMA

Electron microprobe analysis of garnets from samples CS15-49–CS15-51 showed that garnets from all three Leisure Bay samples are consistently almandine-rich (*c.* 64–69 mol%), with lesser amounts of pyrope (*c.* 26–29 mol%) and minor grossular and spessartine contents (*c.* 2.5–3.5 and *c.* 1.5–2.5 mol%, respectively). X-ray maps of Fe, Ca, Mg and Mn show no major element geochemical zoning relating to core or rim textures in any of the three samples (i–iv in Fig. 7a–c).

Garnet trace elements

LA-ICP-MS analysis of garnets from all three Leisure Bay samples (CS15-49–CS15-51) targeted cores (inclusion rich) and rims (inclusion poor). Garnet cores from all three samples showed near flat mid-to-heavy REE (MREE–HREE) slopes (Yb/Gd slopes of *c.* 1–6) with normalized Lu (Lu_N) concentrations between 200 and 1000 chondrite normalized values (v in Fig. 7a–c). Garnet cores in sample CS15-51 had the least amount of scatter to the MREE–HREE with Yb/Gd slopes of 0.45–1.5. Garnet rims from all samples showed near flat to shallowly negative MREE–HREE slopes (0.45–2) with similar to slightly depleted HREE concentrations (Lu_N = 100–600) (v in Fig. 7a–c).

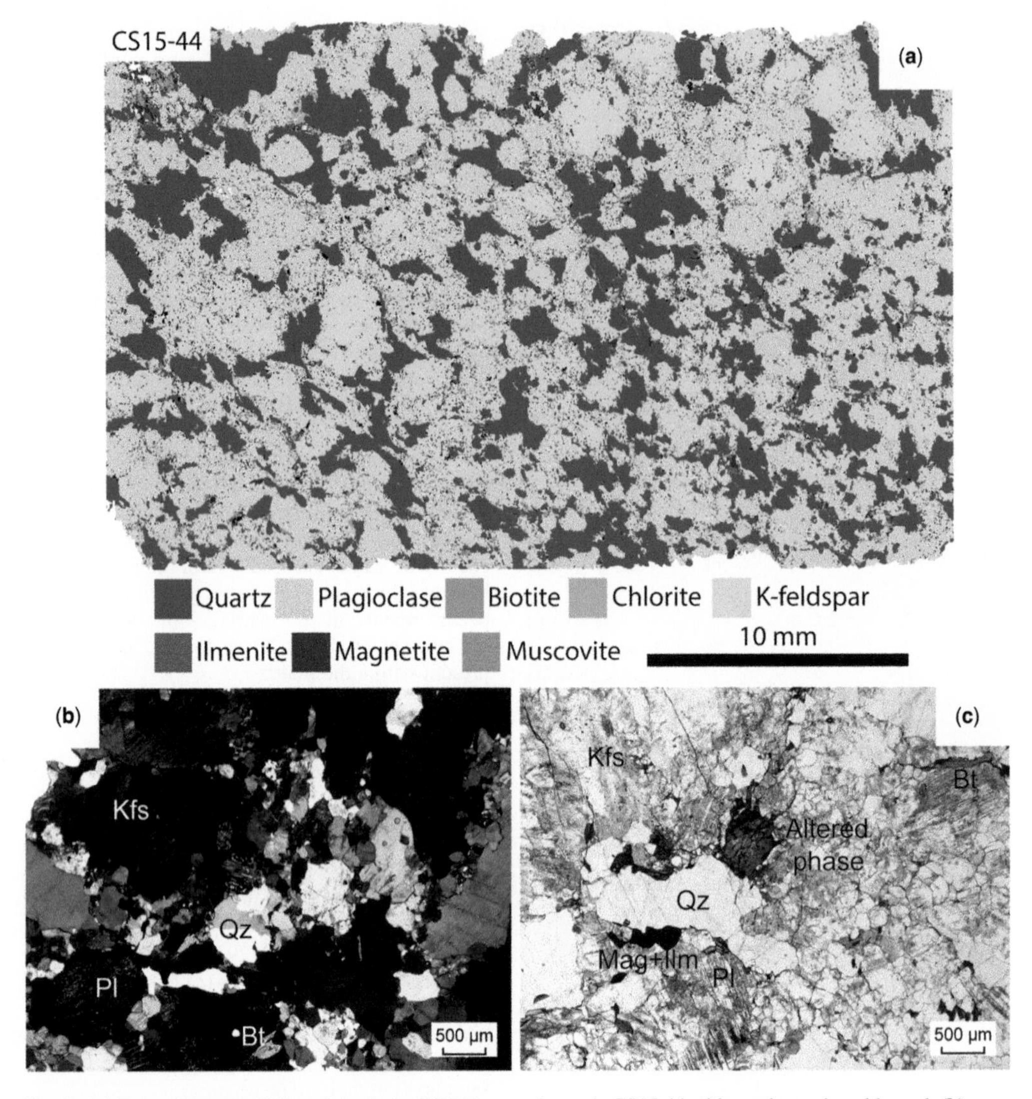

Fig. 3. (**a**) Tescan Integrated Mineral Analysis (TIMA) map of sample CS15-44 with a colour mineral legend. (**b**) and (**c**) Photomicrographs of CS15-44.

Monazite geochronology

Monazite from samples CS15-44 and CS15-49 were investigated *in situ* from within the matrix, within garnet where it was available and large enough to analyse, as well as near the margins of garnet. Monazite from both samples showed no visible zoning under BSE imaging (BSE) (Fig. 7d).

LA-ICP-MS analysis of monazite cores from the Quha and Leisure Bay formations (CS15-44 and CS15-49, respectively) yielded single populations for both samples (Fig. 8a). Monazite from the Quha Formation (CS15-44) had a weighted mean $^{207}Pb/^{206}Pb$ age of 987.4 ± 8.1 Ma (±21.3 Ma systematic uncertainty, $n = 25$, MSWD = 0.8), with monazite from the Leisure Bay Formation (CS15-49) having an older weighted mean age of 1032.7 ± 4.7 Ma (±21.2 Ma systematic uncertainty, $n = 9$, MSWD = 0.8) (Fig. 8b).

Phase equilibrium modelling

Compositional layering was observed within sample CS15-50 (Fig. 4b). However, an investigation of major elements (EPMA) and trace elements (LA-ICP-MS) within garnet from different compositional layers demonstrated that CS15-50, as well as

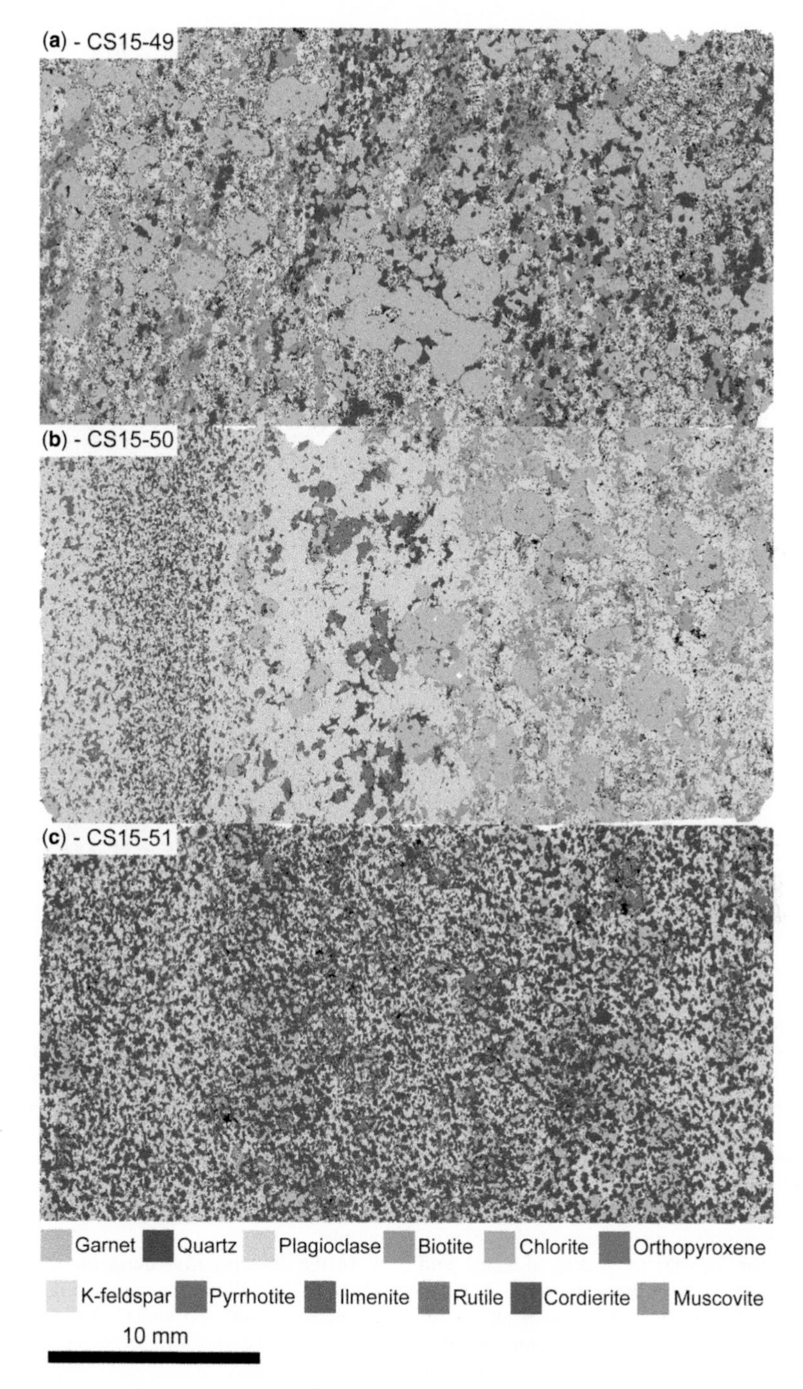

Fig. 4. Tescan Integrated Mineral Analysis (TIMA) map of samples CS15-49–CS15-51 with a colour mineral legend.

CS15-49, showed no significant compositional zoning within garnet or compositional changes across the thin section within all layers, indicating that the equilibration volume and diffusivity of elements within the sample was sufficient for the sample to equilibrate on at least the thin-section scale. As a

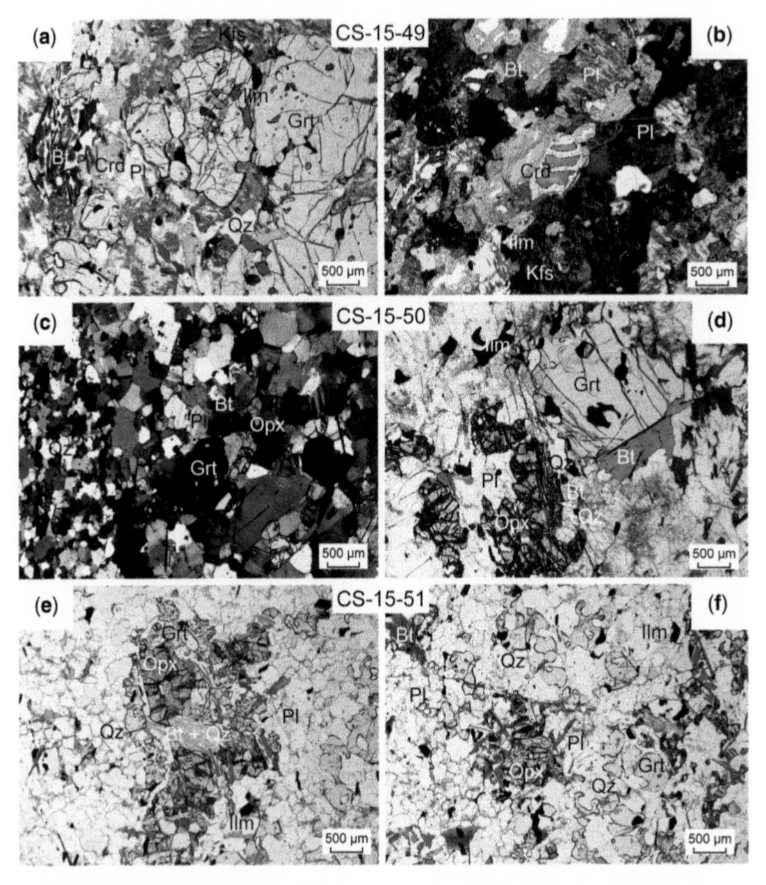

Fig. 5. (**a**) and (**b**) Photomicrographs of sample CS15-49. (**c**) and (**d**) Photomicrographs of sample CS15-50. (**e**) and (**f**) Photomicrographs of sample CS15-51.

result, bulk-rock compositions from material containing all recognized layers were used for the purpose of phase equilibrium modelling, over domain-based compositions.

Mzumbe Terrane. In the *P–T* pseudosection for sample CS15-44 (Fig. 9a) the solidus for the modelled H_2O content lies between 750 and 790°C. The interpreted peak assemblage of garnet, biotite, plagioclase, K-feldspar, quartz, magnetite, ilmenite and melt is predicted to have been between 3.9 and 7.8 kbar at 780–834°C (Fig. 9a). The peak assemblage is limited by the loss of magnetite above *c.* 7 kbar, as well as the growth of orthopyroxene at higher temperatures and cordierite at lower pressures than the peak assemblage.

Margate Terrane. In the *P–T* pseudosection for sample CS15-49 (Fig. 9b) the solidus for the modelled H_2O content lies between 800 and 830°C. The interpreted peak assemblage of garnet, cordierite, K-feldspar, quartz, ilmenite and melt occurred within a restricted *P–T* window between 5.5 and 6.4 kbar at 827–910°C. Sillimanite is stable at higher temperatures and pressures than the peak assemblage, with spinel stable at higher temperatures and magnetite stable at lower pressures. A small proportion of biotite can be grown just before crossing the solidus, consistent with the petrogenesis of the sample.

In the *P–T* pseudosection for sample CS15-50 (Fig. 9c) the solidus for the modelled H_2O content lies between 840 and 876°C. The interpreted peak assemblage of garnet, orthopyroxene, cordierite, biotite, plagioclase, K-feldspar, ilmenite and melt occurred between 5.5 and 6.5 kbar at 841–892°C. Cordierite is consumed up-temperature of the peak field, with the growth of spinel occurring at high temperatures and magnetite at low pressures. A small amount of quartz forms just before crossing the solidus, consistent with the minor amount of quartz in the sample.

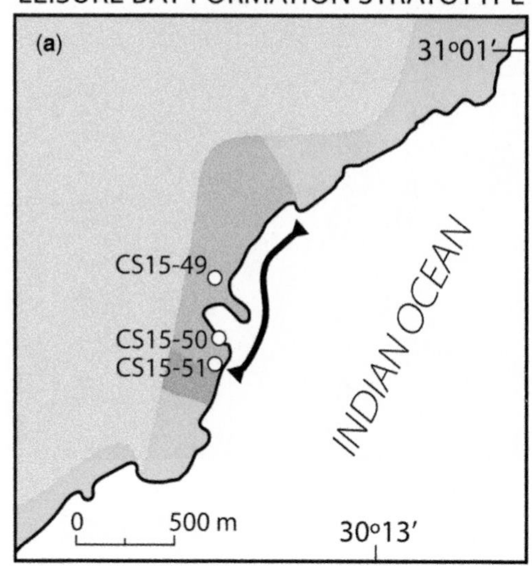

Fig. 6. (**a**) Simplified geological map of the Leisure Bay Formation stratotype showing the sample locations. (**b**) Field photograph of the pelitic gneisses of the Leisure Bay Formation with irregular garnet-bearing leucosomes. Garnet is also present within the melanosome. The hand lens is for scale.

In the *P–T* pseudosection for sample CS15-51 (Fig. 9d) the solidus lies at *c.* 850°C above a pressure of *c.* 6 kbar, with an inflection to *c.* 900°C below *c.* 6 kbar due to the loss of biotite and the presence of cordierite, with some H_2O partitioning into cordierite at lower pressures instead of partial melt. The interpreted peak assemblage of garnet, orthopyroxene, cordierite, plagioclase, K-feldspar, quartz, ilmenite and melt occurs between 5.1 and 6.1 kbar at 850–962°C. The sample grew a small portion of biotite before crossing the solidus, consistent with the petrology of the sample.

Discussion

Revised *P–T* conditions

For the Mzumbe Terrane our new *P–T* work is largely comparable to previous estimates (Evans *et al.* 1987). The psammitic sample from the Quha Formation (CS15-44) yielded *P–T* conditions between 780 and 834°C at 3.9–7.8 kbar. These conditions are *c.* 30°C more elevated than previous estimates (750–800°C at 6–8 kbar: Evans *et al.* 1987) with similar to lower pressures predicted. We saw no textural evidence for the growth of orthopyroxene and/or cordierite or the loss of magnetite that could limit our peak assemblage; however, we are unable to confirm the nature of the *P–T* path at this time due to the simplicity of the mineral paragenesis. Based on previous work (Evans *et al.* 1987), the psammitic assemblages of the Quha Formation are likely to have followed a typical clockwise *P–T* path (Fig. 9e; see also Figs 2 & 9a). We were unable to differentiate between a cooling path with moderate decompression and one with isobaric cooling due to limited pressure constraints (Fig. 9e). Mineral modes did not provide any additional refinement to our *P–T* constraints as none of the minerals stable within our peak assemblage had mineral modes that varied significantly with pressure, which is the broadest variable of our assemblage.

Unlike the Mzumbe Terrane, our *P–T* work in the Margate Terrane significantly refines existing *P–T* data. All three Leisure Bay samples share a field of overlap within the *P–T* space between 850 and 892°C at 5.7–6.1 kbar (Fig. 9e), vastly refined from previous estimates from mineral pair thermometry for M_1 and falling at more elevated pressures than previously recorded for M_1 (434–1100°C at *c.* 4 kbar: Mendonidis and Grantham 2003). Temperatures are similar to those previously reported from *P–T* grids for M_1 but, again, more refined (>850°C: Mendonidis and Grantham 2003). The conditions for our samples also include lower pressures for M_2 (7–9 kbar: Mendonidis and Grantham 2003). Whilst all three of our Leisure Bay samples overlap to some degree in the *P–T* space, there also appears to be a potential relationship between the maximum peak metamorphic temperature recorded and the distance of the sample from the Port Edward pluton (Oribi Gorge Suite, 1034.4 ± 0.6 Ma: Spencer *et al.* 2015) but not the Nicholson's Point granite (Margate Granite Suite, 1084.4 ± 1.7 Ma: Spencer *et al.* 2015), which are located on either side of the sample transect. The recorded peak temperature range increases between samples along the transect, with the most elevated temperatures experienced by

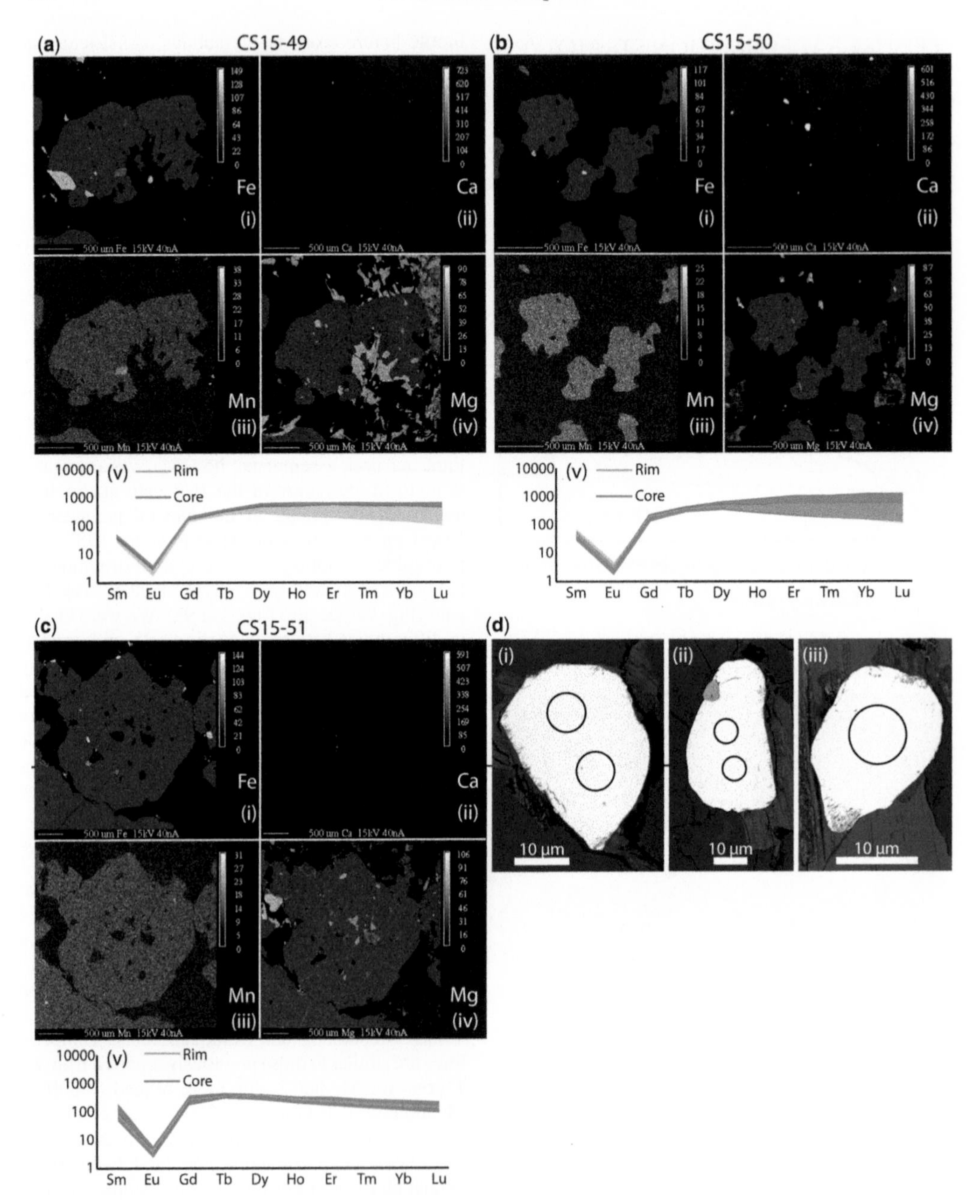

Fig. 7. (**a**) Garnet EPMA X-ray maps for (i–iv) major elements and (v) chondrite-normalized LA-ICP-MS trace elements from sample CS15-49. (**b**) Garnet EPMA X-ray maps for (i–iv) major elements and (v) LA-ICP-MS trace elements from sample CS15-50. (**c**) Garnet EPMA X-ray maps for (i–iv) major elements and (v) chondrite-normalized LA-ICP-MS trace elements from sample CS15-51. (**d**) (i–iii) Backscattered electron (BSE) images of monazite from sample CS15-49 that are representative of monazite in all samples.

sample CS15-51 (850–962°C), which is within 1 m of the Port Edward pluton, and the lowest temperature experienced by CS15-49 (827–910°C), which is the most removed from the Port Edward pluton (*c.* 50 m from the Nicholson's Point granite). The samples, which are up to 500 m apart, seemed to have followed similar *P–T* paths with slight changes in overall maximum temperature along the short

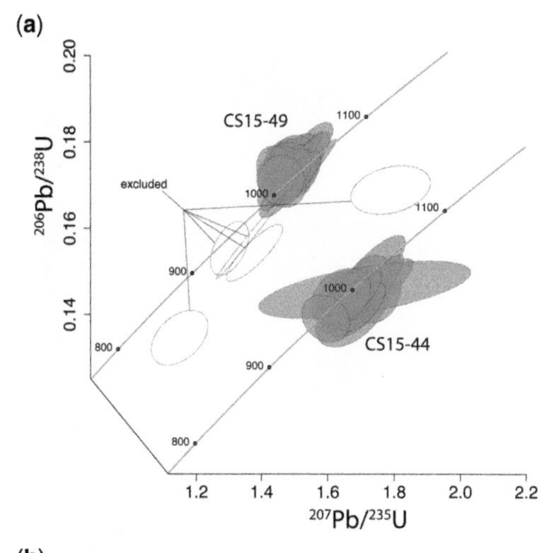

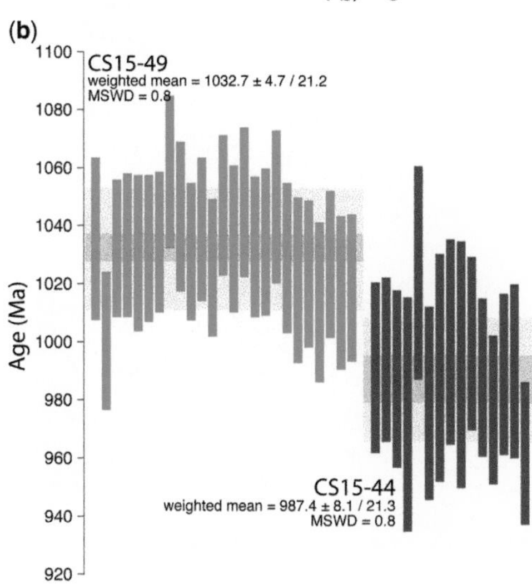

Fig. 8. (**a**) Wetherill concordia plot of laser ablation split-stream (LASS) U–Pb monazite ages from samples CS15-44 (Quha) and CS15-49 (Leisure Bay). (**b**) Weighted mean $^{207}Pb/^{206}Pb$ ages (with 2σ and systematic errors) for samples CS15-44 and CS15-49.

transect. This potentially indicates a relationship between peak metamorphism and the emplacement of the Oribi Gorge Suite, a point we will return to in the following subsection with regard to the age of metamorphism.

All three Leisure Bay samples followed a tight clockwise *P–T* path, followed by effectively isobaric cooling due to the limited stability field of the modelled assemblage as a result of a number of phase changes at higher *P* and/or *T*. These constraints at higher pressures are the loss of cordierite (CS15-50) and orthopyroxene (CS15-51), and the growth of sillimanite (CS15-49) (Fig. 9e), as well as the growth of spinel and magnetite at higher temperatures and lower pressures, respectively, both of which are absent in all samples. The essentially isobaric *P–T* path in this study's samples is consistent with previous studies (Grantham *et al.* 1994; Mendonidis and Grantham 2003) but is much more refined in terms of overall *P–T* conditions. As all of the samples exhibit partial melting, we only infer the prograde history of the *P–T* path.

Unlike previous studies, we did not see any evidence of a second metamorphic event within our samples from the Leisure Bay Formation connected to additional garnet growth (Mendonidis and Grantham 2003), and we did not see a second population, texturally or geochemically (Figs 5 & 7). However, samples CS15-49 and CS15-50 did contain retrograde micas and should have been fertile for later metamorphism, unlike samples described in previous studies. This could potentially indicate local variation in metamorphic conditions across the Margate Terrane. Samples interpreted to contain a secondary metamorphic assemblage should be re-evaluated using modern methods.

Age of metamorphism

The timing of peak metamorphism of the Mzumbe and Margate terranes was constrained using monazite dating from supracrustal metamorphic rocks in the respective terranes. Monazite from the Leisure Bay Formation yielded a late Stenian age (1032.7 $\pm$ 4.7/21.2 Ma, age $\pm$ weighted uncertainty/systematic uncertainty) (Fig. 8), which is in line with previous monazite dating from the Turtle Bay Suite that lies within an enigmatic high-shear zone between the Margate and Mzumbe terranes (Spencer *et al.* 2015; Mendonidis and Armstrong 2016) and overlaps with the *c.* 1050–1030 Ma timing of the Oribi Gorge Suite that spans both terranes (Spencer *et al.* 2015) (Fig. 10). The similar monazite ages of the Leisure Bay Formation and Turtle Bay Suite imply that the Turtle Bay Suite is likely to have experienced metamorphism associated with the intrusion of the Oribi Gorge Suite.

Thomas *et al.* (1999) dated zircon rims from the Quha Formation that were interpreted to be metamorphic in origin with an age of 1065 $\pm$ 15 Ma. This age is a $^{207}Pb/^{206}Pb$ age but, given the age of the grains, this should have been be recorded as a $^{206}Pb/^{238}Pb$ age (Spencer *et al.* 2016), which shows the significant disruption in all of the analyses. The dating of these rims should be reattempted in order to verify and increase the precision of these rims. The earliest Neoproterozoic age of monazite from the Quha Formation of the Mzumbe Terrane (987.4 $\pm$ 8.1/21.3 Ma) (Fig. 8) is the youngest tectonomagmatic event confirmed with robust geochronology (cf. 951 $\pm$ 16 Ma Rb–Sr whole-rock

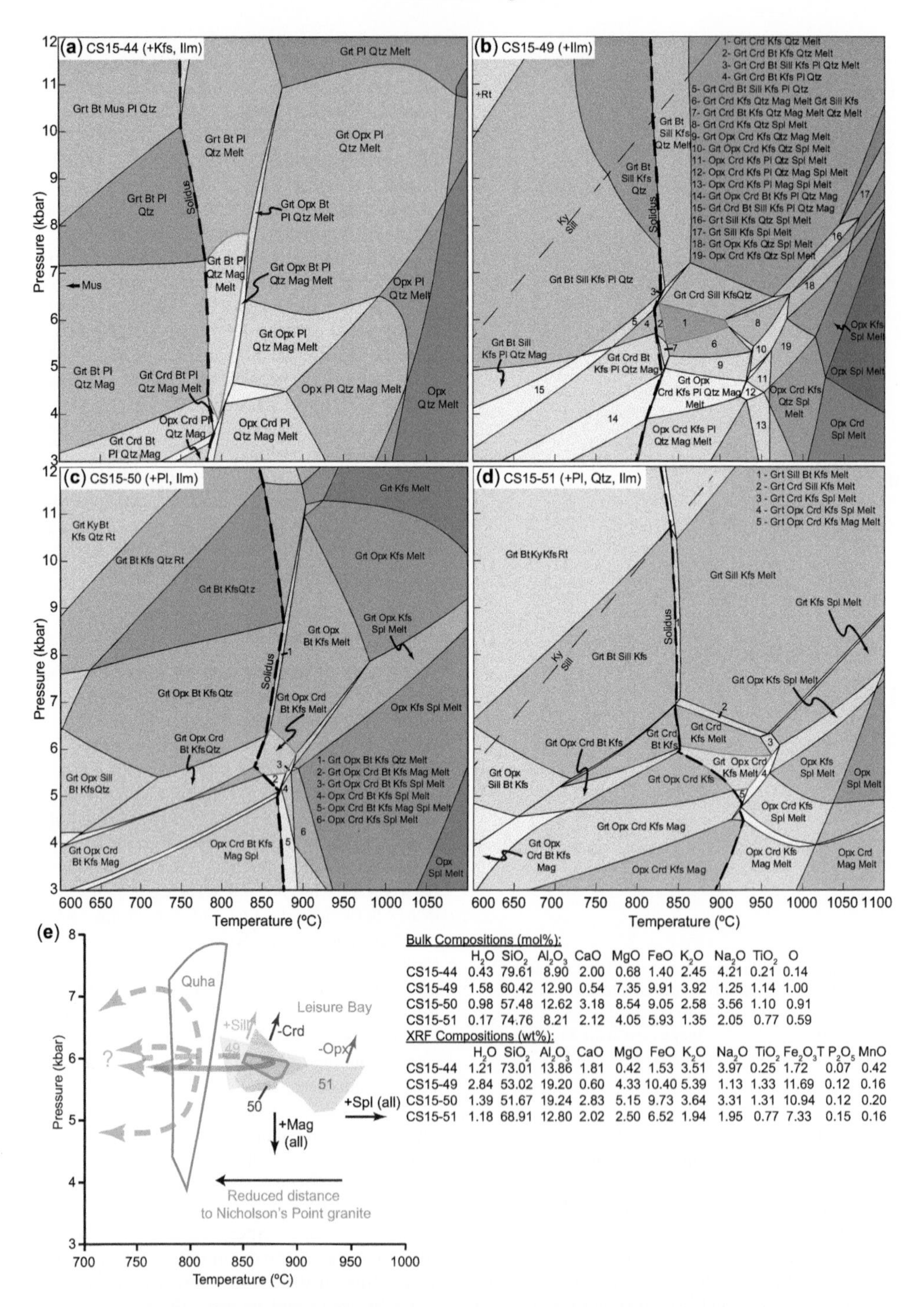

Bulk Compositions (mol%):

	H_2O	SiO_2	Al_2O_3	CaO	MgO	FeO	K_2O	Na_2O	TiO_2	O
CS15-44	0.43	79.61	8.90	2.00	0.68	1.40	2.45	4.21	0.21	0.14
CS15-49	1.58	60.42	12.90	0.54	7.35	9.91	3.92	1.25	1.14	1.00
CS15-50	0.98	57.48	12.62	3.18	8.54	9.05	2.58	3.56	1.10	0.91
CS15-51	0.17	74.76	8.21	2.12	4.05	5.93	1.35	2.05	0.77	0.59

XRF Compositions (wt%):

	H_2O	SiO_2	Al_2O_3	CaO	MgO	FeO	K_2O	Na_2O	TiO_2	Fe_2O_3T	P_2O_5	MnO
CS15-44	1.21	73.01	13.86	1.81	0.42	1.53	3.51	3.97	0.25	1.72	0.07	0.42
CS15-49	2.84	53.02	19.20	0.60	4.33	10.40	5.39	1.13	1.33	11.69	0.12	0.16
CS15-50	1.39	51.67	19.24	2.83	5.15	9.73	3.64	3.31	1.31	10.94	0.12	0.20
CS15-51	1.18	68.91	12.80	2.02	2.50	6.52	1.94	1.95	0.77	7.33	0.15	0.16

Fig. 9. *P–T* pseudosections of (**a**) sample CS15-44 from the Quha Formation, and (**b**) samples CS15-49, (**c**) CS15-50 and (**d**) CS15-51 from the Leisure Bay Formation. The solidus is highlighted by a black dashed line, with the interpreted peak assemblage outlined in red. (**e**) *P–T* summary diagram with potential *P–T* paths for both the Quha Formation (red) and Leisure Bay Formation (orange), and the field of overlap between the three Leisure Bay samples (CS15-49–CS15-51). Bulk compositions used in the modelled pseudosections are also listed in mol%, as well as original XRF bulk compositions in wt%.

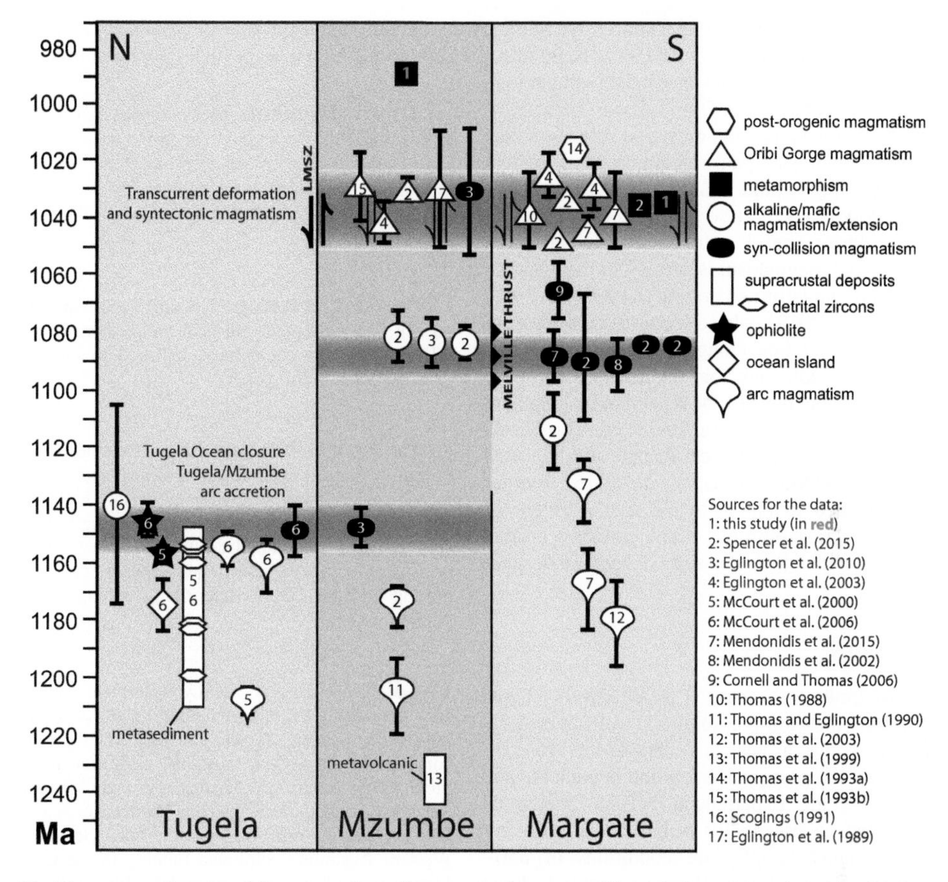

Fig. 10. Time–space diagram of the magmatic, sedimentary and metamorphic evolution of the whole Natal Belt, modified after Spencer *et al.* (2015).

isochron age of the Sezela Suite of alkaline igneous rocks: Eglington *et al.* 1989). Although there is overlap in the individual monazite ages from both of these units, both the weighted means using analytical uncertainties (a less conservative age estimate) and weighted means incorporating systematic uncertainties (a more conservative age estimate) are discrete with no overlapping uncertainties. Depending on whether the zircon rims of Thomas *et al.* (1999) can be verified, it is possible that the *c.* 987 Ma age of monazite could be post-peak crystallization or peak crystallization if the zircon does not represent a discrete population but instead resetting due to later metamorphism. It has been shown that monazite can grow without further or new zircon growth (Morrissey *et al.* 2016), so the lack of a zircon population of this age is possible. However, the difference in metamorphic age of the supracrustals from the two terranes implies that they experienced different metamorphic events in terms of timing and potentially duration.

Impacts on the tectonic history of the Natal Belt

Previous models posit that accretion of the Tugela, Mzumbe and Margate terranes proceeded from north to south with accretion of the Tugela and Mzumbe terranes, with the Kaapvaal cratonic margin occurring at *c.* 1150 Ma (McCourt *et al.* 2006; Spencer *et al.* 2015). The Margate Terrane is then thought to have accreted at *c.* 1100–1090 Ma (Eglington *et al.* 2003; Mendonidis *et al.* 2015; Spencer *et al.* 2015). The young age of the monazite (*c.* 987 Ma) from the Quha Formation provides evidence that tectonism within the Mzumbe Terrane continued during protracted accretion into the earliest Neoproterozoic (Fig. 10). This may have been associated with further crustal thickening via oblique thrusting along the Melville Thrust that lies between the Mzumbe and Margate terranes (Jacobs and Thomas 1994). This hypothesis could be validated through the dating of deformation fabrics associated with the Melville Thrust. This *c.* 990 Ma monazite age of the Quha Formation is not supportive of thermal perturbation due to the Oribi Gorge Suite intrusion as no Oribi Gorge Suite ages exist younger than *c.* 1000 Ma (Spencer *et al.* 2015). The isobaric cooling of the Margate Terrane may also provide a clue as to the duration of peak metamorphism, in that isobaric cooling would prolong high-grade metamorphic conditions more so than isothermal decompression during rapid exhumation. In addition, similarly aged magmatism is present in the Maurice Ewing Bank (undeformed granite at 1006 ± 13 Ma: Chemale *et al.* 2018), whose reconstructed Neoproterozoic position lies along the Natal margin of South Africa, indicating that metamorphism in the Mzumbe Terrane may also be related to magmatism; however, this is yet to be substantiated. The Sezela Suite is too old to explain the monazite ages reported herein (*c.* 1080 Ma: Spencer *et al.* 2015). Further work is needed to elucidate the spatial extent of this younger metamorphic event and any cryptic magmatism with which it may have been associated.

Conclusions

The Mzumbe Terrane experienced high-temperature–medium-pressure (HT–LP) metamorphism, reaching a peak of 780–834°C at 3.9–7.8 kbar but the post-peak evolution is unclear. The Margate Terrane experienced HT–LP metamorphism along an essentially isobaric *P–T* path but achieved higher temperature conditions of 850–892°C at 5.7–6.1 kbar. Previous studies have assumed coeval metamorphic histories for the Mzumbe and Margate terranes. We argue that the thermal peak of metamorphism in the Mzumbe Terrane (987.4 ± 8.1 Ma) post-dates similar grade and style of metamorphism in the Margate Terrane (1032.7 ± 4.7 Ma) by *c.* 40 myr, indicating that the metamorphic evolution of the Natal Belt is more complex than previously thought. Metamorphism in the Margate Terrane appears to have been tied to the *c.* 1030 Ma emplacement of the Oribi Gorge Suite, assumed using the age of metamorphic monazite and potential variation in peak temperature conditions between samples with distance from the Port Edward pluton. Metamorphism at *c.* 987 Ma in the Mzumbe Terrane may have been a product of additional thrusting along the Melville Thrust, although the event could potentially be related to magmatism; further study on this is required. These new data can be explained by advective cooling during orthogonal or oblique transcurrent deformation (Chen *et al.* 2015; Xu *et al.* 2015) that is supported by the oblique structural and tectonic features of the Natal Province (Jacobs and Thomas 1994), these features have yet to be dated directly. This model predicts isobaric cooling while minimizing exhumation and provides a possible explanation for the protracted nature of metamorphism in the NMP.

Acknowledgements Many thanks go to Noreen Evans and Brad McDonald for their assistance with the LA-ICP-MS analysis. We thank Mike Dorais for assistance in the XRF and electron microprobe analyses at Brigham Young University. We would also like to thank the reviewers of this paper.

Competing interests The authors declare that they have no known competing financial interests or personal relationships that could have appeared to influence the work reported in this paper.

Author contributions **EB**: conceptualization (supporting), data curation (equal), formal analysis (equal), investigation (equal), methodology (equal), writing – original draft (lead), writing – review & editing (equal); **CS**: conceptualization (lead), data curation (equal), formal analysis (equal), investigation (equal), methodology (equal), writing – original draft (supporting), writing – review & editing (equal).

Funding GeoHistory Facility instruments (part of the John de Laeter Centre) were funded via an Australian Geophysical Observing System (AGOS) grant provided to AuScope by the AQ44 Australian Education Investment Fund. The Australian Microscopy and Microanalysis Research Facility, AuScope, the Australian Science and Industry Endowment Fund, and the State Government of Western Australia contributed funding to the Centre for Microscopy, Characterization and Analysis at the University of Western Australia. The Tescan Integrated Mineral Analysis (TIMA) instrument at the John de Laeter Centre was funded by a grant from the Australian Research Council (LE140100150) and is operated with the support of the Geological Survey of Western Australia, University of Western Australia and Murdoch University.

Data availability All data generated or analysed during this study are included in this published article (and, if present, its supplementary information files).

References

Aleinikoff, J.N., Schenck, W.S., Plank, M.O., Srogi, L., Fanning, C.M., Kamo, S.L. and Bosbyshell, H. 2006. Deciphering igneous and metamorphic events in high-grade rocks of the Wilmington Complex, Delaware: Morphology, cathodoluminescence and backscattered electron zoning, and SHRIMP U–Pb geochronology of zircon and monazite. *Geological Society of America Bulletin*, **118**, 39–64, https://doi.org/10.1130/B25659.1

Anders, E. and Grevesse, N. 1989. Abundances of the elements: Meteoritic and solar. *Geochimica et Cosmochimica acta*, **53**, 197–214, https://doi.org/10.1016/0016-7037(89)90286-X

Blereau, E., Clark, C., Taylor, R.J.M., Johnson, T.E., Fitzsimons, I.C.W. and Santosh, M. 2016. Constraints on the timing and conditions of high-grade metamorphism, charnockite formation and fluid–rock interaction in the Trivandrum Block, southern India. *Journal of Metamorphic Geology*, **34**, 527–549, https://doi.org/10.1111/jmg.12192

Blereau, E., Johnson, T.E., Clark, C., Taylor, R.J.M., Kinny, P.D. and Hand, M. 2017. Reappraising the *P–T* evolution of the Rogaland–Vest Agder Sector, southwestern Norway. *Geoscience Frontiers*, **8**, 1–14, https://doi.org/10.1016/j.gsf.2016.07.003

Blereau, E., Clark, C. *et al.* 2019. Closed system behaviour of argon in osumilite records protracted high-temperature metamorphism within the Rogaland-Vest Agder Sector, Norway. *Journal of Metamorphic Geology*, **37**, 667–680, https://doi.org/10.1111/jmg.12480

Blereau, E., Clark, C., Kinny, P.D., Sansom, E., Taylor, R.J.M. and Hand, M. 2022. Probing the history of ultra-high temperature metamorphism through rare earth element diffusion in zircon. *Journal of Metamorphic Geology*, **40**, 329–357, https://doi.org/10.1111/jmg.12630

Carvalho, B.B., Bartoli, O. *et al.* 2019. Anatexis and fluid regime of the deep continental crust: New clues from melt and fluid inclusions in metapelitic migmatites from Ivrea Zone (NW Italy). *Journal of Metamorphic Geology*, **37**, 951–975, https://doi.org/10.1111/jmg.12463

Chemale, F., Ramos, V.A., Naipauer, M., Girelli, T.J. and Vargas, M. 2018. Age of basement rocks from the Maurice Ewing Bank and the Falkland/Malvinas Plateau. *Precambrian Research*, **314**, 28–40, https://doi.org/10.1016/j.precamres.2018.05.026

Chen, X., Liu, J., Tang, Y., Song, Z. and Cao, S. 2015. Contrasting exhumation histories along a crustal-scale strike-slip fault zone: The Eocene to Miocene Ailao Shan-River shear zone in southeastern Tibet. *Journal of Asian Earth Sciences*, **144**, 174–187, https://doi.org/10.1016/j.jseaes.2015.05.020

Clark, C., Taylor, R.J.M., Kylander-Clark, A.R.C. and Hacker, B.R. 2018. Prolonged (>100 Ma) ultrahigh temperature metamorphism in the Napier Complex, East Antarctica: A petrochronological investigation of Earth's hottest crust. *Journal of Metamorphic Geology*, **36**, 1117–1139, https://doi.org/10.1111/jmg.12430

Cornell, D.H. and Thomas, R.J. 2006. Age and tectonic significance of the Banana Beach gneiss, KwaZulu-Natal south coast, *South Africa. South African Journal of Geology*, **109**, 335–340, https://doi.org/10.2113/gssajg.109.3.335

Eglington, B.M., Harmer, R.E. and Kerr, A. 1989. Isotope and geochemical constraints on Proterozoic crustal evolution in south-eastern Africa. *Precambrian Research*, **45**, 159–174, https://doi.org/10.1016/0301-9268(89)90037-5

Eglington, B.M., Thomas, R.J., Armstrong, R.A. and Walraven, F. 2003. Zircon geochronology of the Oribi Gorge Suite, KwaZulu-Natal, South Africa: constraints on the timing of trans-current shearing in the Namaqua–Natal Belt. *Precambrian Research*, **123**, 29–46, https://doi.org/10.1016/S0301-9268(03)00016-0

Eglington, B.M., Thomas, R.J. and Armstrong, R.A. 2010. U–Pb SHRIMP zircon dating of Mesoproterozoic magmatic rocks from the Scottburgh Area, central Mzumbe terrane, KwaZulu-Natal, South Africa. *South African Journal of Geology*, **113**, 229–235, https://doi.org/10.2113/gssajg.113.2.229

Engi, M., Lanari, P. and Kohn, M.J. 2017. Significant ages – An introduction to petrochronology. *Reviews in Mineralogy and Geochemistry*, **83**, 1–12, https://doi.org/10.2138/rmg.2017.83.1

Evans, E.P., Eglington, B.M., Kerr, A. and Saggerson, E.P. 1987. The geology of the Proterozoic rocks around Umzinto, southern Natal, South Africa. *South African Journal of Geology*, **90**, 471–488.

Evans, M.J. 1984. *Precambrian Geology West of Scottburgh, Natal*. MSc dissertation, University of Natal, Durban, South Africa.

Grantham, G.H. 1983. *The Tectonic, Metamorphic and Intrusive History of the Natal Mobile Belt between Glenmore and Port Edward, Natal*. MSc dissertation, University of Natal, Durban, South Africa.

Grantham, G.H., Thomas, R.J. and Mendonidis, P. 1991. Leisure Bay Formation. *In*: *Catalogue of South African Lithostratigraphic Units, Volume 3*. Government Printer for the South African Committee for Stratigraphy, Pretoria, 17–18.

Grantham, G.H., Thomas, R.J., Eglington, B.M., De Bruin, D., Atanasov, A. and Evans, M.J. 1993. Corona textures in Proterozoic olivine melanorites of the Equeefa Suite, Natal metamorphic province, South Africa. *Mineralogy and Petrology*, **49**, 91–102, https://doi.org/10.1007/BF01162928

Grantham, G.H., Thomas, R.J. and Mendonidis, P. 1994. Contrasting P–T–t loops from southern East Africa, Natal and East Antarctica. *Journal of African Earth Sciences*, **19**, 225–235, https://doi.org/10.1016/0899-5362(94)90062-0

Grantham, G.H., Mendonidis, P., Thomas, R.J. and Satish-Kumar, M. 2012. Multiple origins of charnockite in the Mesoproterozoic Natal belt, Kwazulu-Natal, South Africa. *Geoscience Frontiers*, **3**, 755–771, https://doi.org/10.1016/j.gsf.2012.05.006

Green, E.C.R., White, R.W., Diener, J.F.A., Powell, R., Holland, T.J.B. and Palin, R.M. 2016. Activity–composition relations for the calculation of partial melting equilibria in metabasic rocks. *Journal of Metamorphic Geology*, **34**, 845–869, https://doi.org/10.1111/jmg.12211

Guiraud, M., Holland, T. and Powell, R. 1990. Calculated mineral equilibria in the greenschist–blueschist–eclogite facies in Na_2O–FeO–MgO–Al_2O_3–SiO_2–H_2O. *Contributions to Mineralogy and Petrology*, **104**, 85–98, https://doi.org/10.1007/BF00310648

Holder, R.M., Hacker, B.R., Kylander-Clark, A.R.C. and Cottle, J.M. 2015. Monazite trace-element and isotopic signatures of (ultra)high-pressure metamorphism: Examples from the Western Gneiss Region, Norway. *Chemical Geology*, **409**, 99–111, https://doi.org/10.1016/j.chemgeo.2015.04.021

Holland, T.J.B. and Powell, R. 2011. An improved and extended internally consistent thermodynamic dataset for phases of petrological interest, involving a new equation of state for solids. *Journal of Metamorphic Geology*, **29**, 333–383, https://doi.org/10.1111/j.1525-1314.2010.00923.x

Horstwood, M.S.A., Košler, J. *et al.* 2016. Community-derived standards for LA-ICP-MS U–(Th–)Pb geochronology – uncertainty propagation, age interpretation and data reporting. *Geostandards and Geoanalytical Research*, **40**, 311–332, https://doi.org/10.1111/j.1751-908X.2016.00379.x

Jacobs, J. and Thomas, R.J. 1994. Oblique collision at about 1.1 Ga along the southern margin of the Kaapvaal continent, south-east Africa. *Geologische Rundaschau*, **83**, 322–333, https://doi.org/10.1007/BF00210548

Johnson, T., Clark, C., Taylor, R., Santosh, M. and Collins, A.S. 2015. Prograde and retrograde growth of monazite in migmatites: An example from the Nagercoil Block, southern India. *Geoscience Frontiers*, **6**, 373–387, https://doi.org/10.1016/j.gsf.2014.12.003

Kohn, M.J. 2016. Metamorphic chronology – a tool for all ages: Past achievements and future prospects. *American Mineralogist*, **101**, 25–42, https://doi.org/10.2138/am-2016-5146

Kylander-Clark, A.R.C., Hacker, B.R. and Cottle, J.M. 2013. Laser-ablation split-stream ICP petrochronology. *Chemical Geology*, **345**, 99–112, https://doi.org/10.1016/j.chemgeo.2013.02.019

Laurent, A.T., Bingen, B., Duchene, S., Whitehouse, M.J., Seydoux-Guillaume, A.-M. and Bosse, V. 2018*a*. Decoding a protracted zircon geochronological record in ultrahigh temperature granulite, and persistence of partial melting in the crust, Rogaland, Norway. *Contributions to Mineralogy and Petrology*, **173**, 29, https://doi.org/10.1007/s00410-018-1455-4

Laurent, A.T., Duchene, S., Bingen, B., Bosse, V. and Seydoux-Guillaume, A.-M. 2018*b*. Two successive phases of ultrahigh temperature metamorphism in Rogaland, S. Norway: evidence from Y-in-monazite thermometry. *Journal of Metamorphic Geology*, **36**, 1009–1037, https://doi.org/10.1111/jmg.12425

Lindsley, D.H. 1983. Pyroxene thermometry. *American Mineralogist*, **68**, 477–493.

McCourt, S., Bisnath, A. *et al.* 2000. Geology and tectonic setting of the Tugela Terrane, Natal Belt, South Africa. *Journal of African Earth Sciences*, **31**, Suppl. 1, 48–49, https://doi.org/10.1016/S0899-5362(00)00050-6

McCourt, S., Armstrong, R.A., Grantham, G.H. and Thomas, R.J. 2006. Geology and evolution of the Natal belt, South Africa. *Journal of African Earth Sciences*, **46**, 71–92, https://doi.org/10.1016/j.jafrearsci.2006.01.013

Mendonidis, P. and Grantham, G.H. 2003. Petrology, origin and metamorphic history of Proterozoic-aged granulites of the Natal Metamorphic Province, southeastern Africa. *Gondwana Research*, **6**, 607–628, https://doi.org/10.1016/S1342-937X(05)71011-X

Mendonidis, P. and Armstrong, R.A. 2016. U–Pb Zircon (SHRIMP) ages of granite sheets and timing of deformational events in the Natal Metamorphic Belt, southeastern Africa: Evidence for deformation partitioning and implications for Rodinia reconstructions. *Precambrian Research*, **278**, 22–33.

Mendonidis, P. and Thomas, R.J. 2019. A review of the geochronology of the Margate Terrane reveals a history of diachronous terrane docking and arc accretion across the Mesoproterozoic Natal belt, southeastern Africa. *Journal of African Earth Sciences*, **150**, 532–545, https://doi.org/10.1016/j.jafrearsci.2018.07.021

Mendonidis, P., Armstrong, R., Eglington, B.M., Grantham, G.H. and Thomas, R.J. 2002. Metamorphic history and U–Pb Zircon (SHRIMP) geochronology of the Glenmore Granite: implications for the tectonic evolution of the Natal Metamorphic Province. *South African Journal of Geology*, **105**, 325–336, https://doi.org/10.2113/1050325

Mendonidis, P., Thomas, R.J., Grantham, G.H. and Armstrong, R.A. 2015. Geochronology of emplacement and charnockite formation of the Margate Granite Suite, Natal Metamorphic Province, South Africa: Implications for Natal–Maud belt correlations. *Precambrian Research*, **265**, 189–202, https://doi.org/10.1016/j.precamres.2015.02.013

Morrissey, L.J., Hand, M., Kelsey, D.E. and Wade, B.P. 2016. Cambrian high-temperature reworking of the Rayner–Eastern Ghats terrane: Constraints from the northern Prince Charles Mountains region, East Antarctica. *Journal of Petrology*, **57**, 53–92, https://doi.org/10.1093/petrology/egv082

Paton, C., Woodhead, J., Hellstrom, J., Hergt, J., Greig, A. and Maas, R. 2010. Improved laser ablation U–Pb zircon and geochronology through robust downhole fractionation correction. *Geochemistry, Geophysics, Geosystems*, **11**, 1–36, https://doi.org/10.1029/2009GC002618

Paton, C., Hellstrom, J., Paul, B., Woodhead, J. and Hergt, J. 2011. Iolite: freeware for the visualisation and processing of mass spectrometric data. *Journal of Analytical Atomic Spectrometry*, **26**, 2508–2518, https://doi.org/10.1039/c1ja10172b

Pearce, N.J.G., Perkins, W.T., Westgate, J.A., Gorton, M.P., Jackson, S.E., Neal, C.R. and Chenery, S.P. 1997. A compilation of new and published major and trace element data for NIST SRM 610 and NIST SRM 612 glass reference materials. *Geostandards Newsletter*, **21**, 115–144, https://doi.org/10.1111/j.1751-908X.1997.tb00538.x

Powell, R. and Holland, T.J.B. 1988. An internally consistent thermodynamic dataset with uncertainties and correlations: 3. Application, methods and worked examples and a computer program. *Journal of Metamorphic Geology*, **6**, 173–204, https://doi.org/10.1111/j.1525-1314.1988.tb00415.x

Rubatto, D. 2017. Zircon: the metamorphic mineral. *Reviews in Mineralogy and Geochemistry*, **83**, 261–295, https://doi.org/10.2138/rmg.2017.83.9

Scogings, A.J. 1991. *Alkaline Intrusives from the Tugela Terrane, Natal Metamorphic Province*. PhD thesis, University of Durban-Westville, Westville, Durban, South Africa, http://hdl.handle.net/10413/11162

Spencer, C.J., Thomas, R.J., Roberts, N.M.W., Cawood, P.A., Millar, I. and Tapster, S. 2015. Crustal growth during island arc accretion and transcurrent deformation, Natal Metamorphic Province, South Africa: new isotopic constraints. *Precambrian Research*, **265**, 203–217, https://doi.org/10.1016/j.precamres.2015.05.011

Spencer, C.J., Kirkland, C.L. and Taylor, R.J.M. 2016. Strategies towards statistically robust interpretations of *in situ* U–Pb zircon geochronology. *Geoscience Frontiers*, **7**, 581–589, https://doi.org/10.1016/j.gsf.2015.11.006

Spencer, C.J., Yakymchuk, C. and Ghaznavi, M. 2017. Visualising data distributions with kernel density estimation and reduced chi-squared statistic. *Geoscience*

Frontiers, **8**, 1247–1252, https://doi.org/10.1016/j.gsf.2017.05.002

Taylor, R.J.M., Clark, C., Fitzsimons, I.C.W., Santosh, M., Hand, M., Evans, N. and McDonald, B. 2014. Post-peak, fluid-mediated modification of granulite facies zircon and monazite in the Trivandrum Block, southern India. *Contributions to Mineralogy and Petrology*, **168**, 1044, https://doi.org/10.1007/s00410-014-1044-0

Taylor, R.J.M., Kirkland, C.L. and Clark, C. 2016. Accessories after the facts: Constraining the timing, duration and conditions of high-temperature metamorphic processes. *Lithos*, **264**, 239–257, https://doi.org/10.1016/j.lithos.2016.09.004

Taylor, R.J.M., Clark, C., Harley, S.L., Kylander-Clark, A.R.C., Hacker, B.R. and Kinny, P.D. 2017. Interpreting granulite facies events through rare earth element partitioning arrays. *Journal of Metamorphic Geology*, **35**, 759–775, https://doi.org/10.1111/jmg.12254

Thomas, R.J. 1988. *The Geology of the Port Shepstone Area. Explanation of Sheet 3030 Port Shepstone*. Geological Survey of South Africa, Johannesburg, South Africa.

Thomas, R.J. 1989. A tale of two tectonic terranes. *South African Journal of Geology*, **92**, 306–321.

Thomas, R.J. 1990. Mzumbe Gneiss Suite. *In*: *Catalogue of South African Lithostratigraphic Units, Volume 2*. Government Printer for the South African Committee for Stratigraphy, Pretoria, 35–36.

Thomas, R.J. 1992*a*. Mapumulo Group. *In*: *Catalogue of South African Lithostratigraphic Units, Volume 4*. Government Printer for the South African Committee for Stratigraphy, Pretoria, 11–14.

Thomas, R.J. 1992*b*. Mzimkulu Group. *In*: *Catalogue of South African Lithostratigraphic Units, Volume 4*. Government Printer for the South African Committee for Stratigraphy, Pretoria, 17–18.

Thomas, R.J. and Eglington, B.M. 1990. A Rb–Sr, Sm–Nd and U–Pb zircon isotopic study of the Mzumbe Suite, the oldest intrusive granitoid in southern Natal, South Africa. *South African Journal of Geology*, **93**, 761–765.

Thomas, R.J., Evans, M.J. and Eglington, B.M. 1991*a*. Equeefa Suite. *In*: *Catalogue of South African Lithostratigraphic Units, Volume 3*. Government Printer for the South African Committee for Stratigraphy, Pretoria, 9–12.

Thomas, R.J., Evans, M.J. and Eglington, B.M. 1991*b*. Quha Formation. *In*: *Catalogue of South African Lithostratigraphic Units, Volume 3*. Government Printer for the South African Committee for Stratigraphy, Pretoria, 41–44.

Thomas, R.J., Mendonidis, P., Grantham, G.H. and Johnson, M.R. 1991*c*. Margate Granite Suite. *In*: *Catalogue of South African Lithostratigraphic Units, Volume 3*. Government Printer for the South African Committee for Stratigraphy, Pretoria, 33–36.

Thomas, R.J., Ashwal, L.D. and Andreoli, M.A.G. 1992*a*. The petrology of the Turtle Bay Suite: a mafic–felsic granulite association from southern Natal, South Africa. *Journal of African Earth Sciences (and the Middle East)*, **15**, 187–206, https://doi.org/10.1016/0899-5362(92)90068-N

Thomas, R.J., Eglington, B.M., Evans, M.J. and Kerr, A. 1992*b*. The petrology of the Proterozoic Equeefa Suite, southern Natal, South Africa. *South African Journal of Geology*, **95**, 116–130.

Thomas, R.J., Eglington, B.M. and Bowring, S.A. 1993*a*. Dating the cessation of Kibaran magmatism in Natal, South Africa. *Journal of African Earth Science (and the Middle East)*, **16**, 247–252, https://doi.org/10.1016/0899-5362(93)90046-S

Thomas, R.J., Eglington, B.M., Bowring, S.A., Retief, E.A. and Walraven, F. 1993*b*. New isotope data from a neoproterozoic porphyritic garnitoid–charnockite suite from Natal, South Africa. *Precambrian Research*, **62**, 83–101, https://doi.org/10.1016/0301-9268(93)90095-J

Thomas, R.J., Cornell, D.H. and Armstrong, R.A. 1999. Provenance age and metamorphic history of the Quha Formation, Natal Metamorphic Province: a U–Th–Pb zircon SHRIMP study. *South African Journal of Geology*, **102**, 83–88.

Thomas, R.J., Armstrong, R.A. and Eglington, B.M. 2003. Geochronology of the Sikombe Granite, Transkei, Natal Metamorphic Province, South Africa. *South African Journal of Geology*, **106**, 403–408, https://doi.org/10.2113/106.4.403

Tomascak, P.B., Krogstad, E.J. and Walker, R.J. 1996. U–Pb monazite geochronology of granitic rocks from Maine: Implications for Late Paleozoic tectonics in the Northern Appalachians. *The Journal of Geology*, **104**, 185–195, https://doi.org/10.1086/629813

Viete, D.R. and Lister, G.S. 2017. On the significance of short-duration regional metamorphism. *Journal of the Geological Society, London*, **174**, 377–392, https://doi.org/10.1144/jgs2016-060

Wang, Y., Zhai, M. *et al.* 2021. Incipient charnockite formation in the Trivandrum Block, southern India: Evidence from melt-related reaction textures and phase equilibria modelling. *Lithos*, **380**, 105825, https://doi.org/10.1016/j.lithos.2020.105825

White, R.W., Powell, R., Holland, T.J.B., Johnson, T. and Green, E.C.R. 2014*a*. New mineral activity–composition relations for thermodynamic calculations in metapelitic systems. *Journal of Metamorphic Geology*, **32**, 261–286, https://doi.org/10.1111/jmg.12071

White, R.W., Powell, R. and Johnson, T. 2014*b*. The effect of Mn on mineral stability in metapelites revisited: new *a–x* relations for manganese-bearing minerals. *Journal of Metamorphic Geology*, **32**, 809–828, https://doi.org/10.1111/jmg.12095

Whitehouse, M.J. 2003. Rare earth elements in zircon: a review of applications and case studies from the Outer Hebridean Lewisian Complex, NW Scotland. *Geological Society, London, Special Publications*, **220**, 49–64, https://doi.org/10.1144/GSL.SP.2003.220.01.03

Xu, Z., Wang, Q. *et al.* 2015. Kinematics of the Tengchong Terrane in SE Tibet from the late Eocene to early Miocene: Insights from coeval mid-crustal detachments and strike-slip shear zones. *Tectonophysics*, **665**, 127–148, https://doi.org/10.1016/j.tecto.2015.09.033

Zircon U–Pb geochronology, Nd isotopes and geochemistry of mafic granulites from the Central Indian Tectonic Zone: isotopic constraints on Proterozoic crustal evolution

Meraj Alam[1]*, Tatiana V. Kaulina[2], Rakhi R. Varma[3] and Talat Ahmad[4]

[1]Department of Geology, Indira Gandhi National Tribal University, Amarkantak 484887, India

[2]Geological Institute of the Kola Science Centre, RAS, Apatity 184209 Russia

[3]ONGC, Subsurface Team, Jorhat 785704, Assam, India

[4]Wadia Institute of Himalayan Geology, Dehradun 248001, India

MA, 0000-0001-6820-9239
*Correspondence: merajdu@gmail.com

Abstract: The Central Indian Tectonic Zone (CITZ) comprises northern and southern Indian cratonic blocks and is a tectonic window that is suitable for investigating the Proterozoic crustal evolution because of the presence of a wide variety of lithologies. Geochemical and geochronological data on mafic granulites by previous workers do not ascertain the possibility of mafic protoliths and their coeval link to other CITZ units. Thus, determining the precise timing of the formation of mafic granulites may indicate a connection between metamorphism and fragmentation of the Columbian supercontinent. This study presents zircon U–Pb ages, Nd isotopes and the geochemistry of mafic granulites to evaluate their genesis and timing of metamorphism. The results show the tholeiitic affinity and primary magmatic differentiation of the parental melt. Depletion of Nb, P, Zr and Ti and positive enrichment of Ba, U and Pb indicate the derivation of mafic granulites from a variably enriched subcontinental lithospheric mantle (SCLM) source. The zircon U–Pb ages (1564 $\pm$ 8 to 1598 $\pm$ 9 Ma) are interpreted as a period of granulite-facies metamorphism. The T_{DM} (depleted-mantle) model ages (2.9–3.4 Ga) of mafic granulites indicate the timing of mafic protolith extraction. The mineral isochron age *c.* 1.0 Ga indicates that these rocks underwent some events during an early Neoproterozoic period. Protolith of mafic granulites could be related to the evolution of melts derived from metasomatized SCLM through fractional crystallization processes.

The proto-Indian continent actively participated in the reconstruction of the Columbian supercontinent assembly (between 2.1 and 1.8 Ga), which is probably defined by the existence of rifting events of similar ages (*c.* 1.6–1.3 Ga) in several distinct continents, including eastern India and the western margin of North America (Rogers and Santosh 2002). Hou *et al.* (2008) emphasized the connection between India and Columbia based on the Paleoproterozoic rifting events. Furthermore, Zhao *et al.* (2003) have also highlighted a Columbia temporally linked with India and the North China Craton (NCC) based on the correlation between the Trans North China Orogen, which was sutured to the eastern and western blocks in the NCC, and the Central Indian Tectonic Zone (CITZ) at *c.* 1.85 Ga (Zhang *et al.* 2011). Since that time, the NCC has recorded events of configuration, followed by a collisional event (*c.* 2.0–1.8 Ga), accretionary events (*c.* 1.8–1.3 Ga) and rifting events (*c.* 1.6–1.3 Ga), which promoted the assembly, outgrowth and break-up of the Columbia supercontinent (Zhao *et al.* 2011). The northern margin of the NCC experienced an intra-continental rifting event at *c.* 1.6–1.2 Ga, which promoted separation amongst the cratonic blocks of the Columbia supercontinent (Zhao *et al.* 2011).

In central India, lamprophyric and alkaline rocks formed at *c.* 1.6 Ga through an extensional tectonic setting, which indicates that central India started undergoing extension after 1.9 Ga and followed fragmentation up to the age of 1.6 Ga (Srivastava and Chalapathi Rao 2007). Srivastava and Chalapathi Rao (2007) also highlighted a similar age of rocks evidently occurring in Australia; therefore, the breakup of the Columbia supercontinent may have begun by 1.6 Ga. The east–west trending CITZ defines the prominent Proterozoic suture zone between the North Indian Block (NIB) and South Indian Block (SIB) (Radhakrishna and Naqvi 1986; Yedekar *et al.* 1990; Jain *et al.* 1991; Eriksson *et al.* 1999; Acharyya and Roy 2000; Mishra *et al.* 2000; Acharyya 2003; Roy and Hanuma Prasad 2003; Figure 1a). The evolutionary history of the CITZ covers the overlapping periods of assembly and breakup of two supercontinents: Columbia,

From: van Schijndel, V., Cutts, K., Pereira, I., Guitreau, M., Volante, S. and Tedeschi, M. (eds) 2024. *Minor Minerals, Major Implications: Using Key Mineral Phases to Unravel the Formation and Evolution of Earth's Crust.* Geological Society, London, Special Publications, **537**, 333–358.
First published online September 13, 2023, https://doi.org/10.1144/SP537-2022-135

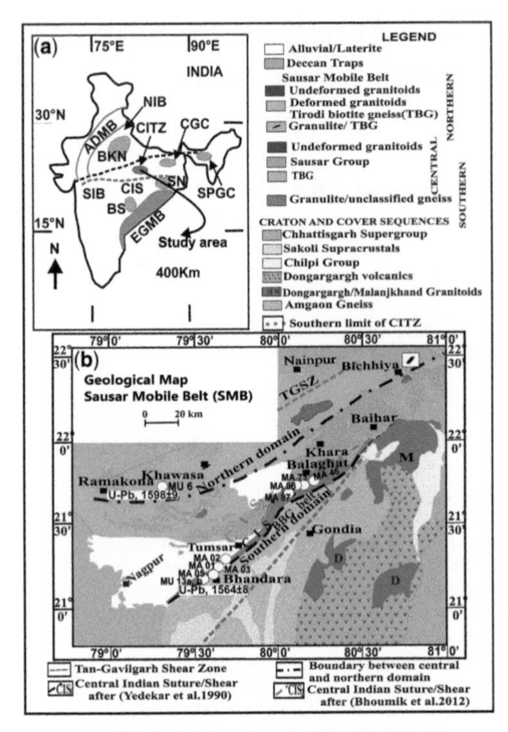

Fig. 1. (**a**) Outline map (inset) of India showing different lithotectonic units, including important blocks (NIB and SIB) with respect to the study area. (**b**) Simplified geological map showing the study area with the location of the studied samples; the map also shows the Sausar mobile belt, including TBG, BBGB, and southern and northern domains, along with major lineaments and supergroups from north to south (modified after Bhowmik 2019) Abbreviations: NIB, North Indian Block; SIB, South Indian Block; CITZ, Central Indian Tectonic Zone; CIS, Central Indian Shear; ADMB, Aravalli Delhi Mobile Fold Belt; CGC, Chhotanagpur Gneissic Complex; EGMB, Eastern Ghat Mobile Belt; SPGC, Shillong Plateau Gneissic Complex; BKN, Bundelkhand; BS, Bastar; SN, Singhbhum; TBG, Tirodi Biotite Gneisses; BBGB, Bhandara–Balaghat Granulite Belt; TGSZ, Tan–Gavilagarh Shear Zone.

assembled at *c.* 2.1–1.8 Ga (Rogers and Santosh 2002; Zhao *et al.* 2002; Zhao *et al.* 2004; Hou *et al.* 2008; Santosh and Kusky 2010; Meert 2012; Nance *et al.* 2014; Meert and Santosh 2017), and Rodinia, at *c.* 1.2 and 0.9 Ga (Meert and Torsvik 2003; Pisarevsky *et al.* 2003; Collins and Pisarevsky 2005; Li *et al.* 2008; Merdith *et al.* 2017). Previous studies have suggested that the activity of subduction–accretion–collision brought out the CITZ from the amalgamation of the NIB and SIB at *c.* 1.8 Ga, which led to the development of the Indian Subcontinent (Yedekar *et al.* 1990; Jain *et al.* 1991; Rogers and Gird 1997; Naganjaneyulu and Santosh 2010; Bhowmik *et al.* 2012). However, this amalgamation is still disputed regarding the direction of the subduction of NIB and SIB units. Two tectonic models have been proposed for the CITZ: one supported the southerly subduction of the NIB beneath the SIB (Yedekar *et al.* 1990; Jain *et al.* 1991; Eriksson *et al.* 1999; Acharyya 2003), and the other was by Roy and Hanuma Prasad (2003), who proposed the northerly subduction of the SIB beneath the NIB (Chattopadhyay *et al.* 2017). The development of a high-grade granulite terrain to the north of the Central Indian Shear (CIS) and a low-grade island-arc tholeiites suite, including Sakoli and Nandgaon volcanics in the Deccan protocontinent to the south, is considered evidence of the southerly subduction of the Bundelkhand protocontinent below the Deccan protocontinent (Yedekar *et al.* 1990). Moreover, a model proposed by Mall *et al.* (2008) suggests that the Bastar craton subducted beneath the Bundelkhand craton, in which the rocks of the Sausar Group, Tirodi gneiss and associated granulites actively participated at a later stage. Furthermore, the formation of the Sausar fold belt and the thrusting of mafic granulites from the north of the CIS were recorded at 1700–1500 Ma (Mohanty 2020), which could be related to the rifting event in the Columbia supercontinent. Zircon U–Pb ages for the Tirodi gneiss (1534 $\pm$ 13 Ma) and granulites (1569 $\pm$ 15 Ma) were interpreted as the ages of peak metamorphism, which facilitated the generation of Tirodi gneiss (Ahmad *et al.* 2009). However, a zircon U–Pb age (1618 $\pm$ 8Ma) is reported by Bhowmik *et al.* (2011) for Tirodi granite gneiss, and Sm–Nd model ages between 2325 and 2494 Ma are considered the protolith ages of Tirodi gneiss (Mishra *et al.* 2009). Moreover, Alam *et al.* (2022) suggested a Sm–Nd Model age (2.20–2.78 Ga) for Tirodi gneiss, indicating the extraction of the protolith from the mantle. Other petrological investigations suggest that the study area had experienced multiple tectonothermal events, including earlier high-grade metamorphism recorded at 2672 $\pm$ 54 Ma, followed by granulite facies metamorphism at 1416 $\pm$ 59 Ma (Sm–Nd age), and a subsequent thermal event occurring at 1380 $\pm$ 28 Ma (Rb–Sr age) (Ramachandra and Roy 2001). However, Alam *et al.* (2017) have also suggested multiple events recorded during *c.* 3.2–1.6 Ga (Sm–Nd model ages) and the long-term evolution of the mafic protolith for the granulites.

Mafic granulite terrains are spatially linked to the supercontinent assembly (Brown 2007*a*, *b*; Sizova *et al.* 2014; Clark *et al.* 2015; Brown and Johnson 2018). They occurred in various high-grade Precambrian terrains that existed on globe scale, including, Fuping Complex, Central China (Zhao *et al.* 2000), Hoggar, Algeria (Ouzegane *et al.* 2001), Saxon, Germany (Romer and Rötzler 2001), Rauer and

Sostrene Islands, Antarctica (Harley 1989), Northern Labrador, Canada (Mengel and Rivers 1991), Varpaisijärvi, Finland (Hölttä *et al.* 2000), Eastern Ghats Mobile Belt (Dasgupta *et al.* 1993), the CITZ (Bhowmik and Roy 2003; Alam *et al.* 2017) and several localities in the Southern Granulite Terrain of India (Kumar and Chacko 1994; Prakash 1999). Mafic granulites are essential constituents of many high-grade metamorphic belts. They preserve their metamorphic history in order to form reaction textures, indicating the condition of high-temperature metamorphism. These granulites are useful for the reconstruction of the crustal evolution history of the belts (McLelland and Whitney 1977; Johnson and Essene 1982; Bohlen *et al.* 1985; Ellis and Green 1985; Sandiford *et al.* 1987; Sandiford *et al.* 1988; Harley 1989; Stüwe and Powell 1989). Thus, it helps to understand the variable nature and composition of the Earth's lower crust (Smithson and Brown 1977; Harley 1989; Bohlen 1991). The granulites are generally formed by underplating mantle-derived mafic magma corresponding to a counter-clockwise evolution path, which is common during the Archean period (Bohlen 1987; Harley 1989; Condie 2002). However, it is a great challenge to determine the link between granulite formation and magmatism, especially in the older and deeper portions of exhumed Proterozoic orogenic systems (Clark *et al.* 2014). The rocks from deeper crustal levels have undergone high-grade metamorphism and deformation (Bohlen and Mezger 1989; Liu and Zhong 1997), and may also have suffered retrograde metamorphism owing to decompression and cooling during exhumation (Ellis 1987; Liu *et al.* 2000). The mafic granulites and their assemblages are considered a well-founded indicator to investigate the Proterozoic crustal evolution of cratons that occurs in the high-pressure metamorphic system (Green and Ringwood 1967; Zhao *et al.* 2001; O'Brien and Rötzler 2003; Pattison *et al.* 2003; Zhang *et al.* 2016). However, significant questions have arisen about the heat source, fluid influence and tectonic setting leading to such extreme crustal conditions. Hence, mafic magmatism could be a reliable source of heat (Collins 2002; Hyndman *et al.* 2005; Brown 2006; Brown 2007*a*; Clark *et al.* 2009; Clark *et al.* 2011).

Previous petrological studies on granulites from CITZ have mainly focused on mineral phase equilibrium characteristics and thermobarometry. However, these studies were not subjected to any detailed geochemical and isotopic studies to put constraints on their geochemical characteristics, geochronology and genesis. These studies will also help to understand the nature and origin of mafic protoliths and the relationship between magmatism and high-grade granulite facies metamorphism.

In this contribution, we have carried out petrography, whole rock (WR) geochemistry, Sm–Nd (WR) isotopic composition, Sm–Nd mineral isochron age and zircon U–Pb ages of the mafic granulite from the CITZ. The aims are to further constrain the nature of protoliths and geochronological and isotopic compositions to discuss the Proterozoic crustal evolution and evaluate the precise timing of the formation of mafic granulites and their relationship with the fragmentation of the Columbian supercontinent.

Geological Background

The NE–SW-trending CITZ has been considered in previous studies as a transitional–continental suture zone (Harris 1993) that emerged after the amalgamation of NIB (comprising Bundelkhand craton and Aravalli craton) and SIB (comprising Bastar craton, Dharwar craton and Singhbhum craton), which is a very important window to understand the crustal evolution of east Gondwanaland (Bhowmik and Roy 2003). The CITZ is linearly extended and lies between the Son-Narmada north fault in the north and a CIS zone in the south (Radhakrishna 1989; Acharyya and Roy 2000). This zone is also blended with at least three distinct supracrustal belts from north to south, namely Mahakoshal belt, Betul belt and Sausar belt. Similarly, the ENE–WSW trending Sausar Mobile Belt (SMB) incorporates highly deformed and metamorphosed non-volcanic and Mn-bearing sediments along with basement gneisses, migmatites (Tirodi biotite gneiss) and granulites (Bhowmik and Roy 2003; Fig. 1b). The opening of the Sausar basin within the CITZ was related to the disintegration of the Proto-Greater Indian Landmass during the mid-Proterozoic crustal expansion. There are three important granulite belts, comprising the Bhandara–Balaghat Granulite Belt (BBGB), Ramakona–Katangi Granulite (RKG) Belt and Makrohar Granulite Belt, which run parallel to the CITZ and are situated in the southern, central and northern domains of the SMB, respectively (Fig. 1). These granulites consist of felsic and aluminous granulites interbedded with norite, gabbro and two pyroxene granulites (Ramachandra and Roy 2001; Bhowmik *et al.* 2005; Alam *et al.* 2017). The gneisses and granulite assemblages from the Sausar supracrustals are shown to have resulted from a single cycle of Sausar orogeny (Brown and Phadke 1983; Sarkar *et al.* 1986). However, Bhowmik *et al.* (2012) proposed that the SMB has experienced two different orogenic events that occurred at 1.62–1.42 and 1.06–0.94 Ga, respectively. However, the southern domain, including the BBGB, records only one orogeny that occurred during Paleoproterozoic to early Mesoproterozoic. Both central and northern domains, including the RKG Belt, have registered evidence of the Grenville-aged orogenic event that reworked the 1.62 Ga crust (Bhowmik *et al.* 2012). These domains record the history of tectono-magmatic, tectono-metamorphic (Bhowmik 2019) and chronological events that occurred in the southern domain (Bhandara domain), northern domain (RKG domain), and western domain, respectively, as cited by the previous studies summarized in Table 1. Later tectono-thermal events are involved to develop granulite-facies metamorphism, followed by exhumation of the BBGB granulite rocks to shallow crustal levels during the Grenville event. Further, such an event was responsible for the formation of the Indian subcontinent through the collision of the Archean Bundelkhand and Bastar cratons (Chetty 2017).

Analytical methods

Fourteen representative samples of mafic granulites were collected from the study area (the GPS coordinates of selected samples are given in Table 2 and Fig. 1) and analysed for major, trace and rare earth elements (REEs). These studied samples were analysed by X-ray fluorescence spectrometry (Panalytical Model Philips Magix Pro Model 2440) at the University of Delhi, India. The precision limits are 1% for SiO_2, 2% for some other major elements, 2–5% for minor elements and better than 10% for trace elements (Longjam and Ahmad 2012). The trace elements and REEs were analysed using inductively coupled plasma mass spectrometry (ICP-MS; Perkin Elmer, Sciex Elan DRC II) at Indian Institute of Technology, Roorkee. Numerous international rock standards (JG-2, G-3, JR-2, JR-3 and DGH) were used for calibrating the instrument. The procedure for trace element analyses by ICP-MS was followed, and the precision limit of the ICP-MS data was ±4.1% RSD (Bhattacharya *et al.* 2012).

Isotope dilution thermal ionization mass spectrometry (ID-TIMS) Sm–Nd and U–Pb analyses were performed at the Laboratory for Geochronology and Isotope Geochemistry of the Geological Institute of the Kola Science Centre of Russian Academy of Science, Apatity, Russia. The Sm–Nd isotopic analyses were carried out using the dissolution of minerals in $HF + HNO_3$ (or $+ HClO_4$) in Teflon beakers at 100°C, followed by the extraction of Sm and Nd by ion-exchange column chromatography. The isotopic dilution technique was used for measuring Nd and Sm with a mixed $^{149}Sm/^{150}Nd$ tracer on double Re + Re filaments. The ratios of $^{143}Nd/^{144}Nd$ were standardized to $^{146}Nd/^{144}Nd = 0.7219$ and the mean values for the Japanese Nd standard material JNdi-1 (Tanaka *et al.* 2000) during analysis. The mean value for the $^{143}Nd/^{144}Nd$ obtained for the La Jolla standard value was 0.511833 ± 6 (2σ, $n = 11$) (Lugmair *et al.* 1983). A minimum error of 0.003% was chosen for the $^{143}Nd/^{144}Nd$ ratio based on the La Jolla standard reproducibility, while the minimum error for the $^{147}Sm/^{144}Nd$ ratio was 0.3% (2σ, $n = 15$) according to BCR standard measurements. Nd blanks were less than 0.3 ng and Sm blanks were less than 0.06 ng.

The zircon ID-TIMS U–Pb analyses were carried out, starting with the separation of zircon using magnetic separation in a Frantz separator and density separation in heavy liquids. Zircons to be analysed were handpicked under a binocular microscope. The U–Pb analytical procedures for zircon followed the method of Krogh (1973). The isotopic dilution technique measured Pb and U with a mixed ^{208}Pb–^{235}U tracer on the multi-collector Finnigan-MAT 262 mass spectrometer. Pb and U were loaded together on outgassed single Re filaments with H_3PO_4 and silica gel. Pb isotope ratios were corrected for mass fractionation with a factor of 0.10% per amu, based on repeat analyses of the NBS SRM 982 standard. Total procedural blanks were 0.1–0.3 ng for Pb and 0.04 ng for U. The U analyses were corrected for mass fractionation with a factor of 0.003% per amu, based on repeat analyses of the NBS U 500 standard. The reproducibility of the U–Pb ratios was determined from the repeated analyses of standard zircon IGFM-87 (Ukraine) and taken as 0.5% for the $^{207}Pb/^{235}U$ and $^{206}Pb/^{238}U$ ratios, respectively, at a 95% confidence level. All calculations were done using the programs PBDAT and ISOPLOT (Ludwig 1991). The chemical composition of zircons, including the REEs, was analysed on a Cameca IMS-4F ion microprobe at the Yaroslavl Branch of the Physical Technological Institute (Yaroslavl, Russia). The primary O^{2-} ion beam spot size was *c.* 20 µm. Each analysis was averaged over five measurement cycles. The relative analytical errors were 10–15%, and the average detection limits were 10 ppb. The respective procedure is described in Fedotova *et al.* (2008).

Results

Lithological and petrographic characteristics

For the petrographic study, 14 representative samples were selected from different locations of the study area (Figs 1b, 2 & 3). The study was carried out in the petrological laboratory of the Department of Geology, Delhi University, under a Leica Orthoplan microscope fitted with an image analyser. The lithological units dominantly comprise meta-igneous and meta-sedimentary rocks of granulite facies. The meta-igneous rocks are dominated by mafic granulites, which are of gabbroic composition (Fig. 2a) and have intruded the granite gneisses. However, metasedimentary rocks are aluminous granulite

Table 1. *Summary of events found in Central Indian Tectonic Zone (CITZ)*

Tectono-magmatic and tectono-metamorphism events in the southern domain (BBG domain) (modified after Bhowmik 2019)			Tectono-metamorphism events in Southern domain (BBG domain) (modified after Bhowmik 2019)		Chronological events in Western domain of CITZ
Metamorphism	Magmatic events	Tectonic environment	Metamorphism	Tectonic environment	Age <900 Ma Granite magmatism in Betul Belt *c.* 850 Ma[c] (W-1); amphibolite facies metamorphism in RKG and BBGB: *c.* 800–900 Ma[b,c] (W-2, 3, 4)
Accretionary oogenesis during late Paleoproterozoic to early Mesoproterozoic Metamorphic events (BM_5–BM_1)			Low-*P* event coincident with mafic dyke emplacement and felsic plutonism (including magmatic charnockite)	Continent– continent collisional orogeny promoting a high-pressure metamorphism	Age- 900–1000 (Ma) Post- tectonic Sausar granite 928 Ma[c] (W-5). Sausar Orogeny closed (*c.* 1000 Ma) and metamorphism and granite intrusion 1000 Ma (W-6). Amphibolite facies metamorphism and syntectonic granite magmatism in Sausar (W-5). Northern and southern domain of SMB formed during final amalgamation of northern and southern Indian block *c.* 955–1062 Ma[d] (W-7)
BM_5: upper greenschist to lower amphibolite facies metamorphism restrained with ductile shear zone	Syn-metamorphic emplacement				Age 1000–1300 Ma Closing of Sausar basin *c.* 1100 Ma[d] (W-1); dolerite dyke in Sausar group 1112 ± 77 Ma[b] (W-4)
Norite Gabbroic emplacement					
BM_4: localized low- to intermediate-pressure, high-temperture granulite facies metamorphism tie up of metamorphic re-working of the BM_3 granulites		Extensional tectonic setting			Age 1300–1400 Ma Mafic granulites of Bhandara craton, thermal resetting age 1407 ± 11 Ma[b] (W-4); Sausar group: metamorphic event *c.* 1415 ± 23 Ma[d] (W-8); Tirodi gneiss 1454 ± 5 Ma[a] (W-9)
BM_3: reburial metamorphic and re-heating of partially exhumed composite BM_1–BM_2 granulites (BM_{3P} at 7.3		Back-arc closure			Age 1400–1500 Ma Gabbroic emplacement in BBGB *c.* 1400 Ma[bc] (W-4)

(*Continued*)

Table 1. *Continued.*

Tectono-magmatic and tectono-metamorphism events in the southern domain (BBG domain) (modified after Bhowmik 2019)			Tectono-metamorphism events in Southern domain (BBG domain) (modified after Bhowmik 2019)	Chronological events in Western domain of CITZ
kbar, 750°C) and closing at isobaric cooling ($BM3_R$) to 580°C at *c.* 1539 Ma				
BM_2: reheating of cooled BM_1 granulites and pushed into a second UHT event (BM_{2P} at 6.8 kbar, 900°C at *c.* 1572 Ma, followed by localized hydration and potash metasomatism as part of a low-*P* >cooling event (BM_{2R} at 2.9 kbar, 680°C)	Emplacement of gabbro-norite dyke in the BM_1 granulite crust	Back-arc extension.		Age 1500–1650 Ma Emplacement of granitoids in Betul belt/deformation of Sausar Group/ MKG gneissesc. 1500 Ma[c] (W-4, 10); granulite facies metamorphism of RKG *c.* 1520 Ma[c] (W-11); Sausar granulite 1553 ± 19, 1569 ± 15 and 1577 ± 19 Ma[a] (W-9); Tirodi gneiss 1534 ± 15 Ma[a] (W-9); 1572 ± 7 and 1584 ± 17 Ma[a] (W-12); Monazite age in BBGB 1525 ± 13 Ma[a] (W-8); 1582 ± 13 Ma[d], 1603 ± 12 Ma[d] (W-13); Betul granite 1550 ± 50 Ma[c] (W-14); granulite metamorphism in BBGB (M4) *c.* 1540–1570 Ma[a,d] (W-15); ultrahigh-temperature granulite metamorphism in BBGB (M3) *c.* 1570–1640 Ma[a] (W-15); Tirodi Gneisses 1618 ± 8 Ma[a] (W-12); thermal imprints on Dongargarh granite 1640 ± 60 Ma[c] (W-16);
BM_1: high-grade event, locally reaching ultrahigh-temperature metamorphic conditions (BM_{1P} at 8–9.5 kbar, 900–1000°C). Post-peak near isobaric cooling (BM_{1R} at 9 kbar, 670°C) during *c.* 1612–1574 Ma	Magmatic events are: (1) psammo-pelitic, pelitic and aluminous granulites had experienced extensive dehydration melting which produced a migmatite banding; (2) syn-metamorphic emplacement of granite and concordant bodies of coarse-grained gabbro-norite suite of rocks in the supracrustal lithopackage	Back-arc extension and closure		

Abbreviations: BM, metamorphism in the Bhandara–Balaghat Granulite (BBG) domain; BBGB, Bhandara–Balaghat Granulite Belt; BMP/R, peak/retrograde metamorphism; MKG, Makrohar granulite belt; RKG, Ramakona–Katangi Gneisses; SMB, Sausar Mobile Belt; W, western domain.

Methods used for age determination: a: U–Pb Zircon age; b: Sm–Nd age; c: Rb–Sr age; d: monazite age.

References for age data: W-1, Roy and Hanuma Prasad (2003); W-2: Bandyopadhyay *et al.* (1990); W-3, Bandyopadhyay *et al.* (1995); W-4, Roy *et al.* (2006); W-5, Chattopadhyay *et al.* (2015); W-6, Acharyya (2003); W-7, Bhowmik *et al.* (2012); W-8, Bhowmik *et al.* (2005); W-9, Ahmad *et al.* (2009); W-10, Phadke (1990); W-11, Sarkar *et al.* (1986); W-12, Bhowmik *et al.* (2011); W-13, Bhandari *et al.* 2010; W-14, Mahakud *et al.* (2000); W-15, Bhowmik *et al.* (2014); W-16, Sarkar *et al.* (1990).

Table 2. *Sample description of selected samples, GPS coordinates of location and their analysis of mafic granulites from the CITZ*

Sample no.	Rock type	GPS coordinates	Rock composition	Model ages (Ma)
MA1	Mafic granulite	N 21°12' 3.8″ E 79°35' 11″	Opx + Cpx + Hbl + Pl + Fe–Ti oxides ± Ap ± Spn	3234 (Sm–Nd)
MA2	Mafic granulite	N 21°12' 2.4″ E 79°35' 7.9″	Opx + Cpx + Hbl + Pl + Qtz + Fe–Ti oxides ± Ap ± Spn ± Zrn	3420 (Sm–Nd)
MA3	Mafic granulite	N 21°12' 1.5″ E 79°35' 5.9″	Opx + Cpx + Hbl + Pl + Qtz + Fe–Ti oxides ± Ap ± Zrn	3103 (Sm–Nd)
MA5	Mafic granulite	N 21°12' 6″ E 79°34' 59.7″	Opx + Cpx + Hbl + Pl + Qtz + Fe–Ti oxides ± Ap ± Spn ± Ilm	2963(Sm–Nd)
MA6	Mafic granulite	N 21°12' 4.2″ E 79°34' 55.7″	Opx + Cpx + Hbl + Pl + Qtz + Mag ± Ap ± Spn ± Ilm	3148(Sm–Nd)
Mu13a	Mafic granulite	N 21°11' 50″ E 79°34' 34.5″	Opx + Cpx + Hbl + Pl + Qtz + Mag ± Zrn ± Ap ± Ilm	1564 ± 16 (zircon U–Pb)
Mu13b	Mafic granulite	N 21°11'50″ E 79°34'34.5″	Opx + Cpx + Hbl + Pl + Qtz + Mag ± Zrn ± Ap ± Ilm ± Spn	–
MA46	Mafic granulite	N 21°42'34.7″ E 80°5' 55.4″	Opx + Cpx + Hbl + Pl + Qtz + Mag ± Grt ± Ap ± Ilm	–
MA71	Mafic granulite	N 21°41'35.1″ E 80° 05' 57.5″	Opx + Cpx + Hbl + Pl + Qtz + Fe–Ti oxides ± Grt ± Ap ± Ilm ± Zrn	–
MA86	Mafic granulite	N 21°41'35.1″ E 80° 05' 57.5″	Opx + Cpx + Hbl + Pl + Qtz + Fe–Ti oxides ± Grt ± Ap ± Ilm ± Zrn	–
MA87	Mafic granulite	N 21°41'35.1″ E 80° 05' 57.5″	Opx + Cpx + Hbl + Pl + Qtz + Fe–Ti oxides ± Grt ± Spn ± Ilm ± Zrn	–
Mu-6	Mafic granulite	N 21°43'21.9″ E 79°26'19.8″	Opx + Cpx + Hbl + Pl + Qtz + Mag ± Zrn ± Ap ± Ilm ± Spn	1599 ± 10 (zircon U–Pb)

Mineral abbreviations used after Whitney and Evans (2010).

Fig. 2. Field photographs of mafic granulites from the CITZ. (**a**) Gabbroic composition, majorly comprising pyroxene and plagioclase; (**b**) variable contact between mafic and felsic granulite; (**c**) quartz–feldspathic leucosome, corresponding to partial melting, which occurred at a later stage; and (**d**) mafic bands that developed by the recrystallization of pyroxene.

with Al-rich mineral (garnet, spinel and sillimanite) assemblages. These two lithologies also appear to coexist as alternating felsic and mafic bands a few centimetres in thickness (Fig. 2b). Although the contacts between mafic and felsic granulites are variable, the quartzo-feldspathic leucosome and mafic bands in these granulites correspond to the partial melting that occurred at a later stage (Fig. 2c, d). Petrographically, they are of gabbroic composition and have medium-grained, inequigranular granoblastic fabric, which dominantly comprises plagioclase, clinopyroxene, orthopyroxene, Fe–Ti oxides and quartz, with some accessory phases of zircon, sphene, ilmenite and garnet (Fig. 3a–e). Hornblende (16–35%) occurs in two generations in the form of green and brown hornblende (Fig. 3a, b). There is a triple junction between orthopyroxene, clinopyroxene and plagioclase, which could suggest annealing processes between pyroxene and plagioclase (Fig. 3a, c, d). Exsolution intergrowth of pyroxene (involving orthopyroxene, clinopyroxene, plagioclase, magnetite and quartz) has also been observed in the laths of orthopyroxene and clinopyroxene, and the small grains of quartz and plagioclase bind these laths (Fig. 3c). The laths of pyroxene may have experienced recrystallization at a later stage in which the plagioclase and quartz were included within the lath (Fig. 3c, d). Plagioclase is fine to coarse grained, surrounded by assemblages of hornblende, orthopyroxene, clinopyroxene and quartz, which show some strong compositional zoning (Fig. 3d). Plagioclases occur as anhedral to subhedral grain and the modal percentage is 12–45%. The modal percentage of orthopyroxene (10–35%) shows an anhedral to subhedral crystal shape and is partially or completely replaced by hornblende (Fig. 3e). The model percentage of clinopyroxene is 15–30%, which occurs both as granules associated with orthopyroxene and sometimes as coronas around the mineral (Fig. 3e). Garnet corona is developed between orthopyroxene and plagioclase (Fig. 3e). The quartz modal percentage is *c.* 11%, which usually forms monomineralic aggregates and displays deformation and undulose extinction (Fig. 3).

Geochemistry. The geochemical analyses of representative rock samples from the southern and northern domains of the SMB within CITZ are presented in Table 3, and their locations are illustrated in Figure 1. The silica content ranges from 48 to 52 wt%, the TiO_2 content from 1.05 to 1.42 wt%, and the Al_2O_3 content from 13 to 16 wt%. The Fe_2O_3 content ranges from 11 to 17 wt%. The MgO content ranges from 4 to 9 wt%, the K_2O content ranges from 0.03 to 3.1 wt%, and the P_2O_5 content ranges from 0.05 to 0.36 wt%. Among trace elements, the Rb content varies from 1 to 6 ppm, the Ba content varies from 13 to 456 ppm, the Sr content varies from 76 to 140 ppm, and the V content varies from 92 to 392 ppm. The concentration of Ni varies

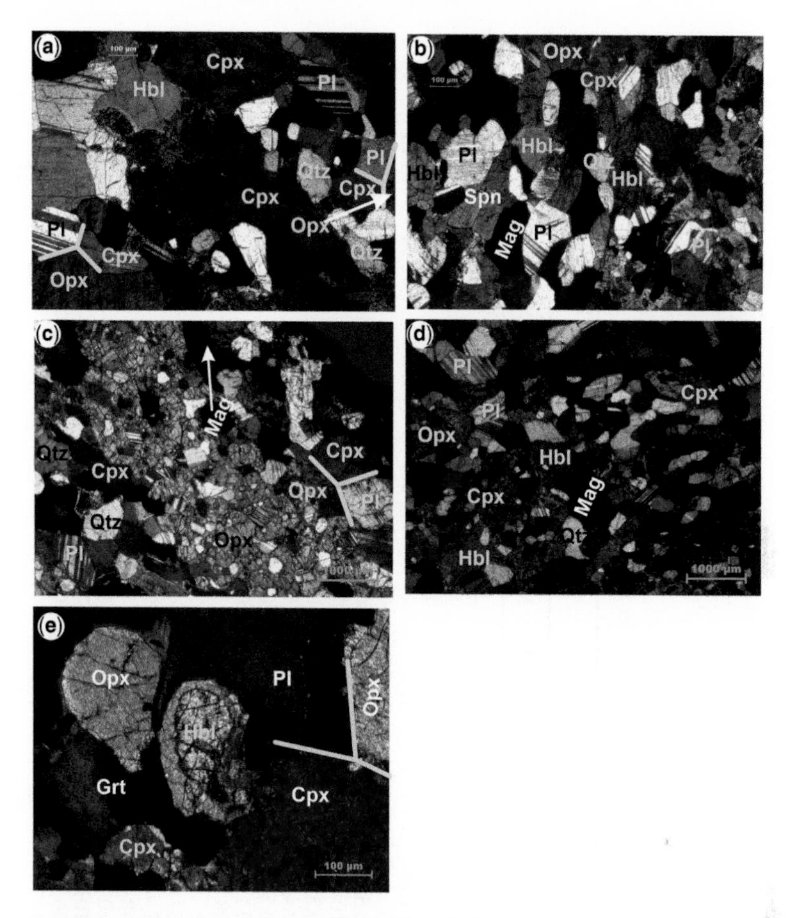

Fig. 3. Microscopic photographs of mafic granulites from the CITZ. (**a**) Triple junction between orthopyroxene, clinopyroxene and plagioclase; (**b**) medium-grained, inequigranular granoblastic fabric, comprising plagioclase, clinopyroxene, orthopyroxene, Fe-oxide and quartz, with some accessory phases of sphene, magnetite, etc.; (**c**) exsolution intergrowth of pyroxene (involving orthopyroxene, clinopyroxene, plagioclase, magnetite and quartz); (**d**) the laths of pyroxene may have experienced recrystallization in the latter stage, in which the plagioclase and quartz are included within the lath; and (**e**) garnet corona developed between orthopyroxene and plagioclase. Mineral abbreviations used after Whitney and Evans (2010).

from 61 to 243 ppm, the Cr content varies from 66 to 707 ppm, the Zr content varies from 28 to 93 ppm and the Y content varies from 14 to 38 ppm.

Major and trace elements for mafic granulite rocks from CITZ have been plotted on various diagrams. In the triangular A–F–M diagram (Na_2O + K_2O)–FeO^T–MgO (Irvine and Baragar 1971) and a binary Zr (ppm) v. Y (ppm) classification diagram (after Barrett and MacLean 1994) (Fig. 4a, b), trends define a tholeiitic affinity. The Nb/Y v. Zr/Y plot (Winchester and Floyd 1977; modified after Pearce (1996), and total alkali v. silica plot (Wilson 1989), plots depict basaltic composition (Fig. 4c, d). The Harker geochemical diagrams (Fig. 5a–d) of Al_2O_3 wt%, Fe_2O_3 wt%, TiO_2 wt%, MgO wt% and CaO wt% are slightly inversely correlated to that of SiO_2 wt% (sample MU6 has relatively high silica (52 wt%), probably suggesting primary magmatic differentiation characteristics of the parental melt for these studied rocks. However, the alkali oxides Na_2O wt%, K_2O wt%, and P_2O_5 wt% show somewhat scattered trends, suggesting either elemental mobility during alteration or remobilization during the granulite-facies metamorphism.

Trace and REE. The trace and REE results are presented in Table 3. The chondrite normalized REE and primitive mantle (PM) normalized multi-element patterns (Sun and McDonough 1989) are very useful to understand the behaviour of trace elements in any magmatic system (Hanson 1980; Ahmad and Tarney 1991; Ahmad and Tarney

Table 3. *Major oxide (wt %) and trace elements (ppm) data of representative samples of mafic granulites from CITZ*

Mafic granulite rocks

Sample	MA01	MA02	MA03	MA05	MA06	Mu13a	Mu13b	MA99	MA100	MA46	MA71	MA86	MA87	MU6
SiO_2	48.77	49.62	49.26	49.33	48.04	48.44	48.37	47.87	47.52	48.33	48.59	48.42	48.17	52.74
TiO_2	1.42	0.93	0.99	0.92	1.32	1.12	1.07	0.84	1.27	1.67	1.33	1.48	1.21	1.05
Al_2O_3	14.59	14.32	14.22	15.16	14.01	13.61	13.36	15.78	14.52	13.79	15.31	14.56	15.35	13.46
Fe_2O_3	15.54	11.11	12.72	11.82	16.52	14.76	14.74	12.21	15.31	13.45	13.72	12.94	13.75	13.72
MnO	0.22	0.18	0.2	0.17	0.23	0.19	0.27	0.18	0.2	0.21	0.2	0.18	0.2	0.03
MgO	5.67	7.77	8.26	7.69	6.86	6.58	6.82	9.25	6.98	7.46	6.23	6.81	6.54	4.12
CaO	13.16	14.51	13.5	12.42	12.24	10.57	10.47	12.39	12.61	11.25	11.6	11.33	12.12	9.33
K_2O	0.05	0.04	0.06	0.06	0.09	0.31	0.28	0.11	0.15	0.62	0.29	0.28	0.17	0.03
Na_2O	1.61	1.73	1.74	1.83	1.68	2.40	2.44	1.62	1.63	2.17	2.41	2.14	1.8	3.58
P_2O_5	0.13	0.08	0.08	0.08	0.12	0.14	0.11	0.08	0.14	0.15	0.13	0.15	0.13	0.36
LOI	0.4	0.9	0.8	0.5	0.6	0.11	0.27	0.33	0.35	0.80	0.50	0.29	0.14	2.59
SUM	101.5	101.1	101.8	99.98	101.71	98.24	98.18	100.66	100.68	99.90	100.31	98.58	99.58	101.0
Rb	6	3	10	3	1	3	4	1	2	12	5	6	3	1
Ba	13	294	456	231	7	63	55	29	49	96	67	99	49	192
Th	1	2	2	1	1	2	1	1	2	0	1	1	0	2
Sc	76	70	68	26	10	48	50	50	13	61	44	55	51	9
Nb	6	5	6	5	1	10	7	3	5	9	6	7	3	6
Sr	128	140	124	118	58	101	118	116	76	259	66	147	157	114
Cu	62	100	86	39	10	76	58	26	93	53	141	70	106	34
Zn	514	363	411	360	40	117	523	109	99	423	108	88	82	66
Ga	13	20	25	18	2	16	15	16	17	7	17	16	17	16
Pb	25	13	20	17	1	7	5	3	1	15	2	6	7	6
V	392	339	368	306	131	274	279	226	314	123	332	312	296	92
Co	72	62	68	69	37	46	51	58	65	70	60	61	62	68
Ni	157	150	179	243	61	71	89	142	74	121	92	78	78	88

Cr	454	707	684	438	116	270	278	436	325	226	288	414	291	66
Zr	81	45	43	50	68	93	58	48	60	103	69	172	173	28
Y	38	25	26	24	18	34	27	14	19	37	20	15	19	47
U	0	0	0	0	0	1	2	1	1	0	1	1	1	1
La	6.00	4.18	4.78	5.44	4.16	16.57	7.36	4.86	7.50	10.21	7.68	9.38	4.35	13.04
Ce	14.91	10.96	12.21	13.09	10.47	37.27	14.90	11.34	18.07	25.92	17.88	21.14	11.08	36.43
Pr	2.37	1.69	1.83	1.86	1.45	4.35	2.05	1.68	2.73	3.87	2.75	2.65	1.54	3.72
Nd	12.89	8.80	9.60	9.30	6.97	20.22	10.04	7.20	11.40	19.78	11.90	14.15	8.95	19.52
Sm	3.92	2.37	2.60	2.41	1.60	5.35	3.34	2.29	3.49	5.24	3.79	3.57	2.92	6.26
Eu	1.26	1.31	1.57	1.26	0.53	1.45	1.13	0.81	1.18	1.83	1.24	1.30	1.07	1.99
Gd	5.30	3.48	3.67	3.38	2.07	4.81	3.15	2.59	4.05	6.40	4.02	3.63	3.29	5.85
Tb	1.14	0.73	0.76	0.70	0.39	1.06	0.62	0.48	0.74	1.27	0.73	0.54	0.57	1.24
Dy	8.43	5.36	5.55	5.07	2.71	6.80	4.20	3.30	4.86	8.48	4.76	3.80	4.45	8.22
Ho	1.32	0.84	0.88	0.80	0.41	1.51	0.95	0.60	0.91	1.31	0.85	0.73	0.94	1.74
Er	4.49	2.88	3.03	2.72	1.44	4.44	2.83	2.05	3.08	4.15	2.82	1.96	2.63	4.77
Tm	0.61	0.39	0.41	0.37	0.19	0.66	0.43	0.35	0.50	0.55	0.46	0.23	0.32	0.64
Yb	4.84	3.21	3.57	3.01	1.54	4.08	2.56	2.09	3.06	4.39	2.81	1.33	1.88	3.47
Lu	0.76	0.53	0.60	0.48	0.23	0.64	0.41	0.34	0.48	0.68	0.43	0.29	0.45	0.48
Mg#	29	44	42	42	32	33	34	46	34	58	53	57	54	25
$(La/Yb)_N$	0.89	0.93	0.96	1.3	1.94	2.91	2.06	1.65	5.82	446	395	514	495	2.70
$(Gd/Yb)_N$	0.91	1.12	1.13	1.14	1.08	0.98	1.02	1.02	1.36	1.67	1.96	1.11	1.88	1.39
$(La/Sm)_N$	1.03	0.85	0.85	1.10	1.46	2.00	1.42	1.37	1.39	1.67	1.31	1.71	0.96	1.34
$(La/Lu)_N$	0.85	0.85	0.85	1.21	1.95	2.77	1.92	1.54	1.67	1.61	1.90	3.47	1.04	2.91

LOI, Loss on ignition; Mg#, MgO/(MgO + FeOt); N chondrite (Cl) normalized to the values of Sun and McDonough (1989).

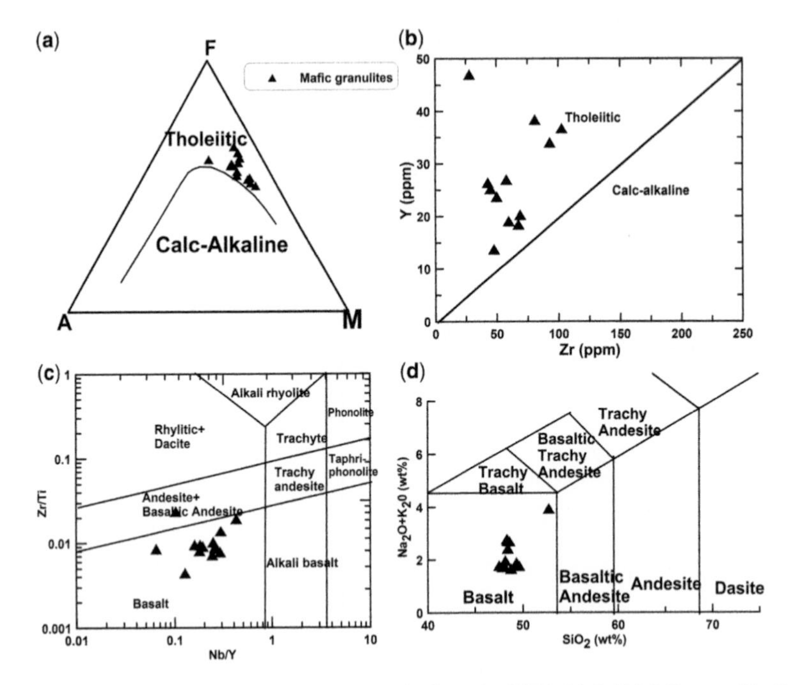

Fig. 4. Major and trace element plots for mafic granulite rocks from the CITZ. (**a**) A–F–M diagram ($Na_2O + K_2O$)–FeO^T–MgO (Irvine and Baragar 1971); (**b**) binary Zr (ppm) v. Y (ppm) classification diagram (after Barrett and MacLean 1994), where the trends define a tholeiitic character; (**c**) Nb/Y v. Zr/Y plot (Winchester and Floyd 1977; modified after Pearce 1996); and (**d**) total alkali v. silica plot (TAS: Wilson 1989). These plots show basaltic composition.

1994). The REE patterns are nearly flat, with light REE (LREE) and high REE (HREE) having similar enrichment to chondrites without any anomalies (Fig. 6a). The samples show relatively flat patterns (LREE enrichment *c.* 1732 times that of chondrite with an abundance ratio $\sum$LREE/HREE of 1.12) with average values of $(La/Sm)_N = 1.18$, $(La/Yb)_N = 1.41$, $(La/Lu)_N = 1.35$ and $(Gd/Yb)_N = 1.05$ (Table 3, Fig. 6a). However, two samples (13a and 6a) depict LREE enrichment (*c.* 50–70 times that of chondrite with an abundance ratio $\sum$LREE/HREE of 1.61) with average values of $(La/Sm)_N = 1.67$, $(La/Yb)_N = 2.80$, $(La/Lu)_N = 2.84$ and $(Gd/Yb)_N = 1.18$.

The PM-normalized multi-element patterns also depict nearly flat patterns with enrichment of about 8–10 times PM (except for two samples). The high field strength elements, including Nb, P, Zr and Ti, show negative anomalies. The U and Pb depict positive anomalies, while Th shows a negative anomaly (Fig. 6b).

Zircon U–Pb systematics. To constrain the period of tectono-magmatic events, we carried out zircon U–Pb dating of mafic granulites (samples MU6 and MU13) from the CITZ (locations shown in the Fig. 1), and the analytical data are illustrated in Table 4. Zircon from sample MU13 is represented by small (70–100 µm) isometric to round pinkish and oval yellowish grains, while the zircon grains from sample MU6 show a colourless to oval yellowish type. Cl images of zircon grains demonstrate sector zoning (Fig. 7). The results show low concentrations of Pb and U in zircon in both samples: Pb = 42–52 ppm in MU13 and 30–156 ppm in MU6; U = 215–268 ppm in MU13 and 198–221 ppm in MU6. These low values are characteristic of metamorphic zircon (Rubatto 2002).

Isotope geochemistry

U–Pb zircon geochronology. Four analysed zircon fractions from sample MU13 yield $^{207}Pb/^{206}Pb$ ages bracketed from 1553 ± 19 to 1569 ± 15 Ma. The upper intercept age calculated for these four data points is 1564 ± 8 Ma (mean square weighted deviation, MSWD = 0.57) (Fig. 7). Three zircon fragments were analysed from sample MU6, and they yielded $^{207}Pb/^{206}Pb$ ages ranging from 1577 ± 89 to 1599 ± 10 Ma. Three fractions with one concordant point determine a discordia line with an upper intercept at 1598 ± 9 Ma (MSWD = 0.74) (Fig. 7).

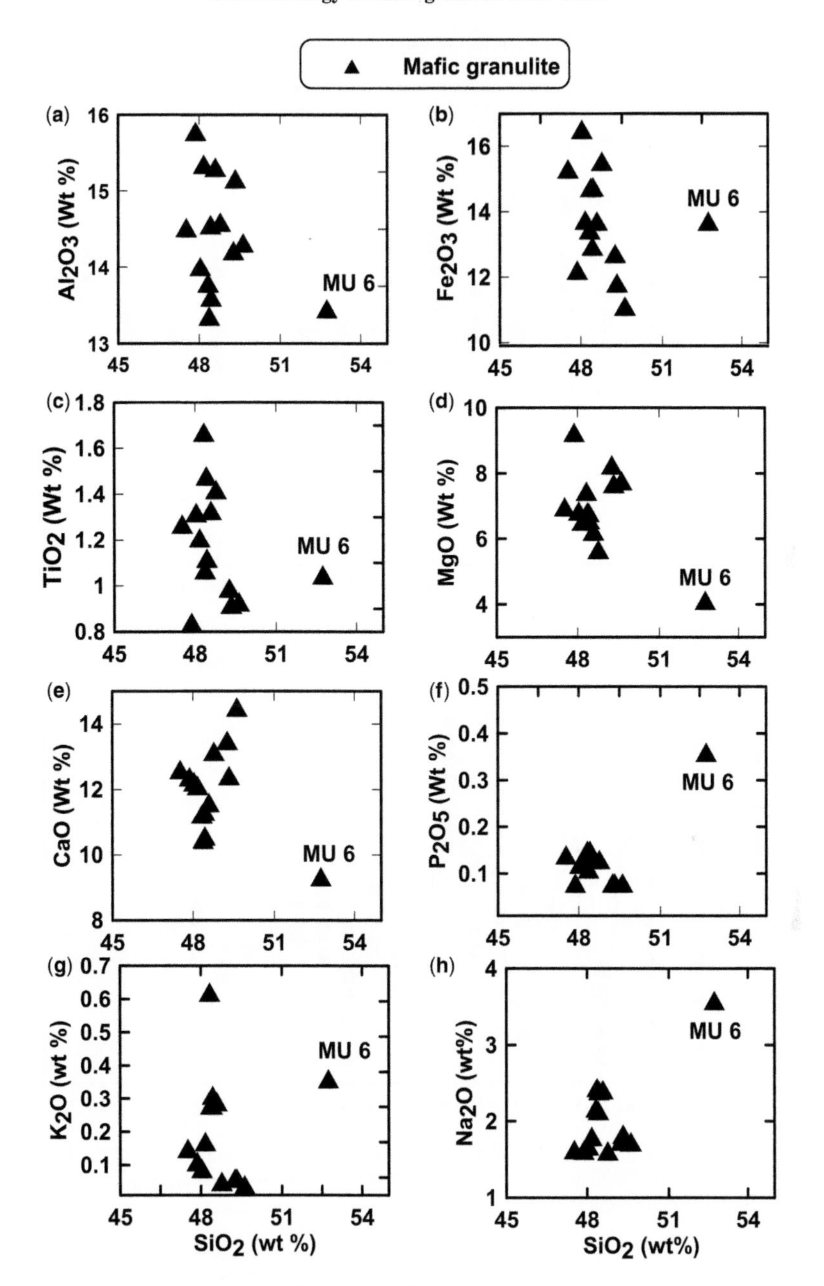

Fig. 5. Harker variation plots for mafic granulites from the CITZ.

Sm–Nd (WR) depleted mantle model and Nd–Sr mineral isochron ages. Sm–Nd (WR) model ages were obtained from the mafic granulites, and the data are presented in Table 5. Nd isotopic diagrams show the evolutionary curves of $^{143}Nd/^{144}Nd$ and ε_{Nd} ($t = 0$) against time (Fig. 8). The data present initial $^{143}Nd/^{144}Nd$ ($t = 1.5$ Ga) ratios (0.509593–0.510011) and the ε_{Nd} ($t = 1.5$ Ga) is (−21.7 to −13.5) with T_{DM} (depleted mantle) model ages ranging from 2.9 to 3.4 Ga, as illustrated in Table 5. As the Tirodi Gneissic Complex (TGC) is associated with granulite rocks, we have also calculated the initial ratios of $^{143}Nd/^{144}Nd$ ($t = 1.5$ Ga) to find out the isotopic relationships between TGC and the granulites. The initial $^{143}Nd/^{144}Nd$ ($t = 1.5$ Ga) ratio varies from 0.510023 to 0.510363, and the ε_{Nd}

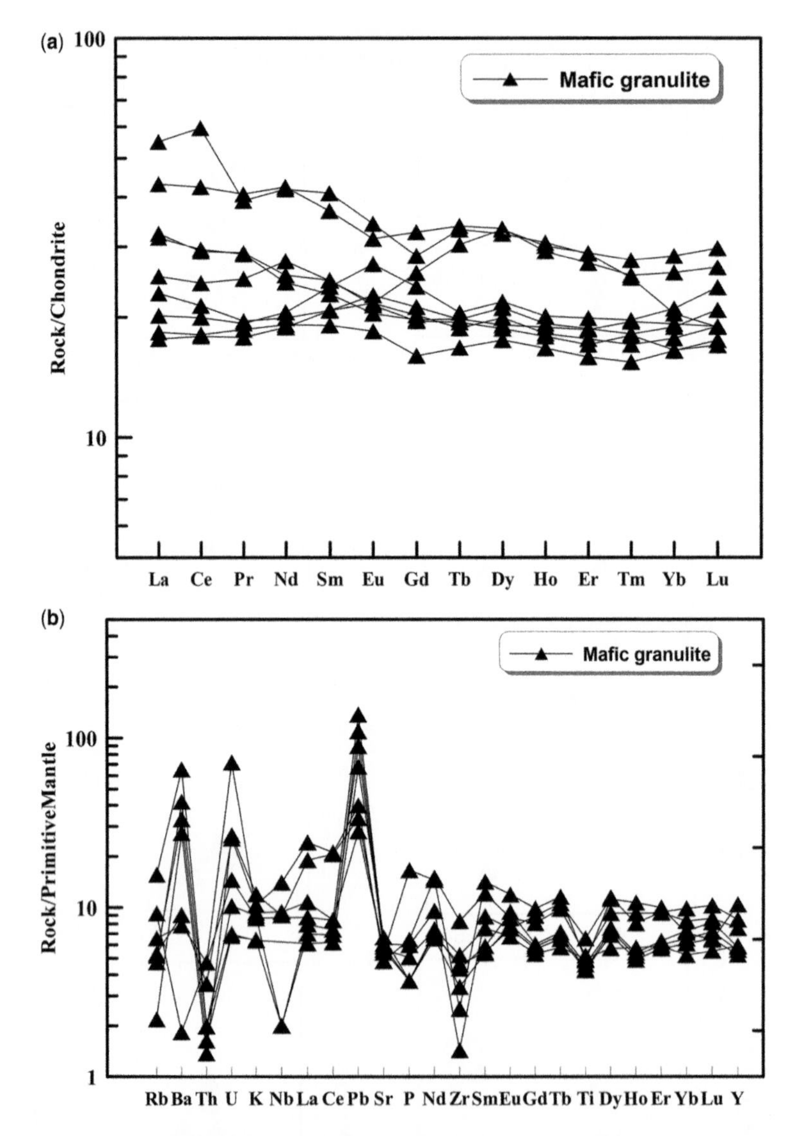

Fig. 6. (**a**) Rock/chondrite normalized rare earth element patterns for mafic granulites in the CITZ (**b**) Rock/primitive mantle normalized multi-element patterns for studied rocks. Normalization values of chondrite and primitive mantle use after Sun and McDonough (1989).

($t = 1.5$ Ga) (-6.6 to -13.2) with depleted mantle model ages (T_{DM}) ranges from 2.36 to 2.78 Ga for TGC granitoid (Alam *et al.* 2022). We consider the U–Pb zircon age of 1.5 Ga for mafic granulites after Ahmad *et al.* (2009) to calculate the initial ratios $^{143}Nd/^{144}Nd$ ($t = 1.5$ Ga) and ε_{Nd} ($t = 1.5$ Ga) for the present set of data (Table 5). The ε_{Nd} ($t = 1.5$ Ga) values from the mafic granulite show a wide range (-21.7 to -13.5), while TGC granitoids depict a narrow range (-6.6 to -13.2) (Fig. 8).

The minerals (plagioclase, apatite and amphibole) and WR of sample MU13 exhibit the Sm–Nd mineral isochron ages of 1024 ± 32 Ma (MSWD $= 1.7$), as illustrated in Table 6, while the Cpx (clinopyroxene) fractions (Cpx + WR) show older ages of 1290–1146 Ma with the highest MSWD, which clearly shows that it is not a good isochron (Fig. 9a). The isochron age of WR, amphibole and garnet without Ap from sample MU6 defines an age of 1030 ± 8 Ma (MSWD $= 0.11$), which is similar to that of MU13 (Fig. 9b).

Table 4. *U–Pb data for zircons of mafic granulites (samples MU-13 and MU-6) from southern and northern domains of the CITZ*

Sample	Size μm	Weight,mg	Concentration, ppm		Pb isotopic composition			Isotopic ratio		Correlation coefficient	Age, Ma
			Pb total	U	$^{206}Pb/^{204}Pb$	$^{206}Pb/^{207}Pb$	$^{206}Pb/^{208}Pb$	$^{207}Pb/^{235}U$ (% error)	$^{206}Pb/^{238}U$ (% error)		$^{207}Pb/^{206}Pb$
MU13/1	<100	0.9	41.5	214.6	1825	9.4996	8.2452	3.0114 (1.1)	0.2255 (0.6)	0.60	1564 ± 16
MU13/2	<100	0.6	51.8	267.7	1309	9.3460	7.6095	2.4615 (1.2)	0.1843 (0.6)	0.58	1564 ± 19
MU13/3	<100	0.8	46.2	238.7	1315	9.3279	7.2439	2.4325 (1.0)	0.1818 (0.5)	0.60	1569 ± 15
MU13/4	<100	0.6	51.5	254.8	1428	9.6899	7.4411	2.3454 (1.3)	0.1767 (0.7)	0.61	1553 ± 19
MU6/5	<100	0.3	90.0	220.5	1477	9.3314	8.8175	3.7067 (1.7)	0.2739 (1.2)	0.75	1589 ± 22
MU6/6	<100	0.2	155.7	183.4	3544	9.7708	9.8134	3.5302 (1.0)	0.2595 (0.9)	0.88	1599 ± 10
MU6/7	<100	0.2	30.4	198.1	1238	9.1640	7.9486	1.5661 (0.9)	0.1160 (0.8)	0.61	1585 ± 14

U mass fractionation is 0.08% per amu. Reproducibility of the U–Pb ratios is 0.7%.

Discussion

Nature and composition of protolith

The mafic granulite presents the exsolution intergrowth of orthopyroxene, clinopyroxene and plagioclase, and triple junction generation occurs between orthopyroxene, clinopyroxene and plagioclase. Hornblende occurs in two generations in the form of green and brown, where the green one is possibly generated at the time of magmatism, while the brown hornblende was generated owing to the recrystallization in the later stage, which led to retrogression. The garnet corona developed between orthopyroxene and plagioclase. Such textural features suggest subsolidus re-equilibration under granulite facies metamorphic conditions. The geochemical characteristics of the studied rocks are of basaltic composition with tholeiitic affinity. Their gabbroic protoliths were emplaced at deep crustal levels, where they underwent slow cooling and recrystallization under granulite-facies conditions. Pressure (0.7–0.8 GPa) and temperature (800–900°C) were estimated for the subsolidus and re-equilibration of the gabbroic protoliths (Marroni and Tribuzio 1996; Montanini 1997). The estimated metamorphic event (BM_1, Bhandara metamorphism) suggested by Bhowmik (2019) showed that the high-grade event locally reached ultra-high temperature (900–1000°C) and pressure (8–9.5 kbar) metamorphic conditions for the gabbroic emplacement in the Bhandara region.

The presence of an inverse geochemical correlation of major oxide against silica correspond to the primary magmatic differentiation of the igneous parental melts (Fig. 5). The variable ratios of Nb/Y (0.06–0.29) and Zr/Ti (0.001–0.1) reflect the index of fractionation and depict tholeiitic affinity that was derived from a variably enriched subcontinental lithospheric mantle (SCLM) source. Enriched REE and multi-element patterns suggest their derivation from the enriched lithospheric mantle source (Drummond and Defant 1990; Martin *et al.* 2005); this enrichment could be made possible by the fluid-driven elemental mobility, especially in REEs, during the metamorphism (Ague 2017). The trace element results show a wide variation in LREEs, as expected to reflect varying degrees of partial melting. Sample 13a with the highest abundances ($\sum$LREE = 303 ppm) and sample MA6 with the lowest abundances ($\sum$LREE = 97 ppm) represent the lowest and highest degrees of partial melting, respectively. The observed variation in depletion in mafic granulite samples could primarily reflect source characteristics and to some extent be related to plagioclase, Ti-magnetite oxide, apatite fractionation, ilmenite–titanite and the stability of titanite in sources; however, the prominent positive Pb anomaly probably suggests crustal influence in their genesis.

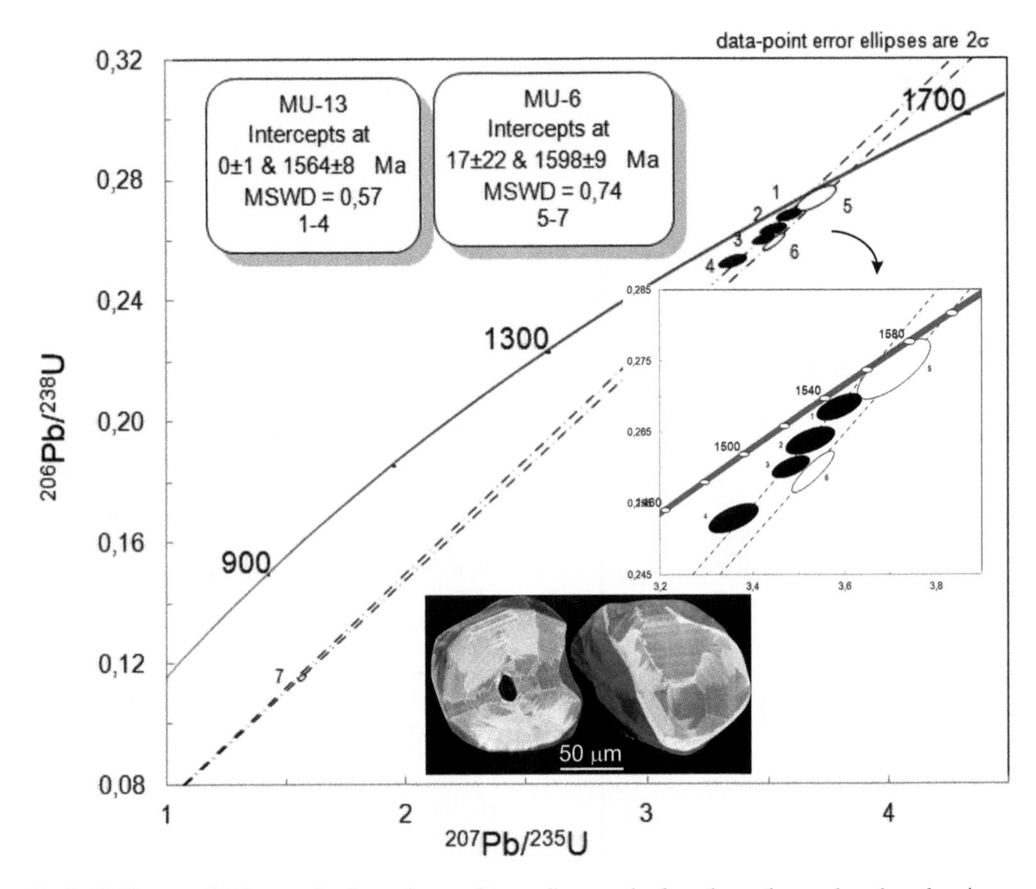

Fig. 7. U–Pb concordia diagrams for zircons from mafic granulites samples from the northern and southern domains of the CITZ.

Geochronological constraints on the Central Indian Tectonic Zone

Previous studies have presented some significant events and recorded relationships with crustal growth and their products in the form of tonalite–trondhjemite–granodiorites and amphibolites (tholeiitic basalts/gabbro and two-pyroxene mafic granulites). Several workers have highlighted the possible genetic link between granites and granulites in the continental crust (Clemens 1990; Vielzeuf and Vidal 1990). It has also been emphasized that the dehydration melting of amphibolitic protoliths has actively participated at lower crustal depth and is capable of generating granitic melts and other forms of granulite like garnet-bearing mafic and ultramafic restites (Johannes and Holtz 1996).

The Tirodi gneissic complex granitoids from the north of the CIS represent basement rock units of gneisses that are closely associated with mafic sequences in the Bhandara region, yielding a T_{DM} modal age range from 2325 to 2494 Ma (Mishra *et al.* 2009). The mafic melts probably trigger the formation of TGC by the fluid flux to the crustal sources during 1.5–1.72 Ga (zircon U–Pb age (Mishra 2011). However, the SMB constraints over magmatic crystallization at *c.* 1.62–1.58 Ga (U–Pb zircon and monazite ages) and overprinting of metamorphic recrystallization were also marked at around 1.57 Ga (Bhowmik *et al.* 2011). The possible extraction of protoliths from the mantle source within the set of BBGB mafic granulites was at c.a. *c.* 3.0 Ga (Sm–Nd model ages), thus having very low Nd (Alam *et al.* 2017). The multiple events varying from 3.2 to 1.6 Ga (Sm–Nd model ages) indicate a longer crustal residence period for the mafic granulites protoliths (Alam *et al.* 2017). Proterozoic collisional events and crustal interaction with lower crustal source rocks are probably responsible for forming mafic granulites through gabbroic emplacement from tholeiitic protoliths (Alam *et al.* 2017). This was produced by 15–40% partial melting in a back-arc setting with a variably enriched mantle source (Alam *et al.* 2017), promoted by adiabatic

Table 5. *Sm–Nd (whole rock) isotopic data of mafic granulite from the CITZ*

Sample	Sm (ppm)	Nd (ppm)	$^{143}Nd/^{144}Nd$	$^{147}Sm/^{144}Nd$	^{f} (Sm/Nd)	$^{143}Nd/_{144}Nd$ ($t = 1.5$Ga)	ε_{Nd} (0)	ε_{Nd} ($t = 1.5$Ga)	Age	T_{DM} (Ma)
MA-01	2.31	13.63	0.5108	0.10272	−0.48	0.50977	−36.3	−18.3	1500	3234
MA-02	2.73	16.55	0.5106	0.10002	−0.49	0.50959	−40.2	−21.7	1500	3420
MA-03	2.38	13.15	0.5110	0.10980	−0.44	0.50994	−31.6	−14.9	1500	3103
MA-05	12.18	68.57	0.5111	0.10740	−0.45	0.51001	−30.6	−13.5	1500	2963
MA-06	7.19	38.88	0.5110	0.11179	−0.43	0.50993	−31.4	−15.1	1500	3148

The ratios of $^{143}Nd/^{144}Nd$ were standardized to $^{146}Nd/^{144}Nd = 0.7219$ and the mean values for the Japanese Nd standard material JNdi-1 (Tanaka *et al.* 2000). $^{143}Nd/^{144}Nd = 0.511833 \pm 6$ (2σ, $n = 11$) (Lugmair *et al.* 1983). Minimum error for $^{143}Nd/^{144}Nd$ ratio was 0.003% and the minimum error for $^{147}Sm/^{144}Nd$ ratio was 0.3% (2σ, $n = 15$) according to BCR standard measurements. Nd blanks were less than 0.3 ng and Sm blanks were less than 0.06 ng. $^{143}Nd/^{144}Nd$ CHUR = 0.512636, $^{147}Sm/^{144}Nd$ CHUR = 0.196700 (Rollinson 1993). The 'Age' column refers to the 1.5 Ga for mafic granulites after Ahmad *et al.* (2009) to calculate the initial ratios $^{143}Nd/^{144}Nd$ ($t = 1.5$ Ga) and ε_{Nd} ($t = 1.5$ Ga) for the present dataset; $\lambda = 6.54 \times 10–12\ a^{-1}$ (Lugmair and Marti 1978) and $\lambda = 1.42 \times 10–11\ a^{-1}$ (Steiger and Jager 1977) for the decay constant for Nd and Sr, respectively. T_{DM} - Depleted-mantle model age.

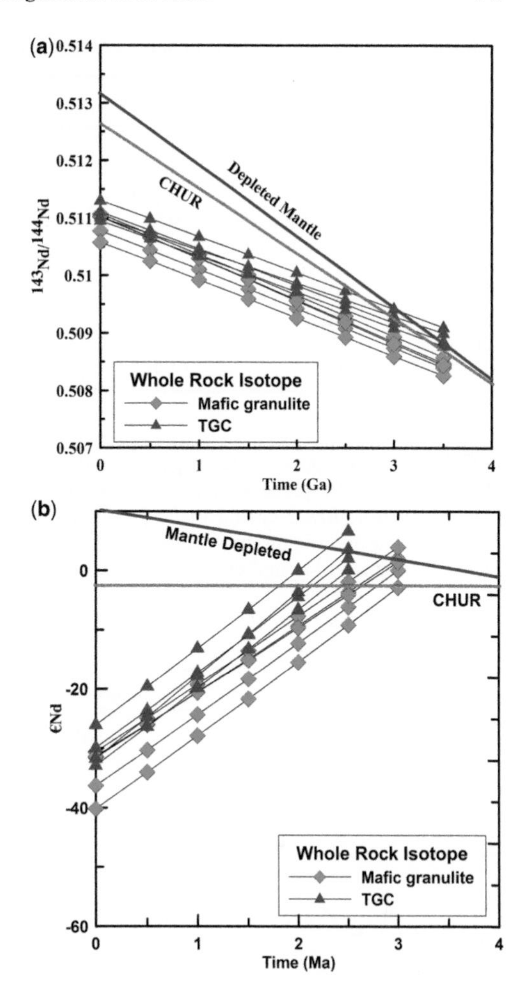

Fig. 8. (**a**) Plot of $^{143}Nd/^{144}Nd$ v. time (Ga). (**b**) Plot of ε_{Nd} v. time (Ma) showing evolution curves for mafic granulites and the Tirodi Gneissic Complex (TGC) from CITZ. The reference lines show the depleted mantle (DM) (Goldstein *et al.* 1984) and the Chondritic Uniform Reservoir (CHUR).

decompression, during syn-metamorphic basaltic emplacement in the Proterozoic BBG domain (Bhowmik 2019).

The zircon U–Pb ages from samples MU13 and MU6 are 1564 ± 8 Ma (MSWD = 0.57) and 1598 ± 9 Ma (MSWD = 0.74), respectively. Similar internal zircon structures for both of these samples visible in CL (cathodoluminescence) show sector zoning, and low U and Pb contents are typical for zircons formed under granulite-facies metamorphic conditions. The two obtained ages do not overlap within error limits, which may indicate some difference in the time of manifestation of granulite-facies metamorphism in the southern and northern domains

Table 6. *Sm–Nd mineral isotopic data of mafic granulite from the CITZ*

Sample	Minerals	Sm (ppm)	Nd (ppm)	$^{143}Nd/^{144}Nd$	$^{147}Sm/^{144}Nd$
MU-13	WR	4.95	18.6	0.51147	0.08021
MU-13a	Ap	59.4	304.6	0.511727	0.11786
MU-13a	Pl	0.158	1.469	0.511385	0.06487
MU-13a	Amf	9.02	34.2	0.512016	0.15957
MU-13a	Cpx-1	1.68	6.0	0.5121	0.17020
MU-13a	Cpx-2	1.72	6.0	0.5123	0.17372
Mu-6	WR	5.99	19.4	0.51194	0,132902
MU-6	Ap	31.1	141.0	0.512095	0.13324
MU-6	Amf	7.5	25.9	0.512225	0.17507
MU-6	Gr	1.58	1.59	0.515119	0.60301

of CITZ. The Mesoproterozoic orogenic event probably facilitated the process of exhumation of the Paleoproterozoic lower crustal mafic granulite. On the basis of sector zoning, these ages may be interpreted as the time of granulite-facies metamorphism (Fig. 7). The temperature of zircon crystallization calculated from the Ti content (‘Ti-in-zircon’ thermometer; Watson 1996) is 730–734°C. The zircon U–Pb ages of 1564 ± 8 and 1598 ± 9 Ma are interpreted as the period of granulite metamorphism of the mafic protoliths, which were emplaced in the form of a gabbroic unit in the Bhandara region of CITZ. These gabbroic melts probably triggered the generation of the TGC (zircon U–Pb ages 1506 ± 11 to 1730 ± 13 Ma) by supplying heat and fluid flux to the crustal sources for the generation of these granitoids.

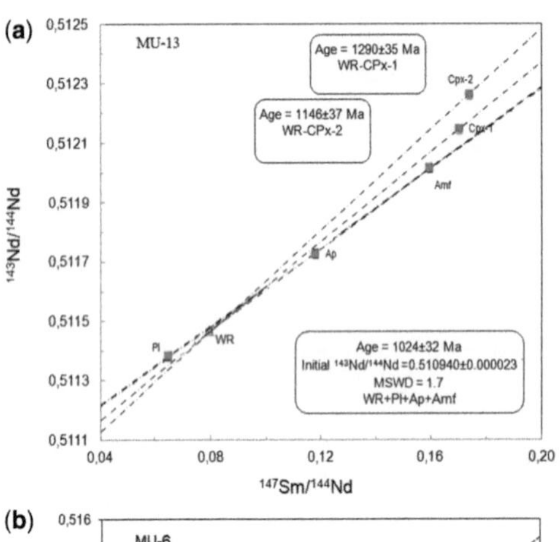

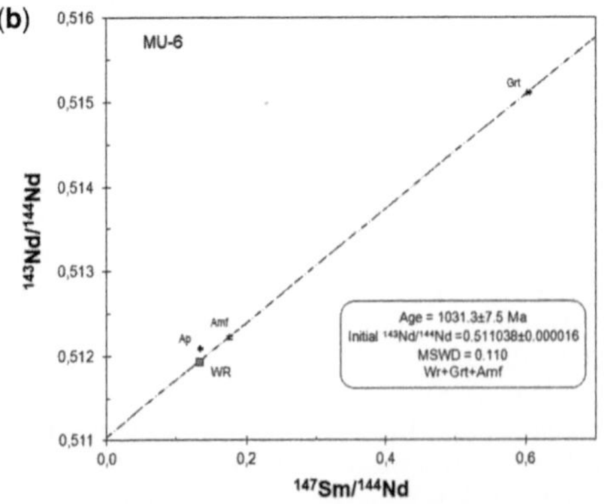

Fig. 9. Sm–Nd mineral isochron diagrams of the mafic granulite rocks from the northern and southern domains of the Sausar mobile belt, CITZ. Cpx, clinopyroxene; Pl, plagioclase; Ap, apatite; Grt, garnet; Amf, amphibole; WR, whole rock.

Formation of protoliths of the mafic granulites and relationship with TGC

The isotopic data of the present study may characterize the protoliths and the period of formation of mafic granulites and their relation to the surrounding granitic units of TGC in the CITZ. The present dataset shows the initial $^{143}Nd/^{144}Nd$ ($t = 1.5$ Ga) ratios (0.509593–0.510011), and the ε_{Nd} ($t = 1.5$ Ga) is -21.7 to -13.5 with T_{DM} (WR) model ages ranging from 2.9 to 3.4 Ga; this span may indicate the extraction of mafic protoliths from the mantle sources. However, the initial $^{143}Nd/^{144}Nd$ ($t = 1.5$ Ga) ratios vary from 0.510023 to 0.510363, ε_{Nd} ($t = 1.5$ Ga) -6.6 to -13.2, with T_{DM} model ages bracketed within 2.20–2.78 Ga for TGC granitoids; this age window indicates the extraction of TGC granitoids protoliths (Alam *et al.* 2022). The lower ratios of $^{147}Sm/^{144}Nd$ (<0.12) for exposed the mafic granulite indicate its interaction with the early crust. The ε_{Nd} ($t = 1.5$ Ga) values from the mafic granulite show a wide range (-21.7 to -13.5), which indicates a longer crustal residence period (since 2.9–3.4 Ga), while TGC granitoids are marked with a narrower range (-6.6 to -13.2), which indicates crustal sources or a relatively shorter residence period for their sources that were mafic (2.2–2.7 Ga; Table 5 and Fig. 8). Therefore, the T_{DM} age of mafic granulites is older than that for the

TGC granitoids, which suggests that the protoliths of mafic granulite are older than the TGC granitoids. As mentioned above, the zircon U–Pb ages of 1564 ± 8 and 1598 ± 9 Ma from the present study correspond to the period of granulite-facies metamorphism of the mafic protolith in the form of gabbroic bodies. These melts could be responsible for the generation of the TGC granitoids during 1506 ± 11 to 1730 ± 13 Ma (zircon U–Pb age; Mishra 2011) by supplying heat and fluid flux. Moreover, the zircon U–Pb ages (1534 ± 13 to 1454 ± 5 Ma (Ahmad *et al.* 2009) of Tirodi biotite gneiss from the northern and central domains of the CITZ are interpreted as the time of magmatic crystallization with two events of metamorphic recrystallization occurring at 1.57–1.56 and 1.42 Ga, respectively (Bhowmik *et al.* 2011). The age bracket (1.57–1.42 Ga) indicates the collision between these two crustal domains within the CITZ.

The obtained Sm–Nd mineral ages of 1024 ± 32 and 1030 ± 8 Ma can be interpreted as some local thermal events with a temperature rise. This is supported by an older Cpx age of 1148–1290 Ma, since Cpx has a higher closure temperature for the Sm–Nd system than other minerals (Pl and Amf). Known Sm–Nd data for the mafic granulites of CITZ are within 1403 ± 77 to 1416 ± 59 Ma (Roy *et al.* 2006), which the authors interpreted as the closure of the Sm–Nd system of minerals during cooling. Preliminary Rb–Sr data for these samples determines two ages: 1369 ± 84 Ma for Ap + Cpx + Amf and 852 ± 130 Ma for Pl and Cpx. Similar Rb–Sr data – 1370 ± 84–1422 ± 104 Ma and 800–973 Ma – were obtained by Roy *et al.* (2006) and Ramachandra and Roy (2001). The interval of 800–900 Ma is considered the time of amphibolite facies metamorphism (675°C and 7 kb), which is described as the last metamorphic event in the CITZ. The rearrangement of the Rb–Sr system is associated with a fluid action, and in rocks (or minerals) that were less accessible to the fluid, the Rb–Sr system was less disturbed.

Summarily, the Sm–Nd model age for WR of 2.9–3.4 Ga suggested that the basaltic/gabbro melts initially separated from the mantle. However, zircon U–Pb ages (1564 ± 8 Ma–1598 ± 9 Ma) reflect the time of granulite-facies metamorphism of mafic granulite protoliths owing to collisional

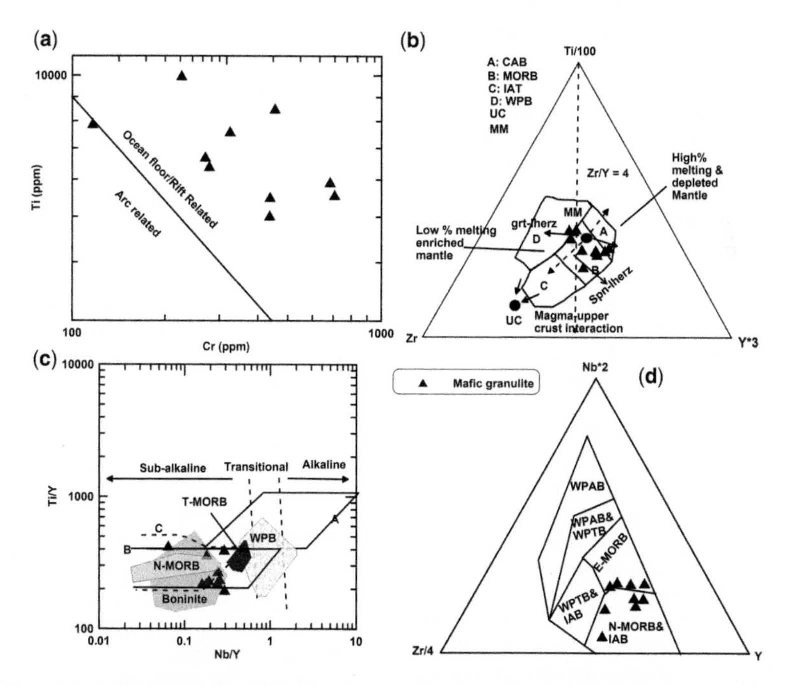

Fig. 10. (**a**) Plot of Cr v. Ti (after Pearce 1975) showing the discrimination between an island arc and ocean floor/rift-related rocks studied. (**b**) Plot of the Ti/100-Zr-Y*3 ternary diagram (after Pearce and Cann 1973). (**c**) Nb/Y v. Ti/Y (modified after Pearce 1982) plot showing subalkaline mid ocean ridge basalt (MORB) affinity with the variable ratios of Nb/Y (0.06–0.29) and Ti/Y (197–427). (**d**) The ternary plot Nb*2-Zr/4-Y (Meschede 1986) shows that most of the samples have been placed in the fields of normal-type MORB (N-MORB) and island-arc basalt (IAB), while a few samples fall within the boundary of incompatible element-enriched MORB (E-MORB). CAB, calc alkaline basalts; IAT, island arc tholeiites; WPB, within plate basalts; UC, upper crust; MM-N, MORB mantle; WPAB, within plate alkali basalts; WPTB, within plate tholeiites basalts; IAB, island arc basalts; T-MORB, transitional mid-oceanic ridge basalts.

orogeny during 1.7–1.5 Ga in the CITZ (Acharyya 2003). The obtained Sm–Nd mineral ages of 1014 ± 34 and 1030 ± 18 Ma reflect thermal perturbation and fluid events during the metamorphism (Fig. 9). Therefore, the precise timing of the formation of mafic granulites through fractional crystallization and moderate crustal contamination, promoted by the rifting event (1.6–1.3 Ga; Rogers and Santosh 2002), infers the fragmentation of the Columbian supercontinent. Thus, we consider that the protolith of the mafic granulite crystallized from an early Mesoproterozoic mafic magma and that it was formed originally at lower crustal depths, which were subducted into the mantle, and underwent metamorphism in the garnet stability field.

Tectonic implications for the development of Proterozoic mafic granulite

The information on regional tectonics, geochronology, magmatism and metamorphism in the CITZ has been provided by previous studies, and it has received significant attention over the last few decades (e.g. Naqvi and Rogers 1987; Yedekar *et al.* 1990; Acharyya 2003; Roy and Hanuma Prasad 2003; Bhowmik *et al.* 2005, 2011; Bhowmik 2006, 2019; Mall *et al.* 2008; Naganjaneyulu and Santosh 2010; Santosh 2012; Chattopadhyay *et al.* 2015; Alam *et al.* 2017, 2022; Chattopadhyay *et al.* 2020; Deshmukh and Prabhakar 2020; Mohanty 2020). Several studies based on geophysical and geological data have addressed the tectonic evolution history and the nature of the deep crust beneath the CITZ hitherto.

Four tectonic diagrams were chosen to constrain the tectonic setting of the protolith for the studied rocks (Fig. 10). Plot Cr v. Ti (after Pearce 1975) shows the discrimination between an island arc and ocean floor/rift-related mafic granulite rocks; all of the samples plot in the ocean floor/rift-related setting (Fig. 10a). In the plot Ti/100-Zr-Y*3 ternary diagram (after Pearce and Cann 1973), it is shown that most samples plot within the mid-ocean ridge basalt (MORB) and very few are straddling the within-plot boundary. It is also suggested that the magma probably derived from the spinel- and garnet-bearing lherzolitic source (Fig. 10b). Furthermore, the Nb/Y v. Ti/Y (modified after Pearce 1982) plot depicts subalkaline MORB affinity with the variable ratios of Nb/Y (0.06–0.29) and Ti/Y (197–427; Fig. 10c). However, the ternary plot Nb*2-Zr/4-Y (Meschede 1986) shows that the majority of samples plot within the normal-type MORB and island-arc basalt fields, with some tendency towards the boundary of incompatible element-enriched MORB (Fig. 10d). Thus, we propose that an arc tectonic setting for the magmatic underplating below the existing continental crust was involved in the formation of the studied mafic granulites, through metamorphic evolution in the CITZ.

Conclusions

The geochemical results of the studied rocks define a tholeiitic affinity with basaltic composition. Major element variation characteristics suggest a primary magmatic differentiation of the parental melt/igneous protoliths of the studied mafic granulites from the southern and northern domains of the SMB within the CITZ. The degree of alkalinity and the fractionation index would also be delineated, as the studied rocks were derived from a variably enriched SCLM. Trace element geochemistry indicate plagioclase, Ti-magnetite and apatite fractionation with a mild crustal influence.

The present data shows that the ε_{Nd} ($t = 1.5$ Ga) is -21.66 to -13.45 with T_{DM} (WR) model ages ranging from 2.9 to 3.4 Ga. Mineral isochrons (Sm–Nd) accompanied by plagioclase, apatite and amphibole exhibit an age of 1024 ± 32 Ma; however, the fractions of apatite, amphibole and garnet describe an age of 1030 ± 8 Ma. These younger ages (*c.* 1.0 Ga) probably reflect that these granulitic rocks have undergone some events during an early Neoproterozoic period.

U–Pb zircon ages of 1564 ± 8 Ma and 1598 ± 9 Ma are interpreted as the period of granulite-facies metamorphism of the mafic protoliths in the form of gabbroic bodies. These gabbroic melts, in turn, triggered the generation of the Tirodi Granite Gneisses during 1506 ± 11 to 1730 ± 13 Ma by supplying heat and fluid flux to the crustal sources for the generation of these granitoids. The protoliths of mafic granulites could be related to the evolution of melts derived from metasomatized SCLM through fractional crystallization processes and have undergone arc/MORB and rift-related settings, which are inferred to the Columbian event.

Acknowledgment We thank the Department of Science and Technology and the Russian Academy of Sciences for financial support. We thank the Department of Geology, University of Delhi, Institute Instrumentation Centre, IIT Roorkee, and the Geological Institute of Kola Science Centre, Russian Academy of Sciences, Apatity, for the analytical facilities. We thank all of the members of the Institute Instrumentation Centre, IIT Roorkee and Isotope Geology Laboratory, Russian Academy of Sciences, Apatity, Russia.

Competing interests The authors declare that they have no known competing financial interests or personal relationships that could have appeared to influence the work reported in this paper.

Author contributions **MA**: conceptualization (equal), investigation (equal), methodology (supporting), writing – original draft (lead); **TVK**: conceptualization (supporting), data curation (lead), formal analysis (supporting), funding acquisition (equal), investigation (supporting), methodology (supporting), project administration (equal); **RRV**: conceptualization (supporting), investigation (supporting), visualization (supporting); **TA**: funding acquisition (lead), investigation (supporting), project administration (lead), resources (lead), supervision (supporting), validation (supporting), visualization (supporting), writing – review & editing (supporting).

Funding TA acknowledges the financial support from the J.C. Bose Fellowship of SERB-DST. The Department of Science and Technology and the Russian Academy of Sciences provided financial support for this work through project no. SR/S4/ES-402/2009 and ILTP project B-2.58, respectively.

Data availability All data generated or analysed during this study are included in this published article, shown for each table.

References

Acharyya, S.K. 2003. The nature of mesoproterozoic Central Indian Tectonic Zone with exhumed and reworked older granulites. *Gondwana Research*, **6**, 197–214, https://doi.org/10.1016/S1342-937X(05)70970-9

Acharyya, S.K. and Roy, A. 2000. Tectonothermal history of the Central Indian Tectonic Zone and reactivation of major faults/shear zones. *Journal of the Geological Society of India*, **55**, 239–256.

Ague, J.J. 2017. Element mobility during regional metamorphism in crustal and subduction zone environments with a focus on the rare earth elements (REE). *American Mineralogist*, **102**, 1796–1821, https://doi.org/10.2138/am-2017-6130

Ahmad, T. and Tarney, J. 1991. Geochemistry and petrogenesis of Garhwal volcanics: implications for evolution of the north Indian lithosphere. *Precambrian Research*, **50**, 69–88, https://doi.org/10.1016/0301-9268(91)90048-F

Ahmad, T. and Tarney, J. 1994. Geochemistry and petrogenesis of late Archaean Aravalli volcanics, basement enclaves and granitoids, Rajasthan. *Precambrian Research*, **65**, 1–23, https://doi.org/10.1016/0301-9268(94)90097-3

Ahmad, T., Kaulina, T.V., Wanjari, N., Mishra, M.K. and Nitkina, E.A. 2009. U–Pb zircon chronology and Sm–Nd isotopic characteristics of the Amgaon and Tirodi Gneissic Complex, Central Indian Shield: constraints on Precambrian crustal evolution. *In*: *Precambrian Continental Growth and Tectonism*. Excel India, New Delhi, 137–138.

Alam, M., Choudhary, A.K., Mouri, H. and Ahmad, T. 2017. Geochemical characterization and petrogenesis of mafic granulites from the central indian tectonic zone (CITZ). *Geological Society, London, Special Publications*, **449**, 207–229, https://doi.org/10.1144/SP449.1

Alam, M., Mishra, M.-K., Kaulina, T.-V., Ahmad, T. and Choudhary, A.-K. 2022. Geochemistry and petrogenesis of Proterozoic granitoids from Central Indian Tectonic Zone (CITZ): elemental and isotopic constraints. *Geochemical Journal*, **56**, 160–176, https://doi.org/10.2343/geochemj.gj22016

Bandyopadhyay, B.K., Bhoskar, K.G. *et al.* 1990. Recent geochronological studies in parts of the Precambrian of central India. *Geological Survey of India Special Publications*, **28**, 199–210.

Bandyopadhyay, B.K., Roy, A. and Huin, A.K. 1995. Structure and tectonics of a part of the Central Indian shield. *In*: Sinha-Roy, S. and Gupta, K.R. (eds) *Continental Crust of Northwestern and Central India*. Geological Society of India, Memoirs, **31**, 433–467.

Barrett, T.J. and MacLean, W.H. 1994. Chemostratigraphy and hydrothermal alteration in exploration for VHMS deposits in greenstones and younger volcanic rocks. *In*: Lentz, D.R. (ed.) *Alteration and Alteration Processes associated with Ore-forming systems*. Geological Association of Canada, Short Course Notes, **11**, 433–467.

Bhandari, A., Bhowmik, S.K., Wilde, S.A. and Pant, N.C. 2010. Inherited monazites and zircons in ~1.6 Ga ultra-high-temperature granulites from the southern margin of the Central Indian Tectonic Zone: implication for sedimentary provenance and Central Indian Orogenesis. *In*: *Proceedings of the Seventh AOGS*, Hyderabad, SE-01, A004.

Bhattacharya, S., Chaudhary, A.K. and Basei, M. 2012. Original nature and source of khondalites in the Eastern Ghats Province, India. *Geological Society, London, Special Publications*, **365**, 147–159, https://doi.org/10.1144/sp365.8

Bhowmik, S.K. 2006. Ultra high temperature-metamorphism and its significance in the Central Indian Tectonic Zone. *Lithos*, **92**, 484–505, https://doi.org/10.1016/j.lithos.2006.03.061

Bhowmik, S.K. 2019. The current status of orogenesis in the Central Indian Tectonic Zone: a view from its Southern Margin. *Geological Journal*, **54**, 2912–2934, https://doi.org/10.1002/gj.3456

Bhowmik, S.K. and Roy, A. 2003. Garnetiferous metabasites from the Sausar Mobile Belt: Petrology, *P–T* path and implications for the tectonothermal evolution of the Central Indian Tectonic Zone. *Journal of Petrology*, **44**, 387–420, https://doi.org/10.1093/petrology/44.3.387

Bhowmik, S.K., Sarbadhikari, A.B., Spiering, B. and Raith, M.M. 2005. Mesoproterozoic reworking of palaeoproterozoic ultrahigh-temperature granulites in the Central Indian Tectonic Zone and its implications. *Journal of Petrology*, **46**, 1085–1119, https://doi.org/10.1093/petrology/egi011

Bhowmik, S.K., Wilde, S.A. and Bhandari, A. 2011. Zircon U–Pb/Lu–Hf and monazite chemical dating of the Tirodi biotite gneiss: Implication for latest Palaeoproterozoic to Early Mesoproterozoic orogenesis in the Central Indian Tectonic Zone. *Geological Journal*, **46**, 574–596, https://doi.org/10.1002/gj.1299

Bhowmik, S.K., Chattopadhyay, A., Gupta, S. and Dsagupta, S. 2012. Proterozoic tectonics: an Indian perspective on the Central Indian Tectonic Zone (CITZ). *Proceedings of the Indian Academy of Sciences, Earth and Planetary Sciences*, **78**, 385–391.

Bhowmik, S.K., Wilde, S.A., Bhandari, A. and Basu Sarbadhikari, A. 2014. Zoned monazite and zircon as monitors for the thermal history of granulite terranes: an example from the Central Indian Tectonic Zone. *Journal of Petrology*, **55**, 585–621, https://doi.org/10.1093/petrology/egt078

Bohlen, S.R. 1987. Pressure–temperature–time paths and a tectonic model for the evolution of granulites. *The Journal of Geology*, **95**, 617–632, https://doi.org/10.1086/629159

Bohlen, S.R. 1991. On the formation of granulites. *Journal of Metamorphic Geology*, **9**, https://doi.org/10.1111/j.1525-1314.1991.tb00518.x

Bohlen, S.R. and Mezger, K. 1989. Origin of granulite terranes and the formation of the lowermost continental crust. *Science (New York)*, **244**, 326–329, https://doi.org/10.1126/science.244.4902.326

Bohlen, S.R., Valley, J.W. and Essene, E.J. 1985. Metamorphism in the adirondacks. I. Petrology, pressure and temperature. *Journal of Petrology*, **26**, 971–992, https://doi.org/10.1093/petrology/26.4.971

Brown, M. 2006. Duality of thermal regimes is the distinctive characteristics of plate tectonics since the Neoarchean. *Geology*, **34**, 961–964, https://doi.org/10.1130/G22853A.1

Brown, M. 2007*a*. Metamorphic conditions in orogenic belts: a record of secular change. *International Geology Review*, **49**, 193–234, https://doi.org/10.2747/0020-6814.49.3.193

Brown, M. 2007*b*. Metamorphism, Plate Tectonics, and the Supercontinent Cycle. *Earth Science Frontiers*, **14**, 1–18, https://doi.org/10.1016/s1872-5791(07)60001-3

Brown, M. and Johnson, T. 2018. Secular change in metamorphism and the onset of global plate tectonics. *American Mineralogist*, **103**, 181–196, https://doi.org/10.2138/am-2018-6166

Brown, M. and Phadke, A.V. 1983. High temperature retrograde reactions in pelitic gneiss from the Precambrian Sausar metasediments of the Ramakona area, Chindwara dist., M.P (India). Definition of the exhumation P-T path and the tectonic implication. *In*: Phadke, A.V. and Phansalkar, V.G. (eds), Prof. K.V. Kelkar, Memorial Volume Indian Society, Earth Scientists, Pune, 61–96.

Chattopadhyay, A., Das, K., Hayasaka, Y. and Sarkar, A. 2015. Syn- and post-tectonic granite plutonism in the Sausar Fold Belt, central India: age constraints and tectonic implications. *Journal of Asian Earth Sciences*, **107**, 110–121, https://doi.org/10.1016/j.jseaes.2015.04.006

Chattopadhyay, A., Chatterjee, A., Das, K. and Sarkar, A. 2017. Neoproterozoic transpression and granite magmatism in the Gavilgarh-Tan Shear Zone, central India: tectonic significance of U–Pb zircon and U–Th–total Pb monazite ages. *Journal of Asian Earth Sciences*, **147**, 485–501, https://doi.org/10.1016/j.jseaes.2017.08.018

Chattopadhyay, A., Bhowmik, S.K. and Roy, A. 2020. Tectonothermal evolution of the Central Indian Tectonic Zone and its implications for Proterozoic supercontinent assembly: the current status. *Episodes*, **43**, 132–144, https://doi.org/10.18814/epiiugs/2020/020008

Chetty, T.R.K. 2017. *Proterozoic Orogens of India: a Critical Window to Gondwana*, https://doi.org/10.1007/s12594-018-0867-0

Clark, C., Collins, A.S., Timms, N.E., Kinny, P.D., Chetty, T.R.K. and Santosh, M. 2009. SHRIMP U–Pb age constraints on magmatism and high-grade metamorphism in the Salem Block, southern India. *Gondwana Research*, **16**, 27–36, https://doi.org/10.1016/j.gr.2008.11.001

Clark, C., Fitzsimons, I.C.W., Healy, D. and Harley, S.L. 2011. How does the continental crust get really hot? *Elements*, **7**, 235–240, https://doi.org/10.2113/gselements.7.4.235

Clark, C., Kirkland, C.L., Spaggiari, C.V., Oorschot, C., Wingate, M.T.D. and Taylor, R.J. 2014. Proterozoic granulite formation driven by mafic magmatism: an example from the Fraser Range Metamorphics, Western Australia. *Precambrian Research*, **240**, 1–21, https://doi.org/10.1016/j.precamres.2013.07.024

Clark, C., Healy, D., Johnson, T., Collins, A.S., Taylor, R.J., Santosh, M. and Timms, N.E. 2015. Hot orogens and supercontinent amalgamation: a Gondwanan example from southern India. *Gondwana Research*, **28**, 1310–1328, https://doi.org/10.1016/j.gr.2014.11.005

Clemens, J.D. 1990. The granulite–granite connexion. *In*: Vielzeuf, D. and Vidal, P. (eds) *Granulites and Crustal Evolution*. NATO ASI Series, **311**. Springer, Dordrecht, 25–36, https://doi.org/10.1007/978-94-009-2055-2_3

Collins, W.J. 2002. Hot orogens, tectonic switching, and creation of continental crust. *Geology*, **30**, 535–538, https://doi.org/10.1130/0091-7613(2002)030<0535:HOTSAC>2.0.CO;2

Collins, A.S. and Pisarevsky, S.A. 2005. Amalgamating eastern Gondwana: the evolution of the Circum-Indian Orogens. *Earth-Science Reviews*, **71**, 229–270, https://doi.org/10.1016/j.earscirev.2005.02.004

Condie, K. 2002. Continental growth during a 1.9-Ga superplume event. *Journal of Geodynamics*, **34**, 249–264, https://doi.org/10.1016/S0264-3707(02)00023-6

Dasgupta, S., Sengupta, P., Mondal, A. and Fukuoka, M. 1993. Mineral chemistry and reaction textures in metabasites from the Eastern Ghats belt, India and their implications. *Mineralogical Magazine*, **57**, 113–120, https://doi.org/10.1180/minmag.1993.057.386.11

Deshmukh, T. and Prabhakar, N. 2020. Linking collision, slab break-off and subduction polarity reversal in the evolution of the Central Indian Tectonic Zone. *Geological Magazine*, **157**, 340–350, https://doi.org/10.1017/S0016756819001419

Drummond, M.S. and Defant, M.J. 1990. A model for trondhjemite–tonalite–dacite genesis and crustal growth via slab melting: Archean to modern comparisons. *Journal of Geophysical Research*, **95**, 21, 503–521, 521, https://doi.org/10.1029/JB095iB13p21503

Ellis, D.J. 1987. Origin and evolution of granulites in normal and thickened crusts. *Geology*, **15**, 167–170, https://doi.org/10.1130/0091-7613(1987)15<167:OAEOGI>2.0.CO

Ellis, D.J. and Green, D.H. 1985. Garnet-forming reactions in mafic granulites from enderby land, Antarctica –

implications for geothermometry and geobarometry. *Journal of Petrology*, **26**, 633–662, https://doi.org/10.1093/petrology/26.3.633

Eriksson, P.G., Mazumder, R., Sarkar, S., Bose, P.K., Altermann, W. and Van Der Merwe, R. 1999. The 2.7–2.0 Ga volcano-sedimentary record of Africa, India and Australia: evidence for global and local changes in sea level and continental freeboard. *Precambrian Research*, **97**, 269–302, https://doi.org/10.1016/S0301-9268(99)00035-2

Fedotova, A.A., Bibikova, E.V. and Simakin, S.G. 2008. Ion-microprobe zircon geochemistry as an indicator of mineral genesis during geochronological studies. *Geochemistry International*, **46**, 912–927, https://doi.org/10.1134/S001670290809005X

Goldstein, S.L., O'Nions, R.K. and Hamilton, P.J. 1984. A SmNd isotopic study of atmospheric dusts and particulates from major river systems. *Earth and Planetary Science Letters*, **70**, https://doi.org/10.1016/0012-821X(84)90007-4

Green, D.H. and Ringwood, A.E. 1967. An experimental investigation of the gabbro to eclogite transformation and its petrological applications. *Geochimica et Cosmochimica Acta*, **31**, 767–833, https://doi.org/10.1016/S0016-7037(67)80031-0

Hanson, G.N. 1980. Rare earth elements in petrogenetic studies of igneous systems. *Annual Review of Earth and Planetary Sciences*, **8**, 371–406, https://doi.org/10.1146/annurev.ea.08.050180.002103

Harley, S. 1989. The origins of granulites – a metamorphic perspective. *Geological Magazine*, **126**, 215–247, https://doi.org/10.1017/S0016756800022330

Harris, L.B. 1993. Correlations of tectonothermal events between the Central Indian Tectonic Zone and the Albany Mobile Belt of Western Australia. *Gondwana Eight: Assembly, Evolution and Dispersal, Proceedings of the 8th Gondwana symposium*, Hobart, 1991, 165–180.

Hölttä, P., Huhma, H., Mänttäri, I. and Paavola, J. 2000. *P–T–t* development of Archaean granulites in Varpaisjärvi, Central Finland. *Lithos*, **50**, 121–136, https://doi.org/10.1016/S0024-4937(99)00055-9

Hou, G., Santosh, M., Qian, X., Lister, G.S. and Li, J. 2008. Configuration of the Late Paleoproterozoic supercontinent Columbia: insights from radiating mafic dyke swarms. *Gondwana Research*, **14**, 395–409, https://doi.org/10.1016/j.gr.2008.01.010

Hyndman, R.D., Currie, C.A. and Mazzotti, S.P. 2005. Subduction zone backarcs, mobile belts, and orogenic heat. *GSA Today*, **15**, 4–10, https://doi.org/10.1130/1052-5173(2005)015<4:SZBMBA>2.0.CO;2

Irvine, T.N. and Baragar, W.R.A. 1971. A guide to the chemical classification of the common volcanic rocks. *Canadian Journal of Earth Sciences*, **8**, 523–548, https://doi.org/10.1139/e71-055

Jain, S.C., Yedekar, D.B. and Nair, K.K.K. 1991. Central Indian Shear Zone – a major Precambrian crustal boundary. *Journal of the Geological Society of India*, **37**, 521–531.

Johannes, W. and Holtz, F. 1996. Petrogenesis and Experimental Petrology of Granitic Rocks. *Minerals and Rocks*, **22**, 115–275, https://doi.org/10.1007/978-3-642-61049-3.

Johnson, C.A. and Essene, E.J. 1982. The formation of garnet in olivine-bearing metagabbros from the Adirondacks. *Contributions to Mineralogy and Petrology*, **81**, 240–251, https://doi.org/10.1007/BF00371301

Krogh, T.E. 1973. A low-contamination method for hydrothermal decomposition of zircon and extraction of U and Pb for isotopic age determinations. *Geochimica et Cosmochimica Acta*, **87**, 485–494, https://doi.org/10.1016/0016-7037(73)90213-5

Kumar, C.R.R. and Chacko, T. 1994. Geothermobarometry of mafic granulites and metapelite from the Palghat Gap, South India: petrological evidence for isothermal uplift and rapid cooling. *Journal of Metamorphic Geology*, **12**, 479–492, https://doi.org/10.1111/j.1525-1314.1994.tb00037.x

Li, Z.X., Bogdanova, S.V. *et al.* 2008. Assembly, configuration, and break-up history of Rodinia: a synthesis. *Precambrian Research*, **160**, 179–210, https://doi.org/10.1016/j.precamres.2007.04.021

Liu, S., Liang, H., Zhao, G., Hua, Y. and Jian, A. 2000. Isotopic chronology and geological events of Precambrian complex in Taihangshan region. *Science in China, Series D: Earth Sciences*, **43**, 386–393, https://doi.org/10.1007/BF02959449

Liu, Y. and Zhong, D. 1997. Petrology of high-pressure granulites from the eastern Himalayan syntaxis. *Journal of Metamorphic Geology*, **15**, 451–466, https://doi.org/10.1111/j.1525-1314.1997.00033.x

Longjam, K.C. and Ahmad, T. 2012. Geochemical characterization and petrogenesis of Proterozoic Khairagarh volcanics: Implication for Precambrian crustal evolution. *Geological Journal*, **47**, 130–143, https://doi.org/10.1002/gj.1312

Ludwig, K.R. 1991. PBDAT Program. US Geological Survey, Open-File Report 88-542.

Lugmair, G.W. and Marti, K. 1978. Lunar initial 143Nd/144Nd: Differential evolution of the lunar crust and mantle. *Earth Planetary Science Letter*, **39**, 349–357, https://doi.org/10.1016/0012-821X(78)90021-3

Lugmair, G., Shimamura, T., Lewis, R.S. and Anders, E. 1983. Samarium-146 in the early solar system: evidence from neodymium in the Allende Meteorite. *Science (New York)*, **222**, 1015–1018, https://doi.org/10.1126/science.222.4627.1015

Mahakud, S.P., Raut, P.K. and Mishra, V.P. 2000. Geological set-up of Kherli Bazar polymetallic deposit, Betul district, Madhya Pradesh. *In*: Gyani, K.C. and Kataria, P. (eds) *Tectonomagmatism, Geochemistry and Metamorphism of Precambrian Terrains*. University Department of Geology, Udaipur.

Mall, D.M., Reddy, P.R. and Mooney, W.D. 2008. Collision tectonics of the Central Indian Suture zone as inferred from a deep seismic sounding study. *Tectonophysics*, **460**, 116–123, https://doi.org/10.1016/j.tecto.2008.07.010

Marroni, M. and Tribuzio, R. 1996. Gabbro-derived granulites from external liguride units (northern Apennine, Italy): implications for the rifting processes in the western Tethys. *Geologische Rundschau*, **85**, 239–249, https://doi.org/10.1007/BF02422231

Martin, H., Smithies, R.H., Rapp, R., Moyen, J.F. and Champion, D. 2005. An overview of adakite, tonalite–trondhjemiten–granodiorite (TTG), and sanukitoid: relationships and some implications for crustal evolution. *Lithos*, **79**, 1–24, https://doi.org/10.1016/j.lithos.2004.04.048

McLelland, J.M. and Whitney, P.R. 1977. The origin of garnet in the anorthosite–charnockite suite of the Adirondacks. *Contributions to Mineralogy and Petrology*, **60**, 161–181, https://doi.org/10.1007/BF00372280

Meert, J.G. 2012. What's in a name? The Columbia (Paleopangaea/Nuna) supercontinent. *Gondwana Research*, **21**, 987–993, https://doi.org/10.1016/j.gr.2011.12.002

Meert, J.G. and Santosh, M. 2017. The Columbia supercontinent revisited. *Gondwana Research*, **50**, 67–83, https://doi.org/10.1016/j.gr.2017.04.011

Meert, J.G. and Torsvik, T.H. 2003. The making and unmaking of a supercontinent: Rodinia revisited. *Tectonophysics*, **375**, 261–288, https://doi.org/10.1016/S0040-1951(03)00342-1

Mengel, F. and Rivers, T. 1991. Decompression reactions and *P–T* conditions in high-grade rocks, northern Labrador: *P–T–t* paths from individual samples and implications for early proterozoic tectonic evolution. *Journal of Petrology*, **32**, 139–167, https://doi.org/10.1093/petrology/32.1.139

Merdith, A.S., Collins, A.S. *et al.* 2017. A full-plate global reconstruction of the Neoproterozoic. *Gondwana Research*, **50**, 84–134, https://doi.org/10.1016/j.gr.2017.04.001

Meschede, M. 1986. A method of discriminating between different types of mid-ocean ridge basalts and continental tholeiites with the Nb1bZr1bY diagram. *Chemical Geology*, **56**, 207–218, https://doi.org/10.1016/0009-2541(86)90004-5

Mishra, D.C., Singh, B., Tiwari, V.M., Gupta, S.B. and Rao, M.B.S.V. 2000. Two cases of continental collisions and related tectonics during the Proterozoic period in India – insights from gravity modelling constrained by seismic and magnetotelluric studies. *Precambrian Research*, **99**, 149–169, https://doi.org/10.1016/S0301-9268(99)00037-6

Mishra, M.K. 2011. *Geochemistry, petrogenesis and U–Pb zircon geochronology of basement granitoids and gneisses of central Indian tectonic zone (CITZ), central Indian Shield.* Unpublished PhD thesis, Delhi University.

Mishra, M.K., Ahmad, T. and Kaulina, T.V. 2009. Geochemical characteristics of Tirodi basement gneisses Central Indian Shield. *In*: Singh, V.K. and Chandra, R. (eds) *Precambrian Continental Growth and Tectonism*. Excel India, New Delhi, 21.

Mohanty, S.P. 2020. Evolution of the 'Central Indian Tectonic Zone': A Critique Based on the Study of the Sausar Belt. *In*: Biswal, T., Ray, S. and Grasemann, B. (eds) *Structural Geometry of Mobile Belts of the Indian Subcontinent.* Society of Earth Scientists Series. Springer, Cham, https://doi.org/10.1007/978-3-030-40593-9_3

Montanini, A. 1997. Mafic granulites in the Cretaceous sedimentary mélanges from the northern Apennine (Italy): petrology and tectonic implications. *Schweizerische Mineralogische und Petrographische Mitteilungen*, **77**, 51–72.

Naganjaneyulu, K. and Santosh, M. 2010. The Central India Tectonic Zone: a geophysical perspective on continental amalgamation along a Mesoproterozoic suture. *Gondwana Research*, **18**, 547–564, https://doi.org/10.1016/j.gr.2010.02.017

Nance, R.D., Murphy, J.B. and Santosh, M. 2014. The supercontinent cycle: a retrospective essay. *Gondwana Research*, **25**, 4–29, https://doi.org/10.1016/j.gr.2012.12.026

Naqvi, S.M. and Rogers, J.J.W. 1987. *Precambrian Geology of India*. Oxford University Press, Oxford.

O'Brien, P.J. and Rötzler, J. 2003. High-pressure granulites: formation, recovery of peak conditions and implications for tectonics. *Journal of Metamorphic Geology*, **21**, 3–20, https://doi.org/10.1046/j.1525-1314.2003.00420.x

Ouzegane, K., Bendaoud, A., Kienast, J.R. and Touret, J.L.R. 2001. Pressure–temperature–fluid evolution in Eburnean metabasitesand metapelites from Tamanrasset (Hoggar, Algeria). *Journal of Geology*, **109**, 247–263, https://doi.org/10.1086/319238

Pattison, D.R.M., Chacko, T., Farquhar, J. and McFarlane, C.R.M. 2003. Temperatures of granulite-facies metamorphism: constraints from experimental phase equilibria and thermobarometry corrected for retrograde exchange. *Journal of Petrology*, **44**, 867–900, https://doi.org/10.1093/petrology/44.5.867

Pearce, J.A. 1975. Basalt geochemistry used to investigate past tectonic environments on Cyprus. *Tectonophysics*, **25**, 41–67, https://doi.org/10.1016/0040-1951(75)90010-4

Pearce, J.A. 1982. Trace element characteristics of lavas from destructive plate boundaries. *In*: Thorpe, R.S. (eds) *Andesites: Orogenic andesites and related rocks*, Wiley and Sons, New York, 525–548.

Pearce, J.A. 1996. A User's Guide to Basalt Discrimination Diagrams. *In*: Wyman, D.A., (eds) *Trace Element Geochemistry of Volcanic Rocks: Applications for Massive Sulphide Exploration*, Geological Association of Canada, Short Course Notes, **12**, 113.

Pearce, J.A. and Cann, J.R. 1973. Tectonic setting of basic volcanic rocks determined using trace element analyses. *Earth and Planetary Science Letters*, **19**, 290–300, https://doi.org/10.1016/0012-821X(73)90129-5

Phadke, A.V. 1990. Genesis of the granitic rocks and the status of the 'Tirodi Biotite Gneiss' in relation to the metamorphites of the Sausar group and the regional tectonic setting. *Geological Society of India, Special Publications*, **28**, 287–302.

Pisarevsky, S.A., Wingate, M.T.D., Powell, C.M., Johnson, S. and Evans, D.A.D. 2003. Models of Rodinia assembly and fragmentation. *Geological Society, London, Special Publications*, **206**, 35–55, https://doi.org/10.1144/GSL.SP.2003.206.01.04

Prakash, D. 1999. Petrology of the Basic Granulites from Kodaikanal, South India. *Gondwana Research*, **2**, 95–104, https://doi.org/10.1016/S1342-937X(05)70130-1

Radhakrishna, B.P. 1989. Suspect tectono-stratigraphic terrane elements in the Indian subcontinent. *Journal – Geological Society of India*, **34**, 1–24.

Radhakrishna, B.P. and Naqvi, S.M. 1986. Precambrian continental crust of India and its evolution. *The Journal of Geology*, **94**, 145–166, https://doi.org/10.1086/629020

Ramachandra, H.M. and Roy, A. 2001. Evolution of the Bhandara–Balaghat granulite belt along the Southern margin of the Sausar mobile belt of Central India.

Proceedings of the Indian Academy of Sciences, Earth and Planetary Sciences, **110**, 351–368, https://doi.org/10.1007/BF02702900

Rogers, J.J.W. and Gird, R.S. 1997. The Indian Shield. *In*: De Wit, and Ashawal, L.D. (eds) *Greenstone Belts*. Clarendon Press, Oxford.

Rogers, J.J.W. and Santosh, M. 2002. Configuration of Columbia, a Mesoproterozoic supercontinent. *Gondwana Research*, **5**, 5–22, https://doi.org/10.1016/S1342-937X(05)70883-2

Rollinson, H.R. 1993. *Using Geochemical Data: Evaluation, Presentation, Interpretation*. Longman Scientific and Technical, Wiley, New York, 352.

Romer, R.L. and Rötzler, J. 2001. *P–T–t* evolution of ultrahigh-temperature granulites from the Saxon Granulite Massif, Germany. Part II: geochronology. *Journal of Petrology*, **42**, 2015–2032, https://doi.org/10.1093/petrology/42.11.2015

Roy, A. and Hanuma Prasad, M. 2003. Tectonothermal events in Central Indian Tectonic Zone (CITZ) and its implications in Rodinian crustal assembly. *Journal of Asian Earth Sciences*, **22**, 115–129, https://doi.org/10.1016/S1367-9120(02)00180-3

Roy, A., Kagami, H. *et al.* 2006. Rb–Sr and Sm–Nd dating of different metamorphic events from the Sausar Mobile Belt, central India: implications for Proterozoic crustal evolution. *Journal of Asian Earth Sciences*, **26**, 61–76, https://doi.org/10.1016/j.jseaes.2004.09.010

Rubatto, D. 2002. Zircon trace element geochemistry: Partitioning with garnet and the link between U–Pb ages and metamorphism. *Chemical Geology*, **184**, 123–138, https://doi.org/10.1016/S0009-2541(01)00355-2

Sandiford, M., Neall, F.B. and Powell, R. 1987. Metamorphic evolution of aluminous granulites from Labwor Hills, Uganda. *Contributions to Mineralogy and Petrology*, **95**, 217–225, https://doi.org/10.1007/BF00381271

Sandiford, M., Powell, R., Martin, S.F. and Perera, L.R.K. 1988. Thermal and baric evolution of garnet granulites from Sri Lanka. *Journal of Metamorphic Geology*, **6**, 351–364, https://doi.org/10.1111/j.1525-1314.1988.tb00425.x

Santosh, M. 2012. India's Palaeoproterozoic legacy. *Geological Society, London, Special Publications*, **365**, 263–288, https://doi.org/10.1144/SP365.14

Santosh, M. and Kusky, T. 2010. Origin of paired high pressure–ultrahigh–temperature orogens: A ridge subduction and slab window model. *Terra Nova*, **22**, 35–42, https://doi.org/10.1111/j.1365-3121.2009.00914.x

Sarkar, S.N., Trivedi, J.R. and Gopalan, K. 1986. Rb–Sr whole-rock and mineral isochron ages of the Tirodi gneiss, Sausar group, Rhandara district, Maharashtra. *Journal of the Geological Society of India*, **27**, 30–37.

Sarkar, G., Paul, D.K., De Leater, J.R., McNaughton, N.J. and Misra, V.P. 1990. A geochemical and Pb, Sr, isotopic study of the Evolution of Granite-Gneisses from the Bastar Craton, Central India. *Journal of the Geological Society of India*, **35**, 480–496.

Sizova, E., Gerya, T. and Brown, M. 2014. Contrasting styles of Phanerozoic and Precambrian continental collision. *Gondwana Research*, **25**, 522–545, https://doi.org/10.1016/j.gr.2012.12.011

Smithson, S.B. and Brown, S.K. 1977. A model for lower continental crust. *Earth and Planetary Science Letters*, **35**, 134–144, https://doi.org/10.1016/0012-821X(77)90036-X

Srivastava, R.K. and Chalapathi Rao, N.V. 2007. Petrology, geochemistry and tectonic significance of Palaeoproterozoic alkaline lamprophyres from the Jungel Valley, Mahakoshal supracrustal belt, Central India. *Mineralogy and Petrology*, **89**, 189–215, https://doi.org/10.1007/s00710-006-0144-6

Steiger, R.H. and Jager, E. 1977. Subcommision on geochronology: Convention on the use of decay constants in geo and cosmochronology. *Earth Planetary Science Letter*. **36**, 359–362, https://doi.org/10.1016/0012-821X(77)90060-7

Stüwe, K. and Powell, R. 1989. Low-pressure granulite facies metamorphism in the Larsemann Hills area, East Antarctica; petrology and tectonic implications for the evolution of the Prydz Bay area. *Journal of Metamorphic Geology*, **7**, 465–483, https://doi.org/10.1111/j.1525-1314.1989.tb00609.x

Sun, S.-S. and McDonough, W.F. 1989. Chemical and isotopic systematics of oceanic basalts: implications for mantle composition and processes. *Geological Society, London, Special Publications*, **42**, 313–345, https://doi.org/10.1144/gsl.sp.1989.042.01.19

Tanaka, T., Togashi, S. *et al.* 2000. JNdi-1: a neodymium isotopic reference in consistency with LaJolla neodymium. *Chemical Geology*, **168**, 279–281, https://doi.org/10.1016/S0009-2541(00)00198-4

Vielzeuf, D. and Vidal, P. 1990. Granulites and crustal evolution. *Granulites and Crustal Evolution*, **311**, https://doi.org/10.1007/978-94-009-2055-2

Watson, E.B. 1996. Surface enrichment and trace-element uptake during crystal growth. *Geochimica et Cosmochimica Acta*, **60**, 5013–5020, https://doi.org/10.1016/S0016-7037(96)00299-2

Whitney, D.L. and Evans, B.W. 2010. Abbreviations for names of rock-forming minerals. *American Mineralogist*, **95**, 185–187, https://doi.org/10.2138/am.2010.3371

Wilson, M. 1989. *Igneous Petrogenesis a Global Tectonic Approach*. Chapman and Hall, London.

Winchester, J.A. and Floyd, P.A. 1977. Geochemical discrimination of different magma series and their differentiation products using immobile elements. *Chemical Geology*, **20**, 325–343, https://doi.org/10.1016/0009-2541(77)90057-2

Yedekar, D.B., Jain, S.C., Nair, K.K.K. and Dutta, K.K. 1990. The central Indian collision suture. *Geological Survey of India, Special Publications*, **28**, 1–43.

Zhang, R.Y., Yang, J.S., Ernst, W.G., Jahn, B.M., Iizuka, Y. and Guo, G.L. 2016. Discovery of in situ super-reducing, ultrahigh-pressure phases in the Luobusa ophiolitic chromitites, Tibet: new insights into the deep upper mantle and mantle transition zone. *American Mineralogist*, **101**, 1285–1294, https://doi.org/10.2138/am-2016-5436

Zhang, Y., Dostal, J., Zhao, Z., Liu, C. and Guo, Z. 2011. Geochronology, geochemistry and petrogenesis of mafic and ultramafic rocks from Southern Beishan area, NW China: implications for crust-mantle interaction. *Gondwana Research*, **20**, 816–830, https://doi.org/10.1016/j.gr.2011.03.008

Zhao, G., Cawood, P.A., Wilde, S.A. and Lu, L. 2001. High-pressure granulites (retrograded eclogites) from the Hengshan Complex, North China Craton: petrology and tectonic implications. *Journal of Petrology*, **42**, 1141–1170, https://doi.org/10.1093/petrology/42.6.1141

Zhao, G., Cawood, P.A., Wilde, S.A. and Sun, M. 2002. Review of global 2.1–1.8 Ga orogens: implications for a pre-Rodinia supercontinent. *Earth-Science Reviews*, **59**, 125–162, https://doi.org/10.1016/S0012-8252(02)00073-9

Zhao, G., Sun, M. and Wilde, S.A. 2003. Correlations between the eastern block of the North China Craton and the South Indian Block of the Indian Shield: an Archaean to Palaeoproterozoic link. *Precambrian Research*, **122**, 201–233, https://doi.org/10.1016/S0301-9268(02)00212-7

Zhao, G., Sun, M., Wilde, S.A. and Li, S. 2004. A Paleo-Mesoproterozoic supercontinent: assembly, growth and breakup. *Earth-Science Reviews*, **67**, 91–123, https://doi.org/10.1016/j.earscirev.2004.02.003

Zhao, G., Li, S., Sun, M. and Wilde, S.A. 2011. Assembly, accretion, and break-up of the Palaeo-Mesoproterozoic Columbia supercontinent: record in the North China Craton revisited. *International Geology Review*, **53**, 1331–1356, https://doi.org/10.1080/00206814.2010.527631

Zhao, G.C., Wilde, S.A., Cawood, P.A. and Lu, L.Z. 2000. Petrology and *P–T* path of the Fuping mafic granulites: Implications for tectonic evolution of the central zone of the North China craton. *Journal of Metamorphic Geology*, **18**, 375–391

The Mutare–Fingeren dyke swarm: the enigma of the Kalahari Craton's exit from supercontinent Rodinia

Ashley P. Gumsley[1,2*], Michiel de Kock[3], Richard Ernst[4], Anna Gumsley[5], Richard Hanson[6], Sandra Kamo[7], Michael Knoper[3], Marek Lewandowski[8], Bartłomiej Luks[8], Antony Mamuse[2] and Ulf Söderlund[9,10]

[1]Institute of Earth Sciences, University of Silesia in Katowice, 41-205, Sosnowiec, Poland

[2]Faculty of Engineering and Geosciences, Midlands State University, Zvishavane, 9005, Zimbabwe

[3]Department of Geology, University of Johannesburg, Johannesburg, 2006, South Africa

[4]Department of Earth Sciences, Carleton University, Ottawa, ON K1S 5B6, Canada

[5]Institute of Geological Sciences, Polish Academy of Sciences, Krakow, 31-002, Poland

[6]Department of Geological Sciences, Texas Christian University, Fort Worth, 76109, USA

[7]Department of Earth Sciences, University of Toronto, Toronto, ON M5S 3B1, Canada

[8]Institute of Geophysics, Polish Academy of Sciences, Warsaw 01-452, Poland

[9]Department of Geology, Lund University, Lund, 223 62, Sweden

[10]Department of Geosciences, Swedish Museum of Natural History, Stockholm, 104 05, Sweden

APG, 0000-0003-1395-3065; MdeK, 0000-0002-5036-3438;
RE, 0000-0001-9474-0314; AG, 0000-0002-0889-3435; MK, 0000-0001-6867-6654;
ML, 0000-0003-1504-3735; BL, 0000-0003-2287-385X
*Correspondence: ashley.gumsley@us.edu.pl

Abstract: The Rodinia supercontinent broke apart during the Neoproterozoic. Rodinia break-up is associated with widespread intraplate magmatism on many cratons, including the *c.* 720–719 Ma Franklin large igneous province (LIP) of Laurentia. Coeval magmatism has also been identified recently in Siberia and South China. This extensive magmatism terminates ~1 myr before the onset of the Sturtian Snowball Earth. However, LIP-scale magmatism and global glaciation are probably related. U–Pb isotope dilution–thermal ionization mass spectrometry (ID-TIMS) baddeleyite dating herein identifies remnants of a new *c.* 724–712 Ma LIP on the eastern Kalahari Craton in southern Africa and East Antarctica: the combined Mutare–Fingeren Dyke Swarm. This dyke swarm occurs in northeastern Zimbabwe (Mutare Dyke Swarm) and western Dronning Maud Land (Fingeren Dyke Swarm). It has incompatible element-enriched mid-ocean ridge basalt-like geochemistry, suggesting an asthenospheric mantle source for the LIP. The Mutare–Fingeren LIP probably formed during rifting. This rifting would have occurred almost ~100 myr earlier than previous estimates in eastern Kalahari. The placement of Kalahari against southeastern Laurentia in Rodinia is also questioned. Proposed alternatives, invoking linking terranes between Kalahari and southwestern Laurentia or close to northwestern Laurentia, also present challenges with no discernible resolution. Nevertheless, LIP-scale magmatism being responsible for the Sturtian Snowball Earth significantly increases.

Supplementary material: Detailed methodology and supplementary tables including mineral chemistry, whole-rock geochemistry and U–Pb ID-TIMS geochronology data are available at https://doi.org/10.6084/m9.figshare.c.6447279

The supercontinent of Rodinia assembled and broke apart during the Meso- to Neoproterozoic (Li *et al.* 2008). Its break-up coincides with extensive magmatism, preserved in large igneous provinces (LIPs), which are short-lived, extensive and predominantly mafic magmatic events in intraplate settings (Ernst 2014). One of the largest of these LIPs is the *c.* 720–719 Ma Franklin (Heaman *et al.* 1992; Denyszyn *et al.* 2009; Pu *et al.* 2022), which occurred approximately ~1 myr before the onset of the Sturtian Snowball Earth (Pu *et al.* 2022). The termination of one event, and the commencement of the other,

From: van Schijndel, V., Cutts, K., Pereira, I., Guitreau, M., Volante, S. and Tedeschi, M. (eds) 2024. *Minor Minerals, Major Implications: Using Key Mineral Phases to Unravel the Formation and Evolution of Earth's Crust.* Geological Society, London, Special Publications, **537**, 359–380.
First published online May 3, 2023, https://doi.org/10.1144/SP537-2022-206

are linked with increased weathering of mafic magmatic material on Rodinia (Cox *et al.* 2016), which was at the equator (Denyszyn *et al.* 2009). After a short-lived episode of volcanic-related carbon dioxide flooding the atmosphere and hydrosphere from the LIPs, the longer-term drawdown of carbon due to weathering probably led to global glaciation (Cox *et al.* 2016; Pu *et al.* 2022), the so-called 'fire and ice' hypothesis (Goddéris *et al.* 2003). These studies have been reinforced by identifying coeval LIPs in Siberia (Ernst *et al.* 2016) and South China (Lu *et al.* 2022).

The Kalahari Craton's position within Rodinia is enigmatic (e.g. Jacobs *et al.* 2008). This uncertainty is primarily due to sparse palaeomagnetic evidence and the lack of crucial precise temporal and spatial piercing points (i.e. LIPs, fossil record) outside of more generic and widespread mobile belts. Various attempts have been made to reconstruct Kalahari's position during this time, mostly placing the craton against the southwestern margin of Laurentia (Dalziel *et al.* 2000; Li *et al.* 2008; Swanson-Hysell *et al.* 2015; Swanson-Hysell 2021). Other possibilities have been proposed by Pisarevsky *et al.* (2003) and Evans (2009), with Kalahari attached to proto-Australia far to the south and west of Laurentia or north of it (present-day coordinates), respectively. Recently, linking terranes between Kalahari and southeastern Laurentia have been proposed (e.g. MARA or Arequipa; Casquet *et al.* 2012; Rapela *et al.* 2016; Hodgin *et al.* 2021). Kalahari is also usually reconstructed with the Grunehogna Craton and the Meso- to Neoproterozoic segments of the Maud Belt. The Grunehogna Craton, together with the Meso- to Neoproterozoic Maud Belt, has long been considered either as a fragment of 'greater' Kalahari (Groenewald *et al.* 1991) or adjacent to the craton (Basson *et al.* 2004) that then broke away in the Jurassic. These are the so-called 'close fit' and 'far fit' scenarios of Basson *et al.* (2004). In addition, the tectonic setting of Kalahari's western, eastern and southern margins during the Neoproterozoic is increasingly being studied (i.e. Jacobs *et al.* 2020). After amalgamation with Laurentia in Rodinia by *c.* 1.0 Ga, a continental arc is interpreted to have developed off eastern Kalahari after *c.* 0.8 Ga. Back-arc extension was then thought to occur by *c.* 650–600 Ma, followed by further metamorphism and deformation at *c.* 590–500 Ma (Jacobs *et al.* 2020). This back-arc extension contrasts with Grosch *et al.* (2007), who interpreted magmatism at *c.* 600 Ma as related to rifting.

This study presents U–Pb isotope dilution–thermal ionization mass spectrometry (ID-TIMS) geochronology on baddeleyite from the Mutare Dyke Swarm in southern Africa and the Fingeren Dyke Swarm in western Dronning Maud Land of East Antarctica (Fig. 1). We identify and document the Fingeren Dyke Swarm for the first time in this study. This geochronology is coupled with petrography and whole-rock major-, minor- and trace-element geochemistry. These two dyke swarms from Kalahari are coeval. They constitute the remnants of a new LIP and, therefore, a new record in the magmatic barcode, as defined by Bleeker and Ernst (2006), for Kalahari and a new piercing point between segments of the craton in western Dronning Maud Land and southern Africa. The dyke swarms may also be genetically related to the Franklin LIP of Laurentia (Heaman *et al.* 1992; Denyszyn *et al.* 2009), along with coeval magmatic events in Siberia and South China (Ernst *et al.* 2016; Lu *et al.* 2022), and therefore probably contributed to the onset of the Sturtian Snowball Earth (Pu *et al.* 2022). The new magmatic barcode also allows us to speculate further on Kalahari's enigmatic placement within Rodinia, the position from which it exited the supercontinent, and its tectonic history during this time.

Regional geology

The bulk of the Kalahari Craton is located in southern Africa. The Grunehogna Craton fragment and its eastward extension into the Maud Belt are in western and central Dronning Maud Land of East Antarctica (Fig. 1). Many definitions exist for the Kalahari Craton (e.g. Jacobs *et al.* 2008). The core of Kalahari consists of the Archean to Paleoproterozoic Kaapvaal Craton, located mainly in South Africa, Eswatini (Swaziland), Lesotho and Botswana; the Zimbabwe Craton, covering most of Zimbabwe and a small part of Botswana and Mozambique; and the Limpopo Belt (Metamorphic Complex) along the South Africa–Zimbabwe border which welded the two cratons together during the Neoarchean to Paleoproterozoic. A later westward extension of Kalahari into Botswana and Namibia is composed of the Paleo- to Mesoproterozoic Rehoboth Block, together with the Paleoproterozoic Kheis and Magondi belts, present west of the Kaapvaal and Zimbabwe craton margins. We follow Jacobs *et al.* (2008) in referring to these combined geologic units as the Proto-Kalahari Craton. The term Kalahari Craton refers to this Archean to Paleoproterozoic core, and the surrounding Mesoproterozoic crust added later (i.e. the Namaqua–Natal Belt in southern Africa, and parts of the Maud Belt in western and central Donning Maud Land, stretching from Heimefrontfjella to the SE of Grunehogna to Kirwanveggan in the south, then H.U. Sverdrupfjella, Gjelsvikfjella and Schirmacher Oasis with increasing distance to the east; Jacobs *et al.* 2008). Most models support the Grunehogna Craton being a direct continuation of Kalahari (Marschall *et al.* 2010).

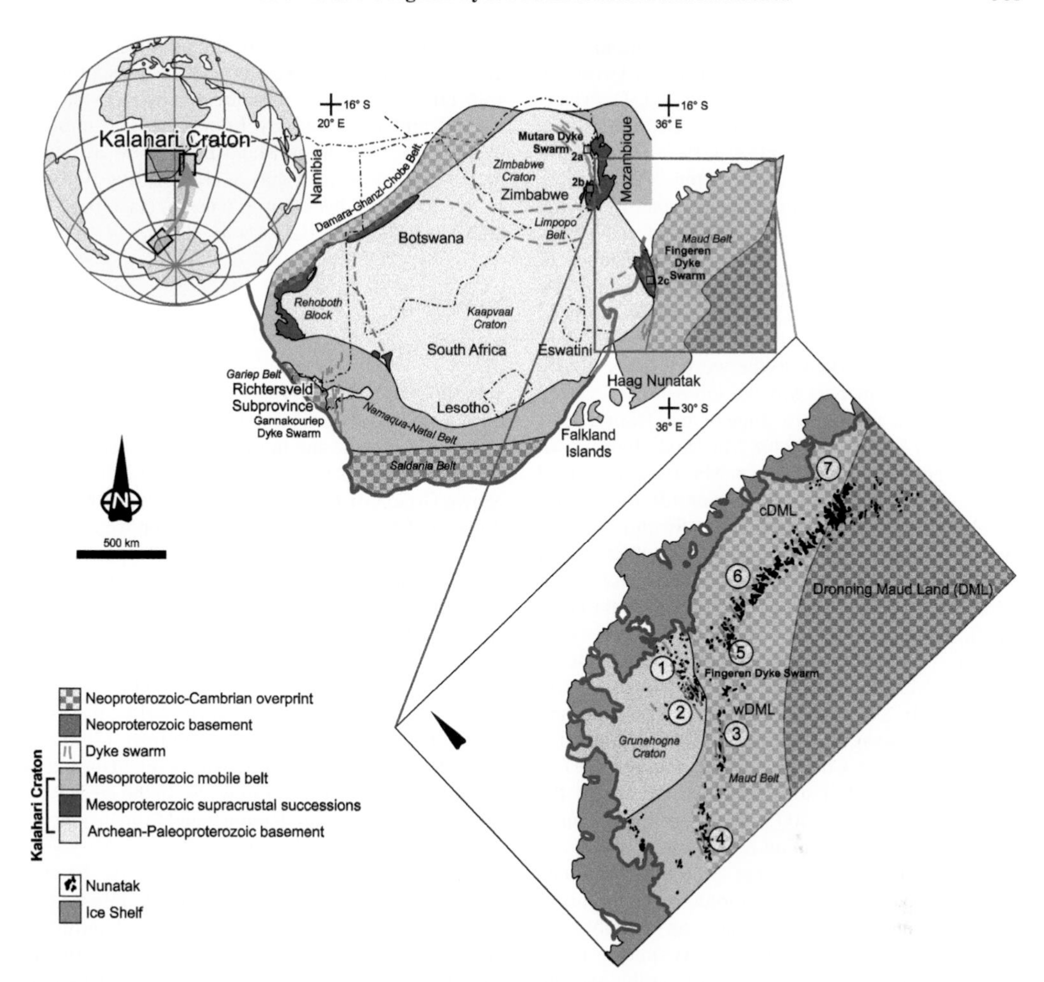

Fig. 1. Geological outline of the Kalahari Craton in southern Africa and East Antarctica, including its core of the Archean to Paleoproterozoic Kaapvaal Craton in South Africa, Eswatini, Lesotho and Botswana, the Zimbabwe Craton in Zimbabwe, Botswana and Mozambique, and the Grunehogna Craton in Western Dronning Maud Land of East Antarctica. Kalahari includes the Limpopo, Kheis and Magondi belts and the Rehoboth Block; Meso- to Neoproterozoic extensions include the Namaqua–Natal Belt and parts of the Maud Belt. The Gariep, Saldania and the Damara–Ghanzi–Chobe belts according to their configuration in the Jurassic (Corner and Durrheim 2018), are also shown. This includes Mesoproterozoic crust in the Falkland Islands and Haag Nunatak. Insert: the geographic position of these crustal fragments today. wDML is western Dronning Maud Land and cDML is central Dronning Maud Land. Numbered localities in Dronning Maud Land: 1, Ahlmanryggen; 2, Borgmassivet; 3, Kirwanveggen; 4, Heimefrontfjella; 5, H.U. Sverdrupfjella; 6, Gjelsvikfjella; and 7, Schirmacher Oasis.

The mostly sedimentary late Mesoproterozoic Umkondo Group of Zimbabwe and the Ritscherflya Supergroup of Grunehogna are near contemporaneous (Marschall *et al.* 2013; Fig. 1). The Umkondo Group was considered to have been deposited in an epicontinental basin by Button (1977). However, as documented for the Ritscherflya Supergroup by Marschall *et al.* (2013), a foreland basin setting is also possible. Both successions are intruded by *c.* 1112–06 Ma Umkondo LIP dykes and sills and contain Umkondo-aged volcanic rocks (Hanson *et al.* 2006; de Kock *et al.* 2014). Later Mesoproterozoic crust accreted onto Kalahari in the Maud Belt is variably overprinted by Ediacaran to Cambrian metamorphism and deformation (Jacobs *et al.* 2008), which increases to the east, and which includes coeval felsic magmatism in H.U. Sverdrupfjella, Gjelsvikfjella and Schirmacher Oasis.

Different generations of Precambrian dyke swarms and sill provinces cut Kalahari ranging from *c.* 2990 to *c.* 795 Ma (de Kock *et al.* 2019). Younger dykes intrude the Umkondo Group and associated Umkondo LIP sills within the Archean Zimbabwe Craton in Zimbabwe and Mozambique, and the Ritscherflya Supergroup and Borgmassivet sills belonging to the Umkondo LIP within Grunehogna in western Dronning Maud Land (Fig. 1). In eastern Zimbabwe, these dykes were called the Mutare Dyke Swarm by Wilson *et al.* (1987). The swarm has an arcuate shape, ranging from NW trending near the northern margin of the Zimbabwe Craton to NNW trending near the margin of the craton south of Mutare into Mozambique, a distance of over ~400 km. The larger dykes in the swarm are well spaced, up to ~28 m in width, and can be traced discontinuously along-strike for more than ~150 km. Approximately ~80 km to the west, this previously undated swarm merges with the sub-parallel Sebanga Dyke Swarm, which has a range of ages from *c.* 2512 to *c.* 2470 Ma and *c.* 2408 Ma (Söderlund *et al.* 2010). However, *c.* 1112–06 Ma Umkondo-aged dykes are also present in the region (de Kock *et al.* 2014). Further, the Mutare Dyke Swarm cuts the Sebanga Dyke Swarm in northern Zimbabwe and the *c.* 1886–72 Ma Mashonaland Sill Province (of the Mashonaland LIP; Stidolph 1977; Söderlund *et al.* 2010; Hanson *et al.* 2011*a*). The Mutare Dyke Swarm was originally considered part of the *c.* 1112–06 Ma Umkondo LIP (Wilson *et al.* 1985). However, based on a whole-rock Rb–Sr date and palaeomagnetism on two dykes NE of Harare (i.e. the Mahumi and Chenjera dykes; Stidolph 1977), Wilson *et al.* (1987) suggested that the age of the swarm was at least *c.* 500 Ma, but not more than *c.* 700 Ma. This interpretation agrees with Leitner and Phaup (1974), who stated that these dykes pre-date late Neoproterozoic to Cambrian metamorphism. Mukwakwami (2004) presented an unpublished U–Pb ID-TIMS baddeleyite weighted mean ^{206}Pb–^{238}U date of 724.0 ± 2.1 Ma on this swarm. Dykes inferred to belong to the Mutare Dyke Swarm were also studied by Ward *et al.* (2000), who assigned three different geochemical groups (M1, M2 and M3) to the swarm; however, as determined herein, only one of the geochemical groups (M1) probably belongs to the Mutare Dyke Swarm.

In the Grunehogna Craton (Fig. 1), NNE-trending dykes intrude into the Ritscherflya Supergroup. They also cut across the Umkondo LIP-aged Borgmassivet sills. The NNE-trending dykes were assigned an Umkondo LIP age in the Borgmassivet area. In contrast, further to the north in the Ahlmanryggen area, both Umkondo LIP-aged and Karoo-aged dykes, following the same trend, have been noted (Riley and Millar 2014). Before this study, these older NNE-trending dykes received only cursory study. In this work, they are termed the Fingeren Dyke Swarm. To the SE, south and east of the Grunehogna Craton, variably NE-trending metamorphosed dykes have been recorded in Heimefrontfjella (Bauer *et al.* 2003) and Kirwanveggan (Grantham *et al.* 1995). In contrast, metamorphosed NNE-trending dykes were recorded in H.U. Sverdrupfjella (Grosch *et al.* 2007). All of these dykes intrude the Mesoproterozoic crust and have subsequently been metamorphosed and deformed by Ediacaran to Cambrian overprinting within the Maud Belt. They are interpreted to have formed during either rifting (Grosch *et al.* 2007) or back-arc extension (Jacobs *et al.* 2020) at *c.* 586 Ma (Bauer *et al.* 2003).

Sampling

Zimbabwe

Sample MD15 (18°00′41″ S, 32°36′5″ E) was taken from a linear subvertical dolerite trending at ~145° that is ~20 m wide and intrusive into Archean quartz diorite and hornblende-bearing tonalite of the Zimbabwe Craton of Kalahari in the vicinity of Nyanga (Figs 2a, d & 3a). Dolerites in the region were generally classified as belonging to two types, based mostly on the palaeomagnetic studies of McElhinny and Opdyke (1964), according to Stocklmayer (1978). The two dolerite types are now included within the *c.* 1886–72 Ma Mashonaland LIP (e.g. Söderlund *et al.* 2010) and the *c.* 1112–06 Ma Umkondo LIP (e.g. de Kock *et al.* 2014), the existence of which was verified by Wingate (2001) in the dolerite sills of the area. However, it is likely that the Mutare Dyke Swarm is also in the area (Wilson *et al.* 1987), as was demonstrated from geochemical evidence by Ward *et al.* (2000).

Sample MD31 (19°56′41″ S, 32°54′19″ E) was taken from a linear subvertical dolerite trending at ~156° that is at least ~40 m wide and intrusive into the low-grade metamorphosed Mesoproterozoic Umkondo Group in eastern Zimbabwe on the eastern margin of the Zimbabwe Craton portion of Kalahari (Fig. 2b, e). This dolerite was interpreted to be older than *c.* 180 Ma Karoo LIP magmatism, according to Watson (1969). The sample was taken from what was originally interpreted as a feeder into the voluminous *c.* 1112–06 Ma Umkondo LIP (e.g. Watson 1969; de Kock *et al.* 2014). However, Mukwakwami (2004) has shown that this dyke belongs to the younger Mutare Dyke Swarm. The sampling site of Mukwakwami (2004) was re-sampled herein.

Dronning Maud Land

Two dyke samples were taken in western Dronning Maud Land (Borgmassivet area) of East Antarctica on the eastern margin of the Grunehogna Craton portion of Kalahari (Fig. 2c, f). Sample 04FN95 (72°38′24″ S, 3°45′18″ W) was taken from a linear subvertical dolerite trending ~011° that is ~80 m wide on the Fingeren Nunatak (Fig. 3b). The dolerite is intrusive into the low-grade metamorphosed Ritscherflya Supergroup (Veten Member, Högfonna

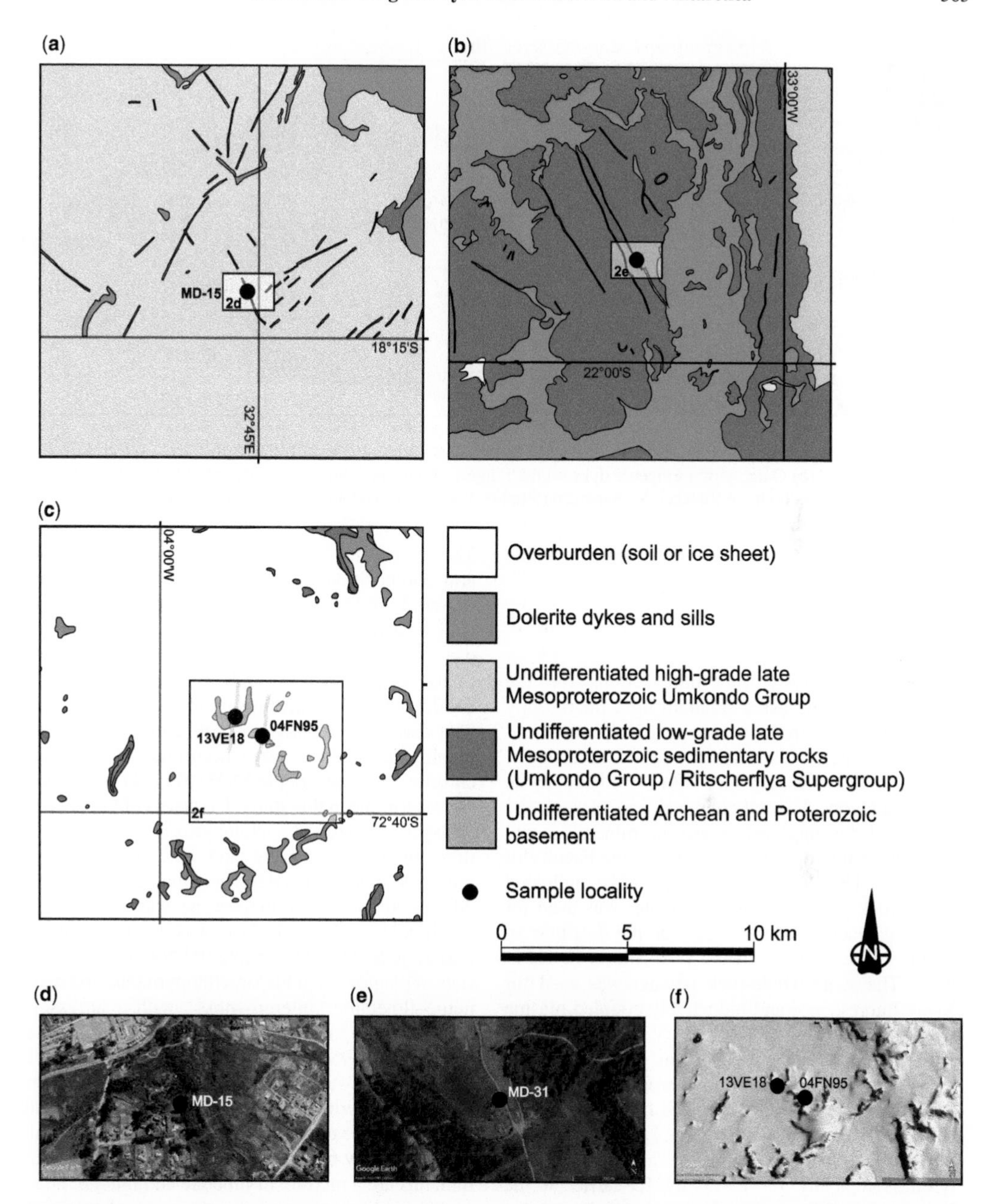

Fig. 2. Geological outline of a part of the Nyanga area (**a**) in eastern Zimbabwe after Stocklmayer (1978) from which sample MD15 was taken. (**b**) Geological outline of a part of the Chimanimani area in eastern Zimbabwe after Watson (1969) from which sample MD31 was taken. Geological outline of a part of the Borgmassivet area (**c**) in western Dronning Maud Land from Eastern Antarctica after Wolmarans and Kent (1982) from which sample 04FN95 and sample 13VE18 were taken. Satellite imagery from the area of (**d**) MD15, (**e**) MD31 and (**f**) 04FN95 and 13VE18.

Formation, Ahlmannryggen Group) and a Umkondo-aged Borgmassivet sill. These dykes were generally considered to be feeders of the Borgmassivet sills in this area, according to the mapping of Wolmarans and Kent (1982).

Sample 13VE18 (72°37′25″ S, 3°49′15″ W) was taken from a linear subvertical dolerite trending ~11° that is ~40 m wide on the Veten Nunatak and intrudes the same part of the Ritscherflya Supergroup (Veten Member, Högfonna Formation,

(a)

(b)

Fig. 3. (**a**) Outcrop of a Mutare dyke and surrounding granitic host rock in the river from Nyanga from which sample MD15 was taken. (**b**) Outcrop of a Fingeren dyke on the Fingeren Nunatak from which sample 04FN95 was taken. The dyke cuts across both the Ritscherflya Supergroup sedimentary rocks and an intrusive Borgmassivet Sill.

Ahlmannryggen Group) as the former sample (Fig. 2c and f; 04FN95).

Methodology

Petrographic analysis of the thin sections from the samples was done using standard light microscopy and scanning electron microscopy (SEM) techniques at the Institute of Earth Sciences, the University of Silesia in Katowice, together with SEM baddeleyite documentation. Mineral-chemical analyses of the main rock-forming and accessory minerals using an electron microprobe were done at the Faculty of Geology, University of Warsaw. The prepared whole-rock powder from the samples was used for X-ray diffraction spectrometry at the Institute of Earth Sciences, the University of Silesia in Katowice. The same whole-rock powder was used for X-ray fluorescence and inductively coupled plasma mass spectrometry at Bureau Veritas to obtain major-, minor- and trace-element geochemistry to assist with petrographic and petrological analysis. Baddeleyite grains were separated from the samples at the Department of Geology, Lund University. Subsequently, these baddeleyite grains were analysed isotopically for U–Pb by ID-TIMS at the Department of Earth Sciences of the University of Toronto and the Department of Geosciences, Swedish Museum of Natural History. Further details of the analytical methodologies can be found in the Supplementary Material (Methodology).

Results

Petrography

The four samples investigated have mineral assemblages typical for dolerite altered at varying degrees. The representative rock-forming minerals and mineral compositions are given in the Supplementary Material (Tables S1–S6). Dolerite sample MD15 (Fig. 4) consists of plagioclase feldspar, clinopyroxene and magnetite–ilmenite intergrowths with minor alkali feldspar, and accessory apatite and baddeleyite. The alteration is characterized by the near-complete replacement of clinopyroxene by amphibole and the common occurrence of secondary chlorite. Dolerite sample MD31 (Fig. 4) is the most altered sample in this study. It consists of plagioclase feldspar, amphibole, clinopyroxene, quartz, an ilmenite–titanite mixture and epidote. Accessory minerals include baddeleyite, apatite and pyrite. Almost all of the clinopyroxene was replaced by amphibole, and plagioclase was strongly altered into epidote. Dolerite sample 04FN95 (Fig. 4) consists of plagioclase feldspar, clinopyroxene and magnetite–ilmenite intergrowths with accessory baddeleyite, apatite, titanite, chalcocite and pyrite. Sample 04FN95 is relatively unaltered, but secondary minerals such as chlorite and epidote were detected. Dolerite sample 13VE18 (Fig. 4) consists of plagioclase, clinopyroxene and opaque minerals with accessory baddeleyite and apatite. The opaque minerals are a mixture of ilmenite and titanite, interpreted as the product of the alteration of titanium-rich magnetite–ilmenite intergrowths. The sample is relatively unaltered, but secondary minerals such as chlorite were detected.

Geochemistry

Analytical results from the four samples are presented in the Supplementary Material (Table S7). The variably metamorphosed dolerites in this study can be classified as sub-alkaline high-Fe tholeiitic

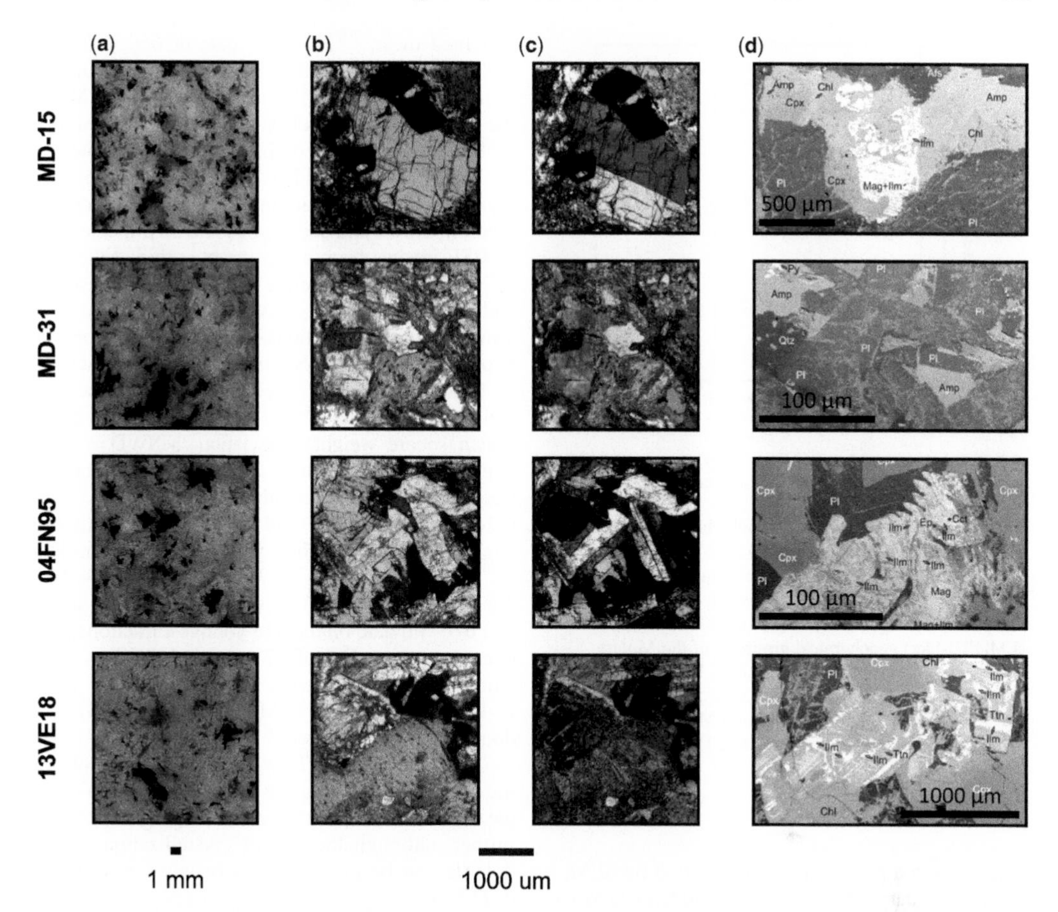

Fig. 4. Petrography of the studied samples in normal light (**a**), plane-polarized light (**b**), cross-polarized light (**c**) and back-scattered electron (**d**) imagery for MD15, MD31, 04FN95 and 13VE18. Mineral abbreviations: amp, amphibole; cct, chalcocite; chl, chlorite; cpx, clinopyroxene; czo, clinozoisite; ep, epidote; ilm, ilmenite; mag, magnetite; pl, plagioclase feldspar; py, pyrite; qtz, quartz; ttn, titanite.

basalts in major-, minor-, and trace-element classification schemes (Fig. 5), indicating limited element mobility owing to alteration. In the primitive mantle-normalized trace element diagram (Fig. 6a; Palme and O'Neil 2014), as well as the chondrite-normalized rare-earth element (REE) diagram (Fig. 6b; Palme *et al.* 2014), the sample patterns are relatively flat, despite being enriched relative to both primitive mantle and chondrite, respectively. The primitive mantle-normalized diagram shows a pronounced negative anomaly in Pb. Additionally, there is a weaker negative anomaly in Sr, indicative of fractionation in plagioclase feldspar, and a slight enrichment in large-ion lithophile elements over high-field-strength elements. In the REE diagram, no anomalies are observed, and only a moderate enrichment of light REE over heavy REE can be seen. The samples, therefore, show no apparent geochemical evidence of a significant subduction signature or crustal contamination. This lack of evidence is further highlighted in the crustal input and residual garnet proxy projections of Pearce *et al.* (2021) after Pearce (2008; Fig. 7a, b), with the samples lying along the mid-oceanic ridge basalt–ocean island basalt array of oceanic basalts close to EMORB (enriched mid-oceanic ridge basalt), with little evidence of crustal/subduction interaction and an intermediate depth of melting. Geochemically, this can be shown in the plume array according to the Pearce *et al.* (2021) two-proxy projection (Fig. 7c).

Geochronology

U–Pb ID-TIMS analytical results from the four dated samples are presented in the Supplementary Material (Tables S8–S9).

Sample MD15 yielded approximately 10 baddeleyite grains. These grains are mottled but clear, with a light brown colour, and are between 20 and 50 µm in length. In back-scattered electron (BSE)

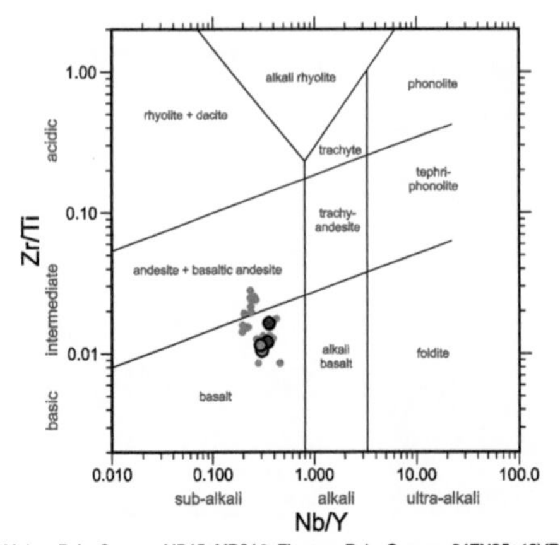

Fig. 5. Trace-element volcanic rock classification using Zr/Ti against Nb/Y after Pearce (1996) modified from Winchester and Floyd (1977). Dark blue denotes samples from the Mutare Dyke Swarm analysed in this study, whereas light blue denotes the M1 (EMORB-like) geochemical grouping of the Mutare Dyke Swarm from Ward *et al.* (2000). Red denotes samples from the Fingeren Dyke Swarm analysed in this study; green denotes the M2 and M3 (arc-like) geochemical groupings of the Mutare Dyke Swarm from Ward *et al.* (2000).

images from a petrographic thin section by SEM, a slight secondary alteration of the baddeleyite to zircon is visible (Fig. 8a). The baddeleyite was arranged in two fractions for dating, composed of three grains in each fraction. Analyses of U–Pb isotopes in the two fractions of baddeleyite by ID-TIMS produced slightly reversely concordant data. A preliminary weighted mean $^{207}Pb/^{206}Pb$ date of 660 ± 33 Ma at the 2σ uncertainty level is not considered further. A preliminary $^{206}Pb/^{238}U$ date of 715 ± 22 Ma at the 2σ uncertainty level (Fig. 9a) is interpreted as a preliminary age of crystallization of the dolerite.

Sample MD31 yielded approximately 60 baddeleyite grains. These grains are variably mottled, medium brown and between 20 and 60 µm in length. In BSE images from a petrographic thin section by SEM (Fig. 8b), moderate secondary alteration of baddeleyite to zircon is visible. Baddeleyite was arranged into six fractions, each composing one or two grains. U–Pb analyses in three of the six fractions of baddeleyite produced concordant results with a weighted mean $^{206}Pb/^{238}U$ date of 712.0 ± 1.5 Ma (mean square weighted deviation, MSWD = 0.27) at the 2σ uncertainty level (Fig. 9b). One discordant fraction produced a slightly older $^{206}Pb/^{238}U$ date (724.1 ± 3.1 Ma; 2σ), and there were two outlying, slightly older and slightly younger concordant data ($^{206}Pb/^{238}U$ dates of 717.1 ± 1.9 and 704.2 ± 2.6 Ma; both at 2σ). Using a weighted mean $^{207}Pb/^{206}Pb$ date on the five youngest fractions produces an older date of 723 ± 19 (MSWD = 0.5), which is within the error of the $^{206}Pb/^{238}U$ date. U–Pb results reported in Mukwakwami (2004) produced a weighted mean $^{206}Pb/^{238}U$ date of 724.0 ± 2.1 Ma (2σ; MSWD = 0.73) from a sample in the same locality along the same dyke. Such a complex isotopic composition is challenging to interpret further, although the age of crystallization of the dolerite can be interpreted as between *c.* 724 and *c.* 712 Ma.

Sample 04FN95 yielded over 80 baddeleyite grains, which are clear, medium brown and between 20 and 80 µm in length. In BSE images from a petrographic thin section by scanning electron microscopy (SEM), no apparent secondary alteration of

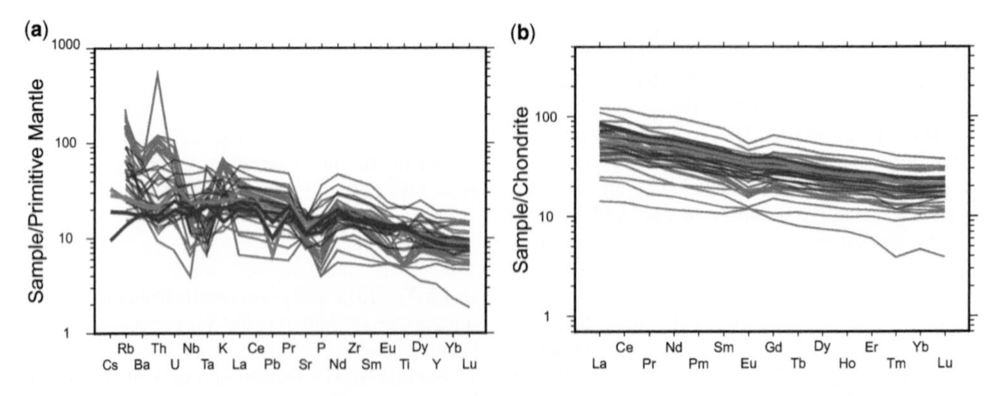

Fig. 6. (**a**) Trace-element primitive mantle normalization diagram using the values of Palme and O'Neil (2014). (**b**) Rare earth element chondrite normalization diagram using the values of Palme *et al.* (2014). Colour coding follows that of Figure 5.

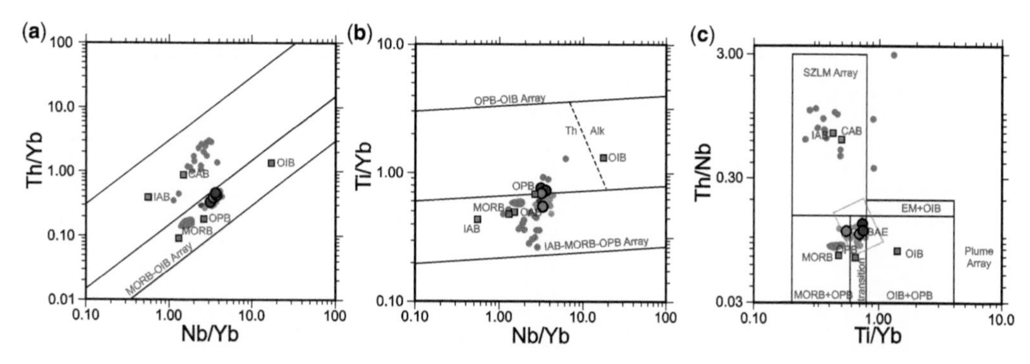

Fig. 7. (**a**) Crustal input proxy projection after Pearce *et al.* (2021), modified from Pearce (2008). (**b**) Residual garnet proxy after Pearce *et al.* (2021), modified from Pearce (2008). (**c**) Two proxy (crustal input and residual garnet) projections after Pearce *et al.* (2021). Colour coding follows that of Figure 5. Abbreviations: Alk, alkali; BAE, back-arc basalts; CAB, continental-arc basalts; EM, enriched mantle; IAB, island-arc basalts; MORB, mid-ocean ridge basalts; OIB, ocean-island basalts; OPB, ocean-plateau basalts; SZLM, subduction-modified lithospheric mantle; and Th, tholeiitic. Green shading denotes the field of dykes from Heimefrontfjella.

the baddeleyite to zircon is visible (Fig. 8c). U–Pb analysis of four baddeleyite fractions, comprising one or two grains in each fraction, produced concordant overlapping results that give a weighted mean $^{206}Pb/^{238}U$ date of 716.6 ± 1.2 Ma (MSWD = 1.43) at the 2σ uncertainty level (Fig. 9c). A distinctly younger concordant result with a $^{206}Pb/^{238}U$ date of 693.7 ± 2.0 Ma (2σ) probably underwent early Pb loss and was therefore excluded from the mean date. A weighted mean $^{207}Pb/^{206}Pb$ date of 717 ± 12 Ma (MSWD = 0.7) on all five fractions is consistent with this interpretation. Therefore, the $^{206}Pb/^{238}U$ date is interpreted as the dolerite's crystallization age.

Sample 13VE18 yielded approximately 40 baddeleyite grains. These grains are pristine and clear, with a dark brown colour, and are between 30 and 100 µm in length. In BSE images from a petrographic thin section by SEM, no secondary alteration of the baddeleyite to zircon was visible (Fig. 8d). Baddeleyite was arranged into four fractions for age dating, composed of one to three grains in each fraction. Analyses of the U–Pb isotopes in all four fractions of baddeleyite by ID-TIMS produced a weighted mean $^{206}Pb/^{238}U$ date of 718.9 ± 5.8 Ma (MSWD = 0.52) at the 2σ uncertainty level (Fig. 9d). This date is consistent with the $^{207}Pb/^{206}Pb$ weighted mean date of 714 ± 13 Ma (MSWD = 0.5). Therefore, the $^{206}Pb/^{238}U$ weighted mean date is interpreted as the dolerite's crystallization age.

Discussion

The Mutare Dyke Swarm

The Mutare Dyke Swarm in eastern Zimbabwe was described previously by Wilson *et al.* (1987), and its age was estimated based on preliminary Rb–Sr whole-rock geochronology, together with palaeomagnetic studies. Here the first U–Pb crystallization ages of the dyke swarm are presented, with ages between 712 and 724 Ma (MD31) and 715 ± 22 Ma (MD15) using U–Pb on baddeleyite by ID-TIMS (Fig. 9). These ages agree with the unpublished 724.0 ± 2.1 U–Pb ID-TIMS age on baddeleyite by Mukwakwami (2004). Two K–Ar whole-rock ages of 715 ± 70 and 757 ± 20 Ma presented by Vail and Dodson (1969) on dolerites from the same region are probably related to the same magmatic event. The two dated samples of the present study are mafic dykes (dolerite) that have experienced deuteric alteration, as recorded by Vail (1966), together with low greenschist facies metamorphism for MD31. This alteration is documented by the partial replacement of pyroxenes and plagioclase feldspar by amphibole, epidote, chlorite and sericite. Probably, Ediacaran to Cambrian deformation and metamorphism on the eastern margin of Kalahari from the Mozambique Belt affected sample MD31.

Ward *et al.* (2000) considered that the dykes of the Mutare Dyke Swarm were composed of three geochemical groupings (i.e. M1, M2 and M3), with modelling failing to indicate any genetic relationship between them. In de Kock *et al.* (2014), the M1 grouping was linked to the *c.* 724 Ma magmatic event based on the unpublished age of Mukwakwami (2004). This study confirms this hypothesis. The two dated samples have high-Fe tholeiitic gabbroic/basaltic EMORB-like compositions, along with other members of the swarm (M1 geochemical group in Ward *et al.* 2000). Therefore, of the three compositions identified by Ward *et al.* (2000), only M1 is considered to belong to the Mutare Dyke Swarm dated herein to 724–712 Ma. Regarding the

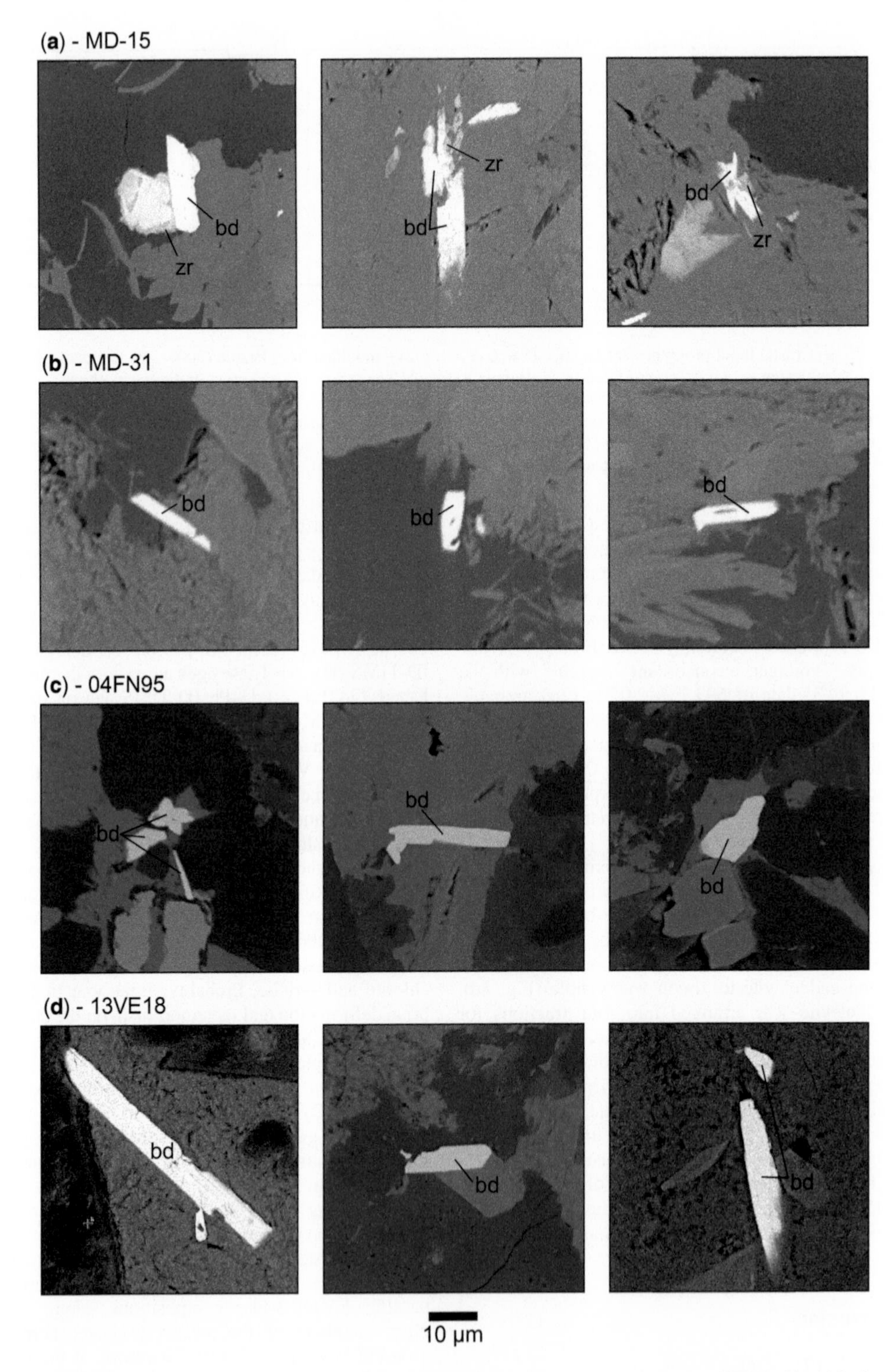

Fig. 8. Representative back-scattered electron (BSE) images by scanning electron microscopy (SEM) of baddeleyite and zircon in samples MD15, MD31, 04FN95 and 13VE18. bd denotes baddeleyite whereas zr denotes duller grey zircon overgrowth on brighter grey baddeleyite.

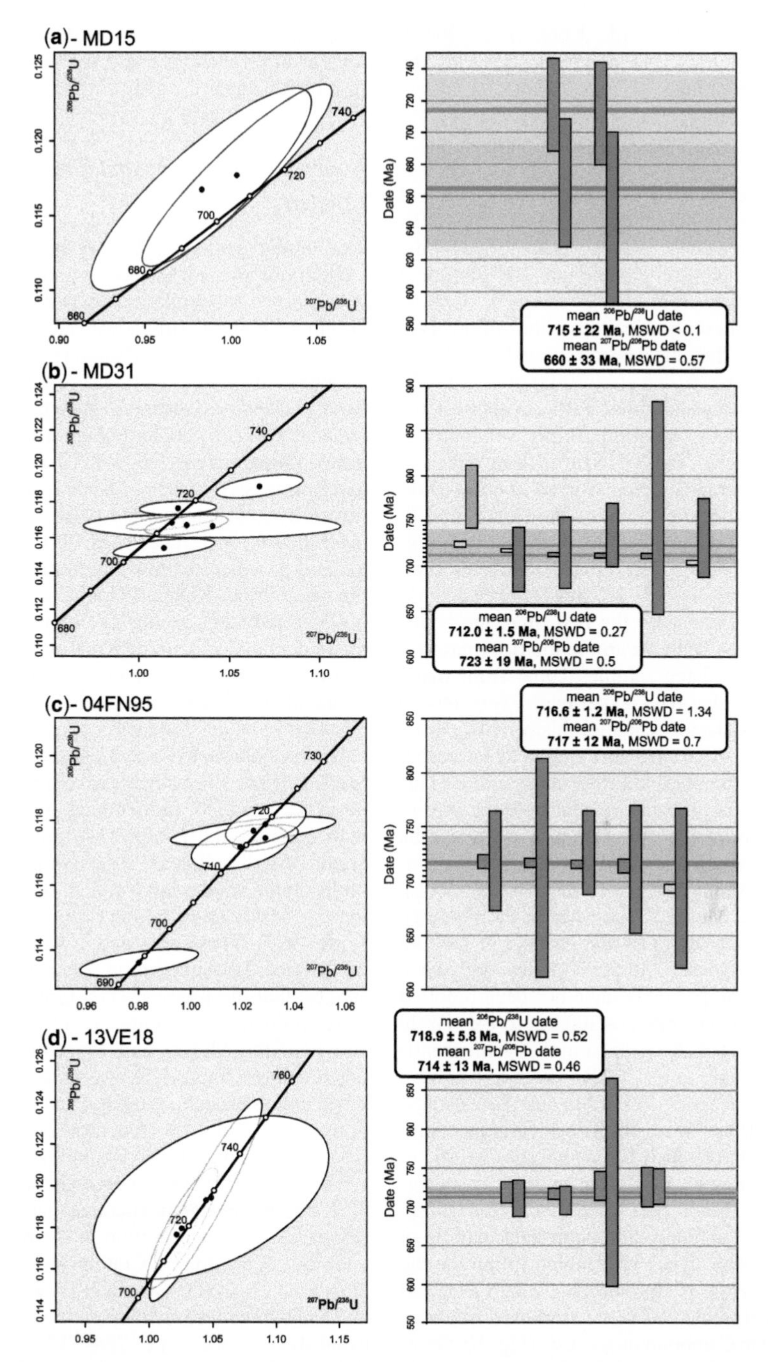

Fig. 9. U–Pb ID-TIMS baddeleyite data presented as Wetherill concordia (left) and (right) weighted mean $^{206}Pb/^{238}U$ dates (blue) and weighted mean $^{207}Pb/^{206}Pb$ dates (green) from samples MD-15 (**a**), MD-31 (**b**), 04FN95 (**c**) and 13VE18 (**d**).

other two compositional groups identified by Ward *et al.* (2000), one (M2) probably belongs to the Sebanga Dyke Swarm (2512–2408 Ma; Söderlund *et al.* 2010), which merges with the Mutare Dyke Swarm to the west. The other (M3) is probably Umkondo age, based on a dated dyke at Devuli

Ranch at 1110 ± 19 Ma (de Kock *et al.* 2014), which has proximity and a similar trend to the Mutare Dyke Swarm.

The Fingeren Dyke Swarm

In East Antarctica, on the Grunehogna Craton of Kalahari in western Dronning Maud Land, the NNE-trending Fingeren Dyke Swarm was identified for the first time in this study. Although NNE-trending dykes are known from the Grunehogna Craton, they have been described mainly in the Ahlmanryggen area, where they have been reported as either *c.* 190–178 Ma dykes (Riley *et al.* 2005) or as *c.* 1112–06 Ma dykes (Riley and Millar 2014). In contrast, the dykes in this study are from further south in the Borgmassivet area. These NNE-trending dykes in the Borgmassivet have been assigned as either Bogmassivet sill feeders belonging to the Umkondo LIP or Jurassic-aged magmatism, as in the Ahlmanryggen area. The new U–Pb ID-TIMS baddeleyite crystallization ages of 716.6 ± 1.2 and 718.9 ± 5.8 Ma are coeval and define, for the first time, the NNE-trending Fingeren Dyke Swarm in the Borgmassivet area. These two dated samples come from mafic dykes (dolerite) that have experienced very low to low greenschist facies metamorphism, with partial replacement of pyroxenes and plagioclase feldspar by amphibole, epidote, chlorite and sericite. This metamorphism is probably related to Ediacaran to Cambrian deformation and metamorphism on the southern and eastern margin of the Grunehogna Craton from the Maud Belt (e.g. Jacobs *et al.* 2003). Like the coeval Mutare Dyke Swarm, the Fingeren Dyke Swarm appears to be the product of melting in the asthenosphere. Although dykes with such EMORB-like compositions have not been reported on the Grunehogna Craton, ~300 km to the SW in the Maud Belt, such dykes exist in the Heimefrontfjella (Fig. 1; Bauer *et al.* 2003). These dykes have been assigned an age of *c.* 586 Ma based on a single zircon grain (Bauer *et al.* 2003), which is inconclusive. Dykes have also been mapped ~100 km south of the Grunehogna Craton in the Kirwanveggan area of the Maud Belt (Grantham *et al.* 1995), but they are NE-trending, although they may be of similar age. These dykes were noted to intrude the late Meso- to early Neoproterozoic Maud Belt and are heavily deformed and metamorphosed in places by Ediacaran to Cambrian orogenesis (Fig. 10; Grantham *et al.* 1995; Bauer *et al.* 2003). Approximately ~200 km east of the Grunehogna Craton, further dykes of similar geochemistry were noted in the Maud Belt from H.U. Sverdrupfjella (Fig. 10; Grosch *et al.* 2007) and were assigned an age of *c.* 800 Ma. These dykes, however, have been metamorphosed at amphibolite facies. Further U–Pb geochronology and geochemical studies are required to determine the full extent of the Fingeren Dyke Swarm in Grunehogna and the surrounding Maud Belt of Kalahari.

Kaapvaal, Zimbabwe and Grunehogna in Kalahari

The similar geological history between the Kaapvaal, Zimbabwe and Grunehogna cratons in Kalahari is confirmed by similarities between the Umkondo Group in eastern Zimbabwe and the Ritscherflya Supergroup in Western Dronning Maud Land (e.g. Groenewald *et al.* 1991; Marschall *et al.* 2013). Both successions preserve lava piles that remain undated by U–Pb geochronology. However, the suggested correlation to the *c.* 1112–06 Ma Umkondo and Borgmassivet sills (de Kock *et al.* 2014) is confirmed by geochemical and palaeomagnetic studies (McElhinny 1966; Jones *et al.* 2003). Dyke swarms and sill provinces of this age intrude across almost the whole Proto-Kalahari Craton and may represent multiple radiating centres or one radiating and one circumferential swarm (de Kock *et al.* 2014; Buchan and Ernst 2019). Further magmatism related to the Umkondo LIP is also preserved along the northwestern margin of the Kalahari Craton in the Damara–Ghanzi–Chobe Belt. Geological similarities between the Kaapvaal, Zimbabwe and Grunehogna cratons have been argued further back into the Archean with ages on the *c.* 3067 Ma Annandagstoppane granite (Marschall *et al.* 2010) compared with batholiths on the southeastern side of the Kaapvaal Craton (i.e. Makhutswi granite, *c.* 3063 Ma; Poujol and Robb 1999). These ages and provenance studies of the Ritscherflya Supergroup (Marschall *et al.* 2013) suggest that the Kaapvaal and Grunehogna cratons were originally contiguous. In such a Kalahari reconstruction, the Mutare and Fingeren dyke swarms define a single linear dyke swarm (Fig. 1). This combined dyke swarm on the African and Antarctic portions of Kalahari shows that these cratons could only have existed adjacent to one another in a 'close fit' scenario (Fig. 11; Basson *et al.* 2004).

Using the palaeomagnetic signature comparison between the African portion of the Kalahari Craton and the Grunehogna Craton at *c.* 1112–06 Ma (Jones *et al.* 2003; Swanson-Hysell *et al.* 2015) and an alignment based on piercing points of the relevant dyke swarms at *c.* 724–712 Ma would place the Grunehogna Craton outboard of the Kaapvaal and Zimbabwe cratons by several hundred kilometres. Therefore, much more of the Grunehogna Craton must be preserved in the ocean down to the Explora Escarpment in the Weddell Sea, which would then match the Lebombo Escarpment along the South Africa–Mozambique Border (Corner and Durrheim 2018). Such a reconstruction would

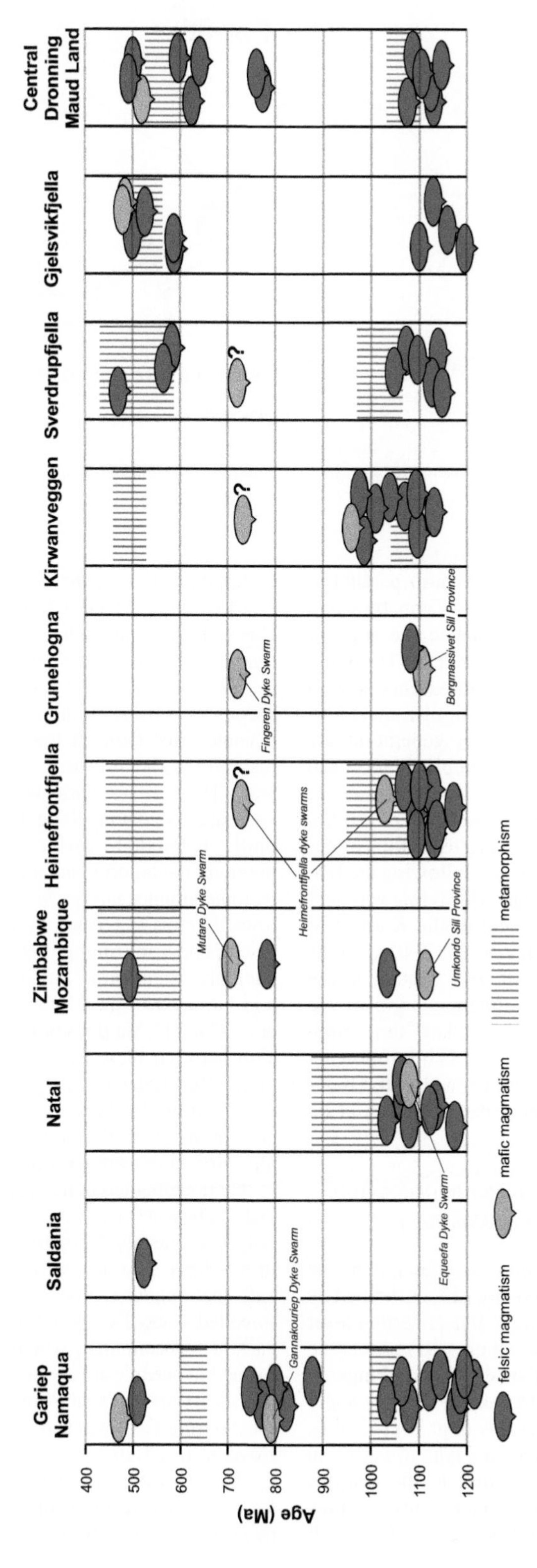

Fig. 10. The Neoproterozoic magmatic and metamorphic evolution of the various marginal terranes of the Kalahari Craton, with pink denoting felsic plutons and grey denoting mafic plutons (as well as dykes and sills). The pink pattern denotes metamorphic and deformational events.

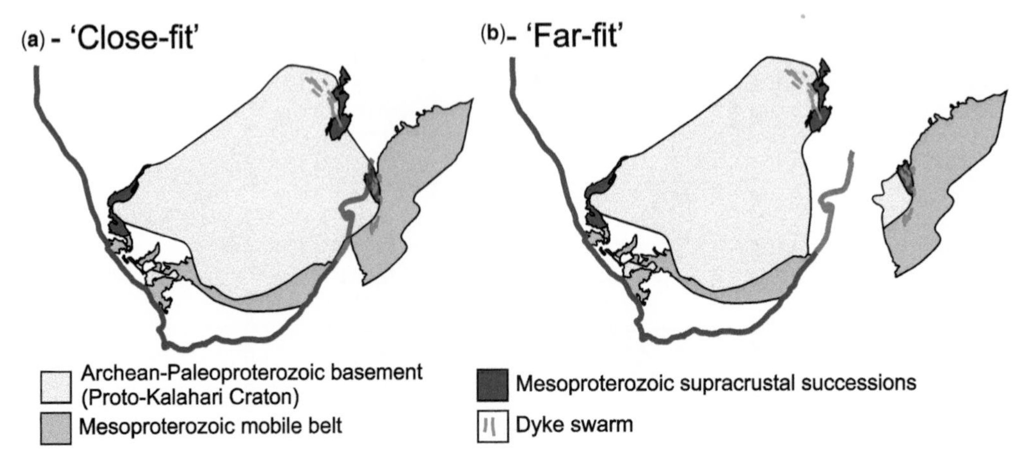

Fig. 11. The 'close-fit' and 'far-fit' scenarios for the Kalahari Craton in (**a**) and (**b**), respectively after Basson *et al.* (2004).

make much of the Mutare Dyke Swarm parallel to the edge of the Kalahari Craton in Zimbabwe and the Fingeren Dyke Swarm parallel to the edge of the Grunehogna Craton in Western Dronning Maud Land. However, as noted above (and shown in Figs 1 & 10), geochemically similar metamorphosed dykes, probably of the same generation, are present within the H.U. Sverderupfjella (Grosch *et al.* 2007), in Kirwanveggen (Grantham *et al.* 1995) and in Heimefrontfjella (Bauer *et al.* 2003), and would run parallel to the margins of the Grunehogna Craton to the east and south following the fabric of the Maud Belt of Kalahari. This fabric was later utilized during the emplacement of the Karoo LIP and subsequent rifting (Riley *et al.* 2005). At its northern terminus, dykes of the Mutare Dyke Swarm terminate near the northern margin of the Zimbabwe Craton. Here, the dykes turn from NNW- to NW-trending and are probably following pre-existing structural anisotropies within the basement of the Zimbabwe Craton itself (Wilson *et al.* 1987).

Possible relationships to tectonic processes along the eastern margin of Kalahari

The combined Mutare–Fingeren Dyke Swarm on the Kalahari Craton is interpreted as a LIP, as defined by Ernst (2014). Definitions vary, but an estimate of >0.1 Mkm^2 and a time interval of <10 myr can be diagnostic, and the combined Mutare–Fingeren Dyke Swarm meets these criteria. LIPs are typically assigned to mantle plume events (e.g. Ernst 2014). However, the Mutare–Fingeren Dyke Swarm, in large parts, occurs relatively close to the margins of the Zimbabwe and Grunehogna cratons and may even intrude into the Mesoproterozoic Maud Belt of Kalahari. East of the Grunehogna Craton in central Dronning Maud Land, near the Schirmacher Oasis, East Antarctica, remnants of a continental magmatic arc have yielded U–Pb zircon ages of *c.* 785–770 Ma, as shown by Jacobs *et al.* (2020). Jacobs *et al.* (2020) inferred that subduction of old oceanic crust beneath the Kalahari Craton margin resulting in slab roll-back leading to back-arc extension. There is no evidence, however, for possible back-arc extension along this part of the margin until *c.* 650–600 Ma, which involved ultra-high temperature metamorphism and the intrusion of syntectonic granites interpreted to record asthenospheric upwelling, as suggested by Baba *et al.* (2010). The discrepancy in timing between arc magmatism/back-arc extension in the Schirmacher Oasis region and intrusion of the Mutare–Fingeren Dyke Swarm at *c.* 724–712 Ma does not support a direct relationship between these events (Fig. 10). Further north along the Kalahari margin, Mesoproterozoic arc rocks within the Mozambique Belt outboard of the eastern margin of the Zimbabwe Craton show generally strong Ediacaran–Cambrian overprinting similar to that documented in the Maud Belt (e.g. Fritz *et al.* 2013; Chaúque *et al.* 2019; Thomas *et al.* 2022). The only documented Neoproterozoic arc-related rocks in this large area are *c.* 820–700 Ma meta-volcanic and meta-plutonic rocks present within the far-travelled Cabo Delgado Nappe Complex in north-eastern Mozambique, which is interpreted to have formed in oceanic arcs or as part of small continental blocks juxtaposed against older Mesoproterozoic arc crust during Gondwana assembly (Viola *et al.* 2008; Boyd *et al.* 2010).

The available data thus provide no clear evidence for a causal relationship between the emplacement of the Mutare–Fingeren Dyke Swarm and tectonomagmatic events along the entire eastern Kalahari margin. We conclude that the dyke swarm results from the decompression melting of upwelling asthenosphere mantle, possibly a mantle plume, unrelated to plate tectonic processes along the eastern Kalahari margin.

The palaeogeography of the Kalahari Craton within the supercontinent Rodinia

Although the Kalahari Craton is usually placed adjacent to the southwestern margin of Laurentia (Dalziel *et al.* 2000; Li *et al.* 2008; Loewy *et al.* 2011; Swanson-Hysell *et al.* 2015), the LIP ages presented in this study on the combined Mutare–Fingeren Dyke Swarm on Kalahari could argue for spatial association with the Franklin and Irkutsk LIPs in a reconstructed Laurentia–Siberia (Ernst *et al.* 2016). However, the geochemical signatures are different, with *c.* 720–719 Ma mafic magmatism in Canada and similary aged magmatism in Siberia being more compositionally enriched relative to coeval mafic magmatism in northeastern Zimbabwe and western Dronning Maud Land, which is more primitive and EMORB-like. This may argue against a direct connection. Palaeomagnetic studies, unfortunately, remain limited from Kalahari. The palaeogeographic model of Dalziel *et al.* (2000) allows for the Grenvillian-aged Llano Uplift in Laurentia to be matched with the Namaqua–Natal Mobile Belt in Kalahari at some point after the emplacement of the unrelated *c.* 1112–06 Ma Umkondo and *c.* 1094–84 Ma southwestern Laurentia LIPs (Bright *et al.* 2014; de Kock *et al.* 2014; Swanson-Hysell *et al.* 2015) in Kalahari and Laurentia, respectively (Fig. 12). The southwestern Laurentia LIP is probably part of the greater Midcontinent Rift/Keweenawan LIP, dated to *c.* 1106–1084 Ma (Bright *et al.* 2014; Swanson-Hysell *et al.* 2019). At that time, the continental blocks of Kalahari and Laurentia were interpreted to be approximately 3000 km apart using palaeomagnetic constraints, with Kalahari probably part of a mega-continent named 'Umkondia' (Swanson-Hysell *et al.* 2015; Choudhary *et al.* 2019; Wang *et al.* 2021). By the Ediacaran–Cambrian boundary, Kalahari is interpreted to have already broken away from Laurentia and have become amalgamated with the Congo and Rio de la Plata cratons in Gondwana along the various Pan African mobile belts (Merdith *et al.* 2017). Against the Llano Uplift in Laurentia, southern Kalahari is proposed to have been an indenter during the later phases of the Grenvillian Orogeny at *c.* 1.0 Ga (Dalziel *et al.* 2000; Fig. 12). In this model, the Namaqua–Natal Belt in Kalahari would align with the Grenville Belt in Laurentia, which is supported by Tonian palaeomagnetic constraints from Kalahari (Swanson-Hysell *et al.* 2015). However, this would place the *c.* 724–712 Ma Mutare–Fingeren Dyke Swarm on the opposite side of Kalahari from Laurentia. These palaeomagnetic constraints would make the *c.* 795 Ma Gannakouriep Dyke Swarm (Rioux *et al.* 2010) a more likely magmatic barcode target and piercing point between Kalahari and Laurentia. However, there is no evidence for the presence of dykes in southern Laurentia (present-day coordinates) equivalent in age to the Gannakouriep Dyke Swarm. Palaeomagnetic studies also appear to show that this model would be untenable for the Gannakouriep Dyke Swarm to intercept southeastern Laurentia (Fig. 12) (Bartholomew 2008). Likewise, the trend of the combined Mutare–Fingeren Dyke Swarm itself would then appear to point to a more eastward neighbour of the Kalahari Craton in Rodinia, such as proto-Australia (Li *et al.* 2008). Alternatively, a piercing point could exist between the Proterozoic Western Basement Terrane of Tasmania and Kalahari, where similarly aged units have also been identified in the Tayatea Dyke Swarm and 733 ± 9 Ma Cooee Sill Province (McGregor *et al.* 2016; Mulder *et al.* 2018*a*), if the palaeogeographic reconstruction of Mulder *et al.* (2018*a*, *b*) is valid. This would mean that both the Western Basement Terrane of Tasmania and Coats Land (Loewy *et al.* 2011) would have been present during Rodinia break-up between proto-Australia, East Antarctica, Laurentia and Kalahari (Fig. 12).

Evidence of *c.* 706 Ma mafic to felsic intraplate volcanism that may relate to rifting along the southern Laurentia margin was presented by Hanson *et al.* (2016) in the Marathon Uplift in Texas. Indeed, further to the east in southeastern Laurentia, drift-related intraplate magmatism occurred sporadically between *c.* 758 and *c.* 702 Ma (Aleinikoff *et al.* 1995; Tollo and Aleinikoff 1996; Tollo *et al.* 2004). This magmatism, however, shows little similarity to southwestern Kalahari, as magmatism there probably began much earlier, at *c.* 890 Ma, and lasted until *c.* 741 Ma (Hanson *et al.* 2011*b*; Frimmel 2018).

Another set of models place Kalahari to the SE of Laurentia and linked to it by a hypothetical ribbon continent, the MARA block, which is now represented by a series of Mesoproterozoic basement inliers in western South America, including the Arequipa Massif in Peru and other inliers further south. In this interpretation, Kalahari first rifted off from the MARA block to form the Clymene Ocean, with the MARA block separating later from Laurentia to form the Iapetus Ocean (e.g. Casquet *et al.* 2012; Rapela *et al.* 2016; Hodgin *et al.* 2021). Detrital zircon age data support aspects of this model (Casquet *et al.* 2018; Hodgin *et al.* 2021), but a fairly complex kinematic scenario is required to account for the disposition of different parts of the MARA block in South America relative

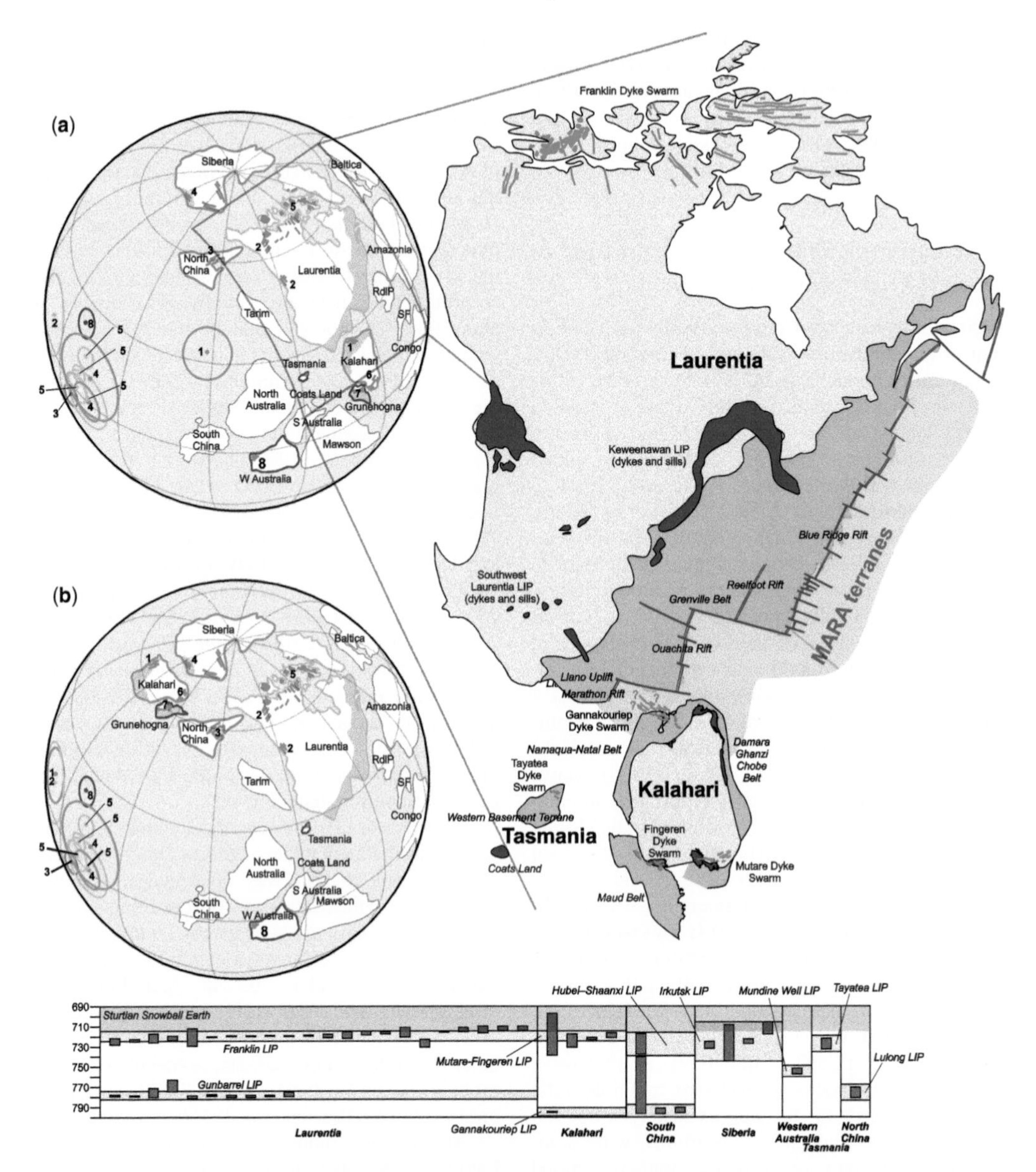

Fig. 12. Palaeogeography of Rodinia after Li *et al.* (2008), showing the two scenarios of 'Kalahari-in-the-south' (**a**) and 'Kalahari-in-the-north' (**b**), together with an enlarged view of Kalahari, Laurentia, Tasmania, Coats Land and Rio de la Plata in the south model. Inset includes the magmatic barcodes for the various cratonic terranes used in the reconstructions. Areas in yellow denote late Meso- to early Neoproterozoic mobile belts associated with the Grenville orogeny. Abbreviations: 1, Gannakouriep Dyke Swarm; 2, Gunbarrel Dyke Swarm; 3, Lulong Dyke Swarm; 4, Yenesei Ridge Dyke Swarm; 5, Franklin Dyke Swarm; 6, Mutare Dyke Swarm; 7, Fingeren Dyke Swarm; 8, Mundine Wells Dyke Swarm.

to the Kalahari Craton within a Gondwana configuration (e.g. Rapela *et al.* 2016).

An alternative interpretation is that the *c.* 724–712 Ma Mutare–Fingeren Dyke Swarm is a distal part of the 721–712 Ma Franklin LIP of Laurentia (and adjacent Irkutsk LIP of Siberia; Heaman *et al.* 1992; Denyszyn *et al.* 2009; Ernst *et al.* 2016). This would allow the Kalahari Craton to be placed near northwestern Laurentia. As argued, it is in this region that Kalahari could form a LIP match with Laurentia following the break-up of the proposed Umkondia mega-continent at *c.* 1112–06 Ma

(Choudhary *et al.* 2019; Wang *et al.* 2021). LIP-related magmatism at *c.* 724–712 Ma between Laurentia and Siberia, and Kalahari could form around a triple-point off northwestern Laurentia. It is also encouraging to see an overlap between the *c.* 795 Ma Gannakouriep and *c.* 780–769 Ma Gunbarrel (of northeastern Laurentia; Mackinder *et al.* 2019) palaeomagnetic poles, producing a possible reconstruction with Kalahari as a more distant neighbour located near northwestern Laurentia, together with North China and Siberia (Fig. 12). However, such a pole may be discrepant by up to ~15 myr. This issue needs to be resolved, with many geodynamic implications.

Implications

The *c.* 724–712 Ma Mutare–Fingeren LIP is coeval with the Franklin LIP. The Franklin LIP, together with coeval LIPs in Siberia (the Irkutsk LIP; Ernst *et al.* 2016) as well as the Hubei–Shaanxi LIP in South China (Lu *et al.* 2022), is widely regarded as responsible for the Sturtian Snowball Earth through weathering of the flood basalt provinces on an equatorial Rodinia (Li *et al.* 2008; Cox *et al.* 2016; Pu *et al.* 2022). This LIP-scale magmatism led to a transient amount of carbon dioxide in the environment, before carbon drawdown from weathering of mafic material led to the sequestration of greenhouse gases and a cooling Earth that resulted in global glaciation at 717–718 Ma – the so-called 'fire and ice' hypothesis (Goddéris *et al.* 2003; Pu *et al.* 2022). The new Mutare–Fingeren LIP adds another magmatic node to this widespread LIP-scale magmatism across Rodinia as it was breaking apart, and therefore probably further contributed to the onset of global glaciation. Any tectonic model to account for all these LIPs will need to invoke either a mantle superplume or mantle plumes impinging on the lithosphere at nearly exactly the same time in widely removed localities. Such a model could use the LLSVPs as proposed by Torsvik (2019), such as TUZO and JASON, in which the edges reflect a plume-generation zone. The Kalahari could exist on the edges of JASON on either the southern or northwestern margins of Laurentia in the Neoproterozoic (present day coordinates).

Another approach is a single plume rising underneath the supercontinent and sliding upward along the asthenosphere–lithosphere boundary to multiple thin spots (two in this scenario; Bright *et al.* 2014), and producing separate LIP events at these widely separated thin spots (i.e. Franklin–Irkutsk and Mutare–Fingeren). Nevertheless, after *c.* 712 Ma, geological similarities end, providing evidence for the rifting and break-up of Kalahari from Laurentia, and Kalahari's migration into Gondwana together with the Congo and Sao Francisco cratons in proto-Africa and South America, supported by palaeomagnetic evidence according to Merdith *et al.* (2017). This would need to occur before the Ediacaran to Cambrian, as the Mozambique and Maud belts would be stitched together as proto-Africa and East Antarctica within Gondwana, with the metamorphosed equivalents of the Umkondo Group thrust up and on to the Kalahari Craton in Zimbabwe. Rifting along the eastern margin of the craton was therefore probably related to rifting between Kalahari and Laurentia and/or Siberia (or various linking terranes), or between Kalahari and Tasmania. During rifting from a Kalahari against southern Laurentia, the Coats Land Block would have remained with Kalahari as part of a rifted fragment from Laurentia, while Tasmania rifted together with proto-Australia.

Conclusion

In this study, the Mutare Dyke Swarm of the Zimbabwe Craton portion of the younger and larger Kalahari Craton is dated at 715 ± 22 and 724–712 Ma by U–Pb ID-TIMS on baddeleyite on two different dykes in eastern Zimbabwe. These dykes show deuteric alteration, have been metamorphosed at low greenschist facies, have EMORB chemistry, probably of asthenosphere affinity, and can be correlated geochemically with the M1 dykes of the Mutare Dyke SWARM recognized by Ward *et al.* (2000). The Mutare Dyke Swarm is coeval with the Fingeren Dyke Swarm on the rifted Grunehogna Craton portion of the Kalahari Craton in Western Dronning Maud Land of Eastern Antarctica, which is, for the first time, identified herein, with ages of 716.6 ± 1.2 and 718.9 ± 5.8 Ma, also by U–Pb ID-TIMS dating on baddeleyite. These Fingeren dykes have been metamorphosed at very low greenschist facies, and also have EMORB chemistry. Thus, they are inferred to be part of the same dyke swarm as the Mutare Dyke Swarm, and can be distinguished from *c.* 1112–06 Ma and *c.* 178 Ma dykes in the region using geochemistry. These dykes run parallel to the craton margin in both regions and are aligned in a close-fit reconstruction. Undated dykes of similar geochemistry and potentially similar age also intrude into the Maud Belt, but were subsequently metamorphosed in the Ediacaran–Cambrian up to amphibolite facies. This may suggest a much larger extent of the dyke swarm, particular in western Dronning Maud Land. This combined Mutare–Fingeren Dyke Swarm was probably associated with rifting, and not back-arc extension, and input of asthenospheric mantle/mantle plume source, with little crustal contamination. Although similar-aged magmatic events occur in northern Laurentia and Siberia in the Franklin LIP and related magmatic units, a direct connection between Kalahari and this region is difficult to reconcile in terms of whole-plate palaeogeography, geochemistry and geology at this stage. However, the traditionally more accepted Kalahari model, with Kalahari against SW Laurentia, also presents significant obstacles, although some can possibly be overcome with the addition of linking terranes existing between Kalahari and southern Laurentia. Without direct evidence of the *c.* 795 Ma Gannakouriep Dyke Swarm in southern Laurentia or on any linking terranes such as Arequipa, combined with a feasible overlapping palaeomagnetic signature, this remains mostly hypothetical However, the Mutare–Fingeren LIP does present another node of coeval LIP-scale magmatism across Rodinia at the Tonian–Cryogenian boundary, which is likely responsible for the onset of global glaciation, which may have been driven by an LLSVP or superplume beneath Rodinia as it broke up.

Acknowledgments The authors wish to thank the editor Valby van Schijndel, as well as Eben Hodgin and an anonymous reviewer, who helped to improve the manuscript. This article is a contribution to International Geoscience Programme 648: Supercontinents and Global Geodynamics. This is publication number 93 of the LIPs-Industry Consortium Project (www.supercontinent.org).

Competing interests The authors declare that they have no known competing financial interests or personal relationships that could have appeared to influence the work reported in this paper.

Author contributions **APG**: conceptualization (lead), data curation (lead), formal analysis (equal), funding acquisition (equal), investigation (lead), methodology (lead), project administration (lead), resources (equal), supervision (lead), validation (lead), visualization (lead), writing – original draft (lead), writing – review & editing (equal); **MDK**: conceptualization (supporting), data curation (supporting), formal analysis (supporting), investigation (supporting), methodology (supporting), validation (supporting), visualization (supporting), writing – review & editing (supporting); **RE**: conceptualization (supporting), funding acquisition (equal), writing – review & editing (supporting); **AG**: data curation (supporting), formal analysis (equal), investigation (supporting), methodology (supporting), validation (supporting), visualization (supporting), writing – review & editing (supporting); **RH**: conceptualization (supporting), methodology (supporting), writing – review & editing (supporting); **SK**: data curation (supporting), formal analysis (equal), investigation (supporting), resources (equal), validation (supporting), writing – review & editing (supporting); **MK**: conceptualization (equal), investigation (equal), methodology (equal), resources (supporting), supervision (supporting), validation (supporting), writing – review & editing (supporting); **ML**: conceptualization (supporting), funding acquisition (supporting), investigation (supporting), methodology (supporting), project administration (supporting), writing – review & editing (supporting); **BL**: investigation (supporting), project administration (supporting), resources (supporting), writing – review & editing (supporting); **AM**: conceptualization (supporting), investigation (supporting), methodology (supporting), project administration (supporting), resources (supporting), supervision (supporting), writing – review & editing (supporting); **US**: data curation (supporting), formal analysis (supporting), investigation (supporting), resources (supporting), validation (supporting), writing – review & editing (supporting).

Funding APG acknowledges financial support through a grant from the National Science Centre (Naradowe Centrum Nauki; NCN), Poland (SONATINA 3 grant no. UMO-2019/32/C/ST10/00238). MdK acknowledges support from the DST-NRF Centre of Excellence for Integrated Mineral and Energy Resource Analysis, South Africa. MdK also acknowledges National Research Foundation, South Africa, incentive funding. Funding for logistical support in East Antarctica was provided by the National Research Foundation of South Africa via the South African National Antarctica Programme to Geoff Grantham (grant nos 80267 and 80915), which are gratefully acknowledged.

Data availability All data generated or analysed during this study are included in this published article (and its supplementary information files) and at Mendeley Data (DOI: 10.17632/6wpm8dshgn.1).

References

Aleinikoff, J.N., Zartman, R.E., Walters, M., Rankin, D.W., Lyttle, P.T. and Burton, W.C. 1995. U–Pb ages of metarhyolites of the Catoctin and Mount Rogers formations, central and southern Appalachians: evidence for two pulses of Iapetan rifting. *American Journal of Science*, **295**, 428–454, https://doi.org/10.2475/ajs.295.4.428

Baba, S., Hokada, T., Kaiden, H., Dunkley, D.J., Owada, M. and Shiraishi, K. 2010. SHRIMP zircon U–Pb dating of sapphirine-bearing granulite and biotite-hornblende gneiss in the Schirmacher Hills, East Antarctica: implications for Neoproterozoic ultrahigh-temperature metamorphism predating the assembly of Gondwana. *Journal of Geology*, **118**, 621–639, https://doi.org/10.1086/656384

Bartholomew, L.T. 2008. *Paleomagnetism of Neoproterozoic intraplate ignrous rocks in the southwest Kalahari Craton, Namibia and South Africa*. MSc thesis, Texas Christian University.

Basson, I.J., Perritt, S., Watkeys, M.K. and Menzies, A.H. 2004. Geochemical correlation between metasediments of the Mfongosi Group of the natal sector of the Namaqua–Natal metamorphic province, South Africa and the Ahlmannryggen Group of the Grunehogna Province, Antarctica. *Gondwana Research*, **7**, 57–73, https://doi.org/10.1016/S1342-937X(05)70306-3

Bauer, W., Fielitz, W., Jacobs, J., Fanning, C.M. and Spaeth, G. 2003. Mafic dykes from Heimefrontfjella and implications for the post-Grenvillian to pre-Pan-African geological evolution of western Dronning Maud Land, Antarctica. *Antarctic Science*, **15**, 371–391, https://doi.org/10.1017/S095410200300 1391

Bleeker, W. and Ernst, R. 2006. Short-lived mantle generated magmatic events and their dyke swarms: the key unlocking Earth's palaeogeographic record back to 2.6 Ga. *In*: Hanski, E., Mertanen, S., Rämö, T. and Vuollo, J. (eds) *Dyke Swarms – Time Markers of Crustal Evolution*. Taylor & Francis, 3–26, https://doi.org/10.1201/NOE0415398992

Boyd, R., Nordgulen, Ø. *et al.* 2010. The geology and geochemistry of the East African Orogen in northeastern Mozambique. *South African Journal of Geology*, **113**, 87–129, https://doi.org/10.2113/gssajg.113.1.87

Bright, R.M., Amato, J.M., Denyszyn, S.W. and Ernst, R.E. 2014. U–Pb geochronology of 1.1 Ga diabase in the southwestern United States: testing models for the origin of a post-Grenville large igneous province. *Lithosphere*, **6**, 135–156, https://doi.org/10.1130/L335.1

Buchan, K.L. and Ernst, R.E. 2019. Giant circumferential dyke swarms: catalogue and characteristics. *In*: Srivastava, R.K., Ernst, R.E. and Peng, P. (eds) *Dyke Swarms of the World: a Modern Perspective*. Springer, 1–44, https://doi.org/10.1007/978-981-13-1666-1_1

Button, A. 1977. *Stratigraphic History of the Middle Proterozoic Umkondo Basin in the Chipinga area, Southeastern Rhodesia*. Economic Geology Research Unit – University of the Witwatersrand, Johannesburg, Information Circulars, **108**.

Casquet, C., Rapela, C.W. *et al.* 2012. A history of Proterozoic terranes in South America: from Rodinia to Gondwana. *Geoscience Frontiers*, **3**, 137–145, https://doi.org/10.1016/j.gsf.2011.11.004

Casquet, C., Dahlquist, J.A. *et al.* 2018. Review of the Cambrian Pampean orogeny of Argentina; a displaced orogen formerly attached to the Saldania Belt of South Africa? *Earth-Science Reviews*, **177**, 209–225, https://doi.org/10.1016/j.earscirev.2017.11.013

Chaúque, F.R., Cordani, U.G. and Jamal, D.L. 2019. Geochronological systematics for the Chimoio–Macossa frontal nappe in central Mozambique: implications for the tectonic evolution of the southern part of the Mozambique belt. *Precambrian Research*, **150**, 47–67, https://doi.org/10.1016/j.jafrearsci.2018.10.013

Choudhary, B.R., Ernst, R.E. *et al.* 2019. Geochemical characterization of a reconstructed 1110 Ma large igneous province. *Precambrian Research*, **332**, 105382, https://doi.org/10.1016/j.precamres.2019.105382

Corner, B. and Durrheim, R.J. 2018. An integrated geophysical and geological interpretation of the Southern African lithosphere. *In*: Siegesmund, S., Basei, M.A.S., Oyhantçabal, P. and Oriolo, S. (eds) *Geology of Southwest Gondwana*. Springer, 19–61, https://doi.org/10.1007/978-3-319-68920-3_2

Cox, G.M., Halverson, G.P. *et al.* 2016. Continental flood basalt weathering as a trigger for Neoproterozoic Snowball Earth. *Earth and Planetary Science Letters*, **446**, 89–99, https://doi.org/10.1016/j.epsl.2016.04.016

Dalziel, I.W.D., Mosher, S., and Gahagan, L.M. 2000. Laurentia–Kalahari collision and the assembly of Rodinia. *Journal of Geology*, **108**, 499–513, https://doi.org/10.1086/314418

de Kock, M.O., Ernst, R. *et al.* 2014. Dykes of the 1.11 Ga Umkondo LIP, Southern Africa: clues to a complex plumbing system. *Precambrian Research*, **249**, 129–143, https://doi.org/10.1016/j.precamres.2014.05.006

de Kock, M.O., Gumsley, A.P., Klausen, M.B., Söderlund, U. and Djeutchou, C. 2019. The Precambrian mafic magmatic record, including large igneous provinces of the Kalahari Craton and its constituents: a paleogeographic review. *In*: Srivastava, R., Ernst, R. and Peng, P. (eds) *Dyke Swarms of the World: a Modern Perspective*. Springer, 155–214, https://doi.org/10.1007/978-981-13-1666-1_5

Denyszyn, S.W., Halls, H.C., Davis, D.W. and Evans, D.A.D. 2009. Paleomagnetism and U–Pb geochronology of Franklin dykes in High Arctic Canada and Greenland: a revised age and paleomagnetic pole constraining block rotations in the Nares Strait region. *Canadian Journal of Earth Sciences*, **46**, 689–706, https://doi.org/10.1139/E09-042

Ernst, R.E. 2014. *Large Igneous Provinces*. Cambridge University Press, Cambridge.

Ernst, R.E., Hamilton, M.A. *et al.* 2016. Long-lived connection between southern Siberia and northern Laurentia in the Proterozoic. *Nature Geoscience*, **9**, 464–469, https://doi.org/10.1038/ngeo2700

Evans, D.A.D. 2009. The palaeomagnetically viable, long-lived and all-inclusive Rodinia supercontinent reconstruction. *Geological Society, London, Special Publications*, **327**, 371–404, https://doi.org/10.1144/SP327.1

Frimmel, H.E. 2018. The Gariep Belt. *In*: Siegesmund, S., Basei, M.A.S., Oyhantçabal, P. and Oriolo, S. (eds) *Geology of Southwest Gondwana*. Springer, Berlin, 353–386, https://doi.org/10.1007/978-3-319-68920-3_13

Fritz, H., Abdelsalam, M. *et al.* 2013. Orogen styles in the East African Orogen: a review of the Neoproterozoic to Cambrian tectonic evolution. *Journal of African Earth Sciences*, **86**, 65–106, https://doi.org/10.1016/j.jafrearsci.2013.06.004

Godderis, Y., Donnadieu, Y. *et al.* 2003. The Sturtian 'snowball' glaciation: fire and ice. *Earth and Planetary Science Letters*, **211**, 1–12, https://doi.org/10.1016/S0012-821X(03)00197-3

Grantham, G.H., Jackson, C., Moyes, A.B., Groenewald, P.B., Harris, P.D., Ferrar, G. and Krynauw, J.R. 1995. The tectonothermal evolution of the Kirwanveggen-H.U. Sverdrupfjella areas, Dronning Maud Land, Antarctica. *Precambrian Research*, **75**, 209–229, https://doi.org/10.1016/0301-9268(95)80007-5

Groenewald, P.B., Grantham, G.H. and Watkeys, M.K. 1991. Geological evidence for a Proterozoic to Mesozoic link between southeastern Africa and Dronning Maud Land, Antarctica. *Journal of the Geological Society*, **148**, 1115–1123, https://doi.org/10.1144/gsjgs.148.6.1115

Grosch, E.G., Bisnath, A., Frimmel, H.E. and Board, W.S. 2007. Geochemistry and tectonic setting of mafic rocks in western Dronning Maud Land, East Antarctica: implications for the geodynamic evolution of the Proterozoic Maud Belt. *Journal of the Geological Society*, **164**, 465–475, https://doi.org/10.1144/0016-76492005-152

Hanson, R.E., Harmer, R.E. *et al.* 2006. Mesoproterozoic intraplate magmatism in the Kalahari Craton: a review.

Journal of African Earth Sciences, **46**, 141–167, https://doi.org/10.1016/j.jafrearsci.2006.01.016

Hanson, R.E., Rioux, M., Gose, W.A., Blackburn, T.J., Bowring, S.A., Mukwakwami, J. and Jones, D.L. 2011*a*. Paleomagnetic and geochronological evidence for large-scale post-1.88 Ga displacement between the Zimbabwe and Kaapvaal cratons along the Limpopo belt. *Geology*, **39**, 487–490, https://doi.org/10.1130/G31698.1

Hanson, R.E., Rioux, M. *et al.* 2011*b*. Constraints on Neoproterozoic intraplate magmatism in the Kalahari craton: geochronology and paleomagnetism of ~890–795 Ma extension-related igneous rocks in SW Namibia and adjacent parts of South Africa. *Geological Society of America Annual Meeting*, 9–12 October, Minneapolis, MN.

Hanson, R.E., Roberts, J.M., Dickerson, P.W. and Fanning, C.M. 2016. Cryogenian intraplate magmatism along the buried southern Laurentian margin: evidence from volcanic clasts in Ordovician strata, Marathon uplift, west Texas. *Geology*, **44**, 539–542, https://doi.org/10.1130/G37889.1

Heaman, L.M., LeCheminant, A.N. and Rainbird, R.H. 1992. Nature and timing of Franklin igneous events, Canada: implications for a Late Proterozoic mantle plume and the break-up of Laurentia. *Earth and Planetary Science Letters*, **109**, 117–131, https://doi.org/10.1016/0012-821X(92)90078-A

Hodgin, E.B., Gutiérrez-Marco, J.C., Colmenar, J., Macdonald, F.A., Carlotto, V., Crowley, J.L. and Newmann, J.R. 2021. Cannibalization of a late Cambrian backarc in southern Peru: new insights into the assembly of southwestern Gondwana. *Gondwana Research*, **92**, 202–227, https://doi.org/10.1016/j.gr.2021.01.004

Jacobs, J., Fanning, C.M. and Bauer, W. 2003. Timing of Grenville-age v. Pan-African medium- to high grade metamorphism in western Dronning Maud Land (East Antarctica) and significance for correlations in Rodinia and Gondwana. *Precambrian Research*, **125**, 1–20, https://doi.org/10.1016/S0301-9268(03)00048-2

Jacobs, J., Pisarevsky, S., Thomas, R.J. and Becker, T. 2008. The Kalahari Craton during the assembly and dispersal of Rodinia. *Precambrian Research*, **160**, 142–158, https://doi.org/10.1016/j.precamres.2007.04.022

Jacobs, J., Mikhalsky, E. *et al.* 2020. Neoproterozoic geodynamic evolution of easternmost Kalahari: constraints from U–Pb–Hf–O zircon, Sm–Nd isotope and geochemical data from the Schirmacher Oasis, East Antarctica. *Precambrian Research*, **342**, 105553, https://doi.org/10.1016/j.precamres.2019.105553

Jones, D.L., Bates, M.P., Li, X.-L., Corner, B. and Hodgkinson, G. 2003. Palaeomagnetic results from the ca. 1130 Ma Borgmassivet intrusions in the Ahlmannryggen region of Dronning Maud Land, Antarctica, and tectonic implications. *Tectonophysics*, **375**, 247–260, https://doi.org/10.1016/S0040-1951(03)00341-X

Leitner, E.G. and Phaup, A.E. 1974. *The Geology of the Country Around Mount Darwin*. Geological Survey of Rhodesia, Salisbury, Bulletins, **73**.

Li, Z.-X., Bogdanova, S.V. *et al.* 2008. Assembly, configuration, and break-up history of Rodinia: a synthesis. *Precambrian Research*, **160**, 179–210, https://doi.org/10.1016/j.precamres.2007.04.021

Loewy, S.L., Dalziel, I.W.D., Pisarevsky, S., Connelly, J.N., Tait, J., Hanson, R.E. and Bullen, D. 2011. Coats land crustal block, East Antarctica: a tectonic tracer for Laurentia? *Geology*, **39**, 859–862, https://doi.org/10.1130/G32029.1

Lu, K., Mitchell, R.N., Yang, C., Zhou, J.-L., Wu, L.-G., Wang, X.-C. and Li, X.-H. 2022. Widespread magmatic provinces at the onset of the Sturtian Snowball Earth. *Earth and Planetary Science Letters*, **594**, 117736, https://doi.org/10.1016/j.epsl.2022.117736

Mackinder, A., Cousens, B.L., Ernst, R.E. and Chamberlain, K.R. 2019. Geochemical, isotopic, and U–Pb zircon study of the central and southern portions of the 780 Ma Gunbarrel Large Igneous Province in western Laurentia. *Canandian Journal of Earth Sciences*, **56**, 738–755, https://doi.org/10.1139/cjes-2018-0083

Marschall, H.R., Hawkesworth, C.J., Storey, C.D., Dhuime, B., Leat, P.T., Meyer, H.-P. and Tamm-Buckle, S. 2010. The Annandagstoppane Granite, East Antarctica: evidence for Archaean intracrustal recycling in the Kaapvaal–Grunehogna Craton from zircon O and Hf isotopes. *Journal of Petrology*, **51**, 2277–2301, https://doi.org/10.1093/petrology/egq057

Marschall, H.R., Hawkesworth, C.J. and Leat, P.T. 2013. Mesoproterozoic subduction under the eastern edge of the Kalahari–Grunehogna Craton preceding Rodinia assembly: the Ritscherflya detrital zircon record, Ahlmannryggen (Dronning Maud Land, Antarctica). *Precambrian Research*, **236**, 31–45, https://doi.org/10.1016/j.precamres.2013.07.006

McElhinny, M.W. 1966. The Palaeomagnetism of the Umkondo Lavas, eastern Southern Rhodesia. *Geophysical Journal of the Royal Astronomical Society*, **10**, 375–381, https://doi.org/10.1111/j.1365-246X.1966.tb03065.x

McElhinny, M.W. and Opdyke, N.D. 1964. The Paleomagnetism of the Precambrian dolerites of eastern Southern Rhodesia, an example of geologic correlation by rock magnetism. *Journal of Geophysical Research*, **69**, 2465–2475, https://doi.org/10.1029/JZ069i012p02465

McGregor, C., Denyszyn, S., Halverson, G., Everard, J., Cumming, G. and Calver, C. 2016. ID-TIMS U–Pb geochronology of the Tayatea Dyke Swarm of Australia. *Paper 38, presented at the Seventh International Dyke Conference*, 18–20 August, Beijing.

Merdith, A.S., Collins, A.S. *et al.* 2017. A full-plate global reconstruction of the Neoproterozoic. *Gondwana Research*, **50**, 84–134, https://doi.org/10.1016/j.gr.2017.04.001

Mukwakwami, J. 2004. *Geological Structure of the Umkondo Group in eastern Zimbabwe and geochronology of associated mafic rocks and possible correlatives in Zimbabwe*. MSc thesis, University of Zimbabwe.

Mulder, J.A., Berry, R.F., Halpin, J.A., Meffre, S. and Everard, J.L. 2018*a*. Depositional age and correlation of the Oonah Formation: refining the timing of Neoproterozoic basin formation in Tasmania. *Australian Journal of Earth Sciences*, **65**, 391–407, https://doi.org/10.1080/08120099.2018.1426629

Mulder, J.A., Karlstrom, K.E., Halpin, J.A., Merdith, A.S., Spencer, C.J., Berry, R.F. and McDonald, B. 2018*b*. Rodinian devil in disguise: correlation of 1.25–1.10 Ga strata between Tasmania and Grand Canyon.

Geology, **46**, 991–994, https://doi.org/10.1130/G45225.1

Palme, H. and O'Neil, H.S.C. 2014. Cosmochemical estimates of mantle composition. *In*: Holland, H.D. and Turekian, K.K. (eds) *Treatise on Geochemistry*. Elsevier, Amsterdam, 1–39, https://doi.org/10.1016/B978-0-08-095975-7.00201-1

Palme, H., Lodders, K. and Jones, A. 2014. Solar system abundances of the elements. *In*: Holland, H.D. and Turekian, K.K. (eds) *Treatise on Geochemistry*. Elsevier, Amsterdam, 15–36, https://doi.org/10.1016/B978-0-08-095975-7.00118-2

Pearce, J.A. 1996. A user's guide to basalt discrimination diagrams. *In*: Wyman, D.A. (ed.) *Trace Element Geochemistry of Volcanic Rocks: Applications for Massive-Sulphide Exploration*. Geological Association of Canada, Ottawa, 79–113.

Pearce, J.A. 2008. Geochemical fingerprinting of oceanic basalts with applications to ophiolite classification and the search for Archean oceanic crust. *Lithos*, **100**, 14–48, https://doi.org/10.1016/j.lithos.2007.06.016

Pearce, J.A., Ernst, R.E., Peate, D.W. and Rogers, C. 2021. LIP printing: use of immobile element proxies to characterize large igneous provinces in the geologic record. *Lithos*, **392–393**, 106068, https://doi.org/10.1016/j.lithos.2021.106068

Pisarevsky, S.A., Wingate, M.T.D., Powell, C., Johnson, S. and Evans, D.A.D. 2003. Models of Rodinia assembly and fragmentation. *Geological Society, London, Special Publications*, **206**, 35–55, https://doi.org/10.1144/GSL.SP.2003.206.01.04

Poujol, M. and Robb, L.J. 1999. New U–Pb zircon ages on gneisses and pegmatite from south of the the Murchison greenstone belt, South Africa. *South African Journal of Geology*, **102**, 93–97.

Pu, J.P., Macdonald, F.A. *et al.* 2022. Emplacement of the Franklin large igneous province and initiation of the Sturtian Snowball Earth. *Science Advances*, **8**, eadc9430, https://doi.org/10.1126/sciadv.adc9430

Rapela, C.W., Verdecchia, S.O. *et al.* 2016. Identifying Laurentian and SW Gondwana sources in the Neoproterozoic to Early Paleozoic metasedimentary rocks of the Sierras Pampeanas: paleogeographic and tectonic implications. *Gondwana Research*, **32**, 193–212, https://doi.org/10.1016/j.gr.2015.02.010

Riley, T.R. and Millar, I.L. 2014. Geochemistry of the 1100 Ma intrusive rocks from the Ahlmannryggen region, Dronning Maud Land, Antarctica. *Antarctic Science*, **26**, 389–399, https://doi.org/10.1017/S0954102013000916

Riley, T.R., Leat, P.T., Curtis, M.L., Millar, I.L., Duncan, R.A. and Fazel, A. 2005. Early-Middle Jurassic Dolerite Dykes from Western Dronning Maud Land (Antarctica): identifying mantle sources in the Karoo Large Igneous Province. *Journal of Petrology*, **46**, 1489–1524, https://doi.org/10.1093/petrology/egi023

Rioux, M., Bowring, S., Dudas, F. and Hanson, R. 2010. Characterizing the U–Pb systematics of baddeleyite through chemical abrasion: application of multi-step digestion methods to baddeleyite geochronology. *Contibutions to Mineralogy and Petrology*, **160**, 777–801, https://doi.org/10.1007/s00410-010-0507-1

Söderlund, U., Hofmann, A., Klausen, M.B., Olsson, J.R., Ernst, R.E. and Persson, P.-O. 2010. Towards a complete magmatic barcode for the Zimbabwe craton: Baddeleyite U–Pb dating of regional dolerite dyke swarms and sill complexes. *Precambrian Research*, **183**, 388–398, https://doi.org/10.1016/j.precamres.2009.11.001

Stidolph, P.A. 1977. *The Geology of the Country around Shamva*. Geological Survey of Rhodesia, Salisbury, Bulletins, **78**.

Stocklmayer, V.R. 1978. *The Geology of the Country around Inyanga*. Geological Survey of Rhodesia, Salisbury, Bulletins, **79**.

Swanson-Hysell, N.L. 2021. The Precambrian paleogeography of Laurentia. *In*: Pesonen, L.J., Salminen, J., Elming, S.-A., Evans, D.A.D. and Veikkolainen, T. (eds) *Ancient Supercontinents and the Paleogeography of Earth*. Elsevier, Amsterdam, 109–153, https://doi.org/10.1016/B978-0-12-818533-9.00009-6

Swanson-Hysell, N.L., Kilian, T.M. and Hanson, R.E. 2015. A new grand mean palaeomagnetic pole for the 1.11 Ga Umkondo large igneous province with implications for palaeogeography and the geomagnetic field. *Geophysical Journal International*, **203**, 2237–2247, https://doi.org/10.1093/gji/ggv402

Swanson-Hysell, N.L., Ramezani, J., Fairchild, L.M. and Rose, I.R. 2019. Failed rifting and fast drifting: midcontinent rift development, Laurentia's rapid motion and the driver of Grenvillian orogenesis. *Geological Society of America Bulletin*, **131**, 913–940, https://doi.org/10.1130/B31944.1

Thomas, R.J., Fullgraf, T. *et al.* 2022. The Mesoproterozoic Nampula Subdomain in southern Malawi: completing the story from Mozambique. *Journal of African Earth Sciences*, **196**, 104677, https://doi.org/10.1016/j.jafrearsci.2022.104667

Tollo, R.P. and Aleinikoff, J.N. 1996. Petrology and U–Pb geochronology of the Robertson River Igneous Suite, Blue Ridge Province, Virginia – evidence for multistage magmatismassociated with an early episode of Laurentian rifting. *American Journal of Science*, **296**, 1045–1090, https://doi.org/10.2475/ajs.296.9.1045

Tollo, R.P., Aleinikoff, J.N., Bartholomew, M.J. and Rankin, D.W. 2004. Neoproterozoic A-type granitoids of the central and southern Appalachians: intraplate magmatism associated with episodic rifting of the Rodinian supercontinent. *Precambrian Research*, **128**, 3–38, https://doi.org/10.1016/j.precamres.2003.08.007

Torsvik, T.H. 2019. Earth history: a journey in time and space from base to top. *Tectonophysics*, **760**, 297–313, https://doi.org/10.1016/j.tecto.2018.09.009

Vail, J.R. 1966. Zones of progressive regional metamorphism across the western margin of the Mozambique Belt in Rhodesia and Mozambique. *Geological Magazine*, **103**, 231–239, https://doi.org/10.1017/S0016756800052808

Vail, J.R. and Dodson, M.H. 1969. Geochronology of Rhodesia. *Transactions of the Geological Society of South Africa*, **72**, 79–113.

Viola, G., Henderson, I.H.C., Bingen, B., Thomas, R.J., Smethurst, M.A. and de Azavedo, S. 2008. Growth and collapse of a deeply eroded orogen: insights from structural, geophysical, and geochronological constraints on the Pan-African evolution of NE Mozambique. *Tectonics*, **27**, TC5009, https://doi.org/10.1029/2008TC002284

Wang, C., Mitchell, R.N., Murphy, J.B., Peng, P. and Spencer, C.J. 2021. The role of megacontinents in the supercontinent cycle. *Geology*, **49**, 402–406, https://doi.org/10.1130/G47988.1
Ward, S.E., Hall, R.P. and Hughes, D.J. 2000. Guruve and Mutare dykes: preliminary geochemical indication of complex Mesoproterozoic mafic magmatic systems in Zimbabwe. *Journal of African Earth Sciences*, **30**, 689–701, https://doi.org/10.1016/S0899-5362(00)00046-4
Watson, R.L.A. 1969. *The Geology of the Cashel, Melsetter and Chipinga Areas*. Geological Survey of Rhodesia, Salisbury, Bulletins, **60**.
Wilson, J.F., Jones, D.L. and Kramers, J.D. 1985. Mafic dyke swarms in Zimbabwe. *Abstracts of the International Conference on Mafic Dyke Swarms*, University of Toronto, Canada. 188.
Wilson, J.F., Jones, D.L. and Kramers, J.D. 1987. Mafic dyke swarms of Zimbabwe. *In*: Halls, H.C. and Fahring, A.F. (eds) *Mafic Dyke Swarms*. Geological Association of Canada, Ottawa, 433–444.
Winchester, J.A. and Floyd, P.A. 1977. Geochemical discrimination of different magma series and their differentiation products using immobile elements. *Chemical Geology*, **20**, 325–343, https://doi.org/10.1016/0009-2541(77)90057-2
Wingate, M.T.D. 2001. SHRIMP baddeleyite and zircon ages for an Umkondo dolerite sill, Nyanga Mountains, Eastern Zimbabwe. *South African Journal of Geology*, **104**, 13–22, https://doi.org/10.2113/104.1.13
Wolmarans, L.G. and Kent, L.E. 1982. Geological investigations in Western Dronning Maud Land, Antarctica – a synthesis. *South African Journal of Antarctic Research*, **2**, 3–93.

Evolution of the neoarchean Kola alkaline granites, northeastern Fennoscandian Shield: insights from SHRIMP-II titanite and zircon U–Pb isotope and rare earth elements data

Andrei A. Arzamastsev[1], Boris V. Belyatsky[2]*, Nickolay V. Rodionov[2], Anton V. Antonov[2], Elena N. Lepekhina[2] and Sergei A. Sergeev[2]

[1]Institute of Precambrian Geology and Geochronology RAS, St Petersburg 199034, Russia

[2]Centre of Isotopic Research, A. P. Karpinsky Russian Geological Research Institute, St Petersburg 199106, Russia

AAA, 0000-0003-4100-9821; BVB, 0000-0002-4022-9366; NVR, 0000-0001-5201-1922
*Correspondence: bbelyatsky@mail.ru

Abstract: An U–Pb isotopic investigation combined with rare earth element data for titanites, zircons and coexisting accessories has been undertaken to gain insight into the formation of the Archean peralkaline granites of the northeastern Fennoscandian Shield and to test the stability of titanite during metamorphic and hydrothermal processes. The obtained set of isotope data shows that whereas more stable zircon retains a memory of the major episodes of granite evolution, the coexisting titanite provides additional information on crystallization, subsequent growth, cooling and alteration of the plutonic complexes. In addition to titanites formed at the magmatic stages of *c.* 2710 and *c.* 2650 Ma, the peralkaline granites contain titanite populations which have undergone major resetting at *c.* 1870 Ma during the burial metamorphism related to the Svecofennian orogeny and the 1760 Ma hydrothermal alteration near contemporaneous with regional metamorphism of *c.* 1780 Ma. The peralkaline granites which contain *c.* 2710 Ma titanites also include inherited titanite grains of an age of 2795 Ma. These data support the concept that the titanite can remain a closed system to Pb diffusion at temperatures of peralkaline granite melt (<800°C) as long as the crystals escape magmatic or metamorphic recrystallization.

Supplementary material: Data tables on the whole rock composition, rare earth element concentrations in titanite and other accessories (S1), U–Pb isotope composition of titanite (S2) and zircon (S4), and BSE images of analysed titanite and zircon grains (S3) are available at https://doi.org/10.6084/m9.figshare.c.6488847

Neoarchean to early Proterozoic alkaline granitoid magmatism played a fundamental role as a marker of geological evolution, with craton-wide events in several Precambrian areas, such as Yilgarn craton in Western Australia (Smithies and Champion 1999; Smithies *et al.* 2018), Superior craton in Canada (Sutcliffe *et al.* 1990; Ducharme *et al.* 2021), Greenland (Blichert-Toft *et al.* 1995), Eastern Dharwar Craton, India (Moyen *et al.* 2003) and Karelian and Kola cratons in the Fennoscandian Shield (Mitrofanov *et al.* 2000; Zozulya *et al.* 2005). In the Kola craton, peralkaline granites occupy a considerable volume (>2500 km^2) of the Keivy terrane. Since these rocks belong to the family of A-type granites (Zozulya *et al.* 2005), they share certain common attributes including enrichment in incompatible lithophile and high-field-strength elements and a broad spatial association with intracratonic rift settings (e.g. Batieva 1976; Batieva and Bel'kov 1984; Mints *et al.* 2015; Vetrin 2018; Zozulya *et al.* 2020; Zakharov *et al.* 2022). Geochronological studies during previous decades have been based essentially on conventional (TIMS – thermal ionization mass spectrometry) zircon U–Pb work, which involved multigrain fractions of several generations of zircon growth and inevitably yielded age data that were difficult to interpret, also with relatively large errors, and provide evidence of an extremely wide time span of peralkaline magmatism in the Kola craton (Vetrin *et al.* 1999; Mitrofanov *et al.* 2000; Bayanova 2004). According to U–Pb isotope zircon data, the early manifestations of alkaline granites in Keivy terrane (Fig. 1) occurred at *c.* 2760–2740 Ma (Vetrin *et al.* 1999; Bayanova 2004), but additional U–Pb isotope sensitive high-resolution ion micro probe (SHRIMP) determinations fixed the maximum igneous activity at 2670 Ma (Vetrin and Rodionov 2009). The finalizing magmatic event is the Sakharjok syenite massif, which was formed at *c.* 2610 Ma (Bayanova 2004; Vetrin

From: van Schijndel, V., Cutts, K., Pereira, I., Guitreau, M., Volante, S. and Tedeschi, M. (eds) 2024. *Minor Minerals, Major Implications: Using Key Mineral Phases to Unravel the Formation and Evolution of Earth's Crust.*
Geological Society, London, Special Publications, **537**, 381–410.
First published online May 4, 2023, https://doi.org/10.1144/SP537-2022-233

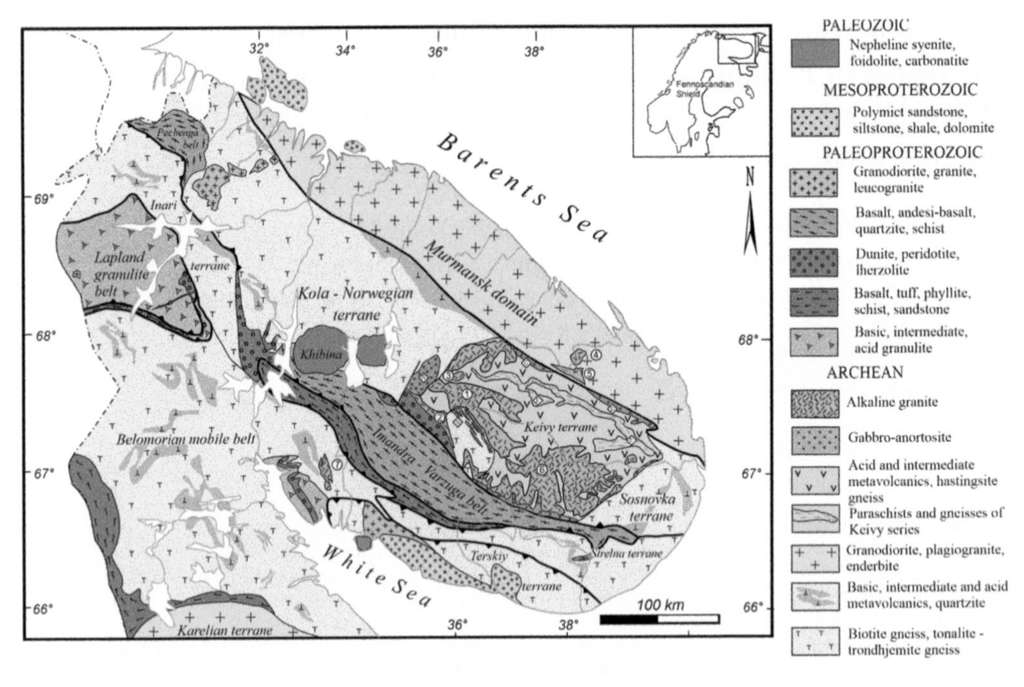

Fig. 1. Main structural and magmatic features of the northeastern Fennoscandian Shield. Numbers in circles denote Neoarchean alkaline granite plutons: 1, Western Keivy; 2, White Tundra; 3, Pessariok–Mariok massif; 4, Iokangskiy; 5, Koutyngskiy; 6, Ponoyskiy; 7, Kanozerskiy. Letters in diamonds denote gabbro-anorthosite massifs: A, Tzaga; B, Medvezhye–Shchuchieozerskiy; C, Patchemvarek; D, Acheriok; E, Lebyazhka volcanic pile. Source: map drafted after Mitrofanov (1996) and Bogdanov *et al.* (2003).

et al. 2014). However, every SHRIMP-analysed alkaline granite sample contained numerous grains whose age was outside that of the major zircon population, thus being indicative of some preceding and subsequent events (Vetrin *et al.* 1999, 2014; Bayanova 2004; Nitkina and Serov 2022).

Zircon may not fully decipher the tectonothermal history of the Precambrian terranes because it does not always participate with metamorphic and hydrothermal fluids (Rubatto *et al.* 2001; Harley *et al.* 2007). In order to more fully illuminate the multi-stage geological history of the Archean provinces, it is necessary to investigate other U- and Th-bearing accessory minerals that more readily participate in metamorphic reactions but still have relatively high closure temperatures. Titanite incorporates significant concentrations of rare earth elements (REE), and owing to significant Pb diffusion at lower-crustal temperatures and recrystallization, it is a more sensitive indicator of petrogenetic processes than zircon. Additionally, combined geochronological data for more stable zircon and titanite U–Pb ages may date not only magmatic events, but also the subsequent growth, cooling, fluid alteration and/or deformation of plutonic rocks (Prowatke and Klemme 2006; Rubatto *et al.* 2009; Garber *et al.* 2017; Olierook *et al.* 2019).

Here, we report sensitive high-resolution ion micro probe (SHRIMP) U–Pb ages of titanite and zircon inclusions and the REE titanite geochemistry of peralkaline meta-granitoids from the Keivy area in the northeastern Fennoscandian Shield, an Archean terrane that has experienced at least two Paleoproterozoic metamorphic and collisional events. Investigation of titanite geochronology and geochemistry data in conjunction with previously obtained geochronological data on zircon allows us to constrain the time of the main episodes to better understand the tectonic and geodynamic processes that operated during the formation of the Kola part of the shield and its subsequent reworking.

Geological background

The major Precambrian provinces of the northeastern Fennoscandian Shield are the Belomorian, Kola and Murmansk domains (Fig. 1; Slabunov *et al.* 2006). The Murmansk domain designated as a Neoarchean craton has been little affected by younger events, while the Belomorian and Kola Provinces both record significant thermal and tectonic reworking and amalgamation related to the Paleoproterozoic Lapland–Kola collisional orogeny (Daly *et al.*

2006). The Murmansk domain is composed dominantly of various granite-gneisses and granitoids, within which supracrustal rocks occur only as enclaves, metamorphosed to amphibolite grade; relics of granulite-facies mineral parageneses have also been described from the central part of the craton (Slabunov *et al.* 2006). The Keivy terrane, which is the main locus of the Neoarchean alkaline granite magmatism, is a fault-isolated fragment of crust in the Kola craton (Fig. 1). In the north it is separated by the detachment zone along which the Murmansk domain overthrust the Keivy terrane. In the south it is surrounded by the Paleoproterozoic (<2500 Ma) Imandra–Varzuga belt (Fig. 1). According to Balagansky *et al.* (2021), three rock complexes are recognized in Keivy.

The first rock complex is represented by biotite and garnet–biotite gneisses and locally hastingsite-bearing and microcline-bearing gneisses displaying relics of volcanic structures, which compose 40% of the terrane. These are combined with minor metasediments into the Archean Lebyazhka volcanic pile (Belolipetsky *et al.* 1980; Rundqvist and Mitrofanov 1993; Mints *et al.* 2015). Zircons from garnet-bearing biotite gneiss interpreted as a rhyodacitic tuff yielded a U–Pb age of 2871 ± 15 Ma (Bayanova 2004). A more recent study by Balagansky *et al.* (2021) determined U–Pb zircon ages of 2678 ± 7 and 2647 ± 35 Ma from the hosting metatrachyrhyolites of the Lebyazhka formation and interpreted them as extrusion ages.

The second rock complex is represented by kyanite, staurolite and garnet schists that fill synforms of the Keivy formation (Belkov 1963; Rundqvist and Mitrofanov 1993; Mitrofanov 1996; Mints *et al.* 2015). These schists are localized only within the Keivy terrane and occupy less than 10% of the area. Isotopic ages from the Keivy schists are lacking and their sedimentation ages are believed to be Archean (Belkov 1963; Zagorodny and Radchenko 1983; Slabunov *et al.* 2006).

The third group of rocks is represented by the gabbroid and peralkaline granite plutonic series (Balagansky *et al.* 2021). Apart from the two largest Tzaga and Medvezhye–Shchuchieozerskiy gabbro-anorthosite massifs, several thin but extended sheet-like bodies are located along southeastern and northeastern boundaries of the Keivy terrane (Fig. 1). Zircon U–Pb data from the Tzaga and Medvezhye–Shchuchieozerskiy massifs yielded ages of 2659 ± 3 and 2663 ± 7 Ma, respectively, whereas gabbro-anorthosites from the Acheriok massif were dated at 2678 ± 16 Ma (Bayanova 2004). Some gabbro-anorthosites occur outside of the Keivy terrane near its boundary with the Murmansk Province, where these rocks compose the Patchemvarek massif (Fig. 1). The TIMS U–Pb determinations of zircon multigrain fractions from these gabbro-anorthosites yielded an age of 2925 ± 7 Ma (Kudryashov and Mokrushin 2011). However, according to SHRIMP determinations by Vrevsky and Lvov (2016), two groups of zircon grains yielded concordia ages of 2938 ± 8 and 2662 ± 7 Ma. The former age of grains is considered to be inherited from the host granitoids; the latter is identical within errors to those from the other gabbro-anorthosite massifs and interpreted as the magmatic crystallization age.

The other plutonic series is represented by the A-type peralkaline granites. The Belye (White) Tundra, Zapadny (Western) Keivskiy and Ponoyskiy massifs together with minor peralkaline granite bodies occupy *c.* 25% of the Keivy terrane (Balagansky *et al.* 2021). The granites intrude the tonalite–trondhjemite gneiss basement of the central Kola terrane and acid-intermediate metavolcanics of the Lebyazhka formation (Fig. 1). The 2678 ± 7 Ma age determination of the hosting Lebyazhka metatrachyrhyolites obtained by Balagansky *et al.* (2021) implies that the gneiss protolith would be near contemporaneous to the peralkaline granites.

According to Zozulya *et al.* (2020), the typical alkali granite is massive with a porphyritic structure, with large (1.5–2.0 cm) subhedral phenocrysts of microcline-perthite (40 vol%) and a fine- to medium grained groundmass composed of xenomorphic albite, quartz, microcline and schlieren-like segregations of arfvedsonite and aegirine, aenigmatite, astrophyllite. The total amount of mafic minerals does not exceed 7–8 vol%. The rock is typical of the White Tundra massif but differs from the Western Keivy granite, which always shows gneissic structure and a subsolvus character (K-feldspar and albite). Variations in the structure of granites from certain massifs in the Keivy province are explained by different degrees of a late metamorphic overprint (Zozulya *et al.* 2020). Recent studies by Vetrin (2018, 2019) indicate the presence of subalkaline rocks among the peralkaline granite series. Subalkaline rocks of the Neoarchean association of latites–monzonites–granites are preserved as xenoliths and outliers among plagiomicrocline granites of the terrane. In addition to Keivy alkaline complexes, two massifs of arfvedsonite-bearing and aegirine-bearing granites are exposed south of the Keivy terrane, one in the southern part of the Imandra–Varzuga Rift and the Kanozerskiy massif just south of this rift (Fig. 1) (Batieva 1976; Nitkina and Serov 2022).

Termination of Neoarchean alkaline magmatism in Kola is recorded by the Sakharjok and Kuliok intrusions. The Sakharjok massif is composed of three intrusive phases: (1) essexite and theralite with an age of 2666 ± 4 Ma (Zozulya and Bayanova 2013); (2) alkaline syenite dated at 2645 ± 7 Ma (Vetrin *et al.* 2014); and (3) nepheline syenite 2613 ± 35 Ma (Bayanova 2004). Alkaline syenites form the southwestern footwall of the massif;

nepheline syenites are confined to its hanging side in the northern and northeastern parts. Alkaline gabbroids occur as xenoliths in nepheline syenites.

As in the most other Precambrian terranes of the Fennoscandian Shield, the Keivy terrane was consolidated by *c.* 2550 Ma and, during the next 600 Ma, underwent a number of episodes of metamorphism and continental rifting with the formation of passive continental margins that were later deformed during the *c.* 1900 Ma Svecofennian orogeny (Rundqvist and Mitrofanov 1993; Slabunov *et al.* 2006; Bushmin *et al.* 2011; Mints *et al.* 2015; Balagansky *et al.* 2021).

Analytical methods

We studied 12 titanite and zircon samples from Western Keivy, White Tundra, Iokangskiy, Ponoyskiy and other alkaline granite massifs from the Keivy terrane (Table 1). Hand-picked under optical binocular microscope, the grains of 0.2–1.0 mm size and broken titanite crystals containing zircon inclusions were mounted in epoxy with reference OLT1 titanite grains (Kennedy *et al.* 2010). After solidification of the epoxy their surface was polished to about half the thickness of the grains and gold coated to eliminate the excess negative ionic charge produced by the primary high-energy beam of the ion microprobe.

Given that titanite is a mineral where distinctive cathodoluminescence is absent, this prevents cathodoluminescence images being used for selecting suitable sites for geochronological dating. Therefore, optical images of the studied grains in both the reflected and transmitted light, as well as backscattered electron (BSE) images, which enable estimations of geochemical heterogeneity of the crystal surface, were taken into account in localizing analytical points for further isotope analysis. The images were used to identify altered and fractured grains, as well as zircon, apatite and other types of inclusions, which were avoided during subsequent SHRIMP analysis. U/Pb relationships in titanite were measured using high-resolution secondary ion microprobe SHRIMP-IIe at the Centre for Isotope Research at Karpinsky Geological Institute, St Petersburg, employing a method similar to that

Table 1. *Location and mineral composition of the studied samples of the Keivy alkaline granites*

Sample	Location*		Rock	Q	Pl	Mi	Aeg	Arf	Aen	Lep	Mus	Other
Western Keivy massif												
1/57	67°52′10″	36°37′45″	Aegirine–arfvedsonite granite with aenigmatite	33	29	26	3	8	1	–	–	1
2/57	67°56′20″	36°30′15″	Aegirine–arfvedsonite granite with aenigmatite	25	30	29	2	11	1	–	–	1
281	67°52′13″	36°03′48″	Aegirine–arfvedsonite granite	39	22	26	4	6	–	–	1	2
282	67°52′13″	36°03′48″	Magnetite–aegirine granite	29	36	25	5	1	–	1	–	4
1086/57	67°52′10″	36°37′45″	Aegirine granite with magnetite	42	19	24	8	1	–	–	–	7
1110/57	67°37′50″	36°24′52″	Aegirine–arfvedsonite granite with aenigmatite	25	32	32	1	8	1	–	–	1
Pessariok–Mariok massif												
90/64	67°42′48″	35°40′08″	Monzonite fine-grained	31	43	17	8	–	–	–	1	1
White Tundra massif												
35/63a	67°29′10″	35°46′25″	Aegirine granite with magnetite	37	29	27	4	1	–	–	–	3
Koyutyngskiy massif												
148/69	67°40′22″	38°38′22″	Augite–lepidomelane granite	23	45	15	–	5	–	9	–	3
Kanozerskiy massif												
458	67°00′35″	34°00′56″	Lepidomelane–ferrohastingsite granite	30	30	26	–	7*	–	5	–	2
Ponoyskiy massif												
79/56	67°17′15″	39°37′10″	Aegirine–arfvedsonite granite	31	28	26	7	3	–	3	–	1
Iokangskiy massif												
143/69	67°48′34″	38°50′26″	Lepidomelane–ferrohastingsite granite	29	31	21	1	6*	–	10	–	2

Q, quartz; Pl, plagioclase; Mi, microcline; Aeg, aegirine; Arf, arfvedsonite; Aen, aenigmatite; Lep, lepidomelane; Mus, muscovite.
*Map datum: UTM, Pulkovo 1942.

described in Kennedy *et al.* (2010). The primary beam of molecular negatively charged oxygen ions (O_2^-) was adjusted so that the secondary emission intensity was maximal, while the current value was 10–15 nA, with an analytical spot (crater) of about 50 μm in diameter. OLT1 titanite, whose grains were obtained by disintegration of a titanite megacryst (31 g) from Ca–Ti-skarn (Otter Lake, Quebec, Canada), was used as a U–Pb reference material. The measured $^{206}Pb/^{238}U$ ratios of OLT1 titanite were normalized by 0.1705, which corresponds to the time of titanite crystallization of 1015 Ma from Kennedy *et al.* (2010: 1014.8 ± 2.0 Ma (isotope dilution TIMS) and 1016.8 ± 3.8 Ma (SHRIMP), see Supplementary Material S2). Given that OLT1 titanite is moderately inhomogeneous with respect to the uranium distribution whose concentrations average between 330 ppm (SHRIMP-II, our data) and 450 ppm (Kennedy *et al.* 2010), the calculated ^{238}U concentration in the studied titanite samples appears tentative. The U contents were calculated by a correlation between the measured $^{238}U^{16}O$ current intensity and the $^{40}Ca^{48}Ti_2^{16}O_4$ reference peak, with the uranium abundances in the sample subsequently normalized to 300 ppm (the value taken for reference OLT1 titanite). The measured reference $CaTi_2O_4^+$ current peak intensity, representing a function of the matrix chemical composition of unknown mineral and reference standard, is sensitive to changes in electrical conductivity and charge, which results in an increase of up to 20% in the uncertainty of the calculated uranium and thorium abundances. When calculating U abundancies (also thorium and radiogenic lead, respectively) in the titanite samples relative to the measured zircon reference sample 91500, the estimated corresponding element abundancies show no more than a two-fold increase compared with the results obtained with reference OLT1 titanite. Mass spectrometric analysis of OLT1 titanite grains as a reference for U–Pb ratios and their concentrations performed in each SHRIMP analytical session showed a slight variance in $^{206}Pb/^{238}U$ ratio values when the linear regression approach was applied to the calibration procedure for correlation ln(Pb/U) v. ln(UO/U). The measurement error for on average 10 analyses per session was ≤1.5% (2σ), which is comparable with the measurement results for the Temora2 standard zircon in the conventional U–Th–Pb analyses when the SHRIMP-II zircon dating method was employed. The calculated U concentrations in OLT1 titanite had moderate variance not exceeding 15% relative to a specified value of 300 ppm. The measured and calculated thorium contents also varied in the same relative range, while the Th/U ratio varied from 3 to 3.5. The affinity of titanite for U and Th is known to be significantly less than in the case of accessory minerals such as zircon and monazite (Cherniak 1993, 2010; Amelin 2009). Whereas ion Pb^{2+} can readily enter the position of bivalent calcium (Ca^{2+}), titanite is characterized by high concentrations of common (non-radiogenic) lead (Frost *et al.* 2001). The share of common Pb thus accounted for 1.3–2.0% of the reference OLT1 titanite, whereas a limit of less than 1% for the geochronological standard was established for U–Pb zircon dating by SHRIMP method. In the OLT1 titanite-based calculations of ages, the measured lead isotopic composition was respectively corrected for the common Pb composition and abundances, according to the model for terrestrial lead isotope evolution at the time of titanite formation of 1015 Ma: $^{206}Pb/^{204}Pb$, 17.0 ± 1; $^{207}Pb/^{206}Pb$, 0.910 ± 0.015; $^{208}Pb/^{206}Pb$, 2.16 ± 0.1 (Stacey and Kramers 1975). Prior to each of the *in situ* U–Th–Pb analyses, the grain surface was subject to ion-rastering for 1 min around the analytical spot. The accumulation of secondary ion pulses for the subsequent calculation of isotopic ratios was carried out on a secondary electron multiplier in a single-collector mode. The registration of isotope ion currents was performed by mass spectrum scanning in the following sequence (in atomic mass units): $^{40}Ca^{48}Ti_2^{16}O_4$ (200), ^{204}Pb (204), 'background' (204.5), ^{206}Pb (206), ^{207}Pb (207), ^{208}Pb (208), ^{238}U (238), $^{232}Th^{16}O$ (248), $^{238}U^{16}O$ (254). The selection of the mass spectrometer input/output slit parameters, as well as adjustment of the secondary ion beam provided a mass resolution of at least 5000 for 0.01 of the UO peak height, which allowed any isobaric overlaps on the measured masses to be ruled out. For the purpose of statistics accumulation and the evaluation of analytical errors, each analysis included repeated measurements (five times per analysis) for this set of masses, by changing the magnetic field and mechanically moving the collector into the focal plane of the secondary beam. Depending on the assumed age of the sample and amounts of the measured isotopes, the timing of signal accumulation, i.e. pulse counts for each mass stop, was determined and documented individually, while the total time of one analysis averaged about 20 minutes.

The raw data were processed using version 1.13 of the SQUID software (Ludwig 2000) while the construction of concordia plots for the calculated isotope ratios and age determinations was performed using the ISOPLOT/EX program (Ludwig 2003, 2009, 2012). The errors calculated for concordant ages, as well as those calculated from the 'mixing lines' with concordia intercepts or weighted average ages, are discussed in the text and listed in Table 1, and the figures are either at the 2σ level or in the 95% confidence interval.

The *in situ* REE analysis of the studied titanites was performed using secondary ion mass spectrometry (SIMS) on a SHRIMP-II instrument according to the procedure described in Hoskin (1998). A 5–6 nA O_2^- primary beam of *c.* 30 μm diameter was employed to sputter positive secondary ions, later extracted using a 10 kV accelerating voltage. A typical MRP of 4500–5000 (1% of the peak height) in combination with energy filtering allowed the separation of possible isobars. Ti was used as an internal standard and the NIST SRM 611 silicate glass calibration standard as a primary standard with 434 ppm of Ti was applied. In each analysis, three mass scans of the isotopes listed below were recorded at the following accumulation times: ^{49}Ti, 4 s; ^{139}La, 3 s; ^{140}Ce, 4 s; ^{141}Pr, 4 s; ^{143}Nd, 4 s; ^{146}Nd, 3 s; ^{147}Sm, 10 s; ^{149}Sm, 10 s; ^{151}Eu, 15 s; ^{153}Eu, 15 s; ^{155}Gd, 15 s; ^{157}Gd, 15 s; ^{159}Tb, 25 s; ^{161}Dy, 20 s; ^{163}Dy, 15 s; ^{165}Ho, 25 s; ^{166}Er, 20 s; ^{167}Er, 25 s; ^{169}Tm, 40 s; ^{171}Yb, 30 s; ^{172}Yb, 30 s; and ^{175}Lu, 90 s. Secondary ions were detected by means of an electron multiplier in the peak-switching mode. The total data acquisition time for one analysis was about 30 min. The raw data were processed using the MS Excel program. Raw data were reduced by normalizing to the reference glass NIST SRM 611 (preferred values from Norman *et al.* 1996). The registration of paired isotopes for several measured elements allowed an additional quality control of the analysis based on the degree of their coincidence (considering the natural abundance). At the beginning of the session and through the analysis, OLT1 titanite was measured every 10 unknowns as a secondary standard. The scatter of measured REE data in OLT1 titanite did not exceed 10–12% during the analytical session and is consistent with the results presented in Ma *et al.* (2019).

Results

Rock descriptions and REE chemistry of accessory minerals

Alkali granites of the Keivy terrane are subdivided into lepidomelane, lepidomelane–arfvedsonite, arfvedsonite–aenigmatite, aegirine–arfvedsonite, aegirine–ferrohastingsite (dominant group in most massifs), aegirine–magnetite and astrophyllite–riebeckite varieties. The transition between rock types is gradational. Geological data indicate that alkali granosyenite is the earliest intrusive phase of these massifs. All peralkaline granites belong to transsolvus type, formerly classified as subsolvus granite (Boily and Williams-Jones 1994; Gysi and Williams-Jones 2013), so named because it contains two separate alkali feldspars (microcline and albite), in addition to recently recognized perthitic alkali feldspar (minor). The transsolvus aegirine–ferrohastingsite granite occupies the bulk of the Western Keivy and Ponoyskiy massifs, whereas the aegirine–magnetite granites tend to be confined to the apical parts of the massifs. Because massive porphyritic granites are also known from some plutons, their gneissic appearance is most likely caused by a later metamorphism. Alkaline granites are medium- to coarse-grained and have equigranular or porphyric textures that are typical of most granite varieties. The whole rock chemistry of the investigated samples is summarized in Supplementary Material S1.

Zircon and titanite are the first high field strength element minerals to have crystallized in the alkali granites, forming euhedral to subhedral crystals interstitial to quartz and feldspar, and in contact with the aegirine and amphibole. Under transmitted light, the titanite grains are colourless to pale honey or brown, with localized domains rich in inclusions. Microinclusions are represented by zircon, apatite, rutile, ilmenite and rock-forming minerals, as well as allanite. Where in contact with other mafic minerals, titanite forms elongated aggregates of euhedral or rounded grains up to 300–900 μm. The BSE imaging revealed the complex internal structure of some grains, with cores enveloped by titanite overgrowths (Supplementary Material S3). The results of REE determinations of titanites from all rock samples are presented in Supplementary Material S1.

Western Keivy massif. Three populations of titanite are identified in aenigmatite–aegirine–arfvedsonite granite. The type WK1 population shows consistent enrichment in light rare earth elements (LREE) relative to medium (MREE) and heavy rare earth elements (HREE), with $(La/Yb)_N$ ratios ranging from 3.6 to 5.8. Abundances for individual REE within a population can range over half an order of magnitude. The WK1 population exhibits a concave-up LREE pattern with a straight MREE and HREE pattern and a weak negative Eu anomaly (Fig. 2a). The REE patterns for the Western Keivy (WK2) type population are similar to those for WK1, but with lower REE concentrations. Chondrite-normalized profiles of the WK3 titanites show the most convex section and a strong negative Eu anomaly Eu/Eu* ($Eu_N/\sqrt{(Sm_N*Gd_N)}$) varying from 0.3 in magnetite-bearing aegirine granites to 0.9 in aenigmatite–aegirine varieties (Fig. 2b). Within the WK3 population there can be half an order of magnitude variation in individual REE abundances, although the shape and slope of individual patterns do not change significantly.

Pessariok–Mariok massif. Two groups of titanite grains are defined in the granodiorites. The REE profiles of the first group (Pessariok–Mariok, PM1) are similar to those in the WK2 population of the adjacent Western Keivy granites, but the total REE concentrations are half an order of magnitude lower (Fig. 2c). The other group of grains (PM2) exhibits

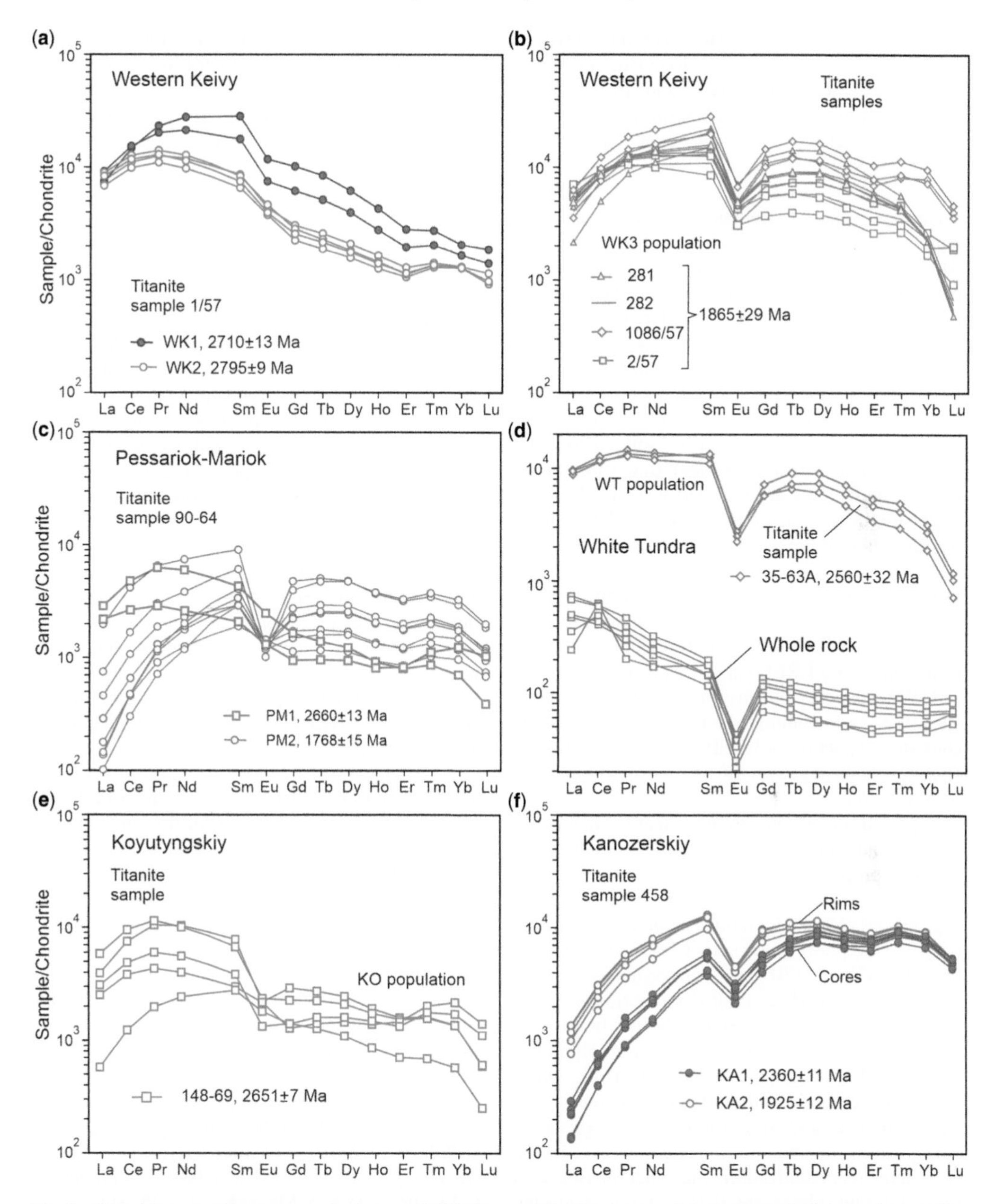

Fig. 2. Chondrite-normalized rare earth element (REE) diagrams for titanite populations from alkaline granite massifs of the Keivy Terrane. Pattern colours mark different titanite populations. Source: in (d) the whole rock REE pattern of the host White Tundra granite is from Vetrin and Kremenetsky (2020). Normalizing values are from Anders and Grevesse (1989). Age data of titanite populations are listed in *Geochronology* section.

a resemblance to the WK3 population as reflected in their convex LREE patterns and strong Eu/Eu* anomalies: 0.19–0.57.

White Tundra massif. Titanite from the aegirine granite with magnetite (WT population) shows concave-up REE patterns combined with a negative Eu anomaly (Eu/Eu* = 0.3) and significant depletion of HREE as reflected in their high $(Gd/Yb)_N$ values (2.1–3.2) (Fig. 2d). The chondrite-normalized profiles of WT are similar to those of the Western Keivy (WK3) population.

Koyutyngskiy massif. The titanite population (Koyutyngskiy, KO) in augite–lepidomelane granite is similar to that in the WK2 and PM1 massifs (Fig. 2e). However, abundances for individual REE within a population can range over half an order of magnitude.

Kanozerskiy massif. Euhedral titanite grains from lepidomelane–ferrohastingsite granite contain BSE-dark cores (KA1 population) and bright rims (KA2 population) which show uniform convex-up REE patterns. All titanites are characterized by strong depletion of LREE – $(La/Sm)_N$, 0.08 ± 0.03 – and a pronounced negative Eu anomaly (Eu/Eu*, 0.3) (Fig. 2f). The relative decrease in LREE abundances probably reflects apatite and allanite fractionation, with allanite strongly influencing the abundances of La and Ce (and La_N/Ce_N) in the melt. Titanite cores are depleted in REE (Table 3), suggesting a significant enrichment of REE during the formation of the rims.

Ponoyskiy massif. Aegirine–arfvedsonite granite (sample 79/56) comprises two populations of titanite grains. The Ponoyskiy1 (PO1) population exhibits a concave-up LREE pattern with straight MREE and HREE patterns and an absence of Eu/Eu* anomalies (Fig. 3). In contrast, the PO2 population shows consistent depletion of LREE relative to MREE and HREE, and a negative Eu anomaly varying from 0.28 to 0.46.

Apart from zircon, titanite, apatite, magnetite, ilmenite and fluorite, alkaline granites contain highly variable proportions of REE-bearing accessories. According to Batieva (1976), the most abundant aegirine–arfvedsonite granites contain chevkinite–(Ce) (<440 ppm), monazite–(Ce) (30–220 ppm), britholite–(Y) (<130 ppm), euxenite (<50 ppm), bastnäsite–(Ce) and rare grains of xenotime–(Y), fergusonite–(Y), allanite–(Ce) and thorite. In contrast, the only rare REE-bearing accessory in lepidomelane–arfvedsonite granites is allanite–(Ce), whose content varies from 30 to 650 ppm. Whereas monazite, xenotime and fergusonite are suggested to have primary magmatic origin and may have undergone later recrystallization, the other phases were formed during subsequent alteration of granites during regional metamorphism and interaction of granites with hydrothermal fluids (Macdonald *et al.* 2015, 2017; Bagiński *et al.* 2016; Zozulya *et al.* 2020). The REE patterns of the above rare accessories exhibit two different types of element distribution which dramatically differ from both the primary and secondary titanites and zircons (Fig. 4). Whereas monazite, chevkinite, britholite, euxenite and allanite are characterized by significant enrichment in LREE ($(La/Yb)_N$, 25.4–91.1), the coexisting xenotime, fergusonite and thorite grains are strongly enriched in HREE ($(La/Yb)_N$, 0.02–0.17) (Table 3 in Supplementary Material S1). The presence of a pronounced negative Eu/Eu* anomaly typical of all the above secondary accessories including recrystallized titanite fractions is notable.

Geochronology

All studied titanites are characterized by significant variations in the uranium content in the range of 5–250 ppm, whereas the Th/U ratio varies from 0 to 13, and the proportion of non-radiogenic ^{206}Pb varies from 0.1% up to 78%, depending on the sample. Nevertheless, all of the calculated ages turned out to be concordant within the errors after correction of the Pb isotopic composition for the share of common (non-radiogenic) lead from the measured ^{204}Pb isotope, and assuming that its plausible composition can be approximated by the Stacey–Kramers model Pb composition for the appropriate ages. The U–Pb isotope data for individual grains are shown in Supplementary Material S2. Additionally, it was of particular interest to obtain the ages of zircon grains and inclusions settled in titanite. The size of inclusions varies in the range of 0.02–0.10 mm, so it was possible to determine the U–Pb isotope characteristics of the zircons using SHRIMP technique (Supplementary Material S3).

Western Keivy massif. One-hundred and twenty-four U–Pb SIMS analyses were performed on the titanite grains from six samples which were previously analysed for REE. Three different age clusters were recognized in aenigmatite–aegirine–arfvedsonite granites. The WK1 and WK2 populations (see REE patterns) of titanites from sample 1/57 are characterized by high U and Th contents of 49 ± 12 and 280 ± 38 ppm, respectively, and a high Th/U ratio of 6.2 ± 0.95. The content of radiogenic ^{206}Pb is 23 ± 6 ppm and unradiogenic lead component is 1.0–2.0% of the total ^{206}Pb (Supplementary Material S1). The WK1 and WK2 populations form two clusters on the Tera–Wasserburg concordia line (Fig. 5a) with ages of 2710 ± 13 Ma (WK1, $n = 5$, MSWD (mean square of weighted deviates) 0.74, probability 0.39) and 2795 ± 9 Ma (WK2, $n = 15$, MSWD 1.6, probability 0.21). The third titanite population (WK3) in the samples 281, 282, 2/57, 1110/57 and 1086/57 is characterized by relatively low U and Th content, 25.3 ± 7.1 and 36.7 ± 9.7 ppm, respectively, and a Th/U ratio of 1.53 ± 0.22. A high proportion of common Pb ($^{206}Pb_{com}$, 63.3 ± 5.2%), combined with a relatively high content of radiogenic ^{206}Pb – 16.1 ± 6.8 ppm – and the unradiogenic component with $^{207}Pb/^{206}Pb$ equal to 0. 9772, resulted in the linear mixing line of data points on a $^{238}U/^{206}Pb$ v. $^{207}Pb/^{206}Pb$ (common Pb/U)

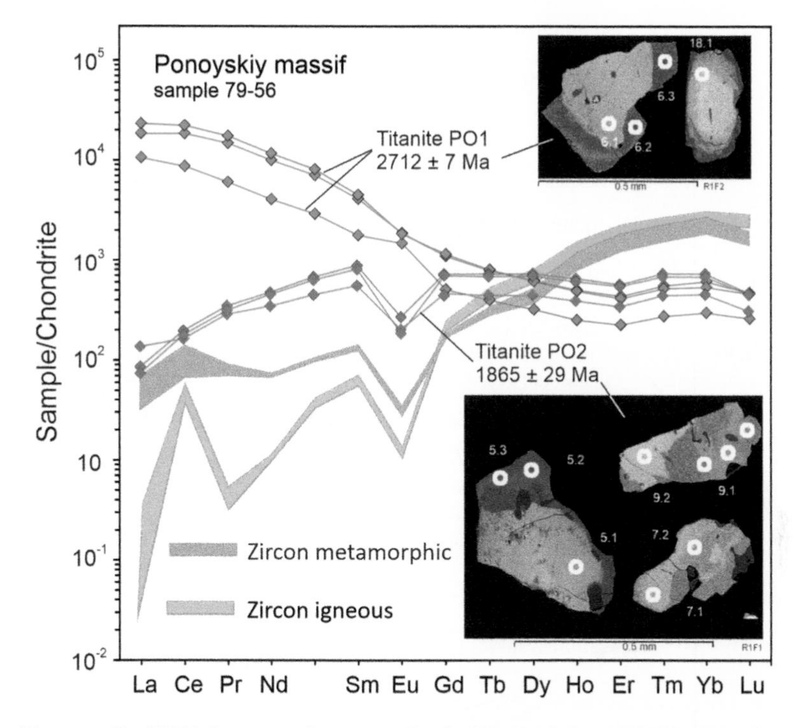

Fig. 3. Chondrite-normalized REE diagrams and corresponding backscattered electron (BSE) images for titanite populations from the Ponoyskiy massif in comparison with the igneous and metamorphic zircon data of the Ponoyskiy massif taken from Balashov and Skublov (2011). Pattern colours mark different titanite populations. Source: normalizing values are from Anders and Grevesse (1989). Age data of titanite populations are listed in *Geochronology* section.

diagram, which corresponds to an age of 1865 ± 29 Ma (Fig. 6). Recalculation of the U/Pb data of the WK3 population for individual samples and correction of the measured ratios for the common Pb composition (^{204}Pb-corrected) according to the Stacey–Kramers model form concordant or near concordant clusters ranging from *c.* 1876 to 1730 Ma (Fig. 5b–f). Titanite analyses of samples 2/57 and 1110/57 (Supplementary Material S2) yield ages of 1765–1730 Ma, which are in agreement with the upper intercept discordia age of 1763 ± 16 Ma obtained for coexisting individual zircon grains from the sample 2/57 (Fig. 7). However, given the very high proportion of common Pb and the uncertainty of the isotopic composition of this component, we assume 1865 ± 29 Ma to be the best approximation for the age of WK3 titanite population.

Pessariok–Mariok massif. U–Pb isotope data on 26 titanite grains from monzonite (sample 90/64, Supplementary Material S2) indicate presence of two isotope-geochemically different systems, which correspond to the PM1 and PM2 populations and can be approximately compared with the outer and inner zones of individual grains. The PM1 population is characterized by the elevated Th content (40. 5 ppm) and low Th/U ratios (0.95 ± 0.4) whereas PM2 titanites are enriched in U (118 ± 62 ppm) and exhibit the lowest Th/U ratios (0.19 ± 0.1). The content of common Pb in both populations varies insignificantly within the range of 0.5–4% (the fraction of the total ^{206}Pb isotope).

The PM1 titanite population in the $^{238}U/^{206}Pb^*$ v. $^{207}Pb^*/^{206}Pb^*$ isotope diagram forms a concordant cluster with an age of 2650 ± 13 Ma (MSWD 5.1, probability 0.02, $n = 13$), whereas the PM2 titanites yield the concordant age of 1745 ± 9 Ma (MSWD 12, $n = 19$) (Fig. 8a). The relatively high MSWD values in both cases reflect the uncertainty of the assumed composition of common Pb when the measured ratios are corrected. However, the weighted mean age calculations based on the radiogenic $^{207}Pb^*/^{206}Pb^*$ ratios corrected for common Pb for the PM1 titanite population give an age of 2660 ± 13 Ma ($n = 14$, MSWD 1.1, probability 0.3), and for the PM2 population an age of 1768 ± 15 Ma ($n = 20$, MSWD 1.1, probability 0.7) (Fig. 8b, c). The latter age estimations are suggested to represent the time of PM1 and PM2 titanite populations.

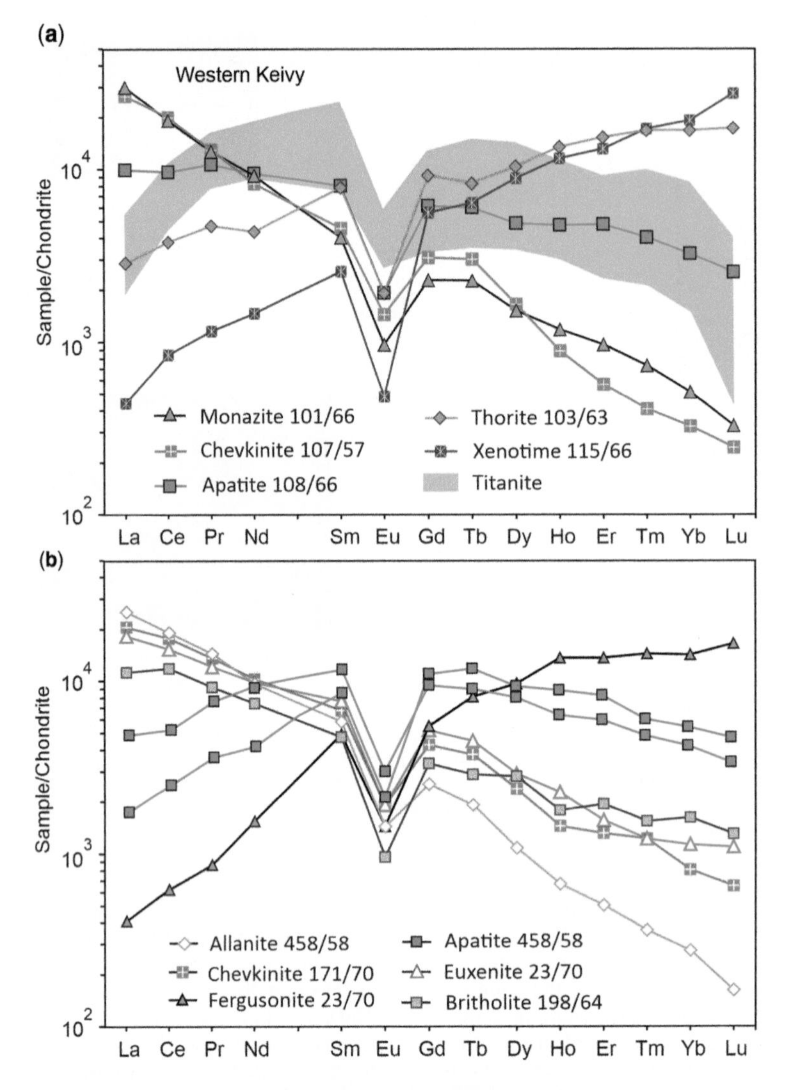

Fig. 4. Chondrite-normalized REE diagrams for: (**a**) monazite, chevkinite, apatite, thorite, xenotime and titanite from peralkaline granite of the Western Keivy massif; (**b**) allanite and apatite from Kanozerskiy massif, chevkinite, euxenite, britholite and fergusonite from the Ponoyskiy massif and apatite from the Iokangskiy and Koutyngskiy massifs. Source: normalizing values are from Anders and Grevesse (1989). The REE concentrations are listed in table 3 of the Supplementary Material S1.

White Tundra massif. Titanite from aegirine granite of the White Tundra massif is characterized by low U and Th contents: 15 ± 4 and 69 ± 15 ppm respectively, and a Th/U ratio of 4.5–7.4. Only two grains (9.2 and 17.1, Supplementary Material S2) out of 22 analysed exhibit an abnormally low Th/U <1.0 and an extremely low Th content, and therefore were excluded from the age calculations. A high but fairly constant proportion of non-radiogenic lead ($^{206}Pb_{com}$ $34.2 \pm 8.7\%$) suggests that the studied titanite represents a homogeneous population formed under uniform conditions. At the same time, the content of radiogenic Pb in the analysed titanites is relatively high ($^{206}Pb_{rad}$ 10 ± 4 ppm after correction of the Pb isotopic composition by the measured ^{204}Pb isotope). Eighteen spot analyses out of 22 form a close group yielding a concordia age of 2574 ± 25 Ma but show a relatively high value of MSWD (4.8) and low probability (0. 01), which are caused by an overestimation of the share of the common Pb. Exclusion of the most discordant analyses exhibiting increased uncertainty data gives a concordia age of 2560 ± 32 Ma (MSWD 4.5, probability 0.04) (Fig. 9a).

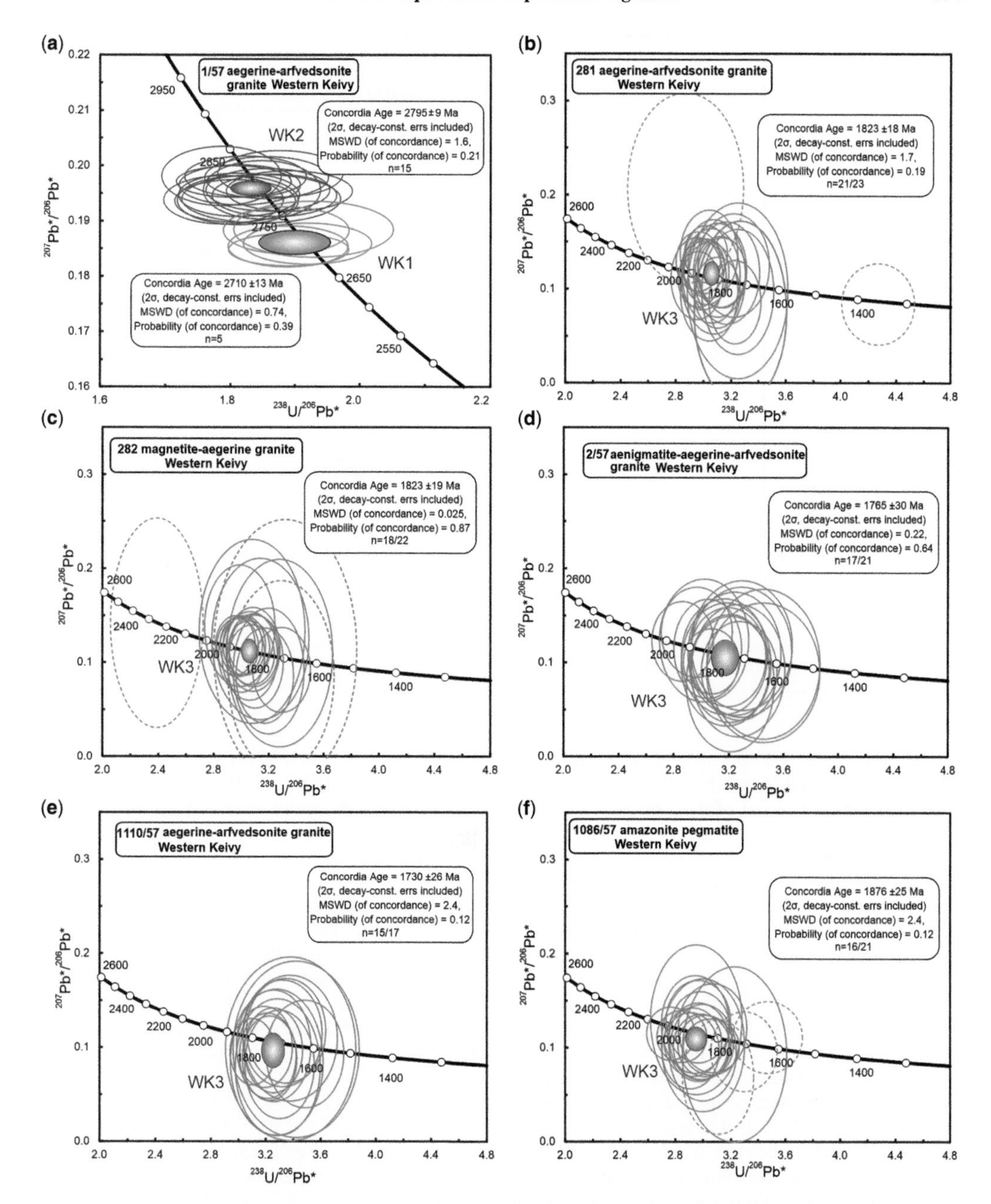

Fig. 5. Tera–Wasserburg diagram of U–Pb dating results for titanites from Western Keivy. Ellipses are coloured according to titanite REE populations (Fig. 2). Solid lines indicate analyses used for age calculation, and dashed lines mark excluded discordant analyses. All data are ^{204}Pb-corrected (for common Pb of the Stacey–Kramers model composition on corresponding ages) and uncertainties are quoted at the 2σ level. The inset text contains the number of analyses (n) used for age calculations.

The tiny irregular shape zircon inclusions in titanite with U up to 1360 ppm and relatively high Th/U ratios of 0.23–0.72 (Fig. 9b, Supplementary Material S4) are characterized by an elevated share of common Pb and a relatively high individual data-point uncertainty of the calculated U/Pb ratios (up to 30%). Of the whole dataset, only one analysis of zircon yields the age of 2556 ± 7 Ma,

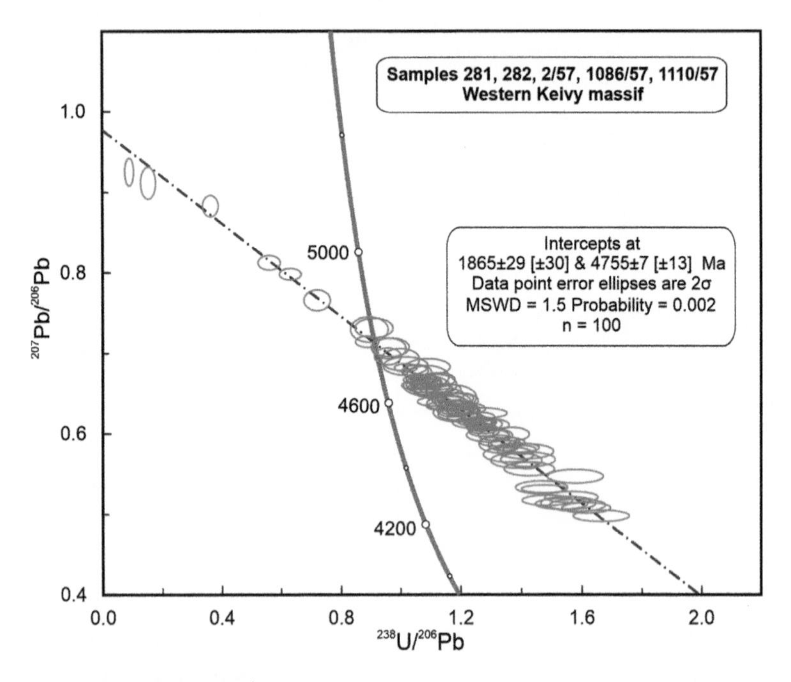

Fig. 6. Total Pb–U diagram (measured U/Pb ratios) for titanite WK1 population represented in the samples 2/57, 281, 282, 1110/57 and 1086/57 from Western Keivy (combined isochron). Uncertainties are quoted at the 2σ level. The inset text contains the number of analyses (n) used for age calculations.

which coincides with the age obtained from the host titanite population (Fig. 9c). Interestingly, the remaining seven out of nine analyses of zircon inclusions of this sample can be approximated by a single discordia line with a significantly younger upper intercept age of 2266 ± 54 Ma (MSWD 0. 1) if compared with the age of the host titanite grains, thus being indicative of the presumably postmagmatic event.

Koyutyngskiy massif. The average contents of both uranium and thorium in 20 studied titanite grains vary from 20 to 60 ppm, but in a few grains concentrations of U and Th may run up to 260 and 360 ppm, respectively. However, the insignificant variations of the Th/U ratio (1.21 ± 0.67), combined with the relatively constant content of the common (non-radiogenic) Pb in the titanite composition (the share of common ^{206}Pb is 1.4 ± 0.9%) suggest the homogeneity of the mineral-forming melt at the time of titanite crystallization.

A histogram $^{238}U/^{206}Pb^*$ v. $^{207}Pb^*/^{206}Pb^*$ based on 20 of 25 analyses shows the most reliable concordia age of 2651 ± 7 Ma (MSWD of concordance 1.1 and probability 0.29) (Fig. 10a). Alternatively, the weighted mean radiogenic $^{207}Pb^*/^{206}Pb^*$ ratio ages based on 22 analyses except for three spots with the lowest contents of radiogenic Pb and U (grains 1, 8, 9, see Supplementary Material S2) yield a similar age of 2652 ± 7 Ma (MSWD, 2.4), whereas all analysed zones of titanite grains ($n = 25$) give a weighted mean $^{206}Pb^*/^{238}U$ age of 2631 ± 16 Ma (MSWD 1.2, probability 0.25).

Kanozerskiy massif. Titanites from lepidomelane–ferrohastingsite granites are separated into two populations on the basis of their REE chemistry (Fig. 2f). These two populations also differ in their U and Th contents, as well as the Th/U ratio and the portion of common Pb (Supplementary Material S2). Whereas the titanite KA2 population exhibits high contents of U (130–210 ppm) and Th (50–98 ppm), and elevated Th/U ratio (*c.* 0.45) and portion of common Pb (9.6 ± 11%), the KA1 population is characterized by significantly lower value for all of the above characteristics (U, 40–110 ppm; Th, 7–11 ppm; Th/U, <0.2; common Pb varies within a narrow range of 2.5–4.5%).

Thirty-seven U–Pb analyses performed on titanites (Fig. 10b) were spread along the concordia line with the oldest age cluster at 2360 ± 11 Ma ($n = 18$) represented by the KA1 population. The KA2 population analyses form a discrete cluster which corresponds to the age of 1925 ± 12 Ma ($n = 14$). The increased MSWD values (9.5 and 16, respectively) for both age clusters indicate an overestimation of the proportion of common Pb when using

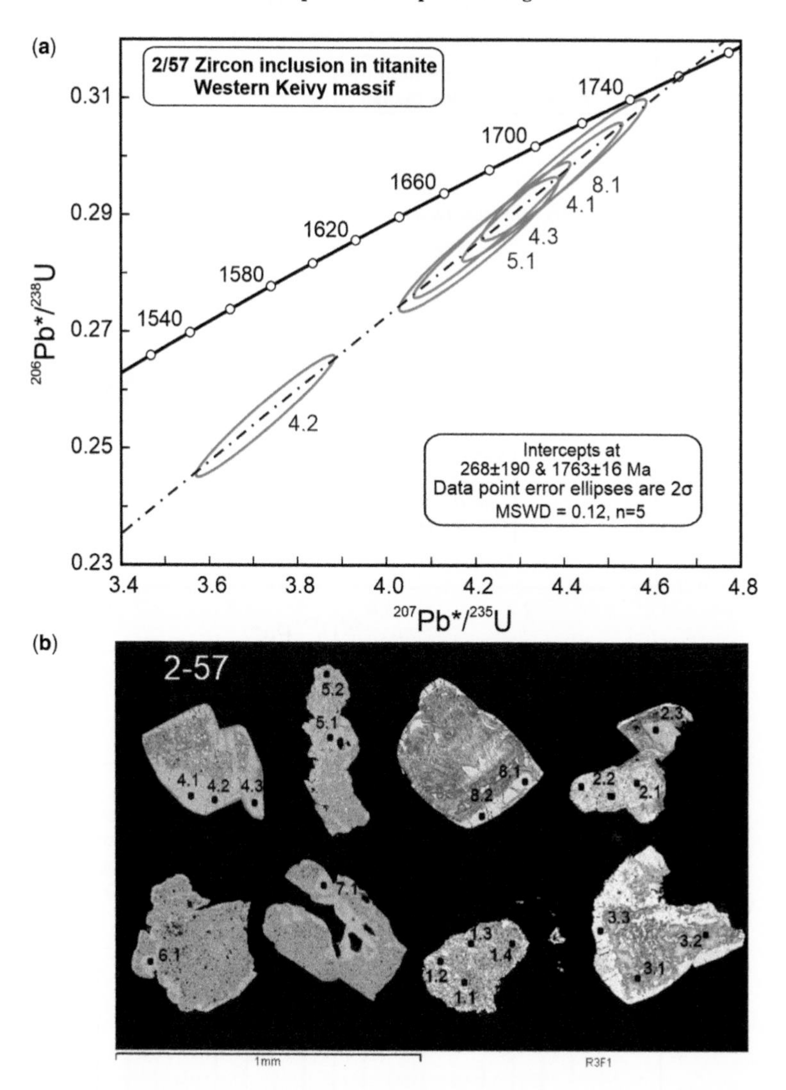

Fig. 7. Diagram with concordia (a) and corresponding BSE images (b) for zircons in WK3 titanite population (sample 2/57) of the Western Keivy massif. All data uncertainties are quoted at 2σ level. Spot numbers correspond to the data from Supplementary Material S4.

the correction of isotopic compositions from the measured ^{204}Pb isotope.

Ponoyskiy massif. The composite titanite grains from aegirine–arfvedsonite granites contain bright BSE cores (PO1 population according to REE patterns) (Fig. 11) with a U content of 75 $\pm$ 11 ppm, a high Th 617 $\pm$ 114 ppm and a relatively low portion of common Pb (0.68 $\pm$ 0.23%). The PO1 population forms a compact cluster with a ^{204}Pb-corrected concordia age of 2712 $\pm$ 7 Ma ($n = 20$, MSWD 2.3, probability 0.13) (Fig. 11a), whereas the weighted mean age (^{207}Pb*/^{206}Pb*) is 2709 $\pm$ 4 Ma (MSWD 0.7, probability 0.78). The PO2 population of dark (in BSE) rims (Fig. 11b) is characterized by highly variable and low Th content (0.2–4.9 ppm), very low Th/U ratio (*c.* 0.03) and a portion of common Pb up to 8.3%. Eighteen of the 20 concordia analyses yield an age of 1758 $\pm$ 11 Ma (MSWD 1.1, probability 0.3), which can be taken as the crystallization age of the PO2 population.

Additionally, SHRIMP analysis of zircon grains coexisting with titanite showed their variable U–Pb isotope characteristics (Supplementary Material S4). Zircon grains (100 × 50 µm) located on the grain boundary of titanite contain U (530–700 ppm), Th (265–470 ppm) and a low portion of common Pb (0. 11–0.55%). The significant degree of discordance

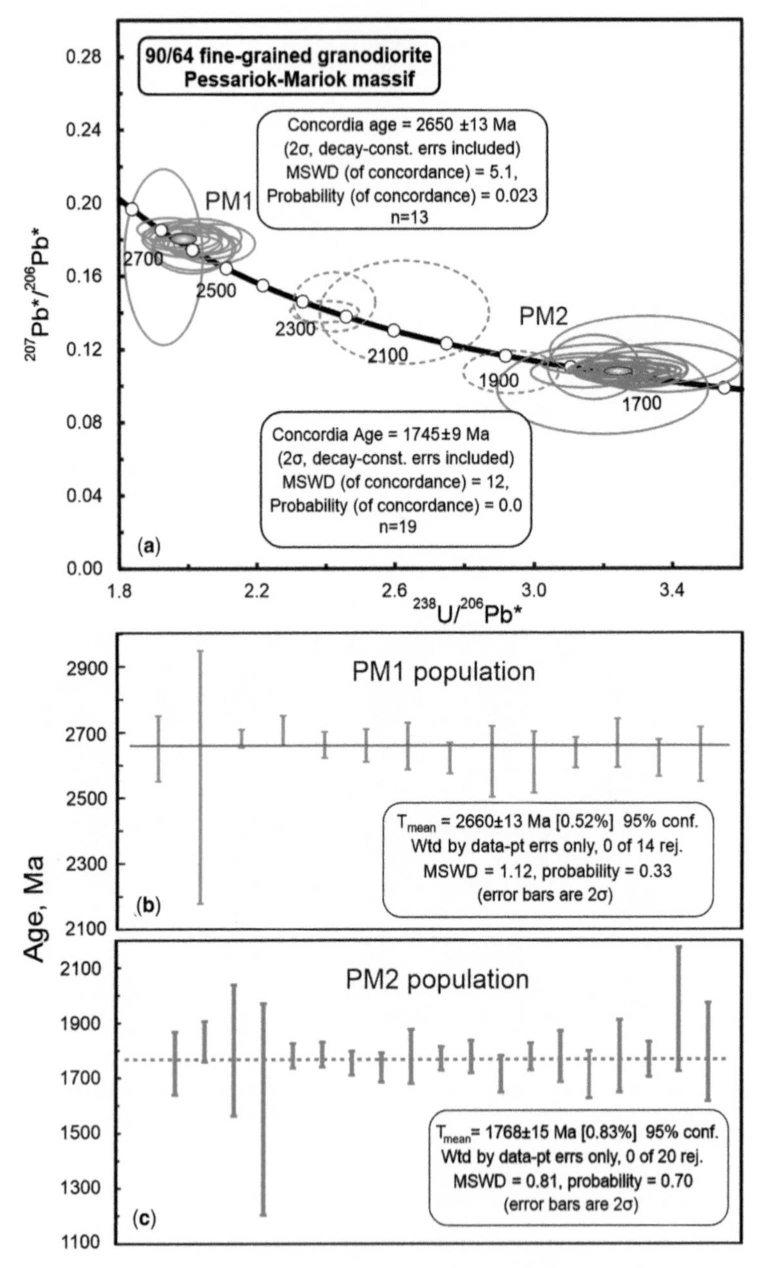

Fig. 8. Tera–Wasserburg diagram (**a**) and weighted mean age calculations based on the radiogenic $^{207}Pb^*/^{206}Pb^*$ ratios corrected for common Pb for the Pessariok–Mariok (PM1) (**b**) and PM2 (**c**) titanite populations from the Pessariok–Mariok massif. Ellipses are coloured according to the titanite REE populations (Fig. 2). Solid lines indicate analyses used for age calculation; dashed lines mark excluded discordant analyses. All data are ^{204}Pb-corrected and uncertainties are quoted at the 2σ level. The inset text contains the number of analyses (*n*) used for age calculations.

(20%) does not allow construction of a discordia-line with satisfactory parameters. However, three analyses of this zircon grain yield a weighted average $^{206}Pb^*/^{238}U$ age of 1959 ± 170 Ma with a high MSWD value of 9.9. Three tiny zircon grains (<20 μm) settled in the titanite matrix at the intersection of intragranular cracks, characterized by elevated U and Th contents (760–1560 and 710–1870 ppm,

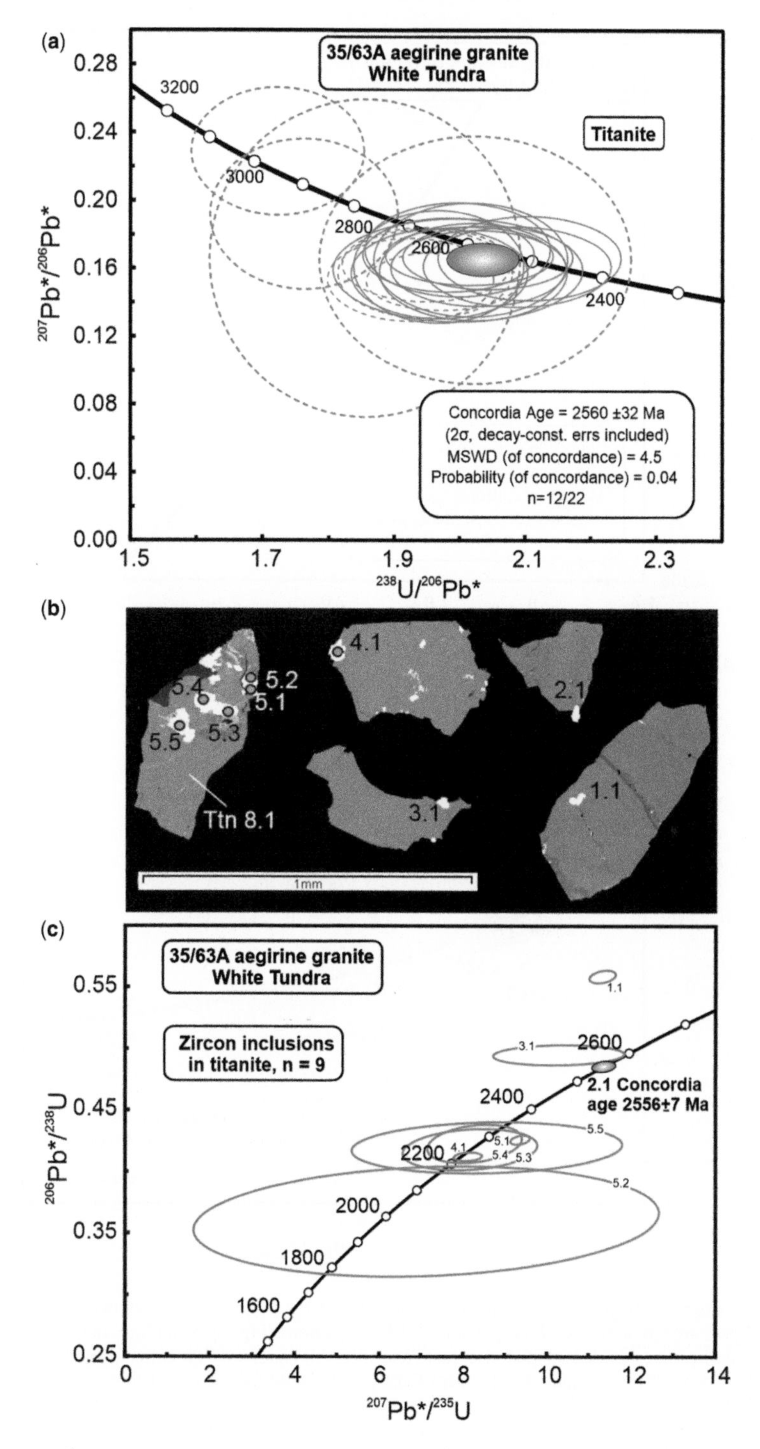

Fig. 9. U–Pb concordia plots for titanite grains and zircon inclusions from the White Tundra massif (**a**, **c**) and corresponding BSE images of zircon inclusions (bright) hosted in titanite (dark) (**b**). Solid lines indicate analyses used for age calculation; dashed lines mark excluded discordant outliers. All data are ^{204}Pb-corrected and uncertainties are quoted at the 2σ level. The inset text contains the number of analyses (*n*) used for age calculations. In (**b**) and (**c**) spot numbers correspond to the data from Supplementary Material S4 respectively.

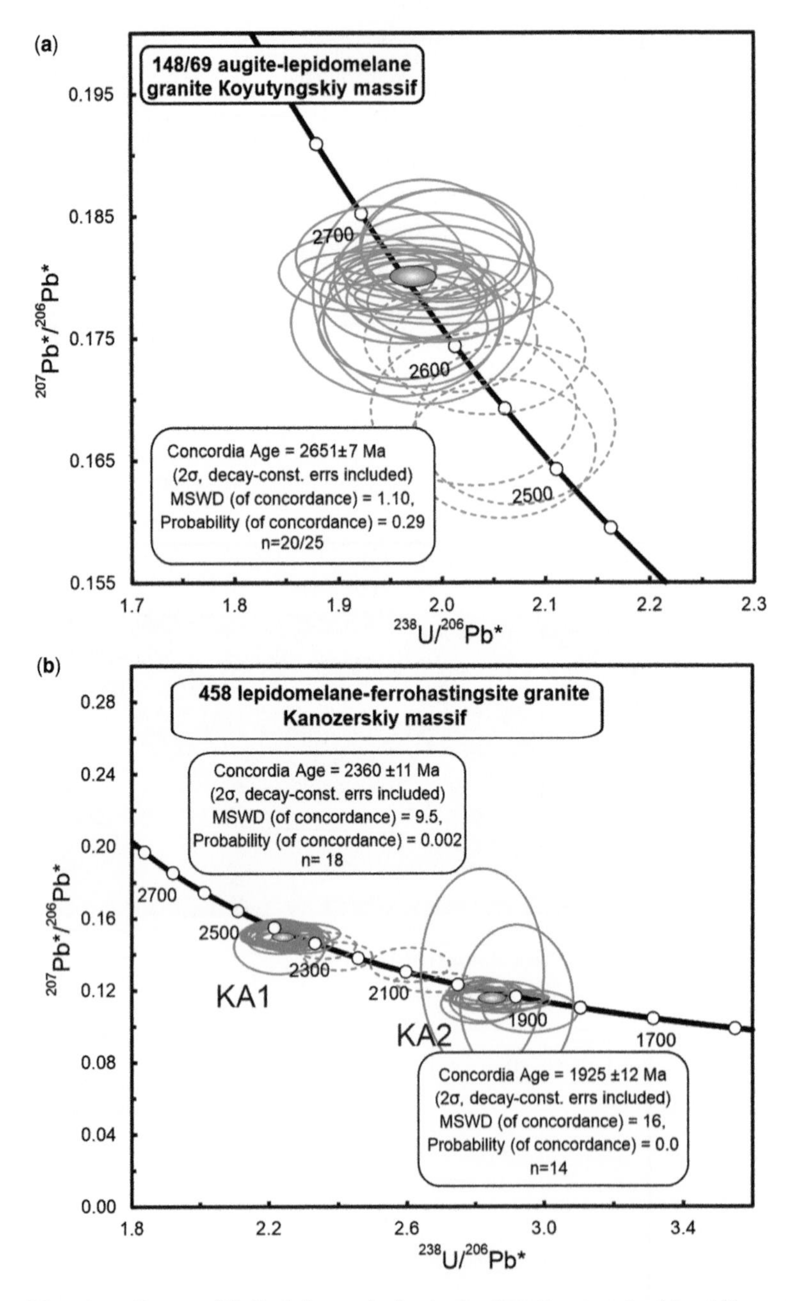

Fig. 10. Tera–Wasserburg diagram of U–Pb dating results for titanites from Koyutyngsky (**a**) and Kanozerskiy (**b**) massifs. Ellipses are coloured according to titanite REE populations (Fig. 2). Solid lines indicate analyses used for age calculation; dashed lines mark excluded discordant analyses. All data are ^{204}Pb-corrected and uncertainties are quoted at the 2σ. The inset text contains the number of analyses (n) used for age calculations.

respectively), have a low Th/U ratio (1.13 ± 0.15) and varying portion of common Pb ranging from 1.1 to 8.3%. The upper intersection of discordia with concordia yield an age of 2208 ± 17 Ma (MSWD, 2.8, lower intercept at 191 ± 75 Ma) (Fig. 11c). This age estimation seems unreliable because of the discordancy of only three analyses of zircon inclusions in different titanite grains.

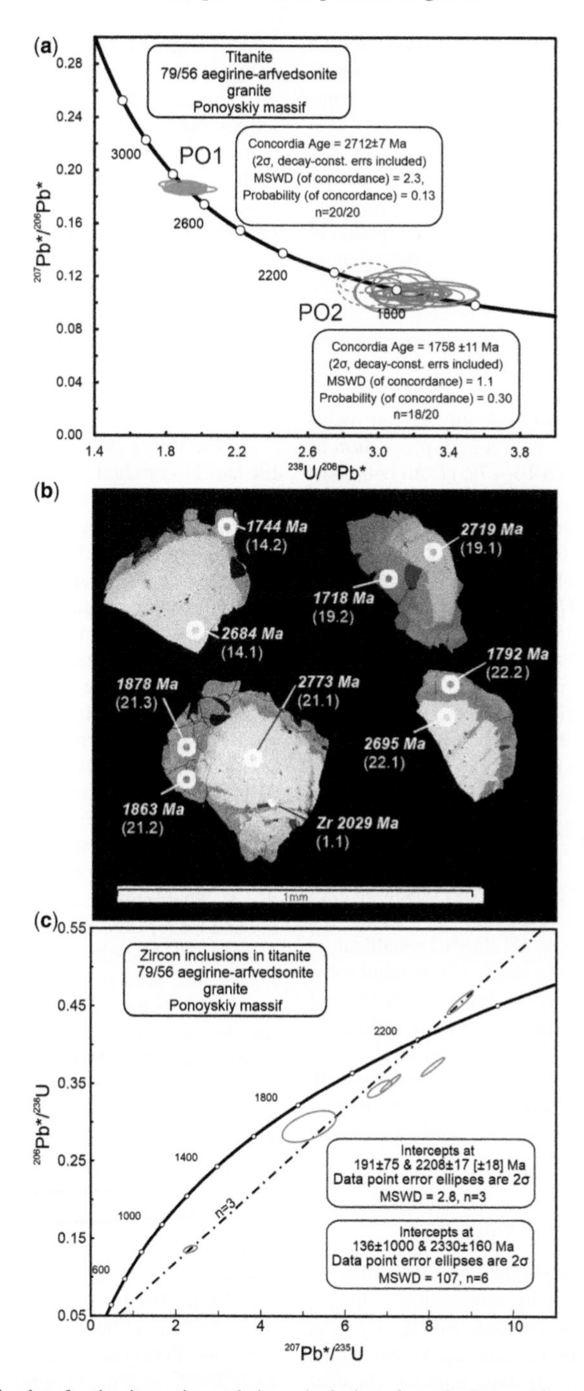

Fig. 11. U–Pb concordia plots for titanite grains and zircon inclusions from the Ponoyskiy massif (**a**, **c**) and corresponding BSE images (**b**) for titanite samples that record different ages for the PO1 and PO2 populations. Ellipses colours are similar to those in Figure 3. Solid lines indicate analyses used for age calculation; dashed lines mark excluded discordant analyses. All data are ^{204}Pb-corrected and uncertainties are quoted at the 2σ level. The inset text contains the number of analyses (n) used for age calculations. In the BSE images spot numbers and ages correspond to analyses from Supplementary Material S2 and S4.

Iokangskiy massif. The U–Pb system of the studied titanite in the lepidomelane–ferrohastingsite granite is characterized by a highly variable share of common Pb varying from 0.6 to 9.4% of ^{206}Pb (with one extreme exception of 65%). The U and Th contents range within 10–170 and 11–132 ppm, respectively, and the Th/U ratio is 0.2–1.9 (average 0.7) (Supplementary Material S2). The relatively high content of radiogenic lead in this titanite (^{206}Pb 35 ± 19 ppm) makes it possible to calculate the concordant age in the coordinates $^{238}U^*/^{206}Pb^*$–$^{207}Pb^*/^{206}Pb^*$ using 24 of 26 analyses: 2647 ± 7 Ma (MSWD of concordance 1.2 and probability 0.3; Fig. 12a).

Local U–Pb analysis of four tiny zircon inclusions (<30 µm, Fig. 12b) with a high proportion of non-radiogenic lead (^{206}Pb 10–27%) from two titanite grains showed a high degree of disturbance of the isotope system (discordance 37–67%). However, the discordia based on all four analyses has an upper intersection with the concordia line corresponding to an age of 2638 ± 370 Ma (MSWD 0.14 and a lower intercept age of 410 Ma; Fig. 12c; Supplementary Material S4). Despite the high errors, the obtained age of zircon inclusions is in accordance with the data on host titanite grains and could be interpreted as a time of the primary zircon and titanite magmatic crystallization.

Discussion

The wide spectra of the obtained U–Pb isotope ages of different populations of titanite can indicate the time of different processes: (1) direct crystallization from the alkaline granite melt; (2) metamorphic reactions involving the breakdown of Zr-bearing minerals; and (3) solid-state *in situ* recrystallization triggered by metamorphic/metasomatic fluids. Coupling titanite and zircon U–Pb isotopic analyses with REE signatures can help to decipher the mineral growth history during magmatic and subsequent metamorphic cycles.

Magmatic titanite v. magmatic zircon

One of the key features of peralkaline silicic magmas is their relatively low liquidus temperatures, despite having elevated bulk iron contents which vary from 2 wt% for incipiently peralkaline up to more than 7 wt% for the most peralkaline granites (Scaillet *et al.* 2016). Petrologic evidence shows that most peralkaline silicic magmas crystallize in invariably water-rich conditions, with H_2O_{melt} >4 wt%, temperatures below 800°C and a redox state at or below NNO-1 (Scaillet and Macdonald 2001). Arfvedsonite occurrence as a near-liquidus mineral indicates H_2O-rich conditions and temperatures below 750°C. Higher fO_2 promotes aegirine crystallization at the expense of amphibole. However, aegirine may also appear at low fO_2 but at lower temperatures than amphibole (Scaillet *et al.* 2016). Magmatic titanite (Ttn) in granites most likely formed because of hydration breakdown processes by the reaction suggested by Broska *et al.* (2007) involving amphibole (Amph), K-feldspar (Kfs), biotite (Bt), titanomagnetite (Ti-Mag) and quarz (Qz) as follows:

$$7\,\text{Amph} + 2\,\text{Kfs} + 3\,\text{Ti-Mag} + 3\,H_2O$$
$$= 3\,\text{Ttn} + 2\text{Bt} + \text{Qz}$$

In the Keivy peralkaline granites Ti-bearing biotite and Ti-enriched magnetite or ilmenite–magnetite composite grains are common (Batieva 1976) and both serve as a source of Ti for the formation of titanite. The presence of titanite in association with Ti-magnetite, feldspar and quartz in the Keivy granites indicates a relatively higher oxygen fugacity (fO_2) close to the FMQ buffer. Under oxidizing conditions, some Ce should exist as Ce^{4+}, and thus Ce anomalies in titanite have the potential to be used to monitor oxygen fugacity. However, none of the Keivy peralkaline granite titanites exhibit Ce/Ce* anomalies, which is typical of the majority of titanites of magmatic, metamorphic and metasomatic origin (e.g. Cao *et al.* 2015; Garber *et al.* 2017; Gros *et al.* 2020), except for varieties in the Yukon granodiorite, which show a weak positive Ce anomalies (Che *et al.* 2013). Alternatively, the pronounced positive anomaly and Ce^{4+}/Ce^{3+} ratio of the Ponoyskiy magmatic zircons (7–29) which is one order of magnitude higher that than in the secondary zircons (1–2), is suggested to be an efficient geochemical indicator for the discrimination of genetic types of zircons (Balashov and Skublov 2011). The flat and even negative slope of the MREE–HREE pattern observed in several titanite rims (Figs 2 & 3) may reflect the growth of titanite after primary magmatic HREE-rich zircon, xenotime, fergusonite and thorite (Fig. 4).

Zircon saturation thermometry performed by Vetrin (2019) for samples which were previously used for geochronological studies (Vetrin and Rodionov 2009) fix the crystallization temperature for White Tundra granites at 870–965°C and the melt crystallization pressure is 1.1–1.2 kbar. Similarly, the Ponoyskiy granites yield temperatures of 880–950°C and a pressure of 0.5 kbar. The given high temperature estimations hardly reflect the real temperature of peralkaline granitoid melt crystallization (Scaillet *et al.* 2016) owing to an underestimation of melt-composition change during plutonic granite formation (Siégel *et al.* 2018; Crisp and Berry 2022). However, as a first approximation, these figures appear to be reconcilable with a

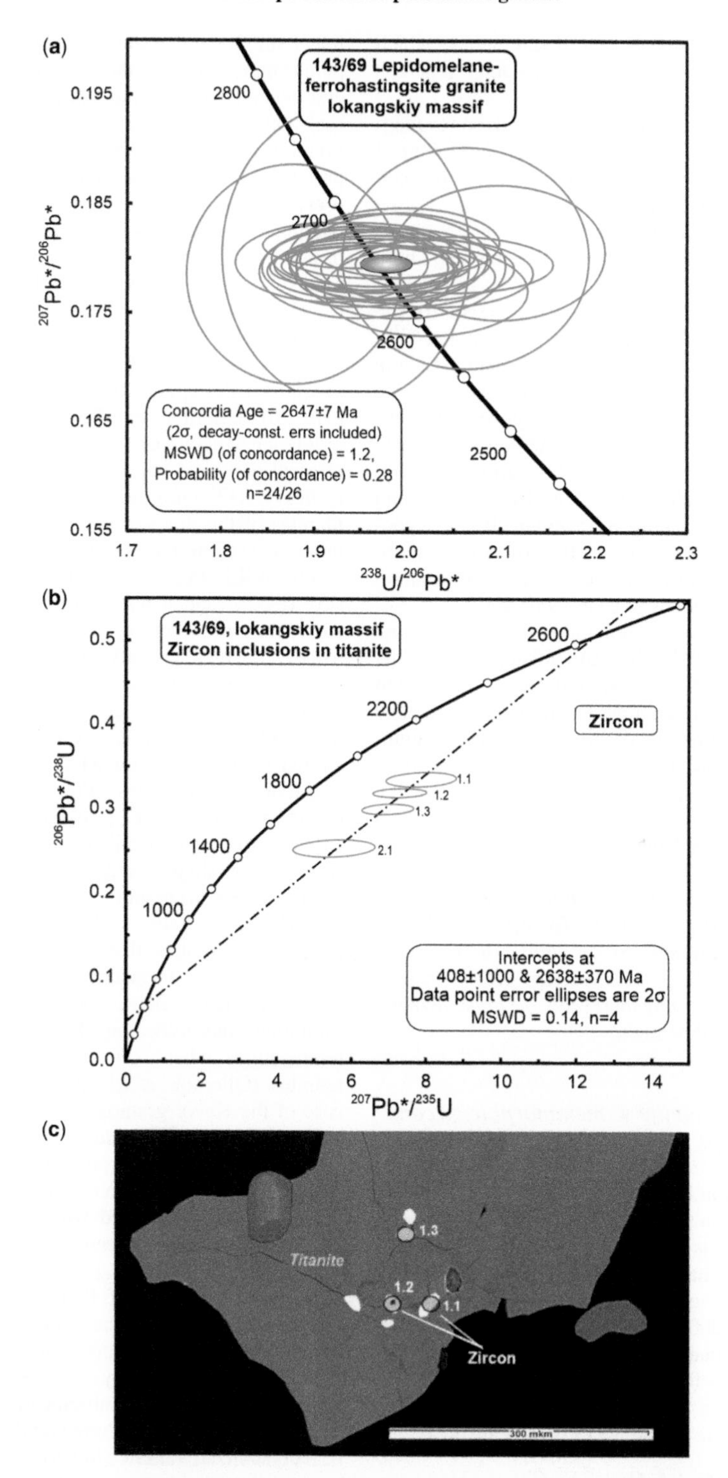

Fig. 12. U–Pb concordia plots (**a**, **c**) and corresponding BSE images (**b**) for zircon inclusions from the Iokangskiy massif. All data are ^{204}Pb-corrected and uncertainties are quoted at the 2σ level. The inset text contains the number of analyses (n) used for age calculations. In (**b**) and (**c**) spot numbers correspond to analyses from Supplementary Material S4.

magmatic growth both of zircons and titanites which fall within the scatter of U–Pb age determinations ranging from 2712 to 2647 Ma.

Two age clusters have been recognized for magmatic titanites. The obtained U–Pb geochronological data for titanite cores from the Ponoyskiy and Western Keivy massifs defined the older ages of 2712 ± 7 and 2710 ± 13 Ma respectively, which are considered as the crystallization ages of the granites. These age estimations are significantly older than the previously published zircon ages 2666 ± 10 Ma for the Ponoyskiy massif (Vetrin and Rodionov 2009). Revisions of the U–Pb data for particular titanite cores in (Fig. 11) show a wide scatter of ages ranging from 2684 to 2719 Ma. Similarly, previously published ages of the same rocks of the Ponoyskiy massif also define the age interval for particular zircon grains between 2661 and 2724 Ma (Vetrin and Rodionov 2009). Taking into account all sets of geochronological data we suggest the probable crystallization age of the primary zircon–titanite association in peralkaline granites to be *c.* 2710 Ma.

The U–Pb geochronological data for titanite cores from the PM1, KO and Iokangskiy massifs that are assigned to the second cluster fall within the *c.* 2650 Ma. In contrast to the older peralkaline massifs of the first cluster, the above massifs comprise subalkaline rocks varying in composition from monzonite to augite–lepidomelane granite. Moreover, whereas the KO and Iokangskiy massifs are located beyond the Keivy terrane in the Murmansk domain, the PM intrusion is settled in the fault zone adjacent to the western contact of the Keivy terrane (Fig. 1). The REE patterns of these titanites are suggested to be similar to the cores from the Ponoyskiy and Western Keivy massifs (Fig. 2), thus indicating their probable crystallization from the peralkaline melt.

Metamorphic titanite v. metamorphic zircon and other REE-bearing minerals

The formation of hydrothermal titanite, especially in the case of titanite rims around titanomagnetite in Keivy granites, is most likely due to hydration reactions involving hastingsite, plagioclase, chlorite and magnetite. The equilibrium reaction proposed by Broska *et al.* (2007) for the formation of hydrothermal titanite explains its presence in the Keivy granites, which is as follows:

$$6\,\text{Fe-Hst} + 12\,\text{Ilm} + 4\,O_2 + 6\,H_2O = 12\,\text{Ttn} + 6\,\text{Ab} + 3\,\text{Fe-Chl} + 8\,\text{Mag}$$

where Ttn is titanite, Fe-Hst is ferrohastingsite, Ilm is ilmenite, Ab is albite, Chl is chlorite and Mag is magnetite. Regardless of the magmatic or metamorphic origin of titanite, the REE balance of this mineral depends both on the composition of the whole rock and on the relative crystallization/recrystallization time of titanite compared with other minerals. In particular, titanite reveals a depleted LREE pattern if it crystallized after an assemblage of rich in LREE-bearing minerals. Since the monazite, chevkinite, bastnaesite and britholite are typical accessories in peralkaline granites, they control the LREE budget to some extent. On the other hand, titanite can be enriched in LREE if it crystallizes after zircon which contains predominantly HREE (Bruand *et al.* 2014; Papapavlou *et al.* 2017; Scibiorski *et al.* 2019). In order to differentiate whether titanite grains are igneous or metamorphic, we used trace element data, and particularly, Th/U, Th/Pb, light to heavy REE ratio and Eu anomalies. Unlike the titanites of the first group already discussed in the previous section, the populations from the Western Keivy (WK3), Pessariok–Mariok (PM2) and Ponoyskiy (PO2) yielded an age cluster of *c.* 1860 Ma. According to Olierook *et al.* (2019), LREE to MREE are preferentially incorporated over HREE at higher temperatures, whereas at lower temperatures MREE to HREE are more dominant. Titanites of 1865 Ma of the Western Keivy cluster are characterized by hump-shaped REE patterns, and strong negative Eu anomalies (Fig. 2b). Similarly, the depletion in LREE influenced the typical LREE-enriched mineral like apatite, which also shows hump-shaped REE patterns in the presence of the above accessories (Fig. 4). The relatively low $(La/Sm)_N$ ratio (<1) of the above titanite populations implies that LREE are more readily transferred to the formation fluids during metamorphism, whereas MREE and HREE are relatively immobile, thus indicating the lower crystallization temperatures when compared with the magmatic titanites (Olierook *et al.* 2019) (Fig. 13). In the case of the Keivy granites this is illustrated by the uniform Gd–Lu REE patterns for both titanite magmatic cores and metamorphic rims (Fig. 3). Furthermore, the coexisting zircon grains display highly variable LREE combined with the stable MREE–HREE patterns, thus indicating the same REE behaviour for both the igneous and metamorphic varieties (Fig. 3). Comparison of REE data for coexisting titanite and zircon indicates opposite REE trends during metamorphic recrystallization of igneous grains. Whereas primary titanites show dramatic loss of LREE during subsequent metamorphism, the coexisting zircons exhibit significant enrichment of recrystallized varieties in LREE, probably owing to mutual redistribution of elements in these two phases (Fig. 3).

Pronounced negative Eu anomalies observed in titanite rims and several cores similar to that in the

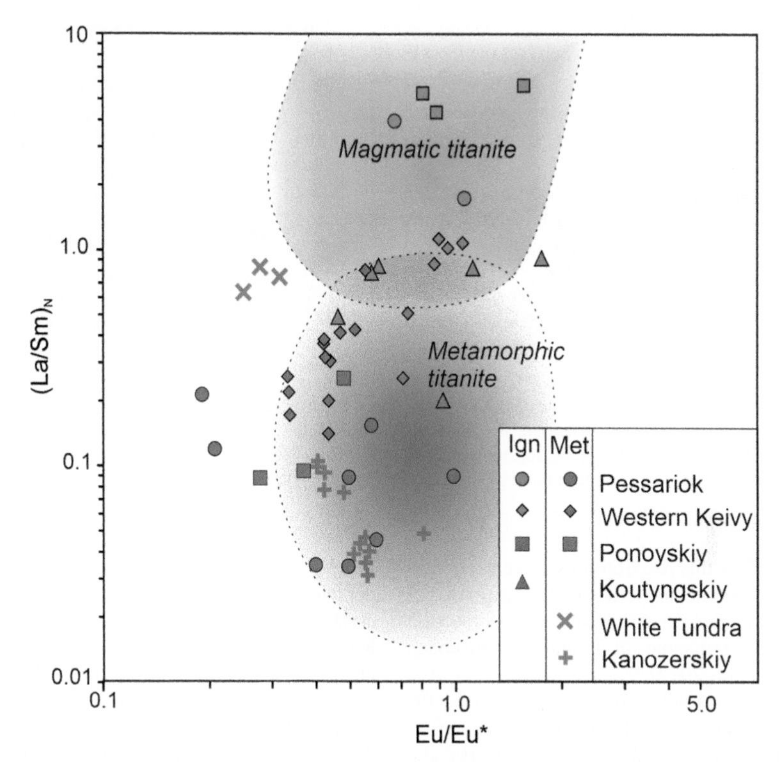

Fig. 13. Titanite trace element discrimination indices $(La/Sm)_N$ v. $(Eu/Eu^*)_N$ of the igneous (Ign, red symbols) and metamorphic (Met, green symbols) titanites from the Keivy alkaline granites. Fields of the igneous and metamorphic titanites from the Precambrian felsic intrusive rocks of Western Australia (Olierook *et al.* 2019) are shown for comparison.

bulk rock REE patterns of the alkaline granites (Fig. 2d) imply their formation in the presence of plagioclase (Schaltegger *et al.* 1999). Thus, the crystallization of plagioclase prior to, or during, titanite saturation would deplete the melt in Eu, generating a negative Eu/Eu* anomaly which would be inherited by titanite crystallizing subsequently. We suggest that the plagioclase effect and/or participation of other REE accessories, such as monazite, xenotime, chevkinite, epidote group minerals (allanite) and zircon (Frei *et al.* 2004; Hawthorne *et al.* 2012), and their breakdown during prograde metamorphism may also affect the REE distribution in titanite. Recrystallized titanite grains can be distinguished by the presence of zircon inclusions, rimming allanite around titanite and/or significant LREE and Th depletions not observed in the primary grains (Fig. 2). Additionally, chevkinite rims around partly substituted titanite grains are common in the Western Keivy granites (Fig. 14a). The wide range of Th/U and Th/Pb ratios in titanites and co-existing zircons does give some indication of possible trace element redistribution at *c.* 1860 Ma. Recrystallization of the primary zircon resulted in a decrease in Th/U and Th/Pb ratios accompanied by a uniform trend observed in the primary titanites (Fig. 15).

Apart from zircon, allanite will also strongly fractionate Th from Pb, and the recrystallization of primary allanite may have decreased the Th/Pb ratio of co-existing titanite. The allanite surrounding the titanite appears to be a reaction product between titanite and the surrounding mineral assemblage, i.e. plagioclase, K-feldspar and quartz, during metamorphic alteration (Fig. 14b). Such a textural and chemical relationship indicates the reaction:

$$\text{Ttn} + \text{Plag} + \text{Kfs} + H_2O = \text{Alla} + \text{Ilm} + \text{Mus}$$

where Ttn is titanite, Plag is plagioclase, Kfs is K-feldspar, Alla is allanite, Ilm is ilmenite and Mus is muscovite. In the case of the Pessariok–Mariok monzonite, Iokangskiy and Ponoyskiy augite–lepidomelane alkaline granites in which allanite is a ubiquitous accessory mineral (Batieva 1976), the elemental losses of LREE combined with low Th/U and Th/Pb can be explained by coeval growth of allanite during recrystallization. The amount of

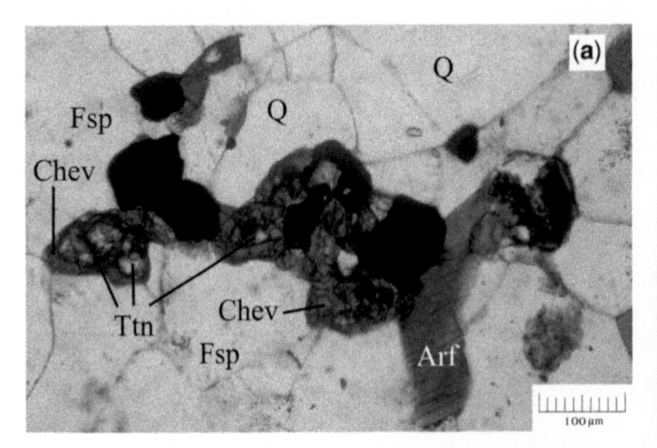

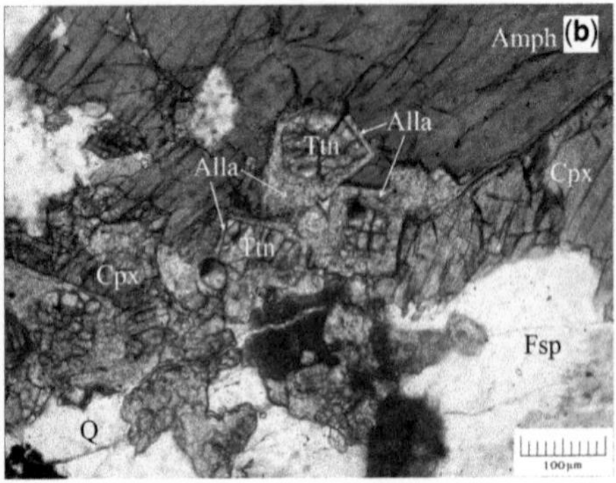

Fig. 14. Photomicrographs: (**a**) of the Western Keivy aegirine–arfvedsonite sample TM-109, showing titanite crystals (Ttn) partially replaced by deep-red chevkinite–(Ce) rims (Chev), set in a matrix of arfvedsonite (Arf), quartz (Q) and feldspar (Fsp); (**b**) lepidomelane–ferrihastingsite granite sample B672 (Pessariok massif) showing titanite crystals (Ttn) rimmed by allanite (Alla), set in a matrix of amphibole (Amph), quartz (Q) and feldspar (Fsp).

allanite formed by this reaction must have been miniscule relative to the amount of titanite involved, because the titanites on either side of the reaction have essentially the same Si, Ti and Ca concentrations (Garber *et al.* 2017).

The other phases which may have affected REE distribution in titanite are chevkinite and monazite. According to mineralogical observations of Batieva (1976), chevkinite, monazite and britholite are the dominant accessory phases in peralkaline granites of the Western Keivy, White Tundra and other massifs, whereas allanite occurs sporadically. At normal crustal pressures, larger negative Eu/Eu* anomalies in monazite, chevkinite and britholite (Fig. 4), as well as in titanite, have been related to the growth of feldspar from melt (Nagy *et al.* 2002; Rubatto *et al.* 2013; Holder *et al.* 2015). However, this interpretation is complicated by the sensitivity of Eu^{2+}/Eu^{3+} ratios to fO_2 (Wilke and Behrens 1999; Aigner-Torres *et al.* 2007), which may not be constant, especially in magmatic and metamorphic environments. For example, fergusonites from the Keivy granite show negative Eu/Eu* anomalies (Fig. 4), whereas the samples from the Keivy metasomatites and amazonite pegmatites (Rova quartzolite, the El'ozero and the Ploskaya) have positive Eu/Eu* anomalies (Zozulya *et al.* 2020). While the above accessories do not incorporate significant Pb during crystallization (e.g. Parrish 1990; Suzuki and Adachi 1991; Holder *et al.* 2015; Zozulya *et al.* 2019), the Th/Pb ratio is also indicative of the coeval crystallization of titanite in the presence of monazite, chevkinite and britholite during subsequent metamorphism (Fig. 15).

Variations in titanite HREE signatures can provide additional information about the conditions during metamorphism operated at the time of titanite (re) crystallization. Titanite that crystallized in a garnet-bearing assemblage is significantly more HREE depleted (higher $(Gd/Yb)_N$ ratios) than titanite that crystallized in the absence of garnet (e.g. Kohn and Kelly 2018, and references therein; Scibiorski *et al.* 2019). In the Keivy terrane the highest $(Gd/Yb)_N$ ratios (3.71–4.95) exhibit igneous titanites (WK1, 2710 Ma) from Western Keivy massif, and are thus diagnostic of growth in a garnet-bearing assemblage of amphibolite facies. However, scarce garnet-bearing granite varieties together with the host Lebyazhka gneisses occur only in intensely sheared zones along the deformed contacts of massifs (Mudruk *et al.* 2013). Titanites in the 2650 Ma population as well as titanites of 1876–1745 Ma populations have a low average $(Gd/Yb)_N$ ratio of 1.77 ± 0.91 (1σ), thus indicating that titanite experienced metamorphic reworking under mid-amphibolite-facies conditions in the absence of garnet.

Implications for the geological history of the Keivy peralkaline granite magmatism

The Neoarchean period. During the Late Neoarchean, the continental crust in the NE of the Fennoscandian Shield was exposed to large-scale and widespread magmatism associated with supercontinental cycles. The well-constrained tectonothermal events in the adjacent areas of the Fennoscandian Shield strongly affected igneous activity within the Keivy terrane. The majority of obtained titanite ages broadly fall into several groups, each of which is associated with a previous event in the adjacent areas of the Fennoscandian Shield (Fig. 16). Available data revealed differences in the U–Pb ages and trace element compositions of the groundmass titanite and zircon inclusions. This suggests that the material that constitutes the Keivy granites crystallized from separate magmas in at least two stages.

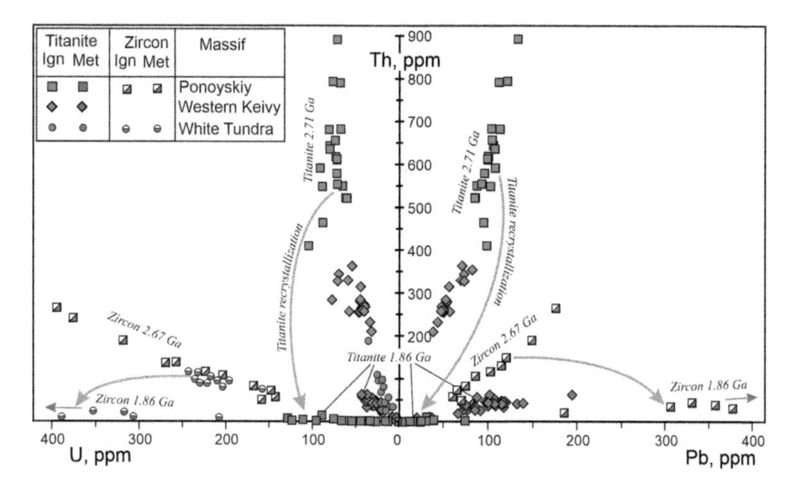

Fig. 15. Plot of U v. Th and Pb v. Th for titanite and zircon populations of the igneous (Ign, red symbols) and metamorphic (Met, green symbols) titanites from the Keivy peralkaline granites. Zircon data are from (Vetrin and Rodionov 2009).

The initial stage involved relatively rapid final emplacement of peralkaline granites from compositionally evolved magmas at *c.* 2710 Ma, triggered by the preceding long-lived endogenous activity in the adjacent Kolmozero–Voron'ya greenschist belt. Titanite U–Pb ages from this study fall into main trends with the data yielding the formation of a wide spectrum of sanukitoids represented by Porosozero and Kolmozero multiphase intrusions located in the northwestern contact of the Keivy terrane. According to Kudryashov *et al.* (2013), monzodiorite–granodiorite of the initial phase of the Porosozero complex yielded a U–Pb zircon age of 2733 ± 6 Ma, whereas the final leucogranite was emplaced at 2712 ± 6 Ma. In addition, anorthosite sills of 2730 Ma age (Vrevskii 2016) are developed in the volcanogenic amphibolite sequence (the Polmos Tundra Formation) on the eastern flank of this structure. Moreover, Vrevskii (2016) reported the presence of plagioclase–microcline granite bodies cutting the komatiite (2697 ± 10 and 2696 ± 9 Ma) in the western part of the belt. Thus, summarizing the above data we suggest that the earliest 2710 Ma pulse of alkaline magmatism records the extinction of endogenous events in the adjacent Kolmozero–Voron'ya belt.

The time interval of the main stage of alkaline magmatism in the Keivy terrane is the subject of discussion. Current models for the geodynamic evolution of the Keivy terrane suggest an extremely broad period of the peralkaline granite igneous activity of 2750–2610 Ma with a maximum constrained to 2670 Ma (Bayanova 2004; Zozulya *et al.* 2005; Vetrin and Rodionov 2009; Nitkina and Serov 2022; Zakharov *et al.* 2022). The Western Keivy, White Tundra, Ponoyskiy and Kanozerskiy granites have been considered to have formed synchronously with the enclosing Lebyazhka acid metavolcanics (Balagansky *et al.* 2021) and gabbro-anorthosite massifs (Bayanova 2004; Vrevsky and Lvov 2016; Kudryashov *et al.* 2019), suggesting that these three types of rocks should have a common origin (Zozulya *et al.* 2005; Balagansky *et al.* 2021). Alternatively, our data for titanite cores and zircon inclusions from the PM1, KO and Iokangskiy massifs fall within the age of *c.* 2650 Ma. Although titanite from granitic rocks may yield younger ages than zircon owing to open system behaviour during slow cooling (Corfu *et al.* 1985; Aleinikoff *et al.* 2002), titanites from the other massifs (except for the Western Keivy and Ponoyskiy) are several million years younger than their previously analysed zircon counterparts (Mitrofanov *et al.* 2000; Zozulya *et al.* 2005; Vetrin and Rodionov 2009; Nitkina and Serov 2022; Zakharov *et al.* 2022), with no apparent spatial control on age distribution. However, the obtained titanite ages are in agreement with the age of the Sakharjok massif (2645 ± 7 Ma; Vetrin and Belousova 2020), the hosting metaporphyres of the Lebyazhka formation (2647 ± 35 Ma; Balagansky *et al.* 2021) and the Tzaga anorthosites (2659 ± 3 Ma; Zozulya *et al.* 2005). In addition, the time span of 2670–2650 Ma is marked by the emplacement of anorthosites of the Patchemvarek massif which defines a concordant U–Pb zircon age 2662 ± 7 Ma (Vrevskii 2016). These data correspond within errors to the ages of the Achinsky (2678 ± 16 Ma) and Tsaginsky (2659 ± 3 and 2660 ± 10 Ma) anorthosite massifs (Bayanova 2004). Whereas the growth temperature of titanite (650–780°C) (Scott and St-Onge 1995; Pidgeon *et al.* 1996; Frost *et al.* 2001; Cherniak 2010) is

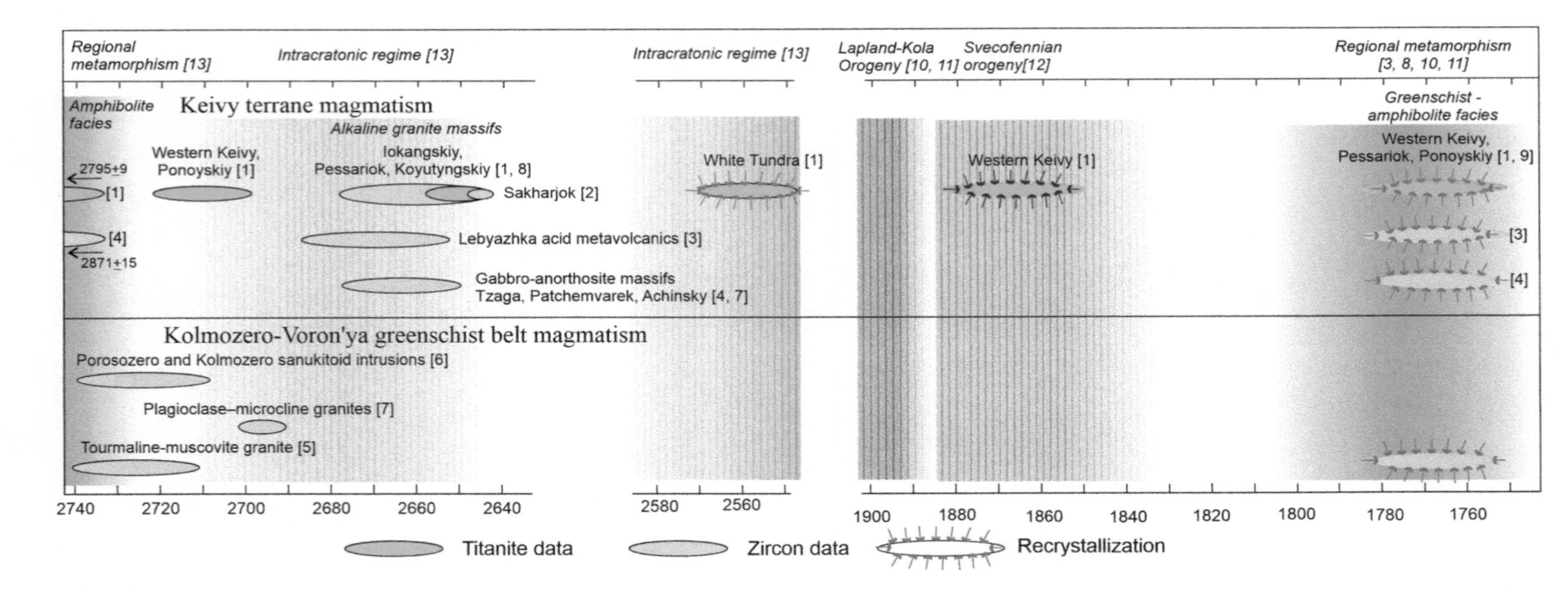

Fig. 16. Schematic chart illustrating relative and approximate absolute timing of magmatic events in the Keivy terrane and adjacent Kolmozero–Voron'ya greenstone belt. Note that the breadth of age bars is based on the reported ages including errors. Data sources: [1] our data; [2] Vetrin and Belousova (2020); [3] Balagansky *et al.* (2021); [4] Bayanova (2004); [5] Kudryashov *et al.* (2019); [6] Kudryashov *et al.* (2013); [7] Vrevskii (2016); [8] Vetrin and Rodionov (2009); [9] Zakharov *et al.* (2022); [10] Lahtinen and Huhma (2019); [11] Balagansky *et al.* (2015); [12] Balagansky *et al.* (2016); [13] Radchenko *et al.* (1992).

consistently higher than the amphibolite facies conditions that occurred during the subsequent metamorphism, precluding partial or total resetting of U/Pb in magmatic titanite, the most plausible explanation for the obtained wide time span of ages is that the Keivy terrane experienced a prolonged cooling history from the crystallization of the granitoids. Recent data from Zakharov *et al.* (2022) support the assumption that the Keivy peralkaline system represents a large and long-lived (several million years) shallow (<5 km depth) magmatic–hydrothermal system. Overall, it can be concluded that the obtained U–Pb titanite ages mark the termination of the *c.* 30 Ma time span of acidic alkaline magmatism and the whole magmatic activity in the Keivy terrane.

A specific group of the titanite cores which shows concordant U–Pb ages of 2795 ± 9 Ma was found in peralkaline granite near the contact of the Western Keivy massif (Fig. 3a). Unlike the other titanite clusters, this population is characterized by a high Th/U ranging from 4.6 to 7.1 and rich LREE patterns. The presence of entrained titanite has never been extracted before owing to previous analyses using whole grain isotope dilution TIMS methods. Therefore, the presence of xenogenic titanites owing to the contamination of primary melts in the intracrustal magmatic chambers is highly possible. Moreover, such xenogenic titanites as well as zircons will possess 'magmatic' geochemical characteristics and crystallization temperatures. The crustal contamination of primary melts by the crustal host rocks is also supported by the $\varepsilon_{Nd}(T)$ (deviation in Nd-isotope composition from the chondrite composition at the time T (in m.y. before present) in parts per 10000) (from +2.2 to −9.0) for the monzonites and peralkaline granites (Zozulya and Bayanova 2003; Vetrin 2018; Balagansky *et al.* 2021). We assume that the most likely candidate for the source of the old 2795 Ma titanite cores might be the rocks of the adjacent Kolmozero–Voron'ya greenschist belt. This Mesoarchean collisional suture zone comprising mantle plume komatiitic rocks, arc-type tholeiitic basaltic–andesitic–dacitic metavolcanic rocks and conglomerate-bearing terrigenous rocks intruded by 2730 Ma. According to Vrevskii (2016), the U–Pb age of zircons from the metadacite pile of the komatiite–tholeiite rocks intercalated with basalts yielded the age 2790 ± 9 Ma, which is in agreement with the obtained U–Pb age of titanite cores in peralkaline granite of the Western Keivy massif.

A titanite sample from the White Tundra records an age of 2560 ± 32 Ma, which is close to the age of 2555 ± 7 Ma determined in the zircon inclusion of this sample. The lower intercept age is not precise, and possible driving events can range from burying under the sediments of the *c.* 2600 Ma Tersky–Allarechenskiy belt to a thermal overprint related to emplacement of the nearby Fedorovo–Panskiy layered ultramafic complex at 2500 Ma (Bayanova 2004).

The Paleoproterozoic period. The major tectonic event in the Paleoproterozoic history of the Fennoscandian Shield was recorded in the Lapland–Kola collisional orogenesis during a time span between 1940 and 1780 Ma, which resulted in the development of Svecofennian and Lapland–Kola distinct orogens (Daly *et al.* 2006; Lahtinen and Huhma 2019) and had an impact on the Keivy terrane and alkaline granite massifs. The two statistically reliable datasets (130 spots) for titanite from Western Keivy, Pessariok–Mariok and Ponoyskiy massifs have yielded two age clusters of *c.* 1870 and *c.* 1770 Ma.

The first cluster is represented by titanites from Western Keivy massif, which have yielded statistically reliable ages of 1865 ± 29 Ma (100 spots). In contrast to the Neoarchean magmatic varieties, hump-shaped REE patterns and strong negative Eu anomalies are typical of these titanites. The obtained geochronological data match the age of metasomatic rocks in the shear zones related to collision-related metamorphism in the Lapland Granulite Belt (1960–1890 Ma) (Bushmin *et al.* 2009) and are roughly consistent with the age obtained for Kanozerskiy alkaline granites (1921 ± 53 Ma; Nitkina and Serov 2022). Overall, it seems more likely that the recorded Proterozoic age of titanite from the Western Keivy massif represents major resetting at *c.* 1870 Ma during the initial burial metamorphism in the period *c.* 1880 Ma related to the Svecofennian orogeny (Lahtinen *et al.* 2009).

The second cluster is represented by titanite samples from the Pessariok–Mariok and Ponoyskiy massifs collected in the marginal zones of the Keivy terrane. Extensive hydrothermal activity and fluid cycling near contemporaneous with regional metamorphism of 1790–1770 Ma resulted in the formation of the Rovozero quartzolite of the Western Keivy (1792–1766 Ma) (Zakharov *et al.* 2022) and muscovite metasomatic rocks of the Bolshie Keivy nappe (Bushmin *et al.* 2011). In addition, this event manifest itself in metaporphyric rocks of the Serpovidnyi Ridge of the Western Keivy (1740 ± 15 Ma) (Myskova *et al.* 2014) and was fixed in hydrothermal alteration zones of 1830–1780 Ma, related to alkaline and nepheline syenites of the Sakharjok massif (Vetrin and Belousova 2020).

The ages derived from core and rim zones of titanite grains of the Kanozerskiy intrusion clearly indicate that the formation and subsequent tectonothermal events of lepidomelane–ferrohastingsite granite were not totally coeval with the other alkaline granite massifs. Whereas the age of 2667 ± 36 Ma of Kanozero alkaline granite obtained by

Nitkina and Serov (2022) is consistent with the time of formation of the Western Keivy and other complexes, the age of aplite (2301 ± 13 Ma) is close to the age of the titanite cores, 2360 ± 11 Ma. Moreover, the obtained ages of titanite rims (1925 ± 12 Ma) exactly fit the age interval of 1921 ± 53 Ma, which is interpreted by Nitkina and Serov (2022) as the time of regional Svecofennian metamorphism.

Conclusions

New U–Pb isotope titanite and trace element data from alkaline granite massifs elucidate the Neoarchean tectonothermal histories of the northeastern Fennoscandian Shield. The obtained data suggest the following.

Two stages of alkaline magmatic activity are identified: Stage 1 (*c.* 2710 Ma) and Stage 2 (*c.* 2650 Ma). Alkaline granite massifs of the 2710 Ma stage are suggested to represent the final pulses of sanukitoid magmatism and thus record the extinction of endogenous events in the adjacent Kolmozero–Voron'ya greenschist belt. Peralkaline complexes of the 2650 Ma stage fall in the time interval of emplacement of the wide spectra of anorthosite and syenite intrusions, thus indicating the major endogenous event that occurred within 2670–2650 Ma time span and mark the termination of magmatic activity in the Precambrian terranes of the central Kola.

The occurrence of 1865 ± 29 Ma titanites demonstrates that the Neoarchean alkaline granite massifs affected significant transformations of granites, predominantly associated with subsequent tectonothermal events, and represent major resetting at *c.* 1880 Ma during the initial burial metamorphism in the period 1880 Ma related to the Svecofennian orogeny. Furthermore, the traces of extensive hydrothermal activity and fluid cycling near contemporaneous with the regional metamorphism of 1790–1770 Ma were revealed in titanite samples from the alkaline granite massifs of the marginal zones of the Keivy terrane.

In addition to geochronological data, which made it possible to distinguish the different stages of granite emplacement, one of the main results was the discovery of primary titanite igneous grains that preserved the crystallization age of granites. The retained U–Pb isotope characteristics of the titanite cores clearly indicate that despite the significant Pb diffusion at lower crustal temperatures and recrystallization, titanite is a more sensitive indicator of petrogenetic processes than zircon. Hence, the complementary U–Pb geochronological data of coexistent titanite and a more stable zircon provide additional information on crystallization, subsequent growth, cooling and alteration of the Precambrian plutonic complexes.

Acknowledgements We thank the Centre for Isotope Research at Karpinsky Geological Institute personnel for their help in the laboratory. A significant part of fieldwork investigations was done by the late Iya Batieva and Igor Belkov, whose mineral collection was used for the current work. We are grateful to Dr Bruno Ribeiro and an anonymous reviewer for their thorough and constructive reviews.

Competing interests The authors declare that they have no known competing financial interests or personal relationships that could have appeared to influence the work reported in this paper.

Author contributions **AAA**: conceptualization (equal), investigation (equal), writing – original draft (lead), writing – review & editing (lead); **BVB**: conceptualization (equal), investigation (equal), writing – original draft (supporting), writing – review & editing (supporting); **NVR**: data curation (equal), investigation (equal), writing – original draft (supporting), writing – review & editing (supporting); **AVA**: data curation (equal), investigation (equal), writing – review & editing (supporting); **ENL**: data curation (supporting), investigation (supporting), writing – review & editing (supporting); **SAS**: conceptualization (equal), supervision (equal), writing – review & editing (supporting).

Funding This work was funded as a part of the project 'Thematic and methodic works within 2018–20 aimed in development of laboratory and analytical methods under the State Geological Survey' performed by the Centre for Isotope Research at Karpinsky Geological Institute and Ministry of Education and Science (project FMUW-2022-0004 for the Institute of Precambrian Geology and Geochronology), and by the Ministry of Natural Resources (state contract 049-00017-20-04).

Data availability All data generated or analysed during this study are included in this published article and its supplementary information files.

References

Aigner-Torres, M., Blundy, J., Ulmer, P. and Pettke, T. 2007. Laser-ablation ICPMS study of trace element partitioning between plagioclase and basaltic melts: an experimental approach. *Contributions to Mineralogy and Petrology*, **153**, 647–667, https://doi.org/10.1007/s00410-006-0168-2

Aleinikoff, J.N., Wintsch, R.P., Fanning, C.M. and Dorais, M.J. 2002. U–Pb geochronology of zircon and polygenetic titanite from the Glastonbury Complex, Connecticut, USA: an integrated SEM, EMPA, TIMS, and SHRIMP study. *Chemical Geology*, **188**, 125–147, https://doi.org/10.1016/S0009-2541(02)00076-1

Amelin, Y.V. 2009. Sm–Nd and U–Pb systematics of single titanite grains. *Chemical Geology*, **261**, 53–61, https://doi.org/10.1016/j.chemgeo.2009.01.014

Anders, E. and Grevesse, N. 1989. Abundances of the elements: meteoritic and solar. *Geochimica et Cosmochimica Acta*, **53**, 197–214, https://doi.org/10.1016/0016-7037(89)90286-X

Bagiński, B., Zozulya, D., Macdonald, R., Kartashov, P.M. and Dzierżanowski, P. 2016. Low-temperature hydrothermal alteration of a rare-metal rich quartz–epidote metasomatite from the El'ozero deposit, Kola Peninsula, Russia. *European Journal of Mineralogy*, **28**, 789–810, https://doi.org/10.1127/ejm/2016/0028-2552

Balagansky, V., Shchipansky, A. *et al.* 2015. Archean Kuru–Vaara eclogites in the northern Belomorian Province, Fennoscandian Shield: crustal architecture, timing and tectonic implications. *International Geology Review*, **57**, 1543–1565, https://doi.org/10.1080/00206814.2014.958578

Balagansky, V., Gorbunov, I. and Mudruk, S. 2016. Palaeoproterozoic Lapland–Kola and Svecofennian Orogens (Baltic Shield). *Herald of the Kola Science Centre of the Russian Academy of Sciences*, **3**, 5–11 [in Russian], http://www.kolasc.net.ru/russian/news/vestnik1/html

Balagansky, V.V., Myskova, T.A., Lvov, P.A., Larionov, A.N. and Gorbunov, I.A. 2021. Neoarchaean A-type acid metavolcanics in the Keivy Terrane, northeastern Fennoscandian Shield: geochemistry, age, and origin. *Lithos*, **380–381**, 105899, https://doi.org/10.1016/j.lithos.2020.105899

Balashov, Y.A. and Skublov, S.G. 2011. Contrasting geochemistry of magmatic and secondary zircons. *Geochemistry International*, **49**, 594–604, https://doi.org/10.1134/S0016702911040033

Batieva, I.D. 1976. *Petrology of Alkaline Granitoids of the Kola Peninsula*. Nauka, Leningrad [in Russian].

Batieva, I.D. and Bel'kov, I.V. 1984. *The Sakharjok Alkaline Massif: Rocks and Minerals*. Kola Branch Academy of Sciences of the USSR, Apatity [in Russian].

Bayanova, T.B. 2004. *Age of Reference Geological Complexes of the Kola Region and the Duration of Magmatic Processes*. Nauka Publishers, St Petersburg [in Russian].

Belkov, I.V. 1963. *Kyanite Schists of the Keivy Formation*. USSR Academy of Sciences Publishers, Moscow–Leningrad [in Russian].

Belolipetsky, A.P., Gaskelberg, V.G., Gaskelberg, L.A., Antonyuk, E.S. and Il'in, Y.I. 1980. *Geology and Geochemistry of Early Precambrian Metamorphic Complexes of the Kola Peninsula*. Nauka, Leningrad [in Russian].

Blichert-Toft, J., Rosing, M.T., Lesher, C.E. and Chauvel, C. 1995. Geochemical constraints on the origin of the Late Archaean Skjoldungen alkaline igneous province, SE Greenland. *Journal of Petrology*, **36**, 515–561, https://doi.org/10.1093/petrology/36.2.515

Bogdanov, Y.B., Jacobson, K.V. and Amantov, A.V. 2003. *Geological Map of Russian Federation (new series). 1:1000000. Q-(35)-37*. Karpinsky Geological Institute (VSEGEI), St Petersburg [in Russian].

Boily, M. and Williams-Jones, A.E. 1994. The role of magmatic and hydrothermal processes in the chemical evolution of the Strange Lake plutonic complex, Québec–Labrador. *Contributions to Mineralogy and Petrology*, **118**, 33–47, https://doi.org/10.1007/BF00310609

Broska, I., Harlov, D., Tropper, P. and Siman, P. 2007. Formation of magmatic titanite and titanite–ilmenite phase relations during granite alteration in the Tribec Mountains, Western Carpathians, Slovakia. *Lithos*, **95**, 58–71, https://doi.org/10.1016/j.lithos.2006.07.012

Bruand, E., Storey, C.D. and Fowler, M. 2014. Accessory mineral chemistry of high Ba–Sr granites from Northern Scotland: constraints on petrogenesis and records of whole-rock signature. *Journal of Petrology*, **55**, 1619–1651, https://doi.org/10.1093/petrology/egu037

Bushmin, S.A., Glebovitskii, V.A., Savva, E.V., Lokhov, K.I., Presnyakov, S.L., Lebedeva, Y.M. and Sergeev, S.A. 2009. The age of HP metasomatism in shear zones during collision-related metamorphism in the Lapland Granulite Belt: the U/Pb SHRIMPII dates on zircons from sillimanite–hypersthene rocks of the Porya Guba Nappe. *Doklady Earth Sciences*, **429**, 1342–1345, https://doi.org/10.1134/S1028334X09080224

Bushmin, S.A., Glebovitskii, V.A., Presnyakov, S.L., Savva, E.V. and Shcheglova, T.P. 2011. New data on the age (SHRIMP II) of protolith and Paleoproterozoic transformations of the Archaean Keivy Terrain (Kola Peninsula). *Doklady Earth Sciences*, **438**, 661–665, https://doi.org/10.1134/S1028334X11050163

Cao, M., Qin, K., Li, G., Evans, N.J. and Jin, L. 2015. In situ LA–(MC)–ICP–MS trace element and Nd isotopic compositions and genesis of polygenetic titanite from the Baogutu reduced porphyry Cu deposit, Western Junggar, NW China. *Ore Geology Reviews*, **65**, 940–954, https://doi.org/10.1016/j.oregeorev.2014.07.014

Che, X.D., Linnen, R.L., Wang, R.C., Groat, L.A. and Brand, A.A. 2013. Distribution of trace and rare earth elements in titanite from tungsten and molybdenum deposits in Yukon and British Columbia, Canada. *The Canadian Mineralogist*, **51**, 415–438, https://doi.org/10.3749/canmin.51.3.415

Cherniak, D.J. 1993. Lead diffusion in titanite and preliminary results on the effects of radiation damage on Pb transport. *Chemical Geology*, **110**, 177–194, https://doi.org/10.1016/0009-2541(93)90253-F

Cherniak, D.J. 2010. Diffusion in accessory minerals: zircon, titanite, apatite, monazite and xenotime. *Reviews in Mineralogy and Geochemistry*, **72**, 827–869, https://doi.org/10.2138/rmg.2010.72.18

Corfu, F., Krogh, T.E. and Ayres, L.D. 1985. U–Pb zircon and sphene geochronology of a composite Archaean granitoid batholith, Favourable Lake area, northwestern Ontario. *Canadian Journal of Earth Science*, **22**, 1436–1451, https://doi.org/10.1139/e85-150

Crisp, L.J. and Berry, A.J. 2022. A new model for zircon saturation in silicate melts. *Contributions to Mineralogy and Petrology*, **177**, 71, https://doi.org/10.1007/s00410-022-01925-6

Daly, J.S., Balagansky, V.V., Timmerman, M.J. and Whitehouse, M.J. 2006. The Lapland–Kola Orogen: Palaeoproterozoic collision and accretion of the northern Fennoscandian lithosphere. *Geological Society, London, Memoirs*, **32**, 579–598, https://doi.org/10.1144/GSL.MEM.2006.032.01.35

Ducharme, T.A., McFarlane, C.R.M., van Rooyen, D. and Corrigan, D. 2021. Petrogenesis of the peralkaline Flowers River Igneous Suite and its significance to the development of the southern Nain Batholith. *Geological Magazine*, **158**, 1911–1936, https://doi.org/10.1017/s0016756821000388

Frei, D., Liebscher, A., Franz, G. and Dulski, P. 2004. Trace element geochemistry of epidote minerals. *Reviews in Mineralogy and Geochemistry*, **56**, 553–605, https://doi.org/10.2138/gsrmg.56.1.553

Frost, B.R., Chamberlain, K.R. and Schumacher, J.C. 2001. Sphene (titanite): phase relations and role as a geochronometer. *Chemical Geology*, **172**, 131–148, https://doi.org/10.1016/S0009-2541(00)00240-0

Garber, J.M., Hacker, B.R., Kylander-Clark, A.R.C., Stearns, M. and Seward, G. 2017. Controls on trace element uptake in metamorphic titanite: implications for petrochronology. *Journal of Petrology*, **58**, 1031–1058, https://doi.org/10.1093/petrology/egx046

Gros, K., Słaby, E., Birski, Ł., Kozub-Budzyń, G. and Sláma, J. 2020. Geochemical evolution of a composite pluton: insight from major and trace element chemistry of titanite. *Mineralogy and Petrology*, **114**, 375–401, https://doi.org/10.1007/s00710-020-00715-x

Gysi, A.P. and Williams-Jones, A.E. 2013. Hydrothermal mobilization of pegmatite-hosted REE and Zr at Strange Lake, Canada: a reaction path model. *Geochimica et Cosmochimica Acta*, **122**, 324–352, https://doi.org/10.1016/j.gca.2013.08.031

Harley, S.L., Kelly, N.M. and Möller, A. 2007. Zircon behaviour and the thermal histories of mountain chains. *Elements*, **3**, 25–30, https://doi.org/10.2113/gselements.3.1.25

Hawthorne, F.C., Oberti, R., Harlow, G.E., Maresch, W.V., Martin, R.F., Schumacher, J.C. and Welch, M.D. 2012. Nomenclature of the amphibole supergroup. *American Mineralogist*, **97**, 2031–2048, https://doi.org/10.2138/am.2012.4276

Holder, R.M., Hacker, B.R., Kylander-Clark, A.R.C. and Cottle, J.M. 2015. Monazite trace-element and isotopic signatures of (ultra)high-pressure metamorphism: examples from the Western Gneiss Region, Norway. *Chemical Geology*, **409**, 99–111, https://doi.org/10.1016/j.chemgeo.2015.04.021

Hoskin, P.W.O. 1998. Minor and trace element analysis of natural zircon ($ZrSiO_4$) by SIMS and laser ablation ICPMS: a consideration and comparison of two broadly competitive techniques. *Journal of Trace and Microprobe Techniques*, **16**, 301–326, https://doi.org/10.1007/BF02719033

Kennedy, A.K., Kamo, S.L., Nasdala, L. and Timms, N.E. 2010. Grenville skarn titanite: potential reference material for SIMS U–Th–Pb analysis. *The Canadian Mineralogist*, **48**, 1423–1443, https://doi.org/10.3749/canmin.48.5.1423

Kohn, M.J. and Kelly, N.M. 2018. Petrology and geochronology of metamorphic zircon. *In*: Moser, D.E., Corfu, F., Darling, J.R., Reddy, S.M. and Tait, K. (eds) *Microstructural Geochronology: Planetary Records Down to Atom Scale*. American Geophysical Union, Washington, DC, 35–61, https://doi.org/10.1002/9781119227250.ch2

Kudryashov, N.M. and Mokrushin, A.V. 2011. Mesoarchaean gabbroanorthosite magmatism of the Kola Region: petrochemical, geochronological, and isotope-geochemical data. *Petrology*, **19**, 167–182, https://doi.org/10.1134/S086959111102007X

Kudryashov, N.M., Petrovsky, M.N., Mokrushin, A.V. and Elizarov, D.V. 2013. Neoarchaean sanukitoid magmatism in the Kola region: geological, petrochemical, geochronological, and isotopic-geochemical data. *Petrology*, **21**, 351–374, https://doi.org/10.1134/S0869591113030041

Kudryashov, N.M., Balagansky, V.V., Udoratina, O.V., Mokrushin, A.V. and Coble, M.A. 2019. The age of gabbro-anorthosites of the Achinsky complex: U–Pb (SHRIMP RG) isotope-geochronological study of zircon. *In*: Kozlov, N.E. (ed.) *The Proceedings of the 16 Fersman Scientific Session*, 7–10 April 2019, Kola Science Centre RAS, Apatity, 318–322 [in Russian], https://doi.org/10.31241/FNS.2019.16.064

Lahtinen, R. and Huhma, H. 2019. A revised geodynamic model for the Lapland–Kola Orogen. *Precambrian Research*, **330**, 1–19, https://doi.org/10.1016/j.precamres.2019.04.022

Lahtinen, R., Korja, A., Nironen, M. and Heikkinen, P. 2009. Palaeoproterozoic accretionary processes in Fennoscandia. *Geological Society, London, Special Publications*, **318**, 237–256, https://doi.org/10.1144/SP318.8

Ludwig, K.R. 2000. *SQUID 1.00, A User's Manual.* BGC, Special Publications, Berkeley, CA, **2**.

Ludwig, K.R. 2003. *User's Manual for Isoplot/Ex 3.0, A Geochronological Toolkit for Microsoft Excel.* BGC, Berkeley, CA, Special Publications, **1a**.

Ludwig, K.R. 2009. *SQUID 2: A User's Manual, Rev. 12.* BGC, Berkeley, CA, Special Publications, **5**.

Ludwig, K.R. 2012. *User's Manual for Isoplot 3.75, A Geochronological Toolkit for Microsoft Excel.* BGC, Berkeley, CA, Special Publications, 4.

Ma, Q., Evans, N.J., Ling, X.-X., Yang, J.-H., Wu, F.-Y., Zhao, Z.-D. and Yang, Y.-H. 2019. Natural titanite reference materials for in situ U–Pb and Sm–Nd isotopic measurements by LA–(MC)–ICP–MS. *Geostandards and Geoanalytical Research*, **43**, 355–384, https://doi.org/10.1111/ggr.12264

Macdonald, R., Bagiński, B., Kartashov, P.M., Zozulya, D. and Dzierżanowski, P. 2015. Hydrothermal alteration of chevkinite-group minerals. Part 2. Metasomatite from the Keivy massif, Kola Peninsula, Russia. *Mineralogical Magazine*, **79**, 1039–1059, https://doi.org/10.1180/minmag.2015.079.5.02

Macdonald, R., Bagiński, B. and Zozulya, D. 2017. Differing responses of zircon, chevkinite–(Ce), monazite–(Ce) and fergusonite–(Y) to hydrothermal alteration: evidence from the Keivy alkaline province, Kola Peninsula, Russia. *Mineralogy and Petrology*, **111**, 523–545, https://doi.org/10.1007/s00710-017-0506-2

Mints, M.V., Konilov, A.N., Kaulina, T.V., Zlobin, V.L. and Bogina, M.M. 2015. Neoarchaean intracontinental areas of sedimentation, magmatism, and high-temperature metamorphism (hot regions) in eastern Fennoscandia. *Geological Society of America, Special Papers*, **510**, 89–124, https://doi.org/10.1130/2015.2510(03)

Mitrofanov, F.P. 1996. *Geological Map of the Kola Region (North-Eastern Part of the Baltic Shield). Scale 1: 500 000.* Kola Science Centre of the Russian Academy of Sciences, Apatity [in Russian].

Mitrofanov, F.P., Zozulya, D.R., Bayanova, T.B. and Levkovich, N.V. 2000. The world's oldest anorogenic alkali granitic magmatism in the Keivy structure on the Baltic Shield. *Doklady Earth Sciences*, **374**, 1145–1148.

Moyen, J.-F., Martin, H., Jayananda, M. and Auvray, B. 2003. Late Archaean granites: a typology based on the Dharwar Craton (India). *Precambrian Research*, **127**, 103–123, https://doi.org/10.1016/S0301-9268(03)00183-9

Mudruk, S.V., Balagansky, V.V., Gorbunov, I.A. and Raevsky, A.B. 2013. Alpine-type tectonics in the Paleoproterozoic Lapland–Kola Orogen. *Geotectonics*, **47**, 251–265, https://doi.org/10.1134/S0016852113040055

Myskova, T.A., Balagansky, V.V., Glebovitsky, V.A., Lvov, P.A., Mudruk, S.V. and Skublov, S.G. 2014. The first isotopic data on the Paleoproterozoic age of the Serpovidnyi Ridge amphibolites, Keivy Terrane, Baltic Shield. *Doklady Earth Sciences*, **459**, 1553–1558, https://doi.org/10.1134/S1028334X14120083

Nagy, G., Draganits, E., Demeny, A., Panto, G. and Arkai, P. 2002. Genesis and transformations of monazite, florencite, and rhabdophane during medium grade metamorphism: examples from the Sopron Hills, Eastern Alps. *Chemical Geology*, **191**, 25–46, https://doi.org/10.1016/S0009-2541(02)00147-X

Nitkina, E.A. and Serov, P.A. 2022. Zircon morphology and isotope U–Pb and Sm–Nd dating the rocks of the Kanozero alkaline granite massif (the Kola region). *Vestnik of the Murmansk State University*, **25**, 50–60 [in Russian], https://doi.org/10.21443/1560-9278-2022-25-1-50-60

Norman, M.D., Pearson, N.J., Sharma, A. and Griffin, W.L. 1996. Quantitative analysis of trace elements in geological materials by laser ablation ICP-MS: instrumental operating conditions and calibration values of NIST glasses. *Geostandards Newsletter*, **20**, 247–261, https://doi.org/10.1111/j.1751-908X.1996.tb00186.x

Olierook, H.K.H., Taylor, R.J.M. *et al.* 2019. Unravelling complex geologic histories using U–Pb and trace element systematics of titanite. *Chemical Geology*, **504**, 105–122, https://doi.org/10.1016/j.chemgeo.2018.11.004

Papapavlou, K., Darling, J.R., Storey, C.D., Lightfoot, P.C., Moser, D.E. and Lasalle, S. 2017. Dating shear zones with plastically deformed titanite: new insights into the orogenic evolution of the Sudbury impact structure (Ontario, Canada). *Precambrian Research*, **291**, 220–235, https://doi.org/10.1016/j.precamres.2017.01.007

Parrish, R.R. 1990. U–Pb dating of monazite and its application to geological problems. *Canadian Journal of Earth Science*, **27**, 1432–1450, https://doi.org/10.1139/e90-152

Pidgeon, R.T., Bosch, D. and Bruguier, O. 1996. Inherited zircon and titanite U/Pb systems in an Archaean syenite from southwestern Australia: implications for U/Pb stability of titanite. *Earth and Planetary Science Letters*, **141**, 187–198, https://doi.org/10.1016/0012-821X(96)00068-4

Prowatke, S. and Klemme, S. 2006. Rare earth element partitioning between titanite and silicate melts: Henry's law revisited. *Geochimica et Cosmochimica Acta*, **70**, 4997–5012, https://doi.org/10.1016/j.gca.2006.07.016

Radchenko, A.T., Balagansky, V.V., Vinogradov, A.N., Golionko, G.B., Petrov, V.P. and Pozhilenko, V.I. 1992. *Precambrian Tectonics of the Northeastern Baltic Shield.* Nauka Publishers, St Peterburg [in Russian].

Rubatto, D., Williams, I.S. and Buick, I.S. 2001. Zircon and monazite response to prograde metamorphism in the Reynolds Range, central Australia. *Contributions to Mineralogy and Petrology*, **140**, 458–468, https://doi.org/10.1007/PL00007673

Rubatto, D., Hermann, J., Berger, A. and Engi, M. 2009. Protracted fluid-induced melting during Barrovian metamorphism in the Central Alps. *Contributions to Mineralogy and Petrology*, **158**, 703–722, https://doi.org/10.1007/s00410-009-0406-5

Rubatto, D., Chakraborty, S. and Dasgupta, S. 2013. Timescales of crustal melting in the Higher Himalayan Crystallines (Sikkim, Eastern Himalaya) inferred from trace element constrained monazite and zircon chronology. *Contributions to Mineralogy and Petrology*, **165**, 349–372, https://doi.org/10.1007/s00410-012-0812-y

Rundqvist, D.V. and Mitrofanov, F.P. (eds) 1993. *Precambrian Geology of the USSR*. Elsevier, Amsterdam.

Scaillet, B. and Macdonald, R. 2001. Phase relations of peralkaline silicic magmas and petrogenetic implications. *Journal of Petrology*, **42**, 825–845, https://doi.org/10.1093/petrology/42.4.825

Scaillet, B., Holtz, F. and Pichavant, M. 2016. Experimental constraints on the formation of silicic magmas. *Elements*, **12**, 109–114, https://doi.org/10.2113/gselements.12.2.109

Schaltegger, U., Fanning, M., Günter, D., Maurin, J.C., Schulmann, K. and Gebauer, D. 1999. Growth, annealing and recrystallization of zircon and preservation of monazite in high-grade metamorphism: conventional and in situ U–Pb isotope, cathodoluminescence and microchemical evidence. *Contributions to Mineralogy and Petrology*, **134**, 186–201, https://doi.org/10.1007/s004100050478

Scibiorski, E., Kirkland, C.L., Kemp, A.I.S., Tohver, E. and Evans, N.J. 2019. Trace elements in titanite: a potential tool to constrain polygenetic growth processes and timing. *Chemical Geology*, **509**, 1–19, https://doi.org/10.1016/j.chemgeo.2019.01.006

Scott, D.J. and St-Onge, M.R. 1995. Constraints on Pb closure temperature in titanite based on rocks from the Ungava orogen, Canada: implications for U–Pb geochronology and *PT–t* path determinations. *Geology*, **23**, 1123–1126, https://doi.org/10.1130/0091-7613(1995)023<1123:COPCTI>2.3.CO;2

Siégel, C., Bryan, S.E., Allen, C.M. and Gust, D.A. 2018. Use and abuse of zircon-based thermometers: a critical review and a recommended approach to identify antecrystic zircons. *Earth-Science Reviews*, **176**, 87–116, https://doi.org/10.1016/j.earscirev.2017.08.011

Slabunov, A.I., Lobach-Zhuchenko, S.B. *et al.* 2006. The Archaean nucleus of the Fennoscandian (Baltic) Shield. *Geological Society, London, Memoirs*, **32**, 627–644, https://doi.org/10.1144/GSL.MEM.2006.035.01.37

Smithies, R.H. and Champion, D.C. 1999. Geochemistry of felsic igneous alkaline rocks in the eastern goldfields, Yilgarn craton, Western Australia: a result of lower crustal delamination? – implications for late Archaean tectonic evolution. *Journal of Geological Society of London*, **156**, 561–576, https://doi.org/10.1144/gsjgs.156.3.0561

Smithies, R.H., Lu, Y., Gessner, K., Wingate, M.T.D. and Champion, D.C. 2018. *Geochemistry of Archaean Granitic Rocks in the South West Terrane of the Yilgarn Craton.* Geological Survey of Western Australia, Record **2018/10**, https://doi.org/10.13140/RG.2.2.35563.85281

Stacey, J.S. and Kramers, J.D. 1975. Approximation of terrestrial lead isotope evolution by a two-stage model. *Earth and Planetary Science Letters*, **26**, 207–221, https://doi.org/10.1016/0012-821X(75)90088-6

Sutcliffe, R.H., Smith, A.R., Doherty, W. and Barnett, R.L. 1990. Mantle derivation of Archaean amphibole-bearing granitoid and associated mafic rocks: evidence from the southern Superior Province, Canada. *Contributions to Mineralogy and Petrology*, **105**, 255–274, https://doi.org/10.1007/BF00306538

Suzuki, K. and Adachi, M. 1991. Precambrian provenance and Silurian metamorphism of the Tsubonasawa paragneiss in the South Kitakami terrane, northwest Japan, revealed by the chemical Th–U–total Pb isochron ages of monazite, zircon and xenotime. *Geochemical Journal*, **25**, 357–376, https://doi.org/10.2343/geochemj.25.357

Vetrin, V.R. 2018. Isotope-geochemical systematics (Sm–Nd, Lu–Hf) of Neoarchaean subalkaline and alkaline rocks of the Keivy structure (Kola Peninsula): their age and genetic relations. *Geology of Ore Deposits*, **61**, 581–588, https://doi.org/10.1134/S1075701519070146

Vetrin, V.R. 2019. Geochemistry and conditions of the crystallization of subalkaline and peralkaline granites of the Keivy megablock. *Harald of the Kola Science Centre of RAS*, **3**, 45–49 [in Russian], https://doi.org/10.25702/KSC.2307-5228.2019.11.3.45-49

Vetrin, V.R. and Belousova, E.A. 2020. Lu–Hf Isotope composition of zircon from syenites of the Sakharjok alkaline massif, Kola Peninsula. *Geology of Ore Deposits*, **62**, 574–583, https://doi.org/10.30695/zrmo/2019.1486.01

Vetrin, V.R. and Kremenetsky, A.A. 2020. Lu–Hf isotope-geochemical zircon systematics and genesis of the Neoarchaean alkaline granites in the Keivy megablock, Kola Peninsula. *Geochemistry International*, **58**, 624–638, https://doi.org/10.1134/S0016702920060129

Vetrin, V.R. and Rodionov, N.V. 2009. Geology and geochronology of Neoarchaean anorogenic magmatism of the Keivy structure, Kola Peninsula. *Petrology*, **17**, 537–557, https://doi.org/10.1134/S0869591109060022

Vetrin, V.R., Kamenskii, I.L., Baynova, T.B., Timmerman, M., Belyatsky, B.V., Levsky, L.K. and Balashov, Y.A. 1999. Melanocratic nodules in alkaline granites of the Ponoiskii Massif, Kola Peninsula: a clue to petrogenesis. *Geochemistry International*, **37**, 1061–1072.

Vetrin, V.R., Skublov, S.G., Balashov, Y.A., Lialina, L.M. and Rodionov, N.V. 2014. The time of formation and genesis of yttrium-zirconium mineralization of the Sakharjok massif, Kola Peninsula. *Geology of Ore Deposits*, **56**, 603–616, https://doi.org/10.1134/S1075701514080011X

Vrevskii, A.B. 2016. Age and sources of the anorthosites of the Neoarchaean Kolmozero–Voron'ya greenstone belt (Fennoscandian Shield). *Petrology*, **24**, 527–542, https://doi.org/10.1134/S0869591116060060

Vrevsky, A.B. and Lvov, P.A. 2016. Isotopic age and heterogeneous sources of gabbro-anorthosites from the Patchemvarek Massif, Kola Peninsula. *Doklady Earth Sciences*, **469**, 716–721, https://doi.org/10.1134/S1028334X16070163

Wilke, M. and Behrens, H. 1999. The dependence of the partitioning of iron and europium between plagioclase and hydrous tonalitic melt on oxygen fugacity. *Contributions to Mineralogy and Petrology*, **137**, 102–114, https://doi.org/10.1007/s004100050585

Zagorodny, V.G. and Radchenko, A.T. 1983. *Tectonics of the Early Precambrian of the Kola Peninsula.* Nauka Publishers, Leningrad [in Russian].

Zakharov, D.O., Zozulya, D.R. and Rubatto, D. 2022. Low-δ^{18}O Neoarchaean precipitation recorded in a 2.67 Ga magmatic–hydrothermal system of the Keivy granitic complex, Russia. *Earth and Planetary Science Letters*, **578**, 117322, https://doi.org/10.1016/j.epsl.2021.117322

Zozulya, D.R. and Bayanova, T.B. 2003. Isotope-geochemical characteristics and the problem of the source of Archaean alkaline rocks of the Keivy terrane, Kola Peninsula. *In*: *Geochemistry of Igneous Rocks*. Proceedings of the XXI Seminar on the Geochemistry of Igneous Rocks, 3–5 September 2003, Apatity, Russia. Kola Science Centre, RAS, 63–64 [in Russian].

Zozulya, D.R. and Bayanova, T.B. 2013. Age and tectonic setting of alkali gabbro from the late Archaean Sakharjok massif (Kola Peninsula). *In*: Voitekhovsky, Y.L. (ed.) *Geology and Geochronology of Rock-forming and Ore Processes in Crystal Rocks*. K&M, Apatity, 60–62 [in Russian].

Zozulya, D.R., Bayanova, T.B. and Eby, G.N. 2005. Geology and age of the Late Archaean Keivy Alkaline Province, Northeastern Baltic Shield. *Journal of Geology*, **113**, 601–608, www.jstor.org/stable/10.1086/431912, https://doi.org/10.1086/431912

Zozulya, D., Lyalina, L., Macdonald, R., Bagiński, B., Savchenko, Y. and Jokubauskas, P. 2019. Britholite group minerals from REE-rich lithologies of Keivy Alkali Granite–Nepheline Syenite Complex, Kola Peninsula, NW Russia. *Minerals*, **9**, 732, https://doi.org/10.3390/min9120732

Zozulya, D.R., Macdonald, R. and Bagiński, B. 2020. REE fractionation during crystallization and alteration of fergusonite–(Y) from Zr–REE–Nb-rich late- to post-magmatic products of the Keivy alkali granite complex, NW Russia. *Ore Geology Reviews*, **125**, 103693, https://doi.org/10.1016/j.oregeorev.2020.103693

Lower crustal hot zones as zircon incubators: Inherited zircon antecryts in diorites from a mafic mush reservoir

Hoseong Lim[1]*, Oliver Nebel[1], Roberto F. Weinberg[1], Yona Nebel-Jacobsen[1], Vitor R. Barrote[1,2], Jongkyu Park[3], Bora Myeong[4] and Peter A. Cawood[1]

[1]School of Earth, Atmosphere and Environment, Monash University, Clayton 3800 Victoria, Australia

[2]Institut für Geologische Wissenschaften, Geochemie, Freie Universität Berlin, Malteserstr. 74-100, 12249 Berlin, Germany

[3]Department of Geology, Kyungpook National University, Daegu 41566, Republic of Korea

[4]GeoZentrum Nordbayern, Friedrich-Alexander-Universität Erlangen-Nürnberg, Schlossgarten 5, 91054 Erlangen, Germany

HL, 0000-0002-0859-7900; ON, 0000-0002-5068-7117; RFW, 0000-0001-9420-8918; YN-J, 0000-0001-5600-7453; VRB, 0000-0001-7442-9748; JP, 0000-0002-9021-5333; BM, 0000-0001-5963-2019; PAC, 0000-0003-1200-3826

*Correspondence: hoseong.lim@monash.edu

Abstract: Continental arcs are key sites of granitic magmatism, yet details of the origins of these magmas, including the role and contribution of mafic magma, the timing and location of initial zircon formation and how zircon isotopic signatures relate to granite formation, remain as challenges. Here we use U–Pb dating, trace elements and Hf isotopic systematics of zircon in mafic microgranular enclaves (MMEs), from the convergent plate margin Satkatbong diorite (SKD) in Korea to understand lower arc magmatism and zircon production. The host granitic body and MMEs display similar major element evolutionary trends and similar ranges of Sr, Nd and Hf isotopes, implying a cognatic relationship. Zircons show a large variability in εHf (*t*) (*c.* 6 units) and age (>30 Ma). We propose that the SKD and MMEs originated from the same, long-lasting, lower crustal mush reservoir, enabling long and variable residence times for zircons. Prolonged zircon ages, combined with the Hf isotope variability within a single pluton (SKD and its MME), indicate that not all zircons were instantaneously crystallized in a rapidly cooling shallow magma chamber but were continuously formed in a long-lasting hot source. A low-melt-fraction mush type reservoir in a deep crustal hot zone provides a viable model for the source setting. Continuous replenishment of mafic magmas acts as the main re-activator of the reservoir, and provide a critical role in spawning zircons that record a long age span, because (1) the magma adds Zr into the reservoir, enabling it to reach zircon saturation and (2) the generated zircon grains are transported upward as antecrysts by flow inside of the reservoir. This means that antecrysts with different ages may mix with each other in the ascending magma body. The significance of this model is that a conclusive time of intrusion cannot be constrained by such zircon ages, as these antecrysts constitute inherited grains.

Supplementary material: The data used in study and analytical settings are available at https://doi.org/10.6084/m9.figshare.c.6365895

Convergent continental margin granitoids range in composition from I-type intrusions, generally of mantle affinity, to S-types granites with a more evolved crustal signature. The chemical spectrum of these intrusions is the result of diverse source compositions within the lower crust and melting reactions during partial melting. An integrated part of the lower arc crust, the regions where convergent margin granitoids originate are the so-called deep crustal hot zones (DCHZ; Annen *et al.* 2006; Solano *et al.* 2012). These are areas within the deeper continental arc crust that act as regions with a prolonged active melt storage (Jackson *et al.* 2018). These mush reservoirs are assumed to be tapped and reactivated by injections of new magma batches from the underlying sub-arc mantle, and range in composition from mafic to felsic reservoirs through igneous differentiation (Jackson *et al.* 2018). The link between these hot zones and magmatic arc granitoids is not clear, but it is most likely that these regions either represent the source or play a key role in the development of parental melts of the high-level plutons.

From: van Schijndel, V., Cutts, K., Pereira, I., Guitreau, M., Volante, S. and Tedeschi, M. (eds) 2024. *Minor Minerals, Major Implications: Using Key Mineral Phases to Unravel the Formation and Evolution of Earth's Crust.* Geological Society, London, Special Publications, **537**, 411–433.
First published online March 13, 2023, https://doi.org/10.1144/SP537-2021-195

The lifespan of some magmatic arc plutons can exceed millions of years (e.g. Idaho Batholith, Gaschnig *et al.* 2013; Ladakh Batholith, Weinberg and Dunlap 2000; Florida Mountains granite, Amato and Mack 2012), and numerous studies have shown that large plutonic bodies do not form in a single event but grow incrementally through repeated injections of new magma batches (Glazner *et al.* 2004; Miller *et al.* 2007). With long-lived DCHZ as possible sources for the magma, it is possible that some of the compositional variety in composite plutons is directly related to melt genesis at depth within the lower arc crust (e.g. Kemp *et al.* 2007; Hammerli *et al.* 2018). Conclusive evidence, however, is yet to be found.

Another prominent feature of many I-type convergent margin plutons is mafic microgranular enclaves (MMEs). These enclaves, which usually exhibit only slightly lower silica content than their host granites (Chen *et al.* 1990; Barbarin 2005), are often interpreted to represent mantle-derived juvenile influx (Barbarin 2005). They have been linked to parental melts from the source of the granites (Collins *et al.* 2001; Barbarin 2005), and thus may hold clues to the genesis and evolution of plutonic bodies.

Zircon is an ideal dating tool of magmatic processes and is a common accessory phase in intermediate to felsic rocks, including MMEs. Age information, combined with Hf in zircon, can help constrain the source characteristics of rocks (Nebel *et al.* 2007; Roberts and Spencer 2015). Furthermore, Hf isotopes are highly stable in the zircon lattice, making them resistant to diffusion, and combined with their relatively high ($> 900°C$) closing temperature for Lu–Hf (Scherer *et al.* 2000), zircons are an ideal tool to study early stages of the magma genesis (Kemp *et al.* 2007).

In this study, we investigate U–Pb ages and Hf isotopes of zircons from MMEs (*c.* 56 wt% SiO_2) and their host dioritic body (*c.* 65 wt% SiO_2; Lim *et al.* 2016; Lim *et al.* 2018), comparing them with the neighbouring Yeongdeok granite (YDG; Cheong *et al.* 2002; Yi *et al.* 2012). Previous studies implied a possible source connection between the Satkatbong diorite (SKD) and the YDG, which differ in age by *c.* 60 Ma, yet exhibit similar Sr and Nd isotopic systematics (Cheong *et al.* 2002; Lim *et al.* 2018). We aim to elucidate the role of lower crust in the genesis of these plutonic bodies and the MMEs.

Geological context and previous work

Geological setting

The southeastern Korean Peninsula lies within the eastern margin of the Eurasian Plate. The basement of the peninsula consists of the Precambrian Yeongnam massif (YM), which resembles the South China Craton (Fig. 1) in radiogenic Sr–Nd–Pb isotope studies (Choi *et al.* 2005). The YM collided with the Gyeonggi massif in the Late Permian to Early Triassic, forming the intervening Okcheon metamorphic Belt along the suture (Fig. 1). During and after the collision, extensive plutonic rocks intruded the YM, forming scattered stocks that are mainly composed of alkaline or medium- to high-K calc-alkaline granite to diorite. These rocks include the YDG and SKD, the study area plutons. Subsequently, the Cretaceous Gyeongsang arc system and associated volcano-sedimentary back arc Gyeongsang Basin developed along the southern segment of the Korean Peninsula (Fig. 1). Plutons in the study area are non-conformably overlain by the Gyeongsang Basin (Lim *et al.* 2016).

Yeongdeok granite

The Yeongdeok granite (YDG) intruded into the Proterozoic Yeongnam massif. Its contacts with surrounding units are masked by a cover of Cretaceous volcano-sedimentary rocks. The granite ranges in composition from a K-feldspar megacryst-bearing biotite granite to a hornblende granite (Woo and Jang 2014). It is mainly composed of quartz, K-feldspar, plagioclase, biotite and hornblende with accessory titanite, zircon and apatite. Moderately enriched Rb (10–100 ppm) and HFSE (high field strength elements; including Y and Nb) deficiency highlight its volcanic arc granite characteristics (Cheong *et al.* 2002; Yi *et al.* 2012; Lim *et al.* 2018), and A/NK and A/CNK (=molar $Al_2O_3/(Na_2O + K_2O)$ and molar $Al_2O_3/(CaO + Na_2O + K_2O)$) relations indicate that the YDG is a metaluminous to peraluminous I-type body (Cheong *et al.* 2002; Yi *et al.* 2012; Cheong *et al.* 2019). Yi *et al.* (2012) first analysed the YDG zircons and reported their age at 257 ± 2 Ma. This age indicates a post-collisional relationship with respect to the formation of the Okcheon metamorphic Belt. The origin of the YDG has been suggested as having formed through mixing between mid-ocean ridge basalt-like mantle melt and Proterozoic basement on the basis of $^{87}Sr/^{86}Sr$ (*t*) and εNd (*t*) modelling (Cheong *et al.* 2002). Lim *et al.* (2018) discussed trace element modelling and showed that the melting of a subducted eclogitic slab followed by crustal assimilation could generate the YDG magma (Cheong *et al.* 2019).

Satkatbong diorite

The SKD is a north–south elongated plutonic body, exposed along the coast. The SKD is a leucocratic, medium- to coarse-grained equigranular tonalite–quartz diorite (Lim *et al.* 2016) composed of plagioclase, hornblende, quartz and K-feldspar with minor titanite, apatite and zircon (Cheong *et al.* 2002; Lim

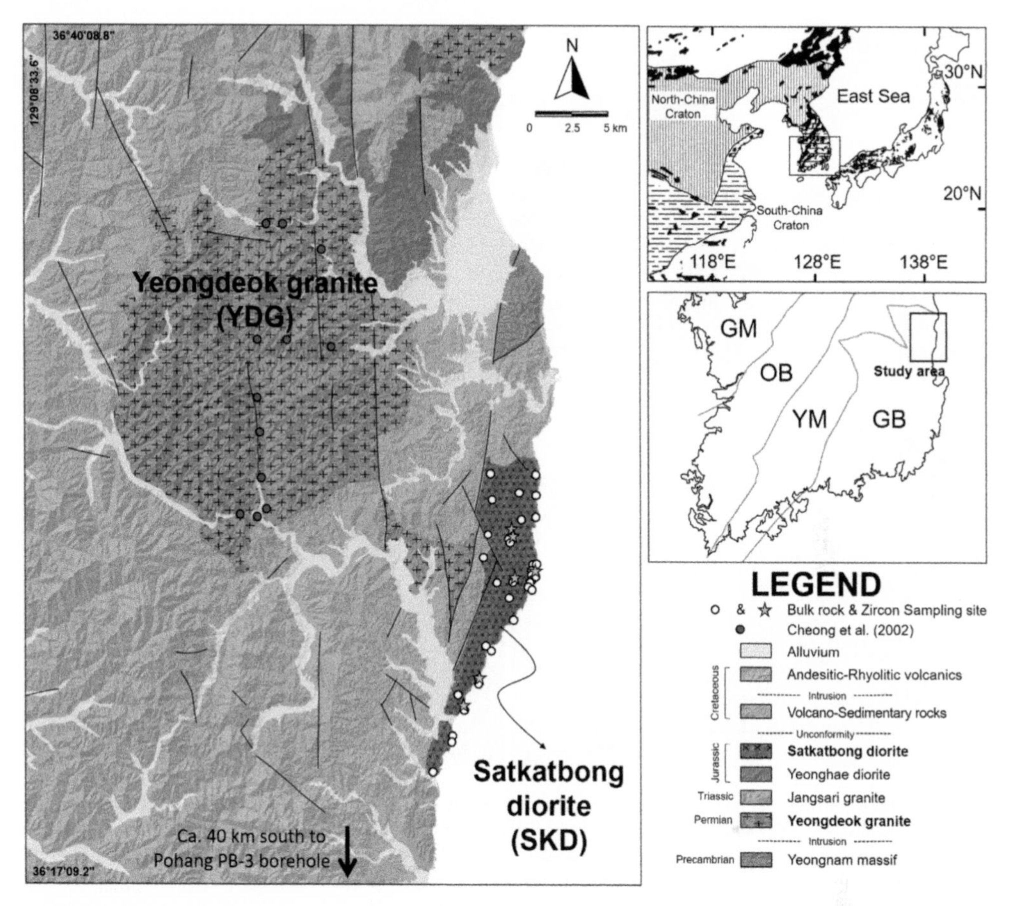

Fig. 1. Simplified geological map of the Yeongdeok area with the Yeongdeok granite (YDG) and Satkatbong diorite (SKD). GM, Gyeonggi Massif; OB, Okcheon Belt; YM, Yeongnam Massif; GB, Gyeongsang Basin (modified after Lim *et al.* 2018).

et al. 2016). Epidote is found as a secondary phase. A U–Pb zircon age of 192 ± 2 Ma was reported for the SKD (Yi *et al.* 2012). This pluton is rich in MMEs, whose composition ranges from quartz diorite to diorite (Fig. 2a; Lim *et al.* 2016). The MMEs are also equigranular, and slightly finer grained than the host diorite and lack cumulate textures (Lim *et al.* 2016). Mafic clots and acicular apatite are found in both host and enclaves on microscopic level (Lim *et al.* 2018). These textures indicate magma mixing between the host and MME magma (Baxter and Feely 2002; Lim *et al.* 2016, 2018), additionally supported by hybrid zones in an outcrop scale. The SKD resembles the YDG in Sr, Nd, and Pb isotope compositions but trace and rare earth element (REE) concentrations are different between the SKD and YDG (Lim *et al.* 2018; Cheong *et al.* 2019). The similar isotopic yet different REE patterns between the YDG and SKD are interpreted to indicate partial melting of similar source materials at different depths (deeper for the YDG; Cheong *et al.* 2002). Lim *et al.* (2018), however, suggested that both plutons could have a similar source with a similar depth if the MME magma mixing altered the source to produce the SKD.

Cumulate

In this study, we report data for an additional unit of the SKD, which has previously not been investigated. The unit, which is exposed in the northern part of the SKD, is distinct from the rest of the body in that there are no MMEs but the unit displays a distinctive magmatic cumulate texture (Fig. 2b, c). The rock suite is characterized by predominantly amphibole–plagioclase that appears aligned on an outcrop scale and also in microstructures (Fig. 2d).

Samples and methodology

From the SKD complex (i.e. main body, MMEs and cumulate), the major and trace elements of the host

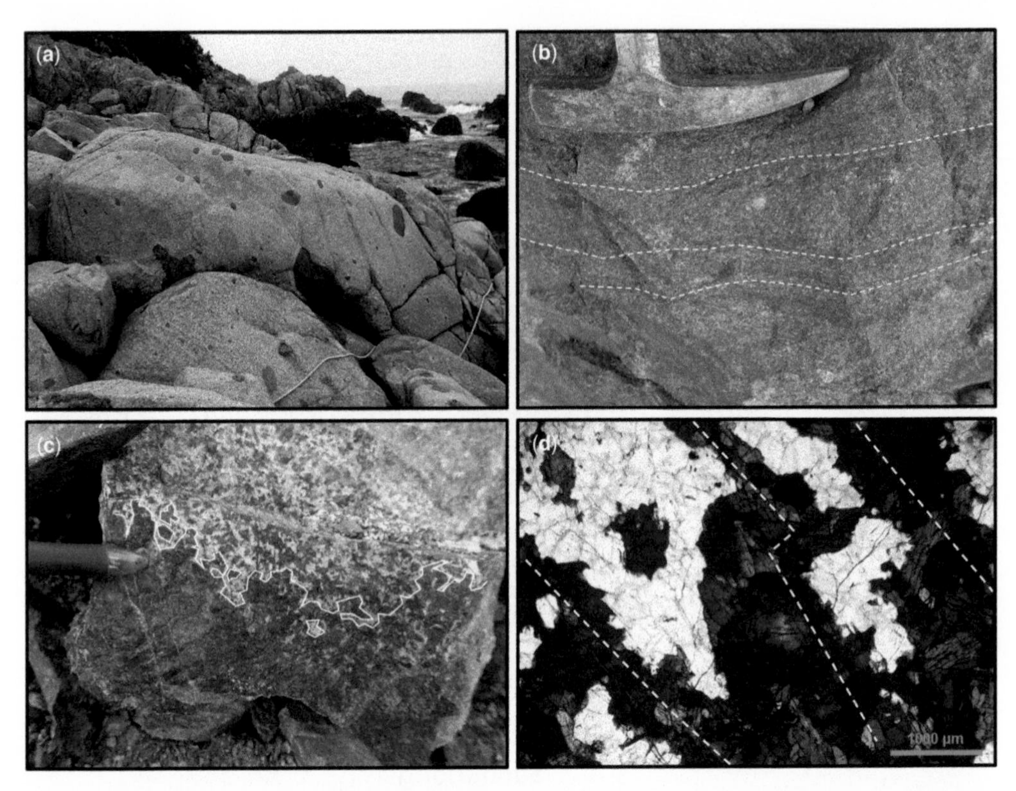

Fig. 2. Nature of the SKD complex including the main body, mafic microgranular enclaves (MMEs) and cumulate. (**a**) The leucocratic host SKD and darker melanocratic MMEs. Note the considerable amount of MMEs. (**b**) Magmatic layers of the cumulate sample shown by the white dashed lines. (**c**) Cumulate texture defined by amphiboles. (**d**) Photomicrograph of aligned amphiboles (white dashed line).

rock and MMEs were analysed by Lim *et al.* (2018). In this study, we present an additional seven analyses of cumulates, four from the host rock and eight from the MMEs. Each sample was visibly fresh. Samples were crushed in a hydraulic press and a jaw crusher, and powdered using a tungsten carbide ball mill and an agate mortar. The samples were mixed with a flux of lithium metaborate and lithium tetraborate and then fused in an induction furnace. The melts were poured into a solution of 5% nitric acid containing an internal standard, and mixed continuously (*c.* 45 min) to fully dissolve potential refractory minerals such as zircon, titanite, monazite and chromite.

The major elements were analysed on a Thermo Jarrell-Ash ENVIROII ICP at Activation Laboratories, Canada. Calibration was performed using certified standard materials from the US Geological Survey and the Canada Centre for Mineral and Energy Technology. The detection limits of MnO and TiO_2 were 0.001%, and those of the other major elements were 0.01%. Trace elements were analysed by the Perkin Elmer Sciex ELAN 9000 ICP-MS (Activation Laboratories, Canada). The precision was estimated to be within $\pm 3\%$ based on repeated analyses of a selected sample (15G) and international rock standards (BIR-1a, DNC-1, JR-1, NIST 694, SY-4 and W-2a).

Zircons were separated from host rock (05G, 2109, 4101) and MME (E09, E11, E12) samples (Fig. 1). After removal of any weathered portions, *c.* 2–3 kg of each sample was crushed by hydraulic press and jaw crusher and then for a short period in a tungsten carbide ball mill (<1 min). The powders were sieved through 250 µm mesh. Water washing, panning, magnetic separation (by a hand magnet and a Frantz) and heavy liquid (tetrabromoethane; 2.97 g cm^{-3}) were used to separate zircons from lighter or magnetic minerals. Zircons were then handpicked under a binocular microscope. Zircon U–Pb and Hf-isotope data were collected using the method developed for laser ablation split stream multi-collector inductively coupled plasma mass spectrometry (LASS-MC-ICPMS) at the Isotopia Facility, Monash University. The isotopic data from both systems were collected simultaneously using a Resonetics S-155-LR 193 nm excimer laser coupled to a Thermo Fisher Neptune *PLUS* multicollector (Hf isotopes) and a Thermo Fisher iCAP-TQ quadrupole mass spectrometer (U–Pb) in single quadrupole mode. The laser aerosol was split evenly between the two instruments, and for enhanced Hf sensitivity, nitrogen gas was added to the sample line of the Neptune only, past the split junction of the ablated material at 6.0 ml min^{-1} flow rate. The laser conditions were set such that a fluence at the sample of *c.* 4.5 J cm^{-2} was obtained, with an 8 Hz frequency and 35 µm spot size. Hafnium isotopes were collected following Fisher *et al.* (2014*a*) with a 1 s integration time and a total of 60 s of ablation (about 30 µm depth) after a 30 s background.

Dwell times for the iCAP TQ were 10 ms for ^{238}U and ^{232}Th, 20 ms for ^{208}Pb, 70 ms for ^{207}Pb, 40 ms for ^{206}Pb and 30 ms for ^{204}Pb and ^{202}Hg. The ablated material was carried to the mass spectrometers by combined He gas, with a flow rate of 0.55 L min^{-1} and Ar gas at *c.* 1 L min^{-1} flow rate. The U–Pb elemental fractionation, down-hole fractionation and calibration drift were corrected by bracketing measurements of unknowns with analyses of the primary zircon reference material Plešovice ($^{206}Pb/^{238}U$ age = 337.13 ± 0.37 Ma; Sláma *et al.* 2008). Mud tank ($^{176}Hf/^{177}Hf = 0.282507 \pm 0.000006$; Woodhead and Hergt 2005) was used as the primary reference material for Hf analysis. The secondary standards for U–Pb and Hf analyses were 91500 (Wiedenbeck *et al.* 1995) and GJ-1 (Jackson *et al.* 2004). Time-resolved data were baseline subtracted and reduced using Iolite 4 (DRS after Paton *et al.* 2011) and the in-built data reduction schemes UPb_Geochron_4 and Hf_isotopes with natural abundance ratios of $^{171}Yb/^{173}Yb = 1.132685$ (Chu *et al.* 2002) and $^{179}Hf/^{177}Hf = 0.7325$ for mass bias correction. The mass bias of Lu was assumed to be identical to that of Yb ($\beta Lu = \beta Yb$; Fisher *et al.* 2014*b*). Age calculations and diagrams were constructed using IsoplotR (Vermeesch 2018). No common lead correction was made in the data reduction scheme, and model-1 discordia (Ludwig 2012) was used if a common lead regression was found. Level 5 of Horstwood *et al.* (2016) does not take into account systematic uncertainties. A complete list of the settings for U/Pb and Hf analyses is provided in the Supplementary Material 3 (Tables S1 and S2). The results for reference material analysis are listed in the Supplementary Material 1.

Results

Whole rock geochemistry

The major and trace element compositions of the YDG and the SKD are listed in Table 1. The total alkali v. SiO_2 (TAS) diagram (Middlemost 1994) indicates that the YDG samples have granodiorite–quartz monzonite–granite compositions, whereas the SKD and MMEs fall within a range of monzogabbroic diorite to granodiorite. Cumulates mostly plot in the gabbro–gabbroic diorite fields (Fig. 3).

The SKD has a wide range of major element values (54.1–74.2 wt% in SiO_2, 0.31–3.61 wt% in MgO, 1.57–4.76 wt% in K_2O, $n = 14$), and the MMEs show a slightly more mafic compositional range in SiO_2 (51.8–64.7 wt%, $n = 17$), plotting on a similar major element trend to the SKD. Compared with their host, the MMEs are slightly more sodic (2.86–4.60 wt%) and potassic (1.46–3.10 wt%) at the same silica content (Fig. 3). A striking feature of the rock suites is that the host rock varies substantially in silica content and extends at lower values into the region of MME silica concentrations, which could be a result of either differentiation (cumulate-melt) or mixing processes (between endmembers). Additionally, the YDG complex defines a similar linear trend to the SKD complex in Al_2O_3, K_2O, MgO and CaO, which may indicate similar differentiation processes. The cumulates show an apparent depletion in SiO_2 and alkali elements, as can be expected from their mafic mineralogy, whereas enriched Al_2O_3, MgO and CaO values correspond to the high proportions of amphibole and plagioclase (Fig. 3).

Together with the YDG (Yi *et al.* 2012), the SKD has relatively high Sr/Y yet straddles between the adakitic and non-adakitic fields in both the Sr/Y v. Y and La/Yb v. Yb plots (Fig. 4). The MMEs do not show Sr/Y enrichment. These trends are also shown in the La/Yb v. Yb diagram (Fig. 4). The MMEs plot outside of the adakitic (in this case, high-Sr/Y granite) field in both diagrams.

In the REE plot, the YDG is characterized by elevated light REEs and depleted heavy REEs and lacks a negative Eu anomaly (Fig. 5), resulting in a relatively steep and linear slope in the REE pattern. The REE pattern for the SKD shows a less steep slope. The MMEs have relatively enriched high REEs compared with their host rock. Negative Eu anomaly ($Eu/Eu^* = 0.75$) of the MME indicates plagioclase fractionation during the MME magma evolution. The cumulates show, on average, lower light REE abundances than the SKD and MMEs and have a positive Eu anomaly ($Eu/Eu^* = 1.17$) indicative of the accumulation of plagioclase (Fig. 5, Table 1).

Zircon

A total of 82 zircons from six samples representative of both SKD and MMEs were systematically analysed for cores and rims (164 analyses in total; Fig. 6). Zircons from the SKD and MMEs, as magmatic zircons, are similar in habit and chemical characteristics. Their morphologies and internal

Table 1. *Major and trace element composition of the Satkatbong diorite (SKD), mafic microgranular enclaves (MMEs) and cumulate*

	Host rock (SKD)									
Sample	02	03	04	12	03G	05G	07G	09G	13G	15G
SiO_2	66.29	65.92	74.22	54.13	65.75	64.67	64.92	64.87	65.13	64.19
Al_2O_3	15.16	15.98	13.59	17.00	16.37	16.20	16.32	16.90	17.03	17.31
Fe_2O_3(T)	3.89	4.06	1.34	8.60	3.99	4.50	4.18	4.45	4.09	4.54
MnO	0.08	0.09	0.03	0.14	0.09	0.10	0.07	0.10	0.09	0.10
MgO	1.63	1.61	0.31	3.38	1.69	1.92	1.95	1.87	1.74	1.87
CaO	3.97	3.83	0.96	6.50	4.60	4.83	4.35	4.76	4.71	4.95
Na_2O	3.70	3.65	3.21	4.28	3.72	3.63	4.12	3.74	3.74	3.84
K_2O	2.69	2.70	4.76	1.57	2.00	2.21	2.07	2.22	2.18	1.97
TiO_2	0.41	0.39	0.14	1.21	0.36	0.43	0.42	0.42	0.40	0.43
P_2O_5	0.09	0.11	0.03	0.23	0.12	0.13	0.12	0.12	0.10	0.13
LOI	1.47	1.86	1.51	1.79	1.99	1.41	2.23	1.27	1.61	1.62
Total (wt%)	99.38	100.20	100.10	98.82	100.70	100.00	100.80	100.70	100.80	100.90
$Na_2O + K_2O$	6.39	6.35	7.97	5.85	5.72	5.84	6.19	5.96	5.92	5.81
FeO(T)	3.50	3.65	1.21	7.74	3.59	4.05	3.76	4.00	3.68	4.09
A/NK	1.68	1.79	1.30	1.95	1.98	1.94	1.81	1.98	2.00	2.05
A/CNK	0.93	1.01	1.12	0.83	0.98	0.94	0.96	0.98	1.00	0.99
Mg #	0.45	0.44	0.31	0.44	0.46	0.46	0.48	0.45	0.46	0.45
K/Na	0.48	0.49	0.98	0.24	0.35	0.40	0.33	0.39	0.38	0.34
Trace elements (ppm)										
V	64	65	11	209	66	73	72	78	66	74
Ba	543	514	756	361	371	550	407	635	580	555
Sr	434	456	146	928	452	430	431	439	465	478
Y	14	18	20	18	13	14	15	13	12	14
Zr	102	120	89	67	100	127	111	123	104	115
Ga	16	15	14	19	17	17	17	17	18	18
Rb	77	71	137	39	58	52	67	49	49	43
Nb	3	3	5	4	3	4	3	3	3	3
Cs	4.8	4.1	4.2	1.7	2.4	2.1	2.1	1.9	2.0	1.6
La	43.1	8.4	26.3	29.2	22.8	14.4	18.2	10.0	9.2	24.2
Ce	77.9	18.6	52.3	55.0	43.5	29.8	34.5	21.9	20.1	45.2
Pr	7.5	2.4	5.5	6.4	4.6	3.5	3.9	2.8	2.5	4.6
Nd	22.8	10.3	18.8	25.0	17.4	13.7	15.1	12.1	10.3	16.5
Sm	3.3	2.4	3.1	5.3	3.0	2.9	3.0	2.7	2.3	3.0
Eu	0.8	0.7	0.5	1.7	0.8	0.8	0.8	0.8	0.7	0.8
Gd	2.4	2.2	2.6	4.2	2.4	2.4	2.5	2.3	1.9	2.4
Tb	0.3	0.4	0.4	0.6	0.4	0.4	0.4	0.4	0.3	0.4
Dy	1.8	2.4	2.4	3.5	2.2	2.3	2.4	2.3	1.9	2.1
Ho	0.4	0.5	0.5	0.7	0.4	0.5	0.5	0.5	0.4	0.4
Er	1.1	1.5	1.5	1.8	1.2	1.3	1.5	1.3	1.2	1.2
Tm	0.2	0.2	0.2	0.3	0.2	0.2	0.2	0.2	0.2	0.2
Yb	1.1	1.7	1.6	1.6	1.4	1.5	1.5	1.4	1.3	1.4
Lu	0.2	0.3	0.3	0.3	0.2	0.2	0.2	0.2	0.2	0.2
Hf	2.3	3.2	2.6	2.0	2.6	3.3	2.9	3.0	2.6	2.9
Ta	0.3	0.4	0.8	0.3	0.5	0.6	0.3	0.3	0.4	0.4
Pb	11	12	29	9	10	10	8	11	11	10
Th	11.4	7.4	14.9	8.2	9.7	6.8	6.8	7.9	7.5	7.9
U	1.4	3.3	2.1	3.3	2.1	2.2	1.7	1.7	2.5	1.9
Eu/Eu*	0.9	1.0	0.6	1.1	0.9	0.9	0.9	1.0	1.1	0.9

structures are similar in back-scattered electron and cathodoluminescence images (Fig. 6, Fig. S2, Supplementary Material 2). The zircons from both SKD and MMEs show sector to weak oscillatory zoning, indicating relative compositional homogeneity (Fig. S2). Both rock types lack obvious inherited zircon cores (Fig. 6). The zircons in both rocks lack thorite microcrysts, which often occur in metamict zircons (Kusiak *et al.* 2009), highlighting their relatively unaltered condition.

The results of U–Pb and Hf isotope analyses are listed in the Supplementary Material 4 (Table S3).

Table 1. *Continued.*

	Host rock (SKD)				MMEs					
Sample	D05	D09	D10	E09	05	06	07	11	03M	09M
SiO_2	62.52	71.25	63.64	57.05	55.11	54.20	54.23	54.40	53.09	57.34
Al_2O_3	15.16	14.48	16.17	17.83	18.25	18.66	19.05	17.86	17.49	14.93
Fe_2O_3(T)	4.50	1.85	4.22	6.97	7.53	7.47	7.80	7.37	9.40	10.58
MnO	0.07	0.03	0.08	0.14	0.21	0.19	0.21	0.22	0.29	0.21
MgO	2.06	0.42	1.78	3.61	4.05	3.70	3.75	4.38	4.85	4.01
CaO	3.05	1.69	4.12	6.86	6.22	5.55	6.68	5.96	6.79	5.18
Na_2O	3.30	3.24	3.62	4.05	4.23	3.78	4.27	4.06	4.03	2.86
K_2O	3.37	4.71	2.77	1.67	2.28	2.39	2.10	2.06	1.84	1.72
TiO_2	0.51	0.15	0.42	0.65	0.77	0.72	0.69	0.77	0.71	0.90
P_2O_5	0.09	0.05	0.12	0.13	0.19	0.16	0.15	0.13	0.10	0.19
LOI	4.17	0.90	2.18	1.69	1.36	2.24	1.75	2.05	1.92	2.77
Total (wt%)	98.79	98.77	99.13	100.70	100.20	99.07	100.70	99.26	100.50	100.70
$Na_2O + K_2O$	6.67	7.95	6.39	5.72	6.51	6.17	6.37	6.12	5.87	4.58
FeO(T)	4.05	1.66	3.80	6.27	6.78	6.72	7.02	6.63	8.46	9.52
A/NK	1.67	1.39	1.81	2.11	1.94	2.12	2.05	2.00	2.03	2.27
A/CNK	1.04	1.07	0.98	0.85	0.88	0.99	0.89	0.90	0.83	0.93
Mg #	0.48	0.31	0.46	0.51	0.52	0.50	0.49	0.54	0.51	0.43
K/Na	0.67	0.96	0.50	0.27	0.35	0.42	0.32	0.33	0.30	0.40
Trace elements (ppm)										
V	93	24	71	146	146	134	140	172	167	181
Ba	407	1245	616	304	560	717	628	631	465	626
Sr	395	247	468	533	429	523	507	511	430	356
Y	16	8	13	15	26	26	35	24	40	25
Zr	130	118	140	98	120	99	111	64	62	211
Ga	14	13	16	17	19	19	20	18	21	20
Rb	107	112	74	65	71	80	68	67	59	43
Nb	3	2	3	2	4	5	6	4	6	6
Cs	4.1	1.4	3.9	5.3	2.8	3.7	2.2	2.1	4.3	1.8
La	22.8	41.0	19.0	11.1	13.0	14.5	13.5	15.2	18.4	12.4
Ce	46.3	80.4	36.6	24.4	33.4	36.6	35.6	37.0	47.6	31.9
Pr	5.1	7.8	3.9	3.0	4.7	5.1	5.3	5.0	6.9	4.5
Nd	17.4	23.6	14.4	12.6	20.3	22.2	24.4	20.7	32.5	20.9
Sm	3.2	3.1	2.8	2.8	4.7	5.0	6.1	5.3	8.0	4.9
Eu	0.6	0.8	0.8	0.9	1.2	1.1	1.3	1.3	1.5	0.9
Gd	2.9	1.8	2.5	2.9	4.3	4.5	5.7	4.7	6.6	4.3
Tb	0.5	0.2	0.4	0.5	0.7	0.7	1.0	0.7	1.1	0.7
Dy	2.9	1.2	2.3	2.7	4.5	4.3	5.9	4.1	6.6	4.3
Ho	0.6	0.2	0.5	0.5	0.9	0.9	1.2	0.8	1.4	0.8
Er	1.7	0.8	1.4	1.6	2.6	2.7	3.6	2.4	3.8	2.5
Tm	0.3	0.1	0.2	0.3	0.4	0.4	0.6	0.4	0.6	0.4
Yb	1.8	1.0	1.5	1.7	2.8	2.7	3.7	2.7	4.2	2.6
Lu	0.3	0.2	0.2	0.3	0.4	0.4	0.6	0.4	0.6	0.4
Hf	3.3	3.1	3.5	2.5	2.9	2.5	2.9	2.0	2.2	5.0
Ta	0.4	0.4	0.4	0.2	0.3	0.3	0.4	0.3	0.6	0.5
Pb	10	23	12	7	11	11	13	10	9	9
Th	6.1	18.6	8.5	4.2	4.0	3.5	2.4	5.0	5.0	8.7
U	2.0	2.3	2.1	1.0	1.0	1.6	1.9	1.5	1.8	1.7
Eu/Eu*	0.6	1.0	0.9	1.0	0.8	0.7	0.7	0.8	0.6	0.6

The U–Pb and the Hf isotope data of the YDG are from Yi *et al.* (2012) and Cheong *et al.* (2019), respectively, and are used here for comparison. Ages with more than 5% discordance lie along a common lead contamination trend and are excluded from this study to avoid influences on either age and/or Hf values. Despite the absence of obvious core–rim textures, ages from both the SKD and MMEs spread by up to *c.* 45 Ma (with mean square weighted deviation (MSWD) = 8.2 and 12 respectively), while weighted mean ages are similar at *c.* 190 Ma (190.6 ± 1.4 Ma for the SKD ($n = 80$, 2 SD = 14 Ma) and 189.7 ± 1.5 Ma for the MMEs ($n = 90$, 2SD = 18 Ma)). Ages extending over tens

Table 1. *Continued.*

	MMEs								
Sample	11M	E04	E10	E03	E07	E08	E11	E12	D06
SiO_2	52.29	54.17	53.94	51.80	55.11	54.26	56.49	54.69	64.66
Al_2O_3	19.28	18.18	18.30	18.73	18.78	18.21	18.23	17.90	15.57
Fe_2O_3(T)	8.83	7.68	7.31	8.49	7.38	6.96	6.75	6.74	4.02
MnO	0.24	0.14	0.20	0.19	0.20	0.18	0.19	0.19	0.08
MgO	4.41	4.08	4.09	4.59	3.43	3.86	3.76	3.77	1.70
CaO	6.76	7.13	6.41	7.98	6.54	6.21	6.41	5.56	4.41
Na_2O	4.42	3.87	3.98	3.70	4.31	4.60	4.01	4.18	4.02
K_2O	2.05	1.46	3.10	1.65	1.94	2.41	2.25	2.11	2.46
TiO_2	0.78	0.78	0.65	0.76	0.66	0.67	0.69	0.72	0.44
P_2O_5	0.14	0.16	0.12	0.14	0.17	0.15	0.13	0.16	0.08
LOI	1.48	1.98	2.13	1.66	1.14	2.73	1.46	3.03	1.72
Total (wt%)	100.70	99.63	100.20	99.69	99.66	100.20	100.40	99.05	99.15
$Na_2O + K_2O$	6.47	5.33	7.08	5.35	6.25	7.01	6.26	6.29	6.48
FeO(T)	7.95	6.91	6.58	7.64	6.64	6.26	6.07	6.06	3.62
A/NK	2.03	2.29	1.85	2.38	2.04	1.79	2.02	1.95	1.68
A/CNK	0.89	0.87	0.85	0.84	0.89	0.85	0.88	0.93	0.90
Mg #	0.50	0.51	0.53	0.52	0.48	0.52	0.52	0.53	0.46
K/Na	0.31	0.25	0.51	0.29	0.30	0.34	0.37	0.33	0.40
Trace elements (ppm)									
V	181	179	139	179	133	152	165	140	82
Ba	520	505	553	389	587	296	551	594	444
Sr	399	544	498	465	494	389	426	435	309
Y	29	17	30	17	22	29	23	23	21
Zr	103	65	20	98	98	60	86	84	129
Ga	21	19	18	20	20	19	19	19	15
Rb	71	43	127	50	71	138	75	66	84
Nb	6	3	5	4	4	5	5	5	3
Cs	2.7	2.5	3.5	1.0	3.9	6.4	2.7	3.9	2.6
La	17.1	11.9	17.9	17.4	18.6	12.7	10.8	16.2	11.6
Ce	42.2	26.6	45.5	39.3	40.5	34.2	27.7	38.2	25.1
Pr	5.9	3.3	6.1	4.7	5.1	5.1	4.1	5.0	3.3
Nd	25.8	13.7	24.1	18.0	21.2	22.3	18.3	21.3	13.2
Sm	6.2	3.3	5.3	3.7	4.9	5.3	4.4	4.9	3.2
Eu	1.3	1.0	1.2	1.2	1.3	1.1	1.1	1.2	0.7
Gd	5.0	3.4	4.9	3.4	4.3	5.0	4.4	4.6	3.2
Tb	0.8	0.5	0.8	0.5	0.7	0.9	0.7	0.7	0.5
Dy	4.8	3.3	5.0	3.3	3.9	5.1	4.3	4.2	3.3
Ho	1.0	0.7	1.0	0.7	0.8	1.0	0.9	0.8	0.7
Er	3.0	1.9	3.2	2.0	2.3	3.2	2.6	2.5	2.0
Tm	0.5	0.3	0.5	0.3	0.4	0.5	0.4	0.4	0.3
Yb	3.1	1.8	3.9	2.2	2.5	3.6	2.7	2.7	2.1
Lu	0.5	0.3	0.6	0.4	0.4	0.6	0.4	0.5	0.3
Hf	2.7	1.7	1.2	2.6	2.4	2.0	2.4	2.3	3.4
Ta	0.5	0.2	0.5	0.3	0.3	0.4	0.4	0.7	0.4
Pb	12	8	8	7	7	7	12	12	7
Th	3.0	4.3	5.6	2.2	3.9	3.0	4.7	7.4	6.8
U	1.0	1.1	2.5	0.7	0.8	2.5	1.3	3.0	2.2
Eu/Eu*	0.7	0.9	0.7	1.0	0.9	0.7	0.7	0.8	0.7

of millions of years are beyond analytical error (*c.* 10 Ma of 2 SE) as confirmed by the Wetherill Concordia plot together with probability density diagrams (Fig. 7).

Sampling locations from northern (192 ± 2 Ma, MSWD = 11, $n = 63$), central (188 ± 2 Ma, MSWD = 9.9, $n = 71$) and southern (193 ± 2 Ma, MSWD = 7.5, $n = 36$) parts of the pluton show no systematic variation in age (Fig. 7). The weighted mean ages of zircon cores are consistently older than the rim (193 ± 1 Ma core with MSWD = 6.3, $n = 80$ and 187 ± 2 Ma rim with MSWD = 12, $n = 89$; Fig. 7). This holds true for both the host rock core (194 ± 0.4 Ma; MSWD = 5.9, $n =$

Table 1. *Continued.*

	MMEs		Cumulate						
Sample	D12	D15	01	08	09	10	D04	E01	E02
SiO_2	64.55	61.19	67.10	58.45	51.46	48.85	48.77	52.64	53.15
Al_2O_3	15.42	16.62	15.19	17.89	18.94	18.94	21.75	19.12	18.75
Fe_2O_3(T)	4.03	6.00	3.84	6.51	8.08	9.59	8.09	8.60	8.29
MnO	0.07	0.10	0.08	0.12	0.17	0.17	0.11	0.15	0.22
MgO	1.79	2.96	1.71	2.99	5.21	5.17	3.64	4.89	5.23
CaO	4.93	5.99	4.51	6.80	8.48	10.52	10.38	9.38	8.72
Na_2O	3.88	3.23	3.84	3.37	2.85	3.64	2.52	2.78	3.77
K_2O	2.33	1.95	1.84	1.33	1.30	0.92	1.00	0.91	0.69
TiO_2	0.45	0.67	0.43	0.65	0.73	0.81	0.82	0.80	0.89
P_2O_5	0.09	0.14	0.09	0.14	0.11	0.24	0.08	0.05	0.14
LOI	2.31	1.38	1.77	1.07	2.48	1.35	1.86	1.35	0.80
Total (wt%)	99.85	100.20	100.40	99.32	99.80	100.20	99.03	100.70	100.70
$Na_2O + K_2O$	6.21	5.18	5.68	4.70	4.15	4.56	3.52	3.69	4.46
FeO(T)	3.63	5.40	3.46	5.86	7.27	8.63	7.28	7.74	7.46
A/NK	1.73	2.24	1.83	2.56	3.11	2.71	4.16	3.44	2.70
A/CNK	0.86	0.91	0.92	0.92	0.88	0.73	0.90	0.85	0.82
Mg #	0.47	0.49	0.47	0.48	0.56	0.52	0.47	0.53	0.56
K/Na	0.40	0.40	0.32	0.26	0.30	0.17	0.26	0.22	0.12
Trace elements (ppm)									
V	82	147	69	146	229	232	315	234	221
Ba	314	402	483	343	174	318	257	222	192
Sr	260	403	544	491	513	854	658	518	483
Y	17	21	13	16	16	11	11	12	13
Zr	115	99	121	95	62	23	46	70	61
Ga	15	16	16	18	18	17	20	18	18
Rb	92	52	54	37	43	22	40	22	14
Nb	3	3	3	2	2	1	1	2	3
Cs	2.3	1.8	1.5	1.8	3.7	3.1	2.5	1.3	1.0
La	12.3	14.3	8.9	14.1	7.5	10.1	6.2	7.8	11.4
Ce	28.5	33.0	17.8	28.5	17.7	21.1	13.9	17.2	22.2
Pr	3.5	4.2	2.1	3.3	2.4	2.5	1.8	2.2	2.6
Nd	13.3	17.0	8.4	12.9	9.8	10.2	7.8	8.8	11.0
Sm	2.9	3.9	1.8	2.8	2.5	2.3	2.0	2.2	2.5
Eu	0.7	0.9	0.8	0.9	0.8	0.9	0.8	0.9	1.0
Gd	2.7	3.8	1.7	2.4	2.6	2.1	2.1	2.3	2.5
Tb	0.4	0.6	0.3	0.4	0.4	0.3	0.3	0.4	0.4
Dy	2.8	3.9	1.6	2.6	2.6	2.0	2.2	2.3	2.6
Ho	0.6	0.8	0.3	0.5	0.5	0.4	0.4	0.5	0.5
Er	1.7	2.2	1.0	1.4	1.5	1.0	1.2	1.3	1.5
Tm	0.3	0.3	0.2	0.2	0.2	0.2	0.2	0.2	0.2
Yb	1.8	2.2	1.0	1.5	1.6	1.0	1.2	1.5	1.4
Lu	0.3	0.3	0.2	0.2	0.3	0.2	0.2	0.3	0.2
Hf	3.1	2.6	2.9	2.5	1.7	0.7	1.3	1.8	1.5
Ta	0.3	0.3	0.4	0.2	0.1	<0.1	0.2	0.2	0.2
Pb	7	8	9	8	<5	11	7	5	7
Th	12.2	7.8	5.7	3.0	1.9	1.6	1.9	3.3	2.7
U	1.4	2.2	2.6	1.2	0.7	1.6	1.2	1.0	0.9
Eu/Eu*	0.7	0.7	1.3	1.1	0.9	1.3	1.2	1.2	1.2

37) v. rim (188 $\pm$ 0.3 Ma rim; MSWD = 9, n = 43) and the MME core (193 $\pm$ 0.3 Ma; MSWD = 6.8, n = 44) v. rim (187 $\pm$ 2 Ma rim; MSWD = 15, n = 46) zircons.

Epsilon Hf values of the SKD and MMEs were calculated from $^{176}Hf/^{177}Hf$ values of each sample at the time of crystallization (zircon $^{206}Pb/^{238}U$ age). The averages and ranges of εHf (t) are: +10.4 for the SKD, +10.9 for the MMEs (ranging between +8.1 and +11.9 for sample 2109, +8.3 and +13.2 for sample 05G and +9.1 and +12.5 for sample 4101 of the SKD; +9.0 and +13.0 for sample E12, +9.1 and +14.0 for sample E11 and +9.2 and +15.6 for sample E09 of the MMEs;

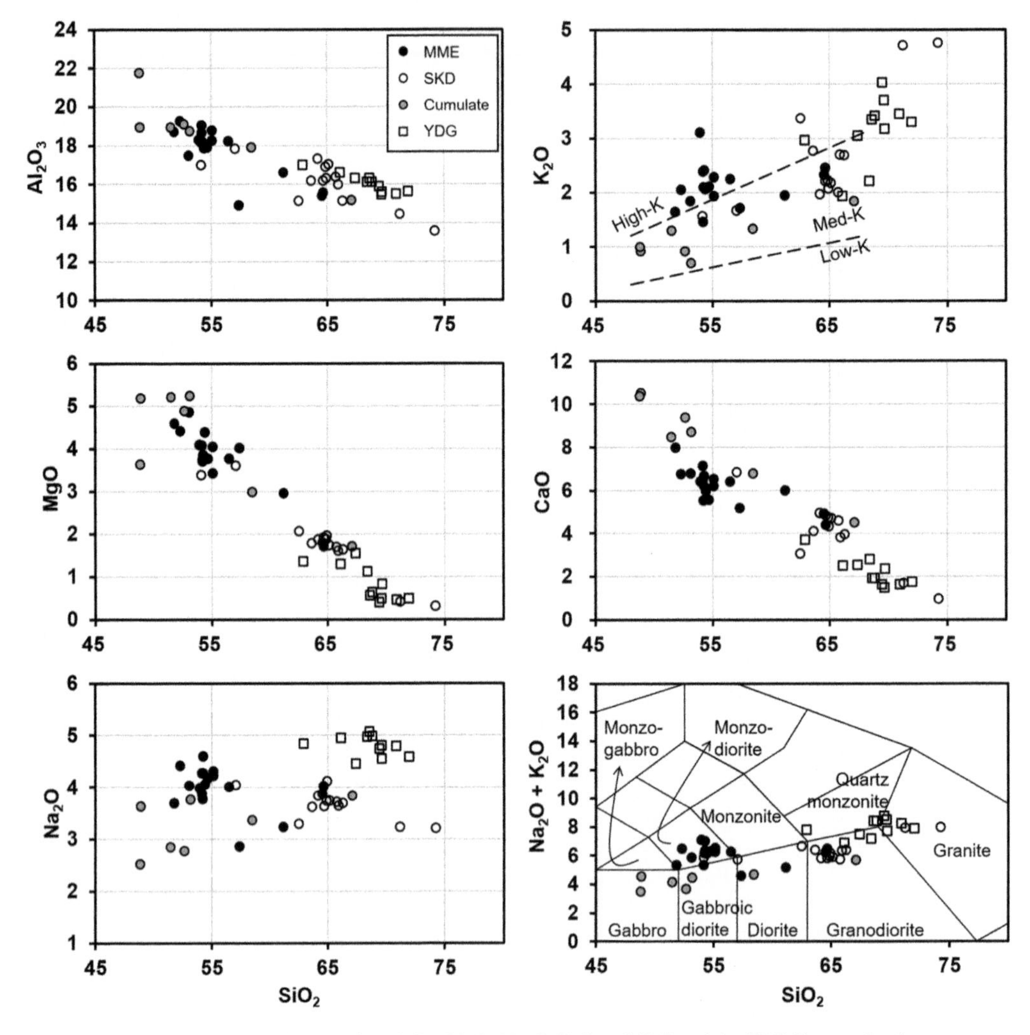

Fig. 3. Major element Harker diagrams for Al_2O_3, MgO, Na_2O, K_2O and CaO and the TAS diagram for the classification of plutonic rocks (Middlemost 1994). Note that the SKD shows a wide range in SiO_2.

Fig. 8). Both zircon populations of the SKD and MMEs show a wide spread of 5–7 epsilon units (Fig. 8). The average value from zircon cores is +10.6, identical to the rim. The weighted average εHf (t) of the MME zircons (10.7 $\pm$ 0.22) is slightly higher than that of the host rock (10.2 $\pm$ 0.21) (Fig. 8). The YDG has an average εHf (t) of +11.1 and also a relatively wide range of *c.* 4 εHf (t) units (Cheong *et al.* 2019).

Discussion

SKD and MME formation

Plutonic systems often show a wide spectrum of zircon ages outside analytical reproducibility (e.g. Idaho Batholith (Gaschnig *et al.* 2013), Ladakh Batholith (Weinberg and Dunlap 2000), Florida Mountains granite (Amato and Mack 2012)). It has been suggested that these bodies grow incrementally and through successive magma injections (Glazner *et al.* 2004; Miller *et al.* 2007), resulting in many millions of years of activity. Age spectra are reported within single units or even hand specimens, and in the absence of evidence for substantial magma mixing, these spectra remain difficult to reconcile with the concept of incremental pluton growth.

Zircon ages of the SKD and its MMEs range from around 210 to 180 Ma. Considering an analytical error of $\pm$5 Ma (*c.* 2.5% of SE), this range of 30 Ma is well outside analytical scatter and thus considered to accurately represent the time span of

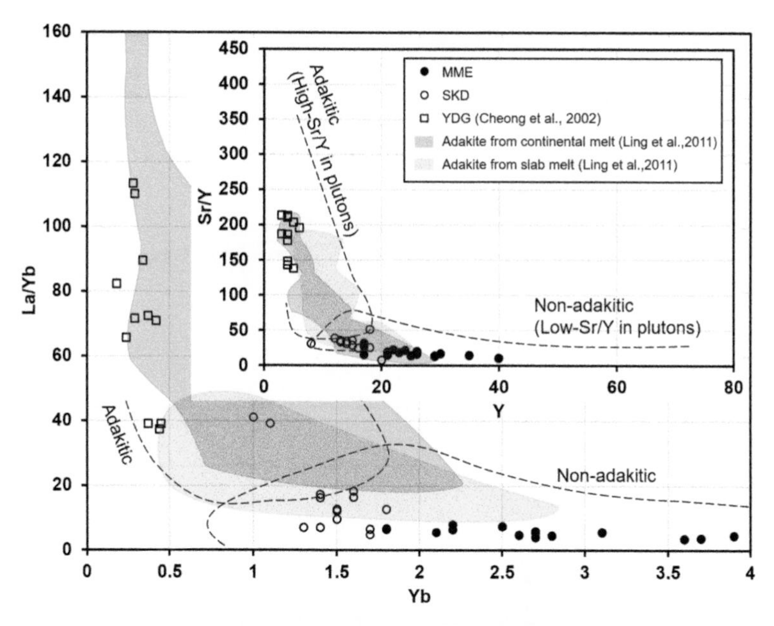

Fig. 4. Sr/Y v. Y and La/Yb v. Yb diagram. Note that the YDG plots in the red area, representing high-Sr/Y granite type. Reference adakite areas are from South China Craton (Ling *et al.* 2011).

magmatic activity. The scattering (*c.* 4% of SD) is larger than the standard error, and is a serious scale compared with lower standard errors (less than *c.* 1.7% in SE) and standard deviation (less than *c.* 1% in SD) in reference materials (Fig. 7c, d).

Considering the systematic difference in zircon core v. rim ages (Δ core–rim = average 6.3 Ma), the *c.* 6 Ma of systematic difference probably indicates the time from zircon growth until the end of the magmatic duration (i.e. emplacement). Notably, however, this span is not related to absolute ages, with some rim ages overlapping with core ages of other zircons, indicating that this process does not equate to a single event. A second critical observation is that these varying core–rim systematics can be observed in zircons from single samples. It is unclear how zircons in close proximity can undergo this core–rim cycle in the presence of other zircons with a similar cycle that is 10–15 Ma older.

Hafnium isotope ratios in the zircons also show a spread well beyond analytical uncertainty. A spread of *c.* 6 εHf (t) units in zircon is an indicator for a diverse source of Hf, in this case with a considerable influence of a juvenile component (εHf (t) = +8 to +14). Such a spread in plutonic rock is not uncommon and is considered a result of juvenile–mature endmember mixing (Hildreth and Moorbath 1988; Dungan and Davidson 2004). This explanation may be applicable to the SKD rocks, given the presence of more mafic MMEs. Petrographic mixing indicators such as mafic clots and acicular apatites in the MMEs and SKD indeed support magma mixing (Lim *et al.* 2018). Mixing here is best explained by chemical- and mechanical-mixing effects between MMEs and their host rock magma (i.e. a physical transfer of minerals between MMEs and host magma; Barbarin 2005). This would imply that some zircons from the MME sample could be derived from the host rock magma and vice versa, and would explain the strong overlap in zircon ages between both suites.

Any such mixing, however, cannot explain the long timespan of zircon ages. If melts remain near their solidus for *c.* 30 Ma, a physical mixing of grains requires a much higher liquid proportion, and therefore higher temperatures. This may be achieved through the incremental injection of new melts, such as the MME melts. However, a viable alternative to a hot magma chamber at upper crustal levels is that the older zircons did not form within the actual magmatic body, so are not phenocrysts *sensu stricto* but are in fact antecrysts. In this scenario, a deep-seated, lower crustal mush reservoir would be the source of the SKD and MME magmas, but at different stages of differentiation. In both cases, zircon would have already grown at depth and not in a shallow magma body.

Both, SKD and MMEs show similar mineralogy and identical age and Hf variabilities, and considerable overlaps in major and trace elements (Figs 3 & 5). In a scenario in which SKD and MME magma formed in a lower crustal hot zone, the difference in magmatic composition can be achieved by

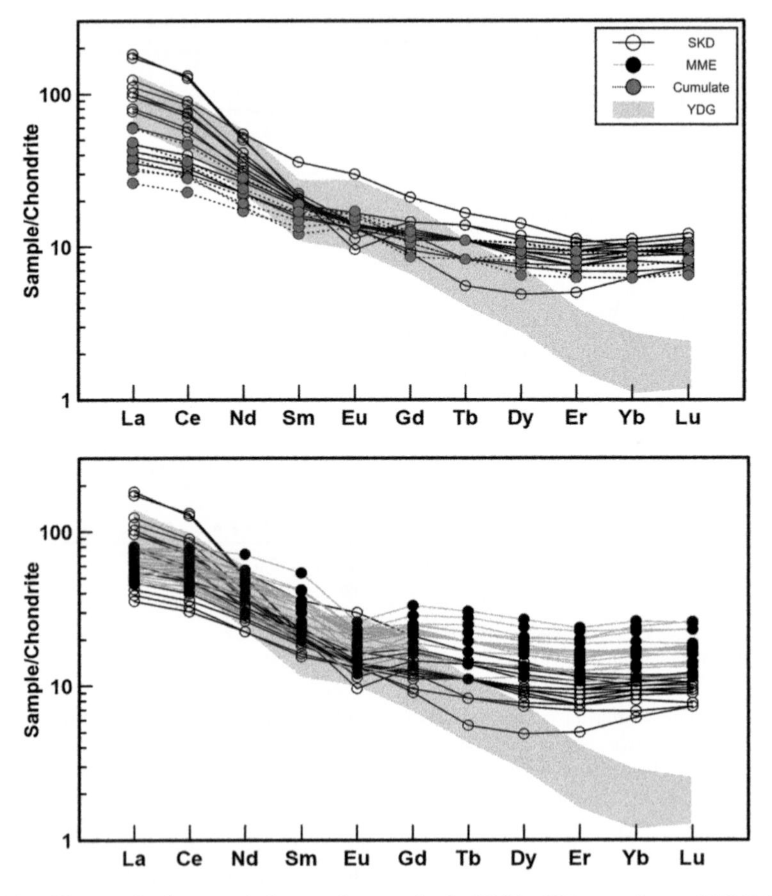

Fig. 5. Chondrite (CI) normalized rare earth element diagrams for the MMEs, SKD, cumulate and YDG. The slope (La/Yb) of the YDG trend is steep as a typical high-Sr/Y granite characteristics, whereas the MME is less steep and the SKD is intermediate. The YDG data are from Cheong *et al.* (2002).

different stages of fractional crystallization (FC). While exact values for each parameter are difficult to constrain, and with this any parental magma composition, it should be possible to calculate a feasible parent to both magma compositions through reverse fractional crystallization (equation (1)):

$$C_O = C_L/F^{D-1} \qquad (1)$$

where C_L is the concentration of the element in the residual liquid and C_0 is the initial concentration of the element, F denotes melt fraction and D is bulk distribution coefficient.

In reality, no such hypothetical magma composition would exist, simply owing to the constant recharge and evacuation mechanisms of melts entering and leaving lower crustal reservoirs. However, it is imperative to test if a hypothetical magma composition can produce both melt reservoirs (SKD and MMEs) for the scenario proposed here to be feasible for producing the zircons within a single reservoir at depth.

From the cumulate sample's mineralogy and chemistry (e.g. Eu* = 1.2; Fig. 5, Table 1), plagioclase and amphibole are considered as the main fractionating phases. Setting an unknown parental liquid composition (C_0), the FC modelling result indicates that the SKD and MME melt could indeed have originated from a similar source material with a simple adjustment of the amount of the amphibole fraction being removed from or added to the system (Fig. 9). The MME melts fractionated less modal amphibole but more pyroxenes (10% orthopyroxene, 80% clinopyroxene, and 10% plagioclase) compared with the host diorite (10% clinopyroxene, 60% amphibole, and 30% plagioclase) fractional phases quote experimental result of Nandedkar *et al.* (2014) (starting material RN7 and RN8). Discrepancies in Ba and Sr might indicate an inaccuracy of this

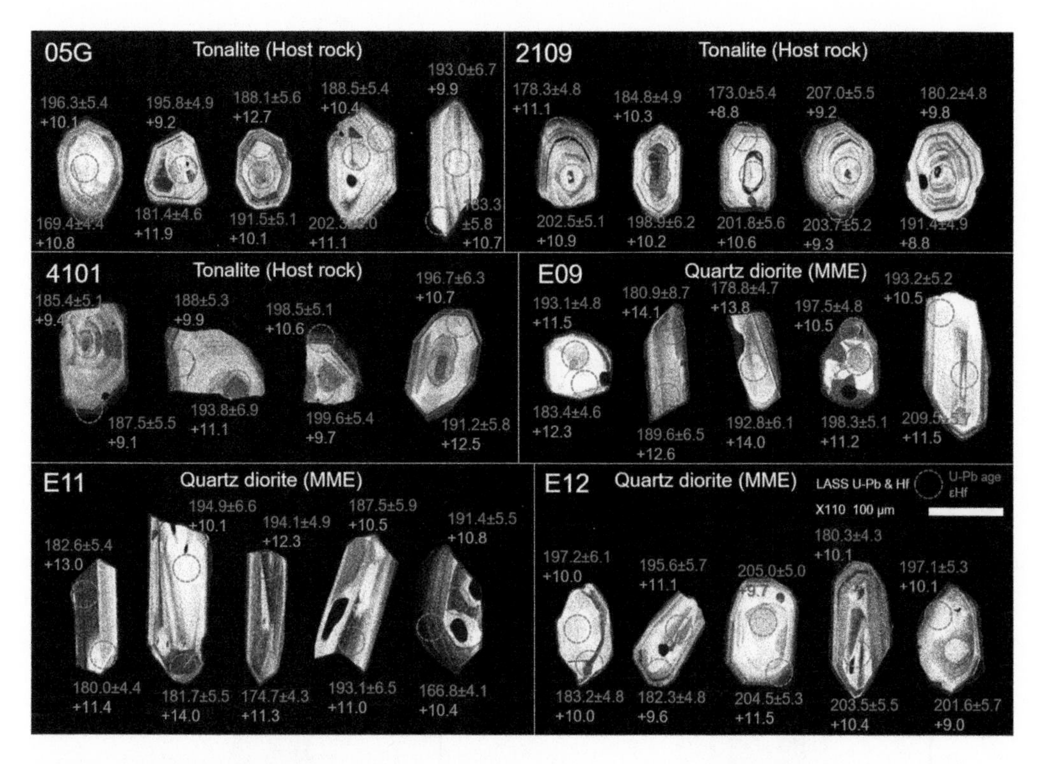

Fig. 6. Representative cathodoluminescence images of analysed zircons and ablated spots (red circle) with *in-situ* ages (red font ± SE) and εHf (t) values (blue font). Note that zircons from the SKD and MMEs record indistinguishable ages.

model in plagioclase effect. Plagioclase can be actively transferred between the MMEs and host rock melt, and also cumulated, which involves complexity. Even though this hypothetical parental melt is not considered a real melt composition here, the successful proof of concept further implies that amphibole is a key phase in the genesis of the melts, which is expected for I-type granitoids (Chappell *et al.* 2012), and a lower arc crust hydrous melt reservoir. This test adequately confirmed the feasibility of a hypothetical parental melt. It is, however, critical to test if any such melt would be saturated in zircon in order to produce and transport the crystals.

Zircon saturation

Experimental zircon saturation in relatively mafic magma compositions is an ongoing debate. Boehnke *et al.* (2013) suggest that an unrealistically high abundance of Zr is required for basaltic melt to have autocrystic zircon, and Siégel *et al.* (2018) conclude that zircon should not be present in a melt with $<$64 wt% SiO_2, apart from some minor crystallization in differentiated melt pockets. Experimental studies on zircon saturation in peralkaline melts demonstrate that zircon would not form because of the unrealistically high levels of Zr enrichment ($>$10 000 ppm) required for this to occur (Gervasoni *et al.* 2016; Shao *et al.* 2019). Borisov and Aranovich (2019) reported that zircon crystallization in the dry, near-solidus evolved basaltic melt is unlikely, while a water-containing condition is more plausible for zircon crystallization. Nevertheless, zircon is found in many mid-ocean ridge environments (basalt or gabbro; Coogan and Hinton 2006; Bortnikov *et al.* 2008; Fu *et al.* 2008; Lissenberg *et al.* 2009; Schmitt *et al.* 2011; Bea *et al.* 2020; Borisova *et al.* 2020; Aitchison *et al.* 2022), and it is often reported that zircons from a granite and MMEs are different from each other in their morphology and/or isotopic characteristics.

The MME zircons from our study appear to have more positive εHf (t) values (Fig. 8), indicating that simple grain mixing between host diorite and MMEs is not the only reason for the distribution of ages between the two reservoirs. However, the MMEs are also more mafic than the host diorite so that any zircon saturation scenario in which zircons either grew or were transported must be tested for these melts. Below, we test if the MME composition could have indeed supported zircon stability through saturation calculations.

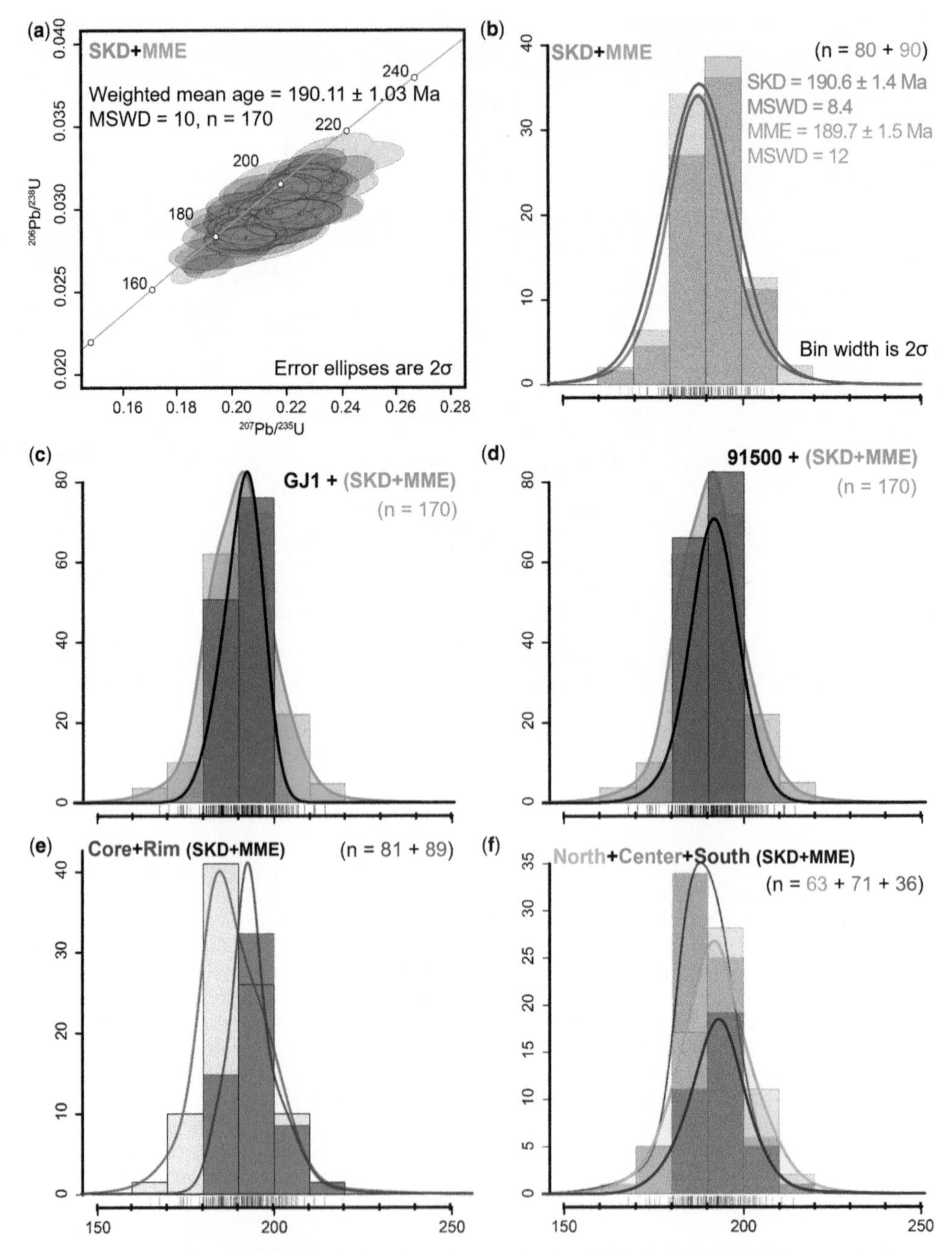

Fig. 7. Concordia and Kernel density plots. (**a**) The weighted mean age is 190 ± 1 Ma ($n = 80$) for the SKD and MMEs. (**b**) The age distributions and variabilities in both the SKD and MMEs are identical. (**c** and **d**) Comparison between the standard materials and the samples in age distribution. The distribution of sample shows larger variability. (**e**) Core v. rim of the SKD rocks showing the most distinct difference in the age range. (**f**) Spatial distribution of the sample does not explain the age variability.

In order to assess whether zircon was stable in the MME melt compositions, we conducted Zr enrichment–zircon saturation modelling using the replenishment–evacuation–fractionation (REFC) model of Lee *et al.* (2014) with the phase equilibrium calibration of Boehnke *et al.* (2013). The modelling

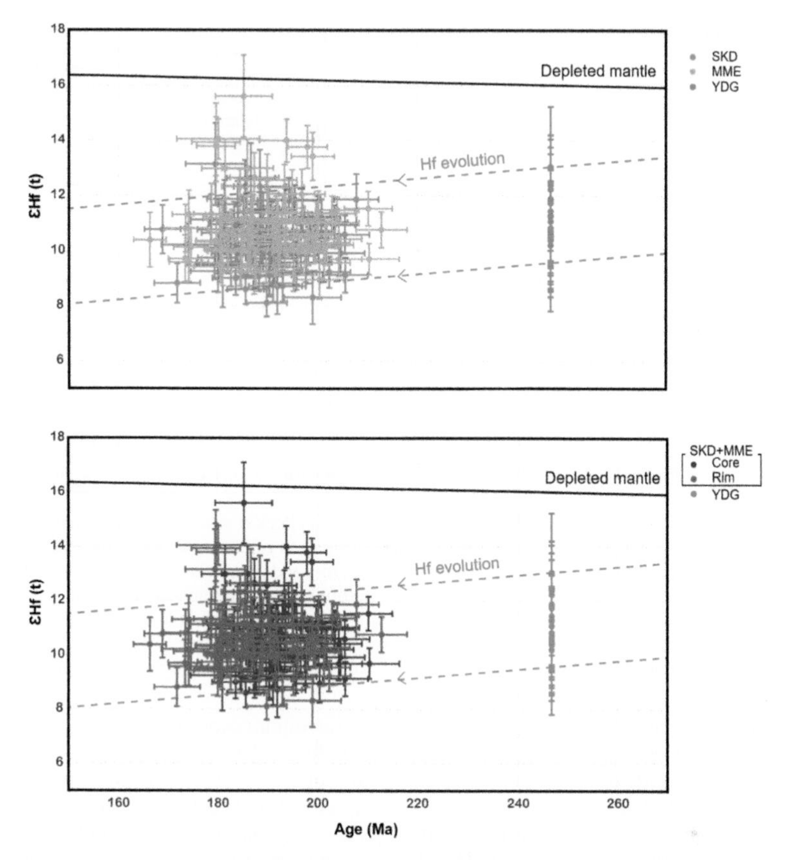

Fig. 8. Epsilon Hf (t) v. age diagram together with Hf isotope evolution lines based on the YDG (red dashed lines). Hafnium isotope evolution lines are calculated with $^{176}Lu/^{177}Hf = 0.0093$ for felsic crust (Amelin *et al.* 1999). Depleted mantle evolution is calculated based on present-day depleted mantle values $^{176}Hf/^{177}Hf = 0.283223$ and $^{176}Lu/^{177}Hf = 0.038$ (Vervoort and Blichert-Toft 1999).

is designed to test if an experimental primitive hydrous arc magma evolution can achieve zircon saturation (from 48.5 wt% in SiO_2, 3.17 wt% in H_2O, and 52 ppm in Zr; after experimental petrology result, Nandedkar *et al.* 2014; and average mantle Zr content, Lee and Bachmann 2014) under mid–low crustal redox-pressure conditions (20–25 km depth and in NiNiO buffer; Nandedkar *et al.* 2014). The model is composed of two parts; the first part of the model investigates Zr enrichment in a recharging melt reservoir and the second part of the model evaluates whether this enrichment is sufficient to achieve zircon saturation.

In the initial stage of any melt reservoir development, source enrichment in Zr was calculated by REFC modelling, which implies that a parental melt undergoes FC, yet is constantly recharged through melt replenishment with volume being controlled through melt evacuation in addition to FC. Mass balance in the magma reservoir is assumed by equal masses of FC and evacuation v. melt replenishment. This relation is expressed in equation (2)

$$\frac{C_{ch}}{C_{ch}^{o}} = \frac{C_{ch}/C_{ch}^{o}}{D\alpha_x + \alpha_e} - \left[\frac{C_{ch}/C_{ch}^{o}}{D\alpha_x + \alpha_e} - 1\right]\exp[-\Delta\bar{M}_{re}(D\alpha_x + \alpha_e)] \quad (2)$$

where C denotes the concentration of a certain element in chamber/initial chamber (C_{ch}/C_{ch}^{o}), α_x equals dM_x/dM_{re} and α_e equals dM_e/dM_{re}, where M represents the mass of fractionation (M_x), replenishment (M_{re}) or evacuation (M_e). Note that $\alpha_x + \alpha_e = 1$ if the mass of magma chamber M_{ch} is at steady state. In this model, number of times that the magma reservoir has recharged ($\Delta\bar{M}_{re}$, cf. 'overturn', Lee *et al.* 2014) is considered.

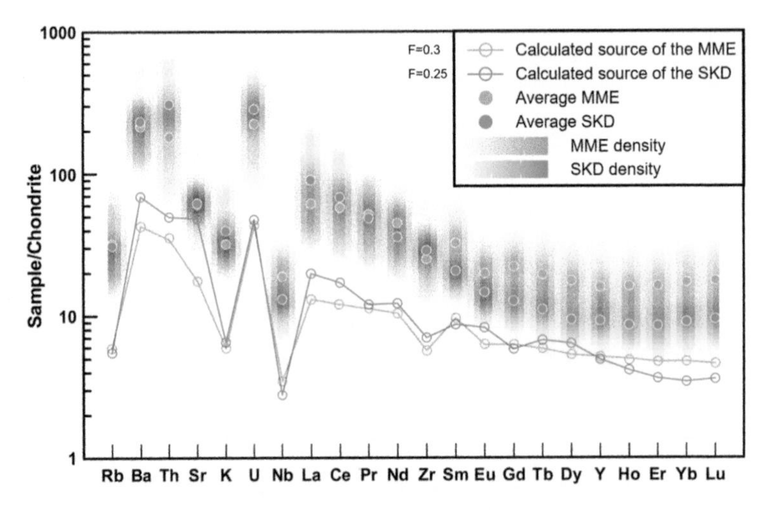

Fig. 9. Fractional crystallization modelling of the MMEs and SKD to check if two different magmas can be generated by different fractional phases. Sources are calculated from each resultant magma composition (Shaw 1970; Kd database (McKenzie and O'Nions 1991) for orthopyroxene, garnet and plagioclase; Foley *et al.* (1996) for clinopyroxene; and Fujimaki *et al.* (1984) for amphibole). Proportions of residual phases (10% orthopyroxene, 80% clinopyroxene and 10% plagioclase for the MMEs; 10% clinopyroxene, 60% amphibole and 30% plagioclase for the host rock) refer the experimental results after Nandedkar *et al.* (2014) (starting materials RN7 and RN8 therein). Note that the calculated concentration implies a nearly identical source composition dictated by the amphibole proportion.

In the REFC model, the concentration of a given element (either incompatible or compatible) converges into a level of 'steady state' as replenishment proceeds. In other words, the maximum enrichment for any incompatible elements ($D < 1$) is finite, similar to the depletion of compatible elements ($D > 1$) (Lee *et al.* 2014; Fig. 10a, b). Another significance of this model is that compatible elements ($D > 1$) quickly achieve a steady state compared with relatively slow enrichment of incompatible elements ($D < 1$) (Lee *et al.* 2014; Fig. 10a, b). This implies that Zr, treated here as an incompatible element ($D < 1$), is more enriched than other major elements, whose concentrations are more or less constant. If the initial Zr concentration in any hypothetical primitive melt is mantle like (i.e. 52 ppm, Lee *et al.* 2014), and replenishing magma is of a similar composition, the maximum enrichment of Zr is 1037 ppm until a steady state is achieved if no evacuation ($\alpha_e = 0$) of the magma chamber is assumed (Fig. 10a). On the other hand, no more than 216 ppm Zr is achieved if 20% evacuation ($\alpha_e = 0.2$) is assumed (Fig. 10b).

The modelling results, and with this any constraints on zircon saturation, are dependent on the bulk distribution coefficient of Zr (D_{Zr}). This D_{Zr}, however, is dependent on (1) magma composition, (2) fractionating phases and (3) their weight fractions, rendering constraints on saturation strongly model dependent. To test for the effect of compatibility, we used a range of D values, from 0.05 to 2 (Fig. 10a, b).

Zirconium concentrations of an evolving magma were then calculated using the fractional crystallization equation (Shaw 1970; equation (1)). In this case, C_L is the Zr concentration in an evolved melt and C_0 is Zr concentration in the starting material and F denotes the melt fraction. Each melt fraction F with corresponding temperature is quoted by experimental results from Nandedkar *et al.* (2014) The starting temperature is 1070°C. The bulk D_{Zr} values tested in this model are 0.1 and 0.3 (Marxer and Ulmer 2019).

In the second part, zircon saturation is calculated by Boehnke *et al.* (2013):

$$\ln D_{Zr} = \frac{10\ 108}{T(\mathrm{K})} - 1.16\ (M - 1) - 1.48 \qquad (3)$$

where M is a compositional parameter given by (Na + K + 2Ca)/(Al·Si). From equation (3) (Boehnke *et al.* 2013), one can create an x–y field of Zr in melt [Zr] v. temperature with variable M values (grey gradient lines in Fig. 10c). Each parabolic line may represent the loss of [Zr] (i.e. zircon crystallization) as temperature drops in a certain melt composition (M). Figure 10c represents a combination of Zr enrichment and zircon crystallization curves simulated by zircon saturation model of Boehnke *et al.* (2013) in mid–lower crustal conditions (Nandedkar *et al.* 2014) when $\alpha_e = 0.2$ (Table S4 in Supplementary Material 5). This hypothetical value 'eruption factor' is not a real value, but an upper limit while

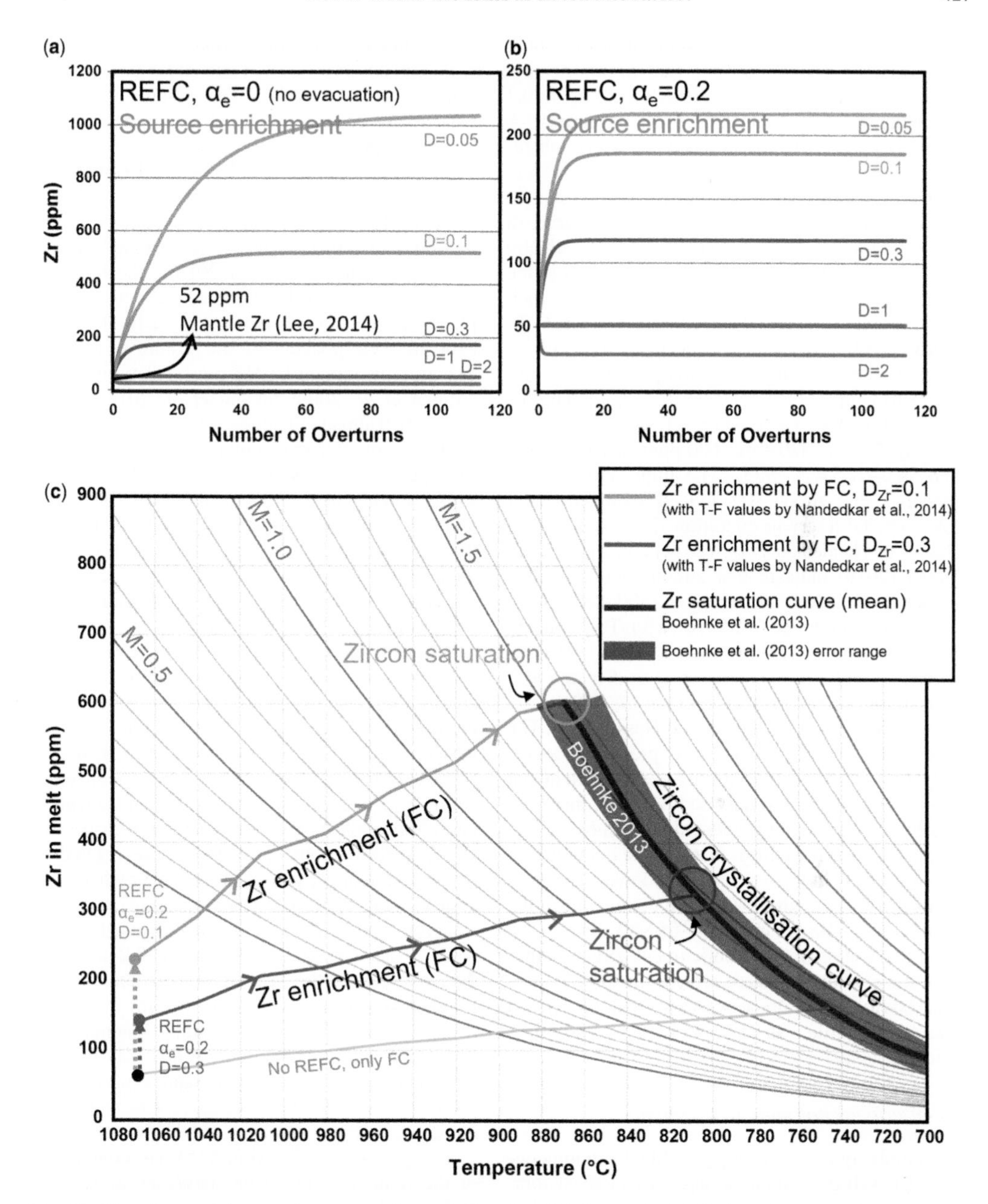

Fig. 10. Zirconium enrichment–zircon saturation modelling for MME-like melt. (**a**) Various source enrichment curves were calculated by the replenishment–evacuation–fractional crystallization (REFC) modelling after Lee *et al.* (2014). $D = 0.05$ is an alternative to $D = 0$. Note that compatible elements take less time to reach steady state than incompatible elements. Concentration change from the beginning to the steady state is not dramatic in compatible elements. (**b**) The same graph as Figure 10a except for $\alpha_e = 0.2$. (**c**) Two Zr enrichment–Zr saturation curves from an average mantle Zr content (52 ppm; Lee and Bachmann 2014), each enriched with $D_{Zr} = 0.1$ and $D_{Zr} = 0.3$. Melt evolution follows the fractional crystallization (FC) curve. Melt fraction (F) and temperature refer to experimental parameters after Nandedkar *et al.* (2014). Zircon saturation is based on Boehnke *et al.* (2013). Whole rock compositions of experimental basaltic hydrous arc melt (RN6–RN17V2 runs in Nandedkar *et al.* 2014) are used to calculate M values $(Na + K + 2Ca)/(Al{\cdot}Si)$ in each temperature. Note that the zircon saturation curve crosses gradient lines of decreasing M values as the temperature drops, which indicates evolving melt composition in the P–T–redox conditions with a lower crust affinity. The concentration of Zr in melt is enriched until a zircon saturation concentration of *c.* 320–600 ppm in a 810–870°C range with 61 wt% SiO_2.

eruption was probably not significant in this study area since no coeval volcanic unit occurs nearby. Thus, it is expected that Figure 10c with $\alpha_e = 0.2$ could constrain the lower limit in Zr enrichment. A Boehnke *et al.* (2013) constrained zircon crystallization model in the evolving basaltic hydrous arc magma (Nandedkar *et al.* 2014) is shown in Figure 10c. The zircon crystallization curve (a thick grey line with a grey area of error) gradually crosses gradient lines of decreasing M values as the temperature drops, which indicates that the melt evolves into a more silicic composition at P–T–redox conditions with a lower crustal affinity (Fig. 10c). According to the example model, evolving melts with $D = 0.1$ and $D = 0.3$ (blue and purple lines in Fig. 10c, Table S4) reach saturation points ranging from *c.* 320 to *c.* 600 ppm in 810–870°C and 59.6–61.5 wt% SiO_2. The implication of this model is that zircon was indeed saturated under lower crustal reservoir condition. The experimental studies of Nandedkar *et al.* (2014) and Marxer and Ulmer (2019) indicate that zircon saturation may be achieved at low temperatures (800–830°C) yet in a higher SiO_2 range (*c.* 65–70 wt%). The model in this study slightly underestimates SiO_2 (*c.* 60 wt%) compared with those experimental studies and also with implications from Siégel *et al.* (2018) (*c.* 64 wt% SiO_2) and Dickinson and Hess (1982). The essential difference is that the previous models only considered fractionation as a Zr-enrichment mechanism, while this study additionally considered basaltic magma replenishment simulated by REFC. The calculation reconfirms that magma fractionation without REFC in this study returns lower saturation temperature and higher SiO_2 (The yellow line in the figure 10; *c.* 750 °C and *c.* 64–65 wt% SiO_2). The main implication of the Zr enrichment–zircon saturation model is that constant replenishment may provide additional Zr to a reservoir and the zircon saturation hurdle can be lower for slightly less silicic magmas.

Deep crustal hot zone magmatism

The similarity between SKD and MMEs in mineralogy, age, εHf (t), and major- and trace element compositions is consistent with their derivation from a common source. Their subtle differences in trace elements are mainly due to the different fractional proportions of amphibole, and this outlines the source depth where magmatic amphiboles are stable (<18 kbar; Gill 1981). Zirconium enrichment plus zircon saturation assessment in this study indicates that this less silicic source may be saturated in zircon. Even though physical mixing of zircon grains between host melt and MMEs cannot be excluded, a viable alternative scenario is that their common magma reservoir already contained zircons. This scenario can successfully explain the long lifespan with the observed systematic older core–younger rim pattern of zircons. The assumed source is expected to be seated at *c.* 7.0 kbar with 810–870° C. The temperature range corresponds to the temperature range of Marxer and Ulmer (2019) with tonalitic melt composition, which has nearly identical lithology to the SKD and MMEs. However, since these experimental studies (e. g., Nandedkar *et al.* 2014; Müntener and Ulmer 2018) indicate a more evolved magma in this temperature range, a more plausible scenario is heterogeneously distributed evolved melt pockets in the primitive reservoir. The aforementioned DCHZ mush reservoir concept (Annen *et al.* 2006; Solano *et al.* 2012; Jackson *et al.* 2018) already includes a similar model of a layered reservoir, where it contains more silicic melts in its upper level (Jackson *et al.* 2018). This is a likely source model for the SKD and MMEs. In analogy this can also explain the Hf isotope variability of the neighbouring (yet older) YDG. For this, no single zircon ages are available, yet the isotope variability in εHf (t) indicates a similar situation to that observed for the SKD (Fig. 8).

In deep-seated (18–30 km depth) reservoirs, the melt fraction is low (<0.2) while fractional crystallization and replenishment occur together (Jackson *et al.* 2018). ‘Early Jurassic magma flare-ups’ suggested by a regional study of Cheong and Jo (2020) could have triggered this replenishment (from Fig. 11a to b). Periodic replenishments of basaltic magma sporadically crank up melt fraction. Higher melt fraction is progressively segregated and moves to the upper part of the reservoir where more evolved magma resides (Jackson *et al.* 2018). The reservoir sporadically evacuates evolved magma when it has a greater melt fraction (0.2–0.35) and thus overcomes the second percolation threshold (Vigneresse *et al.* 1991) to obtain momentum (Fig. 11a, b). These evacuating magmas with basalt-andesitic composition are probably mixtures conveying both relic minerals and evolved liquid pockets which contain zircon antecrysts, and the zircon grains in them should record the timespan of the evolving mush reservoir (Fig. 11b). The composition of the remaining reservoir converges into that of replenishing primitive magma if it is continuously replenished (Lee *et al.* 2014). Evacuating magma in the later stage of the reservoir, therefore, is more juvenile than that in the previous stage, and mingles with the host rock, then forms the MMEs (Fig. 11c, d). Fractional crystallization and mixing may occur at any stages in the evolution, and at any level of the reservoir. This occurs around the solidus temperature of *c.* 800°C, yet under *c.* 900°C (zircon closure temperature), because MMEs mingled at a high temperature and no ‘reset-ages’ but prolonged ages were recorded.

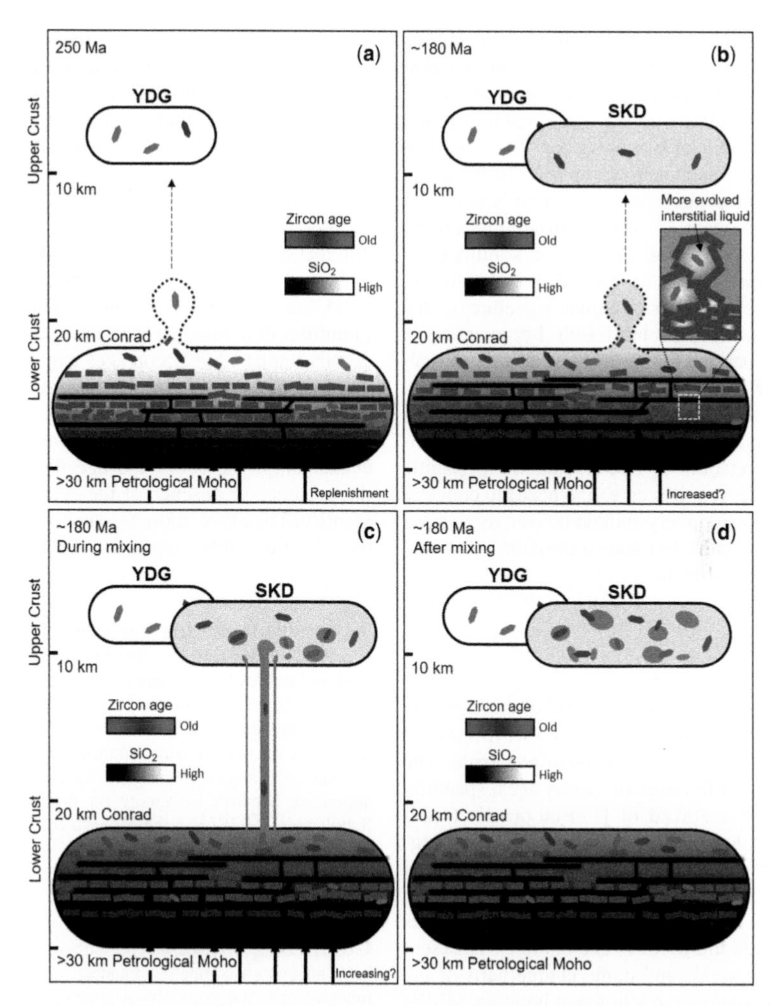

Fig. 11. The cartoon depicting deep crustal hot zone (DCHZ) reservoir model for the study area magmatism (not to strict scale). (**a**) A deep-seated mush reservoir which generates the high-Sr/Y YDG magma. Consistent replenishment progressively changes the composition of the reservoir, while Zr could be keep enriched and saturated in the REFC process. Note the compositional layering/gradation in the reservoir, which facilitates zircon saturation in a more silicic part. (**b**) A compositionally changed mush reservoir could form the less felsic SKD. Previously formed zircon could join the SKD magma. The inset figure illustrates how the zircon-saturated evolved liquids (pockets) can be captured by previously crystallized grains. (**c**) Consistent, or even increasing replenishment kept lowering SiO_2 in the reservoir, where MME magma was generated by a different FC route compared with that for SKD. Zircons in the MME magma also have protracted ages inherited from the reservoir. (**d**) The SKD body approaches mechanical and chemical equilibrium which results in overlaps between the MMEs and host rock. Note that mixing and fractionation may happen any time during these processes.

The significance of the proposed model is that the exact time of an intrusion cannot conclusively be constrained by the ages of this zircon population, as these may include antecrysts, which are basically inherited. The pooling of ages in programs such as Isoplot then indicates a single age and origin with seemingly small total error, where only the elevated MSWD indicates the prolonged age spectrum. Common lead can be responsible for spreading ages. However, this study trimmed data with discordance >5%, and the common lead effect for age-spreading in the surviving spots is not significant, as graphically assumed in Figure 7a. Lead loss, on the other hand, indeed spreads ages in a younger direction, which was reported and warned by different studies (e.g. Spencer *et al.* 2016; Andersen *et al.* 2019). It is difficult to reconcile the presence or absence of lead loss events with post-crystallization events. The dataset in this study might not be enough to fully discuss the lead loss effect on younger ages. However, the Hf isotope variability found in this study implies that the age spreading is not solely due to the lead loss in zircon. The idea that zircons crystallize and achieve their Hf isotopic signature before the shallow crustal emplacement is similar to the view of Hammerli *et al.* (2018). They demonstrated that incomplete solidification of basaltic magmas in a DCHZ (< 35–40 km) leads to the absence of a garnet signature in the example from Lachlan Fold Belt I-type granites, and this may parallel the lack of enriched Sr/Y values in the SKD and MMEs. In addition, they proposed that zircon isotopic variabilities in intrusions probably originated by juvenile magma replenishments. Interpretation of zircon data from granitic rocks, therefore, requires care, and needs to consider conditions of zircon crystallization across various temperatures, depths and source durations, and evidence for replenishment.

Conclusion

We present new LASS-MC-ICP-MS zircon age data and Hf isotope compositions for a diorite (SKD) and associated MMEs from the Korean peninsula. The SKD shows a wide range of zircon ages, spanning *c.* 30 Ma, with a spread of Hf isotopes (*c.* 6 εHf units) far outside analytical reproducibility. We suggest a common parental melt for the SKD and MMEs, which evolved through each fractionation path towards dioritic compositions. Cumulates located towards the north of the complex represent fractionated minerals and mafic counterparts of the intrusion. This genetic relationship between MMEs and ambient diorite, which is reflected in evolutionary plots of major element compositions, can readily explain the formation of the pluton through incremental pluton growth associated with fractional crystallization. This incremental pluton growth scenario, however, fails to explain similar zircon distribution in terms of ages and εHf (t) ranges between both rock types, notably a long timespan (*c.* 30 Ma) of zircon ages.

Based on major trace element distributions, and in the context of proposed deep crustal hot zones, we propose that the source of both intrusion types is a lower arc crust melt reservoir. Parental melts of both intrusions are probably of basalt–andesitic composition, while one of them fractionated more amphibole forming the dioritic SKD, and the other evolved towards MME compositions through fractionation of less amphibole. We propose that a low-melt fraction mush reservoir existed, which is re-activated and tapped when hot, with mafic magma replenished from the mantle. This reservoir was probably a long-lasting zircon spawning mush, where pressure and temperature satisfy near-solidus conditions. Zirconium enrichment–zircon saturation modelling shows that zircon formation in such a lower crustal mush reservoir is viable with consistent replenishments under hydrous (>6 wt% H_2O) conditions.

If correct, such mush reservoirs can serve as a zircon incubator, which can explain the wide range of ages and Hf isotope signatures observed in the SKD, and possibly elsewhere. This scenario for granitoids that originate from lower crustal mush reservoirs implies that zircon ages may not indicate the true age of the intrusion. Instead, zircon Hf-isotope and age data rather record condition of the source. However, a lead loss effect by post-crystallization events might influence the age distribution, while the presence or absence of the events is difficult to identify. Therefore, more research on the zircon system closure will be required.

Acknowledgements The authors would like to thank Jisu Kim for the sampling and field work, Massimo Raveggi and Ashlea Wainwright for the analytical help. Oscar Laurent and an anonymous reviewer are thanked for their useful comments that greatly improved the paper. We also thank the volume editor Martin Guitreau for accepting the manuscript and the smooth editorial procedure. We appreciate Monash University for the Monash Graduate Scholarship (MGS) and the Monash International Tuition Sponsorship (MITS).

Competing interest The authors declare that they have no known competing financial interests or personal relationships that could have appeared to influence the work reported in this paper.

Author contributions **HL**: conceptualization (lead), data curation (lead), formal analysis (lead), investigation (lead), methodology (lead), resources (lead), validation (lead), visualization (lead), writing – original draft (lead), writing – review & editing (lead); **ON**: conceptualization (equal), project administration (equal), supervision (lead), validation (lead), writing – review & editing (equal); **RFW**: conceptualization (equal), investigation (supporting), validation (equal), writing – review & editing (supporting); **YN-J**: investigation (supporting), resources (supporting), supervision (equal), validation (equal), writing – review & editing (supporting); **VRB**: data curation (equal), software (equal), validation (supporting); **JP**: conceptualization (supporting), resources (supporting), validation (supporting), writing – original draft (supporting), writing – review & editing (supporting); **BM**: resources (supporting), validation (supporting), writing – original draft (supporting), writing – review & editing (supporting). **PAC**: Supervision (equal), Funding acquisition (lead), Writing – review & editing (equal).

Funding This work was supported by a grant from the Australian Research Council (FL160100168) to Peter A. Cawood

Data availability Data will be available on request.

References

Aitchison, J.C., Cluzel, D. *et al.* 2022. Solid-phase transfer into the forearc mantle wedge: rutile and zircon xenocrysts fingerprint subducting sources. *Earth and Planetary Science Letters*, **577**, 117251, https://doi.org/10.1016/j.epsl.2021.117251

Amato, J.M. and Mack, G.H. 2012. Detrital zircon geochronology from the Cambrian–Ordovician Bliss Sandstone, New Mexico: evidence for contrasting Grenville-age and Cambrian sources on opposite sides of the Transcontinental Arch. *Bulletin*, **124**, 1826–1840, https://doi.org/10.1130/B30657.1

Amelin, Y., Lee, D.-C., Halliday, A.N. and Pidgeon, R.T. 1999. Nature of the Earth's earliest crust from hafnium isotopes in single detrital zircons. *Nature*, **399**, 252–255, https://doi.org/10.1038/20426

Andersen, T., Elburg, M.A. and Magwaza, B.N. 2019. Sources of bias in detrital zircon geochronology: discordance, concealed lead loss and common lead correction. *Earth-Science Reviews*, **197**, 102899, https://doi.org/10.1016/j.earscirev.2019.102899

Annen, C., Blundy, J. and Sparks, R. 2006. The genesis of intermediate and silicic magmas in deep crustal hot zones. *Journal of Petrology*, **47**, 505–539, https://doi.org/10.1093/petrology/egi084

Barbarin, B. 2005. Mafic magmatic enclaves and mafic rocks associated with some granitoids of the central Sierra Nevada batholith, California: nature, origin, and relations with the hosts. *Lithos*, **80**, 155–177, https://doi.org/10.1016/j.lithos.2004.05.010

Baxter, S. and Feely, M. 2002. Magma mixing and mingling textures in granitoids: examples from the Galway Granite, Connemara, Ireland. *Mineralogy and Petrology*, **76**, 63–74, https://doi.org/10.1007/s007100200032

Bea, F., Bortnikov, N. *et al.* 2020. Zircon xenocryst evidence for crustal recycling at the Mid-Atlantic Ridge. *Lithos*, **354**, 105361, https://doi.org/10.1016/j.lithos.2019.105361

Boehnke, P., Watson, E.B., Trail, D., Harrison, T.M. and Schmitt, A.K. 2013. Zircon saturation re-revisited. *Chemical Geology*, **351**, 324–334, https://doi.org/10.1016/j.chemgeo.2013.05.028

Borisov, A. and Aranovich, L. 2019. Zircon solubility in silicate melts: new experiments and probability of zircon crystallization in deeply evolved basic melts. *Chemical Geology*, **510**, 103–112, https://doi.org/10.1016/j.chemgeo.2019.02.019

Borisova, A.Y., Bindeman, I.N. *et al.* 2020. Zircon survival in shallow asthenosphere and deep lithosphere. *American Mineralogist*, **105**, 1662–1671, https://doi.org/10.2138/am-2020-7402

Bortnikov, N., Sharkov, E., Bogatikov, O., Zinger, T., Lepekhina, E., Antonov, A. and Sergeev, S. 2008. Finds of young and ancient zircons in gabbroids of the Markov Deep, Mid-Atlantic Ridge, 5°54′–5°02.2′ N (results of SHRIMP-II U–Pb dating): implication for deep geodynamics of modern oceans. *In*: *Doklady Earth Sciences*. Springer Nature, 240–248.

Chappell, B.W., Bryant, C.J. and Wyborn, D. 2012. Peraluminous I-type granites. *Lithos*, **153**, 142–153, https://doi.org/10.1016/j.lithos.2012.07.008

Chen, Y., Price, R., White, A. and Chappell, B. 1990. Mafic inclusions from the Glenbog and Blue Gum granite suites, southeastern Australia. *Journal of Geophysical Research: Solid Earth*, **95**, 17757–17785, https://doi.org/10.1029/JB095iB11p17757

Cheong, A.C.-S. and Jo, H.J. 2020. Tectonomagmatic evolution of a Jurassic Cordilleran flare-up along the Korean Peninsula: geochronological and geochemical constraints from granitoid rocks. *Gondwana Research*, **88**, 21–44, https://doi.org/10.1016/j.gr.2020.06.025

Cheong, C.S., Kwon, S.T. and Sagong, H. 2002. Geochemical and Sr–Nd–Pb isotopic investigation of Triassic granitoids and basement rocks in the northern Gyeongsang Basin, Korea: implications for the young basement in the East Asian continental margin. *Island Arc*, **11**, 25–44, https://doi.org/10.1046/j.1440-1738.2002.00356.x

Cheong, A.C.-S., Jo, H.J., Jeong, Y.-J. and Li, X.-H. 2019. Magmatic response to the interplay of collisional and accretionary orogenies in the Korean Peninsula: geochronological, geochemical, and O–Hf isotopic perspectives from Triassic plutons. *Bulletin*, **131**, 609–634, https://doi.org/10.1130/B32021.1

Choi, S.H., Kwon, S.-T., Mukasa, S.B. and Sagong, H. 2005. Sr–Nd–Pb isotope and trace element systematics of mantle xenoliths from Late Cenozoic alkaline lavas, South Korea. *Chemical Geology*, **221**, 40–64, https://doi.org/10.1016/j.chemgeo.2005.04.008

Chu, N., Taylor, R.N. *et al.* 2002. Hf isotope ratio analysis using multi-collector inductively coupled plasma mass spectrometry: an evaluation of isobaric interference corrections. *Journal of Analytical Atomic Spectrometry*, **17**, 1567–1574, https://doi.org/10.1039/b206707b

Collins, W., Richards, S., Healy, B. and Ellison, P. 2001. Origin of heterogeneous mafic enclaves by two-stage hybridisation in magma conduits (dykes) below and in granitic magma chambers. Fourth Hutton Symposium: The Origin of Granites and Related Rocks. *Earth and Environmental Science Transactions of The Royal Society of Edinburgh*, **91**, 27–46, https://doi.org/10.1017/S0263593300007276

Coogan, L.A. and Hinton, R.W. 2006. Do the trace element compositions of detrital zircons require Hadean continental crust? *Geology*, **34**, 633–636, https://doi.org/10.1130/G22737.1

Dickinson, J.E. and Hess, P. 1982. Zircon saturation in lunar basalts and granites. *Earth and Planetary Science Letters*, **57**, 336–344, https://doi.org/10.1016/0012-821X(82)90154-6

Dungan, M.A. and Davidson, J. 2004. Partial assimilative recycling of the mafic plutonic roots of arc volcanoes: An example from the Chilean Andes, *Geology* **32**, 773–776.

Fisher, C.M., Vervoort, J.D. and DuFrane, S.A. 2014*a*. Accurate Hf isotope determinations of complex zircons using the 'laser ablation split stream' method. *Geochemistry, Geophysics, Geosystems*, **15**, 121–139, https://doi.org/10.1002/2013GC004962

Fisher, C.M., Vervoort, J.D. and Hanchar, J.M. 2014*b*. Guidelines for reporting zircon Hf isotopic data by LA-MC-ICPMS and potential pitfalls in the interpretation of these data. *Chemical Geology*, **363**, 125–133, https://doi.org/10.1016/j.chemgeo.2013.10.019

Foley, S.F., Jackson, S.E., Fryer, B.J., Greenouch, J.D. and Jenner, G.A. 1996. Trace element partition coefficients for clinopyroxene and phlogopite in an alkaline lamprophyre from Newfoundland by LAM-ICP-MS. *Geochimica et Cosmochimica Acta*, **60**, 629–638, https://doi.org/10.1016/0016-7037(95)00422-X

Fu, B., Page, F.Z. *et al.* 2008. Ti-in-zircon thermometry: applications and limitations. *Contributions to Mineralogy and Petrology*, **156**, 197–215, https://doi.org/10.1007/s00410-008-0281-5

Fujimaki, H., Tatsumoto, M. and Aoki, K.I. 1984. Partition coefficients of Hf, Zr, and REE between phenocrysts and groundmasses. *Journal of Geophysical Research: Solid Earth*, **89**, B662–B672, https://doi.org/10.1029/JB089iS02p0B662

Gaschnig, R.M., Vervoort, J.D., Lewis, R.S. and Tikoff, B. 2013. Probing for Proterozoic and Archean crust in the northern US Cordillera with inherited zircon from the Idaho batholith. *Bulletin*, **125**, 73–88, https://doi.org/10.1130/B30583.1

Gervasoni, F., Klemme, S., Rocha-Júnior, E.R. and Berndt, J. 2016. Zircon saturation in silicate melts: a new and improved model for aluminous and alkaline melts. *Contributions to Mineralogy and Petrology*, **171**, 21, https://doi.org/10.1007/s00410-016-1227-y

Gill, J.B. 1981. Bulk chemical composition of orogenic andesites. *Orogenic Andesites and Plate Tectonics*. Springer, 97–167, https://doi.org/10.1007/978-3-642-68012-0

Glazner, A.F., Bartley, J.M., Coleman, D.S., Gray, W. and Taylor, R.Z. 2004. Are plutons assembled over millions of years by amalgamation from small magma chambers? *GSA Today*, **14**, 4–12, https://doi.org/10.1130/1052-5173(2004)014<0004:APAOMO>2.0.CO;2

Hammerli, J., Kemp, A.I., Shimura, T., Vervoort, J.D., EIMF and Dunkley, D.J. 2018. Generation of I-type granitic rocks by melting of heterogeneous lower crust. *Geology*, **46**, 907–910, https://doi.org/10.1130/G45119.1

Hildreth, W. and Moorbath, S. 1988. Crustal contributions to arc magmatism in the Andes of central Chile. *Contributions to Mineralogy and Petrology*, **98**, 455–489.

Horstwood, M.S., Košler, J. *et al.* 2016. Community-derived standards for LA-ICP-MS U–(Th–)Pb geochronology – uncertainty propagation, age interpretation and data reporting. *Geostandards and Geoanalytical Research*, **40**, 311–332, https://doi.org/10.1111/j.1751-908X.2016.00379.x

Jackson, S.E., Pearson, N.J., Griffin, W.L. and Belousova, E.A. 2004. The application of laser ablation–inductively coupled plasma–mass spectrometry to in situ U–Pb zircon geochronology. *Chemical Geology*, **211**, 47–69, https://doi.org/10.1016/j.chemgeo.2004.06.017

Jackson, M., Blundy, J. and Sparks, R. 2018. Chemical differentiation, cold storage and remobilization of magma in the Earth's crust. *Nature*, **564**, 405, https://doi.org/10.1038/s41586-018-0746-2

Kemp, A., Hawkesworth, C. *et al.* 2007. Magmatic and crustal differentiation history of granitic rocks from Hf–O isotopes in zircon. *Science (New York)*, **315**, 980–983, https://doi.org/10.1126/science.1136154

Kusiak, M.A., Dunkley, D.J., Słaby, E., Martin, H. and Budzyń, B. 2009. Sensitive high-resolution ion microprobe analysis of zircon reequilibrated by late magmatic fluids in a hybridized pluton. *Geology* , **37**, 1063–1066, https://doi.org/10.1130/G30048A.1

Lee, C.-T.A. and Bachmann, O. 2014. How important is the role of crystal fractionation in making intermediate magmas? Insights from Zr and P systematics. *Earth and Planetary Science Letters*, **393**, 266–274, https://doi.org/10.1016/j.epsl.2014.02.044

Lee, C.-T.A., Lee, T.C. and Wu, C.-T. 2014. Modeling the compositional evolution of recharging, evacuating, and fractionating (REFC) magma chambers: Implications for differentiation of arc magmas. *Geochimica et Cosmochimica Acta*, **143**, 8–22, https://doi.org/10.1016/j.gca.2013.08.009

Lim, H., Kim, J.-H., Woo, H., Do, J. and Jang, Y.-D. 2016. Petrological characteristics of the Satkatbong Pluton, Yeongdeok. Korea. *Journal of Petrological Society of Korea*, **25**, 121–142, https://doi.org/10.7854/JPSK.2016.25.2.121

Lim, H., Woo, H.D., Myeong, B., Park, J. and Jang, Y.-D. 2018. Semi-adakitic magmatism of the Satkatbong diorite, South Korea: Geochemical implications for post-adakitic magmatism in southeastern Eurasia. *Lithos*, **304**, 109–124, https://doi.org/10.1016/j.lithos.2018.01.015

Ling, M.-X., Wang, F.-Y., Ding, X., Zhou, J.-B. and Sun, W. 2011. Different origins of adakites from the Dabie Mountains and the Lower Yangtze River Belt, eastern China: geochemical constraints. *International Geology Review*, **53**, 727–740, https://doi.org/10.1080/00206814.2010.482349

Lissenberg, C.J., Rioux, M., Shimizu, N., Bowring, S.A. and Mével, C. 2009. Zircon dating of oceanic crustal accretion. *Science (New York, NY)*, **323**, 1048–1050, https://doi.org/10.1126/science.1167330

Ludwig, K. 2012. *User's Manual for Isoplot 3.75. A Geological Toolkit for Microsoft Excel*. Berkeley Geochronology Center, Special Publications, **5**.

Marxer, F. and Ulmer, P. 2019. Crystallisation and zircon saturation of calc-alkaline tonalite from the Adamello Batholith at upper crustal conditions: an experimental study. *Contributions to Mineralogy and Petrology*, **174**, 1–29, https://doi.org/10.1007/s00410-019-1619-x

McKenzie, D. and O'Nions, R. 1991. Partial melt distributions from inversion of rare earth element concentrations. *Journal of Petrology*, **32**, 1021–1091, https://doi.org/10.1093/petrology/32.5.1021

Middlemost, E.A. 1994. Naming materials in the magma/igneous rock system. *Earth-Science Reviews*, **37**, 215–224, https://doi.org/10.1016/0012-8252(94)90029-9

Miller, J.S., Matzel, J.E., Miller, C.F., Burgess, S.D. and Miller, R.B. 2007. Zircon growth and recycling during the assembly of large, composite arc plutons. *Journal of Volcanology and Geothermal Research*, **167**, 282–299, https://doi.org/10.1016/j.jvolgeores.2007.04.019

Müntener, O. and Ulmer, P. 2018. Arc crust formation and differentiation constrained by experimental petrology. *American Journal of Science*, **318**, 64–89, https://doi.org/10.2475/01.2018.04

Nandedkar, R.H., Ulmer, P. and Müntener, O. 2014. Fractional crystallization of primitive, hydrous arc magmas: an experimental study at 0.7 GPa. *Contributions to Mineralogy and Petrology*, **167**, 1–27, https://doi.org/10.1007/s00410-014-1015-5

Nebel, O., Nebel-Jacobsen, Y., Mezger, K. and Berndt, J. 2007. Initial Hf isotope compositions in magmatic zircon from early Proterozoic rocks from the Gawler Craton, Australia: a test for zircon model ages. *Chemical Geology*, **241**, 23–37, https://doi.org/10.1016/j.chemgeo.2007.02.008

Paton, C., Hellstrom, J., Paul, B., Woodhead, J. and Hergt, J. 2011. Iolite: freeware for the visualisation and processing of mass spectrometric data. *Journal of Analytical Atomic Spectrometry*, **26**, 2508–2518, https://doi.org/10.1039/c1ja10172b

Roberts, N.M. and Spencer, C.J. 2015. The zircon archive of continent formation through time. *Geological Society, London, Special Publications*, **389**, 197–225, https://doi.org/10.1144/SP389.14

Scherer, E.E., Cameron, K.L. and Blichert-Toft, J. 2000. Lu–Hf garnet geochronology: closure temperature relative to the Sm–Nd system and the effects of trace mineral inclusions. *Geochimica et Cosmochimica Acta*, **64**, 3413–3432, https://doi.org/10.1016/S0016-7037(00)00440-3

Schmitt, A.K., Perfit, M.R. *et al.* 2011. Rapid cooling rates at an active mid-ocean ridge from zircon thermochronology. *Earth and Planetary Science Letters*, **302**, 349–358, https://doi.org/10.1016/j.epsl.2010.12.022

Shao, T., Xia, Y., Ding, X., Cai, Y. and Song, M. 2019. Zircon saturation in terrestrial basaltic melts and its geological implications. *Solid Earth Sciences*, **4**, 27–42, https://doi.org/10.1016/j.sesci.2018.08.001

Shaw, D.M. 1970. Trace element fractionation during anatexis. *Geochimica et Cosmochimica Acta*, **34**, 237–243, https://doi.org/10.1016/0016-7037(70)90009-8

Siégel, C., Bryan, S., Allen, C. and Gust, D. 2018. Use and abuse of zircon-based thermometers: a critical review and a recommended approach to identify antecrystic zircons. *Earth-Science Reviews*, **176**, 87–116, https://doi.org/10.1016/j.earscirev.2017.08.011

Sláma, J., Košler, J. *et al.* 2008. Plešovice zircon – a new natural reference material for U–Pb and Hf isotopic microanalysis. *Chemical Geology*, **249**, 1–35, https://doi.org/10.1016/j.chemgeo.2007.11.005

Solano, J., Jackson, M., Sparks, R., Blundy, J. and Annen, C. 2012. Melt segregation in deep crustal hot zones: a mechanism for chemical differentiation, crustal assimilation and the formation of evolved magmas. *Journal of Petrology*, **53**, 1999–2026, https://doi.org/10.1093/petrology/egs041

Spencer, C.J., Kirkland, C.L. and Taylor, R.J. 2016. Strategies towards statistically robust interpretations of in situ U–Pb zircon geochronology. *Geoscience Frontiers*, **7**, 581–589, https://doi.org/10.1016/j.gsf.2015.11.006

Vermeesch, P. 2018. IsoplotR: A free and open toolbox for geochronology. *Geoscience Frontiers*, **9**, 1479–1493, https://doi.org/10.1016/j.gsf.2018.04.001

Vervoort, J.D. and Blichert-Toft, J. 1999. Evolution of the depleted mantle: Hf isotope evidence from juvenile rocks through time. *Geochimica et Cosmochimica Acta*, **63**, 533–556, https://doi.org/10.1016/S0016-7037(98)00274-9

Vigneresse, J., Cuney, M. and Barbey, P. 1991. Deformation assisted crustal melt segregation and transfer. *Geological Association of Canada/Mineralogical Association of Canada*Abstracts, A128.

Weinberg, R.F. and Dunlap, W.J. 2000. Growth and deformation of the Ladakh Batholith, northwest Himalayas: implications for timing of continental collision and origin of calc-alkaline batholiths. *The Journal of Geology*, **108**, 303–320, https://doi.org/10.1086/314405

Wiedenbeck, M., Alle, P. *et al.* 1995. Three natural zircon standards for U–Th–Pb, Lu–Hf, trace element and REE analyses. *Geostandards Newsletter*, **19**, 1–23, https://doi.org/10.1111/j.1751-908X.1995.tb00147.x

Woo, H.-D. and Jang, Y.-D. 2014. Petrological characteristics of the Yeongdeok granite. *The Journal of the Petrological Society of Korea*, **23**, 31–43, https://doi.org/10.7854/JPSK.2014.23.2.31

Woodhead, J.D. and Hergt, J.M. 2005. A preliminary appraisal of seven natural zircon reference materials for in situ Hf isotope determination. *Geostandards and Geoanalytical Research*, **29**, 183–195, https://doi.org/10.1111/j.1751-908X.2005.tb00891.x

Yi, K., Cheong, C.-S., Kim, J., Kim, N., Jeong, Y.-J. and Cho, M. 2012. Late Paleozoic to Early Mesozoic arc-related magmatism in southeastern Korea: SHRIMP zircon geochronology and geochemistry. *Lithos*, **153**, 129–141, https://doi.org/10.1016/j.lithos.2012.02.007

On the virtues and pitfalls of combined laser ablation Rb–Sr biotite and U–Pb monazite–zircon geochronology: an example from the isotopically disturbed Cape Woolamai Granite, SE Australia

Kaitlin Baggott[1]*, Yona Jacobsen[1,2], Oliver Nebel[1], Jack Mulder[3], Massimo Raveggi[1], Xue-Ying Wang[1], Eric Vandenburg[1], Hoseong Lim[1,4], Angus Rogers[1], Barbara Etschmann[1], Ross Whitmore[1], Alexandra Churchus[1] and Lauren Jennings[1]

[1]School of Earth, Atmosphere & Environment, Monash University, Room 124, 9 Rainforest Walk, Clayton, Victoria 3800, Australia

[2]Faculty of Arts and Education, Deakin University, Burwood, Victoria 3125, Australia

[3]Department of Earth Sciences, University of Adelaide, Adelaide, South Australia 5005, Australia

[4]Department of Earth and Environmental Sciences, James Cook University, 373 Flinders Street, Townsville, Queensland 4810, Australia

KB, 0000-0002-4154-8780; YJ, 0000-0001-5600-7453; ON, 0000-0002-5068-7117; JM, 0000-0001-6197-0439; X-YW, 0000-0001-7071-6555; EV, 0000-0001-7676-2931; HL, 0000-0002-0859-7900; AR, 0000-0001-9990-2191; BE, 0000-0002-7807-2763; RW, 0000-0002-3922-7777; LJ, 0000-0003-2997-3618

*Correspondence: Kaitlin.Baggott@monash.edu

Abstract: Different mineral clocks in granite can provide age information reflecting various aspects of rock formation, including cooling or post-emplacement fluid–rock interaction. However, the dating tool chosen can yield inconclusive age information due to differences in closure temperatures and susceptibility to fluid alteration among chronometers. This has led to an inferred superiority of U–Pb in zircon over U–Pb in monazite or Rb–Sr in mica. Here, we investigate age systematics using Rb–Sr biotite grains, U–Pb in monazite and zircon in a Devonian granite from Australia. Single-grain laser ablation ICP-MS/MS biotite analyses are combined with zircon–monazite U–Pb ages and trace element systematics. Textural and trace element evidence combined with age systematics reveals a Rb–Sr closure age of *c.* 360–330 Ma relative to a putative 364 Ma emplacement age, suggesting hydrothermal alteration of the granite. Trace element systematics and magnetic susceptibility in biotite grains reflect their partial chemical reset and fluid overprint in the granite. However, similar systematics are also observed for zircon and monazite. Our multiple chronometer dating approach, studied with modern laser-ablation methods, highlights the need for detailed investigation of isotope and trace element systematics in single grains and that individual ages should be used cautiously when dating altered granitoids.

Supplementary material: Zircon and monazite U–Pb and trace elements, biotite Rb–Sr and appendix are available at https://doi.org/10.6084/m9.figshare.c.6758747

Accurate and precise dating of plutonic rocks is a complex exercise and often requires a detailed investigation of rocks, minerals and their chemistry (e.g. Schaltegger *et al.* 2015). Obtaining radiometric ages through isotope analyses that confidently define the time of emplacement of a granitoid becomes challenging if rock associations have been subject to fluid–rock interactions. Some isotope systems are known to have different closure temperatures (T_C), the temperature under which isotope exchange of a mineral and its surroundings is no longer possible, e.g. for U–Pb in zircon (*c.* 900°C) (Mezger and Krogstad 1997; Cherniak and Watson 2001) v. Rb–Sr in biotite (*c.* 350–400°C) (Jäger 1979; Glodny *et al.* 2008). Therefore, different minerals will yield different 'ages' for a pluton, depending on the cooling rate of a rock. However, fluid–rock interactions can affect isotope closure at much lower temperatures. It has been shown that closure temperature is affected by factors such as mineral type, composition, matrix, grain size and cooling rate (Jäger 1967; Dodson 1973; Cherniak and

From: van Schijndel, V., Cutts, K., Pereira, I., Guitreau, M., Volante, S. and Tedeschi, M. (eds) 2024. *Minor Minerals, Major Implications: Using Key Mineral Phases to Unravel the Formation and Evolution of Earth's Crust.* Geological Society, London, Special Publications, **537**, 435–454.
First published online September 29, 2023, https://doi.org/10.1144/SP537-2022-320

Watson 2001; Cherniak 2010). Further, closure temperature is only considered valid for thermal cooling, as fluids can induce dissolution–reprecipitation or enhance isotope diffusion between a mineral and its surroundings below closure temperature (Villa 1998; Weinberg *et al.* 2020). When fluids are involved, recrystallization and isotopic redistribution between phases probably have a greater influence than slower thermal/diffusional resetting (Villa and Williams 2013).

Zircon U–Pb ages are considered the most robust measure for geological age determination and thus overwhelmingly dominate our understanding of granite formation ages, among other applications (Spencer *et al.* 2016). Zircon incorporates significant U concentrations (usually *c.*150 to 200 ppm) which, coupled with a high T_C, make them suitable for U–Pb dating, and they are considered robust against thermal overprinting and a wide range of mechanical and chemical processes (Mezger and Krogstad 1997; Kirkland *et al.* 2015). Zircon often survives multiple intrusive events during pluton growth (Miller *et al.* 2007). Therefore, studies of igneous rocks commonly find inherited zircon grain ages that predate magmatic crystallization (Olierook *et al.* 2020*a*). Inherited zircon grains are particularly abundant in S-type granites, which have a higher sedimentary source component and are typically more felsic and formed at lower temperatures (Parrish 1990; Chappell *et al.* 2000; Chew *et al.* 2020). Concordant zircon ages in a pluton do not only represent incremental growth of zircon autocrysts in surrounding melt or fluid: inherited zircon grains can be derived from the source rock, xenocrysts scavenged from country rock or antecrysts derived from a different melt to the one in which they are ultimately emplaced (Miller *et al.* 2007; Siégel *et al.* 2018; Olierook *et al.* 2020*a*).

Additionally, as disruption of zircon U–Pb systematics can occur for a wide range of reasons, for example auto-metasomatism, discordant zircon ages may be produced with challenging and debatable interpretations (Black 1987; Mezger and Krogstad 1997; Spencer *et al.* 2016; Villa and Hanchar 2017). Zircon grains with high U concentrations are subjected to more radiation damage, accumulating over long periods at low temperatures (Mezger and Krogstad 1997). While pioneering work by Mezger and Krogstad (1997) attributed zircon discordance primarily to near-surface diffusional Pb-loss from metamict zircon, among other potential causes, some recent work attributes most discordance to fluid-assisted dissolution–reprecipitation reactions (Geisler *et al.* 2007; Kusiak *et al.* 2009; Villa and Hanchar 2017; Bosse and Villa 2019).

The Rb–Sr system has long been known to be susceptible to thermal and fluid overprint, and whilst this feature resulted in a loss of confidence in this dating scheme, it can reciprocally also be used to determine low-temperature events (Tillberg *et al.* 2021; Redaa *et al.* 2022). Most Rb–Sr ages are based on isotope dilution data of mostly biotite separates as the high Rb/Sr phase (Meißner *et al.* 2002; Challandes *et al.* 2008). These are dissolved as bulk separates, often with a low number of data points because of the complex and time-consuming method of isotope dilution analyses. Age information may be lost or blended in such multi-mineral approaches. However, little is known on individual biotite grain evolution, particularly during fluid–rock interactions, which can be pervasive but also localized in compact plutonic rocks. Development of *in-situ* Rb–Sr dating techniques and refinement of the ^{87}Rb decay constant has revitalized the potential utility of this isotopic system (Nebel *et al.* 2011; Villa *et al.* 2015; Zack and Hogmalm 2016). Single mineral *in-situ* analysis of biotite, combined with initial Sr from other analyses to calculate an isochron model age or inclusion of other minerals to the isochrons, are now possible, painting a much more complex picture of granite genesis (Li *et al.* 2020; Gyomlai *et al.* 2022; Redaa *et al.* 2022). The combined use of chronometers thus bears the potential to elucidate crystallization, subsequent recrystallization or cooling (Ribeiro *et al.* 2020, 2022).

Here, we investigate individual Rb–Sr biotite ages, combined with a multi-mineral, multi-isotope approach in the Cape Woolamai Granite. When combined with textural and trace element information, individual Rb–Sr ages allow us to distinguish between emplacement and reset/overprinting events. We found multiple Rb–Sr ages in biotite grains with different magnetic susceptibility and accessory mineral inclusions within a single granite sample. Multiple mica Rb–Sr ages have previously been found in metamorphic and mineralized terranes (Olierook *et al.* 2020*b*; Gyomlai *et al.* 2022; Redaa *et al.* 2022), but to our knowledge, this is the first reported instance of multiple Rb–Sr biotite ages in a single granite sample.

Geological background

The Cape Woolamai Granite is located on Millowl (aka Phillip Island), *c.* 75 km SE of Melbourne, Victoria, SE Australia (Fig. 1). The geology of Millowl/Phillip Island comprises rare greenstones of probable pre-Ordovician age (Cayley 2011), the Devonian–Carboniferous Cape Woolamai Granite, Cretaceous siliciclastic and volcaniclastic rocks, basalt flows of the Eocene Older Volcanics Province and recent sediments (Edwards 1945; Henry and Birch 1992; Moore *et al.* 2016). The Cape Woolamai Granite may have intruded into country rock comprised of Cambrian or Neoproterozoic greenstones and

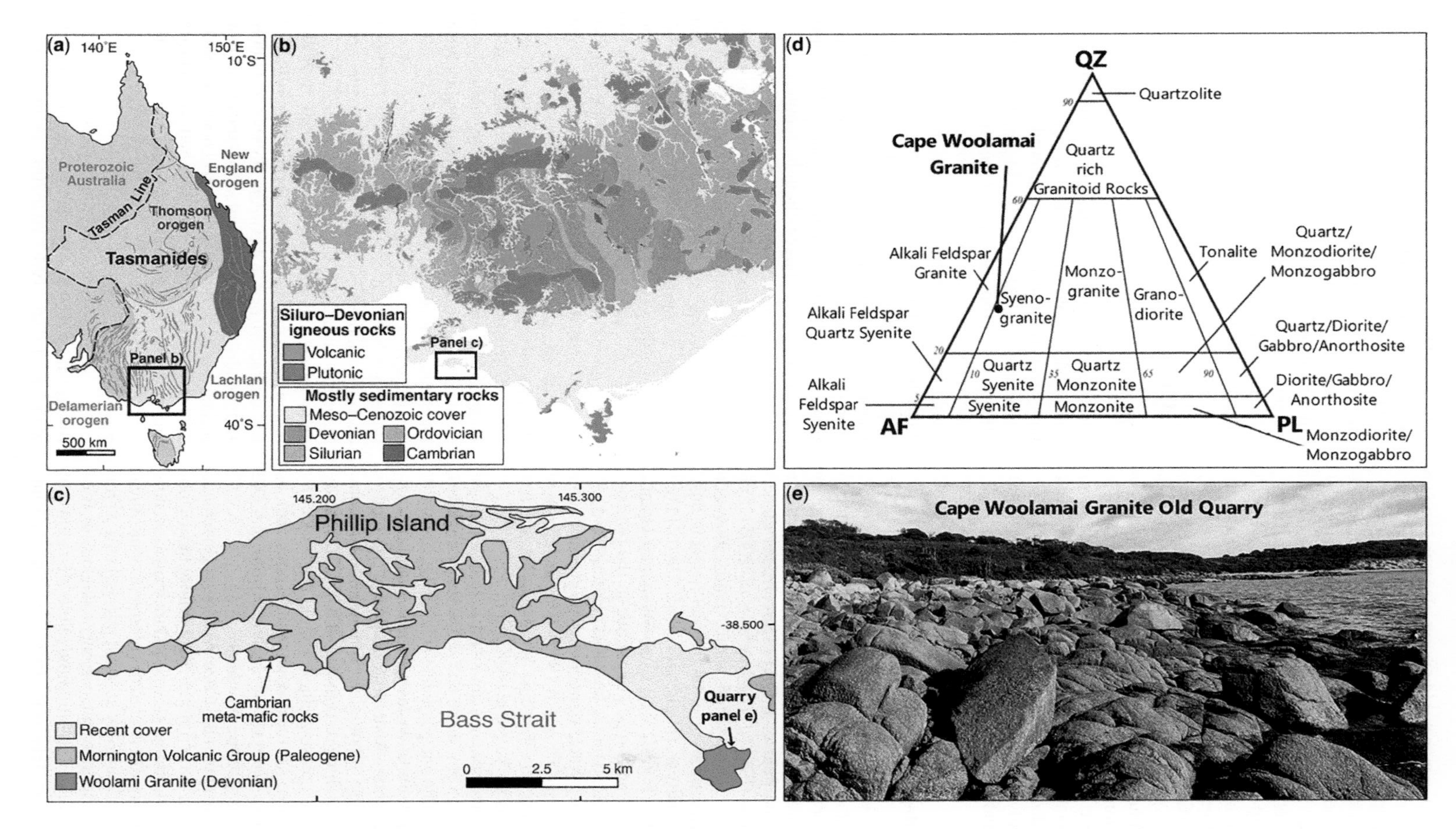

Fig. 1. (**a**) Map of Eastern Australia, showing the Lachlan Orogen in green. (**b**) Geology of central Victoria. (**c**) Map of Millowl (Phillip Island) showing the location of the Cape Woolamai Granite in red (38°33′00.23″ S, 145°21′31.65″ E). (**d**) QAPF Streckeisen quaternary diagram. (**e**) Cape Woolamai Granite Old Quarry, study location. Source: (a) adapted from Gray and Foster (2004); (b and c) adapted from Raymond *et al.* (2012); (d) adapted from Le Maitre *et al.* (2005).

Silurian–Ordovician turbidites, which are not known to crop out on Millowl/Phillip Island but are widespread throughout the Lachlan Fold Belt (LFB).

The Cape Woolamai Granite is part of the Melbourne Zone of the LFB, Fig. 1 (VandenBerg *et al.* 2000). The Melbourne Zone is argued to have a different deformational history to other LFB provinces due to the underlying Selwyn Block, which is interpreted as an unexposed Proterozoic microcontinent accreted to the Gondwanan margin in the Early Paleozoic (Cayley *et al.* 2002; Cayley 2011; Moore *et al.* 2016). The LFB is a sequence of Ordovician–Silurian sediments accreted to the eastern edge of Gondwana (Glen 2005). In the Melbourne Zone, these sedimentary rocks were folded during the Tabberaberran Orogeny (*c.* 385–380 Ma) and then intruded by post-orogenic granites in the Devonian, mostly *c.* 370 Ma (VandenBerg *et al.* 2000; Rossiter and Gray 2008).

Sample description and petrography

The Cape Woolamai Granite is medium- to coarse-grained with the major mineral assemblage comprising K-feldspar (*c.* 52%), quartz (*c.* 30%), plagioclase (*c.* 11%) and biotite (*c.* 7%). The granite plots in the syeno-granite field of a QAPF Streckeisen quaternary diagram (Fig. 1d) (Le Maitre *et al.* 2005). Evidence for post-crystallization fluid alteration includes chloritization of biotite and hematite dusting of K-feldspar (Matheney *et al.* 1990; Putnis *et al.* 2007) (Fig. 2). The Cape Woolamai Granite is considered an S-type granite due to the presence of white mica and biotite but lacks cordierite and garnet, which are typical of many S-type granites in the LFB (White and Chappell 1977). The abundance of inherited zircon in the Cape Woolamai Granite further supports an S-type classification.

The granite contains abundant K-feldspar megacrysts up to 15 cm, mafic and sedimentary enclaves, schlieren, aplite and pegmatite (Fig. 2). Quartz unidirectional solidification textures and miarolitic cavities are common. A small collection of 3D models depicting these lithological and structural features is located at https://skfb.ly/oyOCD. This study focused on a coarse-grained sample that lacks any obvious lithological and structural features such as K-feldspar megacrysts or enclaves from the Old Granite Quarry, Cape Woolamai.

Phyllosilicates in the Cape Woolamai Granite are present as dark brown biotite, light yellow-brown biotite, sericite and chlorite (Fig. 3). Dark brown biotite is frequently partially altered to chlorite or white mica. White mica occurs only as small intergrowths within biotite and as sericite replacement of plagioclase. Biotite with a lighter colour shows little alteration and often contains relatively large (*c.* 150–200 μm) apatite inclusions. Chlorite occurs more commonly as a partial replacement product of some biotite layers, and less commonly as euhedral chlorite grains. Opaque elongated minerals, interpreted as iron oxides, are more frequently associated with white mica or chlorite-altered biotite domains. Zircon grains are often observed as inclusions in biotite grains. Some zircon grains show the growth of tiny zircon grains often parallel to the fabric of host biotite grains.

Methods

Sample preparation

Weathered portions of the sample were removed before the sample was crushed in a jaw crusher and tungsten-carbide ring mill, followed by sieving to 38–63 μm, 63–180 μm, 180–355 μm, 355–600 μm, 600 μm–1.18 mm and >1.18 mm fractions. Magnetic separation was conducted on a Frantz isodynamic magnetic separator with an initial pass at an angle of 15° and 1.4 A. The magnetic fraction was sequentially separated at 0.6, 0.4 and 0.2 A. Paramagnetic minerals in the >1.4 A fraction were further separated at −5° pitch and 1.8 A in a barrier Franz. Magnetic separation was followed by density separation using heavy liquids tetrabromethane (2.96 g cm^{-3}) and diiodomethane (3.3 g cm^{-3}).

These separation processes intended to separate minerals into distinct fractions of feldspar (>1.0 amps, non-paramagnetic), biotite (*c.* 0.3–1.5 amps depending on species), zircon (>1.4 amps, >3.3 g cm^{-3} density) and monazite (*c.* 0.3–1.0 amps, >3.3 g cm^{-3} density) (Rosenblum 1958). However, biotite grains were identified in multiple separated fractions due to natural variations, suspected alteration or accessory mineral inclusions (biotite with ilmenite inclusions was more magnetically susceptible; biotite with large apatite inclusions was less magnetically susceptible). Therefore, biotite grains from different fractions with varying grain size and magnetic susceptibility were mounted separately and are referred to as 'groups' in this paper. We distinguish between three biotite groups (1, 2 and a mix of 1–2). Biotite Group 1 comprises smaller, more magnetic grains; Biotite Group 2 is the least magnetic group and has larger grain sizes with more abundant apatite inclusions; Biotite Group 1–2 Mix has characteristics in between the Biotite Group 1 and 2 end members. Separated minerals were handpicked under a binocular microscope, mounted in 25.4 mm epoxy pucks, and polished to expose the grain interiors for textural analysis and laser ablation.

Textural analysis

Biotite and zircon epoxy pucks were carbon coated with a Leica EM ACE200 carbon coater.

Fig. 2. Field pictures of the Cape Woolamai Granite, Old Granite Quarry, Cape Woolamai, Millowl (38° 33′ 00.23″ S, 145° 21′ 31.65″ E). (**a**) Non-deformed plagioclase, K-feldspar and quartz S-type granite. (**b**) Feldspar megacryst. (**c**) K-feldspar-biotite granite with mafic enclaves intruded by granite. (**d**) Foliated mafic enclave in K-feldspar, quartz and biotite granite. (**e**) Sedimentary enclave with diffuse boundaries intruded by granitic melt in K-feldspar, plagioclase and quartz granite. (**f**) Enclave with surrounding hematite alteration halo in plagioclase, biotite and quartz granite. (**g**) Aplite vein and schlieren in K-feldspar, quartz and plagioclase granite. (**h**) Aplite vein with diffuse boundary on one side and schlieren in plagioclase, K-feldspar and quartz granite. (**i**) Schlieren in plagioclase, K-feldspar and quartz granite. (**j**) Quartz, K-feldspar and biotite pegmatite in plagioclase, K-feldspar and quartz granite. (**k**) Quartz, K-feldspar and biotite pegmatite in plagioclase, quartz and biotite granite. (**l**) Quartz uni-directional solidification textures within plagioclase, K-feldspar and quartz granite.

Fig. 3. Photomicrographs of magmatic biotite crystals of the Cape Woolamai Granite. (**a**) Dark brown biotite partially altered to chlorite and white mica. (**b**) Biotite crystals partially replaced by white mica along the cleavage host zircon inclusions. (**c**) Light yellow-brown biotite with apatite inclusions. (**d**) Biotite with both zircon and apatite inclusions. Ms, muscovite; Bt, biotite; Cl, chlorite; Fsp, feldspar; Ser, sericite; Ap, apatite; Zr, zircon.

Backscattered electron (BSE), energy-dispersive X-ray (EDX) maps and cathodoluminescence (CL) images were produced for biotite and zircon grains using an FEI Quanta 600 tungsten filament MLA SEM at Monash University. BSE images and EDX maps were acquired at 20 kV (EDAX detectors) and CL images at 12.5 kV (Delmic RGB Jolt detector) (see Appendix, Figs 1 to 3).

LA-ICP-MS

Zircon and monazite U–Pb and trace elements. Split-stream laser-ablation inductively-coupled plasma mass-spectrometry (LA-ICP-MS) was used to determine ages, trace element concentrations and isotopic ratios (Kylander-Clark *et al.* 2013; Richter *et al.* 2019; Fisher *et al.* 2020; Weinberg *et al.* 2020; Barrote *et al.* 2022). Analyses were conducted at The Isotopia Labs, Monash University, using an ASI-RESOlution ArF 193 nm excimer laser ablation system, coupled with a Thermo Scientific iCAP Q ICP-MS for trace elements and a Thermo Scientific iCAP TQ ICP-MS for U–Pb, following the method detailed in Ribeiro *et al.* (2020). All other parameters are shown in Table 1. Zircon grains were analysed as fully separated grains and as inclusions in mica. Both cores and rims were analysed where possible.

U–Pb ages and trace elements were analysed for zircon and monazite separated by conventional methods (heavy liquids and picking) and as inclusions in biotite groups. Split stream laser ablation for zircon negated the need for sequential analysis, which can be problematic for zoned minerals, such as igneous zircon which tends to contain heterogeneities, and enabled rapid analysis of samples (Harrison and Zeitler 2005; Nebel-Jacobsen *et al.* 2005; Fisher *et al.* 2020).

Primary reference materials included NIST610 for trace element concentrations and GJ-1 zircon for U–Pb ages (Liu *et al.* 2007), measured every 20 and 40 unknown analyses, respectively. Secondary

Table 1. LA-ICP-MS parameters

	Mica Rb–Sr	Feldspar Rb–Sr	Zircon U–Pb and TE	Monazite U–Pb and TE
Laser Ablation				
Laser energy (J cm^{-2})	3	3	4	4
Laser rep. rate (Hz)	10	10	10	10
Ablation time (sec)	30	30	30	30
Warm up time/ baseline analysis (sec)	20	20	20	20
Washout time (sec)	15	15	15	15
Spot size (μm)	100	100	30	20
He flow	370	370	550	550
ICP-MS				
System and Mode	Triple quadrupole mode of iCAP TQ	Triple quadrupole mode of iCAP TQ	Split stream of single mode iCAP TQ with single quadrupole mode (for U–Pb) and iCAP Q (for TE)	Split stream of single mode iCAP TQ with single quadrupole mode (for U–Pb) and iCAP Q (for TE)
Carrier Gas and Flow (L min^{-2})	Ar 0.9	Ar 0.9	0.8 Ar for iCAP TQ and 1.0 in iCAP Q	0.9924 Ar for iCAP TQ and 0.9945 in iCAP Q
Additional Gas Flow (ml min^{-2})	N_2 2	N_2 2	–	–
Reaction Gas Flow (ml min^{-2})	O_2 0.1	O_2 0.1	–	–

reference materials included glasses NIST612, ATHO-G, BCR-2G and BHVO-2G for trace element analyses, and zircon reference materials Plešovice and 91500 for U–Pb ages. Secondary reference materials were measured every 40 unknowns. ^{29}Si concentrations were used as the trace element internal standard. For secondary reference materials, Plešovice zircon gave a concordia age of 336.50 ± 0.81 Ma (2SE), within error of the recommended age of 337.1 Ma (Sláma *et al.* 2008) or 337.24 Ma (Horstwood *et al.* 2016). The 91500 zircon gave a concordia age of 1068 ± 3 Ma (2SE), also within error of the recommended age of 1065 Ma (Wiedenbeck *et al.* 1995). Secondary reference material details are provided in supplementary materials.

For monazite analyses, we used primary reference materials Madel monazite for U–Pb (Payne 2008) and NIST610 glass for trace elements, which were measured every 20 unknowns. Secondary reference materials were measured every 40 unknowns and included 44069, STK-Monazite and glasses NIST612, ATHO-G, BCR-2G and BHVO-2G (Richter *et al.* 2019). P concentrations of 12.89% were used as the internal standard for the TE data reduction. The 44069 Monazite measured U–Pb concordia age was 422.6 ± 2.2 Ma (2SE), within error of the expected age of 424.9 ± 0.4 Ma (Aleinikoff *et al.* 2006). STK monazite was 1038 ± 3 Ma (2SE), also within error of the expected age of 1047 ± 7 Ma (Liu *et al.* 2012).

Biotite Rb–Sr. Rb–Sr isotopes were analysed in three biotite groups with different magnetic susceptibility, which were separated with a magnetic Franz separator. Biotite Rb–Sr analyses were conducted on a Thermo Scientific iCAP-TQ ICP-MS/MS coupled with an ASI-RESOlution ArF 193 nm excimer at the Isotopia Laboratory, Monash University. Argon gas was used as a carrier with high-purity N_2 to increase the signal. Additionally, O_2 gas was used in the reaction chamber to shift the mass of ^{87}Sr to ^{87}Sr^{16}O+, enabling separation of ^{87}Sr and ^{87}Rb (Zack and Hogmalm 2016). Primary reference materials for biotite analysis were NIST610 (Sr isotopes) and Mica-Mg (Rb/Sr). Secondary reference materials included Mica-Mg flake, BCR-2G, NIST612 and the NBS SRM-607 K-feldspar (Nebel and Mezger 2006). NIST610 was used as the primary reference material for ^{87}Sr/^{86}Sr ratios, producing Mica-Mg ^{87}Sr/^{86}Sr ratios of 1.861 ± 0.05, within error of the expected value of 1.852 ± 0.0024 (Redaa *et al.* 2021). The ^{87}Rb/^{86}Sr ratios were reduced with Mica-Mg as the primary reference material. NIST 610 gave 2.263 ± 0.055, slightly lower than expected (2.33 ± 0.005) (Hogmalm *et al.* 2017).

Data reduction. Data were processed using Iolite 4 software (Paton *et al.* 2011). The zircon U–Pb data reduction scheme (DRS) included baseline, drift and downhole fractionation corrections (Paton *et al.* 2010; Redaa *et al.* 2021). The Rb–Sr DRS included background subtraction, ^{87}Rb calculation derived from ^{85}Rb counts per second and a conversion factor assuming natural abundances, and drift correction of raw $^{87}Rb/^{86}Sr$ and $^{87}Sr/^{86}Sr$ ratios with expected values of primary reference materials (Redaa *et al.* 2021).

Concordia were plotted using IsoplotR (Vermeesch 2018), while Kernel Density Estimation (KDE) were constructed with Density Plotter (Vermeesch 2012) using an adaptive bandwidth based on Botev *et al.* (2010)'s bandwidth selection algorithm, which uses linear diffusion processes to optimize bandwidth according to data density. A discordance filter of 10% concordia distance was used (Vermeesch 2021). U–Pb concordia and isochron ages were calculated using propagated 2SE errors.

Individual spot age calculation and initial Sr correction methods were adapted from methods developed by Rösel and Zack (2022) (see equation 1). Individual ages for KDE plots were calculated using an initial $^{87}Sr/^{86}Sr$ range of 0.706 to 0.710, estimated from nearby Melbourne Zone Granites (Clemens and Elburg 2016).

$$\begin{aligned}&\text{Individual Rb} - \text{Sr Age (Ma)}\\&= \text{Ln}\left(\left(\left(\frac{^{87}\text{Sr}}{^{86}\text{Sr}}\text{sample} - \frac{^{87}Sr}{^{86}\text{Sr}}\text{initial}\right)/\right.\right.\\&\left.\left.\frac{^{87}\text{Rb}}{^{86}\text{Sr}}\text{sample}\right) + 1\right)/(1.3972 * 10^{-11})/\\&1\ 000\ 000\end{aligned} \quad (1)$$

The kernel density plot bandwidth was optimized using Silverman (1986)'s equation to determine the most suitable bandwidth for the period of interest (380 to 330 Ma). The biotite datasets were not smooth and unimodal; therefore, we turned on the adaptive bandwidth function in density plotter, which smoothed areas with lower data density (mostly near-recent ages predominantly derived from biotite magnetic batches 2 and 3) and sharpened areas with higher data density (peaks between 380 and 330 Ma). Isochrons were constructed for the three biotite groups, with data selection boundaries for each group distinguished by KDE relative probability local minima to remove older or younger tails.

Results and discussion

U–Pb systematics in zircon

The severe alteration of the granite is reflected in the U–Pb systematics and trace element content of the zircon and monazite investigated in this study. In order to place the Rb–Sr ages into context, we first evaluate the emplacement age of the granite using U–Pb systematics.

Cape Woolamai Granite zircon grains include pink, light brown and colourless grains, 50–200 µm in length, and often with small cracks. Many zircon grains display oscillatory zoning in CL images (Fig. 4). Zoning is sometimes truncated and overgrown. In some cases, outer oscillatory zoning is preserved whereas inner zoning is patchy and irregular; elsewhere, zonation is completely disrupted. Many zircon grains show evidence of dissolution (dark pores) and reprecipitation (bright zones) along oscillatory zonation within grains or along grain boundaries (Geisler *et al.* 2007). Some zircon grains have an almost skeletal porous structure, locally containing baddeleyite inclusions. Other inclusions include inherited zircon, apatite, biotite, feldspar and composite zones of Si-rich glass with U–Th-rich inclusions, possibly representing breakdown products of zircon (Kozlik *et al.* 2016).

Both individual zircon grains and zircon inclusions within biotite show a high proportion of discordancy, i.e. defined here as >10% distance from concordia. Only 37 zircon spot analyses were concordant from a total of 868 analyses. Concordant zircon analyses were further filtered according to a process adapted from Lim (2022) to remove zircon grains likely affected by secondary processes to restrict U–Pb ages to those most likely to record an igneous event. For this reason, a combination of various geochemical proxies within zircon was employed. Zircon grains which are interpreted to preserve crystallization ages *c.* 380–350 Ma, consistent with nearby Melbourne Zone Granites (Clemens and Elburg 2016; Regmi *et al.* 2016) have LREE depleted patterns, positive Ce anomalies and negative Eu anomalies (Hoskin 2005; Trail *et al.* 2012, Fig. 5a). In contrast, zircon grains yielding younger ages (<350 Ma, which is outside of the LFB age pattern in Victoria, SE Australia) have elevated REE contents, particularly light REEs (LREE), and often do not have a positive Ce anomaly, which are typical features associated with fluid-altered zircon (Bell *et al.* 2019). On a lambda diagram (O'Neill 2016), REE enrichment is notably associated with more positive REE slopes (more LREE-rich) and less-positive quadratic curvature (Fig. 5b). Alteration is greater for zircon rims than cores and more significant for zircon inclusions in biotite. The change in REE patterns is thus correlated with zircon age systematics, as strongly altered zircon grains generally have younger ages (see Appendix, Fig. 4).

Additional filtering based on minor and trace elements in zircon resulted in exclusion of four concordant zircon analyses from the dataset. The filtering stages excluded zircon grains with Al, Fe, Mn, Ca

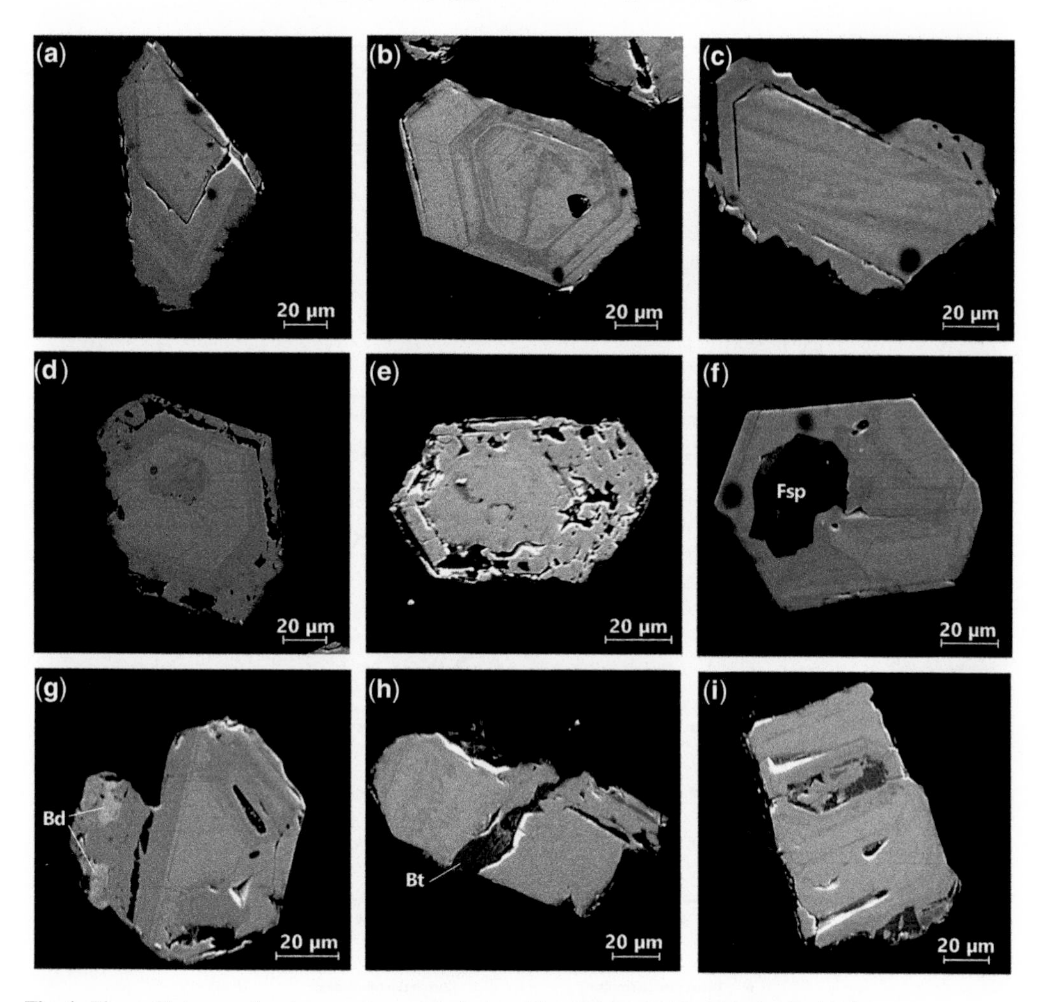

Fig. 4. Zircon CL images showing zonation and inclusions. Fsp, feldspar; Bd, baddeleyite; Bt, biotite.

or K concentrations above 1 wt%, high ΣREE (>3000 ppm) and Ba/Hf *1000>1 (Belousova *et al.* 2002; Geisler *et al.* 2007; Lim 2022). An alteration index or light rare earth element index (LREE-I) was considered to remove altered zircon grains with LREE-I of less than 30 from the dataset. The LREE-I ratio (Dy/Sm + Dy/Nd) can identify altered zircon grains, with LREE-I of less than 30, generally indicating alteration and LREE-I of less than ten associated with metamict zircon or analysis of cracks (Bell *et al.* 2016).

The spread of the ages of the remaining zircon implies that they cannot record a single event. Instead, small groups of concordant zircon analyses gave weighted mean $^{206}Pb/^{238}U$ ages for small zircon populations of 360 ± 3 Ma ($n = 15$), which is our best estimate for the crystallization of the granite based on zircon, 254 ± 13 Ma ($n = 2$), and 112 ± 5 Ma ($n = 2$) when pooled together based on statistical overlap of ages (Fig. 6). The latter ages are interpreted to reflect post-magmatic Pb diffusion or dissolution–reprecipitation (Mezger and Krogstad 1997), and the geological meaning of these ages remains unclear.

U–Pb systematics in monazite

Similar to the approach used for zircon, we separated monazite through conventional techniques and analysed monazite grains as inclusions in larger biotite grains. Firstly, monazite inclusions analysed in individual Biotite Group 1 grains contained variable common Pb proportions, yielding a lower concordia intercept age of 369 ± 12 Ma ($n = 26$, MSWD = 0.9) (Fig. 6m). When filtered for discordancy, the combined age for both conventionally separated monazite and monazite inclusions in Biotite Group 1 gives a concordia age of 365 ± 10 Ma ($n = 7$)

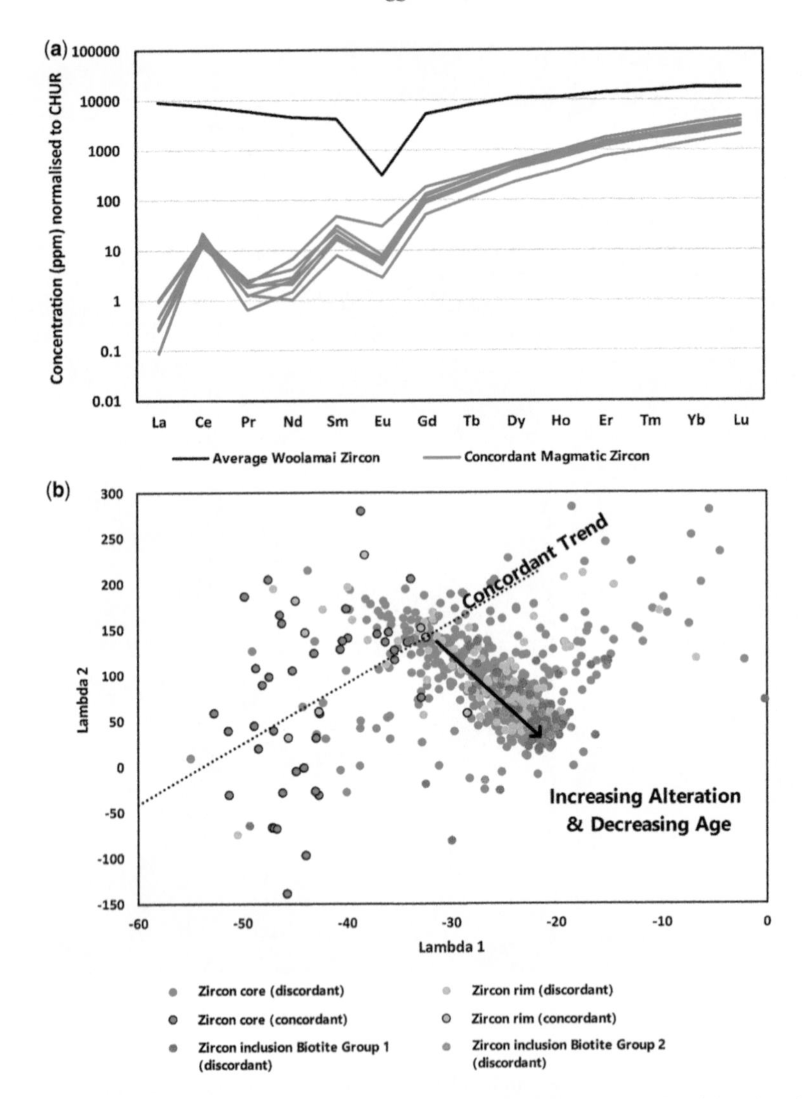

Fig. 5. Zircon REE patterns. The REE concentrations in ppm were normalized to CHUR. Lambda values for REEs were calculated using BLambdaR. (**a**) Average REE pattern of all Woolamai zircons (black) and concordant (15% distance from concordia) inferred magmatic zircons (pink), defined as filtered zircons with U–Pb concordia age between 350 and 380 Ma. (**b**) REE lambdas by zircon sub-sample: conventionally-separated cores (dark pink), conventionally-separated rims (light pink), zircon inclusions in Biotite Group 1 (green) and zircon inclusions in Biotite Group 2 (orange). Concordant zircons indicated by black rims. Source: REE concentrations reported by Nakamura (1974); Lambda values reported by O'Neill (2016), calculated with BLambdaR by Anenburg and Williams (2022).

(Fig. 6i) and a weighted mean of individual $^{206}Pb/^{238}U$ ages of 364 ± 11 (Fig. 6l).

Monazite U–Pb ages are traditionally considered to represent high-temperature cooling after granite emplacement (Engi 2017). The monazite U–Pb closure temperature is still debated, but recent works indicate it is likely to be very high, *c.* 900°C, which is higher than the emplacement temperature of most S-type granites (Copeland *et al.* 1988; Suzuki and Adachi 1994; Cherniak *et al.* 2004; Gardés *et al.* 2006). Therefore, monazite in an S-type granite could approximate the timing of granite crystallization. However, the slightly older age of monazite inclusions in Biotite Group 1 may result from a more effective shielding from resorption due to the large size of biotite grains, or due to timing of monazite growth. Monazite, as a phosphate, is sensitive to dissolution–reprecipitation reactions,

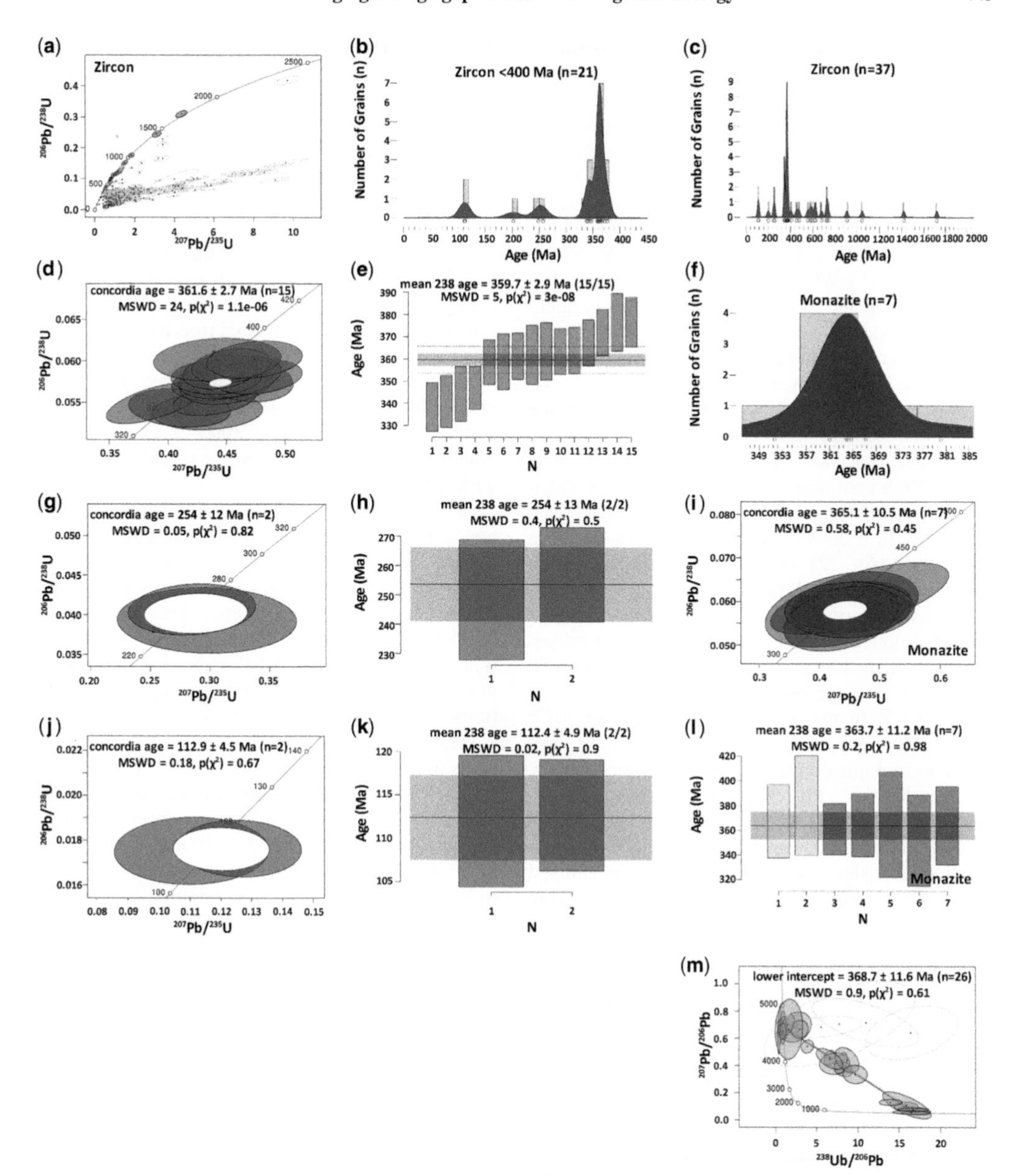

Fig. 6. Zircon and monazite U–Pb results. MSWDs refer to concordance: (**a**) Zircon Concordia, >10% discordant shown as open circles. (**b**) Concordant zircon kernel density plot, limited to ages <400 Ma. (**c**) Concordant zircon kernel density plot showing inherited zircon ages. (**d, g, j**) concordant filtered zircon concordia ages. **(e, h, k)** Concordant zircon weighted mean ^{206}Pb/^{238}U ages. (**f**) Concordant monazite kernel density plot. (**i**) Concordant monazite concordia age. (**l**) Concordant monazite weighted mean ^{206}Pb/^{238}U age, conventionally separated monazites in dark purple, monazite inclusions in Biotite Group 1 in light purple. (**m**) Tera-Wasserburg age of monazite inclusions in Biotite Group 1.

can be resorbed with the introduction of late-stage volatile-rich fluids and might suffer from Pb-loss during interaction with hydrothermal fluids (Engi 2017; Weinberg *et al.* 2020; Budzyń *et al.* 2021). Late-stage fluid interaction is consistent with REE patterns, which show a lower range of values and have higher REE abundances for younger monazite (Appendix, Fig. 6).

Emplacement of the Cape Woolamai Granite based on U–Pb systematics

The small number of concordant zircon that remained after the filtering process produce a weighted mean $^{206}Pb/^{238}U$ age of 360 ± 3 Ma (*n* = 15), which overlaps with the 'oldest' weighted mean $^{206}Pb/^{238}U$ monazite ages of 364 ± 11 Ma (*n* = 7) and the monazite intercept age of 369 ± 12 Ma (*n* = 26). When monazite and zircon are simply combined, and a mean of these analyses is calculated, this yields a weighted mean age of 364 ± 9 Ma. This age is interpreted as the intrusion age of the S-type Cape Woolamai Granite.

Disruption of U–Pb systematics in the Cape Woolamai Granite zircon grains is likely due to two main processes: (1) high U concentrations (1741–5948 ppm) leading to accumulation of radiation damage (Mezger and Krogstad 1997; Anderson *et al.* 2020), followed by multiple recrystallization events; and/or (2) dissolution-reprecipitation of more-soluble zircon domains during late stage magmatic–hydrothermal alteration. Diffusive Pb-loss should produce smooth, bell-shaped U–Pb profiles, which is inconsistent with the sharp, stepped U–Pb profiles of the Cape Woolamai Granite zircon grains (Villa and Williams 2013; Villa and Hanchar 2017; Bosse and Villa 2019). We interpret these 'age steps' as multiple dissolution–reprecipitation events, with radiogenic Pb expelled from the newly-reformed zircon lattice. Radiation-damaged zircon crystals are more susceptible to dissolution–reprecipitation (Geisler 2002). However, recrystallized magmatic zircon in a still-cooling granite is unlikely to have accumulated significant radiation damage; here, the controlling factor may be more readily-dissolved zones with different compositions (Putnis 2002; Tomaschek *et al.* 2003), e.g. greater concentration of substitutions with non-formula elements. In both cases, dissolved zircon did not always reform as zircon in the same position; other minerals such as feldspar, biotite, apatite or baddeleyite filled spaces, while tiny zircon grains nucleated, often along planes in nearby biotite. The granite experienced a substantial amount of alteration, which affected the majority of zircon grains, while monazite appeared to be slightly more robust. Monazite grains that were encapsulated within biotite were apparently the least affected.

Biotite Rb–Sr systematics

The large data base of Rb–Sr analyses in biotite grains facilitated through the laser ablation analyses and ICP-MS/MS allows a completely new way of exploring Rb–Sr systematics in a granite sample. Due to the large number of points but also the relatively large uncertainty (i.e. ±2, 5% on Rb/Sr and Sr isotopes) compared to conventional methods, age relationships are not straightforwardly interpreted. As such, we have followed an approach of separating biotite grains into groups based on their physical and chemical properties. These groups are Biotite Groups 1 and 2, which are distinct and a third group that appears to be a mix of Groups 1 and 2, Biotite Group 1–2 mix.

Biotite BSE images indicate that Biotite Groups 1 and 2 with different magnetic susceptibility also have textural differences. Biotite Group 1 BSE images include dark grey BSE rims associated with REE accessory minerals concentrated along grain boundaries and occasional marble-cake textures at grain edges, possibly related to later chlorite alteration (Fig. 7d). Biotite Group 2 characteristics include large apatite inclusions and ubiquitous dark, patchy zones (Fig. 7g).

The difference in BSE textures and ages is consistent with differences in composition and accessory mineral abundances revealed by mineral liberation analysis (MLA) mapping, (Appendix, Tables 1 and 2). Biotite Group 1 has a small grain size and is more Mg-rich than Biotite Group 2. MLA calculations based on X-ray BSE indicated that zircon grains are more numerous and make up a larger proportion of the area in Biotite Group 1. Biotite Group 1 also contains considerably more ilmenite inclusions (0.18% of the area) compared to ilmenite-poor Biotite Group 2, which possibly affects the magnetic susceptibility. Conversely, there is more apatite in Biotite Group 2 (0.22%) than Biotite Group 1 (0. 10%). Finally, monazite is present in small amounts in Biotite Group 1 and Biotite Group 2 (0.05 and 0. 02%, respectively, Fig. 7j).

A Rb–Sr model age for each individual grain (Rösel and Zack 2022) from the three Biotite Groups was first calculated to create a single-grain-age distribution using the KDE function in Density Plotter (Vermeesch 2012). Individual age distributions were calculated for grains within these different populations using an initial $^{87}Sr/^{86}Sr$ ratio of 0.708 ± 0.002, considered here as a reasonable estimate based on regional granite data (Clemens and Elburg 2016). Biotite Group 1 had a main age peak of 357 Ma (spread of 281–399 Ma), while Biotite Group 2 had a main peak at 343 Ma (spread of 307–381 Ma) (Fig. 7b, c). Biotite Group 1–2 Mix is interpreted to be a mixture of the two main biotite populations with peaks at both 357 and 344 Ma (Fig. 7d); therefore, future discussion will be focusing on Biotite Groups 1 and 2. An important feature of these analyses is the apparent difference in age populations, noting the distinction into groups based on magnetic susceptibility and grain size (Fig. 7). Rb–Sr ages are traditionally considered to represent cooling ages rather than mica growth, except for some muscovite species, which can

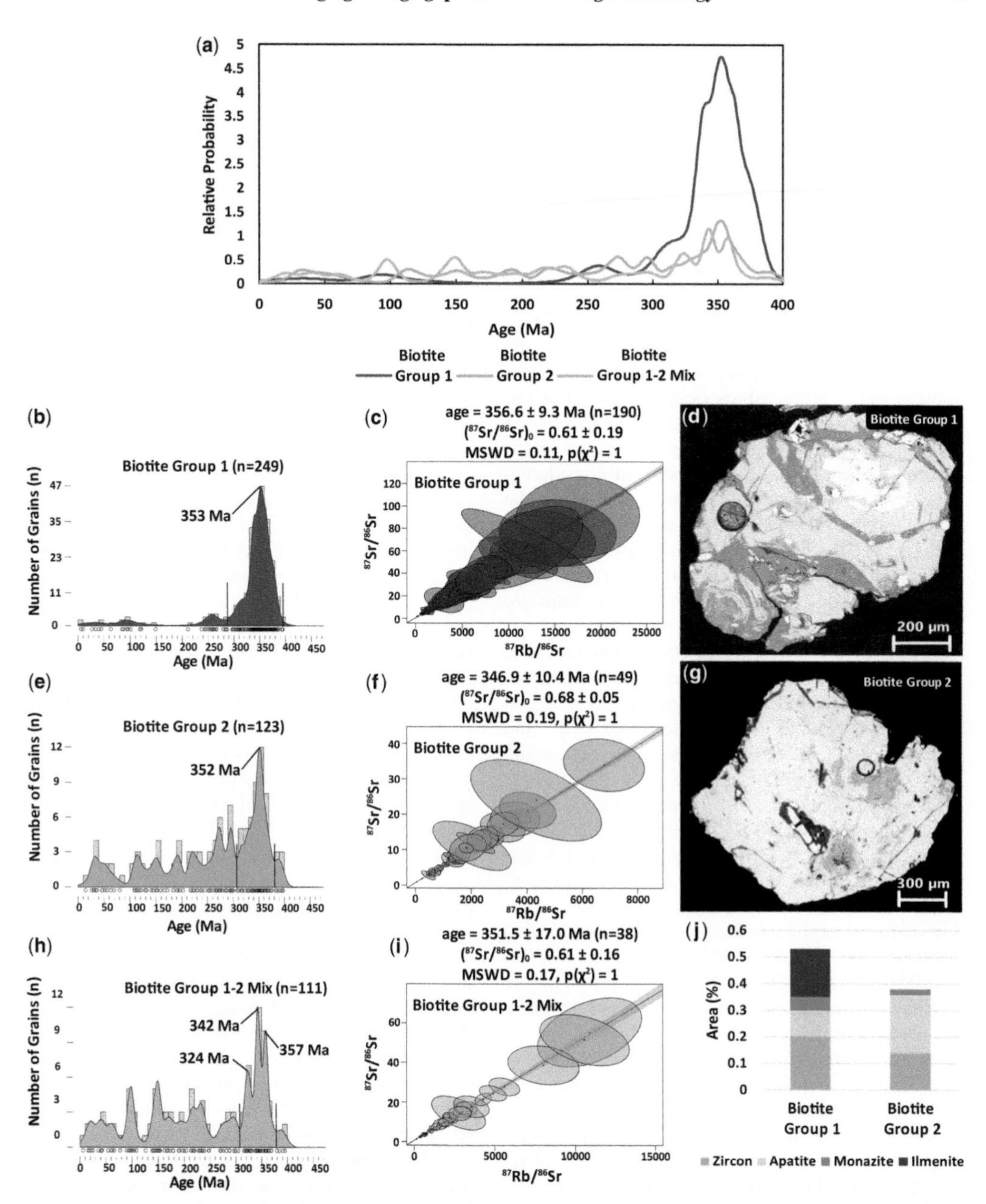

Fig. 7. Biotite Rb–Sr ages. (**a**) Kernel density distribution of Rb–Sr ages for Biotite Groups 1, 2 and 1–2 mix, corrected for initial Sr. Rb–Sr ages calculated with initial Sr of 0.708 $\pm$ 0.002 (2SE). Ages >400 Ma were omitted. (**b**) Biotite Group 1 coloured regions represent kernel density, histograms in grey. (**c**) Biotite Group 1 isochron constructed from Rb–Sr data within the black lines shown in Figure 7b. (**d**) BSE image of a representative grain from Biotite Group 1. (**e**) Biotite Group 2 kernel density plot. (**f**) Biotite Group 2 isochron. (**g**) BSE image of a representative grain from Biotite Group 2. (**h**) Biotite Group 1–2 mix kernel density plot. (**i**) Biotite Group 1–2 mix isochron. (**j**) Abundance of accessory minerals in Biotite Groups 1 and 2.

grow below their closure temperature during metamorphism (Oriolo *et al.* 2018). However, in all mica grains these cooling ages can be altered with fluid-assisted recrystallization at low temperatures (Eberlei *et al.* 2015; Oriolo *et al.* 2018). Considering the textural evidence for post-crystallization fluids alteration of the Cape Woolamai Granite, and the fact that closure temperatures are not considered valid for recrystallized minerals (Villa 1998), many biotite ages reported here do not simply represent a single igneous closure temperature age.

Hence, subsequent to the above evaluations, we pooled biotite analyses from the respective groups together using local minima of kernel density distribution plots with a focus on the oldest peak within a group only for isochron construction. These biotite group isochrons, which have been filtered for apparently younger grains, can then be interpreted in light of the evolution of the granite. Biotite-only isochrons give ages of 357 ± 9 Ma, 347 ± 10 Ma and 351 ± 17 Ma for the three oldest age peak bundles in Biotite Groups 1, 2 and 1–2 mix, respectively. However, these isochron relationships yield unusually low initial Sr ratios (0.61 ± 0.19, 0.680 ± 0.054 and 0.61 ± 0.16, respectively). In the absence of low Rb/Sr anchors for the initial Sr isotopes and considering they are below the lowest value estimated for the solar system of 0.698976 (Papanastassiou and Wasserburg 1968), these initial values are unrealistic. Nevertheless, the isochrons still yield reliable age information due to the high Rb/Sr ratio of the biotites.

Some of the difference in ages between biotite groups could be explained by closure temperature differences, as biotite closure temperatures depend on factors including cooling rate, grain shape and size, rock and mineral composition and surrounding minerals (Jenkin 1997; Willigers *et al.* 2004). Jäger (1967)'s foundational work, which posited closure temperatures of *c.* 300°C for biotite and *c.* 500°C for muscovite for the Rb–Sr system, is still broadly accepted. However, Glodny *et al.* (2008) demonstrated that the Rb–Sr system in coarse-grained biotite in metamorphic rocks can remain closed well above *c.* 300°C. Additionally, Kühn *et al.* (2000) and Willigers *et al.* (2004) show that Mg-rich biotite can have closure temperatures in excess of 650°C in the absence of fluids. Both of these observations are relevant to interpreting the significance of the biotite ages from the Woolamai Granite as the biotites tend to be coarse-grained and Group 1 biotites are notably more Mg-rich (Appendix, Table 2). It is thus suggested that the oldest peak in Biotite Group 1 reflect isotope closure and thus an igneous age.

With reference to the unusually low initial Sr ratio, these are likely due to biotite domains or entire grains being affected by diffusion. Surrounding minerals can affect biotite Rb–Sr systematics through the sub-solidus diffusion of both Rb and Sr. For example, Giletti (1991) estimates that the Rb–Sr closure temperature for biotite can be lowered by *c.* 100°C in the presence of apatite. Similarly, inter-mineral diffusion of Sr between phlogopite and calcite has been observed on scales of *c.* 4 cm (Jenkin *et al.* 1995); this mechanism is only valid if apatite is present in sufficient concentrations. While it may not affect Rb–Sr systematics on a whole rock scale, apatite could locally influence biotite Rb–Sr systematics. The younger, skewed ages that are predominantly observed for Biotite Group 2 may be due to compositional differences or the presence of large apatite inclusions, which increased the biotite's susceptibility to later dissolution–reprecipitation and post-crystallization redistribution of Rb and Sr isotopes. Glodny *et al.* (2008) proposed that Sr isotope redistribution during alteration should be dominated by exchange between biotite and apatite due to their higher diffusivities relative to other minerals. The amount of ^{87}Sr increases quickly in biotite due to high Rb concentrations, but ^{87}Sr/^{86}Sr changes little in apatite over time due to low Rb and high Sr contents (Jenkin *et al.* 1995; Glodny *et al.* 2008; Farina and Stevens 2011). Therefore, the ^{87}Sr/^{86}Sr of apatite will increase during Sr exchange as it inherits radiogenic Sr from cogenetic biotite. Reciprocally, diffusion from high-Sr, low ^{87}Sr/^{86}Sr apatite to low-Sr, high ^{87}Sr/^{86}Sr biotite should lower the original ^{87}Sr/^{86}Sr of biotite, and may even increase the Sr content in individual biotite domains, and thereby lower the Rb/Sr. This diffusion-driven redistribution can thus cause biotite analyses to drift off the isochron (instead of moving along). If used in a combined isochron approach, this can cause the isochron to rotate. The effect on the age is virtually absent here due to the wide spread in Rb–Sr ratios, but the initial Sr is no longer reliable. A slight change in slope will have no effect on the calculated age, but will have a strong, amplified effect on the intercept of the isochron.

Cape Woolamai Granite evolution

The Cape Woolamai Granite emplacement age is inferred to be *c.* 370–360 Ma, reflected by a U–Pb zircon age and the U–Pb monazite (Fig. 8). However, this interpretation is based on a rather small number of concordant analyses, and complex filtering is required to extract this information. The zircon grains show evidence of later dissolution–reprecipitation, including precipitation of tiny zircon grains near larger inclusions within biotite, and precipitation of baddeleyite near skeletal or porous zircon. In future studies, the chemical treatment of zircon may lead to more reliable ages. For monazite, inclusions within biotite grains appear to be the most robust grains.

The biotite Rb–Sr results include a variety of age information, which we interpret as follows. The mica single mineral isochrons yielded two main population, with the older one likely reflecting the igneous mica age. The initial cooling trend of the pluton may be recorded by the Biotite Group 1 age of 357 ± 9 Ma, which is *c.* 10 Myrs younger than the emplacement age. Younger individual age peaks of the

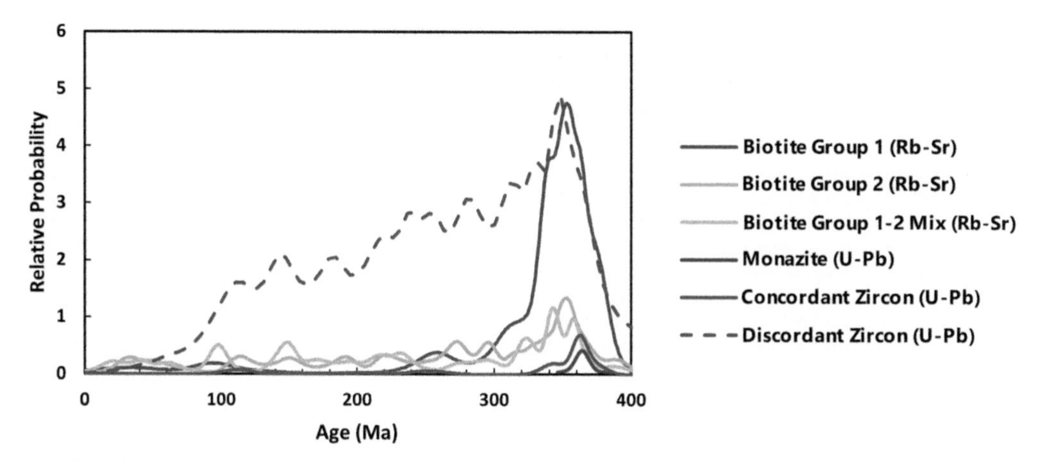

Fig. 8. Multiple-mineral kernel density estimate age distributions. Individual biotite Rb–Sr ages corrected with initial Sr of 0.708 ± 0.002. Zircon and monazite U–Pb weighted mean $^{206}Pb/^{238}U$ ages calculated with discordance filter of 10% distance from concordia.

biotite groups indicate magmatic–hydrothermal alteration *c.*360–330 Ma, during which sub-solidus processes have probably affected Rb–Sr systematics, producing younger ages. These processes include recrystallization below closure temperature, isotope exchange with apatite and chloritization. Subsequent fluid-assisted dissolution–reprecipitation events affected multiple minerals and isotope systems. Emplacement was followed by late-stage alteration and recrystallization, supported by the apatite-rich Biotite Group 2 Rb–Sr isochron age of 347 ± 10 Ma ($n = 49$). These are interpreted as magmatic–hydrothermal ages, possibly associated with aplite/pegmatite/feldspar megacryst-rich dyke intrusions and volatile exsolution in nearly-crystallized granite. This sequence of events shares common features and timing with other granites in the LFB, including the Heemskirk Granite and Native Dog Pluton at Yea (Brooks and Compston 1965; Van Krieken and Wilson 2016).

Conclusion

Our study highlights the possibility of constraining the timing of multiple events in one granite sample, using Rb–Sr analyses of biotite grains. When coupled with U–Pb systematics of other phases, such as zircon and monazite, these ages yield important information about post-emplacement systematics in granitic samples. The emplacement age of the Cape Woolamai Granite is constrained to *c.* 370–360 Ma by combining *in-situ* U–Pb dating of zircon and monazite. However, these chronometers also recorded successive disrupted and reset ages. As such, late-stage magmatic–hydrothermal events disrupted both U–Pb and Rb–Sr isotopes after the emplacement at *c.* 364 Ma. This interpretation is supported by trace element and textural evidence, including REE enrichment and changes in relative incorporation of REE elements in zircon analyses, zircon dissolution and reprecipitation as baddeleyite or small nearby zircon grains, and biotite with patchy BSE textures. Nonetheless, zircon ages of *c.* 113 Ma are interpreted as regional uplift-related recrystallization of strongly metamict zircon grains, consistent with the timing of nearby uplift from previous work by Samsu *et al.* (2019) and Cape Woolamai Granite fission track data (Kohn *et al.* 2020; Boone *et al.* 2022).

Importantly, the multi-mineral and multi-isotope ages that post-date the inferred emplacement by *c.* 364 Ma of the Cape Woolamai Granite cannot be interpreted as a simple cooling trend but requires substantial fluid alteration (Villa 1998; Kühn *et al.* 2000). Cooling is expected to yield a single population when the thermal threshold of a mineral is reached, which is not the case here. Instead, the granite records a complicated history of fluid overprint events inducing recrystallization and isotope redistribution between phases.

A key outcome of this study is that for an altered pluton severely affected by fluid overprint, a simple Rb-Sr biotite age relationship is not observed, which can explain the widely observed whole rock disturbance of this system in granites. Diffusive disturbance of the system on an individual mineral grain scale follows no simple systematics. However, biotite grain selection based on magnetic susceptibility yielded a magmatic cooling age of 357 ± 9 Ma. Biotite with abundant apatite inclusions show a biased age (Biotite Group 2, 348 ± 10 Ma) and should be avoided. It is demonstrated that bulk analyses and even large volume data analyses are insufficient to extract fast reliable ages for an altered granite. This is true for zircon and biotite, with monazite embedded in biotite as inclusions potentially being the most reliable dating tool available.

Acknowledgements This research is supported by an Australian Government Research Training Program (RTP) Scholarship. We thank Dr Hugo Olierook, Professor Marlina Elburg, and an anonymous reviewer for very detailed and constructive comments, and editorial handling by Dr Silvia Volante and Dr Valby van Schijndel.

We acknowledge the Traditional Custodians of the land, waters and sky of Millowl, the Bunurong People of the Kulin Nations. We pay our respects to their Elders past, present, and emerging. We extend this respect to all Aboriginal and Torres Strait Islander Peoples.

Competing interests The authors declare that they have no known competing financial interests or personal relationships that could have appeared to influence the work reported in this paper.

Author contributions **KB**: conceptualization (equal), data curation (lead), formal analysis (lead), investigation (lead), methodology (lead), writing – original draft (lead), writing – review and editing (equal); **YJ**: conceptualization (equal), data curation (supporting), formal analysis (supporting), investigation (supporting), methodology (supporting), supervision (lead), writing – review and editing (supporting); **ON**: conceptualization (supporting), data curation (supporting), formal analysis (supporting), methodology (supporting), supervision (supporting), writing – review and editing (equal); **JM**: conceptualization (supporting), data curation (supporting), formal analysis (supporting), methodology (supporting), writing – review and editing (supporting); **MR**: data curation (supporting), formal analysis (supporting), methodology (supporting); **X-YW**: data curation (supporting), formal analysis (supporting), methodology (supporting), writing – review and editing (supporting); **EV**: data curation (supporting), formal analysis (supporting); **HL**: formal analysis (supporting), methodology (supporting), writing – review and editing (supporting); **AR**: formal analysis (supporting), methodology (supporting), writing – review and editing (supporting); **BE**: data curation (supporting), formal analysis (supporting), methodology (supporting), writing – review and editing (supporting); **RW**: formal analysis (supporting), methodology (supporting), writing – review and editing (supporting); **AC**: formal analysis (supporting), writing – review and editing (supporting); **LJ**: data curation (supporting), writing – original draft (supporting), writing – review and editing (supporting).

Funding This research received no specific grant from any funding agency in the public, commercial or not-for-profit sectors.

Data availability All data generated or analysed during this study are included in this published article (and, if present, its supplementary information files).

References

Aleinikoff, J.N., Schenck, W.S., Plank, M.O., Srogi, L.A., Fanning, C.M., Kamo, S.L. and Bosbyshell, H. 2006. Deciphering igneous and metamorphic events in high-grade rocks of the Wilmington Complex, Delaware: Morphology, cathodoluminescence and backscattered electron zoning, and SHRIMP U–Pb geochronology of zircon and monazite. *Geological Society of America Bulletin*, **118**, 39–64, https://doi.org/10.1130/B25659.1

Anderson, A.J., Hanchar, J.M., Hodges, K.V. and van Soest, M.C. 2020. Mapping radiation damage zoning in zircon using Raman spectroscopy: implications for zircon chronology. *Chemical Geology*, **538**, https://doi.org/10.1016/j.chemgeo.2020.119494

Anenburg, M. and Williams, M.J. 2022. Quantifying the tetrad effect, shape components, and Ceâ Euâ Gd anomalies in rare earth element patterns. *Mathematical Geosciences*, **54**, 47–70, https://doi.org/10.1007/s11004-021-09959-5

Barrote, V.R., Nebel, O., Wainwright, A.N., Cawood, P.A., Hollis, S.P. and Raveggi, M. 2022. Testing the advantages of simultaneous *in-situ* Sm–Nd, U–Pb and elemental analysis of igneous monazite for petrochronological studies. An example from the late Archean, Penzance Granite, Western Australia. *Chemical Geology*, **594**, 120760, https://doi.org/10.1016/j.chemgeo.2022.120760

Bell, E.A., Boehnke, P. and Harrison, T.M. 2016. Recovering the primary geochemistry of Jack Hills zircons through quantitative estimates of chemical alteration. *Geochimica et Cosmochimica Acta*, **191**, 187–202, https://doi.org/10.1016/j.gca.2016.07.016

Bell, E.A., Boehnke, P., Barboni, M. and Harrison, T.M. 2019. Tracking chemical alteration in magmatic zircon using rare earth element abundances. *Chemical Geology*, **510**, 56–71, https://doi.org/10.1016/j.chemgeo.2019.02.027

Belousova, E., Griffin, W., O'Reilly, S.Y. and Fisher, N. 2002. Igneous zircon: trace element composition as an indicator of source rock type. *Contributions to Mineralogy and Petrology*, **143**, 602–622, https://doi.org/10.1007/s00410-002-0364-7

Black, L.P. 1987. Recent Pb-loss in zircon: a natural or laboratory-induced phenomenon? *Chemical Geology: Isotope Geoscience Section*, **65**, 25–33, https://doi.org/10.1016/0168-9622(87)90059-5

Boone, S.C., Dalton, H. *et al.* 2022. AusGeochem: an open platform for geochemical data preservation, dissemination and synthesis. *Geostandards and Geoanalytical Research*, **46**, 245–259, https://doi.org/10.1111/ggr.12419

Bosse, V. and Villa, I.M. 2019. Petrochronology and hygrochronology of tectono-metamorphic events. *Gondwana Research*, **71**, 76–90, https://doi.org/10.1016/j.gr.2018.12.014

Botev, Z.I., Grotowski, J.F. and Kroese, D.P. 2010. Kernel density estimation via diffusion. *The Annals of Statistics*, **38**, 2916–2957, https://doi.org/10.1214/10-AOS799

Brooks, C. and Compston, W. 1965. The age and initial Sr^{87}/Sr^{86} of the Heemskirk granite, Western Tasmania. *Journal of Geophysical Research (1896–1977)*,

70, 6249–6262, https://doi.org/10.1029/JZ070i024p06249

Budzyń, B., Wirth, R. *et al.* 2021. LA-ICPMS, TEM and Raman study of radiation damage, fluid-induced alteration and disturbance of U–Pb and Th–Pb ages in experimentally metasomatised monazite. *Chemical Geology*, **583**, https://doi.org/10.1016/j.chemgeo.2021.120464

Cayley, R.A. 2011. Exotic crustal block accretion to the eastern Gondwanaland margin in the Late Cambrian–Tasmania, the Selwyn Block, and implications for the Cambrian–Silurian evolution of the Ross, Delamerian, and Lachlan orogens. *Gondwana Research*, **19**, 628–649, https://doi.org/10.1016/j.gr.2010.11.013

Cayley, R.A., Taylor, D.H., VandenBerg, A.H.M. and Moore, D.H. 2002. Proterozoic–Early Palaeozoic rocks and the Tyennan Orogeny in central Victoria: the Selwyn Block and its tectonic implications. *Australian Journal of Earth Sciences*, **49**, 225–254, https://doi.org/10.1046/j.1440-0952.2002.00921.x

Challandes, N., Marquer, D. and Villa, I.M. 2008. P-T-t modelling, fluid circulation, and ^{39}Ar-^{40}Ar and Rb–Sr mica ages in the Aar Massif shear zones (Swiss Alps). *Swiss Journal of Geosciences*, **101**, 269–288, https://doi.org/10.1007/s00015-008-1260-6

Chappell, B.W., White, A.J.R., Williams, I.S., Wyborn, D. and Wyborn, L.A.I. 2000. Lachlan Fold Belt granites revisited: high- and low-temperature granites and their implications. *Australian Journal of Earth Sciences*, **47**, 123–138, https://doi.org/10.1046/j.1440-0952.2000.00766.x

Cherniak, D.J. 2010. *Diffusion in Accessory Minerals: Zircon, Titanite, Apatite, Monazite and Xenotime*. Mineralogical Society of America.

Cherniak, D.J. and Watson, E.B. 2001. Pb diffusion in zircon. *Chemical Geology*, **172**, 5–24, https://doi.org/10.1016/S0009-2541(00)00233-3

Cherniak, D.J., Watson, E.B., Grove, M. and Harrison, T.M. 2004. Pb diffusion in monazite: a combined RBS/SIMS study. *Geochimica et Cosmochimica Acta*, **68**, 829–840, https://doi.org/10.1016/j.gca.2003.07.012

Chew, D., O'Sullivan, G., Caracciolo, L., Mark, C. and Tyrrell, S. 2020. Sourcing the sand: accessory mineral fertility, analytical and other biases in detrital U–Pb provenance analysis. *Earth-Science Reviews*, **202**, https://doi.org/10.1016/j.earscirev.2020.103093

Clemens, J.D. and Elburg, M.A. 2016. Possible spatial variability in the Selwyn Block of Central Victoria: evidence from Late Devonian felsic igneous rocks. *Australian Journal of Earth Sciences*, **63**, 187–192, https://doi.org/10.1080/08120099.2016.1158736

Copeland, P., Parrish, R.R. and Harrison, T.M. 1988. Identification of inherited radiogenic Pb in monazite and its implications for U–Pb systematics. *Nature*, **333**, 760–763, https://doi.org/10.1038/333760a0

Dodson, M.H. 1973. Closure temperature in cooling geochronological and petrological systems. *Contributions to Mineralogy and Petrology*, **40**, 259–274, https://doi.org/10.1007/BF00373790

Eberlei, T., Habler, G., Wegner, W., Schuster, R., Körner, W., Thöni, M. and Abart, R. 2015. Rb/Sr isotopic and compositional retentivity of muscovite during deformation. *Lithos*, **227**, 161–178, https://doi.org/10.1016/j.lithos.2015.04.007

Edwards, A. 1945. The geology of Phillip Island. *Proceedings of the Royal Society of Victoria*, **57**, 1–21.

Engi, M. 2017. Chapter 12. Petrochronology Based on REE-Minerals: Monazite, Allanite, Xenotime, Apatite. *Reviews in mineralogy and geochemistry*, **83**, 365–418, https://doi.org/10.2138/rmg.2017.83.12

Farina, F. and Stevens, G. 2011. Source controlled 87Sr/86Sr isotope variability in granitic magmas: the inevitable consequence of mineral-scale isotopic disequilibrium in the protolith. *Lithos*, **122**, 189–200, https://doi.org/10.1016/j.lithos.2011.01.001

Fisher, C.M., Bauer, A.M. *et al.* 2020. Laser ablation split-stream analysis of the Sm–Nd and U–Pb isotope compositions of monazite, titanite, and apatite – improvements, potential reference materials, and application to the Archean Saglek Block gneisses. *Chemical Geology*, **539**, 119493, https://doi.org/10.1016/j.chemgeo.2020.119493

Gardés, E., Jaoul, O., Montel, J.M., Seydoux-Guillaume, A.M. and Wirth, R. 2006. Pb diffusion in monazite: an experimental study of Pb2 + +Th4 + ⇔2 Nd3 + interdiffusion. *Geochimica et Cosmochimica Acta*, **70**, 2325–2336, https://doi.org/10.1016/j.gca.2006.01.018

Geisler, T. 2002. Isothermal annealing of partially metamict zircon: evidence for a three-stage recovery process. *Physics and Chemistry of Minerals*, **29**, 420–429, https://doi.org/10.1007/s00269-002-0249-3

Geisler, T., Schaltegger, U. and Tomaschek, F. 2007. Re-equilibration of zircon in aqueous fluids and melts. *Elements*, **3**, 43–50, https://doi.org/10.2113/gselements.3.1.43

Giletti, B.J. 1991. Rb and Sr diffusion in alkali feldspars with implications for cooling histories of rocks. *Geochimica et Cosmochimica Acta*, **55**, 1331–1343, https://doi.org/10.1016/0016-7037(91)90311-R

Glen, R.A. 2005. The Tasmanides of Eastern Australia. *Geological Society, London, Special Publications*, **246**, 23–96, https://doi.org/10.1144/GSL.SP.2005.246.01.02

Glodny, J., Kühn, A. and Austrheim, H. 2008. Diffusion v. recrystallization processes in Rb–Sr geochronology: isotopic relics in eclogite facies rocks, Western Gneiss Region, Norway. *Geochimica et Cosmochimica Acta*, **72**, 506–525, https://doi.org/10.1016/j.gca.2007.10.021

Gray, D.R. and Foster, D.A. 2004. Tectonic evolution of the Lachlan Orogen, southeast Australia: historical review, data synthesis and modern perspectives. *Australian Journal of Earth Sciences*, **51**, 773–817, https://doi.org/10.1111/j.1400-0952.2004.01092.x

Gyomlai, T., Agard, P. *et al.* 2022. Cimmerian metamorphism and post Mid-Cimmerian exhumation in Central Iran: insights from *in-situ* Rb/Sr and U/Pb dating. *Journal of Asian Earth Sciences*, **233**, 105242, https://doi.org/10.1016/j.jseaes.2022.105242

Harrison, T.M. and Zeitler, P.K. 2005. Fundamentals of Noble Gas Thermochronometry. *Reviews in Mineralogy and Geochemistry*, **58**, 123–149, https://doi.org/10.2138/rmg.2005.58.5

Henry, D.A. and Birch, W.D. 1992. Cambrian greenstone on Phillip Island, Victoria. *Australian Journal of*

Earth Sciences, **39**, 567–575, https://doi.org/10.1080/08120099208728050

Hogmalm, K.J., Zack, T., Karlsson, A.K.O., Sjöqvist, A.S.L. and Garbe-Schönberg, D. 2017. *In-situ* Rb–Sr and K–Ca dating by LA-ICP-MS/MS: an evaluation of N_2O and SF_6 as reaction gases. *Journal of Analytical Atomic Spectrometry*, **32**, 305–313, https://doi.org/10.1039/C6JA00362A

Horstwood, M.S.A., Košler, J. *et al.* 2016. Community-derived standards for LA-ICP-MS U– (Th-)Pb geochronology – uncertainty propagation, age interpretation and data reporting. *Geostandards and Geoanalytical Research*, **40**, 311–332, https://doi.org/10.1111/j.1751-908X.2016.00379.x

Hoskin, P.W.O. 2005. Trace-element composition of hydrothermal zircon and the alteration of Hadean zircon from the Jack Hills, Australia. *Geochimica et Cosmochimica Acta*, **69**, 637–648, https://doi.org/10.1016/j.gca.2004.07.006

Jäger, E. 1967. Die Bedeutung der Biotit-Alterswerte. *In*: Jäger, E., Niggli, E. and Wenk, E. (eds) *Rb–Sr Alterbestmmungen an Glimmern der Zentralalpen Beitr. Geol. Kaarte Schweiz,* swisstopo, *NF*, **134**, 28–31.

Jäger, E. 1979. The Rb-Sr method. *In*: Jäger, E. and Hunziker, J.C. (eds) *Lectures in Isotope Geology*. Springer Berlin Heidelberg, Berlin, Heidelberg, 13–26.

Jenkin, G.R.T. 1997. Do cooling paths derived from mica Rb–Sr data reflect true cooling paths? *Geology*, **25**, 907–910, https://doi.org/10.1130/0091-7613(1997)025<0907:DCPDFM>2.3.CO;2

Jenkin, G.R.T., Rogers, G., Fallick, A.E. and Farrow, C.M. 1995. Rb–Sr closure temperatures in bi-minerallic rocks: a mode effect and test for different diffusion models. *Chemical Geology*, **122**, 227–240, https://doi.org/10.1016/0009-2541(95)00013-C

Kirkland, C.L., Smithies, R.H., Taylor, R.J.M., Evans, N. and McDonald, B. 2015. Zircon Th/U ratios in magmatic environs. *Lithos*, **212–215**, 397–414, https://doi.org/10.1016/j.lithos.2014.11.021

Kohn, B.P., Gleadow, A.J.W., Brown, R.W., Gallagher, K., O'Sullivan, P.B. and Foster, D.A. 2020. *Australian-Wide Apatite Fission Track Data Compilation*. Pangaea, https://doi.org/10.1594/PANGAEA.911861

Kozlik, M., Raith, J.G. and Gerdes, A. 2016. U–Pb, Lu–Hf and trace element characteristics of zircon from the Felbertal scheelite deposit (Austria): new constraints on timing and source of W mineralization. *Chemical Geology*, **421**, 112–126, https://doi.org/10.1016/j.chemgeo.2015.11.018

Kühn, A., Glodny, J., Iden, K. and Austrheim, H. 2000. Retention of Precambrian Rb/Sr phlogopite ages through Caledonian eclogite facies metamorphism, Bergen Arc Complex, W. Norway. *Lithos*, **51**, 305–330, https://doi.org/10.1016/S0024-4937(99)00067-5

Kusiak, M., Dunkley, D., Słaby, E., Martin, H. and Budzyń, B. 2009. Sensitive high-resolution ion microprobe analysis of zircon re-equilibrated by late magmatic fluids in a hybridized pluton. *Geology*, **37**, 1063–1066, https://doi.org/10.1130/G30048A.1

Kylander-Clark, A.R.C., Hacker, B.R. and Cottle, J.M. 2013. Laser-ablation split-stream ICP petrochronology. *Chemical Geology*, **345**, 99–112, https://doi.org/10.1016/j.chemgeo.2013.02.019

Le Maitre, R.W., Streckeisen, A., Zanettin, B., Le Bas, M., Bonin, B. and Bateman, P. 2005. *Igneous Rocks: a Classification and Glossary of Terms: Recommendations of the International Union of Geological Sciences Subcommission on the Systematics of Igneous Rocks*. Cambridge University Press.

Li, S.-S., Santosh, M. *et al.* 2020. Coupled U–Pb and Rb–Sr laser ablation geochronology trace Archean to Proterozoic crustal evolution in the Dharwar Craton, India. *Precambrian Research*, **343**, 105709, https://doi.org/10.1016/j.precamres.2020.105709

Lim, H. 2022. *Timescales of granite infancy: improvement and alternatives on plutonic geochronology*. Doctor of Philosophy, Monash University.

Liu, X., Gao, S., Diwu, C., Yuan, H. and Hu, Z. 2007. Simultaneous *in-situ* determination of U–Pb age and trace elements in zircon by LA-ICP-MS in 20 µm spot size. *Chinese Science Bulletin*, **52**, 1257–1264, https://doi.org/10.1007/s11434-007-0160-x

Liu, Z.-C., Wu, F.-Y., Yang, Y.-H., Yang, J.-H. and Wilde, S.A. 2012. Neodymium isotopic compositions of the standard monazites used in U–Th/Pb geochronology. *Chemical Geology*, **334**, 221–239, https://doi.org/10.1016/j.chemgeo.2012.09.034

Matheney, R.K., Brookins, D.G., Wallin, E.T., Shafiqullah, M. and Damon, P.E. 1990. Incompletely reset Rb–Sr systems from a Cambrian red-rock granophyre terrane, Florida Mountains, New Mexico, USA. *Chemical Geology: Isotope Geoscience Section*, **86**, 29–47, https://doi.org/10.1016/0168-9622(90)90004-V

Meißner, B., Deters, P., Srikantappa, C. and Köhler, H. 2002. Geochronological evolution of the Moyar, Bhavani and Palghat shear zones of Southern India: implications for east Gondwana correlations. *Precambrian Research*, **114**, 149–175, https://doi.org/10.1016/S0301-9268(01)00222-4

Mezger, K. and Krogstad, E.J. 1997. Interpretation of discordant U–Pb zircon ages: an evaluation. *Journal of Metamorphic Geology*, **15**, 127–140, https://doi.org/10.1111/j.1525-1314.1997.00008.x

Miller, J.S., Matzel, J.E.P., Miller, C.F., Burgess, S.D. and Miller, R.B. 2007. Zircon growth and recycling during the assembly of large, composite arc plutons. *Journal of Volcanology and Geothermal Research*, **167**, 282–299, https://doi.org/10.1016/j.jvolgeores.2007.04.019

Moore, D.H., Betts, P.G. and Hall, M. 2016. Constraining the VanDieland microcontinent at the edge of East Gondwana, Australia. *Tectonophysics*, **687**, 158–179, https://doi.org/10.1016/j.tecto.2016.09.009

Nakamura, N. 1974. Determination of REE, Ba, Fe, Mg, Na and K in carbonaceous and ordinary chondrites. *Geochimica et Cosmochimica Acta*, **38**, 757–775, https://doi.org/10.1016/0016-7037(74)90149-5

Nebel-Jacobsen, Y., Scherer, E.E., Münker, C. and Mezger, K. 2005. Separation of U, Pb, Lu, and Hf from single zircons for combined U–Pb dating and Hf isotope measurements by TIMS and MC-ICPMS. *Chemical Geology*, **220**, 105–120, https://doi.org/10.1016/j.chemgeo.2005.03.009

Nebel, O. and Mezger, K. 2006. Reassessment of the NBS SRM-607 K-feldspar as a high precision Rb/Sr and Sr isotope reference. *Chemical Geology*, **233**, 337–345, https://doi.org/10.1016/j.chemgeo.2006.03.003

Nebel, O., Scherer, E.E. and Mezger, K. 2011. Evaluation of the ^{87}Rb decay constant by age comparison against the U–Pb system. *Earth and Planetary Science Letters*, **301**, 1–8, https://doi.org/10.1016/j.epsl.2010.11.004

O'Neill, H.S.C. 2016. The smoothness and shapes of chondrite-normalized Rare Earth Element patterns in basalts. *Journal of Petrology*, **57**, 1463–1508, https://doi.org/10.1093/petrology/egw047

Olierook, H.K.H., Kirkland, C.L. *et al.* 2020*a*. Differentiating between inherited and autocrystic zircon in granitoids. *Journal of Petrology*, **61**, egaa081, https://doi.org/10.1093/petrology/egaa081

Olierook, H.K.H., Rankenburg, K. *et al.* 2020*b*. Resolving multiple geological events using *in-situ* Rb–Sr geochronology: implications for metallogenesis at Tropicana, Western Australia. *Geochronology*, **2**, 283–303, https://doi.org/10.5194/gchron-2-283-2020

Oriolo, S., Wemmer, K., Oyhantçabal, P., Fossen, H., Schulz, B. and Siegesmund, S. 2018. Geochronology of shear zones – a review. *Earth-Science Reviews*, **185**, 665–683, https://doi.org/10.1016/j.earscirev.2018.07.007

Papanastassiou, D.A. and Wasserburg, G.J. 1968. Initial strontium isotopic abundances and the resolution of small time differences in the formation of planetary objects. *Earth and Planetary Science Letters*, **5**, 361–376, https://doi.org/10.1016/S0012-821X(68)80066-4

Parrish, R.R. 1990. U–Pb dating of monazite and its application to geological problems. *Canadian Journal of Earth Sciences*, **27**, 1431–1450, https://doi.org/10.1139/e90-152

Paton, C., Woodhead, J.D., Hellstrom, J.C., Hergt, J.M., Greig, A. and Maas, R. 2010. Improved laser ablation U–Pb zircon geochronology through robust downhole fractionation correction. *Geochemistry, Geophysics, Geosystems*, **11**, https://doi.org/10.1029/2009GC002618

Paton, C., Hellstrom, J., Paul, B., Woodhead, J. and Hergt, J. 2011. Iolite: freeware for the visualisation and processing of mass spectrometric data. *Journal of Analytical Atomic Spectrometry*, **26**, 2508–2518, https://doi.org/10.1039/C1JA10172B

Payne, J.L. 2008. Palaeo- to Mesoproterozoic evolution of the Gawler Craton, Australia: geochronological, geochemical and isotopic constraints. *Doctor of Philosophy, University of Adelaide*, https:/hdl.handle.net/2440/50045

Putnis, A. 2002. Mineral replacement reactions: from macroscopic observations to microscopic mechanisms. *Mineralogical Magazine – MINER MAG*, **66**, https://doi.org/10.1180/0026461026650056

Putnis, A., Hinrichs, R., Putnis, C.V., Golla-Schindler, U. and Collins, L.G. 2007. Hematite in porous red-clouded feldspars: evidence of large-scale crustal fluid–rock interaction. *Lithos*, **95**, 10–18, https://doi.org/10.1016/j.lithos.2006.07.004

Raymond, O.L.L., Gallagher, R., Zhang, W. and Highet, L.M. 2012. *Surface Geology of Australia 1:1 Million Scale Dataset 2012 Edition*. Geoscience Australia.

Redaa, A., Farkaš, J., Hassan, A., Collins, A.S., Gilbert, S. and Löhr, S.C. 2022. Constraints from *in-situ* Rb–Sr dating on the timing of tectono-thermal events in the Umm Farwah shear zone and associated Cu–Au mineralisation in the Southern Arabian Shield, Saudi Arabia. *Journal of Asian Earth Sciences*, **224**, 105037, https://doi.org/10.1016/j.jseaes.2021.105037

Redaa, A., Farkaš, J. *et al.* 2021. Assessment of elemental fractionation and matrix effects during: *in-situ* Rb–Sr dating of phlogopite by LA-ICP-MS/MS: implications for the accuracy and precision of mineral ages. *Journal of Analytical Atomic Spectrometry*, **36**, 322–344, https://doi.org/10.1039/d0ja00299b

Regmi, K.R., Weinberg, R.F., Nicholls, I.A., Maas, R. and Raveggi, M. 2016. Evidence for hybridisation in the Tynong Province granitoids, Lachlan Fold Belt, Eastern Australia. *Australian Journal of Earth Sciences*, **63**, 235–255, https://doi.org/10.1080/08120099.2016.1180321

Ribeiro, B.V., Mulder, J.A. *et al.* 2020. Using apatite to resolve the age and protoliths of mid-crustal shear zones: A case study from the Taxaquara Shear Zone, SE Brazil. *Lithos*, **378-379**, 105817, https://doi.org/10.1016/j.lithos.2020.105817

Ribeiro, B.V., Finch, M.A. *et al.* 2022. From microanalysis to supercontinents: Insights from the Rio Apa Terrane into the Mesoproterozoic SW Amazonian Craton evolution during Rodinia assembly. *Journal of Metamorphic Geology*, **40**, 631–663, https://doi.org/10.1111/jmg.12641

Richter, M., Nebel-Jacobsen, Y., Nebel, O., Zack, T., Mertz-Kraus, R., Raveggi, M. and Rösel, D. 2019. Assessment of five monazite reference materials for U–Th/Pb dating using laser-ablation ICP-MS. *Geosciences (Switzerland)*, **9**, https://doi.org/10.3390/geosciences9090391

Rösel, D. and Zack, T. 2022. LA-ICP-MS/MS Single-Spot Rb–Sr dating. *Geostandards and Geoanalytical Research*, **n/a**, https://doi.org/10.1111/ggr.12414

Rosenblum, S. 1958. Magnetic susceptibilities of minerals in the Frantz Isodynamic magnetic separator1. *American Mineralogist*, **43**, 170–173.

Rossiter, A.G. and Gray, C.M. 2008. Barium contents of granites: key to understanding crustal architecture in the southern Lachlan Fold Belt? *Australian Journal of Earth Sciences*, **55**, 433–448, https://doi.org/10.1080/08120090801888586

Samsu, A., Cruden, A.R., Hall, M., Micklethwaite, S. and Denyszyn, S.W. 2019. The influence of basement faults on local extension directions: insights from potential field geophysics and field observations. *Basin Research*, **31**, 782–807, https://doi.org/10.1111/bre.12344

Schaltegger, U., Schmitt, A.K. and Horstwood, M.S.A. 2015. U–Th/Pb zircon geochronology by ID-TIMS, SIMS, and laser ablation ICP-MS: recipes, interpretations, and opportunities. *Chemical Geology*, **402**, 89–110, https://doi.org/10.1016/j.chemgeo.2015.02.028

Siégel, C., Bryan, S.E., Allen, C.M. and Gust, D.A. 2018. Use and abuse of zircon-based thermometers: a critical review and a recommended approach to identify antecrystic zircons. *Earth-Science Reviews*, **176**, 87–116, https://doi.org/10.1016/j.earscirev.2017.08.011

Silverman, B.W. 1986. *Density Estimation for Statistics and Data Analysis Monographs on Statistics and Applied Probability*. Chapman and Hall, London.

Sláma, J., Košler, J. *et al.* 2008. Plešovice zircon – a new natural reference material for U–Pb and Hf isotopic microanalysis. *Chemical Geology*, **249**, 1–35, https://doi.org/10.1016/j.chemgeo.2007.11.005

Spencer, C.J., Kirkland, C.L. and Taylor, R.J.M. 2016. Strategies towards statistically robust interpretations of *in-situ* U–Pb zircon geochronology. *Geoscience Frontiers*, **7**, 581–589, https://doi.org/10.1016/j.gsf.2015.11.006

Suzuki, K. and Adachi, M. 1994. Middle Precambrian detrital monazite and zircon from the Hida Gneiss on Oki-Dogo Island, Japan: their origin and implications for the correlation of basement gneiss of Southwest Japan and Korea. *Tectonophysics*, **235**, 277–292, https://doi.org/10.1016/0040-1951(94)90198-8

Tillberg, M., Drake, H., Zack, T., Hogmalm, J., Kooijman, E. and Åström, M. 2021. Reconstructing craton-scale tectonic events via *in-situ* Rb–Sr geochronology of poly-phased vein mineralization. *Terra Nova*, **33**, 502–510, https://doi.org/10.1111/ter.12542

Tomaschek, F., Kennedy, A., Villa, I.M., Lagos, M. and Ballhaus, C. 2003. Zircons from Syros, Cyclades, Greece – recrystallization and mobilization of zircon during high-pressure metamorphism. *Journal of Petrology*, **44**, 1977–2002, https://doi.org/10.1093/petrology/egg067

Trail, D., Bruce Watson, E. and Tailby, N.D. 2012. Ce and Eu anomalies in zircon as proxies for the oxidation state of magmas. *Geochimica et Cosmochimica Acta*, **97**, 70–87, https://doi.org/10.1016/j.gca.2012.08.032

Van Krieken, A.T. and Wilson, C.J.L. 2016. Structural and timing constraints on molybdenum and tungsten mineralisation at Yea. *Victoria Australian Journal of Earth Sciences*, **62**, 985–1007, https://doi.org/10.1080/08120099.2015.1121927

VandenBerg, A., Willman, C. *et al.* 2000. *The Tasman Fold Belt System in Victoria*. Geological Survey of Victoria, Melbourne, Victoria.

Vermeesch, P. 2012. On the visualisation of detrital age distributions. *Chemical Geology*, **312–313**, 190–194, https://doi.org/10.1016/j.chemgeo.2012.04.021

Vermeesch, P. 2018. IsoplotR: a free and open toolbox for geochronology. *Geoscience Frontiers*, **9**, https://doi.org/10.1016/j.gsf.2018.04.001

Vermeesch, P. 2021. On the treatment of discordant detrital zircon U–Pb data. *Geochronology*, **3**, 247–257, https://doi.org/10.5194/gchron-3-247-2021

Villa. 1998. Isotopic closure. *Terra Nova*, **10**, 42–47, https://doi.org/10.1046/j.1365-3121.1998.00156.x

Villa, I.M. and Hanchar, J.M. 2017. Age discordance and mineralogy. *American Mineralogist*, **102**, 2422–2439, https://doi.org/10.2138/am-2017-6084

Villa, I.M. and Williams, M.L. 2013. Geochronology of metasomatic events. *In*: Harlov, D.E. and Austrheim, H. (eds) *Metasomatism and the Chemical Transformation of Rock: the Role of Fluids in Terrestrial and Extraterrestrial Processes*. Springer Berlin Heidelberg, Berlin, Heidelberg, 171–202, https://doi.org/10.1007/978-3-642-28394-9_6

Villa, I.M., De Bièvre, P., Holden, N.E. and Renne, P.R. 2015. IUPAC-IUGS recommendation on the half-life of ^{87}Rb. *Geochimica et Cosmochimica Acta*, **164**, 382–385, https://doi.org/10.1016/j.gca.2015.05.025

Weinberg, R.F., Wolfram, L.C., Nebel, O., Hasalová, P., Závada, P., Kylander-Clark, A.R.C. and Becchio, R. 2020. Decoupled U–Pb date and chemical zonation of monazite in migmatites: the case for disturbance of isotopic systematics by coupled dissolution-reprecipitation. *Geochimica et Cosmochimica Acta*, **269**, 398–412, https://doi.org/10.1016/j.gca.2019.10.024

White, A.J.R. and Chappell, B.W. 1977. Ultrametamorphism and granitoid genesis. *Tectonophysics*, **43**, 7–22, https://doi.org/10.1016/0040-1951(77)900 03-8

Wiedenbeck, M., Allé, P. *et al.* 1995. Three natural zircon standards for U–Th/Pb, Lu–Hf, trace element and REE analyses. *Geostandards Newsletter*, **19**, 1–23, https://doi.org/10.1111/j.1751-908X.1995.tb00147.x

Willigers, B.J.A., Mezger, K. and Baker, J.A. 2004. Development of high precision Rb–Sr phlogopite and biotite geochronology; an alternative to ^{40}Ar/^{39}Ar trioctahedral mica dating. *Chemical Geology*, **213**, 339–358, https://doi.org/10.1016/j.chemgeo.2004.07.006

Zack, T. and Hogmalm, K.J. 2016. Laser ablation Rb/Sr dating by online chemical separation of Rb and Sr in an oxygen-filled reaction cell. *Chemical Geology*, **437**, 120–133, https://doi.org/10.1016/j.chemgeo.2016.05.027

Zircon U–Pb geochronology and Hf isotopic compositions of igneous rocks from Sumatra: implications for the Cenozoic magmatic evolution of the western Sunda Arc

Yu-Ming Lai[1]*, Ping-Ping Liu[2], Sun-Lin Chung[3,4], Azman A. Ghani[5], Hao-Yang Lee[3], Long Xiang Quek[6], Shan Li[7], Muhammad Hatta Roselee[5], Sayed Murtadha[8], Lediyantje Lintjewas[1,9] and Yoshiyuki Iizuka[3]

[1]Department of Earth Sciences, National Taiwan Normal University, Taipei 11677, Taiwan

[2]Key Laboratory of Orogenic Belts and Crustal Evolution, School of Earth and Space Sciences, Peking University, Beijing 100871, China

[3]Institute of Earth Sciences, Academia Sinica, Taipei 11529, Taiwan

[4]Department of Geosciences, National Taiwan University, Taipei 10617, Taiwan

[5]Department of Geology, Faculty of Science, University of Malaya, Kuala Lumpur 50603, Malaysia

[6]Institute of Geology, Chinese Academy of Geological Sciences, Beijing 100037, China

[7]College of Earth and Planetary Sciences, University of Chinese Academy of Sciences, Beijing 100049, China

[8]Department of Geology, Syiah Kuala University, Banda Aceh 23111, Indonesia

[9]National Research and Innovation Agency (BRIN), Bandung 40135, Indonesia

Y-ML, 0000-0002-9771-0505
*Correspondence: ymlai@ntnu.edu.tw

Abstract: Sumatra is located at the western end of the Sunda Arc, which resulted from the subduction of the Indo-Australian Plate beneath the Eurasian Plate. In this study, we report detailed zircon U–Pb and Hf isotope data for Cenozoic igneous rocks from the entire island of Sumatra to better constrain the temporal and spatial distribution of arc magmatism. The new dataset, combined with literature information, identifies the following two magmatic stages: (1) Paleocene to Early Eocene (66–48 Ma) and (2) Early Miocene to Recent (23–0 Ma), with a 25 myr-long period of magmatic quiescence in between. The magmatic zircons show predominantly positive and high $\varepsilon_{Hf}(t)$ values, ranging from +19.4 to +7.1 in western Sumatra, +17.1 to +1.6 in central Sumatra and +18.0 to +7.0 in eastern Sumatra, indicating an isotopically juvenile magma source in the mantle wedge along the western Sunda Arc. We explain the negative and low $\varepsilon_{Hf}(t)$ values (+0.5 to −13.1) of young samples around the supervolcano Toba as evidence for the subduction of sediment. We argue for a change in the subduction processes, where the first magmatic stage ceased owing to the termination of the Neo-Tethyan subduction and the following stage corresponded to the modern Sunda subduction.

Supplementary material: Previous dating results, zircon CL images, U–Pb and Hf isotope data are available at https://doi.org/10.6084/m9.figshare.c.6366009

Sundaland is located at the southern edge of the Eurasian Plate and is predominantly a collage of blocks that rifted from Gondwana (Hall 2009). The Sunda Arc (including the islands of Sumatra, Java and West and East Nusa Tenggara) in the southern part of Sundaland is a volcanic arc that was formed by the subduction of the Indo-Australian Plate beneath the Eurasian Plate (Hamilton 1979). This arc system went through the Tethys Ocean closure in the Cenozoic before the current Indian Ocean subduction (Curray 1989; Hall 2012). Sumatra Island, located in the western part of the Sunda Arc and the sixth-largest island on Earth, has Cenozoic magmatism spread along both sides of its Sumatra Fault Zone (SFZ), which trends NW–SE (Katili and Hehuwat 1967).

Van Bemmelen (1949) was the first to describe three volcanic cycles in Sumatra during the Cenozoic Era, excluding the largest-known Quaternary Toba caldera complex (see Chesner (2012) for a review)

From: van Schijndel, V., Cutts, K., Pereira, I., Guitreau, M., Volante, S. and Tedeschi, M. (eds) 2024. *Minor Minerals, Major Implications: Using Key Mineral Phases to Unravel the Formation and Evolution of Earth's Crust.* Geological Society, London, Special Publications, **537**, 455–478.
First published online March 10, 2023, https://doi.org/10.1144/SP537-2022-199

– Old Neogene, Young Neogene and Young Quaternary. After combining the results of the literature on age, Crow (2005) proposed six new volcanic episodes in Sumatra: the Paleocene, Late Middle Eocene, Late Eocene to Late Oligocene, Late Early Miocene, Middle Miocene and Late Miocene to Pliocene. However, the previous dating used various methods, such as mineral or whole-rock ^{40}K–^{40}Ar, $^{40}Ar/^{39}Ar$ and Rb–Sr radiometric dating, and most of the sample localities are untraceable (without GPS data) (Barber *et al.* 2005). Therefore, researchers who want to combine the ages with the geochemical characteristics from these studies might encounter challenges. The K–Ar and Rb–Sr systems in these dating methods are easily influenced by later hydrothermal alteration or weathering (Wilkinson *et al.* 2017). Furthermore, because only a few volcanic samples have both age and geochemical data, studying their petrogenetic evolution is difficult. Although previous research could document some specific volcanoes (e.g. Toba volcano), such studies can provide only a rough view of the petrogenesis or magmatic evolution throughout the Sunda Arc, with no detailed investigation of the magmatic stages combining temporal and spatial geochemical variations along or across the Sumatra Island.

Our paper reports zircon U–Pb ages for 34 samples and Lu–Hf isotopic data for a selection of 28 samples and provides a systematic investigation of the Cenozoic magmatism on Sumatra Island. The results from this study are combined with previous age data to improve our understanding of the distribution of magmatism stages and highlight the temporal and spatial variation of geochemical properties in Sumatra Island since the Paleocene.

Geological background

Sumatra Island, located in the western Sunda Arc in southern Sundaland, is the product of accretion of three terranes: the East Sumatra Block, West Sumatra Block and Woyla Nappe (Hutchison 1994; Metcalfe 1996). The East Sumatra Block is a part of the Sibumasu (Shan–Thai, Bunopas 1982) Block that collided with Indochina during the Middle Triassic period (Hutchison 1994; Metcalfe 2002). The West Sumatra Block arrived on the southern side of Sibumasu after the Mid-Triassic (Barber and Crow 2003; Barber *et al.* 2005) and was separated from West Burma by the opening of the Andaman Sea at around 15 to 13 Ma (Curray 2005). The Jurassic to the Cretaceous Woyla Group is composed of an oceanic island arc and accretionary complex, which collided with southern Sumatra to form the Woyla Nappe (Cameron *et al.* 1980; Wajzer *et al.* 1991). The Cenozoic to present arc-trench system is from the subduction of the Indo-Australian Plate beneath the Sumatra and Java islands and is named the Sunda Arc (Hamilton 1988) (Fig. 1). The motion of the subducting plate is nearly perpendicular to the subduction zone in the Java area; however, subduction is oblique in the Sumatra area (Curray 1989). On the island, the NW–SE SFZ started to develop at 23 to 15 Ma and cut through the entire island (Curray *et al.* 1979). The Wharton Fossil Ridge, an aseismic ridge, is located to the south of Sumatra (Whittaker *et al.* 2007); it forms several transform structures on its eastern part as the Investigator Fracture Zone. This fracture zone was subducted beneath the Toba area by the northward subduction intersecting the SFZ (Crow 2005; Gasparon 2005) from *c.* 15 to 10 Ma (Kopp *et al.* 2008; Lange *et al.* 2010) and might be related to the large eruption at the Toba area (Crow 2005).

Cenozoic volcanic activity was classified by van Bemmelen (1949) into the following three cycles: Old Neogene (Late Oligocene to Middle Miocene), Young Neogene (Middle Miocene to early Quaternary) and Young Quaternary. Several researchers later published detailed geological survey reports and geological maps (Cameron *et al.* 1980; Rock *et al.* 1983; Kusnama *et al.* 1993; McCourt and Cobbing 1993). Following advances in igneous rock dating methods, ^{40}K–^{40}Ar age dates from volcanic and plutonic rocks were released in numerous studies (Kanao *et al.* 1971; De Coster 1974; Hehuwat 1976; Bennett *et al.* 1981; Eubank and Makki 1981; Cameron *et al.* 1982; Rock *et al.* 1982; Koning and Aulia 1985; Wajzer 1986; van Leeuwen *et al.* 1987; JICA 1988; Kallagher 1990; Sato 1991; Kusnama *et al.* 1993; McCourt and Cobbing 1993; Wikarno *et al.* 1993; Amin *et al.* 1994; Gafoer *et al.* 1994; Bellon *et al.* 2004). Furthermore, Imtihanah (2000) provided 16 ages in central and southern Sumatra using $^{40}Ar/^{39}Ar$ and Rb–Sr dating methods. Crow (2005) collated previous ages and classified volcanic episodes that occurred during the Paleocene (60–50 Ma), late Middle Eocene (46–40 Ma), Late Eocene to Late Oligocene (35–30 Ma), late Early Miocene (22–14 Ma), Middle Miocene (12–8 Ma) and Late Miocene through Pliocene (6–1.6 Ma). Some studies concentrated on the supervolcano Toba and gave age data of the Toba tuffs (Haranggaol Dacite Tuff: 1.20 ± 0.16 Ma; Oldest Toba Tuff: 0.84 ± 0.03 Ma; Middle Toba Tuff: 0.501 ± 0.005 Ma; Youngest Toba Tuff: 75.0 ± 0.9 ka) using different methods (Nishimura *et al.* 1977; Ninkovich *et al.* 1978; Yokoyama and Hehanussa 1981; Diehl *et al.* 1987; Chesner *et al.* 1991; Mark *et al.* 2014). Zircon U–Pb dating has been an important dating tool for decades, and has been widely accessible for igneous rocks because of laser ablation coupled plasma spectrometer (LA-ICP-MS) instrumentation. Zhang *et al.* (2018, 2019) and Hu *et al.* (2019) used U–Pb ages from

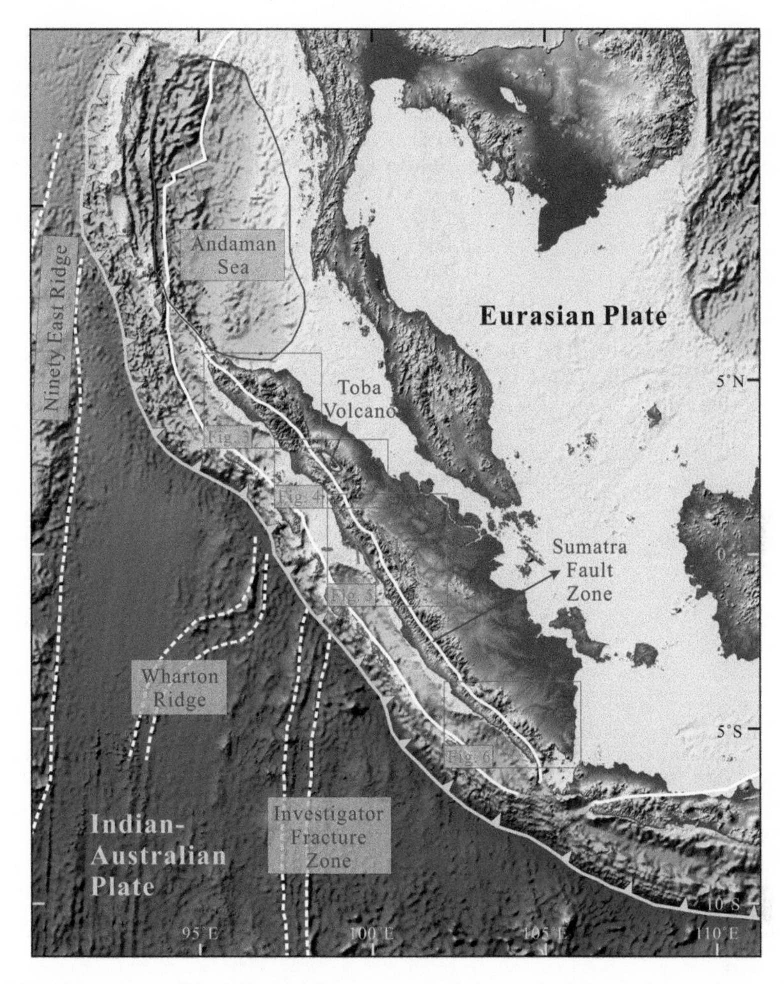

Fig. 1. Simplified tectonic map of SE Asia and adjacent areas. Red squares show four regions of the sample localities in this study, from west to east: western Sumatra (Fig. 3), Toba area (Fig. 4), central Sumatra (Fig. 5) and eastern Sumatra (Fig. 6). Source: after Barber *et al.* (2005) and Metcalfe (2013); the base map is from NOAA's website, https://ngdc.noaa.gov/mgg/global/relief/ETOPO1/image/.

detrital and igneous zircons to investigate the tectonic evolution of southern Asia and the age of Tangse porphyry Cu–Mo deposit, respectively. Furthermore, there has been some research related to the zircon U–Pb geochronology of young volcanic rocks within or near Toba (Ito 2020; Liu *et al.* 2021) and western Sumatra (Lai *et al.* 2021).

Some geochemical research was conducted 20–40 years ago based on a small number of igneous rocks in Sumatra (Whitford 1975; Leo *et al.* 1980; Rock *et al.* 1982; Wajzer 1986; Kallagher 1989; Gafoer *et al.* 1992; Gasparon *et al.* 1994; Gasparon and Varne 1995; Chesner 1998; Wark *et al.* 2000; Turner and Foden 2001). Rock *et al.* (1982) and Gasparon and Varne (1995) analysed and reviewed the previous data, which provided an extensive database for the major and trace elements in volcanic and plutonic rocks. Gasparon and Varne (1995) used the Sr–Nd–Pb method to provide whole-rock isotopic analyses of plutonic rocks. Whitford (1975) and Leo *et al.* (1980) published whole-rock Sr isotopes for some volcanic rocks, whereas Wark *et al.* (2000) published Nd isotopes for others. Previously, researchers combined samples from Sumatra and Java to discuss the petrogenesis in the elongated Sunda Arc (Whitford 1975) and to directly summarize the entire Sunda–Banda Arc petrogenesis (Wheller *et al.* 1987; Turner and Foden 2001) because most samples were from Java Island. Thus, despite a few studies, petrogenesis in Sumatra has remained a difficult subject for discussion.

Methods and samples

Samples

In this study, we collected 34 igneous rocks from the entire Sumatra Island, including 17 volcanic rocks and 17 plutonic rocks, for zircon U–Pb dating and Lu–Hf isotope analyses. In total, 16 zircon U–Pb ages from previous publications (Hu *et al.* 2019; Ito 2020; Lai *et al.* 2021; Liu *et al.* 2021) and 7 ages from Gao *et al.* (2022) were integrated with our data and are shown in Figure 2. In addition, we combined 141 age results dated by ^{40}K–^{40}Ar, $^{40}Ar/^{39}Ar$

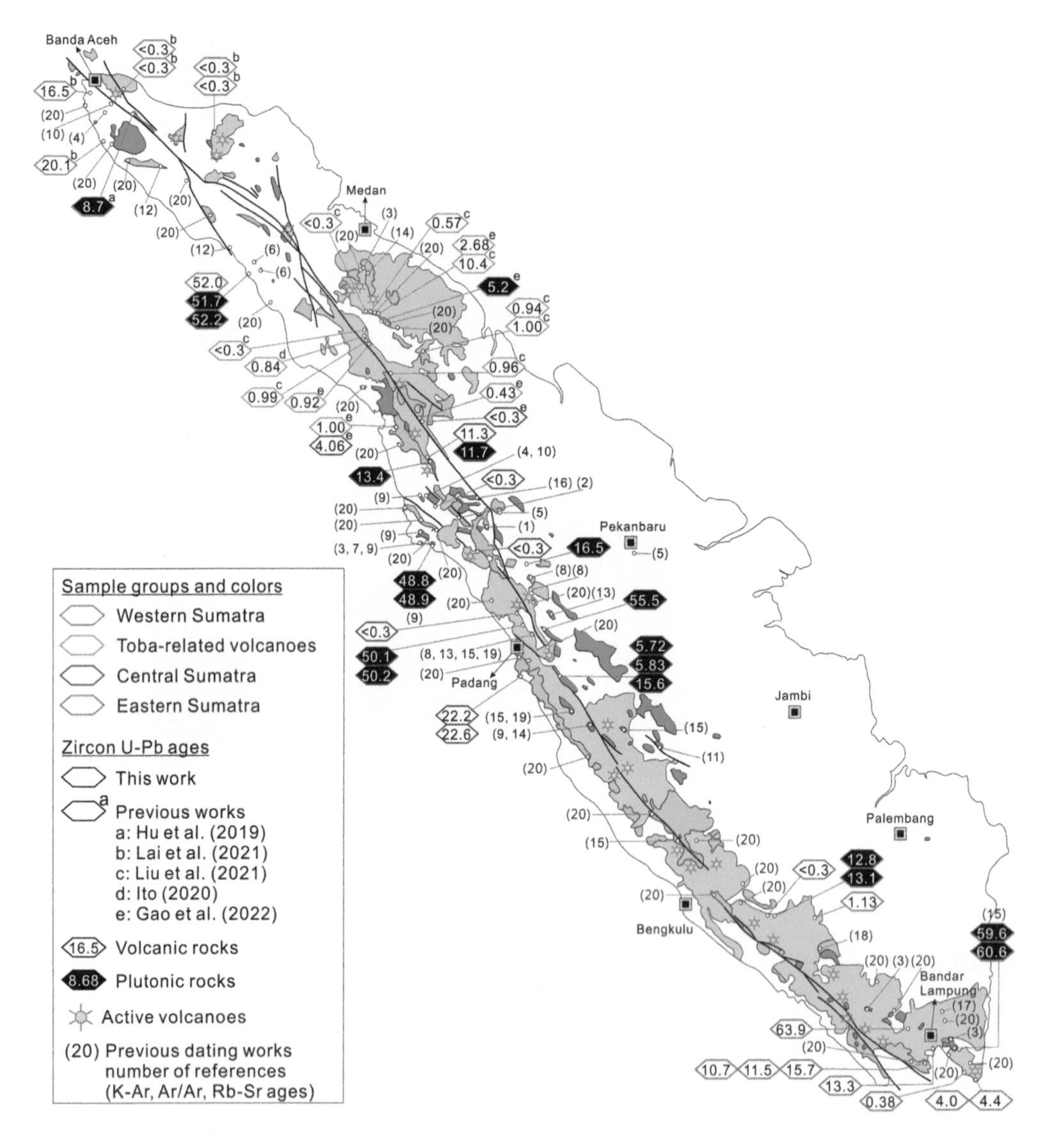

Fig. 2. Cenozoic igneous rocks dating samples from previous works and this study on Sumatra Island. Hexagons show zircon U–Pb dating samples (Hu *et al.* (2019)[a], Lai *et al.* (2021)[b], Liu *et al.* (2021)[c], Ito (2020)[d], Gao *et al.* (2022)[e], and ages without upper index marks are from this study). Different colours of hexagons show four different regions of Sumatra Island; open hexagons are volcanic rocks; solid ones are plutonic rocks. Sun-shape patterns are active volcanoes (Gasparon 2005). Numbers in parentheses show the locations of dating samples from the literature by ^{40}K–^{40}Ar, $^{40}Ar/^{39}Ar$ and Rb–Sr methods (Kanao *et al.* 1971[1]; De Coster 1974[2]; Hehuwat 1976[3]; Bennett *et al.* 1981[4]; Eubank and Makki 1981[5]; Cameron *et al.* 1982[6]; Rock *et al.* 1983[7]; Koning and Aulia 1985[8]; Wajzer 1986[9]; van Leeuwen *et al.* 1987[10]; JICA 1988[11]; Kallagher 1990[12]; Sato 1991[13]; Kusnama *et al.* 1993[14]; McCourt and Cobbing 1993[15]; Wikarno *et al.* 1993[16]; Amin *et al.* 1994[17]; Gafoer *et al.* 1994[18]; Imtihanah 2000[19]; Bellon *et al.* 2004[20], all references are from Barber *et al.* 2005).

and Rb–Sr methods from 20 previous works embodied in a book which was edited by Barber *et al.* (2005) for further discussion (Fig. 2 and Table S1). Furthermore, 28 samples from western, central and eastern Sumatra were analysed for Lu–Hf isotopes in zircon. Table 1 shows 15 Lu–Hf isotopic data from Gao *et al.* (2022) that were combined with our data.

Zircon U–Pb dating

Zircon grains were extracted from *c.* 2 to 3 kg igneous rock samples using standard density and magnetic techniques, mounted in epoxy and polished to reveal the crystal interiors. Cathodoluminescence (CL) images were captured at the Institute of Earth Sciences, Academia Sinica, in Taipei. The internal structures of individual magmatic zircons can be recognized and suitable positions for U–Pb and Lu–Hf isotope determinations selected based on these CL images (Fig. S1).

In this study, all the zircon U–Pb isotopic analyses were performed using the Agilent 7500 s inductively coupled plasma mass spectrometer (ICP-MS) attached to a Photon Machines Analyte G2 (PM193) laser ablation system at the Department of Geosciences, National Taiwan University, in Taipei. The analysis spots have a diameter of approximately 30 µm and analyses were done at a repetition rate of 4 Hz. The analytical procedures herein followed those reported by Chiu *et al.* (2009) and Shao *et al.* (2014). The calibration was performed using the primary zircon standard GJ-1 (608.5 $\pm$ 0.4 Ma, Jackson *et al.* 2004), and data quality was controlled using two secondary zircon standards 91500 (1065.4 $\pm$ 0.4 Ma, Wiedenbeck *et al.* 1995) and Plešovice (337.1 $\pm$ 0.4 Ma, Sláma *et al.* 2008). The two secondary standards yielded average $^{206}Pb/^{238}U$ ages of 1064 $\pm$ 5 Ma ($n = 86$, MSWD = 0.12, 2σ) and 335.2 $\pm$ 2.4 Ma ($n = 43$, MSWD = 0.35, 2σ). The function suggested by Andersen (2002) was used to correct for common lead. GLITTER 4.4 (Griffin *et al.* 2008) and Isoplot v. 4.15 (Ludwig 2008) were used to calculate the U–Th–Pb isotopic ratios and construct-weighted mean U–Pb ages, concordia plots, probability curves and histograms. The weighted means of $^{206}Pb/^{238}U$ ages are used to represent the ages of the Cenozoic zircons in this study because the $^{207}Pb/^{206}Pb$ ratios are only feasible for ancient zircons (>1.4 Ga) (Gehrels *et al.* 2011).

Zircon Lu–Hf isotope analysis

Lu–Hf isotopic analyses were performed using a Nu Plasma HR multi-collector ICP-MS with a Photon Machines Analyte G2 laser ablation system at the Institute of Earth Sciences, Academia Sinica, in Taipei. The analysing spots are *c.* 50 µm in diameter at a rate of 8 Hz. Detailed descriptions of these analytical techniques can be found in Bikramaditya *et al.* (2018). $^{176}Hf/^{177}Hf$ results of the Mud Tank zircon standard during the analysis herein are 0.282496 $\pm$ 0.000022 (2σ, $n = 19$), which are similar to the values reported by Woodhead and Hergt (2005) for solution analysis (0.282507 $\pm$ 0.000006; $n = 5$) and laser ablation microprobe (LAM) (MC) ICP-MS analysis (0.282504 $\pm$ 0.000044; $n = 158$). The chondritic uniform reservoir was used to calculate the initial $^{176}Hf/^{177}Hf$ ratios and $\varepsilon_{Hf}(t)$ values. We used the decay constant for ^{176}Lu of 1.867×10^{-11} a^{-1} (Söderlund *et al.* 2004), the chondritic $^{176}Hf/^{177}Hf$ ratio of 0.282785 and the $^{176}Lu/^{177}Hf$ ratio of 0.0336 (Bouvier *et al.* 2008). We assumed a $^{176}Lu/^{177}Hf$ ratio of 0.015 for average continental rocks (Griffin *et al.* 2002) and used the reference for $^{176}Hf/^{177}Hf$ ratio of 0.28325 (similar to that of average MORB over 4.56 Ga) and a $^{176}Lu/^{177}Hf$ ratio of 0.0384 (Griffin *et al.* 2000).

Results

In total, 34 new zircon U–Pb ages and 28 Lu–Hf isotopic data were analysed (Table 1). We divided Sumatra Island into four segments for further data presentation and discussion based on the ages, isotopic characteristics and geographical distribution of the sample localities: (1) western Sumatra, samples in the NW of Toba volcano from Banda Aceh to Tapaktuan (Fig. 3); (2) Toba-related area, samples within or near the supervolcano Toba between Medan and Sibolga (Fig. 4); (3) central Sumatra, samples to the south of Toba volcano from Sibolga to southern Padang (Fig. 5) and (4) eastern Sumatra, samples from the southeastern Sumatra Island between Lahat and Bandar Lampung (Fig. 6). Table 1 and Figure 2 show the summarized zircon U–Pb results; Table S2 contains the detailed data. Tables 1 and S3 show the results of the zircon Lu–Hf isotope analysis.

Zircon U–Pb dating

In western Sumatra, one basalt (13SU11) and two granites (13SU13 and 13SU15) were collected near Tapaktuan, yielding Eocene ages ranging from 52.2 to 51.7 Ma (Figs 3 & 7). We combined several zircon U–Pb age results from previous studies in west Sumatra (six samples, Lai *et al.* 2021) and the Toba-related area (eight samples from Liu *et al.* (2021) and five samples from Gao *et al.* (2022)) as well (Table 1 and Fig. S2).

Seven volcanic rocks and 10 plutonic rocks were dated in central Sumatra (including two data samples from Gao *et al.* (2022)) (Table 1 and Fig. 5). Near Padang Sidempuan, one andesite (14SU34-2) and

Table 1. *Zircon U-Pb ages and Hf isotopic results in Sumatra Island*

Sample No.	Longitude	Latitude	Elevation	Rock type	SiO_2 (wt.%)	Ages (Ma)	n[c]	MSWD	$\varepsilon_{Hf}(t)$ value	Remark[a]
Western Sumatra										
13-SU-01	N 05°04.523′	E 95°35.948′	32m	Andesite	53.6	16.5±0.5[b]	16	3.2	+12.8 to +16.1	WS
13-SU-04	N 04°51.191′	E 95°25.317′	25m	Andesite	58.5	20.1±0.3[b]	17	1.2	na[d]	WS
13-SU-11	N 03°21.304′	E 97°07.303′	18m	Basalt	48.2	52.0±1.0	21	3.9	+7.1 to +12.5	WS
13-SU-13	N 03°21.304′	E 97°07.303′	18m	Granite	68.2	52.2±0.7	12	0.51	+7.3 to +19.4	WS
13-SU-15	N 03°21.005′	E 97°07.454′	17m	Granite	67.7	51.7±0.4	20	1.12	na	WS
13-SU-28	N 05°01.142′	E 96°41.929′	366m	Andesite	61.8	<0.3[b, e]	22	na	+7.3 to +11.6	WN
13-SU-31	N 05°01.142′	E 96°41.929′	366m	Andesite	na	<0.3[b]	22	na	+7.4 to +11.7	WN
13-SU-38	N 05°26.971′	E 95°41.901′	772m	Andesite	58.1	<0.3[b]	22	na	+10.4 to +14.0	WN
13-SU-39	N 05°26.971′	E 95°41.901′	772m	Andesite	61.2	<0.3[b]	23	na	+10.1 to +13.7	WN
Toba-related										
14SU03	N 02°55.653′	E 98°31.535′	1446m	Andesite	63.7	<0.3[f]	na	na	−4.7 to −8.1[g]	T
14SU05	N 02°53.044′	E 98°39.716′	1244m	Rhyolite	74.6	0.57±0.02[f]	22	1.11	−6.4 to −10.0[g]	T
14SU06	N 02°49.976′	E 98°45.665′	1404m	Andesite	na	10.4±0.2[f]	19	2.1	−13.1 to +0.5[g]	T
14SU07	N 02°46.750′	E 98°49.176′	1139m	Andesite	58.1	2.68±0.29[g]	6	1.3	−6.5 to −9.2[g]	T
14SU08-1	N 02°40.125′	E 98°56.306′	933m	Diorite	60.4	5.2±0.8[g]	18	0.13	−2.2 to −7.4[g]	T
14SU10	N 02°30.310′	E 99°15.773′	752m	Rhyolite	72.8	0.94±0.04[f]	20	4.5	−7.0 to −4.8[g]	T
14SU11-1	N 02°30.564′	E 99°15.339′	895m	Rhyolite	71.4	1.00±0.08[f]	17	8.2	−7.0 to −4.8[g]	T
14SU12-1	N 02°06.674′	E 98°57.488′	1127m	Rhyolite	70.3	0.96±0.04[f]	20	3.7	−8.0 to −5.6[g]	T
14SU23	N 01°34.099′	E 98°54.904′	48m	Baslat	51.3	1.00±0.08[g]	8	20	−6.0 to −4.4[g]	T
14SU48	N 01°36.892′	E 99°14.478′	985m	Rhyolite	69.9	0.43±0.02[g]	18	4.3	−2.4 to +1.7[g]	T
14SU51	N 02°36.983′	E 98°41.008′	945m	Andesite	65.2	< 0.3[f]	na	na	−9.7 to −5.0[g]	T
14SU52	N 02°33.574′	E 98°38.222′	1371m	Rhyolite	71.4	0.92±0.03[g]	19	1.8	−8.3 to −4.8[g]	T
15SU52	N 02°35.405′	E 98°37.371′	1136m	Rhyolite	72.4	0.99±0.04[f]	17	12	−7.4 to −5.4[g]	T
Central Sumatra										
14SU25	N 01°34.100′	E 98°54.905′	49m	Andesite	60.9	4.06±0.04[g]	16	1.3	+4.5 to +7.9[g]	C
14SU34-1	N 01°11.739′	E 99°22.646′	245m	Diorite	na	11.7±0.2	18	0.36	+12.7 to +17.0	C
14SU34-2	N 01°11.739′	E 99°22.646′	245m	Andesite	60.3	11.3±0.3	18	1.3	+10.4 to +13.6	C
14SU39	N 01°11.691′	E 99°22.651′	246m	Diorite	60.3	13.4±0.2	24	0.9	+5.1 to +7.8	C
14SU47	N 01°31.619′	E 99°17.873′	901m	Andesite	59.3	<0.3[g]	na	na	+1.6 to +4.6[g]	C
15SU08	N 00°01.771′	E 100°44.860′	146m	Diorite	58.1	16.5±0.2	24	1	+4.0 to +8.9	C
15SU14-1	S 00°45.928′	E 100°42.173′	394m	Granite	74.4	55.5±0.5	24	1.3	+12.2 to +14.1	C
15SU17-1	S 01°10.521′	E 100°50.253′	1172m	Diorite	63.7	5.72±0.37	15	0.84	na	C
15SU17-2	S 01°10.521′	E 100°50.253′	1172m	Granite	78.5	5.83±0.20	12	4.1	+4.3 to +13.0, +12.9 to +17.1 (n=7, 13-17Ma)[h]	C
15SU17-3	S 01°10.521′	E 100°50.253′	1172m	Granite	72.2	15.6±0.3	22	0.76	na	C

15SU20	S 01°00.898′	E 100°23.398′	9m	Baslat	na	22.2±0.5	21	1.7	+13.0 to +17.0	C
15SU21	S 01°00.898′	E 100°23.398′	9m	Baslat	50.5	22.6±0.5	24	2	na	C
15SU23-1	S 00°37.643′	E 100°21.552′	77m	Diorite	61.8	50.1±0.5	24	0.91	+12.5 to +16.6	C
15SU23-3	S 00°37.643′	E 100°21.552′	77m	Diorite	60.8	50.2±0.7	24	1.6	+12.7 to +16.0	C
15SU24-1	S 00°33.276′	E 100°20.347′	138m	Andesite	60.1	<0.3	na	na	na	C
15SU30-1	N 00°10.343′	E 099°57.058′	383m	Andesite	57.6	<0.3	na	na	+12.6 to +16.0 (n=5, 0−3.2Ma)	C
15SU32	N 00°13.007′	E 099°21.777′	17m	Diorite	63.8	48.8±0.8	24	1.4	na	C
15SU33	N 00°12.449′	E 099°23.588′	10m	Diorite	63.8	48.9±0.8	24	0.84	na	C
15SU46-3	N 00°40.317′	E 099°24.769′	283m	Andesite	57.5	<0.3	na	na	+4.0 to +10.2 (n=10, 0−0.5Ma)	C
Eastern Sumatra										
16SU09-1	S05°27.149′	E105°22.464′	na	Granite	76.9	59.6±0.8	24	0.84	+7.2 to +9.4	E
16SU09-2	S05°27.149′	E105°22.464′	na	Granite	74.7	60.6±0.8	23	0.88	+7.0 to +9.5	E
16SU12-1	S05°45.907′	E105°41.219′	na	Andesite	na	0.38±0.15	17	1.9	+11.6 to +18.0	E
16SU14-2	S05°49.635′	E105°44.351′	na	Rhyolite	70.1	4.4±0.2	23	4.3	+15.9 to +17.6 (n=2)	E
16SU15-2	S05°49.284′	E105°44.423′	na	Rhyolite	66.6	4.0±0.1	21	1.8	na	E
16SU16-2	S05°28.493′	E105°14.794′	na	Andesite	59.7	13.3±0.1	18	0.5	na	E
16SU18-1	S05°38.289′	E105°11.167′	na	Andesite	63.8	10.7±0.2	21	0.95	+12.4 to +16.3	E
16SU18-2	S05°38.289′	E105°11.167′	na	Rhyolite	68.3	11.5±0.2	19	0.48	+11.9 to +14.6	E
16SU18-3	S05°38.289′	E105°11.167′	na	Rhyolite	69.0	15.7±0.3	23	1.2	na	E
16SU20-2	S05°23.062′	E105°01.421′	na	Rhyolite	70.6	63.8±0.7	16	1.7	+13.0 to +15.4	E
16SU41-5	S04°03.571′	E103°51.686′	na	Andesite	60.1	1.13±0.09	22	2.3	+13.9 to +16.1	E
16SU42-5	S03°51.181′	E103°31.570′	na	Granite	82.4	12.8±0.2	25	0.88	+9.9 to +14.7	E
16SU42-7	S03°51.181′	E103°31.570′	na	Granite	75.5	13.1±0.2	24	0.73	+13.9 to +17.1	E
16SU43	S03°59.570′	E103°24.741′	na	Rhyolite	71.4	<0.3	na	na	na	E

[a] WS: South part of the western Sumatra; WN: North part of the western Sumatra
[b] Lai *et al.* (2021)
[c] zircon numbersof analyses
[d] not determined
[e] ages younger than the detection limit (about 0.3 Ma)
[f] Liu *et al.* (2021)
[g] Gao *et al.* (2022)
[h] inherited zircons
T: Toba-related; C: Central Sumatra; E: Eastern Sumatra

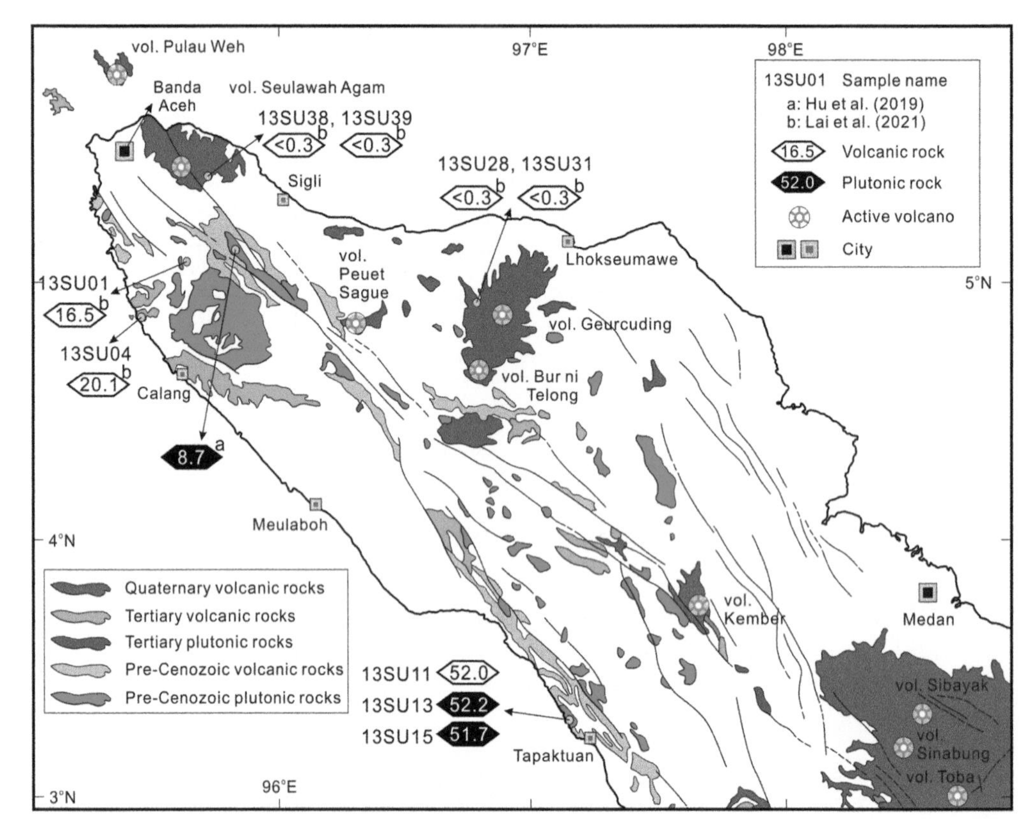

Fig. 3. Zircon U–Pb dating samples from western Sumatra, the symbols are the same as in Figure 2. Source: geological map modified after Gafoer *et al.* (1996).

two dacites (14SU34-1 and 14SU39) yielded Miocene ages ranging from 13.4 to 11.3 Ma (Figs 5 & 7). The andesite from volcano Sorikmerapi (15SU46-3) was too young (less than 0.3 Ma) to yield a mean $^{206}Pb/^{238}U$ age (Fig. 5). Two diorites near Air Bangle showed similar ages of 48.9 Ma (15SU33) and 48.8 Ma (15SU32), respectively (Figs 5 & 8). Zircons of two andesites from volcano Talakmau (15SU30-1) and Tandikat (15SU24-1) showed an age of less than 0.3 Ma (Fig. 5). Moreover, two diorites (15SU23-1 and 15SU23-3) nearby the Tandikat volcano were dated and yielded similar Eocene ages (50.1 and 50.2 Ma, respectively) (Figs 5 & 7). A granite sample (15SU14-1) near Solok also gave an Eocene age of 55.5 Ma (Figs 5 & 7). Two basalts to the south of Padang had enough zircons to yield Miocene ages of 22.6 Ma (15SU21) and 22.2 Ma (15SU20), respectively (Figs 5 & 7). Three plutonic rocks in the SE of Danau Singkarak showed Miocene ages ranging from 15.6 Ma (15SU17-3, a granite) to 5.83 Ma (15SU17-1, a diorite) and 5.72 (15SU17-2, a granite) (Figs 5 & 7).

In eastern Sumatra, 10 volcanic rocks and 4 plutonic rocks were dated (Table 1 and Fig. 6). Two granites near Lahat showed Miocene U–Pb ages of 13.1 Ma (16SU42-7) and 12.8 Ma (16SU42-5) (Figs 6 & 8). Zircons of rhyolite (16SU43) from the west of Airdingin showed an age of less than 0.3 Ma (Figs 6 & 8). Near Pengandonan, an andesite (16SU41-5) yielded a Quaternary age of 1.13 Ma (Figs 6 & 8). Furthermore, a rhyolite sample (16SU20-2) collected from the west of Gedongtataan yielded a Paleocene age of 63.8 Ma (Figs 6 & 8). Two granites from the east of Bandar Lampung showed another two Paleocene ages of 60.6 Ma (16SU09-2) and 59.6 Ma (16SU09-1) (Figs 6 & 8). Three volcanic rocks were dated in the northern Kekatang and yielded the following ages: 15.7 Ma (16SU18-3, rhyolite), 11.5 Ma (16SU18-2, rhyolite) and 10.7 Ma (16SU18-1, andesite) (Figs 6 & 8). In addition, an andesite (16SU16-2) from the south of Bandar Lampung yielded a Miocene age of 13.1 Ma (Figs 6 & 8). From the east of volcano Rajabasa, three volcanic rocks were dated. Two rhyolites among them evinced Pliocene ages of 4.4 Ma

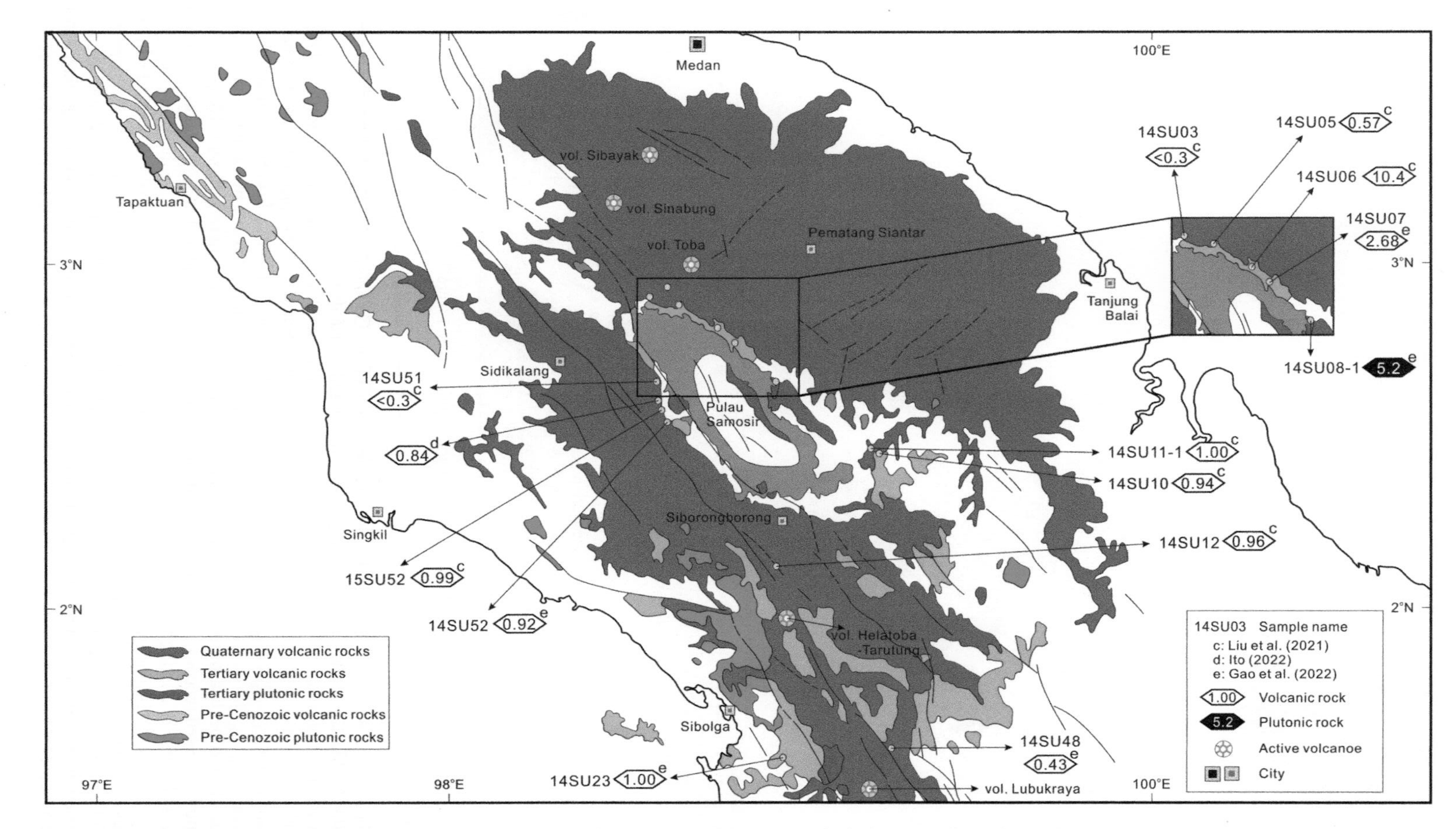

Fig. 4. Zircon U–Pb dating samples from the Toba-related area; symbols as in Figure 2. Source: geological map modified after Gafoer *et al.* (1996).

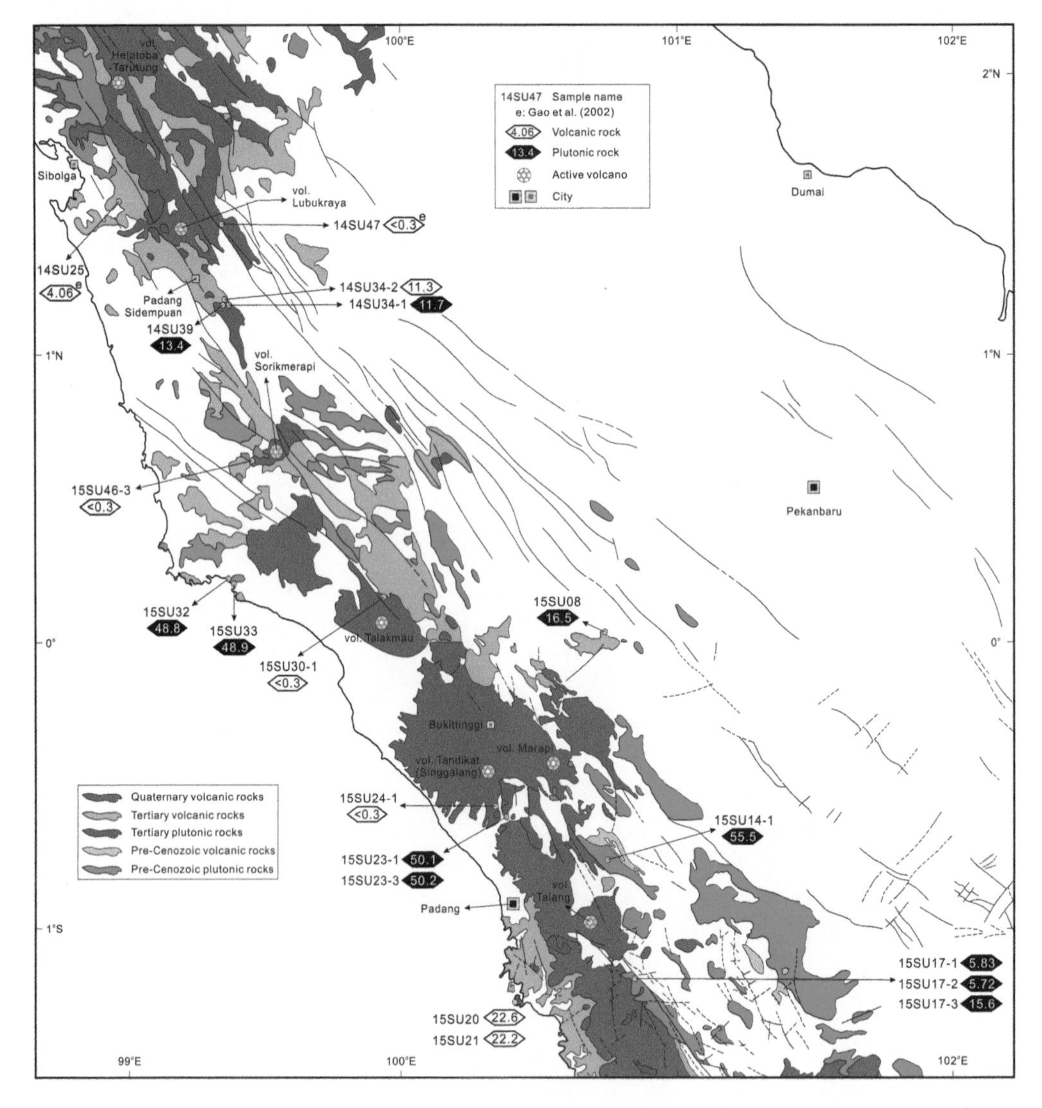

Fig. 5. Zircon U–Pb dating samples from central Sumatra; symbols as in Figure 2. Source: geological map modified after Gafoer *et al.* (1992, 1996).

(16SU14-2) and 4.0 Ma (16SU15-2), and another andesite (16SU12-1) had a Quaternary age of 0.38 Ma (Figs 6 & 8).

Zircon Lu–Hf isotope results

The Lu–Hf isotope data were obtained by analysing 439 grains of dated igneous zircon from 28 samples (Fig. 2 and Table 1). Among the grains dated by Lai *et al.* (2021) and this study are 139 grains from 7 samples in western Sumatra, 163 zircon grains from 11 samples in central Sumatra and the other 137 zircon grains from 10 samples in eastern Sumatra. Table S3 contains the detailed results and some data from Gao *et al.* (2022), all of which are summarized in Table 1.

The samples from the south or north of the SFZ in western Sumatra have $\varepsilon_{Hf}(t)$ values ranging from +19.4 to +7.1 (Table 1). In central Sumatra, the $\varepsilon_{Hf}(t)$ values can be divided into two groups – one group has lower $\varepsilon_{Hf}(t)$ values ranging from +13.6 to +4.0, and another group gave higher $\varepsilon_{Hf}(t)$ values ranging from +17.1 to +12.2; however, neither group was specific to any specific area or time period (Table 1). In eastern Sumatra, excluding two samples (16SU09-1 and 16SU09-2) that showed lower $\varepsilon_{Hf}(t)$ values ranging from +9.5 to +7.0, the samples displayed higher $\varepsilon_{Hf}(t)$ values from +18.0 to +9.9 (Table 1).

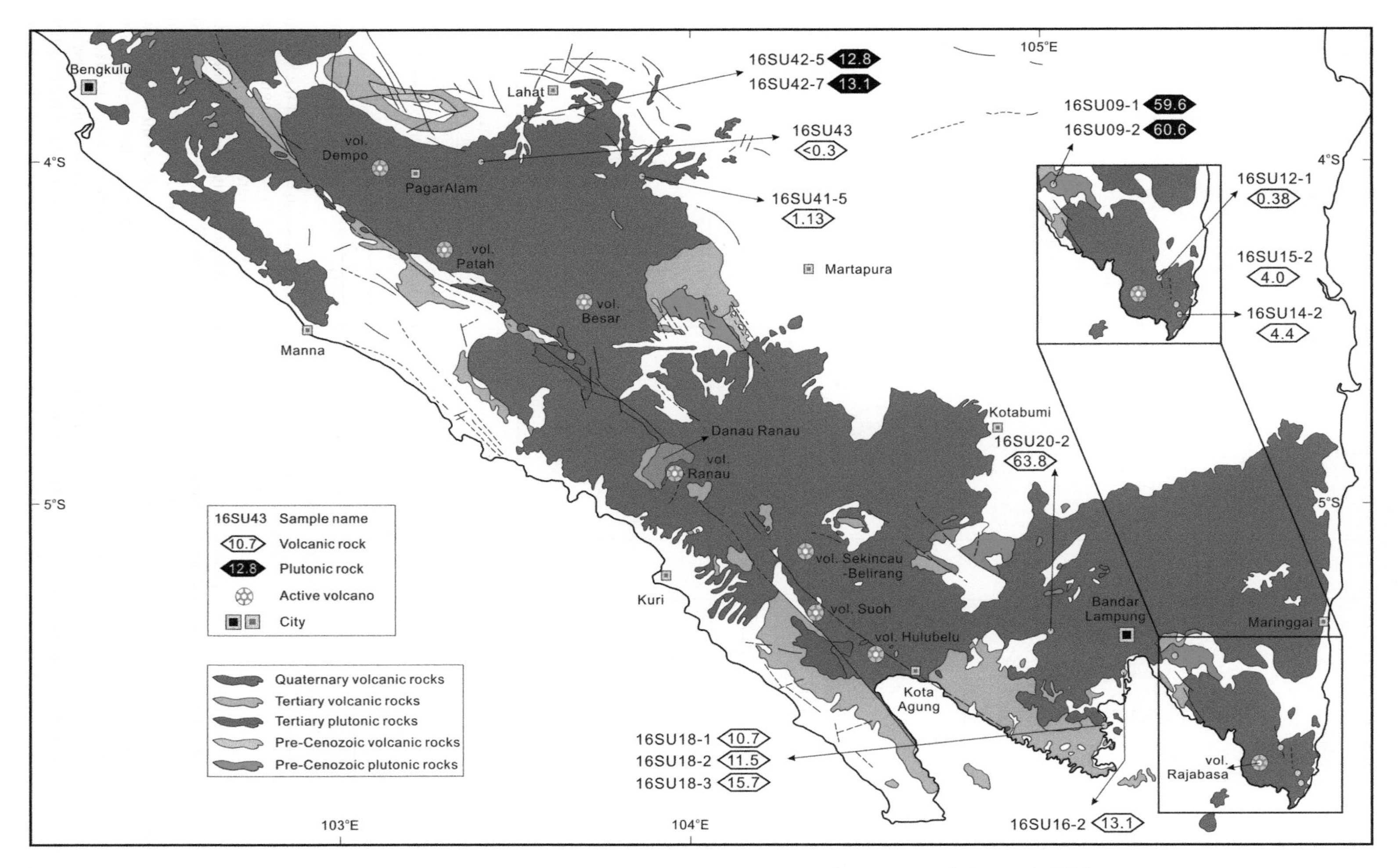

Fig. 6. Zircon U–Pb dating samples from eastern Sumatra; symbols as in Figure 2. Source: geological map modified after Gafoer *et al.* (2012).

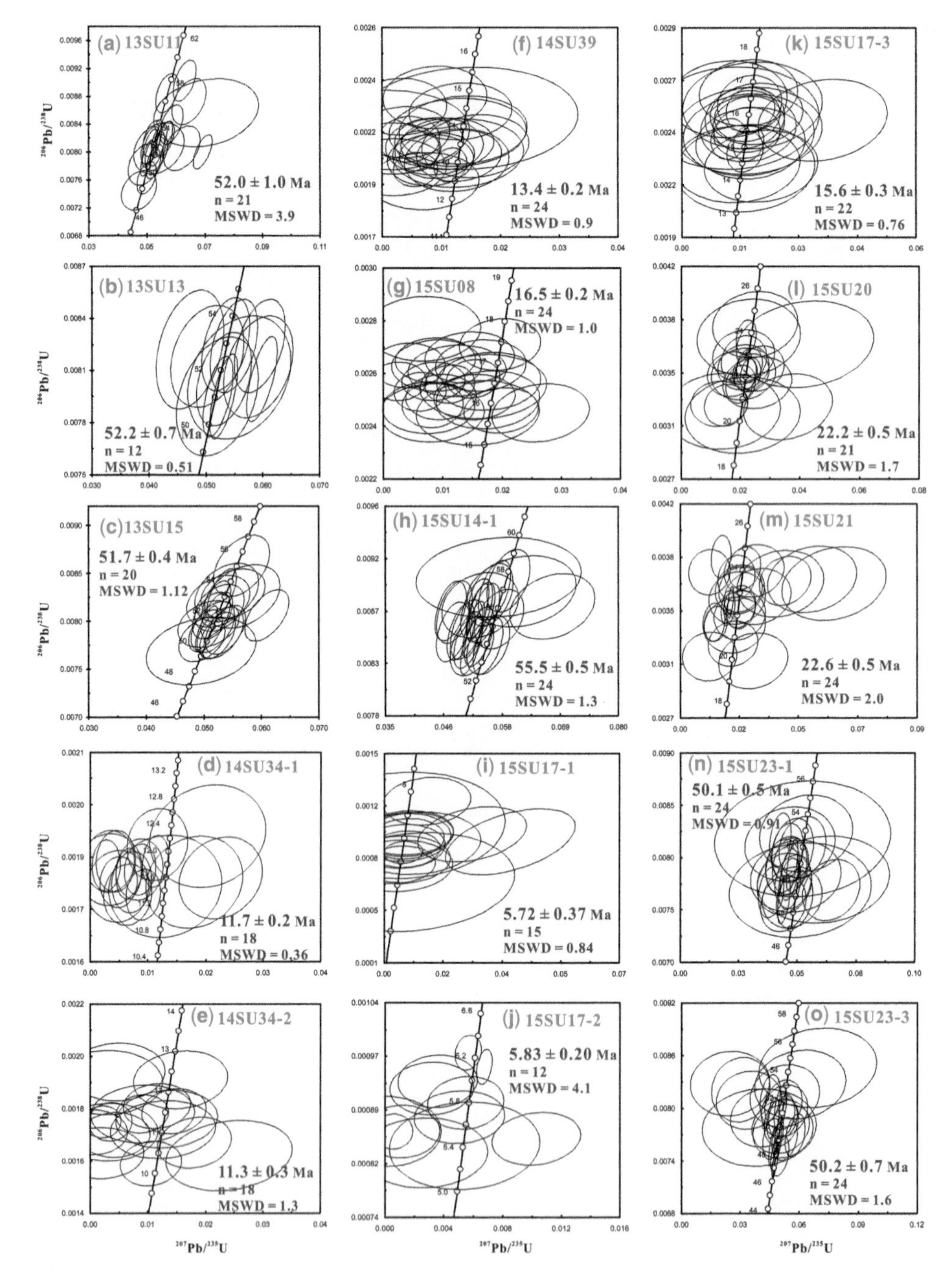

Fig. 7. U–Pb concordia diagrams and age results of igneous rocks; the numeric (*n*) denotes the number of zircon. Sample localities are shown in Figures 3–5 and the detailed age data list in Tables 1 and S2.

Discussion

Cenozoic magmatic stages in Sumatra

Barber *et al.* (2005) gathered over 200 previous dating results of Cenozoic igneous rocks on Sumatra Island, which were dated using the ^{40}K–^{40}Ar, Ar^{40}/Ar^{39} and Rb–Sr methods. Following Barber's work, Crow (2005) classified these ages into six distinct magmatic stages without any specific reason for their division: 65–50, 46–40, 38–24, 22–14, 12–8 and 6–1.6 Ma (Table 2). Several studies have

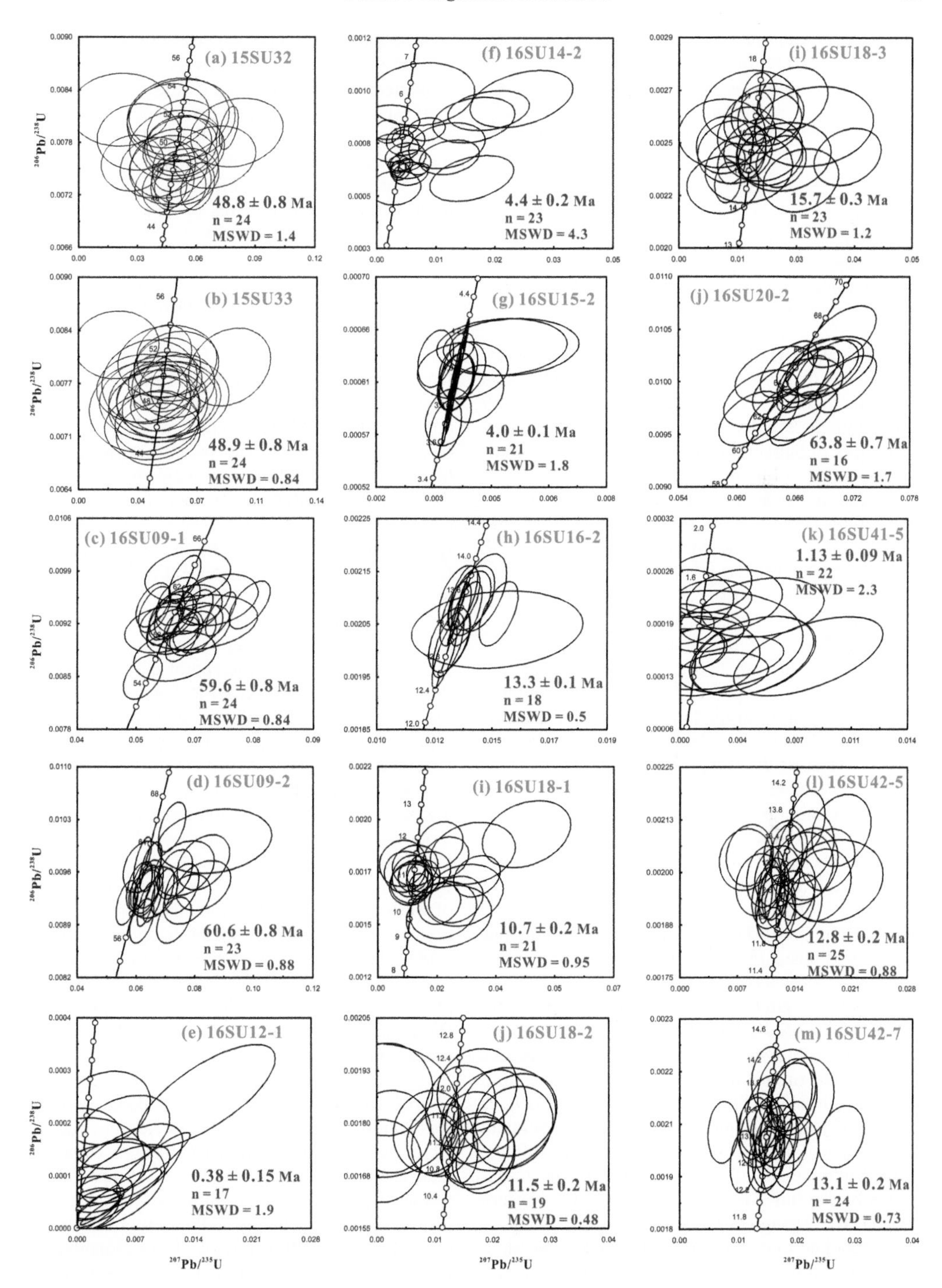

Fig. 8. U–Pb concordia diagrams and age results of igneous rocks; the numeric (n) denotes the number of zircon. Sample localities are shown in Figures 5 and 6 and the detailed age data list in Tables 1 and S2.

recently used zircon U–Pb dating methods for igneous rocks in some specific areas (Hu *et al.* 2019; Ito 2020; Lai *et al.* 2021; Liu *et al.* 2021; Gao *et al.* 2022). To combine all previous ages to better recognize the age spectrum, we selected 157 ages that can be traced to their sample localities in the book that

Table 2. *Magmatism stages proposed by Crow (2005) based on previous dating works*

Stages	Volcanics	Plutonics	Total
6–1.6 Ma	16	10	26
12–8 Ma	1	12	13
22–14 Ma	33	7	40
38–24 Ma	7	3	10
46–40 Ma	3	4	7
65–50 Ma	13	27	40
Total samples	73	63	136

was edited by Barber *et al.* (2005), zircon U–Pb ages from five aforementioned recent studies and 34 new zircon U–Pb ages from this work (Fig. 9 and Table 1). Age distribution of the Cenozoic igneous rocks in Sumatra using various dating methods can be found in Figure 10.

Some analytical biases, such as collecting numerous samples from specific locations, make it easier to show large peaks between 20 and 16 Ma (Fig. 10, ^{40}K–^{40}Ar method). To avoid confusion, Figure 10 excluded 11 samples from this study that were younger than 0.3 Ma. Notably, no ages were obtained from the ^{40}Ar/^{39}Ar, Rb–Sr and zircon U–Pb methods between 48 and 23 Ma. In contrast, there are 18 ages dated using ^{40}K–^{40}Ar methods yielding 48–23 Ma, and we argue that these are not credible owing to weathering or alteration by later magmatic events. Most of the samples which with ^{40}K–^{40}Ar ages between 48 and 23 Ma, can be also found other older ages by using the other dating methods (^{40}Ar/^{39}Ar, Rb–Sr and zircon U–Pb methods) in the same locations, and this can provide strong support for this hypothesis (Fig. 9). Another piece of evidence is the red-coloured sample columns collected from the Air Bangis Granite by both previous works and this study (Fig. 10). Previous ages from the ^{40}K–^{40}Ar method range from 29 to 28 Ma; however, our new zircon U–Pb ages are approximately 49 Ma (Figs 5 & 9). Furthermore, as per our diagram of zircon U–Pb ages v. $\varepsilon_{Hf}(t)$ values (Figs 11 & 12), no inherited zircons for this period can be found in the entire island of Sumatra (Zhang *et al.* 2019). A period of magmatic quiescence existed in Sumatra from 48 to 23 Ma (Fig. 12), consistent with the notion that the Neo-Tethyan subduction was terminated by the India–Eurasia collision in the western Sunda Arc (Zhang *et al.* 2019).

According to the preceding discussion, we identified two distinct magmatic stages during the Cenozoic Era (Table 3 and Fig. 11), namely, Paleocene to Early Eocene (66–48 Ma) and Miocene to Recent (23–0 Ma). Figure 13 shows the reconstructed tectonic model with two magmatic stages, and the map for the older stage is inferred from the slip rate of the SFZ after the Middle Miocene (Genrich *et al.* 2000; Sieh and Natawidjaja 2000). To make the active volcanoes more visible (in particular, those near the Toba area), we divide the second magmatic stage (23–0 Ma) into two graphs: Miocene to Pliocene and Quaternary to Recent. These two magmatic stages are found throughout Sumatra Island; however, some temporal difference remains between these four spatial areas (Figs 11–13), as will be discussed in further detail below.

Temporal and spatial distributions of Cenozoic magmatism in Sumatra

During the first magmatic stage, from 66 to 48 Ma, igneous rocks formed along the coast of southwestern Sumatra Island and erupted in all four regions of Sumatra (Fig. 11). The second magmatic stage began around 23 Ma and coincidentally occurred with the SFZ's initiated movement (*c.* 23 to 15 Ma, Curray *et al.* 1979). Delving into the temporal and spatial variation of the second magmatic stage, different characteristics can be identified between the four regions on Sumatra Island (Figs 11 & 12).

In western Sumatra (Figs 1 & 3), zircon U–Pb ages of igneous rocks show magmatism reignited at *c.* 20 Ma and present a temporal and spatial migration from the southern coast and across the SFZ to the northern portions of this area from 15 Ma to <0.3 Ma (Figs 11 & 13). Lai *et al.* (2021) also observed a migration event between 15 and 10 Ma and suggested that this area was transitioning from the oblique subduction tectonic setting to a dextral motion-governed plate boundary. In Figure 13, there are still some sporadic plutonic rocks with ages ranging from 15 to 8 Ma in the southern part of western Sumatra. Therefore, the temporal migration in this area most likely happened between 8 Ma and <0.3 Ma. It is noteworthy that zircons from a basalt sample show similar ages and Hf isotopic characteristics to granites in this area (Fig. 11). We argue these zircons in basalt are inherited from nearby plutonic rocks.

In Toba-related areas (Figs 1 & 4), zircon U–Pb ages of igneous rocks show that resumed magmatism was continuous from *c.* 13 Ma to Recent (Fig. 11). Here, we can roughly summarize that pre-Toba magmatism during Miocene to Pliocene (Stage 2a) was formed within the overall Sumatra system, and the pre-Toba event in Stage 2b then erupted from 2.2 to 1.2 Ma (Ito 2020; Liu *et al.* 2021). Notably, several inherited zircons from igneous rocks have ages between 58 and 48 Ma (Fig. 11). However, no whole-rock dating results from igneous rocks yielded mean ages in this range (Fig. 13). These results indicate that the inherited zircons were not from main Toba events, and the details will be discussed in the following section.

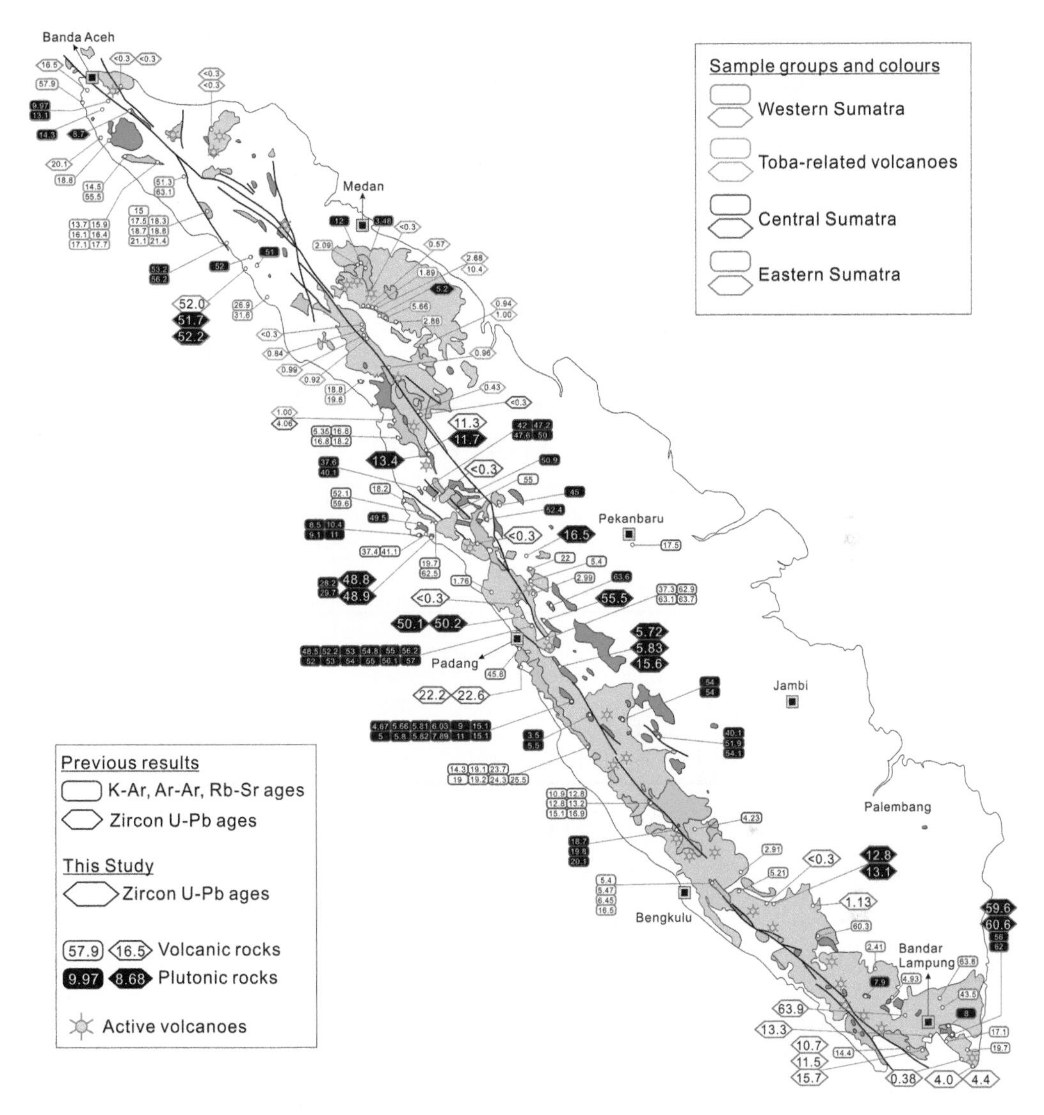

Fig. 9. Age results of igneous rocks from previous data and this study. Larger hexagons show zircon U–Pb ages in this study; smaller ones are data from previous works. Symbols as in Figure 2.

Zircon U–Pb ages of igneous rocks in central Sumatra (Figs 1 & 5) show that revived magmatism continued from *c.* 23 Ma to Recent (Fig. 11). Remarkably, almost all magmatism during 17 to 5 Ma were comprised of plutonic rocks. These results can be attributed to a sampling bias, as there are few fresh volcanic rock outcrops in this area, whereas plutonic outcrops are easier to reach and are better preserved.

In eastern Sumatra (Figs 1 & 6), zircon U–Pb ages of igneous rocks show that magmatism restarted around *c.* 15 Ma, with a significant magmatic gap between 10 and 2 Ma (Fig. 11). However, inspecting the previous age results (Fig. 13), magma erupted continuously from *c.* 20 Ma to Recent without a gap. More evidence is required to prove the differences in zircon U–Pb ages and previous ^{40}K–^{40}Ar ages.

An isotopically juvenile magma source in the mantle wedge along the western Sunda Arc

There have been few isotopic geochemical studies of igneous rocks in Sumatra. Gasparon and Varne (1995) used whole-rock Sr–Nd–Pb isotopic data to

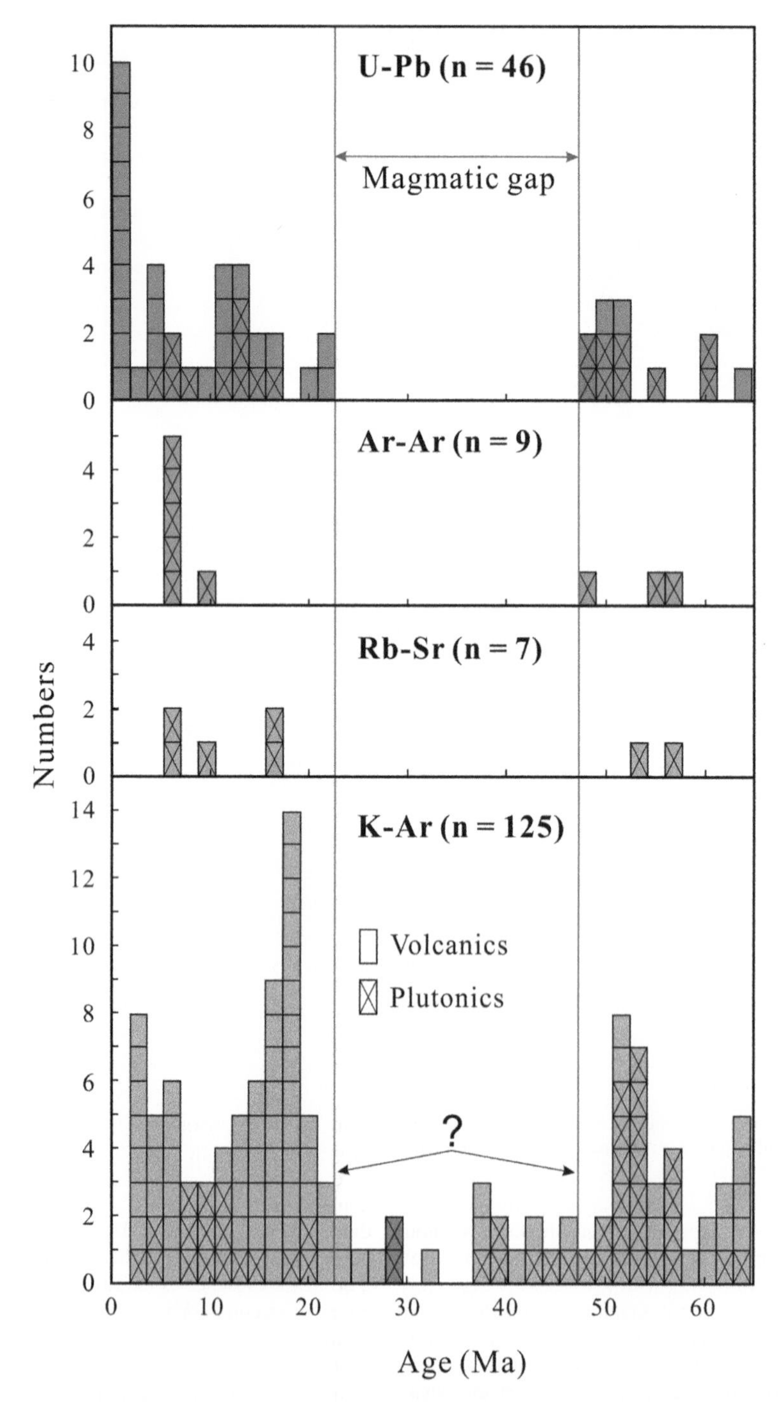

Fig. 10. Histograms of igneous rock ages in Sumatra shown in Figure 9 (*n* is the number of ages analysed in different methods). A magmatic gap during 48–23 Ma can be identified; however, there are still some ^{40}K–^{40}Ar ages that exist in this period, which may be a result of sample weathering or alteration; further discussions can be found in the text. The colours of the columns show different dating methods; the red columns in both U–Pb and ^{40}K–^{40}Ar methods mean samples were collected from the same outcrop but were analysed using two different dating methods.

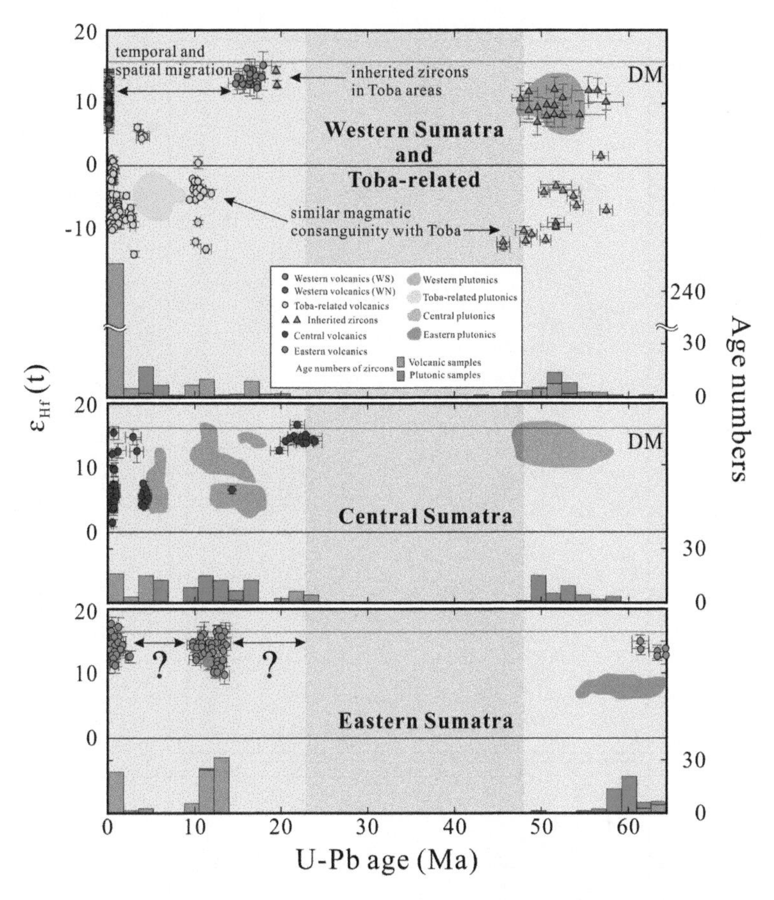

Fig. 11. Diagram of $\varepsilon_{Hf}(t)$ values v. U–Pb ages of magmatic zircons. Four colours indicate different regions of Sumatra. Solid circles show zircons from volcanic rocks; the shaded areas display the data ranges of plutonic rocks. Histograms of volcanic and plutonic rocks are also present here. Note that the age scale is different for samples from the Toba-related area. In addition to the magmatic gap from 48 to 23 Ma, a temporal migration after 15 Ma can also be recognized in western Sumatra. There seem to be two small gaps in eastern Sumatra in the figure. However, some ages within these gaps can be found in previous works using the other dating methods (in Fig. 9), and we use question marks here. Source: data from Lai *et al.* (2021), Liu *et al.* (2021), Gao *et al.* (2022) and this study.

connect the Sumatra granitoids with the Central Granitoid Province in Malaysia. It was a great achievement at that time to acquire multiple isotopes from 15 rock samples; however, the aforementioned authors only focused on the plutonic rocks, without any dating results. Furthermore, Leo *et al.* (1980) used the whole-rock Sr isotopes to study Quaternary volcanic rocks in the Padang area, indicating a crustal source signal. Whitford (1975), Wark *et al.* (2000) and Mucek *et al.* (2017) used whole-rock Sr and Nd isotopic evidence, and preferred crustal materials involved in magmatic processes in the Toba area. In this study, we used zircon Hf isotopes and their U–Pb ages to better recognize the characteristics of igneous rocks in Sumatra (Fig. 11).

According to the zircon Hf isotope data, the magma compositions in western, central and eastern Sumatra have various $\varepsilon_{Hf}(t)$ values (+19.4 to +1.6) (Fig. 11). The $\varepsilon_{Hf}(t)$ ranges in each region are positive and high, e.g. $\varepsilon_{Hf}(t) = +19.4$ to +7.1 in western Sumatra, $\varepsilon_{Hf}(t) = +17.1$ to +1.6 in central Sumatra and $\varepsilon_{Hf}(t) = +18.0$ to +7.0 in eastern Sumatra. Contrasting with the previous study by Leo *et al.* (1980), we propose that an isotopically juvenile magma source exists in the mantle wedge of all of Sumatra (except for Toba-related samples), and this magma source may be the result of two components (the juvenile mantle wedge and old continental crust, i.e. Sundaland) mixing. These compositions can also be traced to the Mesozoic Era since the Neo-Tethys subduction began in the western Sunda Arc (Li *et al.* 2020) (Fig. 12).

Toba-related samples are independent of the general compositions. Except for two zircons with

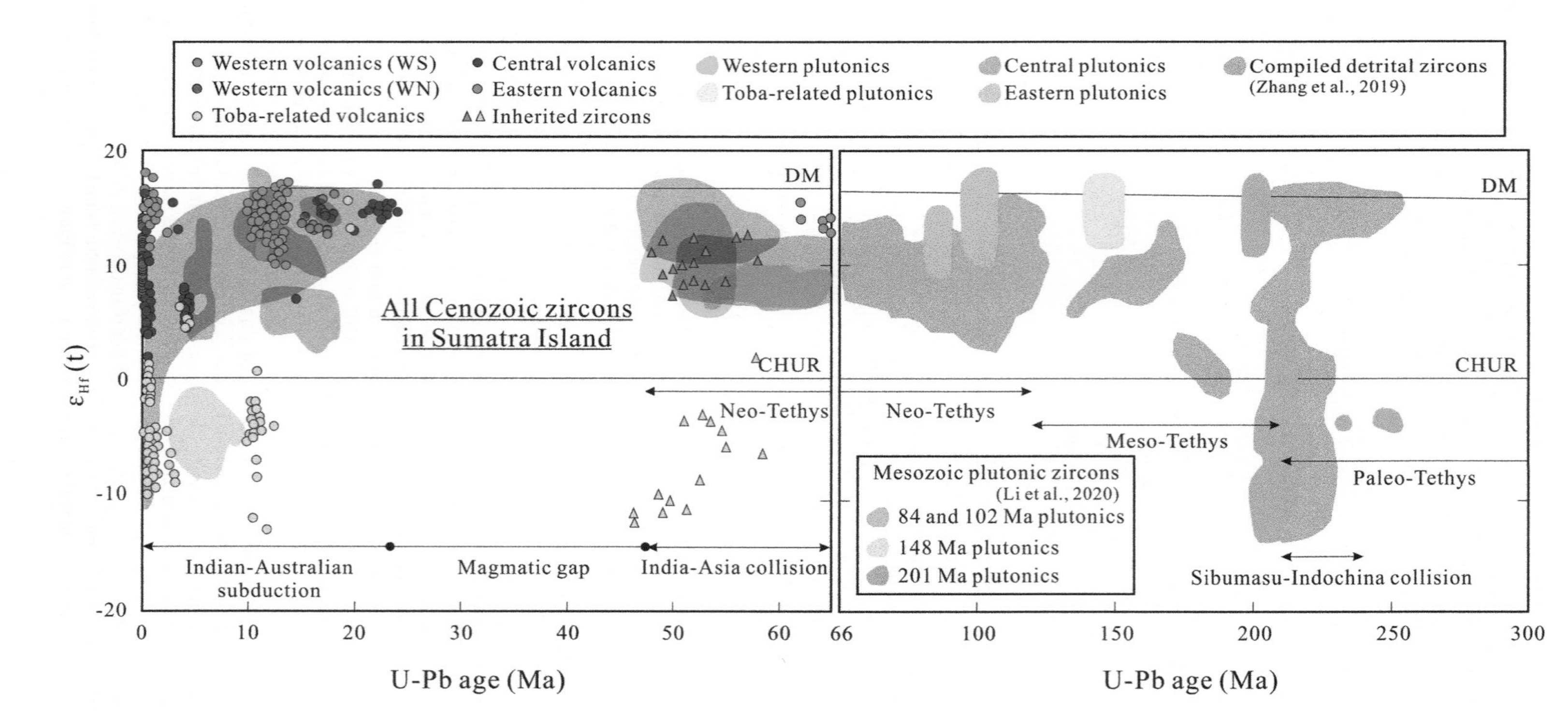

Fig. 12. Plots of $\varepsilon_{Hf}(t)$ values v. U–Pb ages of magmatic zircons since 250 Ma in Sumatra. The tectonic evolution of several subduction stages can be identified in this diagram, e.g. the Meso-Tethys, Neo-Tethys and Indian–Australian subductions. Source: Mesozoic plutonic ages and detrital zircon ages are from Li *et al.* (2020) and Zhang *et al.* (2019), respectively.

Table 3. *Total number of dated igneous rocks in Sumatra Island*

Stages		All dating results			Previous research			This study		
		Volcanics	Plutonics	Total	Volcanics	Plutonics	Total	Volcanics	Plutonics	Total
Miocene to Recent	23–0 Ma	84	37	121	69	29	98	15	8	23
Paleocene to Early Eocene	66–48 Ma	15	36	51	13	27	40	2	9	11
	Total	99	73	172	82	56	138	17	17	34

a high positive $\varepsilon_{Hf}(t)$ value ($> +10$) that could be explained as inherited zircons caught by later magmatism from western Sumatra (Fig. 11), samples from the Toba-related area show abnormally low ε_{Hf} values (Fig. 11, $\varepsilon_{Hf}(t) = +0.5$ to -13.1). Similarly, the other whole-rock isotopic studies of Sr or Nd isotopes also show evident high $^{87}Sr/^{86}Sr$ ratio and low ε_{Nd} values in this area (Whitford 1975; Wark *et al.* 2000; Mucek *et al.* 2017). One possible source is from the same Hf isotope signatures as 214–201 Ma plutonic rocks in Sumatra (Fig. 12). Li *et al.* (2020) suggested that the negative Hf signatures during 214–201 Ma could be explained by two components mixing: the juvenile crust formed from the Meso-Tethys subduction, and the S-type granites from the Malay Peninsula. These Triassic plutonic rocks are not only exposed in Sibolga area, which is near the Toba-related areas, but also exposed in central Sumatra, like Panyambungan and Maura Soma areas (Li *et al.* 2020). Our negative Hf signatures in the Toba-related area during 60–50 Ma and 10–0 Ma could be obtained by either partial melting of these Triassic plutonic rocks, or the contamination between them and Toba-related magmas. However, it is hard to explain the absence of 60–50 Ma igneous rocks in the Toba-related area and these ages only can be found from the inherited zircons. Moreover, this assumption also cannot explain why the Triassic plutonic rocks were merely partial melting in the Toba-related area rather than also occurring in central Sumatra. Crow (2005) and Gasparon (2005) suggest that the subduction of the Investigator Fracture Zone under the Toba area could be a reason for sediment input to the magma source. However, our recent study of the U–Pb–Hf–O isotope in Toba-related volcanic samples also shows a higher $^{87}Sr/^{86}Sr$ ratio (0.71377 to 0.70906) and lower ε_{Nd} (-6.6 to -10.2) than the other samples (Gao *et al.* 2022). This implies that another component is required in the magmatic processes beneath the Toba area to make the magma composition different from the general ones in Sumatra. Gao *et al.* (2022) proposed a notion that magma generation in this particular region involves a significant contribution by subducted sediment from the Himalayas (the formation of S-type granitic rocks) to the source and can be traced to the early Cenozoic Era (58–48 Ma, in inherited zircons) (Fig. 11).

Progressive subduction re-initiation from eastern to western Sunda Arc during the Cenozoic Era

The zircon $\varepsilon_{Hf}(t)$ value v. U–Pb ages on Sumatra Island after 250 Ma can be found in Figure 12. This graph combines the Mesozoic magmatic zircons of granites (Li *et al.* 2020), detrital zircons (Zhang *et al.* 2019) and the results of this study. Magmatism on Sumatra can be traced back to the Jurassic period, between the Palaeo- and Meso-Tethys subduction events and then through Neo-Tethys subduction from the Cretaceous to the Paleogene (Fig. 12).

The first stage of magmatism in the Cenozoic Era (66–48 Ma) can be observed in all of Sumatra (Figs 12 & 13). The India–Asia collision occurred *c.* 60 to 50 Ma, with the demise of the Neo-Tethys Ocean (Rowley 1996; Najman *et al.* 2017; Zheng and Wu 2018). The rate of north-directed drift of the Australian continent decreased dramatically between *c.* 53 and 50 Ma (Sun *et al.* 2020) and caused Sumatra magmatism to halt after 48 Ma (Fig. 12). This decrease and halt in spreading also took place in the Wharton Ridge during 43 Ma in southern Sumatra (Whittaker *et al.* 2007).

After a short period of quiescence of 10 Ma, the Australian continent began moving northward continuously at *c.* 40 Ma (Sun *et al.* 2020) and reactivated the Indian–Australian subduction zone from India to Banda. New magmatism formed in the eastern Sunda Arc in Java Island at *c.* 34 to 32 Ma (Soeria-Atmadja and Noeradi 2005), followed by magmatism in the western Sunda Arc in Sumatra Island after 23 Ma (Figs 12 & 13). Thus, a progressive subduction re-initiation from east (Java Island) to west (Sumatra Island) in the Sunda Arc from the Late Eocene to Early Miocene can be recognized.

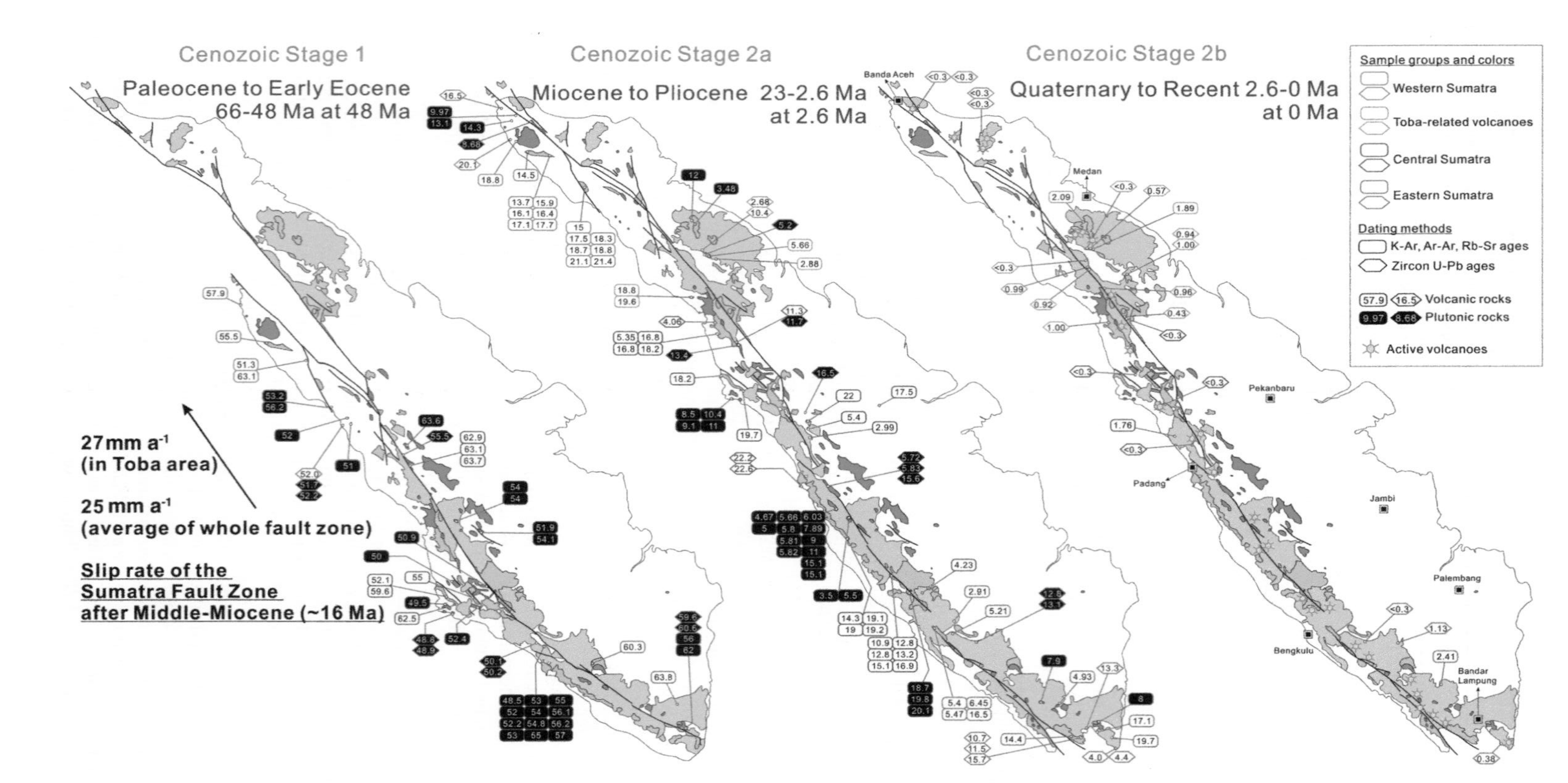

Fig. 13. The reconstructed schematic models of two magmatic stages in the Cenozoic Era from Sumatra. The magma erupted from 66 to 48 Ma in the first magmatic stage along the coast of Sumatra parallel to the Sunda Trench. Following a short rest period of 10 Ma (from 48 to 23 Ma), the SFZ moved with the slip rate of *c.* 25 mm a^{-1} to form the recent Sumatra Island; simultaneously, the second magmatic stage erupted from 23 Ma to Recent along the SFZ in the entire Sumatra Island.

Conclusions

In this study, we performed U–Pb dating and Hf isotopic analysis of magmatic zircons from igneous rocks distributed across the entire island of Sumatra. According to the U–Pb zircon ages and data from previous studies, two distinct Cenozoic Era magmatic stages can be identified in Sumatra (the western Sunda Arc). The first magmatic stage is found throughout the island and ceased at *c.* 48 Ma, following a period of magmatic quiescence from 48 to 23 Ma. The second magmatic stage is first seen at *c.* 20 Ma on the southern coast of western Sumatra and migrated to the north of the SFZ after 10 Ma. The second magmatic stage started in the Toba-related area after 13 Ma and continued to Recent. In central and eastern Sumatra, the second magmatic stage started at *c.* 23 and *c.* 20 Ma, respectively. Combining our data and the age data from the literature, we see that the Sunda Arc subduction recommenced gradually from the east of Java Island to the west of Sumatra Island after the Late Eocene.

The zircon Hf isotopes show the incomplete similarity in magmatic characteristics of samples from the segments of Sumatra. Magmatic compositions in western and eastern Sumatra have shown generally higher $\varepsilon_{Hf}(t)$ values than other areas from +19.4 to +7.1 and +18.0 to +7.0, respectively. The magmatic feature is slightly enriched in central Sumatra, with $\varepsilon_{Hf}(t)$ values ranging from +17.1 to +1.6. These samples indicate an isotopically juvenile magma source in the mantle wedge, formed by the juvenile mantle wedge and the old Sundaland continental crust. However, the enriched isotopic ratios in Toba-related samples ($\varepsilon_{Hf}(t) = +0.5$ to -13.1) show a different petrogenesis, which might result from the contribution of subducted sediment to the source.

Acknowledgements We would like to thank C.-H. Hung, Allie Honda, Y.-J. Hsin and C.-H. Chen for helping with the analyses, W.-Y. Hsia for drawing the geological map and T.-T. Lo for her miscellaneous assistance. We would like to thank Andrew Mitchell and an anonymous reviewer for their considerably helpful and constructive comments and M. Guitreau for his efficient editorial handing of this paper.

Competing interests The authors declare that they have no known competing financial interests or personal relationships that could have appeared to influence the work reported in this paper.

Author contributions **Y-ML**: conceptualization (lead), data curation (lead), formal analysis (lead), funding acquisition (lead), investigation (lead), project administration (lead), resources (equal), supervision (lead), validation (lead), visualization (lead), writing – original draft (lead), writing – review & editing (lead); **P-PL**: data curation (equal), formal analysis (supporting); **S-LC**: conceptualization (equal), funding acquisition (equal), project administration (lead), resources (lead), visualization (equal), writing – original draft (supporting), writing – review & editing (supporting); **AAG**: investigation (supporting), project administration (supporting); **H-YL**: methodology (lead); **LXQ**: formal analysis (supporting), investigation (supporting), writing – original draft (supporting); **SL**: formal analysis (supporting), investigation (supporting); **MHR**: investigation (supporting); **SM**: investigation (supporting); **LL**: formal analysis (supporting); **YI**: methodology (equal).

Funding The Ministry of Science and Technology (MOST) of Taiwan provided financial support for this study (109-2116-M-003-005 and 110-2116-M-003-008 to Y.-M. Lai).

Data availability All data generated or analysed during this study are included in this published article (and its supplementary information files).

References

Amin, T.C., Sidarto, S., Santoso, S. and Gunawan, W. 1994. *Geological Map of the Kotaagung Quadrangle, Sumatera 1:250.000*. Center for Geological Research and Development. Department of Mines and Energy – Indonesia, Bandung.

Andersen, T. 2002. Correction of common lead in U–Pb analyses that do not report ^{204}Pb. *Chemical Geology*, **192**, 59–79, https://doi.org/10.1016/s0009-2541(02)00195-x

Barber, A.J. and Crow, M.J. 2003. An evaluation of Plate Tectonic models for the development of Sumatra. *Gondwana Research*, **20**, 1–28, https://doi.org/10.1016/S1342-937X(05)70642-0

Barber, A.J., Crow, M.J. and Milsom, J.S. (eds) 2005. Sumatra: geology, resources and tectonic evolution. *Geological Society, London, Memoirs*, **31**, 290, https://doi.org/10.1144/GSL.MEM.2005.031

Bellon, H., Maury, R.C., Sutanto Soeria-Atmadja, R., Cotten, J. and Polve, M. 2004. 65 m.y.-long magmatic activity in Sumatra (Indonesia), from Paleocene to Present. *Bulletin de la Societe Geologique de France*, **175**, 61–72, https://doi.org/10.2113/175.1.61

Bennett, J.D., McC, D. *et al.* 1981. *Geologic Map 1:250,000 of Banda Aceh Quadrangle, Sumatra*. Geological Research and Development Centre, Bandung.

Bikramaditya, R.K., Singh, A.K., Chung, S.L., Sharma, R. and Lee, H.Y. 2018. Zircon U–Pb ages and Lu–Hf isotopes of metagranitoids from the Subansiri region, Eastern Himalaya: implications for crustal evolution along the northern Indian passive margin in the early Paleozoic. *Geological Society, London, Special Publications*, **481**, 299–318, https://doi.org/10.1144/SP481.7

Bouvier, A., Vervoort, J.D. and Patchett, P.J. 2008. The Lu–Hf and Sm–Nd isotopic composition of CHUR: constraints from unequilibrated chondrites and implications for the bulk composition of terrestrial planets. *Earth and Planetary Science Letters*, **273**, 48, https://doi.org/10.1016/j.epsl.2008.06.010

Bunopas, S. 1982. *Palaeogeographic history of Western Thailand and adjacent parts of Southeast Asia*. Geological Survey of Thailand Paper 5, Royal Thai Department of Mineral Resources, Thailand.

Cameron, N.R., Clarke, M.C.G., Aldiss, D.T., Aspden, J.A. and Djunuddin, A. 1980. The geological evolution of North Sumatra. *Proceedings of the Indonesian Petroleum Association, Annual Convention, Jakarta, Indonesia*, **9**, 149–187.

Cameron, N.R., Aspden, J.A. *et al.* 1982. *The Geology of the Medan Quadrangle, Sumatra*. Geological Research and Development Centre, Bandung, Indonesia.

Chesner, C.A. 1998. Petrogenesis of the Toba Tufts, Sumatra, Indonesia. *Journal of Petrology*, **39**, 397–438, https://doi.org/10.1093/petroj/39.3.397

Chesner, C.A. 2012. The Toba caldera complex. *Quaternary International*, **258**, 5–18, https://doi.org/10.1016/j.quaint.2011.09.025

Chesner, C.A., Rose, W.I., Deino, A., Drake, R. and Westgate, J.A. 1991. Eruptive history of earth's largest quaternary Caldera (Toba, Indonesia) Clarified. *Geology*, **19**, 200–203, https://doi.org/10.1130/0091-7613(1991)019<0200:EHOESL>2.3.CO;2

Chiu, H.Y., Chung, S.L. *et al.* 2009. Zircon U–Pb and Hf isotopic constraints from eastern Transhimalayan batholiths on the precollisional magmatic and tectonic evolution in southern Tibet. *Tectonophysics*, **477**, 3–19, https://doi.org/10.1016/j.tecto.2009.02.034

Crow, M.J. 2005. Tertiary volcanicity. *Geological Society, London, Memoirs*, **31**, 98–119, https://doi.org/10.1144/GSL.MEM.2005.031.01.08

Curray, J.R. 1989. The Sunda Arc: a model for oblique plate convergence. *Netherlands Journal of Sea Research*, **24**, 131–140, https://doi.org/10.1016/0077-7579(89)90144-0

Curray, J.R. 2005. Tectonics and history of the Andaman Sea region. *Journal of Asian Earth Sciences*, **25**, 187–232, https://doi.org/10.1016/j.jseaes.2004.09.001

Curray, J.R., Moore, D.G., Lawver, L.A., Emmel, F.J., Raitt, R.W., Henry, M. and Kieckhefer, R. 1979. Tectonics of the Andaman Sea and Burma. *American Association of Petroleum Geologists, Memoirs*, **29**, 189–198.

De Coster, G.L. 1974. The geology of the Central and South Sumatra Basins. *Proceedings Indonesian Petroleum Association, 3rd Annual Convention*, Jakarta, 77–110.

Diehl, J.F., Onstott, T.C., Chesner, C.A. and Knight, M.D. 1987. No short reversals of Brunhes age recorded in the Toba tuffs, north Sumatra, Indonesia. *Geophysical Research Letters*, **14**, 753–756, https://doi.org/10.1029/GL014i007p00753

Eubank, R.T. and Makki, A.C. 1981. Structural Geology of the Central Sumatra Back-Arc Basin. *Indonesian Petroleum Association 10th Annual Convention*. https://doi.org/10.29118/IPA.203.153.196

Gafoer, S., Hermanto, B. and Amin, T.C. 1992. *Geological map of Indonesia, Padang sheet (Scale 1:1 000000)*. Directorate General of Geology and Mineral Resources, Geological Research and Development Centre, Bandung.

Gafoer, S., Amin, T.C. and Pardede, R. 1994. *Geological Map of the Baturaja Quadrangle, Sumatera 1:250.000*. Center for Geological Research and Development. Department of Mines and Energy, Indonesia, Bandung.

Gafoer, S., Amin, T.C. and Samodra, H. 1996. *Geologic Map of Indonesia, Medan Sheet, Scale 1:250.000*. Directorate General of Geology and Mineral Resources, Geological Research and Development Centre.

Gafoer, S., Amin, T.C. and Setyogroho, B. 2012. *Geologic Map of Indonesia, Palembang Sheet, Scale 1:1.000.000*. Ministry of Energy and Mineral Resources, Centre for Geological Survey.

Gao, M.H., Liu, P.P. *et al.* 2022. Himalayan zircons resurface in Sumatran arc volcanism through sediment recycling. *Communications Earth & Environment*, **3**, 283, https://doi.org/10.1038/s43247-022-00611-6

Gasparon, M. 2005. Quaternary volcanicity. *Geological Society, London, Memoirs*, **31**, 120–130, https://doi.org/10.1144/GSL.MEM.2005.031.01.09

Gasparon, M. and Varne, R. 1995. Sumatran granitoids and their relationship to Southeast Asian terranes. *Tectonophysics*, **251**, 277–299, https://doi.org/10.1016/0040-1951(95)00083-6

Gasparon, M., Hilton, D.R. and Varne, R. 1994. Crustal contamination processes traced by helium isotopes: examples from the Sunda Arc, Indonesia. *Earth and Planetary Science Letters*, **126**, 15–22, https://doi.org/10.1016/0012-821X(94)90239-9

Gehrels, G., Kapp, P. *et al.* 2011. Detrital zircon geochronology of pre-Tertiary strata in the Tibetan–Himalayan orogen. *Tectonics*, **30**, TC5016, https://doi.org/10.1029/2011TC002868

Genrich, J.F., Bock, Y. *et al.* 2000. Distribution of slip at the northern Sumatran fault system. *Journal of Geophysical Research: Solid Earth*, **105**, 28327–28341, https://doi.org/10.1029/2000JB900158

Griffin, W.L., Pearson, N.J., Belousova, E.A., Jackson, S.E., Achterbergh, E.V., O'Reilly, S.Y. and Shee, S.R. 2000. The Hf isotope composition of cratonic mantle: LAM-MC-ICPMC analysis of zircon megacrysts in kimberlites. *Geochimica et Cosmochimica Acta*, **64**, 133–147, https://doi.org/10.1016/S0016-7037(99)00343-9

Griffin, W.L., Wang, X., Jackson, S.E., Pearson, N.J., O'Reilly, S.Y., Xu, X.S., Zhou, X.M. 2002. Zircon chemistry and magma mixing, SE China: In-situ analysis of Hf isotopes, Tonglu and Pingtan igneous complexes. *Lithos*, 61(3–4), 237–269, https://doi.org/10.1016/S0024-4937(02)00082-8

Griffin, W.L., Powell, W.J., Pearson, N.J. and O'Reilly, S.Y. 2008. GLITTER: Data reduction software for laser ablation ICP-MS. *In*: Sylvester, P. (ed.) *Laser Ablation ICP-MS in the Earth Sciences: Current Practices and Outstanding Issues*. Mineralogical Association of Canada, Short Course Series, **40**, 308–311.

Hall, R. 2009. Hydrocarbon basins in SE Asia: understanding why they are there. *Petroleum Geoscience*, **15**, 131–146, https://doi.org/10.1144/1354-079309-830

Hall, R. 2012. Late Jurassic–Cenozoic reconstructions of the Indonesian region and the Indian Ocean. *Tectonophysics*, **570–571**, 1–41, https://doi.org/10.1016/j.tecto.2012.04.021

Hamilton, W. 1979. Tectonics of the Indonesian region. *US Geological Survey Professional Paper*, **1078**, 345, https://doi.org/10.3133/pp1078

Hamilton, W.B. 1988. Plate tectonics and island arcs. *Geological Society of America Bulletin*, **100**, 1503–1527, https://doi.org/10.1130/0016-7606(1988)100<1503:PTAIA>2.3.CO;2

Hehuwat, F. 1976. Isotopic age determinations in Indonesia: the state of the art. *Proceedings of the Seminar on Isotopic Dating*, May 1975, CCOP, UNDP, Bangkok, Thailand, 135–157.

Hu, P., Cao, L., Zhang, H., Yang, Q., Armin, T. and Cheng, X. 2019. Late Miocene adakites associated with the Tangse porphyry Cu–Mo deposit within the Sunda arc, north Sumatra, Indonesia. *Ore Geology Reviews*, **111**, 102983, https://doi.org/10.1016/j.oregeorev.2019.102983

Hutchison, C.S. 1994. Gondwana and Cathaysian blocks, Palaeotethys sutures and Cenozoic tectonics in South-east Asia. *In*: Giese, P. and Behrmann, J. (eds) *Active Continental Margins – Present and Past*. Springer, Berlin, **83**, 388–405, https://doi.org/10.1007/978-3-662-38521-0_14

Imtihanah. 2000. *Isotopic dating of igneous sequences of the Sumatra Fault System*. MPhil thesis, London University.

Ito, H. 2020. Magmatic history of the Oldest Toba Tuff inferred from zircon U–Pb geochronology. *Scientific Reports*, **10**, 17506, https://doi.org/10.1038/s41598-020-74512-z

Jackson, S.E., Pearson, N.J., Griffin, W.L. and Belousova, E.A. 2004. The application of laser ablation-inductively coupled plasma-mass spectrometry to in situ U–Pb zircon geochronology. *Chemical Geology*, **211**, 47–69, https://doi.org/10.1016/j.chemgeo.2004.06.017

JICA 1988. *Report on the Cooperative Mineral Exploration of Southern Sumatra, Consolidated Report*. Japan International Cooperation Agency, Metal Mining Agency of Japan.

Kallagher, H.J. 1989. *The structural and stratigraphic evolution of the Sunda Forearc Basin, North Sumatra Indonesia*. PhD thesis, University of London.

Kallagher, H.J. 1990. K–Ar dating of selected igneous samples from the Sibolga Basin, Meulaboh and Simeulue Island, western Sumatra. *Lemigas Scientific Contributions on Petroleum Science and Technology, Special Issue*, **13**(1), 99–111.

Kanao, N. *et al.* 1971. Summary report on the survey of Sumatra. *Block No. 5. Japanese Overseas Mineral Development Company Ltd., Bull. N.I.G.M.* **2**, 29–31.

Katili, J.A. and Hehuwat, F. 1967. On the occurrence of large transcurrent faults in Sumatra, Indonesia. *Journal of Geoscience, Osaka City University*, **10**, 5–17.

Koning, T. and Aulia, K. 1985. Petroleum geology of the Ombilin intermontane basin, West Sumatra. *Indonesian Petroleum Association, Proceedings 14th Convention*, Jakarta, **1**, 117–131.

Kopp, H., Weinrebe, W. *et al.* 2008. Lower slope morphology of the Sumatra trench system. *Basin Research*, **20**, 519–529, https://doi.org/10.1111/j.1365-2117.2008.00381.x

Kusnama, Pardede, P., Mangga, S.A. and Sidarto, 1993. *Geological map of the Sungaipenuh and Ketaun quadrangle, Sumatra (1: 250,000)*. Geological Research and Development Centre, Indonesia.

Lai, Y.M., Chung, S.L., Ghani, A.A., Murtadha, S., Lee, H.Y. and Chu, M.F. 2021. Mid-Miocene volcanic migration in the westernmost Sunda arc induced by India-Eurasia collision. *Geology*, **49**, 713–717, https://doi.org/10.1130/G48568.1

Lange, D., Tilmann, F. *et al.* 2010. The fine structure of the subducted investigator ridge in western Sumatra as seen by local seismicity. *Earth and Planetary Science Letters*, **298**, 47–56, https://doi.org/10.1016/j.epsl.2010.07.020

Leo, G.W., Hedge, C.E. and Marvin, R.F. 1980. Geochemistry, strontium isotope data, and potassium–argon ages of the andesite–rhyolite association in the Padang area, West Sumatra. *Journal of Volcanological and Geothermal Research*, **7**, 139–156, https://doi.org/10.1016/0377-0273(80)90024-4

Li, S., Chung, S.L., Lai, Y.M., Ghani, A.A., Lee, H.Y. and Murtadha, S. 2020. Mesozoic juvenile crustal formation in the easternmost Tethys: Zircon Hf isotopic evidence from Sumatran granitoids, Indonesia. *Geology*, **48**, 1002–1005, https://doi.org/10.1130/G47304.1

Liu, P.P., Caricchi, L. *et al.* 2021. Growth and thermal maturation of the Toba magma reservoir. *Proceedings of the National Academy of Sciences of the United States of America*, **118**, e2101695118, https://doi.org/10.1073/pnas.2101695118

Ludwig, K.R. 2008. *User's Manual for Isoplot 4.15*. A Geochronological Toolkit for Microsoft Excel, Berkeley Geochronology Center, Special Publications, Berkeley, CA.

Mark, D.F., Petraglia, M. *et al.* 2014. A high-precision $^{40}Ar/^{39}Ar$ age for the Young Toba Tuff and dating of ultra-distal tephra: forcing of Quaternary climate and implications for hominin occupation of India. *Quaternary Geochronology*, **21**, 90–103, https://doi.org/10.1016/j.quageo.2012.12.004

McCourt, W.J. and Cobbing, E.J. 1993. *The Geochemistry, Geochronology and Tectonic Setting of Granitoid Rocks from Southern Sumatra, Western Indonesia. Southern Sumatra Geological and Mineral Exploration Project. Project Report Series, 9*. Directorate of Mineral Resources/Geological Research and Development Centre, Bandung, Indonesia.

Metcalfe, I. 1996. Pre-Cretaceous evolution of SE Asian terrane. *Geological Society, London, Special Publications*, **106**, 97–122, https://doi.org/10.1144/GSL.SP.1996.106.01.09

Metcalfe, I. 2002. Permian tectonic framework and palaeogeography of SE Asia. *Journal of Asian Earth Sciences*, **20**, 551–566, https://doi.org/10.1016/S1367-9120(02)00022-6

Metcalfe, I. 2013. Gondwana dispersion and Asian accretion: tectonic and palaeogeographic evolution of eastern Tethys. *Journal of Asian Earth Sciences*, **66**, 1–33, https://doi.org/10.1016/j.jseaes.2012.12.020

Mucek, A.E., Danišík, M., de Silva, S.L., Schmitt, A.K., Pratomo, I. and Coble, M.A. 2017. Post-supereruption recovery at Toba Caldera. *Nature Communications*, **8**, Article 15248, https://doi.org/10.1038/ncomms15248

Najman, Y., Jenks, D. *et al.* 2017. The Tethyan Himalayan detrital record shows that India–Asia terminal collision occurred by 54 Ma in the western Himalaya. *Earth and Planetary Science Letters*, **459**, 301–310, https://doi.org/10.1016/j.epsl.2016.11.036

Ninkovich, D., Sparks, R.S.J. and Ledbetter, M.T. 1978. The exceptional magnitude and intensity of the Toba eruption, Sumatra: an example of the use of deep-sea tephra layers as a geological tool. *Bulletin Volcanologique*, **41**, 286–298, https://doi.org/10.1007/BF02597228

Nishimura, S., Abe, E., Yokoyama, T., and Wirasantosa, S. and Dharma,. 1977. Danau Tobae the outline of Lake

Toba, North Sumatra, Indonesia. *Paleolimnology of Lake Biwa and the Japanese Pleistocene*, **5**, 313–332.

Rock, N.M., Syah, H.H., Davis, A.E., Hutchison, D., Styles, M.T. and Lena, R. 1982. Permian to Recent volcanism in northern Sumatra, Indonesia: a preliminary study of its distribution, chemistry, and peculiarities. *Bulletin Volcanologique*, **45**, 127–152, https://doi.org/10.1007/BF02600429

Rock, N.M.S., Aldiss, D.T. *et al.* 1983. *The Geology of the Lubuksikaping Quadrangle (0716), Sumatra, Scale 1: 250.000.* Geological Survey of Indonesia, Directorate of Mineral Resources, Geological Research and Development Centre, Bandung.

Rowley, D.B. 1996. Age of initiation of collision between India and Asia: A review of stratigraphic data. *Earth and Planetary Sciences*, **145**, 1–13, https://doi.org/10.1016/S0012-821X(96)00201-4

Sato, K. 1991. K–Ar ages of granitoids in Central Sumatra, Indonesia. *Bulletin Geological Survey of Japan*, **42**, 111–181.

Shao, W.Y., Chung, S.L. and Chen, W.S. 2014. Zircon U–Pb age determination of volcanic eruptions in Lutao and Lanyu in the northern Luzon magmatic arc. *Terrestrial, Atmospheric and Oceanic Sciences*, **25**, 149–187, https://doi.org/10.3319/TAO.2013.11.06.01(TT)

Sieh, K. and Natawidjaja, D. 2000. Neotectonics of the Sumatran fault, Indonesia. *Journal of Geophysical Research: Solid Earth*, **105**, 28295–28326, https://doi.org/10.1029/2000JB900120

Sláma, J., Košler, J. *et al.* 2008. Plešovice zircon – a new natural reference material for U–Pb and Hf isotopic microanalysis. *Chemical Geology*, **249**, 1–35, https://doi.org/10.1016/j.chemgeo.2007.11.005

Söderlund, U., Patchett, J.P., Vervoort, J.D. and Isachsen, C.E. 2004. The ^{176}Lu decay constant determined by Lu–Hf and U–Pb isotope systematics of Precambrian mafic intrusions. *Earth and Planetary Science Letters*, **219**, 311–324, https://doi.org/10.1016/S0012-821X(04)00012-3

Soeria-Atmadja, R. and Noeradi, D. 2005. Distribution of early Tertiary volcanic rocks in South Sumatra and West Java. *Island Arc*, **14**, 679–686, https://doi.org/10.1111/j.1440-1738.2005.00476.x

Sun, W.D., Zhang, L.P., Li, H. and Liu, X. 2020. The synchronic Cenozoic subduction initiations in the West Pacific induced by the closure of the Neo-Tethys Ocean. *Science Bulletin*, **65**(24), 2068–2071, https://doi.org/10.1016/j.scib.2020.09.001

Turner, S. and Foden, J. 2001. U, Th and Ra disequilibria, Sr, Nd and Pb isotope and trace element variations in Sunda arc lavas: predominance of a subducted sediment component. *Contributions to Mineralogy and Petrology*, **142**, 43–57, https://doi.org/10.1007/s004100100271

van Bemmelen, R.W. 1949. *The Geology of Indonesia.* Government Printing Office, Nijhoff, The Hague.

van Leeuwen, T.M., Taylor, R.P. and Hutagalung, J. 1987. The geology of the Tangse porphyry copper-molybdenum prospect, Aceh, Indonesia. *Economic Geology*, **82**(1), 27–42, https://doi.org/10.2113/gsecongeo.82.1.27

Wajzer, M.R. 1986. *The geology and tectonic evolution of the Woyla Group, Natal area, North Sumatra.* PhD thesis, University of London.

Wajzer, M.R., Barber, A.J., Hidayat, S. and Suharsono, 1991. Accretion, collision and strike-slip faulting: the Woyla Group as a key to the tectonic evolution of North Sumatra. *Journal of Southeast Asian Earth Sciences*, **6**, 447–461, https://doi.org/10.1016/0743-9547(91)90087-E

Wark, D.A., Masturyono, M., McCaffrey, R., Farmer, G.L., Rani, M. and Sukhyar, R. 2000. Plumbing of the Toba magma system; petrologic and geophysical evidence of two shallow reservoirs and their mantle roots. *Eos Transactions American Geophysical Union*, **81–48**, 1387.

Wheller, G.E., Varne, R., Foden, J.D. and Abbott, M.J. 1987. Geochemistry of Quaternary volcanism in the Sunda–Banda arc, Indonesia, and three-component genesis of island-arc basaltic magmas. *Journal of Volcanology and Geothermal Research*, **32**, 137–160, https://doi.org/10.1016/0377-0273(87)90041-2

Whitford, D.J. 1975. Strontium isotopic studies of the volcanic rocks of the Sunda arc, Indonesia, and their petrogenetic implications. *Geochimica et Cosmochimica Acta*, **39**, 1287–1302, https://doi.org/10.1016/0016-7037(75)90136-2

Whittaker, J.M., Muller, R.D., Sdrolias, M. and Heine, C. 2007. Sunda–Java trench kinematics, slab window formation and overriding plate deformation since the Cretaceous. *Earth and Planetary Science Letters*, **255**, 445–457, https://doi.org/10.1016/j.epsl.2006.12.031

Wiedenbeck, M., Alle, P. *et al.* 1995. Three natural zircon standards for U–Th–Pb, Lu–Hf, trace element and REE analyses. *Geostandards and Geoanalytical Research*, **19**, 1–23, https://doi.org/10.1111/j.1751-908x.1995.tb00147.x

Wikarno, R., Hardjono, T. and Graha, D.S. 1993. *Distribution of Radiometric Ages in Indonesia.* Geological Research and Development Centre, Bandung, Indonesia.

Wilkinson, C.M., Ganerød, M., Hendriks, B.W.H. and Eide, E.A. 2017. Compilation and appraisal of geochronological data from the North Atlantic Igneous Province (NAIP). *Geological Society, London, Special Publications*, **447**, 69–103, https://doi.org/10.1144/SP447.10

Woodhead, J.D. and Hergt, J.M. 2005. A preliminary appraisal of seven natural zircon reference materials for in situ Hf isotope determination. *Geostandards and Geoanalytical Research*, **29**, 183–195, https://doi.org/10.1111/j.1751-908X.2005.tb00891.x

Yokoyama, T. and Hehanussa, P.E. 1981. The age of 'Old Toba Tuff' and some problems on the geohistory of Lake Toba, Sumatra, Indonesia. *Paleolimnology of Lake Biwa and the Japanese Pleistocene*, **9**, 177–186.

Zhang, X., Chung, S.L., Lai, Y.M., Ghani, A.A., Murtadha, S., Lee, H.Y. and Hsu, C.C. 2018. Detrital zircons dismember Sibumasu in East Gondwana. *Journal of Geophysical Research: Solid Earth*, **123**, 6098–6110, https://doi.org/10.1029/2018JB015780

Zhang, X., Chung, S.L., Lai, Y.M., Ghani, A.A., Murtadha, S., Lee, H.Y. and Hsu, C.C. 2019. A 6000-km-long Neo-Tethyan arc system with coherent magmatic flare-ups and lulls in South Asia. *Geology*, **47**, 573–576, https://doi.org/10.1130/G46172.1

Zheng, Y.F. and Wu, F.Y. 2018. The timing of continental collision between India and Asia. *Science Bulletin*, **63**, 1649–1654, https://doi.org/10.1016/j.scib.2018.11.022

Index

Page numbers in *italics* refer to Figures. Page numbers in **bold** refer to Tables.